Solar Cell Engineering

PHOTOVOLTAICS BEYOND SILICON
Innovative Materials, Sustainable Processing Technologies, and Novel Device Structures

Edited by

SENTHILARASU SUNDARAM
School of Computing, Engineering and Digital Technologies, Teesside University, Tees Valley, Middlesbrough, United Kingdom

VELUMANI SUBRAMANIAM
Department of Mechanical Engineering, Texas A&M University, College Station, TX, United States

RYNE P. RAFFAELLE
Rochester Institute of Technology, Rochester, NY, United States

MOHAMMAD KHAJA NAZEERUDDIN
EPFL Valais Wallis, Sion, Switzerland

ARTURO MORALES-ACEVEDO
Electrical Engineering Department — SEES, Centro de Investigación y de Estudios Avanzados del Instituto Politécnico Nacional (CINVESTAV), Ciudad de México, México

MARÍA BERNECHEA NAVARRO
Instituto de Nanociencia y Materiales de Aragón (INMA), CSIC-Universidad de Zaragoza, Zaragoza, Aragón, Spain

ALOYSIUS F. HEPP
Nanotech Innovations LLC, Oberlin, OH, United States

ELSEVIER

Elsevier
Radarweg 29, PO Box 211, 1000 AE Amsterdam, Netherlands
125 London Wall, London EC2Y 5AS, United Kingdom
50 Hampshire Street, 5th Floor, Cambridge, MA 02139, United States

Copyright © 2024 Elsevier Inc. All rights are reserved, including those for text and data mining, AI training, and similar technologies.

Publisher's note: Elsevier takes a neutral position with respect to territorial disputes or jurisdictional claims in its published content, including in maps and institutional affiliations.

No part of this publication may be reproduced or transmitted in any form or by any means, electronic or mechanical, including photocopying, recording, or any information storage and retrieval system, without permission in writing from the publisher. Details on how to seek permission, further information about the Publisher's permissions policies and our arrangements with organizations such as the Copyright Clearance Center and the Copyright Licensing Agency, can be found at our website: www.elsevier.com/permissions.

This book and the individual contributions contained in it are protected under copyright by the Publisher (other than as may be noted herein).

Notices

Knowledge and best practice in this field are constantly changing. As new research and experience broaden our understanding, changes in research methods, professional practices, or medical treatment may become necessary.

Practitioners and researchers must always rely on their own experience and knowledge in evaluating and using any information, methods, compounds, or experiments described herein. In using such information or methods they should be mindful of their own safety and the safety of others, including parties for whom they have a professional responsibility.

To the fullest extent of the law, neither the Publisher nor the authors, contributors, or editors, assume any liability for any injury and/or damage to persons or property as a matter of products liability, negligence or otherwise, or from any use or operation of any methods, products, instructions, or ideas contained in the material herein.

ISBN: 978-0-323-90188-8

For Information on all Elsevier publications
visit our website at https://www.elsevier.com/books-and-journals

Publisher: Matthew Deans
Acquisitions Editor: Stephen Jones
Editorial Project Manager: Rafael G. Trombaco
Production Project Manager: Prasanna Kalyanaraman
Cover Designer: Mark Rogers

Typeset by MPS Limited, Chennai, India

Photovoltaics Beyond Silicon

Contents

List of contributors xiii
Preface xvii

Section I Photovoltaics: Background, technologies, and innovation

1. **Photovoltaics overview: Historical background and current technologies**   3
   Aloysius F. Hepp and Ryne P. Raffaelle

   1.1 Introduction and overview   3
   1.2 Short history of photovoltaics   4
   1.3 Background of photovoltaic materials and devices   10
   1.4 Summary and conclusions   55
   References   56

2. **Third-generation photovoltaics: Introduction, overview, innovation, and potential markets**   75
   Sahaya Dennish Babu George, Ananthakumar Soosaimanickam and Senthilarasu Sundaram

   2.1 Introduction   75
   2.2 Features of third-generation solar cells   77
   2.3 Terawatt energy generation challenges   78
   2.4 Summary and overview of third-generation photovoltaics   80
   2.5 Economic assessment and market status of third-generation photovoltaics   83
   2.6 Summary and conclusion   104
   Acknowledgments   105
   References   105

Section II Perovskite materials and devices

3. **Perovskite solar cells: Past, present, and future**   113
   Abhishek Kumar Chauhan, Pankaj Kumar and Shailesh Narain Sharma

   3.1 Introduction to photovoltaic devices   113

3.2	Perovskite solar cells	115
3.3	Working mechanism of perovskite solar cells	118
3.4	Device architectures of perovskite solar cells	122
3.5	Lead-free perovskite solar cells	129
3.6	Tandem architecture of perovskite-silicon solar cells	130
3.7	Stability study of perovskite solar cells	133
3.8	Perovskite film fabrication processes	136
3.9	Different ranges of electron and hole transport layers	140
3.10	Electrodes for perovskite solar cells	145
3.11	Effect of encapsulation on perovskite solar cells	147
3.12	Standard testing protocols	148
3.13	Conclusion and future prospects	149
	References	149

4. Modeling perovskite solar cells — 165

Arturo Morales-Acevedo

4.1	Introduction	165
4.2	Thick versus thin solar cells	166
4.3	Thin perovskite solar cells	169
4.4	Modeling results for thin-film perovskite cells	176
4.5	Conclusion	180
	References	181

5. Optical design of perovskite solar cells — 183

Arturo Morales-Acevedo and Roberto Bernal-Correa

5.1	Introduction	183
5.2	Optical modeling	184
5.3	Multilayered systems	186
5.4	Optical calculations for perovskite solar cells	188
5.5	Conclusion	192
	References	193

6. Organic hole-transporting materials for perovskite solar cells: Progress and prospects — 195

S. Sambathkumar and P. Baby Shakila

6.1	Introduction to perovskite solar cells	195
6.2	Device architecture and mechanism of perovskite solar cells	197
6.3	Organic hole-transporting materials	198

6.4	Conclusions and further perspectives	215
	Acknowledgments	216
	References	216

Section III Alternative photovoltaic materials

7. Advanced fabrication strategies to enhance the performance of dye-sensitized solar cells   223
Anurag Roy, Tapas K. Mallick, Asif Ali Tahir, Mohammad Ashraf Gondal and Senthilarasu Sundaram

7.1	Introduction	223
7.2	Classification of solar cells: Materials and technologies	224
7.3	A brief history of dye-sensitized solar cells	225
7.4	Dye-sensitized solar cells: Working principles	226
7.5	Solar irradiation	228
7.6	Concentrated light irradiation on dye-sensitized solar cells	230
7.7	Strategies to improve interfacial electron kinetics	230
7.8	Photoanode improvement of state-of-the-art dye-sensitized solar cells	234
7.9	Improvement(s) over state-of-the-art dye sensitizers	236
7.10	Improvement(s) over state-of-the-art dye sensitizer electrolytes	237
7.11	Improvement(s) over state-of-the-art dye-sensitized solar cell counter-electrodes	238
7.12	Overall relative scope of performance improvement of dye-sensitized solar cells	239
7.13	Nanostructured materials for performance improvement of dye-sensitized solar cells	241
7.14	Effect of light scattering	243
7.15	Improved dye-sensitized solar cell device performance: Fabrication approaches	245
7.16	Summary and future scope	247
	Acknowledgments	248
	References	248

8. Development of active layer materials for solution-processable organic photovoltaics   255
Geneviève Sauvé

8.1	Introduction	255
8.2	Solution-processed bulk heterojunction organic photovoltaics	260
8.3	Electron donors optimized for fullerene-based acceptors	261

8.4	Nonfullerene electron acceptors	266
8.5	Considerations for commercialization	281
8.6	Summary and conclusions	289
	References	290

9. Copper zinc tin sulfide thin-film solar cells: An overview — 303
Zubair Ahmad Kumar, Towseef Ahmad and Mohd Zubair Ansari

9.1	Introduction	303
9.2	Structure	304
9.3	Synthesis methods of copper zinc tin sulfide nanoparticles and thin films	305
9.4	Copper zinc tin sulfide-based thin-film solar cell deposition techniques	309
9.5	Comparison of performance of vacuum- and nonvacuum-based solar cells	316
9.6	Future scope	318
9.7	Conclusion	319
	References	319

Section IV Green and sustainable aspects of photovoltaics

10. Nature-inspired and computer-aided approaches to enable improved photovoltaic materials, more efficient processing, and novel devices — 325
Aloysius F. Hepp and Ryne P. Raffaelle

10.1	Introduction: Nature-inspired and computer-aided insights	325
10.2	Sustainability, green technologies, and biomimicry	326
10.3	Computer-aided approaches to improved photovoltaics	337
10.4	Metal-organic framework materials for enhanced solar cells	344
10.5	Future paradigms: Novel materials and device fabrication for photovoltaics	358
10.6	Summary and conclusions	375
	References	378

11. Green chemical synthesis of photovoltaic materials — 405
O. Reyes-Vallejo, S. Torres-Arellano, J.L. Aleman-Ramirez and P.J. Sebastian

11.1	Introduction	405
11.2	Green methods for synthesis of semiconductor materials	406

11.3	Biosynthesis	412
11.4	Photovoltaic materials through green synthesis	417
11.5	Considerations for greener fabrication of solar cells	422
11.6	Conclusion: Prospects for green synthesis of solar cells	426
	References	428

12. Sustainable solution-processed solar cells based on environmentally friendly nanocrystals — 437

Sergio Aina, Nichole Scott, M. Pilar Lobera and María Bernechea Navarro

12.1	Introduction	437
12.2	Antimony compounds	438
12.3	Bismuth compounds	445
12.4	Tin compounds	453
12.5	Iron compounds	458
12.6	Conclusions	464
	Acknowledgments	465
	References	465

13. Life cycle assessment of renewable energy from solar photovoltaic technologies — 479

Annick Anctil

13.1	Introduction to life cycle assessment	479
13.2	Life cycle assessment metrics	481
13.3	Life cycle assessment of various photovoltaic technologies	484
13.4	Future need for life cycle assessment of photovoltaic technologies	488
13.5	Conclusion	489
	References	490

Section V Concentrator and multijunction devices

14. Concentrator and multijunction solar cells — 499

Katie Shanks

14.1	Introduction: Classifications of relevant technologies	499
14.2	Concentrator compared to one sun solar cells	503
14.3	Multijunction solar cells for solar concentrator systems	509
14.4	Matching solar cells and concentrator optics for optimum performance	512
14.5	Conclusions	517
	References	518

15. All perovskite tandem solar cells — 523
Arunkumar Prabhakaran Shyma, Nandhakumar Eswaramoorthy, Raja Sellappan, Kamatchi Rajaram, Sowmya Sridharan and Selvakumar Pitchaiya

15.1	Introduction	523
15.2	Light capture and device management	529
15.3	All perovskite tandem architectures	533
15.4	All perovskite flexible tandem solar cells	537
15.5	Device stability	538
15.6	Conclusion	539
	References	540

16. Recent advances in perovskite-containing tandem structures — 545
Maria Khalid, Tapas K. Mallick and Senthilarasu Sundaram

16.1	Introduction	545
16.2	An overview of tandem architectures	548
16.3	Challenges with tandem structures	552
16.4	Power losses in perovskite tandem design	555
16.5	Tandem architecture(s) advantages	556
16.6	Tandem configurations	557
16.7	Perovskite/Si tandem solar cells	565
16.8	Perovskite/copper indium gallium selenide tandem solar cells	569
16.9	Perovskite/perovskite tandem solar cells	573
16.10	Conclusion	575
	References	576

Section VI Practical applications of photovoltaics

17. Transparent photovoltaics: Overview and applications — 585
Ananthakumar Soosaimanickam and Abhirami Murugavel

17.1	Transparent solar cells: An introduction	585
17.2	Solar cell parameters for (semi-)transparent applications	588
17.3	Electrodes in transparent solar cells	590
17.4	Types of transparent solar cells	596
17.5	Applications of transparent solar cells in agriculture	616
17.6	Transparent luminescent solar concentrators	618
17.7	Conclusion and future perspectives	629
	Acknowledgments	630
	References	630

18. Photovoltaic-powered vehicles: Current trends and future prospects   **647**
Jubaer Ahmed

18.1	Introduction	647
18.2	Recent trends in photovoltaic-powered vehicles	649
18.3	Solar cell technologies for photovoltaic-powered electric vehicles	658
18.4	Benefits of photovoltaic-powered vehicles and future directions	666
18.5	Conclusions	669
	Acknowledgment	670
	References	670

19. Space photovoltaics: New technologies, environmental challenges, and missions   **675**
Aloysius F. Hepp, Ryne P. Raffaelle and Ina T. Martin

19.1	A brief history of space solar cells	675
19.2	Space environment: Concerns, challenges, and issues for space photovoltaics	683
19.3	Materials and devices: Connections to space exploration	690
19.4	Space applications and exploration missions: Past, present, and future	716
19.5	Conclusions	748
	References	751

Index   *767*

List of contributors

Towseef Ahmad
Department of Physics, National Institute of Technology Srinagar, Srinagar, Jammu & Kashmir, India

Jubaer Ahmed
Edinburgh Napier University, Edinburgh, United Kingdom

Sergio Aina
Instituto de Nanociencia y Materiales de Aragón (INMA), CSIC-Universidad de Zaragoza, Zaragoza, Aragón, Spain; Department of Chemical and Environmental Engineering (IQTMA), University of Zaragoza, Zaragoza, Aragón, Spain

J.L. Aleman-Ramirez
Renewable Energy Research and Innovation Institute IIIER-UNICACH, Tuxtla Gutiérrez, Chiapas, Mexico

Annick Anctil
Michigan State University, East Lansing, MI, United States

Mohd Zubair Ansari
Department of Physics, National Institute of Technology Srinagar, Srinagar, Jammu & Kashmir, India

Roberto Bernal-Correa
Universidad Nacional de Colombia, Sede Orinoquia, Grupo de Investigación en Ciencias de la Orinoquia, Arauca, Colombia

María Bernechea Navarro
Instituto de Nanociencia y Materiales de Aragón (INMA), CSIC-Universidad de Zaragoza, Zaragoza, Aragón, Spain; Department of Chemical and Environmental Engineering (IQTMA), University of Zaragoza, Zaragoza, Aragón, Spain; CIBER de Bioingeniería, Biomateriales y Nanomedicina, Instituto de Salud Carlos III, Zaragoza, Aragón, Spain; ARAID, Government of Aragon, Zaragoza, Aragón, Spain

Abhishek Kumar Chauhan
Advanced Materials and Devices Metrology Division, CSIR-National Physical Laboratory, New Delhi, India

Nandhakumar Eswaramoorthy
School of Mechanical Engineering, Vellore Institute of Technology, Vellore, Tamil Nadu, India

Sahaya Dennish Babu George
Department of Physics, Chettinad College of Engineering and Technology, Karur, Tamil Nadu, India

Mohammad Ashraf Gondal
Laser Research Group, Physics Department, King Fahd University of Petroleum and Minerals (KFUPM), Dhahran, Saudi Arabia; K.A. CARE Energy Research and Innovation Center, King Fahd University of Petroleum and Minerals (KFUPM), Dhahran, Saudi Arabia

Aloysius F. Hepp
Nanotech Innovations LLC, Oberlin, OH, United States

Maria Khalid
Solar Energy Research Group, Environment and Sustainability Institute, University of Exeter, Penryn Campus, Cornwall, United Kingdom

Pankaj Kumar
Advanced Materials and Devices Metrology Division, CSIR-National Physical Laboratory, New Delhi, India

Zubair Ahmad Kumar
Department of Physics and Astronomy, National Institute of Technology Rourkela, Rourkela, Odisha, India

Tapas K. Mallick
Solar Energy Research Group, Environment and Sustainability Institute, University of Exeter, Penryn Campus, Cornwall, United Kingdom

Ina T. Martin
Case Western Reserve University, Cleveland, OH, United States

Arturo Morales-Acevedo
Electrical Engineering Department – SEES, Centro de Investigación y de Estudios Avanzados del Instituto Politécnico Nacional (CINVESTAV), Ciudad de México, México

Abhirami Murugavel
R&D Division, Intercomet S.L., Madrid, Spain

M. Pilar Lobera
Instituto de Nanociencia y Materiales de Aragón (INMA), CSIC-Universidad de Zaragoza, Zaragoza, Aragón, Spain; Department of Chemical and Environmental Engineering (IQTMA), University of Zaragoza, Zaragoza, Aragón, Spain; CIBER de Bioingeniería, Biomateriales y Nanomedicina, Instituto de Salud Carlos III, Zaragoza, Aragón, Spain

Selvakumar Pitchaiya
Faculty of Engineering and Science, Western Norway University of Applied Sciences, Bergen, Norway

Ryne P. Raffaelle
Rochester Institute of Technology, Rochester, NY, United States

Kamatchi Rajaram
School of Mechanical Engineering, Vellore Institute of Technology, Vellore, Tamil Nadu, India

List of contributors

O. Reyes-Vallejo
Solid State Electronics Section (SEES) CINVESTAV-IPN, Gustavo A. Madero, Mexico City, Mexico; Renewable Energy Institute IER-UNAM, Solar Materials Section, Temixco, Morelos, Mexico

Anurag Roy
Solar Energy Research Group, Environment and Sustainability Institute, University of Exeter, Penryn Campus, Cornwall, United Kingdom

S. Sambathkumar
Department of Chemistry and Biochemistry, Vivekanandha College of Arts and Sciences for Women (Autonomous), Elayampalayam, Tiruchengode, Namakkal, Tamil Nadu, India

Geneviève Sauvé
Department of Chemistry, Case Western Reserve University, Cleveland, OH, United States

Nichole Scott
Smart Material Solutions, Inc., Raleigh, NC, United States

P.J. Sebastian
Renewable Energy Institute IER-UNAM, Solar Materials Section, Temixco, Morelos, Mexico

Raja Sellappan
Center for Nanotechnology Research, Vellore Institute of Technology, Vellore, Tamil Nadu, India

P. Baby Shakila
Department of Chemistry and Biochemistry, Vivekanandha College of Arts and Sciences for Women (Autonomous), Elayampalayam, Tiruchengode, Namakkal, Tamil Nadu, India

Katie Shanks
Solar Energy Research Group, Environment and Sustainability Institute, University of Exeter, Penryn Campus, Cornwall, United Kingdom

Shailesh Narain Sharma
Advanced Materials and Devices Metrology Division, CSIR-National Physical Laboratory, New Delhi, India

Arunkumar Prabhakaran Shyma
Department of Physics, School of Advanced Science, VIT, Vellore, Tamil Nadu, India

Ananthakumar Soosaimanickam
R&D Division, Intercomet S.L., Madrid, Spain

Sowmya Sridharan
School of Mechanical Engineering, Vellore Institute of Technology, Vellore, Tamil Nadu, India

Senthilarasu Sundaram
School of Computing, Engineering and Digital Technologies, Teesside University, Tees Valley, Middlesbrough, United Kingdom

Asif Ali Tahir
Solar Energy Research Group, Environment and Sustainability Institute, University of Exeter, Penryn Campus, Cornwall, United Kingdom

S. Torres-Arellano
Renewable Energy Research and Innovation Institute IIIER-UNICACH, Tuxtla Gutiérrez, Chiapas, Mexico

Preface

The world demand for energy is dramatically increasing, necessitating accelerated research into renewable nonfossil-based sources of electricity. The conversion of light from the Sun into electrical energy using photovoltaics (PV) is a key technology being explored to meet these challenges both on Earth and in space. *Photovoltaics Beyond Silicon* presents the latest innovations in materials, processing, and devices to produce electricity via advanced, sustainable photovoltaic technologies. It provides an overview of the novel materials and device architectures that have been developed to optimize energy conversion efficiencies and minimize environmental impacts. Advances in technologies for harnessing solar energy are extensively discussed, with topics, including materials processing, device fabrication, sustainability of materials and manufacturing, and the current state of the art. Contributions from leading international experts discuss the applications, challenges, and future prospects of research in this increasingly vital field, providing a valuable resource for students and researchers working in this area.

This book includes the following key features: the book offers a comprehensive overview and detailed discussion of solar energy technology options for sustainable energy conversion; provides an understanding of the environmental challenges to be overcome and discusses the importance of efficient materials utilization for clean energy; and investigates how to design materials synthesis processes, optimize device fabrication, and metrics such as power-to-weight ratio, effectiveness at end-of-life (compared to beginning-of-life), and life cycle analysis. The coeditors of this book have selected and organized nineteen chapters authored by experts in the fields of third-generation solar cell materials and devices, green materials and processing, concentrator and multijunction (MJ) devices, as well as PV for terrestrial and aerospace systems. The book is divided into six sections, each devoted to a different aspect of PV: an introductory section, two sections on perovskites and other third-generation materials and devices, green aspects of PV, advanced device designs, and applications of PV with a representative key technology background, materials-focused, and applications-oriented chapters.

The initial section of this book includes two chapters that are coauthored by three coeditors; these overview chapters provide historical background, summaries of current technologies, details of recent innovations, and commentary on applications of PV. Chapter 1 begins with a short account of the early days of photovoltaic solar cell development. Crystalline silicon (Si) technology is a venerable and reliable technology that remains dominant but is being challenged by an ever-growing set of advanced alternatives. In fact, for a majority of space applications, Si has been replaced by MJ gallium arsenide (III-V)-based technologies. Future applications that rely on flexible devices would likely favor lightweight (inorganic, hybrid, or organic) devices. After a historical background and technology background discussion, the chapter explores a series of next (third and beyond) generation materials and device concepts, including dye-sensitized solar cells (DSSCs), perovskite solar cells (PSCs), quantum dot (sensitized) solar cells (QD(S)SCs), intermediate bandgap (concentrator) solar cells, up/down conversion materials, concentrator solar cells, and MJ (including tandem) devices. Chapter 2 focuses on third-generation PV materials and devices; after a brief introduction, an overview of materials, including key innovations and potential markets are addressed. The lead author is Prof. S.D.B. George (Chettinad College of Engineering and Technology, Karur, India) with coauthors Dr. Soosaimanickam (Intercomet S.L., Madrid, Spain) and book coeditor, Prof. Sundaram. Third-generation PV stands out for its higher efficiency, lower-cost manufacturing approach, and applicability for a range of uses. For example, it is feasible to produce solar cells using inexpensive printing methods on flexible substrates. Organic, dye-sensitized, perovskite, and quantum dot solar cells are key examples of these technologies. The advantages and special characteristics of each technology are highlighted in this chapter. These technologies are suited for a variety of applications, including electronic, automotive, and architectural products, because of the promise of third-generation PV for low-cost production, flexibility, and transparency.

Section II includes four chapters on perovskite materials and devices. Mixed organic—inorganic PSC devices are the leading contenders to disrupt terrestrial PV; this is due to their rapid improvement in conversion efficiencies, potential for lower-cost manufacturing, and the potential for large-scale flexible architectures. The first chapter in the section (Chapter 3) is a look back and ahead at PSCs with a historical perspective and current state-of-the-art (SOA) and includes an overview of potential developments, especially enhancing the stability of PSCs to enable eventual commercialization. The contribution, from CSIR-National Physical

Laboratory, New Delhi, India, is coauthored by Drs. Chauhan, Kumar, and S.N. Sharma, the lead author. In particular, the chapter delves into SOA performance, novel materials, designs, processing, electrical and optical properties, degradation, the factors affecting their performance under different storage conditions, large-area printing, and strategies to make them commercially viable. The next two chapters are authored or coauthored by Prof. A. Morales-Acevedo (CINVESTAV-México), another member of the editorial team. Chapter 4 details a simple PSC model, considering both radiative and nonradiative recombination currents. The conversion efficiency dependence on lifetime, perovskite material bandgap, and device thickness is studied. This model allows the estimation of the optimum thickness, and it also helps in determining what should be the required carrier lifetimes for achieving improved efficiencies. It is shown that (nonradiative) lifetimes larger than 5 µs in the perovskite cell material will not cause any efficiency improvement above the limit of 29%. Chapter 5 addresses optical losses that affect the performance of PSCs. The authors detail an optical matrix calculation methodology for planar thin PSCs with both normal and inverted structures. The method is based on the Fresnel equations (and coefficients), which are explained first; a general (optical) matrix equation enables the treatment of multilayer systems. Subsequently, this method is used to determine the spectral reflectance (and transmittance) from planar PSC and it is shown that there is an optimum thickness for the upper transport layer, particularly for the inverted structure. The section concludes with a chapter on organic HTMs for PSCs, reporting on progress and prospects; the chapter is contributed by lead coauthor Prof. Sambathkumar with Prof. Shakila, both of Vivekanandha College of Arts and Sciences for Women, Tamil Nadu, India. This chapter explores the progress of novel design strategies, facile synthetic routes, scale-up, stability, device suitability, and performance of the organic HTMs with carbazole, phenothiazine, triphenylamine, naphthol, thiophene, and pyrene core structure to enable efficient PSCs. The chapter ends with suggested further areas of potential interest to readers.

Section III contains three chapters on alternative photovoltaic materials. Chapter 7, the second chapter coauthored by Prof. Sundaram, explores fabrication strategies to enhance the performance of DSSCs. The author team is led by Prof. A. Roy and also includes Profs. Mallick and Tahir, all from the University of Exeter (Penryn Campus) in Cornwall, UK, and Prof. Gondal, King Fahd University of Petroleum and Minerals, Dhahran, Saudi Arabia. This chapter is focused on alternative methods for

Lansing, MI, United States. LCA is becoming a standard method used for product selection through certification or preferred purchase for commercialized technologies. Existing studies focus on issues that are perceived as being of particular concern, such as toxicity of Pb-containing perovskites and cadmiun telluride (CdTe) or large energy consumption for Si, and are often used to compare alternatives. Emerging technologies can be evaluated using prospective LCA; future studies will need to incorporate other benefits of PV beyond electricity production for product and building-integrated applications.

Section V of the book includes three chapters that focus on concentrator and MJ devices. Chapter 14, the first chapter in the section, addresses single and MJ concentrator cells and systems; it is contributed by Dr. Katie Shanks, University of Exeter (Penryn Campus). Matching cells and optics for optimum cost, efficiency, or carbon footprint are important considerations discussed in this chapter. Although concentrator MJ solar cells are best matched for medium concentration ratios (and upward), single junction solar cells can be used with low concentration ratios with little or zero alterations to cell design, such as the metallization grid layer, potentially leading to more cost-effective systems with little loss in efficiency. Solar cells can be used above their designed working range if adequately cooled but this may affect solar cell lifetime. Optimally matching solar cells with concentrator optics can result in high-efficiency designs that are very space (and carbon) efficient but may not be the most cost-efficient option. Concentrator photovoltaic systems are still being developed with opportunities for customization depending on integration requirements such as building, vehicle, marine, and agriculture-integrated applications. Chapter 15 focuses on all-perovskite tandem solar cells, and it is contributed by Prof. S. Pitchaiya from Western Norway University of Applied Sciences, Bergen, Norway and a team of coauthors (Profs. Shyma, Eswaramoorthy, Sellappan, Sridharan, and Rajaram) from Vellore Institute of Technology, Vellore, India. The chapter considers strategies that can be employed to refine the carrier diffusion length of narrow bandgap bottom cells and voltage loss in wide bandgap top cells. The authors summarize key developments in the area of all perovskite-tandem solar cells. Finally, insights and alternative approaches are considered to overcome existing challenges and enable rapid commercial deployment of all perovskite tandem solar cells. A broader view of perovskite-containing tandem devices is provided by the final chapter (Chapter 16) in the section; it is contributed by Dr. M. Khalid (Exeter University, Penryn, UK) and two coauthors: Profs. Mallick and Sundaram—his third coauthored chapter. The authors stress three advantages that make PSCs ideal candidates for fabricating tandem solar cells: bandgap tunability between 1.48 and 2.3 eV, long-range charge transportation, and high radiative recombination rates. The chapter continues with a summary of recent advances in the tandem device landscape with key motivations and the challenges to fabricating such promising architectures. Finally, an in-depth examination of the current status of the perovskite/perovskite, perovskite/ CIGS, and perovskite/Si tandems utilizing different device technologies with key challenges and related aspects is presented.

The final section of the book includes three chapters that speak to the practical utilization of PV technologies, from terrestrial structures and mobile platforms (both land and sea) through high-altitude airships to space exploration. Chapter 17 provides an overview of transparent PV and provides examples of applications; it is contributed by Drs. A. Soosaimanickam and Murugavel of Intercomet S.L., Madrid, Spain. The chapter addresses recent advances in transparent solar cells and their potential for commercial applications. While semitransparent crystalline Si and PSCs have delivered about 12% efficiency under laboratory conditions, the tandem architecture of these semitransparent solar cells has produced over 20% efficiency. Various fabrication strategies of semitransparent solar cells with different materials and architectures are discussed in detail. Potential alternative outcomes for the growth of future solar cell markets based on this emerging concept are also outlined. Chapter 18, contributed by Prof. Jubaer Ahmed, Edinburgh Napier University (Edinburgh, UK), explores photovoltaic-powered vehicles by addressing current trends and future prospects. Conventional fossil fuel-based vehicles are being challenged in the marketplace by hybrid or electric vehicles (EV). Unlike PV penetration in the energy sector, the direct involvement of PV in vehicles [PV-powered (or PV-integrated) vehicles] has yet to see significant growth. However, this sector has significant growth potential and could revamp the transport sector in the future. In this chapter, a concise overview of the ongoing research and development trend in PV-powered vehicles is presented. Finally, how the adoption of such technologies can shape the future transportation sector is also briefly analyzed. The final chapter (Chapter 19) in the section and book (and the third contributed by coeditors Drs. Hepp and Raffaelle) addresses space PV by examining (new) PV technologies, environmental challenges, and example missions; they are joined by Dr. I.T. Martin of CWRU. A short discussion of the difference between space (air mass zero, AM0) and terrestrial (AM1.0 or AM1.5) solar radiation begins the chapter. Next, the unique stressors in the space environment, varying with location, encountered by photovoltaic devices are summarized. After contrasting crystalline and TFSCs, key components of TFSC devices of the main inorganic absorber materials are considered in the context of overcoming environmental challenges. The chapter briefly addresses legacy (non-PSC or organic-based) thin-film materials (thin silicon, CIGS, and CdTe) for space PV. Issues such as mass-specific power and impact of processing, as well as testing relevant to the use of PV for space missions are examined. A brief summary of past, present, and future applications (missions) for space exploration, including examples of technology transfer from studies of these material systems to other PV systems and devices, concludes the chapter.

The success of this edited book project is the result of the full commitment of each contributing author. Without their availability in sharing their valuable knowledge and providing critical reviews of respective topics, this book could not be published at such a high standard. Our appreciation, indeed gratitude, also goes to Rafael Trombaco, our project manager, who provided tireless support during some challenging times. Stephen Jones, our acquisitions editor, is acknowledged for his constant support throughout. The production staff led by Prasanna Kalyanaraman were very helpful, thorough, and professional throughout the final stages of publication. Finally, we are pleased to honor the work of our colleagues in the PV community through the publication of this book; our sincere aim is to inform current workers in the field and inspire the next generation of PV researchers.

Senthilarasu Sundaram
Velumani Subramaniam
Ryne P. Raffaelle
Mohammad Khaja Nazeeruddin
Arturo Morales-Acevedo
María Bernechea Navarro
Aloysius F. Hepp

SECTION I

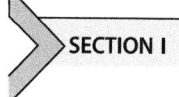

Photovoltaics: Background, technologies, and innovation

CHAPTER ONE

Photovoltaics overview: Historical background and current technologies

Aloysius F. Hepp[1] and Ryne P. Raffaelle[2]
[1]Nanotech Innovations LLC, Oberlin, OH, United States
[2]Rochester Institute of Technology, Rochester, NY, United States

1.1 Introduction and overview

Photovoltaic(s) (PV) technology is one of several critical renewable energy technologies. There are numerous technical overviews and reviews available at a variety of complexities; we call the reader's attention to several recent review articles [1–4] and books [5–7]. Several follow-on chapters in this book review major developments in PV over the past several decades, particularly hybrid, perovskite, and organic materials and advanced device designs [8–11].

Several divergent paradigms that demonstrate the potential to enable further developments in PV beyond current technologies are addressed by us in a follow-on chapter in this book [12]. The topics included in the second half of our overview include green (and sustainable) technologies [13,14]; biomimicry [15,16]; simulation (or modeling) of PV devices [17]; first principles materials modeling, including density functional theory [18]; artificial intelligence- and machine learning-directed discovery [19]; Earth-abundant [20] and nontoxic materials [21]; novel material processing, including single-source precursors [22,23]; and particularly, use of metal-organic frameworks (MOFs) [24–27]. Novel device structures and approaches to be reviewed include integrated power devices and structures [28] and low-cost, novel manufacturing methods [3,5,13,29].

In this chapter, after a historical overview and technology background discussion (Section 1.2), we provide a retrospective of the state-of-the-art of first- and second-generation PV by cell type (Section 1.3). The remainder of the chapter surveys a variety of topics that may be of interest to readers who seek to make contributions to advanced PV systems with potential practical applications. We review a series of next-generation materials and device concepts, including dye-sensitized solar cells (DSSCs), perovskite solar cells (PSCs), quantum dot (sensitized) solar cells (QDSCs), intermediate bandgap (concentrator) solar cells (IB(C)SCs), up/down conversion materials, concentrator solar cells, and multijunction (MJ) (including tandem) devices. Conclusions include key lessons learned and a discussion regarding promising new directions for research that enable novel applications. For example, MJ III−V devices, due to their high power conversion efficiency (PCE) and associated metrics, will quite likely be the predominant nonnuclear space power technology for the foreseeable future [30,31]; however, such concentrator technologies have a potential for terrestrial applications [11].

1.2 Short history of photovoltaics

As we have previously discussed [31], "the history of space PV is in many ways the history of PV." Concurrently, the early development of the silicon (Si) solar cell by its inventors at Bell Laboratories anticipated numerous terrestrial uses for the new source of electrical power in 1954 [32]. The development of space solar power systems drove much of early PV solar cells and array technology. For example, in response to the launch of Sputnik satellites by the Soviet Union, the United States launched Vanguard 1, a Si-powered satellite, on March 17, 1958 [33]. Further improvements in Si efficiency and manufacturing costs led to a burgeoning use of Si PV technology to generate electricity on an industrial (or utility) scale. Finally, the last quarter of the 20th century saw the development of a broad array of new PV materials and devices. The remainder of the chapter lays out the history of PV in a chronological manner, in context.

1.2.1 The first century of photovoltaics: Becquerel, Einstein, and the new physics

Background literature providing an overview of solar cell technologies will often include the discovery of the photovoltaic effect in 1838 by Becquerel [34]. Alexandre Edmond Becquerel was a French physicist who is credited with the discovery of the "Becquerel effect" or as it eventually became known as the "photovoltaic effect." This effect forms the basic operating principle of most solar cells (see detailed discussion in Section 1.3.1). Becquerel was able to demonstrate that when he placed silver chloride in an acidic solution and illuminated it while connected to platinum electrodes, it would generate a voltage and current. Similar demonstrations were soon to follow, most notably with the use of junctions made between platinum (Pt) or gold (Au) and selenium (Se), thus creating a solid-state device, unlike Becquerel's electrochemical cell. The American inventor Charles Fritts made what is considered the first solid-state solar cell using a thin layer of Au on Se [35], forming what is now known as a Schottky solar cell [36]. These cells were able to convert about 1% of the available energy from the sun to usable electricity. These devices were used to make the world's first rooftop solar array, installed on a New York City rooftop in 1884 [37].

Several other interesting phenomena associated with the interactions between light and matter were also being investigated at this time, such as the release of electrons from a metal surface when illuminated (i.e., photoelectric effect) and the changes in the spectrum and intensity of light emitted from a heated surface as its temperature changes (i.e., blackbody radiation), where a blackbody is a perfectly opaque and nonreflective body that is both a perfect absorber and emitter of radiation. From these basic observations, a picture began to emerge where the light was viewed not only as a wave but made up of "quanta" or packets of energy that made up the wave itself [38].

In 1905, Albert Einstein published a paper in which he outlined the concept of an energy quantum (or what G.N. Lewis would eventually call in a 1926 letter to Nature [39], a "photon") to explain the experimental data associated with the photoelectric effect. In 1918, the German physicist Max Planck received the Nobel Prize in physics for turn-of-the-century employment of the quantization of energy to explain the effects of radiation on a "blackbody" substance. Thus the era of modern physics with its quantum theory and mathematical formalism (i.e., quantum mechanics) was born. Einstein eventually received his Nobel Prize in Physics in 1921 [40] "for his services to theoretical physics, and especially for his discovery of the law of the photoelectric effect." Within a decade after Einstein's famous paper explaining the photoelectric effect, William Coblentz received the first US patent for converting sunlight into electricity in 1913 [41].

Niels Bohr, Louis de Broglie, Erwin Schrodinger, Paul M. Dirac, and other scientists advanced quantum theory and led to a probabilistic view of nature, sharply contrasting with classical mechanics, in which all precise properties of objects are, in principle, calculable [40]. Today, the combination of quantum mechanics with Einstein's theory of relativity is the basis of modern physics. Modern physics ultimately led scientists to an understanding of the behavior of semiconducting materials and their junctions and what came to be known as the "photovoltaic effect" or the modern solar cell [42].

1.2.2 The next 20 years: From a practical solar cell device to the dawn of the space race

In 1940, Russell Ohl of Bell Labs in Holmdel, NJ (USA) developed what was the first semiconducting "*p-n* junction;" he received a patent in 1946 for what was called a "light-sensitive device" (Fig. 1.1) [43]. Ohl and colleagues were able to discover the connection between the behavior of silicon and its "purity," or more importantly what "impurities" were present, and how junctions between Si samples with different types of impurities would generate a current when exposed to light (i.e., photocurrent). Ohl and his colleagues discovered that certain impurities

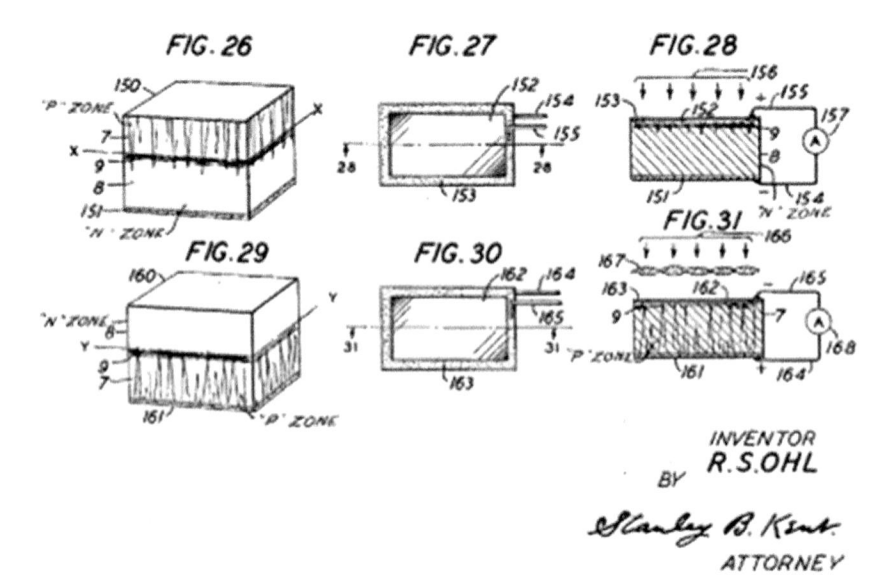

Figure 1.1 Diagram from Russel Ohl's 1946 US Patent (2,402,662) for a "Light sensitive device." *Courtesy: US Patent and Trademark Office.*

(i.e., phosphorus (P), arsenic (As), antimony (Sb), bismuth (Bi), lithium (Li)) would introduce an excess of electrons, negative charge carrier, or a *n*-type materials. Conversely, other impurities (i.e., boron (B), aluminum (Al), gallium (Ga), indium (In)) would do the opposite and yield an excess of positive charge carriers and thus render the Si *p*-type [44].

In 1953, Calvin Fuller and Gerald Pearson from Bell Labs were following in the footsteps of Ohl and Scaff and working on controlling the properties of semiconductors by introducing impurities. They observed a very large photoresponse from a *p-n* junction using Si doped with Ga and Li, respectively. They encouraged Daryl Chapin, another Bell Lab researcher, who was working to develop a "solar battery," a device that could convert light to electricity for remote telephone systems, to consider developing a solar cell from this material as opposed to the selenium he had been using. Eventually due to the instability of the junctions made from lithium- and gallium (Ga-) doped Si, they settled upon the use of As- and B-doped Si. The resulting solar cells delivered a power of about 60 Watts/m^2, which equates to an efficiency of approximately 6%. It was patented in 1957 [45] as a "solar energy converting apparatus" and introduced at a press conference by Bell Labs to the public on April 25, 1954 [32].

There was considerable excitement over the prospect that this new technology might hold for the future. Bell Laboratories Record stated that *"Scientists have long reached for the secret of the Sun. For they have known that it sends us nearly as much energy daily as is contained in all known reserve of coal, oil, and uranium. If this energy could be put to use, there would be enough to turn every wheel and light every lamp that mankind would ever need."* It would however take another 50 years or so to reach the 1 GW mark in terms of deployed terrestrial solar power, and another 20 years after that to reach 100 GW [42,46].

1.2.3 From the space race to the energy crisis

The early days in the development of photovoltaic power were not actually driven by the dream of inexhaustible solar power on Earth, but by the need to harvest energy while in space. The first man-made satellites, Sputnik 1 and 2, were launched by the Soviet Union in October 4 and November 3, 1957, respectively. In response, the US launched Explorer 1 in January 31, 1958, and Vanguard I in March 17, 1958. Vanguard 1, developed by the US Naval Research Laboratory [33,47], was the first solar-powered satellite in space.

Unlike its three predecessors, Vanguard 1 has six body-mounted solar panels (see Fig. 1.2) made with eighteen 2×0.5 cm individual solar cells, protected by 0.16 cm quartz windows. These solar panels provided the power to transmitter to operate for over 6 years, as compared to Sputnik, which transmitted for only 3 weeks until its three silver–zinc batteries died. Also, unlike its predecessors, which lasted only a few months before aerodynamic drag caused them to fall back into and burn within the Earth's atmosphere, Vanguard 1 continues to orbit the Earth making the oldest man-made object in space. It is expected to continue doing so for approximately 240 years. The success of Vanguard demonstrated the potential of solar-powered satellites [47–49].

Less than 5 years after the launch of Vanguard, a USA/UK/France coalition launched the Telstar 1 satellite on July 10, 1962, to attempt to establish communications over the Atlantic Ocean [50]. Telstar 1 was covered with a number of solar panels consisting of 3600 solar cells that were capable of producing 14 W of electrical power. Telstar I successfully relayed through space the first television pictures, telephone calls, and telegraph images, and it provided the first live transatlantic television feed. Although no longer functional, like Vanguard 1, it too continues to orbit the Earth. However, Telstar 1 has been followed over the years by twenty subsequent Telstar communication satellites. With each successive generation of Telstar and the myriad of other satellites developed over the years, the trend was for ever-larger satellites with more capability and consequently greater power demands [51,52].

This demand drove much-continued improvement of PV over the next several decades, but the promise of terrestrial solar power continued

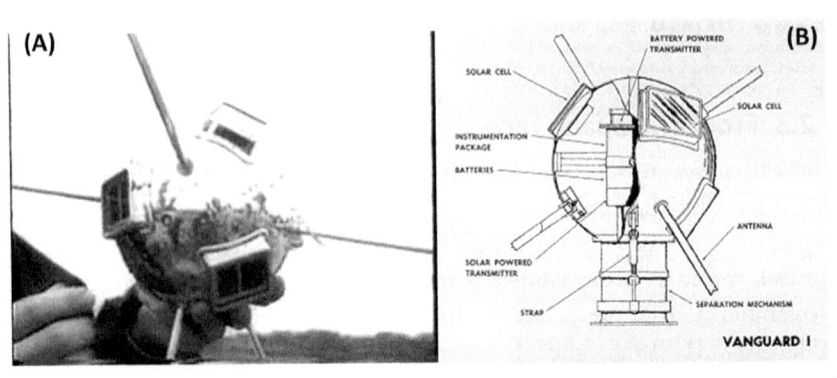

Figure 1.2 (A) Photograph and (B) schematic of Vanguard 1. *Courtesy: US Naval Research Laboratory.*

to garner attention during this time, especially during the oil crisis in 1973 when the Organization of Petroleum Exporting Countries proclaimed an oil embargo and after the Three Mile Island accident in 1979. Silicon solar cell technology continued to improve and manufacturing costs decreased. It was during this period that the Si solar cell "learning curve" was clearly demonstrated. A "learning curve" is a visual method to portray the relationship that exists between the cost of solar power and the deployment of solar power. In other words, a symbiotic relationship exists between the price and the number of installations. The Si solar cell learning factor was 21.5% from 1976 to 2015. Crystalline Si solar cell efficiencies went from approximately 14% to over 20% from 1976 to 1986. In that same period, their cost went from $80/W to below $10/W. The widespread use of terrestrial solar power was much more viable in areas that had good solar insolation and high electricity costs. By 2015, total annual deployment had reached the 100 GW scale [40,42,46].

1.2.4 End of the 20th century: New materials and new challenges

During the 1970s, notorious for the 1973 oil embargo and 1979 Three Mile Island nuclear accident, U.S. President Jimmy Carter opened the Solar Research Institute (SERI) in Golden, Colorado on July 5, 1977. After 14 years as a solar institute, SERI achieved national laboratory status in 1991 under President George H.W. Bush; SERI became the US Department of Energy's (DOE) National Renewable Energy Laboratory (NREL) [53]. It was during this period that the landscape of PV began to not only broaden but also to diversify from a materials perspective [3,4,6,42,46].

New thin-film approaches to photovoltaic energy conversion were starting to achieve some impressive results. By 1990, polycrystalline thin-film devices based upon cadmium telluride (CdTe) and $CuInGa(S,Se)_2$ (CIGS) had both exceeded the 10% conversion efficiency threshold [54–56]. These materials presented the promise of less reactive materials, low-cost deposition methods, and even flexible substrates [3,5,13,22,23]. Although there are still some environmental stability challenges to address, these materials pose a challenge to the dominance of crystalline and polycrystalline Si for terrestrial solar deployment [57–61].

However, perhaps the most impressive new solar cell system to emerge at this time was group 13–group 15 (III–V) solar cells. Gallium arsenide (GaAs) has been a material of interest to the solar cell community for many years [3,4,62]. In 1955, a Radio Corporation of America group was funded by the US Army Signal Corps, and later by the United States Air Force (USAF), to work on the development of GaAs-based solar cells; this binary alloy has a nearly ideal direct bandgap of 1.42 eV for operation in a terrestrial solar spectrum [62,63]. It also has favorable thermal stability and radiation resistance as compared to silicon. However, it took nearly 30 years of efficiency improvements until the use of GaAs-based cells could be justified due to its much higher costs. The arena in which this premium could most easily be justified was in space power. At present, the vast majority of cells being launched for space solar power generation are MJ III–V devices [30,62–64].

The lattice-matched heterojunction cells developed in the early 1970s led to the acceptance of GaAs as a viable photovoltaic material. The USAF launched the manufacturing technology for GaAs Solar Cells (or ManTech) program in 1982. This program was designed to develop metal-organic chemical vapor deposition (MOCVD) techniques necessary for the large-scale production of large quantities of GaAs solar cells [31,62–64]. The other significant development in GaAs technology was in the use of alternative substrates. In 1986, the Air Force supported work by the Applied Solar Energy Corp. on the growth of GaAs cells on Ge substrates. This was possible due to the similarity in the lattice constants and thermal expansion coefficients of the two materials. This resulted in an improvement in the mechanical stability of the cells and a lowering of the production costs [30,62–64].

The development of GaAs-based cells continued throughout the late 1980s with the primary focus being on the development of MJ approaches to photovoltaic conversion. The initial work focused on mechanically stacked cells; however, this quickly transitioned into epitaxially grown dual-junctions with tunnel junctions in between. Laboratory cell efficiencies doubled over nearly two decades as liquid phase epitaxy was replaced by MOCVD and molecular beam epitaxy (MBE) processing (see Fig. 1.3) [63–65].

1.3 Background of photovoltaic materials and devices

This section delves into details of a variety of materials and devices for solar cell applications. We briefly review key developments of Si,

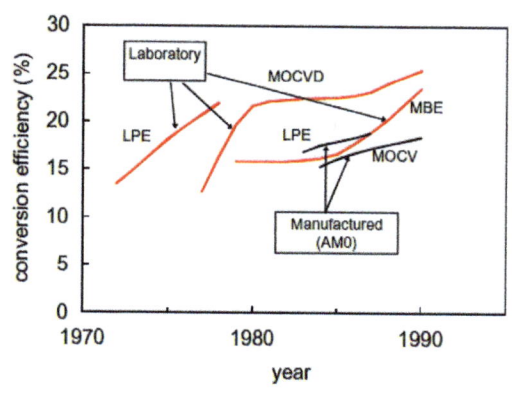

Figure 1.3 Chronological improvements in the efficiencies of GaAs solar cells fabricated by liquid phase epitaxy (LPE), metal-organic chemical vapor deposition (MOCVD), and molecular beam epitaxy (MBE). *Reproduced from M. Yamaguchi, High-efficiency GaAs-based solar cells, in: M.M. Rahman, A.M. Asiri, A. Khan, I. Inamuddin, and T. Tabbakh (Eds.), Post-Transition Metals, InTech, 2021. Open, DOI:10.5772/intechopen.94365 under the terms of the Creative Commons Attribution 3.0 Unported license (CC BY 3.0).*

GaAs, and related III−V materials, as well as thin-film inorganic materials. Recently-developed hybrid technologies including DSSCs, organic (including polymer) PV or solar cells (OPV or OSCs), and PSCs will be briefly surveyed given that the majority of the book provides coverage of these materials. We will then survey advanced concepts such as concentrators, MJ, and nanowires/quantum dots (QDs). Fig. 1.4 provides a visual summary of the four generations of PV technologies currently in use or under development [4].

1.3.1 Silicon solar cells: A short review of photovoltaic devices and recent technology advances

When a junction is created between the *p*-type layer of silicon with an excess of positive carriers or "holes" and an *n*-type layer with excess electrons, the dominant carrier type on each side will diffuse across the junction to the concentration gradient present [4,42]. This will continue until the original charge imbalance is rectified leaving a "built-in" potential difference across the junction. If this junction is subsequently illuminated, any photons that have sufficient energy to be absorbed (i.e., energy greater than the bandgap energy of the semiconductor) and which are absorbed in or near the junction where the potential difference exists will be driven by the built-in field, one way or another, based on their charge

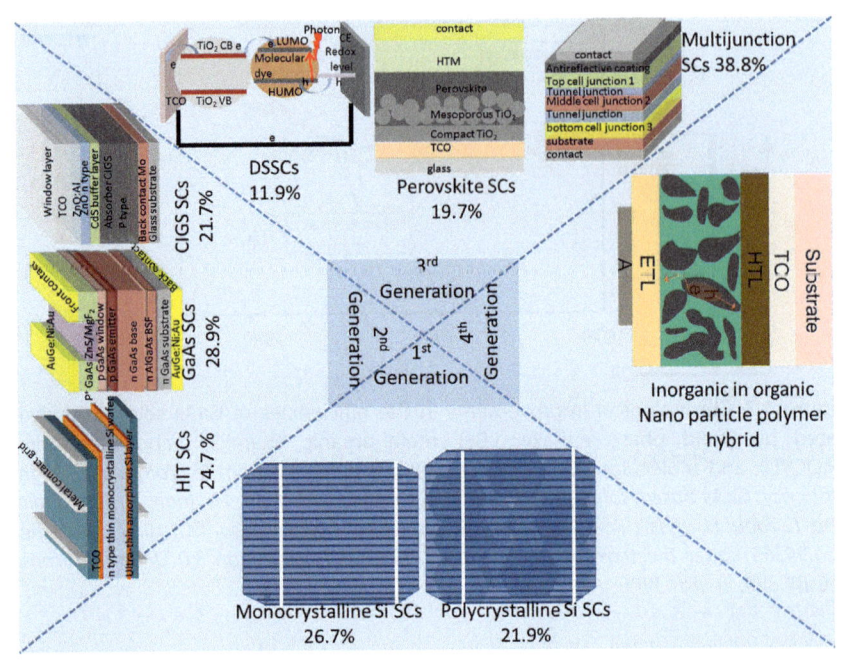

Figure 1.4 Schematic diagram showing the evolution of different generations of solar cells with respective device structures and current photovoltaic efficiencies for first through third generations. *Reprinted with permission from A. Sahu, A. Garg, A. Dixit, A review on quantum dot sensitized solar cells: Past, present and future towards carrier multiplication with a possibility for higher efficiency, Solar Energy 203 (June 2020) 210–239, copyright (2020) Elsevier.*

(i.e., electrons toward the n-side and holes toward the *p*-side). This resulting flow of charges is called the photocurrent (Fig. 1.5).

The electrical behavior of the basic *p-n* junction solar cell is characterized by its nonlinear current versus applied voltage characteristics or $I-V$ curve. For a solar cell, the changes in this $I-V$ curve under illumination are what is used to measure its basic properties such as short-circuit current (density) (I_{sc} (J_{sc})), open-circuit (or built-in) voltage (V_{oc}), and its fill factor (FF) or rectangular area beneath the curve that is related to its photovoltaic efficiency (η) or PCE (Fig. 1.6). Shockley and Queisser published a determination of the maximum theoretical efficiency (i.e., Shockley–Queisser limit or detailed balance limit) of single *p-n* junction solar cell in 1961 [66]. This calculation placed the maximum solar conversion efficiency of approximately 33.7% for a *p-n* junction; it assumed a 1 sun intensity (not concentrated) and an air mass (AM) 1.5 spectrum; see discussion in the next paragraph.

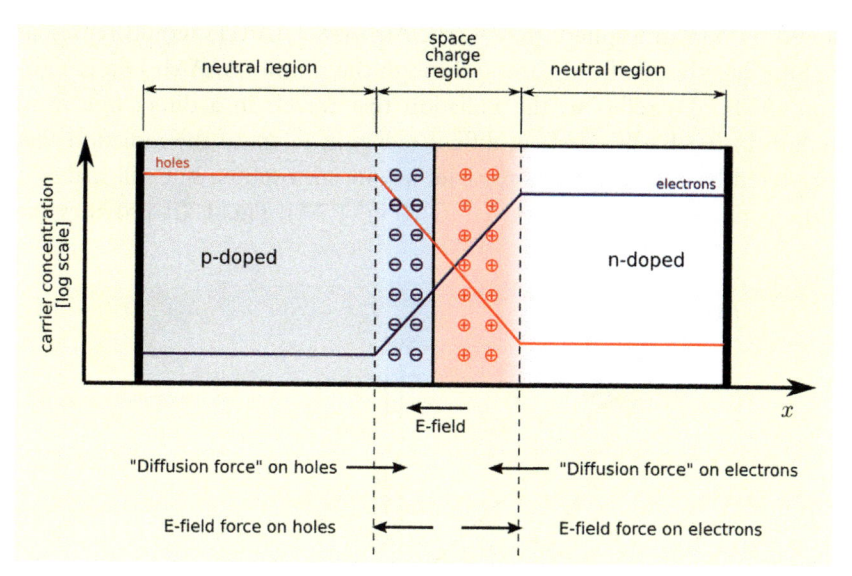

Figure 1.5 A *p-n* junction in thermal equilibrium with zero bias voltage applied. Electron and hole concentrations are indicated by blue and red lines, respectively. Gray regions are charge neutral. Light red zone is positively charged. Light blue zone is negatively charged. *From The Noise at https://commons.wikimedia.org/wiki/File:Pn-junction-equilibrium.png under the terms of the Creative Commons Attribution 3.0 Unported license (CC BY 3.0).*

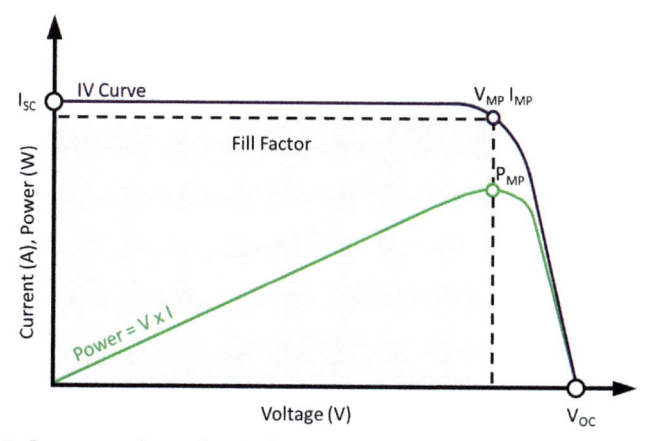

Figure 1.6 Current–voltage (*I*–*V*) characteristics of a solar cell with the figures of merit illuminated. *Courtesy: NASA.*

As sunlight passes through the atmosphere it is filtered due to both absorption and scattering. The more "air mass" it passes through the more it is filtered [67]. Space radiation (AM0) indicates that sunlight is not filtered by our atmosphere [62,64,68]. Air mass 1 (AM1) would represent the Sun directly overhead passing through the minimum AM. Direct radiation (AM1.5D) refers to the radiation that travels in a direct line from the Sun to the Earth's surface; diffuse radiation refers to the radiation that is scattered by molecules and particles in our atmosphere but still makes it to the surface of the Earth (AM1.5G) (Fig. 1.7A). Fig. 1.7B provides the

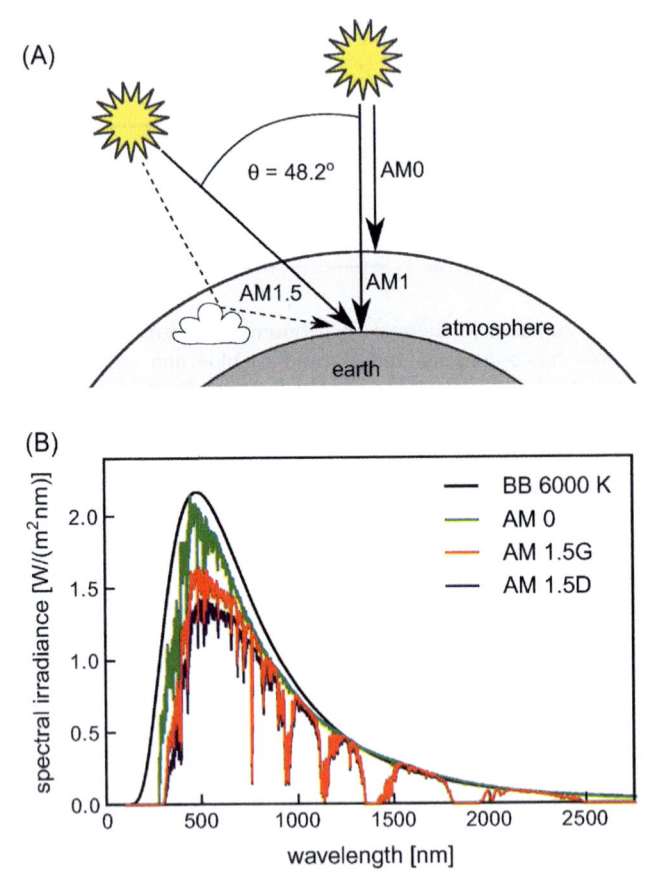

Figure 1.7 (A) Schematic representation of the spectral irradiance outside the Earth's atmosphere (AM0) and on the Earth's surface for direct sunlight shown by a solid arrow (AM1); direct sunlight (AM1.5D) together with the scattered contribution from atmosphere integrated over a hemisphere (AM 1.5G), solid and dashed arrow, respectively; (B) Spectral irradiance according to ASTM G173-03 (https://www.astm.org/standards/g173, accessed June 11, 2023) in comparison to the spectrum used by Shockley and Queisser of a blackbody with a surface temperature of 6000 K (BB 6000 K). *Reproduced with permission from S. Rühle, Tabulated values of the Shockley−Queisser limit for single junction solar cells, Solar Energy 130 (June 2016) 139−147, copyright (2016) Elsevier.*

spectral irradiance as a function of wavelength for each type of solar radiation [67].

A reasonable approximation of the solar spectrum can be achieved using a blackbody radiation spectrum with a temperature of 6000 K (Fig. 1.7B); this is the Shockley−Queisser limit for an ideal solar cell (i.e., bandgap of 1.4 eV) under AM1.5 illumination with a maximum electricity production of 337 W/m^2 [66,67]. Fig. 1.8 shows two different versions of maximum PCE of several single p-n junction solar cells compared to a 6000 K blackbody spectrum calculated by the Shockley−Queisser model [66], as a function of bandgap; see discussion above. Fig. 1.8A includes calculated efficiencies that include nonradiative recombination loss, characterized by an external radiative efficiency (ERE), the ratio of radiatively recombined carriers and all recombined carriers [63]. At the Shockley−Queisser limit, ERE = 1 [63,65,67]. Fig. 1.8B includes a greater variety of solar cell devices in a simpler plot. Note the difference scales in each subfigure's x-axis (bandgap) [67]. Detailed discussions about the various solar cell technologies represented in Fig. 1.8 are included in the remainder of this section.

Silicon PV is the most mature of solar cell technologies; it is still used for the vast majority (>90%) of consumer, utility, and aerospace applications. Silicon is the second most abundant element in the Earth's crust, is nontoxic, and has a practical bandgap (1.1 eV) [3,69]. As discussed above, by the early 1960s, silicon solar cells were approximately 11% efficient, relatively inexpensive, and well suited for low power and limited lifetime application [62,69]. There have been many enhancements to silicon cells over the years to improve their efficiency. Textured front surfaces for better light absorption, extremely thin cells with back surface reflectors for internal light trapping, and passivated cell surfaces to reduce losses due to recombination effects are just a few examples [69]. The present single-junction efficiency record for monocrystalline Si solar cells is 26.3% [2,70], about 5% higher (27.6%) for concentrator cell technologies [2]. This can be compared to the intrinsic limit of approximately 30%, the maximum thermodynamic efficiency of a single-junction solar cell, with a bandgap of 1.1 eV, calculated by Shockley and Queisser [2,67].

A lower-cost alternative, although less efficient than standard crystalline Si (c-Si) technology is polycrystalline- or multicrystalline-Si (p-Si or mc-Si) [71,72]. This material is manufactured by pouring liquid silicon into a mold. Upon solidification, multicrystallites form with associated grain boundaries. The resulting blocks of material are sliced into

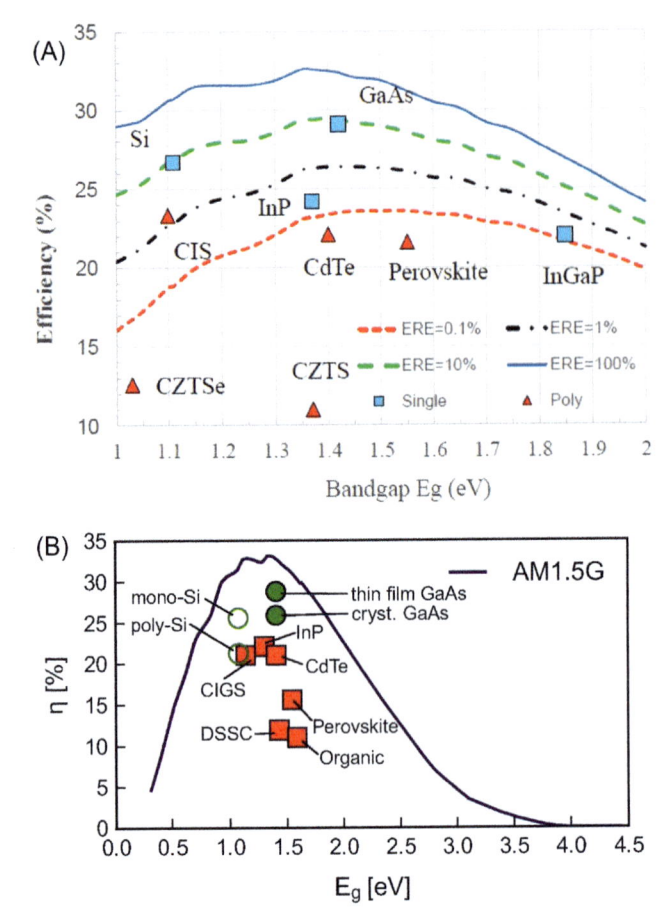

Figure 1.8 (A) Calculated and recently obtained efficiencies of single-junction: Single-crystalline (blue squares) and polycrystalline (red triangles) solar cells. The nonradiative recombination loss is characterized by external radiative efficiency (ERE), which is the ratio of radiatively recombined carriers against all recombined carriers; ERE = 1 at the Shockley–Queisser limit; (B) The maximum PCE for a solar cell operating at 298.15 K and illuminated with the AM1.5G spectral irradiance (ASTM 173-03, see Fig. 1.7B caption) in accordance with standard solar test conditions as a function of band gap energy. Record efficiencies of laboratory cells differentiating between homojunction (circles) and heterojunction devices (squares) with indirect band gaps (empty symbols) and absorbers with direct optical transitions (full symbols). *(A) Reproduced from M. Yamaguchi, High-efficiency GaAs-based solar cells, in: M. M. Rahman, A.M. Asiri, A. Khan, I. Inamuddin, and T. Tabbakh (Eds.), Post-Transition Metals, InTech, 2021. Open, DOI:10.5772/intechopen.94365 under the terms of the Creative Commons Attribution 3.0 Unported license (CC BY 3.0). (B) With permission from S. Rühle, Tabulated values of the Shockley-Queisser limit for single junction solar cells, Solar Energy 130 (June 2016) 139–147, copyright (2016) Elsevier.*

suitable wafers. The light absorption properties along with its simple manufacture and broadly tunable morphology have made it an attractive photovoltaic material for some time now [62,71,72]. Due to the defects associated with the grain boundaries, the best polycrystalline Si solar cell efficiencies stand at 20.3%, 25% less than its monocrystalline Si counterpart [67,71,72].

1.3.2 Gallium arsenide and III−V materials: Higher efficiency and multijunction devices

The highest efficiency solar cells are based upon III−V materials such as GaAs, AlGaAs, and GaInP [30,63]. Like silicon, these crystalline materials are grown on wafers, but they have distinct advantages: (1) They are direct-band gap materials, as opposed to silicon's indirect bandgap; (2) they are heterojunction devices (the *p* and *n* sides consisting of different semiconductor compounds), as opposed to Si homojunction devices, which introduces some advantages with regard to the open-circuit voltage; and (3) they are capable of having multiple *p-n* junctions on top of one another with different stoichiometries or elemental concentrations, which can be connected in series via tunnel junctions. This allows for a much higher η or PCE due to the use of a larger portion of the solar spectrum than could be achieved by a single junction. This new multijunction approach pushed III−V cell efficiencies to a whole new level. By the year 2000, triple-junction III−V solar cells had broken the 30% barrier under 1-sun illumination and would soon go even higher under solar concentration, see below. Continuing on this path, by 2020, 6-junction cells under concentration have exceeded the 47% threshold [3,63].

The drawback of the III−V material system is that they are very expensive to produce. The best cells require an ultrahigh vacuum to produce and involve the use of very toxic metal-organic precursors. As these III−V multijunction solar cells began to emerge, they very quickly displaced silicon as the solar cell material system of choice for space power systems. Unlike in a terrestrial application where solar energy must compete with a wide variety of other options in terms of cost, for space power producing electricity from the Sun's radiation is generally the best, if not the only option; thus, spacecraft developers will pay a premium for the highest efficiency solar cells, in this arena where watts per kilogram rules the day. It is an important feature that III−V materials have proven themselves to be remarkably radiation tolerant, particularly for use in space [31,62−64].

Investigations into further efficiency improvement toward the end of the 20th century turned toward the development of multiple junction cells and concentrator cells. Much of the development of MJ GaAs-based PV was supported by a cooperative program funded by the USAF (ManTech program, Space Vehicles Directorate, and the Space Missile Center) and NASA [30,31,64]. This work resulted in the development of a "dual-junction" cell that incorporates a high-bandgap GaInP cell grown on a GaAs low-bandgap cell. The 1.85 eV GaInP converts higher energy photons and GaAs converts lower energy photons. Fig. 1.9 illustrates the bandgap energy versus lattice constant for various semiconductors, emphasizing the utility of lattice-matched growth of III–V materials for MJ PV cells to produce optimal efficiencies [3,63].

MJ technology has provided the most dramatic increase in performance; many groups have reported laboratory efficiencies well over 30% under AM1.5 radiation. Industrial lot averages are already above 28% [73]. The move toward a quadruple-junction solar cell had been

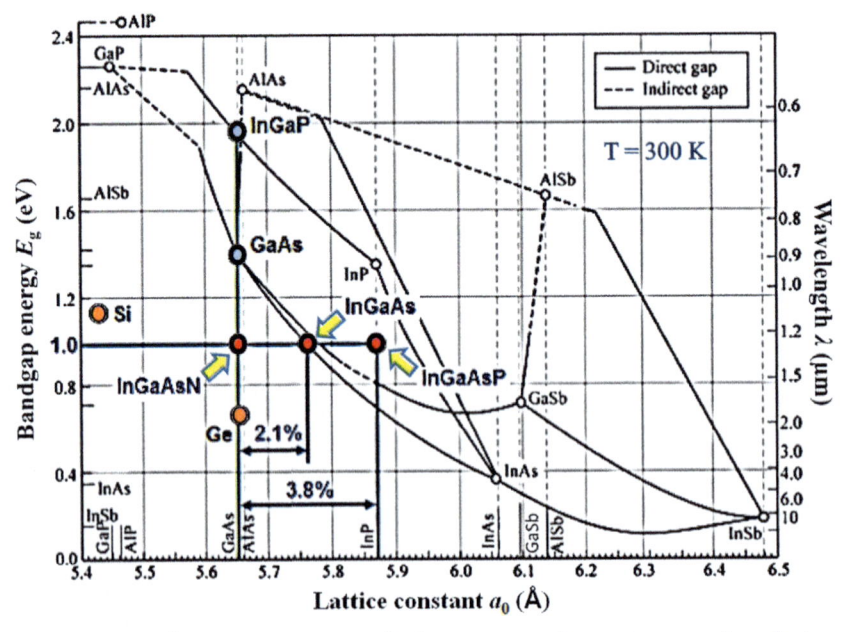

Figure 1.9 Bandgap energy versus lattice constant for various semiconductors, highlighting the utility of lattice-matched growth of III–V materials for PV cells. Reproduced with permission from A.K.-W. Chee, On current technology for light absorber materials used in highly efficient industrial solar cells, Renew. Sustain. Energy Rev. 173 (March 2023) 113027, copyright (2023) Elsevier.

problematic due to lattice constant constraints. However, there were several successful efforts looking at lattice mismatch approaches to MJ III−V, which can give a better match to the solar spectrum than what is available in an ordinary lattice-matched triple-junction cell. In addition, there is a considerable amount of attention being paid to the use of nanostructures (i.e., QDs, wires, and wells) to improve the efficiencies of III−V devices. Theoretical results and experimental advances have shown that the use of such structures may afford dramatic improvement in cell efficiency [74,75].

The development of MJ III−V devices is illustrated in Fig. 1.10. Commercially available dual-junction GaAs/GaInP cells have an AM0 efficiency of 22% with a V_{oc} of 2.06 V [2,3,63,65,73]. Beginning with an AM0 efficiency for InGaP/GaAs tandem solar cells (31%), calculated idealized efficiencies of 47% and 52% were reported for InGaP/GaAs/InGaNAs (triple junction) and InGaP/GaAs/InGaNAs/Ge (quadruple junction) solar cells, respectively, utilizing terrestrial concentrator photovoltaic (CPV) systems (500 suns) [76].

The high-quality epitaxial growth of active layer materials, such as InGaP and InGaAs, lattice-matched to Ge (see Fig. 1.11), allows for current matching between the subcells to ensure maximum light harvesting

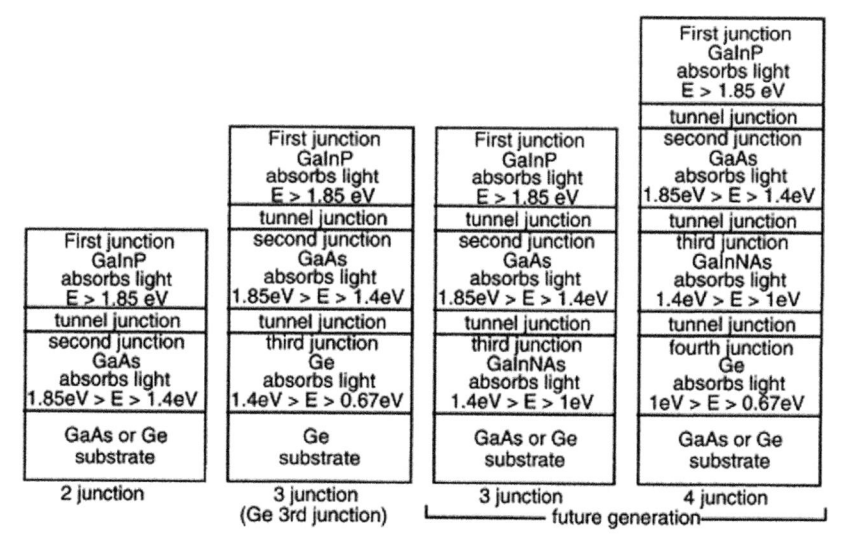

2 junction	3 junction (Ge 3rd junction)	3 junction	4 junction
			First junction GaInP absorbs light E > 1.85 eV
			tunnel junction
	First junction GaInP absorbs light E > 1.85 eV	First junction GaInP absorbs light E > 1.85 eV	second junction GaAs absorbs light 1.85eV > E > 1.4eV
	tunnel junction	tunnel junction	tunnel junction
First junction GaInP absorbs light E > 1.85 eV	second junction GaAs absorbs light 1.85eV > E > 1.4eV	second junction GaAs absorbs light 1.85eV > E > 1.4eV	third junction GaInNAs absorbs light 1.4eV > E > 1eV
tunnel junction	tunnel junction	tunnel junction	tunnel junction
second junction GaAs absorbs light 1.85eV > E > 1.4eV	third junction Ge absorbs light 1.4eV > E > 0.67eV	third junction GaInNAs absorbs light 1.4eV > E > 1eV	fourth junction Ge absorbs light 1eV > E > 0.67eV
GaAs or Ge substrate	Ge substrate	GaAs or Ge substrate	GaAs or Ge substrate

future generation

Figure 1.10 Schematic representation of the structure of dual-, triple-, and quadruple-junction solar cells. *Reproduced with permission from D.J. Friedman, J.F. Geisz, S.R. Kurtz, J.M. Olson, 1-eV solar cells with GaInNAs active layer, J. Cryst. Growth 195 (1998) 409−415, copyright (1998) Elsevier.*

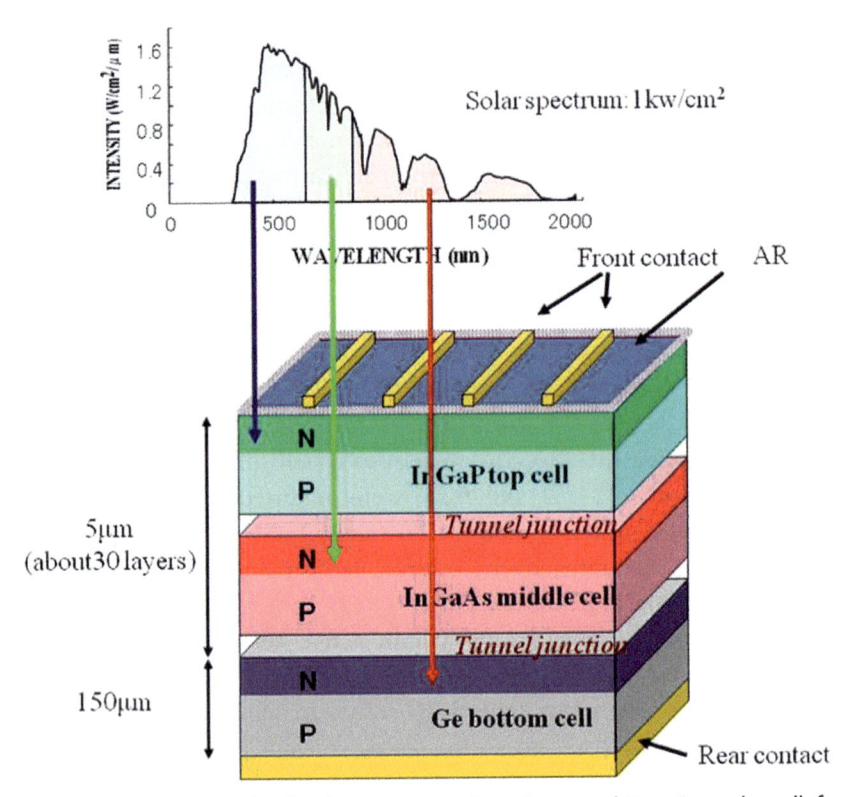

Figure 1.11 Principle of wide photo response by using a multijunction solar cell, for the case of an InGaP/GaAs/Ge triple-junction solar cell. *Reproduced from M. Yamaguchi, High-efficiency GaAs-based solar cells, in: M.M. Rahman, A.M. Asiri, A. Khan, I. Inamuddin, and T. Tabbakh (Eds.), Post-Transition Metals, InTech, 2021. Open, DOI:10.5772/intechopen.94365 under the terms of the Creative Commons Attribution 3.0 unported license (CC BY 3.0).*

over the solar spectrum [63]. They are grown in series connected layers and have been produced with a 28.3% efficiency with a V_{oc} of 2.66 V in production lots, with laboratory cells of 29%. Emcore, Inc., and Spectrolab, Inc., currently produce cells that are commercially available in the 25%–27% range. Their high efficiencies are due to their ability to effectively convert a larger portion of the available sunlight [30,63,64].

Given the need to increase both the efficiency and radiation resistance of GaAs and related III−V materials, a significant increase in the world-wide research effort resulted in a number of III−V materials available to researchers, cell designers, solar cell vendors, and mission planners [12,31,63]. The most straightforward (ternary) materials include indium phosphide (InP), aluminum gallium arsenide (AlGaAs), and indium gallium phosphide (InGaP). In keeping with the paradigm of looking toward the future, Fig. 1.12 plots world record efficiencies of single-junction devices looking back to approximately 1960 and 15 years into the future. The solid and dashed lines in Fig. 1.12 are the fitted trajectories using Eq. (1.1) from the work of Goetzberger et al. [77].

$$\eta(t) = \eta_L \cdot \left\{ 1 - \exp\left(\frac{(t_0 - t)}{c}\right) \right\} \tag{1.1}$$

In Eq. (1.1), $\eta(t)$ is the time-dependent efficiency, η_L is the limiting asymptotic maximum efficiency, t_0 is the year for which $\eta(t)$ is zero, t is the calendar year, and c is a characteristic development time. Fitting of the curve was done with three parameters (η_L, t_0, c) for each of the four materials as follows: GaAs (30.5, 1953, 20), InP (28, 1965, 17), AlGaAs (22, 1972, 15), and InGaP (23, 1975, 12). It should be noted that in the original report, Goetzberger et al. stated that multiple functions could be utilized; also, the function was employed to predict the future development

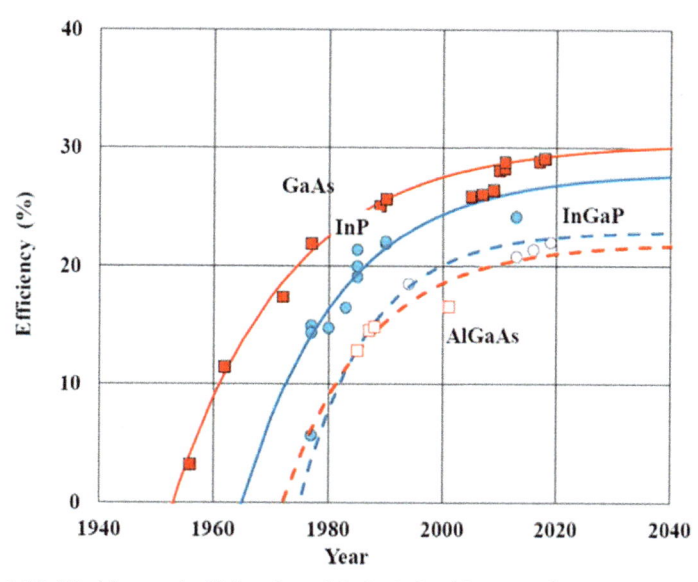

Figure 1.12 World record efficiencies of GaAs, InP, AlGaAs, and InGaP single-junction solar cells over 80 years. *Reproduced from M. Yamaguchi, High-efficiency GaAs-based solar cells, in: M.M. Rahman, A.M. Asiri, A. Khan, I. Inamuddin, and T. Tabbakh (Eds.), Post-Transition Metals, InTech, 2021. Open, DOI:10.5772/intechopen.94365 under the terms of the Creative Commons Attribution 3.0 Unported license (CC BY 3.0).*

of a number of non-III–V materials including c-Si, multiple thin-film materials, organic cells, and a "new material" [77]; these are now described as emerging materials (Section 1.3.4).

1.3.3 Thin-film solar cells: Traditional inorganic materials

1.3.3.1 Thin-film solar cells: Introduction and background

Thin-film cells have been investigated for over fifty years. Early work began with Cu_2S-CdS cells that were developed in the mid-1970s [3,4,42,58,59,62]. As mentioned above, the traditional inorganic thin-film solar cell (TFSC) materials that have received the most attention are CdTe, CIGS, and amorphous Si (a-Si or a-Si:H) [54–61,78,79]. Thin-film cell technologies are quite attractive due to the fact that many of the proposed fabrication methods are relatively inexpensive and lend themselves well to mass production [54,58,62]. These cells can be made to be extremely lightweight and flexible [57,80], especially if produced on polymeric [60,81] or metal foil substrates [82–85]. In fact, an important metric for space solar arrays is mass-specific power (MSP); lightweight substrates give all lightweight solar cell technologies an advantage [30,48,62,64,80–88]. While we are considering high MSP traditional inorganic materials technologies in this section, organic and hybrid solar cells for space are covered in other published sources [89–91].

As discussed above, the founding of the DOE National Laboratory, which is currently known as NREL [53], coincided with the beginning of a significant diversification in solar cell materials and device technologies. The steady progress of efficiency improvements and the advent of new device technologies since 1976 is illustrated by NREL's ubiquitous best research cell efficiency chart (Fig. 1.13). Throughout the remainder of this chapter, we will refer back to this figure for context.

1.3.3.2 Thin-film silicon solar cells

Unlike c-Si devices, where Si has a well-ordered crystalline lattice structure [92], a-Si:H does not exhibit long-range order and thus contains a much higher defect concentration. As there is no long-range crystalline order in a-Si, not all the atoms are fourfold coordinated [57,58]. The unused valence orbitals are dangling bonds, which act as defect sites. Thus, the material is passivated with H, resulting in a-Si:H, which has a low enough defect density to make solar cells. The change in coordination of the amorphous material results in very different absorption properties; a-Si:H is a direct bandgap semiconductor, with a bandgap of 1.7 eV. Due to this material's defect density, it takes orders of magnitude higher

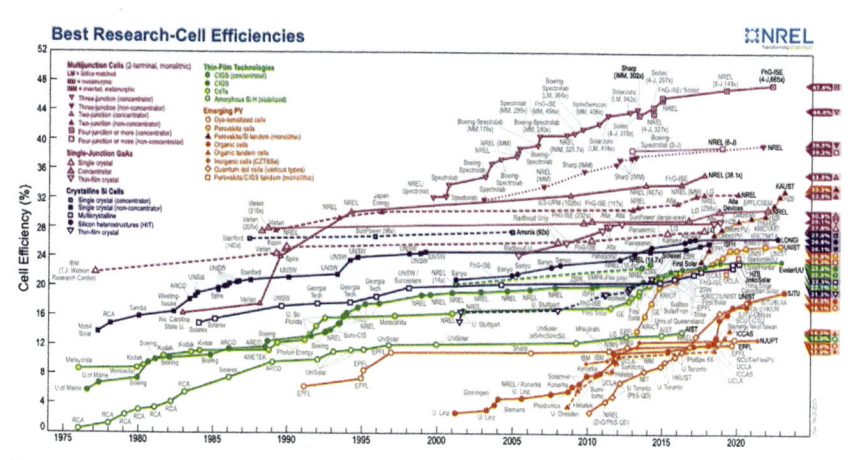

Figure 1.13 National Renewable Energy Laboratory champion solar cell efficiency chart. *From US DOE NREL. https://www.nrel.gov/pv/cell-efficiency.html (accessed September 16, 2023), public domain as a US Government agency work product.*

dopant levels to render the material *p*- or *n*-type, on the order of several percent. The majority of the dopants are not electrically active, which makes *p-n* junction devices unsuitable. Instead, a-Si:H devices are made in a *p-i-n* or *n-i-p* configuration, where the intrinsic (undoped) layer acts as the absorber, creating the electron-hole pairs that can be swept away in the electric field created by the adjacent *p*- and *n*-layers [58].

Although inexpensive and easy to deposit, a-Si:H efficiencies are quite low; the record single-junction cell has an efficiency of 10.2% [93]. The device performance degrades under operation conditions (i.e., under illumination and current injection), in what is known as the Staebler–Wronski effect [94]; thicker films experience enhanced degradation. To mitigate this, various improvement strategies have been utilized such as limiting layer thickness, fabricating MJ cells where many layers in series are used to absorb different portions of the solar spectrum, and replacing amorphous Si:H with nano- or microcrystalline silicon, which is more resistant to degradation. Despite these advances, the record efficiencies for double and triple-junction cells with amorphous and nanocrystalline absorbers are only 12.7% [95] and 14.0% [96], respectively.

1.3.3.3 Chalcogenide thin-film solar cells

The two most inorganic important TFSC absorbers (CdTe and CIGS) are metal chalcogenides, materials with one or (often) more heavier-than-oxygen chalcogen (e.g., S, Se, and Te) anions. Extensive research has

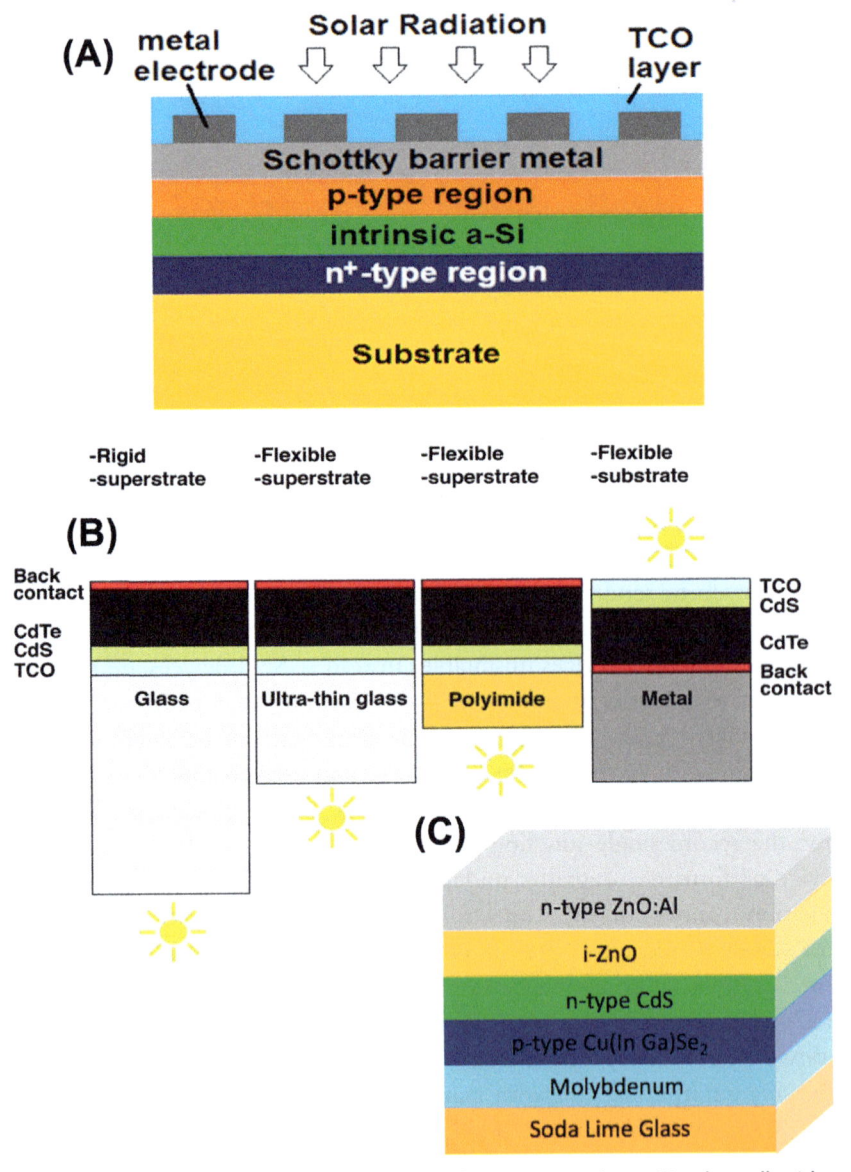

Figure 1.14 (A) An illustration of a Schottky barrier amorphous Si solar cell with a highly doped *p*-type region adjacent to the Schottky barrier high work function metal. (B) Schematic of superstrate and substrate CdTe devices (the substrates' width is not proportional but indicative of the differences between the devices). Both in substrate and superstrate configurations, light enters through wide bandgap window material (CdS). In the substrate structure, opaque substrates are used, and CdS does
(Continued)

been conducted into terrestrial TFSCs based on CdTe [57,59] and other chalcogenide materials, in particular, chalcopyrite (I−III−VI$_2$) [4,55,57]. Kesterite (I$_2$−II−IV−VI$_4$) materials, composed of Earth-abundant elements, are also being explored as alternative TFSC absorbers [97]. Overall results utilizing these materials for PV devices have produced varying success in terms of efficiency, stability, and ramping up actual production; typical inorganic thin-film device structures are depicted in Fig. 1.14 [57,58].

From a terrestrial perspective, CdTe solar cells are arguably the most successful thin-film PV technology. First Solar, a CdTe manufacturer based in Toledo, Ohio, USA, is the only thin-film PV company to regularly appear on the annual top ten list of solar manufacturers [98]. As an absorber layer, CdTe is a strongly absorbing, thermodynamically stable, and direct bandgap material, with a near-ideal 1.44 eV bandgap. Over four decades of research have resulted in the evolution of the device from a 4.1% CdTe/CdS heterojunction [99] to a >22% efficient device (19% large area) produced by First Solar [59], enabled by a graded CdTe/CdSe$_x$Te$_{1-x}$ junction and a high transparency MZO buffer layer [100]. This performance is 35% higher than the flex glass superstrate device [101].

CuIn$_{1-x}$Ga$_x$Se$_2$ is another widely studied chalcogenide thin-film absorber. The quaternary material is commonly deposited via vacuum coevaporation of all constituents on a substrate heated to 400°C−600°C [57,102,103]; alternate deposition methods will typically include one or more steps at higher temperature to optimize device performance [60,85]. Table 1.1 provides examples of CIGS devices on glass that held terrestrial efficiency records for a polycrystalline thin-film absorber that approached 23% (AM1.5) efficiency [103,104]. To provide a broader device comparison, Table 1.1 also includes examples of higher(est)-efficiency a-Si, organic, and hybrid devices on glass as well as recent reports on flexible, lightweight substrates (or superstrates) for CIGS and CdTe [59,93,101,103−113]. Finally, Solar Frontier, a leading Japanese CIS

◀ not need to undergo high-temperature processing. (C) Structure of a CIGS solar cell. *(A and C) Reproduced with permission from T.D. Lee, A.U. Ebong, A review of thin film solar cell technologies and challenges, Renew. Sust. Energ. Rev. 70 (April 2017) 1286−1297, copyright (2017) Elsevier; (B) Reproduced with permission from J. Ramanujam, D.M. Bishop, T.K. Todorov, O. Gunawan, J. Rath, R. Nekovei, E. Artegiani, A. Romeo, Flexible CIGS, CdTe and a-Si:H based thin film solar cells: A review, Prog. Mater. Sci. 110 (2020) 100619, copyright (2020) Elsevier.*

Table 1.1 Representative CIGS and CdTe devices on flexible substrates (or superstrates[a]); efficiencies on glass substrates and others are included for comparison.

Material	Substrate or superstrate[a]	Eff. (h) (%)	V_{oc} (mV)	J_{sc} (mA/cm^2)	FF (%)	Year	Alkali metals/ buffer layer	References
CIGS	Glass	22.6	741	37.8	80.6	2016	KF + RbF	[104]
		22.9	746	38.5	79.7	2019	Cs	[103]
	Polyimide	20.4	736	35.1	78.9	2013	KF	[105]
		20.8	734	36.7	77.2	2019	RbF	[106]
	Stainless steel	17.5	601	40	72.5	2020	–	[107]
		18.0	692	34.5	75.5	2018	RbF + Ni/Cr	[108]
CdTe	Glass[a]	22.1	887	31.7	78.5	2015	–	[59]
	Flex glass[a]	16.4	831	25.5	77.4	2015	–	[101]
	Polyimide[a]	13.6	846	22.3	73.4	2012	–	[109]
	Mo foil	11.5	821	21.8	63.9	2013	–	[110]
a-Si	Glass	10.2	896	16.4	69.8	2015	–	[93]
Perovskite	Glass	20.7[b]	1060	26.0	75.1	2019	–	[111]
Organic	Glass	11.9	744	22.4	71.2	2012	–	[112]
DSSC	Glass	12.5	860	23.2	76.1	2020	–	[113]

[a]Superstrate.
[b]Small area device aged 500 h – large area (1 cm^2) device as prepared = 22.3%.
Source: Adapted from J. Ramanujam, D.M. Bishop, T.K. Todorov, O. Gunawan, J. Rath, R. Nekovei, E. Artegiani, A. Romeo, Flexible CIGS, CdTe and a-Si:H based thin film solar cells: A review, Prog. Mater. Sci. 110 (2020) 100619 and other literature sources.

company, in partnership with Japan's National Research and Development Agency's New Energy and Industrial Technology Development Organization, announced a new record of 23.35% in 2019 [114], an increase of 0.45% over the previous record [103].

1.3.4 Dye-sensitized, organic, and perovskite solar cells

There are multiple later chapters in this book that address emerging inorganic, hybrid, and organic devices (DSS-Cs, PSCs, and OPV) [8–10,115–118], as well as improvements from the utilization of MOFs [12]. Therefore, in this subsection, we will merely provide a broad overview of these third (and fourth) generation devices; see Fig. 1.4 for a diagram of generations of solar cell technologies. The reader is directed to one of many excellent sources including later chapters in this book [8–10,115–118], archival journal publications [1,3,12,72,119–129], and books that address these topics [4–6,130–136]. Fig. 1.15 provides a schematic of typical architectures and mechanisms of DSSCs, OSCs (OPV), and PSCs [129].

Given the eclectic mix of topics, relevant to the development of new PV technologies, the NREL best efficiency chart (Fig. 1.13) provides a

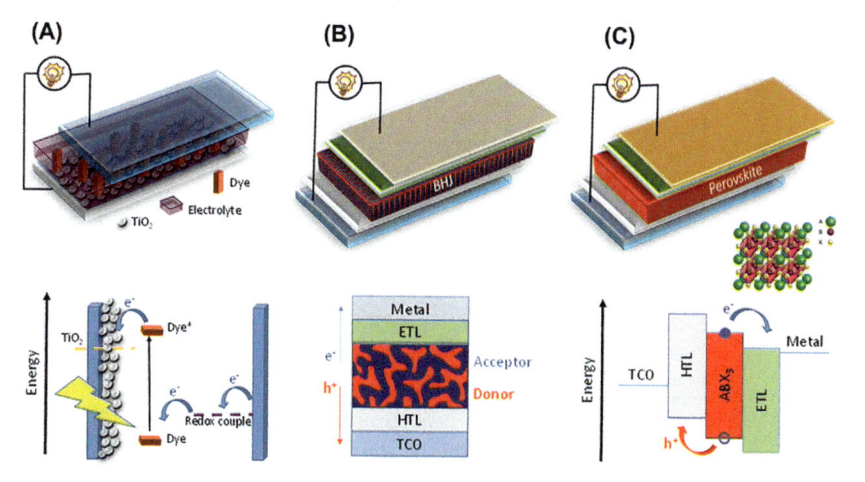

Figure 1.15 Schematic views of the standard architectures (*top*) and working mechanisms (*bottom*) in (A) DSSCs (*left*), (B) OSCs (*center*), and (C) PSCs (*right*). Inset: Generic ABX_3 perovskite structure. *Reprinted from L.M. Cavinato, E. Fresta, S. Ferrara, R.D. Costa, Merging biology and photovoltaics: how nature helps sun-catching, Adv. Energy Mater. 11 (43) (November 2021) 2100520, this article is distributed under the terms of the Creative Commons (CC-BY 4.0) license (http://creativecommons.org/licenses/by/4.0/), which permits unrestricted use.*

convenient (albeit busy) snapshot of the development and improvement of emerging hybrid/organic solar cell technologies. Furthermore, when considering sustainable or green alternative PV devices, the interested reader is asked to consider Fig. 1.16 [129], a record of the efficiency improvements of various device technologies that include bio-based materials. Several observations can be made about the improvements in efficiencies (updated with results from Fig. 1.13) of these technologies that emerged roughly every 10 years going back to the early 1990s for DSSCs, followed by OSCs in the early 2000s and PSCs in the early 2010s.

While the efficiencies of champion DSSCs doubled within 5 years since their first report, performance improvement has essentially remained flat over the past 25 years to approximately 13%. Organic-based devices have experienced a near-order of magnitude improvement from initial reports over 20 years ago to nearly 20% today. The most recent entry into the emerging hybrid/organic family, PSCs, has steadily increased

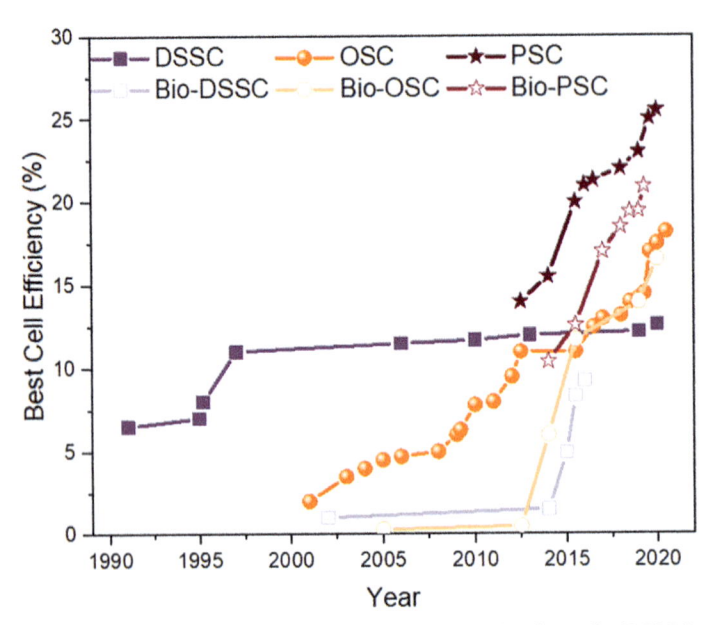

Figure 1.16 Chart of the best performing dye-sensitized solar cells (DSSCs), organic solar cells (OSCs), and perovskite solar cells (PSCs), as well as their bio-based processed versions since 1991. *Reprinted from L.M. Cavinato, E. Fresta, S. Ferrara, R.D. Costa, Merging biology and photovoltaics: How nature helps sun-catching, Adv. Energy Mater. 11 (43) (November 2021) 2100520, this article is distributed under the terms of the Creative Commons (CC-BY 4.0) license (http://creativecommons.org/licenses/by/4.0/), which permits unrestricted use.*

from a quite respectable 13% to nearly 26% today—a near doubling in a decade. Other device technologies that are categorized by NREL as emerging (see Fig. 1.13) will be addressed in follow-on subsections (quantum dot (sensitized) solar cells—QDSCs) (Section 1.3.5.1) and PSC tandems (Section 1.3.6) and later chapters in this book [11,118,119]. Finally, an interesting observation to make regarding PV devices that are bio-inspired [12] or contain biomaterials is that bio-OSCs have nearly matched OSCs for record efficiencies while bio-versions of DSSCs and PSCs are approximately 5% below the all-synthetic devices [129]; this could be explained, in part, by the preponderance of organic moieties in biological materials.

1.3.5 Advanced device concepts and structures

Over the past several decades, significant cell efficiency improvements have resulted from the move toward MJ devices and/or concentrator cells, see Fig. 1.13 [3,4,11,31,63,76]. However, as mentioned above (Section 1.3.2), researchers continue to push the envelope by looking toward new approaches, such as the use of nanotechnology [4–6,22,23,74,75,134,136], to further improve device efficiencies. Fig. 1.17A summarizes the evolution of three generations of PV technology and illustrates several key innovations leading to improvement in device efficiency [4,137].

The subsection begins with an overview of QDSCs (Section 1.3.5.1), another emerging PV technology. Fig. 1.18 provides an overview of the progress over time of PCE for this very broad and heterogeneous class of PV technologies [137]. Several remaining topics—concentrators (Section 1.3.5.2), IB(C)SCs (Section 1.3.5.3), up- and down-conversion (DC) materials (Section 1.3.5.4), and MJ cells (Section 1.3.5.5) are discussed in follow-on subsections; as several of these topics are discussed extensively by other chapters in this book [11,118,119], subsections discussing such technologies will be merely summarized in brief.

1.3.5.1 Quantum dot (sensitized) solar cells

Quantum dots (or nanocrystallites) have been used successfully to improve the performance of devices such as lasers, light-emitting diodes, and photodetectors [138–140]. Quantum dots behave essentially as a three-dimensional potential well for electrons (i.e., the quantum mechanical "particle in a box"). Luque and Marti calculated a theoretical efficiency of 63.2% for a photovoltaic device [141]; an intermediate band (IB) of states,

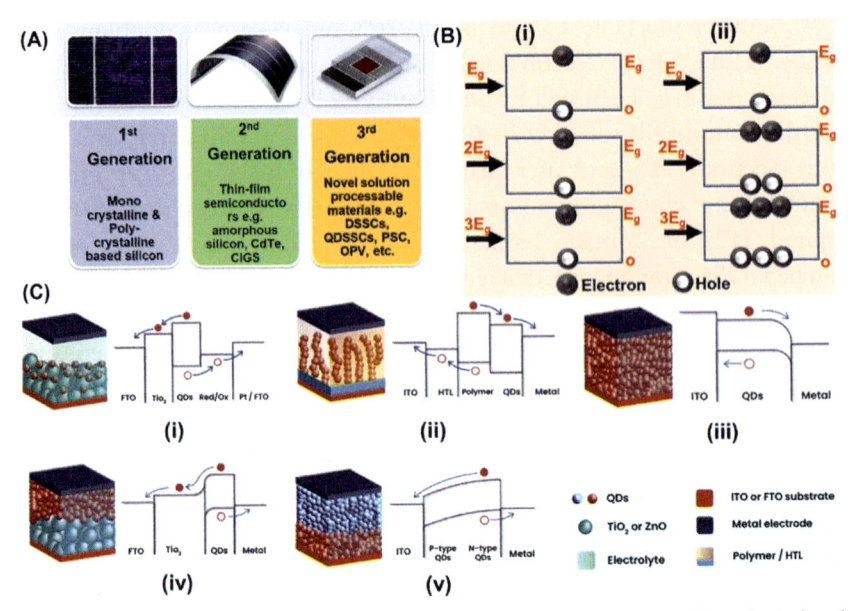

Figure 1.17 (A) Three generations of photovoltaic technologies based on their development and innovation; (B) graphic drawing illustrates generation of two types of carriers [(i) traditional and (ii) multiple exciton generation)] when photon energy is excited; (C) illustrative drawings of various device configurations in the quantum dot (sensitized) solar cells. *Reprinted with permission from A.S. Rasal, S. Yadav, A.A. Kashale, A. Altaee, J.-Y. Chang, Stability of quantum dot-sensitized solar cells: A review and prospects, Nano Energy 94 (April 2022) 106854, copyright (2022) Elsevier.*

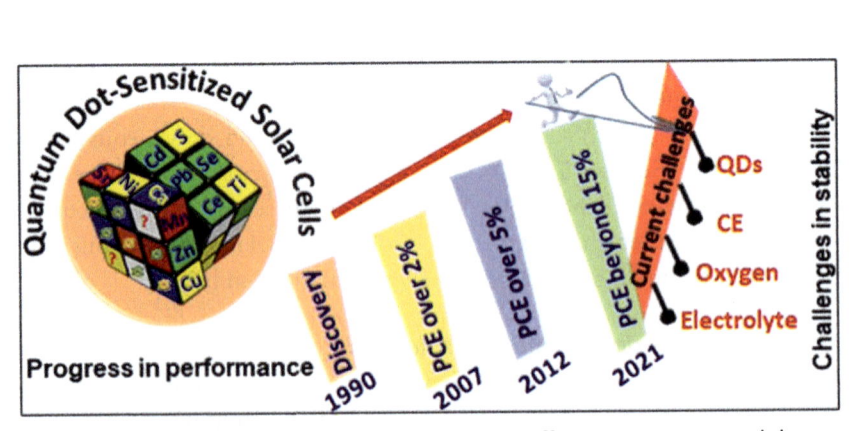

Figure 1.18 Schematic progress in power conversin efficiency over time and the current challenges (on right) affecting quantum dot (sensitized) solar cells device stability. *Reprinted with permission from A.S. Rasal, S. Yadav, A.A. Kashale, A. Altaee, J.-Y. Chang, Stability of quantum dot-sensitized solar cells: A review and prospects, Nano Energy 94 (April 2022) 106854, copyright (2022) Elsevier.*

resulting from the introduction of QDs, results in predicted efficiencies exceeding the Shockley—Queisser model efficiency [66] of not only a single junction but also a tandem cell device [141]. Semiconductor QDs are currently a subject of great interest mainly due to their size-dependent electronic structures, in particular the increased band gap and therefore tunable optoelectronic properties [138–140]. However, the unique properties, such as the size-dependent increase in oscillator strength due to the strong confinement exhibited in QDs and the blue shift in the band gap energy of QDs, are properties that can be exploited for developing photovoltaic devices, offering advantages over conventional solar cells [4,137,139,140,142–149].

As mentioned above, theoretical studies predict a potential efficiency of 63.2% for a single-size QD cell, which is approximately twice as efficient as a typical SOA single-junction, nonconcentrated device available today. For the most general case, a system with an infinite number of sizes of QDs has the same theoretical efficiency as an infinite number of band gaps or 86.5% [12,13,141]. A collection of different-sized QDs can be regarded as an array of semiconductors that are individually size-tuned for optimal absorption at their band gaps throughout the solar energy emission spectrum. This is in contrast to a bulk material where photons are absorbed at the band gap; energies above the band gap result in less efficient photogeneration of carriers. In addition, bulk materials used in photovoltaic cells suffer from reflective losses for convertible photons near the band gap, whereas for individual QDs, reflective losses are minimized. Some recent work has shown that QDs may also offer some additional radiation resistance and favorable temperature coefficients [137].

Another potential approach to improve efficiency would rely on multiexciton generation processes (Fig. 1.17B), this would enable QDSCs to approach a theoretical efficiency of 44% [4,145,150–152]. Typically, based on their working mechanism, QD-based solar cells have been categorized in slightly different manners by multiple authors [4,137,142,145]. For the sake of simplicity, we will utilize the categories of Rasal et al. [137], shown in Fig. 1.17C, to review five types of QD(S)SCs: (1) QDSSCs—a QD-sensitized version of a DSSC [4,136,142,146,149,153] (Fig. 1.17C(i)); see Fig. 1.19 for a more detailed schematic of a typical QDSSC working mechanism; (2) hybrid QD—polymer solar cells [154–156] (Fig. 1.17C(ii)); (3) QD Schottky junction cells [4,149–159] (Fig. 1.17C(iii)); (4) quantum $p-n$ heterojunction [160–162] (Fig. 1.17C(iv)); and (5) $p-n$ homojunction solar cells [163–165] (Fig. 1.17C(v)).

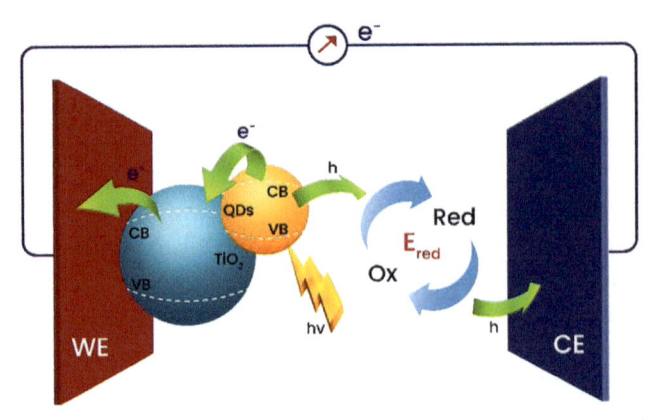

Figure 1.19 Schematic and mechanism of a quantum dot (QD)-sensitized nanocrystalline TiO_2 solar cells with key components labeled: Working (oxide substrate) electrode (WE), QD, nanocrystalline TiO_2, counter (metal) electrode (CE), and electrolyte (not shown). *Reprinted with permission from A.S. Rasal, S. Yadav, A.A. Kashale, A. Altaee, J.-Y. Chang, Stability of quantum dot-sensitized solar cells: A review and prospects, Nano Energy 94 (April 2022) 106854, copyright (2022) Elsevier.*

Examination of Figs. 1.13 and 1.18 show that after more than a decade since initial reports of QDSCs [166,167], PCEs had only reached the low single digits [168,169] in efficiency. Nozik and coworkers sensitized TiO_2 with InAs QDs for a 0.3% PCE [168]; in the following year, Toyoda and coworkers produced a device with a PCE of 2.7% by introducing a passivation layer of wide bandgap zinc sulfide (ZnS) over CdSe-sensitized TiO_2 followed by fluoride anion deposition [169]. Subsequently, doping or alloying of QDs with a suitable metal has been proven to enhance both the PCE and stability of QDSCs [4,137,139,140,143,144,146,149]. For example, in 2012, Santra and Kamat demonstrated a near doubling of PCE to over 5% with Mn-doped CdS QDs that also extended the performance stability of devices [170]. Within 5 years, Du et al. reported QDSCs with a PCE of 11.6%, another doubling of efficiency, comprised of alloyed QDs of Cu−In−Se and ZnSe; the devices utilized porous TiO_2 photoanodes with Zn−Cu−In−Se QDs (In/Cu = 1, Zn/Cu = 0.4, and size of ca. 4 nm) [171]. In 2021, Song et al. reported efficiencies that reached 15.31%, employing Zn − Cu − In − S − Se QD-sensitized films on an enhanced QD-loading on a metal oxyhydroxide-treated TiO_2 film, resulting in decreased recombination of corresponding QDSSCs [172]. A recent review explores the synthesis properties and applications (i.e., QDSC) of I−III−VI$_2$ QDs; it compares binary (i.e., CdSe and PbS) and ternary (i.e., CuInS$_2$ and CuGaSe$_2$) chalcogenide QDs for enhancing the performance of photofunctional devices [173]. While legacy efforts for well-studied binaries give them a performance edge currently, ternaries have superior long-term potential; a visual summary is provided in Fig. 1.20 [173]. Finally, the right-hand side of Fig. 1.18 lists some key challenges and stability issues to be resolved to enable further efficiency and stability improvements for QDSCs [4,137,139,146,148,149,156,161].

Based upon a reading of results summarized by the various review articles referenced in this subsection, it is clear that QDSSC devices, the QD analogy to DSSCs, are the most frequently researched device as fabrication is relatively straightforward [4,6,136,142,146,149,153,173−177]. A related nanotechnology alternative to QD(S)SCs is the intermediate bandgap solar cell (IBSC) [75]. While the long-term potential for PCE of IBSCs is well beyond the SQ limit [66] and toward a theoretical efficiency of 63.2% is intriguing [141], the fabrication technologies reported thus far are quite

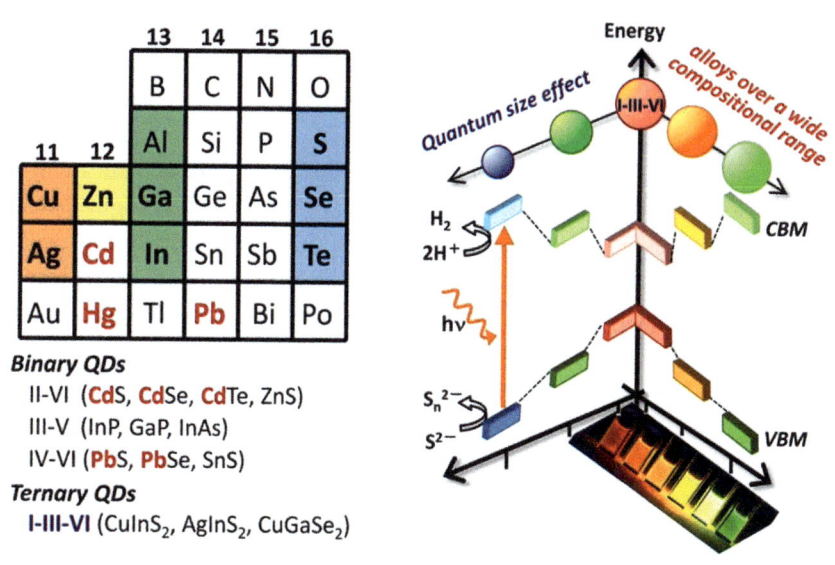

Figure 1.20 (A) Representative binary and ternary semiconductors investigated for solution-phase preparation of quantum dots (QDs). (B) Schematic illustration of the control of the electronic energy structure of I−III−VI QDs for various applications. The energy levels of the conduction band (CB) minimum (CBM) and the valence band maximum (VBM) are tuned by the particle size and chemical composition of QDs. *Reproduced with permission from T. Torimoto, T. Kameyama, T. Uematsu, S. Kuwabata, Controlling optical properties and electronic energy structure of I−III−VI semiconductor quantum dots for improving their photofunctions, J. Photochem. Photobiol. C: Photochem. Rev. 54 (2023) 100569, copyright (2023) Elsevier.*

complex. Such devices will have a prohibitive cost in $/W for some time to come for terrestrial applications [1,3,4,42,77,167,178,179]. We will discuss IBSCs in one of the following subsections (Section 1.3.5.3) and provide examples of IBSC concentrator devices.

1.3.5.2 Concentrator solar cells: Multiple photovoltaics technologies for higher efficiency

As mentioned in the subsection's introduction, the topic of CPV is covered in some depth in a later chapter in this book [11]; we will confine this discussion to a few key points that connect to other aspects of this chapter. Concentrator PV systems can be categorized by the concentration ratio of solar radiation onto the cell. This ratio (known as "suns") indicates the number of times that solar light is concentrated. A previous review [180] has defined three different CPV regimes:
- Low concentration (LCPV): Refers to systems that concentrate light between 1 and 40 times (1−40×) or 1 and 40 suns;
- Medium concentration (MCPV): Systems that concentrate sunlight between 40 and 300 times (40−300×) or 40 and 300 suns; and
- High concentration (HCPV): The concentration level of these systems is greater than 300 suns (>300×).

The majority of solar cells that have been deployed to date are not only made of Si but have also been optimized for direct sunlight (AM1 and AM1.5); see Fig. 1.7. These cells are optimized for maximum solar energy conversion for direct radiation that enters perpendicularly to their surface [3,4,11,69]. However, some diffuse light will still be converted, contributing to the overall electricity generated. In this way, we experience what is referred to as "cosine loss" from the path of the Sun across the sky throughout the day. Only once during the day (i.e., noon), is the Sun directly overhead such that the direct rays from the sun strike the cells in a perpendicular fashion; we must also mention the change in angle due to the change in path of the Sun throughout the seasons [67].

The very high materials and manufacturing costs of III−V solar cells have limited them from transitioning from space [4,31,48,62−64,181,182] to terrestrial applications, with the possible exception of some concentrator technologies [3,4,11,31,42,48,62,63,180,183,184]. In a concentrator device, sunlight is focused onto a small area using lenses or mirrors, thus the necessary area given up to the most expensive part of a solar array (i.e., the cells) is reduced [3,11,180−188]; see Fig. 1.21 for examples of concentrator technologies for both devices and systems [186−188]. This

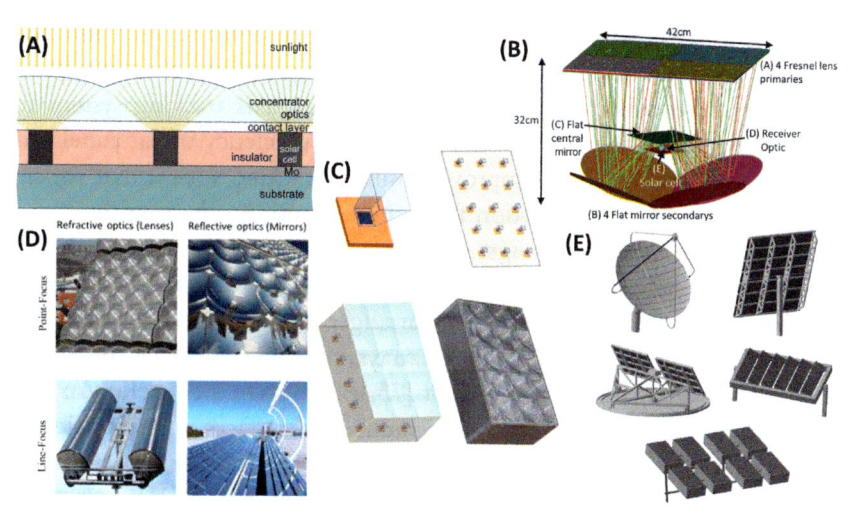

Figure 1.21 Examples of technologies relevant to concentrator photovoltaics (CPV) (A) Concept of CuInSe$_2$ microconcentrator solar cells; (B) Ray trace diagram of ultra-high concentrator showing four square Fresnel lenses focusing toward four angled flat reflectors, the light it reflected again by a central flat mirror onto a central refractive optic made of four spherical filled lenses completing concentration onto the solar cell receiver, the green and red rays represent 400 and 1600 nm light, respectively; (C) Structure of a CPV module-based on point-focus Fresnel lenses; (D) Examples of CPV systems classified as a function of their strategy for concentrating sun rays; (E) Types of two-axis sun trackers for high-CPV systems, i.e., point-focus dish, pedestal-mounted, Carousel (rotate-and-roll), and tilt-and-roll. *(A) S. Sadewasser, P.M.P. Salomé, H. Rodriguez-Alvarez, Materials efficient deposition and heat management of CuInSe$_2$ micro-concentrator solar cells, Solar Energy Mater. Solar Cells 159 (January 2017) 496−502, copyright (2017) Elsevier; (B) K. Shanks, J.P. Ferrer-Rodriguez, E. F. Fernández, F. Almonacid, P. Pérez-Higueras, S. Senthilarasu, T. Mallick, A >3000 suns high concentrator photovoltaic design based on multiple Fresnel lens primaries focusing to one central solar cell, Solar Energy 169 (July 2018) 457−467, copyright (2018) Elsevier; (C-E) E.F. Fernández, F. Almonacid, P.M. Rodrigo, P. Pérez-Higueras, Chapter II-4-A - CPV Systems, in: S. Kalogirou (Ed.), McEvoy's Handbook of Photovoltaics: Fundamentals and Applications, third edition, Academic Press, 2018, pp. 931−985, copyright (2018) Elsevier.*

technology has the added advantage of improving the PCE, which tends to increase with increased intensity [11,62,63,182−193]. However, this does come at the cost of also generating more heat in the cell, which will decrease efficiency and lifetime, thus creating a new set of problems to be confronted [31,62−64,182,184,186−193].

Concentrator solar collectors use devices such as Fresnel lenses (Fig. 1.21B and C) and parabolic mirrors to concentrate light onto solar

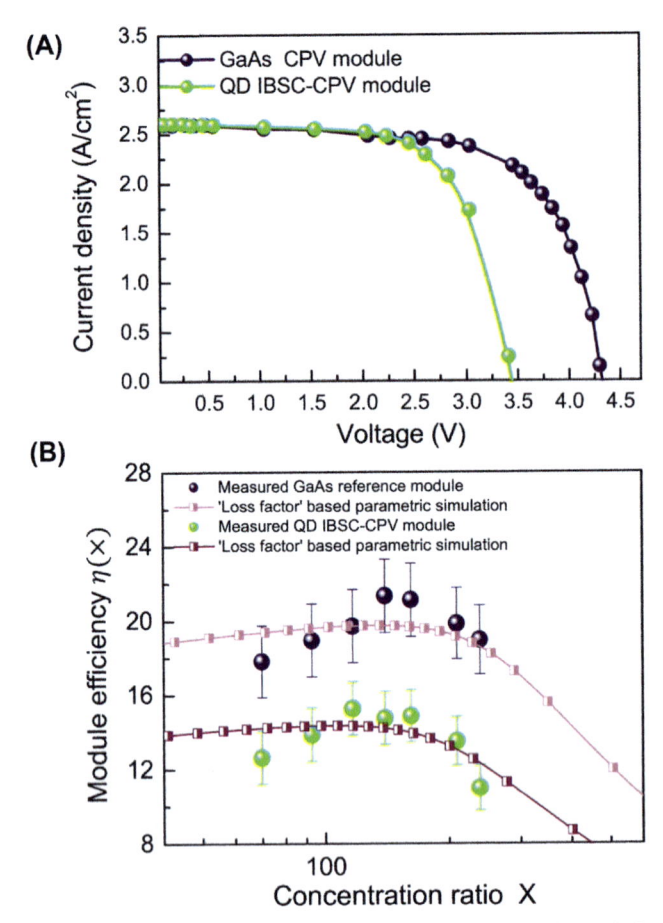

Figure 1.23 (A) Current–voltage curve of the IBSC-CPV module and GaAs control module. (B) Conversion efficiency η(X) under different concentration ratios, X. Fitted results, based on "loss factor" parametric method, are also presented, see details in the original reference. *Reproduced from T. Sogabe, Y. Shoji, M. Ohba, K. Yoshida, R. Tamaki, et al., Intermediate band dynamics of quantum dots solar cell in concentrator photovoltaic modules. Sci Rep 4 (April 2014) 4792, this article is distributed under the terms of the Creative Commons (CC-BY 4.0) license (http://creativecommons.org/licenses/by/4.0/), which permits unrestricted use.*

devices reported by Beattie et al. [197] surpassed the earlier results of Sogabe et al. for both IBSC and single-junction devices at > 100 suns concentration [195].

Quite recently, Sogabe et al. reported a theoretical study and experimental demonstration of an InGaP top cell-integrated QD-IBSC with a

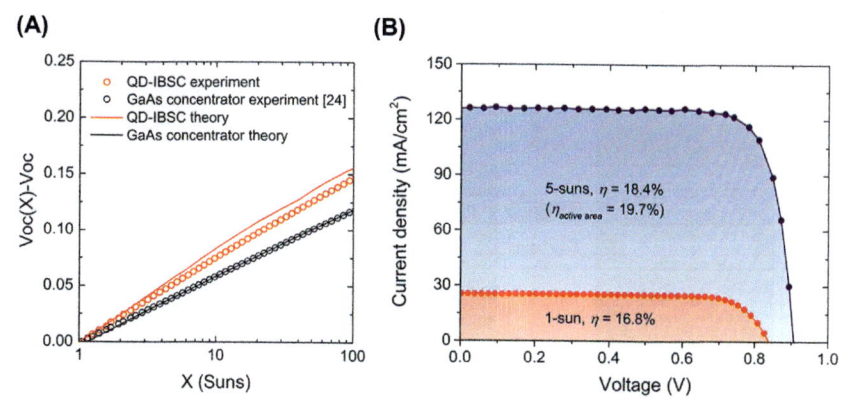

Figure 1.24 (A) Variation of the open-circuit voltage offset, $V_{oc}(X) - V_{oc}(X=1)$ as a function of the light concentration (X), for a quantum-engineered QD-IBSC and a single gap conventional GaAs solar cell. Open symbols represent experiment ([24] for GaAs data refers to: H.F. MacMillan, H.C. Hamaker, N.R. Kaminar, J.M. Gee, M.S. Kuryla, et al., 28% efficient GaAs concentrator solar cells, in: *Proc. of the 20th IEEE Photovoltaics Specialists Conference*, Vol. 1, pp. 462–468) and solid lines theory (see: R. Strandberg, T.W. Reenaas, Photofilling of intermediate bands, J. Appl. Phys. 105 (2009) 124512); (B) JV characteristic for an optimized quantum-engineered QD-IBSC at 1 and 5 suns irradiance. *Reproduced from N.S. Beattie, P. See, G. Zoppi, P.M. Ushasree, M. Duchamp, et al., Quantum engineering of InAs/GaAs quantum dot based intermediate band solar cells, ACS Photon. 4 (11) (2017) 2745–2750, this article is distributed under the terms of the Creative Commons (CC-BY 4.0) license (http://creativecommons.org/licenses/by/4.0/), which permits unrestricted use.*

"current constraint (CC)" feature—a CC-QD-IBSC [204]. A hybrid MBE and MOCVD shuttle growth of the proposed CC-QD-IBSC device was utilized to better preserve QD-induced IB features and overcome practical growth difficulties (Fig. 1.25A). The advantages of the proposed device concepts, such as increased efficiency, improved operation under concentrated sunlight, and efficiency stability against IB variation, were identified by comparing them with a conventional IBSC (Fig. 1.25C). Furthermore, by introducing an InGaP top cell for constraining the current of InAs/GaAs QD-IBSC, the short-circuit current density of CC-QD-IBSC decreased to 13.0 mA/cm^2 under 1 sun illumination (Fig. 1.25B), a near 50% reduction compared to the conventional InAs/GaAs QD-IBSC. The utility of the current constraint effect was described in the detailed discussion of the simulation reported in this study [204]. However, as it is beyond the scope of this review, interested readers are encouraged to consult the original source. By adjusting the regrowth

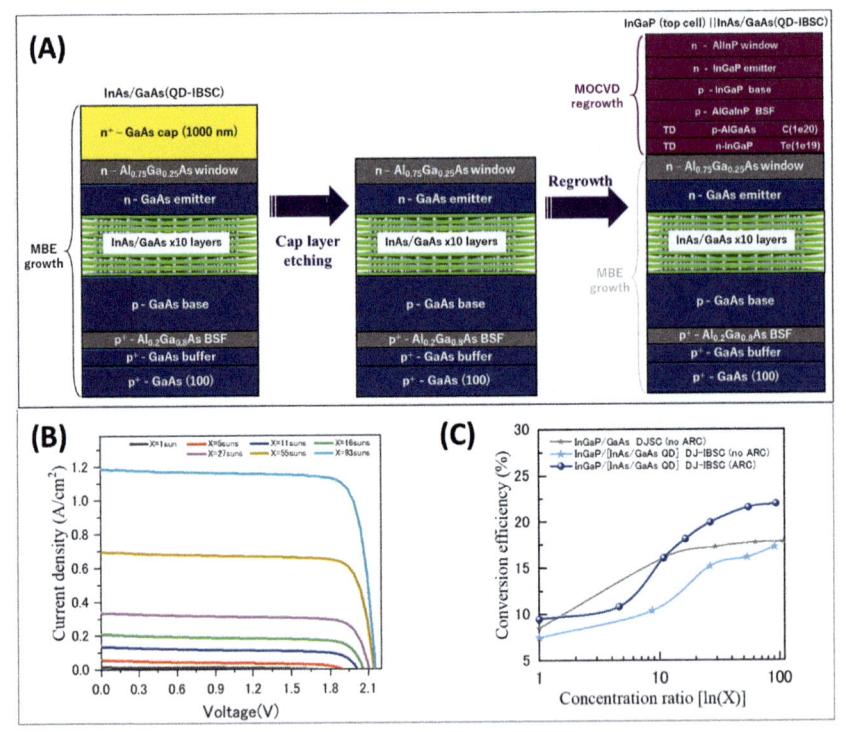

Figure 1.25 (A) Sketch of the proposed "shuttle" growth scheme and the detailed device structure information for CC-QD-IBSC grown by MBE and an InGaP layer grown by MOCVD; (B) J–V curves of CC-QD-IBSC under various light concentration ratios; (C) Dependence of device efficiency on the light concentration ratio. Reproduced from T. Sogabe, Y. Shoji, N. Miyashita, D.J. Farrell, K. Shiba, H.-F. Hong, Y. Okada, High-efficiency InAs/GaAs quantum dot intermediate band solar cell achieved through current constraint engineering, Next Mater. 1 (2) (2023) 100013, this article is distributed under the terms of the Creative Commons (CC-BY 4.0) license (http://creativecommons.org/licenses/by/4.0/), which permits unrestricted use.

interface between the MBE and MOCVD-processed parts of the device, Sogabe et al. achieved a PCE of 22.8% measured under a light concentration ratio of 93 suns (Fig. 1.25C); this was attributed to contributions of both QD and IB features to the CC-QD-IBSC. In the low concentration regions between 4 and 30 suns, the InAs QD CC-QD-IBSC displayed an accelerated increase rate of efficiency compared to the DJ InGaP/GaAs control cell. A similar higher increase rate of efficiency as a function of relation was reported by Sogabe et al. in previous work [195] (see above); this has been attributed (in part) to the effect due to intermediate levels formed in the InAs/GaAs QD-IBSC.

1.3.5.4 Rare earth-containing up- and down-conversion materials to improve solar cell efficiency

Thus far, we have discussed several mechanisms for enhancing the utilization of the solar spectrum via QD(S)SCs (section(s)), MJ solar cells (section(s)), and IBSCs (section(s)). An alternative approach known as DC [205–209] or upconversion (UC) [206,208,210–216] typically utilizes luminescent rare earth-containing materials [212–217] to either split high-energy photons or combine low-energy photons into visible light photons, respectively. Fig. 1.26 illustrates the utility of these phenomena for accessing the energy available in parts of the AM1.5 spectrum that are not utilized by a typical Si solar cell (green zone); the extended fraction of light that could potentially be converted to electricity by DC and UC processes is colored blue with added lines [206].

The DC/UC processes are a specific application of a theoretical study described in 1953 by Dexter [218]. A summary schematic of both mechanisms is provided in Fig. 1.27 for a $CdSe_{0.4}S_{0.6}$ QDSSC [219] enhanced with a bidirectional (UC/DC) material system. An in-depth discussion of the complex chemical physics of DC and UC mechanisms is beyond the scope of this abbreviated summary. Interested readers are encouraged to consult the original literature or a suitable review [217] referenced in this subsection to appreciate the complexity of these intriguing phenomena.

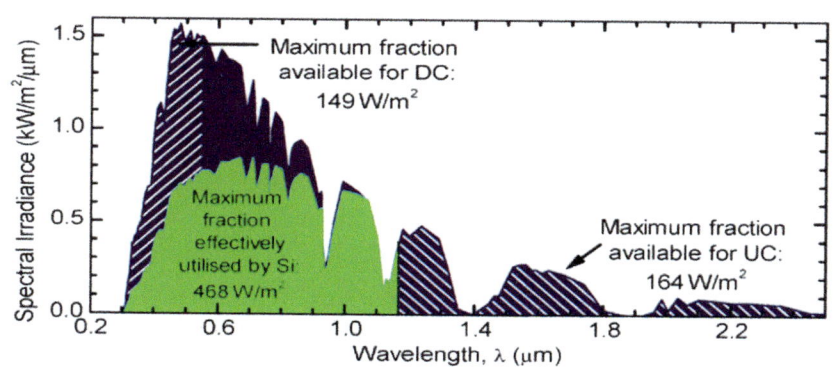

Figure 1.26 Regions of the AM1.5G spectrum that can be used in a solar cell through DC and UC (blue with lines). The fraction absorbed by a thick solar silicon device is shown in green. *Reproduced with permission from B.S. Richards, Enhancing the performance of silicon solar cells via the application of passive luminescence conversion layers, Sol. Energy Mater. Sol. Cells 90 (15) (September 2006) 2329–2337, copyright (2006) Elsevier.*

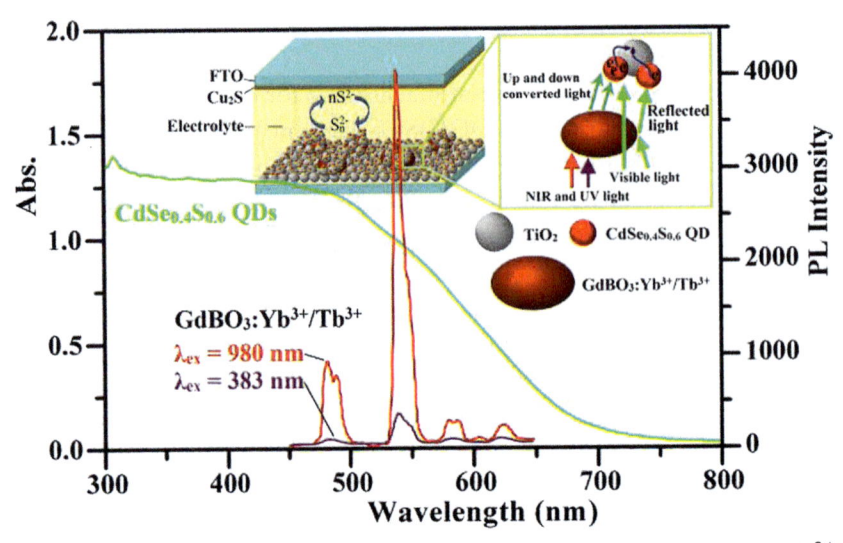

Figure 1.27 A novel bidirectional-conversion luminescence agent (GdBO$_3$:Yb^{3+}/Tb^{3+}), capable of converting ultraviolet photons and near-infrared photons into visible light, is employed to enhance the light capture capability of CdSe$_{0.4}$S$_{0.6}$ QDSSCs. The GdBO$_3$:Yb^{3+}/Tb^{3+}@TiO$_2$ photoanode is fabricated by a sol–gel drop coating method; subsequently, the CdSe$_{0.4}$S$_{0.6}$ QDs are deposited on the prepared photoanode surface by successive ionic layer adsorption and reaction (SILAR) technique at room temperature. *Reproduced with permission from D. Fang, X. Zhang, C. Zhao, X. Liu, X. Shu, J. Wang, Application of bidirectional (up and down)-conversion luminescence material (GdBO$_3$:Yb^{3+}/Tb^{3+}) in CdSe$_{0.4}$S$_{0.6}$ quantum dot-sensitized solar cells, Opt. Mater. 88 (February 2019) 80–90, copyright (2019) Elsevier.*

Up- and down-conversion materials have been studied for enhancing the efficiency of a variety of solar cell devices over the past several decades, including Si [206–208,211,216,220,221], a-Si [208,211], III–V solar cells [206,208,211,212,222,223], DSSCs [208,211,224,225], QD(S)SCs [219], and PSCs [208,226,227]. Fig. 1.28 outlines the fabrication process of a flexible DSSC utilizing Er^{3+}-Yb^{3+} (EY) codoped TiO$_2$ nanoparticle (NP) UC enhancement. The PCE of devices composed of an uncoated TiO$_2$ nanowire assembly is 4.8%; this is increased by nearly 70% to 8.10% when fully enhanced by UC codoped TiO$_2$ NPs that are Nb$_2$O$_5$ coated [225].

In fact, UC or DC enhancement results in a range of increased efficiencies across example devices; however, the enhancement was greatest for low-efficiency devices, somewhat mitigating the beneficial impact of UC/DC materials. The PCE of a QD(S)SC increased fivefold (0.1% to

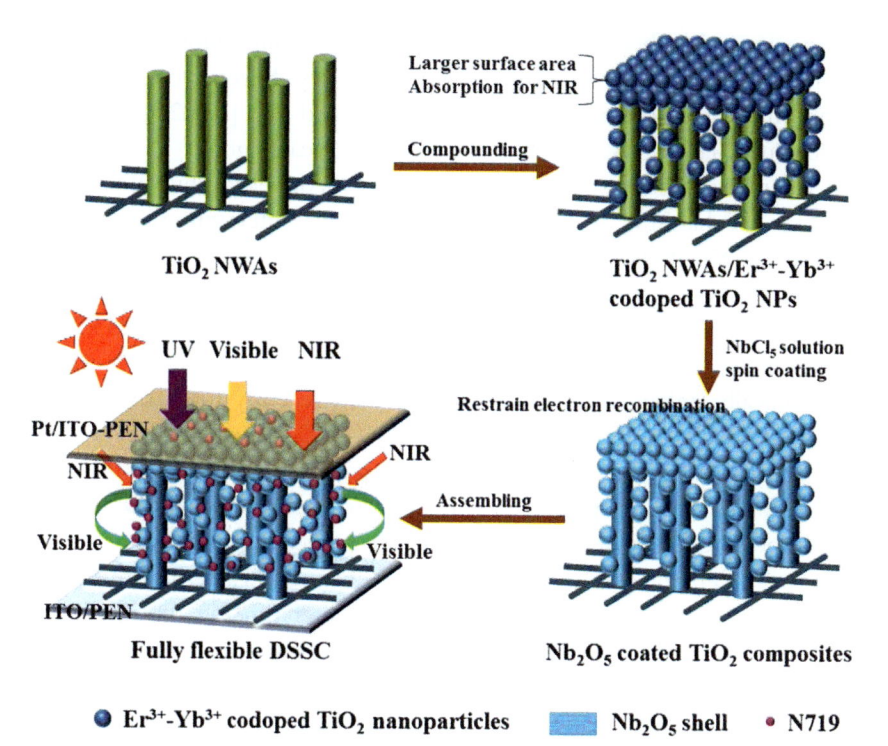

Figure 1.28 The assembly diagram of the fully flexible DSSCs based on Nb_2O_5 coated TiO_2 nanowire assembly/UC-EY-TiO_2 NPs composite micro–nano structures. *Reproduced with permission from W. Liu, H. Zhang, H.-G. Wang, M. Zhang, M. Guo, Titanium mesh supported TiO_2 nanowire arrays/upconversion luminescence Er^{3+}-Yb^{3+}-codoped TiO_2 nanoparticles novel composites for flexible dye-sensitized solar cells, Appl. Surf. Sci. 422 (November 2017) 304–315, copyright (2017) Elsevier.*

0.52%) when adding 1% $GdBO_3$:Yb^{3+}/Tb^{3+} NPs [219]. The PCE of a hybrid perovskite device was increased by 10% from 15.5% to 16.9% [225]. A DC Tb/Yb-doped tellurite glass coated onto commercial Si cells improved the PCE by 7% but only increased a low (<1%) PCE GaP cell by 1% [207]. The use of molybdate ($La_2Mo_2O_9$:Yb, R (R = Er, Ho)) phosphors for UC enhancement of PCE of Si solar cells resulted in a 1.5%–2.7% improvement [212]. A recently published study compared an antireflective coated (ARC) GaAs (22.9%) baseline cell to a device with one or both Eu-doped and Yb/Er-doped phosphors (23.2% - 23.8%). This resulted in 1% to 5% PCE enhancement [223].

Finally, in order to place this subsection in context, it must be noted that there are other approaches to UC and DC for solar cells that employ other approaches besides rare earth-containing luminophores (or lumophores) to enhance light absorption and efficiency. Several recent reviews nicely summarize alternative methods for DC and UC [228,229]. Triplet-triplet annihilation UC was first described 60 years ago by Parker [230] and recently detailed by Cheng et al. [231]. Upconversion by QDs occurs via two mechanisms, as detailed in the original literature [232] and a review by Nozik [142]. Another mechanism involves UC via thermal radiation, where light is used to raise a substance's temperature, shifting its blackbody emission spectrum to shorter wavelengths [233,234]. Interestingly, the performance of UC rare-earth nanocrystals (NC) can be plasmonically enhanced as has been described recently for GaAs devices [235]. The PCE of GaAs solar cells on a UCNC-incorporated plasmonic substrate [Ag nanostructure (NS)] is increased by approximately 6% and approximately 12%, respectively, compared to those on an Ag NS reflector without UCNC and on a plain silver reflector with UCNC; thus, each enhancement results in an approximately 6% PCE improvement. Lastly, Day et al. provide a summary of the literature on the impact of UC, DC, and luminescent downshifting—a less-often, recently described mechanism [236–238]—on a variety of solar cells [228]. The results mirror the PCE enhancements for various solar cells discussed above [207,212,215,219,223].

1.3.6 Multijunction devices: Focus on third-generation tandem solar cells

As mentioned above (Section 1.3.5), two topics [concentrators (Section 1.3.5.2) and MJ cells (Section 1.3.2)] are addressed in a prior subsection and/or later chapters in this book [11,118,119]. Thus, these technologies will be merely summarized in brief in this subsection. Clearly, MJ III–V devices that already approach 40% efficiency at one sun (3 J/NREL—39.5% and 6 J/NREL—39.2%) will continue to be the leading option to provide nonnuclear power for future space missions [31,63–65,68,182,239]. In fact, concentrated (302X) 3 J (Sharp IMM) cells have hit 44.4% PCE; concentrated (865X) 4 J (FhG-ISE) have achieved 47.6% PCE. Fig. 1.29 provides a historical perspective and a forward look, utilizing Eq. (1.1) and the approach of Goetzberger et al. [77], as illustrated in Fig. 1.12. It compares numerous inorganic device technologies to concentrated MJ III–V cells, including SJ GaAs for some perspective.

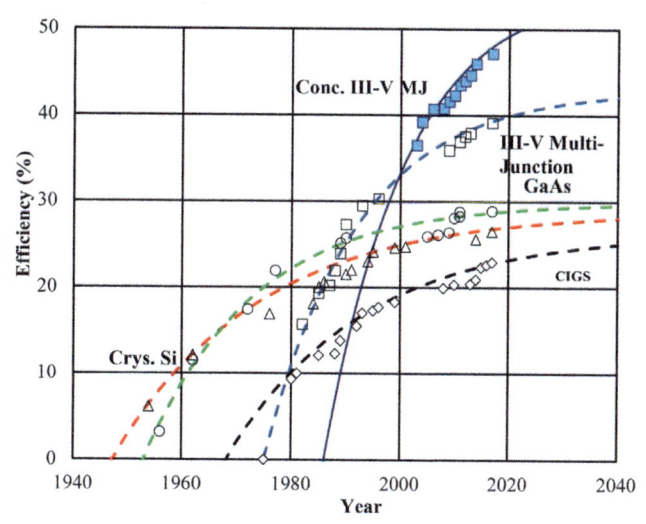

Figure 1.29 Historical record-efficiency of III–V multijunction (MJ) and concentrator MJ solar cells in comparison with 1-sun efficiencies of GaAs and crystalline Si solar cells, along with their extrapolations. *Reproduced from M. Yamaguchi, High-efficiency GaAs-based solar cells, in: M.M. Rahman, A.M. Asiri, A. Khan, I. Inamuddin, and T. Tabbakh (Eds,), Post-Transition Metals, InTech, 2021. Open, DOI:10.5772/intechopen.94365 under the terms of the Creative Commons Attribution 3.0 Unported license (CC BY 3.0).*

Having addressed MJ III–V devices in detail, the remainder of this subsection will mainly focus on third (and fourth) generation subcell-containing tandems. Fig. 1.30A provides the spectral response for a generic tandem solar cell and Fig. 1.30B illustrates the essential utility of a tandem where lower energy photons can be utilized by the lower band-gap (bottom) subcell [240]. The three most common dual-junction configurations are shown schematically in Fig. 1.30C-E. Monolithic 3 T devices are not addressed in this chapter, for the sake of simplicity; these devices are reviewed in the literature and/or later chapters in this book [118,119,241]. Interested readers are directed to these sources. Consulting Fig. 1.13 and efficiency tables from the literature [2], the PCE of a perovskite/CIGS tandem (24.2%) is currently below the record PCE of a champion SJ-PSC (25.8%) but slightly (<5%) higher than concentrated (23.3%) or one sun CIGS (23.6%) champion devices.

In the remainder of this subsection, we highlight trends, identify key findings, and briefly review the literature; the interested reader is encouraged to delve further in several later chapters in the book [11,63,65,118,119] and the literature referenced in this chapter

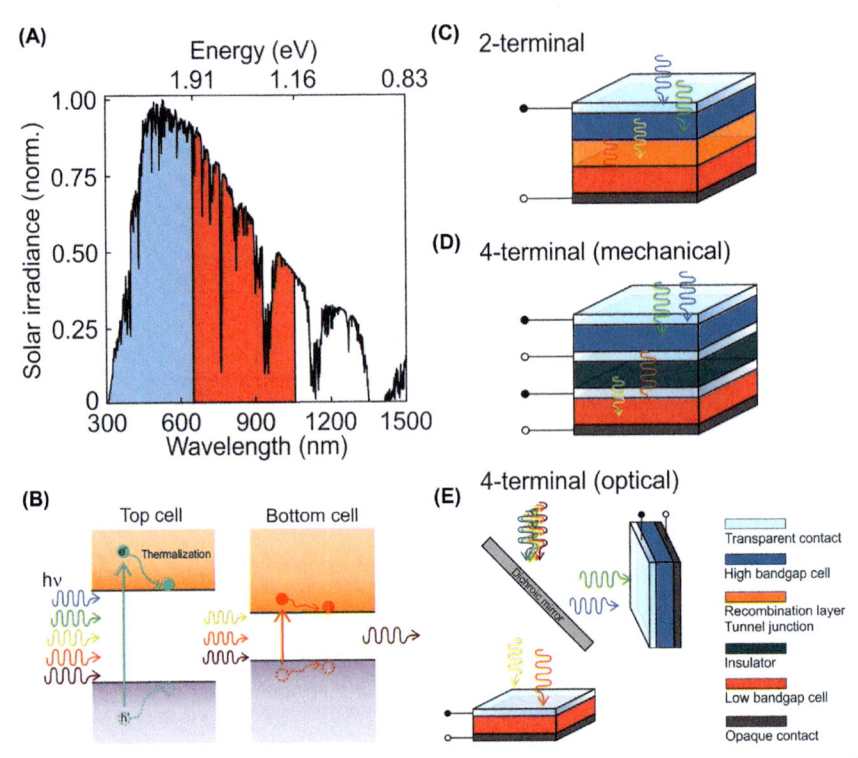

Figure 1.30 Working principles of tandem solar cells; (A) Spectral response of generic top (blue) and bottom (red) subcells comprising a tandem device working in different regions of the reference solar spectral irradiance AM1.5; (B) Energy diagram for a dual-junction tandem solar cell; (C–E) Scheme of the three most common dual-junction configurations: (C) two-terminal, (D) mechanically stacked four-terminal, and (E) optically coupled four-terminal. *Reprinted with permission from M. Anaya, G. Lozano, M.E. Calvo, H. Míguez, ABX₃ perovskites for tandem solar cells, Joule 1 (4) (December 2017) 769–793, copyright (2017) Elsevier.*

[183,222,239–275] for further details. We consulted numerous review articles on PSC-containing tandem devices in order to present a representative overview of the current status and future prospects for these device technologies [241–253]. Table 1.2 was assembled from several reviews [2,248,249,252] that included performance details of a large number of prior results on tandem SC studies [254–274]. A cross-section of results provides examples of PSC-containing tandem devices with a comparison to more traditional all-inorganic III–V/Si devices [252,254–257]. The bandgap tunability of PSCs used as the top subcell or bottom subcell in an all-perovskite tandem is clearly demonstrated in Fig. 1.31B. Fig. 1.31A

Table 1.2 Overview of PSC-containing tandems compared to III–V/Si tandem devices.

Top cell	Bottom cell	Device details[a]	Area (cm²)	J_{sc} (mA/cm²)	V_{oc} (V)	FF (%)	PCE (%)	References
Al$_{0.15}$Ga$_{0.85}$As	Si	Mono. 2-term	0.25	23	1.51	77.2	19.9 (AM0)	[254]
GaAs$_{0.75}$P$_{0.25}$	Si	Mono. 2-term	1.03	17.34	1.73	77.7	23.4	[255]
GaAsP	Si	Mono. 2-term	0.123	18.8	1.65	81	25.0	[256]
GaAs	Si	MS 4-term	1.00	28.9 (III–V) 11.1 (Si)	1.09 (III–V) 0.683 (Si)	85.0 (III–V) 79.2 (Si)	32.8	[257]
Perovskite[b]	Organic[b]	2-term	0.062	13.05	1.902	83.1	20.6	[258]
CsPbI$_2$Br	Organic[c]	2-term	1	12.42	2.08	74.06	19.18	[259]
CsPbI$_2$Br	Organic[c]	Elec. Iso.	1	16.5 (PSC) 10.3 (OPV)	1.28 (PSC) 0.84 (OPV)	73.3 (PSC) 78.9 (OPV)	22.34	[259]
Perovskite[d]	Organic[e]	4-term	0.08	14.83	2.06	77.2	23.60	[260]
CsPbI$_2$Br	Organic[f]	2-term	0.04	12.68	2.22	76	21.4	[261]
Perovskite[g]	Perovskite[h]	2-term Flexible	0.09	15.6	2.017	76.2	24.1	[262]
Perovskite[g]	Perovskite[h]	2-term Rigid	0.09	15.6	2.049	78.7	25.3	[262]
Perovskite[i]	Perovskite[j]	2-term Flexible	0.058	15.6	1.82	75.0	21.3	[263]
Perovskite[i]	Perovskite[j]	2-term Rigid	0.058	16.0	1.88	77.0	23.1	[262]
Perovskite[k]	Perovskite[l]	2-term 4 SCs	20.3	3.55	8.04	75.9	21.7	[264]
Perovskite[k]	Perovskite[l]	2-term	1.05	15.4	2.025	79.4	24.8	[264]
Perovskite[m]	Perovskite[l]	Elec. Iso. 4-term	0.1188	19.9 (WB) 11.5 (NB)	1.146 (WB) 0.817 (NB)	81.6 (WB) 78.9 (NB)	26.3	[265]

(Continued)

Table 1.2 (Continued)

Top cell	Bottom cell	Device details[u]	Area (cm²)	J_{sc} (mA/cm²)	V_{oc} (V)	FF (%)	PCE (%)	References
Perovskite[n]	CIGS	2-term	0.042	17.3	1.774	73.1	22.43	[266]
Perovskite[n]	CIGS	2-term	0.52	16.67	1.780	70.2	20.83	[266]
Perovskite[o]	CIGS	2-term	1.04	18.8	1.77	71.2	24.2	[267]
Perovskite[p]	CIGS	MS Flex.	0.07	20.45 (PVK)	1.015 (PVK)	73.74 (PVK)	21.56	[268]
		4-term		12.43 (CIGS)	0.656 (CIGS)	76.65 (CIGS)		
CH₃NH₃PbI₃	Si	Mono. SHJ	4	16.1	1.68	78.0	21.0	[269]
Perovskite[q]	Si	Mono. SHJ	16	16.2	1.65	78.0	21.9	[270]
Perovskite[r]	Si	Mono. SHJ	1.00	17.8	1.76	78.1	24.5	[271]
Perovskite[s]	Si	Mono. SHJ	0.5	16.99	1.92	77.95	25.42	[272]
Perovskite[t]	Si	Mono. SHJ	1.064	19.26	1.90	79.52	29.15	[273]
Perovskite	Si	2-term	1.00	20.47	1.9131	79.8	31.25	[274]

Note: Details of the organic structures (2,3,5,6) can be found by consulting the original literature [258–261].

[a] $FA_{0.8}MA_{0.02}Cs_{0.18}PbI_{1.8}Br_{1.2}$.
[b] $PBDBT-2F:Y6:PC_{71}BM$ (1:1.2:0.2 w/w).
[c] $D18-Cl-B:N3:PC_{61}BM$ (1:1.4:0.2).
[d] $FA_{0.75}Cs_{0.25}Pb(I_{0.6}Br_{0.4})_3$.
[e] $PM6:Y6:PC_{71}BM$.
[f] $PM6:Y6$ (1:1.2 w/w).
[g] $FA_{0.8}Cs_{0.2}Pb(I_{0.6}Br_{0.4})_3$.
[h] $FA_{0.7}MA_{0.3}Pb_{0.5}Sn_{0.5}I_3$.
[i] $DMA_{0.1}FA_{0.6}Cs_{0.3}PbI_{2.4}Br_{0.6}$.
[j] $FA_{0.75}Cs_{0.25}Sn_{0.5}Pb_{0.5}I_3$.
[k] $Cs_{0.35}FA_{0.65}PbI_{1.8}Br_{1.2}$.
[l] $MA_{0.3}FA_{0.7}Pb_{0.5}Sn_{0.5}I_3$.
[m] $FA_{0.75}Cs_{0.25}Pb(I_{0.8}Br_{0.2})_3$.
[n] $Cs_{0.09}FA_{0.77}MA_{0.14}Pb(I_{0.86}Br_{0.14})_3$.
[o] $Cs_{0.05}(FA_{0.77}MA_{0.23})Pb_{1.1}(I_{0.77}Br_{0.23})_3$.
[p] $Cs_{0.05}MA_{0.10}FA_{0.85}Pb(I_{0.97}Br_{0.03})_3$.
[q] $(FAPbI_3)_{0.83}(MAPbBr_3)_{0.17}$.
[r] $(Cs_{0.05}Rb_{0.05}FA_{0.765}MA_{0.135})PbI_{2.55}Br_{0.45}$.
[s] $FA_{0.83}Cs_{0.17}Pb(I_{0.80}Br_{0.20})_3$.
[t] $Cs_{0.05}(FA_{0.77}MA_{0.23})_{0.95}Pb(I_{0.77}Br_{0.23})_3$.; see Fig. 1.30A for details of the organic cations found in perovskites, DMA = dimethylammonium⁺.
[u] Device details: *Mono.*, Monolithic; *MS*, Mechnically Stacked; *Elec. Iso.*, Electrically isoloated; *SHJ*, Si heterojunction.

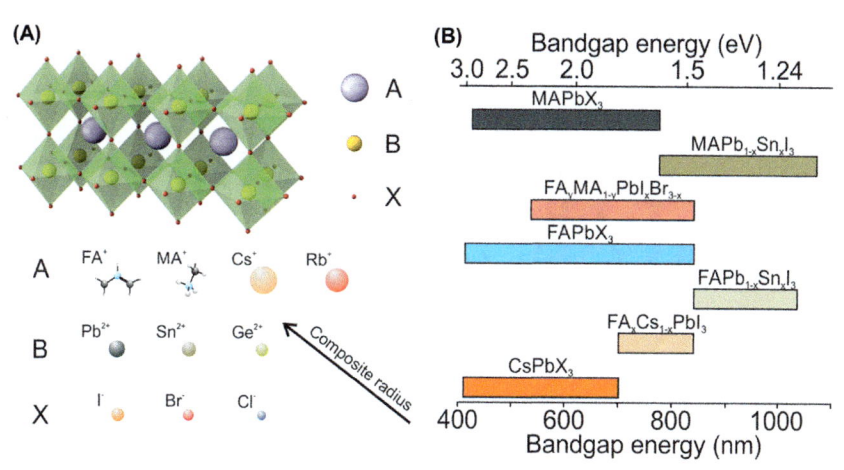

Figure 1.31 Bandgap tunability of inorganic or hybrid perovskites (ABX$_3$) (A) Schematic illustration of an ABX$_3$ perovskite and typical cations and anions occupying the different positions in the structure; (B) Bandgap energies that can be attained with the different combinations of elements in an ABX$_3$. *Reprinted with permission from M. Anaya, G. Lozano, M.E. Calvo, H. Míguez, ABX$_3$ perovskites for tandem solar cells, Joule 1 (4) (December 2017) 769–793, copyright (2017) Elsevier.*

also includes details of the locations of organic or inorganic cations (A and B) and anions (X) in the perovskite unit cell [240].

Several options that were considered when selecting devices to be included in Table 1.2 were: (a) 4-terminal (4-T) (mechanically stacked or optically coupled but electrically independent) versus monolithic (currently matched) 2-terminal (2-T) devices; (b) size of the device active area; and (c) rigid versus flexible substrate. As expected, smaller area and rigid substrate devices tend to produce higher measured PCEs [2,260,262,263,266]. Electrically independent (mechanically stacked or optically linked) 4-T devices, particularly for III−V/Si tandems, demonstrate a significant (approximately 30%) PCE advantage [257]. A recent study of CsPbI$_2$Br/OPV tandems also demonstrated an approximately 30% PCE advantage of 4-T over 2-T device for a 1 cm^2 active area [259]. An all-PSC tandem assembled in a 4-T configuration recently exceeded 26% PCE, albeit for a small (0.1188 cm^2) active area device [265]. Even with a small device area (0.07 cm^2), a recently reported [268] flexible 4-T perovskite/CIGS tandem does not demonstrate a better PCE than a 1 cm^2 2-T device [267]; see Table 1.2. The situation for PSC/Si tandems is more complex; see Fig. 1.32 [244]. Up to 5 years ago, 4-T devices had predominated higher PCE reports [242−244,252]; however, over the past

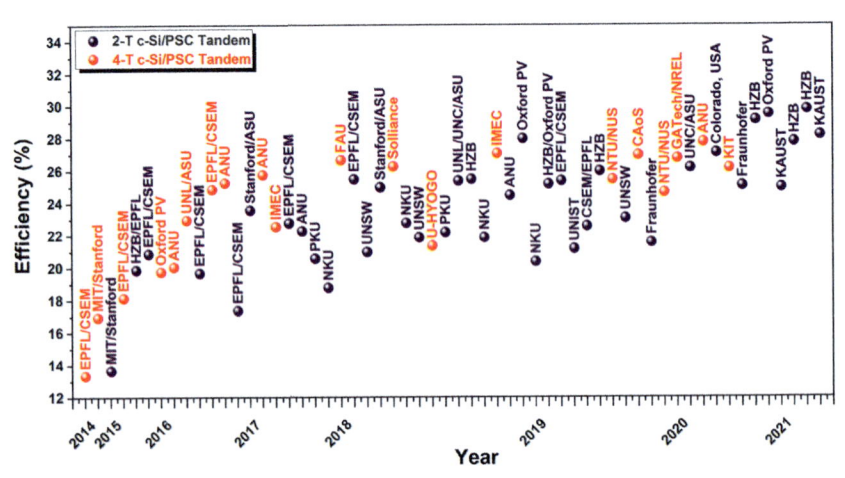

Figure 1.32 Variation in efficiency values of 2-T and 4-T c-Si/PSC tandem devices with respect to publication date includes the identity of universities and/or research Institutes [Note that this is the same approach adopted by NREL (Fig. 1.13)]. In order to save significant space, the interested reader is directed to the original (open access) source to identify the contributing institutions and references. *Reprinted from E. Raza, Z. Ahmad, Review on two-terminal and four-terminal crystalline-silicon/perovskite tandem solar cells: Progress, challenges, and future perspectives, Energy Reports 8 (November 2022) 5820–5851, this article is distributed under the terms of the Creative Commons (CC-BY 4.0) license (http://creativecommons.org/licenses/by/4.0/), which permits unrestricted use.*

3 years, monolithic 2-T devices have demonstrated their ability to achieve equally high PCE [244,252,272–274]. The reader is directed to the original literature for more in-depth discussions [244].

When considering future prospects for the five types of tandem devices considered plus the MJ devices covered above (Section 1.3.5.2), it is clear that III–V/Si tandems have been superseded by 3 J (and beyond) devices. MJ III–V devices have a promising future, mainly for space applications, where array cost is not a factor [30,31,63–65,68,182,239]. Concentrators utilizing III–V-containing tandems may be viable, provided that both electrical and thermal energies are utilized [183,184,186,189–193]. While PSC/OPV [248,253,258–261] (Fig. 1.33D) and PSC/CIGS [266–268] (Fig. 1.33A–C) tandems may have utility in applications where flexibility, higher MSP, and possibly cost are important, thus far these technologies have not demonstrated efficiency advantages compared to all-PSC [262–265] and PSC/Si tandems [241–245,252,269–274], see Table 1.2 and Fig. 1.13.

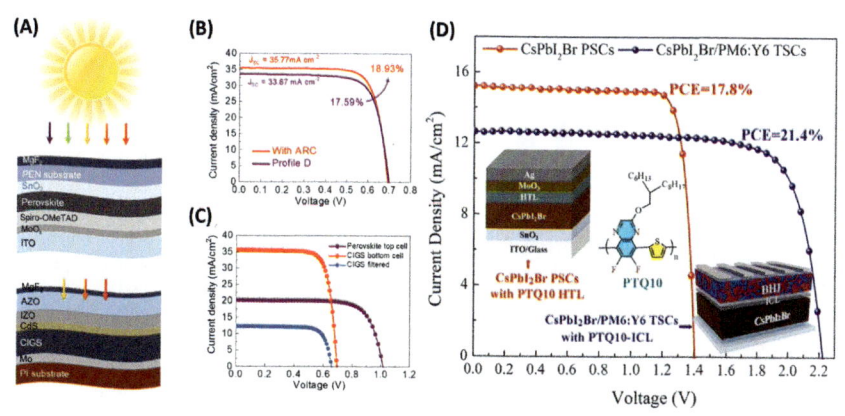

Figure 1.33 (A) Schematic diagram of the all-flexible 4-T perovskite/CIGS solar cells; (B) J–V curves of flexible champion CIGS solar cell with and without a 100 nm MgF$_2$ antireflective coating; (C) J–V curves of CIGS, filtered CIGS, and PSC; (D) (*left*) device structure of CsPbI$_2$Br PSC and molecular structure of PTQ10 (low-cost HTL) and (*right*) device structure of CsPbI$_2$Br/organic TSC. These are plotted on a J–V curve of the PSC and TSC devices. *(A–C) Reprinted with permission from J. Luo, L. Tang, S. Wang, H. Yan, W. Wang, et al., Manipulating Ga growth profile enables all-flexible high-performance single-junction CIGS and 4 T perovskite/CIGS tandem solar cells, Chem. Eng. J. 455 (Part 2) (January 2023) 140960, copyright (2023) Elsevier and (D) Reprinted with permission from Y. Ding, Q. Guo, Y. Geng, Z. Dai, Z. Wang, et al., A low-cost hole transport layer enables CsPbI$_2$Br single-junction and tandem perovskite solar cells with record efficiencies of 17.8% and 21.4%, Nano Today 46 (October 2022) 101586, copyright (2022) Elsevier.*

Both all-PSC and PSC/Si tandems provide significant flexibility due to the options available for both wide bandgap and narrow bandgap perovskites (Fig. 1.31B). While all-perovskite TSCs have the advantage of flexibility in bandgap and chemistry, Si bottom cells have the advantage of decades of research, a deep understanding of optimal processing methods, stable device options, and cost. Fig. 1.34 illustrates a deeper dive into the chronological scheme of 2-T and 4-T PSC/Si TSCs of Fig. 1.32. The evolution of both types of devices is illustrated including details of the PSC upper subcell chemistry and the bottom subcell Si device(s) employed. In their excellent review, Raza and Ahmad summarize the findings of recent studies that have established four approaches to enhance Si cell-based device efficiency and/or efficacy, according to each parameter-dependent cell metric, which are carrier recombination analysis, optical analysis, resistive loss, and cost reduction [244]. Given the greater opportunity for device degradation in a monolithic device, recent efforts

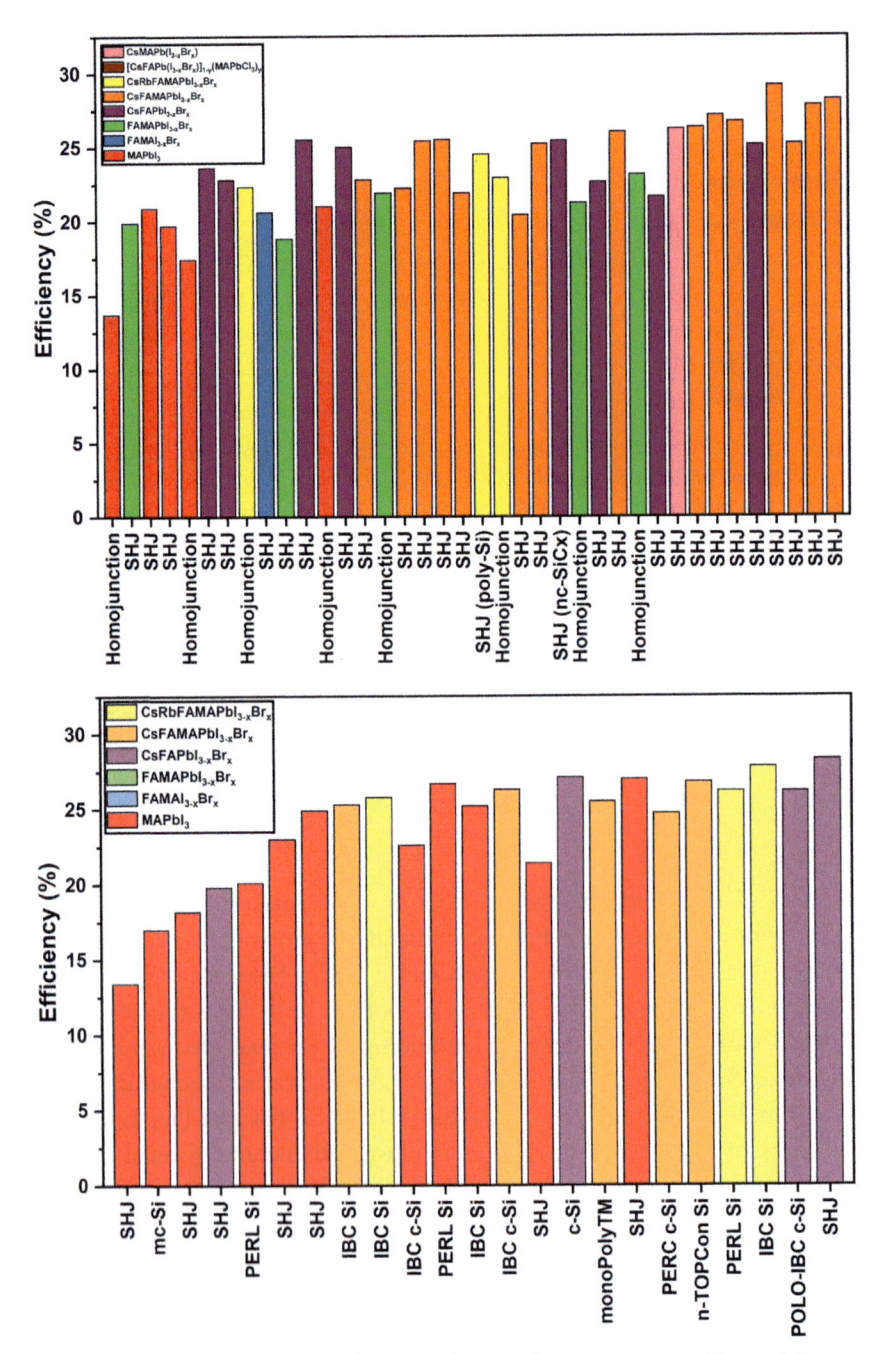

Figure 1.34 (*Top*) Evolution of efficiency values with various compositions of the top perovskite and bottom silicon cells in 2-T c-Si/perovskite TSC; (*Bottom*) Evolution of
(*Continued*)

in 2-T PSC/Si tandems have focused on (CsFAMA)PbI$_3$Br$_{3-x}$ top cells and SHJ bottom cells to progress toward higher efficiency [241–245, 269–273], whereas 4-T tandems have a greater diversity of PSC and Si cell technologies; see Fig. 1.34. Despite the overall theme of this chapter and the book, which looks "beyond Si," further work in improving Si PV technology will continue, especially for TSCs that employ Si as a lower bandgap bottom cell [241–245,252,254–257,269–275].

This subsection has focused on the details of TSC devices and the evolution of increased performance. In order to transition into the commercial world as MJ III–V devices have for space applications, there are numerous challenges to be confronted and overcome. In general, issues common to any MJ or tandem devices must be addressed when contemplating scale-up, manufacturing, and eventual commercialization, including: (1) stability, (2) light trapping/surface texturing, (3) cost-effectiveness, (4) encapsulation and reliability testing, and (5) material toxicity [241–246,252,258,269–271,275]. Further, for every TSC that includes PSC or organic constituents, the major concern is the environmental and processing stability of the PSC subcell [112,242–246] (Fig. 1.33A and D, Fig. 1.35 bottom); this is further complicated by similar concerns regarding the OPV subcell [248,253,258–261] (Fig. 1.33D). These stability issues include device damage and performance reduction due to moisture, oxygen, UV-light, thermal instability, hole transport layer (HTL) and electron transport layer (ETL) interface instability, instability based on metal electrode layer, and processing and postprocessing issues. The complex mechanism of the PSC subcell of a PSC/Si tandem device (Fig. 1.35 top) as well as the complicated structure of both PSC-containing devices (Fig. 1.35 middle) results in numerous potential opportunities for performance reduction due to the issues enumerated above [241,243–246]. In addition, morphological issues or process compatibility make it difficult to fabricate solution-processed perovskite top cells [246]. Other issues specific

◀ efficiency values with various compositions of the top perovskite and bottom silicon cells in 4-T c-Si/perovskite tandem solar cell (TSC); definition of abbreviations: Passivated emitter and rear cell (PERC), PERC + (PERC solar cells with rear Al grid), Si heterojunction (SHJ), interdigitated back contact (IBC) silicon (IBC)-SHJ. *Reprinted with permission from E. Raza, Z. Ahmad, Review on two-terminal and four-terminal crystalline-silicon/perovskite tandem solar cells; Progress, challenges, and future perspectives, Energy Reports 8 (November 2022) 5820–5851, this article is distributed under the terms of the Creative Commons (CC-BY 4.0) license (http://creativecommons.org/licenses/by/4.0/), which permits unrestricted use.*

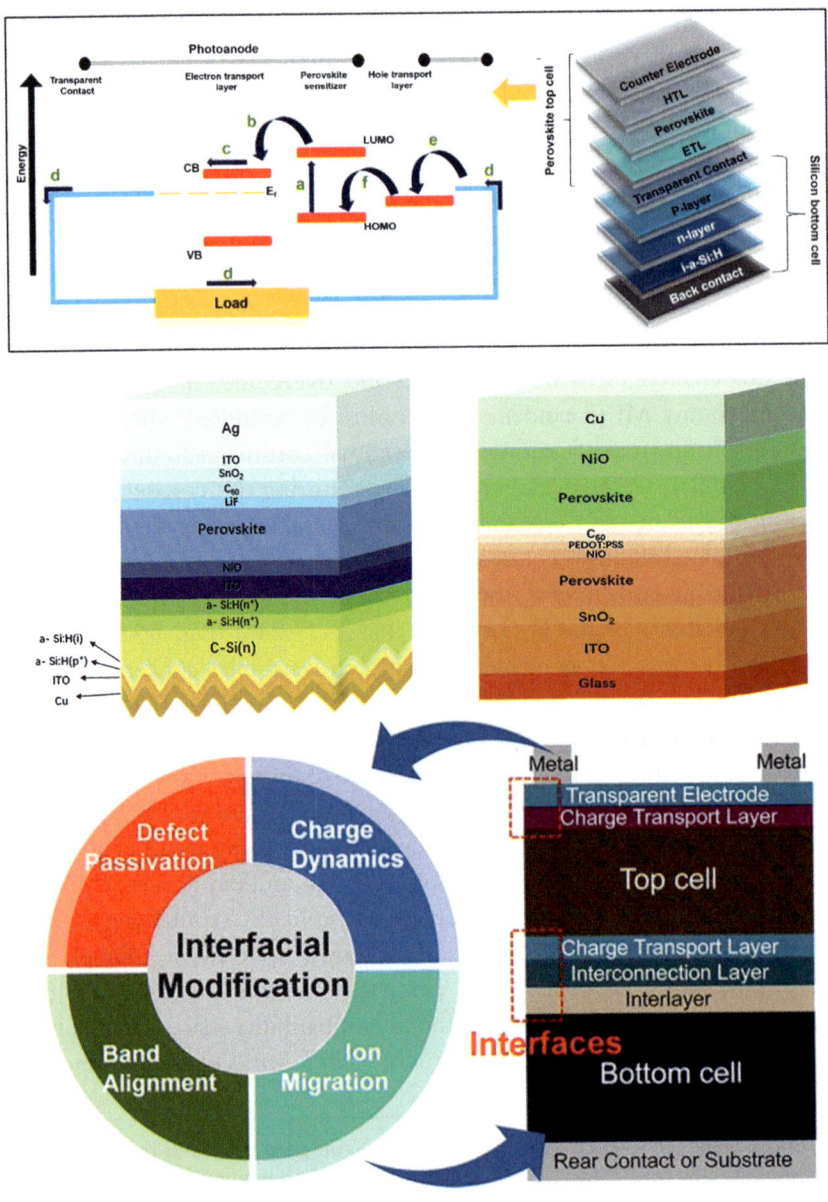

Figure 1.35 (*Top*) *Left*: Schematic representation of energy levels and functional layers in the perovskite subcell of a 2-T tandem cell, *Right*: Schematic view of a conventional (2 T) perovskite silicon tandem solar cell; (*Middle*) *Left*: Schematic diagrams of a 2-T silicon/perovskite tandem cell, *Right*: Schematic diagram of a 2-T perovskite/perovskite tandem cell; (*Bottom*) schematic of interfaces in a generic 2-T monolithic

(*Continued*)

to PSC devices include perovskite hysteresis effects, ion migration and segregation, and perovskite free surface states. Fig. 1.35 at the bottom illustrates the numerous interfaces (listed above and others, i.e., interconnecting layers (ICLs)) and related issues that arise, as well as potential solutions. This topic is explored in much greater depth in the relevant literature, covering creative approaches to improved performance via novel processing approaches [241,243−245,251,275].

1.4 Summary and conclusions

It is important to keep in mind that solar cells, regardless of the material(s) or device structure, are designed to generate electricity. While highlighting important terrestrial applications, where the price per watt will be a critical driver, the challenges of aerospace and defense have spurred the development of advanced technologies, particularly for first and second-generation (crystalline and inorganic thin-film) technologies. After a historical background and technology background discussion, we progressed through a series of next-generation materials and device concepts, including OPV, DSSCs, PSCs, and QD(S)SCs, as well as concentrator solar cells, IBSCs, up/downconversion materials, and MJ devices. We noted that multijunction III−V devices will quite likely be the predominant nonnuclear space power technology for the foreseeable future due to their high PCE and radiation resistance. The cost of PV arrays is essentially irrelevant, given the minimal impact on overall spacecraft and mission expenses. A key structure with promise for future terrestrial applications is the tandem device (both 2-T and 4-T), particularly with PSC top layers.

◂ tandem cells with corresponding interfacial issues to be resolved to enable high efficiency and stability. *(Top) S. Akhil, S. Akash, A. Pasha, B. Kulkarni, M. Jalalah, et al., Review on perovskite silicon tandem solar cells: Status and prospects 2 T, 3 T and 4 T for real world conditions, Materials & Design 211 (December 2021) 110138, copyright (2021) Elsevier; (Middle) Z. Li, Y. Zhao, X. Wang, Y. Sun, Z. Zhao, et al., Cost analysis of perovskite tandem photovoltaics, Joule 2 (8) (2018) 1559−1572, copyright (2018) Elsevier; (Bottom) I.J. Park, H.K. An, Y. Chang, J.Y. Kim, Interfacial modification in perovskite-based tandem solar cells, Nano Convergence 10 (2023) 22, this article is distributed under the terms of the Creative Commons (CC-BY 4.0) license which permits unrestricted use (http://creativecommons.org/licenses/by/4.0/).*

The authors would like to emphasize a number of lessons to be imparted to the reader after considering this omnibus solar cell overview chapter. From the point of view of cost and Earth abundance, MJ III−V technology will almost certainly dominate space applications due to the quite small market share (< 1%) and routine PCE approaching 40%. While the main focus of this book is PV beyond Si, Si-containing devices are the premier Earth-abundant element; therefore, tandem PV with Si bottom cells is the most attractive from this perspective. PSCs, either as single-junction or as the top cell in a tandem device, or all-perovskite devices have a promising future due, in large part to the enormous available materials and processing technologies. Concentrator PV technologies that generate heat should be utilized in systems that exploit thermal energies.

A further lesson to be learned from a more practical consideration of solar cells is that a wide array of power generation applications exists each with its own set of challenges. These challenges can be addressed by a variety of technologies that overcome specific issues involving available area, efficiency, reliability, and specific power at an optimal cost. The particular application may be in aerospace, defense, utility, consumer, grid-based, off-grid (housing), recreational, vehicular or industrial settings. The PV community often relies on PCE (or efficiency) as an essential metric; however, for practical applications, significant cost savings during manufacture can enable the use of lower-efficiency technologies. Finally, we encourage our colleagues in the worldwide energy conversion community to continue their impressive level of effort toward new PV materials, devices, and technologies as applications for other energy or electronic devices may be discovered.

References

[1] S. Almosni, A. Delamarre, Z. Jehl, Daniel Suchet, Ludmila Cojocaru, et al., Material challenges for solar cells in the twenty-first century: directions in emerging technologies, Sci. Technol. Adv. Mater. 19 (1) (2018) 336−369. Available from: https://doi.org/10.1080/14686996.2018.1433439.

[2] M.A. Green, E.D. Dunlop, J. Hohl-Ebinger, M. Yoshita, N. Kopidakis, et al., Solar cell efficiency tables (Version 60), Prog. Photovolt. Res. Appl. 30 (7) (2022) 687−701. Available from: https://doi.org/10.1002/pip.3595.

[3] A.K.-W. Chee, On current technology for light absorber materials used in highly efficient industrial solar cells, Renew. Sustain. Energy Rev. 173 (2023) 113027. Available from: https://doi.org/10.1016/j.rser.2022.113027.

[4] A. Sahu, A. Garg, A. Dixit, A review on quantum dot sensitized solar cells: past, present and future towards carrier multiplication with a possibility for higher efficiency,

Sol. Energy 203 (2020) 210−239. Available from: https://doi.org/10.1016/j.solener.2020.04.044.
[5] S. Kalogirou (Ed.), McEvoy's Handbook of Photovoltaics: Fundamentals and Applications, Third Edition, Academic Press, 2018, p. 1340.
[6] M.A. Fraga, D. Amos, S. Sonmezoglu, V. Subramaniam (Eds.), Sustainable Material Solutions for Solar Energy Technologies: Processing Techniques and Applications, 1st Edition, Elsevier, 2021, p. 647.
[7] V.K. Khanna (Ed.), Nano-Structured Photovoltaics: Solar Cells in the Nanotechnology Era, CRC Press, Boca Raton, FL, USA, 2022, p. 494.
[8] A. Roy, T.K. Mallick, A.A. Tahir, M.A. Gondal, S. Sundaram, Advanced fabrication strategies to enhance the performance of dye-sensitized solar cells, in: S. Sundaram, V. Subramaniam, R.P. Raffaelle, M.K. Nazeeruddin, A. Morales-Acevedo, M. Bernechea Navarro, A.F. Hepp (Eds.), Photovoltaics Beyond Silicon, Elsevier, Cambridge, MA, 2024, pp. 223−254.
[9] G. Sauvé, Development of active layer materials for solution-processable organic photovoltaics, in: S. Sundaram, V. Subramaniam, R.P. Raffaelle, M.K. Nazeeruddin, A. Morales-Acevedo, M. Bernechea Navarro, et al. (Eds.), Photovoltaics Beyond Silicon, Elsevier, Cambridge, MA, 2024, pp. 255−302.
[10] A.K. Chauhan, P. Kumar, S.N. Sharma, Perovskite solar cells: Past, present and future, in: S. Sundaram, V. Subramaniam, R.P. Raffaelle, M.K. Nazeeruddin, A. Morales-Acevedo, M. Bernechea Navarro, A.F. Hepp (Eds.), Photovoltaics Beyond Silicon, Elsevier, Cambridge, MA, 2024, pp. 113−164.
[11] K. Shanks, Concentrator and multijunction PV structures, in: S. Sundaram, V. Subramaniam, R.P. Raffaelle, M.K. Nazeeruddin, A. Morales-Acevedo, M. Bernechea Navarro, et al. (Eds.), Photovoltaics Beyond Silicon, Elsevier, Cambridge, MA, 2024, pp. 499−522.
[12] A.F. Hepp, R.P. Raffaelle, Nature-inspired and computer-aided approaches to enable improved photovoltaic materials, more efficient processing and novel devices, in: S. Sundaram, V. Subramaniam, R.P. Raffaelle, M.K. Nazeeruddin, A. Morales-Acevedo, M. Bernechea Navarro, A.F. Hepp (Eds.), Photovoltaics Beyond Silicon, Elsevier, Cambridge, MA, 2024, pp. 325−404.
[13] K.Y. Cheong, A. Apblett (Eds.), Sustainable Materials and Green Processing for Energy Conversion, Elsevier, Cambridge, MA, USA, 2021, 504 pp.
[14] J. Ajayan, D. Nirmal, P. Mohankumar, M. Saravanan, M. Jagadesh, L. Arivazhagan, A review of photovoltaic performance of organic/inorganic solar cells for future renewable and sustainable energy technologies, Superlatt. Microstruct. 143 (2020) 106549. Available from: https://doi.org/10.1016/j.spmi.2020.106549.
[15] M. Eggermont, V. Shyam, A.F. Hepp (Eds.), Biomimicry for Materials, Design and Habitats: Innovations and Applications, Elsevier, Cambridge, MA, USA, 2022, 590 pp.
[16] R.J. Martín-Palma, A. Lakhtakia, Engineered biomimicry for harvesting solar energy: A bird's eye view, Int. J. Smart Nano Mater. 4 (2) (2013) 83−90. Available from: https://doi.org/10.1080/19475411.2012.663812.
[17] A. Gholami, M. Ameri, M. Zandi, R.G. Ghoachani, Electrical, thermal and optical modeling of photovoltaic systems: step-by-step guide and comparative review study, Sustain. Energy Technol. Assess. 49 (2022) 101711. Available from: https://doi.org/10.1016/j.seta.2021.101711.
[18] F. Oba, Y. Kumagai, Design and exploration of semiconductors from first principles: A review of recent advances, Appl. Phys. Exp. 11 (6) (2018) 060101. Available from: https://iopscience.iop.org/article/10.7567/APEX.11.060101.
[19] A. Mellit, S.A. Kalogirou, Machine learning and deep learning for photovoltaic applications, Artificial Intelligence for Smart Photovoltaic Technologies, AIP

Publishing (online), Melville, New York, NY, USA, 2022. Available from: https://doi.org/10.1063/9780735424999_001.
[20] D. Shin, B. Saparov, D.B. Mitzi, Defect engineering in multinary Earth-abundant chalcogenide photovoltaic materials, Adv. Energy Mater. 7 (11) (2017) 1602366. Available from: https://doi.org/10.1002/aenm.201602366.
[21] M. Ikram, R. Malik, R. Raees, M. Imran, F. Wang, et al., Recent advancements and future insight of lead-free non-toxic perovskite solar cells for sustainable and clean energy production: A review, Sustain. Energy Technol. Assess. 53 (A) (2022) 102433. Available from: https://doi.org/10.1016/j.seta.2022.102433.
[22] A. Apblett, A.R. Barron, A.F. Hepp (Eds.), Nanomaterials via Single-Source Precursors: Synthesis, Processing and Applications, Elsevier, Cambridge, MA, USA, 2022, 628 pp.
[23] S.L. Castro, S.G. Bailey, R.P. Raffaelle, K.K. Banger, A.F. Hepp, Synthesis and characterization of colloidal $CuInS_2$ nanoparticles from a molecular single-source precursor, J. Phys. Chem. B 108 (33) (2004) 12429−12435.
[24] O.M. Yaghi, J.L.C. Rowsell, Metal-organic frameworks: A new class of porous materials, Microporous Mesoporous Mater. 73 (1−2) (2004) 3−14.
[25] H. Furukawa, K.E. Cordova, M. O'Keeffe, O.M. Yaghi, The chemistry and applications of metal-organic frameworks, Science 341 (6149) (2013) 1230444.
[26] X. Liang, X. Zhou, C. Ge, H. Lin, S. Satapathi, Q. Zhu, et al., Advance and prospect of metal-organic frameworks for perovskite photovoltaic devices, Org. Electron. 106 (2022) 106546. Available from: https://doi.org/10.1016/j.orgel.2022.106546.
[27] B. Chen, Z. Yang, Q. Jia, R.J. Ball, Y. Zhu, Y. Xia, Emerging applications of metal-organic frameworks and derivatives in solar cells: Recent advances and challenges, Mater. Sci. Eng. R: Rep. 152 (2023) 100714. Available from: https://doi.org/10.1016/j.mser.2022.100714.
[28] L. Fagiolari, M. Sampò, A. Lamberti, J. Amici, C. Francia, S. Bodoardo, et al., Integrated energy conversion and storage devices: Interfacing solar cells, batteries and supercapacitors, Energy Storage Mater. 51 (2022) 400−434. Available from: https://doi.org/10.1016/j.ensm.2022.06.051.
[29] T.-Y. Yang, Y.Y. Kim, J. Seo, Roll-to-roll manufacturing toward lab-to-fab-translation of perovskite solar cells, APL. Mater. 9 (2021) 110901. Available from: https://doi.org/10.1063/5.0064073.
[30] A.F. Hepp, R.P. Raffaelle, I.T. Martin, Space photovoltaics: new technologies, environmental challenges, and missions, in: S. Sundaram, V. Subramaniam, R.P. Raffaelle, M.K. Nazeeruddin, A. Morales-Acevedo, M. Bernechea Navarro, A.F. Hepp (Eds.), Photovoltaics Beyond Silicon, Elsevier, Cambridge, MA, 2024, pp. 675−766.
[31] R.P. Raffaelle, An introduction to space photovoltaics: Technologies, issues, and missions, in: S.G. Bailey, A.F. Hepp, D.C. Ferguson, R.P. Raffaelle, S.M. Durbin (Eds.), Space Photovoltaics: Materials, Missions and Alternative Technologies, Elsevier, Cambridge, MA, USA, 2022, pp. 3−27.
[32] D.M. Chapin, C.S. Fuller, G.L. Pearson, New silicon p-n junction photocells for conversion of solar radiation into electrical power, J. Appl. Phys. 25 (1954) 676−677.
[33] <https://nssdc.gsfc.nasa.gov/nmc/spacecraft/display.action?id=1958-002B> (accessed 13.02.23).
[34] E. Becquerel, On electron effects under the influence of solar radiation, Comptes Rendus, Academie des Sciences, Paris 9 (1839) 561−567.
[35] C.E. Fritts, On a new form of selenium photocell, Am. J. Sci. 26 (1883) 465−472. Available from: https://doi.org/10.2475/ajs.s3-26.156.465.

[36] R.T. Tung, The physics and chemistry of the Schottky barrier height, Appl. Phys. Rev. 1 (2014) 011304. Available from: https://doi.org/10.1063/1.4858400.
[37] A brief history of solar panels. <https://www.smithsonianmag.com/sponsored/brief-history-solar-panels-180972006/> (accessed 23.05.23).
[38] K. Krane, Modern Physics, 4th Edition, John Wiley & Sons, Hoboken, NJ, 2020, 592 pp.
[39] G.N. Lewis, The conservation of photons, Nature 118 (1926) 874−875.
[40] J.Z. Buchwald, R. Fox (Eds.), The Oxford Handbook of the History of Physics, Oxford University Press, 2013, 945 pp.
[41] W.W. Coblenz, Thermal generator, U.S. Patent 1.077,219 (Oct. 28, 1913).
[42] A. Luque, S. Hegedus (Eds.), Handbook of Photovoltaic Science and Engineering, John Wiley & Sons, Ltd, 2010, 1132 pp.
[43] R.S. Ohl, Light-sensitive electric device, U.S. Patent 2,402,662, June 25, 1946.
[44] G.L. Pearson, J. Bardeen, Electrical properties of pure silicon and silicon alloys containing boron and phosphorus, Phys. Rev. 75 (5) (1949) 865−883. Available from: https://journals.aps.org/pr/abstract/10.1103/PhysRev.75.865.
[45] D.M. Chapin, C.S. Fuller, G.L. Pearson, Solar energy converting apparatus, U.S. Patent 2,780,765, February 5, 1957.
[46] F.M. Smits, History of silicon solar cells, IEEE Trans. Electron. Devices 23 (7) (1976) 640−643. Available from: https://ieeexplore.ieee.org/document/1478479.
[47] R.L. Easton, M.J. Votaw, Vanguard I, Satellite IGY (1958 Beta), Rev. Sci. Instrum. 30 (2) (1959) 70−75. Available from: https://doi.org/10.1063/1.1716492.
[48] S. Bailey, R. Raffaelle, Space solar cells and arrays, in: A. Luque, S. Hegedus (Eds.), Handbook of Photovoltaic Science and Engineering, John Wiley & Sons, Ltd, 2010, pp. 365−401. Available from: https://doi.org/10.1002/9780470974704.ch9.
[49] A.A. Siddiqi, Beyond Earth: A Chronicle of Deep Space Exploration, 1958−2016, second edition, National Aeronautics and Space Administration, Office of Communications, NASA, Washington, DC. History Division, 2018, NASA SP2018-4041, The NASA history series, report available at <https://www.nasa.gov/connect/ebooks/beyond_earth_detail.html> (accessed 24.05.23).
[50] <https://www.nasa.gov/topics/technology/features/telstar.html> (accessed 24.05.23).
[51] D.F. Hoth, E.F. O'Neill, I. Welber, The Telstar satellite system, Bell Syst. Tech. J. 42 (4) (1963) 765−799. Available from: https://doi.org/10.1002/j.1538-7305.1963.tb04019.
[52] D.R. Glover, NASA experimental communications satellites, 1958−1995, in: A.J. Butrica (Ed.), Beyond the Ionosphere: Fifty Years of Satellite Communication SP-4217, Washington, D.C. <https://history.nasa.gov/SP-4217/ch6.htm>, 1997.
[53] <https://www.nrel.gov/index.html> (accessed 24.05.23).
[54] P. Sheldon, Process integration issues in thin-film photovoltaics and their impact on future research directions, Prog. Photovolt. Res. Appl. 8 (2000) 77−91.
[55] H.-W. Schock, R. Noufi, CIGS-based solar cells for the next millennium, Prog. Photovolt. Res. Appl. 8 (2000) 151−160.
[56] D. Bonnet, P.V. Meyers, Cadmium telluride − material for thin film solar cells, J. Mater. Res. 13 (1998) 2740−2753.
[57] J. Ramanujam, D.M. Bishop, T.K. Todorov, O. Gunawan, J. Rath, R. Nekovei, et al., Flexible CIGS, CdTe and a-Si:H based thin film solar cells: A review, Prog. Mater. Sci. 110 (2020) 100619. Available from: https://doi.org/10.1016/j.pmatsci.2019.100619.
[58] T.D. Lee, A.U. Ebong, A review of thin film solar cell technologies and challenges, Renew. Sust. Energ. Rev. 70 (2017) 1286−1297. Available from: https://doi.org/10.1016/j.rser.2016.12.028.

[59] A. Romeo, E. Artegiani, CdTe-based thin film solar cells: past, present and future, Energies 14 (6) (2021) 1684. Available from: https://doi.org/10.3390/en14061684.
[60] B.M. Başol, V.K. Kapur, C.R. Leidholm, A. Halani, K. Gledhill, Flexible and light weight copper indium diselenide solar cells on polyimide substrates, Sol. Energy Mater. Sol. Cell 43 (1) (1996) 93—98.
[61] N.G. Dhere, V.S. Gade, A.A. Kadam, A.H. Jahagirdar, S.S. Kulkarni, S.M. Bet, Development of $CIGS_2$ thin film solar cells, Mater. Sci. Eng.: B 116 (3) (2005) 303—309.
[62] A.F. Hepp, S.G. Bailey, R.P. Raffaelle, Inorganic photovoltaic materials and devices: past, present, and future, in: S.-S. Sun, N.S. Sariciftci (Eds.), Organic Photovoltaics: Mechanisms, Materials and Devices, CRC Press, Boca Raton, FL, USA, 2005, pp. 19—36.
[63] M. Yamaguchi, High-efficiency GaAs-based solar cells, in: M.M. Rahman, A.M. Asiri, A. Khan, I. Inamuddin, T. Tabbakh (Eds.), Post-Transition Metals, InTech Open, 2021. Available from: https://doi.org/10.5772/intechopen.94365.
[64] S. Bailey, R. Raffaelle, Chapter II-4-B — Operation of solar cells in a space environment, in: S. Kalogirou (Ed.), McEvoy's Handbook of Photovoltaics: Fundamentals and Applications, Third Edition, Academic Press, 2018, pp. 987—1003.
[65] M. Yamaguchi, Multi-junction solar cells: present and future, in: Proceedings of 12th International Photovoltaics Science and Engineering Conference, Jeju, ROK, June 11—15, 2001, Kyung Hee Information Printing Co. Ltd., Seoul, Korea, 2001, pp. 291—295.
[66] W. Shockley, H.J. Queisser, Detailed balance limit of efficiency of p-n junction solar cells, J. Appl. Phys. 32 (1961) 510—519. Available from: https://doi.org/10.1063/1.1736034.
[67] S. Rühle, Tabulated values of the Shockley-Queisser limit for single junction solar cells, Sol. Energy 130 (2016) 139—147. Available from: https://doi.org/10.1016/j.solener.2016.02.015.
[68] R. Verduci, V. Romano, G. Brunetti, N.Y. Nia, A. Di Carlo, G. D'Angelo, et al., Solar energy in space applications: Review and technology perspectives, Adv. Energy Mater. 12 (29) (2022) 2200125. Available from: https://doi.org/10.1002/aenm.202200125.
[69] M.A. Green, The future of crystalline silicon solar cells, Prog. Photovolt. Res. Appl. 8 (2000) 127—139.
[70] K. Yoshikawa, H. Kawasaki, W. Yoshida, T. Irie, K. Konishi, et al., Silicon heterojunction solar cell with interdigitated back contacts for a photoconversion efficiency over 26%, Nat. Energy 2 (2017) 17032. Available from: https://doi.org/10.1038/nenergy.2017.32.
[71] J. Zhao, A. Wang, M.A. Green, 19.8 percent efficient, 'honeycomb' textured, multicrystalline and 24.4 percent monocrystalline silicon solar cells, Appl. Phys. Lett. 73 (1998) 1991—1993.
[72] L. El Chaar, L.A. lamont, N. El Zein, Review of photovoltaic technologies, Renew. Sust. Energy Rev. 15 (5) (2011) 2165—2175. Available from: https://doi.org/10.1016/j.rser.2011.01.004.
[73] D.J. Friedman, S.R. Kurtz, K.A. Bertness, A.E. Kibbler, C. Kramer, et al., 30.2 percent efficient GaInP/GaAs monolithic two-terminal tandem concentrator cells, Prog. Photovolt. Res. Appl. 3 (1995) 47—50.
[74] R.P. Raffaelle, S.L. Castro, A.F. Hepp, S.G. Bailey, Quantum dot solar cells, Prog. Photovolt. Res. Appl. 10 (2002) 433—439.
[75] A. Luque, A. Martí, C. Stanley, Understanding intermediate-band solar cells, Nat. Photon. 6 (3) (2012) 146—152. Available from: https://doi.org/10.1038/nphoton.2012.1.

[76] D.J. Friedman, J.F. Geisz, S.R. Kurtz, J.M. Olson, 1-eV solar cells with GaInNAs active layer, J. Cryst. Growth 195 (1998) 409−415. Available from: https://doi.org/10.1016/S0022-0248(98)00561-2.
[77] A. Goetzberger, J. Luther, G. Willek, Solar cells: Past, present, future, Sol. Energy Mater. Sol. Cell 74 (1) (2002) 1−11. Available from: https://doi.org/10.1016/S0927-0248(02)00042-9.
[78] S. Guha, J. Yang, A. Banerjee, Amorphous silicon alloy photovoltaic research − present and future, Prog. Photovolt. Res. Appl. 8 (2000) 141−150.
[79] K.K. Banger, J.A. Hollingsworth, J.D. Harris, J. Cowen, W.E. Buhro, A.F. Hepp, Ternary single-source precursors for polycrystalline thin-film solar cells, Appl. Organomet. Chem. 16 (2002) 617−627. Available from: https://doi.org/10.1002/aoc.362.
[80] J.E. Dickman, A.F. Hepp, D.L. Morel, C.S. Ferekides, J.R. Tuttle, D.J. Hoffmann, et al., Utility of thin film solar cells on flexible substrates for space power, in: Proceedings of the 1st International Energy Conversion Engineering Conference, Portsmouth, VA, August 2003, AIAA 2003−5922, 10 pp. Available from: https://doi.org/10.2514/6.2003-5922.
[81] A.F. Hepp, J.S. McNatt, J.E. Dickman, D.L. Morel, C.S. Ferekides, et al., Thin-film solar cells on polymer substrates for space power, in: Proceedings of 2nd International Energy Conversion Engineering Conference, Providence, August 2004, RI, AIAA-2004-5537, 18 pp. Available from: https://doi.org/10.2514/6.2004-5737.
[82] N.G. Dhere, S.R. Ghongadi, M.B. Pandit, A.H. Jahagirdar, D. Scheiman, $CIGS_2$ thin-film solar cells on flexible foils for space power, Prog. Photovolt.: Res. Appl. 10 (6) (2002) 407−416.
[83] B.M. Başol, V.K. Kapur, A. Halani, C. Leidholm, Copper indium diselenide thin film solar cells fabricated on flexible foil substrates, Sol. Energy Mater. Sol. Cell 29 (2) (1993) 163−173.
[84] J.A. AbuShama, J. Wax, T. Berens, J. Tuttle, Progress toward improved device performance in large-area $Cu(In,Ga)Se_2$ thin film solar cells, in: 2006 IEEE 4th World Conference on Photovoltaic Energy Conference, 2006, pp. 487−490. <https://ieeexplore.ieee.org/document/4059670>.
[85] R.P. Raffaelle, A.F. Hepp, J.E. Dickman, D.J. Hoffman, N. Dhere, J.R. Tuttle, et al., Thin-film solar cells on metal foil substrates for space power 18 (2004, AIAA-2004-5735). Available from: https://doi.org/10.2514/6.2004-5735.
[86] G.A. Landis, A.F. Hepp, Applications of thin-film photovoltaics for space, in: Proceedings of the 26th Intersociety Energy Conversion Engineering Conference, Vol. 2, August 4−9, 1991, pp. 256−261.
[87] D.J. Hoffman, T.W. Kerslake, A.F. Hepp, M.K. Jacobs, D. Ponnusamy, Thin-film photovoltaic solar array parametric assessment, in: Proceedings of the 35th Intersociety Energy Conversion Engineering Conference, vol. 1, July 24−28, 2000, pp. 670−680, AIAA-2000-2919, also: NASA/TM-2000-210342. https://ntrs.nasa.gov/api/citations/20000081753/downloads/20000081753.pdf.
[88] S.G. Bailey, A.F. Hepp, R.P. Raffaelle, Thin film photovoltaics for space applications, in: Proceedings of the 36th Intersociety Energy Conversion Engineering Conference, vol. 1, July 29−August 2, 2001, pp. 235−238.
[89] L.K. Reb, M. Böhmer, B. Predeschly, S. Grott, C.L. Weindl, et al., Perovskite and organic solar cells on a rocket flight, Joule 4 (9) (2020) 1880−1892. Available from: https://doi.org/10.1016/j.joule.2020.07.004.
[90] L. McMillon-Brown, T.J. Peshek, Perovskite solar cells: background and prospects for space power applications, in: R.P. Raffaelle, S.G. Bailey, D.C. Ferguson, S. Durbin, A.F. Hepp (Eds.), Space Photovoltaics: Materials, Missions and Alternative Technologies, Elsevier, Cambridge, MA USA, 2023, pp. 129−156.

Eu-doped phosphor species, Thin Solid. Films 618 (2016) 141−145. Available from: https://doi.org/10.1016/j.tsf.2016.03.063.
[239] P. Chiu, Space applications of III-V single- and multijunction solar cells, in: S.G. Bailey, A.F. Hepp, D.C. Ferguson, R.P. Raffaelle, S.M. Durbin (Eds.), Space Photovoltaics: Materials, Missions and Alternative Technologies, Elsevier, Cambridge, MA, USA, 2023, pp. 79−127.
[240] M. Anaya, G. Lozano, M.E. Calvo, H. Míguez, ABX_3 perovskites for tandem solar cells, Joule 1 (4) (2017) 769−793. Available from: https://doi.org/10.1016/j.joule.2017.09.017.
[241] S. Akhil, S. Akash, A. Pasha, B. Kulkarni, M. Jalalah, et al., Review on perovskite silicon tandem solar cells: Status and prospects 2T, 3T and 4T for real world conditions, Mater. & Des. 211 (2021) 110138. Available from: https://doi.org/10.1016/j.matdes.2021.110138.
[242] M. Jošt, L. Kegelmann, L. Korte, S. Albrecht, Monolithic perovskite tandem solar cells: a review of the present status and advanced characterization methods toward 30% efficiency, Adv. Energy Mater. 10 (26) (2020) 1904102. Available from: https://doi.org/10.1002/aenm.201904102.
[243] M.I. Elsmani, N. Fatima, M.P.A. Jallorina, S. Sepeai, M.S. Su'ait, Recent issues and configuration factors in perovskite-silicon tandem solar cells towards large scaling production, Nanomaterials 11 (12) (2021) 3186. Available from: https://doi.org/10.3390/nano11123186.
[244] E. Raza, Z. Ahmad, Review on two-terminal and four-terminal crystalline-silicon/perovskite tandem solar cells: Progress, challenges, and future perspectives, Energy Rep. 8 (2022) 5820−5851. Available from: https://doi.org/10.1016/j.egyr.2022.04.028.
[245] F. Fu, J. Li, T.C.-J. Yang, H. Liang, A. Faes, et al., Monolithic perovskite-silicon tandem solar cells: From the lab to fab? Adv. Mater. 34 (24) (2022) 2106540. Available from: https://doi.org/10.1002/adma.202106540.
[246] P. Wu, D.T. Gangadharan, M.I. Saidaminov, H. Tan, Roadmap for efficient and stable all-perovskite tandem solar cells from a chemistry perspective, ACS Cent. Sci. 9 (1) (2023) 14−26. Available from: https://doi.org/10.1021/acscentsci.2c01077.
[247] L. Fara, L. Chilibon, D. Craciunescu, A. Diaconu, S. Fara, Review: Heterojunction tandem solar cells on Si-based metal oxides, Energies 16 (2023) 3033. Available from: https://doi.org/10.3390/en16073033.
[248] S. Wu, M. Liu, A.K.-Y. Jen, Prospects and challenges for perovskite-organic tandem solar cells, Joule 7 (3) (2023) 484−502. Available from: https://doi.org/10.1016/j.joule.2023.02.014.
[249] T. Nie, Z. Fang, X. Ren, Y. Duan, S.F. Liu, Recent advances in wide-bandgap organic−inorganic halide perovskite solar cells and tandem application, Nano-Micro Lett. 15 (2023) 70. Available from: https://doi.org/10.1007/s40820-023-01040-6.
[250] N.E.I. Boukortt, C. Triolo, S. Santangelo, S. Patanè, All-perovskite tandem solar cells: From certified 25% and beyond, Energies 16 (8) (2023) 3519. Available from: https://doi.org/10.3390/en16083519.
[251] I.J. Park, H.K. An, Y. Chang, J.Y. Kim, Interfacial modification in perovskite-based tandem solar cells, Nano Convergence 10 (2023) 22. Available from: https://doi.org/10.1186/s40580-023-00374-6.
[252] O.M. Saif, A.H. Zekry, M. Abouelatta, A. Shaker, A comprehensive review of tandem solar cells integrated on silicon substrate: III/V vs perovskite, Silicon (2023). Available from: https://doi.org/10.1007/s12633-023-02466-8.
[253] R. Wang, M. Han, Y. Wang, J. Zhao, J. Zhang, et al., Recent progress on efficient perovskite/organic tandem solar cells, J. Energy Chem. 83 (2023) 158−172. Available from: https://doi.org/10.1016/j.jechem.2023.04.036.

[254] M. Umeno, T. Kato, T. Egawa, et al., High efficiency AlGaAs/Si tandem solar cell over 20%, Sol. Energy Mater. Sol Cell 41–42 (1996) 395–403. Available from: https://doi.org/10.1016/0927-0248(95)00123-9.

[255] T.J. Grassman, D.J. Chmielewski, S.D. Carnevale, J.A. Carlin, S.A. Ringel, et al., GaAs$_{0.75}$P$_{0.25}$/Si dual-junction solar cells grown by MBE and MOCVD, IEEE J. Photovolt. 6 (1) (2016) 326–331. Available from: https://doi.org/10.1109/JPHOTOV.2015.249336546.

[256] S. Fan, Z.J. Yu, R.D. Hool, P. Dhingra, W. Weigand, et al., Current-matched III–V/Si epitaxial tandem solar cells with 25.0% efficiency, Cell Rep. Phys. Sci. 1 (9) (2020) 100208. Available from: https://doi.org/10.1016/j.xcrp.2020.100208.

[257] S. Essig, C. Allebé, T. Remo, J.F. Geisz, M.A. Steiner, et al., Raising the one-sun conversion efficiency of III–V/Si solar cells to 32.8% for two junctions and 35.9% for three junctions, Nat. Energy 2 (9) (2017) 17144. Available from: https://doi.org/10.1038/nenergy.2017.144.

[258] X. Chen, Z. Jia, Z. Chen, T. Jiang, L. Bai, et al., Efficient and reproducible monolithic perovskite/organic tandem solar cells with low-loss interconnecting layers, Joule 4 (7) (2020) 1594–1606. Available from: https://doi.org/10.1016/j.joule.2020.06.006.

[259] L. Liu, H. Xiao, K. Jin, Z. Xiao, X. Du, et al., 4-Terminal inorganic perovskite/organic tandem solar cells offer 22% efficiency, Nano-Micro Lett. 15 (2023) 23. Available from: https://doi.org/10.1007/s40820-022-00995-2.

[260] W. Chen, Y. Zhu, J. Xiu, G. Chen, H. Liang, Monolithic perovskite/organic tandem solar cells with 23.6% efficiency enabled by reduced voltage losses and optimized interconnecting layer, Nat. Energy 7 (3) (2022) 229–237. Available from: https://doi.org/10.1038/s41560-021-00966-8.

[261] Y. Ding, Q. Guo, Y. Geng, Z. Dai, Z. Wang, et al., A low-cost hole transport layer enables CsPbI$_2$Br single-junction and tandem perovskite solar cells with record efficiencies of 17.8% and 21.4%, Nano Today 46 (2022) 101586. Available from: https://doi.org/10.1016/j.nantod.2022.101586.

[262] Y. Wang, R. Lin, X. Wang, C. Liu, Y. Ahmed, et al., Oxidation-resistant all-perovskite tandem solar cells in substrate configuration, Nat. Commun. 14 (2023) 1819. Available from: https://doi.org/10.1038/s41467-023-37492-y.

[263] A.F. Palmstrom, G.E. Eperon, T. Leijtens, R. Prasanna, S.N. Habisreutinger, et al., Enabling flexible all-perovskite tandem solar cells, Joule 3 (9) (2019) 2193–2204. Available from: https://doi.org/10.1016/j.joule.2019.05.009.

[264] K. Xiao, Y.H. Lin, M. Zhang, R.D.J. Oliver, X. Wang, et al., Scalable processing for realizing 21.7%-efficient all-perovskite tandem solar modules, Science 376 (6594) (2022) 762–767. Available from: https://doi.org/10.1126/science.abn7696.

[265] X. Hu, J. Li, C. Wang, H. Cui, Y. Liu, et al., Antimony potassium tartrate stabilizes wide-bandgap perovskites for inverted 4-T all-perovskite tandem solar cells with efficiencies over 26%, Nano-Micro Lett. 15 (2023) 103. Available from: https://doi.org/10.1007/s40820-023-01078-6.

[266] Q.F. Han, Y.T. Hsieh, L. Meng, J.L. Wu, P.Y. Sun, et al., High-performance perovskite/Cu(In, Ga)Se$_2$ monolithic tandem solar cells, Science 361 (6405) (2018) 904–908. Available from: https://doi.org/10.1126/science.aat5055.

[267] M. Jošt, E. Köhnen, A. Al-Ashouri, T. Bertram, Š. Tomšič, et al., Perovskite/CIGS tandem solar cells: From certified 24.2% toward 30% and beyond, ACS Energy Lett. 7 (4) (2022) 1298–1307. Available from: https://doi.org/10.1021/acsenergylett.2c00274.

[268] J. Luo, L. Tang, S. Wang, H. Yan, W. Wang, et al., Manipulating Ga growth profile enables all-flexible high-performance single-junction CIGS and 4 T perovskite/CIGS tandem solar cells, Chem. Eng. J. 455 (Part 2) (2023) 140960. Available from: https://doi.org/10.1016/j.cej.2022.140960.

[269] J. Zheng, C.F.J. Lau, H. Mehrvarz, F.-J. Ma, Y. Jiang, et al., Large area efficient interface layer free monolithic perovskite/homo-junction-silicon tandem solar cell with over 20% efficiency, Energy Environ. Sci. 11 (2018) 2432−2443. Available from: https://doi.org/10.1039/C8EE00689J.

[270] J. Zheng, H. Mehrvarz, F.-J. Ma, C.F.J. Lau, M.A. Green, et al., 21.8% efficient monolithic perovskite/homo-junction-silicon tandem solar cell on 16 cm^2, ACS Energy Lett. 3 (9) (2018) 2299−2300. Available from: https://doi.org/10.1021/acsenergylett.8b01382.

[271] H. Shen, S.T. Omelchenko, D.A. Jacobs, S. Yalamanchili, Y. Wan, et al., In situ recombination junction between p-Si and TiO_2 enables high-efficiency monolithic perovskite/Si tandem cells, Sci. Adv. 4 (12) (2018) eaau9711. Available from: https://doi.org/10.1126/sciadv.aau971.

[272] B. Chen, P. Wang, R. Li, N. Ren, Y. Chen, et al., Composite electron transport layer for efficient *n-i-p* type monolithic perovskite/silicon tandem solar cells with high open-circuit voltage, J. Energy Chem. 63 (2021) 461−467. Available from: https://doi.org/10.1016/j.jechem.2021.07.018.

[273] A. Al-Ashouri, E. Köhnen, B. Li, A. Magomedov, H. Hempel, et al., Monolithic perovskite/silicon tandem solar cell with >29% efficiency by enhanced hole extraction, Science 370 (6522) (2020) 1300−1309. Available from: https://doi.org/10.1126/science.abd4016.

[274] <https://www.csem.ch/press/new-world-records-perovskite-on-silicon-tandem-solar?pid = 172296> (accessed 23.06.23).

[275] Z. Li, Y. Zhao, X. Wang, Y. Sun, Z. Zhao, et al., Cost analysis of perovskite tandem photovoltaics, Joule 2 (8) (2018) 1559−1572. Available from: https://doi.org/10.1016/j.joule.2018.05.001.

CHAPTER TWO

Third-generation photovoltaics: Introduction, overview, innovation, and potential markets

Sahaya Dennish Babu George[1], Ananthakumar Soosaimanickam[2] and Senthilarasu Sundaram[3]
[1]Department of Physics, Chettinad College of Engineering and Technology, Karur, Tamil Nadu, India
[2]R&D Division, Intercomet S.L., Madrid, Spain
[3]School of Computing, Engineering and Digital Technologies, Teesside University, Tees Valley, Middlesbrough, United Kingdom

2.1 Introduction

Advanced photovoltaic (PV) technologies, which began development approximately 30 years ago, are commonly known as third (3rd)-generation solar cells [1,2]. The various PV technologies are being pursued with the aim(s) of enhanced efficiency, reliability, and affordability of PV. These advanced solar cells are based on new materials and technologies, such as nanotechnology, quantum dots (QDs), and multijunction devices; see Fig. 2.1 [2]. The overall goal of 3rd-generation PV is to create solar cells that enable increased energy production at a lower cost and to be used in a wider range of applications [1,2].

Some of the key technological advances of 3rd-generation PV include the following:

1. Organic photovoltaics (OPV): OPV are made of organic materials and are lightweight, flexible, and easy to produce. They have the potential to be used in a variety of applications, including portable electronic devices, wearable technology, and building-integrated photovoltaics (BIPV) [2].
2. Dye-sensitized solar cells (DSSCs): These cells absorb light using dyes, which then transform it into electrical energy. They are a viable

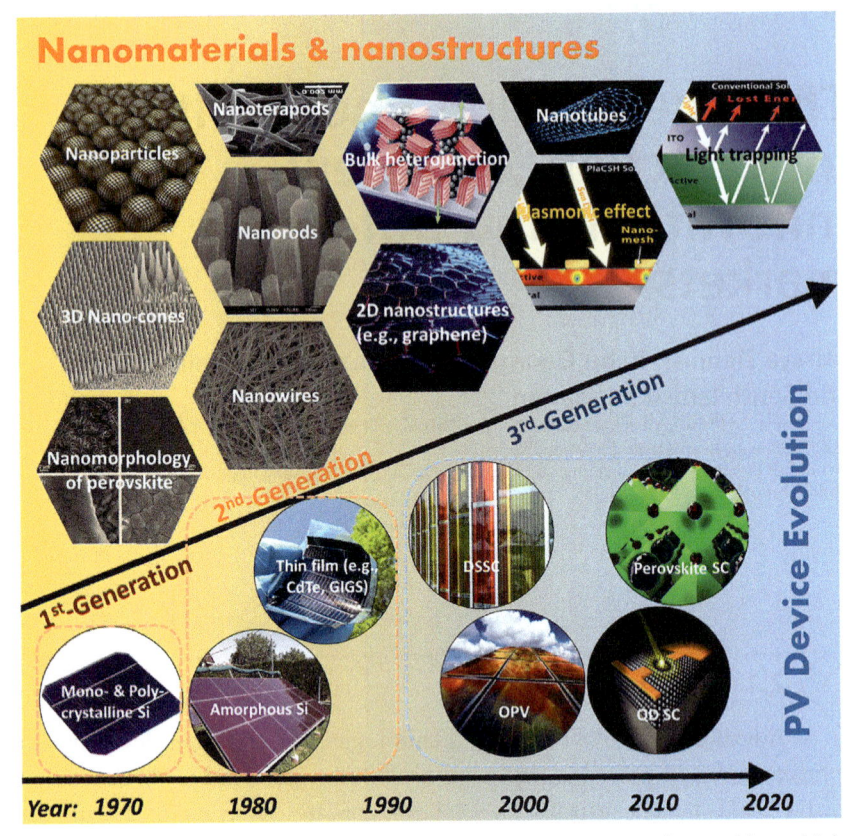

Figure 2.1 Timeline of the three photovoltaic (PV) generations along with multiple nanomaterials and nanostructures that have been successfully employed in the 3rd-generation PV, including dye-sensitized solar cells, organic photovoltaics, perovskite solar cells, and quantum dot (sensitized) solar cells. *Reproduced with permission from C. Wu, K. Wang, M. Batmunkh, A.S.R. Bati, D. Yang, et al., Multifunctional nanostructured materials for next generation photovoltaics, Nano Energy 70 (April 2020) 104480, copyright (2020) Elsevier.*

   option for broad solar energy generation because they are constructed from inexpensive materials and can be produced using straightforward procedures [3].
3. Quantum dot (sensitized) solar cells (QD(S)SCs): QDs contain bandgaps that can be adjusted in size to function at a variety of energies. The material selection determines the bandgap in bulk materials. This characteristic makes QDs appealing for multijunction solar cells, which increase efficiency by collecting various parts of the solar spectrum using a variety of materials [4].

4. Perovskite solar cells (PSCs): These modern emerging solar cells collect light and transform it into electrical energy using a unique family of substances known as perovskites. They have shown to be very effective and to have low-cost manufacturing processes, which makes them a potential technology for the generation of solar energy in the future [5].

Several new prospects for the advancement of solar energy technology are presented by 3rd-generation PV. Compared to conventional PV, they are more effective, adaptable, and affordable, which makes them a desirable alternative for the generation of renewable energy. Overall, the development of 3rd-generation solar cells is an area that is growing quickly and has a great deal of promise to increase solar energy production's effectiveness, performance, and cost [4,5]. To be made commercially available and widely used, however, they still need to go through a lot of research and development.

2.2 Features of third-generation solar cells

Devices that convert photons into energy with great efficiency may be made more affordably with 3rd-generation PV technology [5–7]; the key features and expectations of 3rd-generation solar cells include the following:

1. High efficiency: 3rd-generation solar cells are intended to eventually enhance energy conversion rates, leading to better levels of efficiency than most (except for III–V materials) 1st and 2nd-generation solar panels.
2. Thin-film technology: Thin-film technology (like second-generation inorganic materials) is used in this new generation of solar cells, making the panels lighter, thinner, and more flexible.
3. DSSCs: This type of solar cell uses a dye to absorb light and convert it into electricity. This allows for a more efficient use of light, increasing the energy conversion rate.
4. Hybrid solar cells: 3rd-generation solar cells also include hybrid solar cells, which combine multiple technologies to improve performance.
5. OPV cells: This type of solar cell uses organic materials instead of inorganic materials, making them more environmentally friendly.
6. Reduced cost: 3rd-generation solar cells can be fabricated with less expensive materials, reducing the overall cost of production, and making them more accessible for a variety of applications.

7. Customizable design: 3rd-generation solar cells can be customized to fit specific needs and requirements, making them more versatile and adaptable for a range of applications.

2.3 Terawatt energy generation challenges

To address the global energy issue and facilitate the transition to a greener, more sustainable energy future, the Terawatt Challenge is an international initiative [8]. The mission is a call to action for people, groups, and governments to collaborate to come up with answers that would lower greenhouse gas emissions, boost energy efficiency, and encourage the use of renewable sources of energy. To accomplish a global shift to renewable energy sources, including solar, wind, hydro, and geothermal power, is one of the key objectives of the Terawatt Challenge. This would help lessen the effects of climate change by reducing reliance on fossil fuels, which are the main source of greenhouse gas emissions [9]. The Terawatt Challenge calls for a considerable investment in renewable energy technologies, including research and development (R&D), production, and infrastructure, to meet this objective. Increasing energy efficiency is one of the Terawatt Challenge's main goals. By utilizing energy more effectively, we may cut down on energy waste and lower our overall energy usage. Energy-efficient solutions, such as insulation, LED lighting, and energy-efficient appliances, can be used to achieve this.

The Terawatt Challenge also asks for the creation of smart grid technologies that can facilitate the utilization of renewable energy sources and more effectively control energy use [10]. The Terawatt Challenge also acknowledges the value of education and public awareness in advancing a future powered by sustainable energy. We can contribute to the development of a more knowledgeable and involved populace that can embrace and propel the migration to a greener energy future by informing individuals and groups about the advantages of clean technology. Collaboration and coordination between individuals, groups, and governments are necessary to successfully complete the Terawatt Challenge [11]. The creation and implementation of new technology as well as financial investments in renewable energy all depend heavily on the private sector. Incentives for the advancement of renewable energy, energy efficiency requirements, and the establishment of renewable energy regulations are just a few of

Figure 2.2 Challenges of terawatt energy generation.

the policies and laws that governments must play a significant role in developing to promote the move to a future of clean energy.

The Terawatt Challenge also calls for international cooperation and collaboration. By working together across borders, we can share knowledge and resources, reduce the costs of renewable energy technologies, and accelerate the transition to a sustainable energy future. This collaboration should include both developed and developing countries, as all nations have a role to play in addressing the global energy crisis and promoting a cleaner energy future [12,13]. By transitioning to renewable energy sources, increasing energy efficiency, and promoting education and awareness, we can achieve a cleaner, more sustainable energy future. As depicted in Fig. 2.2, the Terawatt Challenges is an opportunity for the global community to work together to build a better future for this and future generations; the six primary challenges are detailed below [9,12–15]:

1. Renewable energy storage: To generate terawatt-level energy, renewable energy sources need to be stored efficiently to ensure that energy is available whenever required. Currently, the storage of renewable energy is limited and requires significant investments to improve the technology.
2. Grid infrastructure: To support the integration of large amounts of renewable energy into the power grid, there is a need to upgrade and expand the grid infrastructure. This will require significant investments and the development of smart grid technology.

3. Integration of multiple energy sources: To achieve terawatt-level energy generation, a mix of renewable energy sources such as solar, wind, hydro, and geothermal need to be integrated into the grid. This requires complex planning and coordination between different energy sources.
4. Investment costs and returns: The development of renewable energy projects and the associated infrastructure require significant investments. Governments and private companies need to work together to provide funding for these projects. The private sector needs to see credible, reliable, and compelling return on investment projections.
5. Public acceptance: Renewable energy projects often face opposition from local communities due to concerns about unreliability, the impact on the (local) environment, and the visual impact of large installations. It is important to address these concerns through objective studies, public education, and engagement.
6. Regulatory framework: A clear and stable regulatory framework is necessary to provide a predictable environment for the development of renewable energy projects. The absence of clear regulations can hinder investment and development in this field.

2.4 Summary and overview of third-generation photovoltaics

Third-generation cells are "emerging" technologies that are yet-to-be commercially viable. As discussed above, OPV, DSSCs, QD(S)SCs, and PSCs are grouped together as 3rd generation or emerging technologies; see Fig. 2.1. In the past fifteen years, both organic and PSCs have advanced to the point where large-scale deployment is feasible. For this reason, we choose to concentrate on these four developing technologies because they have promising futures. In a typical PV cell, the conversion begins with the creation of charges caused by light, which are then transported and collected by the electrodes [12,15]. Due to the fundamentally different properties of the active layer components, organic semiconductors and hybrid organic-inorganic perovskites, which will collect light and convert it into electricity, organic and PSCs vary in this context in the process of charge production. The 3rd-generation solar cell's manufacturing costs are lower, and the cells' efficiency is acceptable [1–3,15].

3rd-generation solar cell technologies cover a wide range of technologies. They include DSSCs, QD(S)SCs, and perovskite-sensitized solar cells. Like other solar cells, these are made up of a photoanode, a counter electrode (CE), and a medium enabling charge transmission. The operating theory is also comparable. In the 1970s, work on sensitized PV began using organic dyes as sensitizers [2,6,7]. Organic dyes can indeed be synthetic or natural. Plant sources can be used to make natural organic dyes; however, their effectiveness and performance are poor. QD(S)SCs, which make use of QDs or particle semiconductor nanocrystals with a tiny band gap as well as a high extinction coefficient, have emerged as DSSCs technology continues to advance [2–4,16,17].

Initially, scientists used perovskite materials as photosensitizers for DSSC [16]. However, perovskites must be manufactured under extremely sterile and controlled settings because they are highly moisture-sensitive compounds [1,2,5–7]. The photoelectrode is a highly important component in sensitized solar cells because it is where the sensitizer generates the electrons. When lighted, photoanodes will absorb energy, excite, and transport electrons. A hole conductor or a redox mediator in the electrolyte will be used to transfer the electrons from the photoanode to the electrode and back to the sensitizer. The dye sensitizer, a mesoporous semiconductor oxide layer, and a transparent conducting oxide make up the photoanode for DSSCs [17,18]. The redox mediator utilized in the electrolyte is another distinction between them. The development of 3rd-generation PV is still in its infancy, and significant technological advancements are required for them to become economically feasible [19]. Fig. 2.3 illustrates

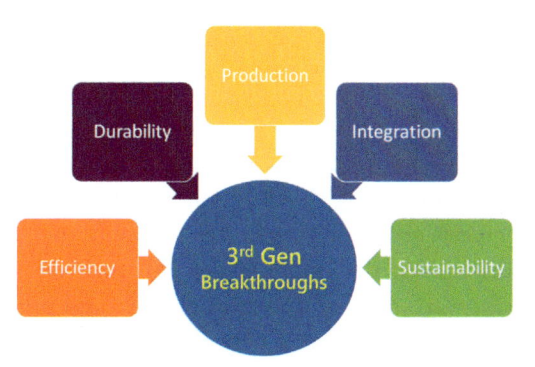

Figure 2.3 Major areas where breakthroughs are needed to enable commercial applications of 3rd-generation solar cell technologies.

the five most important areas where technical breakthroughs [2,6, 7,9,10,16–19] are needed, a detailed explanation of each metric is as follows:
1. Efficiency: To compete, 3rd-generation PV must outperform current technology in terms of efficiency. While certain innovations, such as PSCs, have already shown encouraging results, further study is still required to improve their stability and efficacy.
2. Durability: Solar cells must be strong enough to survive challenging environmental factors, including UV radiation, high humidity, and severe temperatures. For 3rd-generation PVs new coatings and materials must be created that can increase durability.
3. Production: For 3rd-generation PV to be commercially feasible, production costs must be lower than those of previous technology. For them to be commercialized, low-cost, scalable production procedures would be crucial.
4. Integration: To provide successful and effective energy management, PV must be integrated with some other energy system components, such as power storage systems as well as smart grids. The effectiveness of 3rd-generation PV will depend heavily on the creation of technology that can make this integration easier.
5. Sustainability: From the extraction of raw materials to the final disposal, 3rd-generation PV must be ecologically sustainable during their entire existence. For PV to be widely used, it will be crucial to develop technologies that reduce the negative environmental effects of its usage and manufacture.

For a broader context and alternative perspective, it is useful to consider a recently introduced metric of cumulative publications by Dale and Scarpulla as an indicator of total R&D efforts [20]. Interestingly, a 1st generation (Si), two 2nd generation (Cu(In,Ga)Se$_2$ (CIGS)) and CdTe, and a 3rd-generation (halide PSCs) technologies have each achieved 20%–24% efficiency (also termed: following the same learning curve) within 10,000 publications (Fig. 2.4A), with a consistent marginal rate of 5% efficiency increase order of magnitude in publications. The authors of the study acknowledge that insights gleaned from non-PV technologies, cross-pollination from other PV technologies, and proprietary (commercial) efforts are not accounted for by this metric. This analysis yields an interesting alternative perspective and novel insights into PV technology development trajectories [20].

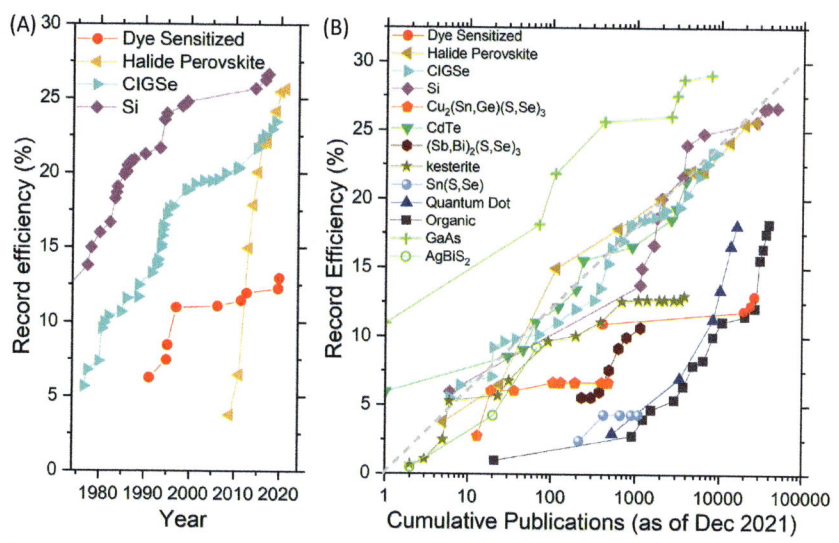

Figure 2.4 (A) Record certified efficiency versus time for four selected single-junction photovoltaic technologies in three (Si (1st), CIGS (2nd), and DSSC and PSC (3rd)) generations exhibiting very different trajectories; see NREL record cell efficiency chart (https://www.nrel.gov/pv/cell-efficiency.html, accessed November 3, 2023). (B) Record efficiencies for a larger set of technologies as a function of cumulative publications since the first working solar cell was announced. The dashed line indicates a learning slope of 5% absolute efficiency per each factor of 10 in publications, taken as a proxy for effort. *Reproduced from P.J. Dale, M.A. Scarpulla, Efficiency versus effort: A better way to compare best photovoltaic research cell efficiencies? Sol. Energy Mater. Sol. Cells 251 (March 2023) 112097, this article is distributed under the terms of the Creative Commons (CC-BY 4.0) license (http://creativecommons.org/licenses/by/4.0/), which permits unrestricted use.*

2.5 Economic assessment and market status of third-generation photovoltaics

This section of the chapter examines business or commercial prospects for the major types of 3rd-generation solar cell technologies. We cover OPV, DSSCs, QD(S)SCs, and PSCs. For each solar cell technology, we address applications and potential markets at present and into the future.

2.5.1 Organic solar cells

Organic solar cells (OSCs) (or OPV) have been the subject of intense study over the past few decades; several key publications are listed that explore interesting results [21–24]. An OSC consists of a sandwich structure where the active layer is fabricated in between electron-transport layer and hole-transport layer (HTL). The bandgap of the active layer materials should be smaller than the energy range of the visible region. In this aspect, low-bandgap organic semiconducting polymers play a vital role. Because of the large absorption coefficient, there are several types of semiconducting polymers studied for the fabrication of high-efficiency OSCs through cost-effective strategies. Advantages of OSCs include the ability to structurally tune the conjugated polymers to alter their photo and electronic properties, ease of deposition using simple coating and printing techniques, and mechanical flexibility for fabricating flexible devices [25]. The features also include the feasibility of large-area coating, easy integration in different devices, cost-reduction compared with other technologies, and ecological and economic advantages [26]. The carrier dissociation at the interface of the donor/acceptor junction is crucially analyzed with different types of organic acceptor and donor molecules.

Given its lower bandgap of 1.1 eV, Si solar cells can harvest a larger fraction of the solar spectrum, compared with OSCs. This is because the best-performing polymers have a bandgap of nearly 2 eV, for example, poly-3-hexyl thiophene (P3HT). In this case, P3HT could harvest only 22% of the visible photons and this limitation hinders the achievement of highly efficient OSCs. However, through tremendous efforts, with the ternary donor-acceptor morphology formation, the efficiency of the OSCs has reached over 19% in single-junction [27,28]. This is still high in the tandem configuration when OSC is coupled with PSCs as n-i-p monolithic type [29–31]. Also, fully transparent OSCs have been fabricated using low-bandgap polymeric donor and acceptor molecules, which is a major breakthrough in this area [32,33]. Furthermore, growing plants under the greenhouse atmosphere, for other indoor lighting purposes, highlights the importance of OSCs for future generation applications. The evolution of OSCs from opaque to transparent is quite impressive; significant milestones have been achieved in terms of new material development and device fabrication. For large-scale production, this technology has been successfully realized by means of slot-die coating, screen printing, and spray deposition. However, several manufacturing challenges currently exist in OSCs, which limit the fabrication technologies within a production scale for industrial and commercial applications, such as how far we could manage the radiative and nonradiative recombination, phase separation of the

Table 2.1 Typical materials used for the fabrication of an organic solar cell.

Layer	Example materials
Substrate	PET, other flexible plastic, and glass
Front electrode	ITO, Ag, and graphite
Electron carrier	ZnO
Active layer	P3HT and PCBM
Hole carrier	PEDOT:PSS and PEDOT:PET
Back electrode	LiF/Ag and Al
Back laminate	PET and PET/PVF

Source: Reproduced with permission from A. Gambhir, P. Sandwell, J. Nelson, The future costs of OPV - A bottom-up model of material and manufacturing costs with uncertainty analysis, Solar Energy Mater. Solar Cells 156 (2016) 49−58, copyright (2016) Elsevier.

donor and acceptor blends, physical degradation, and effective encapsulation strategies. Along with all these efforts, the cost of materials used for the fabrication of OSCs is another important criterion to assess how far OSC modules are economically viable for commercialization.

The commercial use of OSCs is extended toward PV windows, façade applications for the BIPV, semitransparent greenhouse gas application, indoor PV, semitransparent floating PV, underwater PV, PV driven water-splitting, vehicle-integrated PV, and space PV [34]. The annual capacity of the OSCs for the year range 2018−25 has been predicted as 60 GW [35]. Because the global PV market is expected to reach US$223.3 billion by the year 2026 [36], the application of OSCs in these segments is expected to capture a considerable percentage. The typical raw materials used in the fabrication of OSCs are listed in Table 2.1 [37].

The position of any kind of solar cell technology is fixed through three important factors, namely, efficiency, lifetime of the solar cell, and cost per W_p (watt-peak, maximum output in watts for the 1 m^2 solar panel at 25°C) [38]. The fabrication cost includes the procurement, manpower, deposition of each layer using these materials, completion of the module with suitable sealants, and additional costs associated with the completion of the module. Simply put, the balance of module cost is included with materials cost, production cost, and overhead costs, and balance of installation cost is included with area-related costs and energy-related costs [39]. According to Mulligan et al., the material cost in a solar cell fabrication can be estimated using the following assumptions [40]:

1. solution preparation, the printing of anode and active layers, and deposition of the contact electrode;

2. fabrication of donor/acceptor film (P3HT/PCBM), HTL (PEDOT: PSS) in the presence of a solvent (chloroform), and deposition of electrode and encapsulant (PET);
3. consideration of current bulk commodities prices;
4. a production rate of 60 m^2/minute;
5. the lifetime of a plant is 20 years;
6. the operation of the plant is 8 hours (per day), 5 days (per week), and 44 weeks (per year);
7. labor and production cost estimates are based on a developed country; and
8. units' money is estimated based on the US dollar (USD).

The mathematical expression of a module cost is given by Eq. (2.1) [40]:

$$\text{Module cost}\left(\frac{\$}{m^2}\right) = \text{Material cost}\left(\frac{\$}{m^2}\right) + \frac{\text{Total capital cost}(\$)}{\text{Annual production rate}\left(\frac{m^2}{y}\right) \times \text{plant life time}(Y)} + \frac{\text{Annual operating cost}\left(\frac{\$}{y}\right)}{\text{Annual production rate}\left(\frac{m^2}{y}\right)} \quad (2.1)$$

Because of the higher absorption coefficient value of organic semiconducting polymers, the thickness of the OSCs is lower than the other technologies such as CdTe and DSSC. According to Kalowekamo and Baker, the fabrication cost of a solar cell module, which provides 5% efficiency, is $1.0 and $2.83/W$_p$ [41,42]. In the same analysis, the authors have pointed out that for an OSC with 7% efficiency and 67% module active area, the module cost remains between $0.20 and $0.30/Wp. Using space-charge-limited current analysis, Yang et al. found that electron and hole mobility values of the aged thick films are lower than the freshly prepared films, which means that the donor/acceptor blend morphology is a transport-limited one [43]. This indicates the careful protection of different layers for the long-term life of a module. The cost of the fabrication of a module is found to be dependent on the efficiency of the solar cell and it is found to be decreasing with the increase of efficiency [44].

Considering the architecture, Guo and Min carried out a detailed cost-wise analysis for the fabrication of an ITO-free OSC module [45]. According to the authors, the material cost is the major part of the cost of a solar cell module (70.75%) and the cost of the active layer alone acquires about 36.07%. Also, the minimum sustainable price of an OSC module could come down from $0.27/W$_p$ when the efficiency is increased above 20%. It is found that the cost of the substrate is accounted to be a higher percentage of the cost if the solar module is fabricated using a plastic substrate like PET [42]. Lee et al. indicated that the fabrication cost of the semitransparent organic module could be reduced to $0.47/W$_p$, which is a positive aspect for the future generation of OSC devices [46]. Another estimation predicts that a low-cost organic module with 5–10 years lifetime with 5%–10% efficiency could produce at $0.10/k/Wh in the central Europe landscape. This shows that producing electricity using organic solar modules can be achieved using cost-effective fabrication technologies. However, compared with the other thin-film solar cell technologies, the fabrication cost of the OSC is always high owing to the protection of organic components in the solar cell and also having an ultra-thin structure, which makes a higher packing cost. Also, the flexible type OSCs in the research sector are focused on exceeding 5 years lifetime and the glass-based OSCs are focused on exceeding 10 years, which is quite lower than the lifetime of existing silicon-based technology [47]. In the commercial market, the product development (i.e., OSC) would mostly depend on two factors, namely, competitive cost and moderate efficiency to withstand in the market. The possible market sections related to the OSC are given in Fig. 2.5 [26].

The predicted polymer solar cell market between the years 2009 and 2017 was from US$0.2 million to US$462.3 million, which was approximately 5000% growth [35]. However, commercial companies such as Konarka, Solarmer, and Heliatek can produce organic solar panels, which

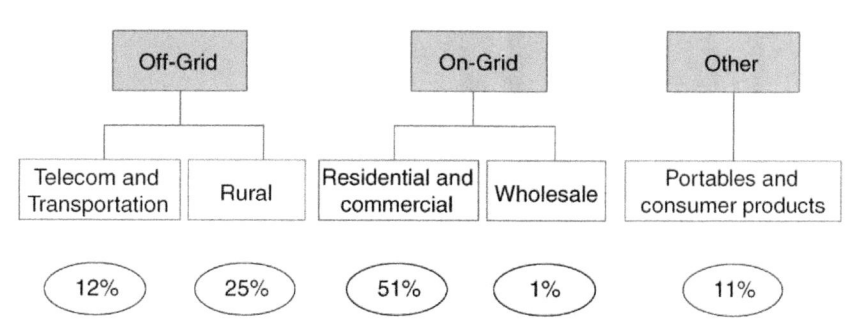

Figure 2.5 The expected possible market segments of an organic solar cell. *Adapted with permission from C. J. Brabec, Organic photovoltaics: Technology and market, Sol. Energy and Sol. Cells 83 (2004) 273–292, copyright (2004) Elsevier.*

have less than 10% efficiency, not suitable to compete with the other existing potential technologies in the market. Although recently developed donor/acceptor molecules are providing nearly 20% efficiency under laboratory conditions, large-scale production of these compounds and maintaining their stability under atmospheric conditions remain as a bottleneck in this field. Despite this, the global indoor PV market is expected to grow toward US$1 billion by 2024 [36,48]. This is still very high for the BIPV application, which is approximately US$5 billion; see Fig. 2.6 [48]. There is considerable development in PV installation in the European Union and the countries in the EU intend to increase renewable energy resources to 27% by 2030 [49]. The current industrial value chain should be strengthened to attract a large number of companies to invest and to investigate OSCs. It is expected that fully transparent OSCs and semitransparent OSCs would possibly capture the future market in this field.

In terms of innovation, the materials, device structure, and patent publications are considered the main sources to achieve the state of art. Specifically, the growing number of patents related to OSCs strongly envisages the importance of this field for future generations of optoelectronic applications. However, the interest related to polymer solar cells is quite low because consumers do not prefer OSCs specifically. Despite this, the current growing trends of OSCs in different directions can be

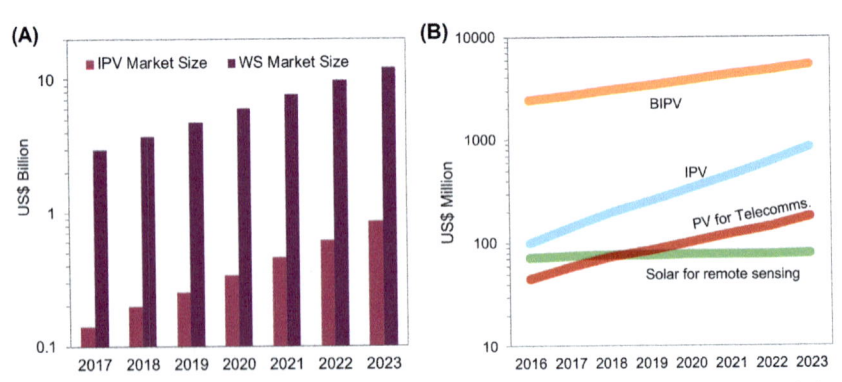

Figure 2.6 Indoor photovoltaics (PV) market size(s): (A) The projected size of the wireless sensor and indoor PV market in billions of dollars and (B) the expected size of alternative markets for PV technologies over the coming years. *Reproduced with permission from I. Mathews, S.N. Kantareddy, T. Buonassisi, I.M. Peters, Technology and market perspective for indoor photovoltaic cells, Joule 3 (6) (2019) 1415–1426, copyright (2019) Elsevier.*

visually seen and several kinds of innovative approaches have led this field into a new arena [50]. Fully transparent solution-processed OSCs have been fabricated with 4% efficiency and this has made a revolution in this field [33]. These kinds of transparent solar cells can be used as power-generating windows in households and buildings, which could selectively block UV or NIR radiation.

Fabricating solar glass is another useful application for mobile sensing applications. Here, two semitransparent solar cells are harvesting energy where the prohibition of UV light is achieved through modifying the absorber layers. These semitransparent solar cells could serve as glass facades in houses and buildings, and could also function as heat-insulation windows [51]. These solar cells have been successfully demonstrated as floating bodies in the water sources for the growth of a plant, for example, algae [52]. It is expected that crop production under semitransparent solar panels reach a new dimension in the near future. Interestingly, Landerer et al. fabricated a semitransparent solar cell with the structure glass/ITO/ZnO/absorber(PBTZT-stat-BDTT-8/PC61BM/PC71BM)/PEDOT:PSS, and this was further used in the glass lens to work as the solar glass [53]. Here, the thickness of the active layer and PEDOT:PSS CE was maintained as 300 and 150 nm. Under 1 sun illumination, this lens-fitted solar cell delivered 0.06% efficiency. These kinds of innovative ideas could help to enlarge the diversity of this field. Construction of "solar tree" is another concept that has several solar cells, which are like a branch of a tree that are coupled through one junction, and this is helpful to harvest a large number of photons. Moreover, the fabrication of an organic solar concentrator together with an OSC is another potential approach to achieve higher efficiency.

Integration of polymer solar cells with another kind of potential device is an emerging technology to utilize the concept for multipurpose applications. For example, OSCs coupled with electrochromic devices also could potentially function as future generation self-powered smart windows applications [54]. Similarly, a semitransparent solar cell coupled with a steam generator can work as solar powered water-electricity generator [55], and a semitransparent polymer solar cell (ST-PSC)/triboelectric nanogenerator (TENG) hybrid system could function for the weather-dependent electricity production [56]; see Fig. 2.7. The dual-energy conversion and optical and heat-insulating properties of the ST-PSC/TENG hybrid devices can be correlated to their various structures [56].

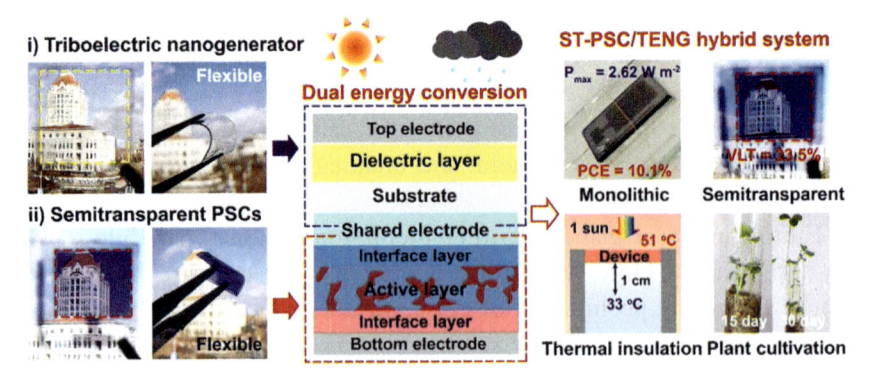

Figure 2.7 Schematic illustration and digital photographs of the semitransparent polymer solar cells (ST-PSC) and triboelectric nanogenerators (TENG), and ST-PSC/TENG hybrid systems developed in the study by Liu et al. *Reproduced with permission from T. Liu, Y. Zheng, Y. Xu, X. Liu, C. Wang, et al., Semitransparent polymer solar cell/triboelectric nanogenerator hybrid systems: Synergistic solar and raindrop energy conversion for window-integrated applications, Nano Energy 103 (2022) 107776, copyright (2022) Elsevier.*

Recently, fabricating OSCs through an artificial intelligence guided machine learning approach has attained popularity among researchers and several attempts have been made to fabricate highly efficient solar cells. The machine learning process is consisted of three important steps, namely, (1) collection and preprocessing of the sample data (input), (2) training and updating the model (learning), and (3) usage of an optimized model to predict the unknown data (output) [57,58]. Fabrication of OSCs using these novel ideas would definitely lead this field in another direction. The energy payback period of OSCs is comparatively high and so all these approaches are getting interested to explore new possibilities.

2.5.2 Dye-sensitized solar cells

DSSCs are considered as a precursor technology for advanced solar cell concepts such as QD(S)SCs and perovskite QDs solar cells (PQDSCs). A typical DSSC consists of four major components, namely, a working electrode (typically a metal oxide), sensitizer, electrolyte, and CE. The sensitizer molecules are effectively anchored on the metal oxide electrode. A charge transfer process in the sensitizer under the exposure of light results extraction of carriers in the respective electrodes in the presence of electrolyte. The theoretical efficiency value of a DSSC is higher than 20%, which makes promising hope to develop this technology to the next level. Since the first report during the year 1991 [59], numerous achievements have been made in this technology to fabricate low-cost, highly stable efficient solar cells. Until now, about approximately 12% efficiency has been achieved using ruthenium-based sensitizer (single dye sensitizing approach), and several other organic and inorganic compound sensitizers are lifting the efficiency to a higher level. The photo-excitation characteristics of several molecular dye compounds have been investigated to explore the possibilities of absorption of a large number of photons. These dye compounds also include several kinds of classifications such as organic dyes—which include cyanine dyes, squaraine dyes, natural dyes, etc.—and metal-free organic dyes, and several kinds of inorganic metal complex dyes such as ruthenium, cobalt, and osmium complex dyes [60–62].

For the extraction of electrons and holes, redox electrolytes in DSSCs play an important role. Although liquid electrolytes are resulting higher efficiency on the laboratory scale, solid-state electrolytes are preferred for the industrial-scale production of DSSCs. In advance of this, cosensitization using different classes of molecular dyes has emerged as a promising approach for enhancing efficiency, and with this, 15.2% efficiency with 500 hours operational stability has been achieved [63]. Although liquid electrolytes initially led in this field, the emergence of solid-state electrolytes has made a revolution to achieve highly stable, higher-performance DSSCs. For example, a quasisolid-state electrolyte-based DSSC has delivered a laboratory-measured (not AM1.5 certified) power conversion efficiency of approximately 25% efficiency, quite promising for large-scale fabrication in industrial applications [64]. In this, polymer-based electrolytes play a crucial role owing to their ionic conductivity property. In the case of CE, because of the highly expensive nature of platinum (Pt), highly electrocatalytic elements are investigated to extract the carriers and considerable efficiency has been achieved with different types of materials, composites, and alloys. Nature and thickness of the working electrodes and CEs, the adsorbing ability of the dye molecules, organic and inorganic binders and additives, and solvents and composition of electrolytes used in the device are playing a key role in the charge transport and consequence efficiency enhancement of the DSSCs.

Compared with c-Si-, CdTe-, and CIGS-based solar cells, DSSCs have several advantages: (1) They can be fabricated using roll-to-roll (R2R) processes without vacuum and high temperature, (2) they can be fabricated with lightweight and flexible substrates, (3) they are mostly fabricated using nontoxic, Earth-abundant elements, (4) they provide less than one-year pay-back period, (5) they are easily available materials and require less investment, (6) they offer different shape, size, and transparent (semitransparent, bifacial) conditions, and (7) they can be fixed to walls, windows, and other BIPV, indoor applications [48]. Furthermore, the toxicity associated with per kg/m^2 of a DSSC is 0.021, which is lower than mc-Si, CIGS, and CdTe solar cell technologies [65]. Also, the material cost to fabricate a module with 10% efficiency should fall between $15/m^2 and $30/m^2 [66]. A reference model of a DSSC with tungsten disulfide (WS_2) CE indicates that the manufacturing cost is US$22.4/m^2 and the processing cost is 5.9/m^2 [67]. Smestad et al. calculated the cost analysis to fabricate a 10% efficient DSSC and the cost per peak watt per unit module area in this measurement was $0.48–$0.80/W_p [68]. This is quite less than the cost of a silicon solar cell, despite having recycling issues. A recent report by Syed and Wei reveals that levelized cost of energy of the DSSC technology is calculated as 0.0438/kWh and the module price is US$0.8 (80 cents)/W [67]. It is predicted that to compete with Si-based solar cell module (approximately $0.20–$0.30/W_p) for niche applications, the module cost of a DSSC should be $0.6/$W_p$ [66]. When the consumption is high, it is expected that the cost of fabricating a module will be even lower.

The global DSSC market was estimated at $49.6 million, a growth rate of over 12% between the years 2015 and 2022 was anticipated at the time of the study's publication [69]. Although the recently emerging PSC is slowly capturing the market, there is still a demand for DSSCs owing to the toxic-free fabrication. The major expected field of applications of the future generation DSSCs includes the following: (1) Internet of things, (2) energy-independent Wi-Fi subcomponents, (3) wireless surveillance camera), (4) (sports) performance monitors, (5) portable medical devices, (6) sensors, and (7) smartwatches [69]. Private companies (i.e., Oxford Photovoltaics, Dye Solar, Solaronix, Sharp Corporation) have established a solid platform to fabricate DSSC modules to apply to different sectors [70]. In particular, considerable attention is paid to applying DSSCs for BIPV applications. Also, other applications such as rooftop greenhouses, supporting the growth of plants, wearable devices, and biomedical sensors are of commercial interest. Perhaps the lifetime of liquid-based electrolytes is quite low for indoor applications, but still the development of solid-state electrolytes will help to lift this issue. An experimental analysis using nine modules of DSSC connected through a series connection reveals that the average power conversion efficiency of the sloped DSSC BIPV window lay at 3.01%, whereas it was 3.40% for the vertically aligned DSSC module for the period of two years [71].

Similar to OSCs, DSSC also can be integrated with other potential energy devices. Integration of electrochromic devices with DSSCs is a potential idea to fabricate future energy devices in which the decoloration/coloration in electrochemical deposition takes place when it receives the voltage while DSSC works [72,73]. The integration of other kinds of devices includes lithium-ion batteries, nanogenerators, OSCs, photoelectrosynthetic cells, PSC, etc. [71,74]. Furthermore, research into the fabrication of DSSCs on flexible substrates is also accelerating utilizing new concepts [75]. The low power conversion efficiency of DSSC is hindering the establishment of a big platform in the field of the industrial scale of DSSC. Although challenges such as charge transfer at the interface, highly conductive substrates for flexible applications, incompetent light scattering in the electrodes, enhancing the stability using novel compounds and composites, and replacement of highly cost Pt electrodes still remain, the future efforts in these areas are expected to solve many issues [72–75]. Recent exploration has focused on the growth rate of the DSSC market in the United States for different potential applications [70].

2.5.3 Quantum-dot-sensitized solar cells

A technological advance called QD solar cells can completely revolutionize the solar PV sector. These solar cells absorb sunlight and transform it into power using tiny semiconductor particles known as QDs. Fig. 2.8 illustrates the complex details of electron transfer in a QD-doped OSC [23]. QD solar cells have several benefits over conventional solar cells, including higher efficiency, reduced cost, and better performance in low light [75]. The capacity of QDs solar cells to absorb a wider variety of sunlight wavelengths, including both visible and infrared light, is one of their main advantages.

Due to their ability to produce more power than conventional solar cells, they are a desirable option for large-scale energy generation. Moreover, lead sulfide (PbS), zinc sulfide (ZnS), and cadmium selenide (CdSe), which are inexpensive and widely accessible, although Pb and Cd are problematic from an environmental perspective, can be used to make QD solar cells, making them more accessible and less expensive than other solar technologies. They, therefore, can become a commonplace energy source with substantially lower carbon emissions [71]. Although the QDs solar cell market is still in its infancy, it is anticipated to expand quickly

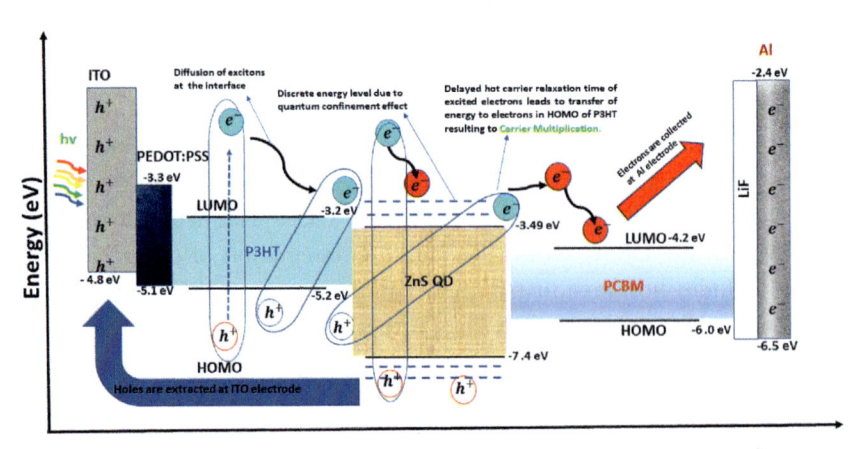

Figure 2.8 Schematic representation of electron-transport energy cascade in zinc sulfide (ZnS) quantum dot-doped P3HT:PCBM blend (organic) solar absorber. *Reproduced with permission from A.K. Ogundele, G.T. Mola, Semiconductor quantum dots as a mechanism to enhance charge transfer processes in polymer solar cells, Chemosphere 45 (December 2023) 140453, copyright (2023) Elsevier.*

over the next several years. The international demand for QDSCs is anticipated to reach $7.6 billion by 2030, expanding at a compound annual growth rate (CAGR) of approximately 25%, beginning in 2022, according to a recently published proprietary report [76].

QD solar cells have enormous market potential, and several businesses have already begun to invest in this technology [75,77]. To increase the effectiveness and scalability of QD solar cells, numerous academic institutes, including the National Renewable Energy Laboratory (NREL) and the Massachusetts Institute of Technology, are conducting considerable research in this area [78,79]. The economic viability and commercial position of QD solar cell manufacturing are currently being assessed as it is still in its infancy. Nevertheless, there seems to be a strong interest in this new PV technology; many researchers and scientists are working on improving its performance and scalability [75–79]. Due to the use of cheap and readily accessible components, it is anticipated that the production costs of QD solar cell technology will be less than those of conventional solar cells. This might enable QD PV cells a more affordable option for generating vast amounts of energy [77,78]. QD solar cells' capacity to absorb a wider spectrum of light wavelengths can also lead to greater energy yields and lower per-watt prices [77–79]. The market growth of QDSCs and its driving forces, chances and opportunities, and challenges from the industrialization perspective are given in Fig. 2.9 and their details follow.

Quantum Dots Solar Cell Market

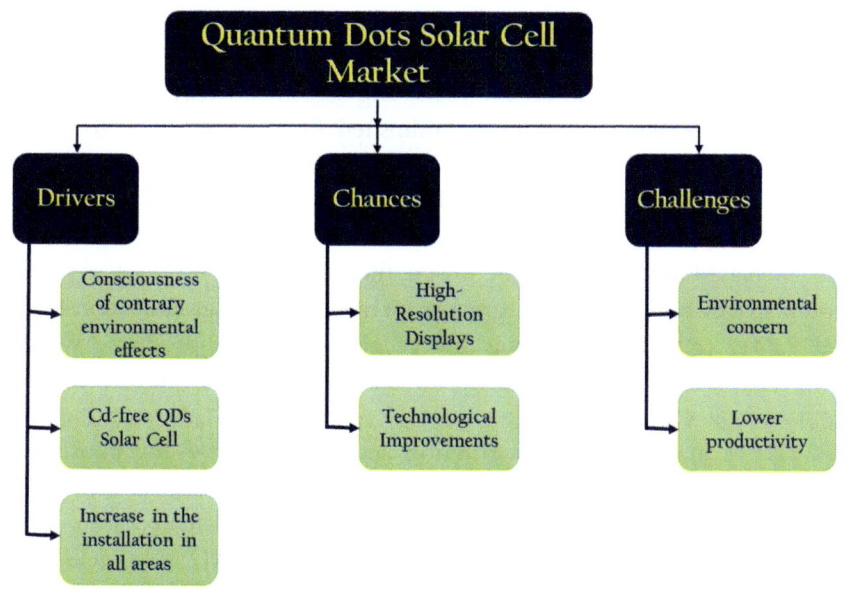

Drivers
- Consciousness of contrary environmental effects
- Cd-free QDs Solar Cell
- Increase in the installation in all areas

Chances
- High-Resolution Displays
- Technological Improvements

Challenges
- Environmental concern
- Lower productivity

Figure 2.9 Details of the quantum dot solar cell market.

Because QD solar cells are a sustainable and eco-friendly energy source, their demand has increased as consumers and industry participants become more conscious of the negative environmental effects of other solar cells. Moreover, the market is positively impacted by the growing reliance on unconventional energy sources, which is anticipated to fuel the market for QD solar cells [75–77]. Consumers and manufacturers have a more environmentally friendly and secure choice in cadmium-free QDs, which gives them the technology's color benefits without any toxicity or significant regulatory risks. These solar cells can be used for displays and a variety of applications, including solar power, lighting, and healthcare [74–78]. The market for QD solar cells is therefore anticipated to develop at a significant CAGR, due to the cadmium-free QD segment [75,77,80].

The market for QD solar cells is expanding because of the increase in the number of renewable systems being installed in homes and businesses. The need to lower solar cell module prices is predicted to increase the desire for QDs in solar cells, which will fuel market expansion. Also, the use of products in the building industries is expanding, there are more government programs to use renewable energy sources, and funding levels for energy availability in numerous locations are increasing [81]. Over the forecast period of 2022–30, multiple aspects are anticipated to boost the QD solar cell market's growth [76].

Due to falling LCD prices and consumer demand for better visual experiences, there has been a noticeable growing market for elevated and large-size displays in recent years. Big-screen showcases are increasingly plausible because of their diminished LCD prices. Due to QDs' capacity to enhance display quality, the attention of display producers is shifting more and more toward the production of 4 and 8 K screens [82–84]. The use of QDs in screens offers a number of benefits, including increased bit depth, faster image frames per second, and a wider color spectrum, which will greatly increase market growth potential. QD solar cells are becoming more inexpensive to produce, which increases the lucrative potential for the big industry players in the approaching years. The market for QD solar cells will also continue to increase as infrastructure construction activities improve and consumer electronics demand rises [76,81–85].

Several governments around the world have implemented strict laws that are anticipated to impede market demand for QD solar cells and restrain market expansion due to the adverse effects of elements such as Cd and Pb on the environment. Over the projected period of 2022–29, the worldwide QD PV cell market could see growth challenges caused by the QD solar cell's lower productivity compared to competing products [76,77,82–86]. This market research study on the QD solar cell industry details recent developments, trade laws, shipping assessment, business model, value chain optimization, customer base, and the effects of domestic and local market participants. It also examines potential revenue pockets, regulatory changes, strategic market expansion analysis, market growth, classification market expansions, application niches and supremacy, product approvals, and product launches.

2.5.4 Perovskite solar cells

A promising new materials technology for producing sustainable energy is PSCs. These solar cells make use of a substance called perovskite; a particular kind of crystalline structure that has proven to be quite effective in converting solar energy into electrical energy. PSCs provide several advantages over other solar energy generation options. Perovskite PV cells have already surpassed regular Si solar cells in conversion efficiency of over 25% [87]. As a result, PSCs are a viable choice for producing solar energy because they can generate a great deal of power from a small surface area. The adaptability of solar cells made of perovskite is another significant quality. They can be produced utilizing a broad spectrum of supplies and methods, which enables a variety of applications [88].

Its tremendous effectiveness, which has already approached 25% in some situations, is one of its most remarkable characteristics. As a result, PSCs are a viable choice for solar power because they can produce a substantial quantity of electricity from a very tiny surface area. PSCs' versatility in terms of the materials and methods used in production is another intriguing quality [88]. Due to their adaptability, they can be used in a variety of contexts, such as the construction of portable electronics and building materials. Furthermore, portable and flexible perovskite solar panels are simple to install and environment adaptable. They are a desirable alternative for supplying electricity to rural places and powering mobile gadgets because of this capability. Fig. 2.10 provides important features of PSCs for possible applications in industrial, commercial, or other practical settings.

PSCs do have significant drawbacks, though, such as their sensitivity to humidity and temperature variations, which over time can reduce their

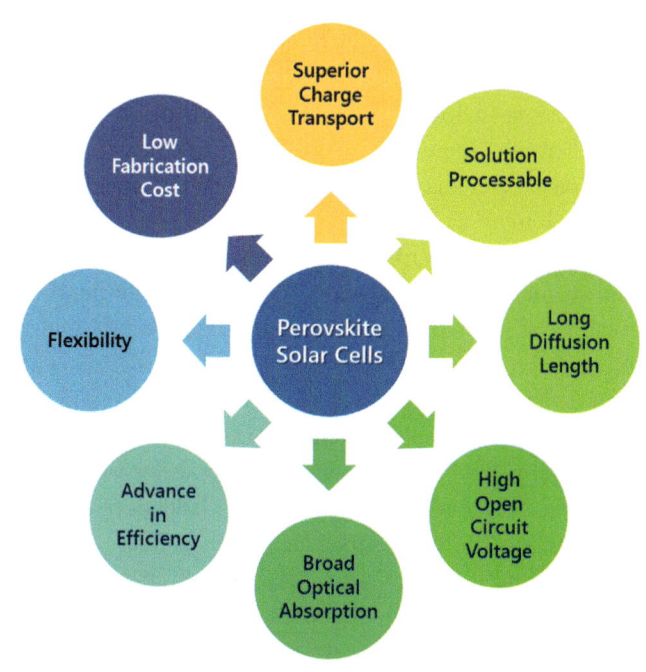

Figure 2.10 Features of perovskite solar cells.

performance [89–92]. The following is a list of some variables that may have an impact on PSCs' market expansion; despite these obstacles, PSCs are a promising technology that merits more research and development:
1. Efficiency: The efficiency of the cells is one of the major aspects that will impact the market for PSCs. Although PSCs have already exceeded 25% efficiency, there is still room for growth [93]. PSCs will be more competitive with conventional silicon-based solar cells, thanks to higher efficiency.
2. Cost: The price of producing PSCs is another crucial element. In comparison to solar cells made of silicon, PSCs are still currently more expensive. The cost is anticipated to go down as the technology advances and scale savings are realized [94].
3. Stability and reliability: Consumers may be concerned about PSCs' lack of stability and dependability compared to silicon-based solar cells. PSCs must become more stable and reliable if they are to be widely used [92].
4. Regulations: Government laws may also influence the market expansion for PSCs. PSC usage can be encouraged by policies that support renewable energy and encourage their use [94].
5. Research and development: PSC technology will need to be developed further to increase its effectiveness, stability, and dependability. This will also assist in lowering prices and enhancing the technology's scalability, all of which are essential for its wider adoption [93].

The main drivers of media attention are improvements in the stated efficiency of limited-area technologies and an increase in publications. It should be highlighted, nevertheless, that the exponential rise in publications and improvement in effectiveness do not necessarily transfer into success in the real world, nor does it ensure that the product will indeed be produced. As of now, the PSC business is still in its infancy, with most of the research and development taking place in university and commercial labs [95]. However, over the past several years, there has been a substantial advancement in the technology, and some businesses have started to market PSC products. PSC economic evaluation is a challenging task because it necessitates examining both possible revenue sources and production expenses. PSCs' cheap manufacturing costs, which may be achieved by using straightforward and affordable fabrication techniques, are one of their key advantages [96,97]. This indicates that compared to conventional silicon-based solar cells, PSCs have much reduced production costs. Until perovskites become a widely used commercial product, however, several technological obstacles still need to be overcome as highlighted in this chapter subsection.

The stability and longevity of the perovskite layer, which is susceptible to deterioration over time, is one of the major difficulties. The solar cells' lifespan and efficiency may be shortened as a result, which may have an impact on their economic feasibility. The adaptability of perovskite PV manufacturing is another problem. Although PSCs can be made using very straightforward techniques, there is still a lack of readily accessible facilities and specialized equipment [93]. This might make it more difficult to scale up to higher production volumes and raise the cost of scaling up PSC production. Notwithstanding these obstacles, PSCs have a large market potential, notably in the areas of portable electronics and BIPV. BIPV, which involves directly integrating solar cells into the building surface, has the potential to significantly lower buildings' energy usage and advance the creation of sustainable cities [94]. PSCs are already being produced by several businesses, and some of these items are commercially available.

Nowadays, the preferred electrode material for PV devices is gold. The presence of the problematic element Pb is a significant barrier to the perovskite solar panel industry. The toxic material used in PSCs, which has an adverse effect on the environment, is limiting the market for these products [94]. Nevertheless. PSCs have a smaller market share than conventional silicon-based solar cells. This is caused in part by the technical difficulties noted earlier, but it is also a result of the dominance of well-established solar cell producers and a shortage of facilities for the mass production of PSCs. In conclusion, because of their excellent performance and low production costs, PSCs get the potential to play a large role in the market for renewable energy [95−98]. However, there are still technological difficulties to be solved, especially about stability and scalability. With continued development and research, it is anticipated that PSCs would become extremely competitive in the market for renewable energy in the future years. In the realm of PV, PSCs have garnered significant interest because of its quick advancement in efficiency, inexpensive manufacturing, and scaling potential. Numerous contemporary PSC varieties have been produced via ongoing research and development to solve issues with stability, efficiency, and commercial viability.

2.5.4.1 Single-junction perovskite solar cells

The fundamental building block of perovskite-based PV devices is represented by single-junction PSCs [5,87−90,98]. Usually, a single perovskite layer is encased in layers of electron and hole transport in these cells. The main goals of single-junction PSC development have been to increase the cells' stability over time and power conversion efficiency. Scholars have investigated many approaches aimed at improving material quality, streamlining device topologies, and alleviating concerns associated with heat stability and moisture absorption. The vulnerability of single-junction PSCs to environmental factors, especially moisture and light exposure, is one of the main problems with them [7]. To increase device stability, attempts have been undertaken to design more stable perovskite formulations, implement efficient encapsulation methods, and create strong interfacial layers [88]. Additionally, improvements in scalable manufacturing techniques like R2R and slot-die printing have increased the possibility of single-junction PSCs being commercialized.

2.5.4.2 Flexible perovskite solar cells
The potential use of flexible PSCs in lightweight, conformable PV systems has attracted a lot of attention. Because of their flexible and bendable form, these cells may be integrated into wearable technology, portable electronics, and curved surfaces. In order to allow for mechanical deformation of the devices without sacrificing their optoelectronic performance, flexible substrates such as metal foils and polymer films have been the focus of research into the creation of flexible PSCs [92]. Achieving high efficiency and mechanical flexibility simultaneously has been a key challenge in the development of flexible PSCs. Researchers have investigated the mechanical properties of various substrates, as well as the impact of mechanical stress on the perovskite layers and charge transport materials [93]. Strategies such as the development of novel encapsulation techniques, introduction of stretchable interfacial layers, and optimization of electrode materials have contributed to the advancement of flexible PSCs toward practical applications in lightweight and portable energy harvesting systems [5–7,29–31,89,98].

2.5.4.3 Tandem perovskite solar cells: An introduction
In comparison to single-junction solar cells, tandem PSCs attain greater efficiencies by combining two or more subcells with complementing light absorption qualities. By using a wider range of the solar spectrum, these tandem structures improve the overall efficiency of power conversion. Perovskite materials are frequently combined with other well-known PV

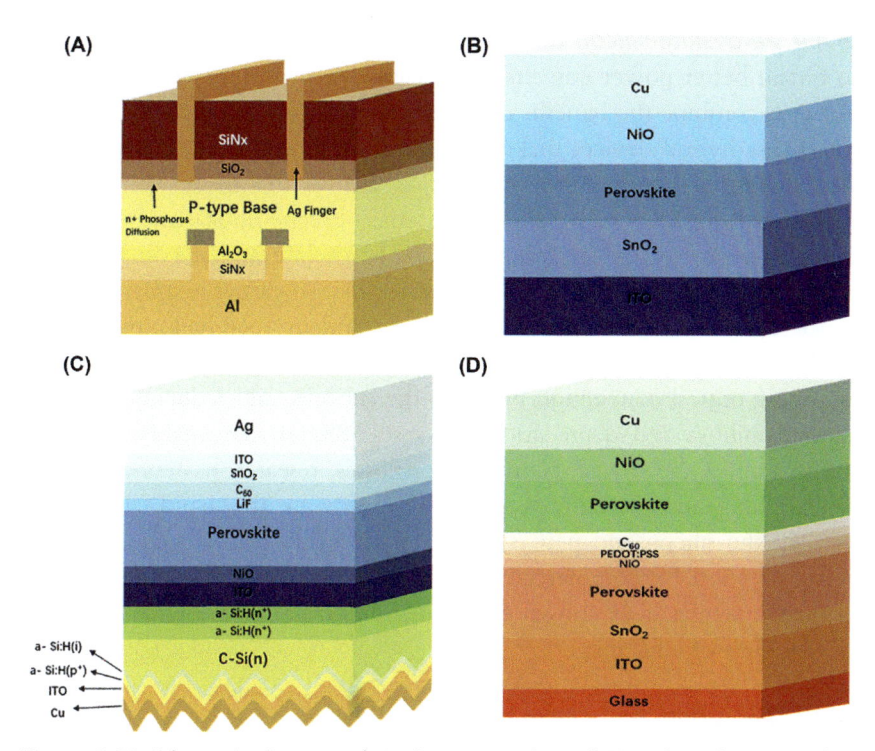

Figure 2.11 Schematic diagram of single-junction (A and B) and tandem (C and D) photovoltaic devices: (A) Traditional silicon cell; (B) planar perovskite cell; (C) silicon/perovskite tandem cell; (D) perovskite/perovskite tandem cell. *Reproduced with permission from Z. Li, Y. Zhao, X. Wang, Y. Sun, Z. Zhao, et al., Cost analysis of perovskite tandem photovoltaics, Joule 2 (8) (August 2018) 1559–1572, copyright (2018) Elsevier.*

materials, such as Si, CIGS, or OPV materials, in tandem PSCs [29–31,97,99,100]. In order to optimize light absorption and charge carrier extraction, tandem PSCs have required complex engineering of device designs, connecting layers, and optical control techniques [90,91]. Innovative material combinations and sophisticated manufacturing processes have been used to overcome issues with interfacial compatibility, optical losses, and current matching between subcells. There is hope for tandem PSCs to be included into next-generation PV technologies because they have proven they can attain efficiencies higher than that of single-junction devices. Fig. 2.11 illustrates two generic device structures of single-junction Si (Fig. 2.11A) and PSC (Fig. 2.11B), as well as SI/PSC (Fig. 2.11C) and PSC/PSC (Fig. 2.11D) tandem devices [97]; Fig. 2.11 provides a visual guide for the reader.

2.5.4.4 Perovskite-silicon tandem solar cells

To obtain better power conversion efficiencies, perovskite-silicon tandem solar cells combine the benefits of both perovskite and silicon PV technology. These hybrid devices make use of silicon solar cells' well-established infrared light absorption infrastructure as well as the better visible light absorption capabilities of perovskite materials. Precise optical and electrical control is required for the integration of perovskite and silicon layers in tandem solar cells in order to maximize charge carrier extraction and reduce interface losses. Perovskite-silicon tandem solar cells provide a number of challenges, such as the creation of strong connecting layers, improving optical connection between the subcells, and reducing material incompatibilities and degradation processes [88,94]. To enable effective charge transfer and lower recombination losses, research has concentrated on designing intermediate layers and improving the band alignment between the silicon and perovskite materials. High-performance PV systems may soon be commercialized, thanks to perovskite-silicon tandem solar cells, which have shown they can outperform single-junction solar cells in terms of efficiency.

2.5.4.5 Perovskite-perovskite tandem solar cells

A breakthrough development in PV technology, perovskite-perovskite tandem solar cells have attracted a lot of attention because of their high efficiency and intriguing properties. These cutting-edge gadgets integrate two distinct perovskite-based materials for increased light absorption and better performance. Perovskite-perovskite tandem solar cells are noteworthy for their ability to outperform conventional single-junction solar cells in terms of efficiency. These tandem solar cells' main advantage is their capacity to combine the complimentary absorption qualities of two separate perovskite materials, each of which is designed to absorb a certain range of the solar spectrum. Overall efficiency is raised by this design's ability to effectively use solar energy over a wider range of wavelengths. [89-93]. Perovskite materials also make it easier to create flexible, lightweight solar cells, which increases the range of possible uses for them, such as in BIPV and portable electronics. Moreover, the absorption spectra may be optimized to more closely match the solar spectrum, thanks to the variable bandgap of perovskite materials. Tandem solar cells may be tailored using this functionality to match unique environmental and energy requirements [95]. Furthermore, the large-scale production of perovskite-based solar cells is made feasible by their low-cost processing methods and ease of fabrication, which might possibly lower the overall cost of solar energy generation. Additionally, the tandem structure improves power conversion efficiency by reducing thermalization and photon energy losses [96]. Perovskite-perovskite tandem solar cells are a viable option for next-generation PV technologies due to their high efficiency potential and the process's scalability and affordability. This will greatly advance the development of renewable energy sources and the worldwide effort to combat climate change.

2.5.4.6 Overview of perovskite single-junction and tandem cells

To summarize, the advancement of contemporary PSC types has greatly increased the potential for scalable, economical, and effective PV technologies. While tandem PSCs have the potential to achieve better efficiencies through wider light absorption, single-junction solar cells have made progress in terms of stability and production scalability. While perovskite-silicon tandem solar cells and perovskite-perovskite tandem solar cells show the potential to exceed the efficiency limits of conventional single-junction devices, flexible PSCs allow the integration of PV systems into various

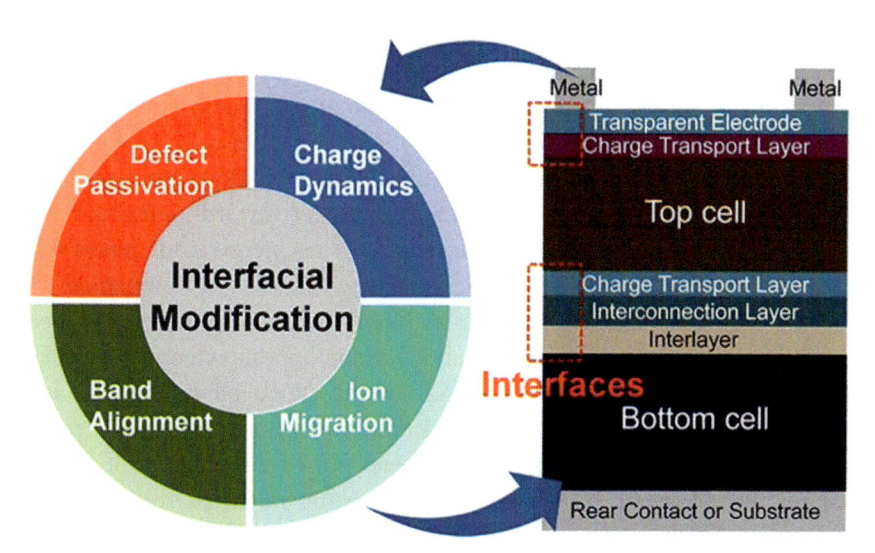

Figure 2.12 Visual summary of issues and concerns arising from interfacial modification while processing perovskite solar cell-top cell tandem photovoltaic devices. *Reproduced from I.J. Park, H.K. An, Y. Chang, J.Y. Kim, Interfacial modification in perovskite-based tandem solar cells. Nano Convergence 10 (1) (2023) 22, this article is distributed under the terms of the Creative Commons (CC-BY 4.0) license (http://creativecommons.org/licenses/by/4.0/), which permits unrestricted use.*

flexible and portable applications. Tandem solar cells can potentially achieve high efficiency using first (Si), second (CIGS), and 3rd-generation (PSC) combined with perovskite top subcells via facile solution fabrication processes [31,95−97,99,100]. However, as photovoltages of the subcells are combined and the structures necessarily contain multiple layers, interfacial issues that cause V_{oc} deficit need to be analyzed and addressed. Other concerns, such as morphological issues or process compatibility, may complicate processing of perovskite top cells; see Fig. 2.12 for a visual summary of interfacial issues and concerns [96]. Persistent investigation and advancement in material science, device optimization, and manufacturing methods will help make PSCs a practical and long-term option to contribute to the world's energy requirements [1,2,7,87−90,92−100].

2.6 Summary and conclusion

Third-generation PV technologies have the potential to outperform current PV technologies in terms of efficiency, robustness, and adaptability. OSCs, PSCs, and QD solar cells are some instances of 3rd-generation PV technology. Compared to current PV technologies, these newer devices have several advantages: Lower potential manufacturing costs, improved efficiency, enhanced flexibility, and a broader application space. BIPV, integrating PV materials into the facades, windows, and roofs of buildings, represent a potentially lucrative market for 3rd-generation PV technologies. By supplying a resource of on-site energy from renewable sources, BIPV can result in significant energy savings. In comparison to conventional PV materials, 3rd-generation PV technologies are expected to provide better design freedom and esthetic appeal, which may make them more appealing to architects and builders. Wearable technology and portable electronics are prospective markets for 3rd-generation PV technologies. With the use of these technologies, solar cells that may be incorporated into clothing, bags, and other accessories might be thin, flexible, and extremely efficient.

In conclusion, because of their high effectiveness, low cost, and minimal impact on the environment, 3rd-generation PV technologies have demonstrated considerable potential for future uses. By offering a dependable, environmentally friendly, and economically advantageous source of energy, these innovations are poised to completely transform the energy sector. 3rd-generation PV have enormous potential for innovation and growth. For instance, it has been shown that DSSCs can be used in interior illumination, portable electronics, and BIPV. Potential uses for OPV include wearable technologies and flexible solar panels. QDSCs have shown promise in products like electricity-producing windows and tandem solar cells. PSCs can be employed in consumer electronics, large-scale power generation, and PV incorporated into buildings. As the demand for renewable energy keeps rising, the market for 3rd-generation PV is anticipated to expand quickly in the future years. In conclusion, 3rd-generation PV methods can completely transform the energy sector by offering a dependable, long-lasting, and affordable source of electricity. These technologies have a huge amount of promise for innovation and growth in the years to come.

Acknowledgments

G.S.D. Babu is grateful to the administrators of Chettinad College of Engineering and Technology, Karur for their constant support and encouragement. A. Soosaimanickam acknowledges the management of Intercomet S.L. for their consistent support during this work.

References

[1] G. Conibeer, Third-generation photovoltaics, Mater. Today 10 (11) (2007) 42–50.
[2] C. Wu, K. Wang, M. Batmunkh, A.S.R. Bati, D. Yang, et al., Multifunctional nanostructured materials for next generation photovoltaics, Nano Energy 70 (2020) 104480. Available from: https://doi.org/10.1016/j.nanoen.2020.104480.
[3] V. Sugathan, E. John, K. Sudhakar, Recent improvements in dye sensitized solar cells: A review, Renew. Sustain. Energy Rev. 52 (2015) 54–64.
[4] D. Sharma, R. Jha, S. Kumar, Quantum dot sensitized solar cell: Recent advances and future perspectives in photoanode, Sol. Energy Mater. Sol. Cell 155 (2016) 294–322.
[5] J.-P. Correa-Baena, M. Saliba, T. Buonassisi, M. Grätzel, A. Abate, W. Tress, et al., Promises and challenges of perovskite solar cells, Science 358 (6364) (2017) 739–744.
[6] S. Khalid, M. Sultan, E. Ahmed, W. Ahmed, Third-generation solar cells, in: W. Ahmed, M. Booth, E. Nourafkan (Eds.), in: Emerging Nanotechnologies for Renewable Energy, Elsevier, 2021, pp. 3–35.
[7] M. Seri, F. Mercuri, G. Ruani, Y. Feng, M. Li, Z.-X. Xu, et al., Toward real setting applications of organic and perovskite solar cells: A comparative review, Energy Technol. 9 (5) (2021) 2000901. Available from: https://doi.org/10.1002/ente.202000901.
[8] R.E. Smalley, Future global energy prosperity: The terawatt challenge, MRS Bull. 30 (6) (2005) 412–417.
[9] P.J. Verlinden, Future challenges for photovoltaic manufacturing at the terawatt level, J. Renew. Sustain. Energy 12 (5) (2020) 053505.
[10] S. Huang, Grand challenges and opportunities in photovoltaic materials and devices, Front. Photonics 2 (2021) 651766.

[11] V. Fthenakis, Sustainability metrics for extending thin-film photovoltaics to terawatt levels, MRS Bull. 37 (4) (2012) 425−430.
[12] T. Meng, Terawatt Solar Photovoltaics Roadblocks and Opportunities, Springer, 2014.
[13] N.S. Lewis, Research opportunities to advance solar energy utilization, Science 351 (6271) (2016) aad1920.
[14] D.S. Ginley, D. Cahen (Eds.), Fundamentals of Materials for Energy and Environmental Sustainability, Cambridge University Press, 2011.
[15] R.G. Newell, D. Raimi, G. Aldana, Global energy outlook 2019: The next generation of energy, Resource for the Future, Report 19-06 Washington D.C., July 2019 46 pp. <https://www.rff.org/publications/reports/global-energy-outlook-2019/>.
[16] S. Rühle, M. Shalom, A. Zaban, Quantum-dot-sensitized solar cells, ChemPhysChem 11 (11) (2010) 2290−2304.
[17] B.E. Hardin, H.J. Snaith, M.D. McGehee, The renaissance of dye-sensitized solar cells, Nat. Photonics 6 (3) (2012) 162−169.
[18] S. Kumar, S. Kumar, R.N. Rai, Y. Lee, T.H.C. Nguyen, et al., Recent development in two-dimensional material-based advanced photoanodes for high-performance dye-sensitized solar cells, Sol. Energy 249 (2023) 606−623.
[19] J. Zheng, W. Liu, T. Cui, H. Wang, F. Chen, et al., A novel domino-like snow removal system for roof PV arrays: Feasibility, performance, and economic benefits, Appl. Energy 333 (2023) 120554.
[20] P.J. Dale, M.A. Scarpulla, Efficiency versus effort: A better way to compare best photovoltaic research cell efficiencies? Sol. Energy Mater. Sol. Cell 251 (March 2023) 112097.
[21] C. Zhao, J. Wang, X. Zhao, Z. Du, R. Yang, J. Tang, Recent advances, challenges and prospects in ternary organic solar cells, Nanoscale 13 (2021) 2181−2208.
[22] Y. Li, W. Huang, D. Zhao, L. Wang, Z. Jiao, et al., Recent progress in organic solar cells: A review on materials from acceptor to donor, Molecules 27 (6) (1800) 2022.
[23] A.K. Ogundele, G.T. Mola, Semiconductor quantum dots as a mechanism to enhance charge transfer processes in polymer solar cells, Chemosphere 45 (Dec. (2023) 140453.
[24] G. Sauvé, Development of active layer materials for solution-processable organic photovoltaics, in: S. Sundaram, V. Subramaniam, R.P. Raffaelle, M.K. Nazeeruddin, A. Morales-Acevedo, M. Bernechea Navarro, et al. (Eds.), Photovoltaics Beyond Silicon, Elsevier, Cambridge, MA, 2024, pp. 255−302.
[25] J.-T. Chen, C.-S. Hsu, Conjugated polymer nanostructures for organic solar cell applications, Polym. Chem. 2 (2011) 2707−2722.
[26] C.J. Brabec, Organic photovoltaics: Technology and market, Sol. Energy Sol. Cell 83 (2004) 273−292. Available from: https://doi.org/10.1016/j.solmat.2004.02.030.
[27] L. Zhu, M. Zhang, J. Xu, C. Li, G. Zhou, et al., Nat. Mater. 21 (2022) 656−663.
[28] W. Gao, F. Qi, Z. Peng, F.R. Lin, K. Jang, et al., Achieving 19% power conversion efficiency in planar-mixed heterojunction organic solar cells using a pseudosymmetric electron acceptor, Adv. Mater. 34 (32) (2022) 2202089.
[29] X. Gu, X. Lai, Y. Zhang, T. Wang, W.L. Tan, et al., Organic solar cell with efficiency over 20% and V_{oc} exceeding 2.1V enabled by tandem with all-inorganic perovskite and thermal annealing-free process, Adv. Sci. 9 (2022) 2200445.
[30] Y.-M. Xie, Q. Yao, H.-L. Yip, Y. Cao, Influence of component properties on the photovoltaic performance of monolithic perovskite/organic tandem solar cells: Subcell, interconnecting layer, and photovoltaic parameters, Small Methods 7 (4) (April 2023) 2201255. Available from: https://doi.org/10.1002/smtd.202201255.
[31] S.-Q. Sun, X. Xu, Q. Sun, Q. Yao, Y. Cai, et al., All-inorganic perovskite-based monolithic perovskite/organic tandem solar cells with 23.21% efficiency by dual-interface engineering, Adv. Energy Mater. (2023) 2204347.

[32] F. Guo, P. Kubis, T. Stubhan, N. Li, D. Baran, et al., Fully solution-processing route toward highly transparent polymer solar cells, ACS Appl. Interfaces 6 (20) (2014) 18251–18257.
[33] C.-C. Chen, L. Dou, R. Zhu, C.-H. Chung, T.-B. Song, et al., Visibly transparent polymer solar cells produced by solution processing, ACS Nano 6 (8) (2012) 7185–7190.
[34] H. Meddeb, M. Gotz-Kohler, N. Neugebohrn, U. Banik, N. Osterthun, et al., Tunable photovoltaics: Adapting solar cell technologies to versatile applications, Adv. Energy Mater. 12 (28) (2022) 2200713.
[35] T.D. Nielsen, C. Cruickshank, S. Foged, J. Thorsen, F.C. Krebs, Business, market and intellectual property analysis of polymer solar cells, Sol. Energy Mater. Sol. Cell 94 (2010) 1553–1571.
[36] F.M. van der Staaij, I.M. van Keulen, E. von Hauff, Organic photovoltaics: Where are we headed? Sol. RRL 5 (8) (2021) 2100167.
[37] A. Gambhir, P. Sandwell, J. Nelson, The future costs of OPV - a bottom-up model of material and manufacturing costs with uncertainty analysis, Sol. Energy Mater. Sol. Cell 156 (2016) 49–58. Available from: https://doi.org/10.1016/j.solmat.2016.05.056.
[38] C.J. Brabec, J.A. Hauch, P. Schilinsky, C. Waldauf, Production aspects of organic photovoltaics and their impact on the commercialization of devices, MRS Bull. 30 (2005) 50–52.
[39] G. Dennler, M.C. Scharber, C.J. Brabec, Polymer-fullerene bulk-heterojunction solar cells, Adv. Mater. 21 (13) (2009) 1323–1338.
[40] C.J. Muligan, M. Wilson, G. Bryant, B. Vaughan, X. Zhou, W.J. Belcher, et al., A projection of commercial-scale organic photovoltaic module costs, Sol. Energy Mater. Sol. Cell 120 (2014) 9–17.
[41] J. Kalowekamo, E. Baker, Estimating the manufacturing cost of purely organic solar cells, Sol. Energy 83 (8) (2009) 1224–1231.
[42] B. Azzopardi, C.J.M. Emmott, A. Urbina, F.C. Krebs, J. Mutale, J. Nelson, Economic assessment of solar electricity production from organic-based photovoltaic modules in a domestic environment, Energy Environ. Sci. 4 (2011) 3741–3753.
[43] W. Yang, W. Wang, Y. Wang, R. Sun, J. Guo, et al., Balancing the efficiency, stability, and cost potential for organic solar cells via a new figure of merit, Joule 5 (5) (2021) 1209–1230.
[44] S. Rasool, J.Y. Kim, Prospects of glove-box versus air-processed organic solar cells, Phys. Chem. Chem. Phys. 25 (2023) 19,337–19,357.
[45] J. Guo, J. Min, A cost analysis of fully solution-processed ITO-free organic solar modules, Adv. Energy Mater. 9 (3) (2019) 1802521.
[46] B. Lee, L. Lahann, Y. Li, S.R. Forrest, Cost estimates of production scale semitransparent organic photovoltaic modules for building integrated photovoltaics, Sustain. Energy Fuels 4 (2020) 5765–5772.
[47] C.J. Brabec, A. Distler, X. Du, H.-J. Egelhaaf, J. Hauch, T. Heumueller, et al., Material strategies to accelerate OPV technology toward a GW technology, Adv. Energy Mater. 10 (43) (2020) 2001864.
[48] I. Mathews, S.N. Kantareddy, T. Buonassisi, I.M. Peters, Technology and market perspective for indoor photovoltaic cells, Joule 3 (6) (2019) 1415–1426.
[49] P. Borawski, J. Holden, A. Beldycka-Borawska, Perspectives of photovoltaic energy market development in the European union, Energy 270 (2023) 126804.
[50] A. Colsmann, H. Rohm, C. Sprau, Shining light on organic solar cells, Sol. RRL 4 (6) (2020) 2000015.
[51] C. Sun, R. Xia, H. Shi, H. Yao, X. Liu, et al., Heat-insulating multifunctional semitransparent polymer solar cells, Joule 2 (9) (2018) 1816–1826.

[52] L. Yin, Y. Zhou, T. Jiang, Y. Xu, T. Liu, et al., Semitransparent polymer solar cells floating on water: Selected transmission windows and active control of algal growth, J. Mater. Chem. C. 9 (2021) 13132−13143.
[53] D. Landerer, D. Bahro, H. Rohm, M. Koppitz, A. Mertens, et al., Solar glasses: A case study on semitransparent organic solar cells for self-powered, smart, wearable device, Energy Tech. 5 (11) (2017) 1936−1945.
[54] X. Jia, E.C. Baird, J. Blochwitz-Nimoth, S. Reineke, K. Vandewal, D. Spoltore, Selectively absorbing small-molecule solar cells for self-powered electrochromic windows, Nano Energy 89 (2021) 106404.
[55] Q. Ji, N. Li, S. Wang, S. Li, F. Li, et al., Synergistic solar-powered water-electricity generation via rational integration of semitransparent photovoltaics and interfacial steam generators, J. Mater. Chem. A 9 (2021) 21197−21208.
[56] T. Liu, Y. Zheng, Y. Xu, X. Liu, C. Wang, et al., Semitransparent polymer solar cell/triboelectric nanogenerator hybrid systems: Synergistic solar and raindrop energy conversion for window-integrated applications, Nano Energy 103 (2022) 107776.
[57] A. Mahmood, J.-L. Wang, Machine learning for high performance organic solar cells: Current scenario and future prospects, Energy Environ. Sci. 14 (2021) 90−105.
[58] Y. Miyake, A. Saeki, Machine learning-assisted development of organic solar cell materials: Issues, analyses and outlooks, J. Phys. Chem. Lett. 12 (51) (2021) 12391−12401.
[59] B. O'Regan, M. Gratzel, A low-cost, high efficiency solar cell based on dye-sensitized colloidal TiO_2 films, Nature 353 (1991) 737−740.
[60] A.S. Polo, M.K. Itokazu, N.Y.M. Iha, Metal complex sensitizers in dye-sensitized solar cells, Coord. Chem. Rev. 248 (2004) 1343−1361.
[61] A. Mishra, M.K.R. Fischer, P. Bauerle, Metal-free organic dyes for dye-sensitized solar cells: From structure: property relationships to design rules, Angew. Chem. 48 (14) (2009) 2474−2499.
[62] B. Pashaei, H. Shahroosvand, P. Abbasi, Transition metal complex redox shuttles for dye-sensitized solar cells, RSC Adv. 5 (2015) 94814−94848.
[63] Y. Ren, D. Zhang, J. Suo, Y. Cao, F.T. Eickemeyer, et al., Hydroxamic acid pre-adsorption raises the efficiency of sensitized solar cells, Nature 613 (2023) 60−65.
[64] I.-P. Liu, Y.-S. Cho, H. Teng, Y.-L. Lee, Quasi-solid-state dye-sensitized indoor photovoltaics with efficiencies exceeding 25%, J. Mater. Chem. A 8 (2020) 22423−22433.
[65] R.D. McConnell, Assessment of the dye-sensitized solar cell, Renew. Sus. Energy Rev. 3 (6) (2002) 271−293.
[66] H.S. Jung, J.-K. Lee, Dye sensitized solar cells for economically viable photovoltaic systems, J. Phys. Chem. Lett. 4 (10) (2013) 1682−1693.
[67] T.H. Syed, W. Wei, Technoeconomic analysis of dye sensitized solar cells (DSSCs) with WS_2/carbon composite as counter electrode material, Inorganics 10 (11) (2022) 191.
[68] A. Maalouf, T. Okoroafor, Z. Jehl, V. Babu, S. Resalati, A comprehensive review on life cycle assessment of commercial and emerging thin-film solar cell systems, Renew. Sustain. Energy Rev. 186 (2023) 113652.
[69] A. Aslam, U. Mehmood, M.H. Arshad, A. Ishfaq, J. Zaheer, et al., Dye-sensitized solar cells (DSSCs) as a potential photovoltaic technology for the self-powered internet of things (IoTs) applications, Sol. Energy 207 (2020) 874−892.
[70] Grand View Research, Dye sensitized solar cell market size, share & trends analysis report by application and segment forecasts, 2020−2027, Report ID: 978-1-68038-644-8, San Francisco, CA, 2019, 105 pp. <https://www.grandviewresearch.com/industry-analysis/dye-sensitized-solar-cell-market>.
[71] H.M. Lee, J.H. Yoon, Power performance analysis of a transparent DSSC BIPV window based on 2-year measurement data in a full-scale mock-up, Appl. Energy 225 (2018) 1013−1021.

[72] A. Fakharuddin, R. Jose, T.M. Brown, F. Fabregat-Santiago, J. Bisquert, A perspective on the production of dye-sensitized solar modules, Energy Environ. Sci. 7 (2014) 3952–3981.
[73] Z. Xie, X. Jin, G. Chen, J. Xu, D. Chen, G. Shen, Integrated smart electrochromic windows for energy saving and storage applications, Chem. Commun. 50 (2014) 608–610.
[74] S. Yun, Y. Qin, A.R. Uhl, N. Vlachopoulos, M. Yin, et al., New-generation integrated devices on dye-sensitized and perovskite solar cells, Energy Environ. Sci. 11 (2018) 476–526.
[75] A.K.-W. Chee, On current technology for light absorber materials used in highly efficient industrial solar cells, Renew. Sustain. Energy Rev. 173 (2023) 113027.
[76] Report Linker, Global quantum dot solar cells industry, Global Industry analysts, Report ID: 6032716, April 2023, 89 pp. <https://www.reportlinker.com/p06032716/Global-Quantum-Dot-Solar-Cells-Industry.html>.
[77] A. Sahu, A. Garg, A. Dixit, A review on quantum dot sensitized solar cells: Past, present and future towards carrier multiplication with a possibility for higher efficiency, Sol. Energy 203 (June 2020) 210–239.
[78] A.J. Nozik, M.C. Beard, J.M. Luther, M. Law, R.J. Ellingson, J.C. Johnson, Semiconductor quantum dots and quantum dot arrays and applications of multiple exciton generation to third-generation photovoltaic solar cells, Chem. Rev. 110 (11) (2010) 6873–6890.
[79] C.-H.M. Chuang, P.R. Brown, V. Bulović, M.G. Bawendi, Improved performance and stability in quantum dot solar cells through band alignment engineering, Nat. Mater. 13 (2014) 796–801. Available from: https://doi.org/10.1038/nmat3984.
[80] M. Kouhnavard, S. Ikeda, N.A. Ludin, N.B. Ahmad Khairudin, B.V. Ghaffari, et al., A review of semiconductor materials as sensitizers for quantum dot-sensitized solar cells, Renew. Sustain. Energy Rev. 37 (2014) 397–407.
[81] K. Khaled, U. Berardi, Current and future coating technologies for architectural glazing applications, Energy Build. 244 (2021) 111022.
[82] A.R. Kirmani, J.M. Luther, M. Abolhasani, A. Amassian, Colloidal quantum dot photovoltaics: Current progress and path to gigawatt scale enabled by smart manufacturing, ACS Energy Lett. 5 (9) (2020) 3069–3100.
[83] F.P. García de Arquer, D.V. Talapin, V.I. Klimov, Y. Arakawa, M. Bayer, E.H. Sargent, Semiconductor quantum dots: Technological progress and future challenges, Science 373 (6555) (2021) eaaz8541.
[84] A.S. Rasal, S. Yadav, A.A. Kashale, A. Altaee, J.-Y. Chang, Stability of quantum dot-sensitized solar cells: A review and prospects, Nano Energy 94 (2022) 106854.
[85] P.V. Kamat, Quantum dot solar cells. The next big thing in photovoltaics, J. Phys. Chem. Lett. 4 (6) (2013) 908–918.
[86] M.V. Kovalenko, Opportunities and challenges for quantum dot photovoltaics, Nat. Nanotech 10 (12) (2015) 994–997.
[87] M.A. Green, A. Ho-Baillie, H.J. Snaith, The emergence of perovskite solar cells, Nat. Photonics 8 (7) (2014) 506–514.
[88] C. Zuo, H.J. Bolink, H. Han, J. Huang, D. Cahen, L. Ding, Advances in perovskite solar cells, Adv. Sci. 3 (7) (2016) 1500324. Available from: https://doi.org/10.1002/advs.201500324.
[89] T. Wu, Z. Qin, Y. Wang, Y. Wu, W. Chen, et al., The main progress of perovskite solar cells in 2020–2021, Nano-Micro Lett. 13 (1) (2021) 18.
[90] Y. Zhou, L.M. Herz, A.K.-Y. Jen, M. Saliba, Advances and challenges in understanding the microscopic structure–property–performance relationship in perovskite solar cells, Nat. Energy 7 (9) (2022) 794–807.

[91] X. Li, Q. Xie, T. Daim, L. Huang, Forecasting technology trends using text mining of the gaps between science and technology: The case of perovskite solar cell technology, Technol. Forecast. Soc. Change 146 (2019) 432–449.

[92] N.-G. Park, K. Zhu, Scalable fabrication and coating methods for perovskite solar cells and solar modules, Nat. Rev. Mater. 5 (5) (2020) 333–350.

[93] M. Cai, Y. Wu, H. Chen, X. Yang, Y. Qiang, L. Han, Cost-performance analysis of perovskite solar modules, Adv. Sci. 4 (1) (2017) 1600269.

[94] A. Alberti, E. Smecca, S. Valastro, I. Deretzis, G. Mannino, et al., Perovskite solar cells from the viewpoint of innovation and sustainability, Phys. Chem. Chem. Phys. 24 (2022) 21549–21566.

[95] R. Pandey, S. Bhattarai, K. Sharma, J. Madan, A.K. Al-Mousoi, M.K.A. Mohammed, et al., Halide composition engineered a non-toxic perovskite–silicon tandem solar cell with 30.7% conversion efficiency, ACS Appl. Electron. Mater. 5 (10) (2023) 5303–5315.

[96] I.J. Park, H.K. An, Y. Chang, J.Y. Kim, Interfacial modification in perovskite-based tandem solar cells, Nano Convergence 10 (1) (2023) 22.

[97] Z. Li, Y. Zhao, X. Wang, Y. Sun, Z. Zhao, et al., Cost analysis of perovskite tandem photovoltaics, Joule 2 (8) (August 2018) 1559–1572.

[98] A.K. Chauhan, P. Kumar, S.N. Sharma, Perovskite solar cells: Past, present, and future, in: S. Sundaram, V. Subramaniam, R.P. Raffaelle, M.K. Nazeeruddin, A. Morales-Acevedo, M. Bernechea Navarro, A.F. Hepp (Eds.), Photovoltaics Beyond Silicon, Elsevier, Cambridge, MA, 2024, pp. 113–164.

[99] M. Khalid, Recent advances in perovskite-containing tandem structures, in: S. Sundaram, V. Subramaniam, R.P. Raffaelle, M.K. Nazeeruddin, A. Morales-Acevedo, M. Bernechea Navarro, et al. (Eds.), Photovoltaics Beyond Silicon, Elsevier, Cambridge, MA, 2024, pp. 545–582.

[100] A.S. Shyma, N. Eswaramoorthy, R. Sellappan, S. Sridharan, K. Rajaram, S. Pitchaiya, All perovskite tandem solar cells, in: S. Sundaram, V. Subramaniam, R.P. Raffaelle, M.K. Nazeeruddin, A. Morales-Acevedo, M. Bernechea Navarro, A.F. Hepp (Eds.), Photovoltaics Beyond Silicon, Elsevier, Cambridge, MA, 2024, pp. 523–544.

SECTION II

Perovskite materials and devices

CHAPTER THREE

Perovskite solar cells: Past, present, and future

Abhishek Kumar Chauhan, Pankaj Kumar and
Shailesh Narain Sharma

Advanced Materials and Devices Metrology Division, CSIR-National Physical Laboratory, New Delhi, India

3.1 Introduction to photovoltaic devices

Energy plays a crucial role in developing society or country at large; the rapid increment in energy demand and insufficient supply have become a global problem. Most of the energy is drawn from conventional fossil fuels, which face many challenges such as security concerns (most countries are dependent on the limited number of countries with significant production), price hikes, and other severe concerns like climate change. As an alternative to these challenges, government, industry, and research organizations support the development of renewable energy sources. Conventional sources may eventually become quite expensive or unavailable, but renewable energy sources are unlimited. Solar power is one of the most important renewable energies, and it is easily harvested using suitable techniques. Electricity generation from solar power has emerged as one of the most promising and rapidly growing power generation areas. Solar power is sufficient to meet the country's energy demand if it is effectively collected. Electricity generation from solar power will provide all the countries with energy security and boost the country's economy. Solar energy has a yearly forecasted power output potential of 23 PW, although total world energy consumption in 2050 is projected to be just 27 TW [1].

Solar energy is generated by using solar cells, which convert sunlight directly into electric energy. This process is also known as the photovoltaic effect. In 1839, French physicist Edmond Becquerel identified the first photovoltaic effect when he noticed that an electrolytic cell generated more power when exposed to light [2]. Later advancements in semiconductor materials, as well as significant progress in the fabrication of solar cells. Daryl Chapin, Calvin Fuller, and Gerald Pearson demonstrated the first silicon semiconductor-based photovoltaic solar cells at Bell Labs in 1954 [3]. Subsequently, several different semiconductor materials were used to fabricate solar cells with high power conversion efficiency (PCE), cost-effectiveness, and ease of processing. Such photovoltaic technology has been classified with their generation. First-generation solar cell technology consists of wafers of single-crystal and multicrystalline [4].

Single-crystal silicon (Si) solar cells, demonstrating over 20-year stability, have a certified 26.1% PCE; Si solar cell heterostructures have a slightly higher certified PCE of 26.8% [5,6]. However, because of higher production costs and a complicated fabrication procedure, this technology has not reached every part of the globe. Second-generation solar cells are thin film-based solar cell technologies such as cadmium telluride (CdTe), copper indium gallium selenide (CIGS), and amorphous silicon (a-Si) solar cells with PCEs of 22.1%, 23.6%, and 14%, respectively [6—9]. Although these solar cells are lighter in weight than first-generation solar cells, the fabrication process is still complicated; handling and installation issues persist. Third-generation solar cell technology, with leading candidates being organic solar cells and dye-sensitized solar cells, suffers from moderate efficiency of 19.2% and 13%, respectively, and very little stability. However, they are quite inexpensive, being solution (nonvacuum) processable.

The new generation perovskite solar cell (PSC) is a currently prominent photovoltaic technology with 25.8% certified PCE; it is the most economical solar cell technology known thus far [6]. The fabrication process is straightforward and solution-processable with relatively inexpensive materials; installation and deinstallation operations are quite facile and appear to have a very low maintenance cost. The devices can be deposited on plastic substrates, by adopting printing technology over a large area, which enables quite low-cost processing. Perovskite materials have the ability to tune their bandgap properties, making them suitable for tandem perovskite-silicon solar cells. Perovskite-silicon tandem solar cells significantly improve PCE to 33.2% and have the potential to exceed 40% efficiency. However, the bottleneck toward its commercialization is the lack of enough stability of the perovskite devices in ambiance environmental conditions.

3.2 Perovskite solar cells

The very first Si solar cell was fabricated at Bells lab with an efficiency (PCE) of 6% by Pearson et al. in 1954 [10]. It took more than 30 years to reach the efficiency of 25%, and then it remained almost constant [6]. Hence whenever we talk about solar cell technology, a feeling of plodding progress comes to the mind. Fig. 3.1. shows the progress of the efficiency chart of the silicon solar cells and PSCs development. Perhaps this trend was broken with the advent of the PSC technology by rapid growth in the PCE from 3.8% to 25.8% in less than a decade and which is expected to boost further and may go beyond 30% in the next few years [11,12]. Perovskite fabrication processes are simple and versatile, allowing for extensive and widespread research. This section provides a contextualized review of the fundamental properties and community developments.

PSCs are thin, lightweight, and mechanically flexible and promise high throughput via roll-to-roll (R2R) fabrication. The light absorbers in these devices are the semiconductors with perovskite structure having a molecular formula ABX_3, where A is an organic cation, B is an inorganic cation, and X is a halide anion such as I^-, Cl^-, Br^-, or a mixture of these [13–15]. Methyl ammonium lead iodide ($CH_3NH_3PbI_3$) is one of the examples of perovskite semiconductors that have shown high PCE in solar cells [16,17]. Fig. 3.2 shows the crystal structure of a synthetic

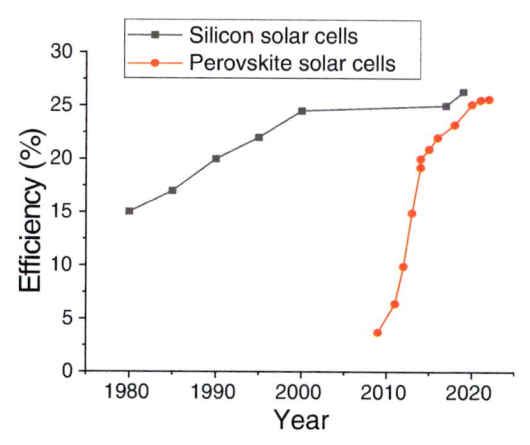

Figure 3.1 Chart illustrates the yearly progress in efficiency of silicon and perovskite solar cells.

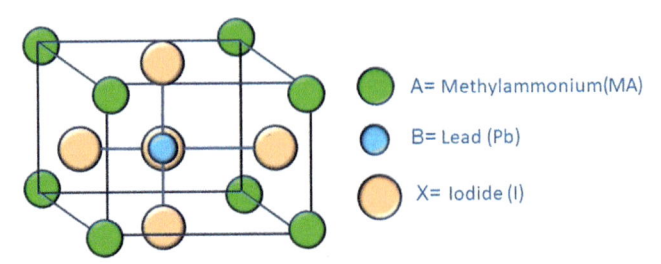

Figure 3.2 Crystal structure of an example perovskite (ABX$_3$) material.

perovskite methylammonium lead iodide. The high power conversion performance of the perovskites comes from their excellent charge transport properties, wide absorption spectrum, high charge carrier mobilities, and large charge carrier diffusion lengths [18–21], which generate high current and voltage [20].

A typical PSCs possess several hundred-nanometer thick film of perovskite light absorber sandwiched between two electrodes with suitable hole and electron transport layers. Methylammonium lead iodide (CH$_3$NH$_3$PbI$_3$), methylammonium lead iodide chloride (CH$_3$NH$_3$PbI$_{3-x}$Cl$_x$), and formamidinium lead iodide (CH(NH$_2$)$_2$PbI$_3$) are some examples of synthetic perovskites used as light absorbers in PSCs. The structure of a typical PSC is shown in Fig. 3.3.

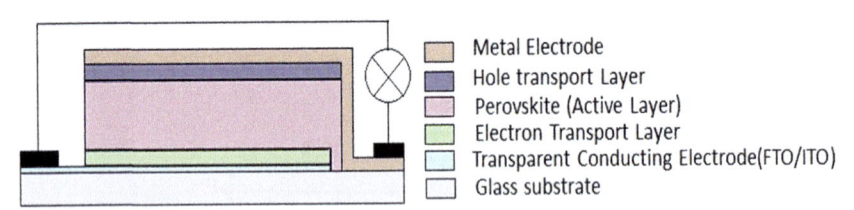

Figure 3.3 Schematic structure of a typical perovskite solar cell.

Different layers in PSCs could be deposited by thermal evaporation in vacuum or solution casting using different techniques such as spin coating, dip coating, and printing. The growth of perovskite materials and their crystalline properties tremendously affect cell performance [22]. Significant efforts have been made toward ease in processing perovskite films and improved efficiency in solar cells. Researchers have used various methods and processes to grow high-quality perovskite thin films; one-step [23], two-step sequential deposition [24], hot casting [25], vapor-assisted growth [26], and mixed solvent approaches [27] are some examples. The underlying charge transport layer could be compact and smooth, giving a planer heterojunction solar cell, or a mesoporous with an underlying compact layer, giving a mesoporous structured solar cell. Both structures have shown high PCE [28–32].

In 2009, Miyasaka et al. used organo-metal halide perovskite semiconductors as light absorbers in solar cells and achieved PCE approximately 3.8% using liquid electrolytes [33]. The devices were stable only for a few minutes because the perovskite layer dissolved in the liquid electrolyte. Further research was carried out to overcome this issue until Gratzel and Park replaced the liquid electrolyte with a solid-state hole-transporting material (HTM) [19,34], specifically spiro-OMeTAD, in 2012. This modification resulted in a PCE of 9.7%, with the stability of 500 hours stored in air at room temperature without encapsulation, measured under one sun illumination [19]. In 2013, Burschka et al. demonstrated a sequential deposition technique for the development of perovskite film to obtain a PCE of 15% [20]. Further modification of perovskite composition to form mixed-halide perovskite ($CH_3NH_3PbI_{3-x}Cl_x$) and reverse scan efficiency reported at 19.3% by Zhou et al., in the year 2014 [21].

A team comprised of researchers from the South Korea Research Institute of Chemical Technology and South Korea Ulsan National Institute of Science and Technology (UNIST) reported a newly recorded efficiency of 22.1% in single-junction PSCs in March 2016 [6]. Five years later, researchers from the Swiss Federal Institute of Technology, École Polytechnique Fédérale de Lausanne (EPFL) and UNIST reported single-junction PSCs with 25.6% efficiency [35]. Though PSCs have shown very high PCE, they face many challenges, such as stability, reproducibility, and hysteresis behavior in device operation, which are considered critical roadblocks for the commercial viability of these devices [28,29,36,37].

To compete with other solar cell technologies in the market, PSCs need to be highly stable; unfortunately, they degrade very rapidly when exposed to ambient conditions both in the dark and under illumination [38,39]. For example, $CH_3NH_3PbI_3$ is very sensitive to moisture, and when it is exposed, it tends to hydrolyze and break into CH_3NH_3I and PbI_2. Further, CH_3NH_3I breaks into CH_3NH_2 and HI [40]. Frost et al. proposed that in $CH_3NH_3PbI_3$, water molecules remove protons from ammonium to form $[(CH_3NH_3^+)_{n-1}(CH_3NH_2)_n PbI_3][H_3O^+]$ an intermediate, which decomposes into CH_3NH_2, PbI_2 and HI [41]. Smith et al. used a two-dimensional (2D) $(C_6H_5(CH_2)_2NH_3)(CH_3NH_3)_2Pb_3I_{10}$ perovskite semiconductor in solar cells that showed better stability against moisture compared to $CH_3NH_3PbI_3$ [42,43]. Not only are the perovskite molecules responsible for the solar cell degradation, but the electrode materials, charge transport materials, and their interfaces with perovskite films also play an important role in deciding solar cell stability [44]. Sanehira et al. studied the effect of electrode interfaces on the stability of PSCs and reduced their degradation using MoO_x as a hole buffer layer with an Al electrode [45]. The degradation issue in PSCs is still not completely understood, and to seek their commercial viability, this issue needs to be studied and understood in more detail. To improve the stability of these devices, more work is needed to be performed in both the areas of molecular engineering and device engineering.

3.3 Working mechanism of perovskite solar cells

Solar cells are electronic devices that convert light energy into electrical energy. The process of converting light into electrical energy through cells demonstrates the PCE of a device. The device consists of three crucial sections that contribute to the device's efficiency: The light absorber layer, the carrier collection layer, and the highly conducting electrode contact. The light absorber layer, also known as the main active layer of PSCs, is responsible for converting light into electrical energy. The carrier collection layer supports transporting the generated charge carriers (electrons and holes) from the light absorber layer and prevents their recombination. Lastly, the highly conducting electrode contact, made of metal, transfers the charge carriers to the electrical circuit.

Excellent absorption capability and low exciton binding energy are the most important characteristics that distinguish PSCs. Sunlight absorption is the primary physical mechanism for generating solar energy. An exciton or electron-hole pair is generated when the active layer (perovskite) is activated by a more energetic photon than its bandgap. Metal halide perovskite has a very high extinction coefficient ($>10^4$/cm) [46], incident photon absorption, and is easy to tune its bandgap via compositional engineering in perovskite material [47], despite the fact that the highest PSC efficiencies to date have been achieved with a bandgap (1.6 eV) that is slightly higher than the optimal one for photovoltaics (1.3 eV) [48]. The most efficient PSCs have an absorber layer in the 300–700 nm range, which is approximately three orders of magnitude thinner than silicon solar cells [49,50]. The thinner films are useful, and light absorption necessitates shorter charge transport routes. As a result, the possibility for recombination is reduced, leading to fewer energy losses and a higher device final voltage.

In perovskite photovoltaics, the ability to change the composition of the perovskite bandgap is advantageous. Variations of the original $CH_3NH_3PbI_3$ solar cell are used in single-junction solar cells. Multication perovskites are formed by combining $CH_3NH_3^+$ with other organic or inorganic cations like cesium [13], formamidinium [51], and rubidium [52] etc. As a result, the bandgap ranges from 1.5 to 1.7 eV. An increase in Br concentration results in a wider bandgap [53], which may be particularly appealing for multijunction tandem solar cell applications. Perovskites show a high population of charge carriers (electrons and holes) inside the same absorber material due to their low exciton binding energy. This is crucial in determining the PSC's working mechanism. The energy required to separate an exciton is determined by the dielectric properties of a material, which indicate its charge screening capabilities. Perovskites such as $CH_3NH_3PbI_3$ and $CH_3NH_3PbBr_3$ have extremely high dielectric constants [54], which are dependent on illumination [55], and may exhibit polaronic behavior [56,57]. The electrons and holes must arrive at the contacts to be retrieved after being photogenerated in the perovskite. This mechanism is related to the charge transport properties of halide perovskite. Long charge carrier diffusion lengths (L) of more than approximately 5 μm and an associated lifespan of approximately 1 μs have been reported in both single-crystal and polycrystalline films of metal halide PSCs [58,59]. The generated charge may be extracted with such diffusion lengths even in thicker layers that absorb all incident light.

Metal halide perovskites have diffusion lengths that are orders of magnitude longer than most organic solution-processed materials (such as phenyl-C61-butyric acid methyl ester (PCBM) and poly(3-hexylthiophene) (P3HT)), which are generally less than 10 nm [60]. The carrier mobility μ determines the diffusion length ($L = \sqrt{D\tau}$), where D is the diffusion coefficient, $D = \mu\ k_B\ T/q$, where q is the elementary charge, k_B is the Boltzmann constant, and T is the temperature. Theoretical calculations show that perovskite mobilities are relatively high, comparable to crystalline inorganic semiconductors. On the other hand, experimental values are difficult to come by and are rarely consistent. Despite the fact that Hall-effect tests provide reasonably high mobility values (ranging from 0.5 to 60 cm^2/V/second depending on the nature of the material) [61].

The electric field can be very significant in charge extraction; the perovskite absorber layer is sandwiched between two charge extraction layers (*n*-type for the electron transport layer and *p*-type for the HTL) in the solar cell, and the band diagram of the system can then be approximated by an $n-i-p$ model, in which the intrinsic perovskites distribute the electric field homogeneously. Metal halide perovskites benefit from the remarkable crystallinity achieved through various fabrication procedures (including low-temperature solution processes). Changes in trap density were observed in single crystals ($109-1010$ cm^3 for $CH_3NH_3PbI_3$ and $CH_3NH_3PbBr_3$) [58,62]. Several researchers have focused on increasing the grain size of perovskite films to enhance solar cell performance [63], citing increased grain-to-grain charge transfer as the reason [64].

The experimental results indicated that the flaws were intrinsically benign [28], reducing the necessity for a large grain size. Grain sizes greater than 100 nm have been shown to play a minor effect in recombination [65]. Similarly, extremely high-efficiency ($>20\%$) PSCs with grain sizes as small as 10 nm have been reported, with high FFs ($>80\%$) suggesting nearly perfect charge transport [66]. In PSCs, three types of recombination processes occur (i) shunt pathways directly contacting the hole and electron transport layers (HTL and ETL), (ii) recombination inside the bulk of the perovskite film, and (iii) at the perovskite/ETL or perovskite/HTL interface [67]. Because charge recombination competes with charge extraction to an external circuit, the kinetics of both processes eventually limit solar cells' total photovoltaic conversion efficiency.

Charge recombination through shunt pathways indicates that the electron and hole extraction layers are in direct electrical contact. Direct contact between ETL and HTL at perovskite pinholes was most likely one of the primary sources of charge loss via recombination, resulting in low open-circuit voltages, V_{oc}, and poor performance of PSCs [68]. The amount of charge lost is determined not only by the material qualities of the perovskite but also by the architecture or morphology of the layer. Bulk recombination, on the other hand, accounts for electron and hole recombination inside the perovskite bulk as well as loss rates. According to studies, the number of trap states is directly related to the quality of the perovskite crystal lattice, such as ion vacancies or interstitials, as well as the number of defects at grain boundaries [69]. Exciton dissociation into free charges occurs relatively fast in hybrid organic–inorganic perovskite crystals. As a result, the majority of these charges are lost by electron-hole radiative recombination [70]. In low-illumination conditions, where the electron population is fewer than the hole population (due to the confinement of certain photoexcited electrons), recombination is monomolecular [71]. Nonradiative trap-assisted recombination might have an adverse influence on perovskite bulk. Illumination, on the other hand, is found to lead the perovskite to form (reversible) metastable deep-level trap states, which enhance bulk recombination over time. The majority of trap states appear to be the result of lattice flaws. Biasing PSCs have also been shown to change recombination processes [72]. As a result, the operating parameters of a PSC may influence how charges recombine and, more significantly, the region in which recombination happens. According to recent research, recombination is dominated by the bulk of the film at low light intensities and when the internal electric field is moderate (voltages near V_{oc}) [73].

After absorption and charge transfer, the final mechanism required for the photovoltaic process is extracting the photogenerated charge. As a result, a good extraction interface should ensure that voltage, current, and FF are all lost to a minimum. To relate these needs to physical processes, it is essential to understand the role of interfaces in PSCs. The optimum selective contact should not absorb light to avoid a drop in light intensity in the perovskite. Similarly, no energy losses should occur during carrier injection from the absorber; there is no interfacial recombination. These contact materials must be selective and have a low series resistance (Rs) to allow only one type of carrier to be injected. As a result, the active layer and the charge extraction layers must be balanced [67]. While introducing any interfacial material it may introduce some losses, which eventually will affect the photovoltaic performance of the solar cell in terms of— short-circuit current density (J_{sc}) (through variation in charge extraction and optical absorption), V_{oc} (through losses such as interfacial recombination), and FF (through interfacial resistance for extraction of the photogenerated charge). Therefore reduced interfacial (surface) recombination, is a critical characteristic when selecting a charge carrier material to yield good outcomes. Nonradiative recombination is reduced by using large bandgap intrinsic materials at the interface. As a result, increasing the contact's charge carrier selectivity directly increases V_{oc} [66], similar to using buffer layers to minimize recombination [74,75]. Decreasing the electron-transporting material's conductivity can improve current-voltage hysteresis by creating a charge carrier imbalance [76].

3.4 Device architectures of perovskite solar cells

PSCs have been researched and manufactured in various configurations using various materials, with success and the potential for mass production in the near future. To better understand how this technology evolved and developed, it is worthwhile to go back to the beginning of this fascinating and remarkable period of organic–inorganic perovskite materials and their application in solar cells. In 2009, Professor Miyasaka's group at Tokyo Polytechnique University in Japan published an article in the Journal of the American Chemical Society on organometal halide perovskites solar cells. In this groundbreaking study, the scientists investigated two compounds, $CH_3NH_3PbI_3$ and $CH_3NH_3PbBr_3$, and were the first to use them in photovoltaics. Miyasaka et al. used lead halide perovskite as the light-absorbing material in a liquid-based dye-sensitized solar cell instead of dye molecules (DSSCs) [33]. Perovskites were deposited on a transparent conducting electrode sandwiched between a mesoporous film containing titanium dioxide (TiO_2) nanoparticles and a platinized counter electrode. A liquid electrolyte comprising redox couple I^-/I_3^- was poured between the two electrodes to transport electrons back and forth from the counter electrode to the oxidized sensitizer that had injected its excited electron into the TiO_2 conduction band. In comparison to traditional DSSCs, the functioning mechanism of this device was believed to be similar. The primary drawback is the stability of the perovskite material in the electrolyte, which is generally thought to be very corrosive, resulting in the device only operating for a few minutes and rapidly degrading after measurement [77].

In 2012, Park's group chose to replace the liquid electrolyte with a solid-state HTM to create solid–state solar cells. In collaboration with Grätzel's group, they successfully developed all-solid-state thin-film solar cells using perovskite as an active layer and spiro-OMeTAD as the HTM. Compared to other device architecture perovskite-based solar cells in previous publications, this PSC had a much higher efficiency of over 9%. It was much more environmentally stable over 500 hours [19]. This game-changing paper began a new era for PSCs. This new device architecture works with the extremely thin absorber (ETA) cell structure, which is commonly used for inorganic materials or quantum dots with a high absorption coefficient and does not require a thick TiO_2 coating [78,79]. The most recent generation of PSCs can be divided into two device

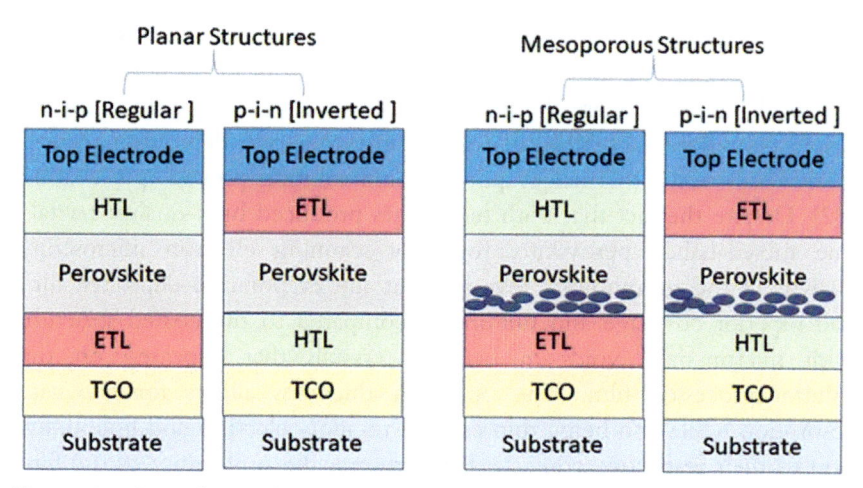

Figure 3.4 Perovskite solar cell device architecture in planar and mesoporous structures.

architecture categories: Planar PSCs, which have a thin layer of perovskite sandwiched between two selective contacts, and Mesoscopic PSCs, which have a layer of nanostructured materials that serve as a scaffold for the perovskite layer; Fig. 3.4 illustrates the structures.

3.4.1 Planar architecture of perovskite solar cells

The perovskite layer is a thin film and a planar configuration without the scaffold was proposed. Despite this, due to the unequal diffusion length [80], separating charges via selective connections and blocking opposing charges is essential to decrease recombination [81]. The planar PSC configuration is similar to a thin film or heterojunction solar cell, with a layer of perovskite sandwiched between two selective contacts. This is a substantial advantage for developing low-temperature flexible electronics on PET substrates using printable manufacturing techniques [61]. It also simplifies the incorporation of perovskite into the top or bottom tandem solar cells. Different techniques of deposition for organometal halide perovskites can also be investigated and applied. Because nanostructures are no longer necessary for planar arrangement, the choices for electrons or HTMs are broadened. Planar heterojunction PSCs can have either an $n-i-p$ or an inverted $p-i-n$ design. In this section, each structure will be thoroughly detailed.

In an $n-i-p$ design, the bottom contact collects negative charges, whereas the top contact collects positive charges. The most common structure for a classic $n-i-p$ planar PSC is FTO/TiO$_2$/CH$_3$NH$_3$PbI$_3$/spiro-OMeTAD/Ag. In planar devices, the quality of the perovskite layer is critical. Liu et al. were the first to construct this structure with mixed-halide perovskite CH$_3$NH$_3$PbI$_x$Cl$_{3-x}$, comparing two deposition techniques: Dual-source thermal evaporation and solution process spin-coating [82]. Despite the fact that both techniques produced high-quality crystalline mixed-halide perovskite, top-view scanning electron microscopy analysis of the morphology revealed that the evaporation-deposited film had superior coverage and uniformity compared to the coated structure with micron-sized voids in between crystals that appeared on the solution-processed film. The optimum thickness allows for adequate absorption while also being thin enough to allow electron and hole diffusion to their respective contacts. Furthermore, the uniformity of the film across length scales prevents shunting caused by direct contact between the HTL and the electron transport compact layer.

The ETL and HTL are critical in developing planar PSCs because they extract charges, block opposite charge carriers, and reduce recombination. Because planar PSCs significantly benefit low-temperature processability, several research groups have become interested in creating a TiO$_2$ compact layer at low temperatures. Conings et al. established a technique for forming TiO$_2$ layers at temperatures below 150°C. This layer was used as the electron-transporting layer in the architecture of ITO/TiO$_2$/CH$_3$NH$_3$PbI$_x$Cl$_{3-x}$/P3HT/Ag (with P3HT poly (3-hexylthiophene) as the HTL), resulting in a 13.6% efficiency [83].

Other oxides, such as ZnO, have been used as the ETL in planar $n-i-p$ PSC. The electron mobility of ZnO is high, and its energy band values are similar to those of TiO$_2$. Kumar et al. presented planar ZnO-based PSC for the first time in 2013, utilizing the electrodeposition technique. When nanostructured ZnO and planar ZnO were compared, the planar devices had 5.54% and 2.18% efficiency on FTO and ITO/PET flexible substrates, respectively [84].

Replacing TiO$_2$ (bandgap 3.3 eV) with a larger band gap metal oxide such as SnO$_2$ (3.6–4.2 eV) aims to decrease ETL ultraviolet (UV) instability and lower processing temperature. The greater bandgap also reduces parasitic absorption [20,85]. When compared to TiO$_2$, crystalline SnO$_2$ has almost two orders of magnitude greater electron mobility [86], indicating that it might be a powerful ETL. Two notable papers on planar SnO$_2$-only ETL deposition techniques have reported efficiencies of greater than 20%, with Jiang et al. utilizing an off-the-shelf nanoparticle

SnO$_2$ product (np-SnO$_2$) resulting in a record-breaking planar $n-i-p$ device PCE of 23.3% [87–91].

The initial few investigations on the $p-i-n$ structure were motivated by the design of organic photovoltaics (OPV) heterojunction concept. This concept was inspired by the fact that perovskite has an ambipolar semiconductor property, which means it can produce and transport both electrons and holes. FTO/PEDOT:PSS/CH$_3$NH$_3$PbI$_3$/PC61BM (or C60 fullerene)/Al (where PEDOT:PSS is poly(3,4-ethylene dioxythiophene)-poly (styrene sulfonate) and PC61BM is fullerene derivative [6,6]-phenyl-C61-butyric acid methyl ester) was the first known $p-i-n$ PSCs [92]. The actual result was just 3.9%, but with the ITO/PEDOT:PSS/CH$_3$NH$_3$PbI$_3$/PC61BM/Au combination, device performance was quickly improved to almost 18% in a short amount of time [93]. The high efficiency of inverted planar PSCs was attributed to the balanced rate of electron and hole transport in each layer instead of the imbalance rate in regular cells, where the hole was carried faster than the electron. It is worth mentioning that the distinctions between $p-i-n$ and $n-i-p$ devices are not just about the design (normal or inverted) but also the materials used. Because organometal halide perovskite is susceptible to humidity and oxygen, selecting the materials that will be put on top of it is crucial. In contrast, the layer underneath it is more flexible. PEDOT:PSS, for example, is water-based and cannot be deposited on top of perovskite through the solution, but it may be spin-coated on the fluorine-doped tin oxide (FTO) (or ITO) substrate before perovskite deposition. As a result, while PEDOT:PSS may be used in an inverted $p-i-n$ solar cell, it cannot be utilized in a typical conventional solar cell ($n-i-p$). PEDOT:PSS's work function changes considerably with the ratio of its ionomers (4.9–5.2 eV), and it may not perfectly match the valence band of perovskite, despite its benefit of being a low-temperature easy technique [94]. Furthermore, because of the corrosive and hygroscopic nature of PEDOT:PSS, which compromises the stability of PSCs, graphene oxide has been proposed as a substitute for PEDOT:PSS in bulk heterojunctions and, more recently, in perovskite heterojunctions. By quenching the photoluminescence of CH$_3$NH$_3$PbI$_x$Cl$_{3-x}$, the charge transfer between the materials can be observed, and the perovskite layer grown on graphene oxide (GO) displayed greater homogeneity and lower roughness than PEDOT:PSS, making it a potential choice. Wu et al. developed an inverted planar PSC using graphene oxide as the HTL that had a 12.4% efficiency, which was higher than the 9.3% of the PEDOT:PSS equivalent [95,96].

Polythiophene (PT) was chosen as the HTL in an ITO/PT/$CH_3NH_3PbI_3$/C60/BCP/Ag structure with optimal efficiency of 11.8% [97] because its highest occupied molecular orbital (HOMO) level (5.2 eV) matches the maxima of the valence band of perovskites. Although the conductivity of PT rises with thickness, it is restricted by its light absorption, which causes photons to be lost before reaching the perovskite layer when used as the front electrode.

Inverted PSCs employ a range of *p*-type materials for the HTL, including organic semiconductors such as PT [97], spiro-OMeTAD, and inorganic materials, such as NiO_x [98−100], MoO_x [101], CuI [101−103], Cu_2O [103,104], or CuSCN [105]. NiO_x, for example, has long been employed as a *p*-type material in solid-state DSSC and has become an attractive option for PSCs [106]. Because of the proper energy alignment between NiO_x and perovskite, NiO_x could be deposited through sol-gel methods with enhanced hole-transporting mobility and higher V_{oc} [107]. UV-ozone treatment enhances the surface wettability of perovskite on NiO_x film, resulting in a greater active area quality as well as improved charge extraction and transport.

3.4.2 Mesoscopic architecture of perovskite solar cells

In the mesoscopic PSCs arrangement, a nanostructured material serves as a scaffold for the deposition of perovskite [108]. This nanostructured material is often a semiconducting metal oxide with a wide bandgap, although it can also be an insulator. Depending on the function of the metal oxide scaffold, some contain charge carriers while others do not. Many metal oxide nanostructures, including mesoporous TiO_2 nanoparticles, have been studied as active scaffolding for mesoscopic PSCs. An example of a typical mesoscopic PSC configuration is included: FTO/TiO_2 compact layer/TiO_2 mesoporous layer/$CH_3NH_3PbI_3$/spiro-OMeTAD/Ag [109]. The existence of a mesoscopic metal oxide layer (mesoporous network of nanoparticles) in this device design is frequently referred to as $n-i-p$ mesoscopic, and the physical bottom layer (closest to the FTO substrate) is the electron transporter and is composed of an *n*-type semiconductor. The TiO_2 compact film acts as a hole-blocking layer, reducing recombination at the anode. The purpose of the TiO_2 mesoporous film is still up for discussion. It can operate as an ETL and a scaffold to assist the development of the perovskite $CH_3NH_3PbI_3$ film. The hole collector or counter electrode is Ag, and the HTM is spiro-OMeTAD.

Spin-coating is commonly used to produce a blocking layer on an FTO substrate. However, other chemical and physical deposition processes, such as spray pyrolysis or sputtering, have also been reported. The blocking layer should be dense and without pinholes to avoid recombination between the perovskite and the FTO substrate. Although this layer is often made of the same materials as the mesoscopic layer, it can also be made of different materials with comparable energy levels. Doping has been widely utilized to improve conductivity. The mesoscopic layer is deposited on the compact layer using spin coating or other synthetic processes. Single or sequential deposition of the perovskite layer on top of the mesoporous layer is possible. Both precursors (PbI_2 or $PbCl_2$ and CH_3NH_3I) are combined in the solution before spin coating and postannealing in the single deposition technique. Burschka et al. proposed a sequential deposition technique in which PbI_2 is spin-coated on the mesoscopic layer first, then dipped in CH_3NH_3I solution [20]. PbI_2 was transformed to $CH_3NH_3PbI_3$ during the dipping process. In the sequential deposition approach, the benefit of the mesoscopic layer was demonstrated to have a substantial impact. Because it is thought that only a thin layer of PbI_2 can be converted to perovskite using this technique, keeping the PbI_2 coating thickness to a few tens of nanometers is critical. The tiny pore size of the mesoporous TiO_2 film allows homogeneous PbI_2 coating to be confined to the pore size.

Leijitens et al. investigated the pore-filling fractions of perovskite and observed that the thickness of the TiO_2 layer had a significant impact on them [110]. The perovskite layer acts as a barrier between the holes and the TiO_2 nanoparticle to avoid recombination with the HTM. The concentration and solvent used in the perovskite precursor solution, as well as the deposition process, all have an impact on the quality of the perovskite layer. According to Leijitens et al., with a perovskite concentration of 40%, the TiO_2 layer thickness should be thin (below 440 nm) to achieve 100% pore filling and optimal device performance. The effectiveness of pore filling is determined not only by TiO_2 thickness but also by pore size and porosity. Although the most widely employed nanostructure for mesoscopic PSCs is TiO_2 nanoparticles producing a mesoporous film, other materials and architectures have also been studied. The size and shape of various TiO_2 nanocrystals, such as nanorods [111], nanowires [112], and nanofibers [113], have been studied. Other *n*-type semiconductors with similar bandgaps and energy levels to TiO_2, such as ZnO [84], aluminum oxide (Al_2O_3), and WO_3 [114], have also been studied as scaffolds for mesoscopic PSCs.

Aluminum oxide is an electrically insulating substance with a large bandgap (7−9 eV). Replacing the mesoporous TiO_2 layer with a passive scaffold of mesoporous Al_2O_3 film produces a device with a greater PCE, especially a higher V_{oc} (0.98 V vs 0.80 V in TiO_2). Because Al_2O_3 is an insulator, electrons are carried through the perovskite layer rather than injected into it, preventing recombination between the injected electron in the metal oxide and the hole in the perovskite [109].

The performance of passive scaffold devices is influenced by film thickness, particle size, porosity, and pore size. Ball et al. investigated the performance of Al_2O_3 mesoscopic PSCs by altering the Al_2O_3 layer thickness from 0 to 140 nm. Both V_{oc} and FF increased rapidly with increasing Al_2O_3 thickness until they reached the optimal point of 40−80 nm, whilst J_{sc} obtained an optimum of 16.9 ± 1.9 mA/cm^2 with a thin layer of just 80 nm [115]. According to optical property analysis, internal quantum efficiency peaked at 80 nm and dropped as the film thickness increased.

Other materials with high bandgaps, such as ZrO_2 or SiO_2, have been employed as passive scaffolding in PSCs. The working mechanism of passive scaffold PSCs was hypothesized to be comparable to that of ETA solar cells based on the first Science paper that outlined the use of Al_2O_3. Later, Kim et al. supported it with electrochemical impedance spectroscopy (EIS) studies of ZrO_2-based PSCs [116]. In the EIS spectrum, charge accumulation may be understood as capacitance in an equivalent electrical circuit. This was the first charge accumulation detected in a perovskite layer, which was previously thought to be a light-absorbing material. It also demonstrated that the perovskite layer functions as ambipolar electron and hole transporters in addition to being a light absorber. This discovery sheds new light on perovskite as a potential energy conversion material and the PSCs' working mechanism and architecture. Furthermore, EIS's analysis revealed a decrease in chemical capacitance in the insulating oxide. Chemical capacitance is a measure of charge buildup linked to the material's subbandgap or density of states. Unlike semiconductors such as TiO_2, insulators such as Al_2O_3, SiO_2, and ZrO_2 lack subbandgap states, lowering the chemical capacitance of the cell substantially.

The influence of porosity and particle size on infiltration and, ultimately, the performance of passive scaffold-based PSCs was investigated using SiO_2 monodispersed particles ranging in size from 15 to 100 nm. Pores larger than 50 nm were shown to be sufficient for perovskite penetration, although 100 nm SiO_2 may result in scattering of either SiO_2 or perovskite, resulting in enhanced absorption at longer wavelengths.

However, because of the complicated scattering, the overall absorption was reduced, and devices with particles bigger than 50 nm showed a moderate drop in J_{sc}, whilst those with particles smaller than 50 nm had a steeper fall in J_{sc} due to lower perovskite loading. Poor penetration also increased resistances and reduced FF with particles smaller than 50 nm.

3.5 Lead-free perovskite solar cells

Over the last several years, there has been an increase in the number of studies aimed at developing ecologically acceptable lead-free PSCs that would eliminate the use of lead in metal halide perovskite layers. In PSCs, a variety of lead-free perovskites based on tin (Sn), bismuth (Bi), antimony (Sb), germanium (Ge), titanium (Ti), copper (Cu), and germanium (Ge) has been utilized [117−121]. Among the lead-free perovskite materials, tin perovskite appears to be the most promising. Because Sn and Pb have comparable outer electronic structures and ionic radius, tin may be utilized to replace lead in perovskite lattices without causing phase separation. Tin-based perovskites material also has the following benefits: (i) Low exciton binding energy (29 meV for $MASnI_3$), (ii) an ideal bandgap near the Shockley−Queisser limit (1.3−1.4 eV), and (iii) high charge carrier mobility (electron mobility = 2000 cm^2/v/second for $MASnI_3$) [122,123]. The oxidation of Sn^{2+} to Sn^{4+} significantly compromises the stability of tin-based PSCs [124]. In this approach, partial substitution of lead with divalent metal-ion can improve the performance of PSCs while causing no harm to the environment. Furthermore, adding bulky organic ligands changes grain orientation and development, boosting out-of-plane photoinduced bulk polarization and spin-orbit coupling (SOC). They are responsible for photovoltaic performance in 2D/3D PSCs. SOC, in particular, enhances photovoltaic activity; increased SOC improves spin-conversion from optically induced states. Spin-permitted recombination causes bright states, whereas spin-forbidden recombination generates dark levels. Grains orientated out of a plane in 2D/3D-perovskites with condensed traps, on the other hand, are helpful for aligning optical-transition dipoles, which increases photovoltaic activity.

Antisolvent engineering approaches have recently been utilized. Carefully designing antioxidant additives, controlling crystallization rate, designing suitable A-site cations such as phenethyl ammonium (PEA) and fuorophen-oxyethylammonium (FOE), and constructing the oriented low-dimensional perovskite structure, the efficiency of tin-based PSCs increased rapidly from 6% to 13% within a few years of development [125–127]. To reduce V_{oc} loss in tin PSCs, researchers are currently focusing on defect passivation and the selection of optimal ETLs. Nishimura et al. employed a combination of bulk ethyl ammonium (EA) doping and surface examine passivation to reduce the defect density of formamidinium tin iodide (FASnI$_3$) perovskite films by up to one order of magnitude, resulting in a device with high V_{oc} of 0.84 V [128]. The research group also designed a template-growth approach supported by an *n*-propylammonium iodide (PAI) salt to decrease the bulk defect density in solution-processed FASnI$_3$ films. Because of increased crystal orientation along with directions, the PAI post-treatment increased electron diffusion length from 70 to 180 nm, lowering recombination loss in FASnI$_3$ absorbers and resulting in a significant V_{oc} enhancement of 200 mV and a record-certified PCE of 11.22% for tin PSCs [129].

Other lead-free PSCs based on wide-bandgap bismuth (Bi) and antimony (Sb) perovskites demonstrated significant efficiency increases. Hu et al. created bulk-heterojunction perovskite active layers, Cs$_3$Bi$_2$I$_9$ and Ag$_3$Bi$_2$I$_9$, to optimize grain orientation and interface band alignment, attaining a record efficiency of roughly 3.6% and a high V_{oc} of 0.89 V for Bi-based PSCs [130]. Singh et al. improved the morphology of wide-bandgap Cs$_3$Sb$_2$I$_9$ perovskites and increased electron extraction capacity in Sb-based PSCs using an indacenodithiophene-based organic acceptor with Lewis base groups, obtaining a PCE of 3.25% for the inverted-structure device [131].

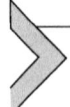

3.6 Tandem architecture of perovskite-silicon solar cells

The ABX$_3$ perovskite absorber's tunable bandgap from 1.2 to 3.0 eV enables the fabrication of silicon-perovskite tandem devices (Fig. 3.5), thereby increasing the PCE above the Shockley–Queisser limit for single-junction solar cells [132]. It is crucial to find suitable wide-bandgap perovskite materials (1.6–1.7 eV) with a spectrum response that matches the 1.1 eV-bandgap silicon absorber to obtain high efficiency in silicon-perovskite tandems [116].

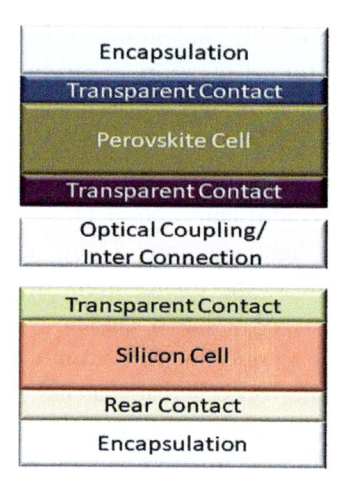

Figure 3.5 Tandem perovskite-silicon device structure.

The efficiency of silicon–perovskite tandems has grown rapidly in the last two years, from around 25% to over 33.2%, mainly due to the decreased V_{oc} deficit of wide-bandgap perovskite top cells enabled by compositional engineering for a stable phase and the selection of charge transport layers [133]. Based on their device construction, tandem solar cells are classified as monolithic two-terminal [134,135], mechanical four-terminal [136], and optical four-terminal devices [137] (shown in Fig. 3.6).

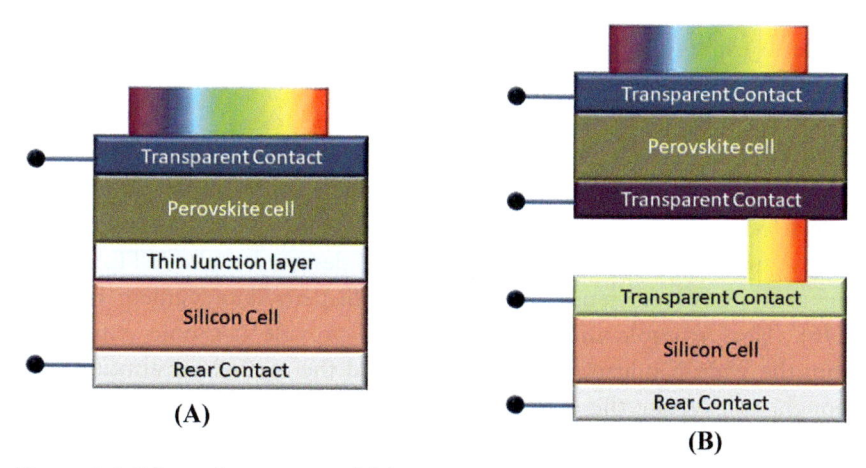

Figure 3.6 Schematic structure of (A) Two-terminal tandem perovskite-silicon solar cells, (B) Four-terminal tandem perovskite-silicon solar cells.

In a two-terminal monolithic tandem solar cell, two subcells are directly coupled in series with a tunnel junction or recombination layer, with the entire tandem current limited by the subcell with the lowest photocurrent [138]. It is vital that each subcell generates the same photocurrent by properly sharing the incoming photons for the tandem solar cell to reach a high matching photocurrent. Controlling optical properties such as layer bandgap/transmittance [139], light scattering [140], and incoming light up/downconversion [141] allows the photocurrent in the subcells to be correctly matched. Most perovskite-based tandem solar cells, including the champion device, employ a crystalline-Si bottom cell. The optimum bandgap energy of a perovskite top cell that matches Si (1.12 eV) perfectly is 1.68 eV, which is somewhat higher than that of an optimal PSC and needs the usage of a mixed-halide system that contains Br. Each subcell's photovoltage must be independently enhanced to increase the photovoltage of tandem solar cells. As a result, several strategies for improving the photovoltage of single-junction PSCs, such as passivation of defects with additives, may be used for wide-bandgap perovskites [116,142].

To adjust carrier lifetime and avoid light-induced phase segregation in wide-bandgap perovskite films, Xu et al. employed triple-halide alloys made up of chlorine, bromine, and iodine [143]. Direct integration of large amounts of Cl (more than 15%) into the double-halide perovskite resulted in a uniform halide distribution across the film, with an optical bandgap (1.67 eV) that matched the spectrum of the bottom 1.12 eV Si absorber. This action significantly decreased phase separation in most mixed-halide perovskite compositions [144,145]. Expanding the double-halide to triple-halide components with a V_{oc} of over 100 mV raised the efficiency of opaque single-junction PSCs from 18.15% to 20.42%. Scientists were able to attain a PCE of 27.04% in 1 cm^{-2} two-terminal monolithic tandems by merging enhanced perovskite top cells with silicon bottom cells [116]. To decrease nonradiative recombination loss at the hole-selective contact in wide-bandgap PSCs, a unique self-assembled, methyl-substituted carbazole monolayer was designed as the HTL, resulting in a record efficiency of 29.15% for Si-perovskite tandem devices [146]. The carbazole unit was thought to provide fast hole extraction from the wide-bandgap perovskite layer, and the methyl substitution was thought to passivate the defects on the perovskite surface, minimizing the FF loss induced by Rs (transport loss) and interfacial nonradiative recombination in PSCs, according to intensity-dependent absolute photoluminescence measurements. The Me-4PACz hole-selective layer increased the FF of single-junction PSCs from 79.8% (control cell-based on PTAA) to 84.0%, and the device V_{oc} increased from 1.19 to 1.25 V. For Si-perovskite tandem devices, textured crystalline-Si subcells have also been reported, with the purpose of minimizing device reflectance and boosting photon utilization rate. Hou et al. used a solution-processed micrometer-thick perovskite top cell deposited directly over a textured Si-heterojunction bottom cell to make tandem devices [147]. Expanding the depletion width at silicon's bases to 26.0% [148] for textured Si-perovskite tandem systems solved the carrier extraction problem in micrometer-thick perovskite layers [149].

3.7 Stability study of perovskite solar cells

The most challenging aspect of the perovskite cell is its instability when exposed to moisture at high temperatures, light (especially UV or direct sunlight), and electrical bias, especially when moisture is present. Despite the abundance of research on perovskite stability, including methods to improve cell durability, there is still a lack of a unified reporting system or a standardized testing process. This is because the technology is still in its early stages of development, perovskite solar cells exhibit multiple facets of instability in terms of device performance and operation, and different behavior and failure mechanisms in cells with different architectures.

In mid-2012, the first report of a stability study on the first efficient solid-state perovskite cell with a 9.7% was reported [19]. The unencapsulated FTO/c-TiO$_2$/mp-TiO$_2$/CH$_3$NH$_3$PbI$_3$/spiro-MeOTAD/Au device was stored at room temperature with no mention of light or humidity levels. Electrical characteristics were monitored irregularly over a 500-hour reporting period, revealing a 9% decrease in short-circuit current, relative fluctuations in open-circuit voltage, and an improvement in fill factor (FF), resulting in a 14% increase in energy conversion efficiency. Longer-term stability tests were not performed in that study.

Researchers reported on the influence of humidity during air exposure and mixed-halide concentration on the durability of FTO/c-TiO$_2$/mp-TiO$_2$/CH$_3$NH$_3$PbI$_{3-x}$Br$_x$/PTAA/Au solar cells [150]. The conversion efficiency of CH$_3$NH$_3$PbI$_3$ and CH$_3$NH$_3$PbI$_{2.94}$Br$_{0.06}$ is drastically reduced when humidity is increased from 35% to 55%. The conversion efficiency of

$CH_3NH_3PbI_{2.8}Br_{0.2}$ and $CH_3NH_3PbI_{2.7}Br_{0.3}$ cells with increased bromide content was maintained during a 20-day (480-hour) reporting period at room temperature with periodic current-voltage measurements. The environmental conditions during storage were not described.

In a nitrogen-filled glove box, Leijtens et al. encased FTO/c-TiO_2/mp-TiO_2/$CH_3NH_3PbI_{3-x}Cl_x$/spiro-MeOTAD/Au. After that, the devices were exposed to a 100 mW/cm^{-2} light (Xenon arc lamp) at 40°C with and without UV filtering [151]. The efficiency of the encapsulated TiO_2-based device declined by 80% after 4 hours, but the unencapsulated device lost just 50% of its original efficiency in the same time frame, which was unexpected. The device's performance was improved by using a UV filter. Light-induced desorption of surface-adsorbed oxygen in the mp-TiO_2 layer was thought to cause UV-instability, leaving unoccupied, deep surface trap sites and a free electron per site, which recombined readily with holes on the spiro-OMeTAD hole transporter. In a similar experiment performed by replacing mp-TiO_2 with mp-Al_2O_3, while keeping other device materials the same, the author reported that under UV, the devices were more stable [152,153]. Under continuous illumination at 40°C, these encapsulated devices were claimed to have a lifetime of 1000 hours, despite a 50% loss in efficiency in the first 200 hours of the illumination test [154].

Zhou H. et al. reported that the device structure, which consists of FTO/c-TiO_2/mp-TiO_2/$CH_3NH_3PbI_3$/Carbon, shows significant device stability of more than 480 hours when the unencapsulated device is kept in the dark and exposed to air, with its electrical properties being checked on a regular basis throughout the reporting period [155]. The method used to measure the temperature of the cells is unclear, and the humidity levels during the reporting period were not provided.

More recently, systematic investigations regarding the degradation of the $CH_3NH_3PbI_3$ solar cell have been published [156,157]. Yang et al. investigated in situ absorption spectroscopy and grazing incidence X-ray diffraction (XRD) at 22.9°C ± 0.5°C and controlled relative humidity (RH). It was observed that at a RH of 20%, it takes 10,000 hours for the absorbance of the perovskite film to degrade half of its initial value, 1000 hours at RH = 50%, 34 hours at RH = 80%, and 4 hours at RH = 98%. The film's absorbance remained constant in 0% RH dry oxygen-containing air during a 2-week reporting period [156].

HTMs with and without lithium bis (trifluoromethanesulfonyl) imide (Li-TFSI) additions were also studied. When HTLs, 2,2′,7,7′-tetrakis (N,N-di-4-methoxyphenylamino)-9, 90-spirobifluorene (spiro-OMeTAD), come into contact with the perovskite layer, the $\tau_{1/2}$ lifespan ($\tau_{1/2}$ or half-life is defined as the time it takes for half of the initial amount of the material to deteriorate or degrade) or remains the same depending on the presence of Li-TFSI. The $\tau_{1/2}$ improves when poly[bis(4-phenyl)(2,4,6-trimethylphenyl)-amine] (PTAA) or poly(3-hexylthiophene) (P3HT) were utilized as a polymer-based HTM. According to the reports, polymer-based HTMs operate as moisture barriers by collecting and storing moisture inside the HTM, which slows water percolation to the perovskite layer. It was proposed that the inclusion of Li-TFSI, which makes the HTM hydrophilic, aids moisture absorption and trapping in the HTM.

Perovskites with smaller dimensions have recently been found to have improved moisture resistance and stability [158—162]. Because of steric effects, replacing the small MA^+ or FA^+ cation with a considerably larger organic primary spacer cation confines the perovskite in two dimensions. The improved moisture resistance is thought to be related to the long cation chain's hydrophobicity and the highly orientated and dense structure of the perovskite films, which prohibit direct contact with water [158,160,161]. Multilayered perovskite structures might change the processes of electronic and ionic mobility across the film, suppressing hysteresis [162]. Thermal stability is still a challenge with these devices, with the best-reported stability for an encapsulated cell, 250 hours under steady lighting. Other strategies, such as adding a phosphonic acid ammonium additive, which is reported to cross-link the perovskite grains, have been shown to improve stability [163]. The addition of RbI to a mixed-cation [$(FA_{0.83}MA_{0.17})Pb(I_{0.83}Br_{0.17})$]perovskite was recently found to improve the long-term stability and cell performance [164]. Other than environmental instabilities, reversible (ion migration) and irreversible (decomposition) behavior have been observed under various operating circumstances such as light, electrical bias, and moisture [165]. Their effects, which are more obvious at grain boundaries and interfaces, may be observed using characterization techniques such as photoluminescence and Kelvin probe force microscopy. Hysteresis becomes more intensified as the cell degrades. Long-term electrical stability and no hysteresis in perovskite cells are still being investigated.

3.8 Perovskite film fabrication processes

Photovoltaic performance is influenced by crystallinity, thickness, the morphology of the perovskite film, and the material purity. Several factors, such as annealing time optimization, temperature modulation, selected material thickness, and various deposition techniques, have contributed to forming perovskite thin films. The solution-based method is simple and inexpensive; however, synthetic crystals produce more internal flaws. The HTL may come into direct contact with the electron transport layer, decreasing the device's FF and open-circuit voltage. The vapor deposition approach creates perovskite films with a high surface density and fewer flaws, resulting in a higher FF and open-circuit voltage. This method, however, demands a high-vacuum environment and uses a significant amount of energy. In this section, we will discuss, in detail, several types of film deposition techniques.

3.8.1 Spin coating method

This is a solution-based approach in which a drop of the solution is spun and dried onto the substrate using a spin coating system. The excess solution is removed by spinning the solution at a certain rpm (revolutions per minute), followed by evaporation. One-step and two-step approaches to perovskite material production via solution techniques can be distinguished (shown in Fig. 3.7).

In a one-step approach, PbI_2 and methylammonium iodide (MAI) are dissolved in solvents at a certain stoichiometric ratio, then spin-coated on a substrate, and annealed at an optimum temperature to make the perovskite light-absorbing layer. The quality of the crystals and perovskite layer properties are affected by the solvent used, annealing temperature, and annealing time [166]. The one-step deposition procedure is simple, but the synthetic crystals' form and size are challenging to control. Liang et al. [167] initially proposed the two-step sequential deposition process. This approach uses a saturated solution of PbI_2 as the precursor for spin-coating on the substrate. The PbI_2-coated substrate is then immersed for a suitable period of time in a 2-propanol solution containing MAI before being rinsed in 2-propanol. After annealing at a high enough temperature, the PbI_2 film interacts with MAI and creates a perovskite layer. Dipping duration and MAI solution concentration affect the form and optoelectronic properties of final $MAPbI_3$ films. To make PSCs with a high PCE of 15%, Burschka et al. employed a slightly modified two-step procedure that involves spin-coating MAI solution on the PbI_2 film rather than

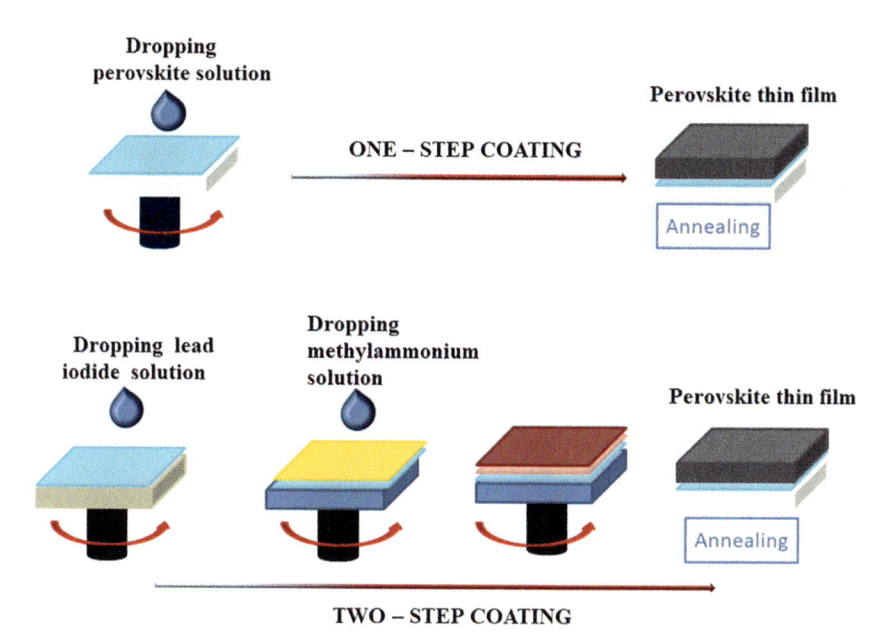

Figure 3.7 One-step and two-step spin-coating procedures for perovskite thin-film formation.

dipping [168]. The two-step procedure outperforms the one-step solution method for producing perovskite films in high humidity conditions; RH of less than 60% did not influence overall performance [168].

3.8.2 Vacuum thermal deposition methods

Thermal evaporation is a physical vapor deposition method used to form thin films, or coatings, on a variety of surfaces. The solid material is heated in a high-vacuum chamber, causing the substance to evaporate and form a vapor stream, which is subsequently deposited onto the substrate (shown in Fig. 3.8). A high-vacuum atmosphere is typically used to create the perovskite absorbent layer. PbI_2 and MAI are deposited on the substrate simultaneously or alternatively through heat evaporation from dual sources of PbI_2 and MAI. After being created at the correct temperature and environment, $MAPbI_3$ crystallizes into a perovskite film. Snaith et al. were the first to reveal coevaporation deposited $MAPbI_3$ film and utilize it in planar heterojunction PSCs with PCE of 15.7% and J_{sc} of 21.5 mA/cm^2 [82,169,170].

The problems in regulating the MAI deposition rate in the codeposition strategy led to the creation of sequential deposition. PbI_2 was deposited first by thermal evaporation, followed by MAI via vapor deposition.

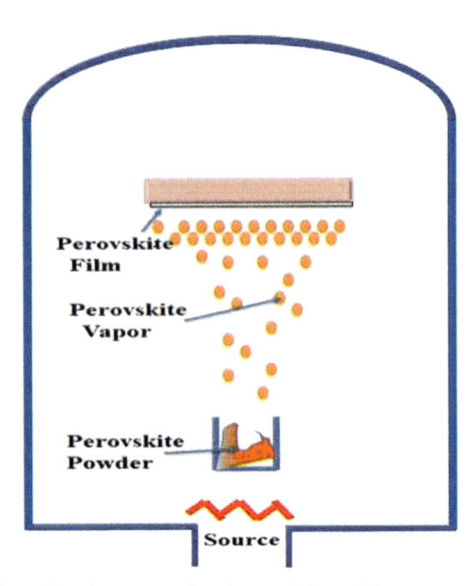

Figure 3.8 Perovskite film formation by thermal deposition.

The temperature of the substrate has a significant impact on the photovoltaic performance of sequentially deposited devices. For devices based on MAPbI$_3$ perovskite thin films made by sequential vapor deposition, Chen et al. obtained a PCE of 15.4% [171]. Zhu et al. reported the fabrication of pinhole-free cesium substitution perovskite thin films using this process [172]. The researchers showed 20.13% efficient planar PSCs, which they obtained by lowering trap-state density and boosting carrier longevity. Forgacs et al. demonstrated the potential of sequentially built multilayer structures to boost the efficiency of single-junction perovskite devices by presenting a perovskite/perovskite tandem solar cell made by sequential vapor deposition with a maximum PCE of 18% [173]. The technique requires a high temperature to convert solid PbI$_2$ to vapor. The procedure demands a large amount of machinery and may emit harmful pollutants. The drawbacks of the perovskite layer synthesis might be eliminated if a new metal halide with a low evaporation temperature and low toxicity could be developed and chosen to replace PbI$_2$.

3.8.3 Spray coating method

Spray coating of perovskite films is a standard approach for producing large-scale, high-throughput films. It may be used on any substrate material because it is a contact-free process. Furthermore, the approach is suitable for low-temperature processing. Barrows and colleagues used ultrasonic spray coating techniques to create planar heterojunction $CH_3NH_3PbI_xCl_{3-x}$ PSCs with a PCE of 11% [174]. This low-cost approach has also been used to make perovskite thin films on a glass substrate for solar cells with a PCE of 13% [175]. P. Kanjanaboos et al. found that layer-by-layer spray coating of mixed-halide quasi-2D perovskite on top of 3D perovskite increases film stability and preserves 3D perovskite performance in a humid environment (40%–50%RH). Sequential spray deposition opens new opportunities for a wide range of stacking patterns and large-scale production in a cost-effective manner [176].

3.8.4 Printing method

A printing technique has recently been used to fabricate PSCs. Screen printing of TiO_2 is the initial step, followed by imprinting ZrO_2 and carbon electrodes. Because it is done in a layer-by-layer step process, the method is sometimes referred to as a layer-by-layer printing procedure. The perovskite solution is applied to the porous carbon electrode layer and passes through the mesoporous TiO_2 and ZrO_2 layers. ZrO_2 is often employed as a porous insulating layer to prevent direct contact between the carbon electrode and the FTO/TiO_2 substrate. The PCE of these devices, on the other hand, remains low (12%) due to inadequate perovskite infiltration [177]. In PSCs, Seo's research group showed complete R2R manufacture of all layers except electrodes on a trial scale[178]. Tert-butyl alcohol (tBuOH) is a new environmentally safe antisolvent with a broad processing window. High PCEs of 23.5% for glass-based spin-coated PSCs and 19.1% for gravure-printed flexible PSCs were achieved by tBuOH:EA(ethyl acetate) bathing confirmed highly crystalline, uniform formamidinium (FA)-based perovskite production [179]. In Fig. 3.9, a R2R fabrication process of PSCs is illustrated.

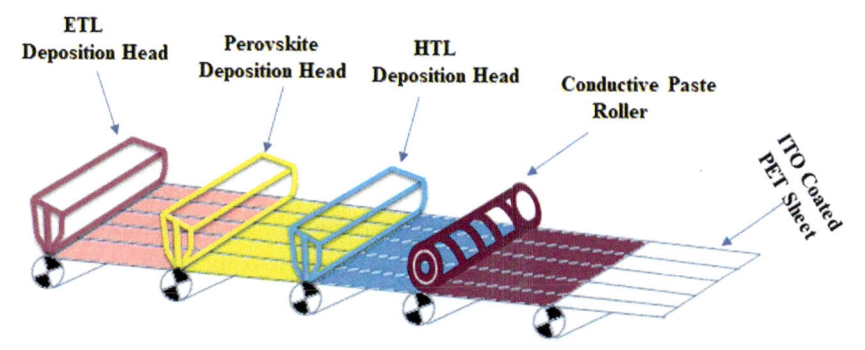

Figure 3.9 Roll-to-roll printing of perovskite solar cells.

3.9 Different ranges of electron and hole transport layers

There are several different materials that can be used as electron and HTL in PSCs. Electron transport materials (ETMs) and HTMs are crucial components in PSCs that play a significant role in determining the efficiency of the device. ETMs are responsible for transporting the negative charges (electrons) from the perovskite layer to the electrodes and for preventing the recombination of electrons and holes. Some common ETMs used in PSCs include titanium(IV) oxide (TiO_2), tin(IV) oxide (SnO_2), and PCBM (phenyl-C61-butyric acid methyl ester).

HTMs are responsible for transporting the positive charges (holes) from the perovskite layer to the electrodes and for preventing the recombination of electrons and holes. Some common HTMs used in PSCs include Spiro-OMeTAD (2,2′,7,7′-tetrakis(N,N-di-p-methoxyphenylamine)-9,9′-spirobifluorene), Nickel oxide (NiO_x), poly(triaryl amine) (PTAA), poly(3-hexylthiophene-2,5-diyl) (P3HT), and poly(3,4-ethylenedioxythiophene) (PEDOT). The choice of HTMs and ETMs depends on several factors, including cost, efficiency, stability, and compatibility with other components of the device. A combination of different materials may also be used to optimize the performance of the perovskite solar cell and some of such materials are shown in Fig. 3.10.

3.9.1 Electron transport layer

Electrons are transported from the active layer (perovskite layer) to the electrode through the electron transport layer. TiO_2 is widely utilized as an electron transport layer in mesoporous-based PSCs. However, nonstoichiometry defects in the TiO_2 layer, such as oxygen vacancies and titanium interstitial, can cause defects [180,181]. The performance of solar cells is harmed by these flaws. Although oxygen is necessary to passivate these flaws, other layers of the device require oxygen and moisture protection. Pathak et al. found a solution by doping aluminum into TiO_2. The conductivity of the electron transport layer increases with a low degree of Al doping in TiO_2. The number of trap states is reduced when Al is doped, according to photothermal deflection spectroscopy measurements. The addition of aluminum to TiO_2 improved device performance and stability [180].

The use of tin(IV) oxide (SnO_2) ETLs has been extensively investigated as an alternative to TiO_2 due to its excellent conduction band alignment

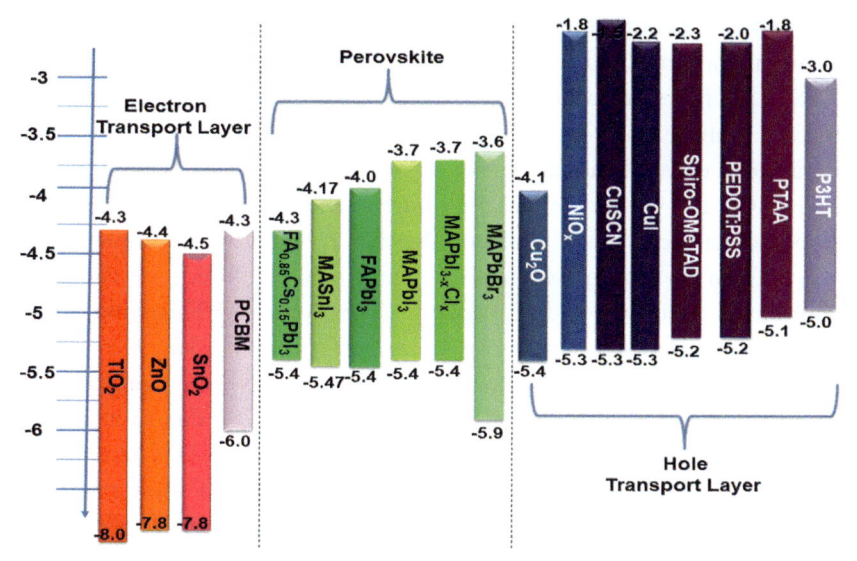

Figure 3.10 Different types of electron transport, perovskite, and hole transport material with their HOMO and LUMO energy levels.

with perovskites, high electron mobility, broad bandgap, high transmittance, good stability, and ease of processing [182]. In terms of processing temperatures, I—V hysteresis, and photostability, SnO_2 surpasses TiO_2 substantially. The majority of high-performance SnO_2 ETLs has been created at temperatures below 200°C, using a variety of synthetic processes such as spin-coating of nanoparticles or precursor solutions [183,184], atomic layer deposition (ALD) [185,186] and chemical bath deposition. In a planar PSC, Song et al. used SnO_2 as the electron transport layer [187]. This layer was deposited by spin coating SnO_2 nanoparticles and annealing them for 1 hour at 150°C, which is a far lower temperature than that utilized to make TiO_2 layers. To create the perovskite layer, they used a two-step deposition approach, first growing crystalline PbI_2 using solvent vapor annealing and then depositing methyl ammonium iodide solution to make $CH_3NH_3PbI_3$, resulting in a champion cell efficiency of 13%. After 700 hours, SnO_2-based devices were more stable than TiO_2-based devices. When exposed to light for 8 hours, both the SnO_2- and TiO_2-based devices degraded. The conversion efficiency of PSCs made using SnO_2-based ETL has gradually increased related to recent optimization strategies such as bilayering (film two electron transport layer). SnO_2 is stable in a variety of processes, including oxygen-rich and humid environments, which aids in the material's long-term stability [188], interfacial cross-linking [91], and ethylene diamine tetraacetic acid (EDTA) complexation [88], with a maximum confirmed efficiency of >23% [142].

PSCs are more environmentally friendly than their organic counterparts, and ETL is made SnO_2-based. SnO_2 is extremely stable under visible light illumination, especially in the presence of UV light [189], because SnO_2, unlike TiO_2, is a very stable material under visible light illumination because the photocatalytic activity is absent. Several techniques for employing a SnO_2 ETL in $n-i-p$ type devices have been reported, which are considerably more advantageous to device stability. The inorganic ETL on top of the perovskite layer acts as an encapsulating layer effectively blocking the passage of water and/or oxygen molecules. Tin(IV) oxide ETLs are typically deposited on another ETL, such as C60 [190] or PCBM [178], due to interfacial adhesion issues, although devices with top-layer SnO_2 ETLs have much superior long-term stability. As replacements for the two most prevalent inorganic ETL materials, many oxide materials with acceptable electrical/optical properties (e.g., ZnO [191], Zn_2SnO_4 [192,193], and $BaSnO_3$ [194]) have been investigated. Their ETL performance, however, has to be enhanced to compete with TiO_2 and SnO_2.

PCBM is the most often used organic ETL, particularly for $p-i-n$ type and/or flexible PSCs. PSCs manufactured from C60 or PCBM have high FF values and minimal IV hysteresis due to their high electron mobility and powerful electron-extracting abilities. Furthermore, because they have been widely used in organic solar cells, their manufacturing conditions, which include vacuum deposition or solution deposition, are well-known. They can be dissolved in a nonpolar solvent orthogonal to the perovskite layer and then directly deposited on top of it during the solution-phase method. One downside of fullerene-based ETLs is that stable noble metals such as Ag cannot be used due to their misaligned energy. The energy barrier for electron transfer from PCBM (or C60) to the Ag metal electrode can be successfully reduced by introducing interfacial layers such as bathocuproine (BCP) [195], polyethyleneimine ethoxylated (PEIE) [196,197], and other small [198]/macro-molecules [199] between PCBM (or C60) and the metal electrode, resulting in significantly improved FF values for the $p-i-n$-type configuration.

3.9.2 Hole transport layer

Spiro-OMeTAD is a widely utilized material in PSCs for hole transport. Additives such as 4-tert-butylpyriding and bis(trifluoromethane) sulfonamide lithium (Li-TFSI) salt are used to increase conductivity. On the other hand, these additives break the perovskite layer, lowering device stability. Han et al. used Spiro-OMeTAD as the HTL to characterize the mechanism of device deterioration [200]. Gradual increment in Rs, measured at 55°C and 50% RH under continuous one sunlight, led the researcher to observe that voids had formed in the Spiro-OMeTAD due to the release of volatile constituents as gaseous by-products [201]. These by-products passed through the Spiro-OMeTAD layer, resulting in an increase in Rs and a decrease in device performance.

Yang et al. investigated the effectiveness of HTL in preventing moisture-induced deterioration of perovskite films [156]. They used poly (3-hexylthiophene) (P3HT), poly[bis(4-phenyl)(2,4,6-trimethylphenyl)] (PTAA), and SpiroOMeTAD HTLs to study degradation in PSCs. Under constant RH, both P3HT and PTAA reduced the rate of deterioration (RHs). Under constant RH, the Spiro-OMeTAD layer accelerated the deterioration and breakdown of perovskite ($CH_3NH_3PbI_3$). Habisreutinger et al. investigated the influence of the HTL on PSC deterioration caused by heat and moisture [202]. Spiro-OMeTAD, PTAA, and P3HT are the HTLs that have proved to be the most effective in PSCs. Absorption and XRD studies confirmed that all HTLs inhibited fast deterioration. PSCs' performance and stability were improved by using polycarbonate (PC) and single-wall carbon nanotubes (SWNTs). With a 96-hour exposure duration, the device's PCE grew from 13.3% to 14.3%. A further increase in cell temperature to over 80°C resulted in a decrease in PCE. The author conducted an experiment to test the P3HT/SWNT-PC layer's water resistance properties of a P3HT/SWNT-PC HTL-based PSCs subjected to flowing tap water. After being exposed to water, the cell's performance was remarkably comparable to that of the original cell. Before and after water exposure, the solar cell's PCE was 12.9% and 12.7%, respectively. In a planar structure of PSCs, Zheng et al. substituted the unstable Spiro-OMeTAD with oligothiophene, Dr3TBDTT, as HTL [203]. Without further doping, this material has good hole mobility, conductivity, and hydrophobic nature. However, Dr3TBDTT has low surface coverage on perovskite film. Adding a tiny quantity of polydimethylsiloxane (PDMS) to Dr3TBDTT solved the problem. The performance of the planar device construction with PDMS-doped Dr3TBDTT as HTL yielded a PCE of 8.8%. The nonencapsulated devices based on Spiro-OMeTAD deteriorated after 3 days when exposed to simulated light in high humidity ($>50\%$). In contrast, those based on Dr3TBDTT:

PDMS HTL only showed a slight drop from 8.8% to 8%. The hydrophobic character of the PDMS-doped Dr3TBDTT film contributed to the increase in stability. Christians. et al. reported that when spiro-OMeTAD was replaced with EH44 (9-(2-ethylhexyl)-N,N,N,N-tetrakis(4-methoxyphenyl)-9H-carbazole-2,7-diamine), the operating stability of the unencapsulated PSCs was increased by more than 1000 hours [189].

To solve the instability problems that organic HTLs present, many inorganic p-type semiconducting materials have been investigated as HTLs in PSCs. NiO is one of the most investigated p-type materials because of its large bandgap, high transmittance, and deep valence band [204]. Spin coating [205], electrodeposition [196], sputtering [206], e-beam evaporation [207], and ALD [208] are some of the industrial procedures that may be utilized to deposit it. However, NiO HTLs have some disadvantages, such as low hole conductivity and poor electrical/physical contact with the perovskite layer, which have been successfully improved by doping with monovalent acceptor ions such as Li^+ [209] and Cu^+ [210], interfacial treatment with PEDOT: PSS [205], and the introduction of nanoscale roughness on the NiO HTL surface [196].

Copper thiocyanate (CuSCN) [211] is another inorganic HTL material with a deep valence band, good carrier mobility, thermal stability, and high transparency. It dissolves quickly in a wide range of solvents and crystallizes without the need for postannealing; as a result, it has been used in both $p-i-n$ and $n-i-p$-type devices. However, in $p-i-n$ devices, it was revealed that the underlying CuSCN HTL reacts with the perovskite layer during annealing, crystallizing the perovskite layer and reducing device performance [212].

Copper oxide (CuO_x) has been investigated as an HTL material in PSCs because of its deep valence band and excellent carrier mobility. Solution-phase [213], evaporation [214], and sputtering [215] have all been used to produce CuO_x HTLs. CuO_x has a low bandgap energy (1.2 eV for cupric oxide, CuO; 2.1 eV for cuprous oxide, CuO_2), which results in significant optical loss, especially in $p-i-n$ devices where incoming light should pass through the HTL. With $n-i-p$ devices, optical loss is not an issue; however, due to the breakdown of the perovskite layer during processing, only a limited number of synthetic techniques (e.g., evaporation) are available.

$CuCrO_2$ (3.0 eV) [216] and $CuGaO_2$ (3.6 eV) [217] are two examples of innovative p-type delafossite oxides that have been produced to raise the bandgap energy of CuO_x. Delafossite-based HTLs, which may be used in both $p-i-n$ and $n-i-p$ device configurations, have predominantly been created using a low-temperature-based spin-coating deposition. Using $CuCrO_2$-based HTLs, the conversion efficiency of the $p-i-n$ and $n-i-p$ devices is 20.54% [218] and 16.25% [219], respectively. The $n-i-p$ device has a conversion efficiency of 16.25% and significantly improved long-term stability because the undoped $CuCrO_2$ HTL was directly placed onto the perovskite layer, suggesting that by utilizing a proper doping technique during nanoparticle manufacturing, device performance might be increased much further. $CuGaO_2$-based HTLs, such as NiO HTL, performed well in the $p-i-n$ device with Zn doping and a bi-layered structure (20.15%) [220].

3.9.3 Buffer layer in perovskite solar cells

The stability of PSCs can also be improved by adding buffer layers between the HTL and the top metal electrode. It keeps humidity and oxygen out of the perovskite film. Guarnera et al. utilized Al_2O_3 nanoparticles as a buffer layer between the HTL and the perovskite absorber [221]. Compared to the control device, adding this buffer layer enhanced device performance. It has been discovered that adding a buffer layer to the HTL reduces the HTL's thickness, resulting in lower Rs and better solar cell performance. The stability of solar cells with and without a buffer layer was measured by the researcher. In a nitrogen-filled glove box, the devices were encapsulated with UV epoxy. Both devices were exposed to constant simulated sunshine illumination. Solar cells with no buffer layer deteriorated quickly, whereas those with an Al_2O_3 buffer layer kept 95% of their initial PCE after 350 hours of exposure. A buffer layer between the metal electrode and the perovskite film prevented direct contact and reaction.

3.10 Electrodes for perovskite solar cells

Gold (Au) is commonly utilized as the top electrode for high-efficiency PSCs [222–224], although it is a costly material. The use of silver (Ag) as the top electrode for PSCs was shown to be problematic by Kato et al. [225]. When compared to the Au electrode, using Ag as the top electrode exhibited poor environmental stability. The alterations in electrodes were investigated using XRD and XPS analyses. Ag interacted with perovskite and generated silver iodide (AgI) at high humidity (>50%), which reduced the effectiveness of PSCs. The five steps that can lead to the development of AgI include (i) water molecule diffusion through the Spiro-OMeTAD HTL via pinholes [30], (ii) moisture-induced breakdown of perovskite layer into MAI and HI, (iii) transfer of iodine-containing materials to Ag electrode through Spiro-OMeTAD pinholes, (iv) surface diffusion of iodine compound, and (v) production of AgI. The formation of AgI alters the color of the Ag electrode from reflecting metal to yellowish. This revealed that in the deterioration of PSCs, all layers are connected.

Pankaj et al. reported the use of copper (Cu) as an electrode [226]; in comparison to traditional Al and Ag electrodes, Cu has been shown to be a more appropriate electrode material. They reported the results of a study into the photostability of PSCs based on top Cu electrodes. The PSCs were produced in standard geometry and kept in the dark, with constant illumination from a white LED bulb inside the lab and direct sunshine outdoors, and tested according to the worldwide summit on OPV stability procedures. The encapsulated solar cells were most stable in dark storage, but their performance degraded under illumination, with the deterioration occurring most quickly in direct sunshine. The photo-oxidation of the perovskite film has been blamed for the degradation under the light. Cu was seen to diffuse into and react with the underlying perovskite film, and the photo-oxidation and chemical interactions between Cu and the perovskite film were enhanced by the ultraviolet and infrared concentrations in direct sunshine. Under natural working conditions, the Cu electrode's chemical interactions with perovskite components caused it to vanish after a while. According to these studies, Cu appears to be a less stable electrode material for PSCs. Schmidt et al. established the first PSCs on flexible surfaces utilizing a printing and coating method [227]. The performance of flexible devices was severely lowered when compared to conventional devices manufactured by spin coating and thermal evaporation of materials on glass substrates, dropping from 9.4% to 4.9%.

Carbon ink is the best choice for electrodes in printing techniques because of its low cost [228] and minimal environmental effect [229]. The solvents used to make the ink are a major issue in printing electrodes. Li et al. manufactured flexible PSCs with a 3.03% efficiency [230]. Carbon nanotube (CNT) fibers were used as the top electrode in this device. Even after 1000 bending cycles, the cells remained highly stable. The CNT network on the perovskite layer eliminated the requirement for an HTL and metal electrode, yielding a PCE of 6.29%. After 20 days of storage in the dark and ambient conditions, the PCE of PSCs based on carbon electrodes decreased from 13.53% to 12.87%.

3.11 Effect of encapsulation on perovskite solar cells

PSCs are extremely sensitive to their surroundings, and they begin to degrade when exposed to oxygen and moisture. Encapsulation protects them from the surrounding environment. Encapsulation of solar cells on glass substrates is usually done with a glass cover slip and UV-curable epoxy [221,231,232]. Pankaj. et al. performed degradation studies of PSCs and observed the degradation effect on photovoltaic parameters [233], including the degradation of PSCs stored in various open-air environmental conditions. The solar cells were kept in dark in the ambient environment of the laboratory (RH 47.5%, temperature 23.4°C), under compact fluorescent lamp (CFL) lighting (irradiance 10 mW/cm^2), and outside the laboratory under natural sunlight. The environmental conditions in the outdoor storage scenario changed throughout the day (irradiance 100 mW/cm^2, RH 18%, and temperature 45°C at midday) and at night (irradiance 0 mW/cm^2, RH 66%, temperature 16°C at midnight). Photovoltaic parameters were measured regularly in the lab, following the guidelines of the International Summit on Organic Photovoltaic Stability (ISOS). Solar cells stored outside degraded quicker than those placed under CFL or in the dark in terms of all photovoltaic parameters, including short circuit current density (J_{sc}), open-circuit voltage (V_{oc}), FF, and PCE. The most stable solar cells were those that were kept in the dark. Solar cells retained under CFL light maintained >60% of their initial efficiency after 1100 hours, but encapsulated solar cells placed outside entirely deteriorated after around 560 hours. The solar cells that were held in the dark for 1100 hours showed no loss in PCE but rather a slight improvement, owing to an increase in their V_{oc}. The rapid deterioration in the open air outside the laboratory under direct sunlight has been linked to high temperature during the day, high humidity at night, high solar illumination intensity, and the presence of ultraviolet and infrared radiation in incident solar light. During CFL storage, the top Ag electrode broke down and interacted with the active layer. The degradation and reaction of the Ag electrode were increased when it was exposed to direct sunlight.

Han et al. focus on two different PSC encapsulation methods. Between the simple glass cover and the silver contact, UV-curable epoxy resin was applied in the first technique. The second technique involved filling a space between the electrode and the cover slip with desiccant material and sealing it with UV epoxy glue [200]. Krebs et al. devised an encapsulation technique for R2R organic solar cells in 2009 [234]; devices are manufactured on a flexible polyethlyeneterephthalate (PET) substrate and enclosed by adhering a 25 μm- thick PET barrier foil to the substrate with an acrylic adhesive. The stability of solar cells under ambient environmental conditions was enhanced by this technique. However, there was an issue with the edges when devices were cut off the roll. By sealing the edges, Tanenbaum and colleagues enhanced the encapsulation process. The samples were sliced and then laminated with an adhesive around the edges [235]. Schmidt et al. reported the first comprehensive R2R manufacturing of flexible PSCs with encapsulation to protect them from the environment [227]. Shao et al. presented a new, low-cost device manufacturing and encapsulation technique in which commercially available conductive copper tape is used to replace the top contact. In basic planar construction, the tape was deposited by rolling it with a glass rod to attach the active region, resulting in a PCE of 12.7% [236]. Devices employing conductive copper tape as an electrode offered a barrier to external entities and enhanced device stability when compared to control devices.

3.12 Standard testing protocols

Stability studies of PSCs from different research groups yield conflicting results because of the differences in testing conditions, including a wide range of RH values. For rapid technological advancement, the progress made in the area by diverse groups must be similar. The OPV community attended the ISOS with the same goals in mind and created certain testing procedures. Several research groups, including Roesch et al. [237], Reese et al. [238], and others, have reported on the performance of their solar cells using these procedures. For testing, safety, and dependability of thin-film solar modules, the International Electrotechnical Commission 61646 standards were created. To some degree, previously accepted methods and standards can be used for PSCs, but they must be adjusted to account for (i) perovskite film degradation processes owing to moisture and increasing temperature, (ii) the existence of a mesoporous scaffold, and (iii) measurement induced hysteresis. PSCs need a re-evaluation of the ISOS protocols.

3.13 Conclusion and future prospects

In just a few years, PSCs have experienced a sixfold increase in PCE from 3.8% to 25.7% [6]. PSCs have the potential to outperform conventional solar cells in terms of PCE; also, they are suitable for low-cost processing methods such as flexible and R2R inkjet printing. To produce high-efficiency tandem solar cells, additional perovskite absorber materials might be used with silicon solar cells [170]. Perovskite materials are highly popular in current photovoltaic communities, and a lot of effort has been put into addressing a number of PSCs-related problems. However, they are unable to approach the market because of poor stability.

This chapter studied the primary concern for perovskite materials, charge transport layers, and different PSC device architectures. Chemical engineering and modification in the composition of the constituent elements in ABX_3 structure have been attempted to find more stable perovskite materials [117,150,170,239]. Rather than addressing a single layer's stability, we must view all of the materials in PSCs as a unified system. Moisture-induced breakdown of perovskite materials is the main cause of instability. The electrode material should be low-cost and long-lasting. Several research papers emphasize the problems with the silver (Ag) electrode [225]. Carbon-based ink is shown to be the best choice for this purpose, but more work is needed in this area for improved efficiency [240]. In order to increase the stability of PSCs in the future, two methods must be considered: (i) A suitable encapsulation procedure and (ii) the intrinsic stability of perovskite materials in a humid environment. Encapsulation gives the devices some stability and speeds up the commercialization of perovskite solar technology. Finally, because various research organizations use a wide variety of testing protocols, we require standardized measuring procedures for these devices. Overall, the fact that substantial progress has already been made in PSCs, particularly in terms of reaching high efficiencies, is quite promising.

References

[1] N.S. Lewis, D.G. Nocera, Powering the planet: Chemical challenges in solar energy utilization, Proc. Natl Acad. Sci. 103 (43) (2006) 15729−15735. Available from: https://doi.org/10.1073/pnas.0710683104. later corrected.
[2] D. Lincot, The new paradigm of photovoltaics: From powering satellites to powering humanity, Comptes Rendus Phys. 18 (7) (2017) 381−390.
[3] R.P. Raffaelle, An introduction to space photovoltaics: Technologies, issues, and missions, in: S.G. Bailey, A.F. Hepp, D.C. Ferguson, R.P. Raffaelle, S.M. Durbin (Eds.), Photovoltaics for Space: Key Issues, Missions and Alternative Technologies, Elsevier, 2022, pp. 3−27.

[4] O.K. Simya, P. Radhakrishnan, A. Ashok, Engineered nanomaterials for energy applications, in: C.M. Hussain (Ed.), Handbook of Nanomaterials for Industrial Applications, Elsevier, 2018, pp. 751−767.

[5] M. Green, E. Dunlop, J. Hohl-Ebinger, M. Yoshita, N. Kopidakis, X. Hao, Solar cell efficiency tables (version 57), Prog. Photovolt. 29 (1) (2021) 3−15.

[6] U.S. Department of Energy, National Renewable Energy Laboratory, Solar cell efficiencies, 2022. <https://www.nrel.gov/pv/assets/pdfs/best-research-cell-efficiencies.pdf>.

[7] M. Laul, Financing firm level research & development in the United States' solar photovoltaic industry: The case of First Solar, Inc., 2002−2016, Master's Thesis, Tallinn University of Technology, Estonia, 2017, 45 pp.

[8] Y.B.M. Yusoff, Copper indium gallium selenide solar cells, in: M. Akhtaruzzaman, V. Selvanathan (Eds.), in: Comprehensive Guide on Organic and Inorganic Solar Cells, Academic Press, 2022, pp. 85−113.

[9] H. Kang, Crystalline silicon vs. amorphous silicon: The significance of structural differences: In Photovoltaic Applications, IOP Conf. Series: Earth Environ. Sci. 726 (1) (2021) 012001.

[10] D.M. Chapin, C.S. Fuller, G.L. Pearson, A new silicon p-n junction photocell for converting solar radiation into electrical power, J. Appl. Phys. 25 (5) (1954) 676−677.

[11] T. Wu, Z. Qin, Y. Wang, Y. Wu, W. Chen, et al., The main progress of perovskite solar cells in 2020−2021, Nano-Micro Lett. 13 (1) (2021)152.

[12] C. Ma, N.-G. Park, A realistic methodology for 30% efficient perovskite solar cells, Chem 6 (6) (2020) 1254−1264.

[13] M. Saliba, T. Matsui, J.-Y. Seo, K. Domanski, J.-P. Correa-Baena, et al., Cesium-containing triple cation perovskite solar cells: Improved stability, reproducibility and high efficiency, Energy Environ. Sci. 9 (6) (2016) 1989−1997.

[14] R.L.Z. Hoye, P. Schulz, L.T. Schelhas, A.M. Holder, K.H. Stone, et al., Perovskite-inspired photovoltaic materials: Toward best practices in materials characterization and calculations, Chem. Mater. 29 (5) (2017) 1964−1988.

[15] D. Zhou, T. Zhou, Y. Tian, X. Zhu, Y. Tu, Perovskite-based solar cells: Materials, methods, and future perspectives, J. Nanomaterials 2018 (2018) 8148072.

[16] C. Momblona, O. Malinkiewicz, C. Roldán-Carmona, A. Soriano, L. Gil-Escrig, et al., Efficient methylammonium lead iodide perovskite solar cells with active layers from 300 to 900 nm, APL. Mater. 2 (8) (2014)081504.

[17] Y. Hu, J. Schlipf, M. Wussler, M.L. Petrus, W. Jaegermann, et al., Hybrid perovskite/perovskite heterojunction solar cells, ACS Nano 10 (6) (2016) 5999−6007.

[18] M.M. Lee, J. Teuscher, T. Miyasaka, T.N. Murakami, H.J. Snaith, Efficient hybrid solar cells based on meso-superstructured organometal halide perovskites, Science 338 (6107) (2012) 643−647.

[19] H.-S. Kim, C.-R. Lee, J.-H. Im, K.-B. Lee, T. Moehl, et al., Lead iodide perovskite sensitized all-solid-state submicron thin film mesoscopic solar cell with efficiency exceeding 9%, Sci. Rep. 2 (1) (2012)591.

[20] J. Burschka, N. Pellet, S.-J. Moon, R. Humphry-Baker, P. Gao, M.K. Nazeeruddin, et al., Sequential deposition as a route to high-performance perovskite-sensitized solar cells, Nature 499 (7458) (2013) 316−319.

[21] H. Zhou, Q. Chen, G. Li, S. Luo, T.-B. Song, et al., Interface engineering of highly efficient perovskite solar cells, Science 345 (6196) (2014) 542−546.

[22] J. Li, Q. Wang, A. Abate, Perovskite solar cells, in: S. Thomas, E.H.M. Sakho, N. Kalarikkal, S.O. Oluwafemi, J. Wu (Eds.), Nanomaterials for Solar Cell Applications, Elsevier, 2019, pp. 417−446.

[23] B.G. Krishna, G.S. Rathore, N. Shukla, S. Tiwari, Perovskite solar cells: A review of architecture, processing methods, and future prospects, in: I. Khan, A. Khan, M.A. Khan, S. Khan, F. Verpoort, A. Umar (Eds.), Hybrid Perovskite Composite Materials, Woodhead Publishing, 2021, pp. 375−412.

[24] F. Shao, L. Xu, Z. Tian, Y. Xie, Y. Wang, P. Sheng, et al., A modified two-step sequential deposition method for preparing perovskite $CH_3NH_3PbI_3$ solar cells, RSC Adv. 6 (48) (2016) 42377−42381.

[25] K. Liao, C. Li, L. Xie, Y. Yuan, S. Wang, Z. Cao, et al., Hot-casting large-grain perovskite film for efficient solar cells: Film formation and device performance, Nano-Micro Lett. 12 (1) (2020)156.

[26] Q. Chen, H. Zhou, Z. Hong, S. Luo, H.-S. Duan, et al., Planar heterojunction perovskite solar cells via vapor-assisted solution process, J. Am. Chem. Soc. 136 (2) (2014) 622−625.

[27] A.D. Taylor, Q. Sun, K.P. Goetz, Q. An, T. Schramm, et al., A general approach to high-efficiency perovskite solar cells by any antisolvent, Nat. Commun. 12 (1) (2021) 1878.

[28] W.-J. Yin, T. Shi, Y. Yan, Unique properties of halide perovskites as possible origins of the superior solar cell performance, Adv. Mater. 26 (27) (2014) 4653−4658.

[29] P.P. Boix, S. Agarwala, T.M. Koh, N. Mathews, S.G. Mhaisalkar, Perovskite solar cells: Beyond methylammonium lead iodide, J. Phys. Chem. Lett. 6 (5) (2015) 898−907.

[30] W. Chen, X. Yin, M. Que, H. Xie, J. Liu, et al., A comparative study of planar and mesoporous perovskite solar cells with printable carbon electrodes, J. Power Sources 412 (2019) 118−124.

[31] K.P. Bhandari, R.J. Ellingson, An overview of hybrid organic−inorganic metal halide perovskite solar cells, in: T.M. Letcher, V.M. Fthenakis (Eds.), A Comprehensive Guide to Solar Energy Systems, Academic Press, 2018, pp. 233−254.

[32] N. Islam, M. Yang, K. Zhu, Z. Fan, Mesoporous scaffolds based on TiO_2 nanorods and nanoparticles for efficient hybrid perovskite solar cells, J. Mater. Chem. A 3 (48) (2015) 24315−24321.

[33] A. Kojima, K. Teshima, Y. Shirai, T. Miyasaka, Organometal halide perovskites as visible-light sensitizers for photovoltaic cells, J. Am. Chem. Soc. 131 (17) (2009) 6050−6051.

[34] N.-G. Park, Perovskite solar cells: An emerging photovoltaic technology, Mater. Today 18 (2) (2015) 65−72.

[35] J. Jeong, M. Kim, J. Seo, H. Lu, P. Ahlawat, Pseudo-halide anion engineering for α-$FAPbI_3$ perovskite solar cells, Nature 592 (7854) (2021) 381−385.

[36] J. Ling, P.K.K. Kizhakkedath, T.M. Watson, I. Mora-Seró, L. Schmidt-Mende, T. M. Brown, et al., A perspective on the commercial viability of perovskite solar cells, Sol. RRL 5 (11) (2021) 2100401.

[37] E.L. Unger, L. Kegelmann, K. Suchan, D. Sörell, L. Korte, S. Albrecht, Roadmap and roadblocks for the band gap tunability of metal halide perovskites, J. Mater. Chem. A 5 (2017) 11401−11409.

[38] D. Bryant, N. Aristidou, S. Pont, I. Sanchez-Molina, T. Chotchunangatchaval, et al., Light and oxygen induced degradation limits the operational stability of methylammonium lead triiodide perovskite solar cells, Energy Environ. Sci. 9 (5) (2016) 1655−1660.

[39] C.C. Boyd, R. Cheacharoen, T. Leijtens, M.D. McGehee, Understanding degradation mechanisms and improving stability of perovskite photovoltaics, Chem. Rev. 119 (5) (2019) 3418−3451.

[40] W.S. Yang, B.-W. Park, E.H. Jung, N.J. Jeon, Y.C. Kim, et al., Iodide management in formamidinium-lead-halide-based perovskite layers for efficient solar cells, Science 356 (6345) (2017) 1376−1379.
[41] J.M. Frost, K.T. Butler, F. Brivio, C.H. Hendon, M. van Schilfgaarde, A. Walsh, Atomistic origins of high-performance in hybrid halide perovskite solar cells, Nano Lett. 14 (5) (2014) 2584−2590.
[42] K. Zheng, T. Pullerits, Two dimensions are better for perovskites, J. Phys. Chem. Lett. 10 (19) (2019) 5881−5885.
[43] Z. Wang, Q. Lin, F.P. Chmiel, N. Sakai, L.M. Herz, H.J. Snaith, Efficient ambient-air-stable solar cells with 2D−3D heterostructured butylammonium-caesium-formamidinium lead halide perovskites, Nat. Energy 2 (9) (2017)17135.
[44] S. Shao, M.A. Loi, The role of the interfaces in perovskite solar cells, Adv. Mater. Interf. 7 (1) (2020)1901469.
[45] E.M. Sanehira, B.J. Tremolet de Villers, P. Schulz, M.O. Reese, S. Ferrere, et al., Influence of electrode interfaces on the stability of perovskite solar cells: Reduced degradation using MoOx/Al for hole collection, ACS Energy Lett. 1 (1) (2016) 38−45.
[46] X. Ziang, L. Shifeng, Q. Laixiang, P. Shuping, W. Wei, et al., Refractive index and extinction coefficient of $CH_3NH_3PbI_3$ studied by spectroscopic ellipsometry, Optical Mater. Express 5 (1) (2015) 29−43.
[47] D. Ju, Y. Dang, Z. Zhu, H. Liu, C.-C. Chueh, et al., Tunable band gap and long carrier recombination lifetime of stable mixed $CH_3NH_3Pb_xSn_{1-x}Br_3$ single crystals, Chem. Mater. 30 (5) (2018) 1556−1565.
[48] T.C.-J. Yang, P. Fiala, Q. Jeangros, C. Ballif, High-bandgap perovskite materials for multijunction solar cells, Joule 2 (8) (2018) 1421−1436.
[49] Y. Kuang, Y. Ma, D. Zhang, Q. Wei, S. Wang, X. Yang, et al., Enhanced optical absorption in perovskite/Si tandem solar cells with nanoholes array, Nanoscale Res. Lett. 15 (1) (2020)213.
[50] Q.G. Du, G. Shen, S. John, Light-trapping in perovskite solar cells, AIP Adv. 6 (6) (2016)065002.
[51] J.-W. Lee, D.-J. Seol, A.-N. Cho, N.-G. Park, High-efficiency perovskite solar cells based on the black polymorph of $HC(NH_2)_2PbI_3$, Adv. Mater. 26 (29) (2014) 4991−4998.
[52] M. Saliba, T. Matsui, K. Domanski, J.-Y. Seo, A. Ummadisingu, et al., Incorporation of rubidium cations into perovskite solar cells improves photovoltaic performance, Science 354 (6309) (2016) 206−209.
[53] N.J. Jeon, J.H. Noh, W.S. Yang, Y.C. Kim, S. Ryu, J. Seo, et al., Compositional engineering of perovskite materials for high-performance solar cells, Nature 517 (7535) (2015) 476−480.
[54] K. Tanaka, T. Takahashi, T. Ban, T. Kondo, K. Uchida, N. Miura, Comparative study on the excitons in lead-halide-based perovskite-type crystals $CH_3NH_3PbBr_3$ $CH_3NH_3PbI_3$, Solid. State Commun. 127 (9) (2003) 619−623.
[55] E.J. Juarez-Perez, R.S. Sanchez, L. Badia, G. Garcia-Belmonte, Y.S. Kang, I. Mora-Sero, et al., Photoinduced giant dielectric constant in lead halide perovskite solar cells, J. Phys. Chem. Lett. 5 (13) (2014) 2390−2394.
[56] T.M. Brenner, D.A. Egger, A.M. Rappe, L. Kronik, G. Hodes, D. Cahen, Are mobilities in hybrid organic−inorganic halide perovskites actually "high"?, J. Phys. Chem. Lett. 6 (23) (2015) 4754−4757.
[57] X.Y. Zhu, V. Podzorov, Charge carriers in hybrid organic−inorganic lead halide perovskites might be protected as large polarons, J. Phys. Chem. Lett. 6 (23) (2015) 4758−4761.

[58] D. Shi, V. Adinolfi, R. Comin, M. Yuan, Erkki Alarousu, et al., Low trap-state density and long carrier diffusion in organolead trihalide perovskite single crystals, Science 347 (6221) (2015) 519−522.
[59] W. Nie, H. Tsai, R. Asadpour, J.-C. Blancon, A.J. Neukirch, et al., High-efficiency solution-processed perovskite solar cells with millimeter-scale grains, Science 347 (6221) (2015) 522−525.
[60] P.M. Beaujuge, J.M.J. Fréchet, Molecular design and ordering effects in π-functional materials for transistor and solar cell applications, J. Am. Chem. Soc. 133 (50) (2011) 20009−20029.
[61] K. Frohna, S.D. Stranks, Hybrid perovskites for device applications, in: O. Ostroverkhova (Ed.), in: Handb. Org. Mater. Electron. Photonic Devices (Second. Ed.), Woodhead Publishing, 2019, pp. 211−256.
[62] Q. Dong, Y. Fang, Y. Shao, P. Mulligan, J. Qiu, L. Cao, et al., Electron-hole diffusion lengths > 175 μm in solution-grown CH3NH3PbI3 single crystals, Science 347 (6225) (2015) 967−970.
[63] M. Long, T. Zhang, W. Xu, X. Zeng, F. Xie, et al., Large-grain formamidinium PbI$_{3-x}$Br$_x$ for high-performance perovskite solar cells via intermediate halide exchange, Adv. Energy Mater. 7 (12) (2017) 1601882.
[64] J.-P. Correa-Baena, M. Anaya, G. Lozano, W. Tress, K. Domanski, et al., Unbroken perovskite: Interplay of morphology, electro-optical properties, and ionic movement, Adv. Mater. 28 (25) (2016) 5031−5037.
[65] J.-P. Correa-Baena, W. Tress, K. Domanski, E.H. Anaraki, S.-H. Turren-Cruz, et al., Identifying and suppressing interfacial recombination to achieve high open-circuit voltage in perovskite solar cells, Energy Environ. Sci. 10 (5) (2017) 1207−1212.
[66] C. Momblona, L. Gil-Escrig, E. Bandiello, E.M. Hutter, M. Sessolo, et al., Efficient vacuum deposited p-i-n and n-i-p perovskite solar cells employing doped charge transport layers, Energy Environ. Sci. 9 (11) (2016) 3456−3463.
[67] A. Marchioro, J. Teuscher, D. Friedrich, M. Kunst, R. van de Krol, T. Moehl, M. Grätzel, J.-E. Moser, Unravelling the mechanism of photoinduced charge transfer processes in lead iodide perovskite solar cells, Nat. Photonics 8 (3) (2014) 250−255.
[68] T. Salim, S. Sun, Y. Abe, A. Krishna, A.C. Grimsdale, Y.M. Lam, Perovskite-based solar cells: Impact of morphology and device architecture on device performance, J. Mater. Chem. A 3 (17) (2015) 8943−8969.
[69] A. Senocrate, I. Moudrakovski, G.Y. Kim, T.-Y. Yang, G. Gregori, M. Grätzel, et al., The nature of ion conduction in methylammonium lead iodide: A multimethod approach, Angew. Chem. Int. Ed. Engl. 56 (27) (2017) 7755−7759.
[70] Y. Yamada, T. Nakamura, M. Endo, A. Wakamiya, Y. Kanemitsu, Photocarrier recombination dynamics in perovskite CH3NH3PbI3 for solar cell applications, J. Am. Chem. Soc. 136 (33) (2014) 11610−11613.
[71] S.D. Stranks, V.M. Burlakov, T. Leijtens, J.M. Ball, A. Goriely, H.J. Snaith, Recombination kinetics in organic-inorganic perovskites: Excitons, free charge, and subgap states, Phys. Rev. Appl. 2 (3) (2014) 034007.
[72] W. Yang, Y. Yao, C.-Q. Wu, Mechanism of charge recombination in mesostructured organic-inorganic hybrid perovskite solar cells: A macroscopic perspective, J. Appl. Phys. 117 (15) (2015) 155504.
[73] J.-P. Correa-Baena, S.-H. Turren-Cruz, W. Tress, A. Hagfeldt, C. Aranda, et al., Changes from bulk to surface recombination mechanisms between pristine and cycled perovskite solar cells, ACS Energy Lett. 2 (3) (2017) 681−688.
[74] Y. Lin, B. Chen, F. Zhao, X. Zheng, Y. Deng, et al., Matching charge extraction contact for wide-bandgap perovskite solar cells, Adv. Mater. 29 (26) (2017) 1700607.

[75] H. Tan, A. Jain, O. Voznyy, X. Lan, F.P. García de Arquer, et al., Efficient and stable solution-processed planar perovskite solar cells via contact passivation, Science 355 (6326) (2016) 722–726.
[76] C. Wang, C. Xiao, Y. Yu, D. Zhao, R.A. Awni, et al., Understanding and eliminating hysteresis for highly efficient planar perovskite solar cells, Adv. Energy Mater. 7 (17) (2017) 1700414.
[77] A. Hagfeldt, G. Boschloo, L. Sun, L. Kloo, H. Pettersson, Dye-sensitized solar cells, Chem. Rev. 110 (11) (2010) 6595–6663.
[78] P.V. Kamat, J.A. Christians, E.J. Radich, Quantum dot solar cells: Hole transfer as a limiting factor in boosting the photoconversion efficiency, Langmuir 30 (20) (2014) 5716–5725.
[79] S. Emin, S.P. Singh, L. Han, N. Satoh, A. Islam, Colloidal quantum dot solar cells, Sol. Energy 85 (6) (2011) 1264–1282.
[80] G. Xing, N. Mathews, S. Sun, S.S. Lim, Y.M. Lam, et al., Long-range balanced electron- and hole-transport lengths in organic-inorganic $CH_3NH_3PbI_3$, Science 342 (6156) (2013) 344–347.
[81] E.J. Juarez-Perez, M. Wußler, F. Fabregat-Santiago, K. Lakus-Wollny, E. Mankel, et al., Role of the selective contacts in the performance of lead halide perovskite solar cells, J. Phys. Chem. Lett. 5 (4) (2014) 680–685.
[82] M. Liu, M.B. Johnston, H.J. Snaith, Efficient planar heterojunction perovskite solar cells by vapour deposition, Nature 501 (7467) (2013) 395–398.
[83] B. Conings, L. Baeten, T. Jacobs, R. Dera, J. D'Haen, J. Manca, et al., An easy-to-fabricate low-temperature TiO_2 electron collection layer for high efficiency planar heterojunction perovskite solar cells, APL. Mater. 2 (8) (2014) 081505.
[84] M.H. Kumar, N. Yantara, S. Dharani, M. Grätzel, S. Mhaisalkar, P.P. Boix, et al., Flexible, low-temperature, solution processed ZnO-based perovskite solid state solar cells, Chem. Commun. 49 (94) (2013) 11089–11091.
[85] M. Cheng, C. Zuo, Y. Wu, Z. Li, B. Xu, Y. Hua, et al., Charge-transport layer engineering in perovskite solar cells, Sci. Bull. 65 (15) (2020) 1237–1241.
[86] P. Tiwana, P. Docampo, M.B. Johnston, H.J. Snaith, L.M. Herz, Electron mobility and injection dynamics in mesoporous ZnO, SnO_2, and TiO_2 films used in dye-sensitized solar cells, ACS Nano 5 (6) (2011) 5158–5166.
[87] J. You, Efficient and stable of perovskite optoelectronic devices, The International Photonics and Optoelectronics Meeting (POEM), OSA Technical Digest (2018) paper PF1B.2. <https://doi.org/10.1364/PFE.2018.PF1B.2>.
[88] D. Yang, R. Yang, K. Wang, C. Wu, X. Zhu, et al., High efficiency planar-type perovskite solar cells with negligible hysteresis using EDTA-complexed SnO_2, Nat. Commun. 9 (1) (2018) 3239.
[89] L. Xiong, Y. Guo, J. Wen, H. Liu, G. Yang, P. Qin, et al., Review on the application of SnO_2 in perovskite solar cells, Adv. Funct. Mater. 28 (35) (2018) 1802757.
[90] W. Ke, G. Fang, Q. Liu, L. Xiong, P. Qin, et al., Low-temperature solution-processed tin oxide as an alternative electron transporting layer for efficient perovskite solar cells, J. Am. Chem. Soc. 137 (21) (2015) 6730–6733.
[91] J. Chen, X. Zhao, S.-G. Kim, N.-G. Park, Multifunctional chemical linker imidazoleacetic acid hydrochloride for 21% efficient and stable planar perovskite solar cells, Adv. Mater. 31 (39) (2019) 1902902.
[92] J.-Y. Jeng, Y.-F. Chiang, M.-H. Lee, S.-R. Peng, T.-F. Guo, P. Chen, et al., $CH_3NH_3PbI_3$ perovskite/fullerene planar-heterojunction hybrid solar cells, Adv. Mater. 25 (27) (2013) 3727–3732.
[93] J.H. Heo, H.J. Han, D. Kim, T.K. Ahn, S.H. Im, Hysteresis-less inverted $CH_3NH_3PbI_3$ planar perovskite hybrid solar cells with 18.1% power conversion efficiency, Energy Environ. Sci. 8 (5) (2015) 1602–1608.

[94] C.T. Howells, S. Saylan, H. Kim, K. Marbou, T. Aoyama, et al., Influence of perfluorinated ionomer in PEDOT:PSS on the rectification and degradation of organic photovoltaic cells, J. Mater. Chem. A 6 (33) (2018) 16012−16028.

[95] J. Niu, D. Yang, X. Ren, Z. Yang, Y. Liu, X. Zhu, et al., Graphene-oxide doped PEDOT:PSS as a superior hole transport material for high-efficiency perovskite solar cell, Org. Electron. 48 (2017) 165−171.

[96] S. Wang, X. Huang, H. Sun, C. Wu, Hybrid UV-ozone-treated rGO-PEDOT: PSS as an efficient hole transport material in inverted planar perovskite solar cells, Nanoscale Res. Lett. 12 (1) (2017) 619.

[97] W. Yan, Y. Li, Y. Li, S. Ye, Z. Liu, S. Wang, et al., Stable high-performance hybrid perovskite solar cells with ultrathin polythiophene as hole-transporting layer, Nano Res. 8 (8) (2015) 2474−2480.

[98] J.H. Park, J. Seo, S. Park, S.S. Shin, Y.C. Kim, et al., Efficient $CH_3NH_3PbI_3$ perovskite solar cells employing nanostructured p-type NiO electrode formed by a pulsed laser deposition, Adv. Mater. 27 (27) (2015) 4013−4019.

[99] X. Yin, P. Chen, M. Que, Y. Xing, W. Que, C. Niu, et al., Highly efficient flexible perovskite solar cells using solution-derived NiO_x hole contacts, ACS Nano 10 (3) (2016) 3630−3636.

[100] U. Kwon, B.-G. Kim, D.C. Nguyen, J.-H. Park, N.Y. Ha, et al., Solution-processible crystalline NiO nanoparticles for high-performance planar perovskite photovoltaic cells, Sci. Rep. 6 (1) (2016) 30759.

[101] Z.-L. Tseng, L.-C. Chen, C.-H. Chiang, S.-H. Chang, C.-C. Chen, C.-G. Wu, Efficient inverted-type perovskite solar cells using UV-ozone treated MoO_x and WO_x as hole transporting layers, Sol. Energy 139 (2016) 484−488.

[102] G.A. Sepalage, S. Meyer, A. Pascoe, A.D. Scully, F. Huang, U. Bach, et al., Leone Spiccia, Copper(I) iodide as hole-conductor in planar perovskite solar cells: Probing the origin of J−V hysteresis, Adv. Funct. Mater. 25 (35) (2015) 5650−5661.

[103] C. Zuo, L. Ding, Solution-processed Cu_2O and CuO as hole transport materials for efficient perovskite solar cells, Small 11 (41) (2015) 5528−5532.

[104] L.-C. Chen, C.-C. Chen, K.-C. Liang, S.H. Chang, Z.-L. Tseng, et al., Nanostructured $CuO-Cu_2O$ complex thin film for application in $CH_3NH_3PbI_3$ perovskite solar cells, Nanoscale Res. Lett. 11 (1) (2016) 402.

[105] S. Ye, W. Sun, Y. Li, W. Yan, H. Peng, Z. Bian, et al., CuSCN-based inverted planar perovskite solar cell with an average PCE of 15.6%, Nano Lett. 15 (6) (2015) 3723−3728.

[106] F. Odobel, Y. Pellegrin, E.A. Gibson, A. Hagfeldt, A.L. Smeigh, L. Hammarströmd, Recent advances and future directions to optimize the performances of p-type dye-sensitized solar cells, Coord. Chem. Rev. 256 (21) (2012) 2414−2423.

[107] L. Hu, J. Peng, W. Wang, Z. Xia, J. Yuan, et al., Sequential deposition of $CH_3NH_3PbI_3$ on planar NiO film for efficient planar perovskite solar cells, ACS Photonics 1 (7) (2014) 547−553.

[108] M. Gabski, S. Ostendorp, M. Peterlechner, G. Wilde, Stability and performance of nanostructured perovskites for light-harvesting applications, Small Methods 3 (8) (2019) 1800404.

[109] D. Ramirez, K. Schutt, J.F. Montoya, S. Mesa, J.C. Lim, H.J. Snaith, et al., Meso-superstructured perovskite solar cells: revealing role mesoporous layer, J. Phys. Chem. C. 122 (37) (2018) 21239−21247.

[110] T. Leijtens, B. Lauber, G.E. Eperon, S.D. Stranks, H.J. Snaith, The importance of perovskite pore filling in organometal mixed halide sensitized TiO_2-based solar cells, J. Phys. Chem. Lett. 5 (7) (2014) 1096−1102.

[111] H.-S. Kim, J.-W. Lee, N. Yantara, P.P. Boix, S.A. Kulkarni, S. Mhaisalkar, et al., High efficiency solid-state sensitized solar cell-based on submicrometer rutile TiO_2

nanorod and $CH_3NH_3PbI_3$ perovskite sensitizer, Nano Lett. 13 (6) (2013) 2412−2417.
[112] Q. Jiang, X. Sheng, Y. Li, X. Feng, T. Xu, Rutile TiO_2 nanowire-based perovskite solar cells, Chem. Commun. 2014 50 (94) (2014) 14720−14723.
[113] S. Dharani, H.K. Mulmudi, N. Yantara, P.T.T. Trang, N.G. Park, et al., High efficiency electrospun TiO_2 nanofiber based hybrid organic−inorganic perovskite solar cell, Nanoscale 6 (3) (2014) 1675−1679.
[114] J. Werner, J. Geissbühler, A. Dabirian, S. Nicolay, M. Morales-Masis, et al., Parasitic absorption reduction in metal oxide-based transparent electrodes: Application in perovskite solar cells, ACS Appl. Mater. & Interfaces 8 (27) (2016) 17260−17267.
[115] J.M. Ball, M.M. Lee, A. Hey, H.J. Snaith, Low-temperature processed meso-superstructured to thin-film perovskite solar cells, Energy Environ. Sci. 6 (6) (2013) 1739−1743.
[116] D. Kim, H.J. Jung, I.J. Park, B.W. Larson, S.P. Dunfield, et al., Efficient, stable silicon tandem cells enabled by anion-engineered wide-bandgap perovskites, Science 368 (6487) (2020) 155−160.
[117] B. Saparov, F. Hong, J.-P. Sun, H.-S. Duan, W. Meng, et al., Thin-film preparation and characterization of $Cs_3Sb_2I_9$: A lead-free layered perovskite semiconductor, Chem. Mater. 27 (16) (2018) 5622−5632.
[118] D. Cortecchia, H.A. Dewi, J. Yin, A. Bruno, S. Chen, et al., Lead-free $MA_2CuCl_xBr_{4-x}$ hybrid perovskites, Inorg. Chem. 55 (3) (2016) 1044−1052.
[119] T. Krishnamoorthy, H. Ding, C. Yan, W.L. Leong, T. Baikie, et al., Lead-free germanium iodide perovskite materials for photovoltaic applications, J. Mater. Chem. A 3 (47) (2015) 23829−23832.
[120] Z. Shao, T.L. Mercier, M. Madec, T. Pauporté, Exploring $AgBi_xI_{3x+1}$ semiconductor thin films for lead-free perovskite solar cells, Mater. Des. 141 (2018) 81−87.
[121] S. Öz, J.-C. Hebig, E. Jung, T. Singh, A. Lepcha, et al., Zero-dimensional $(CH_3NH_3)_3Bi_2I_9$ perovskite for optoelectronic applications, Sol. Energy Mater. Sol. Cell 158 (2016) 195−201.
[122] T. Wu, X. Liu, X. He, Y. Wang, X. Meng, et al., Efficient and stable tin-based perovskite solar cells by introducing π-conjugated Lewis base, Sci. China Chem. 63 (1) (2020) 107−115.
[123] T. Nakamura, S. Yakumaru, M.A. Truong, K. Kim, J. Liu, et al., Sn(IV)-free tin perovskite films realized by in situ Sn(0) nanoparticle treatment of the precursor solution, Nat. Commun. 11 (1) (2020) 3008.
[124] L. Lanzetta, T. Webb, N. Zibouche, X. Liang, D. Ding, et al., Degradation mechanism of hybrid tin-based perovskite solar cells and the critical role of tin (IV) iodide, Nat. Commun. 12 (1) (2021) 2853.
[125] J. Qiu, Y. Lin, X. Ran, Q. Wei, X. Gao, et al., Efficient and stable Ruddlesden-Popper layered tin-based perovskite solar cells enabled by ionic liquid-bulky spacers, Sci. China Chem. 64 (9) (2021) 1577−1585.
[126] X. Meng, Y. Wang, J. Lin, X. Liu, X. He, J. Barbaud, et al., Surface-controlled oriented growth of $FASnI_3$ crystals for efficient lead-free perovskite solar cells, Joule 4 (4) (2020) 902−912.
[127] X. Liu, Y. Wang, T. Wu, X. He, X. Meng, et al., Efficient and stable tin perovskite solar cells enabled by amorphous-polycrystalline structure, Nat. Commun. 11 (1) (2020) 2678.
[128] K. Nishimura, M.A. Kamarudin, D. Hirotani, K. Hamada, Q. Shen, et al., Lead-free tin-halide perovskite solar cells with 13% efficiency, Nano Energy 74 (2020) 104858.

[129] X. Liu, T. Wu, J.-Y. Chen, X. Meng, X. He, et al., Templated growth of FASnI$_3$ crystals for efficient tin perovskite solar cells, Energy Environ. Sci. 13 (9) (2020) 2896−2902.

[130] W. Hu, X. He, Z. Fang, W. Lian, Y. Shang, et al., Bulk heterojunction gifts bismuth-based lead-free perovskite solar cells with record efficiency, Nano Energy 68 (2020) 104362.

[131] A. Singh, P.-T. Lai, A. Mohapatra, C.-Y. Chen, H.-W. Lin, Y.-J. Lu, et al., Panchromatic heterojunction solar cells for Pb-free all-inorganic antimony based perovskite, Chem. Eng. J. 419 (2021) 129424.

[132] Z. Wang, Z. Song, Y. Yan, S.F. Liu, D. Yang, Perovskite—a perfect top cell for tandem devices to break the S−Q limit, Adv. Sci. 6 (7) (2019) 1801704.

[133] F. Fang, Q. Zeng, C. Zuo, L. Zhang, H. Xiao, et al., Perovskite-based tandem solar cells, Sci. Bull. 66 (6) (2021) 621−636.

[134] Z. Qiu, Z. Xu, N. Li, N. Zhou, Y. Chen, et al., Monolithic perovskite/Si tandem solar cells exceeding 22% efficiency via optimizing top cell absorber, Nano Energy 53 (2018) 798−807.

[135] I.Y. Choi, C.U. Kim, W. Park, H. Lee, M. Song, et al., Two-terminal mechanical perovskite/silicon tandem solar cells with transparent conductive adhesives, Nano Energy 65 (2019) 104044.

[136] H.H. Park, J. Kim, G. Kim, H. Jung, S. Kim, et al., Transparent electrodes consisting of a surface-treated buffer layer based on tungsten oxide for semitransparent perovskite solar cells and four-terminal tandem applications, Small, Methods 4 (5) (2020) 2000074.

[137] E.R. Martins, A. Martins, B.-H.V. Borges, J. Li, T.F. Krauss, Photonic intermediate structures for perovskite/c-silicon four terminal tandem solar cells, IEEE J. Photovolt. 7 (5) (2017) 1190−1196.

[138] J. Zheng, C.F.J. Lau, H. Mehrvarz, F.-J. Ma, Y. Jiang, et al., Large area efficient interface layer free monolithic perovskite/homo-junction-silicon tandem solar cell with over 20% efficiency, Energy Environ. Sci. 11 (9) (2018) 2432−2443.

[139] J. Tong, Z. Song, D.H. Kim, X. Chen, C. Chen, et al., Carrier lifetimes of of >1 μs in Sn-Pb perovskites perovskites enable efficient all-perovskite tandem solar cells, Science 364 (6439) (2019) 475−479.

[140] M. Jošt, E. Köhnen, A.B. Morales-Vilches, B. Lipovšek, K. Jäger, et al., Textured interfaces in monolithic perovskite/silicon tandem solar cells: Advanced light management for improved efficiency and energy yield, Energy Environ. Sci. 11 (12) (2018) 3511−3523.

[141] J. Zheng, H. Mehrvarz, C. Liao, J. Bing, X. Cui, et al., Large-area 23%-efficient monolithic perovskite/homojunction-silicon tandem solar cell with enhanced UV stability using down-shifting material, ACS Energy Lett. 4 (11) (2019) 2623−2631.

[142] Q. Jiang, Y. Zhao, X. Zhang, X. Yang, Y. Chen, et al., Surface passivation of perovskite film for efficient solar cells, Nat. Photonics 13 (7) (2019) 460−466.

[143] J. Xu, C.C. Boyd, Z.J. Yu, A.F. Palmstrom, D.J. Witter, et al., Triple-halide wide-band gap perovskites with suppressed phase segregation for efficient tandems, Science 367 (6482) (2020) 1097−1104.

[144] I.L. Braly, R.J. Stoddard, A. Rajagopal, A.R. Uhl, J.K. Katahara, et al., Current-induced phase segregation in mixed halide hybrid perovskites and its impact on two-terminal tandem solar cell design, ACS Energy Lett. 2 (8) (2017) 1841−1847.

[145] M. Jaysankar, B.A.L. Raul, J. Bastos, C. Burgess, C. Weijtens, et al., Minimizing voltage loss in wide-bandgap perovskites for tandem solar cells, ACS Energy Lett. 4 (1) (2019) 259−264.

[146] A. Al-Ashouri, E. Köhnen, B. Li, A. Magomedov, H. Hempel, et al., Monolithic perovskite/silicon tandem solar cell with > 29% efficiency by enhanced hole extraction, Science 370 (6522) (2020) 1300−1309.

[147] Y. Hou, E. Aydin, M. De Bastiani, C. Xiao, F.H. Isikgor, et al., Efficient tandem solar cells with solution-processed perovskite on textured crystalline silicon, Science 367 (6482) (2020) 1135−1140.

[148] K.A.L. Abdullah, B. Bakour, Influence of depletion region width on performance of solar cell under sunlight concentration, Energy Procedia 6 (2011) 36−45.

[149] K. Jäger, S. Albrecht, Perovskite-based tandem solar cells, in: H. Fujiwara (Ed.), Hybrid Perovskite Solar Cells: Characteristics and Operation, Wiley-VCH, 2021, pp. 463−508.

[150] J.H. Noh, S.H. Im, J.H. Heo, T.N. Mandal, S.I. Seok, Chemical management for colorful, efficient, and stable inorganic−organic hybrid nanostructured solar cells, Nano Lett. 13 (4) (2013) 1764−1769.

[151] T. Leijtens, G.E. Eperon, S. Pathak, A. Abate, M.M. Lee, H.J. Snaith, Overcoming ultraviolet light instability of sensitized TiO_2 with meso-superstructured organometal tri-halide perovskite solar cells, Nat. Commun. 4 (1) (2013) 2885.

[152] G. Mathiazhagan, L. Wagner, S. Bogati, K.Y. Ünal, D. Bogachuk, et al., Double-mesoscopic hole-transport-material-free perovskite solar cells: Overcoming charge-transport limitation by sputtered ultrathin Al_2O_3 isolating layer, ACS Appl. Nano Mater. 3 (3) (2020)) 2463−2471.

[153] G. Niu, W. Li, F. Meng, L. Wang, H. Dong, Y. Qiu, Study on the stability of $CH_3NH_3PbI_3$ films and the effect of post-modification by aluminum oxide in all-solid-state hybrid solar cells, J. Mater. Chem. A 2 (3) (2014) 705−710.

[154] Y. Wang, I. Ahmad, T. Leung, J. Lin, W. Chen, et al., Encapsulation and stability testing of perovskite solar cells for real life applications, ACS Mater. Au 2 (3) (2022) 215−236.

[155] H. Zhou, Y. Shi, Q. Dong, H. Zhang, Y. Xing, et al., Hole-conductor-free, metal-electrode-free $TiO_2/CH_3NH_3PbI_3$ heterojunction solar cells based on a low-temperature carbon electrode, J. Phys. Chem. Lett. 5 (18) (2014) 3241−3246.

[156] J. Yang, B.D. Siempelkamp, D. Liu, T.L. Kelly, Investigation of $CH_3NH_3PbI_3$ degradation rates and mechanisms in controlled humidity environments using in situ techniques, ACS Nano 9 (2) (2015) 1955−1963.

[157] J.A. Christians, P.A. Miranda Herrera, P.V. Kamat, Transformation of the excited state and photovoltaic efficiency of $CH_3NH_3PbI_3$ perovskite upon controlled exposure to humidified air, J. Am. Chem. Soc. 137 (4) (2015) 1530−1538.

[158] H. Tsai, W. Nie, J.-C. Blancon, C.C. Stoumpos, R. Asadpour, et al., High-efficiency two-dimensional Ruddlesden−Popper perovskite solar cells, Nature 536 (7616) (2016) 312−−316.

[159] I.C. Smith, E.T. Hoke, D. Solis-Ibarra, M.D. McGehee, H.I. Karunadasa, A layered hybrid perovskite solar-cell absorber with enhanced moisture stability, Angew. Chem. Int. Ed. Engl. 53 (42) (2014) 11232−11235.

[160] D.H. Cao, C.C. Stoumpos, O.K. Farha, J.T. Hupp, M.G. Kanatzidis, 2D homologous perovskites as light-absorbing materials for solar cell applications, J. Am. Chem. Soc. 137 (24) (2015) 7843−7850.

[161] K. Yao, X. Wang, Y. Xu, F. Li, L. Zhou, Multilayered perovskite materials based on polymeric-ammonium cations for stable large-area solar cell, Chem. Mater. 28 (9) (2016) 3131−3138.

[162] L.N. Quan, M. Yuan, R. Comin, O. Voznyy, E.M. Beauregard, et al., Ligand-stabilized reduced-dimensionality perovskites, J. Am. Chem. Soc. 138 (8) (2016) 2649−2655.

[163] X. Li, M.I. Dar, C. Yi, J. Luo, M. Tschumi, et al., Improved performance and stability of perovskite solar cells by crystal crosslinking with alkylphosphonic acid ω-ammonium chlorides, Nat. Chem. 7 (9) (2015) 703–711.
[164] Y. Hu, M.F. Aygüler, M.L. Petrus, T. Bein, P. Docampo, Impact of rubidium and cesium cations on the moisture stability of multiple-cation mixed-halide perovskites, ACS Energy Lett. 2 (10) (2017) 2212–2218.
[165] T. Duong, H.K. Mulmudi, H. Shen, Y.L. Wu, C. Barugkin, et al., Structural engineering using rubidium iodide as a dopant under excess lead iodide conditions for high efficiency and stable perovskites, Nano Energy 30 (2016) 330–340.
[166] J.L. Barnett, V.L. Cherrette, C.J. Hutcherson, M.C. So, Effects of solution-based fabrication conditions on morphology of lead halide perovskite thin film solar cells, Adv. Mater. Sci. Eng. 2016 (2016) 4126163.
[167] K. Liang, D.B. Mitzi, M.T. Prikas, Synthesis and characterization of organic − inorganic perovskite thin films prepared using a versatile two-step dipping technique, Chem. Mater. 10 (1) (1998) 403–411.
[168] Y. Xu, L. Zhu, J. Shi, X. Xu, J. Xiao, et al., The effect of humidity upon the crystallization process of two-step spin-coated organic−inorganic perovskites, ChemPhysChem 17 (1) (2016) 112–118.
[169] H.J. Snaith, Perovskites: The emergence of a new era for low-cost, high-efficiency solar cells, J. Phys. Chem. Lett. 4 (21) (2013) 3623–3630.
[170] M.A. Green, A. Ho-Baillie, H.J. Snaith, The emergence of perovskite solar cells, Nat. Photonics 8 (7) (2014) 506–514.
[171] C.-W. Chen, H.-W. Kang, S.-Y. Hsiao, P.-F. Yang, K.-M. Chiang, H.-W. Lin, Efficient and uniform planar-type perovskite solar cells by simple sequential vacuum deposition, Adv. Mater. 26 (38) (2014) 6647–6652.
[172] X. Zhu, D. Yang, R. Yang, B. Yang, Z. Yang, et al., Superior stability for perovskite solar cells with 20% efficiency using vacuum co-evaporation, Nanoscale 9 (34) (2017) 12316–12323.
[173] D. Forgács, L. Gil-Escrig, D. Pérez-Del-Rey, C. Momblona, J. Werner, et al., Efficient monolithic perovskite/perovskite tandem solar cells, Adv. Energy Mater. 7 (8) (2017) 1602121.
[174] A.T. Barrows, A.J. Pearson, C.K. Kwak, A.D.F. Dunbar, A.R. Buckley, D.G. Lidzey, Efficient planar heterojunction mixed-halide perovskite solar cells deposited via spray-deposition, Energy Environ. Sci. 7 (9) (2014) 2944–2950.
[175] S. Das, D. Pandey, J. Thomas, T. Roy, The role of graphene and other 2D materials in solar photovoltaics, Adv. Mater. 31 (1) (2019) 1802722.
[176] K. Amratisha, J. Ponchai, P. Kaewurai, P. Pansa-Ngat, K. Pinsuwan, et al., Layer-by-layer spray coating of a stacked perovskite absorber for perovskite solar cells with better performance and stability under a humid environment, Optical Mater. Express 10 (7) (2020) 1497–1508.
[177] X. Li, M. Tschumi, H. Han, S.S. Babkair, R.A. Alzubaydi, et al., Outdoor performance and stability under elevated temperatures and long-term light soaking of triple-layer mesoporous perovskite photovoltaics, Energy Tech. 3 (6) (2015) 551–555.
[178] S. Seo, S. Jeong, C. Bae, N.-G. Park, H. Shin, Perovskite solar cells with inorganic electron- and hole-transport layers exhibiting long-term (≈ 500 h) stability at 85 °C under continuous 1 sun illumination in ambient air, Adv. Mater. 30 (29) (2018) 1801010.
[179] Y.Y. Kim, T.-Y. Yang, R. Suhonen, A. Kemppainen, K. Hwang, N.J. Jeon, et al., Roll-to-roll gravure-printed flexible perovskite solar cells using eco-friendly antisolvent bathing with wide processing window, Nat. Commun. 11 (1) (2020) 5146.

[180] S.K. Pathak, A. Abate, P. Ruckdeschel, B. Roose, K.C. Gödel, et al., Performance and stability enhancement of dye-sensitized and perovskite solar cells by Al doping of TiO_2, Adv. Funct. Mater. 24 (38) (2014) 6046−6055.
[181] N. Ueoka, T. Oku, A. Suzuki, H. Sakamoto, M. Yamada, S. Minami, et al., Effects of TiO_2 nanoparticles with different sizes on the performance of $CH_3NH_3PbI_{3-x}Cl_x$ solar cells, AIP Conf. Proc. 2067 (2019) 020001. Available from: https://doi.org/10.1063/1.5089434. 10 pp.
[182] Q. Jiang, X. Zhang, J. You, SnO_2: A wonderful electron transport layer for perovskite solar cells, Small 14 (31) (2018) 1801154.
[183] G.S. Han, J. Kim, S. Bae, S.-H. Han, Y.J. Kim, et al., Spin-coating process for 10 cm × 10 cm perovskite solar modules enabled by self-assembly of SnO_2 nanocolloids, ACS Energy Lett. 4 (8) (2019) 1845−1851.
[184] Q. Jiang, L. Zhang, H. Wang, X. Yang, J. Meng, et al., Enhanced electron extraction using SnO_2 for high-efficiency planar-structure $HC(NH_2)_2PbI_3$-based perovskite solar cells, Nat. Energy 2 (1) (2016) 16177.
[185] S. Jeong, S. Seo, H. Park, H. Shin, Atomic layer deposition of a SnO_2 electron-transporting layer for planar perovskite solar cells with a power conversion efficiency of 18.3%, Chem. Commun. 55 (17) (2019) 2433−2436.
[186] E.H. Anaraki, A. Kermanpur, L. Steier, K. Domanski, T. Matsui, et al., Highly efficient and stable planar perovskite solar cells by solution-processed tin oxide, Energy Environ. Sci. 9 (10) (2016) 3128−3134.
[187] J. Song, E. Zheng, J. Bian, X.-F. Wang, W. Tian, Y. Sanehira, et al., Low-temperature SnO_2-based electron selective contact for efficient and stable perovskite solar cells, J. Mater. Chem. A 3 (20) (2015) 10837−10844.
[188] S. Song, G. Kang, L. Pyeon, C. Lim, G.-Y. Lee, T. Park, et al., Systematically optimized bilayered electron transport layer for highly efficient planar perovskite solar cells ($\eta = 21.1\%$), ACS Energy Lett. 2 (12) (2017) 2667−2673.
[189] J.A. Christians, P. Schulz, J.S. Tinkham, T.H. Schloemer, S.P. Harvey, et al., Tailored interfaces of unencapsulated perovskite solar cells for >1,000 hour operational stability, Nature Energy 3 (2018) 68−74.
[190] Z. Zhu, Y. Bai, X. Liu, C.-C. Chueh, S. Yang, A.K.-Y. Jen, Enhanced efficiency and stability of inverted perovskite solar cells using highly crystalline SnO_2 nanocrystals as the robust electron-transporting layer, Adv. Mater. 28 (30) (2016) 6478−6484.
[191] D.-Y. Son, K.-H. Bae, H.-S. Kim, N.-G. Park, Effects of seed layer on growth of ZnO nanorod and performance of perovskite solar cell, J. Phys. Chem. C 119 (19) (2015) 10321−10328.
[192] L.S. Oh, D.H. Kim, J.A. Lee, S.S. Shin, J.-W. Lee, et al., Zn_2SnO_4-based photoelectrodes for organolead halide perovskite solar cells, J. Phys. Chem. C 118 (40) (2014) 22991−22994.
[193] J. Dou, Y. Zhang, Q. Wang, A. Abate, Y. Li, M. Wei, Highly efficient Zn_2SnO_4 perovskite solar cells through band alignment engineering, Chem. Commun. 55 (97) (2019) 14673−14676.
[194] S.S. Shin, E.J. Yeom, W.S. Yang, S. Hur, M.G. Kim, et al., Colloidally prepared La-doped $BaSnO_3$ electrodes for efficient, photostable perovskite solar cells, Science 356 (6334) (2017) 167−171.
[195] N. Shibayama, H. Kanda, T.W. Kim, H. Segawa, S. Ito, Design of BCP buffer layer for inverted perovskite solar cells using ideal factor, APL. Mater. 7 (3) (2019) 031117.
[196] I.J. Park, G. Kang, M.A. Park, J.S. Kim, S.W. Seo, et al., Highly efficient and uniform 1 cm^2 perovskite solar cells with an electrochemically deposited NiO_x hole-extraction layer, ChemSusChem 10 (12) (2017) 2660−2667.

[197] S. Song, B.J. Moon, M.T. Hörantner, J. Lim, G. Kang, et al., Interfacial electron accumulation for efficient homo-junction perovskite solar cells, Nano Energy 28 (2016) 269–276.
[198] J. Ciro, S. Mesa, J.I. Uribe, M.A. Mejía-Escobar, D. Ramirez, et al., Optimization of the Ag/PCBM interface by a rhodamine interlayer to enhance the efficiency and stability of perovskite solar cells, Nanoscale 9 (27) (2017) 9440–9446.
[199] X. Meng, C.H.Y. Ho, S. Xiao, Y. Bai, T. Zhang, et al., Molecular design enabled reduction of interface trap density affords highly efficient and stable perovskite solar cells with over 83% fill factor, Nano Energy 52 (2018) 300–306.
[200] Y. Han, S. Meyer, Y. Dkhissi, K. Weber, J.M. Pringle, et al., Degradation observations of encapsulated planar $CH_3NH_3PbI_3$ perovskite solar cell high temperature humidity, J. Mater. Chem. A 3 (15) (2015) 8139–8147.
[201] E. Kasparavicius, M. Franckevičius, V. Malinauskiene, K. Genevičius, V. Getautis, T. Malinauskas, Oxidized spiro-OMeTAD: Investigation of stability in contact with various perovskite compositions, ACS Appl. Energy Mater. 4 (12) (2021) 13696–13705.
[202] S.N. Habisreutinger, T. Leijtens, G.E. Eperon, S.D. Stranks, R.J. Nicholas, H.J. Snaith, Enhanced hole extraction in perovskite solar cells through carbon nanotubes, J. Phys. Chem. Lett. 5 (23) (2014) 4207–4212.
[203] L. Zhang, C. Liu, J. Zhang, X. Li, C. Cheng, Y. Tian, et al., Intensive exposure of functional rings of a polymeric hole-transporting material enables efficient perovskite solar cells, Adv. Mater. 30 (39) (2018) 1804028.
[204] J.-Y. Jeng, K.-C. Chen, T.-Y. Chiang, P.-Y. Lin, T.-D. Tsai, et al., Nickel oxide electrode interlayer in $CH_3NH_3PbI_3$ perovskite/PCBM planar-heterojunction hybrid solar cells, Adv. Mater. 2014 26 (24) (2014) 4107–4113. 4.
[205] I.J. Park, M.A. Park, D.H. Kim, G.D. Park, B.J. Kim, et al., New hybrid hole extraction layer of perovskite solar cells with a planar p–i–n geometry, J. Phys. Chem. C. 119 (49) (2015) 27285–27290.
[206] X. Yin, Z. Yao, Q. Luo, X. Dai, Y. Zhou, et al., High efficiency inverted planar perovskite solar cells with solution-processed NiO_x hole contact, ACS Appl. Mater. Interfaces 9 (3) (2017) 2439–2448.
[207] T. Abzieher, S. Moghadamzadeh, F. Schackmar, H. Eggers, F. Sutterlüti, et al., Electron-beam-evaporated nickel oxide hole transport layers for perovskite-based photovoltaics, Adv. Energy Mater. 9 (12) (2019) 1802995.
[208] S. Seo, I.J. Park, M. Kim, S. Lee, C. Bae, et al., An ultra-thin, un-doped NiO hole transporting layer of highly efficient (16.4%) organic–inorganic hybrid perovskite solar cells, Nanoscale 8 (22) (2016) 11403–11412.
[209] S. Usami, Y. Ando, A. Tanaka, K. Nagamatsu, M. Deki, et al., Correlation between dislocations and leakage current of p-n diodes on a free-standing GaN substrate, Appl. Phys. Lett. 112 (18) (2018) 182106.
[210] J.W. Jung, C.-C. Chueh, A.K.-Y. Jen, A low-temperature, solution-processable, Cu-doped nickel oxide hole-transporting layer via the combustion method for high-performance thin-film perovskite solar cells, Adv. Mater. 27 (47) (2015) 7874–7880.
[211] M. Lyu, J. Chen, N.-G. Park, Improvement of efficiency and stability of CuSCN-based inverted perovskite solar cells by post-treatment with potassium thiocyanate, J. Solid. State Chem. 269 (2019)) 367–374.
[212] J. Liu, S.K. Pathak, N. Sakai, R. Sheng, S. Bai, Z. Wang, et al., Identification and mitigation of a critical interfacial instability in perovskite solar cells employing copper thiocyanate hole-transporter, Adv. Mater. Interf. 3 (22) (2016) 1600571.
[213] H. Rao, S. Ye, W. Sun, W. Yan, Y. Li, et al., A 19.0% efficiency achieved in CuO_x-based inverted $CH_3NH_3PbI_{3-x}Cl_x$ solar cells by an effective Cl doping method, Nano Energy, 27, 2016, pp. 51–57.

[214] Y. Guo, H. Lei, L. Xiong, B. Lia, G. Fang, An integrated organic—inorganic hole transport layer for efficient and stable perovskite solar cells, J. Mater. Chem. A 6 (5) (2018) 2157—2165.
[215] I. van Scodeller, K. De Oliveira Vigier, E. Muller, C. Ma, F. Guégan, R. Wischert, et al., A combined experimental—theoretical study on Diels-Alder reaction with bio-based furfural: Towards renewable aromatics, ChemSusChem 14 (1) (2021) 313—323.
[216] S. Jeong, S. Seo, H. Shin, p-Type $CuCrO_2$ particulate films as the hole transporting layer for $CH_3NH_3PbI_3$ perovskite solar cells, RSC Adv. 8 (49) (2018) 27956—27962.
[217] Y. Chen, J. Yang, S. Wang, Y. Wu, N. Yuan, W. Zhang, Interfacial contact passivation for efficient and stable cesium-formamidinium double-cation lead halide perovskite solar cells, iScience 23 (1) (2020) 100762.
[218] B. Yang, D. Ouyang, Z. Huang, X. Ren, H. Zhang, W.C.H. Choy, Multifunctional synthesis approach of In:$CuCrO_2$ nanoparticles for hole transport layer in high-performance perovskite solar cells, Adv. Funct. Mater. 29 (34) (2019) 1902600.
[219] S. Akin, Y. Liu, M.I. Dar, S.M. Zakeeruddin, M. Grätzel, S. Turan, et al., Hydrothermally processed $CuCrO_2$ nanoparticles as an inorganic hole transporting material for low-cost perovskite solar cells with superior stability, J. Mater. Chem. A 6 (41) (2018) 20327—20337.
[220] Y. Chen, Z. Yang, X. Jia, Y. Wu, N. Yuan, et al., Thermally stable methylammonium-free inverted perovskite solar cells with Zn^{2+} doped $CuGaO_2$ as efficient mesoporous hole-transporting layer, Nano Energy 61 (2019) 148—157.
[221] S. Guarnera, A. Abate, W. Zhang, J.M. Foster, G. Richardson, A. Petrozza, et al., Improving the long-term stability of perovskite solar cells with a porous Al_2O_3 buffer layer, J. Phys. Chem. Lett. 6 (3) (2015) 432—437.
[222] J. Xu, A. Buin, A.H. Ip, W. Li, O. Voznyy, et al., Perovskite—fullerene hybrid materials suppress hysteresis in planar diodes, Nat. Commun. 6 (1) (2015) 7081.
[223] M.A. Mutalib, N.A. Ludin, N.A.N.A. Ruzalman, V. Barrioz, S. Sepeai, et al., Progress towards highly stable and lead-free perovskite solar cells, Mater. Renew. Sustain. Energy 7 (2) (2018) 7.
[224] G. Sfyri, C.V. Kumar, D. Raptis, V. Dracopoulos, P. Lianos, Study of perovskite solar cells synthesized under ambient conditions and of the performance of small cell modules, Sol. Energy Mater. Sol. Cell 134 (2015) 60—63.
[225] Y. Kato, L.K. Ono, M.V. Lee, S. Wang, S.R. Raga, Y. Qi, Silver iodide formation in methyl ammonium lead iodide perovskite solar cells with silver top electrodes, Adv. Mater. Interf. 2 (13) (2015) 1500195.
[226] A.K. Chauhan, P. Kumar, Photo-stability of perovskite solar cells with Cu electrode, J. Mater. Science: Mater. Electron. 30 (10) (2019) 9582—9592.
[227] T.M. Schmidt, T.T. Larsen-Olsen, J.E. Carlé, D. Angmo, F.C. Krebs, Upscaling of perovskite solar cells: Fully ambient roll processing of flexible perovskite solar cells with printed back electrodes, Adv. Energy Mater. 5 (15) (2015) 1500569.
[228] F. Machui, M. Hösel, N. Li, G.D. Spyropoulos, T. Ameri, et al., Cost analysis of roll-to-roll fabricated ITO free single and tandem organic solar modules based on data from manufacture, Energy Environ. Sci. 7 (9) (2014) 2792—2802.
[229] N. Espinosa, L. Serrano-Luján, A. Urbina, F.C. Krebs, Solution and vapour deposited lead perovskite solar cells: Ecotoxicity from a life cycle assessment perspective, Sol. Energy Mater. Sol. Cell 137 (2015) 303—310.
[230] R. Li, X. Xiang, X. Tong, J. Zou, Q. Li, Wearable double-twisted fibrous perovskite solar cell, Adv. Mater. 27 (25) (2015) 3831—3835.

[231] T. Leijtens, S.D. Stranks, G.E. Eperon, R. Lindblad, E.M.J. Johansson, et al., Electronic properties of meso-superstructured and planar prganometal halide perovskite films: Charge trapping, photodoping, and carrier mobility, ACS Nano 8 (7) (2014) 7147–7155.
[232] J.-W. Lee, D.-H. Kim, H.-S. Kim, S.-W. Seo, S.M. Cho, N.-G. Park, Formamidinium and cesium hybridization for photo- and moisture-stable perovskite solar cell, Adv. Energy Mater. 5 (20) (2015) 1501310.
[233] A.K. Chauhan, P. Kumar, Degradation in perovskite solar cells stored under different environmental conditions, J. Phys. D: Appl. Phys. 50 (32) (2017) 325105.
[234] F.C. Krebs, S.A. Gevorgyan, J. Alstrup, A roll-to-roll process to flexible polymer solar cells: Model studies, manufacture and operational stability studies, J. Mater. Chem. 19 (30) (2009) 5442–5451.
[235] D.M. Tanenbaum, H.F. Dam, R. Rösch, M. Jørgensen, H. Hoppe, F.C. Krebs, Edge sealing for low cost stability enhancement of roll-to-roll processed flexible polymer solar cell modules, Sol. Energy Mater. Sol. Cell 97 (2012) 157–163.
[236] Y. Shao, Q. Wang, Q. Dong, Y. Yuan, J. Huang, Vacuum-free laminated top electrode with conductive tapes for scalable manufacturing of efficient perovskite solar cells, Nano Energy 16 (2015) 47–53.
[237] M.O. Reese, S.A. Gevorgyan, M. Jørgensen, E. Bundgaard, S.R. Kurtz, et al., Consensus stability testing protocols for organic photovoltaic materials and devices, Sol. Energy Mater. Sol. Cell 95 (5) (2011) 1253–1267.
[238] R. Roesch, T. Faber, E. von Hauff, T.M. Brown, M. Lira-Cantu, H. Hoppe, Procedures and practices for evaluating thin-film solar cell stability, Adv. Energy Mater. 5 (20) (2015) 1501407.
[239] G.E. Eperon, V.M. Burlakov, P. Docampo, A. Goriely, H.J. Snaith, Morphological control for high performance, solution-processed planar heterojunction perovskite solar cells, Adv. Funct. Mater. 24 (1) (2014) 151–157.
[240] G.A. dos Reis Benatto, B. Roth, M.V. Madsen, M. Hösel, R.R. Søndergaard, M. Jørgensen, et al., Carbon: The ultimate electrode choice for widely distributed polymer solar cells, Adv. Energy Mater. 4 (15) (2014) 1400732.

CHAPTER FOUR

Modeling perovskite solar cells

Arturo Morales-Acevedo
Electrical Engineering Department—SEES, Centro de Investigación y de Estudios Avanzados del Instituto Politécnico Nacional (CINVESTAV), Ciudad de México, México

4.1 Introduction

High-purity crystalline silicon solar cells have achieved power conversion efficiencies exceeding 26% [1] and long-term durability, making them an ideal product for the photovoltaic (PV) market. However, emerging PV technologies based on thin films (<4 μm) and simple deposition methods promise to reduce production costs and produce high-quality semiconductors for solar cells, rivaling other established ones such as Si and GaAs [2,3]. Lead halide perovskite solar cells (PSCs) have emerged as one such candidate. In just a few years, PSCs have achieved conversion efficiencies similar to those established for thin CdTe and CuInGaSe$_2$ (CIGS) solar cells, surpassing the 22% mark [4].

As for other solar cell device technologies, modeling perovskite cells is fundamental for understanding their measured $J-V$ characteristics, so that improved performing cells can be designed. In this regard, two different approaches can be followed: Firstly, using conventional solar cell models, which do not take into account the specific properties of perovskite materials, or secondly, proposing complex numerical models that do not allow a clear understanding of these new devices. Such complex models have already been developed [5,6], and they can be used to have a more precise behavior prediction, but they do not help for a physical understanding and quick evaluation of the cells. On the other hand, the first approach, that is, using cell models developed for thick, diffusion-limited semiconductors, such as silicon, is not useful for very thin PSCs because carrier transport may be limited by electric field drift and for which radiative recombination may become comparable to the nonradiative recombination rates, as will be explained later.

Photovoltaics Beyond Silicon.
DOI: https://doi.org/10.1016/B978-0-323-90188-8.00003-8
© 2024 Elsevier Inc. All rights reserved, including those for text and data mining, AI training, and similar technologies.

In this chapter, it will be shown that a simple analytical model can be developed for PSCs, using classical expressions already developed for describing relevant phenomena in semiconductor devices, but built explicitly into a new model for taking into account the specific characteristics of perovskite cells. For this purpose, a brief explanation of the differences between these solar cells and the conventional ones will be discussed first. Then, the fundamental properties of perovskite materials will be considered for developing the appropriate model for PSCs. Finally, this model will be used to better understand and predict the behavior of the cells, in such a way that it will be possible to visualize the importance of parameters such as nonradiative recombination lifetimes for solar cells with a specific absorber thickness.

4.2 Thick versus thin solar cells

The simplest and most basic model for a solar cell is based on the Shockley equations for a single semiconductor diode. Carrier transport occurs by ambipolar diffusion in the quasineutral regions of a $p-n$ junction. When the diode is illuminated, some excess photo-generated carriers move by diffusion in the emitter and base regions of the solar cell until the electrons and holes that do not recombine are collected at the junction's space-charge region. Carrier collection occurs in this region because the electrons and holes are separated by the high electric field existing there. Typically, the space—charge region thickness is very small compared with the total cell thickness, and therefore in a first-order approximation model, the total recombination in the space—charge region is considered negligible as compared to the recombination that occurs in the quasineutral regions. The photo-generated carriers that do not recombine on their way toward the space—charge region are separated and collected by the electric field there, generating the cell photo-current (due to the hole and electrons being drifted in opposite directions). It is assumed that both photo-generated and injected carriers do not recombine within the very thin space—charge region. Under these assumptions, applying the superposition principle, the current density versus voltage ($J-V$) curve under illumination [Eq. (4.1)], for a thick solar cell, should be:

$$J = J_L - J_0\left[\exp(V/V_T) - 1\right] \qquad (4.1)$$

where J_L is the photo-current density due to the collection of the photo-generated carriers at the junction. J_0 represents the dark saturation current density due to the carriers injected at this junction to support the hole and electron recombination within the quasineutral regions, given a specific bias voltage V. The thermal voltage at the absolute temperature T is V_T (kT/q).

This is the simplest and basic model for a solar cell. Analytical expressions can be obtained for J_L and J_0 by solving the ambipolar transport equation for homogeneous semiconductors. Within the quasineutral regions, carrier transport is assumed to be limited by diffusion, and therefore the most important transport parameter is the so-called diffusion length in each of the semiconductors forming the p–n junction. An improved model is obtained if carrier recombination at defect traps within the space–charge region is taken into account. These traps cause very active recombination levels at the semiconductor mid-gap, causing an additional dark recombination current density. Thus, a more realistic J–V characteristic [Eq. (4.2)] for a solar cell under illumination is:

$$J = J_L - J_{0d}\left[\exp(V/V_T) - 1\right] - J_{0r}\left[\exp(V/2V_T) - 1\right] \quad (4.2)$$

where J_{0d} is now the diffusion-limited dark saturation current density, and J_{0r} is the dark saturation current density due to the recombination within the space–charge region. Notice that a different ideality factor ($n = 2$) affects the exponential voltage dependence in this case.

Unfortunately, the cell behavior described by the two exponential terms in Eq. (4.2) has been replaced by a single exponential term, as in Eq. (4.1), but with an ideality factor n between 1 and 2, depending upon the real contribution of each of the two phenomena related to the exponentials in Eq. (4.2). In other words, typically the J-V curve is fitted [Eq. (4.3)] to the following expression:

$$J = J_L - J_{0e}\left[\exp(V/nV_T) - 1\right] \quad (4.3)$$

with J_{0e} being an effective dark saturation current density and n the respective (effective) ideality factor.

Often the extracted J_{0e} is interpreted as J_{0d} or J_{0r}, but this is not physically correct. This is correct only in the case that the solar cell behavior is limited by either the quasineutral region recombination or by the recombination at the junction's space–charge region. Unfortunately, it is very common to see many reports where J_{0e} is taken as a real dark saturation current density for solar cells. Conclusions about how to improve the solar cells are taken based on this extracted (fitted) parameter, without taking care of determining the real values of the two important saturation current densities (J_{0d} or J_{0r}) mentioned above. The correct procedure should also be important for perovskite cells, as will be explained below. In general, we shall consider that a single exponential (i.e., single diode) $J-V$ curve is not correct and should be dismissed as a model for a solar cell, except for the cell where only one carrier transport phenomenon dominates the dark behavior. This is the case for high-efficiency solar cells; for example, highly efficient silicon solar cells have a dark saturation current density dominated by carrier diffusion and recombination in the cell's quasineutral regions, having large recombination lifetimes.

Based on the above discussion, it is important here to mention one of the main differences between thick and thin solar cells and their respective electrical behavior. The so-called first-generation silicon solar cells require very thick wafers (200–300 μm) due to the relatively small sunlight absorption coefficient. On the other hand, semiconductors with direct bandgap having large sunlight absorption coefficients require only thin layers for absorbing the solar radiation. This is the case for second-generation solar cells based on CdTe, CIGS, etc., where only thin layers of less than 4 μm are required to achieve high conversion efficiencies. In some cases, good efficiencies are achieved with even thinner absorber layers, below 2 μm. This is a relevant difference because recombination in the quasineutral and the space–charge regions will contribute with a different magnitude for each kind of solar cell. The space–charge region thickness is typically around or below 1–2 μm, so the phenomena occurring in this region become unimportant for very thick cells (200–300 μm). However, this becomes very important for very thin cells for which, in some cases, the space–charge region extends along the whole thickness of the absorbing layer. In other words, for good thick solar cells, the dark current will be dominated by recombination in the quasineutral region ($n = 1$), while for very thin solar cells, the dominant dark current might be caused by recombination in the space–charge region ($n = 2$). Then, the physics involved is also different in each case, but usually, thin solar cells are analyzed without understanding this important difference with thick cells.

Moreover, not only the dark recombination current is affected by this fact but also the photo-current density (J_L). In thick cells, the space–charge region contribution to the photo-generated current is also small and it is considered negligible, so it is seldom mentioned. However, for very thin cells, the photo-current generated within this region becomes the largest contribution because most of the photo-generated carriers are collected there. Electrons and holes photo-generated in the space−charge region are immediately drifted and separated causing the photo-current in the cell. Then, we can assume a high collection probability (approximately 100%) for the charge carriers generated within this region. In addition, it must be noticed that the space-charge region thickness depends on the operating voltage V so both J_L and J_0 will also depend on the operating voltage. This voltage dependence is usually not taken into account for thin solar cells as should be. Hence, a more realistic general model [Eq. (4.4)] for this kind of solar cell would be:

$$J = J_L(V) - J_{0d}(V)\left[\exp(V/V_T) - 1\right] - J_{0r}(V)\left[\exp(V/2V_T) - 1\right] \quad (4.4)$$

with $J_L(V)$, $J_{0d}(V)$, and $J_{0r}(V)$ being the photo-current density, $J_{0d}(V)$ the dark quasineutral saturation current density, and $J_{0r}(V)$ the space−charge recombination dark current density at voltage V, respectively. In our prior work [7], the complete equations for an analytical model of thin solar cells are described.

In summary, in this section, it was shown that the behavior and physics of thick and thin solar cells are different, but many experimentalists and solar cell professionals do not take all the above effects into account, sometimes reaching incorrect conclusions about the functioning of thin solar cells under study. It will be shown below that these aspects are also very important for understanding and modeling thin perovskite cells (<1 μm).

4.3 Thin perovskite solar cells

PSCs are intrinsically very thin-film devices, and therefore the aspects discussed above for thin-film solar cells (TFSCs) are very important to understand their behavior. Below, it will be shown that for extremely thin solar cells, such as the PSC, a simplified model can be used to calculate the $J-V$ characteristics. In this model, in addition to the nonradiative recombination, we must include the radiative recombination because by making the absorber extremely thin, the nonradiative recombination current may become comparable to the radiative recombination current, so both current components must be taken into consideration in the model.

The reader is reminded that under these conditions, there is an optimum thickness for the absorber layer because the total nonradiative recombination will be smaller as the absorber becomes thinner, but the total absorbance for very thin layers might be less than 100%. In other words, there is a compromise: Larger thickness is required for a larger illumination current density (J_L), but a smaller thickness is required for a reduced nonradiative recombination dark saturation current density (J_{0nr}). It will be shown that for perovskite light absorbers, the optimum thickness is below 1 μm.

Good cells made with small absorption length materials such as the hybrid perovskites will be more efficient if the carrier transport is determined by electric field drift instead of diffusion. The characteristic lengths in each of these cases are the drift length [$L_D = \mu\tau E$] and the diffusion length [$L_d = (\mu\tau V_T)^{1/2}$], where E is the electric field and V_T is the thermal potential (kT/q). Typically, the product of carrier mobility by lifetime ($\mu\tau$) is small for polycrystalline materials and, therefore, the diffusion length is also small, but if the electric field is large enough ($> 10^4$ V/cm), the drift length will be several orders of magnitude larger than the corresponding diffusion length. For example, for a material where $\mu\tau = 10^{-8}$ cm^2/V, the diffusion length will be of the order of 158 nm, at room temperature, while in the presence of an electric field of the order of 10^4 V/cm, the drift length becomes larger than 1 μm. For a cell with an absorber layer (500 nm thick) of this material, the carrier collection probability will be around 100% for the case of drift transport, but it will be much less than 100% when carriers diffuse in it.

For materials where the absorption length is small, and where the required thickness is less than 1 μm, it is feasible to have internal electric fields above 10^4 V/cm. One way of having these electric fields is making $n-i-p$ (or $p-i-n$) device structures, so that—as was explained before—all the photo-generated carriers are collected, generating the cell photocurrent. Therefore, very thin cells should be made with a *n-i-p* device structure. In PSCs, the absorber layers have a very high absorption coefficient for the solar spectrum, requiring less than 1 μm for the full absorption of the sunlight photons (with $E > E_g$). It has been confirmed that planar perovskite cells made with very thin absorber layers, between appropriate selective contacts (Fig. 4.1), behave as *n-i-p* devices [8].

Hence, in the following sections, it will be considered that present PSCs are $n-i-p$ (or $p-i-n$) structures. Therefore, because of the high electric field in the perovskite absorber, when the cell is operating,

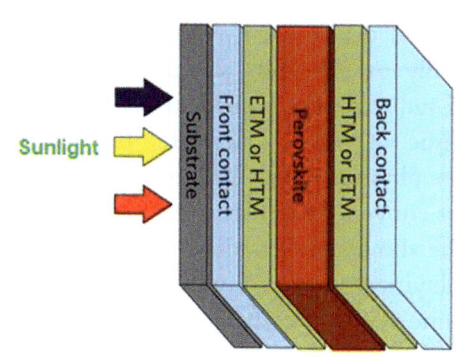

Figure 4.1 Schematics of a typical planar perovskite solar cell structure. HTM and ETM refer to Hole Transport Material (e.g., Spiro-OMeTAD) and Electron Transport Material (e.g., TiO_2), respectively. The front contact can be a transparent conducting oxide [e.g., Fluorine Tin Oxide (FTO), Indium Tin Oxide (ITO)] while the back contact is usually a highly reflective metal such as gold or silver. The ETM layer acts as an n + region and the HTM layer acts as a p^+ region. The perovskite layer can be considered a highly resistive intrinsic region, forming a $n-i-p$ ($p-i-n$) structure, where a high electric field exists. Source authors; no permission required.

complete photo-generated carrier collection can be assumed, and the photo-current density will be determined by the total electron–hole pair generation in this region. In addition, it will be assumed that this layer will cause nonradiative recombination due to the very active Shockley–Read–Hall traps at the middle of the bandgap. In other words, the dark current due to nonradiative recombination will behave similarly to the current that occurs in devices where this is limited by space–charge recombination. Finally, the radiative recombination must also be considered in the model, both because the radiative recombination has been shown to be important for perovskite cells and because for very thin perovskite layers, the nonradiative recombination can become small, depending upon the carrier lifetimes in the material. Under certain conditions, these phenomena can be comparable, and both recombination mechanisms must be considered in a complete model.

Despite its mathematical simplicity, this new model allows the estimation of the optimum perovskite thickness, and it also helps in determining what should be the required carrier lifetime for having increased efficiencies, above the present 25% record. It is predicted that carrier recombination lifetimes above 5 μs in the perovskite cell material will not cause large efficiency improvements, as in this case the total recombination will be already limited by the radiative recombination. It should be noted that

PSC models have been developed by other authors [9,10], but they are not as mathematically or physically straightforward as the present one, including both the radiative and nonradiative recombination mechanisms that are characteristic of PSCs. Therefore, the model presented here can help to provide physical insight into the cell behavior, similar to the Shockley model for conventional solar cells.

According to the above considerations, for developing a simple model, the following simplifying assumptions will be made:

- Inside the intrinsic perovskite absorber layer, a high electric field will exist. The magnitude of the electric field can be estimated to be of the order of 10^4 V/cm. The photo-generated electrons and holes will be separated and drifted by this high electric field.
- Low-efficiency solar cells will be limited by high recombination velocities at the interfaces. Here, it will be assumed that interface recombination velocities are well below 10^3 cm/s so that interface effects are small and can be neglected.
- No hysteresis due to ion movement in the perovskite material will be considered.
- Optical and series resistance losses will also be considered small.

4.3.1 Dark radiative recombination in thin cells

For ideal solar cells, the limiting efficiency occurs when all the absorbed photons generate electron—hole pairs that are collected at the selective contacts, and the recombination is just limited by the radiative recombination, that is, without any other nonradiative carrier recombination sink. In thermal equilibrium, the absorbed energy flux should be equal to the emitted energy flux in the cell. Hence, in equilibrium with the surroundings [Eq. (4.5)], the cell will emit photons at a rate equal to the rate of photons absorbed from the surroundings:

$$\varphi_{em,0}(E) = \varphi_{abs}(E) = a(E)\varphi_{bb}(E) \quad (4.5)$$

φ_{bb} is the flux of photons with energy E due to the surroundings at temperature T. Ideally, φ_{bb} will have the blackbody energy spectrum at temperature T. $a(E)$ is the cell absorbance of photons with energy E.

For very thick cells, the ideal absorbance $a(E)$ is $a(E) = 0$ for $E < Eg$ and $a(E) = 1$ for $E > Eg$. However, for very thin cells, the total absorbance will vary as a function of the absorber thickness because some photons from the surroundings will not be absorbed. In this case, the total absorbance for an absorber layer with a given thickness d is described in [Eq. (4.6)] as:

$$a(E) = [1-\exp(-2\alpha(E)d)] \quad (4.6)$$

where $\alpha(E)$ is the absorption coefficient for photons with energy $E > E_g$; the factor 2 in this expression assumes a 100% reflection at the back contact.

The emitted photons from the cell are due to the radiative recombination of electrons (at the conduction band) and holes (at the valence band) in the absorber material. When a voltage V is applied to the cell (in dark conditions), it becomes out of equilibrium, and we expect the electron and hole concentration to increase, causing an exponential (in the Boltzmann approximation) enhancement of the radiative recombination rate:

$$\varphi_{em}(E) = \varphi_{em,0}(E)\exp\left(\frac{qV}{kT}\right) = a(E)\varphi_{bb}(E)\exp\left(\frac{V}{V_T}\right) \quad (4.7)$$

Hence, when there is no other recombination process under dark conditions, the total current density should be:

$$J_{rad}(V) = J_{0,rad}\left[\exp\left(\frac{V}{V_T}\right) - 1\right] \quad (4.8)$$

V_T is the thermal potential (kT/q) at temperature T, q is the electron's charge magnitude, and k is the Boltzmann constant. Then, considering the blackbody spectrum at temperature T, the saturation current density, when limited by the radiative recombination, is obtained from Eq. (4.7):

$$J_{0,rad} = \left(\frac{2\pi q}{h^3 c^2}\right)\int_{E_g}^{\infty}\frac{[1-\exp(1-2\alpha(E)d)]E^2}{\exp\left(\frac{E}{kT}\right) - 1}dE \quad (4.9)$$

In real cells, in addition to this current due to radiative recombination, there will be nonradiative recombination. For very thin n-i-p devices, the Shockley–Read–Hall nonradiative recombination due to the traps in the intrinsic semiconductor (perovskite) has also to be considered, as will be explained below.

4.3.2 Dark nonradiative recombination

At equilibrium, the intrinsic region in n–i–p devices will be totally depleted of free carriers, similarly as in the space–charge region of n–p junctions. Then, strong carrier generation-recombination due to the traps in the bandgap of the intrinsic semiconductor is expected. The most active traps will be those in the middle of the bandgap (trapping electrons and holes with similar probabilities). Then, an analysis analogous to the one made for the depleted region of $p-n$ junctions [11] should give us the expected dark current [Eq. (4.10)] due to first-order nonradiative recombination in a n-i-p diode:

$$J_{\text{nonrad}}(V) = J_{0,\text{nonrad}} \left[\exp\left(\frac{V}{2V_T}\right) - 1 \right] \quad (4.10)$$

and

$$J_{0,\text{nonrad}} = q\left(\frac{n_i d}{\tau}\right) \quad (4.11)$$

n_i is the intrinsic carrier concentration in the perovskite semiconductor, d is the layer thickness, and τ is the carrier (effective SRH) lifetime in this region. No second-order recombination mechanisms are considered here. For calculating n_i, an additional assumption must be made for the effective conduction band (N_c) and valence band (N_v) densities of states at room temperature, so that the intrinsic carrier concentration can be calculated in each case. In this work, we have assumed $N_c = N_v = 1 \times 10^{18}$ cm^{-3}; note that the ideality factor for this current component is $n = 2$.

4.3.3 Photo-generated current

When the cell is illuminated (by sunlight), it will generate a photo-current. As explained before, due to the high electric field in the intrinsic absorber region, all the photo-generated carriers will be collected to produce the cell photo-current. Therefore, the photo-current density (J_L) [Eq. (4.12)] will be:

$$J_L = q \int_{E_g}^{\infty} a(E) * N_0(E) dE \quad (4.12)$$

where $N_0(E)$ is the photon flux density from the solar spectrum (global AM1.5). For simplicity, here we have assumed a very small optical reflectivity for all the solar spectrum wavelengths, but this may not be true in real perovskite cells. In the future, this optical loss should be considered for a more realistic calculation. In general, the recombination and the photo-generated currents depend upon the absorber layer thickness, as seen in Eqs. (4.2), (4.5), (4.7), and (4.8). Then, an optimum thickness is expected because, as the perovskite film thickness is increased, the photo-current would increase, but the total nonradiative recombination would also increase, and vice-versa.

4.3.4 Total current density in perovskite solar cells

In summary, now we have an approximate model for a perovskite thin TFSC, involving two recombination processes, the radiative and the nonradiative ones, so that the total current density [Eq. (4.13)] for a given applied voltage V is:

$$J(V) = J_L - J_{0,\text{rad}}\left[\exp\left(\frac{V}{V_T}\right) - 1\right] - J_{0,\text{nonrad}}\left[\exp\left(\frac{V}{2V_T}\right) - 1\right] \quad (4.13)$$

The open circuit voltage (V_{oc}) and the fill factor (FF) can be determined from the respective $J-V$ curve [Eq. (4.13)] so that the expected conversion efficiency (η) can be estimated using Eq. (4.14):

$$\eta = \left(\frac{J_L V_{oc}}{P_{\text{in}}}\right) FF \quad (4.14)$$

where P_{in} is the total incident power from the sun radiation (100 mW/cm^2 for the AM1.5 global solar spectrum). The same model can be applied to other very thin n-i-p solar cells, made on direct bandgap semiconductors, for example, CdTe or CIGS cells, using the appropriate parameters (E_g, α, τ, and d) for each material.

For simplifying the calculations [Eq. (4.15)], the absorption coefficient $\alpha(E)$ for direct bandgap semiconductors will be assumed to have the form:

$$\alpha(E) = A(E - E_g)^{1/2} \text{ for } E > E_g \quad (4.15)$$

The constant A depends upon the effective densities of states at the conduction and valence bands. For highly absorbent semiconductors, such as the hybrid perovskites, at room temperature, it has values around 1×10^5 cm^{-1}/eV$^{1/2}$.

We note that Eq. (4.13) describes a cell with an equivalent electrical circuit based on two diodes in parallel with one current source that represents the illumination current density (Fig. 4.2). One of the diodes is associated with the direct band-to-band radiative recombination and the other diode is associated to the nonradiative recombination mechanism. Then, when these dark current components are comparable, an effective ideality factor between 1 and 2 could be expected when the $I-V$ dark characteristic of the cell is modeled with a single exponential. However, to avoid errors and inappropriate interpretations, the extraction of an

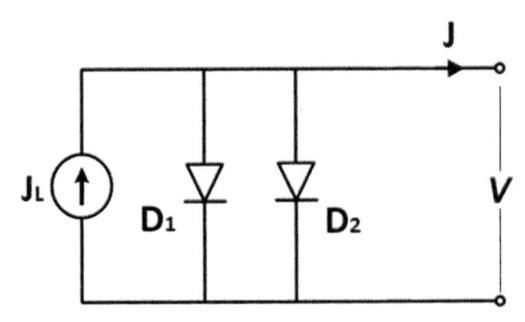

Figure 4.2 Equivalent circuit for a perovskite solar cell. D1 is a diode associated with the first exponential in Eq. (4.13), and it is related to the radiative carrier recombination. D2 is a second diode related to the nonradiative carrier recombination, and it is associated with the second exponential in Eq. (4.13). A more complete equivalent circuit should include both series and shunt resistances (not included here).

effective ideality factor from the experimental $J-V$ curve should be avoided, and the two exponential models should be used instead [12].

The above model is novel, in the sense that both radiative and nonradiative recombination phenomena are present. It avoids any confusion about how the carrier transport occurs within these cells. For example, it is common to assume an effective lifetime associated with both recombination mechanisms, but as the above model implicitly explains, each mechanism should be associated with different carrier statistics, so that the corresponding diodes have different ideality factors. The ideal cell will be limited by the radiative recombination, with an ideality factor $n = 1$. On the opposite, for a cell where the nonradiative recombination dominates, the ideality factor would become $n = 2$. In the more general case, when both recombination processes are of the same order, both diodes must be included in the model.

4.4 Modeling results for thin-film perovskite cells

Present technology of hybrid perovskite materials, such as $CH_3NH_3PbI_3$, produce materials with a bandgap above 1.55 eV. When Cl_2 or Br_2 is included in the compound, the bandgap increases above this value. For this reason, our calculations will be made in the range between 1.45 and 1.8 eV, using the AM1.5 global solar spectrum (normalized to 100 mW/cm^2).

The first calculation refers to the ideal (Shockley-Queisser) efficiency [13] for a bulk hybrid perovskite with a bandgap in this range.

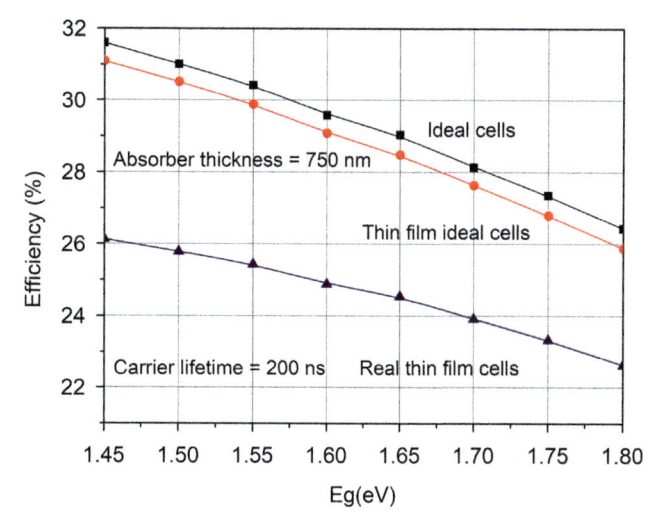

Figure 4.3 Conversion efficiency as a function of the bandgap for ideal thick and thin perovskite cells, compared to real thin-film cells.

The second calculation is made for very thin films of the same materials (i.e., absorbance less than one), but without nonradiative carrier recombination. Then, a third calculation is made for cells with very thin perovskite materials and nonradiative carrier lifetimes (τ) around 200 ns, as has been reported for this kind of materials. For this comparison, the absorber material thickness will be 750 nm, which is a typical thickness. In Fig. 4.3, the results for the efficiency as a function of the bandgap can be seen for all the three cases mentioned above. As expected, the efficiency will be reduced as the bandgap is increased above 1.45 eV, which is around the optimum ideal value, according to the Shockley-Queisser limit [13]. The reader will note that there is only a small difference between the expected ideal efficiencies for *bulk* and *thin-film perovskite cells*. This small difference is due to the noncomplete absorbance of the incident photons (from the solar spectrum and the surroundings), causing both a smaller illumination current density and a smaller radiative recombination current density. Note in Fig. 4.3 that, for present perovskite bandgaps (around 1.6 eV), the maximum efficiency of thin-film (750 nm) solar cells would be around 29%.

The curve, labeled as "real thin-film cells," includes both the radiative and the nonradiative dark recombination currents for *n-i-p* solar cells, as explained above. For a bandgap of around 1.6 eV (typical for perovskite materials), the expected efficiency is around 25%. The effect due to the radiative carrier recombination is small in this case (for nonradiative lifetimes around 200 ns). In this case, almost all the recombination current is due to nonradiative recombination. This is opposite to the ideal cases for which all the recombination is due to radiative recombination. Therefore, it is important to determine how the efficiency would change as the nonradiative lifetime is increased (reducing the nonradiative recombination rate). This is shown in Fig. 4.4.

It is clear in this figure that as the lifetime is increased from 100 ns to 1 μs, in finished perovskite cells, the efficiency might increase by around 3%, from around 24% to 27%. Therefore, increased efficiencies can be expected for PSCs because lifetimes above 1 μs have already been observed in these materials (for pure perovskite layers, that is, not in terminated solar cells). Above this carrier nonradiative lifetime value, the radiative recombination becomes important in comparison with the nonradiative recombination, and this is the reason why there is a slope change in the curve of Fig. 4.4 for lifetimes higher than 750 ns. For nonradiative lifetimes above 5 μs, the Shockley-Queisser ideal efficiency [13] should be achieved, and no further efficiency increase would be expected, in agreement with previous results by Tress [14,15].

Finally, to see if there is an optimum perovskite thickness in the cells, in Fig. 4.5, the efficiency calculation as a function of thickness, for two

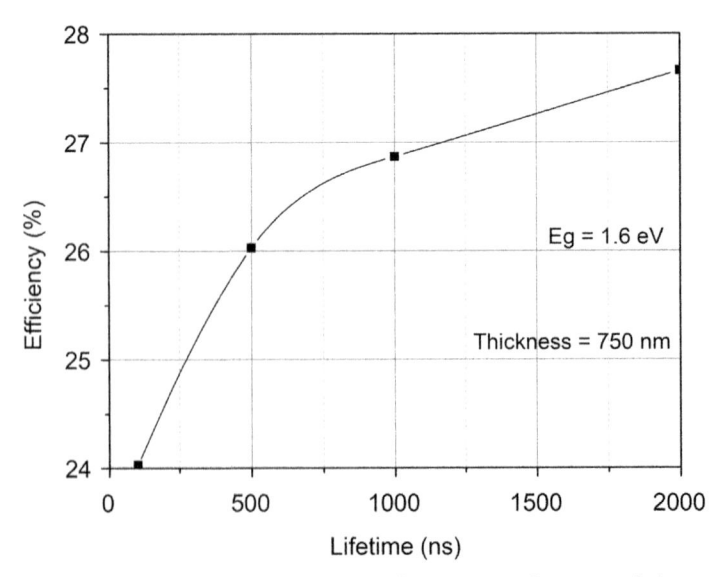

Figure 4.4 Perovskite solar cell conversion efficiency as a function of the nonradiative carrier lifetime.

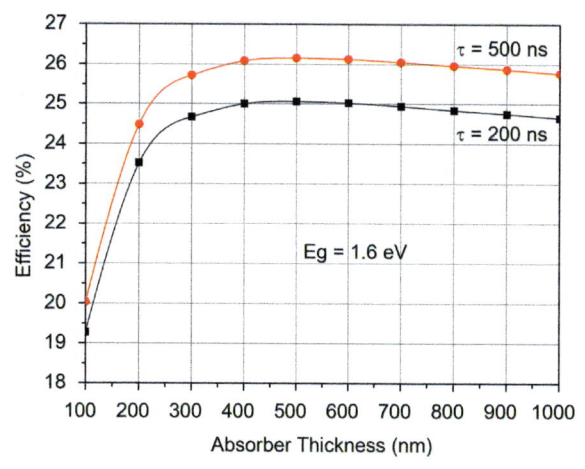

Figure 4.5 Expected conversion efficiency as a function of the perovskite thickness.

Table 4.1 Calculation results of perovskite cells.

Lifetime (ns)	J_L (mA/cm^2)	V_{oc} (V)	FF	Efficiency (%)
200	24.88	1.21	0.83	25.06
500	24.88	1.25	0.84	26.17

different carrier lifetimes, is shown. It can be observed that, in fact, there is an optimum value, around 500 nm, but the efficiencies are almost constant for perovskite thickness between 300 and 1000 nm. In other words, the thickness must be larger than 300 nm with an optimum of around 500 nm. The above thickness values agree with those used for real cells.

Besides the efficiency, other parameters such as V_{oc}, J_L, and FF can be calculated with the help of the model explained before. In Table 4.1, these parameters are given for the cells shown in Fig. 4.5, when the perovskite thickness is optimum (500 nm). When comparing the experimental open circuit voltages (V_{oc}) for high-efficiency solar cells, which are around 1.25 V [16,17], with the V_{oc} values given in Table 4.1, we can conclude that the carrier lifetimes in the perovskite materials used in the current terminated cells is at least 200 ns.

The calculated open-circuit voltages (V_{oc}) for PSCs (500 ns case), made from a material with a bandgap of approximately 1.6 eV, reach 1.25 V (experimentally reported by Liu et al.) [16]. According to our model, this V_{oc} value is only possible if the nonradiative recombination lifetime is of the order of 500 ns in the terminated device. It is remarkably high, particularly taking into account that the perovskite films are fabricated by solution processing, which is a method that comes along with various challenges such as inhomogeneous films, remnants of solvents, impurities in the precursors, and contamination introduced during fabrication under no clean room conditions. A key factor for the progress was better control of the film morphology by engineering the deposition techniques and the precursors' composition. State-of-the-art devices comprise very compact and smooth films.

The values for J_L and V_{oc} in Table 4.1 are very close to those reported experimentally for high-efficiency PSCs [18]; however, the experimental fill factors are smaller than those in Table 4.1, due to the shunt and series resistance effects (not included here). These effects depend on the perovskite material and the selective contact fabrication methods. The present technology for high-efficiency cells seems to cause no problem with leakage currents, but it still produces some fill factor degradation associated with series resistance effects.

Therefore, the above calculations demonstrate that it is feasible to achieve conversion efficiencies above 25%—26% because nonradiative recombination lifetimes of the order of 500—1000 ns have already been reported for single perovskite layers. Keeping these lifetimes in the terminated devices would allow efficiencies as high as 26% if there are no large losses associated with the cell's reflectance and the series resistance effect.

A final consideration must be made here regarding the model described above. Some expressions are well known, but great physical knowledge of the solar cell behavior is required to establish the whole set of equations, from (4.5)—(4.13), as a simple model for thin PSCs. These equations, together with the explained device physics, can provide insight into what can be expected from optimized perovskite cell designs. These equations can be modified to include spectral reflectance, series and shunt resistances, and even interface (perovskite selective contacts) recombination effects, allowing for more realistic calculations in the future.

4.5 Conclusion

In this chapter, a new complete analytical model for PSCs has been developed. This model gives results that are comparable with experimental results, in some cases better than numerical, or other more sophisticated analytical models that do not take into account the device physics in a simple way. The model includes both radiative and nonradiative carrier recombination mechanisms and assumes that the solar cell is a $n-i-p$ (or $p-i-n$) planar device so that a high electric field is established in the perovskite region. This electric field should cause a large carrier drift length so that the photo-carrier collection probability shall be 100%, particularly for very thin perovskite layers. This model allows the calculation of the $J-V$ curve and all the important electrical parameters such as J_{sc}, V_{oc}, FF, and conversion efficiency. The cell reflectance is assumed to be small, and the series and shunt resistance affect the fill factor also, but except for these simplifications, the model agrees well with reported experimental results.

For example, for cells with perovskites having a bandgap around 1.6 eV and nonradiative recombination lifetimes above 100 ns, the expected efficiency would be above 24%. For solar cells in which the perovskite, in the finished device, has a carrier lifetime larger than 500 ns, the expected efficiency will become between 25% and 26%. Above these efficiencies, increasing the lifetime to 5 μs should cause a diminishing nonradiative recombination current as compared to the limiting radiative recombination rate. Therefore, increasing the nonradiative recombination lifetime just by one order of magnitude (from 500 ns to 5 μs) would be enough to have these cells behave close to the ideal Shockley-Queisser limit. This is a remarkable fact for a micro-crystalline material, but it is also a consequence of an appropriate device structure ($n-i-p$ or $p-i-n$) design, where carriers transport by electrical drift, together with a very small absorption length (high absorption coefficient), reducing in this way the total photo-generated carrier loss.

References

[1] K. Yoshikawa, H. Kawasaki, W. Yoshida, T. Irie, K. Konishi, K. Nakano, et al., Silicon heterojunction solar cell with interdigitated back contacts for a photoconversion efficiency over 26%, Nat. Energy 2 (2017) 17032.
[2] D.B. Needleman, J.R. Poindexter, R.C. Kurchin, I.M. Peters, G. Wilson, T. Buonassisi, Economically sustainable scaling of photovoltaics to meet climate targets, Energy Environ. Sci. 9 (6) (2016) 2122−2129.
[3] Z. Song, C.L. McElvany, A.B. Phillips, L. Celik, P.W. Krantz, S.C. Watthage, et al., A technoeconomic analysis of perovskite solar module manufacturing with low-cost materials and techniques, Energy Environ. Sci. 10 (6) (2017) 1297−1305.
[4] NREL, Best research-cell efficiency chart <https://www.nrel.gov/pv/cell-efficiency.html> (accessed 12.12.2023).
[5] X. Ren, Z. Wang, W.E.I. Sha, W.C.H. Choy, Exploring the way to approach the efficiency limit of perovskite solar cells by drift-diffusion model, ACS Photon. 4 (4) (2017) 934−942.
[6] W.E.I. Sha, X. Ren, L. Chen, W.C.H. Choy, The efficiency limit of $CH_3NH_3PbI_3$ perovskite solar cells, Appl. Phys. Lett. 106 (22) (2015) 221104.

[7] A. Acevedo-Luna, R. Bernal-Correa, J. Montes-Monsalve, A. Morales-Acevedo, Design of thin film solar cells based on a unified simple analytical model, J. Appl. Res. Technol. 15 (6) (2017) 599–608.

[8] J. Cui, H. Yuan, J. Li, X. Xu, Y. Shen, H. Lin, et al., Recent progress in efficient hybrid lead halide perovskite solar cells, Sci. Technol. Adv. Mater 16 (3) (2015) 036004.

[9] X. Sun, R. Asadpour, W. Nie, A.D. Mohite, M.A. Alam, A physics-based analytical model for perovskite solar cells, IEEE J. Photovolt. 5 (5) (2015) 1389–1394.

[10] Y. Zhoua, A. Gray-Weale, A numerical model for charge transport and energy conversion of perovskite solar cells, Phys. Chem. Chem. Phys. 18 (2016) 4476–4486.

[11] S.M. Sze, M.-K. Lee, Semiconductor Devices – Physics and Technology, 3rd ed., John Wiley & Sons, 2012, p. 592.

[12] A. Acevedo-Luna, A. Morales-Acevedo, Study of validity of the single-diode model for solar cells by I–V curves parameters extraction using a simple numerical method, J. Mat. Sci.: Mat. Electron. 29 (2018) 15284–15290.

[13] W. Shockley, H.J. Queisser, Detailed balance limit of efficiency of p-n junction solar cells, J. Appl. Phys. 32 (3) (1961) 510–519.

[14] W. Tress, Maximum efficiency and open-circuit voltage of perovskite solar cells, in: N.-G. Park, M. Gratzel, T. Miyasaka (Eds.), Organic-Inorganic Halide Perovskite Photovoltaics: From Fundamentals to Device Architectures, Springer International Publishing Switzerland, 2016, pp. 53–77.

[15] W. Tress, Perovskite solar cells on the way to their radiative efficiency limit – Insights into a success story of high open-circuit voltage and low recombination, Adv. Energy Mater 7 (14) (2017) 1602358.

[16] Z. Liu, L. Krückemeier, B. Krogmeier, B. Klingebiel, J.A. Márquez, S. Levcenko, et al., Open-circuit voltages exceeding 1.26 V in planar methylammonium lead iodide perovskite solar cells, ACS Energy Lett. (2019) 110–117.

[17] M. Saliba, T. Matsui, K. Domanski, J.-Y. Seo, A. Ummadisingu, S.M. Zakeeruddin, et al., Incorporation of rubidium cations into perovskite solar cells improves photovoltaic performance, Science 354 (6309) (2016) 206–209.

[18] W.S. Yang, B.-W. Park, E.H. Jung, N.J. Jeon, Y.C. Kim, D.U. Lee, et al., Iodide management in formamidinium-lead-halide–based perovskite layers for efficient solar cells, Science 356 (6345) (2017) 1376–1379.

CHAPTER FIVE

Optical design of perovskite solar cells

Arturo Morales-Acevedo[1] and Roberto Bernal-Correa[2]

[1]Electrical Engineering Department – SEES, Centro de Investigación y de Estudios Avanzados del Instituto Politécnico Nacional (CINVESTAV), Ciudad de México, México
[2]Universidad Nacional de Colombia, Sede Orinoquia, Grupo de Investigación en Ciencias de la Orinoquia, Arauca, Colombia

5.1 Introduction

The scientific community has become increasingly interested in theoretical models that allow the prediction of physical parameters and performance characteristics so that the required experimental work is shortened when developing new semiconductor devices. In the case of optoelectronic devices, such as solar cells, their performance can be affected by the optical properties of the materials used in their interaction with light. Therefore, interest in this type of study has increased [1–3].

For solar cells, it has been demonstrated that optical losses are one of the important limitations that might circumvent the achievement of high efficiencies [4,5]. Thus, it is important to use mathematical models to design solar cells so that their optical losses are minimized [2–8]. In this chapter, one model for determining the optical reflectance and transmittance of multilayered solar cells is explained.

Among the existing technologies, perovskite-based solar cells stand out due to their high performance achieved in a short time in conjunction with a low production cost [9,10]. However, it is necessary to identify possible ways to improve their performance further, one of them being their appropriate optical design.

Some of the most studied ($MAPbI_3$) perovskite solar cells (PSCs) can be classified by arrangement into one of two groups: normal $n-i-p$ or inverted $p-i-n$ cells (depending on the front material where the photons flow) [11]. In the first case, Spiro-OMeTAD is used as the back hole selective layer (HSL), while for the so-called inverted cells, PEDOT:PSS

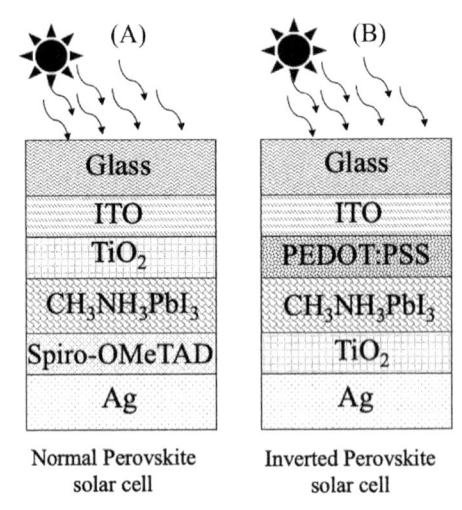

Figure 5.1 Schematics of perovskite solar cell (PSC) structures: (A) normal and (B) inverted cells.

is used as the front HSL. This difference in HSLs is due to some technological problems when making the devices [12,13].

One of the strategies considered in the design of solar cells is to reduce the optical losses in the solar spectrum (terrestrial—AM1.5). Therefore, estimating the spectral and total average reflectance (and transmittance in the upper layers above the absorber) should allow these devices' design to be improved [14,15]. In this chapter, calculations for normal and inverted perovskite solar cells (Fig. 5.1) are made using a model to determine the expected reflectance and transmittance for both cell structures. An optical comparison of these structures is made for determining the best one in this regard, by evaluating their optical losses.

5.2 Optical modeling

Based on the optical phenomena and properties of multilayer systems and the materials that compose them, an optical model can be developed using expressions that relate the transmission and reflection wave-vectors to the incident one. Such a model is a useful tool for the analysis and design of different cell structures [16—18]. If a multilayer system without structural defects is considered, homogeneous, isotropic, and in a state of equilibrium (i.e., the properties of interest for each layer are indistinct as a function of

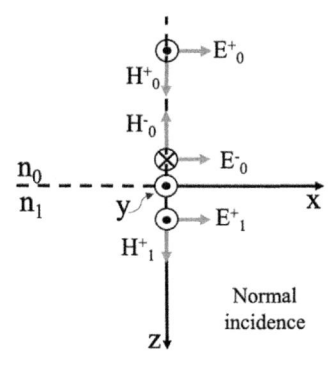

Figure 5.2 Electromagnetic field diagram for normal incidence, reflection, and transmission of waves in a system made up of two different media.

thickness), the transfer matrix model can be applied, using the Fresnel coefficients, giving the relationship between electric field vectors when a wave suffers a change in the propagation direction [19].

In the simple case, where the materials are considered nonabsorbent and the ray incidence on the layered system is normal to the surface, it is possible to calculate the transmitted and reflected wave power fractions due to the properties of the materials. Considering the total components of the electric and magnetic field vectors (E and H, respectively, shown schematically in Fig. 5.2), one can write a set of expressions by relating the electric field vectors of the reflected wave (E_o^-) and the transmitted wave (E_1^+) to the wave vector arriving at the surface (E_o^+).

These expressions (5.1) are known as Fresnel equations, which also define the Fresnel coefficients r and t [20]. For a two-layer system:

$$\begin{cases} \dfrac{n_0 - n_1}{n_0 + n_1} = r = \dfrac{E_0^-}{E_0^+} \\ \dfrac{2n_0}{n_0 + n_1} = t = \dfrac{E_1^+}{E_0^+} \end{cases} \quad (5.1)$$

where n_0 and n_1 correspond to the refractive indices of each of the media. In general, the refractive index n is defined as the ratio of the velocity of light in free space (c) and the velocity of light in the medium (v), and it can also be written (5.2) [21] as

$$n = (\varepsilon * \mu)^{1/2} \rightarrow n = (\varepsilon)^{1/2}$$
(at optical frequencies $\mu = 1$)

(5.2)

If, in addition to refraction, the absorption in the system to be studied is considered, the complex refractive index (5.3) can be written as follows:

$$\mathbf{n} = n + ik \tag{5.3}$$

where the imaginary part k is the extinction coefficient. The complex refraction index $\mathbf{n}$ can be related (5.4) to the complex relative dielectric constant:

$$\varepsilon = \varepsilon_r + i\varepsilon_i \tag{5.4}$$

in which ε_r and ε_i are the real and imaginary parts. From expressions (5.2) to (5.4), the following relations (5.5) can be obtained:

$$\begin{cases} \varepsilon_r = n^2 - k^2 \\ \varepsilon_i = 2nk \end{cases} \tag{5.5}$$

Additionally, there is a relation (5.6) of k with the absorption coefficient α:

$$k = \frac{\lambda \alpha}{4\pi} \tag{5.6}$$

where λ is the wavelength in vacuum. Now, for the system shown in Fig. 5.1, with **medium 1** having a complex refractive index $\mathbf{n_1} = n_1 - ik_1$, according to the equations (5.1)–(5.5), the Fresnel coefficients can be rewritten as

$$\begin{cases} \dfrac{n_0 - n_1 + ik_1}{n_0 + n_1 + ik_1} = r \\ \dfrac{2n_0}{n_0 + n_1 + + ik_1} = t \end{cases} \tag{5.7}$$

5.3 Multilayered systems

Following this methodology for the electric field vectors, a system of equations for "m-layers" can be obtained. In this case, for each layer (j) the complex refractive indices $\mathbf{n_j} = n_j - ik_j$ are included, taking into account that the multilayer system includes absorbing layers. Then, the Fresnel coefficients (5.8) involving these layers are:

$$\begin{cases} \dfrac{n_{j-1} - n_j}{n_{j-1} + n_j} \equiv g_j + ih_j = r_j \\ 1 + g_j + ih_j = t_j \end{cases} \quad (5.8)$$

where g_j and h_j depend on the refraction index and extinction coefficient of the respective (j) layer. The recurrence relation for the incident and reflected electric field vectors (5.9) at each interface between layers may be written as

$$\begin{pmatrix} E^+_{j-1} \\ E^-_{j-1} \end{pmatrix} = \frac{1}{t_j} \begin{pmatrix} e^{i\delta_{j-1}} & r_j e^{i\delta_{j-1}} \\ r_j e^{-i\delta_{j-1}} & e^{-i\delta_{j-1}} \end{pmatrix} \begin{pmatrix} E^+_j \\ E^-_j \end{pmatrix} \quad (5.9)$$

For a system with m layers, the relation (5.10) can be defined as

$$\begin{pmatrix} E^+_0 \\ E^-_0 \end{pmatrix} = (M_1)\ (M_2)\ \cdots\ (M_{m+1}) \begin{pmatrix} E^+_{m+1} \\ E^-_{m+1} \end{pmatrix} \quad (5.10)$$

where M_j is given by the following expression (5.11); the elements of that matrix (M_j) are associated with the Fresnel coefficients, and $\delta_j = 2\pi n_j d_j/\lambda$ for normal incidence ($j = 1, \ldots, m$).

$$M_j = \frac{1}{t_j} \begin{pmatrix} e^{i\delta_{j-1}} & r_j e^{i\delta_{j-1}} \\ r_j e^{-i\delta_{j-1}} & e^{-i\delta_{j-1}} \end{pmatrix} \quad (5.11)$$

Then, the optical matrix (M_{opt}) can be obtained (5.12) by the product of this set of matrices with $j = 1, \ldots, m$.

$$M_{opt} = (M_1)\ (M_2)\ \cdots\ (M_{m+1}) = \begin{pmatrix} A_{11} & A_{12} \\ A_{21} & A_{22} \end{pmatrix} \quad (5.12)$$

From this matrix multiplication resulting in elements A_{11} through A_{22}, it is possible to obtain the necessary elements to determine the reflectance and transmittance in the layer system to be studied. Thus, it is important to use the set of expressions associated with M_1 to M_{m-1}, as a function of wavelength. Finally, the reflectance (5.13) and transmittance (5.14) for each wavelength λ are

$$R(\lambda) = \frac{(E^-_0)(E^-_0)^*}{(E^+_0)(E^+_0)^*} = \frac{|A_{21}|^2}{|A_{11}|^2} \quad (5.13)$$

$$T(\lambda) = \frac{(E^-_{m+1})(E^-_{m+1})^*}{(E^+_0)(E^+_0)^*} = \frac{|t_1 t_2 ... t_{m+1}|^2}{|A_{11}|^2} \qquad (5.14)$$

To determine the cell weighted average reflectance $<R>$ (5.15) and transmittance $<T>$ (5.16) in the solar spectrum, the following expressions can be used:

$$<R> = \frac{\int_{\lambda\min}^{\lambda\text{material}} R(\lambda)N(\lambda)d\lambda}{\int_{\lambda\min}^{\lambda\text{material}} N(\lambda)d\lambda} \qquad (5.15)$$

$$<T> = \frac{\int_{\lambda\min}^{\lambda\text{material}} T(\lambda)N(\lambda)d\lambda}{\int_{\lambda\min}^{\lambda\text{material}} N(\lambda)d\lambda} \qquad (5.16)$$

where $N(\lambda)$ is the incident spectral photon flux density as a function of the wavelength. $\lambda_{\min}$ is the minimum wavelength for which there are photons in the solar spectrum, and $\lambda_{\text{material}}$ is the maximum wavelength for which there is absorption by the absorbent material in the device. It is possible to determine the internal spectral transmittance $T(\lambda)$ and the total average transmittance due to the upper layers above the perovskite absorber material. The cell photocurrent density due to sunlight absorption in the perovskite absorber layer should be proportional to the average $<(1-R)^*T>$ product for each of the structures. Hence, this product is an important parameter to be determined as a way of understanding the optical behavior of solar cells.

5.4 Optical calculations for perovskite solar cells

The optical losses and thickness optimization of the layers of the solar cells studied here can be determined using the model described above. For this purpose, the values of the refractive index n and the extinction coefficient k of each layer involved were used (see Fig. 5.3) [22,23]. Two cases were considered for the normal and inverted solar cell calculations: In the first case, the layer thickness of the perovskite-type material ($CH_3NH_3PbI_3$) was varied in a range from 50 to 1000 nm, and the thickness of the TiO_2, PEDOT:PSS, and Spiro-OMeTAD layers was kept constant at 50 nm. In the second case, the thickness of the

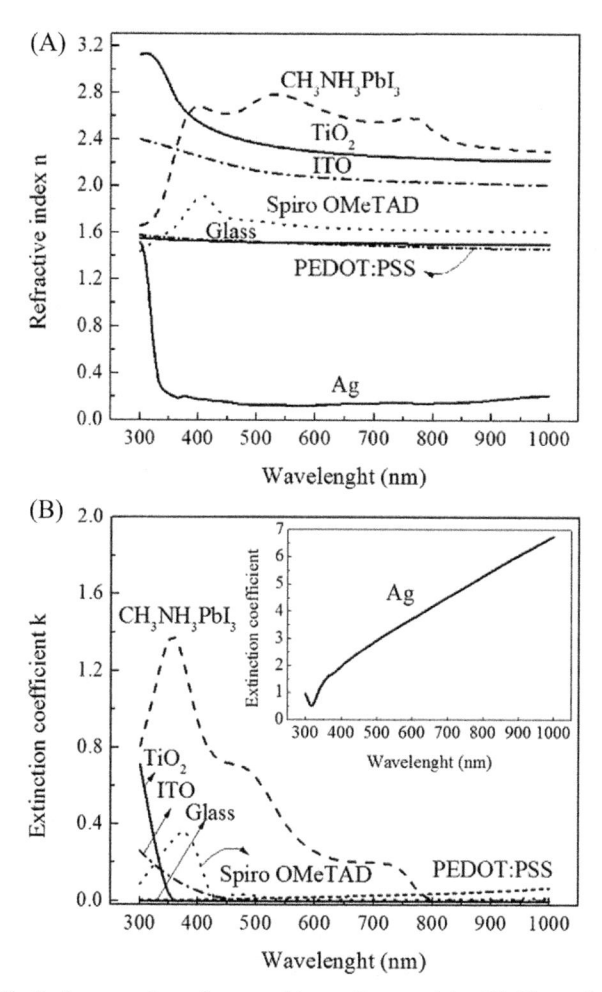

Figure 5.3 Optical properties of perovskite cell materials: (A) The refractive index and (B) the extinction coefficient for the different layers of solar cells.

perovskite-type material was fixed at 800 nm, and that of the top TiO_2 and PEDOT:PSS layers was varied between 50 and 1000 nm for normal and inverted solar cells, respectively. Other thicknesses were kept constant in all cases (glass 3 mm and ITO 500 nm).

The weighted average reflectance (Eq. 5.15) for PSCs as a function of the absorber layer thickness is shown in Fig. 5.4 for the two cases of interest. Note that for perovskite layer thicknesses greater than 500 nm, the reflectance tends to become almost constant for both types of solar cells. However, it is still relatively large, around 10%, for a normal perovskite-type

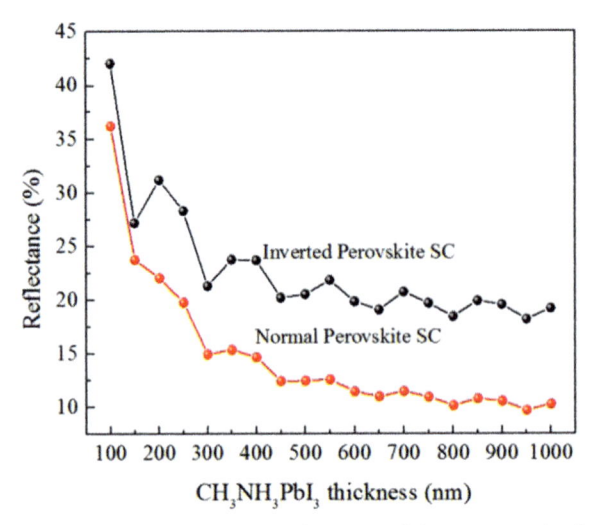

Figure 5.4 Average reflectance <R> as a function of the $CH_3NH_3PbI_3$ layer thickness for normal and inverted solar cells, respectively.

solar cell (14-Roberto Bernal-Correa & Morales-Acevedo, 2021) [14], and double in percentage for an inverted perovskite-type solar cell (20%).

To determine the dependence of the average reflectance on the TiO_2 and PEDOT:PSS layer thicknesses in both cells, 800 nm perovskite layers were selected. Fig. 5.5 shows that the average reflectance values for a normal perovskite solar cell do not depend significantly on the TiO_2 layer thickness. However, a different behavior is observed for an inverted perovskite solar cell when the PEDOT:PSS layer thickness is varied. In this case, the increase in the PEDOT:PSS thickness causes a reduced average reflectance, but as it will be explained below this is due to the increased absorption in this layer.

The average upper-layer transmittance is shown in Fig. 5.6. This indicates that for a normal cell, there is also no strong dependence of the transmittance as a function of the TiO_2 layer thickness because it is highly transparent in the whole solar spectrum. In the case of the inverted solar cell, the transmittance is decreased as the PEDOT:PSS layer thickness is increased due to the sunlight absorption by this layer.

As a way of optimizing the TiO_2 or PEDOT:PSS layers for each cell structure, the behavior of the $<(1-R)*T>$ product as a function of TiO_2 and PEDOT:PSS thickness is shown in Fig. 5.7. In a conventional $n-i-p$ perovskite cell, the TiO_2 layer thickness can be changed from very small (50 nm) up to large thickness (900 nm) without much optical

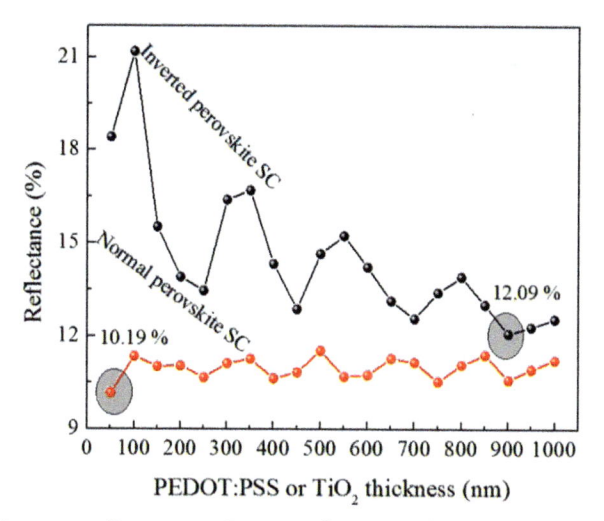

Figure 5.5 Average reflectance <R> as a function of the front layer thickness of TiO$_2$ and PEDOT:PSS layers for normal and inverted perovskite solar cells, respectively. The minimum values are indicated.

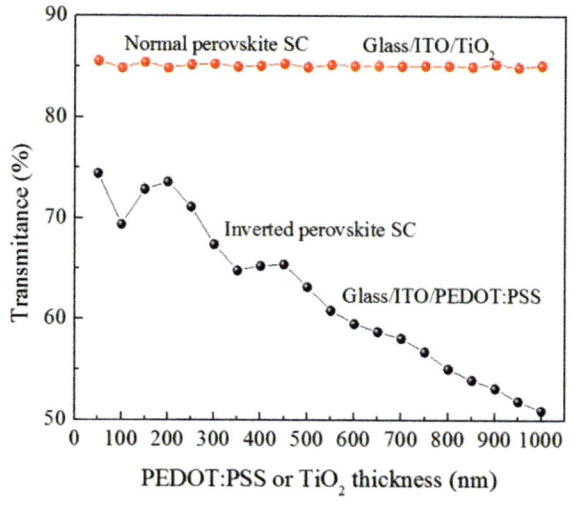

Figure 5.6 Average transmittance <T> as a function of the front layer thickness of TiO$_2$ and PEDOT:PSS layers for normal and inverted perovskite solar cells, respectively.

variation. On the other hand, for an inverted $p-i-n$ solar cell, there is a PEDOT:PSS optimum thickness of approximately 200 nm, for which the $<(1-R)*T>$ product becomes maximum as required for achieving a high cell photocurrent density.

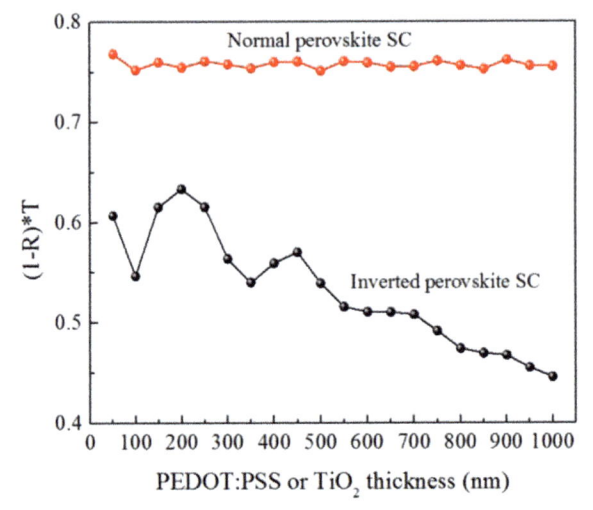

Figure 5.7 Average $<(1 - R)*T>$ as a function of the TiO$_2$ and PEDOT:PSS layer thickness for normal and inverted perovskite solar cells, respectively.

The reader should also note that there is a big difference in the optimum $<(1 - R)^*T>$ products for these two kinds of structures. A larger photocurrent density is typically expected for normal $n-i-p$ solar cells, as compared to inverted $p-i-n$ solar cells. Therefore, a higher photon flux is available to be absorbed by the perovskite material in the normal $n-i-p$ cell structure, so that a larger photo-current density (and efficiency) can be expected in this case.

5.5 Conclusion

Using an optical matrix method, it is possible to calculate the external spectral reflectance and internal transmittance due to multilayered planar perovskite solar cells. This calculation method has been applied to both normal (conventional) and inverted perovskite solar cells to compare their optical behavior. It can be demonstrated that the optical performance of normal $n-i-p$ cells, with TiO$_2$ as the electron selective layer, is superior to inverted $p-i-n$ cells, with PEDOT:PSS as the HSL, due to the sunlight absorption in this layer, reducing the available photons to be absorbed by the perovskite material. For the inverted cell it was shown that there is an optimum PEDOT:PSS layer thickness of approximately

200 nm while the TiO_2 layer thickness for normal cells can be selected in a broad range (100−1000 nm) without much variation of the optical behavior for the cell. Therefore, from an optical perspective, for a high cell efficiency, the normal $n-i-p$ structure is recommended as the preferred device structure.

References

[1] A. Berkhout, A.F. Koenderink, A simple transfer-matrix model for metasurface multilayer systems, Nanophotonics 9 (12) (2020) 3985−4007. Available from: https://doi.org/10.1515/nanoph-2020-0212.

[2] O.D. Iakobson, O.L. Gribkova, A.R. Tameev, J.M. Nunzi, A common optical approach to thickness optimization in polymer and perovskite solar cells, Sci. Rep. 11 (1) (2021) 5005. Available from: https://doi.org/10.1038/s41598-021-84452-x.

[3] Q. Lv, J. Cui, H. Jarimi, H. Lv, Z. Zhai, Y. Su, S. Riffat, S. Dong, Theoretic analysis and experimental evaluation of the spectrum transmission coefficient of a multilayer photovoltaic vacuum glazing, Int. J. Low-Carbon Technol. 15 (4) (2020) 574−582. Available from: https://doi.org/10.1093/ijlct/ctaa026.

[4] S. Manzoor, J. Häusele, K.A. Bush, A.F. Palmstrom, J. Carpenter, Z.J. Yu, S.F. Bent, et al., Optical modeling of wide-bandgap perovskite and perovskite/silicon tandem solar cells using complex refractive indices for arbitrary-bandgap perovskite absorbers, Opt. Exp. 26 (21) (2018) 27441−27460. Available from: https://doi.org/10.1364/OE.26.027441.

[5] A. Purkayastha, A.T. Mallajosyula, Optical modelling of tandem solar cells using hybrid organic-inorganic tin perovskite bottom sub-cell, Solar Energy 218 (2021) 251−261. Available from: https://doi.org/10.1016/j.solener.2021.01.054.

[6] J. Hossain, B.K. Mondal, S.K. Mostaque, S.R.A. Ahmed, H. Shirai, Optimization of multilayer anti-reflection coatings for efficient light management of PEDOT:PSS/c-Si heterojunction solar cells, Mater. Res. Exp. 7 (2020) (2020) 015502. Available from: https://doi.org/10.1088/2053-1591/ab5ac7.

[7] T. Rahman, S.A. Boden, Optical modeling of black silicon for solar cells using effective index techniques, IEEE J. Photovolt. 7 (6) (2017) 1556−1562. Available from: https://doi.org/10.1109/JPHOTOV.2017.2748900.

[8] R. Sharma, Effect of obliquity of incident light on the performance of silicon solar cells, Heliyon 5 (7) (2019) e01965. Available from: https://doi.org/10.1016/j.heliyon.2019.e01965.

[9] M. Green, E. Dunlop, J. Hohl-Ebinger, M. Yoshita, N. Kopidakis, X. Hao, Solar cell efficiency tables (version 57), Prog. Photovolt.: Res. Appl. 29 (1) (2021) 3−15. Available from: https://doi.org/10.1002/pip.3371.

[10] M. Jeong, I.W. Choi, E.M. Go, Y. Cho, M. Kim, B. Lee, S. Jeong, et al., Stable perovskite solar cells with efficiency exceeding 24.8% and 0.3-V voltage loss, Science 369 (6511) (2020) 1615−1620. Available from: https://doi.org/10.1126/science.abb7167.

[11] A.N. Cho, N.G. Park, Impact of interfacial layers in perovskite solar cells, ChemSusChem 10 (19) (2017) 3687−3704. Available from: https://doi.org/10.1002/cssc.201701095.

[12] C. Duan, Z. Liu, L. Yuan, H. Zhu, H. Luo, K. Yan, PEDOT:PSS-metal oxide composite electrode with regulated wettability and work function for high-performance inverted perovskite solar cells, Adv. Opt. Mater. 8 (17) (2020) 2000216. Available from: https://doi.org/10.1002/adom.202000216.

[13] Z. Yang, W. Chen, A. Mei, Q. Li, Y. Liu, Flexible $MAPbI_3$ perovskite solar cells with the high efficiency of 16.11% by low-temperature synthesis of compact anatase TiO_2 film, J. Alloys Comp. 854 (2021) 155488. Available from: https://doi.org/10.1016/j.jallcom.2020.155488.

[14] R. Bernal-Correa, A. Morales-Acevedo, Spectral reflectance optimization for planar perovskite solar cells, Optik 227 (2021) 165973. Available from: https://doi.org/10.1016/j.ijleo.2020.165973.
[15] M.J. Taghavi, M. Houshmand, M.H. Zandi, N.E. Gorji, Modeling of optical losses in perovskite solar cells, Superlatt. Microstruct. 97 (2016) 424–428. Available from: https://doi.org/10.1016/j.spmi.2016.06.031.
[16] A. Acevedo-Luna, R. Bernal-Correa, J. Montes-Monsalve, A. Morales-Acevedo, Design of thin film solar cells based on a unified simple analytical model, J. Appl. Res. Technol. 15 (6) (2017) 599–608. Available from: https://doi.org/10.1016/j.jart.2017.08.002.
[17] R. Bernal-Correa, A. Morales-Acevedo, A. Pulzara Mora, J. Montes Monsalve, M. López López, Design of $Al_xGa_{1-x}As/GaAs/In_yGa_{1-y}As$ triple junction solar cells with anti-reflective coating, Mater. Sci. Semicond. Process. 37 (2015) 57–61. Available from: https://doi.org/10.1016/j.mssp.2015.01.020.
[18] R. Bernal-Correa, A. Morales-Acevedo, J. Montes-Monsalve, A. Pulzara-Mora, Design of the TCO (ZnO:Al) thickness for glass/TCO/CdS/CIGS/Mo solar cells, J. Phys. D: Appl. Phys. 49 (12) (2016) 125601. Available from: https://doi.org/10.1088/0022-3727/49/12/125601.
[19] M. Sadiku, Elements of Electromagnetics, 7th Ed., Oxford University Press, Oxford, UK, 2018, p. 920.
[20] O.S. Heavens, Optical Properties of Thin Solid Films, 2nd Ed., Dover Books, 2011, p. 272.
[21] M. Fox, Optical Properties of Solids, 2nd Ed., Oxford University Press, Oxford, UK, 2010, p. 416.
[22] Filmetrics, Refractive index database. <http://www.filmetrics.com/refractive-index-database> (accessed 06.12.22).
[23] PV Lighthouse, Refractive index library. https://www.pvlighthouse.com.au/refractive-index-library (accessed 06.12.22).

CHAPTER SIX

Organic hole-transporting materials for perovskite solar cells: Progress and prospects

S. Sambathkumar and P. Baby Shakila
Department of Chemistry and Biochemistry, Vivekanandha College of Arts and Sciences for Women (Autonomous), Elayampalayam, Tiruchengode, Namakkal, Tamil Nadu, India

6.1 Introduction to perovskite solar cells

Considering the issues surrounding fossil fuels, researchers are working toward renewable energy sources that are clean and green. Among the various photovoltaic technologies, dye-sensitized solar cells (DSSCs) have gained more attention due to cost effectiveness, easy fabrication, eco-friendliness, better stability, flexibility, and high power conversion efficiencies (PCEs). A breakthrough was made by O'Regan and Gratzel in 1991 with the PCE of 7% [1]. Soon after, it reached 12.3% PCE with a zinc porphyrin photosensitizer and cobalt-based electrolyte system [2–13]. The DSSC device was facing stability issues because of the liquid phase electrolyte, which does not allow for commercialization and leads the research toward developing PSCs. During the past decades, PSCs have attracted significant interest in photovoltaic research, leading to the outstanding enhancement in the PCE, presently exceeding 25.5% [14–25]. Since the seminal work of Miyasaka et al., PSCs consist of a conductive oxide layer [fluorine doped tin oxide (FTO), conductive substrate], hole-transporting material (HTM), a light-absorbing layer (perovskite), electron-transporting material (ETM), and metal electrode (Au) [17].

The HTM layer plays a key role in PSCs, as it facilitates the hole transfer from perovskite to the electrode suppressing recombination, and also protects the perovskite surface against degradation due to moisture/oxygen. The HTM plays a vital role in transmitting holes from the perovskite layer to the electrode because it has low charge carrier loss and exciton binding energy. Moreover, one of the critical factors resulting in the rapid development of PSCs is the existence of HTMs in devices because they could facilitate the photogenerated hole transfer from the perovskite layer to the back electrode efficiently, which dramatically enhances the performance of the devices [26–29]. It is noted that a huge enhancement in PCE is gained within a decade (3.8%–24.2%), which is competitive with the commercially available silicon solar cells [30]. Solid state HTMs can be categorized into three main classes, such as organic, inorganic, and hybrid. Each class has its own advantages and disadvantages, however, the organic HTM based on small molecules served as a suitable candidate for efficient PSCs, which will be discussed in the following sections.

Traditionally, spiro-OMeTAD (2,2′,7,7′-tetrakis-(N,N-di p-methoxyphenylamine)-9,9′-spirobifluorene) has been used as standard HTMs. However, the spiro-OMeTAD is not sustainable for large-scale applications [31] due to its (i) high price, (ii) complicated synthesis routines and laborious purification process, (iii) demand of many dopants due to low hole mobility, and (iv) sufficient thickness to obtain high performance, typically 200 nm, leading to more material usage. Therefore, it is necessary to find new efficient and facile candidates to replace spiro-OMeTAD [32–38]. Thus, there is a need to develop alternative HTMs, which should have suitable energy levels; good solubility in the organic solvent; high thermal stability; comparable hole mobility; and low cost, eco-friendly, facile, and scalable synthesis to allow commercialization of PSCs.

Small organic molecules based on carbazole, phenothiazine, triphenylamine, pyrene, and several others [39–43] have been proposed as HTMs

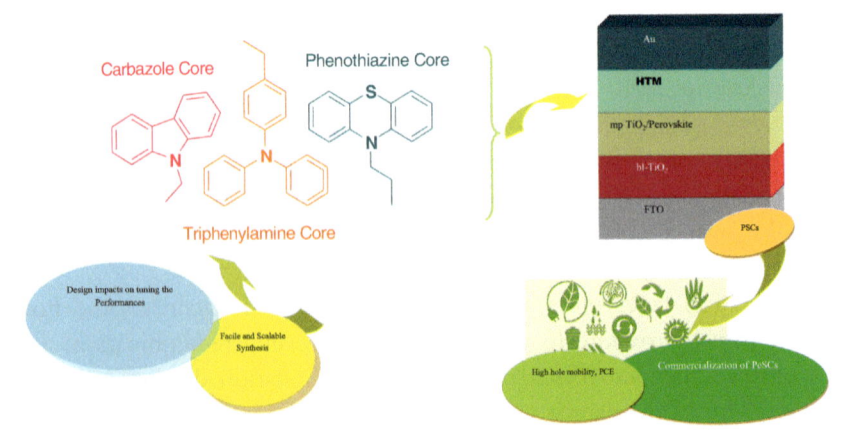

Figure 6.1 Impact of design of hole-transporting materials on the performance of perovskite solar cells.

to sort out the major drawbacks of benchmark spiro-OMeTAD for highly efficient PSCs. Moreover, a comparative study of linear or starburst shape of HTMs on the photovoltaic performance of the PSCs is rarely studied [44,45]. The importance of tuning the performance of PSCs based on the design prospects HTMs has been illustrated in Fig. 6.1. It is necessary to further reveal the relationship between the molecular structure and the photovoltaic performance of PSCs for seeking new efficient HTMs.

6.2 Device architecture and mechanism of perovskite solar cells

Generally, the device architecture consists of an FTO/c-TiO$_2$/mp-TiO$_2$/Perovskite/Organic HTM/Metal electrode (Au or Ag). In the *n-i-p* device structure, a thin and compact TiO$_2$ layer is applied as an electron-selective contact to facilitate the collection of photogenerated electrons from the metal halide perovskite. The c-TiO$_2$ layer also works as a blocking layer to prevent charge recombination by reducing the contact between the FTO substrate with the HTM or the FTO with the metal halide perovskite, which otherwise tends to cause shunting and leakage of current in the PSCs. The most common device architecture of the meso-structured planar PSCs is outlined in Fig. 6.2A.

The simplified operating principle with different components of conventional meso-structured PSCs and key charge transfer process are outlined in Fig. 6.2B. Incident photons excite the perovskite light-harvesting layer through the transparent electrode, leading to the photo-generation of electron−hole pairs in the materials. Electrons are (1) separated from the holes; (2) injected into the conduction band of the ETM before migrating to the anode (FTO); (3) the holes generated in the perovskite are transferred to the highest occupied molecular orbital (HOMO) level of the HTM, and (4) then transferred to the cathode (metallic contact). Electron and hole injection occur quite efficiently due to the high diffusion length of charge carriers. Undesirable charge transfer such as electron recombination of the charge carriers at ETM/perovskite/HTM interfaces may also happen including nonradiative recombination. Meanwhile, it is expected that the kinetic process of intrinsic radiative recombination, contact characteristics of the electrode, charge accumulation, and diffusion lengths of the charge carriers greatly control the limit of photovoltaic conversion efficiency of the perovskite devices [46].

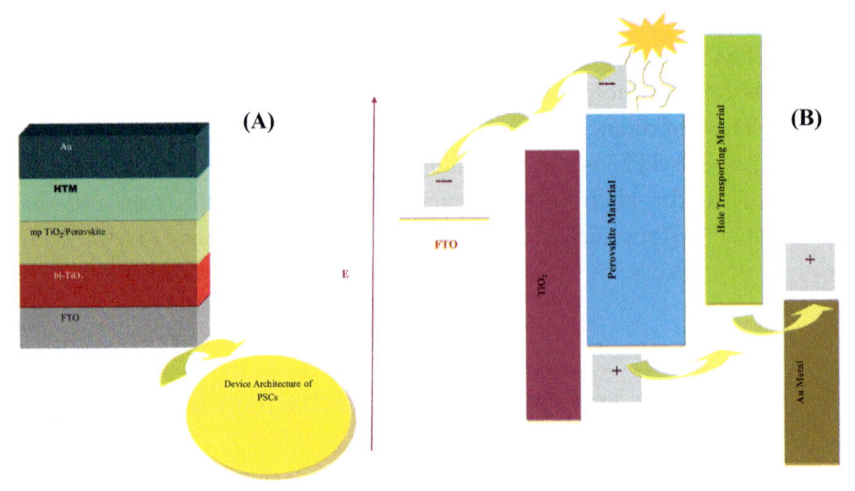

Figure 6.2 Schematic illustration of most common device architecture of PeSCs; (A) n-i-p mesoporous structure; (B) energy level diagram of the different components of a conventional meso-structured PSCs.

6.3 Organic hole-transporting materials

Synthetic organic chemistry offers a wide variety of tools and techniques for synthesizing an extensive assortment of molecules, which can be employed as a central scaffold for HTMs in PSCs. The designed molecules should possess the following requirements to be considered prominent materials for HTM: (i) They must be thermally and photochemically stable; (ii) the HOMO must be slightly higher in energy than the valance band edge of the perovskite material to ensure efficient extraction of the photogenerated holes [47–49]; (iii) they must possess an optimal or at least a reasonable hole mobility for transporting the holes to the counter electrode [50]; (iv) a reduced tendency to crystallize ($T_g > 100°C$) is required for avoiding phase transition during device operation; (v) the solubility of the synthesized materials should be very high, which enables the better film-forming properties; (vi) to scale up and commercialization, the processes must be clear and easily attainable; (vii) strong absorption is not an important requirement for obtaining highly efficient HTMs.

HTMs make several critical contributions to PSC devices, including (i) efficient extraction of photogenerated holes improving their transport through the counter electrode and acting as an energy barrier to suppress the electron transfer to the anode material; (ii) the stability of the device has improved by preventing the direct contact between the perovskite and the metal electrode, preferably Au, Ag, and Al, and protecting from diffusion of the electrode into the perovskite layer [51–53]; (iii) suppressing the charge recombination losses at the interface between the perovskite and HTM by obtaining fully and uniform coating of the perovskite layer [54]; and (iv) affecting the open circuit voltage, (V_{oc}) not only by the presence of dopants of the hole-transporting layer but also with the reduced energy disorder with respect of the chemical structure [55] and by the different mobility and conductivity values.

Generally, carbazole, phenothiazine, triphenylamine, pyrene, naphthalene, phthalocyanin, coumarin, imidazole incorporated motifs, and several others have been proposed so far as small organic molecule-based HTMs, enabling high hole mobility and improving the performance of the PSCs [26]. Among the small organic molecule-based HTMs proposed so far, many reports found on carbazole-based scaffolds, due to their specific nature of enhanced hole mobility. This study deals with the progress and prospects of small organic molecule-based HTMs used in standardized PSCs. As mentioned above, in order to overcome the drawbacks of spiro-OMeTAD and many HTMs have been introduced with the same features as offered by spiro-OMeTAD, which are discussed below in detail.

6.3.1 Carbazole and triphenylamine core-based hole-transporting materials

Park and coworkers recently reported simple molecular-designed HTMs such as the carbazole derivative (**1**) and a novel fluorinated analog (**2**) (shown in Scheme 6.1) used in planar PSCs [31]. A comparison of the

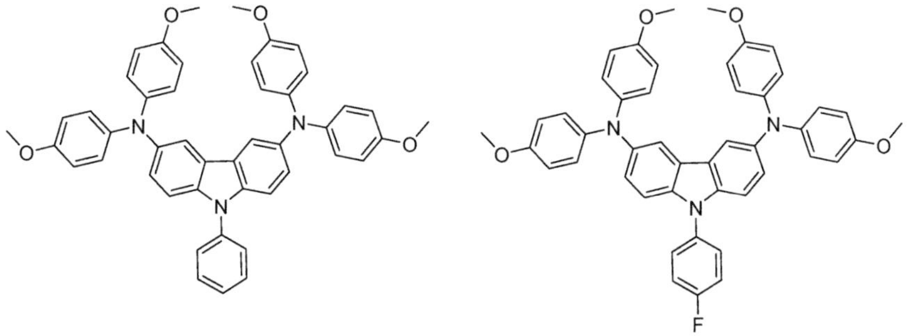

Scheme 6.1 Chemical structure of two carbazole-based HTMs (**1** and **2**).

properties of **1** and **2** clearly demonstrates that the incorporation of a fluorine atom (**2**) increases the hole mobility. A PCE closer to 13% in MAPbI$_3$ planar structure devices was obtained using these amorphous carbazole-based HTMs. Using a conventional perovskite (MAPbI$_3$) in planar structure photovoltaic devices, it is also demonstrated that both carbazole-based HTMs have similar photovoltaic performances to that of spiro-OMeTAD. The strategy was also recently applied to increase the hole mobility of donor−acceptor (D-A) molecular HTMs in PSCs [56]. Since fluorine atoms are usually incorporated into acceptor units, examples of fluorination on electron-rich HTMs remain scarce [57].

A simply designed and easily synthesized star-shaped carbazole core-based organic HTM with triphenylamine side arms (**3**), see Scheme 6.2, has been used in mesoscopic PSCs [58]. Compound **3** displays excellent optoelectronic, electrochemical, and high carrier mobility properties with a deep HOMO level; high hole mobility characteristics; a relatively high glass transition temperature; and good film-forming ability. Promisingly, PSCs using the dopant-free HTM (**3**) have a PCE of 14.29% and a higher PCE of 18% using a doped analog of **3** [58]; the latter PSC performance is comparable to the conventional doped spiro-OMeTAD. The cost of synthesis cost of **3** is considerably lower than that of spiro-OMeTAD [59]. This study shows the important step forward to the commercialized application of PSCs. Moreover, these results prove that the new HTM has significant potential to replace the spiro-OMeTAD standard.

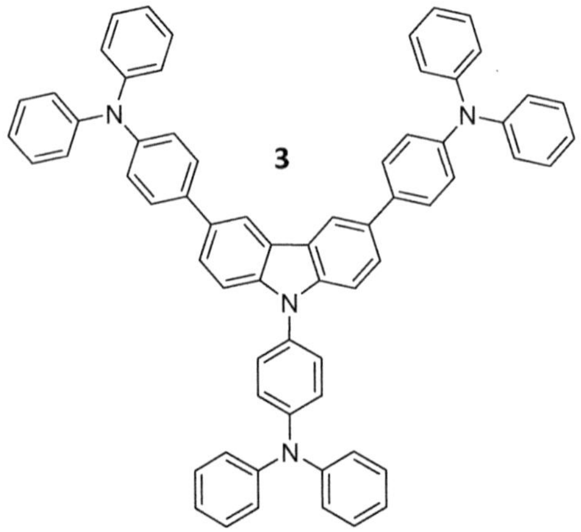

Scheme 6.2 Chemical structure of star-shaped carbazole core-based organic HTM (**3**).

One-step facile synthesis of carbazole-based HTMs has been developed by Yin et al. [60]. Solution-processed planar PSCs with a 50 nm thickness hole-transporting layer delivers a PCE of 18.32%, which is comparable to the most commonly used spiro-OMeTAD (PCE = 18.28%) hole-transporting layer [60]. Notably, the unit cost for the synthesis of such carbazole-based HTMs is much lower than that of spiro-OMeTAD, indicating that these compounds could be promising candidates for the commercialization of low-cost PSC technology.

A series of HTMs with a carbazole core and one to four dimethoxytriphenylamine side chains (4–7) were synthesized and characterized; see Scheme 6.3. By modifying the triphenylamine components in the HTM, the photovoltaic performances of PSCs could be controlled [61]. From the observed results, compound **5**, with two triphenylamine side units, provides better performances than compounds **4**, **6**, and **7**, which have one, three, and four triphenylamine units in their structure, respectively. The incorporation of one or two triphenylamine units appeared to improve device performance; however, additional branches introduced to the carbazole core did not lead to higher efficiency. Overall, the introduction of triphenylamine units was found to play an important part in HTMs; the number of triphenylamine groups has an impact on the device performance [61]. These results will offer further consideration of the molecular engineering of HTMs for the fabrication of efficient and stable PSCs.

More recently, triphenylamine-based donor (D)-π-D HTMs (peripheral groups is triphenylamine unit) have been widely developed by varying the π-linker such as biphenyl, fluorene (fused tricyclic hydrocarbon unit), or carbazole derivative [62], namely H-Bi (**8**), H-F (**9**), and H-Ca (**10**), are illustrated in Scheme 6.4. The impacts of π-linkers on the performance of optical, electrochemical, thermal, photovoltaic, and hole mobility were systematically analyzed [63]; molecules with biphenyl (**8**) and carbazole (**10**) π-linkers clearly display better hole mobility than the fluorene-containing analog (**9**), leading to better hole extraction ability at the perovskite/HTM interface. When these molecules are applied to PSCs, the carbazole-based counterpart exhibits higher efficiency than that of biphenyl and fluorene-based molecules, which is even comparable with traditional spiro-OMeTAD. Hence, carbazole units show great potential for efficient HTMs.

Generally, the film morphology of the hole-transporting layer has proven to be the key element to charge transfer. Liu et al. presented a new dibenzo [a,c]−carbazole (DBC) core with multiple reaction sites with a Y-shape shown in Scheme 6.5, in which the phenanthrene group has been

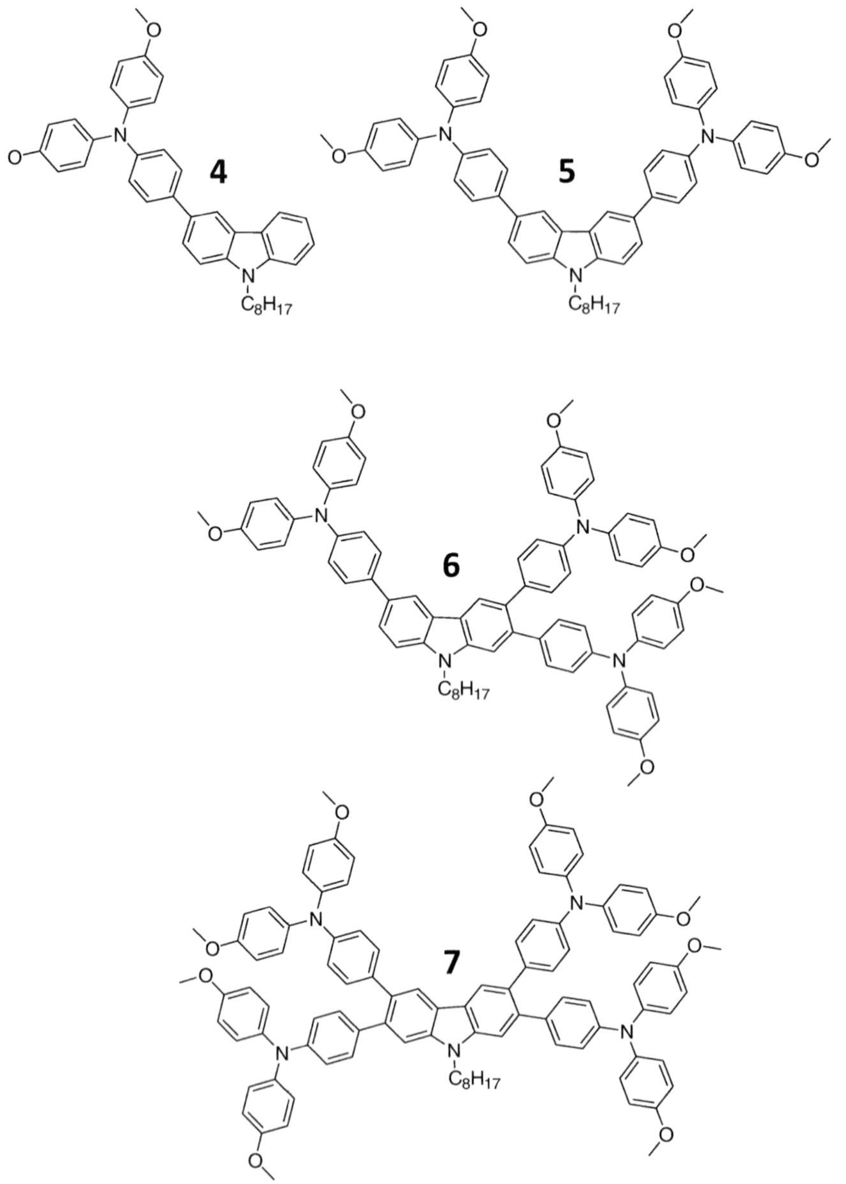

Scheme 6.3 Chemical structure of synthesized carbazole-based HTMs (**4–7**); TPA$_n$C ($n = 1$–4).

integrated as a plane π-structure into the common carbazole moiety [64]. Accordingly, three DBC-based HTMs with two (**11**) or three (**12** and **13**) side groups were synthesized by the introduction of N-(4-methoxyphenyl)-9,9-dimethyl-9H-fluoren-2-amine (F(Me)NPh) as the periphery group at

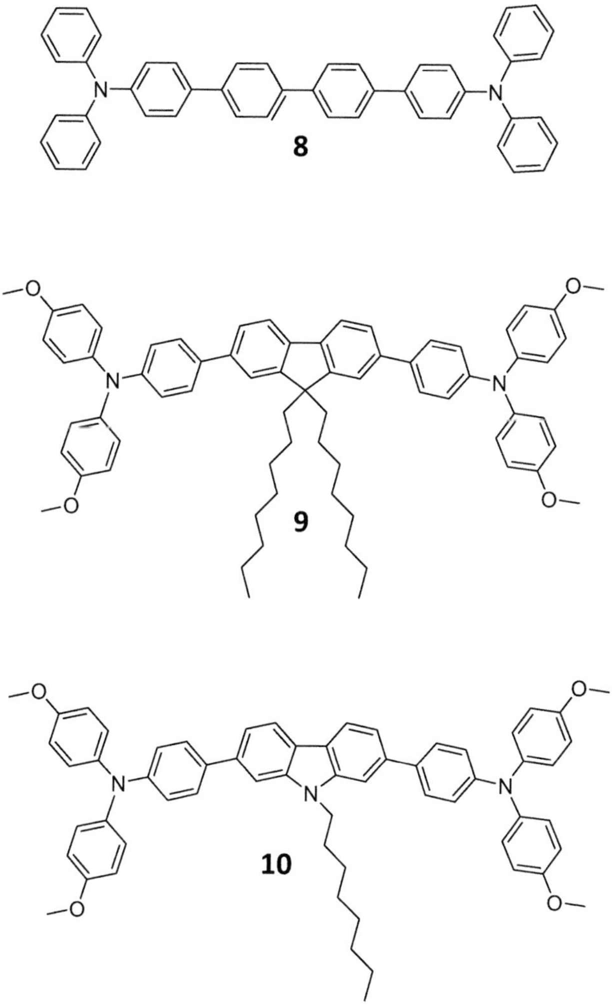

Scheme 6.4 Molecular structure of facile donor (D)-π-D triphenylamine-based organic hole-transporting material with different π-linkers (H-Bi (**8**), H-F (**9**), and H-Ca (**10**)).

different linkage positions [64]. Among the proposed configurations, compound **12** with a twisted and asymmetric structure linked at position C_{12} has better hole-transporting ability than the other two structures with an efficiency of 20.02% [64].

Two simple biphenyl or carbazole derivatives with four di(anisyl)amino substituents were synthesized and applied successfully to $FA_xMA_{1-x}PbI_{3-y}Br_y$ PSCs; increased PCEs of 13.6% and 11.5% were achieved [65]. The simple design, easy synthesis, and good photovoltaic performances of these HTMs make them promising for practical applications toward commercialization.

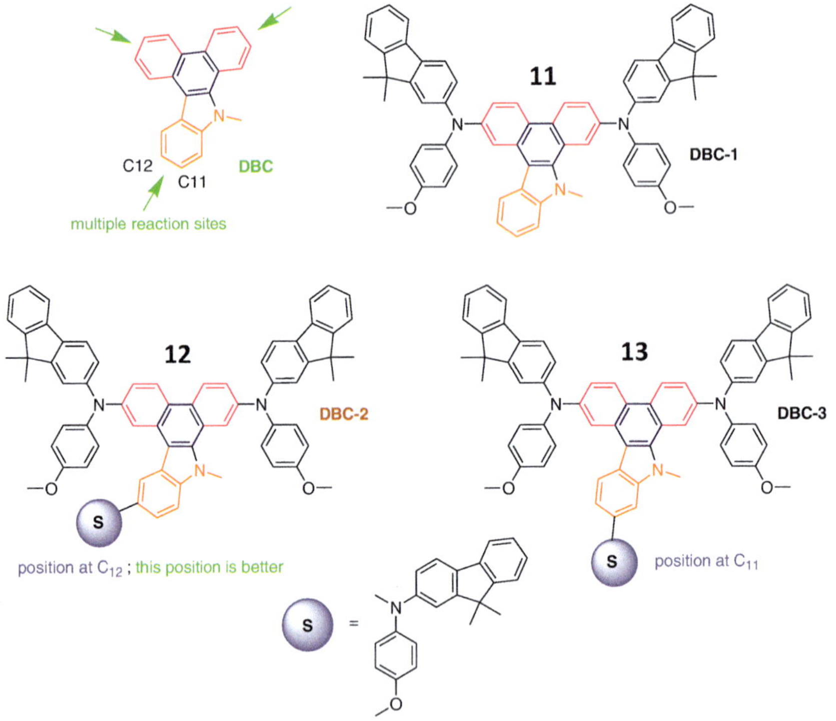

Scheme 6.5 Molecular design feature of target HTMs (**11–13**).

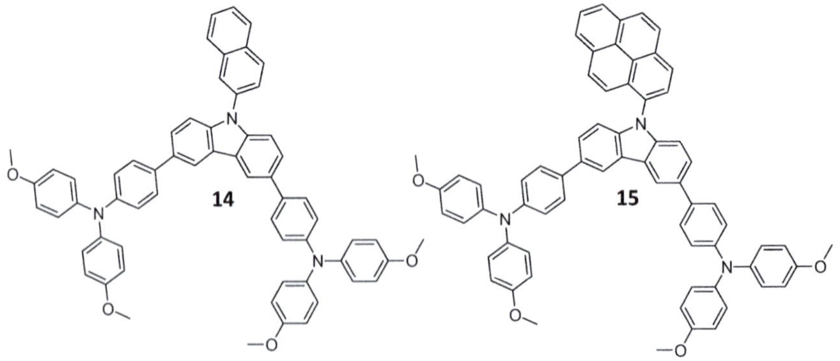

Scheme 6.6 Structures of hole-transporting materials with fused aromatic rings: Cz-N (**14**) and Cz-Pyr (**15**).

Further, two simple carbazole-based organic HTMs substituted with fused aromatic rings (Cz-N (**15**) and Cz-Pyr (**16**)) were designed and synthesized with maximum yield and successfully applied for the PSCs, which ensure the ease of scale-up toward the commercialization [66]. The synthesized compounds are shown in Scheme 6.6. Compound **15** (Cz-Pyr) enables better performance and is considered to be a promising candidate as HTM, owing to high hole mobility, ease synthesis, relatively high T_g, and high PCE, outperforming the reference HTM spiro-OMeTAD. Thus, the present study highlights the interest in extended fused benzene rings as substituents on carbazole-based scaffolds for the efficient PSCs.

Triphenylamine and carbazole core-based HTMs (**16** and **17**) were designed (Scheme 6.7) and applied in planar-structured PSCs, producing a PCE of 18.2%. The initial performances of the HTMs were retained for up to 50 days, indicating the stability of the materials toward commercialization [67]. This work demonstrates the potential alternatives for efficient PSCs with improved stability.

The development of simply designed and easily synthesized dopant-free organic HTMs with high efficiency is of great significance for the potential application of PSCs. Yang et al. reported two novel indolo[3,2-b]carbazole (ICZ)-based small molecules obtained via a facile synthetic route with high yield without any catalysts. The synthesized molecules shown in Scheme 6.8 (**18** and **19**) were systematically characterized with various analytical techniques and successfully applied for PSCs as dopant-free HTMs. The two biphenylamino groups of **19** significantly improve their spatial configuration [68]. It is observed that the interplay between

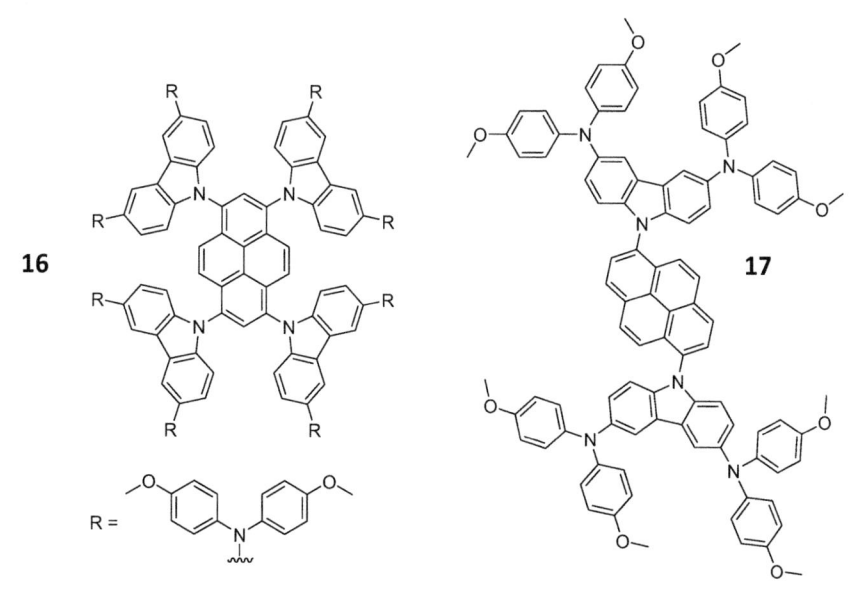

Scheme 6.7 Molecular structures of triphenylamine- and carbazole-based HTMs (**16** and **17**).

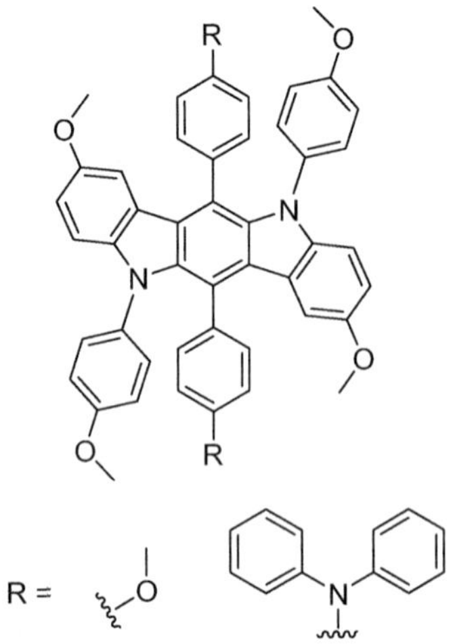

Scheme 6.8 Molecular structures of indolo[3,2-b]carbazole-based HTMs (**18** and **19**).

the molecular geometry and the aggregation behavior can exert a great impact on the film formation capability and thus on the device performance. The device with **19** as an HTM produced a high efficiency, up to 17.7%, which is approximately double that of a device containing **18** (8.7%) under 100 mW/cm^{-2} illumination AM1.5G [68]. Yang and coworkers discovered that the capping layer containing **19** exhibits a more homogeneous and uniform surface morphology as compared to that of **18**; this effectively reduces the charge recombination losses and facilitates charge extraction, leading to a much enhanced PV performance. This study is the first report of ICZ core-based small molecule as dopant-free organic HTM in efficient PSCs. Furthermore, the device consisting of **19** has extraordinary long-term stability under ambient conditions (40% relative humidity) as compared to the standard device with doped spiro-OMeTAD, due to the hydrophobic nature of **19**, which prevents moisture from destroying the perovskite film. This study offers a new avenue for developing cost-effective and stable dopant-free HTMs for efficient PSCs and also other optoelectronic devices [68].

Getautis and coworkers reported a set of novel HTMs based on π-extension through carbazole units via a facile synthetic procedure and applied them successfully in PSCs [16], yielding promising efficiencies of up to almost 18% under standard 100 mWcm^{-2} global AM1.5G and showing better stability, outperforming 2,2′,7,7′-tetrakis-(N,

N-dimethoxyphenylamine)-9,9′-spirobifluorene. This work provides guidance for the molecular design strategy of effective HTMs for perovskite photovoltaics and other similar electronic devices.

Recently, many researchers focused on the design and development of linear (**20**) and starburst (**21**) shaped HTMs as a potential alternative to conventional spiro-OMeTAD in PSCs [69]. Zhou et al. reported that two facile ethane-based HTMs were designed and prepared with a linear or starburst shape shown in Scheme 6.9 [69]. The molecular structure variations in the photophysical, electrochemical, hole mobility, hydrophobicity, film-forming, and photovoltaic properties in PSCs are comprehensively investigated. Interestingly, the linear molecule (**20**) exhibits bathochromic-shift spectra, slightly lower HOMO energy level, higher hole mobility, and better hydrophobicity than its starburst (**21**) counterpart. Moreover, a photoluminescence study confirmed that linear **20** can efficiently extract holes at the perovskite/HTM interface better than its starburst-shaped (**21**) counterpart. Consequently, a higher efficiency of 16.85% for a PSC with a doped linear HTM (**20**) is obtained compared to that of its starburst counterpart (**21**) with a lower efficiency of 14.87%. Furthermore, the linear HTM (**20**) also shows better stability. These results indicate the better application potential of linear HTMs (**20**) than that of the starburst-shaped molecules (**21**) [69].

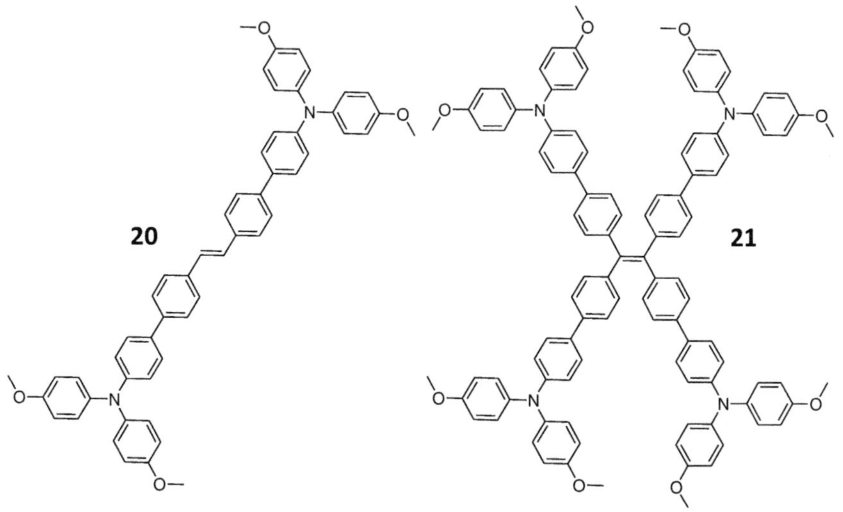

Scheme 6.9 Molecular structure of the investigated linear (**20**) and starburst-shaped (**21**) hole-transporting materials.

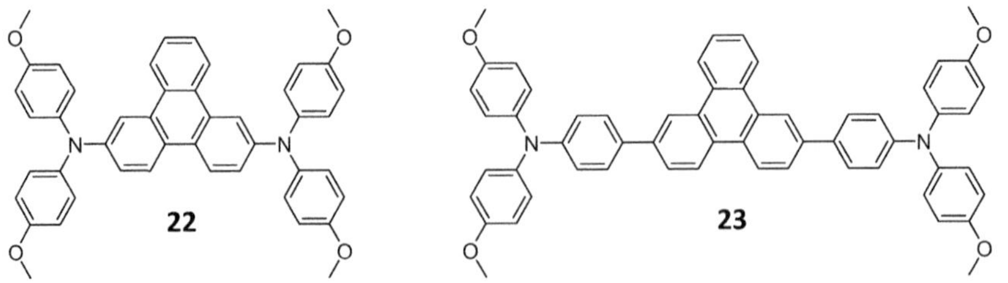

Scheme 6.10 Molecular structures of two hole-transporting materials (**22** and **23**) with a triphenylene core.

Chen et al. designed and developed two organic HTMs, comprised of a two-dimensional triphenylene core and methoxyl-arylamine terminal units (**22** and **23**), see Scheme 6.10, and applied in PSCs [39]. Enhanced photovoltaic and stability performance has been obtained with **23**, with an added benzene ring between the core and terminal units, as compared to PSCSs using a traditional spiro-OMeTAD HTM [39].

6.3.2 Phenothiazine core-based organic hole-transporting materials

Organic HTMs, shown in Scheme 6.11, were synthesized via Schiff base chemistry by functionalizing a phenothiazine (Pz) core with triarylamines integrated with azomethane bridges [42]. Substantial enhancement in the PCE and stability of PSCs were achieved when switching from mono- (**24**) to disubstituted (**25**) HTMs. Use of mono-substituted HTM (**24**) produced a PCE of 12.6%, whereas the disubstituted (**25**) compound resulted in a 14% efficiency; after 60 days, use of **24** and **25** resulted in 68% and 91% PCE (measure of stability), respectively. Also, the extremely low production costs together with the Pd-catalyst-free synthesis make these materials excellent candidates for low-cost and eco-friendly PSCs. Further modification of the phenothiazine core can be conducted via N-substitution, in order to tune the energy levels, improve solubility in green solvents, and enhance the charge carrier mobility, while retaining the low cost of production, hence paving the way toward green, phenothiazine-based PSCs [30] with ever-improving performance.

A series of HTMs was designed and developed based on a phenothiazine core comprising carbazole and triphenylamine as peripheral groups connected via double bonds: **26** (4,4-dimenthyltriphenylamine as a peripheral group), **27** (N-ethylcarbazole as a peripheral group), and **28** (dimethoxytriphenylamine as a peripheral group); see Scheme 6.12. Among these, **28** produced the best results: A PCE of 19.17% with very

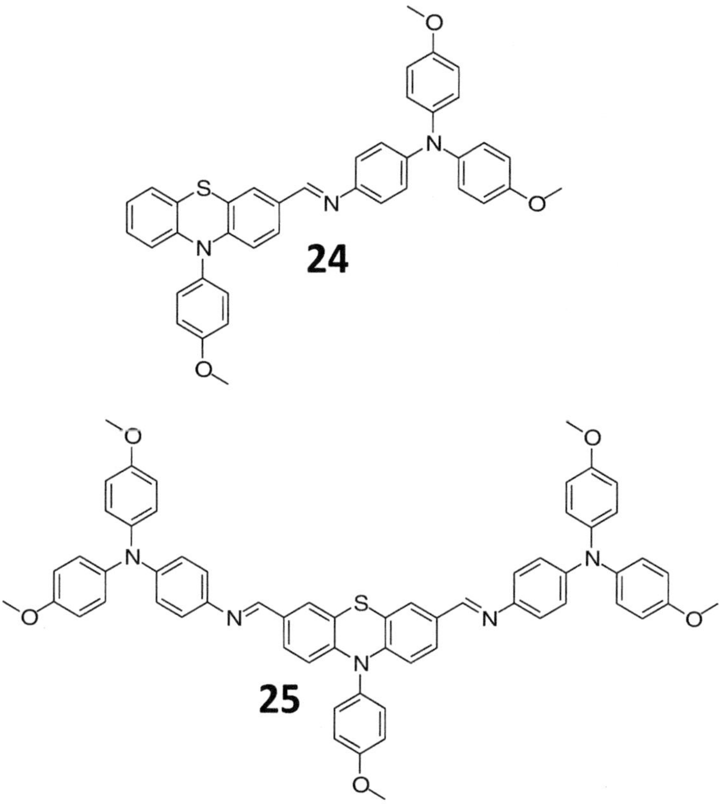

Scheme 6.11 Molecular structures of **24** and **25**, mono- and di-substituted hole-transporting materials.

good stability [40]. These results revealed that the 4,4-dimethoxytriphenylamine is a promising peripheral group in combination with the phenothiazine core, providing another alternative pathway to developing small molecule HTM for more efficient and stable PSCs.

Two triphenylbenzene (TPB) derivatives, 1,3,5-*tris*(2′-((N,N-di(4 methoxyphenyl) amino)phenyl))benzene (TPB(2-MeOTAD)) and 1,3,5-*tris*(2′-(N-phenothiazylo) phenyl)benzene (TPB(2-TPTZ)), have been synthesized via cost-efficient two-step processes. For the first time, phenothiazine had been introduced in PSCs, as a low-cost substituent to replace commonly used dimethoxydiphenylamine, which constitutes almost 90% of the HTM cost [70]. The use of a more flexible central core than a state-of-the-art spirobifluorene lowers the HOMO energy level, increases solubility, and decreases the glass transition temperature. The best devices show an optimized PCE of 12.14% and 4.32% for cells based on 4,4′-dimethoxydiphenylamine and phenothiazine substituents, respectively.

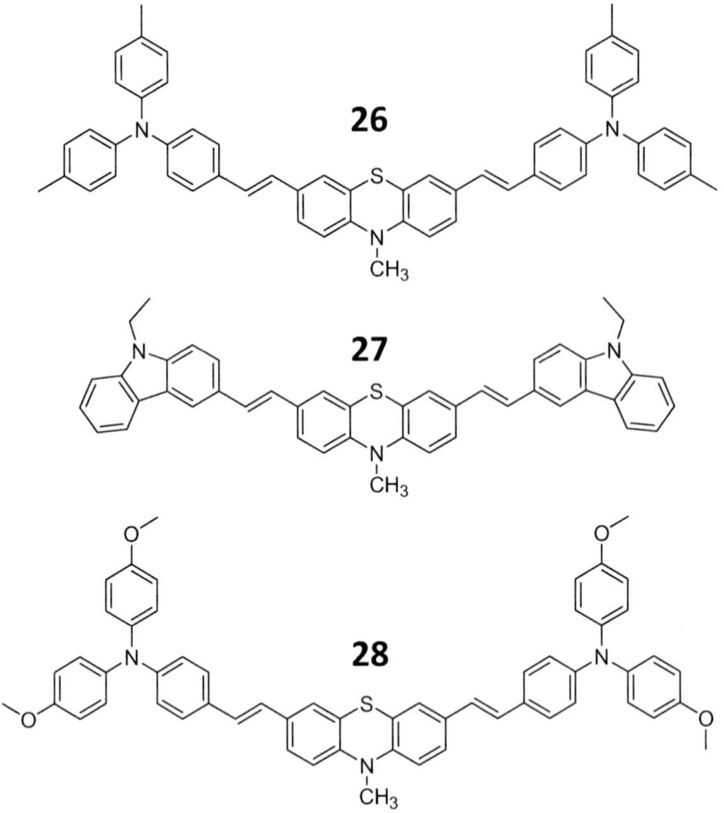

Scheme 6.12 Chemical structure of three promising hole-transporting materials: **26**, **27**, and **28**.

This shows that the approach is commercially viable with potential to deliver HTMs with a cost contribution to the final module of as little as 1%. These results underline the necessity of a deeper insight into designing materials as hole conductors for PSCs taking into account trade-offs between conversion efficiency, scalability, and cost to deliver materials for large-scale production, that is, commercially viable. Thus, it seems likely that this approach will be of broad interest as it is the first work to identify the use of phenothiazine to reduce the cost of hole-transport materials in PSCs.

Phenothiazine is one of the most extensively investigated S, N heterocyclic aromatic hydrocarbons due to its unique optical and electronic properties, flexibility of functionalization, low cost, and commercial availability [71]. Hence, phenothiazine and its derivative materials have been attractive in various optoelectronic applications in the last few years.

Giribabu and coworkers focused on the most significant characteristics of phenothiazine and reported a perspective highlighting how the structural modifications, such as different electron donors or acceptors, length of the π-conjugated system or spacers, polar or nonpolar chains, and other functional groups, influence the optoelectronic properties. They provided a recent account of the advances in phenothiazine derivative materials as active layers for optoelectronic (viz. DSSCs, PSCs, organic solar cells, and organic light-emitting diodes).

6.3.3 Naphthol, thiophene, and pyrene core-based small molecule hole-transporting materials

Chen and coworkers have recently reported the design and facile synthesis of new star-shaped HTMs based on a 1,1′-bi-2-naphthol central scaffold (**29**, **30**, and **31**) illustrated in Scheme 6.13 [72]. This study is the first report

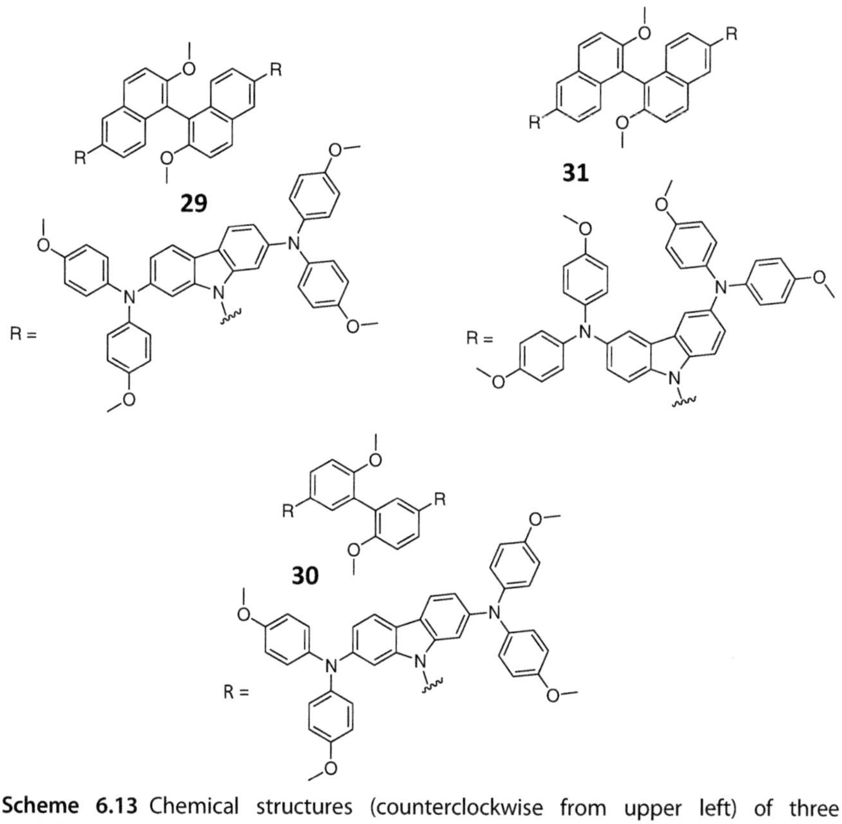

Scheme 6.13 Chemical structures (counterclockwise from upper left) of three naphthol-core hole-transporting materials: **29**, **30**, and **31**; see text for the relative effectiveness for enhancing PSC performance.

on the use of molecules with a 1,1′-bi-2-naphthol central core for efficient PSCs. The 2,7 carbazole *bis*(4-methoxy phenyl) amine and 3,6 carbazole *bis* (4-methoxy phenyl) amine groups were used as branches. The new materials have suitable HOMO levels to match well with the valence band of $CH_3NH_3PbI_3$ and they all possess good glass transition temperatures of greater than 160°C. A device has been constructed with the HTM based on 1,1′-bi-2-naphthol and 2,7 carbazole *bis*(4-methoxy phenyl) amine in combination with the carbon counter electrode yielding high PCE of 8.38%, which is comparable to that of commercial spiro-OMeTAD HTM. The introduction of 2,7 carbazole *bis*(4-methoxy phenyl) amine and 3,6 carbazole *bis*(4-methoxy phenyl) amine groups tends to enhance the performance of the device, in particular J_{sc} and V_{oc}, respectively. $J-V$ characteristics show that PSCs with an HTM using **29** display a higher J_{sc} and the highest PCE (8.38%), greater than **30** (7.08%) and **31** (6.51%). Devices enhanced by **29** are comparable to PSCs fabricated by commercial spiro-OMeTAD (8.73%). This new series of naphthol-core HTMs is yet another pathway to enable economic and efficient PSCs [72].

Three HTMs (**32**, **33**, and **34**) were prepared (structures in Scheme 6.14) via a straightforward synthetic route by cross-linking arylamine-based ligands with a simple thieno- [3,2-b]thiophene (TbT) central core [73]. The synthesized novel HTMs were fully characterized via various analytical techniques to gain insight into their optical and electrochemical properties and were incorporated in solution-processed mesoporous $(FAPbI_3)_{0.85}$ $(MAPbBr_3)_{0.15}$ perovskite-based solar cells. The similar molecular structure of the synthesized HTMs has leveraged to investigate the role that the bridging units between the conjugated TbT central core and the peripheral arylamine units play on their properties and, thereby, on the photovoltaic response. Notably, a remarkable PCE exceeding 18% has been achieved for one of the TbT derivatives, which was slightly higher than the value of the benchmark spiro-OMeTAD HTM. Moreover, extending the central unit by using π-conjugated thiophene bridges results in an improved solubility of the HTMs and, therefore, in better processability from solution, which eventually leads to better surface coverage and film-forming ability.

A novel 4,4′-dimethoxy triphenylamine (donor), thiophene (bridge), and fluorine substituted benzothiadiazole (FBTD, acceptor) based molecule with a D-A-D architecture (**35**), shown in Scheme 6.15, was synthesized via two steps facile route from commercial precursors [74]. The absorption, photoluminescence, thermal, and electrochemical properties of the prepared HTM were examined. The compound possesses a planar

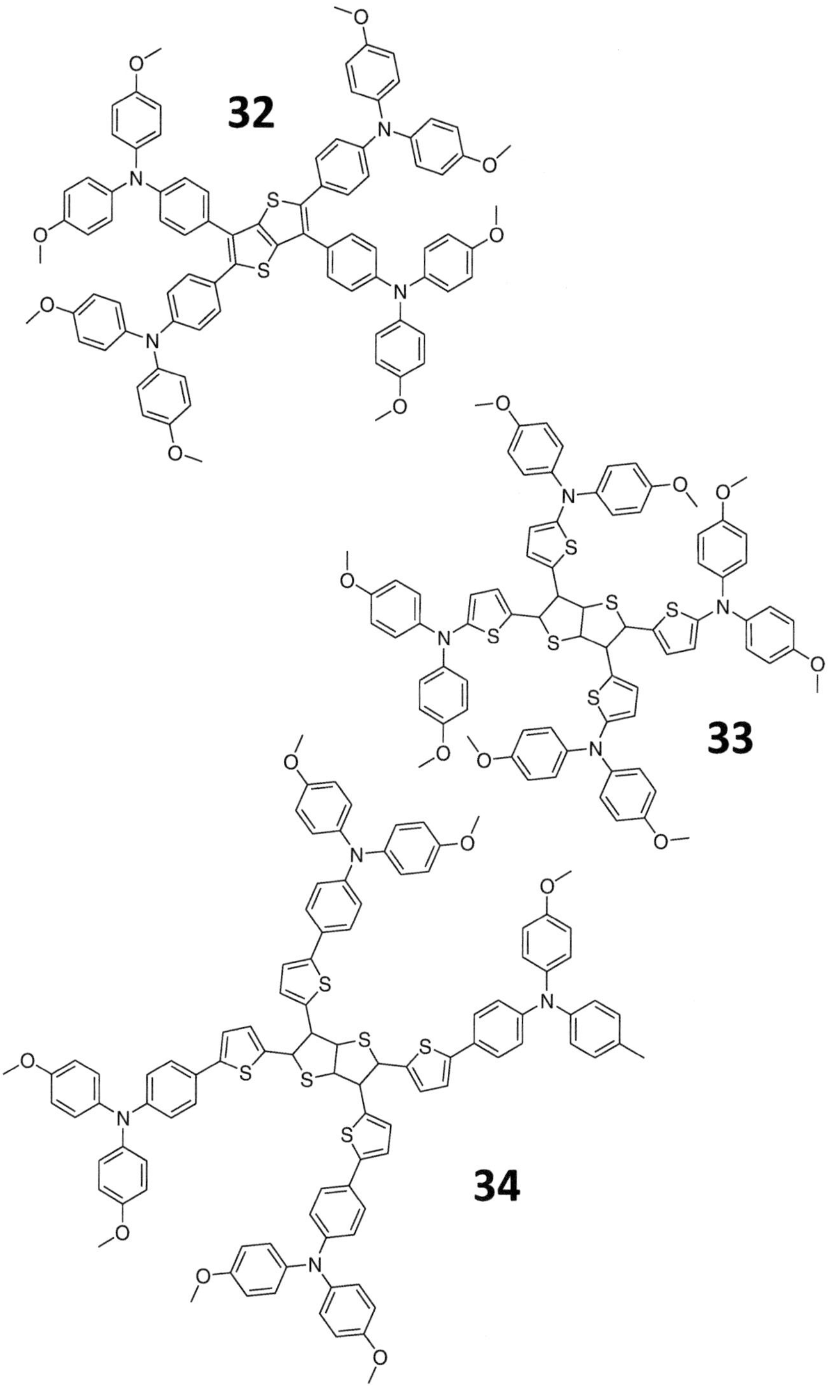

Scheme 6.14 Molecular structures of star-shaped HTMs: **32**, **33**, and **34**.

molecular configuration, suitable energy level, high hole mobility, and excellent thermal stability. When used as dopant-free hole transport material in planar PSCs, the devices offer a high PCE of 18.9% with good ambient stability by maintaining 87% of the initial efficiency after 480 hours degradation. Those results, observed in this present case, enable the readers to further understand the potential of FBTD as a promising building block of HTM for highly efficient PSCs.

A set of three N,N-di-p-methoxyphenylamine-substituted pyrene derivatives (**36**, **37**, and **38**), illustrated in Scheme 6.16, have been synthesized and characterized by $^1H/^{13}C$ NMR spectroscopy, mass spectrometry, and elemental analysis [43]. These pyrene derivatives were employed as HTMs in mesoporous $TiO_2/CH_3NH_3PbI_3$/HTM/Au-based devices. The performance of the materials based on pyrene core is comparable to that of the standard spiro-OMeTAD, even though the V_{oc} is slightly lower. In short, this type of newly synthesized pyrene derivative holds promise as HTMs for highly efficient PSCs.

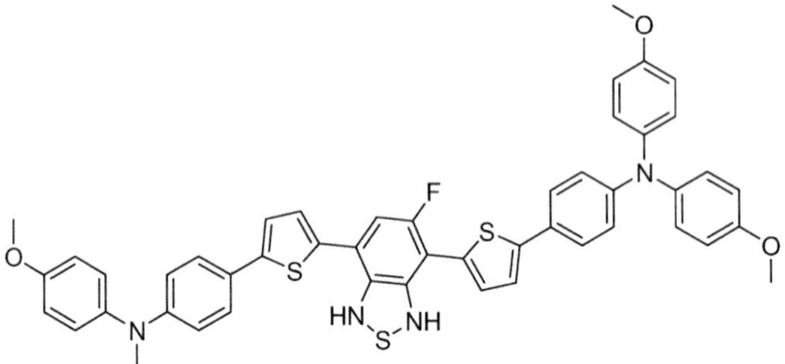

Scheme 6.15 Molecular structure of thiophene bridged hole-transporting materials (**35**).

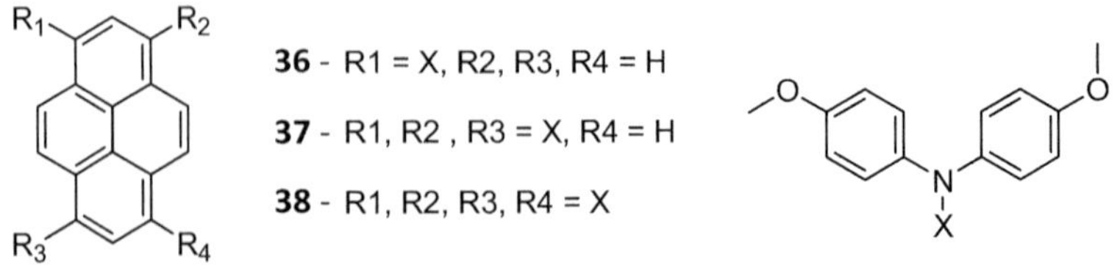

36 - R1 = X, R2, R3, R4 = H

37 - R1, R2 , R3 = X, R4 = H

38 - R1, R2, R3, R4 = X

Scheme 6.16 Chemical structure of three pyrene-based hole-transporting materials (**36**, **37**, and **38**).

6.4 Conclusions and further perspectives

Since the advent of perovskites for energy conversion devices in 2009, tremendous advances have been made in every aspect of PSCs such as design, synthetic route, low cost, eco-friendly, stability, processability, scale-up toward commercialization and high PCE, from fundamental knowledge and material properties to device configurations that enhance stability. PCEs have also rapidly increased, exceeding 25.5%, revolutionizing the photovoltaic field and delivering one of the hottest topics in science. In PSCs, the role of the hole-transporting layer has received considerable attention for enabling higher efficiency.

This crucial role is attributed to spiro-OMeTAD, which is typically the HTM of choice and represents the benchmark for PSCs. However, the use of spiro-OMeTAD has some major drawbacks. Alternatively, a wide variety of chemical structures has been designed with small organic molecules such as carbazole or phenothiazine, which have been investigated as p-type materials, opening new avenues to boost the PCE of the perovskite device. This chapter mainly addressed the progress of novel design strategies, facile synthetic route, scale-up, stability, suitability, and performance of the organic HTMs comprising of carbazole, phenothiazine, triphenylamine, naphthol, thiophene, and pyrene toward the efficient perovskite platform.

Additionally, herein the authors would like to suggest a few perspectives based on the cited literature to interest the readers to consider the following: (i) A structural viewpoint, basic designs of new HTM rely on a variety of different central scaffolds, mainly decorated with arylamine derivatives. Their three-dimensional propeller structures are beneficial for improving solution processability. (ii) Strong electron-donating properties and excellent hole injection render an appropriate band alignment with the valence band edge of the perovskite. (iii) The presence of heteroatoms (i.e., O, S, N, and Si) and their combinations determine their influence in the search for optimized and more stable HTMs at lower costs and with the ultimate goal of producing PSCs at the commercial level. (iv) The D−π−A approach tends to successfully prepare highly efficient HTMs, in which the π-extended conjugation allows face-on organization via strong π−π interactions, affording remarkable charge-transport properties. Also, the strategy was recently applied to increase the hole mobility of D-A molecular HTMs in PSCs by incorporating the fluorine atom. (v) Linear hole-transporting molecule-based device delivered higher PCE compared to that of a starburst counterpart. Furthermore, the linear HTM also shows better stability. (vi) Another chemical approach that could attract a lot of attention is the exploration of new cross-linkable HTMs with various central scaffolds and side arms. (vii) The presence of extended fused benzene rings as substituents on central core, for example, carbazole-based scaffolds, highlights the interest toward the efficient PSCs

In conclusion, the authors have endeavored, in this chapter, to systematically gather the most significant progress carried out on HTMs over the past several years, organized according to the chemical structures. Our goal for the chapter was to be a stimulus for the chemical community working in the ever-expanding area of the design and development of small organic compounds for energy conversion. Finally, we hope to facilitate the discovery of low-cost organic small molecule-based HTMs for large-scale application of PSCs.

Acknowledgments

The authors thank the Indian National Science Academy, New Delhi for INSA Visiting Scientist Fellowship Program 2021 (FY 2021−22; INSA/SP/VSP-25/2021−22; Dt. 31.05.2021) and Department of Science and Technology (DST), Fund for Improvement of S&T Infrastructure (FIST), Government of India for the financial support under DST FIST level "A" PG Colleges (SR/FIST/COLLEGE-/2022/1290, dt. 19.12.2022).

References

[1] B. O'Regan, M. Grätzel, A low-cost, high-efficiency solar cell based on dye-sensitized colloidal TiO_2 films, Nature 353 (1991) 737−740.
[2] M. Grätzel, Recent advances in sensitized mesoscopic solar cells, Acc. Chem. Res. 42 (2009) 1788−1798.
[3] N. Robertson, Optimizing dyes for dye-sensitized solar cells, Angew. Chem. Int. Ed. 45 (2006) 2338−2345.
[4] I. Chung, B. Lee, J. He, R.P.H. Chang, M.G. Kanatzidis, All solid-state dye-sensitized solar cells with high efficiency, Nature 485 (2012) 486−489.
[5] M.K. Nazeeruddin, P. Péchy, M. Grätzel, Efficient panchromatic sensitization of nanocrystalline TiO_2 films by a black dye based on a trithiocyanato−ruthenium complex, Chem. Commun. (1997) 1705−1706.
[6] Y. Hua, S. Chang, J. He, C. Zhang, J. Zhao, et al., Molecular engineering of simple phenothiazine-based dyes To modulate dye aggregation, charge recombination, and dye regeneration in highly efficient dye-sensitized solar cells, Chem. Eur. J. 20 (2014) 6300−6308.
[7] S. Sambathkumar, D. Zych, E. Kavitha, P. Ramesh, R. Jagatheesan, Theoretical investigations on the electronic absorption properties of phenothiazine based organic materials for dye sensitized solar cells, Mater. Today: Proceed 47 (9) (2021) 1937−1941. Available from: https://doi.org/10.1016/j.matpr.2021.03.715.
[8] S. Shalini, R.B. Prabhu, S. Prasanna, T.K. Mallick, S. Senthilarasu, Review on natural dye sensitized solar cells, Renew. Sustain. Energy Rev. 51 (2015) 1306−1325.

[9] A. Hagfeldt, G. Boschloo, L. Sun, L. Kloo, H. Pettersson, Dye-sensitized solar cells, Chem. Rev. 110 (2010) 6595−6663.
[10] T. Chen, W. Hu, J. Song, G.H. Guai, C.M. Li, Interface functionalization of photoelectrodes with graphene for high performance dye-sensitized solar cells, Adv. Funct. Mater. 22 (2012) 5245−5250.
[11] S. Chang, Q. Li, X. Xiao, K.Y. Wong, T. Chen, Enhancement of low energy sunlight harvesting in dye-sensitized solar cells using plasmonic gold nanorods, Energy Environ. Sci. 5 (2012) 9444−9448.
[12] A. Yella, H.W. Lee, H.N. Tsao, C.Y. Yi, A.K. Chandiran, et al., Porphyrin-sensitized solar cells with cobalt (II/III)−based redox electrolyte exceed 12 percent efficiency, Science 334 (2011) 629−634.
[13] S. Zhang, X. Yang, Y. Numata, L. Han, Highly efficient dye-sensitized solar cells: Progress and future challenges, Energy Environ. Sci. 6 (2013) 1443−1464.
[14] C. Chen, J.-Y. Liao, Z. Chi, B. Xu, X. Zhang, et al., Metal-free organic dyes derived from triphenylethylene for dye-sensitized solar cells: Tuning of the performance by phenothiazine and carbazole, J. Mater. Chem. 22 (2012) 8994−9005.
[15] J. Urieta-Mora, L. Zimmerann, J. Arago, A. Molina-Ontoria, E. Orti, N. Martin, et al., Dibenzoquinquethiophene- and dibenzosexithiophene-based hole transporting materials for perovskite solar cells, Chem. Mater. 31 (2019) 6435−6442.
[16] K. Rakstys, S. Paek, A. Drevilkauskaite, H. Kanda, S. Daskeviciute, et al., Carbazole terminated isomeric hole transporting materials for perovskite solar cells, ACS Appl. Mater. Interfaces 12 (2020) 19710−19717.
[17] A. Kojima, K. Teshima, Y. Shirai, T. Miyasaka, Organometalhalide perovskites as visible-light sensitizers for photovoltaic cells, J. Am. Chem. Soc. 131 (17) (2009) 6050−6051.
[18] M.M. Lee, J. Teuscher, T. Miyasaka, T.N. Murakami, H.J. Snaith, Efficient hybrid solar cells based on meso-super structured organometal halide perovskites, Science 338 (2012) 643−647.
[19] H.-S. Kim, C.-R. Lee, J.-H. Im, K.-B. Lee, T. Moehl, et al., Lead iodide perovskite sensitized all-solid-state submicron thin film mesoscopic solar cell with efficiency exceeding 9%, Sci. Rep. 2 (2012) 591.
[20] J.-H. Im, C.-R. Lee, J.-W. Lee, S.-W. Park, N.-G. Park, 6.5% efficient perovskite quantum-dot-sensitized solar cell, Nanoscale (2011) 4088−4093.
[21] A. Jena, A. Kulkarni, T. Miyasaka, Halide perovskite photovoltaics status, and future prospects, Chem. Rev. 119 (2019) 3036−3103.
[22] J. Prakash, A. Singh, G. Sathiyan, R. Ranjan, A. Singh, A. Garg, et al., Progress in tailoring perovskite based solar cells through compositional engineering: Materials properties, photovoltaic performance and critical issues, Mater, Today Energy 9 (2018) 440−486.
[23] U.S. Dept. of Energy, National Renewable Energy Laboratory, Best Research Cell Efficiency Chart. <https://www.nrel.gov/pv/cell-efficiency.html> (accessed 11.01.23).
[24] S. Khanna, S. Senthilarasu, K.S. Reddy, T.K. Malick, Performance analysis of perovskite and dye-sensitized solar cells under varying operating conditions and comparison with monocrystaline cell, Appl. Therm. Eng. 127 (2017) 559−565.
[25] G. Sathiyan, J. Prakash, R. Ranjan, A. Singh, A. Garg, R.K. Gupta, Recent progress on hole-transporting materials for perovskite-sensitized solar cells, in: B.A. Bhanvase, V.B. Pawade, S.J. Dhoble, S.H. Sonawane, M. Ashokkumar (Eds.), Nanomaterials for Green Energy, Elsevier, 2018, pp. 279−324.
[26] P. Vivo, J.K. Salunke, A. Priimagi, Hole-transporting materials for printable perovskite solar cells, Materials 10 (9) (2017) 1087.
[27] W. Tress, Perovskite solar cells on the way to their radiative efficiency limit - insights into a success story of high open-circuit voltage and low recombination, Adv. Energy Mater. 7 (14) (2017) 1602358.

[28] Z.H. Bakr, Q. Wali, A. Fakharuddin, L. Schmidt-Mende, T.M. Brown, R. Jose, Advances in hole transport materials engineering for stable and efficient perovskite solar cells, Nano Energy 34 (2017) 271–305.
[29] P. Yan, D. Yang, H. Wang, S. Yang, Z. Ge, Recent advances in dopant-free organic hole-transporting materials for efficient, stable and low-cost perovskite solar cells, Energy. Env. Sci. 15 (2022) 3630–3669.
[30] S. Thokala, S.P. Singh, Phenothiazine based hole transporting materials for perovskite solar cells, ACS Omega 5 (11) (2020) 5608–5619.
[31] S. Benhattab, A.-N. Cho, R. Nakar, N. Berton, F. Tran-Van, N.G. Park, et al., Simply designed carbazole-based hole transporting materials for efficient perovskite solar cells, Org. Electron. 56 (2018) 27–30.
[32] M. Saliba, S. Orlandi, T. Matsui, S. Aghazada, M. Cavazzini, et al., A molecularly engineered hole-transporting material for efficient perovskite solar cells, Nat. Energy 1 (2016) 15017.
[33] L. Yang, B. Xu, D. Bi, H. Tian, G. Boschloo, L. Sun, et al., Initial light soaking treatment enables hole transport material to outperform spiro-OMeTAD in solid-state dye-sensitized solar cells, J. Am. Chem. Soc. 135 (2013) 7378–7385.
[34] U.B. Cappel, T. Daeneke, U. Bach, Oxygen-induced doping of spiro-MeOTAD in solid-state dye-sensitized solar cells and its impact on device performance, Nano Lett. 12 (2012) 4925–4931.
[35] K. Rakstys, C. Igci, M.K. Nazeeruddin, Efficiency vs. stability: Dopant-free hole transporting materials towards stabilized perovskite solar cells, Chem. Sci. 10 (2019) 6748–6769.
[36] F. Zhang, X. Zhao, C. Yi, D. Bi, X. Bi, et al., Dopant-free star-shaped hole-transport materials for efficient and stable perovskite solar cells, Dye. Pigment. 136 (2017) 273–277.
[37] R. Nakar, F.J. Ramos, C. Dalinot, P.S. Marques, C. Cabanetos, et al., Cyclopentadithiophene and fluorene spiro-core based hole-transporting materials for perovskite solar cells, J. Phys. Chem. C. 123 (2019) 22767–22774.
[38] X.-J. Ma, X.-D. Zhu, K.-L. Wang, F. Igbari, Y. Yuan, et al., Planar starburst hole-transporting materials for highly efficient perovskite solar cells, Nano Energy 63 (2019) 103865.
[39] W. Chen, H. Zhang, H. Zheng, H. Li, F. Guo, et al., Two-dimensional triphenylene cored hole-transporting materials for efficient perovskite solar cells, Chem. Commun. 56 (2020) 1879–1882.
[40] F. Zhang, S. Wang, H. Zhu, X. Liu, H. Liu, et al., Impact of peripheral groups on phenothiazine-based hole-transporting materials for perovskite solar cells, ACS Energy Lett. 3 (2018) 1145–1152.
[41] X. Liu, F. Kong, R. Ghadari, S. Jin, W. Chen, et al., Thiophene-arylamine hole transporting materials in perovskite solar cells: Substitution position effect, Energy Technol. 5 (2017) 1788–1794.
[42] J. Salunke, X. Guo, Z. Lin, J.R. Vale, N.R. Candeias, et al., Phenothiazine-based hole-transporting materials toward ecofriendly perovskite solar cells, ACS Appl. Energy Mater. 2 (2019) 3021–3027.
[43] N.J. Jeon, J. Lee, J.H. Noh, M.K. Nazeeruddin, M. Grätzel, S.I. Seok, Efficient inorganic-organic hybrid perovskite solar cells based on pyrene arylamine derivatives as hole transporting materials, J. Am. Chem. Soc. 135 (2013) 19087–19090.
[44] H. Choi, K. Do, S. Park, J.S. Yu, J. Ko, Efficient holetransporting materials with two or four N, N-di (4-methoxyphenyl)aminophenyl arms on an ethene unit for perovskite solar cells, Chem.-Eur. J. 21 (2015) 15919–15923.
[45] W. Ke, P. Priyanka, S. Vegiraju, C.C. Stoumpos, L. Spanopoulos, et al., Dopant-free tetrakis-triphenylamine hole transporting materialfor efficient tin-based perovskite solar cells, J. Am. Chem. Soc. 140 (2018) 388–393.

[46] X. Ren, Z. Wang, W.E.I. Sha, W.C.H. Choy, Exploring the way to approach the efficiency limit of perovskite solar cells by drift-diffusion model, ACS Photonics 4 (2017) 934–942.
[47] S.M. Park, S.M. Mazza, Z. Liang, A. Abtahi, A.M. Boehm, S.R. Parkin, et al., Processing dependent influence of the hole transport layer ionization energy on methyl ammonium lead iodide perovskite photovoltaics, ACS Appl. Mater. Interfaces 10 (2018) 15548–15557.
[48] R.A. Belisle, P. Jain, R. Prasanna, T. Leijtens, M.D. McGehee, Minimal effect of the hole-transport material ionization potential on the open-circuit voltage of perovskite solar Cells, ACS Energy Lett. 1 (2016) 556–560.
[49] L.E. Polander, P. Pahner, M. Schwarze, M. Saalfrank, C. Koerner, K. Leo, Hole-transport material variation in fully vacuum deposited perovskite solar cells, APL. Mater. 2 (2014) 081503.
[50] M. Neophytou, J. Griffiths, J. Fraser, M. Kirkus, H. Chen, C.B. Nielsen, et al., High mobility, hole transport materials for highly efficient PEDOT:PSS replacement in inverted perovskite solar cells, J. Mater. C. 5 (2017) 4940–4945.
[51] L. Zhao, R.A. Kerner, Z. Xiao, Y.L. Lin, K.M. Lee, J. Schwartz, et al., Redox chemistry dominates the degradation and decomposition of metal halide perovskite optoelectronic devices, ACS Energy Lett. 1 (2016) 595–602.
[52] E.M. Sanehira, B.J. Tremolet de Villers, P. Schulz, M.O. Reese, S. Ferrere, et al., Pure Cs_4PbBr_6: Highly luminescent zero-dimensional perovskite solids, ACS Energy Lett. 1 (2016) 840–845.
[53] G. Nazir, S.-Y. Lee, J.-H. Lee, A. Rehman, J.-K. Lee, S.I. Seok, et al., Stabilization of perovskite solar cells: Recent developments and future perspectives, Adv. Mater. 34 (50) (2022) 2204380.
[54] J.-P. Correa-Baena, W. Tress, K. Domanski, E.H. Anaraki, S.-H. Turren-Cruz, et al., Identifying and suppressing interfacial recombination to achieve high open-circuit voltage in perovskite solar cells, Energy Environ. Sci. 10 (2017) 1207–1212.
[55] Y. Shao, Y. Yuan, J. Huang, Correlation of energy disorder and open-circuit voltage in hybrid perovskite solar cells, Nat. Energy (2016) 15001.
[56] J.H. Yun, S. Park, J.H. Heo, H.S. Lee, S. Yoon, J. Kang, et al., Enhancement of charge transport properties of small molecule semiconductors by controlling fluorine substitution and effects on photovoltaic properties of organic solar cells and perovskite solar cells, Chem. Sci. 7 (2016) 6649–6661.
[57] H. Chen, D. Bryant, J. Troughton, M. Kirkus, M. Neophytou, X. Miao, et al., One-step facile synthesis of a simple hole transport material for efficient perovskite solar cells, Chem. Mater. 28 (2016) 2515–2518.
[58] X. Liu, X. Ding, Y. Ren, Y. Yang, Y. Ding, et al., A star-shaped carbazole-based hole-transporting material with triphenylamine side arms for perovskite solar cells, J. Mater. Chem. C. 6 (2018) 12912–12918.
[59] A. Mogomedov, S. Peak, P. Gratia, E. Kasaparavicius, M. Daskeviciene, et al., Diphenylamine-substituted carbazole-based hole transporting materials for perovskite solar cells: Influence of isomeric derivatives, Adv. Funct. Mater. 28 (2018) 1704351.
[60] X. Yin, L. Guan, J. Yu, D. Zhao, C. Wang, et al., One-step facile synthesis of a simple carbazole-cored hole transport material for high-performance perovskite solar cells, Nano Energy 40 (2017) 163–169.
[61] Y. Wang, X. Chen, J. Shao, S. Pang, X. Xing, et al., Influence of dimethoxytriphenylamine groups on carbazole-based hole transporting materials for perovskite solar cells, Sol. Energy190 (2019) 361–366.
[62] H.D. Pham, L. Gil-Escrig, K. Feron, S. Manzhos, S. Albrecht, H.J. Bolink, et al., Boosting inverted perovskite solar cell performance by using 9,9-bis (4-diphenylamino-phenyl) fluorene functionalized with triphenylamine as a dopant-free hole transporting material, J. Mater. Chem. A 7 (2019) 12507–12517.

[63] M. Li, S. Ma, M. Mateen, X. Liu, Y. Ding, et al., Facile donor (D)-π-D triphenylamine-based hole transporting materials with different π-linker for perovskite solar cells, Sol. Energy 195 (2020) 618–625.
[64] F. Liu, F. Wu, W. Ling, Z. Tu, J. Zhang, et al., Facile-effective hole-transporting materials based on dibenzo[a,c]carbazole: The key role of linkage position to photovoltaic performance of perovskite solar cells, ACS Energy Lett. 4 (2019) 2514–2521.
[65] J.-Y. Shao, D. Li, K. Tang, W.-W. Zhong, Q. Meng, Simple biphenyl or carbazole derivatives with four di(anisyl)amino substituents as efficient hole-transporting materials for perovskite solar cells, RSC Adv. 6 (2016) 92213–92217.
[66] F. Al-Zohbi, Y. Jouane, S. Benhattab, J. Faure-Vincent, F. Tran-Van, et al., Simple carbazole-based hole transporting materials with fused benzene rings substituents for efficient perovskite solar cell, New. J. Chem. 43 (2019) 12211–12214.
[67] D. Li, J.-Y. Shao, Y. Li, Y. Li, L.-Y. Deng, W.-W. Zhong, et al., New hole transporting materials for planar perovskite solar cells, Chem. Commun. 54 (2018) 1651–1654.
[68] B. Cai, X. Yang, X. Jiang, Z. Yu, A. Hagfeldt, L. Sun, Boosting the power conversion efficiency of perovskite solar cells to 17.7% with an indolo[3,2-b]carbazole dopant-free hole transporting material by improving its spatial configuration, J. Mater. Chem. A 7 (2019) 14835–14841.
[69] Z. Zhou, X. Zhang, S. Ma, C. Liu, X. Liu, et al., Comparative study of linear and starburst ethane-based hole transporting materials for perovskite solar cells, J. Phys. Chem. C. 124 (2020) 2886–2894.
[70] M. Maciejczyk, R. Chen, A. Brown, N. Zheng, N. Robertson, Beyond efficiency: Phenothiazine, a new commercially viable substituent for hole transport materials in perovskite solar cells, J. Mater. Chem. C. 7 (2019) 8593–8598.
[71] P.S. Gangadhar, G. Reddy, S. Prasanthkumar, L. Giribabu, Phenothiazine functional materials for organic optoelectronic applications, Phys. Chem. Chem. Phys. 23 (2021) 14969–14996.
[72] W. Qiao, Y. Chen, F. Li, X. Zong, Z. Sun, M. Liang, et al., Novel efficient hole-transporting materials based on a 1,1′-bi-2-naphthol core for perovskite solar cells, RSC Adv. 7 (2017) 482–492.
[73] J. Urieta-Mora, I. Garcia-Benito, L. Zimmerann, J. Arago, A. Molina-Ontoria, E. Orti, et al., Tetrasubstituted thieno[3,2-b]thiophenes as hole-transporting materials for perovskite solar cells, J. Org. Chem. 85 (2020) 224–233.
[74] Y. Tian, L. Tao, C. Chen, H. Lu, H. Li, X. Yang, et al., Facile synthesized fluorine substituted benzothiadiazole based dopant-free hole transport material for high efficiency perovskite solar cells, Dye. Pigment. 184 (2021) 108786.

SECTION III

Alternative photovoltaic materials

CHAPTER SEVEN

Advanced fabrication strategies to enhance the performance of dye-sensitized solar cells

Anurag Roy[1], Tapas K. Mallick[1], Asif Ali Tahir[1], Mohammad Ashraf Gondal[2,3] and Senthilarasu Sundaram[4]

[1]Solar Energy Research Group, Environment and Sustainability Institute, University of Exeter, Penryn Campus, Cornwall, United Kingdom
[2]Laser Research Group, Physics Department, King Fahd University of Petroleum and Minerals (KFUPM), Dhahran, Saudi Arabia
[3]K.A. CARE Energy Research and Innovation Center, King Fahd University of Petroleum and Minerals (KFUPM), Dhahran, Saudi Arabia
[4]School of Computing, Engineering and Digital Technologies, Teesside University, Tees Valley, Middlesbrough, United Kingdom

7.1 Introduction

Continuous energy supply is a significant requirement to enable a technologically advanced lifestyle. The ever-increasing insistence on energy depends on naturally stored traditional energy sources (nonrenewable) such as oil, petroleum, natural gas, and coal [1]. However, the availability of those sources is limited and thus creates scarcity for this futuristic user demand-supply. In this regard, researchers are concentrating on various renewable energy sources: Sunlight, wind, geothermal, biogas, etc., which can secure an unlimited supply of sources and surpass environmental pollution-related issues. The imminent depletion has spurred the advancement of renewable energy technologies. European Union has targeted achieving at least 27% of energy originating from renewable energy sources by 2030 [2]. The amount of energy from the sun that hits the Earth's surface every year is 10,000 times bigger than the overall energy consumption. In other words, the sun delivers 120 PJ of energy to the Earth every second. In 1 hour, the sun gives more energy to the Earth than the energy consumed by humans in an entire year [3–5]. This natural phenomenon deserves a crucial strategy for renewable energy production. Thus, photovoltaic (PV) technology exhibits a potential alternative energy resource where the direct conversion of sunlight into high-quality electricity energy is highly desirable.

Silicon (Si) solar cells play a leading role in commercially accessing the current PV technology. However, silicon-based solar cells are restricted to the terrestrial PV market due to their relatively high production and environmental costs as well as energy utilization and chemical by-products, respectively. Unlike high-cost conventional silicon solar cells, dye-sensitized solar cells (DSSCs) are cost-effective PV devices because of their inexpensive materials and simple fabrication processes. DSSC can be assumed to be a mesoscopic heterojunction solar cell or a mesoscopic electron injection solar cell with significant advantages over other PV technologies regarding cost-effectiveness, shorter energy payback time, performance under diffuse light conditions, and novel architectures. Considering several important DSSC device high-performance output parameters, it is necessary to understand the fundamental strategies to enhance its noticeable performance. A comprehensive overview of the DSSC covering the fundamental principles followed by possible ways of performance improvement area involved with the essential components has been demonstrated in this chapter. The basic strategies include various aspects of physics, chemistry, and additional materials science, where a combination of all these fields can be found.

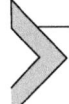

7.2 Classification of solar cells: Materials and technologies

The first, second, and third generations PV devices have undergone significant evolution across three generations: the first, second, and third generations of solar cells. This classification has been established based on considerations of the device's working principle, fabrication protocols, and cost analysis. Fundamental requirements for the operation of a PV cell include the absorption of a broad spectrum of solar radiation, the generation of electron-hole pairs (and, in some cases, excitons), the separation of these generated charge carriers (sometimes involving exciton ionization and carrier separation), and the extraction of these carriers to an external circuit. Depending on factors such as the absorbing material utilized, the manufacturing techniques or processes employed, and the type of junction formed, solar cell technologies can be broadly categorized as outlined in Fig. 7.1. The first-generation solar cells typically consist of p–n junctions led by silicon. However, the

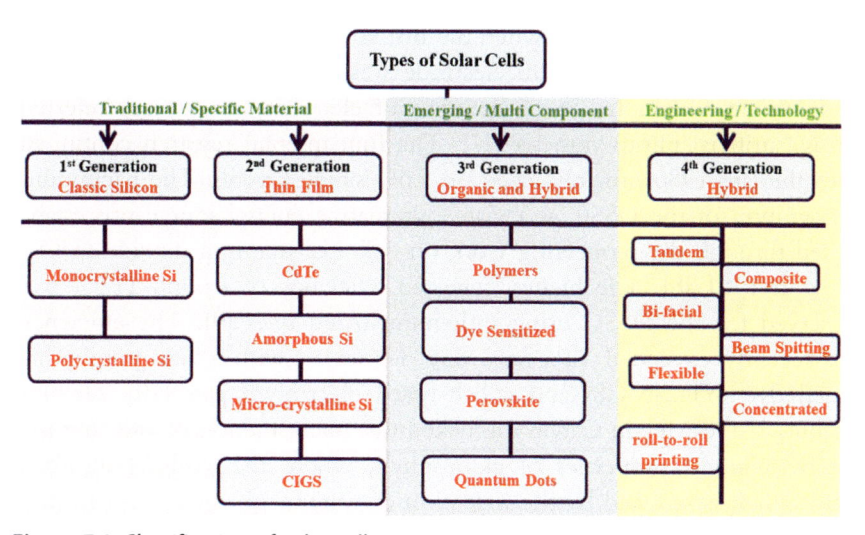

Figure 7.1 Classification of solar cells.

fabrication process for silicon solar cells is complex, leading to high processing costs and increased manufacturing throughput. Consequently, as second-generation solar cells, cadmium telluride (CdTe) and copper indium gallium selenide (CIGS) were developed, offering comparatively lower costs and simpler fabrication processes as potential replacements for silicon solar cells. However, second-generation cells face challenges such as production scarcity and toxicity, hindering their commercialization. In this context, solution-processed solar technologies, such as third-generation solar cells, become appealing options for reducing costs while maintaining high power conversion efficiencies (PCEs) from solar energy [6]. growing immensely and has furthermore, they offer the additional benefit of facilitating the construction of large surface areas and flexible structures through roll-to-roll coating or printing techniques. These technologies encompass DSSCs, organic PV, solution-processed bulk inorganic PV, and colloidal quantum dot (QD) solar cells. While many third-generation devices have yet to achieve widespread commercial implementation, research in this field is rapidly expanding, indicating a promising future.

7.3 A brief history of dye-sensitized solar cells

The capability of organic dyes to generate electricity has been well-known since the late 1960s. The history of DSSC technology, on the other hand, started in 1972 when the first attempt with a chlorophyll sensitized zinc oxide (ZnO) film electrode was taken by Helmut Tributsch and Melvin Calvin for the generation of electricity, a concept referred to as "artificial photosynthesis" [7]. The fundamental research continued into the late 1980s, mainly based on ZnO single crystals. The paramount recognition of the DSSC as a viable alternative energy source was established through the pioneering work on colloidal titanium dioxide (TiO_2) films in 1991 [8]. The highest reported efficiency of around 11.5% was observed for the DSSC using ruthenium-based dyes [9]. The efficiency has recently exceeded 12% for DSSC fabricated with a zinc porphyrin-based dye (YD2-o-C8) and cobalt-based electrolyte pair [10]. Helmut Tributsch and Melvin Calvin mimicked this phenomenon by utilizing the electrochemical properties of chlorophyll, where the sunlight-absorbed chlorophyll is reduced by the oxidation of water with subsequent formation of molecular oxygen (O_2) and protons (H^+) and further converted to chemical energy. Similarly, DSSCs with a different cellular environment for electricity generation from sunlight brought the DSSC device.

7.4 Dye-sensitized solar cells: Working principles

DSSCs have received widespread attention in recent years due to their specific advantages concerning other conventional PV devices owing to their low production cost, shorter energy payback time, compatibility with flexible substrates, feedstock availability to reach terawatt scale, superior performance under diffuse light conditions, novel architectures, and most importantly the environmentally benign nature [11,12]. The fundamental advantage of DSSCs lies in greater solar-to-electricity conversion efficiencies with lower production costs, flexibility, and ecofriendliness. The fundamental mechanism of photoelectrochemical technology closely resembles the natural photosynthesis process. Like chlorophyll in plants, a monolayer of dye molecules (sensitizer) absorbs the incident light, generating positive and negative charge carriers in the cell. Therefore, a DSSC provides a technically and economically credible alternative to present-day $p-n$ junction PV devices. Nanostructured semiconductor films are the framework of the photoanodes in DSSC. The photoanode serves dual functions to support sensitizer loading and transport photoexcited electrons from the sensitizer to the external circuit. In contrast to the conventional systems, the semiconductor assumes light absorption and charge carrier transport.

A DSSC is, in essence, a photochemical cell with two electrodes and an electrolyte that generates an electrical current via redox reactions [13]. The cell's working electrode or photoanode comprises a nanostructured, porous semiconducting oxide layer deposited on a transparent conducting oxide (TCO) glass or plastic substrate. The glass coated with fluorine-doped tin oxide (FTO) or indium-doped tin oxide is commonly used. A light-sensitive charge transfer dye monolayer has been attached to the semiconducting oxide film. Similarly, the electrically conducting glass substrate coated with a thin layer of platinum catalyst acts as the counter electrode (CE). This has been placed parallel with a face-to-face configuration of the working electrode. The interspacing has been filled with the liquid or solid electrolyte that plays the role of conducting media. On light irradiation, the dye molecules absorb photons of a wavelength corresponding to the energy difference between its highest occupied molecular orbital (HOMO) and lowest unoccupied molecular orbital (LUMO). Electrons from the ground electronic state of the dye get promoted to the excited state, that is photoexcitation of dye followed by the electron injection to the conduction band (CB) of the semiconducting n-type metal oxide. After being injected into the CB of the n-type oxide, the electrons are transported through the semiconductor layer by diffusion to reach the transparent conducting layer on a glass substrate (TCO). The iodide ion (I^-) then donates an electron to the oxidized dye and regenerates the dye molecules at the anode. The oxidized species of electrolyte, that is, triiodide (I_3^-) in the iodide triiodide (I_2/I_3^-) complex, is reduced to an iodide cathode. The above processes go in a cycle; as a consequence, current flow continues through the external circuit as long as the light is incident on the cell [14]. Fig. 7.2 provides a schematic representation of various steps occurring in a DSSC with the help of an energy level diagram. The primary pathways for electron transfer processes in a DSSC device are shown in Fig. 7.2.

The operating cycle can be summarized as discussed previously [5,13]. Sunlight absorption by the dye sensitizer followed by excitation (7.1):

$$S + h\nu \rightarrow S^* \rightarrow S^+ + e^- \tag{7.1}$$

The electrons (e^-) are injected into the CB of the n-type semiconducting oxide, where they are transported between oxide with diffusion toward the back contact (TCO) and external circuit. The oxidized photosensitizer (S^+) accepts electrons from the I^- ion redox mediator, leading to

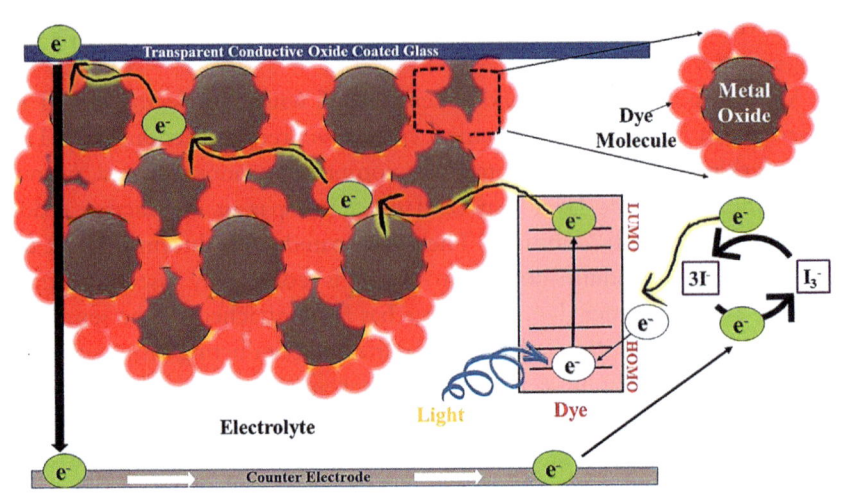

Figure 7.2 Schematic of a dye-sensitized solar cell device and corresponding operational principle governed by photoelectrochemical redox process.

the regeneration of the ground state (S) (7.2). Iodide ions (I^-) are oxidized to elemental iodine that reacts with I^- to form the redox mediator, I_3^-,

$$S^+ + e^- \rightarrow S \tag{7.2}$$

Upon formation, oxidized I_3^- diffuses toward the CE and is reduced to I^- ions (7.3).

$$I_3^- + 2e^- \rightarrow 3I^- \; I_3^- + 2e^- \rightarrow 3I^- \tag{7.3}$$

The redox mediator of the electrolyte transports between the working and CEs following a diffusion process. The diffusion coefficient crucially depends on the conductivity, ionic strength, concentration of the redox mediator, and the effective distance between the two electrodes. The electron is transferred to the redox couple faster than the recombination at the semiconductor/electrolyte interface at the CE. The reduction reaction is further catalyzed by the thin layer of Pt on the CE and completes the overall regenerative cycle.

7.5 Solar irradiation

A characteristic spectrum with intensity peaks in the visible and infrared (IR) range is noticeable once incident sunlight hits the Earth's surface. However, it does not have uniform irradiance across wavelengths; hence, specific wavelengths have poor irradiance even though a solar device may be able to absorb them. The American Society for Testing and Materials has set a global reference standard for solar spectral distribution. It provides tables for spectral irradiance distributions used to compare the performances of products and technologies [15]. The total integrated irradiance for the so-called hemispherical tilted spectrum is 1000.4 W/m^2 (Fig. 7.3A). In effect, the attenuation of sunlight increases with the distance photons have to travel through the atmosphere because of photon absorption and scattering. The ratio of a solar photon's path length through the atmosphere divided by the thickness d of the atmosphere is defined as the air mass (AM) coefficient. AM0 corresponds to a solar radiation spectrum above the Earth's atmosphere, and AM1.5G designates the total global hemispherical irradiance commonly used when measuring solar cells. This aspect can be best illustrated by Fig. 7.3. A $J-V$ measurement is carried out under natural sunlight in the outdoor environment or a closed laboratory environment with the help of a solar simulator.

To evaluate the parameters of a DSSC device rated under the so-called Standard Test Conditions corresponds to an irradiance of 1000 W/m^2, an AM1.5 spectrum, and a device temperature of 25°C [16]. Generally, the sun is directly overhead (at zenith), which is considered for the path length is 1.0 (AM1.0). The sun's angle from the zenith increases by secθ, changing

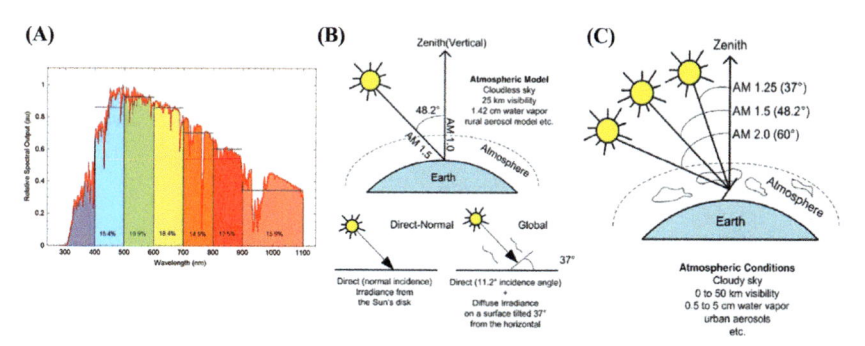

Figure 7.3 (A) Solar radiation spectrum for direct light at the top of the atmosphere (AM0) and sea level (AM1.5G), (B) schematic of AM1.5 reference spectral conditions, and (C) schematic representation of other realistic atmospheric and air mass conditions compared to AM1.5. *Reprinted with permission from V. Esen, S. Sağlam, B. Oral, Light sources of solar simulators for photovoltaic devices: A review, Renew. Sust. Energ. Rev. 77 (2017) 1240–1250, copyright (2017) Elsevier.*

the AM as well. For example, about 48 degrees from the vertical, the AM is 1.5, and when the sun is at 60 degrees, the AM is reflected 2.0. These AM values are shown in Figs. 7.3B and 7.3C.

7.6 Concentrated light irradiation on dye-sensitized solar cells

Concentrating sunlight is another effective approach to increasing DSSC devices' efficiency by concentrating incoming sunlight and focusing it on the solar cell. Optical lens-based solar concentrator systems mounted on top of DSSCs still pose several challenges in terms of efficiency, the cost-effectiveness of optical design, and the provision of uniform and concentrated illumination on a DSSC [17]. Light conditioning techniques such as concentrating photovoltaic (CPV) can address the scalability problem. Choi et al. used a condenser lens for a vertically stacked cell, with an 8 mm separation distance between the lens and the cell, and the device efficiency increased from 2.5% to 8.3% [18]. Barber et al. proposed a concentrator for a hybrid silicon-DSSC system with two different optical filters for visible and IR absorption to achieve about 20% efficiency [19]. Selvaraj et al. observed a 67% PCE increment and maximum J_{sc} of 25.55 mA/cm^2 for the concentrator-coupled DSSC device under 1000 W/m^2; the coupled devices offered >50% efficiency enhancement using low-concentrator PV compared to the bare DSSCs [20]. Interestingly, the CPV system was applied in indoor and outdoor working conditions for a solar cell device. The concentrator factor rose to 1.5 for the outdoor result, exhibiting a linear behavior [21]. Overall, CPV integration provides valuable insights into solving the scaling-up problem of DSSCs by simplifying the concentrated sunlight into a small area, generating concentrated sun intensity, a critical light management strategy of a DSSC device.

7.7 Strategies to improve interfacial electron kinetics

The light-harvesting efficiency of the dye governs the short-circuit photocurrent density (J_{sc}), its capacity to inject electrons into the CB of the metal oxide photoanode, and the ability of the semiconductor to transport them to the collecting electrode. At the same time, the open-circuit voltage (V_{oc}) originates from the difference between the Fermi level of electrons in the metal oxide and the redox potential of the corresponding electrolyte. The fill factor (FF) is controlled mainly by the total series resistance of the device, although it also contributes to recombination resistance [22]. Swift electron transfer and charge transport processes are the fundamental working parameters of a DSSC to achieve a high PCE device. Therefore, the fabrication strategy associated with a DSSC device mainly corresponds to this feature only. It is challenging to maintain continuous electron transport and regeneration throughout the device because a DSSC device experiences several loss pathways, shown as black arrows in Fig. 7.4. The dye excited state to ground and charge recombination of injected electrons are two significant loss pathways, including decay of the dye cations and the redox couple.

Fig. 7.4 illustrates a photochemical view of the function of DSSCs, depicting the sequence of electron transfer and charge transport processes that result in PV device function. It is significant that to improve the lifetime of the excited electron, the charge-separated state must increase the spatial separation of electrons and holes [23,24]. The spectral response of the solar cell to the incoming light in terms of current is called incident photon to current efficiency (IPCE). It is defined as the ratio of incident photons to the number of charge carriers generated and the excitation wavelength function to ascertain the photocurrent's sensitization by the dyes under investigation [25]. The efficiency was calculated at each excitation wavelength, resulting in a sigmoidal curve from the values of short-circuit

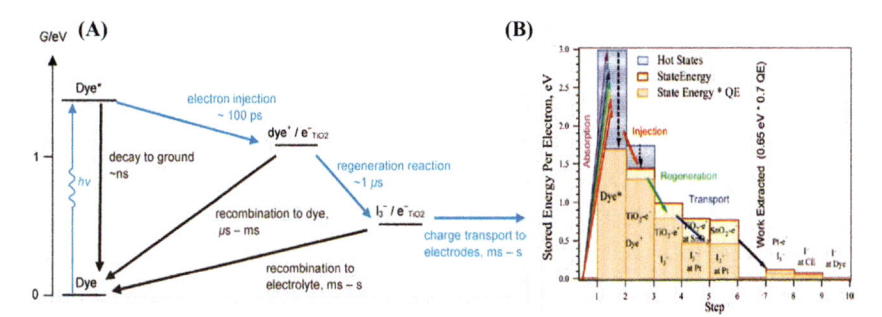

Figure 7.4 State diagram representation of the kinetics of the dye-sensitized solar cell function. *Adapted with permission from A. Listorti, B. O'Regan, J.R. Durrant, Electron transfer dynamics in dye-sensitized solar cells, Chem. Mater. 23 (2011) 3381−3399, copyright (2011) American Chemical Society.*

current density and the intensity of the corresponding monochromatic light. Therefore, generating high IPCE is another crucial factor in retaining electrons' continuous flow instead of enhancing the excitation rate. Besides, the fundamental kinetics of electron injection in DSSCs also depend on the magnitude of these injection kinetics relative to excited-state decay to the ground. Intersystem crossing from the singlet to triplet state can be as fast as 1013/second for inorganic coordination dyes, such as ruthenium bipyridine analogues. In general, singlet state lifetimes range from 10^{13} to 10^{19}/second, and triplet state lifetimes range from nanoseconds to milliseconds (dependent on the solvent/electrolyte environment) [26]. Electron injection efficiency requires careful consideration of the rate of electron injection and the kinetics of excited-state decay. Considering the electron injection rate from the dye excited state, this depends upon the electronic coupling between the dye LUMO orbital and accepting states in the TiO_2 and on the number density of these states at the surface near the dye [27,28]. The time constant for electron transport to the collection electrode is faster than the charge recombination of injected electrons, which allows efficient charge collection by the external circuit requirement.

In Fig. 7.4, the successive shaded bars exhibit reduced IPCE of each step's costs in an actual cell under operation. The excess energy generated by the absorption of photons at shorter wavelengths than the optical bandgap signifies excess energy lost to excited-state thermalization processes (blue colored). Each functional step is associated with a free energy loss driving this step. While maintaining high quantum efficiencies, minimizing these free energy losses is critical to achieving advances in PCE. Introducing thiophene forms a molecular complex with iodine molecules in the electrolyte solution, increasing the J_{sc} from 4.18 to 8.23 mA.cm^{-2} [29]. A high molar extinction coefficient of monosubstituted-by Ru (II) complexes sensitization also facilitates strong dye absorption for better light harvesting, resulting in a high J_{sc} of 17.61 mA/cm^{-2} [30].

Cosensitizer with a high V_{oc} offers a promising way to augment the performance of DSSCs by slowing charge recombination dynamics. Recently, Swiss Federal Institute of Technology Lausanne has developed a cosensitized solar cell made with the bulky donor N-(2′,4′-bis(dodecyloxy)-[1,1′-biphenyl]-4-yl)-2′,4′-*bis*(dodecyloxy)-N-phenyl-[1,1′-biphenyl]-4-amine and the electron acceptor 4-(benzo[*d*] [1,2,5]thiadiazol-4-yl)benzoic acid. Employing Ms5 with a copper (II/I) electrolyte enables a DSSC to achieve a strikingly high V_{oc} of 1.24 V [31].

Gopalraman et al. have observed no V_{oc}–J_{sc} trade-off while using oleic acid and C106 dye cosensitization employing Förster resonance energy transfer (FRET) [32]. FRET is an excellent photophysical scheme for enhancing the PCE of the device. There was a robust spectral overlap between the absorption spectrum of oxalic acid and the emission spectrum of the acceptor (C106), leading to effective FRET between oxalic acid and C106 and suggesting an excellent opportunity to improve the PV performances of DSSCs [32]. FRET assays based on organic dyes have the advantages of simple preparation and low cost. However, a high photo-bleaching rate and pH sensitivity are the major pitfalls of traditional dyes. Much effort has been devoted to looking for new alternatives for both donors and acceptors to replace conventional organic dyes to improve FRET efficiency and sensitivity.

Nanoparticles, including semiconductor QDs, graphene QDs, and rare-earth-doped upconversion nanoparticles, have been used as FRET donors, providing higher efficiency, better stability, and performance for high performance of DSSC. Using H-NIM molecules as a fluorophore for DSSC, H-NIM (FRET donor) to N719 (FRET acceptor) via in situ FRET promotes light-harvesting of N719 molecules spectral range of ultraviolet light improved the PCE from 16.46% to 19.92% under 600 lux illumination [33].

Nath et al. describe an electrochemical approach [34]; interhalogen-based binary-redox couples were examined, varying the [I$_2$Br-] enhances the V_{oc} of the device. The degree of enhancement of V_{oc} was controllable by varying the electrolyte composition. It is typically due to the higher the contribution of [I$_2$Br$^-$], the more positive shift in the energy level of the binary redox couples [34].

Scuto et al. observed the effect of electrical stresses as a function of stress bias voltage, electrolyte type, illumination, and temperature conditions, improving DSSC performance [35]. A V_{oc} of 1.4 V was remarkably achieved using a novel silyl-anchor coumarin dye with alkyl-chain substitutes, a Br$_3^-$/Br$^-$ redox electrolyte solution containing water, and an Mg^{2+}-doped anatase-TiO$_2$ electrode with twofold surface modification by MgO and Al$_2$O$_3$ DSSC components [36]. Besides, Chen et al. have developed a dye called 2-cyano acrylic acid-4-(bis-dimethylfluoreneaniline) thiophene, which resulted in >1 V of V_{OC} due to the shift of TiO$_2$ conduction band edge (CBE) rather than slow recombination as observed by the transient photovoltage decay [37]. Therefore, the selection of candidates with the advantages of intense absorption in a wide spectral range, tunable photophysical and electrochemical properties, and long-lived excited states facilitating electron injection. These results represent a perspective for designing new efficient additives to fabricate low-cost and high-efficiency DSSCs.

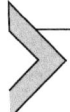

7.8 Photoanode improvement of state-of-the-art dye-sensitized solar cells

In a DSSC, a thick film of a three-dimensional (3D) network of randomly dispersed TiO_2 nanoparticles is typically employed as a photoanode. Although the large surface area of nanoparticles enables a high dye-loading capacity, the disordered network with numerous grain boundaries weakens electron mobility. It results in slow transport and recombination of photoexcited electrons. This greatly restricts the overall efficiency of such devices. The inherent problems associated with the standard photoanode construction necessitate searching for more effective nanostructured photoanode materials and morphologies. In DSSC, the overall PCE strictly depends on the competition between the charge collection and electron transport efficiency through the photoanode and the recombination of electrons with electrolytes on the photoanode/electrolyte interface. Hence, a photoanode should possess a high surface area for better interaction with the photosensitized dye molecules, sufficient light-harvesting capacity, and improved electron collection and transportation properties compared to charge recombination [38–40]. One of the most critical factors that affect the PCE of DSSCs is the electron transport across the oxide electrode. It is established that the greater the electron mobility, the higher the device's PCE. Regarding the development of the photoanode metal oxide for DSSC, considerable R&D has been performed emphasizing mesoporous TiO_2 nanoparticles as the photoanode. Suitable bandgap adjustment for electron injection from most popular dyes; high surface area; robust, stable, and ideal scaffold for dye anchoring; and high electron mobility for photogenerated electron collection establish TiO_2 as the most suitable photoanode DSSC. The PCE reported in Table 7.1 indicates the maximum efficiency recorded for TiO_2-based photoanodes [9,10,38,39,41–50].

Besides, the charge recombination processes that generally inhibit the injected electrons from oxide material to the conducting glass substrate can diminish the performance of DSSCs. Therefore, to ensure a higher conversion efficiency of DSSC, the fast photo-induced electron transport in the

Table 7.1 Performance of TiO$_2$-based photoanode in dye-sensitized solar cell devices.

Types	Oxide	Cell construction (dye/electrolyte/counter)	PCE (%)	References
Binary	TiO$_2$	ADEKA-1. LEG-4 dye/F:[Co (phen)]$^{3+/2+]/}$graphene	14.3	[41]
Binary	TiO$_2$	SM315 dye/Co^{2+}-Co^{3+}/graphene	13.0	[42]
Binary	TiO$_2$	Designed dye/ Co^{2+}-Co^{3+}/Pt	12.0	[10]
Binary	TiO$_2$	N719 dye/I$^-$-I$_3^-$/Pt	11.1	[9]
Modified	BaTiO$_3$ coated TiO$_2$	N719 dye/I$^-$-I$_3^-$/Pt	7.52	[38]
Modified	Er, Yb-CeO$_2$/TiO$_2$	N719 dye/I$^-$-I$_3^-$/Pt	7.05	[39]
Modified	Au@TiO$_2$	N719 dye/I$^-$-I$_3^-$/Pt	8.13	[43]
Doped	Nb-doped TiO$_2$	N719 dye/I$^-$-I$_3^-$/Pt	4.10	[44]
Doped	Sn-doped TiO$_2$	N719 dye/I$^-$-I$_3^-$/Pt	9.43	[45]
Doped	Cu/N-TiO$_2$	N719 dye/I$^-$-I$_3^-$/Pt	11.35	[46]
Doped	Zr/N-doped TiO$_2$	N719 dye/I$^-$-I$_3^-$/Pt	12.62	[47]
Mixed	CeO$_2$-TiO$_2$	N719 dye/I$^-$-I$_3^-$/Pt	8.92	[48]
Mixed	Al$_2$O$_3$-TiO$_2$	N719 dye/I$^-$-I$_3^-$/Pt	6.55	[49]
Mixed	TiO$_2$-GeO$_2$	N719 dye/I$^-$-I$_3^-$/Pt	7.91	[50]

Table 7.2 Performance of various types of oxide-based photoanode in dye-sensitized solar cell devices.

Types	Oxide	Cell construction (dye/electrolyte/counter)	PCE (%)	References
Binary	ZnO	N719 dye/I$^-$-I$_3^-$/Pt	8.03	[51]
Binary	SnO$_2$	N719 dye/I$^-$-I$_3^-$/Pt	3.20	[52]
Ternary	Zn$_2$SnO$_4$	N719 dye/I$^-$-I$_3^-$/Pt	6.30	[53]
Ternary	SrTiO$_3$	N719 dye/I$^-$-I$_3^-$/Pt	1.80	[54]
Ternary	CdSnO$_3$	N719 dye/I$^-$-I$_3^-$/Pt	1.42	[55]
Ternary	BiFeO$_3$	N719 dye/I$^-$-I$_3^-$/Pt	3.02	[56]
Ternary	BaSnO$_3$	N719 dye/I$^-$-I$_3^-$/Pt	6.20	[57]
Doped	B doped ZnO	N3 dye/I$^-$-I$_3^-$/Pt	7.20	[58]
Mixed	SnO$_2$-ZnO	N719 dye/I$^-$-I$_3^-$/Pt	6.31	[59]

working electrode and the suppression of charge recombination processes are prerequisites. In addition to TiO$_2$, different types of photoanode have been studied, and their respective performance in DSSC devices are summarized in Table 7.2 [51–59].

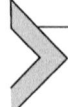

7.9 Improvement(s) over state-of-the-art dye sensitizers

The sensitizer is the central component of DSSCs because it harvests sunlight and produces photoexcited electrons at the semiconductor interface. For efficient performance, the sensitizer component has several requirements: Functional group to anchor on the semiconducting material, appropriate LUMO and HOMO levels for effective charge injection into the semiconductor and dye regeneration from the electrolyte, high molar extinction coefficients in the visible and near-IR region for light-harvesting, and good photostability and solubility.

The most widely used photosensitizers are Ru(II) polypyridyl dyes [14,30], where the light absorption originates due to a metal ligand charge transfer process with a molar extinction coefficient (ε) of 10,000–20,000 $M^{-}cm^{-}$. The development of dyes for DSSCs has evolved to include functionalizing the ancillary polypyridyl ligand through various substituents, including alkyl, aryl, alkoxy, phenyl, and heterocycles. Additionally, the enhancement process involves replacing the chelating anions with popular dyes such as N719, N3, N749 (black dye), Z907, K51, K60, N886, C103, and JK56 to improve sensitization efficiency. At the same time, natural dyes derived from bark, leaves, fruit peel, juice, and flower petal extraction have become a viable alternative to expensive and rare organic sensitizers because of their low cost, easy attainability, abundance in the supply of raw materials and environmentally friendly nature [50,60,61]. In Table 7.3,

Table 7.3 Various sensitizers used in TiO_2-based dye-sensitized solar cell devices.

Dye	Type of dye	PCE (%)	References
SM315	Donor-π-acceptor	16.7	[62]
ADEKA-1	Silyl-anchor dye	14.7	[41]
Dye R6	Polycyclic aromatic hydrocarbon	12.6	[63]
YD-2-o-C8	Porphyrin-based	11.9	[10]
YKP-88	Benzothiadiazole-based	8.7	[64]
Coumarin containing thiophene and indole moieties	Cosensitization	7.09	[65]
Butea monosperma	Natural	1.8	[66]
SGT-149 with a SGT-021	Cosensitization	14.2	[67]

different types of dyes used as sensitizers in TiO_2-based DSSC devices are presented [10,41,62–67], including cosensitization of ruthenium(II) DSSCs by coumarin-based dyes [65].

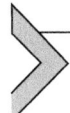

7.10 Improvement(s) over state-of-the-art dye sensitizer electrolytes

The electrolyte is one of the most crucial components in DSSCs; it is responsible for the inner charge carrier transport between electrodes and continuously regenerates the dye during DSSC operation. The electrolyte significantly influences the PCE and long-term stability of the devices. An electrolyte provides pure ionic conductivity between the positive and the negative electrodes in an electrochemical device. However, the potential problems caused by liquid electrolytes, such as the leakage and volatilization of solvents, possible desorption and photodegradation of the attached dyes, and the corrosion of Pt CE, are considered some of the critical factors limiting the long-term performance and practical use of DSSCs. Therefore, much attention has been given to improving the liquid electrolytes or replacing the liquid electrolytes with solid-state or quasisolid-state electrolytes. Over the years, this type of electrolyte has been optimized by including various additives, affecting both electrode surface electronics and structure and the electrolyte chemistry itself. This type of electrolyte provides cells with the highest conversion efficiency of 14.2% [67]. All alternatives, so far, lag in terms of PCEs, although they may offer advantages in terms of long-term endurance.

Table 7.4 illustrates the PCE observed for different electrolytes and their components. The electrolytes must transport the charge carriers between the photoanode and CE, provide fast diffusion of charge carriers (higher conductivity), and produce good interfacial contact with the mesoporous semiconductor layer and the CE. The electrolytes must have long-term stabilities, including chemical, thermal, optical, electrochemical, and interfacial stability with sensitized dye. The electrolytes should not exhibit a significant absorption in the visible light range. Because the I^-/I_3^- redox couple in the electrolyte shows color and reduces visible light absorption, I_3^- ions can react with the injected electrons and increase the dark current. The concentration of I^-/I_3^- must be optimized. The choice of the electrolyte should consider the redox potential and regeneration of the dye and itself [42,68–77].

Table 7.4 Performance of different electrolytes in dye-sensitized solar cell devices.

Type	Composition of electrolyte	Dye	PCE (%)	References
IE	0.25M Co(bpy)$_3$(TFSI)$_2$, 0.06M Co(bpy)$_3$(TFSI)$_3$, 0.1 M LiTFSI, and 0.5 M 4-TBP in acetonitrile	SM315	13.00	[42]
	Poly(vinylidene fluoride-cohexafluoropropylene)-based	N719	7.18	[68]
	[Co((MeIm-Bpy)PF$_6$)$_3$]$^{2+/3+}$) (MeIm-Bpy)PF$_6$)$_3$ = 3,3′-(2,2′-bipyridine-4,4′-diyl-bis(methylene))bis(1-methyl-1H-imidazol-3-ium)hexafluorophosphate	N719	8.29	[69]
	0.5 M I$_2$ and 0.1 M CuI in HMImI	N3	4.54	[70]
	0.4 M LiBr + 0.04 M Br$_2$ in acetonitrile	TC306	5.22	[71]
	1 M DMII, 0.15 M I$_2$, 0.5 M NBB, 0.1 M GNCS, and 50 mM NaI in BN	C106	10.00	[72]
	0.8 M PMImI, 0.15 M I$_2$, 0.1 M GSCN, and 0.5 M NMBI in MPN	K19	8.00	[73]
	1 M PMII, 0.1 M [MeIm-TEMPO][TFSI], 0.01 M NOBF$_4$, 0.2 M LiTFSI, and 0.5 M NBB in MPN	D205	8.20	[74]
QSIE	I$_2$, TBAI, EMImI, EC/PC (4:1), and PEG	N719	7.13	[75]
	PMAPII (16 wt.%) 0.05 M I$_2$ and 0.05 M TBP in GBL	TBA	8.12	[76]
	DMII/EMImI/EMImB(CN)4/I$_2$/NBB/GNCS (molar ratio 12/12/16/1.67/3.33/0.67)	C103	8.50	[77]

IE, Ionic electrolyte; *QSIE*, quasi ionic electrolyte; *EY*, Eosin Y; *HMImI*, 1-hexyl-3-metyl-imidazolium iodide; *NMB*, 1-methylbenzimidazole; *MPN*, methoxypropionitrile; *TBAI*, tetrabutylammonium iodide; *TBP*, tert-but-pyridine; *NBB*, *N*-butylbenzimidazole; *PMMI*, 1-methyl-3-propylimidazolium iodid; *BN*, butyronitrile; EMINCS, 1-ethyl-3- methylimidazolium thiocyanate; GNCS, 0.1 M guanidinium thiocyanate; TFSI, *N*-methyl-*N*-propylpiperidinium bis(trifluoromethanesulfonyl)imide; GBL, γ-butyrolactone; PEG, polyethelene glycol.

7.11 Improvement(s) over state-of-the-art dye-sensitized solar cell counter-electrodes

The CE contribution emphasizes a significant impact on the enhancement of DSSC performance, and Pt-based CE sets an essential benchmark in this field. The active responsibility is reducing the redox electrolyte and collecting electrons from the external load to the electrolyte. Noble metals are more acceptable as CE's due to their superior catalytic activity and faster redox. Also, high conductivity for charge transport, good electrocatalytic activity for reducing the redox couple, and excellent stability significantly contribute to choosing noble metals such as Pt, Au, and Ag-based CE. From theoretical considerations, the catalytic activity toward the I_3^- reduction was calculated, and it was found to be in the order of Pt (111) > Pt (411) > Pt (100) [78]. The main drawback of Pt-based CE is that the noble metals are expensive and corroded in the liquid electrolyte presence. Table 7.5 illustrates the DSSC performance of Pt along with Pt-based alloy and hybrid Pt CE [79–85].

7.12 Overall relative scope of performance improvement of dye-sensitized solar cells

Approaches for third-generation PV aim to circumvent the Shockley-Queisser limit for single-bandgap devices fabricated at low cost [1]. Impressive progress on PCEs is being made in the solution-processed solar technologies (see Section 7.15). However, the technological challenge is to utilize the entire solar spectrum by using better sensitizers and overcoming the fundamental power-loss mechanism, as shown in Fig. 7.5.

As mentioned in Section 7.6, all the modification strategies enhance the effectivity of the PV parameters such as J_{sc}, V_{oc}, and FF. Fig. 7.6 exhibits mechanisms for these improved parameter modifications. Electrochemical impedance spectroscopy has been widely employed to study the kinetics of electrochemical and photoelectrochemical processes, including elucidating salient electronic and ionic processes occurring in the DSSC device [86].

Table 7.5 Performance of Pt and Pt-based alloys and hybrid counter-electrodes in dye-sensitized solar cell devices.

Material	J_{sc} (mA/cm^2)	V_{oc} (V)	FF	PCE (%)	References
Pt–Co	18.53	0.73	0.75	10.23	[79]
Pt$_3$–Ni	17.05	0.72	0.72	8.78	[80]
Pt–Pd	16.88	0.73	0.64	7.45	[81]
Pt–Au	16.50	0.65	0.31	3.40	[82]
Pt–Cu–Ni	18.31	0.76	0.70	7.66	[83]
Pt–Co–Ni	17.01	0.75	0.69	8.85	[84]
Pt/NiO/Ag	30.10	0.81	0.46	11.27	[85]

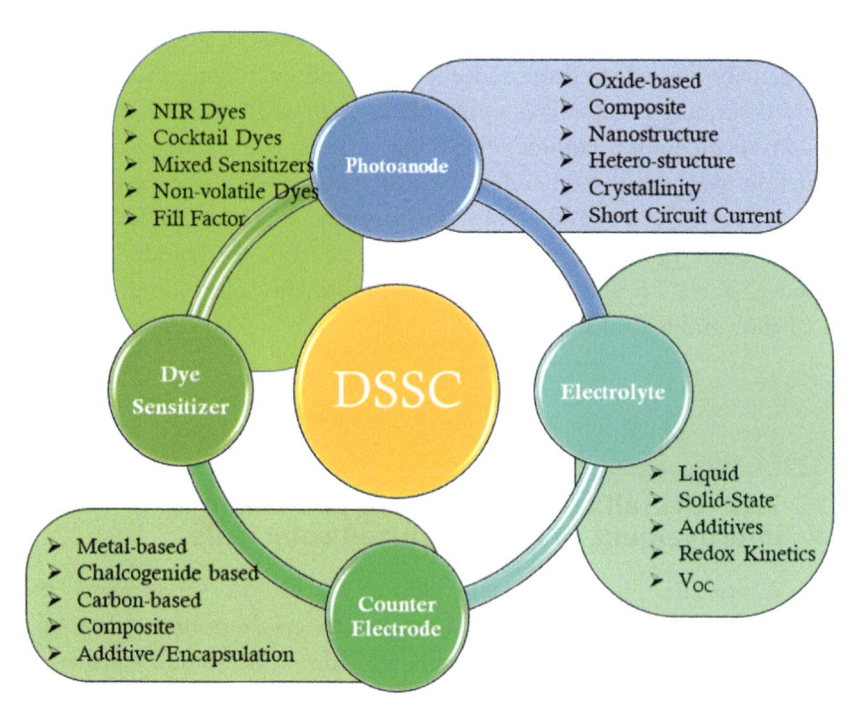

Figure 7.5 Schematic representation of the significant component of a dye-sensitized solar cell device, current solar cell challenges, and complementary strategies to address them.

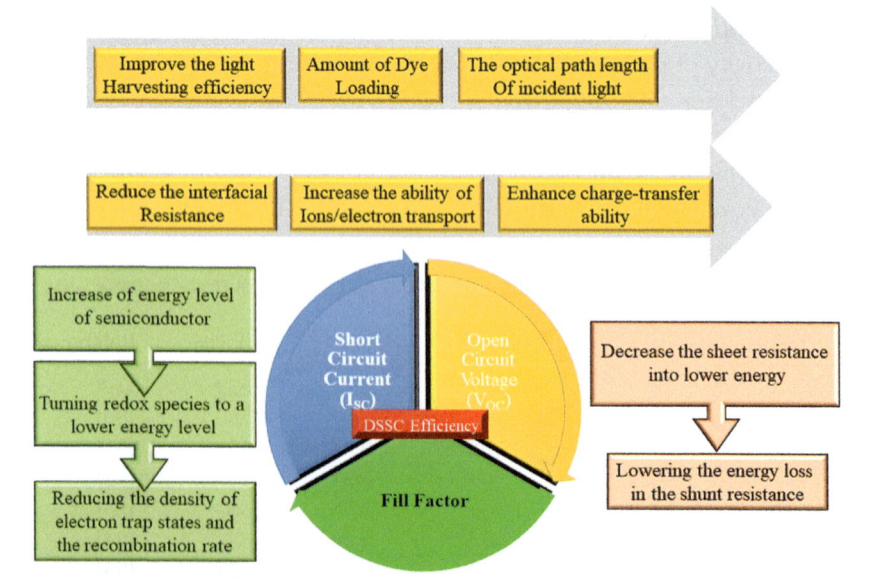

Figure 7.6 A summary of various strategies to enhance the power conversion efficiency of dye-sensitized solar cell by regulating the three crucial photovoltaic parameters.

The impedance of DSSC has been described effectively and informatively by an equivalent circuit model.

The equivalent circuit is constructed by assigning different cell components and their interface resistances that hinder charge transport in the materials and charge transfer across their interfaces. At the same time, capacitances represent a charge accumulation at these interfaces. The ohmic serial resistance is associated with the series resistance of the electrolyte and electric contacts in the DSSC. The Nyquist diagram typically featured three semicircles in increasing frequency attributed to the Nernst diffusion within the electrolyte, the electron transfer at the oxide/electrolyte interface, and the redox reaction CE fitted against an equivalent circuit. The high-frequency intercepts on the real axis represent the series resistance (R_S), mainly composed of CE material's bulk resistance, FTO glass substrates' resistance, and contact resistance. The slope of the $J-V$ curve at the V_{oc} point roughly determines the series resistance of a DSSC. The shunt resistance (R_{SH}) is mainly generated from the current leakage around the edge of the device and across the semiconductor/electrolyte interface in particular. Low shunt resistance offers an alternative path for the photogenerated current. This reduces FF and V_{oc} and affects maximum power output. The slope of the $J-V$ curve at the J_{sc} point further defines the shunt resistance approximately. All the cells exhibit similar semicircles in the higher frequency region, emphasizing an equivalent charge transfer resistance (R_{CT1}) for CE and corresponding capacitance called C_1. The high-frequency semicircle indicates that the charge transfer resistance is attributed to the metal oxide/dye/electrolyte interface (R_{CT2}) and the corresponding capacitance denoted as C_2 [87,88].

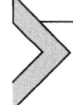

7.13 Nanostructured materials for performance improvement of dye-sensitized solar cells

Considerable attention has been given to the fact that morphological features can impact the performance of photoelectrodes in DSSCs, mainly in electron transport effectiveness and electron recombination with electrolytes. Particular interest is given to 1D nanostructures such as rods, tubes, fibers, and wires that are believed to improve electron transport by providing direct pathways throughout the structure of the photoelectrode to the collecting substrate. Consisting of an ordering pattern, 1D morphology has the advantage of higher diffusion coefficients simplifying that generated electrons can move and reach the collecting substrate faster than nonordered nanostructures. This indicates that 1D structures give electron diffusion lengths more considerable than their film thickness [89–92]. However, due to their low internal surface area, the adsorption of dye molecules exhibits a lesser tendency to the 1D structures. In these circumstances, an increment of the surface area is highly desirable without compromising the 1D structure feature, and as a result, the concept of synthesizing nanowires and nanofibers has developed. This will be a fantastic strategy for the high conductive 1D nanostructure by keeping a high surface area of the material. Recently, Shayi et al. (2021) showed a very high surface area (363.15 m^2/g) of TiO$_2$ nanostructures, resulting in 41% better PCE improvement than the P25 nanoparticles [93]. Besides, Zhong et al. showed that it is possible to improve the PV performance of DSSCs by blending different concentrations of TiO$_2$ nanotubes into a TiO$_2$ mesoporous film [94].

This is mainly attributed to the comparatively low internal surface area available for the adsorption of dye molecules. A common strategy to achieve high electron mobility and increased surface area is blending 1D nanostructures, such as TiO$_2$ nanotubes, with TiO$_2$ nanoparticles. Therefore, the photoelectrode can benefit from the more conductive 1D nanostructure and the increased surface area available for dye adsorption; see Fig. 7.6. The figure depicts a plausible correlation between the device's efficiency and the electron diffusion coefficient, electron collection efficiency, and electron diffusion length (Fig. 7.7) [89].

Recent studies show that 2D or 3D-shaped TiO$_2$ such as nanobelt, nanocomb, hollow hierarchical microsphere, nanoflower, and nanoparticles decorated nanorods also exert excellent properties as compared to the nanoparticles owing to the synergistic effect of optimized surface area, better light-harvesting and directional property for charge migration [95,96]. Along with the morphological variation, the dye adsorbing capability and the charge transfer phenomenon at the TiO$_2$/electrolyte interface are also sensitive to the exposed crystal lattice planes/facets on the TiO surface. According to Wulff construction, the less reactive and thermodynamically most stable {101} facets constitute more than $\sim$ 90% of a naturally occurring anatase-TiO$_2$. In contrast, the higher energy facets diminish rapidly during growth [97]. Consequently, synthesizing TiO$_2$ nanocrystals with exposed high-energy facets becomes a significant challenge to the researchers for potential applications. Chen et al. reported

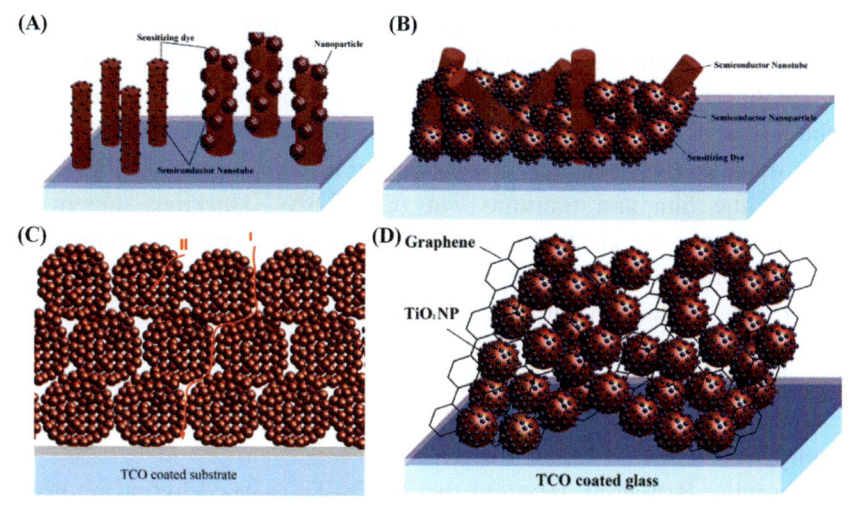

Figure 7.7 Schematic of one-dimensional nanostructures with (A) sensitizing dye adsorbed in the surface and semiconductor nanoparticles, (B) with metal oxide particles, respectively. (C) Schematic representation of a hierarchical nanoparticle film, having two different scale pores: (I) Meso/macropores and (II) smaller nanopores. (D) Schematic representation of aTiO$_2$/graphene hybrid photoanode. *Reprinted with permission from J. Maçaira, L. Andrade, A. Mendes, Review on nanostructured photoelectrodes for next generation dye-sensitized solar cells. Renew. Sustain. Energy Rev. 27 (2013) 334–349, copyright (2013) Elsevier.*

that {010}-faceted anatase nanocrystals exhibit high J_{sc} due to strong adsorption of sensitizer dye molecules on the {010}-facet [98]. The posttreatment can further enhance the energy conversion efficiency of photoanode comprised of native single-phase TiO$_2$ with P25 layer or TiCl$_4$. The increased surface roughness leads to more dye adsorbing capability onto the photoanode surface and synergistically improves the PCE.

7.14 Effect of light scattering

Nanocrystallite aggregation is quite a natural phenomenon while using them in the DSSC. Sometimes, due to the aggregate formation, electron recombination occurs in the device, reducing the device's performance. In contrast, sometimes, the aggregates serve as scatterers causing light scattering within the photoelectrode film. The absorbed light scattering experiences the extension of light propagation within the device, which secures the photon interaction with the dye molecules on the nanocrystallite's surface [99–102]. The photoanode film's light scattering ability thus facilitates the film's optical absorption property. Therefore, light propagation in a nanocrystalline film occurs along a straight path. As a result, except for the absorbed portion, the incident light flows through the film and transmits out (Fig. 7.8A). Thickness increment does not imply the electron mobilization process. Therefore, improving optical absorption through the light scattering effect passively promotes a higher rate of electron mobilization and improves efficiency. Fig. 7.8A exhibits a schematic representation of the light scattering effect within the aggregates. It is straightforward; once the scattering phenomena occur, it increases the light diffusion path length, thus reducing the electron recombination rate.

Recently, Roy et al. reported that CeO_2-TiO_2-based hybrid photoanode DSSCs exhibited approximately 46% more efficiency enhancement than only TiO_2-based DSSCs [48]. A correlation was reported between the light scattering effect of CeO_2, the dye-loading effect, and the heterojunction formation with the TiO_2 responsible for this massive improvement. The heterojunction formed within the hybrid layer interface leads to charge carriers' spatial separation with enhanced electron mobility [48]. Besides, the CeO_2-based scattering approach to the efficiency improvement of DSSCs was initially implemented by Yu et al. by utilizing the

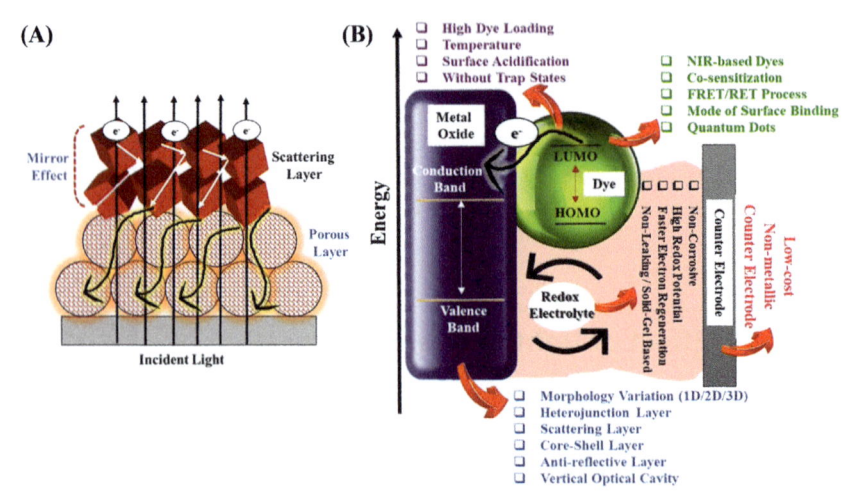

Figure 7.8 (A) Schematic representation of the light scattering effect and (B) various approaches for each component of a dye-sensitized solar cell device for efficiency enhancement.

mirror-like light scattering effect of CeO_2, which produced a noticeable PCE improvement of 17.8% in the case of a TiO_2-based DSSC [103]. Implementing an atomic layer-deposited TiO_2 photoanode modified with a flower-like CeO_2 scattering layer resulted in a ∼31% PCE improvement compared to a conventional TiO_2 photoanode. Mustafa et al. demonstrated that polyvinyl-alcohol/titanium dioxide nanofibers as a scattering layer over the TiO_2 nanoparticles exhibited 33% PCE improvement from their unmodified form [104]. Furthermore, cadmium-doped TiO_2 nanofibers as a light scattering layer in DSSC were successfully prepared by Motlak et al. via electrospinning [105]. Similarly, a mixed phase of copper oxide nanoparticles (CuO-Cu_2O) prepared by microwave heating technique was used as a scattering layer in DSSC, and a higher PCE of 2.31% was achieved compared with pure DSSC without a scattering layer (1.76%) along with the dye-loading capacity of the device [106].

7.15 Improved dye-sensitized solar cell device performance: Fabrication approaches

Fig. 7.8B represents possible ways for a DSSC device improvement for each component. This also indicates the strategies required to maintain high-efficiency device performance. Among the current challenges, this chapter has focused on demonstrating materials that are alternatives to TiO_2 and Pt, modified synthesis techniques, and improved physicochemical properties to achieve enhanced efficiency. Remarkable efforts have been endeavored to establish a series of semiconducting oxides from TiO_2 to other binary oxides. Their suitable electronic configuration is analogous to providing superior electronic properties for developing better photoanode materials for the DSSCs device. Apart from the extensively acclaimed photoanode TiO_2, various ternary oxides such as $BaSnO_3$, $CdSnO_3$, and $BiFeO_3$ have been projected as alternatives to TiO_2 photoanode [57–59].

To overcome the drawback of binary oxides, such as dye-metal oxide complex formation at low pH, degraded stability in acidic dye solutions, and a prolonged period of dye loading, introducing ternary oxides as photoanodes were explored to address the mentioned problems of binary oxides. Accordingly, efforts to design new nanostructured architectures for the next-generation solar cells have been a significant area of research. Implementing new materials can be attributed to modifying photoanode materials, scattering layers or active layers over TiO_2, deposition of conductive polymer layers, introducing defects or doping other elements, or using bi-layer oxides TiO_2. These are some significant alternative ways to improve the performance of TiO_2-based DSSC devices. Optical optimization becomes increasingly essential to overtake the electrical one to enhance their PCE further to approach the theoretical limit. Developing an adequate physicochemical understanding of photoanode sensitization is another parameter where temperature plays a crucial role. Recently, Roy et al. reported that a ZnO-based device exhibited maximum J_{sc} during dye sensitization at 60°C. This is entirely governed by the mode of surface interaction and further fruitful connection of N719-ZnO to promote better electron conduction. A TiO_2 surface plays a significant role in determining the efficiency of the DSSCs [107]. Treatment with various acids such as HCl, HNO_3, H_3PO_4, and H_2SO_4 is reported to affect the binding mode of the dye to the TiO_2 surface, affecting the device's PV performance. A shift of the TiO_2 CBE during an acid treatment enhances dye absorption and increases electron injection dynamics [108].

Benefitting from the advantages of high extinction coefficient, broad and tunable light absorption, low-cost solution-processed preparation, and the potential of multielectron generation, QDs have been more and more employed in solar cells as photosensitizers, to construct quantum dot-sensitized solar cells (QDSSCs) [109]. QDs are nanosized semiconductor particles whose physical and chemical properties are size-dependent. Among the notable characteristics of QDs include tunability of bandgap energy, narrow emission spectrum, good photostability, broad excitation spectra, high extinction coefficient, and multiple exciton generation. With these advantages, fabrication of Zn−Cu−In−S−Se (ZCISSe) QD-sensitized TiO_2 film electrode devices with up to 7% efficiency has been achieved [110]. The evolution of DSSC to QDSSC has not taken a giant leap. The only physical difference between the DSSC and QDSSC is the sensitizing materials. In the QDSSC, the dye is replaced by inorganic nanoparticles of QDs. Many experiments have been carried out with various QDs for QDSSCs to achieve high-performance devices. Therefore, in past years, all the materials used in QDSCs have experienced rational design and careful modification, including the fabrication of nanostructured semiconductor photoanodes with effective dye loading and fast electron transport, the synthesis of versatile QDs with high absorption coefficients and appropriate bandgaps to maximize light harvesting, the preparation of redox electrolytes and long-term stability, and the formation of various CEs with high electrocatalytic ability in reduction of the oxidized electrolyte. Optical management in the device architecture provides solutions complementary or alternative to the ones mentioned above, for example:

- Antireflection coatings are implemented at the air/glass interface [111]. Various antireflection designs also incorporate optical scattering capabilities for focusing or increasing the optical path of light in the photoanode material, thus increasing the optical absorption in this medium.
- Vertical optical cavity design. It allows tuning the electromagnetic field in-depth profile in planar solar cells to optimize the absorption in the photoanode material and (or) the cell reflectance/transmittance.
- Employment of plasmonic nanostructure in the different layers of the materials of the cell also exhibits light scattering ability. This mainly improves the electromagnetic energy in their surrounding region (near-field enhancement), resulting in higher optical absorption of the photoanode.

7.16 Summary and future scope

In summary, this chapter has addressed approaches to improve the design and further development of a DSSC device capable of producing maximum PCE. Various fabrication strategies and possible improvement areas were discussed to improve the possibility of large-scale implementation of the DSSC concept, reducing the time and cost of making these devices. For this purpose, several approaches are underway to understand many critical issues related to the fundamental physicochemical characteristics of each component in DSSC and enhance the stability of the photoanode and kinetics of the electron transfer process, especially at the interfacial junctions of photoanode-dye-electrolyte. Due to its inherent cost reduction potential, DSSC has appeared to be a promising future PV technology. This is due to their inexpensive components, relatively simple production technology, and broad applicability features ease. Concerning the present status of PV technologies, there is enough scope for improvements in several areas, such as conversion efficiency, cost, durability, and sustainability for a DSSC-related research field. Implementing CPV, morphology tuning, various nanostructure material synthesis, applying scattering layer, introducing plasmonic materials, high temperature-dependent dye-sensitization, near-IR absorption ability of dye molecule design, and incorporating potential material for more extended electron diffusion pathway are discussed throughout the chapter, which can pave way fundamental understanding about the PCE improvement strategies and encourage for future development in this field.

Given the progress achieved thus far, further research studies on some underexplored issues deserve more attention to achieve high-efficiency DSSCs in the future. With appropriate band structure configuration, geometrical morphology, photoelectrochemical stability, and broad spectral coverage are novel devices for commercialization. Although some improvements have been achieved in structural modification in the fabricated photoanodes, there is a long way to go to achieve enhanced efficiency from the known systems. Besides, several approaches are underway to understand many critical issues related to each component's fundamental physicochemical characteristics in DSSC and enhance the stability of the photoanode and kinetics of the electron transfer process, especially at the interfacial junctions of the electron transfer process photoanode-dye-electrolyte.

Acknowledgments

A.R. acknowledges the DST-INSPIRE fellowship, Government of India, for supporting his Ph.D. A.R., A.T., and T.K.S acknowledge the Engineering and Physical Sciences Research Council (EPSRC), U.K., under research grant number EP/T025875/1. A.R. and M.A.G. acknowledge the UK-Saudi Challenge Fund 2022, funded by the British Council. The project's funders were not directly involved in the writing of this chapter.

References

[1] G.M. Wilson, M. Al-Jassim, W.K. Metzger, S.W. Glunz, P. Verlinden, G. Xiong, et al., The photovoltaic technologies roadmap, J. Phys. D: Appl. Phys. 53 (2020) (2020) 493001.
[2] European Commission, Delivering the European Green Deal. <https://commission.europa.eu/strategy-and-policy/priorities-2019-2024/european-green-deal/delivering-european-green-deal_en> (accessed 19.12.22).
[3] A.S. Sánchez, E.A. Torres, R.A. Kalid, Renewable energy generation for the rural electrification of isolated communities in the amazon region, Renew. Sust. Energ. Rev. 49 (2015) 278−290.
[4] O. Ellabban, H. Abu-Rub, F. Blaabjerg, Renewable energy resources: CURRENT status, future prospects and their enabling technology, Renew. Sust. Energ. Rev. 39 (2014) 748−764.
[5] K. Kalyanasundaram, Dye-sensitized Solar Cells, EPFL Press, 2010.
[6] N. Rathore, N.L. Panwar, F. Yettou, A. Gama, A comprehensive review of different types of solar photovoltaic cells and their applications, Int. J. Amb. Ener. 42 (10) (2019) 1200−1217.

[7] H. Tributsch, Reaction of excited chlorophyll molecules at electrodes and in photosynthesis, Photochem. Photobiol. 16 (1972) 261–269.
[8] B. O'Regan, M. Grätzel, A low-cost, high-efficiency solar cell based on dye-sensitized colloidal TiO_2 films, Nature 353 (1991) 737–740.
[9] Y. Chiba, A. Islam, Y. Watanabe, R. Komiya, N. Koide, L. Han, Dye-sensitized solar cells with conversion efficiency of 11.1%, Jpn. J. Appl. Phys. 45 (2006) L638–L640.
[10] A. Yella, H.W. Lee, H.N. Tsao, C. Yi, A.K. Chandiran, M.K. Nazeeruddin, et al., Porphyrin-sensitized solar cells with cobalt (II/III)-based redox electrolyte exceed 12 percent efficiency, Science 334 (2011) 629–634.
[11] J. Gong, J. Liang, K. Sumathy, Review on dye-sensitized solar cells (DSSCs): Fundamental concepts and novel materials, Renew. Sust. Energ. Rev. 16 (2012) 5848–5860.
[12] L.L. Li, E.W.G. Diau, Porphyrin-sensitized solar cells, Chem. Soc. Rev. 42 (2013) 291–304.
[13] M. Grätzel, Dye-sensitized solar cells, J. Photochem. Photobiol. C: Photochem. Rev. 4 (2003) 145–153.
[14] N. Tomar, A. Agarwal, V.S. Dhaka, P.K. Surolia, Ruthenium complexes based dye sensitized solar cells: Fundamentals and research trends, Sol. Energy 207 (1) (2020) 59–76.
[15] V. Esen, S. Sağlam, B. Oral, Light sources of solar simulators for photovoltaic devices: A review, Renew. Sust. Energ. Rev. 77 (2017) 1240–1250.
[16] A. Guechi, M. Chegaar, M. Aillerie, Air mass effect on the performance of organic solar cells, Energy Procedia 36 (2013) 714–721.
[17] K.J. Moon, S.W. Lee, Y.H. Lee, J.H. Kim, J.Y. Ahn, S.J. Lee, et al., Effect of TiO_2 nanoparticle-accumulated bilayer photoelectrode and condenser lens-assisted solar concentrator on light harvesting in dye-sensitized solar cells, Nanoscale Res. Lett. 8 (2013) 283.
[18] S. Choi, E. Cho, S. Lee, Y. Kim, D. Lee, Development of a high-efficiency laminated dye-sensitized solar cell with a condenser lens, Opt. Express 19 (4) (2011) A818–A823.
[19] G.D. Barber, P.G. Hoertz, S.H.A. Lee, N.M. Abrams, J. Mikulca, T.E. Mallouk, et al., Utilization of direct and diffuse sunlight in a dye-sensitized solar cell - silicon photovoltaic hybrid concentrator system, J. Phys. Chem. Lett. 2 (2011) 581–585.
[20] P. Selvaraj, H. Baig, T.K. Mallick, J. Siviter, A. Montecucco, W. Li, et al., Enhancing the efficiency of transparent dye-sensitized solar cells using concentrated light, Sol. Energy Mater. Sol. Cell 175 (2018) 29–34.
[21] A. Sacco, M. Gerosa, S. Bianco, L. Mercatelli, R. Fontana, L. Pezzati, et al., Dye-sensitized solar cell for a solar concentrator system, Sol. Energy 125 (2016) 307–313.
[22] S.R. Raga, E.M. Barea, F. Fabregat-Santiago, Analysis of the origin of open circuit voltage in dye solar cells, J. Phys. Chem. Lett. 3 (12) (2012) 1629–1634.
[23] A. Listorti, B. O'Regan, J.R. Durrant, Electron transfer dynamics in dye-sensitized solar cells, Chem. Mater. 23 (2011) 3381–3399.
[24] H.J. Snaith, A.J. Moule, C. Klein, K. Meerholz, R.H. Friend, M. Gratzel, Efficiency enhancements in solid-state hybrid solar cells via reduced charge recombination and increased light capture, Nano Lett. 7 (2007) 3372–3376.
[25] J.A. Christians, J.S. Manser, P.V. Kamat, Best practices in perovskite solar cell efficiency measurements. avoiding the error of making bad cells look good, J. Phys. Chem. Lett. 6 (5) (2015) 852–857.
[26] B.C. O'Regan, J.R. Durrant, Kinetic and rnergetic paradigms for dye-sensitized solar cells: Moving from the ideal to the real, Acc. Chem. Res. 42 (2009) 1799–1808.
[27] J.R. Durrant, M. Gratzel, Dye-sensitised mesoscopic solar cells, in: M.D. Archer, A.J. Nozik (Eds.), Nanostructures Photoelectrochemical Systems Solar and Photonic Converion, Imperial College Press, London, UK, 2008, pp. 503–536.

[28] A. Mishra, M.K.R. Fischer, P. Bauerle, Metal-free organic dyes for dye-sensitized solar cells: From structure: property relationships to design rules, Angew. Chem. 48 (2009) 2474–2499.
[29] M. Afrooz, H. Dehghani, Significant improvement of photocurrent in dye-sensitized solar cells by incorporation thiophene into electrolyte as an inexpensive and efficient additive, Org. Elec. 29 (2016) 57–65.
[30] I.M. Abdellah, A.I. Koraiem, A. El-Shafei, Structure-property relationship of novel monosubstituted Ru (II) complexes for high photocurrent and high efficiency DSSCs: Influence of donor versus acceptor ancillary ligand on DSSCs performance, Sol. Energy 177 (2019) 642–651.
[31] D. Zhang, M. Stojanovic, Y. Ren, Y. Cao, F.T. Eickemeyer, E. Socie, et al., A molecular photosensitizer achieves a V_{oc} of 1.24 V enabling highly efficient and stable dye-sensitized solar cells with copper(II/I)-based electrolyte, Nat. Commun. 12 (2021) 1777.
[32] A. Gopalraman, S. Karuppuchamy, S. Vijayaraghavan, High efficiency dye-sensitized solar cells with V_{OC}-J_{SC} trade off eradication by interfacial engineering of the photoanode|electrolyte interface, RSC Adv. 9 (2019) 40292–40300.
[33] Y.-J. Lin, J.-W. Chen, P.-T. Hsiao, Y.-L. Tung, C.-C. Chang, C.-M. Chen, Efficiency improvement of dye-sensitized solar cells by *in situ* fluorescence resonance energy transfer, J. Mater. Chem. A 5 (2017) 9081–9089.
[34] N.C.D. Nath, H.J. Lee, W.-Y. Choi, J.-J. Lee, Electrochemical approach to enhance the open-circuit voltage (V_{OC}) of dye-sensitized solar cells (DSSCs), Electrochem. Acta 109 (2013) 39–45.
[35] A. Scuto, S. Lombardo, G.D. Marco, G. Calogero, I. Citro, F. Principato, et al., Improvement of DSSC performance by voltage stress application, in: 2016 IEEE International Reliability Physics Symposium (IRPS), 2016, P-V-3-1 – P-V-3-6. <https://doi.org/10.1109/IRPS.2016.7574635>.
[36] K. Kakiage, H. Osada, Y. Aoyama, T. Yano, K. Oya, S. Iwamoto, et al., Achievement of over 1.4 V photovoltage in a dye-sensitized solar cell by the application of a silyl-anchor coumarin dye, Sci. Rep. 6 (2016) 35888.
[37] P. Chen, J.H. Yum, F.D. Angelis, E. Mosconi, S. Fantacci, S.-J. Moon, et al., High open-circuit voltage solid-state dye-sensitized solar cells with organic dye, Nano Lett. 9 (6) (2009) 2487–2492.
[38] L. Zhang, Y. Shi, S. Peng, J. Liang, Z. Tao, J. Chen, Dye-sensitized solar cells made from $BaTiO_3$-coated TiO_2 nanoporous electrodes, J. Photochem. Photobiol. A 197 (2008) 260–265.
[39] W. Song, Y. Gong, J. Tian, G. Cao, H. Zhao, C. Sun, Novel photoanode for dye-sensitized solar cells with enhanced light-harvesting and electron-collection efficiency, ACS Appl. Mater. Interfaces 8 (2016) 13418–13425.
[40] R.S. Dubey, S.R. Jadkar, A.B. Bhorde, Synthesis and characterization of various doped TiO_2 nanocrystals for dye-sensitized solar cells, ACS Omega 6 (2021) 3470–3482.
[41] K. Kakiage, Y. Aoyama, T. Yano, K. Oya, J. Fujisawab, M. Hanaya, Highly-efficient dye-sensitized solar cells with collaborative sensitization by silyl-anchor and carboxy-anchor dyes, Chem. Commun. 51 (2015) 15894–15897.
[42] S. Mathew, A. Yella, P. Gao, R.H. Baker, B.F.E. Curchod, N.A. Astani, et al., Dye-sensitized solar cells with 13% efficiency achieved through the molecular engineering of porphyrin sensitizers, Nat. Chem. 6 (2014) 242.
[43] J. Du, J. Qi, D. Wang, Z. Tang, Facile synthesis of Au@TiO_2 core–shell hollow spheres for dye-sensitized solar cells with remarkably improved efficiency, Energy Environ. Sci. 5 (2012) 6914–6918.

[44] T. Nikolay, L. Larina, O. Shevaleevskiy, B.T. Ahn, Electronic structure study of lightly Nb-doped TiO_2 electrode for dye-sensitized solar cells, Energy Environ. Sci. 4 (2011) 1480–1486.

[45] S. Ni, F. Guo, D. Wang, S. Jiao, J. Wang, Y. Zhang, et al., Modification of TiO_2 nanowire arrays with sn doping as photoanode for highly efficient dye-sensitized solar cells, Crystals 9 (2019) 113.

[46] J.-Y. Park, C.-S. Kim, K. Okuyama, H.-M. Lee, H.-D. Jang, S.-E. Lee, et al., Copper and nitrogen doping on TiO_2 photoelectrodes and their functions in dye-sensitized solar cells, J. Power Sources 306 (29) (2016) 764–771.

[47] J.-Y. Park, K.H. Lee, B.S. Kim, C.S. Kim, S.-E. Lee, K. Okuyama, et al., Enhancement of dye-sensitized solar cells using Zr/N-doped TiO_2 composites as photoelectrodes, RSC Adv. 4 (2014) 9946–9952.

[48] A. Roy, S. Bhnadari, S. Sundaram, T.K. Mallick, Intriguing CeO_2-TiO_2 hybrid nanostructured photoanode resulting up to 46% efficiency enhancement for dye-sensitized solar cells, Mater. Chem. Phys. 272 (1) (2021)125036.

[49] Z. Salam, E. Vijayakumar, A. Subramania, Influence of Al_2O_3 nanoparticles embedded-TiO_2 nanofibers based photoanodes on photovoltaic performance of a dye sensitized solar cell, RSC Adv. 4 (2014) 52871–52877.

[50] Y. Duan, Q. Tang, Z. Chen, B. He, H. Chen, Enhanced dye illumination in dye-sensitized solar cells using TiO_2/GeO_2 photo-anodes, J. Mater. Chem. A 2 (2014) 12459–12465.

[51] Y. He, J. Hu, Y. Xie, High-efficiency dye-sensitized solar cells of up to 8.03% by air plasma treatment of ZnO nanostructures, Chem. Commun. 51 (2015) 16229–16232.

[52] A. Birkel, Y.G. Lee, D. Koll, X.V. Meerbeek, S. Frank, M.J. Choi, et al., Highly efficient and stable dye-sensitized solar cells based on SnO_2 nanocrystals prepared by microwave-assisted synthesis, Energy Environ. Sci. 5 (2012) 5392–5400.

[53] D. Hwang, J.S. Jin, H. Lee, H.J. Kim, H. Chung, D.Y. Kim, et al., Hierarchically structured Zn_2SnO_4 nanobeads for high-efficiency dye-sensitized solar cells, Sci. Rep. 4 (2014) 7353.

[54] S. Burnside, J.E. Moser, K. Brooks, M. Grätzel, Nanocrystalline mesoporous strontium titanate as photoelectrode material for photosensitized solar devices: Increasing photovoltage through flatband potential engineering, J. Phys. Chem. B 103 (1999) 9328–9332.

[55] G. Natu, Y. Wu, Photoelectrochemical study of the ilmenite polymorph of $CdSnO_3$ and its photoanodic application in dye-sensitized solar cells, J. Phys. Chem. C. 114 (2010) 6802–6807.

[56] G.S. Lotey, N.K. Verma, Synthesis and characterization of $BiFeO_3$ nanowires and their applications in dye-sensitized solar cells, Mater. Sci. Semicond. Process. 21 (2014) 206–211.

[57] S.S. Shin, J.S. Kim, J.H. Suk, K.D. Lee, D.W. Kim, J.H. Park, et al., Improved quantum efficiency of highly efficient perovskite $BaSnO_3$-based dye-sensitized solar cells, ACS Nano 7 (2013) 1027–1035.

[58] K. Mahmood, H.J. Sung, A dye-sensitized solar cell based on a boron-doped ZnO (BZO) film with double light-scattering-layers structured photoanode, J. Mater. Chem. A 2 (2014) 5408–5417.

[59] W. Chen, Y. Qiu, Y. Zhong, K.S. Wong, S. Yang, High-efficiency dye-sensitized solar cells based on the composite photoanodes of SnO_2 nanoparticles/ZnO nanotetrapods, J. Phys. Chem. A 114 (2010) 3127–3138.

[60] S. Shalini, R. Balasundaraprabhu, T.S. Kumar, N. Prabavathy, S. Senthilarasu, S. Prasanna, Status and outlook of sensitizers/dyes used in dye sensitized solar cells (DSSC): A review, Int. J. Energy Res. 40 (10) (2016) 1303–1320.

[61] P. Chawla, M. Tripathi, Novel improvements in the sensitizers of dye-sensitized solar cells for enhancement in efficiency - a review, Int. J. Energy Res. 39 (2015) 1579−1596.
[62] Y. Sheng, M. Li, M.M. Flores-Leoner, W. Lu, J. Yang, Y. Hu, Rational design of SM315-based porphyrin sensitizers for highly efficient dye-sensitized solar cells: A theoretical study, J. Mol. Struct. 1205 (2020)127567.
[63] Y. Ren, D. Sun, Y. Cao, H.N. Tsao, Y. Yuan, S.M. Zakeeruddin, et al., A stable blue photosensitizer for color palette of dye-sensitized solar cells reaching 12.6% efficiency, J. Am. Chem. Soc. 140 (2018) 2405−2408.
[64] M. Godfroy, J. Liotier, V.M. Mwalukuku, D. Joly, Q. Huaulme, L. Cabau, et al., Benzothiadiazole-based photosensitizers for efficient and stable dye-sensitized solar cells and 8.7% efficiency semi-transparent mini-modules, Sustain. Energy Fuels 5 (2021) 144−153.
[65] A. Babu, A.S. Thangaraj, S. Kalaiyar, Co-sensitization of ruthenium(II) dye-sensitized solar cells by coumarin based dyes, Chem. Phys. Lett. 699 (2018) 32−39.
[66] S.A. Agarkar, R.R. Kulkarni, V.V. Dhas, A.A. Chinchansure, P. Hazra, S.P. Joshi, et al., Isobutrin from *Butea Monosperma* (flame of the forest): A promising new natural sensitizer belonging to chalcone class, ACS Appl. Mater. Interfaces 3 (2011) 2440−2444.
[67] J.M. Ji, H. Zhou, Y.K. Eom, C.H. Kim, H.K. Kim, 14.2% Efficiency dye-sensitized solar cells by co-sensitizing Novel Thieno [3,2-b]indole-based organic dyes with a promising porphyrin sensitizer, Adv. Energy Mater. 10 (15) (2020)2000124.
[68] Z. Huo, S. Dai, K. Wang, F. Kong, C. Zhang, X. Pan, et al., Nanocomposite gel electrolyte with large enhanced charge transport properties of an I_3^-/I^- redox couple for quasi-solid-state dye-sensitized solar cells, Sol. Energy Mater. Sol. Cell 91 (2007) 1959−1965.
[69] D. Xu, H. Zhang, X. Chen, F. Yan, Imidazolium functionalized cobalt tris(bipyridyl) complex redox shuttles for high efficient ionic liquid electrolyte dye-sensitized solar cells, J. Mater. Chem. A 1 (2013) 11933−11941.
[70] L. Chen, B. Xue, X. Liu, K. Li, Y. Luo, Q. Meng, et al., Efficiency enhancement of dye-sensitized solar cells: Using salt CuI as an additive in an ionic liquid, Chin. Phys. Lett. 24 (2007) 555−558.
[71] C. Teng, X. Yang, C. Yuan, C. Li, R. Chen, H. Tian, et al., Two novel carbazole dyes for dye-sensitized solar cells with open-circuit voltages up to 1V based on Br^-/Br_3^- electrolytes, Org. Lett. 11 (2009) 5542−5545.
[72] F. Sauvage, S. Chhor, A. Marchioro, J. Moser, M. Grätzel, Butyronitrile-based electrolyte for dye-sensitized solar cells, J. Am. Chem. Soc. 133 (2011) 13103−13109.
[73] P. Wang, C. Klein, R. Humphry-Baker, S. Zakeeruddin, M. Grätzel, Stable ≥ 8% efficient nanocrystalline dye-sensitized solar cell based on an electrolyte of low volatility, Appl. Phys. Lett. 86 (2005) 123508.
[74] X. Chen, D. Xu, L. Qiu, S. Li, W. Zhang, F. Yan, Imidazolium functionalized TEMPO/iodide hybrid redox couple for highly efficient dye-sensitized solar cells, J. Mater. Chem. A 1 (2013) 8759−8765.
[75] H. Lee, W. Kim, S. Park, W. Shin, S. Jin, J. Lee, et al., Effects of nanocrystalline porous TiO_2 films on interface adsorption of phthalocyanines and polymer electrolytes in dye-sensitized solar cells, Macromol. Symp. 235 (1) (2006) 230−236.
[76] N. Jeon, D. Hwang, Y. Kang, S. Im, D. Kim, Quasi-solid-state dye-sensitized solar cells assembled with polymeric ionic liquid and poly(3,4-ethylenedioxythiophene) counter electrode, Electrochem. Commun. 34 (2013) 1−4.
[77] D. Shi, N. Pootrakulchote, R. Li, J. Guo, Y. Wang, S. Zakeeruddin, et al., New efficiency records for stable dye-sensitized solar cells with low-volatility and ionic liquid electrolytes, J. Phys. Chem. C. 112 (2008) 17046−17050.

[78] B. Zhang, D. Wang, Y. Hou, S. Yang, X.H. Yang, J.H. Zhong, et al., Facet-dependent catalytic activity of platinum nanocrystals for triiodide reduction in dye-sensitized solar cells, Sci. Rep. 3 (2013) 1836.
[79] B. He, X. Meng, Q. Tang, Low-cost counter electrodes from CoPt alloys for efficient dye-sensitized solar cells, ACS Appl. Mater. Interfaces 6 (2014) 4812–4818.
[80] J. Wan, G. Fang, H. Yin, X. Liu, D. Liu, M. Zhao, et al., Pt–Ni Alloy nanoparticles as superior counter electrodes for dye-sensitized solar cells: Experimental and theoretical understanding, Adv. Mater. 26 (2014) 8101–8106.
[81] Q. Yang, P. Yang, J. Duan, X. Wang, L. Wang, Z. Wang, et al., Ternary platinum alloy counter electrodes for high-efficiency dye-sensitized solar cells, Electrochim. Acta 190 (2016) 85–91.
[82] J. Han, K. Yoo, M. Ko, B. Yu, Y. Noh, O. Song, Effect of the thickness of the Ru-coating on a counter electrode on the performance of a dye-sensitized solar cell, Met. Mater. Int. 18 (2012) 105–108.
[83] P. Yang, Q. Tang, A branching NiCuPt alloy counter electrode for high-efficiency dye-sensitized solar cell, Appl. Surf. Sci. 362 (2016) 28–34.
[84] P. Yang, Q. Tang, Alloying of Pt with Ni microtubes and Co nanosheets for counter electrode of dye-sensitized solar cell, Mater. Lett. 164 (2016) 206–209.
[85] Z. Lan, L. Que, W. Wu, J. Wu, High-performance Pt-NiO nanosheet-based counter electrodes for dye-sensitized solar cells, J. Solid. State Electrochem. 20 (2016) 759–766.
[86] M.M. Rahman, N.C.D. Nath, J.-J. Lee, Electrochemical impedance spectroscopic analysis of sensitization-based solar cells, Isr. J. Chem. 55 (9) (2015) 990–1001.
[87] A. Omar, M.S. Ali, N.A. Rahim, Electron transport properties analysis of titanium dioxide dye-sensitized solar cells (TiO_2-DSSCs) based natural dyes using electrochemical impedance spectroscopy concept: A review, Sol. Energy 207 (1) (2020) 1088–1121.
[88] A.R.C. Bredar, A.L. Chown, A.R. Burton, B.H. Farnum, Electrochemical impedance spectroscopy of metal oxide electrodes for energy applications, ACS Appl. Energy Mater. 3 (1) (2020) 66–98.
[89] J. Maçaira, L. Andrade, A. Mendes, Review on nanostructured photoelectrodes for next generation dye-sensitized solar cells, Renew. Sustain. Energy Rev. 27 (2013) 334–349.
[90] C. Cavallo, F.D. Pascasio, A. Latini, M. Bonomo, D. Dini, Nanostructured semiconductor materials for dye-sensitized solar cells, J. Nanomater. (2017)5323164.
[91] M.-E. Yeoh, K.-Y. Chan, Recent advances in photo-anode for dye-sensitized solar cells: A review, Int. J. Energy Res. 41 (2017) 2446–2467.
[92] J.S. Shaikh, N.S. Shaikh, S.S. Mali, J.V. Patil, K.K. Pawar, P. Kanjanaboos, et al., Nanoarchitectures in dye-sensitized solar cells: Metal oxides, oxide perovskites and carbon-based materials, Nanoscale 10 (2018) 4987–5034.
[93] H. Sayahi, K. Aghapoor, F. Mohsenzadeh, M.M. Morad, H.R. Darabi, TiO_2 nanorods integrated with titania nanoparticles: Large specific surface area 1D nanostructures for improved efficiency of dye-sensitized solar cells (DSSCs), Sol. Energy 215 (2021) 311–320.
[94] P. Zhong, W. Que, J. Zhang, Q. Jia, W. Wang, Y. Liao, et al., Charge transport and recombination in dye-sensitized solar cells based on hybrid films of TiO_2 particles/ TiO_2 nanotubes, J. Alloy. Compd. 509 (29) (2011) 7808–7813.
[95] Y. Du, Q. Feng, C. Chen, Y. Tanaka, X. Yang, Photocatalytic and dye-sensitized solar cell performances of {010}-faceted and [111]-faceted anatase TiO_2 nanocrystals synthesized from tetratitanate nanoribbons, ACS Appl. Mater. Interfaces 6 (2014) 16007–16019.
[96] J.-Y. Liao, J.-W. He, H. Xu, D.-B. Kuang, C.-Y. Su, Effect of TiO_2 morphology on photovoltaic performance of dye-sensitized solar cells: Nanoparticles, nanofibers, hierarchical spheres and ellipsoid spheres, J. Mater. Chem. 22 (2012) 7910–7918.

[97] Z. Lai, F. Peng, Y. Wang, H. Wang, H. Yu, P. Liu, et al., Low temperature solvothermal synthesis of anatase TiO_2 single crystals with wholly {100} and {001} faceted surfaces, J. Mater. Chem. 22 (2012) 23906−23912.

[98] C. Chen, L. Xu, G.A. Sewvandi, T. Kusunose, Y. Tanaka, S. Nakanishi, et al., Microwave-assisted topochemical conversion of layered titanate nanosheets to {010}-faceted anatase nanocrystals for high performance photocatalysts and dye-sensitized solar cells, Cryst. Growth Des. 14 (2014) 5801−5811.

[99] M.N. Mustafa, Y. Sulaiman, Review on the effect of compact layers and light scattering layers on the enhancement of dye-sensitized solar cells, Sol. Energy 215 (2021) 26−43.

[100] J. Gong, K. Sumathy, Q. Qiao, Z. Zhou, Review on dye-sensitized solar cells (DSSCs): Advanced techniques and research trends, Renew. Sustain. Energy Rev. 68 (2017) 234−246.

[101] Q. Zhang, D. Myers, J. Lan, S.A. Jenekhe, G. Cao, Applications of light scattering in dye-sensitized solar cells, Phys. Chem. Chem. Phys. 14 (2012) 14982−14998.

[102] T.G. Deepak, G.S. Anjusree, S. Thomas, T.A. Arun, S.V. Nair, A.S. Nair, A review on materials for light scattering in dye-sensitized solar cells, RSC Adv. 4 (2014) 17615−17638.

[103] H. Yu, Y. Bai, X. Zong, F. Tang, G.Q.M. Lu, L. Wang, Cubic CeO_2 nanoparticles as mirror-like scattering layers for efficient light harvesting in dye-sensitized solar cells, Chem. Commun. 48 (2012) 7386−7388.

[104] M.N. Mustafa, S. Shafie, M.H. Wahid, Y. Sulaiman, Light scattering effect of poly-vinyl-alcohol/titanium dioxide nanofibers in the dye-sensitized solar cell, Sci. Rep. 9 (2019) 14952.

[105] M. Motlak, A.M. Hamza, M.G. Hammed, N.A.M. Barakat, Cd-doped TiO_2 nanofibers as effective working electrode for the dye sensitized solar cells, Mater. Lett. 246 (2019) 206−209.

[106] D. Wongratanaphisan, K. Kaewyai, S. Choopun, A. Gardchareon, P. Ruankham, S. Phadungdhitidhada, CuO-Cu_2O nanocomposite layer for light-harvesting enhancement in ZnO dye-sensitized solar cells, Appl. Surf. Sci. 474 (2019) 85−90.

[107] A. Roy, S. Sundaram, T.K. Mallick, Effect of dye sensitization's temperature on ZnO-based solar cells, Chem. Phys. Lett. 776 (2021) 138668.

[108] J. Singh, A. Gusain, V. Saxena, A.K. Chauhan, P. Veerender, S.P. Koiry, et al., XPS, UV−Vis, FTIR, and EXAFS studies to investigate the binding mechanism of N719 dye onto oxalic acid treated TiO_2 and its implication on photovoltaic properties, J. Phys. Chem. 117 (2013) 21096−21104.

[109] A. Sahu, A. Garg, A. Dixit, A review on quantum dot sensitized solar cells: Past, present and future towards carrier multiplication with a possibility for higher efficiency, Sol. Energy 203 (2020) 210−239.

[110] H. Song, Y. Lin, Z. Zhang, H. Rao, W. Wang, Y. Fang, et al., Improving the efficiency of quantum dot sensitized solar cells beyond 15% via secondary deposition, J. Am. Chem. Soc. 143 (2021) 4790−4800.

[111] N. Shanmugam, R. Pugazhendhi, R.M. Elavarasan, P. Kasiviswanathan, N. Das, Anti-reflective coating materials: A holistic review from PV perspective, Energies 13 (2020) 2631.

CHAPTER EIGHT

Development of active layer materials for solution-processable organic photovoltaics

Geneviève Sauvé
Department of Chemistry, Case Western Reserve University, Cleveland, OH, United States

8.1 Introduction

Organic photovoltaics (OPV) use organic semiconductors to convert sunlight into electricity. Organic photovoltaics are attractive because they can be fabricated using roll-to-roll fast throughput methods on flexible substrates to enable low-cost, lightweight, flexible, and colored or semitransparent devices, greatly expanding photovoltaic applications, for example, Fig. 8.1 [1–9]. Their lightweight and flexible properties enable portable and wearable power, such as using OPV on tents, canopies, backpacks, or clothing [10]. The ability to tailor the color enables a plethora of designs for integrated power generation where you need it, such as urban trees, tents, and backpacks [11]. They can efficiently harvest indoor ambient light and self-power Internet of Things devices [12]. By selecting active layer materials that selectively harvest UV and NIR light while transmitting visible light, one can make semitransparent OPV that could be integrated into the windows of buildings [13] or create self-powered greenhouses with better plant growth efficiencies [2,14,15]. Polymer-based OPV that are solution-processed promise to have very low energy payback time of less than a few months [16] and have a very low power CO_2 footprint, at less than 10 g/kWh [11].

Organic semiconductors are typically π-conjugated molecules or polymers made from abundant elements (C, H, O, and N), whose optoelectronic properties can be tuned by altering the chemical composition [17]. They typically have high extinction coefficients such that very thin layers

Photovoltaics Beyond Silicon.
DOI: https://doi.org/10.1016/B978-0-323-90188-8.00007-5
© 2024 Elsevier Inc. All rights reserved, including those for text and data mining, AI training, and similar technologies.

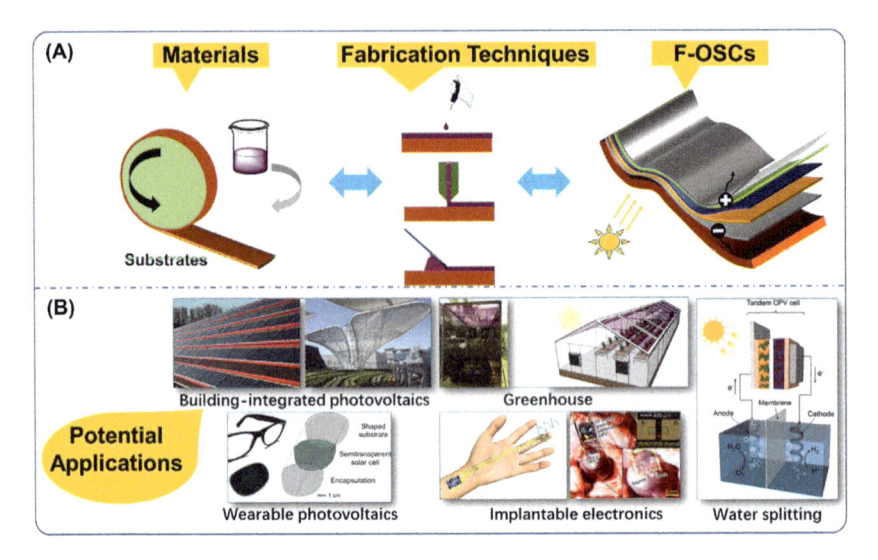

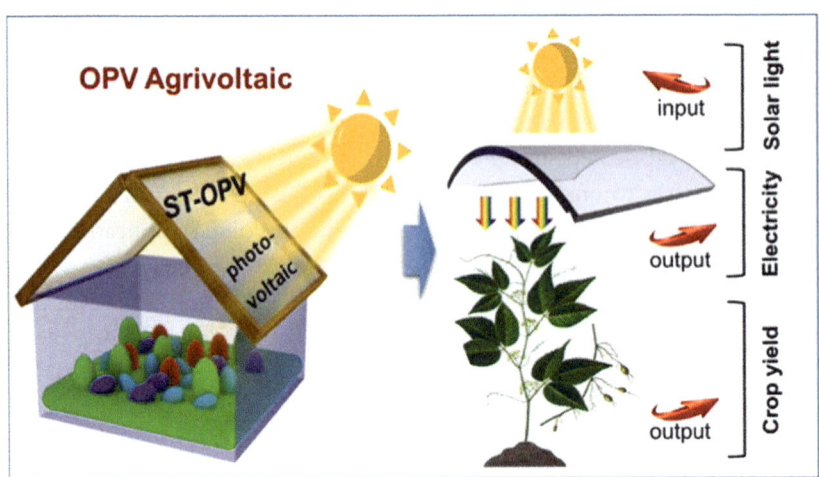

Figure 8.1 (*Top*) Schematic diagram of source materials and fabrication techniques of flexible organic solar cells and potential types of applications. (*Bottom*) Schematic depiction of a greenhouse agrivoltaic system with production of both electricity and crops. *(Top) Reprinted with permission from C. Liu, C. Xiao, C. Xie, W. Li, Flexible organic solar cells: Materials, large-area fabrication techniques and potential applications, Nano Energy 89 (B) (2021) 106399, copyright (2021) Elsevier. (Bottom) Reproduced with permission from W. Song, J. Ge, L. Xie, Z. Chen, Q. Ye, et al., Semi-transparent organic photovoltaics for agrivoltaic applications, Nano Energy 116 (2023) 108805, copyright (2023) Elsevier.*

(100–200 nm) can absorb all incident light. Unlike inorganic semiconductors, organic semiconductors typically have a low dielectric constant and thus a high exciton binding energy [18]. As a result, an interface between an electron donor (or *p*-type) and an electron acceptor (or *n*-type) with energy offsets is required to split the excitons into free charges [19]. The energy offset between the lowest occupied molecular orbital (LUMO) of the donor and acceptor (ΔLUMO) must be larger than the exciton binding energy, often at least 0.3 eV, to "drive" electron transfer from donor to acceptor, see Fig. 8.2A.

Similarly, an energy offset between the highest occupied molecular orbital (HOMO) of the donor and acceptor (ΔHOMO) is beneficial for hole transfer. These energy offsets reduce the maximum open circuit voltage (V_{oc}) that can be obtained. Because organic semiconductors typically have low exciton diffusion lengths (10–30 nm), it is also necessary to have a large interfacial area. To obtain a large interfacial area, OPV research has focused on donor−acceptor bulk heterojunction (BHJ), where the donor and acceptor are mixed and processed to obtain a favorable film morphology for OPV [3,19]. The ideal film morphology is complex [20,21], but at a minimum, the donor and acceptor materials must phase separate on the nanoscale to allow efficient exciton splitting, and each phase must form an interconnected network to transport charges to the collecting electrodes. The basic working principles are as follows: first light is absorbed to form a bound exciton. If light is absorbed by the donor, it is called "channel I" and if light is absorbed by the acceptor, it is called "channel II," see Fig. 8.2B.

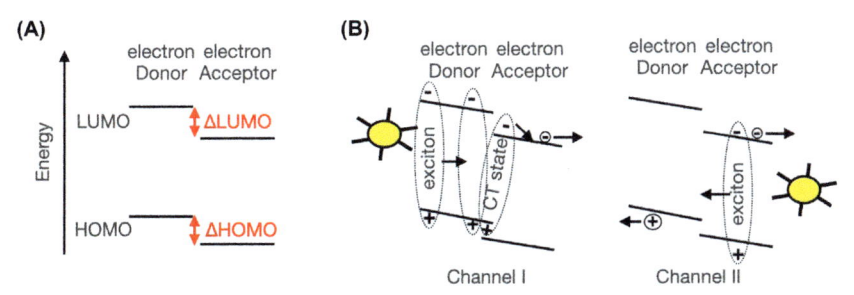

Figure 8.2 (A) Energy level diagram illustrating the energy offsets ΔHOMO and ΔLUMO; (B) Channel I is the most well-understood. Here, upon light absorption, a bound exciton is formed in the donor, which then diffuses to the interface. At the interface, it forms a charge transfer (CT) state which then dissociates into free charges. In Channel II, the electron acceptor absorbs the light.

Ideally, both donor and acceptor light absorption should be optimized. The exciton must then diffuse to a donor—acceptor interface for charge separation, which occurs first by creating a bound charge transfer (CT) state, and then free charges. The free charges can then be transported to the collecting electrodes and extracted using selective contacts. To ensure efficient free charge collection, drift is obtained by choosing electrodes with different work functions. Additionally, electron-transport (hole blocking) and hole transport (electron blocking) layers are used at the cathode and anode, respectively. Depending on the order of the layers, two common device geometries are commonly used: the conventional and inverted device structures, shown in Fig. 8.3 [19].

Considerable research has been devoted to OPV development over the past three decades, which has resulted in steady increases in power conversion efficiency (PCE) over time. Fig. 8.4 depicts the highest confirmed PCE for OPV between 2001 and 2023; data utilized are obtained from the US Department of Energy National Renewable Energy Laboratory's interactive best research-cell efficiency chart [7]. While performance improvements result from a combination of device engineering, processing optimization, and materials development, we observe three major stages that roughly coincide with three major active layer material

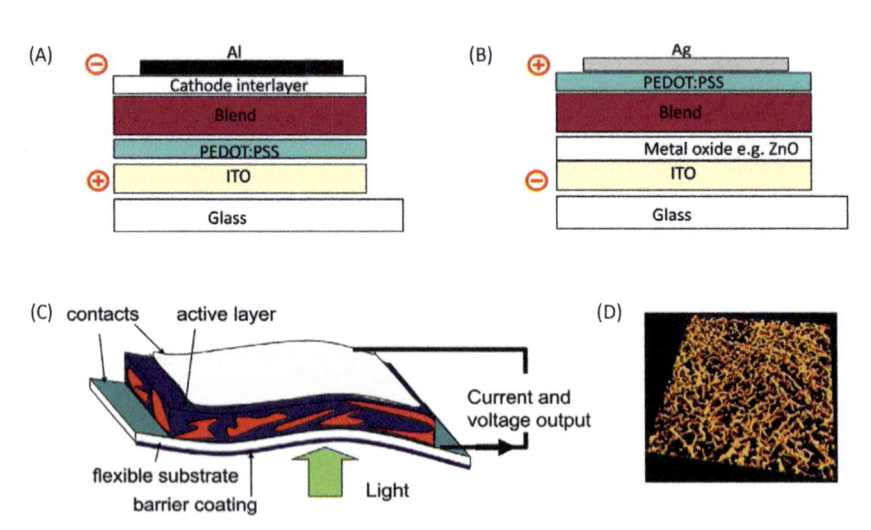

Figure 8.3 Layer structure of an organic solar cell in the traditional (A) and inverted (B) polarity; (C) schematic of the layers in an OPV module on a flexible substrate; (D) image of the microstructure within the active layer, obtained with electron tomography. *Reprinted with permission from J. Nelson, Polymer: fullerene bulk heterojunction solar cells, Mater. Today 14 (2011) 462—470, copyright (2011) Elsevier.*

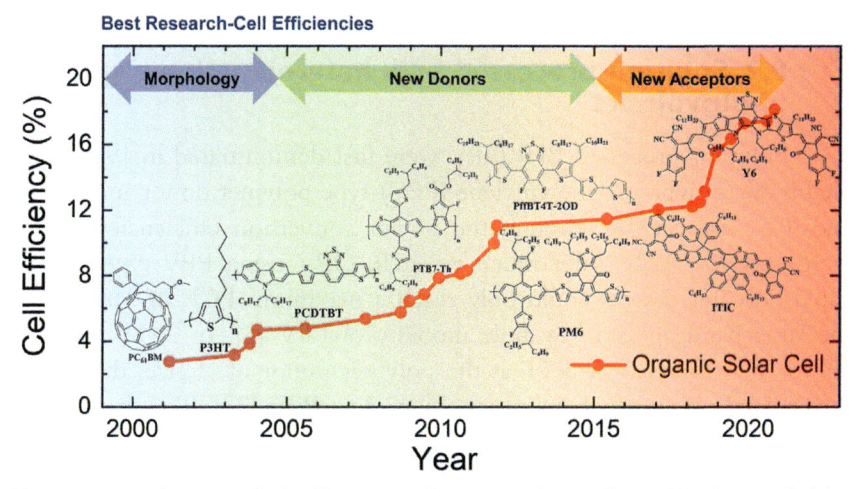

Figure 8.4 Highest certified efficiencies of organic photovoltaics (OPV) recorded by the National Renewable Energy Laboratory (NREL) with depiction of the three main stages of OPV material development. *Reproduced with permission from G. Zhang, F.R. Lin, F. Qi, T. Heumuller, A. Distler, et al., Renewed prospects for organic photovoltaics. Chem. Rev. 122 (2022) 14180–14274, copyright (2022) American Chemical Society.*

developments: (1) OPV that uses regioregular poly(3-hexylthiophene) (P3HT) as the polymer electron donor and a fullerene derivative such as [6,6]-phenyl C_{61} butyric acid methyl ester (PCBM) as the electron acceptor. This system was used to learn about BHJ morphology; (2) the electron donor is optimized while keeping fullerene derivatives as the electron acceptor; and (3) the discovery of new high-performance nonfullerene acceptors (NFA). Active layer materials are designed either for vacuum or solution processability.

Material systems that are solution-processed dominate in Fig. 8.4 [8], mostly due to much greater interest and research in solution-processable material development. Additionally, while solution-processed OPV containing both molecular donor and acceptor have been published, it is more common to have a least one of the two materials be a polymer because polymers are more viscous and thus have much better film formation properties than small molecules. This chapter will thus discuss active layer material development for solution-processed OPV where at least one component is a polymer. We note that this chapter is not intended to be a complete account of all the active layer material development for OPV. Rather, this chapter focuses on some of the major developments in the field as well as considerations for commercialization and future opportunities.

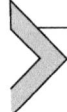

8.2 Solution-processed bulk heterojunction organic photovoltaics

Solution-processed BHJ OPV were first demonstrated in 1995 with blends of a polyphenylene vinylene (PPV)-type polymer donor and fullerene (C_{60}) as acceptor, though the power conversion efficiencies (PCE) were very low [22]. After much research, OPV using PPV-type conjugated polymer donors could only reach a maximum PCE of 3% due to the large bandgap and low hole mobility of PPV [3]. In 2002, research efforts then turned to P3HT as the polymer donor of choice, due to its lower bandgap and higher hole mobility than PPV [23]. The first breakthrough for P3HT blended with [6,6]-phenyl C_{61}-butyric acid methyl ester (PCBM) OPV was in 2003, where a PCE was of 3.5% with postproduction treatment (annealing) [24]. Since then, P3HT:PCBM blends (shown in Fig. 8.5) have been heavily studied for BHJ OPV, with 1033 publications between 2002 and 2010 alone [25]. Because these materials are readily available, they have been used as a benchmark and used to study how to control and image film morphology [26,27], optimize device architectures, interlayers, and electrodes [25], as well as in manufacturing larger area OPV and measuring device stability [28,29].

Optimizing P3HT:PCBM BHJ OPV has provided some of the basics in using processing conditions to optimize film morphology and PCE. In the first demonstration of a 3.5% PCE, Sariciftci and coworkers demonstrated that annealing the OPV devices while simultaneously applying an electric field greater than V_{oc} increased PCE from 2.5% to 3.5%, presumably by enhancing crystallization and orientation of the polymer [24]. In 2005, Yang and coworkers demonstrated that controlling the active layer growth rate increased hole mobility and balanced charge transport of holes and electrons, resulting in a high PCE of 4.4% [30]. This high efficiency was

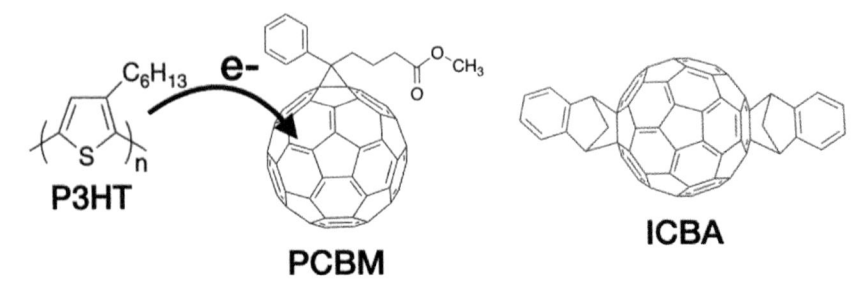

Figure 8.5 Chemical structure of P3HT and the fullerene derivatives PCBM and ICBA.

obtained through the combination of: (1) slowing the active layer growth rate by replacing a low boiling point solvent (chloroform) with a high boiling point solvent (1,2-dichlorobenzene), which encourages ordering of the P3HT and (2) thermal annealing to further improve crystallization and reduce traps. Kroeze and coworkers systematically studied the effect of thermal annealing of a chloroform-cast film [31]. Thermal annealing at 80°C allowed P3HT to form crystalline nanofibrils, whereas further annealing at 130°C resulted in phase separation and formation of nanometer-sized PCBM clusters. In 2006, Kowalewski and coworkers demonstrated that P3HT nanofibrils are π-stacked P3HT sheets where the polymer backbone is perpendicular to the nanofibril axis [32]. In 2009, Loos and coworkers used electron tomography to demonstrate the formation of three-dimensional nanoscale organization of P3HT:PCBM blends upon either thermal or solvent annealing, see Fig. 8.3D [33]. They clearly image the formation of micrometer-long 15 nm wide P3HT nanofibrils and a favorable concentration gradient across the active layer thickness.

Overall, most reported PCEs for P3HT:PCBM are between 3% and 4%, with certified cells by Plextronics, Konarka, and Sharp between 3.5% and 4% [25]. This relatively low PCE has limited commercial appeal. To increase PCE, it is critical to optimize the active layer materials. For example, replacing PCBM with [6,6]-phenyl C71-butyric acid methyl ester (PC71BM) slightly increases the efficiency to approximately 4.7% due to enhanced absorption [34]. To increase V_{oc}, PCBM can be replaced with fullerene derivatives that have higher LUMO energy levels. For example, an indene-C60 bis-adduct (ICBA, Fig. 8.3C), which has a LUMO level 0.17 eV higher than PCBM, gave a V_{oc} of 0.84 V and PCE of 6.5% [35], while the corresponding indene-C70 bisadduct (IC70BA) reached a PCE as high as 7.5% [36,37]. Alternatively, one can replace P3HT (bandgap approximately 1.9 eV) with π-conjugated polymers that have lower bandgap to absorb more of the visible light. This will be examined in the next section.

8.3 Electron donors optimized for fullerene-based acceptors

As mentioned in the previous section, P3HT:PCBM OPV PCE is limited to 4%—5%, which is too low for practical applications. These relatively low PCEs are mainly due to poor overlap with the solar spectrum and large energy losses [3]. To overcome these limitations, the community shifted to developing new π-conjugated p-type materials to be blended with fullerene derivatives. Much of the work was guided by a key paper published in 2006 by Brabec and coworkers [38]. They showed how the V_{oc} of a BHJ polymer-PCBM OPV is determined by the HOMO level of the electron donor and the LUMO level of the electron acceptor following Eq. (8.1):

$$V_{oc} = \frac{1}{e}\left(\left|E^{Donor}_{HOMO}\right| - \left|E^{PCBM}_{LUMO}\right|\right) - 0.3 \tag{8.1}$$

where e is the elementary charge, E_{LUMO} of PCBM is -4.3 eV, and 0.3 V is an empirical number that depends on dark current−voltage loss and field-driven current loss. They were then able to relate the bandgap and LUMO level of the donor to OPV performance, depicted in Fig. 8.6 [38].

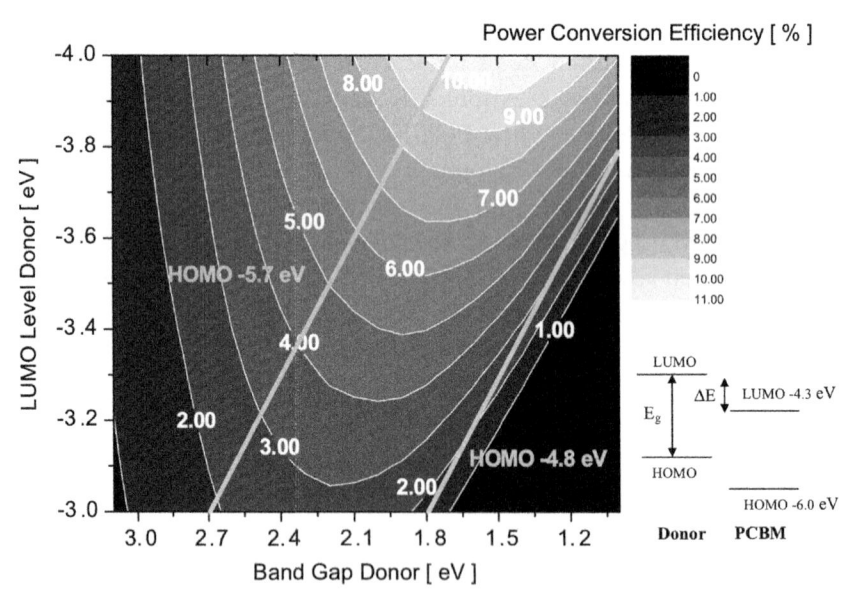

Figure 8.6 Contour plot showing the calculated energy-conversion efficiency (contour lines and gray shades) versus the bandgap and the LUMO level of the donor polymer according to the model described in reference (see below); straight lines starting at 2.7 and 1.8 eV indicate HOMO levels of −5.7 and −4.8 eV, respectively. A schematic energy diagram of a donor PCBM system with the bandgap energy (E_g) and the energy difference (ΔE) is also shown. *Reproduced with permission from M. Scharber, D. Muehlbacher, M. Koppe, P. Denk, C. Waldauf, A. Heeger, C. Brabec, Design rules for donors in bulk-heterojunction solar cells-towards 10% energy-conversion efficiency, Adv. Mater. 18 (2006) 789−794, copyright (2006) John Wiley & Sons, Inc.*

The model assumes an FF and average EQE of 0.65, which were typical high values experimentally observed for OPV at the time, and a LUMO energy offset of 0.3 eV. The model predicted that to reach a PCE above 10%, the bandgap and LUMO energy level of the donor must be <1.74 and < − 3.92 eV, respectively.

To synthesize a new conjugated polymer with optimal bandgap and LUMO energy level, the OPV community turned to alternate D-A copolymers, where D is an electron-rich moiety and A is an electron-poor moiety [39—41]. The basic approach is that the choice of D mainly tunes the HOMO level, and the choice of A mainly tunes the LUMO level of the polymer, resulting in a reduced band gap, see Fig. 8.7A. Electron-rich units (D) often have heteroatoms with lone electron pairs that can donate electrons to the π-system, such as S in thiophene, N in carbazole, or alkyloxy (−OR) groups. Electron-poor units (A) typically have electron-withdrawing groups such as sp^2-hybridized N (−N═C−), halogens such as F and Cl, carbonyls (−C═O), and cyano (−C≡N) groups. Representative examples of D-A copolymers with lower band gap designed for OPV are given in Fig. 8.7B, including cyclopentadithiophene-based copolymers such as PCPDTBT, carbazole-based copolymers such as PCDTBT, and benzodithiophene-based copolymers such as PTB7.

Designing polymers with optimal energy levels is not sufficient to obtain an efficient OPV. For high performance, a favorable film morphology between the polymer and PCBM is required. However, D-A copolymers did not form nanowires like P3HT does, making it difficult to

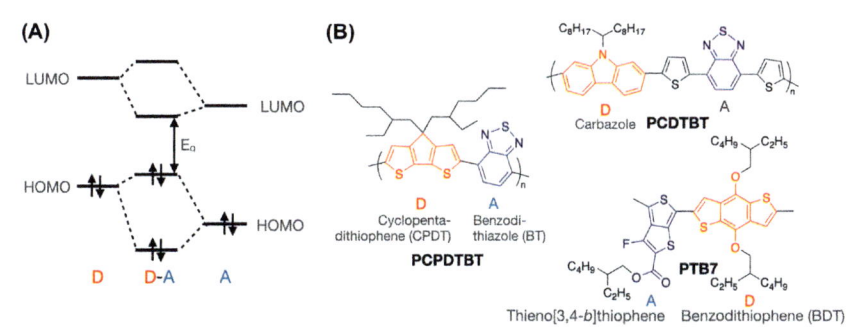

Figure 8.7 (A) Orbital interactions of electron-rich D and electron-poor A units lead to a smaller band gap when they are chemically linked (A−D). (B) Example of D-A copolymers with lower band gaps. Red depicts electron-rich units D, and blue depicts electron-poor units A. In PCDTBT, Thiophene (in black) is electron-rich but is used as a pi-bridge between A and D here.

obtain a nanoscale blend phase separation. In 2007, Heeger and coworkers published a major breakthrough to improve film morphology of D-A copolymer:PCBM blends with the use of processing additives [42]. In this contribution, they discovered that 1,8-diiodooctane (DIO) enabled improved morphology due to its high boiling point and selective solubility of the fullerene component; By keeping the fullerene component in solution, it encourages the conjugated polymer donor to precipitate and crystallize before the fullerene derivative, see Fig. 8.8. The PCPDTBT:$PC_{71}BM$ cell improved from 3.4% without additive to 5.1% with DIO. Using this type of processing additive enabled D-A copolymer donors to form favorable morphologies with fullerene derivatives. For example, both carbazole-based donors (e.g., PCDTBT) and benzodithiophene-based donors (e.g., PTB7) blends reached PCEs above 7% [43,44].

Further polymer donor optimization led to several conjugated polymer donor chemical structure designs that encourage favorable blend morphologies, propelling PCEs of fullerene-based OPV to the 10%−12% range. The following are a few selected examples. High PCE can be obtained by controlling the polymer aggregation in the blend. Yan and coworkers designed conjugated polymer donors whose aggregation could be temperature-controlled to give highly crystalline yet small polymer domains [45]. This was achieved by combining side chains with branches away from the polymer backbone with selective backbone fluorination, for example, PffBT4T-2OD in Fig. 8.9. Interestingly, high PCE >10%

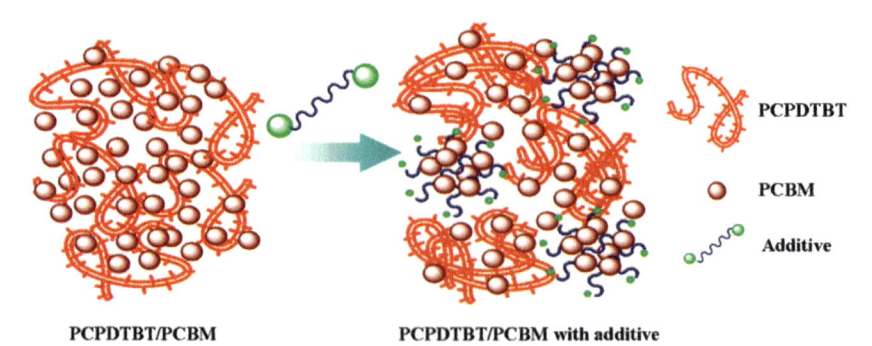

Figure 8.8 Schematic depiction of the role of the processing additive in the self-assembly of bulk heterojunction blend materials and structures of PCPDTBT, $PC_{71}BM$, and additive. *Reproduced with permission from J. Lee, W. Ma, C. Brabec, J. Yuen, J. Moon, et al., Processing additives for improved efficiency from bulk heterojunction solar cells. J. Am. Chem. Soc., 130 (2008) 3619−3623, copyright (2008) American Chemical Society.*

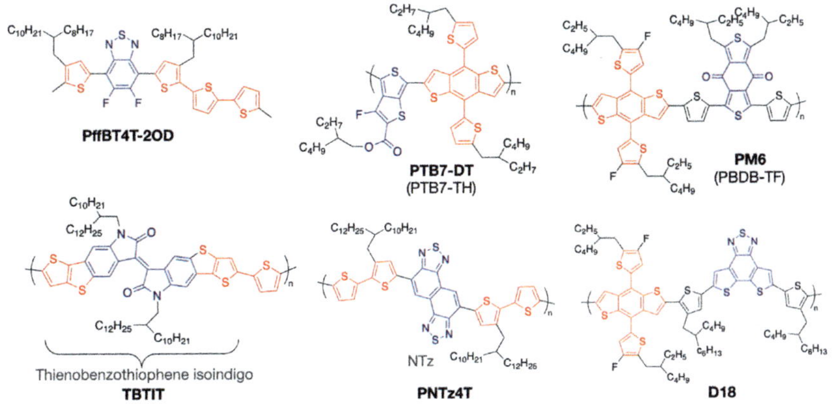

Figure 8.9 Chemical structures of polymer donors discussed in the text.

was achieved irrespective of the type of fullerene derivative used for several polymer:fullerene combinations.

Another polymer design that has led to better PCE is the so-called "2D polymers" where nonconjugated side chains are replaced with conjugated 5-alkylthiophene-2-yl side chains; see for example PTB7-DT (also called PTB7-TH) in Fig. 8.9 [46]. This extends conjugation perpendicular to the polymer backbone, narrows the bandgap, tightens π−π stacking, and increases hole mobility, resulting in a PCE slightly over 10%. Hou and coworkers added F on the alkylthiophene side chains to lower the HOMO of the copolymer in their design of the high-performance polymer PM6 [47], also known as PBDB-TF and PBDB-T-2F.

The development of large planar fused ring units has resulted in D-A polymers with tighter π−π stacking of the conjugated polymer backbone and higher crystallinity. For example, McCulloch and coworkers designed a conjugated polymer donor based on an eight-ring fused isoindigo moiety copolymerized with thiophene, called TBTIT [48]. The high molecular weight version of this polymer combined with $PC_{71}BM$ gave a high PCE of 9.1% without the use of additives or annealing due to the new polymer's high degree of crystallinity. In another example, Cao and coworkers replaced the common A unit benzodithiazole (BT) with a naphthyl fused ring analog, the naphthothiadiazole (NTz) unit [49]. A copolymer that uses the NTz unit, PNTz4T, showed a high PCE of approximately 10% when blended with $PC_{71}BM$, in inverted OPV with thick active layers of 300 nm thick [50]. The high PCE was attributed to a large population of crystallite in the "face-on" orientation, facilitating hole

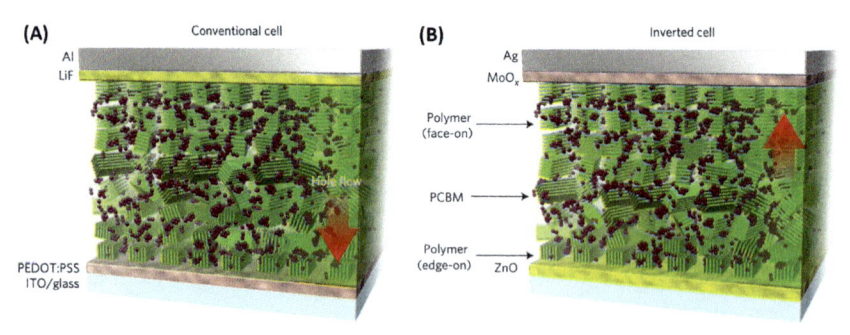

Figure 8.10 Schematic illustrations of PNTz4T:PCBN blend films. (A) Conventional cell with PEDOT:PSS as the bottom and LiF as the top interlayer. (B) Inverted cell with ZnO as the bottom and MoO_x as the top interlayer. The population of face-on crystallite is larger in the inverted cell than in the conventional cell. In both cases, the population of edge-on crystallites is large at the bottom interface and the population of face-on crystallites is large in the bulk through the top interface. Note that the amount of PCBM shown is markedly reduced compared with real cells, and the distribution of the orientation is exaggerated in order to better visualize the polymer orientation. *Reproduced with permission from V. Vohra, K. Kawashima, T. Kakara, T. Koganezawa, I. Osaka, K. Takimiya, H. Murata, Efficient inverted polymer solar cells employing favourable molecular orientation. Nat. Photonics, 9 (2015) 403–408, copyright (2015) Springer Nature.*

transport across the film through $\pi-\pi$ stacking. Also, the PCE was higher in the inverted geometry than the conventional geometry because the top of the blend film mainly consists of the polymer donor with face-on orientation, which is more beneficial for the inverted cell where holes must flow to the top interface, see Fig. 8.10 [50].

8.4 Nonfullerene electron acceptors

Fullerene-derivatives were initially the electron acceptor of choice in OPV because they can accept up to six electrons in its three degenerate LUMOs, have low reorganization energy for electron transfer, high electron mobility, and form favorable blend morphologies with conjugated polymer donors [51–53]. Moreover, their large spherical shape facilitates isotropic electron transport across the active layer [54,55]. However, fullerene-derivatives have several disadvantages. Fullerene derivatives have weak absorption in the visible range, which limits photocurrent. Their energy levels cannot be tuned easily without breaking the conjugated buckyball and usually require large energy offsets with the polymer donor to work well, both limiting V_{oc}. Their spherical shape also limits commercialization because they tend to aggregate over time and are mechanically brittle [56]. To reach higher PCEs and have stable OPV, it is therefore important to seek alternatives to fullerene-based acceptors.

Prior to 2012, the PCE for OPV based on NFA blended with a conjugated polymer donor rarely reached above 2% due to unfavorable blend morphology [57–59]. Low PCEs were obtained despite extensive optimization of the processing conditions, such as choice of solvent, thermal or solvent annealing, and additives. To further increase PCE, the community turned to optimizing the NFA chemical structure [60]. NFAs usually fall into two groups: molecular and polymeric electron acceptors. Research on polymeric NFA has been more limited than molecular NFA because it is generally harder to obtain favorable morphology by blending two conjugated polymers. That is because polymer–polymer blends are more likely than polymer–molecule blends to phase separate on a large scale, which would limit interfacial area. Additionally, n-type conjugated polymers that can accept and transport charges efficiently are rarer than p-type conjugated polymers. Nevertheless, several examples have shown PCE greater than 2%. Notably, OPV using NFA copolymers based on naphthalene diimide have reached PCEs greater than 8% [61]. We are not aware of devices that use a molecular donor blended with a conjugated polymer acceptor with a high(er) PCE; this combination is thus an opportunity for further discovery.

Most research has been devoted to the development of molecular NFAs [62–68]. The shape of the molecular acceptor is critical for BHJ morphology and optimizing device performance. Small molecular NFAs do not work well because they tend to be too miscible with polymer donors, which leads to poor electron transport across the blend film. In addition, small planar molecules perform particularly poorly in mixed phases, as has been observed in active layer blends. Imaging of P3HT: PCBM blends showed that film morphology may have three phases: pure P3HT, pure PCBM, and a mixed P3HT and PCBM, see Fig. 8.11A [69]. This mixed phase can be beneficial as it provides more interfaces for exciton splitting. In addition, excitons formed in the mixed phases may be more likely to split into free charges, as there is an energetic driving force for electrons to go into the pure PCBM phase and for holes to go to the pure P3HT phase, provided charges can find a connected path out of the mixed phase before recombining. Kopidakis and coworkers demonstrated

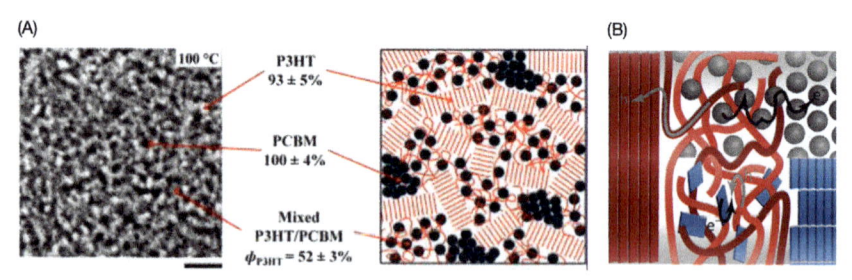

Figure 8.11 (A) Sulfur elemental map generated through EFTEM with schematic illustration for P3HT:PCBM blend annealed at 100°C for 30 min. The image intensity is proportional to the sulfur concentration. The scale bar is 100 nm. (B) Schematic illustration showing that a spherical acceptor shuttles electrons out of the mixed phase whereas a small planar acceptor results in poor electronic coupling between acceptor molecules, thus hindering electron transport. *Reproduced with permission from: (A) C. Guo, D.R. Kozub, S.V. Kesava, C. Wang, A. Hexemer, E.D. Gomez, Signatures of multiphase formation in the active layer of organic solar cells from resonant soft x-ray scattering, ACS Macro Lett. 2 (2013) 185−189, copyright (2013) American Chemical Society; (B) A.M. Nardes, A.J. Ferguson, P. Wolfer, K. Gui, P.L. Burn, P. Meredith, N. Kopidakis, Free carrier generation in organic photovoltaic bulk heterojunctions of conjugated polymers with molecular acceptors: Planar versus spherical acceptors. ChemPhysChem 15 (2014) 1539−1549, copyright (2014) John Wiley & Sons, Inc.*

that the acceptor molecular shape is critical to enable this [70]. They compared the photophysical performance of P3HT:PCBM with P3HT:K12 where K12 is a small planar NFA. The photophysical results from both systems were correlated with film structure. In the case of P3HT:PCBM, the mixed phase is efficient at creating free charges and transporting them out of the mixed phase, whereas in the P3HT:K12 blend, approximately 85% of the excitons fail to produce free charge carriers. This is attributed to the poor electron-transport properties in the mixed P3HT:K12 phase (subpercolation) coupled with enhanced charge recombination due to strong intermolecular interactions between the donor and acceptor (Fig. 8.11B) [70]. Designing large and planar acceptors would solve this issue, but such molecules tend to aggregate strongly, resulting in unfavorable large blend phase separation. To balance miscibility and aggregation and maximize charge generation in mixed phases, the community turned to designing large nonplanar molecular acceptors [60].

The concept of using a large nonplanar structure to optimize blend morphology was first illustrated using perylene dimiides [71]. Perylene diimides (PDIs, Fig. 8.12A) are well-known stable n-type organic semiconductors that have been used as high-performance pigments [72]. These large

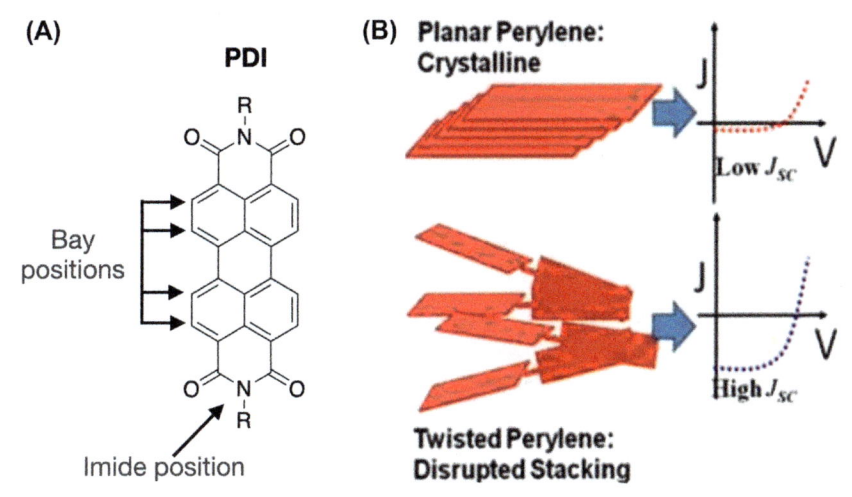

Figure 8.12 (A) Chemical structure of PDI. (B) Schematic illustration showing that planar PDI are crystalline and give low J_{sc} whereas the twisted PDI dimers have disrupted crystallinity and high J_{sc}. (B) Reproduced with permission from S. Rajaram, R. Shivanna, S.K. Kandappa, K.S. Narayan, Nonplanar perylene diimides as potential alternatives to fullerenes in organic solar cells. J. Phys. Chem. Lett. 3 (2012) 2405−2408, copyright (2012) American Chemical Society.

planar molecules have a strong tendency to aggregate, making it hard to obtain a favorable morphology when blended with a conjugated polymer donor. In 2012, Rajaram and coworkers showed that by linking two PDI molecules together at the imide position, PCE improved from 0.13% for the unsubstituted PDI to 2.8% for the dimer, see Fig. 8.12B [71]. The dimer is nonplanar because the two PDI planes are perpendicular to each other due to electronic repulsions between the carbonyl groups. While the planar PDI blends showed micrometer-sized crystals, the PDI dimer blends were featureless, indicating that the donors and acceptors are intimately mixed, yielding a large interfacial area for better exciton harvesting. Interface engineering to improve extraction increased the PCE to 4.6% [73] and tailoring the alkyl chains on the PDI dimer and using a 2D conjugated polymer donor further increased PCE to 5.4% [74].

Following this success, researchers have synthesized and tested a variety of PDI dimers using a linker at the bay positions, where the nature of the linker varied the dihedral angle between the two PDI planes from 40 to 80 degrees, see Fig. 8.13 [60,62]. All these acceptors resulted in optimized OPV with PCE of about 6% by 2015. Similar results were also obtained with other large planar acceptors. For example, Jenekhe and coworkers

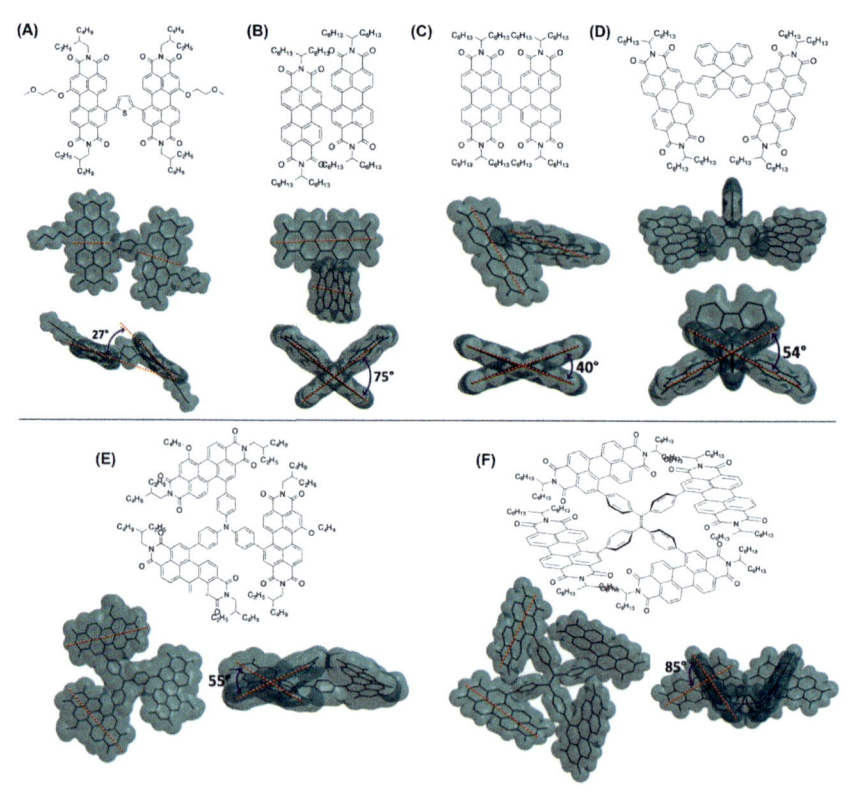

Figure 8.13 Chemical structure and electron density surfaces with chemical structure overlay of the optimized geometry for two orthogonal views for PDI-based acceptors. (A) PDI dimer with a thiophene linker; (B) PDI dimer with a single bond linker at bay positions; (C) PDI dimer with a two-carbon bridge; (D) PDI dimer with spirobifluorene linker; (E) PDI trimer with triphenylamine core; and (F) PDI tetramer with tetraphenylethylene core. *Reproduced with permission from G. Sauvé, R. Fernando, Beyond fullerenes: Designing alternative molecular electron acceptors for solution-processable bulk heterojunction organic photovoltaics, J. Phys. Chem. Lett. 6 (2015) 3770–3780, copyright (2015) American Chemical Society.*

demonstrated the concept using tetraazabenzo-difluororanthene diimide (BFI) molecules [75,76]. Further increasing size and 3D nature by using PDI trimers and tetramers were also heavily explored, with PCEs of approximately 4%–5% [60]. In 2016, Yan and coworkers optimized the conjugated polymer donor to be paired with the PDI dimer in Fig. 8.13D, resulting in a high PCE of 9.5% [77]. They observed ultrafast and efficient charge separation with a large internal quantum efficiency of nearly 90%, despite negligible driving force for charge separation. An impressive low voltage loss of 0.61 eV was observed, which is lower than the best polymer donor:PCBM OPV (0.87 V), a big advantage of using NFAs to reach higher OPV PCEs.

The CWRU OPV group also demonstrated the advantages of nonplanar 3D geometry using azadipyrromethenes derivatives as electron acceptors [60,78,79]. Azadipyrromethenes are conjugated bidentate ligands with intense absorption in the visible to near-IR regions, high electron affinity, and tunable properties through chemical substitutions and coordination with boron or transition metals [80–82]. The tetraphenyl version, named ADP in Fig. 8.14A, is especially easy to synthesize [80]. It is typically coordinated with boron difluoride (BF_2^+) (aza-BODIPY in Fig. 8.14B) to give a planar molecule with an absorption max approximately 650 nm and a high quantum yield. It has been mostly studied for photodynamic and imaging applications. ADP ligands can also be coordinated with transition metals. For example, coordination with M(II) metals such as zinc(II) gives "dimers" or complexes with distorted tetrahedral structures, shown in Fig. 8.14C [81]. Interestingly, each proximal phenyl of one ADP ligand π-stack with a pyrrole group of the other ADP ligand. This leads to a highly rigid and conjugated nonplanar structure. The Zn(II) complex, Zn(ADP)$_2$, absorbs strongly from 550–700 nm, and its absorption is ligand centered where the Zn(II) does not participate in the HOMO and

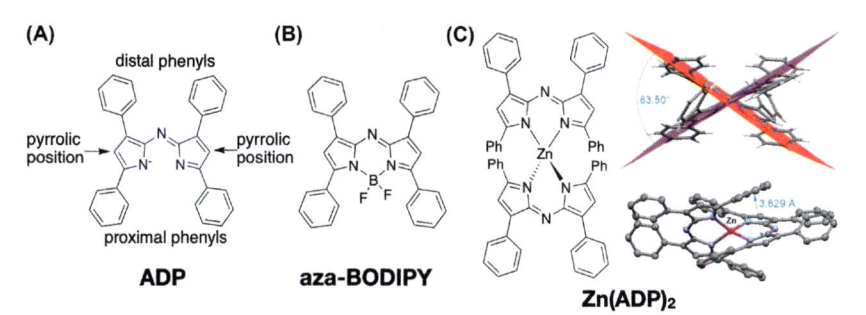

Figure 8.14 (A) Chemical structure of 1,3,7,9-tetraphenylazadipyrromethene (ADP). The ADP core is shown in blue. (B) Chemical structure of the BF_2^+ chelate, aza-BODIPY. (C) The chemical structure of the Zn(II) dimer, Zn(ADP)$_2$. Top right is the crystal structure of Zn(ADP)$_2$ illustrating the dihedral angle between best-fit planes of the two six-membered chelate rings. The bottom right shows a view of the crystal structure showing π-stacking between the proximal phenyl group of one ligand and a pyrrole moiety of the opposite ligand. *Reproduced with permission from T.S. Teets, D.V. Partyka, J.B. Updegraff, T.G. Gray, Homoleptic, four-coordinate azadipyrromethene complexes of d(10) zinc and mercury, Inorg. Chem. 47 (2008) 2338–2346, copyright (2008) American Chemical Society.*

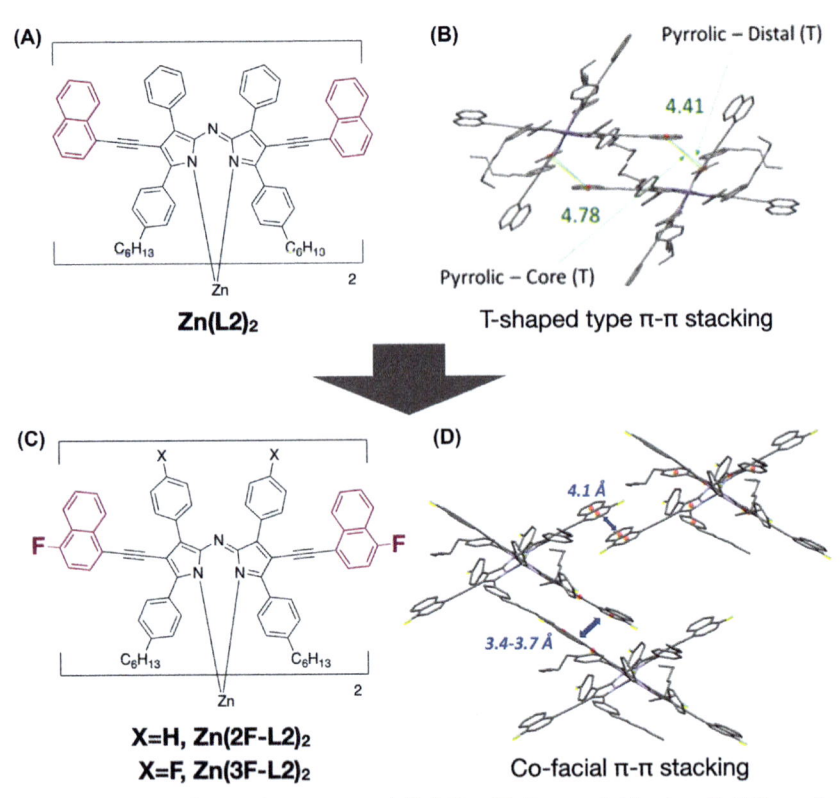

Figure 8.16 (A) Chemical structure of Zn(L2)$_2$. (B) Two neighboring Zn(L2)$_2$ molecules from the crystal packing, showing T-shaped intermolecular π—π stacking. (C) Chemical structure of fluorinated Zn(L2)$_2$. (D) Three neighboring Zn(2F-L2)$_2$ obtained from the crystal packing, showing intermolecular cofacial intermolecular π—π stacking between the fluorinated naphthyls of adjacent molecules and between the fluorinated naphthylethynyls of adjacent molecules.

why PCEs are limited to approximately 6%. Experiments with variable light intensity show that bimolecular charge recombination is not an issue in these thin cells, but trap recombination may limit PCE. Further, unbalanced charge carrier mobility may also play a factor in limiting the fill factor. Interestingly, our fluorinated molecules have a very high hole mobility of approximately 1×10^{-3} cm^2/V•s and have relatively high frontier energy levels for an electron acceptor [93], which have limited the choices of polymer donor to P3HT. These complexes may thus be better suited as molecular electron donors to be paired with a polymeric electron acceptor. Research is needed to discover a polymeric electron acceptor with suitable energy levels to match our zinc complexes.

Around the same time (2015) that nonplanar acceptors were discovered, an alternative strategy for the design of NFA also emerged: the use of highly planar fused-ring A-D-A acceptors with bulky solubilizing groups on the D moiety to tune down self-aggregation, see Fig. 8.17A. McCulloch and coworkers reported a molecule where D is fluorene and A is BTs and 3-ethylrhodanines, FBR in Fig. 8.17B [94]. The center D unit has two alkyl substituents to tune solubility and crystallinity. They reported a PCE of 4.1% when blended with P3HT. They found that device performance was limited by bimolecular recombination because FBR was too miscible with P3HT. To decrease miscibility with P3HT, they replaced the fluorene with the larger fused ring indacenodithiophene (IDT), giving IDTBR [95]. The fused ring center planarized the molecule and increased the tendency to crystallize. They reported a PCE of 6.4% with P3HT, a high PCE at the time. Zhan and coworkers reported IEIC (Fig. 8.17C), which showed a PCE of 6.3% when combined with PTB7-TH (shown in Fig. 8.9) [96]. They also increased planarity and rigidity of the center unit by replacing the center unit with a large 7-fused ring system, called ITIC [97]. Blended with

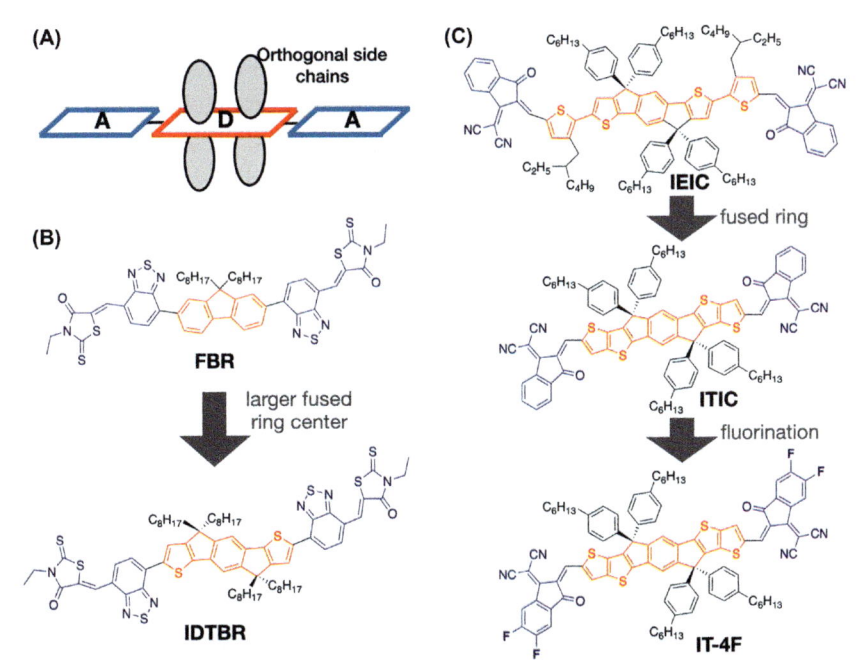

Figure 8.17 (A) Schematic of A-D-A type NFA with bulky side chains on the center D unit; (B) chemical structures of FBR and IDTBR; and (C) chemical structures of IEIC, ITIC, and IT-4F.

PTB7-TH, ITIC showed a high PCE of 6.8%. Optimization of the conjugated polymer donor to be paired with ITIC has led to PCEs in the 10%–12% range [98–101]. Unlike fullerene-based OPV, efficient hole transfer was observed for very small HOMO energy offsets (0.04 eV), leading to low energy losses [101]. It was postulated that ITIC may be forming charge transfer states with lower energetic disorder than fullerene based acceptors. Many variations of ITIC have been reported [66,67]. The best performance for an ITIC derivative was obtained for fluorinated ITIC, called IT-4F, as shown in Fig. 8.17C. It achieved a high PCE of 13.1% when blended with PM6 [102].

A breakthrough occurred in 2019 with the report of Y6, a similar fused ring system but with the A-D-A'-D-A motif, as shown in Fig. 8.18. Zou and coworkers reported a PCE of 15.7% when blended with the conjugated polymer donor PM6, breaking the 15% threshold for the first time [103]. They first combined the D-A'-D center of BZTP-1 with the end group A of ITIC to create BZIC in Fig. 8.18 [104]. The A' was then changed to benzothiadiazole, fluorine was installed on the A end groups, and alkyl groups were added on the periphery of the central unit to increase solubility without making the pyrrole groups bulkier, highlighted in yellow in Fig. 8.18 [104]. Interestingly, Y6 combines both strategies discussed above to tune miscibility while limiting self-aggregation; it is rigid with bulky alkyl substituents, and a side view shows that the molecule has a slight twist originating from the central A' unit [103]. These highly efficient OPV have high charge generation efficiency despite having very low energy offsets, fast charge extraction, low charge carrier recombination, and very low energy losses. These exciting results triggered significant interest in optimizing OPV based on Y6 derivatives, yielding PCEs in the 15%–19% range [8,104,105]. Such rapid development is impressive and PCEs for single-layer OPV are now reaching the predicted limit of 19.4% for single-layer OPV [106]. Incorporating these Y6-type acceptors into tandem solar cells has propelled PCEs above 20% [107].

It is important to understand how these systems can achieve such high PCE to further advance the OPV field. Central to the success of the Y6 derivatives is their unique crystal structure, illustrated in Fig. 8.19 [8,108,109]. Like the previous generation of fused ring NFA, Y6 shows strong $\pi-\pi$ stacking between the terminal groups to form J-aggregates. These are illustrated in Fig. 8.19D with the pink and blue shaded areas. Interestingly, the addition of the central BT in Y6 also promotes strong pi-core interactions that form H-aggregates, Fig. 8.19D green-shaded areas.

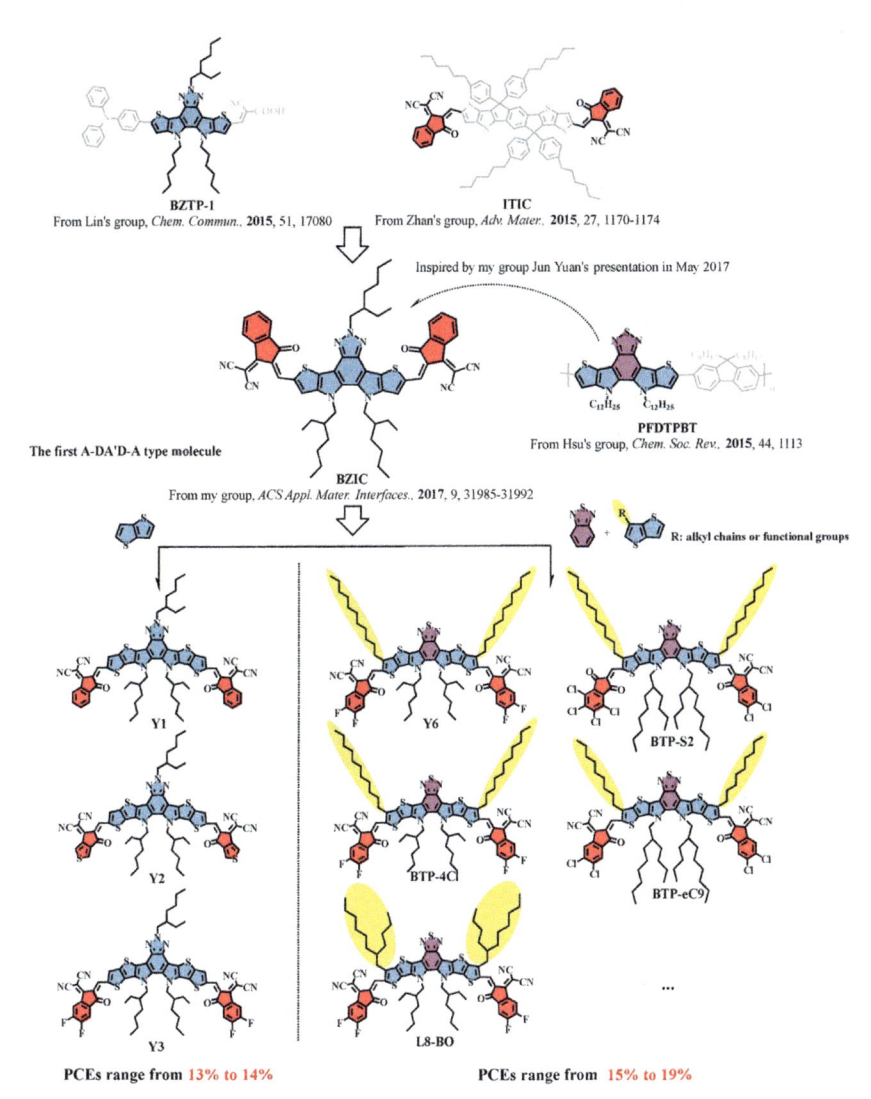

Figure 8.18 Diagram showing the progress of A-DA'D-A type nonfullerene small molecule acceptors. *Reproduced with permission from J. Yuan, Y. Zou, The history and development of Y6, Org. Electron. 102 (2022) 106346, copyright (2022) Elsevier.*

The BT unit is involved in S-N core interactions. The combination of strong intermolecular interactions and the curved geometry of the A-D-A'-D-A motif results in honeycomb-like 3D interpenetrating networks, Fig. 8.19C. Importantly, this distinctive 3D molecular packing structure is preserved in spin-coated films of blends with PM6. Additionally, both Y6 and PM6 prefer to adopt a favorable face-on orientation in films [109]. These combine to result in very favorable morphology and high PCEs.

The strong π—π interactions found in the unique Y6 packing structure are key to achieving high efficiency for several reasons. The strong π—π stacking interaction between adjacent Y6 molecules combined with the 3D network structure results in high charge carrier mobility for both electrons (1.8×10^{-4} cm^2/Vs) and holes (5.6×10^{-4} cm^2/Vs) [108]. Such a balanced high ambipolar charge transport is unusual and may help with facilitating hole transport with low HOMO energy offsets. The strong π—π interactions are also beneficial for exciton diffusion via energy transfer. Fused-ring NFAs, including Y6, have been found to have very high exciton diffusion lengths (L_D) in the 20 nm—50 nm range (compared to 5—10 nm for typical organic semiconductors), which explains why PM6:Y6

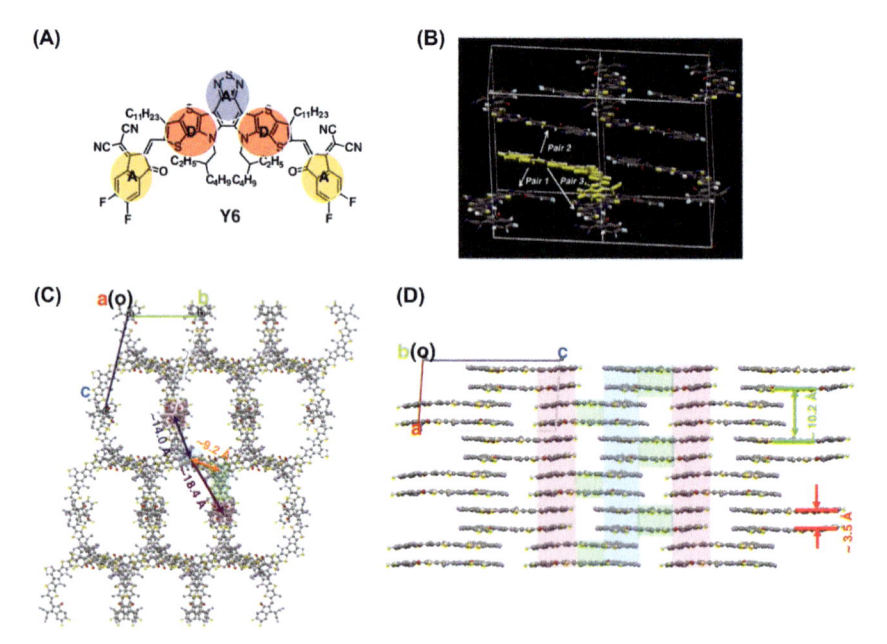

Figure 8.19 (A) Molecular structure of Y6; (B) Molecular pairs in the Y6 single crystal; (C) Top and (D) side views of the extended-crystal structure (the blue column is the stack of end groups in the b direction, the pink column is the stack of end groups in the c direction, and the green one is a molecular packing pair of D—A' fragment). *Reproduced from G. Zhang, X.K. Chen, J. Xiao, P.C.Y. Chow, M. Ren, et al., Delocalization of exciton and electron wavefunction in non-fullerene acceptor molecules enables efficient organic solar cells. Nat. Commun. 11 (2020) 3943, this article is distributed under the terms of the Creative Commons (CC-BY 4.0) license (http://creativecommons.org/licenses/by/4.0/), which permits unrestricted use.*

cells are very efficient even with large 20—50 nm scale phase separation [110,111]. The high L_D is beneficial as OPV performance becomes less sensitive to the BHJ morphology and processing conditions. The high π—π stacking interactions favor the formation of Y6 clusters near donor: acceptor interfaces. The clusters were shown to delocalize the CT state, which reduces the Coulomb interaction between the electron-hole pair and makes it easier to separate the CT state into free charges [108]. The strong π—π stacking interactions also contribute to a very low energetic disorder that has been observed for Y6-based OPV [112—114]. As a result, PM6:Y6 solar cells have unusually low charge trapping for an OPV, enabling high charge carrier mobility and efficient charge generation despite the low energy offsets [112]. The low energetic disorder is thought to be due to the conformational uniformity and rigidity of Y6. This conformational uniformity is encouraged by the strong intramolecular $C=O\cdots S$ interactions (also found in the ITIC), as well as the presence of β-alkyl substituents on the outer thiophene (unique to Y6 derivatives) that restrict rotation of the end groups [8,112].

Of similar importance is the observation of barrier-less charge generation more akin to inorganic semiconductors. Due to the Y6 molecular shape, it has a high quadrupole moment that can cancel the electron-hole Coulomb interactions, thus enabling high charge generation with low energy offsets [111]. Zhang and coworkers identified an intra-moiety excimer (i-EX) in Y6 domains, Fig. 8.20, an intermediate hole-electron state that can separate into free charges, similar to the CT but in Y6 domains instead of at the donor-acceptor interface [115]. This provides a pathway for a hole transfer channel without using the bound CT state and could explain the high generation efficiency independent of HOMO offsets. This is different from the electron transfer from PM6 to Y6, which goes through the bound CT state at the donor/acceptor interface (Fig. 8.20B). Other researchers believe this i-EX-mediated charge generation pathway cannot be a main channel for hole transfer because it is inconsistent with the observed high photoluminescence quantum yields of Y6 films [8]. More research is required to fully understand this system.

It is worth mentioning that the PM6:Y6 high performance greatly depends on the PM6 batch. Depending on the percentage of low molecular weight fraction (LMWF) of PM6 in one batch, PCE can vary from approximately 15% for a batch containing 1% LMWF to 5% for a batch containing 52% LMWF [116]. Interestingly, the percentage of LMWF greatly affected both the bulk and interfacial morphology of the blend, thus

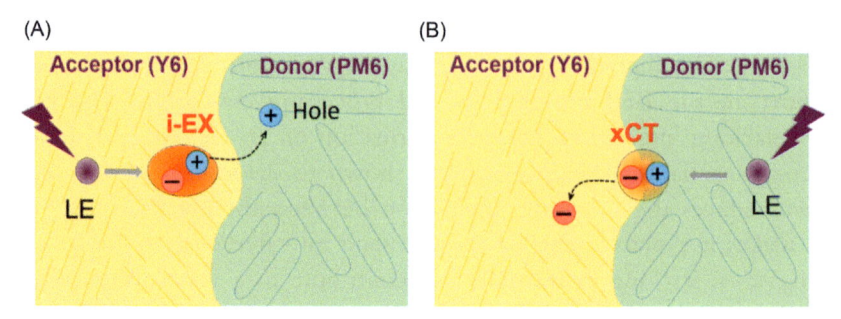

Figure 8.20 Schematic diagrams of the (A) hole transfer and (B) electron transfer channels of charge separation. When a photo is absorbed by acceptor Y6, the LE state at the acceptor side is separated into free polarons through the i-EX state as its intermediate step; when a photo is absorbed by donor PM6, the LE state at the donor side is separated into free polarons through the interfacial xCT state as its intermediate step. *Reproduced with permission from R. Wang, C. Zhang, Q. Li, Z. Zhang, X. Wang, M. Xiao, Charge separation from an intra-moiety intermediate state in the high-performance PM6:Y6 organic photovoltaic blend, J. Am. Chem. Soc. 142 (2020) 12751−12759, copyright (2020) American Chemical Society.*

giving insights into the importance of morphology to explain the high performance of PM6:Y6. Not surprisingly, the higher-performing blend (1% LMWF PM6 batch) had better long-range ordering and a more optimal nanophase phase separation than the less-performing blend. More importantly, they also demonstrated that the higher performing blend had much closer donor−acceptor interactions that facilitate ultrafast electron and hole transfer at the donor−acceptor interface, greatly improving charge generation. The high-performing blend also has less trap-assisted recombination, as well as better charge transport and extraction.

While polymer:small molecule blends have been successful, there is still a strong interest in all polymer solar cells, as they promise better mechanical properties and long-term stability. To create polymers that have the advantages of successful small molecular acceptors, Li et al. introduced the idea of polymerizing small molecule acceptors (PMSAs) [117]. Optimization of PMSA chemical structure using Y6 derivatives as building blocks quickly increased all polymer solar cell PCE to the 17%−18% range [117]. However, the polymers tend to have low MW and low yields and consequently suffer from batch-to-batch reproducibility. To improve the long-term stability of small molecule acceptor-based OPV without making polymers, Kim and coworkers dimerized a Y-6 type acceptor called MYBO via a covalent link to create DYBO [118]. This increases thermal stability significantly, due to a higher glass transition temperature and improved blend miscibility. The DYBO gave a high PCE of approximately 18%.

8.5 Considerations for commercialization

For OPV to be successful, efficiency, lifetime, and production costs are essential. To evaluate the potential for electron donor and acceptor materials, an industrial figure of merit (i-FoM) was introduced, defined by Eq. (8.2) [119]:

$$i - \text{FoM} = \frac{\text{PCE} \times \text{photostability}}{\text{synthetic complexity}} \quad (8.2)$$

where the photostability is the stability of a device under illumination in an inert atmosphere (scale of 0–1.0), and the synthetic complexity (SC, %) is an estimate of how difficult a material's synthesis is, as a percentage relative to the most complex conjugated polymer synthesis (assigned 100%) [120]. Synthetic complexity takes into account: (1) the number of synthetic steps, (2) yields, (3) the number of operations, and (4) the use (and number) of hazardous chemicals. While not perfect, synthetic complexity can be related to scalability and the cost of a material [120]. The i-FoM is thus maximized for large PCE, high photostability, and low SC. Assuming that a large PCE (module efficiency >10%) and good photostability can be achieved, the SC becomes critical [106]. Many of the high-performance materials, including PM6 and Y6, have very high SC, which will limit their scalability [121]. A target SC for the donor:acceptor blend may be that of P3HT: PCBM, which is the cheapest material combination so far with a blend SC of 14% [106]. For comparison, the PM6:Y6 blend has an SC of 62.0% [121]. An i-FoM value greater or equal to 0.7 is suggested as a desirable target for industrial viability [106]. From this perspective, there is work to be done because most i-FoM reported to date are below 0.25.

Of all the polymer donors, P3HT has by far the lowest SC, at 7%. As a result, it is one of the least expensive conjugated polymer donors. For comparison, PM6 costs approximately 14 times more than P3HT (Ossila website, September 2023). P3HT, however, does not give the best performance due to its high oxidation potential (high HOMO).

Alternative high-performance polymers with simple chemical structures are being developed for OPV. For example, a simpler polymer called PTQ10, shown in Fig. 8.21, was developed to be cheaper to make while having lower HOMO than P3HT [122,123]. This polymer achieved a PCE of 17.7% when blended with a Y6 derivative [124]. Alternatively, simple polythiophene backbones with electron-withdrawing groups have also shown promise. For example, polythiophene with electron-withdrawing carbonyl groups, PDCBT, formed a favorable nanoscale morphology with ITIC to give a PCE of 10.2% [125]. This polymer has a SC of about 34% [126]. A fluorinated polythiophene derivative, P4T2F-HD, was miscible with a Y6 derivative, Y6-BO, forming a favorable blend morphology and a high PCE of 13.6% [126]. This polymer has a calculated SC of approximately 42%. Recently, polythiophene P5TCN-Fx was developed with simple design, using cyano groups to lower the HOMO and fluorine groups to increase interchain interactions [127]. This random copolymer has an SC of approximately 49% and achieved a PCE of 16.6% in blends with Y6 and 17% in a ternary blend with Y6 and $PC_{71}BM$.

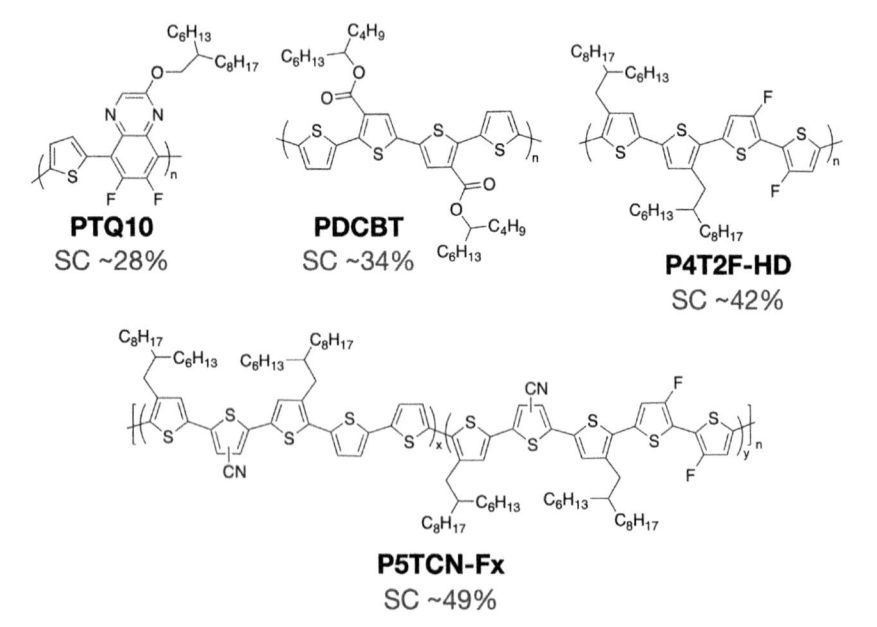

Figure 8.21 Chemical structures of some polymer donors with relatively low synthetic complexity.

It is also important to develop high-performance NFAs with lower SC. Our Zn(L2)$_2$ NFA discussed above has a low SC of 22%, and the P3HT:Zn(L2)$_2$ blend has an SC of only 15%, similar to P3HT:PCBM [90]. Combining this with a PCE of 5.5% and a photostability of 0.81 gave an i-FoM value of 0.30, which is amongst the highest reported i-FoM to date. Nevertheless, the performance and stability are still not high enough for commercialization. There are also efforts to develop high-performance NFAs that are inspired by ITIC and Y6 but reduce the number of fused rings because fused rings are difficult to synthesize [128,129]. Here, the central 7-fused ring system is replaced with smaller fused-ring systems or no fused ring. This introduces single bonds that can rotate. To prevent rotation and favor planarity, molecular designs use noncovalent intramolecular interactions and strategically placed bulky alkyl side chains. Extensive research over the past few years has led to several high-performance nonfused ring acceptors, with maximum PCE now reaching 15%−16% [129].

The first successful NFA reported using this approach was reported in 2018 by Chen and coworkers [130]. They developed and replaced the 7-fused ring center of ITIC with two cyclopentadithiophene moieties separated by one 2,5-difluorobenzene group, see DF-PCIC in Fig. 8.22. The molecule is nearly planar due to noncovalent intramolecular F···H interactions and a PCE of 10.1% was reported when blended with PBDB-T. Later on, unfused NFAs inspired by Y6 were also reported, including BFC-4F, which has a central fluorinated benzothiadiazole moiety [131]. This NFA achieved a PCE of 12% when blended with PTQ10. Huang and coworkers replaced the benzothiadiazole with the benzotriazole moiety in BTzO-4F [132]. Planarity is obtained from the combination of N···H and S···O noncovalent interactions. This NFA achieved a PCE of 13.8% when blended with PBDB-T. Bo and coworkers explored using thienothiophene moieties with two bulky bis(4-butylphenyl)amino substituents, 2BTh-2F. The single crystal structure shows that the molecule is nearly planar due to the S···N and S···O interactions. Interestingly, it formed a 3D network packing with close π−π stacking distances. As a result, this NFA gave a PCE of 14.5% when blended with PBDB-T and 15.4% when blended with D8. Li and coworkers later added halogen substituents on the aromatic side chains to further tune the properties of molecules L1 and L2 [133]. They showed that the substituted aromatic side chains promote an interlocked tic-tac-toe 3D network. When blended with PM6, L1 and L2 gave a high PCE of

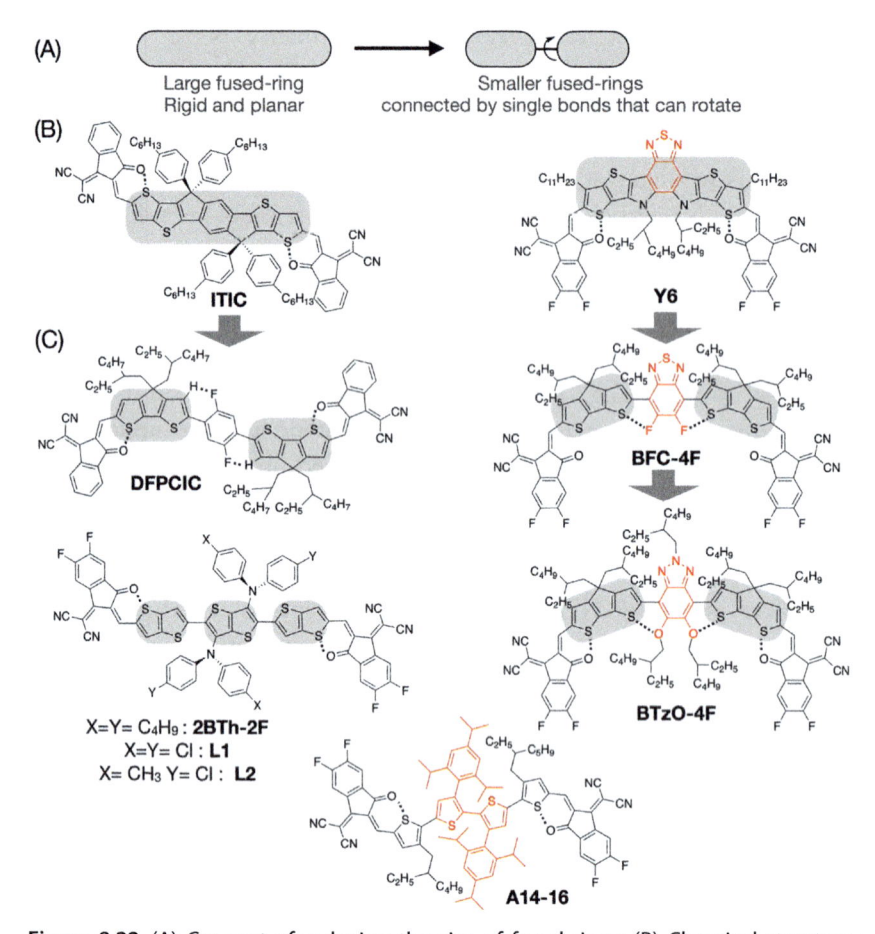

Figure 8.22 (A) Concept of reducing the size of fused rings; (B) Chemical structures of ITIC and Y6, which both have a 7-fused-ring center; (C) Chemical structures of NFA with centers consisting of smaller fused-rings. The dotted line represents a noncovalent intramolecular interaction.

15.4% and 16.2%, respectively. The unsymmetrical L2 had better photostability, with an extrapolated T_{80} lifetime of 17,247 hours. Hou and coworkers developed a completely nonfused NFA, A4T-16 [134]. This design relies on steric hindrance to lock the thiophenes into one plane. DFT calculations show that this method is more effective than using the common noncovalent intramolecular interactions. A4T-16 also forms favorable 3D interpenetrated networks with short $\pi-\pi$ distance. Blends of A4T-16 with PM6 achieved a PCE of 15.2%. Chang and coworkers published a stable highly efficient OPV (PCE approximately 16%) based on undisclosed proprietary active materials with a blend SC of approximately 40%, lower than PM6:Y6 (~ 62%) [121]. Assuming a photostability of 100%, the corresponding i-FoM is improved to 0.39. These results show that, with time, we will find high-performance donor—acceptor systems, where both active layer materials have low SC.

Long-term stability is of critical importance for commercialization. Research into the complex topic of stability has significantly increased over the past decade, leading to a better understanding of degradation mechanisms and strategies to enhance OPV stability [8,11,135—137]. Stability is complex, as there are several external and internal factors to consider: External factors are light, heating, moisture, oxygen, electrical load, and mechanical stress, whereas internal factors include electrode diffusion, unstable buffer layers, and unstable morphology [136]. Ideally, one isolates a single stress factor, but this can be difficult. Further, it is not just the active layer that can fail; encapsulation, electrodes, and interfacial layers can also fail. In terms of active layer material development, it is most important to focus on intrinsic degradation [138]. As such, research has focused on the thermal and photostability of OPV measured under an inert atmosphere. Thermal stability is done in the dark to avoid light effects, and photo-stability is done at room temperature to avoid heat effects. To facilitate comparison between laboratories, the international summit on OPV stability published stability testing protocols for OPV materials and devices [139]. Progress toward improved stability is promising, as summarized below.

Early stability research focused on polymer-fullerene BHJ, which has degradation mechanisms that are unique to these systems. For example, air and moisture cause photochemical reactions of fullerenes and polymers, leading to photooxidation [140]. Such issues can be alleviated with encapsulation. Polymer:fullerene BHJ typical lifetime curve has two regimes: a burn-in degradation where performance degrades relatively quickly and nonlinearly, followed by a linear and slower degradation. The main cause for the burn-in degradation is thought to be light-induced fullerene dimerization and/or initial BHJ morphology changes [137]. For example, an OPV based on PCDTBT:$PC_{71}BM$ showed an initial burn-in of nearly 40%, followed by stable PCE over continuous light exposure under an inert atmosphere with an extrapolated lifetime exceeding 15 years [141].

Initial BHJ morphology changes are an issue that arises because the optimal BHJ morphology is usually not the thermodynamic equilibrium case. Upon heating, fullerenes can diffuse easily and self-aggregate, causing demixing and large fullerene aggregates. This reduces donor−acceptor interfaces and free charge generation. Several strategies have been explored to successfully stabilize the BHJ morphology of polymer:fullerene OPV, such as introducing additives or cross-linkers to "freeze" the morphology [142−144]. Modifying the chemical structure of the donor or acceptor material to increase donor−acceptor miscibility also plays a major role, where increasing miscibility increases morphology stability [145]. Since we need some phase separation to obtain a BHJ, a balance between miscibility and photovoltaic performance must thus be achieved. Using active layer materials that do not diffuse as easily, such as replacing fullerenes with appropriate NFAs, also reduces BHJ morphology changes. Differences in NFA chemical structure were shown to affect the diffusion coefficient in films [146]. A lower diffusion coefficient will provide kinetic morphology stabilization and thus improve thermal stability. A popular approach to optimize NFA-based OPV has been to use the ternary approach, where a third active layer component is added to the binary blend [147]. This approach was originally used to optimize light-harvesting ability but has also been used to optimize and even stabilize blend morphology. One recent example tunes the ratio of two similar acceptors (L8BO and Y6) blended with PTzBI donor to achieve long-term stability of over 1400 hours under 85°C of continuous heating [148]. In addition to the improved thermal stability, the ternary blend had a higher PCE, approximately 18%, higher than the binary blends, at approximately 16%.

Given a stable morphology, it is also critical that the donor−acceptor blend has intrinsic photochemical stability [68]. It was recently shown that all conjugated molecules and polymers can undergo photodegradation under an inert atmosphere, which leads to crosslinking and the accumulation of defects and traps [138]. This photodegradation can be slowed significantly by using UV filters. Because photostability depends on the chemical structure, understanding the relationship between chemical structure and stability of a variety of compounds is necessary to design new materials with improved photostability [149]. Other factors that can improve photostability are to increase polymer crystallinity and avoid the usage of unstable additives [136].

In 2017, Brabec and coworkers compared the photostability of P3HT: PCBM and P3HT:IDTBR BHJ (under an inert atmosphere and continuous light with a UV filter) and showed that the P3HT:IDTBR cell had no burn-in region and lost only 5% of the original PCE (of 5%), compared to 34% loss for the P3HT:PCBM cell. The improvements were attributed to the better photostability of the NFA over PCBM. Stability was further improved in 2021 by replacing the PEDOT:PSS hole-transport layer (HTL) with MoO_x [68], and a P3HT:IDTBR OPV processed from THF and using MoO_x as HTL has now shown excellent photostability over 20,000 hours of continuous exposure, equivalent to about 20 years lifetime, see Fig. 8.23 [8]. While PEDOT:PSS is one of the most used HTLs in conventional OPV, it is known to cause stability issues due to its acidity and hydrophilic nature [136,150]. It is thus not surprising that stability improved by replacing it. Note that the P3HT:IDTBR blend SC is approximately 26% [106], which makes this donor–acceptor combination scalable, but the PCE is limited to about 5%. Stable scalable systems with higher PCE are needed. A study comparing OPV of PBDB-T blended with various ITIC derivatives, all with PCE of 8%–10%, showed that end groups and side chain modifications strongly affect photo- and thermal stability [151]. The most stable was PBDB-T:ITIC-2F, with a PCE of approximately 8% and a T_{80} of over 11,000 hrs (lifetime nearing 10 years), but the i-FoM was only 0.123, pointing again to the need to lower the materials SC. OPV-based PM6 blended with nonfused NFA L2 (Fig. 8.22) has a high PCE of 16% with very good photostability over 1600 hours of continuous illumination

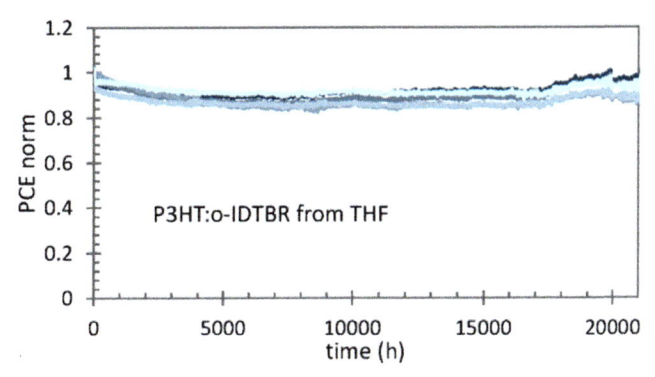

Figure 8.23 P3HT:o-IDTBR solar cells fabricated from a green solvent (tetrahydrofuran) and exposed to metal halide light source illumination. Excellent stability has been observed for more than 20,000 hours. *Reproduced with permission from G. Zhang, F.R. Lin, F. Qi, T. Heumuller, A. Distler, et al., Renewed prospects for organic photovoltaics. Chem. Rev. 122 (2022) 14180–14274, copyright (2022) American Chemical Society.*

without UV-filtration [133]. An extrapolated T_{80} lifetime of 17,241 hours was reported.

Electrodes, interlayers, and the interface between the active layer and interlayers can also be a source of instability [68,137]. The inverted geometry is usually more stable than the conventional geometry because it avoids the use of low work-function electrodes, which oxidize easily and avoids the use of the common PEDOT:PSS HTL. One common ETL used in inverted OPV is ZnO, but it can also lower stability due to interfacial traps. As such, modifying the ETL layer [14,150,152], or adding buffer layers between the transport layers and the active layers [153], have shown superior stability. Intrinsic operational lifetimes exceeding 30 years have now been demonstrated [153], and thermally evaporated OPV has even shown high stability under high-intensity illumination, equivalent to 27,000 years outdoors [154]. Stable OPV with PCEs greater than 15% are also emerging [148,150,155]. For example, Che, Wan, and coworkers showed that modifying ZnO ETL with a modifying agent called NMA significantly improved photo-stability of high-performance OPV [150]. An inverted device using D18:N3 as the active layer had an initial PCE of 18.2% and a T80 extrapolated lifetime of 7572 hours, equivalent to approximately 5 years of outdoor exposure. Note that the active layer materials D18 (donor shown in Fig. 8.9) and N3 (a Y6 derivative) have high SC, but nevertheless demonstrate that highly efficient and stable OPV can be made. Developments will be accelerated with the advent of high-throughput strategies [8]. In time, materials, interfaces, and device structure will be optimized to create stable, scalable, and highly efficient OPV.

The PCEs reported so far are usually for small-area devices. As the solar cell area is increased, there are challenges that limit the efficiency. According to the latest solar cell efficiency tables (version 62) [156], the highest OPV PCE is 19.2% for a 0.0325 cm^2 cell. As the cell area increases to 1.015 cm^2, the maximum PCE drops to 15.2%. For a module with 1475 cm^2 area, the maximum PCE is 13.1%. Some of the challenges with upscaling to large-area modules include area-related losses and device-related losses. From an active layer development perspective, a main area-related loss is that it is difficult to obtain a uniform BHJ. As the cell area increases, there are greater chances for defects and inhomogeneities. The maximum PCE for a large device is therefore no larger than the average PCE for small devices. Developing active layer materials that form more homogeneous BHJ and films under wider processing windows is thus important as we scale up.

Other factors are also important when considering module fabrication [11]. One must consider stability during manufacturing. To take full advantage of solution-processing, materials must be printed under ambient conditions. As such, all materials must be stable toward oxygen, humidity, and visible light. They must also be processed from nontoxic solvents, so popular chlorinated solvents such as chlorobenzene are not acceptable. As a result, there is significant interest in using alternative "greener" solvents [157,158]. To form uniform films using roll-to-roll printing technology, it is necessary to print films with thicknesses of at least 200 nm [159,160]. Most optimal PCEs are reported for thin films of approximately 100 nm. There is thus interest in developing materials with better charge transport properties and less defects, to minimize opportunities for charge recombination. Additionally, materials must be thermally stable under module printing conditions [11]. Drying and curing during R2R fabrication expose materials to a higher thermal load than small-scale laboratory fabrication conditions. Lamination may require heating at 120°C for several hours; it is critical that materials and the active layer morphology be thermally stable. Other types of stability are also required depending on the application. This includes physical appearance, for example, the encapsulation must not turn yellow over time, and stability to mechanical stress, especially for flexible applications. The final product must also be safe to use, for example, if used as a building façade, it must past safety tests. For flexible applications, it must withstand bending tests. Ultimately, OPV modules must also be fully recyclable for long-term sustainability.

Despite all these challenges, there is hope, as the number of reported stable modules with PCE > 10% is increasing [8,161,162], including finding ways to improve the uniformity of thick active layer films [159]. Research in all aspects of product development, including active layer material development, is key to the future success of OPV.

8.6 Summary and conclusions

Over the past 20 years, there has been tremendous progress in the field of OPV, with single-layer devices improving from PCEs of approximately 3% in 2001 to approximately 19% in 2023. Critical to this progress is the development of active layer materials. This chapter discusses the development of solution-processed materials starting with the first successful BHJ OPV using a blend of P3HT as donor and PCBM (a fullerene derivative) as acceptor. The field then focused on optimizing the electron donors to be paired with fullerene-based acceptors. When PCEs plateaued at approximately 12%, the community shifted its focus to developing NFA, which could overcome some of the limitations of fullerene derivatives. Through structure-property studies, the community learned to optimize the NFA chemical structure, enabling PCEs nearing 20%. These high-performance OPV typically use donors that were originally optimized for fullerene derivatives. It is also necessary to develop new donors for high-performance NFAs. Through understanding structure-property relationships, the field is learning how to optimize the active material's chemical structure for high performance.

Current efforts to commercialize OPV point to other important factors to consider, such as material cost, solvent toxicity, and materials and device stability. Research to enable simpler and cheaper high-performance materials is on-going. In the literature, active layer materials with lower SC are typically paired with an expensive material to maximize PCE. In the future, it will be important to optimize material combinations where all components have low SC. It is now evident that OPV can be stable when properly encapsulated. It will be essential to combine interface engineering with the development of active layer materials to not only improve stability but also to enable scale-up with minimal loss. To scale-up, for example, it is important to form uniform films from nontoxic solvents. To form uniform films with a large processing window, one must not rely heavily on processing conditions. Instead, the interfaces and the active layer material's chemical structure must be optimized to strongly favor a desirable morphology. The emerging high-throughput methods combined with machine learning will certainly accelerate research in optimizing these multivariable complex systems. In conclusion, this author believes that the future for the commercialization of OPV is bright.

References

[1] J. Wu, M. Gao, Y. Chai, P. Liu, B. Zhang, J. Liu, et al., Towards a bright future: the versatile applications of organic solar cells, Mater. Reports: Energy 1 (2021) 100062. Available from: https://doi.org/10.1016/j.matre.2021.100062.
[2] W. Song, J. Ge, L. Xie, Z. Chen, Q. Ye, et al., Semi-transparent organic photovoltaics for agrivoltaic applications, Nano Energy 116 (2023) 108805. Available from: https://doi.org/10.1016/j.nanoen.2023.108805.
[3] C. Brabec, S. Gowrisanker, J. Halls, D. Laird, S. Jia, S. Williams, Polymer-fullerene bulk-heterojunction solar cells, Adv. Mater. 22 (2010) 3839–3856. Available from: https://doi.org/10.1002/adma.200903697.

[4] C. Liu, C. Xiao, C. Xie, W. Li, Flexible organic solar cells: materials, large-area fabrication techniques and potential applications, Nano Energy 89 (B) (2021) 106399. Available from: https://doi.org/10.1016/j.nanoen.2021.106399.
[5] L. Lu, T. Zheng, Q. Wu, A.M. Schneider, D. Zhao, L. Yu, Recent advances in bulk heterojunction polymer solar cells, Chem. Rev. 115 (2015) 12666−12731. Available from: https://doi.org/10.1021/acs.chemrev.5b00098.
[6] H. Kang, G. Kim, J. Kim, S. Kwon, H. Kim, K. Lee, Bulk-heterojunction organic solar cells: five core technologies for their commercialization, Adv. Mater. 28 (2016) 7821−7861. Available from: https://doi.org/10.1002/adma.201601197.
[7] <https://www.nrel.gov/pv/interactive-cell-efficiency.html> (accessed 27.09.23).
[8] G. Zhang, F.R. Lin, F. Qi, T. Heumuller, A. Distler, et al., Renewed prospects for organic photovoltaics, Chem. Rev. 122 (2022) 14180−14274. Available from: https://doi.org/10.1021/acs.chemrev.1c00955.
[9] V.V. Brus, J. Lee, B.R. Luginbuhl, S.J. Ko, G.C. Bazan, T.Q. Nguyen, Solution-processed semitransparent organic photovoltaics: from molecular design to device performance, Adv. Mater. 31 (2019) e1900904. Available from: https://doi.org/10.1002/adma.201900904.
[10] M.B. Schubert, J.H. Werner, Flexible solar cells for clothing, Mater. Today 9 (2006) 42−50. Available from: https://doi.org/10.1016/s1369-7021(06)71542-5.
[11] S.V. Subramaniam, D. Kutsarov, T. Sauermann, S.B. Meier, Commercialization of organic photovoltaics and its product integration: a perspective focusing on durability, Energy Technol. 8 (2020) 2000234. Available from: https://doi.org/10.1002/ente.202000234.
[12] M. Jahandar, S. Kim, D.C. Lim, Indoor organic photovoltaics for self-sustaining IoT devices: progress, challenges and practicalization, ChemSusChem 14 (2021) 3449−3474. Available from: https://doi.org/10.1002/cssc.202100981.
[13] C. Sun, R. Xia, H. Shi, H. Yao, X. Liu, J. Hou, et al., Heat-insulating multifunctional semitransparent polymer solar cells, Joule 2 (2018) 1816−1826. Available from: https://doi.org/10.1016/j.joule.2018.06.006.
[14] Y. Zhao, Z. Li, C. Deger, M. Wang, M. Peric, et al., Achieving sustainability of greenhouses by integrating stable semi-transparent organic photovoltaics, Nat. Sustain. 6 (2023) 539−548. Available from: https://doi.org/10.1038/s41893-023-01071-2.
[15] Y. Zhao, Y. Zhu, H.-W. Cheng, R. Zheng, D. Meng, Y. Yang, A review on semitransparent solar cells for agricultural application, Mater. Today Energy 22 (2021) 100852. Available from: https://doi.org/10.1016/j.mtener.2021.100852.
[16] A.L. Roes, E.A. Alsema, K. Blok, M.K. Patel, Ex-ante environmental and economic evaluation of polymer photovoltaics, Progress in Photovoltaics: Research and Applications 17 (2009) 372−393. Available from: https://doi.org/10.1002/pip.891.
[17] Handbook of Conducting Polymers, 3rd edition, CRC Press, New York, 2007 (Vol. 2), 1680 pp.
[18] M. Dvorak, S.-H. Wei, Z. Wu, Origin of the variation of exciton binding energy in semiconductors, Phys. Rev. Lett. 110 (2013) 016402. Available from: https://doi.org/10.1103/PhysRevLett.110.016402.
[19] J. Nelson, Polymer:Fullerene bulk heterojunction solar cells, Mater. Today 14 (2011) 462−470. Available from: https://doi.org/10.1016/S1369-7021(11)70210-3.
[20] F. Liu, Y. Gu, J.W. Jung, W.H. Jo, T.P. Russell, On the morphology of polymer-based photovoltaics, J. Polym. Sci., Part B Polym. Phys. 50 (2012) 1018−1044. Available from: https://doi.org/10.1002/polb.23063.
[21] L.J. Richter, D.M. DeLongchamp, A. Amassian, Morphology development in solution-processed functional organic blend films: an in situ viewpoint, Chem. Rev. 117 (2017) 6332−6366. Available from: https://doi.org/10.1021/acs.chemrev.6b00618.

[22] G. Yu, J. Hummelen, F. Wudl, A. Heeger, Polymer photovoltaic cells: enhanced efficiencies via a network of internal donor-acceptor heterojunctions, Science 270 (1995) 1789−1791.
[23] P. Schilinsky, C. Waldauf, C.J. Brabec, Recombination and loss analysis in polythiophene based bulk heterojunction photodetectors, Appl. Phys. Lett. 81 (2002) 3885−3887. Available from: https://doi.org/10.1063/1.1521244.
[24] F. Padinger, R.S. Rittberger, N.S. Sariciftci, Effects of postproduction treatment on plastic solar cells, Adv. Funct. Mater. 13 (2003) 85−88. Available from: https://doi.org/10.1002/adfm.200390011.
[25] M.T. Dang, L. Hirsch, G. Wantz, P3HT:PCBM, best seller in polymer photovoltaic research, Adv. Mater. 23 (2011) 3597−3602. Available from: https://doi.org/10.1002/adma.201100792.
[26] S.S. van Bavel, M. Bärenklau, G. de With, H. Hoppe, J. Loos, P3HT:PCBM bulk heterojunction solar cells: impact of blend composition and 3d morphology on device performance, Adv. Funct. Mater. 20 (2010) 1458−1463. Available from: https://doi.org/10.1002/adfm.200902247.
[27] B.A. Collins, F.A. Bokel, D.M. DeLongchamp, in: C. Brabec, U. Scherf, V. Dyakonov (Eds.), Organic Photovoltaic Morphology, In Organic Photovoltaics: Materials, Device Physics, and Manufacturing Technologies, Wiley-VCH Verlag GmbH & Co. KGaA, 2014, pp. 377−420.
[28] F.C. Krebs, T. Tromholt, M. Jorgensen, Upscaling of polymer solar cell fabrication using full roll-to-roll processing, Nanoscale 2 (2010) 873. Available from: https://doi.org/10.1039/b9nr00430k.
[29] S.A. Gevorgyan, I.M. Heckler, E. Bundgaard, M. Corazza, M. Hosel, et al., Improving, characterizing and predicting the lifetime of organic photovoltaics, J. Phys. D Appl. Phys. 50 (2017) 103001. Available from: https://doi.org/10.1088/1361-6463/50/10/103001.
[30] J. Huang, G. Li, Y. Yang, Influence of composition and heat-treatment on the charge transport properties of poly(3-hexylthiophene) and [6,6]-phenyl c61-butyric acid methyl ester blends, Appl. Phys. Lett. 87 (2005) 112105.
[31] T. Savenije, J. Kroeze, X. Yang, J. Loos, The effect of thermal treatment on the morphology and charge carrier dynamics in a polythiophene-fullerene bulk heterojunction, Adv. Funct. Mater. 15 (2005) 1260−1266.
[32] R. Zhang, B. Li, M. Iovu, M. Jeffries-El, G. Sauve, J. Cooper, et al., Nanostructure dependence of field-effect mobility in regioregular poly(3-hexylthiophene) thin film field effect transistors, J. Am. Chem. Soc. 128 (2006) 3480−3481. Available from: https://doi.org/10.1021/ja055192i.
[33] S.S. van Bavel, E. Sourty, G. de With, J. loos, Three-dimensional nanoscale organization of bulk heterojunction polymer solar cells, Nano Lett. 9 (2009) 507−513. Available from: https://doi.org/10.1021/nl8014022.
[34] A.M. Al-Harthi, W.E. Mahmoud, Influence of the fullerene acceptor derivative on the structure, molecular interaction and optoelectronic properties of poly(3-hexylthiophene) blend films: elucidating key factors, Opt. Mater. 124 (2022) 112009. Available from: https://doi.org/10.1016/j.optmat.2022.112009.
[35] G. Zhao, Y. He, Y. Li, 6.5% efficiency of polymer solar cells based on poly(3-hexylthiophene) and indene-c(60) bisadduct by device optimization, Adv. Mater. 22 (2010) 4355−4358. Available from: https://doi.org/10.1002/adma.201001339.
[36] Y. Sun, C. Cui, H. Wang, Y. Li, Efficiency enhancement of polymer solar cells based on poly(3-hexylthiophene)/indene-c70 bisadduct via methylthiophene additive, Adv. Energy Mater. 1 (2011) 1058−1061. Available from: https://doi.org/10.1002/aenm.201100378.

[37] X. Guo, C. Cui, M. Zhang, L. Huo, Y. Huang, J. Hou, et al., High efficiency polymer solar cells based on poly(3-hexylthiophene)/indene-c70 bisadduct with solvent additive, Energy Environ. Sci. 5 (2012) 7943–7949. Available from: https://doi.org/10.1039/c2ee21481d.

[38] M. Scharber, D. Muehlbacher, M. Koppe, P. Denk, C. Waldauf, A. Heeger, et al., Design rules for donors in bulk-heterojunction solar cells-towards 10% energy-conversion efficiency, Adv. Mater. 18 (2006) 789–794. Available from: https://doi.org/10.1002/adma.200501717.

[39] Y.-J. Cheng, S.-H. Yang, C.-S. Hsu, Synthesis of conjugated polymers for organic solar cell applications, Chem. Rev. 109 (2009) 5868–5923. Available from: https://doi.org/10.1021/cr900182s.

[40] H. Zhou, L. Yang, W. You, Rational design of high performance conjugated polymers for organic solar cells, Macromolecules 45 (2012) 607–632. Available from: https://doi.org/10.1021/ma201648t.

[41] T. Xu, L. Yu, How to design low bandgap polymers for highly efficient organic solar cells, Mater. Today 17 (2014) 11–15. Available from: https://doi.org/10.1016/j.mattod.2013.12.005.

[42] J. Lee, W. Ma, C. Brabec, J. Yuen, J. Moon, et al., Processing additives for improved efficiency from bulk heterojunction solar cells, J. Am. Chem. Soc. 130 (2008) 3619–3623.

[43] S. Beaupre, M. Leclerc, Pcdtbt: en route for low cost plastic solar cells, J. Mater. Chem. A 1 (2013) 11097–11105. Available from: https://doi.org/10.1039/c3ta12420g.

[44] H.J. Son, F. He, B. Carsten, L. Yu, Are we there yet? Design of better conjugated polymers for polymer solar cells, J. Mater. Chem. 21 (2011) 18934–18945. Available from: https://doi.org/10.1039/c1jm12388b.

[45] Y. Liu, J. Zhao, Z. Li, C. Mu, W. Ma, H. Hu, et al., Aggregation and morphology control enables multiple cases of high-efficiency polymer solar cells, Nat. Commun. 5 (2014) 5293. Available from: https://doi.org/10.1038/ncomms6293.

[46] C. Liu, C. Yi, K. Wang, Y. Yang, R.S. Bhatta, M. Tsige, et al., Single-junction polymer solar cells with over 10% efficiency by a novel two-dimensional donor-acceptor conjugated copolymer, ACS Appl. Mater. Interfaces 7 (2015) 4928–4935. Available from: https://doi.org/10.1021/am509047g.

[47] M. Zhang, X. Guo, W. Ma, H. Ade, J. Hou, A large-bandgap conjugated polymer for versatile photovoltaic applications with high performance, Adv. Mater. 27 (2015) 4655–4660. Available from: https://doi.org/10.1002/adma.201502110.

[48] W. Yue, R.S. Ashraf, C.B. Nielsen, E. Collado-Fregoso, M.R. Niazi, S.A. Yousaf, et al., A thieno[3,2-b][1]benzothiophene isoindigo building block for additive- and annealing-free high-performance polymer solar cells, Adv. Mater. 27 (2015) 4702–4707. Available from: https://doi.org/10.1002/adma.201501841.

[49] M. Wang, X. Hu, P. Liu, W. Li, X. Gong, F. Huang, et al., Donor–acceptor conjugated polymer based on naphtho[1,2- c:5,6- c]bis[1,2,5]thiadiazole for high-performance polymer solar cells, J. Am. Chem. Soc. 133 (2011) 9638–9641. Available from: https://doi.org/10.1021/ja201131h.

[50] V. Vohra, K. Kawashima, T. Kakara, T. Koganezawa, I. Osaka, K. Takimiya, et al., Efficient inverted polymer solar cells employing favourable molecular orientation, Nat. Photonics 9 (2015) 403–408. Available from: https://doi.org/10.1038/nphoton.2015.84.

[51] Y. He, Y. Li, Fullerene derivative acceptors for high performance polymer solar cells, Phys. Chem. Chem. Phys. 13 (2011) 1970–1983.

[52] J. Peet, M.L. Senatore, A.J. Heeger, G.C. Bazan, The role of processing in the fabrication and optimization of plastic solar cells, Adv. Mater. 21 (2009) 1521–1527. Available from: https://doi.org/10.1002/Adma.200802559.

[53] A.B. Sieval, J.C. Hummelen, Fullerene-based acceptor materials, in: C. Brabec, U. Scherf, V. Dyakonov (Eds.), , in Organic Photovoltaics: Materials, Device Physics and Manufacturing Technologies, Wiley-VCH Verlag GmbH & Co. KGaA, 2014, pp. 209−238.

[54] R.D. Pensack, C. Guo, K. Vakhshouri, E.D. Gomez, J.B. Asbury, Influence of acceptor structure on barriers to charge separation in organic photovoltaic materials, J. Phys. Chem. C 116 (2012) 4824−4831. Available from: https://doi.org/10.1021/jp2083133.

[55] D.M. Guldi, Fullerenes: three dimensional electron acceptor materials, Chem. Commun. (2000) 321−327. Available from: https://doi.org/10.1039/a907807j.

[56] B. Roth, S. Savagatrup, N. V. de los Santos, O. Hagemann, J.E. Carle, M. Helgesen, et al., Mechanical properties of a library of low-band-gap polymers, Chem. Mater. 28 (2016) 2363−2373. Available from: https://doi.org/10.1021/acs.chemmater.6b00525.

[57] K.B. Krueger, P.E. Schwenn, K. Gui, A. Pivrikas, P. Meredith, P.L. Burn, Morphology dependent electron transport in an n-type electron accepting small molecule for solar cell applications, Appl. Phys. Lett. 98 (2011). Available from: https://doi.org/10.1063/1.3556280. 083301-1-083301-3.

[58] P. Sonar, J.P. Fong Lim, K.L. Chan, Organic non-fullerene acceptors for organic photovoltaics, Energy Environ. Sci. 4 (2011) 1558−1574. Available from: https://doi.org/10.1039/c0ee00668h.

[59] G. Sauve, Designing alternative non-fullerene molecular electron acceptors for solution-processable organic photovoltaics, Chem. Rec. 19 (2019) 1078−1092. Available from: https://doi.org/10.1002/tcr.201800157.

[60] G. Sauvé, R. Fernando, Beyond fullerenes: designing alternative molecular electron acceptors for solution-processable bulk heterojunction organic photovoltaics, J. Phys. Chem. Lett. 6 (2015) 3770−3780. Available from: https://doi.org/10.1021/acs.jpclett.5b01471.

[61] C. Lee, S. Lee, G.U. Kim, W. Lee, B.J. Kim, Recent advances, design guidelines, and prospects of all-polymer solar cells, Chem. Rev. 119 (2019) 8028−8086. Available from: https://doi.org/10.1021/acs.chemrev.9b00044.

[62] Aa.F. Eftaiha, J.-P. Sun, I.G. Hill, G.C. Welch, Recent advances of non-fullerene, small molecular acceptors for solution processed bulk heterojunction solar cells, J. Mater. Chem. A 2 (2014) 1201−1213. Available from: https://doi.org/10.1039/c3ta14236a.

[63] C.B. Nielsen, S. Holliday, H.-Y. Chen, S.J. Cryer, I. McCulloch, Non-fullerene electron acceptors for use in organic solar cells, Acc. Chem. Res. 48 (2015) 2803−2812. Available from: https://doi.org/10.1021/acs.accounts.5b00199.

[64] Z. Liu, Y. Wu, Q. Zhang, X. Gao, Non-fullerene small molecule acceptors based on perylene diimides, J. Mater. Chem. A 4 (2016) 17604−17622. Available from: https://doi.org/10.1039/C6TA06978A.

[65] W. Chen, Q. Zhang, Recent progress in non-fullerene small molecule acceptors in organic solar cells (oscs), J. Mater. Chem. C 5 (2017) 1275−1302. Available from: https://doi.org/10.1039/c6tc05066b.

[66] F. Shen, J. Xu, X. Li, C. Zhan, Nonfullerene small-molecule acceptors with perpendicular side-chains for fullerene-free solar cells, J. Mater. Chem. A 6 (2018) 15433−15455. Available from: https://doi.org/10.1039/C8TA04718A.

[67] G. Zhang, J. Zhao, P.C.Y. Chow, K. Jiang, J. Zhang, Z. Zhu, et al., Nonfullerene acceptor molecules for bulk heterojunction organic solar cells, Chem. Rev. 118 (2018) 3447−3507. Available from: https://doi.org/10.1021/acs.chemrev.7b00535.

[68] A. Armin, W. Li, O.J. Sandberg, Z. Xiao, L. Ding, et al., A history and perspective of non-fullerene electron acceptors for organic solar cells, Adv. Energy Mater. 11 (2021) 2003570. Available from: https://doi.org/10.1002/aenm.202003570.

[69] C. Guo, D.R. Kozub, S. Vajjala Kesava, C. Wang, A. Hexemer, E.D. Gomez, Signatures of multiphase formation in the active layer of organic solar cells from resonant soft X-ray scattering, ACS Macro Lett. 2 (2013) 185−189. Available from: https://doi.org/10.1021/mz300547x.

[70] A.M. Nardes, A.J. Ferguson, P. Wolfer, K. Gui, P.L. Burn, P. Meredith, et al., Free carrier generation in organic photovoltaic bulk heterojunctions of conjugated polymers with molecular acceptors: planar versus spherical acceptors, Chemphyschem 15 (2014) 1539−1549. Available from: https://doi.org/10.1002/cphc.201301022.

[71] S. Rajaram, R. Shivanna, S.K. Kandappa, K.S. Narayan, Nonplanar perylene diimides as potential alternatives to fullerenes in organic solar cells, J. Phys. Chem. Lett. 3 (2012) 2405−2408. Available from: https://doi.org/10.1021/jz301047d.

[72] C. Li, H. Wonneberger, Perylene imides for organic photovoltaics: yesterday, today, and tomorrow, Adv. Mater. 24 (2012) 613−636. Available from: https://doi.org/10.1002/adma.201104447.

[73] R. Shivanna, S. Rajaram, K.S. Narayan, Interface engineering for efficient fullerene-free organic solar cells, Appl. Phys. Lett. 106 (2015) 123301. Available from: https://doi.org/10.1063/1.4916216.

[74] L. Ye, K. Sun, W. Jiang, S. Zhang, W. Zhao, H. Yao, et al., Enhanced efficiency in fullerene-free polymer solar cell by incorporating fine-designed donor and acceptor materials, ACS Appl. Mater. Interfaces 7 (2015) 9274−9280. Available from: https://doi.org/10.1021/acsami.5b02012.

[75] H. Li, T. Earmme, G. Ren, A. Saeki, S. Yoshikawa, N.M. Murari, et al., Beyond fullerenes: design of nonfullerene acceptors for efficient organic photovoltaics, J. Am. Chem. Soc. 136 (2014) 14589−14597. Available from: https://doi.org/10.1021/ja508472j.

[76] H. Li, Y.J. Hwang, B.A. Courtright, F.N. Eberle, S. Subramaniyan, S.A. Jenekhe, Fine-tuning the 3d structure of nonfullerene electron acceptors toward high-performance polymer solar cells, Adv. Mater. 27 (2015) 3266−3272. Available from: https://doi.org/10.1002/adma.201500577.

[77] J. Liu, S. Chen, D. Qian, B. Gautam, G. Yang, J. Zhao, et al., Fast charge separation in a non-fullerene organic solar cell with a small driving force, Nat. Energy 1 (2016) 16089. Available from: https://doi.org/10.1038/nenergy.2016.89.

[78] W. Senevirathna, G. Sauve, Introducing 3d conjugated acceptors with intense red absorption: homoleptic metal(ii) complexes of di(phenylacetylene) azadipyrromethene, J. Mater. Chem. C 1 (2013) 6684−6694. Available from: https://doi.org/10.1039/c3tc31377h.

[79] Z. Mao, W. Senevirathna, J.Y. Liao, J. Gu, S.V. Kesava, C. Guo, et al., Azadipyrromethene-based zn(ii) complexes as nonplanar conjugated electron acceptors for organic photovoltaics, Adv. Mater. 26 (2014) 6290−6294. Available from: https://doi.org/10.1002/adma.201400647.

[80] A. Loudet, K. Burgess, Bodipy dyes and their derivatives: syntheses and spectroscopic properties, Chem. Rev. 107 (2007) 4891−4932. Available from: https://doi.org/10.1021/cr078381n.

[81] T.S. Teets, D.V. Partyka, J.B. Updegraff, T.G. Gray, Homoleptic, four-coordinate azadipyrromethene complexes of d^{10} zinc and mercury, Inorg. Chem. 47 (2008) 2338−2346. Available from: https://doi.org/10.1021/ic701190g.

[82] A. Palma, J.F. Gallagher, H. Mueller-Bunz, J. Wolowska, E.J.L. McInnes, D.F. O'Shea, Co(ii), ni(ii), cu(ii) and zn(ii) complexes of tetraphenylazadipyrromethene, Dalton Trans. (2009) 273−279. Available from: https://doi.org/10.1039/b811764k.

[83] D.R. Kozub, K. Vakhshouri, L.M. Orme, C. Wang, A. Hexemer, E.D. Gomez, Polymer crystallization of partially miscible polythiophene/fullerene mixtures controls morphology, Macromolecules 44 (2011) 5722−5726. Available from: https://doi.org/10.1021/ma200855r.

[84] W. Senevirathna, C.M. Daddario, G. Sauvé, Density functional theory study predicts low reorganization energies for azadipyrromethene-based metal complexes, J. Phys. Chem. Lett. 5 (2014) 935–941. Available from: https://doi.org/10.1021/jz402735c.
[85] R. DiScipio, G. Sauvé, C.E. Crespo-Hernández, Photodynamics in metal-chelating tetraphenylazadipyrromethene complexes: implications for their potential use as photovoltaic materials, J. Phys. Chem. C 122 (2018) 13579–13589. Available from: https://doi.org/10.1021/acs.jpcc.7b12657.
[86] F.S. Etheridge, R.J. Fernando, S. Pejić, M. Zeller, G. Sauvé, Synthesis and characterization of fluorinated azadipyrromethene complexes as acceptors for organic photovoltaics, Beilstein J. Org. Chem. 12 (2016) 1925–1938. Available from: https://doi.org/10.3762/bjoc.12.182.
[87] S. Pejić, A.M. Thomsen, F.S. Etheridge, R. Fernando, C. Wang, G. Sauvé, Fluorination increases the electron mobility of zinc azadipyrromethene-based electron acceptors and enhances the performance of fullerene-free organic solar cells, J. Mater. Chem. C 6 (2018) 3990–3998. Available from: https://doi.org/10.1039/c7tc05820a.
[88] C. Wang, C. Daddario, S. Pejić, G. Sauvé, Synthesis and properties of azadipyrromethene-based complexes with nitrile substitution, Eur. J. Org. Chem. 2020 (2020) 714–722. Available from: https://doi.org/10.1002/ejoc.201901736.
[89] C. Wang, M. Zhao, A.L. Rheingold, G. Sauvé, Structure–property study of homoleptic zinc(ii) complexes of di(arylethynyl) azadipyrromethene as nonfullerene acceptors for organic photovoltaics: effect of the aryl group, J. Phys. Chem. C 124 (2020) 8541–8549. Available from: https://doi.org/10.1021/acs.jpcc.0c00401.
[90] C. Wang, P. Wei, J.H.L. Ngai, A.L. Rheingold, T.G. Gray, Y. Li, et al., A zinc(ii) complex of di(naphthylethynyl)azadipyrromethene with low synthetic complexity leads to opv with high industrial accessibility, J. Mater. Chem. A 7 (2019) 24614–24625. Available from: https://doi.org/10.1039/c9ta08654d.
[91] J.D. Mottishaw, H. Sun, Effects of aromatic trifluoromethylation, fluorination, and methylation on intermolecular π-π interactions, J. Phys. Chem. A 117 (2013) 7970–7979. Available from: https://doi.org/10.1021/jp403679x.
[92] H. Liu, S. Kang, J.Y. Lee, Electronic structures and charge transport of stacked annelated β-trithiophenes, J. Phys. Chem. B 115 (2011) 5113–5120. Available from: https://doi.org/10.1021/jp1045595.
[93] M. Zhao, J.C. Jimenez, C. Wang, G. Rui, T. Ma, C. Lu, et al., Monofluorination of naphthyls promotes the cofacial π–π stacking and increases the electron mobility of non-planar zinc(ii) complexes of di(naphthylethynyl)azadipyrromethene, J. Phys. Chem. C 126 (2022) 6543–6555. Available from: https://doi.org/10.1021/acs.jpcc.1c09734.
[94] S. Holliday, R.S. Ashraf, C.B. Nielsen, M. Kirkus, J.A. Rohr, C.H. Tan, et al., A rhodanine flanked nonfullerene acceptor for solution-processed organic photovoltaics, J. Am. Chem. Soc. 137 (2015) 898–904. Available from: https://doi.org/10.1021/ja5110602.
[95] S. Holliday, R.S. Ashraf, A. Wadsworth, D. Baran, S.A. Yousaf, C.B. Nielsen, et al., High-efficiency and air-stable p3ht-based polymer solar cells with a new non-fullerene acceptor, Nat. Commun. 7 (2016) 11585. Available from: https://doi.org/10.1038/ncomms11585.
[96] Y. Lin, Z.-G. Zhang, H. Bai, J. Wang, Y. Yao, Y. Li, et al., High-performance fullerene-free polymer solar cells with 6.31% efficiency, Energy Environ. Sci. 8 (2015) 610–616. Available from: https://doi.org/10.1039/C4EE03424D.
[97] Y. Lin, J. Wang, Z.-G. Zhang, H. Bai, Y. Li, D. Zhu, et al., An electron acceptor challenging fullerenes for efficient polymer solar cells, Adv. Mater. 27 (2015) 1170–1174. Available from: https://doi.org/10.1002/adma.201404317.

[98] H. Bin, Z.-G. Zhang, L. Gao, S. Chen, L. Zhong, L. Xue, et al., Non-fullerene polymer solar cells based on alkylthio and fluorine substituted 2d-conjugated polymers reach 9.5% efficiency, J. Am. Chem. Soc. 138 (2016) 4657−4664. Available from: https://doi.org/10.1021/jacs.6b01744.

[99] W. Zhao, D. Qian, S. Zhang, S. Li, O. Inganas, F. Gao, et al., Fullerene-free polymer solar cells with over 11% efficiency and excellent thermal stability, Adv. Mater. 28 (2016) 4734−4739. Available from: https://doi.org/10.1002/adma.201600281.

[100] H. Bin, L. Gao, Z.-G. Zhang, Y. Yang, Y. Zhang, C. Zhang, et al., 11.4% efficiency non-fullerene polymer solar cells with trialkylsilyl substituted 2d-conjugated polymer as donor, Nat. Commun. 7 (2016) 13651. Available from: https://doi.org/10.1038/ncomms13651.

[101] Z. Zheng, O.M. Awartani, B. Gautam, D. Liu, Y. Qin, W. Li, et al., Efficient charge transfer and fine-tuned energy level alignment in a thf-processed fullerene-free organic solar cell with 11.3% efficiency, Adv. Mater. 29 (2017) 1604241. Available from: https://doi.org/10.1002/adma.201604241.

[102] W. Zhao, S. Li, H. Yao, S. Zhang, Y. Zhang, B. Yang, et al., Molecular optimization enables over 13% efficiency in organic solar cells, J. Am. Chem. Soc. 139 (2017) 7148−7151. Available from: https://doi.org/10.1021/jacs.7b02677.

[103] J. Yuan, Y. Zhang, L. Zhou, G. Zhang, H.-L. Yip, T.-K. Lau, et al., Single-junction organic solar cell with over 15% efficiency using fused-ring acceptor with electron-deficient core, Joule 3 (2019) 1140−1151. Available from: https://doi.org/10.1016/j.joule.2019.01.004.

[104] J. Yuan, Y. Zou, The history and development of Y6, Org. Electron. 102 (2022) 106346. Available from: https://doi.org/10.1016/j.orgel.2022.106436.

[105] Z.-C. Wen, H. Yin, X.-T. Hao, Recent progress of pm6:Y6-based high efficiency organic solar cells, Surf. Interfaces 23 (2021) 100921. Available from: https://doi.org/10.1016/j.surfin.2020.100921.

[106] N. Li, I. McCulloch, C.J. Brabec, Analyzing the efficiency, stability and cost potential for fullerene-free organic photovoltaics in one figure of merit, Energy Environ. Sci. 11 (2018) 1355−1361. Available from: https://doi.org/10.1039/c8ee00151k.

[107] Z. Zheng, J. Wang, P. Bi, J. Ren, Y. Wang, Y. Yang, et al., Tandem organic solar cell with 20.2% efficiency, Joule 6 (2022) 171−184. Available from: https://doi.org/10.1016/j.joule.2021.12.017.

[108] G. Zhang, X.K. Chen, J. Xiao, P.C.Y. Chow, M. Ren, et al., Delocalization of exciton and electron wavefunction in non-fullerene acceptor molecules enables efficient organic solar cells, Nat. Commun. 11 (2020) 3943. Available from: https://doi.org/10.1038/s41467-020-17867-1.

[109] W. Zhu, A.P. Spencer, S. Mukherjee, J.M. Alzola, V.K. Sangwan, S.H. Amsterdam, et al., Crystallography, morphology, electronic structure, and transport in non-fullerene/non-indacenodithienothiophene polymer:Y6 solar cells, J. Am. Chem. Soc. 142 (2020) 14532−14547. Available from: https://doi.org/10.1021/jacs.0c05560.

[110] Y. Firdaus, V.M. Le Corre, S. Karuthedath, W. Liu, A. Markina, W. Huang, et al., Long-range exciton diffusion in molecular non-fullerene acceptors, Nat. Commun. 11 (2020) 5220. Available from: https://doi.org/10.1038/s41467-020-19029-9.

[111] L. Perdigon-Toro, H. Zhang, A. Markina, J. Yuan, S.M. Hosseini, C.M. Wolff, et al., Barrierless free charge generation in the high-performance pm6:Y6 bulk heterojunction non-fullerene solar cell, Adv. Mater. 32 (2020) e1906763. Available from: https://doi.org/10.1002/adma.201906763.

[112] J. Wu, J. Lee, Y.-C. Chin, H. Yao, H. Cha, J. Luke, et al., Exceptionally low charge trapping enables highly efficient organic bulk heterojunction solar cells, Energy Environ. Sci. 13 (2020) 2422−2430. Available from: https://doi.org/10.1039/d0ee01338b.

[113] S. Liu, J. Yuan, W. Deng, M. Luo, Y. Xie, Q. Liang, et al., High-efficiency organic solar cells with low non-radiative recombination loss and low energetic disorder, Nat. Photonics 14 (2020) 300−305. Available from: https://doi.org/10.1038/s41566-019-0573-5.

[114] L. Perdigón-Toro, L.Q. Phuong, F. Eller, G. Freychet, E. Saglamkaya, J.I. Khan, et al., Understanding the role of order in y-series non-fullerene solar cells to realize high open-circuit voltages, Adv. Energy Mater. 12 (2022) 2103422. Available from: https://doi.org/10.1002/aenm.202103422.

[115] R. Wang, C. Zhang, Q. Li, Z. Zhang, X. Wang, M. Xiao, Charge separation from an intra-moiety intermediate state in the high-performance PM6:Y6 organic photovoltaic blend, J. Am. Chem. Soc. 142 (2020) 12751−12759. Available from: https://doi.org/10.1021/jacs.0c04890.

[116] A. Karki, J. Vollbrecht, A.J. Gillett, S.S. Xiao, Y. Yang, Z. Peng, et al., The role of bulk and interfacial morphology in charge generation, recombination, and extraction in non-fullerene acceptor organic solar cells, Energy Environ. Sci. 13 (2020) 3679−3692. Available from: https://doi.org/10.1039/d0ee01896a.

[117] X.-J. Li, G.-P. Sun, Y.-F. Gong, Y.-F. Li, Recent research progress of n-type conjugated polymer acceptors and all-polymer solar cells, Chin. J. Polym. Sci. 41 (2023) 640−651. Available from: https://doi.org/10.1007/s10118-023-2944-0.

[118] C. Sun, J.-W. Lee, C. Lee, D. Lee, S. Cho, S.-K. Kwon, et al., Dimerized small-molecule acceptors enable efficient and stable organic solar cells, Joule 7 (2023) 416−430. Available from: https://doi.org/10.1016/j.joule.2023.01.009.

[119] J. Min, Y.N. Luponosov, C. Cui, B. Kan, H. Chen, X. Wan, et al., Evaluation of electron donor materials for solution-processed organic solar cells via a novel figure of merit, Adv. Energy Mater. 7 (2017) 1700465. Available from: https://doi.org/10.1002/aenm.201700465.

[120] R. Po, G. Bianchi, C. Carbonera, A. Pellegrino, "All that glisters is not gold": an analysis of the synthetic complexity of efficient polymer donors for polymer solar cells, Macromolecules 48 (2015) 453−461. Available from: https://doi.org/10.1021/ma501894w.

[121] C.-Y. Liao, Y.-T. Hsiao, K.-W. Tsai, N.-W. Teng, W.-L. Li, J.-L. Wu, et al., Photoactive material for highly efficient and all solution-processed organic photovoltaic modules: study on the efficiency, stability, and synthetic complexity, Sol. RRL 5 (2021) 2000749. Available from: https://doi.org/10.1002/solr.202000749.

[122] C. Sun, F. Pan, H. Bin, J. Zhang, L. Xue, B. Qiu, et al., A low cost and high performance polymer donor material for polymer solar cells, Nat. Commun. 9 (2018) 743. Available from: https://doi.org/10.1038/s41467-018-03207-x.

[123] J.J. Rech, J. Neu, Y. Qin, S. Samson, J. Shanahan, R.F. Josey III, et al., Designing simple conjugated polymers for scalable and efficient organic solar cells, ChemSusChem 14 (2021) 3561−3568. Available from: https://doi.org/10.1002/cssc.202100910.

[124] G. Chai, Y. Chang, J. Zhang, X. Xu, L. Yu, X. Zou, et al., Fine-tuning of side-chain orientations on nonfullerene acceptors enables organic solar cells with 17.7% efficiency, Energy Environ. Sci. 14 (2021) 3469−3479. Available from: https://doi.org/10.1039/d0ee03506h.

[125] Y. Qin, M.A. Uddin, Y. Chen, B. Jang, K. Zhao, Z. Zheng, et al., Highly efficient fullerene-free polymer solar cells fabricated with polythiophene derivative, Adv. Mater. 28 (2016) 9416−9422. Available from: https://doi.org/10.1002/adma.201601803.

[126] J. Xiao, X.E. Jia, C. Duan, F. Huang, H.-L. Yip, Y. Cao, Surpassing 13% efficiency for polythiophene organic solar cells processed from nonhalogenated solvent, Adv. Mater. 33 (2021) 2008158. Available from: https://doi.org/10.1002/adma.202008158.

[127] X. Yuan, Y. Zhao, D. Xie, L. Pan, X. Liu, C. Duan, et al., Polythiophenes for organic solar cells with efficiency surpassing 17%, Joule 6 (2022) 647–661. Available from: https://doi.org/10.1016/j.joule.2022.02.006.

[128] A. Mishra, G.D. Sharma, Harnessing the structure-performance relationships in designing non-fused ring acceptors for organic solar cells, Angew. Chem. Int. Ed Engl. 62 (2023) e202219245. Available from: https://doi.org/10.1002/anie.202219245.

[129] Q. Shen, C. He, S. Li, L. Zuo, M. Shi, H. Chen, Design of non-fused ring acceptors toward high-performance, stable and low-cost organic photovoltaics, Acc. Mater. Res. 3 (2022) 644–657. Available from: https://doi.org/10.1021/accountsmr.2c00052.

[130] S. Li, L. Zhan, F. Liu, J. Ren, M. Shi, C.Z. Li, et al., An unfused-core-based non-fullerene acceptor enables high-efficiency organic solar cells with excellent morphological stability at high temperatures, Adv Mater 30 (2018) 1705208. Available from: https://doi.org/10.1002/adma.201705208.

[131] H. Yu, Z. Qi, X. Li, Z. Wang, W. Zhou, H. Ade, et al., Modulating energy level on an a-d-a′-d-a-type unfused acceptor by a benzothiadiazole core enables organic solar cells with simple procedure and high performance, Sol. RRL 4 (2020) 2000421. Available from: https://doi.org/10.1002/solr.202000421.

[132] X. Liu, Y. Wei, X. Zhang, L. Qin, Z. Wei, H. Huang, An a-d-a′-d-a type unfused nonfullerene acceptor for organic solar cells with approaching 14% efficiency, Sci. China Chem. 64 (2020) 228–231. Available from: https://doi.org/10.1007/s11426-020-9868-8.

[133] D.L. Ma, Q.Q. Zhang, C.Z. Li, Unsymmetrically chlorinated non-fused electron acceptor leads to high-efficiency and stable organic solar cells, Angew. Chem. Int. Ed Engl. 62 (2023) e202214931. Available from: https://doi.org/10.1002/anie.202214931.

[134] L. Ma, S. Zhang, J. Zhu, J. Wang, J. Ren, J. Zhang, et al., Completely non-fused electron acceptor with 3d-interpenetrated crystalline structure enables efficient and stable organic solar cell, Nat. Commun. 12 (2021) 5093. Available from: https://doi.org/10.1038/s41467-021-25394-w.

[135] M. Joergensen, K. Norrman, S.A. Gevorgyan, T. Tromholt, B. Andreasen, F.C. Krebs, Stability of polymer solar cells, Adv. Mater. 24 (2012) 580–612. Available from: https://doi.org/10.1002/adma.201104187.

[136] X. Xu, D. Li, J. Yuan, Y. Zhou, Y. Zou, Recent advances in stability of organic solar cells, EnergyChem 3 (2021) 100046. Available from: https://doi.org/10.1016/j.enchem.2020.100046.

[137] L. Duan, A. Uddin, Progress in stability of organic solar cells, Adv. Sci. 7 (2020) 1903259. Available from: https://doi.org/10.1002/advs.201903259.

[138] O.R. Yamilova, I.V. Martynov, A.S. Brandvold, I.V. Klimovich, A.H. Balzer, A.V. Akkuratov, et al., What is killing organic photovoltaics: light-induced crosslinking as a general degradation pathway of organic conjugated molecules, Adv. Energy Mater. 10 (2020) 1903163. Available from: https://doi.org/10.1002/aenm.201903163.

[139] M.O. Reese, S.A. Gevorgyan, M. Jørgensen, E. Bundgaard, S.R. Kurtz, D.S. Ginley, et al., Consensus stability testing protocols for organic photovoltaic materials and devices, Sol. Energy Mater. Sol. Cells 95 (2011) 1253–1267. Available from: https://doi.org/10.1016/j.solmat.2011.01.036.

[140] S. Rafique, S.M. Abdullah, N. Badiei, J. McGettrick, K.T. Lai, N.A. Roslan, et al., An insight into the air stability of the benchmark polymer:Fullerene photovoltaic films and devices: a comparative study, Org. Electron. 76 (2020) 105456. Available from: https://doi.org/10.1016/j.orgel.2019.105456.

[141] W.R. Mateker, I.T. Sachs-Quintana, G.F. Burkhard, R. Cheacharoen, M.D. McGehee, Minimal long-term intrinsic degradation observed in a polymer solar cell illuminated in an oxygen-free environment, Chem. Mater. 27 (2015) 404–407. Available from: https://doi.org/10.1021/cm504650a.

[142] C. Zhang, T. Heumueller, S. Leon, W. Gruber, K. Burlafinger, X. Tang, et al., A top-down strategy identifying molecular phase stabilizers to overcome microstructure instabilities in organic solar cells, Energy Environ. Sci. 12 (2019) 1078–1087. Available from: https://doi.org/10.1039/c8ee03780a.

[143] J.W. Rumer, I. McCulloch, Organic photovoltaics: crosslinking for optimal morphology and stability, Mater. Today 18 (8) (2015) 425–435. Available from: https://doi.org/10.1016/j.mattod.2015.04.001. Ahead of Print.

[144] D. Landerer, C. Sprau, D. Baumann, P. Pingel, T. Leonhard, D. Zimmermann, et al., Thermal stabilization of the bulk-heterojunction morphology in polymer: Fullerene solar cells using a bisazide cross-linker, Sol. RRL 3 (2019) 1800266. Available from: https://doi.org/10.1002/solr.201800266.

[145] C. Zhang, A. Mumyatov, S. Langner, J.D. Perea, T. Kassar, J. Min, et al., Overcoming the thermal instability of efficient polymer solar cells by employing novel fullerene-based acceptors, Adv. Energy Mater. 7 (2016) 1601204. Available from: https://doi.org/10.1002/aenm.201601204.

[146] H. Hu, M. Ghasemi, Z. Peng, J. Zhang, J.J. Rech, W. You, et al., The role of demixing and crystallization kinetics on the stability of non-fullerene organic solar cells, Adv. Mater. 32 (2020) 2005348. Available from: https://doi.org/10.1002/adma.202005348.

[147] R. Suthar, H. Dahiya, S. Karak, G.D. Sharma, Ternary organic solar cells: recent insight on structure-processing-property-performance relationships, Energy Technol. (Weinheim, Ger.) 11 (2023) 2201176. Available from: https://doi.org/10.1002/ente.202201176.

[148] K. An, W. Zhong, F. Peng, W. Deng, Y. Shang, H. Quan, et al., Mastering morphology of non-fullerene acceptors towards long-term stable organic solar cells, Nat. Commun. 14 (2023) 2688. Available from: https://doi.org/10.1038/s41467-023-38306-x.

[149] Z.X. Liu, Z.P. Yu, Z. Shen, C. He, T.K. Lau, et al., Molecular insights of exceptionally photostable electron acceptors for organic photovoltaics, Nat. Commun. 12 (2021) 3049. Available from: https://doi.org/10.1038/s41467-021-23389-1.

[150] S. Li, Q. Fu, L. Meng, X. Wan, L. Ding, et al., Achieving over 18% efficiency organic solar cell enabled by a ZnO-based hybrid electron transport layer with an operational lifetime up to 5 years, Angew. Chem. Int. Ed Engl. 61 (2022) e202207397. Available from: https://doi.org/10.1002/anie.202207397.

[151] X. Du, T. Heumueller, W. Gruber, A. Classen, T. Unruh, N. Li, et al., Efficient polymer solar cells based on non-fullerene acceptors with potential device lifetime approaching 10 years, Joule 3 (2019) 215–226. Available from: https://doi.org/10.1016/j.joule.2018.09.001.

[152] X. Xu, J. Xiao, G. Zhang, L. Wei, X. Jiao, H.L. Yip, et al., Interface-enhanced organic solar cells with extrapolated t(80) lifetimes of over 20 years, Sci. Bull. (Beijing) 65 (2020) 208–216. Available from: https://doi.org/10.1016/j.scib.2019.10.019.

[153] Y. Li, X. Huang, K. Ding, H.K.M. Sheriff Jr., L. Ye, et al., Non-fullerene acceptor organic photovoltaics with intrinsic operational lifetimes over 30 years, Nat. Commun. 12 (2021) 5419. Available from: https://doi.org/10.1038/s41467-021-25718-w.

[154] Q. Burlingame, X. Huang, X. Liu, C. Jeong, C. Coburn, S.R. Forrest, Intrinsically stable organic solar cells under high-intensity illumination, Nature 573 (2019) 394–397. Available from: https://doi.org/10.1038/s41586-019-1544-1.

[155] Y. Li, B. Huang, X. Zhang, J. Ding, Y. Zhang, et al., Lifetime over 10000 hours for organic solar cells with Ir/IrO$_x$ electron-transporting layer, Nat. Commun. 14 (2023) 1241. Available from: https://doi.org/10.1038/s41467-023-36937-8.

[156] M.A. Green, E.D. Dunlop, M. Yoshita, N. Kopidakis, K. Bothe, G. Siefer, et al., Solar cell efficiency tables (version 62), Prog. Photovolt.: Res. Appl. 31 (2023) 651−663. Available from: https://doi.org/10.1002/pip.3726.

[157] J. Lee, S.A. Park, S.U. Ryu, D. Chung, T. Park, S.Y. Son, Green-solvent-processable organic semiconductors and future directions for advanced organic electronics, J. Mater. Chem. A 8 (2020) 21455−21473. Available from: https://doi.org/10.1039/d0ta07373c.

[158] M. Mooney, A. Nyayachavadi, S. Rondeau-Gagne, Eco-friendly semiconducting polymers: from greener synthesis to greener processability, J. Mater. Chem. C 8 (2020) 14645−14664. Available from: https://doi.org/10.1039/d0tc04085a.

[159] S. Yoon, S. Park, S.H. Park, S. Nah, S. Lee, et al., High-performance scalable organic photovoltaics with high thickness tolerance from 1 cm^2 to above 50 cm^2, Joule 6 (2022) 2406−2422. Available from: https://doi.org/10.1016/j.joule.2022.07.014.

[160] N. Camaioni, C. Carbonera, L. Ciammaruchi, G. Corso, J. Mwaura, R. Po, et al., Polymer solar cells with active layer thickness compatible with scalable fabrication processes: a meta-analysis, Adv. Mater. 35 (2023) 2210146. Available from: https://doi.org/10.1002/adma.202210146.

[161] Q. Wu, Y. Yu, X. Xia, Y. Gao, T. Wang, et al., High-performance organic photovoltaic modules using environmentally-friendly solvents for various indoor application scenarios to improve photovoltaic efficiency and stability, Joule 6 (2022) 2138−2151. Available from: https://doi.org/10.1016/j.joule.2022.07.001.

[162] F. Qin, L. Sun, H. Chen, Y. Liu, X. Lu, et al., 54 cm^2 large-area flexible organic solar modules with efficiency above 13, Adv. Mater. 33 (2021) e2103017. Available from: https://doi.org/10.1002/adma.202103017.

CHAPTER NINE

Copper zinc tin sulfide thin-film solar cells: An overview

Zubair Ahmad Kumar[1], Towseef Ahmad[2] and Mohd Zubair Ansari[2]

[1]Department of Physics and Astronomy, National Institute of Technology Rourkela, Rourkela, Odisha, India
[2]Department of Physics, National Institute of Technology Srinagar, Srinagar, Jammu & Kashmir, India

9.1 Introduction

The energy that powers the globe comes primarily from finite resources such as fossil fuels. The combustion of fossil fuels contributes to pollution in the Earth's atmosphere and increases the planet's average temperature. Significant exhaust gases such as carbon dioxide (CO_2), nitrogen oxides (NO_x), carbon monoxide (CO), sulfur oxides (SO_x), and many others are responsible for the greenhouse gases or pollutants that result from burning fossil fuels. As a result, a significant amount of energy will be required to switch from traditional sources of energy to renewable, lower-pollution, and environment-friendly sources. Over the past ten years, the availability of a variety of sources of renewable energy, such as wind, hydro, and biofuels, has increased [1]. However, geothermal energy has several disadvantages, such as the requirement for a sizable amount of space, the need for more confidence regarding water availability for hydrothermal energy, and the high expenses associated with its development. Solar energy, among all other forms of renewable energy, is now the only alternative technology with abundant, clean energy and sources that are not harmful to the environment. One of the Earth's most plentiful energy sources is solar radiation, but the high cost of existing solar technologies has made it difficult to use. The current photovoltaic (PV) industry derives 80% of its electricity from solar cells based on crystalline silicon (c-Si) and polycrystalline silicon (p-Si) wafer technologies [1–3]. Silicon solar cells use an indirect bandgap absorber material; therefore, a thick layer is required to absorb higher percentages of the incidence radiation. Perfect crystals are also necessary to produce high-efficiency solar modules, which raises the price of PV devices. Therefore, recent research on PV innovations has focused on thin-film PV technologies. These thin-film technologies are based on direct-bandgap materials such as copper indium gallium diselenide (CIGS), cadmium telluride (CdTe), and copper indium diselenide (CIS). The availability of In and Te, as well as the toxicity of Cd and Se, limits the number of PV devices that can be produced using these absorber layers [4]. In this context, alternative direct-bandgap absorber materials with a high absorption coefficient, low cost, and less toxicity are sought [1,5].

Copper zinc tin sulfide (Cu_2ZnSnS_4 or CZTS) is an alternative, nontoxic, affordable, and accessible absorber material that has received a lot of recent interest from the scientific community. It is a quaternary compound [6−8] and the most stable phases for this p-type semiconductor adopt a tetragonal structure (see discussion below). Although CZTS is a relatively new material, encouraging results suggest that it could be used as a direct-bandgap solar cell absorber material [1−3]. The optimal semiconductor bandgap needed for solar cells is 1.35 eV, which is extremely close to the roughly 1.50 eV bandgap of CZTS [3]. Moreover, CZTS films show a high absorption coefficient in the range of 10^4/cm [5,9]. Additionally, CZTS is seen as a strong candidate for the upcoming thin-film solar cells due to its 32% potential conversion efficiency [10].

All of the CZTS's ingredients are inexpensive, less harmful, and widely available on Earth. Fig. 9.1 depicts Earth's content (crust) of the elements utilized in Sn(S, Se), $Cu_2ZnSn(S, Se)_4$, $Cu(In, Ga)(S, Se)_2$, and $Sb_2(S, Se)_3$ light absorbers. In the Earth's crust, Zn (790 ppm) and Sn (22 ppm) concentrations are approximately 500 and 13 times higher than those of In (1.6 ppm), the element replaced to yield CZTS [11].

9.2 Structure

A very brief overview of the structural aspects of Cu_2ZnSnS_4 (CZTS) has been published [12]. Cubic sphalerite [Fig. 9.2A—space group (SG) #216] and hexagonal wurtzite (Fig. 9.2B—SG #186) phases are metastable, with randomly distributed cations with different stacking order; the two most prevalent (and stable) structures of CZTS are tetragonal phases: kesterite (Fig. 9.2C—SG #82) and stannite (Fig. 9.2D—SG #121); the caption of Fig. 9.2 details the crystal structure space group details [12].

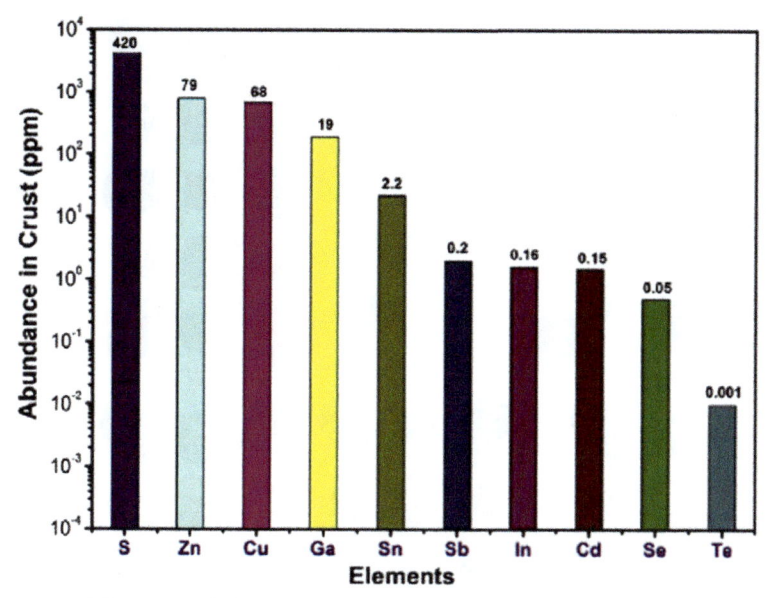

Figure 9.1 Schematic of the relative (ppm/10) Earth abundance of elements that comprise some semiconductor solar cell materials such as Sn(S, Se), $Cu_2ZnSn(S, Se)_4$, $Cu(In, Ga)(S, Se)_2$, $Sb_2(S, Se)_3$, and CdTe; note that abundance is plotted in log scale. Reproduced with permission from B. Pejjai, V.R.M. Reddy, S. Gedi, C. Park, Status review on Earth-abundant and environmentally green Sn-X (X = Se, S) nanoparticle synthesis by solution methods for photovoltaic applications, Int. J. Hydrogen Energy 42 (5) (February 2017) 2790–2831, copyright (2017) Elsevier.

The tetragonal CZTS phases can be derived from the very similar chalcopyrite phase of $CuInS_2$ (CIS) (SG #122, $I\bar{4}2d$): Zn from Group 12 (or IIB) and tin (Sn) from Group 14 (or IVA) can replace indium (In) from Group 13 (or IIIA) to produce the more sustainable and Earth-abundant CZTS, a quaternary group $I_2-II-IV-VI_4$ compound [12]. However, the Cu and Zn atom configurations in the two structures are different (see Fig. 9.2C and D). CZTS materials are usually grown in the kesterite phase because it is more thermodynamically stable than stannite [2,12,13].

9.3 Synthesis methods of copper zinc tin sulfide nanoparticles and thin films

In this section, we overview five methods to produce CZTS nanomaterials and thin films. Sol–gel, chemical bath deposition (CBD), spray pyrolysis,

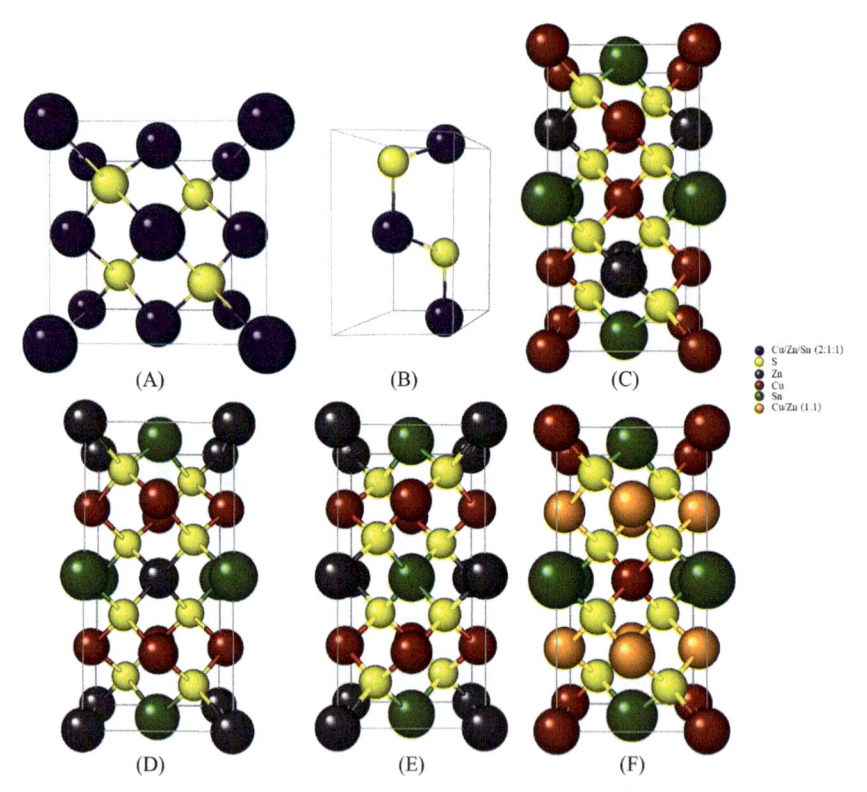

Figure 9.2 Illustrations of copper zinc tin sulfide (CZTS) structural phase: (A) cubic-CZTS (space group (SG) #216—$F\bar{4}3m$), (B) hexagonal-CZTS (SG #186—$P6_3mc$), (C) (ordered) tetragonal kesterite-CZTS (SG #82—$I\bar{4}$), (D) tetragonal stannite-CZTS (SG #121—$I\bar{4}2m$), (E) tetragonal (primitive-mixed CuAu (PMCA))-CZTS (SG #111—$P\bar{4}2m$), and (F) (disordered) tetragonal kesterite-CZTS (SG #121—$I\bar{4}2m$); the following colors are used to designate elements: blue (Cu/Zn/Sn—2:1:1), yellow (S), silver (Zn), red (Cu), green (Sn), and orange (Cu/Zn—1:1). *Reproduced from U. Syafiq, N. Ataollahi, P. Scardi, Progress in CZTS as hole transport layer in perovskite solar cell, Sol. Energy 196 (2020) 399—408, copyright (2020) Elsevier.*

spin coating, solvothermal, and thermal evaporation processes are briefly discussed. Several of these methods (spray pyrolysis, spin coating, and CBD) are better suited for thin-film deposition. Solvothermal, sol—gel, and thermal evaporation are more amenable for the production of nanomaterials.

9.3.1 Sol—gel method

Cu_2ZnSnS_4 (CZTS) films can be made using thiourea $SC(NH_2)_2$, tin (II) chloride dihydrate, zinc (II) acetate dihydrate, and copper (II) acetate monohydrate as precursors, monoethanolamine (MEA) and 2-methoxyethanol (2-ME) were also utilized [14]. After dissolving the precursor solution in 2-ME and stirring at a temperature of 45°C for 1 hour, the metal complexes were completely dissolved; MEA was added as a stabilizer. After this step, the solution was dropped onto the silicon and glass, spun at 2000 rpm for 60 seconds to prepare the CZTS film, and the film was annealed for one hour at 500°C. After thermal annealing, using vacuum-thermal evaporation (4.5×10^{-5} Torr), metal contacts were produced. The diode's contact area was determined to be 3.14×10^{-2} cm^2, and was tested utilizing halogen lamp tests [6].

9.3.2 Chemical bath deposition

Rana et al. [15] used the CBD approach to deposit CZTS thin films. They created two different stacking sequences: SLG/SnS/CuS/ZnS (stack 1) and SLG/ZnS/CuS/SnS (stack 2); note that "SLG" is an abbreviation for soda lime glass. The appropriate metal ions were present as a CuS layer was deposited in between the ZnS and SnS layers. After that, the "ZnS" and "SnS" layers either at the top or bottom were swapped to facilitate alloying. An SnS thin-film was applied to the glass substrate using a CBD precursor solution containing tin-chloride ($SnCl_2$) (0.2 M), thioacetamide (C_2H_5NS) (0.4 M), and tartaric acid (0.01 M) at pH 2 for the preparation of stack 1 sample. For 3 hours, the CBD bath was held at 70°C. In the following step 2, copper sulfate ($CuSO_4$) (0.1 M), acetic acid (CH_3COOH) (2 N), and thioacetamide (C_2H_5NS) at pH 2 solutions are used to deposit a thin layer of CuS over SnS. CuS was deposited over a period of 1 hour at a temperature of 60°C. Finally, in step three, a top layer of ZnS was deposited using a CBD solution (at pH = 8) of zinc acetate, thiourea [$SC(NH_2)_2$], tartaric acid, and aqueous ammonium (NH_3). Zinc sulfide (ZnS) had a 4-hour CBD period and an 80°C CBD temperature; however, deposition of CZTS was not successful. Alternatively, the aforementioned deposition conditions for the stack 2 sample (SLG/ZnS/CuS/SnS) were tried; however, CZTS deposition in a stack 2 film was unsuccessful. In view of this, they changed the CuS layer deposition conditions to shield the ZnS thin-film layer from being pulled away from the substrate surface. The CuS bath chemical composition was altered: copper chloride ($CuCl_2$) (0.1 M), thiourea (0.2 M), and aqueous ammonium (NH_3) added drop by drop to achieve a pH of approximately 12. The bath was then heated for 1 hour at 80°C. SnS was similarly deposited in the last stage as it was in the stack 1 multilayer. Then, for 45 minutes at 500°C, stacks 1 and 2 of the CBD thin-film (TF) CZTS samples, SLG/ZnS/CuS/SnS, and SLG/ZnS/CuS/SnS, respectively, were sulfurized, producing pure CZTS films with a bandgap of 1.48−1.51 eV [15].

9.3.3 Spray pyrolysis

Making use of a perfume atomizer and the spray pyrolysis method, Cu_2ZnSnS_4 thin films were successfully deposited. Thiourea (0.24 M) and cupric chloride (0.06 M) were dissolved either in 25 mL or 50 mL of DI water. The aqueous solution also contained stannic chloride (0.03 M), thiourea (0.06 M), and zinc acetate (0.03 M). Sulfur is a volatile element at high temperatures; it needed to be added in excess to the solution to prevent precipitation and to make up for the sulfur that was lost during pyrolysis. Following vigorous stirring for 30 minutes with a magnetic stirrer, these combined solutions were manually sprayed in tiny droplets for 5 minutes at 350°C on a preheated glass substrate. The sprayed substrates go through an hour-long 350°C thermal treatment [16]. A net balanced chemical reaction Eq. (9.1) that accounts for all of the reactant atoms can be written as follows:

$$2CuCl + ZnCl_2 + SnCl_4 + 4SC(NH_2)_2 + 8H_2O + heat \rightarrow Cu_2ZnSnS_4 + 4CO_2 + 8NH_4Cl$$

(9.1)

9.3.4 Spin-coating

The sol−gel spin-coating process used to make thin films combines several novel methods [1,3,16]. For example, a nonaqueous solution containing thiourea (1.4 M), stannic chloride (0.14 M), cupric chloride (0.34 M), and zinc chloride (0.14 M) combined in 20 mL of 2-methoxy-ethanol and 2 mL of MEA. MEA and 2-methoxy-ethanol are used as solvents and stabilizers, respectively. MEA and CZTS are held at a constant molar ratio of one. The creation of a yellow sol-gel was achieved after the solution was vigorously agitated at 60°C for 1 hour. Filtered sol−gel solution is then applied to a glass substrate, which then rotates by a spin-coater for 45 s at 3600 rpm. The films are then immediately warmed for 15 minutes at 150°C with the purpose of evaporating the solvent. Three and five times each of the coating and drying processes are used; the CZTS thin films that are obtained are annealed for an hour at 350°C [17].

9.3.5 Solvothermal method

The preparation of CZTS thin films using the solvothermal method was reported by Guan et al. [18]. They used 40 mL of ethylene glycol as a precursor solvent; $Cu(NO_3)\cdot 3H_2O$ (0.06 M), $Zn(CH_3COO)_2\cdot 2H_2O$ (0.03 M), $SnCl_2\cdot 2H_2O$ (0.03 M), H_2NCSNH_2 (0.15 M) and appropriate amounts of polyvinylpyrrolidone (PVP) were added while stirring for 3 hours at room temperature. A 50 mL Teflon liner was then filled with the precursor solution, and the well-cleaned glass substrates were submerged inside of it. For 24 hours, the autoclave was closed and kept at 200°C, and after that, it was cooled until it reached room temperature. The glass substrates were cleaned with deionized water and dried. The entire process is straightforward and at a relatively low reaction temperature, which may help make the final products more reproducible.

9.3.6 Thermal evaporation technique

Shi et al. utilized a hydrothermal approach to prepare CZTS semiconducting material powder from a mixture of freshly synthesized CuS, ZnS, and SnS powder precursors. For characterizing CZTS thin films and creating PV devices, SLG and $F:SnO_2$ conductive glass (FTO) were employed as substrates, respectively. The substrates were fastened to a revolving disk that rotated at a speed of 60 rpm, and they were positioned parallel and 30 cm apart from the source in order to produce more homogeneous and smooth thin films. The vacuum thermal evaporator's chamber had a baseline pressure of 2 mPa before the deposition. A molybdenum boat was used to evaporate the CZTS semiconducting material powder while the current was steadily increased from 100 A to 260 A. During the deposition, the evaporation process was sustained for 10 minutes at a working pressure of 0.01 Pa. Formation of a CZTS thin-film (approximately 1 μm) is followed by annealing for 40 minutes at 300°C under a high-purity N_2 environment [19].

9.4 Copper zinc tin sulfide-based thin-film solar cell deposition techniques

Deposition is the process of coating a substrate with a thin layer of material to improve the quality or functionality of synthesized materials.

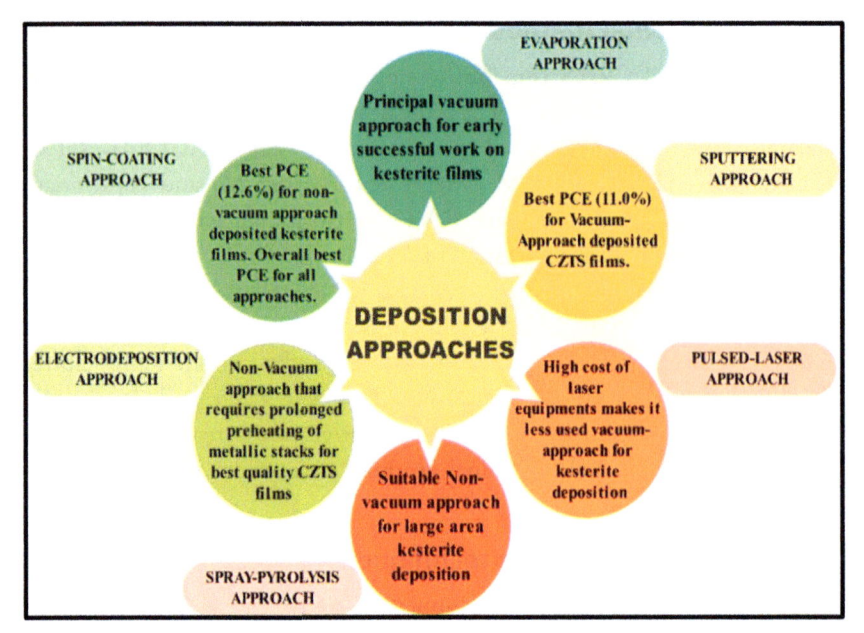

Figure 9.3 Schematic overview of various thin-film deposition approaches to produce kesterite, including comments relative to solar cell performance. *Reproduced with permission by Elsevier from K.C. Nwambaekwe, V.S. John-Denk, S.F. Douman, P. Mathumba, S.T. Yussuf, et al., Crystal engineering and thin-film deposition strategies towards improving the performance of kesterite photovoltaic cell, J. Mater. Res. Tech. 12 (May–June 2021) 1252–1287, published by Elsevier under a Creative Commons (BY-NC-ND 4.0) license.*

CZTS thin films can be deposited using a variety of vacuum- and nonvacuum-based processes. These methods and comments regarding CZTS solar cell performance are illustrated in Fig. 9.3 [20].

9.4.1 Vacuum-based approaches

Vacuum-based deposition procedures offer superior opportunities for the creation of high-quality thin-film devices due to their high homogeneity. These methods include depositing the CZTS compound's constituent atoms on a substrate either through sputtering or through the evaporation or coevaporation of the target sources at optimal pressure and temperature. These methods also offer good reproducibility and the benefit of simple chemical composition control in thin films. These vacuum-based methods can be further divided into sputtering, evaporation, and pulsed laser deposition (PLD) methods, among other categories. The vacuum-based approaches are expensive.

9.4.1.1 Sputtering techniques
Many researchers have employed the sputtering deposition technique to manufacture high-quality thin films. RF, ion beam, DC, argon beam, hybrid, and reactive-magnetron sputtering are some of the sputtering methods adopted to deposit CZTS thin films [1–3,19–22]. Sputtering of a single quaternary target for CZTS thin-film deposition will result in compositional issues and relatively poor performance [21]. Tanaka et al. were one of the first groups to report a hybrid vacuum-based sputtering and S flux annealing process to produce CZTS on a quartz glass substrate [22]. Two sputtering sources were used for Sn and Cu; two effusion cells were used for Zn and S deposition. Tin was deposited by DC-sputtering for 80 minutes at 650 V and 40 Pa of Ar pressure. Zn was first vaporized in a quartz-crucible at 300°C, and then Cu was added by RF-sputtering at 100 W for 220 minutes under 2 Pa of atmospheric pressure. Stannite-phase CZTS TFs were produced by annealing the resulting copper (Cu), zinc (Zn), and tin (Sn) precursor films under S flux for 60 minutes. The resultant CZTS films demonstrated good optical and suitable electrical properties for producing solar cells [22]. Zhang used a similar method of successive deposition employing metallic Cu, Zn, and Sn sputtering targets, followed by sulfurization. CZTS thin films are formed by stacking Cu/Sn/Zn precursors [23]. Katagiri created these CZTS solar cells by employing a similar approach of sequential metal layer deposition, but with changing concentrations of H_2S in the sulfurization [24]. With a 5% H_2S concentration, they found the solar cell's efficiency to be 4.1%, and with a 20% H_2S concentration, it was 3.8%.

9.4.1.2 Evaporation techniques
In 1997, Katagiri reported the first functioning evaporated CZTS-based kesterite device. Electron-beam evaporation was applied to deposit precursors, such as Cu/Sn/Zn stacked layers on SLG substrates heated to 150°C [25]. First, a 1600-micron-thick Zn layer on the substrate was applied. Then, an Sn layer with a thickness of 230 nm was placed on top of the Zn layer. In the end, the precursor was created by depositing an 1800 Å thick Cu layer on top of the Sn layer, and then the precursors were sulfurized in an electric furnace heated to 500°C in an environment of 5% (N_2 + H_2S) to produce the CZTS films. The CZTS thin-film based

solar cell demonstrated an efficiency of 0.66% with $J_{sc} = 6.0$ mA/cm^2, $V_{oc} = 400$ mV, and fill factor (FF) = 0.277.

An efficiency of 4.3% and a $V_{oc} = 608$ mV were obtained by Zhi et al. [26] by employing the single-step coevaporation method. Dual-layer molybdenum (Mo) contacts are plated above the glass substrate using DC sputtering at chamber pressures of 1.2 and 0.1 Pa, respectively. The Mo/glass substrate is then moved inside the elemental evaporation chamber without coming into contact with the air. When the Knudsen cells simultaneously evaporate the elemental sources of Cu, Zn, Sn, and S, the substrate temperature is changed from 500°C to 550°C. Without baking out, a base pressure of less than 0.1 mPa and a working pressure of 0.05 Pa were used to create CZTS thin films. Then, using cadmium acetate, thiourea, and ammonia solution as the reactant, the CBD method for 20 minutes at a temperature of 80°C is used to create the 60 nm-thick n-type CdS buffer layer. RF sputtering is used to deposit 200 nm Al:ZnO thin films with a sputtering power of 125 W and a working pressure of 0.1 Pa Ar, as well as 50 nm i-ZnO thin films with a sputtering power of 50 W, a working pressure of 0.1 Pa Ar, and 2 mPa O$_2$. E-beam deposition is used to create around 3-micron-high, tens-of-micron-wide Al grids with an evaporation rate of 15 m/second. Less than 5% of each cell is covered by the grid shade.

9.4.1.3 Pulsed laser deposition methods

An effective method for producing high-quality films with complex compositions is PLD. An intense laser beam pulse is employed in this method to vaporize the target material's surface and turn it into plasma, which typically turns back into gas before it reaches the substrate. A thin material film is created on the substrate. The epitaxial development of CZTS thin films using the PLD method was initially described by Sekiguchi et al. [27]. A laser beam was directed onto a sintered target, and then ablated pieces were deposited on an n-type (100) GaP (n-GaP) substrate. The solid-state reaction approach was utilized to create CZTS pellets, which were made by mixing Cu$_2$S, ZnS, and SnS$_2$ powders in a (1:1:1 mol.) ratio. The mixture powder was formed into a pellet, placed inside an evacuated quartz ampoule, and heated to 750°C in a furnace for 24 hours. The temperature changed at a rate of 2 degrees/minute, both up and down. The substrate temperature was changed between 300°C, 350°C, and 400°C, while the pulsed laser energy density on the target was fixed at 0.85 JC/m^2. Cazzaniga et al. developed a CZTS thin-film that was

5.2% efficient [28]. When Mohlkar et al. varied the Cu and (Zn/Sn) proportions from 0.8 to 1.2 while maintaining a constant Zn/Sn concentration, they obtained 4.13% efficiency [29].

9.4.1.4 Nonvacuum-based approaches

Most studies on CZTS thin-film production and its application as a PV device have been published over the past few decades using various vacuum approaches, as was previously stated. However, modern research has shown that the majority of nonvacuum techniques enable low-cost thin-film solar cells (TFSCs). While nonvacuum procedures are less expensive, use a lesser amount of energy, and produce better and more uniform thin films, vacuum-based approaches are typically expensive and consume a very large amount of energy. Thus, taking into account the advantages of low-cost nonvacuum techniques that are extremely straightforward and produce high output, they are frequently employed for the plentiful and easily accessible constituents of CZTS materials.

9.4.1.5 Spray pyrolysis deposition technique

The creation of CZTS thin films of the stannite phase was initially described by Nakayama in 1996 [30] by utilizing the spray pyrolysis process. CuCl, $ZnCl_2$, $SnCl_4$, and thiourea solutions were dissolved in 50:50 DW/ethanol and sprayed over SLG substrates at temperatures ranging from 280°C to 360°C. After annealing at 550°C, stoichiometric films with a stannite structure having a bandgap energy of 1.46 eV were produced. Rajeshmon [31] created CZTS thin films for solar cell applications utilizing the spray pyrolysis process. They used two tin precursors, namely stannous chloride ($SnCl_2$) and stannic chloride ($SnCl_4$). An aqueous solution comprising thiourea 0.15 M, zinc acetate 0.0125 M, stannous chloride ($SnCl_2$) or stannic chloride ($SnCl_4$) 0.0125 M, and cuprous chloride 0.02 M was deposited at a rate of 8 mL/minute on the substrate maintained at a temperature of 623 K. In order to create *p-n* junctions, films of CZTS were first applied to glass substrates coated with tin-doped indium oxide. This was followed by the chemical spray pyrolysis deposition of an In_2S_3 buffer layer at 350°C. The solar cell prepared with stannic chloride displayed $V_{oc} = 380$ mV and $J_{sc} = 2.4$ mA/cm^2. The aforementioned findings demonstrated that CZTS thin films created using stannic-chloride as a precursor produced the desired characteristics for the solar cell absorber layer.

9.4.1.6 Electrodeposition technique

One of the most affordable and environmentally friendly methods for depositing a thin coating over a large area is electrodeposition. The two types of deposition are sequential electrodeposition and single-step electrodeposition. To create the final film, multiple layers of metal salts are successively electrodeposited, annealed in a sulfur atmosphere (H_2S/S vapor), or constituent metal precursors are done so. In both small-scale research applications and large-scale industrial applications, diverse semiconductor thin films can be prepared at low cost using the electrodeposition technique. It has been widely used to create thin-film absorber layers for solar cells such as CdTe and CIGS.

Scragg et al. used a traditional three-electrode cell with a platinum counter-electrode, an Ag-AgCl reference electrode, and 4 N purity (or above) metal salts that were deposited on Mo-coated glass substrates in the following order: Cu, Sn, and Zn [32]. At -1.14 V versus Ag/AgCl, bright and firmly adhering copper layers were formed from an alkaline solution containing 0.1 M sorbitol, 50 mM $CuCl_2$, and 1.5 M NaOH. Then, taking an alkaline solution made up of 0.1 M sorbitol, 55 mM $SnCl_2$, and 2.25 M NaOH, they applied tin layers to the copper film at -1.21 V versus Ag/AgCl. At 0.15 M $ZnCl_2$ buffered to pH 3, the final zinc layer was deposited at -1.20 V versus Ag/AgCl. Subsequently, samples made up of a three-layer stack of Zn, Sn, and Cu were annealed in a sulfur environment. After the samples cooled to room temperature, they were removed once the tube had been evacuated with dry nitrogen. In order to construct solar cells, a CdS buffer layer was chemically deposited directly over the CZTS layer. Using magnetron sputtering, a transparent-conductive bilayer of indium tin oxide and zinc oxide was sputtered, and Ni or Al contacts were used to facilitate current collection. The efficiency observed was 0.8% with an open circuit voltage (V_{oc}) of 295 mV, a short circuit current density of J_{sc} 8.7 mA/cm^2, and an FF of 0.32.

9.4.1.7 Sol−gel deposition technique

For the fabrication of thin films, the spin coating method based on sol−gel is very straightforward and inexpensive. Precursor sol−gel solution preparation and spin coating of the solution on the substrate are the two stages in the production of thin films. The cost-effective and low-temperature sol−gel technology is used for the deposition procedure. Quantitative doping and homogenous, multicomponent chalcogenide films are simple to obtain. There are also some drawbacks to this technology, such as the lengthy process and the difficulty in regulating the porosity of the material. Tanaka et al. were the first to report the use of the sol—gel method to create CZTS thin films [33]. MEA, 2-methoxyethanol, tin chloride, zinc acetate, and copper acetate monohydrate were used as precursors to produce CZTS thin films. The solution was agitated at 45°C for 1 hour before being spin-coated for 30 second at a speed of 4000 rpm. They were repeatedly coated and kept to dry at 300°C. The precursors were annealed for 1 hour at 500°C in a N_2/H_2S environment. The CZTS thin films that were produced have a bandgap of 1.49 eV. Su et al. created a CZTS thin-film absorber using the sol—gel process, and they then investigated how doping and etching affected the functionality of the device [34]. They noted that when there was no Na doping, the CZTS-based device demonstrated 4.10% efficiency along with 558 mV open circuit voltage (V_{oc}), 12 mA/cm^2 short circuit current density (J_{sc}), and an FF of 53%, whereas when Na doping was 1%, the device performance improved to 5.01% efficiency with improvements in V_{oc} (650 mV), J_{sc} (12.90 mA/cm^2), and FF of (59%).

9.4.1.8 Solvothermal method

The solvothermal method conducts the reaction in a closed container with high pressure that is kept at a constant temperature. The types of the reaction's reagents, solvents, and temperatures have an impact on the film's quality. We describe the synthesis of CZTS using the solvothermal technique, which is modified to create CZTS nanostructures that are friendly to the environment. In a typical synthesis, 20 mL of ethylene glycol:water (70:30) were used to dissolve 1.1639 g of copper chloride ($CuCl_2 \cdot 2H_2O$), 1.1966 g of tin chloride ($SnCl_2 \cdot 5H_2O$), and 0.465 g of zinc chloride ($ZnCl_2$). The resulting ratios are 0.59 for Cu/(Zn + Sn), 1.9 for Cu/Zn, and 0.43 for Zn/Sn. Additionally, 20 mL of ethyleneglycol-water (which contains 70% ethylene glycol and 30% water) was used to dissolve 1.0392 g of CH_4N_2S. Then, while stirring, it was slowly dropped into the metal chloride solution. 0.5 g of polyvinyl alcohol (PVA) was then added to the mixture after 30 minutes of stirring. The mixture was added to a Teflon-lined stainless steel autoclave, sealed, and heated to 180°C for 12 hours. The autoclave then gradually cooled down to room temperature through natural processes. The resulting black product was centrifuged and repeatedly cleaned with distilled water and pure ethanol [35,36]. Screen-printed CZTS on an n-Si substrate produced a device with a η of 6.72% [35].

9.5 Comparison of performance of vacuum- and nonvacuum-based solar cells

We have discussed two approaches for the synthesis of CZTS thin-film solar cells (Fig. 9.4), which are vacuum- and nonvacuum-based, both having their own advantages and disadvantages; see Fig. 9.3 [20]. The vacuum-based methods are expensive; however, they also have benefits, as higher-homogeneity materials are produced by this method. As a result, the final product is of much higher quality [1–3,20].

9.5.1 Copper zinc tin sulfide vacuum-processed solar cells

An example of a vacuum-processed CZTS solar cell was reported by Yan et al. [37]. A cosputter deposition technique was employed with Cu, SnS, and ZnS precursors under the following environmental conditions: sulfur and SnS were combined at 560°C for 3 minutes with Al_2O_3 at the back, which improved efficiency to 11%. The improvement in efficiency was facilitated by heat treatment at 270°C [37]. Different vacuum-based techniques, as indicated in Table 9.1, are discussed in this book chapter, including sputtering, evaporation, and PLD [13,25,28,37–41]. It has been discovered, after extensive research and implementation over a long

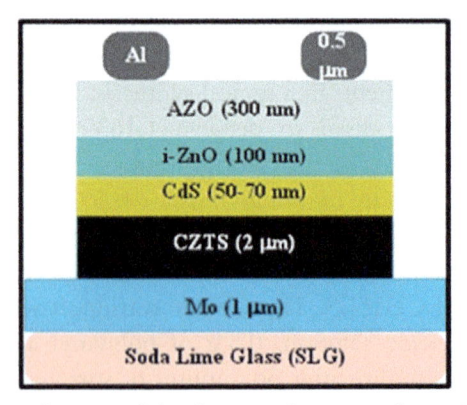

Figure 9.4 Schematic drawing of the device architecture of a typical kesterite photovoltaic cell showing different layers and their thickness. *Reproduced with permission by Elsevier from K.C. Nwambaekwe, V.S. John-Denk, S.F. Douman, P. Mathumba, S.T. Yussuf, et al., Crystal engineering and thin-film deposition strategies towards improving the performance of kesterite photovoltaic cell, J. Mater. Res. Tech. 12 (May–June 2021) 1252–1287, published by Elsevier under a Creative Commons (BY-NC-ND 4.0) license.*

Table 9.1 Examples of vacuum-based approach results.

Technique	V_{oc} (mV)	FF(%)	J_{sc} (mA/cm^2)	Efficiency	References
Cosputtering	730.6	69.27	21.76	11.01	[37]
Evaporation	661	65.8	19.5	8.4	[38]
Sputtering	610	62	17.9	6.77	[39]
RF cosputtering	662	55	15.7	5.74	[40]
Pulsed laser deposition	616	47.9	17.6	5.2	[28]
RF cosputtering	612	62	10.6	4.1	[13]
Evaporation	522.4	36	14.11	2.62	[41]
Evaporation	400	28	6.0	0.66	[25]

Table 9.2 Examples of non-vacuum based approach results.

Technique	V_{oc} (mV)	FF(%)	J_{sc} (mA/cm^2)	Efficiency	References
Coelectrodeposition	691	62.9	20.2	8.7%	[42]
Coelectrodeposition	636	58.7	19.2	7.23%	[43]
Solvothermal method	900	33.5	8.9	6.72%	[35]
Sol–gel	650	59	12.90	5.01%	[34]

period of time, that vacuum methods are more expensive, more complicated, and use less composition material. This prompts researchers to look for a less expensive or cheaper method that is nevertheless effective for preparing CZTS on an industrial scale.

9.5.2 Copper zinc tin sulfide nonvacuum-processed solar cells

A different approach to material preparation was required as the demand for TFSCs grew. As a result, it appeared that the nonvacuum methods indicated in Table 9.2 [34,35,42,43] were preferable because they allow for greater utilization of the elements used in the related compositions while using less energy and doing so at a lower cost [18,44]. The nonvacuum-based methods discussed in this chapter are spray pyrolysis, electrodeposition, sol–gel, and the solvothermal method. Zhang et al. used coelectrodeposition and postsulfurization to deposit CZTS thin films [42]. They varied pressure from 10–60 Torr to get the highest efficiency of 8.7% at 40 Torr. In summary, there are different parameters that can affect the device's efficiency, such as conditions for annealing, slight variations in composition, and doping of elements.

9.6 Future scope

Several researchers have made strides toward commercially viable CZTS-based solar cells, but further developments are required, for example, doping in the CZTS layer to boost efficiency or V_{oc} (or both) have been reported [45]. Because of its p-type conductivity, inherently lower cost, and stability, relative to organic compounds, such as spiro-MeOTAD, CZTS can be utilized to improve third-generation solar cell devices [1–3,12,46–48]. For example, CZTS has been considered as an alternative hole transport material in perovskite-based solar cells [12,46]. Also, due to its enhanced stability, lower cost, and more facile handling during cell preparation, CZTS is being studied as an alternative counter electrode for dye-sensitized solar cells [47]; see Fig. 9.5. Therefore, a study into the preparation, optimization, and use of CZTS films is urgently needed, especially for solar cell applications [48]. Finally, CZTS has shown promise as a visible-light photocatalyst in recent years. For instance, it exhibits significantly better degradation of methyl orange, exceeding the performance of commercial P25 TiO_2 powder by more than five times under the same conditions [49]. There is a substantial window of opportunity for additional study into photocatalysis and related applications using CZTS [4,49–51].

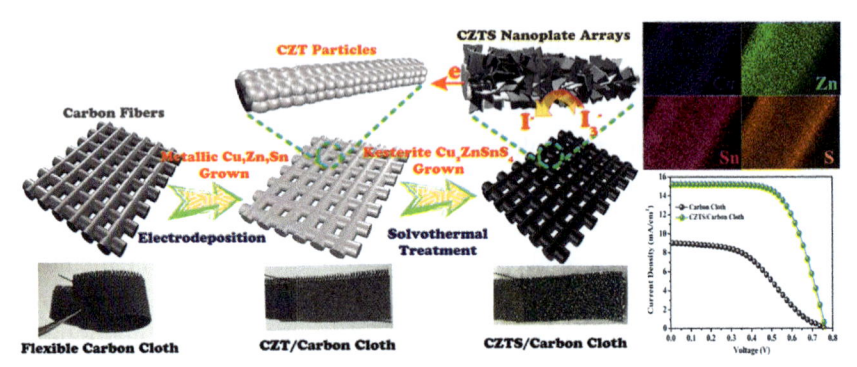

Figure 9.5 (*Left*) Schematic illustrating the growth procedure of Cu_2ZnSnS_4 nanoplate arrays on flexible carbon cloth and the corresponding sample photos; (*upper right*) energy dispersive spectroscopy element (Cu, Zn, Sn, and S) map of CZTS nanoplate arrays grown on flexible carbon cloth; (*lower right*) $J-V$ characteristics of DSSCs with carbon cloth substrate and CZTS grown on carbon cloth as counter electrodes under a light intensity of 100 mW/cm^2 (1 sun). *Reproduced with permission from S.L. Chen, J. Tao, H.-J. Tao, Y.-Z. Shen, T. Wang, L. Pan, High-performance and low-cost dye sensitized solar cells based on kesterite Cu_2ZnSnS_4 nanoplate arrays on a flexible carbon cloth cathode, J. Power Sources 330 (October 2016) 28–36, copyright (2016) Elsevier.*

9.7 Conclusion

While it is an intriguing Earth-abundant alternative, the power conversion efficiency of CZTS TFSCs needs further improvement to replace conventional c-Si solar cells. Silicon solar cells now available on the market are bulky and have a long energy payback period. On the other hand, the CZTS TFSCs can be produced using less energy and materials; thus it has a shorter energy payback period. In comparison to third-generation solar cells, it has lower manufacturing costs and greater stability. The CZTS TFSC has a higher efficiency value because there are fewer material faults and better band alignment with the buffer layer. By reducing the copper vacancy and antisite defects, CZTS material can achieve the desired absorber properties. CZTS is a serious candidate for the production of low-cost, high-efficiency thin-film solar cells. It highlights the PV industry's huge potential to create products that are profitable. Devices produced using nonvacuum-based deposition techniques are frequently affordable and efficient. Metal-doped CZTS films outperformed Na and Ge in terms of efficiency, carrier concentration, shape, and device performance.

References

[1] A.K.-W. Chee, On current technology for light absorber materials used in highly efficient industrial solar cells, Renew. Sustain. Energy Rev. 173 (2023)113027. Available from: https://doi.org/10.1016/j.rser.2022.113027.

[2] T. Amrillah, Enhancing the value of environment-friendly CZTS compound for next generation photovoltaic device: a review, Sol. Energy 263 (2023)111982. Available from: https://doi.org/10.1016/j.solener.2023.111982.

[3] X. Song, X. Ji, M. Li, W. Lin, X. Luo, H. Zhang, A review on development prospect of CZTS based thin film solar cells, Int. J. Photoenergy (2014)613173. Available from: https://doi.org/10.1155/2014/613173. May 2014.

[4] M.Z. Ansari, M. Faraz, S. Munjal, V. Kumar, N. Khare, Highly dispersible and uniform size Cu_2ZnSnS_4 nanoparticles for photocatalytic application, Adv. Powder Technol. 28 (9) (2017) 2402−2409. Available from: https://doi.org/10.1016/j.apt.2017.06.023.

[5] H. Katagiri, Cu_2ZnSnS_4 thin film solar cells, Thin Solid. Films 480−481 (2005) 426−432. Available from: https://doi.org/10.1016/j.tsf.2004.11.024.

[6] M.Z. Ansari, N. Khare, Structural and optical properties of CZTS thin films deposited by ultrasonically assisted chemical vapour deposition, J. Phys. D. Appl. Phys. 47 (18) (2014) 185101. Available from: https://doi.org/10.1088/0022-3727/47/18/185101.

[7] M.Z. Ansari, N. Khare, Thermally activated band conduction and variable range hopping conduction in Cu_2ZnSnS_4 thin films, J. Appl. Phys. 117 (2) (2015)025706. Available from: https://doi.org/10.1063/1.4905673.

[8] M.Z. Ansari, N. Khare, Mott variable range hopping conduction mechanism in single-phase CZTS thin film, AIP Conf. Proc. 1665 (1) (June 2015) 110012. <https://doi.org/10.1063/1.4918068>.

[9] S. Kumar, M.Z. Ansari, N. Khare, Influence of compactness and formation of metallic secondary phase on the thermoelectric properties of Cu_2ZnSnS_4 thin films, Thin Solid. Films 645 (2018) 300−304. Available from: https://doi.org/10.1016/j.tsf.2017.11.001.

[10] S.S. Fouad, I.M. El Radaf, P. Sharma, M.S. El-Bana, Multifunctional CZTS thin films: structural, optoelectrical, electrical and photovoltaic properties, J. Alloy. Compd. 757 (2018) 124−133. Available from: https://doi.org/10.1016/j.jallcom.2018.05.033.

[11] B. Pejjai, V.R.M. Reddy, S. Gedi, C. Park, Status review on Earth-abundant and environmentally green Sn-X (X = Se, S) nanoparticle synthesis by solution methods for photovoltaic applications, Int. J. Hydrog. Energy 42 (5) (2017) 2790−2831. Available from: https://doi.org/10.1016/j.ijhydene.2016.11.084.

[12] U. Syafiq, N. Ataollahi, P. Scardi, Progress in CZTS as hole transport layer in perovskite solar cell, Sol. Energy 196 (2020) 399−408. Available from: https://doi.org/10.1016/j.solener.2019.12.016.

[13] M.Z. Ansari, S. Munjal, V. Kumar, N. Khare, Electrical conduction noise and its correlation with structural properties of Cu_2ZnSnS_4 thin films, Mater. Res. Express 3 (7) (2016) 076404. Available from: https://doi.org/10.1088/2053-1591/3/7/076404.

[14] F. Yakuphanoglu, Nanostructure Cu_2ZnSnS_4 thin film prepared by sol-gel for optoelectronic applications, Sol. Energy 85 (10) (2011) 2518−2523. Available from: https://doi.org/10.1016/j.solener.2011.07.012.

[15] T.R. Rana, N.M. Shinde, J. Kim, Novel chemical route for chemical bath deposition of Cu_2ZnSnS_4 (CZTS) thin films with stacked precursor thin films, Mater. Lett. 162 (2016) 40−43. Available from: https://doi.org/10.1016/j.matlet.2015.09.100.

[16] M.Z. Ansari, S. Munjal, N. Khare, Intrinsic strain dependent redshift in optical band gap of Cu_2ZnSnS_4 nanostructured thin films, Thin Solid. Films 657 (2018) 95−100. Available from: https://doi.org/10.1016/j.tsf.2018.05.003.

[17] T. Chtouki, L. Soumahoro, B. Kulyk, H. Bougharraf, H. Erguig, K. Ammous, et al., Comparative study on the structural, morphological, linear and nonlinear optical properties of CZTS thin films prepared by spin-coating and spray pyrolysis, Mater. Today: Proc. 4 (4C) (2017) 5146−5153. Available from: https://doi.org/10.1016/j.matpr.2017.05.020.

[18] H. Guan, H. Hou, F. Yu, L. Li, Synthesis of wurtzite Cu_2ZnSnS_4 thin films directly on glass substrates by the solvothermal method, Mater. Lett. 159 (2015) 200−203. Available from: https://doi.org/10.1016/j.jiec.2019.03.035.

[19] C. Shi, G. Shi, Z. Chen, P. Yang, M. Yao, Deposition of Cu_2ZnSnS_4 thin films by vacuum thermal evaporation from single quaternary compound source, Mater. Lett. 73 (2012) 89−91. Available from: https://doi.org/10.1016/j.matlet.2012.01.018.

[20] K.C. Nwambaekwe, V.S. John-Denk, S.F. Douman, P. Mathumba, S.T. Yussuf, et al., Crystal engineering and thin-film deposition strategies towards improving the performance of kesterite photovoltaic cell, J. Mater. Res. Tech. 12 (2021) 1252−1287. Available from: https://doi.org/10.1016/j.jmrt.2021.03.047.

[21] M.T. Ferdaous, P. Chelvanathan, S.A. Shahahmadi, M.M.I. Sapeli, K. Sopian, N. Amin, Compositional disparity in Cu_2ZnSnS_4 (CZTS) thin film deposited by RF-sputtering from a single quaternary compound target, Mater. Lett. 221 (2018) 201−205. Available from: https://doi.org/10.1016/j.matlet.2018.03.098.

[22] T. Tanaka, T. Nagatomo, D. Kawasaki, M. Nishio, Q. Guo, et al., Preparation of Cu_2ZnSnS_4 thin films by hybrid sputtering, J. Phys. Chem. Solids 66 (11) (2005) 1978−1981. Available from: https://doi.org/10.1016/j.jpcs.2005.09.037.

[23] J. Zhang, L. Shao, Y. Fu, E. Xie, Cu_2ZnSnS_4 thin films prepared by sulfurization of ion beam sputtered precursor and their electrical and optical properties, Rare Met. 25 (6−Suppl. 1) (2006) 315−319. Available from: https://doi.org/10.1016/S1001-0521(07)60096-5.

[24] H. Katagiri, K. Jimbo, M. Tahara, H. Araki, K. Oishi, The influence of the composition ratio on CZTS-based thin film solar cells, MRS Online Proc. 1165 (2009) 401. Available from: https://doi.org/10.1557/PROC-1165-M04-01.
[25] H. Katagiri, N. Sasaguchi, S. Hando, S. Hoshino, J. Ohashi, T. Yokota, Preparation and evaluation of Cu_2ZnSnS_4 thin films by sulfurization of E-B evaporated precursors, Sol. Energy Mater. Sol. Cell 49 (1−4) (1997) 407−414. Available from: https://doi.org/10.1016/S0927-0248(97)00119-0.
[26] Z. Zhi, S. Wang, L. Huang, J. Li, X. Xiao, Fabrication of CZTS thin films and solar cells via single-step co-evaporation method, Mater. Sci. Semicond. Proc. 144 (2022) 106592. Available from: https://doi.org/10.1016/j.mssp.2022.106592.
[27] K. Sekiguchi, K. Tanaka, K. Moriya, H. Uchiki, Epitaxial growth of Cu_2ZnSnS_4 thin films by pulsed laser deposition, Phys. Stat. Solidi C. 3 (8) (2006) 2618−2621. Available from: https://doi.org/10.1002/pssc.200669603.
[28] A. Cazzaniga, A. Crovetto, C. Yan, K. Sun, X. Hao, et al., Ultra-thin Cu_2ZnSnS_4 solar cell by pulsed laser deposition, Sol. Energy Mater. Sol. Cell 166 (2017) 91−99. Available from: https://doi.org/10.1016/j.solmat.2017.03.002.
[29] A.V. Moholkar, S.S. Shinde, G.L. Agawane, S.H. Jo, K.Y. Rajpure, et al., Studies of compositional dependent CZTS thin film solar cells by pulsed laser deposition technique: an attempt to improve the efficiency, J. Alloy. Compd. 544 (2012) 145−151. Available from: https://doi.org/10.1016/j.jallcom.2012.07.108.
[30] N. Nakayama, K. Ito, Sprayed films of stannite Cu_2ZnSnS_4, Appl. Surf. Sci. 92 (1996) 171−175. Available from: https://doi.org/10.1016/0169-4332(95)00225-1.
[31] V.G. Rajeshmon, C.S. Kartha, K.P. Vijayakumar, C. Sanjeeviraja, T. Abe, Y. Kashiwaba, Role of precursor solution in controlling the opto-electronic properties of spray pyrolysed Cu_2ZnSnS_4 thin films, Sol. Energy 85 (2) (2011) 249−255. Available from: https://doi.org/10.1016/j.solener.2010.12.005.
[32] J.J. Scragg, P.J. Dale, L.M. Peter, G. Zoppi, I. Forbes, New routes to sustainable photovoltaics: evaluation of Cu_2ZnSnS_4 as an alternative absorber material, Phys. Stat. Solidi 1778 (9) (2008) 1772−1778. Available from: https://doi.org/10.1002/pssb.200879539.
[33] K. Tanaka, N. Moritake, H. Uchiki, Preparation of Cu_2ZnSnS_4 thin films by sulfurizing sol−gel deposited precursors, Sol. Energy Mater. Sol. Cell 91 (13) (2007) 1199−1201. Available from: https://doi.org/10.1016/j.solmat.2007.04.012.
[34] Z. Su, K. Sun, Z. Han, H. Cui, F. Liu, et al., Fabrication of Cu_2ZnSnS_4 solar cells with 5.1% efficiency via thermal decomposition and reaction using a non-toxic sol−gel route, J. Mater. Chem. A 2 (2014) 500−509. Available from: https://doi.org/10.1039/c3ta13533k.
[35] N.M. Shaalan, A.Z. Mahmoud, D. Hamad, Synthesis of CZTS without sulfurization process and its performance evaluation on n-Si substrate as ITO-free photovoltaic cell, Mater. Sci. Semicond. Proc. 120 (2020)105318. Available from: https://doi.org/10.1016/j.mssp.2020.105318.
[36] M.Z. Ansari, N. Khare, V. Kumar, Synthesis of Cu2ZnSnS4 nanoparticles by solvothermal route, AIP Conf. Proc. 1731 (1) (May 2016) 120024. <https://doi.org/10.1063/1.4948096>.
[37] C. Yan, J. Huang, K. Sun, S. Johnston, Y. Zhang, et al., Cu_2ZnSnS_4 solar cells with over 10% power conversion efficiency enabled by heterojunction heat treatment, Nat. Energy 3 (2018) 764−772. Available from: https://doi.org/10.1038/s41560-018-0206-0.
[38] B. Shin, O. Gunawan, Y. Zhu, N.A. Bojarczuk, S.J. Chey, S. Guha, Thin film solar cell with 8.4% power conversion efficiency using an earth-abundant Cu_2ZnSnS_4 absorber, Prog. Photovolt. 21 (1) (2013) 72−76. Available from: https://doi.org/10.1002/pip.1174.

[39] H. Katagiri, K. Jimbo, S. Yamada, T. Kamimura, W.S. Maw, et al., Enhanced conversion efficiencies of Cu_2ZnSnS_4-based thin film solar cells by using preferential etching technique, Appl. Phys. Express 1 (4) (2008) 041201. Available from: https://doi.org/10.1143/apex.1.041201.

[40] E. Kask, T. Raadik, M. Grossberg, R. Josepson, J. Krustok, Deep defects in Cu_2ZnSnS_4 monograin solar cells, Energy Proc. 10 (2011) 261−265. Available from: https://doi.org/10.1016/j.egypro.2011.10.188.

[41] H. Katagiri, K. Saitoh, T. Washio, H. Shinohara, T. Kurumadani, S. Miyajima, Development of thin film solar cell based on Cu_2ZnSnS_4 thin films, Sol. Energy Mater. Sol. Cell 65 (2001) 141−148. Available from: https://doi.org/10.1016/S0927-0248(00)00088-X.

[42] C. Zhang, J. Tao, J. Chu, An 8.7% efficiency co-electrodeposited Cu_2ZnSnS_4 photovoltaic device fabricated via a pressurized post-sulfurization process, J. Mater. Chem. C. 6 (2018) 13275−13282. Available from: https://doi.org/10.1039/c8tc05058a.

[43] J. Tao, L. Chen, H. Cao, C. Zhang, J. Liu, et al., Co−electrodeposited Cu_2ZnSnS_4 thin−film solar cells with over 7% efficiency fabricated via fine−tuning of the Zn content in absorber layers, J. Mater. Chem. A 4 (2016) 3798−3805. Available from: https://doi.org/10.1039/C5TA09636G.

[44] M.Z. Ansari, N. Khare, Effect of intrinsic strain on the optical band gap of single phase nanostructured Cu_2ZnSnS_4, Mater. Sci. Semicond. Proc. 63 (2017) 220−226. Available from: https://doi.org/10.1016/j.mssp.2017.02.011.

[45] D.S. Dhawale, A. Ali, A.C. Lokhande, Impact of various dopant elements on the properties of kesterite compounds for solar cell applications: a status review, Sustain. Energy Fuels 3 (6) (2019) 1365−1383. Available from: https://doi.org/10.1039/C9SE00040B.

[46] S.Z. Haider, H. Anwar, M. Wang, Remarkable performance optimization of inverted *p-i-n* architecture perovskite solar cell with CZTS as hole transport material, Phys. B: Condens. Matter 620 (2021)413270. Available from: https://doi.org/10.1016/j.physb.2021.413270.

[47] S.L. Chen, J. Tao, H.-J. Tao, Y.-Z. Shen, T. Wang, L. Pan, High-performance and low-cost dye sensitized solar cells based on kesterite Cu_2ZnSnS_4 nanoplate arrays on a flexible carbon cloth cathode, J. Power Sources 330 (2016) 28−36. Available from: https://doi.org/10.1016/j.jpowsour.2016.08.134.

[48] M. Ravindiran, C. Praveenkumar, Status review and the future prospects of CZTS based solar cell − a novel approach on the device structure and material modeling for CZTS based photovoltaic device, Renew. Sust. Energy Rev. 94 (2018) 317−329. Available from: https://doi.org/10.1016/j.rser.2018.06.008.

[49] X. Hou, Y. Li, J.-J. Yan, C.-W. Wang, Highly efficient photocatalysis of *p*-type Cu_2ZnSnS_4 under visible-light illumination, Mater. Res. Bull. 60 (2014) 628−633. Available from: https://doi.org/10.1016/j.materresbull.2014.09.041.

[50] R. Mudike, A. Bheemaraju, T. Rashid, N. Singh, S.B. Dhage, et al., Enhanced photocatalysis and photodetection using highly crystalline CZTS thin films optimized using stabilizers, Ceram. Int. 48 (23, Part B) (2022) 35666−35675. Available from: https://doi.org/10.1016/j.ceramint.2022.05.169.

[51] M.Z. Ansari, S. Singh, N. Khare, Visible light active CZTS sensitized CdS/TiO_2 tandem photoanode for highly efficient photoelectrochemical hydrogen generation, Sol. Energy 181 (2019) 37−42. Available from: https://doi.org/10.1016/j.solener.2019.01.067.

SECTION IV

Green and sustainable aspects of photovoltaics

CHAPTER TEN

Nature-inspired and computer-aided approaches to enable improved photovoltaic materials, more efficient processing, and novel devices

Aloysius F. Hepp[1] and Ryne P. Raffaelle[2]
[1]Nanotech Innovations LLC, Oberlin, OH, United States
[2]Rochester Institute of Technology, Rochester, NY, United States

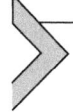

10.1 Introduction: Nature-inspired and computer-aided insights

This chapter completes an omnibus overview of solar cell or photovoltaic(s) (PV) technology; the first half is the initial chapter of this book [1]. Solar cells are semiconductor devices that generate electrical energy from solar (or more generally electromagnetic) radiation. While crystalline silicon (Si) technology is a reliable technology that remains dominant for terrestrial applications, it is being challenged by an ever-growing set of advanced alternative materials and devices [1].

We briefly introduce two pairs of seemingly divergent paradigms that have evolved into significant approaches to enable improved PV (and numerous related technologies) over the past several decades. Green (or environmentally benign) materials technologies and biomimicry or bionics can be characterized as sustainable and/or nature-inspired approaches to technology [2–7]. An alternative paradigm relies on a number of recent advances in theory, computer programming, and numerical methods [8–10] for device modeling, materials simulation (including density functional theory (DFT) [9]), and artificial intelligence (AI) [including machine learning (ML)] for materials discovery and processing as well as advances in device technology [8,10]. This is quite an active area of investigation that often includes technically dense and mathematically complex (equations and derivations) information. We will simply survey interesting and instructive examples, including AI/ML as a paradigmatic follow-on to biomimicry, in fact, some methods include biomimetic mathematical methods. The section then addresses several interesting applications of nature-inspired and computer-enabled approaches to developing and producing novel, sustainable, lower cost, and enhanced-performance PV (-containing) technologies.

10.2 Sustainability, green technologies, and biomimicry

While it is possible to trace the origins of the environmental movement to numerous influences and events, the work of Rachel Carson and the aftermath of her pivotal call to action, *Silent Spring*, in 1962 [11] have been credited by many as inspiring the beginning of the environmental movement [12]. Practical applications of environmentalism began in earnest in the 1990s through the green chemistry and engineering paradigms and related approaches [3,13–15]. This approach has blossomed today into an active multidisciplinary effort; several chapters in this book address this important concern for green(er) PV technologies [16–18].

10.2.1 Background and overview

The desire to adapt lessons of nature to practical systems can be said to date back to the study of the flight of birds by Leonardo Da Vinci in 1505 [19,20]. More recently, Otto Schmidt in the 1950s coined the term "biomimetics" [21] to describe the use of biological analogs to enhance technology [22]. Col. J.E. Steele, a US Air Force doctor, first formally used the term "bionics" at a technical meeting in 1960 [21,23]; he defined it as the science of systems having "*...some function copied from nature, or which represent characteristics of natural systems or their analogues.*" The term biomimicry [4,24,25] in this chapter will be used and thus will include a broader scope of scientific disciplines (biomimetics, bionics, and bio-inspired design) that seek to employ biological knowledge and insights to solve human problems [4,5,20–25].

Table 10.1 summarizes ten principles of chemistry, design, and engineering that can be traced back to green chemistry/engineering and/or

Table 10.1 Summary of connections between green and biomimicry principles and topics of relevance to advanced PV topics addressed in this chapter [importance of principle to the subject (estimated by authors)].

Green and/or biomimicry principles[a]	Earth-abundant; nontoxic	Metal-organic frameworks	Single-source precursors	Power-integrated devices	AI/ML and materials science	Low-cost manufacture
Atom efficiency (GC)		X	XXX			
Multifunctional materials (RE)		XX	X	XXX	XX	
Low energy processes (RE) Energy efficiency (GC) (GE)		X	XX		XX	
Readily available materials (LR)	XXX	XX	X			XXX
Modular and nested components (IDG)		XXX	XX	XX		X
Minimal material diversity (GE)						
Self-organize (IDG)		XX	X			XX
Benign constituents (L-FC)	XXX	X	X		XX	XXX
Reduced toxicity (GC) (GE)						
Computational methods (SMD)				X	XXX	XX
Reshuffle information (ES)						
Renewable feedstocks (GC) (GE)	XX	X	X		XXX	X
Improve, innovate, and invent (new) materials and technologies (SC)	X	XX	X	XXX	XXX	XX

Notes: Critical (XXXX), Important (XXX), Minor (X). Table assembled from text references [3,13–15,24] and literature germane to topics included in column headings. Bold columns are discussed in more detail in the text or in prior publications of authors.

[a]Topics included from several sources: Green Chemistry (GC) [13]; Green Engineering (GE) [14]; Resource Efficient (RE), Evolve to Survive (ES), Life-Friendly Chemistry (L-FC), Integrate Development with Growth (IDG), Locally Responsive (LR) [24]; Sustainable Molecular Design (SMD) [3]; Sandestin Conference (SC) [15]. *Italics describe principles that coincide.*

As is abundantly clear from the content of this book as well as the abundance of original research reviews, PSCs and perovskite materials are currently being intensely researched both as single-junction and tandem devices [1,34,57−59]. While substituting Sn for Pb in perovskites seems to be a straightforward approach to reducing toxicity [41−45], analogous to kesterites replacing CdTe, there are other considerations such as stability to consider [36−40]. In an earlier review, Abate identified a number of approaches to potentially be explored to reduce toxicity and enhance performance and stability in PSCs [41]. He astutely observed that a combination of these approaches would be more likely to succeed: (1) Novel perovskite formulations including doping, multiple cations, and halides (including pseudohalides, i.e., chalcogenides); (2) control of perovskite film morphology to minimize grain boundary performance reduction; (3) minimizing moisture and oxygen impact during material and device preparation; (4) passivating the perovskite surface, postprocessing; and (5) encapsulation to reduce environmental impact(s) during the entire device lifetime. A variety of alternative materials that adopt the first approach suggested by Abate [41] are illustrated in Fig. 10.2, along with accompanying advantages and disadvantages [44]. Presently, the best Sn-based PSCs have demonstrated a PCE of 13%−14% [44,55]; this is slightly more than half the efficiency (25.7%) of a champion small-area (0.1 cm^2) Pb-containing PSC [58]. However, as pointed out by Wu et al., there are other approaches to mitigating Pb issues, including Pb sequestration during operations and recycling [55]. The interested reader is encouraged to consult the PSC literature referenced in this chapter, including tandems [41−45,54−56,60−67], later chapters in this book [16,17,68], and other books [34,59,69−71] for more in-depth discussions.

An alternative approach to simply learning from nature is to employ actual natural products in PV devices. The literature referenced in this subsection of the chapter is focused on natural dyes and pigments from a variety of sources in DSSCs [46−50]. Fig. 10.3 provides an overview of the use of natural or "bio-dyes" in DSSCs; a listing of key issues (2) related to dyes is on the left side of Fig. 10.3; the right side includes a typical device structure (1) and donor−acceptor physics (3) [47]. Typically, the efficiency of natural DSSCs is lower than synthetic dye devices [46−48]. However, a recent study reported that the use of an optimized mixture of green (16.5%) and green and violet (83.5%) dyes from a *Lactuca sativa* leaf resulted in a researcher-reported (nonverified) 24% PCE device with an active area of 3.4 cm^2 [49]; a novel approach that enabled the optimization of the PCE of this DSSC.

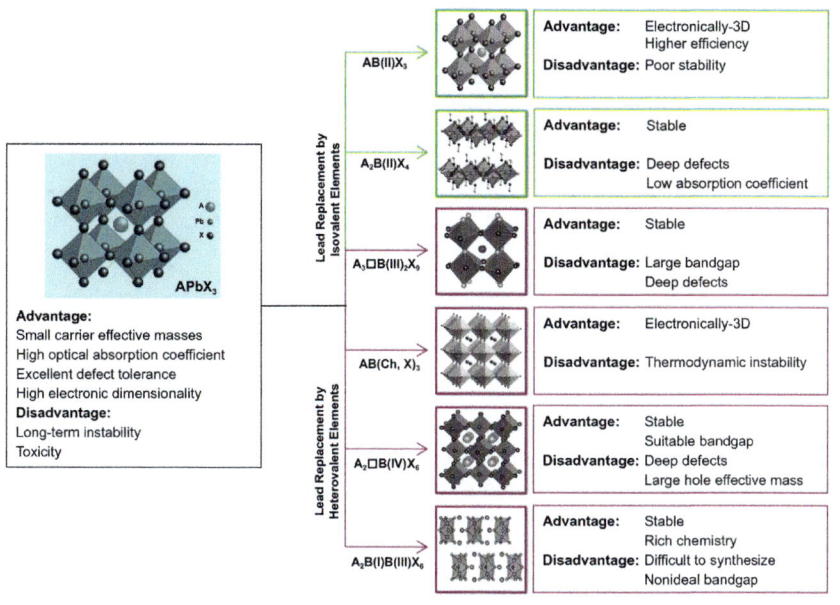

Figure 10.2 Schematic illustration of the approaches and consequences of potential Pb replacement in perovskite solar cells. *Note*: A is an organic (or inorganic or hybrid) cation, usually includes methylammonium (MA), formamidinium (FA), Cs, or their mixture; B includes isovalent elements consisting of Sn (II), Ge (II), and divalent transition metals (e.g., Cu, Fe, Zn), and the heterovalent elements containing Bi(III), Sb (III), Sn (IV), Ti (IV), and double cations of Ag(I)Bi(III) are considered; X is a halide and Ch is a chalcogenide (Group 16). Note that a box sign (☐) indicates vacancies. The inclusion of vacancies reduces the electronic dimension, thus affecting the optoelectronic performance; for a more in-depth discussion of this phenomenon, please see Z. Xiao, Z. Song, Y. Yan, from lead halide perovskites to lead-free metal halide perovskites and perovskite derivatives. Adv. Mater. 31 (47) (2019) 1803792–1803813. *Reproduced from M. Wang, W. Wang, B. Ma, W. Shen, L. Liu, et al., Lead-free perovskite materials for solar cells, Nano-Micro Lett. 13 (2021) 62, this article is distributed under the terms of the Creative Commons (CC-BY 4.0) license which permits unrestricted use, (http://creativecommons.org/licenses/by/4.0/).*

Completing this subsection would ideally embody the broad-brush approach adopted in assembling Table 10.1. There are numerous green processing and eco-friendly technologies that are addressed later in this chapter, other chapters in this book, and the literature [2,3,13–18,51–56]. Fig. 10.4 attempts to visualize the overarching issues and methods that must be addressed to optimize the sustainability of PV technologies, including life cycle analysis [56]. Utilizing a cradle-to-grave approach to sustainability for any PV technology includes a thorough consideration of materials,

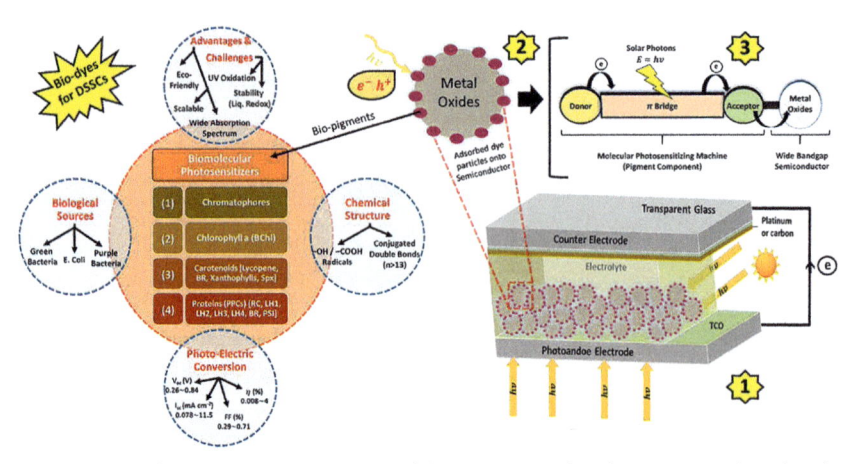

Figure 10.3 Schematic representation of key aspects of a dye-sensitized molecular device: (1) typical components and architecture including photosensitizer for electron injection, a photoanode electrode for charge separation/transport, a counter electrode for electron collection, and redox electrolyte for dye neutralization; (2) detailed look at metal oxide-dye component including various considerations of the sources and issues related to natural dyes from a broad spectrum of sources; (3) donor−acceptor (D−A) structure of an organic dye in a DSSC with a wide bandgap semiconductor photoanode. *Reproduced with permission from H.A. Maddah, V. Berry, S.K. Behura, Biomolecular photosensitizers for dye-sensitized solar cells: Recent developments and critical insights, Renew.Sustain. Energy Rev. 121 (April 2020) 109678, copyright (2020) Elsevier.*

processing, operation, and end-of-life disposal of materials, devices, and arrays [56]. A similar but broader perspective is outlined by Mulvaney et al. [72], which includes other technologies as well as applications.

10.2.3 Lessons learned from nature: Biomimicry and bio-inspired approaches

The principles of biomimicry (see Table 10.1) have the potential to introduce an intriguing new set of paradigms for developing improved PV technologies from a performance and sustainability perspective [4,5,20−25]. A recent book chapter that reviews biomimicry for PV [73], with possible aerospace applications [25], provides several examples of the literature that directly or indirectly apply various biomimicry principles to PV [74−86]. A straightforward approach to characterizing the literature results in three significant direct (or indirect) applications of biomimicry principles to PV materials, components, and systems: Multifunctional materials (and benign constituents) for biomimicking structures, interfaces,

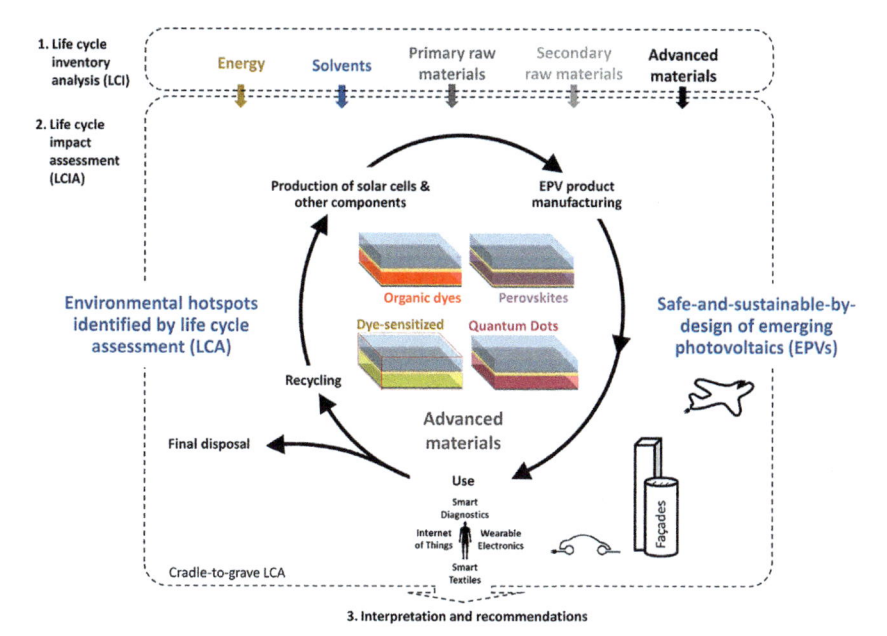

Figure 10.4 Graphic illustration of a safe- and sustainable-by-design approach, where not only environmental compatibility but also recyclability should already be considered in the design phase of emerging (third generation and beyond) PV (EPV) technology. The challenge in developing EPV technology is to find a compromise between the highest power conversion efficiency, best stability, and longest service life under real-life conditions, economic interests, and sustainability of all materials and chemicals applied during the entire life cycle of an EPV system. *Reproduced with permission from S. Gressler, F. Part, S. Scherhaufer, G. Obersteiner, M. Huber-Humer, Advanced materials for emerging photovoltaic systems — Environmental hotspots in the production and end-of-life phase of organic, dye-sensitized, perovskite, and quantum dots solar cells, Sust. Mater. Tech. 34 (December 2022) e00501, copyright (2022) Elsevier.*

and surfaces [73–76,79,80,82,84]; computational methods for simulation of biological systems [49,76–78,81,83]; and improving, innovating, and inventing (new) technologies [78,83,85,86]. It should be noted at this juncture that many of these biomimicry for PV references involve more than one biomimicry principle (or heuristic) [24].

Several examples of biomimicry applied to PV systems from the literature cited in this chapter that involve modeling [49,82,85] biomimicking materials and structures [82,85], and novel technologies [82,85] are instructive. As discussed above (10.2.2), Concepcion et al. determined an optimal mix of dyes to maximize PCE in a DSSC system [82]. Interestingly, this study utilized two bio-inspired optimization methods, namely genetic algorithm and evolutionary strategy, to obtain the exact green and violet concentrations to produce the highest PCE [82]. In fact, this paradigm of utilizing simultaneous approaches (both experimental and theoretical) to improve various PV parameters and stability [83] is being adopted for PV and related devices going forward.

Zhu et al. demonstrated that a moth eye-inspired structure (MEIS) enabled more efficient, robust semitransparent-PSCs (ST-PSCs) for building-integrated PV (BIPV) [85]. The combination of average visible transmittance (AVT) (32.5%), PCE (10.5%), and near-neutral color compared favorably to prior efforts with PSCs (PCE = 5.2%; AVT = 28%) [87] as well as parallel efforts with unmodified PSCs by the authors (PCE = 8.8%; AVT = 35%) [85]. The incorporation of MEISs combines enhanced light-trapping and benefits carrier collection at the interface. This was attributed to the MEIS-enhanced light manipulation, which benefits the light-harvesting in the wavelength range where the human eye is less sensitive. The alteration of the absorption spectra of perovskites based on MEIS resulted in the significantly enhanced J_{sc} while sacrificing >10% AVT, as well as a modification of desired near-color neutrality. Additionally, the ST-PSCs exhibited appealing resistance against moisture and solar irradiation after engineering the halide blend ratio in perovskites [85]. The power-integrated nature of this application is relevant to other valuable applications of solar cell devices, discussed below (10.5.2). Fig. 10.5 includes a schematic device structure of MEIS ST-PSCs (Fig. 10.5A), a device fabrication process overview (Fig. 10.5B), images of as-assembled polystyrene (PS) spheres and MEIS-ITO (Fig. 10.5C1, C2, D1, and D2), and views of the assembled device (Fig. 10.5E and F), confirming conformal coating of SnO_2 on the perovskite [85].

Meng et al. reported on the experimental characterizations and theoretical simulations of a bionic interface layer biomimicking articular cartilage that protects vertebrae from stress damage [82]. One significant reason for this protection is that cartilage functions as a force absorber, by distributing force evenly, thus enlarging the bearing surface due to the elasticity and adhesiveness. Fig. 10.6 demonstrates that biomimicry of a vertebrae mechanism can be accomplished via a soft poly(3,4-ethylenedioxythiophene):poly(ethylene-co-vinyl acetate) (PEDOT:EVA) layer [82]; the multifunctional material provides for both bionic oriented crystallization (Fig. 10.6A) and bionic structure (Fig. 10.6B). The PEDOT: EVA layer is applied between the brittle perovskite and ITO films, which acts as the cartilage between vertebrae to improve the mechanical

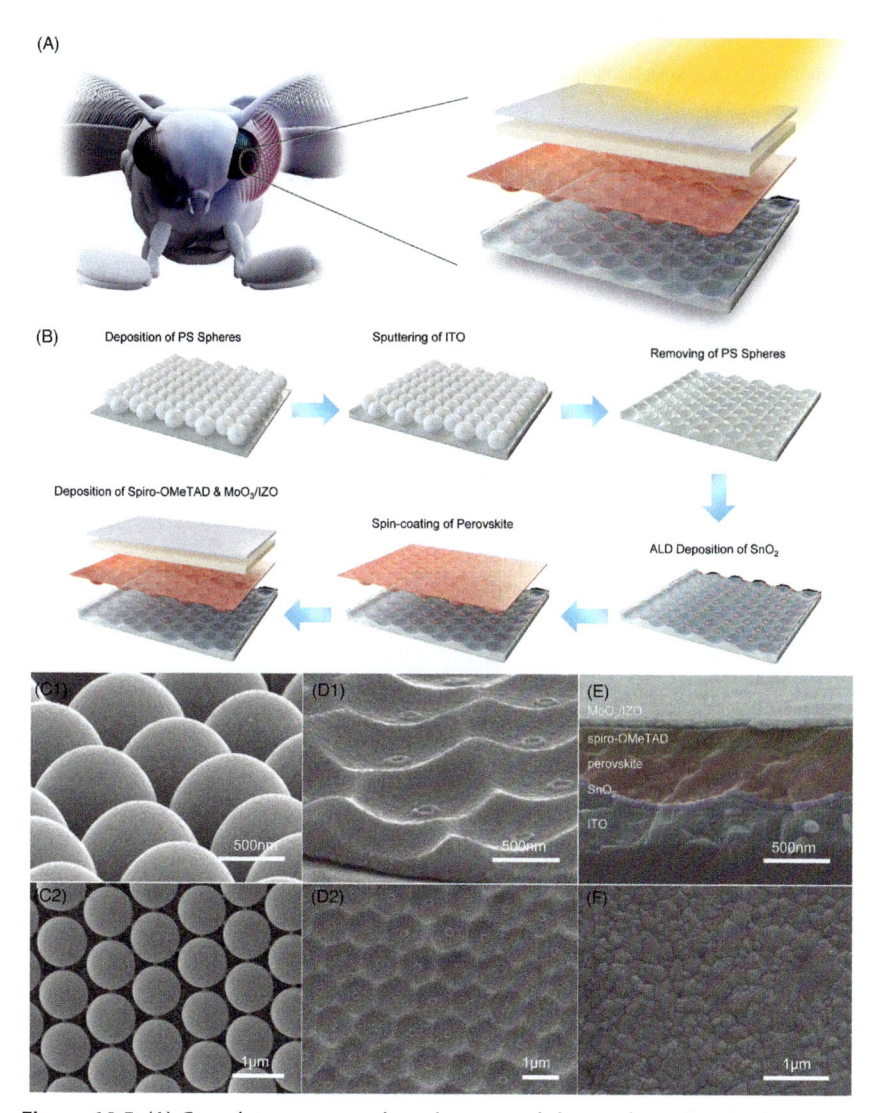

Figure 10.5 (A) Complete compound moth eyes and the moth-eye-inspired structure (MEIS) device structure diagram; (B) Schematic illustration of the fabrication procedure for the MEIS ST-PSCs; (C1) The tilted-view and (C2) top-view scanning electron micrograph (SEM) images of polystyrene spheres on indium tin oxide (ITO); (D1) The tilted-view and (D2) top-view SEM images of MEIS-ITO; (E) Cross-sectional SEM image of the whole device based on MEIS, and (F) top-view SEM of perovskites on the SnO_2/MEIS-ITO. *Reproduced from Y. Zhu, L. Shu, Q. Zhang, Y. Zhu, S. Poddar, et al., Moth eye-inspired highly efficient, robust, and neutral-colored semitransparent perovskite solar cells for building-integrated photovoltaics, EcoMat 3 (4) (August 2021) e12117, this article is distributed under the terms of the Creative Commons (CC-BY 4.0) license which permits unrestricted use, (http://creativecommons.org/licenses/by/4.0/).*

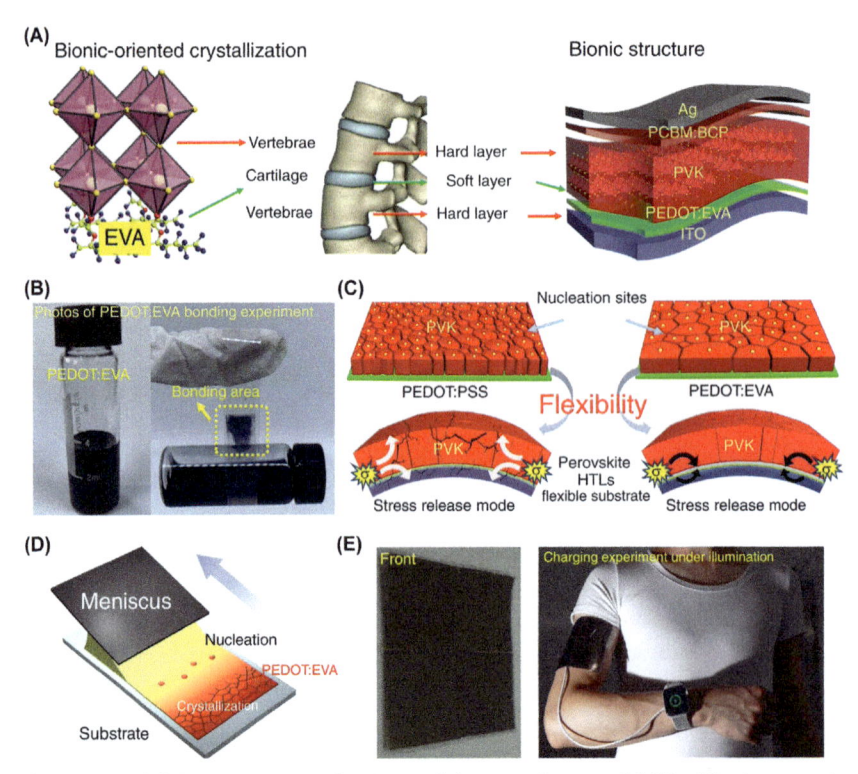

Figure 10.6 (A) Biomimetic mechanisms of the vertebrae and PSCs; (B) photographs of a PEDOT:EVA bonding experiment; (C) scheme of stress release for the PEDOT:EVA structure; (D) schematic of PSCs by meniscus-coating; (E) photographs of flexible perovskite solar modules as a wearable power source. *Reproduced from X. Meng, Z. Cai, Y. Zhang, X. Hu, Z. Xing, et al., Bio-inspired vertebral design for scalable and flexible perovskite solar cells, Nat. Commun. 11 (2020) 3016, this article is distributed under the terms of the Creative Commons (CC-BY 4.0) license which permits unrestricted use, (http://creativecommons.org/licenses/by/4.0/).*

flexibility of PSCs; the adhesiveness of PEDOT:EVA is shown in Fig. 10.6B. From the perspective of bionic crystallization, the PEDOT:EVA layer apparently controls nucleation sites and crystal-oriented growth for high-quality flexible perovskite films (Fig. 10.6C). Thus, given the optimal synthetic conditions, improved device efficiency is enabled via the meniscus-coating technology (Fig. 10.6D); meanwhile, the PEDOT:EVA layer effectively absorbs and releases the stress to reduce the bending resistance for the flexible devices.

To summarize, the multifunctional PEDOT:EVA layer both controls crystallization and acts as an adhesive; the use of this bio-inspired material improved both the perovskite device performance and mechanical stability. The flexible PSCs achieved PCEs of 19.87% and 17.55% at effective areas of 1.01 and 31.20 cm^2, respectively. Furthermore, they retained over 85% of original efficiency after 7000 narrow bending cycles with negligible angular dependence. Finally, the modules were assembled into a wearable solar-power-integrated device (see Section 10.5.2), enabling the upscaling of flexible electronics [82]. As a result of improved PSC performance and mechanical resilience, high-quality large-area perovskite films can readily be fabricated; a perovskite module, containing four devices (36 cm^2), shown in Fig. 10.6E, demonstrates the feasibility of low-power solar-integrated devices [82].

10.3 Computer-aided approaches to improved photovoltaics

As noted at the beginning of this section, recent advances in theory, computer programming, and numerical methods have resulted in a quite significant level of activity in device simulation, molecular modeling, and AI methods [8−10,88−126], beginning over 60 years ago [88]. A Scopus search utilizing terms such as DFT, first principles, AI, ML, simulation, and modeling (abstract, title, and keywords—ATK) paired with PV or solar cells (titles only) resulted in over 47,000 results; a Scopus search utilizing the same terms related to computer-enabled methods (title only) paired with PV or solar cells (titles only) resulted in approximately 8600 results. These search results attest to the prominence in the literature (both original research and reviews) that includes both experimental and theoretical components—nearly six times more prevalent than literature that addresses theoretical methods alone. In fact, many of the references included in this chapter include both experiments and theory. Given that the details and complexity of this topic are well outside the purview of this review chapter, we will simply provide a context for the various theoretical methods and their value when applied in concert with experimental methods, both synthesis and characterization. The term "artificial intelligence" provides a connection back to biomimicry (Section 10.2). A recent overview of AI, ML, and deep learning (DL) provides a concise graphical explanation (Fig. 10.7) of the connection of these (theoretical) tools [110].

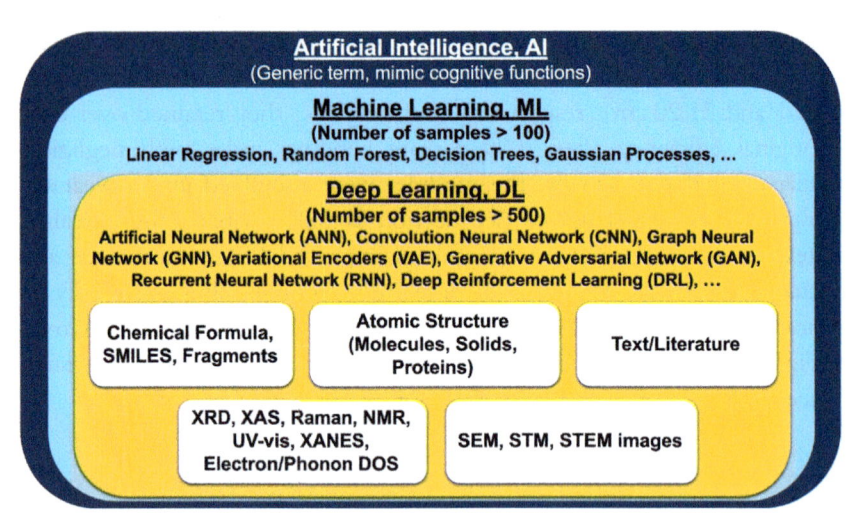

Figure 10.7 Schematic showing an overview of artificial intelligence (AI), machine learning (ML), and deep learning (DL) methods and their applications in materials science and engineering; DL is considered a part of ML, which is contained in an umbrella term AI. *Reproduced from K. Choudhary, B. DeCost, C. Chen, A. Jain, F. Tavazza, et al., Recent advances and applications of deep learning methods in materials science, NPJ Comput. Mater. 8 (2022) 59, this article is distributed under the terms of the Creative Commons (CC-BY 4.0) license which permits unrestricted use, (http://creativecommons.org/licenses/by/4.0/).*

Beginning in the early 1960s [88], a large number of studies on simulation or modeling of PV devices have been published that are primarily theoretical or combine experimental and theoretical results; a representative cross-section of references is included in this chapter [88−102]. Two of the key modeling programs that have come to prominence over the past several decades include the analysis of microelectronic and photonic structures (AMPS) [91,100] and solar cell capacitance simulator (SCAPS) [93,98,101]. The work of Profs. Morel and Ferekides [91,125] on AMPS modeling and experimental efforts toward a tandem chalcogenide device, from the perspective of space applications, is described in more detail in a chapter from our recent book that is focused on space PV [127].

The use of DFT and other first principles methods to gain insights into PV materials began in earnest at the beginning of this century [9]; this approach has evolved to often include AI/ML and DL methods [10]. A sampling of recent general studies and reviews published over the past 5 years is provided in the references [21,103−109]; these can be consulted by the interested reader. As expected, significant literature is available, specifically for perovskites [110−120], while the literature specific to OPV materials is less prominent [121−124].

As we have previously delved into nontoxic or Pb-free PSC materials, we will address two approaches for discovering hybrid organic−inorganic perovskite (HOIP) materials: Solely (or primarily) theoretical [44,111,113,114] and combined theoretical and experimental methods [112]. Fig. 10.8 illustrates a framework devised by Wu and Wang [113] that utilized both ML and DFT methods to discover nontoxic [Pb, Cd, and mercury (Hg) free] HOIPs. Beginning with 230,808 candidates, 686 candidate PSC materials remained after ML method screening. Density functional calculations produced 204

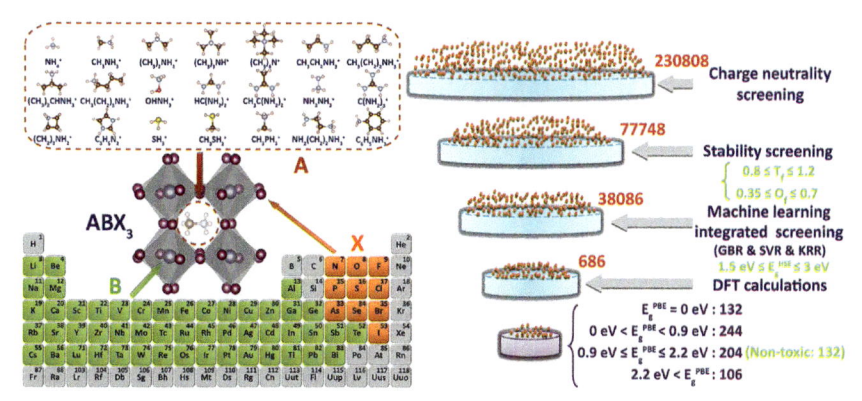

Figure 10.8 Schematic design framework for predicting novel hybrid organic−inorganic perovskites (HOIPs) based on the combination of ML and DFT calculations for photovoltaic applications. Here the combinations of 21 monovalent organic molecular cations (A-site), 50 metallic cations, and 10 typical anions across the periodic table constitute the unexplored HOIPs components giving rise to 230808 initial HOIPs candidates. After the screening using the charge neutrality condition, the stability condition, and the ML method, 686 HOIPs suitable for solar cells are selected. Finally, electronic properties and chemical stability of these selected candidates are further verified by DFT calculations, and 204 ideal HOIPs (132 nontoxic HOIPs) with proper bandgap are finally selected. Note that T_f and O_f are structural criteria or metrics; ML models utilized include gradient boosting regression (GBR), supporting vector regression (SVR), and kernel ridge regression (KRR); the Perdew−Burke−Ernzerhof (PBE) exchange-correlation functional was utilized to improve the accuracy of the calculated bandgaps (please see the referenced publication for further details: M. Ernzerhof, G.E. Scuseria, Assessment of the Perdew−Burke−Ernzerhof exchange-correlation functional, J. Chem. Phys. 110 (11) (March 1999) 5029−5036, https://doi.org/10.1063/1.478401). *Reproduced with permission from T. Wu, J. Wang, Global discovery of stable and nontoxic hybrid organic-inorganic perovskites for photovoltaic systems by combining machine learning method with first principle calculations, Nano Energy 66 (2019) 104070, copyright (2019) Elsevier.*

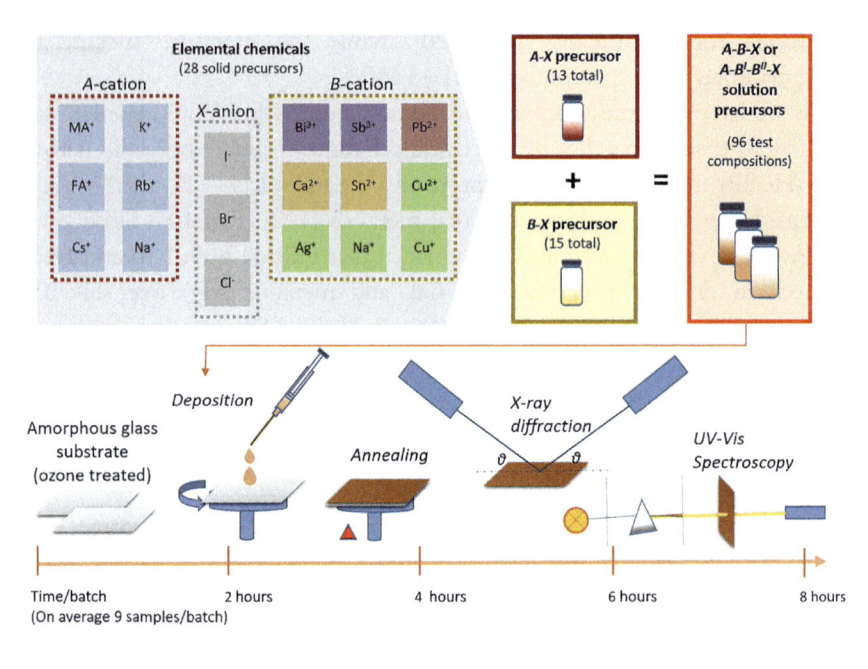

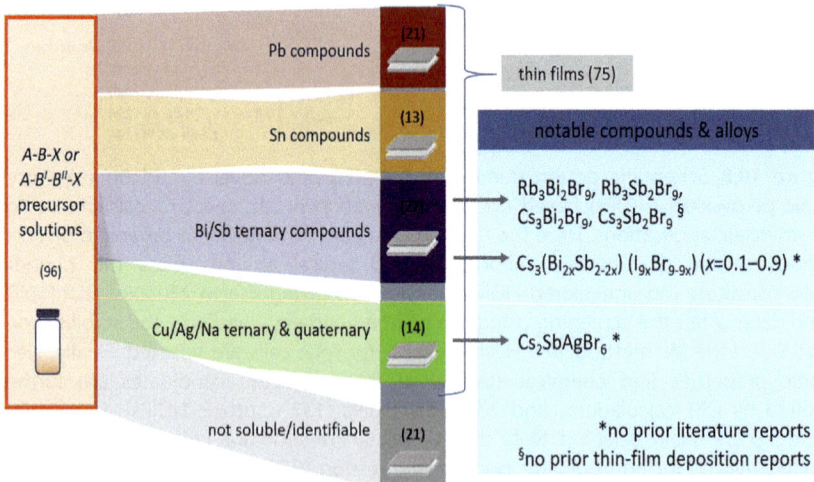

Figure 10.9 (*Top*) Sketch of the optimized experimental workflow employed by Sun et al.: precursor solutions of 96 perovskite-inspired target compositions (ABX$_3$) were prepared by mixing commercially available A-X and B-X solid precursors in stoichiometric ratios. Six cations from the periodic table were dissolved as A-site cations in the target compounds, together with nine cations for the B-site and three halide ions for the X-site; 28 solid precursors, which include 13 A-X (e.g., CsI) and 15 B-X (e.g., BiI$_3$), were used in this study. Through one cycle of learning, each successfully

(*Continued*)

promising candidate HOIPs, 132 of these were nontoxic (Fig. 10.8) [113]. Fifteen candidate HOIPs were subjected to a more detailed analysis; six were found to have indirect bandgaps. The six most promising candidates from the point of view of predicted bandgap (near 1.3−1.4 eV) with Earth abundance as a secondary consideration include $(CH_3)_3NHSnBrI_2$ (1.081), $[NH_2NH_3]$ $InSeI_2$ (1.196 eV), $[CH_3CH_2NH_3]SnClI_2$ (1.28 eV), $[HC(NH_2)_2]SnCl_2I$ (1.31 eV), $[C(NH_3)_3]SnClBrI$ (1.51 eV), and $[CH_3PH_3]GeClI_2$ (1.62) [113]. Lu et al. adopted a similar approach with a much smaller candidate pool of 5158 compounds; two potential candidates identified that were most promising were $[NH_4]InBr_3$ (1.13 eV) and $[C_2H_6N]SnBr_3$ (1.14 eV) [111]. Resulting compounds discovered by these DFT/ML searches would still need to be synthesized; of course, as discussed above (Section 10.2.2), numerous potential doping strategies could be employed to alter the bandgap of candidate compounds. Tandems would employ wider and narrower bandgap materials; thus, several of these and other compounds identified during both studies could be considered [111,113].

An alternative approach adopted by Sun et al. relies on both theoretical and experimental methods; see Fig. 10.9 [112]. A synthetic approach was to combine AX and BX solid precursors in stoichiometric mixtures; the specific A, B, and X ions are shown in the top of Fig. 10.9. The standard perovskite stoichiometry (ABX_3) was the target composition; however, thin-film materials produced included the following five perovskite-related crystal systems: ABX_3 (3D—conventional perovskite), ABX_4 (2D—layered perovskite), $A_3B_2X_9$ (0D—molecular dimer), $A_3B_2X_9$ (2D), and a double perovskite $A_2B^IB^{II}X_6$ (3D) (A = MA, FA, Cs, Rb, K, or Na; B = Pb, Sn, Ag, Cu, Na, Bi, or Sb; and X = Cl, Br, or I); see Fig. 10.2 for a series of crystal structure diagrams. The bottom of Fig. 10.9 summarizes the results of this work; of the 96 compositions prepared, 75 produced thin films [112].

◀ prepared precursor solution followed the three-step experimental process, thin-film deposition, X-ray diffraction, and UV-visible (UV-vis) spectroscopy, to examine the structure and optical properties. Candidate thin films are screened based on their bandgap and the formation of crystalline phases. (*Bottom*) Sankey (flow visualization) diagram demonstrating the material synthesis flow of the 96 precursor compositions attempted, of which 75 resulted in thin films. *Reproduced with permission from S. Sun, N.T.P. Hartono, Z.D. Ren, F. Oviedo, A.M. Buscemi et al., Accelerated development of perovskite-inspired materials via high-throughput synthesis and machine-learning diagnosis, Joule 3 (6) (June 2019) 1437−1451, copyright (2019) Elsevier.*

When focusing on the Bi/Sb ternary and Cu/Ag/Na ternary and quaternary compounds, six novel results were obtained. Four compounds had not previously been successfully deposited as thin films: $Rb_3Bi_2Br_9$, $Rb_3Sb_2Br_9$, $Cs_3Bi_2Br_9$, and $Cs_3Sb_2Br_9$ (all 2D; measured bandgaps range = direct 2.6−2.8, indirect 2.4−2.9). Two compounds were previously unknown: $Cs_2AgSbBr_6$ (3D; measured bandgap of direct = 2.27, indirect = 1.89) and $Cs_3(Bi_{2x}Sb_{2-2x})(I_{9x}Br_{9-9x})$ ($x = 0-0.9$); note that $Cs_3Sb_2Br_9$ (2D layered perovskite) and Cs_3BiI_9 (0D dimer) are at the end points of the stoichiometric scale; the quaternary structure type shifts to 2D when $x = 0.8$ (measured bandgaps range = direct 2.2−2.5, indirect 1.9−2.2). In the final analysis, while this study provided for actual thin films of potential PSC materials, several issues should be evident; the measured bandgap types were not distinguished between direct and indirect; even the narrowest bandgap compound ($Cs_2AgSbBr_6$) had a bandgap that is too wide even for the top cell of a tandem device [1,62−67]. Finally, a valuable review of ML-driven design of promising PSC materials, including predicted PCEs, was published in late 2022 [120]. Several dozen 2D and 3D HOIPs and inorganic perovskites reported by multiple groups in the recent literature are discussed. The calculated bandgaps typically range from 1.2 to 1.6 eV with several estimated compounds having PCEs over 25% [120].

A very-recently published study addresses several concerns regarding differentiating between bandgap types, as well as identifying potential PSC materials with bandgaps in a more useful bandgap range [126]. Prof Sargent and coworkers developed an ML framework with the apt title of DARWIN, derived from the phrase "deep adaptive regressive weighted intelligent network." The three components of DARWIN are illustrated in Fig. 10.10: A surrogate model, a search algorithm, and a program to capture knowledge in a user-friendly way. A key part of the approach generates multiple candidates that meet the desired target properties such as stability and bandgap. Subsequently, statistical techniques and supervised ML are utilized to generate and identify relevant statistically significant chemical rules.

A list of unreported compounds with bandgaps in the range of 1.2−1.6 eV was provided in this report; the five unreported, promising (Pb-free) compounds (with ABX_3 stoichiometry) listed by increasing bandgap are $CsGaBr_3$ (1.16 eV), $KGaBr_3$ (1.17 eV), $CsSn_{0.67}Sb_{0.33}I_3$ (1.51 eV), $CsInBr_3$ (1.54 eV), and $CsGe_{0.67}Sn_{0.33}I_3$ (1.56 eV) [126]. A further interesting aspect of this study was the use of DARWIN to convert semiconductors from indirect to direct bandgap materials through alloying

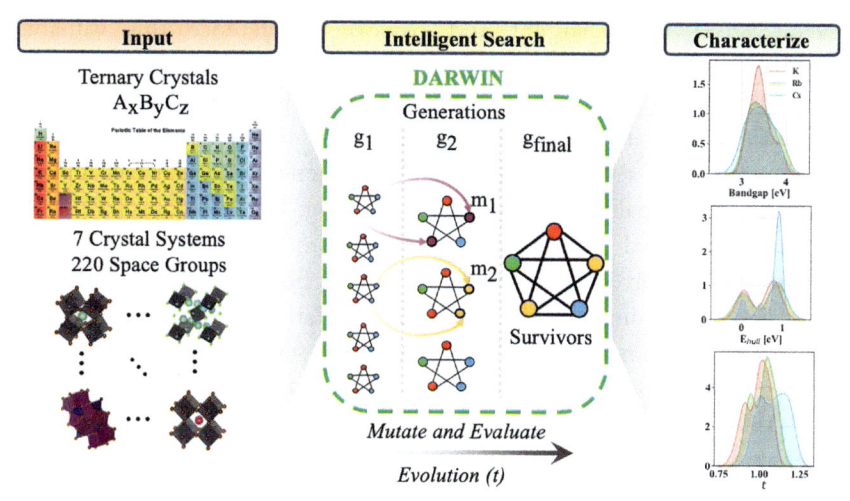

Figure 10.10 Different steps of DARWIN: (1) input crystals were generated using substitutions in prototype structures, spanning 7 crystal systems and 220 space groups; (2) DARWIN then uses trained Graph Networks as surrogate models and mutations to find new candidates that meet the target specifications; (3) both the negative and positive pools of candidates generated are then characterized using different chemical featurization and subjected to supervised learning combined with statistical testing; (4) statistically significant rules derived in this way enable the discovery of new compounds and uncover new chemical trends, which are then explicable to experimentalists. *Reproduced from H. Choubisa, P. Todorović, J.M. Pina, D.H. Parmar, Z. Li, et al., Interpretable discovery of semiconductors with machine learning, NPJ Comput. Mater. 9 (2023) 117, this article is distributed under the terms of the Creative Commons (CC-BY 4.0) license which permits unrestricted use, (http://creativecommons.org/licenses/by/4.0/).*

or doping; examples include SiC to $Si_{0.5}S_{0.5}C$, BP to $B_{0.67}In_{0.33}P$, SnS to $(SnSb)_{0.5}(ClS)_{0.5}$, BiO_2 to BiOSe, and InSe to $In_{0.75}Tl_{0.25}Se$ [126]; of course, the same issues discussed above regarding synthesis, toxicity, and Earth abundance remain for these five proposed compounds. The potential utility and rapid pace of progress in ML methods are further illustrated by the fact that this study [126] appeared online (June 29, 2023) during the writing of this section of the chapter.

Finally, it is important to cover ways that a nonspecialist is interested in applying ML and related methods in their own work. Many of the references in this chapter to AI and ML, as well as other computer-enabled methods, are open acces published or lengthy reviews, in order to facilitate access to this intriguing topic for interested readers [8,10,126,128–165]. The Materials Project is an open-access database that can be utilized in several ways [166]; Sargent and coworkers reference this valuable resource (among others) in their recent study [126]. The journal Nature Review Materials has an open-access collection published online on the topic of ML in materials science [167]. The Materials Research Society, for example, has several symposia on the general topic of computer methods in materials science at every meeting; a link to the Fall 2023 meeting is included here [168]. There are numerous public and private organizations involved in this rapidly growing field, several examples are listed here, including DOE Brookhaven National Laboratory [169], Penn State University [170], and Sunthetics, an ML-focused start-up company [171]. Several universities, such as M.I.T. [172] and Purdue [173], offer free (audit only) online courses, including those on ML and materials science. The Purdue course is part of NanoHUB.org, an open and free online platform for computational education, research, and collaboration in nanotechnology, materials science, and related fields, founded in 2002 [174]. The authors anticipate that after reading this chapter section, a number of interested readers find a suitable route to facilitate their own research and technology efforts through computer-enabled methods.

10.4 Metal-organic framework materials for enhanced solar cells

Earlier in the chapter (Section 10.2 and Table 10.1), MOF materials were introduced. Another related structural class that is used often synonymously with MOFs is "MCP," sometimes shortened to coordination polymer. The two material structures are indeed similar but not actually identical, nor are their applications [175–177]. A search of the two terms (MOF and MCP) on Scopus finds a 4:1 prominence of MOF over MCP of the approximately 83,000 results that are returned; we will, therefore, primarily use the term "MOF" going forward. The earliest archival publications that mention the term "coordination polymer" date back more than 50 years ago [178,179]. Subsequently, Yaghi and Li published a seminal paper in 1995 that described the hydrothermal synthesis of an MOF [180]. Fig. 10.11 illustrates examples of metal clusters (or secondary building units (SBUs)) and organic linkers that are essential components of MOF structures [181].

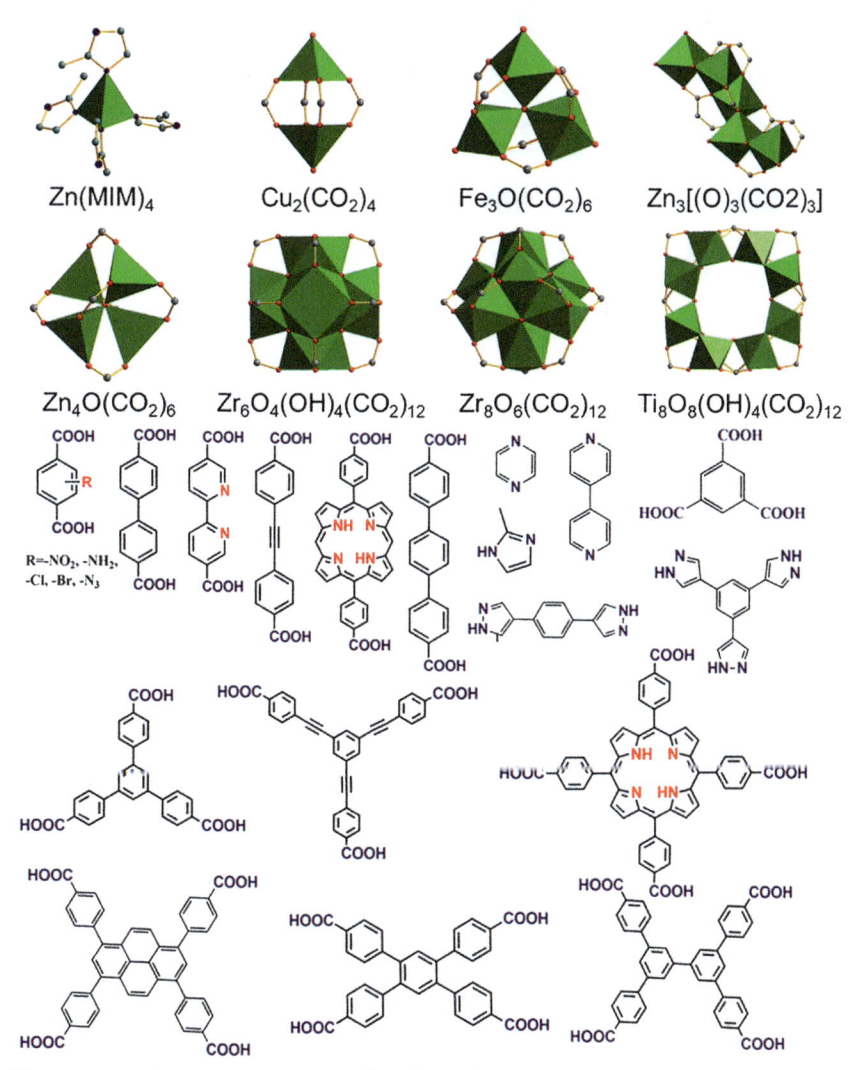

Figure 10.11 Representative examples of metal clusters (or secondary building units) and organic linkers of metal-organic frameworks. *Reproduced with permission from L. Jiao, J.Y.R. Seow, W.S. Skinner, Z.L. Wang, H.-L. Jiang, Metal−organic frameworks: Structures and functional applications, Mater. Today 27 (July−August 2019) 43−68, copyright (2019) Elsevier.*

10.4.1 Metal-organic framework materials: Introduction and synthesis

Fig. 10.12 provides an illustrative overview of common MOF materials reported in the literature [181,182]. All four key construction elements,

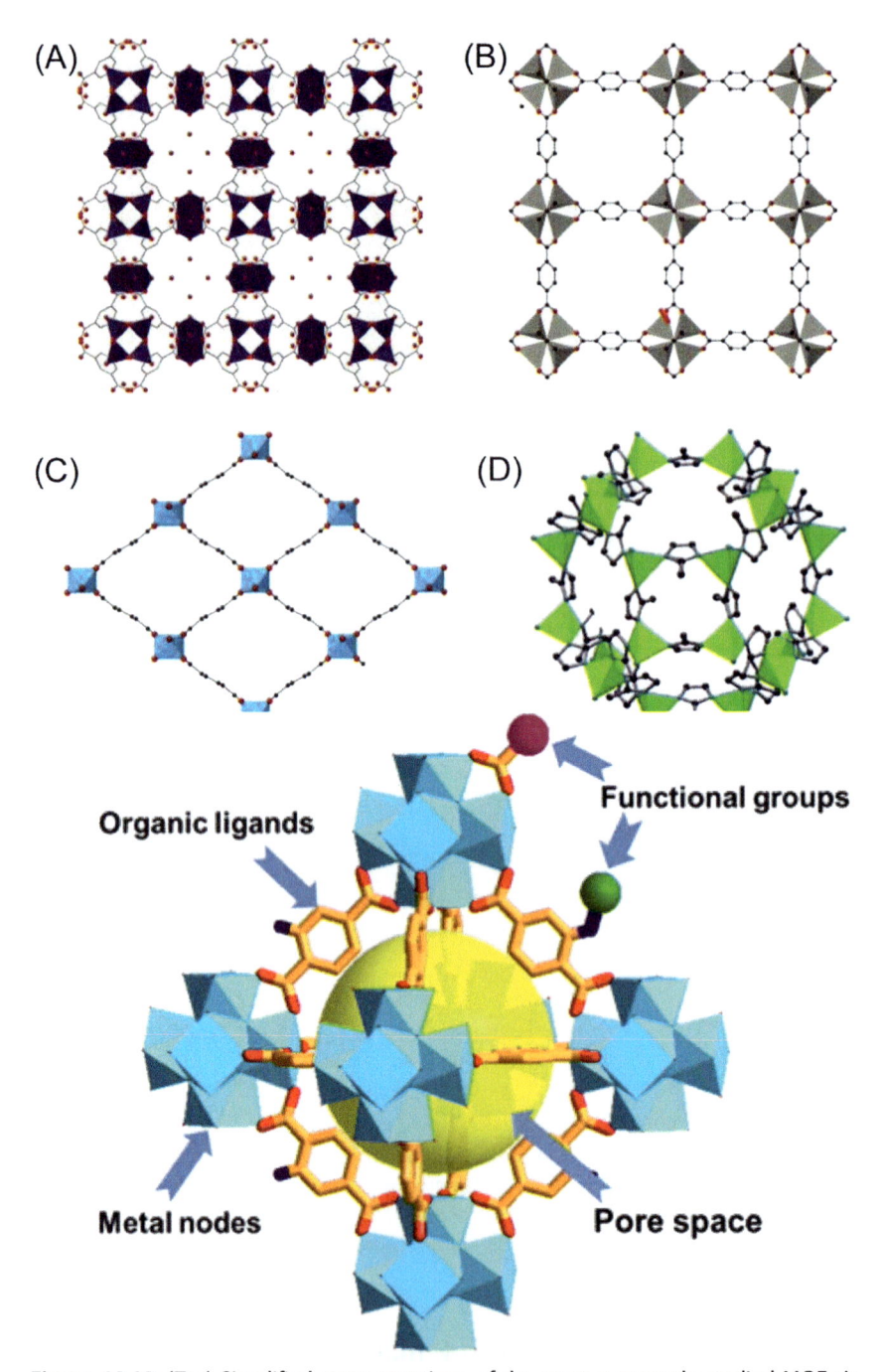

Figure 10.12 (*Top*) Simplified representations of the most commonly studied MOFs in the literature (with first reported metal ion and publication date): (A) HKUST-1 (or MOF-199, Cu^{2+}—1999), (B) MOF-5 (Zn^{2+}—2003), (C) MIL-53 (Cr^{3+}—2002), and (D) ZIF-8 (Zn^{2+}—2006); many MOF structures can readily accommodate isovalent (sometimes allovalent) metal ions. (*Bottom*) Illustration of the construction elements of functional

(*Continued*)

including pore spaces and functional groups that are usually not important for PV devices, are detailed in Fig. 10.12 [181,182]. Several reviews and retrospectives have been published over the past 20 years because MOF and MCP materials have shown great promise for practical applications. A few examples of the literature are noted here [181–188].

Yaghi's group reported success in producing Cu-MOF by a hydrothermal method in 1995 [180]. There have subsequently been reports of successful syntheses using numerous other methods [185,188]. Fig. 10.13 illustrates

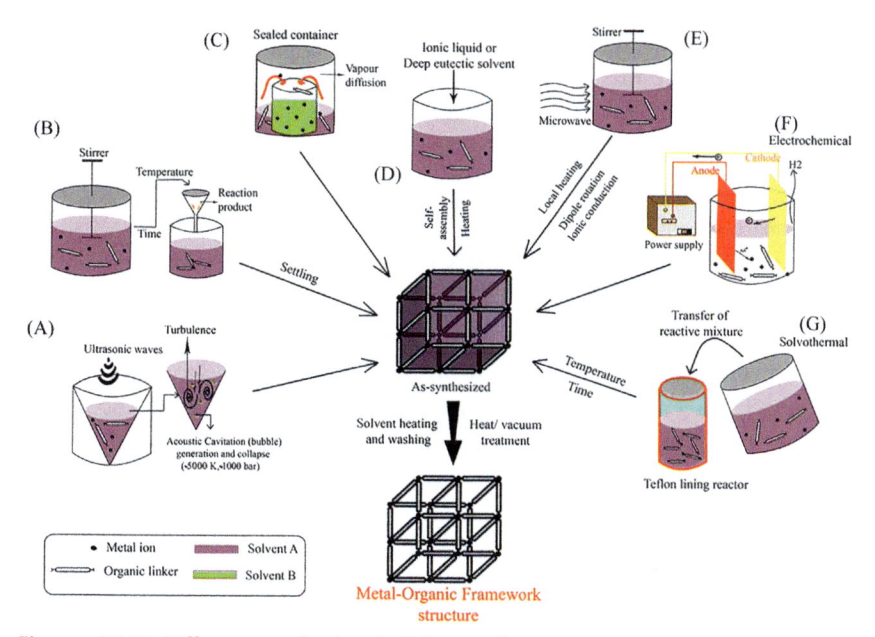

Figure 10.13 Different methods of MOF synthesis: (A) Sono-chemical synthesis method, (B) Conventional solution method, (C) Diffusion synthesis method, (D) Ionothermal process method, (E) Microwave synthesis method, (F) Electrochemical synthesis method, and (G) Solvothermal synthesis method. *Reproduced with permission from P.S. Sharanyakanth, M. Radhakrishnan, Synthesis of metal-organic frameworks (MOFs) and its application in food packaging: A critical review, Trends Food Sci. Technol. 104 (October 2020) 102–116, copyright (2020) Elsevier.*

◀ MOFs, including pore spaces and functional groups, that are critical for numerous applications but less important for PV devices. *(Top) Reproduced with permission from G.E. Gomez, F. Roncaroli, Photofunctional metal-organic framework thin films for sensing, catalysis and device fabrication, Inorganica Chimica Acta 513 (December 2020) 119926. (Bottom) Reproduced with permission from L. Jiao, J.Y.R. Seow, W.S. Skinner, Z. L. Wang, H.-L. Jiang, Metal–organic frameworks: Structures and functional applications, Mater. Today 27 (July–August 2019) 43–68, copyright (2019) Elsevier.*

multiple example methods of producing MOFs, including a sonochemical (Fig. 10.13A) and electrochemical methods (Fig. 10.13F), as well as a more generalized version of (water-based) hydrothermal processing (solvothermal method) (Fig. 10.13G) [185].

10.4.2 Metal-organic framework materials: Further details and photovoltaic applications

While there are numerous applications involving MOF (or MCP) materials, particularly where the pores and cavities that result from the innate structural properties of the organic linkers, a Scopus search of the terms (MCP or MOF) + (PV or solar cells) (ATK for all) results in 666 publications; (MOF or MCP − ATK) + (PV or solar cell) (Title) produces 266 results; the same search will all titles produces 116 publications. The three most cited articles resulting from such searches [189−191], including one of the earliest references involving MOFs and PV, explore the use of MOFs to enhance the V_{oc} of DSSCs [189]. The other publications report on MOF-derived platinum (Pt)-free counter electrodes (CEs) for DSSCs [190] and a 2D MOF for a PSC to minimize Pb leakage [191]. Examples of the (mainly) review literature in this evolving and rapidly growing area of research on enhanced third-generation PV are included in the references [191−206]. Fig. 10.14 provides an illustrated overview of the application of MOFs or MOF-derived materials for improving the performance of third-generation PV devices [193], given the primary focus on DSSCs and PSCs in the literature, these devices will be the focus for the remainder of the subsection.

As we will soon discover, the names for these materials are diverse and often are abbreviations or acronyms for where the materials were first synthesized (MIL) or structural families (ZIF). As seen in Figs. 10.11 and 10.12, there are three main aspects (besides the pore) of a MOF: The metal unit, or SBU, the organic linker or organic framework (Figs. 10.11 and 10.12), and functional groups (Fig. 10.12), which are located on aromatic rings that typically comprise the organic linkers [181,182,184−188,193,200,202,204,206−211]. Some compounds are specific to one metal or SBU, others are structural types that can accommodate a number of different metals. Given that there are well over 10,000 organic linkers and >80 metals currently used in MOFs. It has recently been estimated that there are nearly 100,000 MOFs known, with 500,000 potential MOFs theoretically estimated [207,209,211]. A final consideration is the form of MOFs when utilized in solar cell

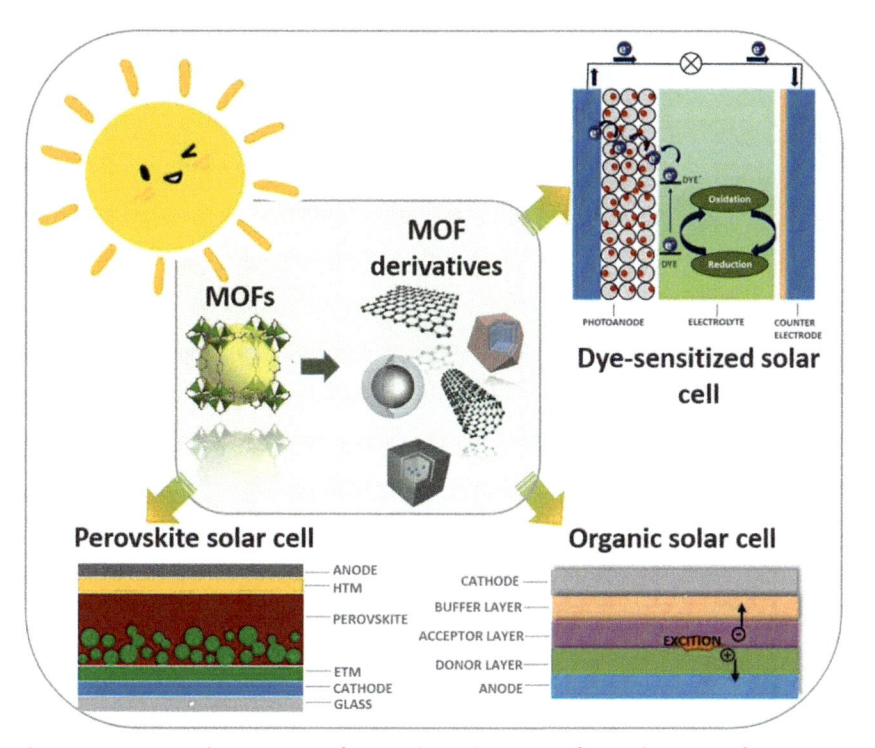

Figure 10.14 Visual summary of several applications of metal-organic frameworks and their derivatives in three third-generation solar cell devices (dye-sensitized solar cells, perovskite solar cells, and organic solar cells), device components, or layers include electrodes, photoactive materials, charge carriers, and additives. *Reproduced from B. Chen, Z. Yang, Q. Jia, R.J. Ball, Y. Zhu, Y. Xia, Emerging applications of metal-organic frameworks and derivatives in solar cells: Recent advances and challenges, Mater. Sci. Eng. R: Rep. 152 (February 2023) 100714, this article is distributed under the terms of the Creative Commons (CC-BY 4.0) license which permits unrestricted use, (http://creativecommons.org/licenses/by/4.0/).*

devices (see Fig. 10.14) either as intact MOFs or MOF-derived materials [26,182,184,193,208].

10.4.3 Use of metal-organic framework materials in dye-sensitized solar cells

Table 10.2 provides a diverse set of examples of DSSC devices (mostly) enhanced by the inclusion of MOFs or MOF-derived materials as either photoanodes or CEs [194,195,197,199,201,203,212−225]. The example devices included in the table are primarily liquid junction (LJ) cells, either I^-/I_3^- or Co^{2+}/Co^{3+}. The most common structural types added to

Table 10.2 Representative dye-sensitized solar cell devices enhanced with metal-organic framework (MOF) or MOF-derived materials.

MOF (see references for structural details of compounds)	MOF (or MOF-derived) materials in device structure[a]	Device type[b]	J_{sc} (mA/cm^2)	V_{oc} (V)	FF (%)	PCE[c] (%)	References
Co-DAPV	MOF photoanode[#]	SS	4.92	0.67	57	2.1	[212]
ZIF-8 (Zn)	ZnO aggregate photoanode[*]	LJ (I$^-$/I$_3^-$)	9.13	0.67	55.4	3.37	[213]
MOF-5 (Zn)	ZnO parallelepiped photoanode[*]	LJ (I$^-$/I$_3^-$)	8.13	0.66	68	**3.67**	[214]
[InK(ox)$_2$(H$_2$O)$_4$]$_n$	MOF photoanode[#]	LJ (I$^-$/I$_3^-$)	16.54	0.73	62	**7.42**	[215]
MIL-125 (Ti)	Mesoporous–TiO$_2$ photoanode[#]	QS	16.51	0.74	61	**7.45**	[201]
[Ni(bpp)(H$_2$O)$_2$]$_3$[P$_2$W$_{18}$O$_{62}$]·24 H$_2$O	P$_2$W$_{18}$·NiO@TiO$_2$ photoanode[#]	LJ (I$^-$/I$_3^-$)	17.68	0.75	57.2	**7.56**	[216]
UiO-66 (Zr)	MOF photoanode[#]	LJ (I$^-$/I$_3^-$)	18.6	0.68	60.8	**7.67**	[203]
Ni-BTC	MOF photoanode[#]	QS	27.32	0.62	51.6	**8.85**	[217]
MIL-125 (Ti)	MoS$_2$/dual-phase TiO$_2$ photoanode[#]	—	17.72	0.60	77.9	**8.96**	[218]
ZIF-8 (Zn)	MOF photoanode[#]	LJ (Co^{2+}/Co^{3+})	14.39	0.90	73	**9.42**	[219]
ZIF-8 (Zn)	ZIF-8/PEDOT:PSS counter electrode	LJ (Co^{2+}/Co^{3+})	11.46	0.85	70	6.84	[220]
CuTCPP	Cu$_{2-x}$Se@N-doped C counter electrode	LJ (Co^{2+}/Co^{3+})	13.12	0.89	65.4	7.61	[221]
ZIF-8 (Zn)	ZnO-NC/PEDOT:PSS counter electrode	LJ (Co^{2+}/Co^{3+})	13.58	0.84	71.5	**8.12**	[222]
Cu-MOF	Cu-MOF/PEDOT counter electrode	LJ (I$^-$/I$_3^-$)	16.36	0.78	65	**8.26**	[223]

ZIF-67 (Co)	CoS$_2$/N-doped C counter electrode	LJ (I$^-$/I$_3^-$)	18.51	0.72	65	**8.74**	[224]
MOF-525 (Zr-porphyrin)	MOF-525/s-PT counter electrode	LJ (I$^-$/I$_3^-$)	16.14	0.80	70	**8.91**	[194]
Ni–Co-MOF	NiCo$_{0.2}$@C counter electrode	LJ (I$^-$/I$_3^-$)	17.80	0.78	67	**9.30**	[195]
ZIF-67 (Co)	Co–MoS$_x$ counter electrode	LJ (I$^-$/I$_3^-$)	17.4	0.80	69	**9.64**	[197]
ZIF-8 (Zn)	m-SiO$_2$@NC-PEDOT:PSS counter electrode	LJ (I$^-$/I$_3^-$)	18.88	0.76	70	**10.0**	[225]
ZIF-67 (Co)	Co embedded carbon counter electrode	LJ (I$^-$/I$_3^-$)	21.44	0.73	68	**10.6**	[199]

Note: Table assembled by the authors from the literature.
[a]Photoanode contains either TiO$_2$[#] or ZnO[*].
[b]Solid-state (SS), quasisolid (QS) and liquid-junction (LJ) solar cell, respectively.
[c]Bold PCE indicates higher PCE for modified device.

DSSCs were zeolitic imidazolate framework materials, both ZIF-8 (Zn) [213,219,220,222,225] and ZIF-67 (Co) [197,199,224]. Aside from the addition of ZIF-8 to TiO_2 [219], the Zn- or Ti-MOF-enhanced photoanodes matched the primary oxide material, ZnO or TiO_2, respectively. The other metals added via MOFs were Ni, Co, Zr, Cu, and In. Devices modified with MOFs that resulted in higher PCEs are indicated by bold PCE numbers - 16 of twenty entries.

While a majority of the DSSC PCEs reported were improved by the addition of MOFs or MOF-derived materials as either photoanodes or CEs, the lone solid-state cell with the addition of a Co–MOF to a TiO_2 photoanode, with a PCE of 2.1%, was not improved [212]. The two quasisolid devices were both improved by the addition of Ti- and Ni-MOFs to TiO_2 photoanodes, respectively [201,217]. There were two primary mechanisms reported for the increased PCE of DSSC devices, depending upon whether the inclusion of MOFs (or MOF-derived materials) was used in photoanodes or CEs. ZIF-8 improves the dye-loading and PCE by approximately 20% when coating the TiO_2 layer of a photoanode [219]; UiO-66 more than doubles the dye-loading capacity when coating TiO_2 and increases PCE by 55% (ZIF-8 performs nearly as well in the same study) [203]. Mesoporous-TiO_2 produced from MIL-125 also enhanced both dye-loading and PCE by approximately 50% [201]. A composite TiO_2-Ni-MOF (0.5%) aerogel quasisolid DSSC [217] demonstrated a 30% higher PCE compared to a pure aerogel photoanode; dye-loading was not a primary factor in this system. CEs composed of Co nanoparticle (NP)-doped carbon rhombic dodecahedra [199] (top of Fig. 10.15) (or yolk-shell m-SiO_2@NC-PEDOT: PSS [225]) demonstrated higher electroactivity in catalyzing the I^-/I_3^- redox couple and a subsequent 15% improvement in PCE. A flexible MOF-525 (Zr-porphyrin) 3%-conducting polymer composite carbon CE demonstrated superior (electroactivity) performance to Pt (8.9% vs 8.2%); the efficiency increased to 9.75% when the light intensity was decreased from 100 to 10 mW/cm^2; see bottom of Fig. 10.15 [194].

10.4.4 Use of metal-organic framework materials in perovskite solar cells

Table 10.3 includes a similar cross-section of MOFs (or MOF-derived materials) intended to enhance both the PV performance but also the stability of PSCs [226–240]. While MOFs are involved as enhancements to DSSC photoanodes and CEs, there are actually four different aspects of

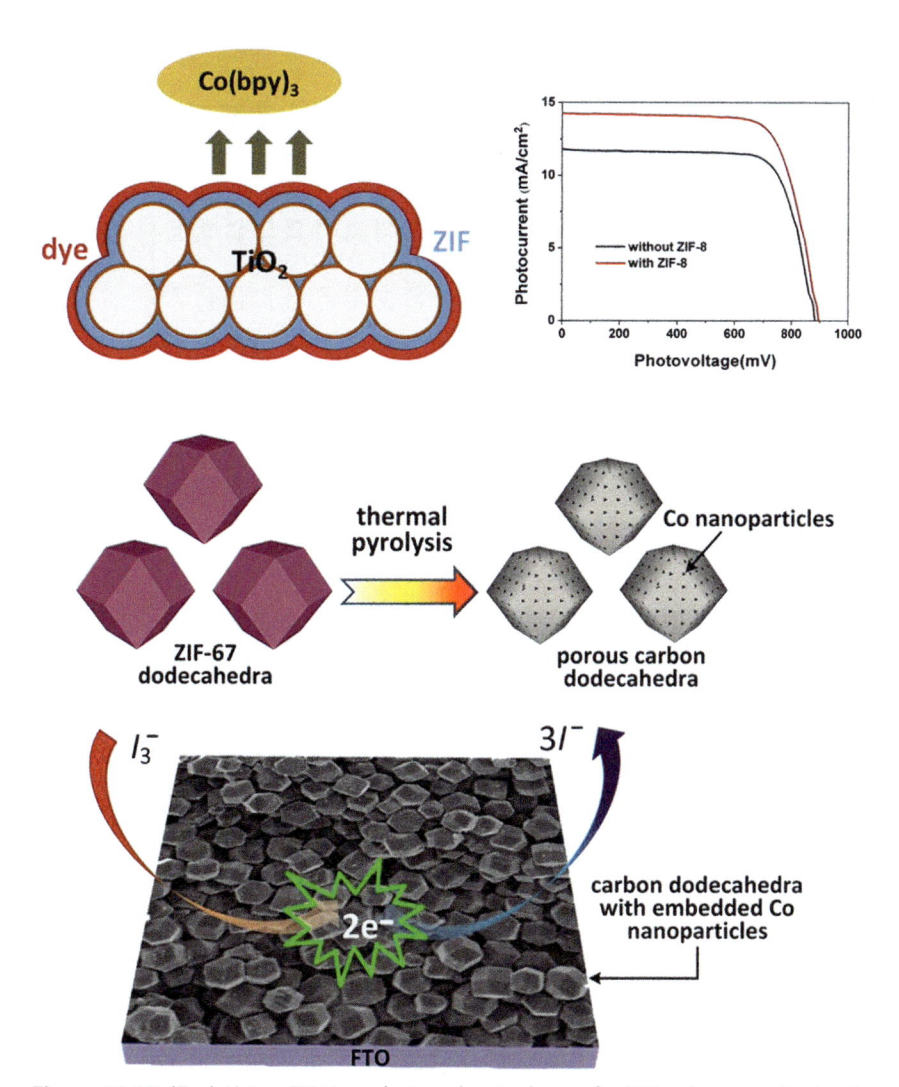

Figure 10.15 *(Top)* Using ZIF-8 as electron barrier layers for TiO$_2$ photoanodes in tris (2,2′-bipyridine)cobalt(II)/(III) based DSSCs has improved (by approximately 20%) both the dye-loading and PCE (to 9.42%) under 100 mW/cm^{-2} simulated sunlight. *(Bottom)* (A) The core-shell structure of MOF-525/s-PT composite film coated on a carbon cloth; (B) SEM images of 3 wt.% MOF-525/s-PT (inset is higher magnification); (C) photocurrent density−voltage curves of the DSSC with 3 wt.% MOF-525/s-PT, illuminated under 10 and 100 mW/cm^{-2}. *(Top)* Reproduced with permission from A. Gu, W. Xiang, T. Wang, S. Gu, X. Zhao, Enhance photovoltaic performance of tris(2,2′-bipyridine) cobalt(II)/(III) based dye-sensitized solar cells via modifying TiO$_2$ surface with metal-organic frameworks, Sol. Energy 147 (2017) 126−132, copyright (2017) Elsevier. *(Bottom)* Reproduced with permission from T.Y. Chen, Y.J. Huang, C.T. Li, C.W. Kung, R. Vittal, K.C. Ho, Metal-organic framework/sulfonated polythiophene on carbon cloth as a flexible counter electrode for dye-sensitized solar cells, Nano Energy 32 (2017) 19−27, copyright (2017) Elsevier.

Table 10.3 Representative PSC devices enhanced with metal-organic framework (MOF) or MOF-derived materials.

MOF (see references for structural details of compounds)	MOF (or MOF-derived) application in device structure	J_{sc} (mA/cm^2)	V_{oc} (V)	FF (%)	PCE^a (%)	References
Co-doped Ti-MIL-100	Co-TiO$_2$ (MOF-derived) modified ETL	24.08	1.03	65.0	**15.75**	[226]
MIL-125 (Ti)	TiO$_2$ (MOF-derived) modified ETL	22.81	1.01	71.8	**16.56**	[227]
ZIF-8 (ZnO)	ZnO (MOF-derived) modified ETL	22.1	1.11	74	**18.1**	[228]
MIL-125 (Ti)	MOF-modified ETL	23.18	1.08	75.5	**18.94**	[229]
Li-TFSI@ NH$_2$-MIL-101 (Fe)	MOF-modified HTL	23.41	1.07	75.7	19.01	[230]
[In(HPyia)Cl$_2$] · CH$_3$CN (In-Pyia)	MOF-modified HTL	23.53	1.09	79	**20.26**	[231]
FJU-17 (In) {[(CH$_3$)$_2$NH$_2$][In(L)] · 4.5DMF · 16H$_2$O}	MOF-modified HTL	25.09	1.05	77	**20.34**	[232]
Zn-CBOB	MOF-modified HTL	23.17	1.14	78.4	**20.64**	[233]
[Cu$_2$(BTC)$_4$/3(H$_2$O)$_2$]$_6$ [H$_3$PMo$_{12}$O$_{40}$]$_2$ (POM@Cu-BTC)	MOF-modified HTL	23.9	1.11	80	21.44	[234]
ZIF-8@FAI (FAI intercalated in Zn-MOF)	MOF-involved interface engineered	23.93	1.06	75.6	**19.13**	[235]
ZnL-MOF	MOF-involved interface engineered	23.86	1.12	79.3	**21.15**	[205]
2D-MOF ZrL$_3$	MOF-involved interface engineered	22.58	1.20	81.3	**22.02**	[191]
Eu-MOF	MOF-involved interface engineered	23.71	1.14	82	**22.16**	[236]
PEIE-2D MOF (Cu)	MOF-involved interface engineered	25.36	1.11	79	**22.22**	[237]
MOF-525 (Zr-porphyrin)	MOF/perovskite heterojunction	23.04	0.93	60	**12.0**	[238]
Cr-MOF	MOF/perovskite heterojunction	16.51	1.30	79	17.02	[239]
UiO-66 (Zr)	MOF/perovskite heterojunction	21.85	1.07	76.9	**17.81**	[198]
MOF-808 (Zr)	MOF/perovskite heterojunction	21.01	1.06	79.8	**18.01**	[198]
[In$_{12}$O(OH)$_{16}$(H$_2$O)$_5$(btc)$_6$]$_n$	MOF/perovskite heterojunction	23.55	1.12	79	**20.87**	[240]

Note: Table assembled by the authors from the literature.
^{a}Bold PCE indicates higher PCE for modified device.

PSC devices where enhancement by MOFs (or MOF-derived materials) have been reported: ETLs [226−229], HTLs [230−234], interfaces [191,205,235−237], and perovskite heterojunctions [198,238−240]. Given the greater diversity of applications, not surprisingly the structural diversity of MOF (and MOF-derived) materials is greater, including eight different metal ions and over 15 different structures, including novel (i.e., not MOF, MIL, etc.) structures [191,205,230−237,240]. Devices modified with MOFs that resulted in higher PCEs are indicated by bold PCE numbers - 16 of 19 devices.

MOF-derived materials [26,182,184,193,208] have been shown to be effective at enhancing the performance of PSCs by improving the efficiency of ETLs, typically ZnO or TiO_2 [226−229]. The improvement of PCE varies from 10% [226] to 50% [227]. Furthermore, porous hierarchical TiO_2 produced from a MIL-125 precursor used as an ETL for a $CH_3NH_3PbI_3$ PSC demonstrated a >250% improved PCE over a conventional device and also demonstrated enhanced stability over both a conventional and np-enhanced TiO_2; see top of Fig. 10.16. Organic hole transport materials with improved performance are enabled by addition of FJU-17 (In) to form an enhanced hybrid HTL [232]. The PCE improvement of 10% is accompanied by improved stability—90% of the original PCE after 1000 hours of storage at ambient conditions versus 25% for a device with an unmodified HTL [232]. A combined theoretical (Vienna Ab Initio Simulation Package (VASP) [241]) and experimental study examined the passivation ability of a ZnL-MOF ($\{[(Me_2NH_2)_3(SO_4)]_2[Zn_2(ox)_3]\}_n$ [242]) and whether amine-containing cations could effectively passivate negatively charged defects, as well as if the anionic framework (with dense oxygen sites) could form strong interaction with the perovskite film, thus anchoring the atoms of the perovskite (Fig. 10.16 (bottom A−E)). The results of the experimental aspect demonstrate that a ZnL MOF-incorporated PSC gains approximately 10% in PCE from 19.75% to 21.15% with enhanced stability (factor of two greater) during 3000 hours (Fig. 10.16, bottom F).

A recent study examined the effectiveness of perovskite/Zr-MOF heterojunctions, comparing a bilayer architecture and hybrid materials, integrated into inverted $p-i-n$ PSCs with a baseline device [198]. Two different Zr-MOFs, UiO-66 and MOF-808, were used as surface modifiers for the NiO_x HTL in the device, enhancing the crystallization of the perovskite film while at the same time facilitating charge-extraction at the

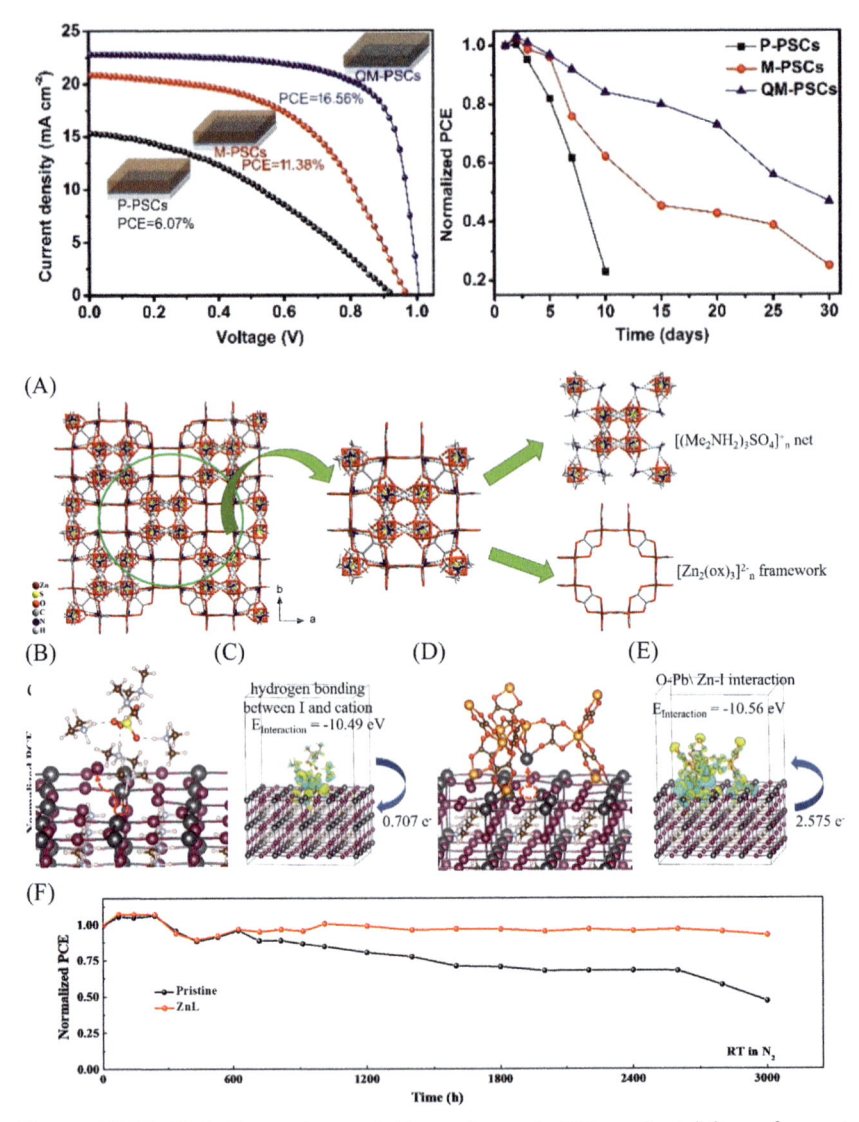

Figure 10.16 (*Top*) Comparison of the enhanced PCE and stability of quasi-mesoscopic PSCs (QM-PSCs) with porous hierarchical TiO$_2$ (hier-TiO$_2$) nanostructures versus conventional nanoparticles of TiO$_2$ and conventional PSCs with compact TiO$_2$ layer. (*Bottom*) Visualization of results from first-principles calculations using Vienna Ab Initio Simulation Package (VASP) [241]. (A) structure of ZnL-MOF; (B) and (C) [(Me$_2$NH$_2$)$_3$(SO$_4$)]$^+$ cation interacts with PbI$_2$ terminated surface with I$_{Pb}$ antisite defect and the corresponding charge-density difference; (D) and (E) [Zn$_2$(ox)$_3$]$_3^{2-}$ anion framework interacts with PbI$_2$ terminated surface with Pb$_I$ antisite defect and
(*Continued*)

interface. The UiO-66 and MOF-808-modified PSCs increased PCEs to 17.01% and 16.55%, respectively, compared to the baseline device (15.79%); while PCEs of the UiO-66/MOF-808-hybrid PSCs were found to be further improved to 18.01% and 17.81%, respectively [198]. The hybrid MOF/perovskite materials were proposed to passivate the perovskite grain boundary, providing a "grain-locking effect." It not only mitigates potential impact of defects but also reinforces the film's robustness against moisture invasion; in fact, approximately 70% and approximately 80% of initial PCE, respectively, were retained for the UiO-66 hybrid and MOF-808-hybrid devices after two weeks, while the PCE of the baseline device dropped precipitously to less than 5% [198].

As stated in several recent publications, there are minimal reports of the use of MOFs to enhance the performance of OSCs [193,206]. Metal-organic nanosheets composed of the 2D-MOF $Zn_2(ZnTCPP)$ (where TCPP is tetrakis(4-carboxyphenyl) porphyrin [243,244]) were blended into the bulk heterojunction layers of six different OSCs and found to have a significant impact on the PCE, up to a doubling of efficiency [206]. As discussed above (10.4.4), it is difficult to delineate the impact of MOF (or MOF-derived) materials in OSCs. However, as discussed previously for bio-OSC devices [1,245], the preponderance of organic moieties in a MOF (or MON) material should facilitate blending into OSC device structures. Finally, given the tremendous promise demonstrated thus far of PSCs and tandems containing PSCs, it is clear that given the impressive diversity of MOF materials and derived materials, this aspect of PV device processing will certainly witness increasing research activity going forward, particularly as not only PCE but also stability enhancement has been demonstrated. However, further (increasing levels) research on the performance enhancement of DSSCs and OSCs is expected.

◀ the corresponding charge-density difference; (F) Long-term stability test of unencapsulated PSCs stored at room temperature under natural day/night lighting in N_2. *(Top) Reproduced with permission from X. Hou, L. Pan, S. Huang, O.-Y. Wei, X. Chen, Enhanced efficiency and stability of perovskite solar cells using porous hierarchical TiO_2 nanostructures of scattered distribution as a scaffold, Electrochim. Acta 236 (2017) 351–358, copyright (2017) Elsevier. (Bottom) reproduced with permission from C. Li, J. Qiu, M. Zhu, Z. Cheng, J. Zhang, et al., Multifunctional anionic metal–organic frameworks enhancing stability of perovskite solar cells, Chem. Eng. J. 433 (2) (April 2022) 133587, copyright (2022) Elsevier.*

10.5 Future paradigms: Novel materials and device fabrication for photovoltaics

To conclude the main topics of this chapter, it is useful to reconnect with connections to biomimicry and green technologies (Table 10.1). There are several concepts covered in this subsection that capture the promise of more efficient, sustainable, and/or nature-inspired processing approaches and paradigms. Given our prior publications and reviews on SSPs, PIDs, or I(M)PSs [26,27,30−33,127,246,247] and prior discussions in this chapter, involving materials discovery and low-cost processing (via numerous approaches including ML and AI), as much as practicable, we will make references to prior literature and reviews and include connections to other parts of this chapter and book.

10.5.1 Single-source precursor routes to solar cell materials

A key focus of the primary author's efforts at the NASA Glenn Research Center (GRC) was the development of high mass-specific power PV technologies for future space applications to minimize launch mass (and costs) [28,29,127,247−254]. Thin-film solar cells composed of chalcogenide materials were chosen as the technology of choice, by our group and our research partners in academia and industry [26−33,91,125,127,251,254−265], because at that time (late 1980s [1,28,58,59,247]), third-generation materials had not yet been discovered or at least developed to a level where a mission-oriented research organization would invest significant resources [28,59,247]. In order to enable the deposition of CIGS and related chalcogenide PV materials (typically processed by PVD at higher temperatures [251,254,256,266−271]) on polymer substrates at temperatures below 400°C, our group, with some work involving collaborations with academic colleagues, relied on a SSP approach to produce thin films and quantum dots by spray- or atmosphere-assisted CVD (AACVD) [30,33,127,257−259,261−264] and solution processing methods [27,29,31,33], respectively. A simplified representation of this approach is depicted at the top of Fig. 10.17 [265]. In fact, over the past several decades, the SSP approach has grown to include a wide variety of materials for numerous applications [26], chalcogenide materials for PV being one prominent example [27,29,31,33,127,258−265].

Elsevier's recently published book on nanomaterials via SSPs [26] was dedicated by the coeditors (Apblett, Barron, and Hepp) to Prof. Paul O'Brien, who sadly passed away in 2018 [272]. The dedication of this book was in part a tribute to our friend and colleague as well as a well-deserved acknowledgment of his many important contributions to the production of semiconductor materials via SSP processing [273–277]. Of relevance to this chapter, Prof. O'Brien and his colleagues produced numerous new metal chalcogenide complexes, particularly dithiocarbamates (DTCs) and related compounds, and explored their use for the fabrication of metal sulfide and selenides for energy (including PV) applications [273,276,277]; see Fig. 10.17 (Middle) for an illustration of the synthesis of an SSP and the subsequent production of Sb_2S_3 [277]. The literature on SSP-processed metal sulfides for PV utilizing sulfur-coordinated complexes is quite extensive as demonstrated by another chapter [278] in the above-discussed SSP book [26] that is limited to the main group (binary and ternary) metal complexes.

Ahmet et al. detail the production of tin sulfide (α-SnS and π-SnS) solar cells via AACVD of a dimeric Sn precursor with a SnN_3S coordination environment [279]; the study includes a comparison to the performance of SnS solar cells produced using other methods [280]. In fact, a few further examples of recent work on SSP processing involving metal-sulfur-containing complexes by other groups evidence the ubiquity of this approach [280–284]; these publications also describe efforts to produce [PSC, hybrid organic−inorganic solar cell (HOISC) [285], and DSSC] solar cells, including discussion of solar cell performance with PCE (or η). Fan et al. [281] and Kou et al. [282] describe PVD processing that utilizes an SSP to produce PSCs as well as device performance. Dowland et al. produced solar cells of blended CdS:P3HT films, where CdS is produced via spin-on processing using a Cd dithiolate $(Cd(S_2COC_2H_5)_2(C_5H_5N)_2)$ SSP heated to 160°C under N_2 [283]. Lastly, Ehsan et al. describe the processing and performance of DSSCs with a Pt CE produced using two different Pt dialkyldithiocarbamate SSPs via AACVD (Fig. 10.17, Bottom), comparing them both with a DSSC containing a traditional Pt CE; interestingly, the SSP-deposited Pt films do not show any measurable sulfur contamination [284]. This curious result coincides with earlier work by O'Brien and coworkers demonstrating that the use of somewhat similar ME_4-coordinated [M{N(EP^{i}Pr)$_2$}$_2$] (M = Pd, Pt; E = S, Se) SSPs for deposition of metal chalcogenide films via low-pressure metal-organic CVD at temperature between 425°C and 500°C produced metal selenides ($PdSe_2$, $PtSe_2$). However, only PdS_2 as deposition using the Pt-S analog produced metallic Pt films [286].

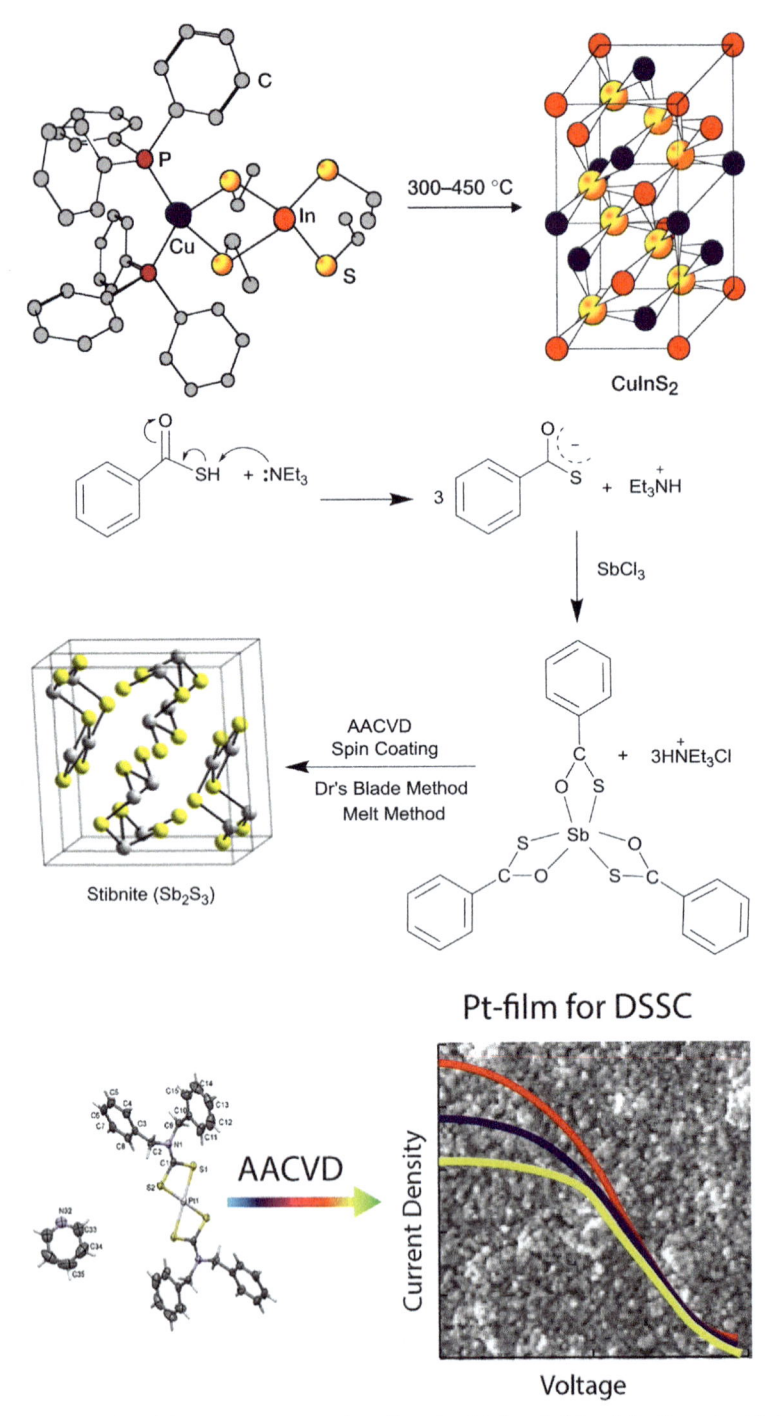

Figure 10.17 (*Top*) Decomposition of a dimeric thiolate-bridged In-Cu tetrathiolate (bisphosphine) complex to form the (tetragonal) chalcopyrite phase of $CuInS_2$—it should be noted that there are numerous potential variations of this structural family

(*Continued*)

A key issue that has not been addressed thus far in this subsection is the PV performance of devices processed using SSPs. At NASA GRC, we worked assiduously to produce device-quality I-III-VI$_2$ materials while optimizing η by working with our academic partner at the University of Delaware's Institute for Energy Conversion [287]. Unfortunately, we were never able to produce devices with a PCE over 1% [33,262,264]. Further, as documented in our recent book chapter, NASA reprioritized funding to mission-oriented research versus base research and technology [33]. Therefore, we terminated our efforts in 2005, aside from publishing results from our efforts, ending with a structural report on several SSP compounds in 2015 [265]. We were able to capture some intellectual property [288] and eventually commercialized an updated version of our AACVD reactor to produce multiwalled carbon nanotubes via the formation of a small company in 2008 [289]; this is detailed in the SSP book chapter [33].

Tin sulfide TFSCs, reported by Ahmet et al., produced by AACVD of an SSP demonstrated a maximum of 0.9% PCE [280]; the maximum efficiency reported for SnS solar cells produced using a variety of methods in ten other studies was 4.4% for atomic layer deposited films [290]. The hybrid PSC devices reported by Fan et al. demonstrated a 4.5% PCE for a 2D $(BA)_2(MA)_3Pb_4I_{13}$ material [281]; while Kou et al. achieved a PCE of 7% with SSP evaporation and inclusion of QDs of $CsPbBr_3$ [282]. While both of these reports are improvements over typical PCEs of SSP-processed devices, they are well below even routine

◀ (Ag replaces Cu; Ga replaces In; Se replaces S; N- or As-containing ligands replacing P (E); different organic groups bonded to C or E); (Middle) Schematic of the synthesis of an SSP to produce (orthorhombic) phase of stibnite (Sb$_2$S$_3$). *(Bottom)* Decomposition of a Pt SSP ([Pt(S$_2$CNBz$_2$)$_2$]·py) to produce Pt thin films; superimposed on an SEM of Pt films is a plot of the current-voltage plots of fabricated DSSCs with the Pt SSP with various deposition times and efficiencies (10 min. (yellow)—4.2%, 20 min. (red) —5.0%, 30 min. (blue) —4.3%). *(Top) Courtesy: NASA; reproduced with permission from G. Murtaza, M. Akhtar, M.A. Malik, P. O'Brien, N. Revaprasadu, Aerosol assisted chemical vapor deposition of Sb$_2$S$_3$ thin films: Environmentally benign solar energy material, Mater. Sci. Semicon. Proc. 40 (December 2015) 643—649, copyright (2015) Elsevier. (Bottom) Reproduced with permission from M. A. Ehsan, M. Younas, A. Rehman, M. Altaf, M.Y. Khan, et al., Synthesis and utilization of platinum(II) dialkyldithiocarbamate precursors in aerosol assisted chemical vapor deposition of platinum thin films as counter electrodes for dye-sensitized solar cells, Polyhedron 166 (July 2019) 186—195, copyright (2019) Elsevier.*

efficiencies of traditionally processed PSCs [34,58,59,68,291,292]. The HOISCs reported by Dowland et al. achieved a maximum efficiency of 2.1% at a 50/50 mix (by volume) of CdS and P3HT [283]. To provide some context, a report by Fradler et al. appearing a year later demonstrated a PCE of 1.6% for a small-area CIS/polymer (9:1 w/w) HOISC on a poly(ethylene terephthalate) (PET) substrate, annealed at 140°C [293]. Devices on glass substrates annealed at higher temperatures (180°C) with the addition of n-hexylamine were improved by approximately 50% to a PCE of 2.45%. Finally, the DSSC with a Pt CE reported by Ehsan et al. using a Pt (dibenzyldithiocarbamate)$_2$ SSP (bottom of Fig. 10.17) essentially matched the PCE of a traditional Pt CE (4.3 vs 4.4%). The reported PCE for a DSSC with an SSP-processed Pt CE is less than half of the PCE of the top five DSSCs with MOF-derived CEs listed in Table 10.2 [194,195,197,199,225]; see bottom of Fig. 10.15.

While our efforts at NASA GRC were not able to unambiguously determine the issues that prevented PCEs higher than 1%, we noted that V_{oc} (and FF) seemed to be an issue for cells prepared with AACVD-deposited films in general [264]. Further studies on other problems, such as defects, junction alignment, recombination losses, and undetected (and unanticipated) process issues, were also identified when the overview was published in 2008 [264]; at that time, the insightful study of defects in CuInSe$_2$, a related material, by Kazmerski and coworkers served as an important guide [294]. In order to broaden the consideration of materials and device challenges that must be overcome to expedite further progress on second and third-generation solar cells, the interested reader is encouraged to consult the series of references considering issues for examples of the PV technologies addressed in this chapter [295–299].

An intriguing recent study reported on solar cells produced using antimony selenoiodide (SbSeI), an orthorhombic metal chalcohalide. A spin-on and anneal process involving an SSP was used to deposit Sb$_2$Se$_3$ followed by SbI$_3$ spin-on/anneal fabrication of SbSeI [300]. This two-step process is somewhat analogous to our previously reported AACVD deposition of indium sulfide from In(arylalkylDTC)$_3$ on a Cu substrate to produce CuInS$_2$ [263]. Going from 8 to 10 to 12 deposition cycles resulted in an improvement and then degradation of PCE from 2.8% to 3.8% to 1.2%, with a drastic reduction in J_{sc} for 12 cycles. The authors determined that insufficient absorption (8 cycles) and inefficient charge transfer (12 cycles) diminished the PCE when compared to 10 cycles where the absorption and charge transfer were determined to be acceptable [300]. An earlier result for Pt CEs in DSSCs gives an analogous result of improvement to an optimal PCE being reduced with thicker films (Fig. 10.17C) [284]. In fact, this trend in optimal film thickness followed by degraded cell performance for thicker films is quite common.

Clearly, using SSP-processed materials for the primary light-absorbing component or film in a device is fraught with issues related to high defect densities and other challenges, as discussed above [264]. However, given the improvement of SSP-processed MOF-derived materials for third-generation PV devices detailed above (Sections 10.4.3 and 10.4.4), there are definite advantages to potentially be realized by processing device components using this intriguing, green approach. However, until device challenges are solved for the primary solar cell light absorbers, secondary device components may be the best option for SSP-processed materials.

10.5.2 Photovoltaics-integrated devices and integrated power sources

Our own experience with an integrated micro-power source (IMPS) dates back to a successful demonstration of a III-V SC/Li ion battery tested in low-Earth orbit on board a student experiment space (Starshine 3) launched in late 1999 from Kodiak, AK. The IMPS consisted of an in-house fabricated 7 V monolithic integrated module-GaAs array, commercial control electronics, and a commercial 45 mAh lithium-ion battery (LIB). Further details of this successful space experiment are provided in multiple book chapters (including references to the original literature) from the perspective(s) of the device concept and fabrication (focusing on the battery) as well as flight data [32], technology transfer [33], solar cell performance [127], and a low-Earth space mission [247]. During the first 100 days in orbit, all five IMPS units on Starshine 3 maintained battery voltages between 2.5 and 2.9 V, even the three units that relied on albedo (light reflected off the Earth) and operating at an average temperature of $-10°C$; to our knowledge, this remains the only successful flight demonstration of an IMPS [32].

The level of research activity in this intriguing area of device technology has increased dramatically since 2000. As we (and others) have pointed out, combining multiple devices is analogous to producing a tandem PV device [28,32,127,246,301,302]. Numerous reviews (or overviews) have recently been published that highlight the PID and IPS literature, including studies that combine PV with either energy storage or devices (i.e., biomedical,

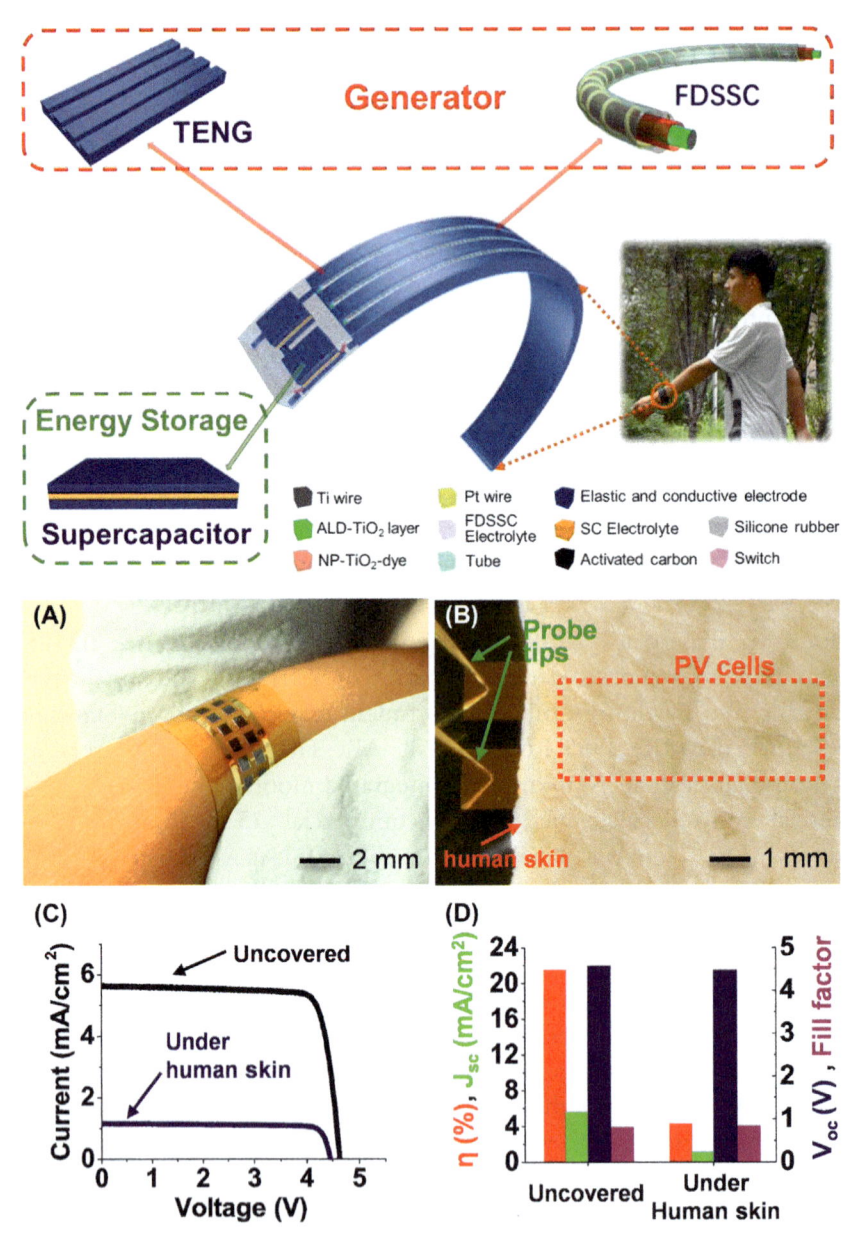

Figure 10.18 (*Top*) Schematic of a flexible DSSC (FDSSC) and triboelectric nanogenerator (TENG) that harvest solar and mechanical energy, respectively, charging a carbon black @silicone rubber supercapacitor within a silicone bracelet. (*Bottom*) Electrical properties of the flexible implantable PV (IPV) device under human skin: (A) optical image of the thin flexible IPV device bent on a forearm with a radius of approximately 3 mm; (B) image of fixed human hand dorsum skin (thickness: (*Continued*)

sensor, electronic) [32,302−308]. Energy conversion and storage devices can be combined and/or integrated into a larger system that includes other functions to enhance efficiency or accomplish multiple tasks; these PV-integrated systems have been the subject of numerous reviews but aside from providing relevant literature will not be further detailed [308−316]. The remainder of this subsection examines a series of example PIDs that power (portable) electronics [317−320] and provide power for biomedical purposes [306,318,321,322], IPSs utilizing batteries and/or supercapacitors [323−328], and wearable electronics and power sources [82,302,303,318,320,329,330].

Figs. 10.5 and 10.6 illustrate bio-inspired solar cell devices that serve multiple functions for building-integrated PV [85], essentially part of a larger system [310], and wearable power devices [82], respectively. These solar PIDs and IPSs [32,33,82,85,127,247,302−330] are a relatively recently developed approach to realizing novel technologies involving multifunctional materials that improve efficiency from a power, mass, and volume perspective, mimicking the efficiency of nature [331]; see Table 10.1. There are essentially two main types of PIDs and IPSs, analogous to tandem PV:monolithic (or integrated, 2-junction or 3-junction) and electrically connected (3-junction or 4-junction) devices; 4-junction integrated devices with redox flow batteries (RFBs) have been studied with some frequency over the past decade, but aside from providing relevant references, they will not be further detailed [32,302,303,305,307]. Another example of a wearable PID (or IPS) [82,318,320] is shown at the top of Fig. 10.18, which provides a schematic of a flexible DSSC (FDSSC) and triboelectric nanogenerator (TENG) that harvest solar and

◀ approximately 0.7 mm), which covers the flexible IPV device (red dotted area); (C) Current−voltage ($I-V$) curves of the IPV device when not covered (black line) and when covered (blue line) with the human hand dorsum skin under AM 1.5 G illumination; (D) Electrical characteristics of the IPV device when uncovered and when covered, the efficiency (η) and short circuit current density (J_{sc}) of the IPV device decrease from 21.5% to 4.3% and from 5.63 mA/cm^2 to 1.17 mA/cm^2, respectively. The V_{oc} (4.6 to 4.5 V) and FF (0.83 to 0.84) do not show significant change under the human skin. *(Top) Reproduced with permission from W. Song, X. Yin, D. Liu, W. Ma, M. Zhang, et al., A highly elastic self-charging power system for simultaneously harvesting solar and mechanical energy, Nano Energy 65 (November 2019) 103997, copyright (2019) Elsevier. (Bottom) Reproduced with permission from K. Song, J.H. Han, H.C. Yang, K.I. Nam, J. Lee, Generation of electrical power under human skin by subdermal solar cell arrays for implantable bioelectronic devices, Biosens. Bioelectron. 92 (June 2017) 364−371, copyright (2017) Elsevier.*

mechanical energy, respectively, charging a carbon black @silicone rubber supercapacitor within a silicone bracelet [329]. The FDSSC has a PCE of 4.85% and is an electrically connected device that in conjunction with the TENG can charge the supercapacitor to 1.8 V in 43 second; devices such as LEDs, watches, and sensors can be powered by the flexible IPS.

For biomedical electronics [306] within a body, a much more challenging environment requires complex systems that rely on much higher power per unit area (also called "area-specific power") to minimize space. The bottom of Fig. 10.18 shows a series of images relating to an implantable PV (IPV) device [321]; while not strictly speaking an IPS, IPV devices could enable numerous implantable biomedical devices or systems. Song et al. studied the performance of tandem GaInP/GaAs devices grown epitaxially on GaAs buried under approximately 0.7 mm of human cadaver hand skin; Fig. 10.18 (bottom A) is an optical image of the thin flexible IPV device bent on a forearm with a radius of approximately 3 mm; Fig. 10.18 (bottom B) is an image of a human cadaver hand (dorsum) skin covering a flexible IPV device (red dotted). The electrical properties were measured by probing the exposed square metal pads connected to the IPV device; Fig. 10.18 (bottom C) are current−voltage (I−V) curves of the IPV device when not covered (black line) and when covered (blue line) with approximately 0.7 mm of cadaver hand skin under AM1.5G illumination. Fig. 10.18 (bottom D) illustrates the electrical characteristics of the IPV device when uncovered and when covered, that is, under the human skin. The efficiency (η) and short circuit current density (J_{sc}) of the IPV device decreased from 21.5% to 4.3% and from 5.63 to 1.17 mA/cm^2, respectively; both V_{oc} and FF did not show any significant change under the human skin. As with the Starshine 3 IMPS flight experiment, when the area was at a premium, a III-V device is currently the best option to maximize efficiency and power [321].

A recent example of a highly efficient IPS is shown in Fig. 10.19 [326]. The top of Fig. 10.19 illustrates the three-terminal integrated photocapacitor (IPC) including a PSC and a symmetrical supercapacitor (SSC) (left) and an equivalent circuit diagram on the right. The bottom of Fig. 10.19 provides visual performance information of the PSC and IPC (Fig. 10.19A and B) as well as performance metrics during charging (Fig. 10.19C and D) and discharge (Fig. 10.19E and F). The efficiency terms in Fig. 10.19 include η_{L-E} (or PCE) light-to-electric energy conversion efficiency; η_{E-C} (electric-to-chemical energy storage efficiency) for PSC photo-charge to SSC; η_{C-E} is the Joule efficiency, dependent upon

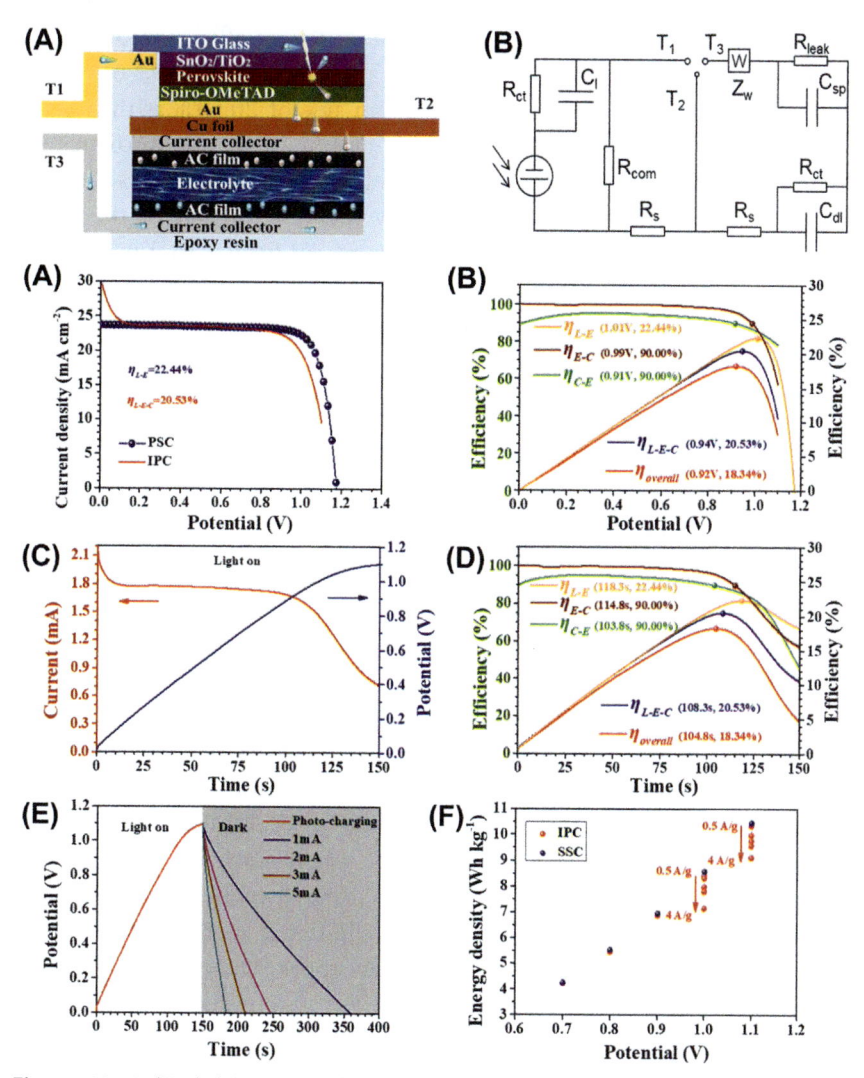

Figure 10.19 (*Top*) (A) Structural schematic and (B) Equivalent circuit of the integrated photocapacitor (IPC). (*Bottom*) (A) Photo-charging current−voltage curves of PSC and IPC; (B) efficiency-voltage curves of IPC; (C) photovoltage-time and photocurrent-time profiles of IPCs during photo-charging; (D) efficiency-time profiles of the IPCs during photo-charging; (E) photo-charging (white sections) and galvanostatic discharge (gray sections) curves of IPCs at different discharge current densities; (F) energy densities of the SSCs and IPCs at different voltage windows in galvanostatic discharge. *Reproduced from Z. Song, J. Wu, L. Sun, T. Zhu, C. Deng, et al., Photocapacitor integrating perovskite solar cell and symmetrical supercapacitor generating a conversion storage efficiency over 20%, Nano Energy 100 (September 2022) 107501, copyright (2022) Elsevier.*

the properties of the SSC; η_{L-E-C} (conversion storage efficiency or η_{CS}) is the product of η_{L-E} and η_{E-C}; and finally, $\eta_{overall}$ is the product of η_{L-E-C} and η_{C-E}. In fact, IPC achieves a maximum η_{CS} of 20.53% at the maximum efficiency point (MEP) of 0.944 V and an η_{C-E} of approximately 90% at 0.913 V, resulting in an $\eta_{overall}$ of 18.34% at 0.919 V. As the discharge performance of the IPCs is nearly identical to that of the SSCs (Fig. 10.19E), as the discharge current is increased from 1 to 5 mA in the dark, the complete discharge time is reduced from 212 to 34 seconds; the energy densities of the IPCs and the SSCs in different potential windows are shown in Fig. 10.19F [326].

In order to estimate comparable performance for IPSs, a general overall device or system efficiency, equivalent to the η_{CS} discussed above, computes system or device efficiency (η_{SD}) as the product of η_{PV} and η_{ES} [32,303−305,307]. However, as recently discussed [307], given different optimal current densities for PIDs or IPSs and the broad range of applications for hybrid devices, reporting η_{SD} values are of questionable utility unless comparing a common PV or storage sub-system or component, or perhaps a specific application. For example, a recent insightful publication by Andrés et al. proposes that a more nuanced figure of merit be utilized for IPSs as a result of their in-depth study of three-junction (monolithically-integrated) IPCs consisting of a large-area OSC and a supercapacitor [328]. When taking into account actual systems performance, the photoelectrochemical energy conversion efficiency of 17.4% (with a discharge of 0.2 mA/cm^{-2}) is reduced to a full cycle efficiency (with losses) of 2% (30 seconds and 1 mA/cm^{-2}) [328]; this is a much more dramatic reduction than the case considered above and illustrated in Fig. 10.19 [326]. Thus, η_{SD} (or η_{CS}) should be considered an upper bound of the practical system efficiency that will be reduced to reflect system performance under conditions that closely resemble working conditions.

Nevertheless, as a first-order heuristic for comparing various IPS technologies, such a crude figure-of-merit (uncorrected η_{SD} or $\eta_{overall}$) may have some merit. In our review chapter [32], the critical review of Vega-Garita [305], the review of Wu et al. [307], and lists of $\eta_{overall}$ (or η_{SD}) reported in the literature are assembled including details such as the PV technology, energy storage device, device structure, and some rough description of the IPS morphology. The three junction (or 3 T) IPC reported by Song et al. [326], even using the more meaningful ($\eta_{overall}$) system efficiency of 18.34%, is the highest that was reported at the time of the study's publication. Reports of devices with a system efficiency ($\eta_{overall}$) of approximately 10% or more, culled from the above-mentioned sources [32,305,307], follow. A GaAs discrete electrically connected (4 junctions or 4 T) IPC device was reported with an $\eta_{overall}$ of 17.6%, while an analogous 4 T p-Si device was reported by the same group with an $\eta_{overall}$ of 12% [332]. Two other notable IPCs are a 2 T p-Si/porous Si SSC with an $\eta_{overall}$ of 12.4% reported by Westover et al. [333] and a discrete 4 T PSC/polypyrrole-based SSC with a reported $\eta_{overall}$ of 10% reported by Xu et al. [334]. An integrated 3 T PSC/aluminum-iron battery was reported by Hu et al. with an $\eta_{overall}$ of 12% [335]; a 4 T solar battery IPS consisting of a PSC and a LIB with advanced electronics was reported by Kin et al. to have an $\eta_{overall}$ of 9.9% [336]. Finally, two other solar battery IPSs using traditional PV reported system efficiencies of over 10%: A PV component of a solar charging station using commercial Si SCs and lithium iron phosphate batteries (discrete or 4 T), with $\eta_{overall}$ of 10.5%–14.5% (dependent upon a number of connected cells at maximum power point) [337]. Additionally, a 3 J GaInP/GaAs/Ge solar cell integrated with an RFB with an $\eta_{overall}$ of 14.1% [338].

Of the higher-system efficiency IPSs (of PIDs) detailed above, the PCE of the SC is clearly critical. With two exceptions [328,329], every system discussed or illustrated above contains either a Si device [332,333,337], a III-V-based device [32,127,321,332,338], or a PSC [128,324–327,334–336]. Lau et al. analyzed the IPS literature and highlighted reports cited more than 50 times [304]. Of the 25 reports that were heavily cited, three were DSSCs in a 2E-monolithic structure, 14 were 3E-monolithic devices (9 DSSCs and 2 PSCs), and eight were 4E-discrete devices (5 DSSC and 1 PSC); the remaining five PV technologies were split between 2 OSCs, a HOISC, and 2 Si SCs [304]. Each PV technology presents advantages in an integrated device: Traditional technologies such as Si and III-V are reliable, highly efficient, exhaustively studied, and are suitable for both monolithic and discrete rigid devices [32,127,321,332,333,337,338]; PSCs are also highly efficient, intensively researched, and suitable for either rigid or flexible devices, either in a monolithic or discrete device [128,301,326,334–336]; and OSCs and DSSCs are lower in efficiency but more suitable in flexible devices that can be assembled in a monolithic (preferred) or discrete structure [304,305,317,323,324,328–330]. In the final analysis, it must be stressed that a device, be it an IPS or PID powering a sensor or biomedical implant, must be able to accomplish the task for which it was designed. Performance in situ is more important than a metric such as η_{SD}. The IMPS on Starshine 3 was designed with more power capacity than needed, each component was tested as was the overall system and successfully accomplished its mission when tested in LEO [246,247,270]. The five integrated devices illustrated (Figs. 10.5, 10.6, 10.18, and 10.19) throughout this chapter span a range of applications [82,85,321,326,329]. Notably, only one of them emphasizes $\eta_{overall}$ [326].

Finally, practical concerns need to be addressed. In our recent book chapter [32], we assembled guidance from numerous literature sources to inform readers who may consider the application of such technologies [32,302—308,321]. These concerns include durability and autonomous (continuous) operations [291,302,304,305,308]; the need to tailor device design to specific applications [303,304,306,308,321]; materials and device structure [302,304—306,308,318]; and chemical corrosion of energy storage or devices [303,304,306—308,321]. Monolithic or integrated 2E devices can be straightforward to assemble, thus minimizing costs while minimizing volume, footprints, and power loss, but potentially presenting the most difficult issues with undesirable reactions [303—305,307,333]. Discrete (4E) devices separate solar energy harvesting and either energy storage or other functions, minimizing chemical reaction issues and enabling numerous potential applications [246,302—305,308,332,334,337]. The prominence of 3E (integrated) devices evidence a useful compromise as space and volume can be minimized while also limiting decomposition issues and providing a wide range of potential applications. This compromise of conflicting metrics has resulted in the intense interest in this device format given the number of highly cited (>50) reports [302—305,307,326—328]. It is clear that in order to move forward toward practical application of IPS or PID technologies, what is required is more research and development focused on integrated devices in terms of sharing components between energy harvesting and storage (or active devices), development of reliable (and common) fabrication processes, and manufacturing via scaling up this emerging field of advanced PV applications [128,302—308,317—321,326—330].

10.5.3 Low-cost manufacturing: Materials, methods, modeling, and practical concerns

A final topic that was previewed in Section 10.1 [2,3,13,59,291,339] and called out in Table 10.1 is (novel) low-cost manufacturing. In fact, the USAF ManTech program was, in its day, such an effort [1,28,247,248,340]. The cost of III-V technologies, while not an issue for space applications, remains a significant barrier to large-scale terrestrial utilization, with the possible exception of concentrator technologies [28,59,340−347]. Our prior discussion of tandem (and MJ) devices [1], particularly those including third-generation technologies, included a discussion of issues to consider when manufacturing MJ devices [63,348−357]. We include references related to three topics germane to manufacturing to enable utilization of (mainly) third-generation(-containing) PV technologies: Material considerations and alternative (large-scale) processing approaches [129,358−369], theoretical methods to further explore and enable discovery of optimal manufacturing approaches [130−134], and practical considerations, including logistics and economics [135−140]. It should be noted that numerous references include multiple perspectives [131,132,134,136−139,359,364−366].

As discussed previously [1], crystalline and thin-film (first and second generation) PV are produced using vacuum processing [28,59,141−149,291,340]. Low(er)-cost manufacturing will likely involve solution- or nonvacuum processing, either in part or exclusively [16,17,34,37,38,53,59,150,264,359−368]; these processes will enable the use of lightweight, nontraditional substrates such as polymers, metal foils, and thin glass [1,26,27,59,151−157, 251−255,269,291,293]. A key enabling technology is the development of raw materials or feedstocks for greener, hopefully lower cost, manufacturing approaches such as MOFs (Section 10.4), SSPs (Section 10.5.1), or inks [2,3,16,17,34,35,52,53,55,158−165]. Several candidate or prototype manufacturing approaches for producing PSCs have recently been reported; they are illustrated in Fig. 10.20. Fig. 10.20A−C provides three views of a traditional vacuum approach for a manufacturing prototype specifically designed for codepositing organic−inorganic perovskites via deposition and characterization of MAPbI$_3$ films for PSCs [129]; Fig. 10.20D provides a schematic for a novel high-throughput open-air processing method, rapid spray plasma processing (RSPP), for the scalable production of PSCs [133,137,363]; Fig. 10.20E and F are schematic drawings of two approaches to roll-to-roll (R2R) production of PSCs or modules: Module patterns are directly formed without additional striping procedures (Fig. 10.20E) and module patterns are produced by mechanical or laser scribing (Fig. 10.20F) [365]. While there are definitely intriguing aspects (and legacy) of vacuum processing [129] and open-air manufacturing enabled by RSPP [133,137,363], we will focus on R2R processing for more in-depth consideration.

R2R manufacturing has been recently projected as a viable manufacturing process for not only PSCs but also OPV [136,359,360, 365,367,368]; previous reports of R2R processing of earlier generation

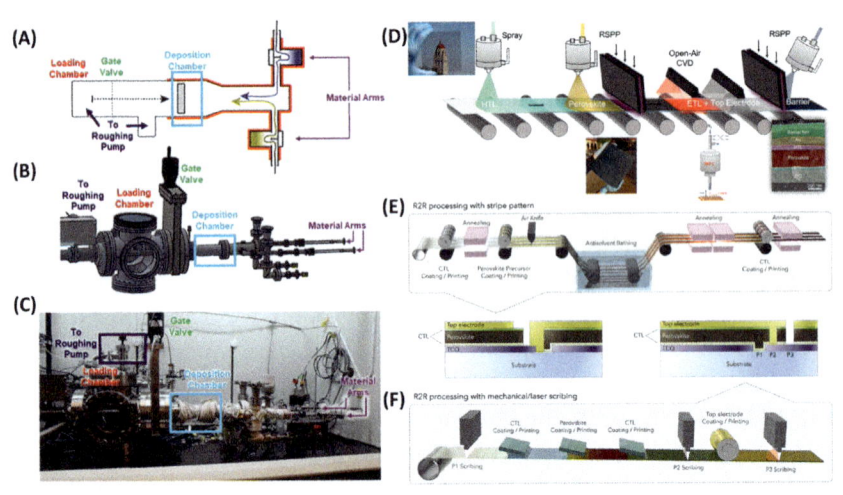

Figure 10.20 Schematics or illustrations of three different manufacturing processing approaches: 1. Design of a vacuum processing technology system; (A) diagram of significant locations within the chamber, including, loading chamber, gate valve, deposition chamber, lines to roughing pump, and material arms; (B) computer-aided design diagram of the constructed system; (C) image of the constructed system highlighting the same significant locations as in the above diagram (A). 2. Schematic of high-throughput open-air production for in-line, fast, and scalable manufacturing of perovskite solar modules using rapid spray plasma processing (RSSP) (D). 3. Roll-to-roll processing method for fabrication of PSCs (and the monolithic structure of a module), using two different approaches to module definition, stripe patterning (E) or scribing (F). *Reproduced from: (A—C) E. Wassweiler, A. Panda, T. Kadosh, T. Nguyen, W.-J. Hsu, et al., Design of a custom vapor transport co-deposition system for scalable production of perovskite solar cells, J. Vac. Sci. Technology A 41 (July 2023) 052801; (D) N. Rolston, A. Sleugh, J.P. Chen, O. Zhao, T.W. Colburn, A.C. Flick, R.H. Dauskardt, Perspectives of open-air processing to enable perovskite solar cell manufacturing, Front. Energy Res. 9 (May 2021) 684082; (E and F) T.-Y. Yang, Y.Y. Kim, J. Seo, Roll-to-roll manufacturing toward lab-to-fab-translation of perovskite solar cells, APL Mater. 9 (2021) 110901, all three articles are distributed under the terms of the Creative Commons (CC-BY 4.0) license which permits unrestricted use, (http://creativecommons.org/licenses/by/4.0/).*

(thin-film) PV technologies had also been reported [370,371]. Table 10.4 includes a brief overview of the compatibility of various deposition methods to R2R processing [359,360,365]; of the seven methods listed, the bottom three (gravure [top of Fig. 10.21] [365], screen, and inkjet printing) are most compatible with R2R processing and are also "green" from the perspective of efficiency of precursor utilization [13]. The bottom of Fig. 10.21 includes five concerns for R2R manufacturing, both process-detail oriented as well as overarching issues [365]. While there are

Table 10.4 Comparison of deposition methods and compatibility with roll-to-roll processing.

Deposition methods	Precursor consumption	Process speed	Compatible with R2R process	Resolution	Layer thickness uniformity
Spin coating	Very high	Low	No	N/A	Good
Blade coating	Low	High	Yes	N/A	Good
Spray coating	High	Low	Yes	Poor	Bad
Slot-die coating	Low	High	Yes	Poor	Very good
Gravure printing	Very low	High	Yes	Very good	Good
Screen printing	Very low	High	Yes	Good	Moderate
Inkjet printing	Very low	Low	Yes	Very good	Good

Source: Adapted from T.-Y. Yang, Y.Y. Kim, J. Seo, Roll-to-roll manufacturing toward lab-to-fab-translation of perovskite solar cells, APL Mater. 9 (2021) 110901, this article is distributed under the terms of the Creative Commons (CC-BY 4.0) license (http://creativecommons.org/licenses/by/4.0), which permits unrestricted use.

certainly numerous options for film deposition methods (Table 10.4) that vary with respect to compatibility with R2R processing [365] as well as multiple precursor chemistries [16,17,34,55,162–165] and device materials and designs, developing a reliable, economically viable overall process will be quite challenging.

As discussed above (10.3), ML/AI methods have made a significant impact on the design and processing of PV materials and devices [10,103–117]. These methods are also being utilized to potentially optimize individual process steps (Fig. 10.22, left) [134], the overall process design (Fig. 10.22, right) [133], and for developing optimized logistics [136,138,139]. Thus, it is likely that a successful manufacturing and sales operation for advanced PV technologies to compete successfully in the marketplace will require advanced computer methodologies that are informed by practical considerations such as raw materials availability and electricity demand [135–140].

Given prior discussions involving future prospects for a variety of PV technologies for large-scale terrestrial applications, we see tandem device technologies utilizing PSCs as the top layer with one of several bottom cell

Figure 10.21 (*Top*) Sophisticated pattern of a perovskite layer fabricated via gravure printing: (A) Master pattern engraved on a printing plate, (B) gravure-printed perovskite layer on a flexible substrate via gravure printing; R2R processing of layers

(*Continued*)

technologies (Si, PSC, OPV, or CIGS) [59,62−68,71,99,132,140, 348−357,372−386]; a Si-based tandem device seems to be the leading candidate for an economically viable manufacturable technology [62,140,348, 350−353,355−357,372,374,375,385]. While developing manufacturable processes as well as considering overarching issues for other technologies such as OPV [131,135,366,367] and CIGS [269,362,376,379] (or kesterite alternatives [36−38]) is certainly worthwhile; these other PV technologies would more likely be viable candidates for use as bottom cells in a tandem device or in a PID or an IPS (Section 10.5.2).

A final but by no means inconsequential consideration is the issue of financial investment required to bring a process (or device) from the laboratory to the marketplace. In our recent chapter on SSPs and commercialization [33], we discussed in some detail the issues surrounding technology transfer and our own experience. We delved into the various metrics of technology readiness level (TRL—NASA term [387−389]) and manufacturing readiness level (MRL—US Dept. of Defense term [390,391]). Transitioning any technology from the "lab-to-fab" (MRL 5/6 to 9) [33,135−140] requires a significant capital investment [392,393], as well as buy-in from multiple stakeholders [391,393], and, most importantly, one or more committed individuals or champions to see the technology process through to a successful conclusion [33].

10.6 Summary and conclusions

A major theme of this chapter has been to address novel concepts through a paradigm that integrates green engineering and materials (processing) principles with biomimicry. We then examined and considered six recent initiatives and advanced concepts into two overarching themes:

◄ into PSCs, (C) Printing of SnO_2 NPs as an ETL, (D) printing of a perovskite layer, (E) printing of P3HT as a HTL, and (F) photograph of fully R2R-processed PSCs. (*Bottom*) Considerations for the successful R2R production of PSCs; Causes (or issues) are shaded gray on top and solutions are on the bottom. *Reproduced from T.-Y. Yang, Y.Y. Kim, J. Seo, Roll-to-roll manufacturing toward lab-to-fab-translation of perovskite solar cells, APL Mater. 9 (2021) 110901; distributed under the terms of the Creative Commons (CC-BY 4.0) license which permits unrestricted use, (http://creativecommons. org/licenses/by/4.0/).*

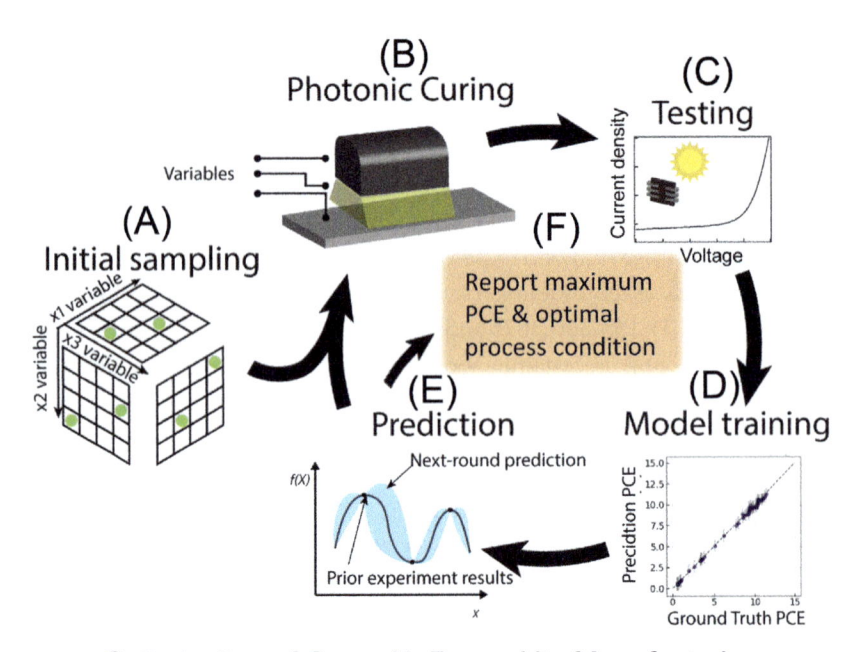

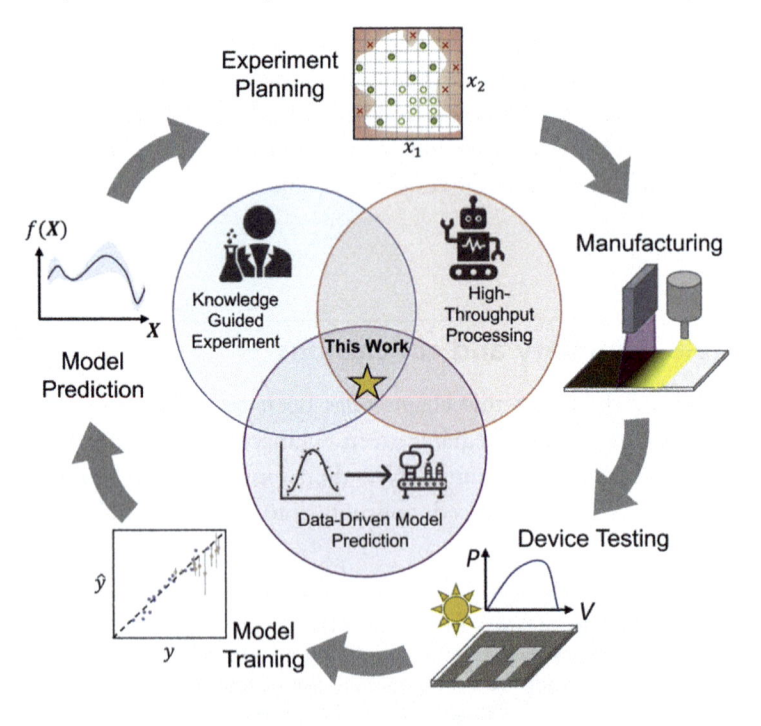

Figure 10.22 (*Top*) Schematic of the Bayesian optimization of PSCs on indium tin oxide-coated willow glass, including (A) initial sampling, (B) photonic curing, (C) testing, (D) model training, and (E) next-round prediction. This workflow iterates until (F) reporting maximum PCE and optimal process conditions. (*Bottom*) A schematic of
(*Continued*)

Nature-inspired and/or sustainable and computer-enabled. Earth-abundant and nontoxic materials, SSPs, and low-cost manufacturing seek to employ efficiency in several domains to minimize energy and materials utilization (reduce waste). ML/AI (a bio-inspired computer application), MOFs, and PIDs/IPSs are relatively recent discoveries that emphasize enabling novel (or optimized) devices and processing through innovation enabled by deep insights into chemistry, materials science, and device physics. Of course, there is an overlap of green/biomimicry principles and the connection to all six of these advanced (computer) concepts, (green) materials, and advanced devices.

The authors would like to emphasize a number of lessons to be imparted to the reader after considering this chapter addressing new directions to explore post-Si solar cell technology. Multijunction III-V technology will almost certainly dominate space applications due to the quite small market share (<1%) and routine PCE approaching 40%; cost is essentially irrelevant given the minimal impact on overall spacecraft and mission costs. While the main focus of this book is PV beyond Si, Si-containing devices are the premier Earth-abundant element; therefore, tandem PV with Si bottom cells is the most attractive from this perspective. PSCs, either as single-junction or as the top cell in a tandem device, or all-perovskite devices, have a very promising future due, in large part, to the enormous available materials and processing space.

Green technologies, biomimicry (nature-inspired), and sustainability (including recycling) provide a useful perspective to inform manufacturing from the perspective of intrinsically lower cost and minimal environmental impacts, including energy efficiency. Novel processing technologies that employ SSPs, particularly MOFs, show great promise to enhance PCE and

◂ the optimization of open-air perovskite manufacturing. The overall five-step workflow includes planning, manufacturing, testing, model training, and prediction. This workflow iterates until the target efficiency is achieved or the maximum experimental budget is reached. The authors demonstrated an efficiency improvement of 18.5% from a device fabricated by RSPP with a limited experimental budget of screening 100 process conditions. *(Left) Reproduced with permission from W. Xu, Z. Liu, R.T. Piper, J.W.P. Hsu, Bayesian Optimization of photonic curing process for flexible perovskite photovoltaic devices, Sol. Energy Mater. Sol. Cells 249 (January 2023) 112055, copyright (2023) Elsevier. (Right) Reproduced with permission from Z. Liu, N. Rolston, A.C. Flick, T. W. Colburn, Z. Ren, A. Sleugh, R.H. Dauskardt, T. Buonassisi, Machine learning with knowledge constraints for process optimization of open-air perovskite solar cell manufacturing, Joule 6 (4) (April 2022) 834–849, copyright (2022) Elsevier.*

long-term durability, particularly of PSCs. Inspired by tandems, PIDs or IPSs have been successfully tested in situ and provide an excellent opportunity for a number of advanced niche applications, with a high-profit margin, using technologies such as OPV, DSSCs, and QD(S)SCs that may not be suitable for large-scale utilization. ML and AI methods will be essential for optimizing device performance and enabling lower cost, reliable manufacturing approaches for advanced PV technologies. Given the criticality of computer-enabled (ML/AI) methods, it is important for research teams to either partner with expert groups (public or private sector or academia), integrate such expertise into their ongoing efforts, or organically develop the expertise. Finally, we encourage our colleagues in the worldwide energy conversion community to continue their impressive level of effort toward new PV materials, devices, and technologies, as applications for other energy or electronic devices may be discovered.

References

[1] A.F. Hepp, R.P. Raffaelle, Photovoltaics overview: Historical background and current technologies, in: S. Sundaram, V. Subramaniam, R.P. Raffaelle, M.K. Nazeeruddin, A. Morales-Acevedo, M. Bernechea Navarro, A.F. Hepp (Eds.), Photovoltaics Beyond Silicon, Elsevier, Cambridge, MA, 2024, pp. 3−74.

[2] K.Y. Cheong, A. Apblett (Eds.), Sustainable Materials and Green Processing for Energy Conversion, Elsevier, Cambridge, MA, USA, 2021, 504 pp.

[3] M.A. Gonzalez, Introduction to green and sustainable chemistry, in: M.A. Abraham (Ed.), Encyclopedia of Sustainable Technologies, Volume 3, Elsevier, 2017, pp. 487−495. Available from: http://doi.org/10.1016/B978-0-12-409548-9.10249-0.

[4] M. Eggermont, V. Shyam, A.F. Hepp (Eds.), Biomimicry for Materials, Design and Habitats: Innovations and Applications, Elsevier, Cambridge, MA, USA, 2022, 590 pp.

[5] R.J. Martín-Palma, A. Lakhtakia, Engineered biomimicry for harvesting solar energy: A bird's eye view, Int. J. Smart Nano Mater. 4 (2) (2013) 83−90. Available from: https://doi.org/10.1080/19475411.2012.663812.

[6] D. Shin, B. Saparov, D.B. Mitzi, Defect engineering in multinary Earth-abundant chalcogenide photovoltaic materials, Adv. Energy Mater. 7 (11) (2017) 1602366. Available from: https://doi.org/10.1002/aenm.201602366.

[7] M. Ikram, R. Malik, R. Raees, M. Imran, F. Wang, et al., Recent advancements and future insight of lead-free non-toxic perovskite solar cells for sustainable and clean energy production: A review, Sustain. Energy Technol. Assess. 53 (A) (2022) 102433. Available from: https://doi.org/10.1016/j.seta.2022.102433.

[8] A. Gholami, M. Ameri, M. Zandi, R.G. Ghoachani, Electrical, thermal and optical modeling of photovoltaic systems: Step-by-step guide and comparative review study, Sustain. Energy Technol. Assess. 49 (2022) 101711. Available from: https://doi.org/10.1016/j.seta.2021.101711.

[9] F. Oba, Y. Kumagai, Design and exploration of semiconductors from first principles: A review of recent advances, Appl. Phys. Express 11 (6) (2018) 060101. Available from: https://iopscience.iop.org/article/10.7567/APEX.11.060101.

[10] A. Mellit, S.A. Kalogirou, Machine learning and deep learning for photovoltaic applications, Artif. Intell. Smart Photovolt. Technol, AIP Publishing (online),

Melville, New York, 2022, pp. 1-1–1-20. Available from: https://doi.org/10.1063/9780735424999.
[11] R. Carson, Silent Spring, 26, Houghton Mifflin Company, 1962, 368 pp.
[12] R.H. Lutts, Chemical fallout: Rachel Carson's *Silent Spring*, radioactive fallout, and the environmental movement, Environ. Rev. 9 (3) (1985) 211–225.
[13] P. Anastas, N. Eghbali, Green chemistry: Principles and practice, Chem. Soc. Rev. 39 (2010) 301–312. Available from: https://doi.org/10.1039/B918763B.
[14] P.T. Anastas, J.B. Zimmerman, Design through the twelve principles of green engineering, Environ. Sci. Technol. 37 (5) (2003) 94A–101A.
[15] M.A. Abraham, N. Nguyen, Green engineering: Defining the principles—results from the Sandestin Conference, Environ. Prog. 22 (4) (2004). Available from: https://doi.org/10.1002/ep.670220410.
[16] S. Aina, N.C. Scott, M.P. Lobera, M. Bernechea, Sustainable solution-processed solar cells based on environment friendly nanocrystals, in: S. Sundaram, V. Subramaniam, R.P. Raffaelle, M.K. Nazeeruddin, A. Morales-Acevedo, M. Bernechea Navarro, A.F. Hepp (Eds.), Photovoltaics Beyond Silicon, Elsevier, Cambridge, MA, 2024, pp. 437–478.
[17] O. Reyes-Vallejo, S. Torres-Arellano, J.L. Aleman-Ramirez, P. Sebastian, Green chemical synthesis of photovoltaic materials, in: S. Sundaram, V. Subramaniam, R.P. Raffaelle, M.K. Nazeeruddin, A. Morales-Acevedo, M. Bernechea Navarro, et al. (Eds.), Photovoltaics Beyond Silicon, Elsevier, Cambridge, MA, 2024, pp. 405–436.
[18] A. Anctil, Life cycle assessment of renewable energy from photovoltaic technologies, in: S. Sundaram, V. Subramaniam, R.P. Raffaelle, M.K. Nazeeruddin, A. Morales-Acevedo, M. Bernechea Navarro, et al. (Eds.), Photovoltaics Beyond Silicon, Elsevier, Cambridge, MA, 2024, pp. 479–496.
[19] J.D. Anderson Jr., A History of Aerodynamics, Cambridge University Press, 1977, pp. 14–27. Available from: https://doi.org/10.1017/CBO9780511607158.
[20] A. du Plessis, C. Broeckhoven, Functional synergy of biomimicry and additive manufacturing: Toward a bio-enhanced engineering approach, in: M. Eggermont, V. Shyam, A.F. Hepp (Eds.), Biomimicry for Materials, Design and Habitats: Innovations and Applications, Elsevier, 2022, pp. 269–289.
[21] J.F.V. Vincent, O.A. Bogatyreva, N.R. Bogatyrev, A. Bowyer, A.-K. Pahl, Biomimetics: Its practice and theory, J. R. Soc. Interface 3 (9) (2006) 471–482. Available from: https://doi.org/10.1098/rsif.2006.0127, https://royalsocietypublishing.org/.
[22] R. Romanofsky, Parallels in communication technology and natural phenomena, in: V. Shyam, M. Eggermont, A.F. Hepp (Eds.), Biomimicry for Aerospace: Technologies and Applications, Elsevier, 2022, pp. 81–101.
[23] Bionics symposium: living prototypes—the key to new technology, Wright-Patterson Air Force Base, Dayton, OH, USA, September 13–15, 1960.
[24] Biomimicry 3.8, DesignLens: Life's principles. <https://biomimicry.net/the-buzz/resources/designlens-lifes-principles/> (accessed 23.03.23).
[25] V. Shyam, M. Eggermont, A.F. Hepp (Eds.), Biomimicry for Aerospace: Technology and Applications, Elsevier, Cambridge, MA, USA, 2022, 504 pp.
[26] A. Apblett, A.R. Barron, A.F. Hepp (Eds.), Nanomaterials via Single-Source Precursors: Synthesis, Processing and Applications, Elsevier, Cambridge, 2022, 628 pp.
[27] S.L. Castro, S.G. Bailey, R.P. Raffaelle, K.K. Banger, A.F. Hepp, Synthesis and characterization of colloidal $CuInS_2$ nanoparticles from a molecular single-source precursor, J. Phys. Chem. B 108 (33) (2004) 12429–12435.
[28] A.F. Hepp, S.G. Bailey, R.P. Raffaelle, Inorganic photovoltaic materials and devices: Past, present, and future, in: S.-S. Sun, N.S. Sariciftci (Eds.), Organic Photovoltaics: Mechanisms, Materials and Devices, CRC Press, Boca Raton, FL, USA, 2005, pp. 19–36.
[29] R.P. Raffaelle, S.L. Castro, A.F. Hepp, S.G. Bailey, Quantum dot solar cells, Prog. Photovolt. Res. Appl. 10 (2002) 433–439.

[30] K.K. Banger, J.A. Hollingsworth, J.D. Harris, J. Cowen, W.E. Buhro, A.F. Hepp, Ternary single-source precursors for polycrystalline thin-film solar cells, Appl. Organomet. Chem. 16 (2002) 617−627. Available from: https://doi.org/10.1002/aoc.362.

[31] S.L. Castro, S.G. Bailey, R.P. Raffaelle, K.K. Banger, A.F. Hepp, Nanocrystalline chalcopyrite materials (CuInS$_2$ and CuInSe$_2$) via low-temperature pyrolysis of molecular single-source precursors, Chem. Mater. 15 (2003) 3142−3147.

[32] A.F. Hepp, P.N. Kumta, O.I. Velikokhatnyi, R.P. Raffaelle, Batteries for integrated power and CubeSats: Recent developments and future prospects, in: P.N. Kumta, A. F. Hepp, M.K. Datta, O.I. Velikokhatnyi (Eds.), Silicon Anode Systems for Lithium-Ion Batteries, Elsevier, Cambridge, MA, USA, 2022, pp. 457−508.

[33] A.F. Hepp, J.D. Harris, A.W. Apblett, A.R. Barron, Commercialization of single-source precursors: Applications, intellectual property, and technology transfer, in: A. Apblett, A.R. Barron, A.F. Hepp (Eds.), Nanomaterials via Single-Source Precursors: Synthesis, Processing and Applications, Elsevier, 2022, pp. 563−600.

[34] M.A. Fraga, D. Amos, S. Sonmezoglu, V. Subramaniam (Eds.), Sustainable Material Solutions for Solar Energy Technologies: Processing Techniques and Applications, 1st Edition, Elsevier, 2021, 647 pp.

[35] J. Ajayan, D. Nirmal, P. Mohankumar, M. Saravanan, M. Jagadesh, L. Arivazhagan, A review of photovoltaic performance of organic/inorganic solar cells for future renewable and sustainable energy technologies, Superlattices Microstructures 143 (2020) 106549. Available from: https://doi.org/10.1016/j.spmi.2020.106549.

[36] A. Wang, M. He, M.A. Green, K. Sun, X. Hao, A critical review on the progress of kesterite solar cells: Current strategies and insights, Adv. Energy Mater. 13 (2) (2023) 2203046. Available from: https://doi.org/10.1002/aenm.202203046.

[37] D.B. Mitzi, O. Gunawan, T.K. Todorov, K. Wang, S. Guha, The path towards a high-performance solution-processed kesterite solar cell, Sol. Energy Mater. Sol. Cell 95 (6) (2011) 1421−1436. Available from: https://doi.org/10.1016/j.solmat.2010.11.028.

[38] S. Abermann, Non-vacuum processed next generation thin film photovoltaics: Towards marketable efficiency and production of CZTS based solar cells, Sol. Energy 94 (2013) 37−70. Available from: https://doi.org/10.1016/j.solener.2013.04.017.

[39] M. Sahu, V.R.M. Reddy, C. Park, P. Sharma, Review article on the lattice defect and interface loss mechanisms in kesterite materials and their impact on solar cell performance, Sol. Energy 230 (2021) 13−58. Available from: https://doi.org/10.1016/j.solener.2021.10.005.

[40] H.S. Nugroho, G. Refantero, N.L.W. Septiani, M. Iqbal, S. Marno, A progress review on the modification of CZTS(e)-based thin-film solar cells, J. Ind. Eng. Chem. 105 (2022) 83−110. Available from: https://doi.org/10.1016/j.jiec.2021.09.010.

[41] A. Abate, Perovskite solar cells go lead free, Joule 1 (4) (2017) 659−664. Available from: https://doi.org/10.1016/j.joule.2017.09.007.

[42] T. Handa, A. Wakamiya, Y. Kanemitsu, Photophysics of lead-free tin halide perovskite films and solar cells, APL Mater. 7 (8) (2019) 080903. Available from: https://doi.org/10.1063/1.5109704.

[43] K. Nishimura, M.A. Kamarudin, D. Hirotani, K. Hamada, Q. Shen, et al., Lead-free tin-halide perovskite solar cells with 13% efficiency, Nano Energy 74 (2020) 104858. Available from: https://doi.org/10.1016/j.nanoen.2020.104858.

[44] M. Wang, W. Wang, B. Ma, W. Shen, L. Liu, et al., Lead-free perovskite materials for solar cells, Nano-Micro Lett. 13 (2021) 62. Available from: https://doi.org/10.1007/s40820-020-00578-z.

[45] M. Zhang, Z. Zhang, H. Cao, T. Zhang, H. Yu, et al., Recent progress in inorganic tin perovskite solar cells, Mater. Today Energy 23 (2022) 100891. Available from: https://doi.org/10.1016/j.mtener.2021.100891.

[46] S. Shalini, R.B. Prabhu, S. Prasanna, T.K. Mallick, S. Senthilarasu, Review on natural dye sensitized solar cells: Operation, materials and methods, Renew. Sustain. Energy Rev. 51 (2015) 1306–1325. Available from: https://doi.org/10.1016/j.rser.2015.07.052.

[47] H.A. Maddah, V. Berry, S.K. Behura, Biomolecular photosensitizers for dye-sensitized solar cells: Recent developments and critical insights, Renew. Sustain. Energy Rev. 121 (2020) 109678. Available from: https://doi.org/10.1016/j.rser.2019.109678.

[48] A. Orona-Navar, I. Aguilar-Hernández, K.D.P. Nigam, A. Cerdán-Pasarán, N. Ornelas-Soto, Alternative sources of natural pigments for dye-sensitized solar cells: Algae, cyanobacteria, bacteria, archaea and fungi, J. Biotechnol. 332 (2021) 29–53. Available from: https://doi.org/10.1016/j.jbiotec.2021.03.013.

[49] R. Concepcion II, J. Alejandrino, C.H. Mendigoria, E. Dadios, A. Bandala, E. Sybingco, et al., Lactuca sativa leaf extract concentration optimization using evolutionary strategy as photosensitizer for TiO$_2$-filmed Grätzel cell, Optik 242 (2021) 166931. Available from: https://doi.org/10.1016/j.ijleo.2021.166931.

[50] A.S. Teja, A. Srivastava, J.A.K. Satrughna, A. Kanwade, M.K. Tiwari, et al., Optimal processing methodology for futuristic natural dye-sensitized solar cells and novel applications, Dye. Pigment. 210 (2023) 110997. Available from: https://doi.org/10.1016/j.dyepig.2022.110997.

[51] M.-G. Ju, M. Chen, Y. Zhou, J. Dai, L. Ma, N.P. Padture, et al., Toward eco-friendly and stable perovskite materials for photovoltaics, Joule 2 (7) (2018) 1231–1241. Available from: https://doi.org/10.1016/j.joule.2018.04.026.

[52] N.P. Holmes, H. Munday, M.G. Barr, L. Thomsen, M.A. Marcus, et al., Unravelling donor–acceptor film morphology formation for environmentally-friendly OPV ink formulations, Green. Chem. 21 (18) (2019) 5090–5103. Available from: https://doi.org/10.1039/C9GC02288K.

[53] B.F. Gonçalves, S. Sadewasser, L.M. Salonen, S. Lanceros-Méndez, Y.V. Kolen'ko, Merging solution processing and printing for sustainable fabrication of Cu(In,Ga)Se$_2$ photovoltaics, Chem. Eng. J. 442 (Part 2) (2022) 136188. Available from: https://doi.org/10.1016/j.cej.2022.136188.

[54] A. Chakraborty, N. Pai, J. Zhao, B.R. Tuttle, A.N. Simonov, V. Pecunia, Rudorffites and beyond: perovskite-inspired silver/copper pnictohalides for next-generation environmentally friendly photovoltaics and optoelectronics, Adv. Funct. Mater. 32 (36) (2022) 2203300. Available from: https://doi.org/10.1002/adfm.202203300.

[55] X. Wu, D. Zhang, X. Wang, X. Jiang, B. Liu, et al., Eco-friendly perovskite solar cells: From materials design to device processing and recycling, EcoMat 5 (7) (2023) e12352. Available from: https://doi.org/10.1002/eom2.12352.

[56] S. Gressler, F. Part, S. Scherhaufer, G. Obersteiner, M. Huber-Humer, Advanced materials for emerging photovoltaic systems – environmental hotspots in the production and end-of-life phase of organic, dye-sensitized, perovskite, and quantum dots solar cells, Sust. Mater. Tech. 34 (2022) e00501. Available from: https://doi.org/10.1016/j.susmat.2022.e00501.

[57] S. Almosni, A. Delamarre, Z. Jehl, D. Suchet, L. Cojocaru, et al., Material challenges for solar cells in the twenty-first century: Directions in emerging technologies, Sci. Technol. Adv. Mater. 19 (1) (2018) 336–369. Available from: https://doi.org/10.1080/14686996.2018.1433439.

[58] M.A. Green, E.D. Dunlop, J. Hohl-Ebinger, M. Yoshita, N. Kopidakis, et al., Solar cell efficiency tables (Version 60), Prog. Photovolt. Res. Appl. 30 (7) (2022) 687–701. Available from: https://doi.org/10.1002/pip.3595.

[59] A.K.-W. Chee, On current technology for light absorber materials used in highly efficient industrial solar cells, Renew. Sustain. Energy Rev. 173 (2023) 113027. Available from: https://doi.org/10.1016/j.rser.2022.113027.

[60] J. Chen, D. Jia, E.M.J. Johansson, A. Hagfeldt, X. Zhang, Emerging perovskite quantum dot solar cells: Feasible approaches to boost performance, Energy Environ. Sci. 14 (2021) 224–261. Available from: https://doi.org/10.1039/D0EE02900A.

[61] Q. Zhao, A. Hazarika, X. Chen, S.P. Harvey, B.W. Larson, et al., High efficiency perovskite quantum dot solar cells with charge separating heterostructure, Nat. Commun. 10 (1) (2019) 2842. Available from: https://doi.org/10.1038/s41467-019-10856-z.

[62] M. Anaya, G. Lozano, M.E. Calvo, H. Míguez, ABX_3 perovskites for tandem solar cells, Joule 1 (4) (2017) 769–793. Available from: https://doi.org/10.1016/j.joule.2017.09.017.

[63] P. Wu, D.T. Gangadharan, M.I. Saidaminov, H. Tan, Roadmap for efficient and stable all-perovskite tandem solar cells from a chemistry perspective, ACS Cent. Sci. 9 (1) (2023) 14–26. Available from: https://doi.org/10.1021/acscentsci.2c01077.

[64] N.E.I. Boukortt, C. Triolo, S. Santangelo, S. Patanè, All-perovskite tandem solar cells: From certified 25% and beyond, Energies 16 (8) (2023) 3519. Available from: https://doi.org/10.3390/en16083519.

[65] A.F. Palmstrom, G.E. Eperon, T. Leijtens, R. Prasanna, S.N. Habisreutinger, et al., Enabling flexible all-perovskite tandem solar cells, Joule 3 (9) (2019) 2193–2204. Available from: https://doi.org/10.1016/j.joule.2019.05.009.

[66] K. Xiao, Y.H. Lin, M. Zhang, R.D.J. Oliver, X. Wang, et al., Scalable processing for realizing 21.7%-efficient all-perovskite tandem solar modules, Science 376 (6594) (2022) 762–767. Available from: https://doi.org/10.1126/science.abn7696.

[67] X. Hu, J. Li, C. Wang, H. Cui, Y. Liu, et al., Antimony potassium tartrate stabilizes wide-bandgap perovskites for inverted 4-T all-perovskite tandem solar cells with efficiencies over 26%, Nano-Micro Lett. 15 (2023) 103. Available from: https://doi.org/10.1007/s40820-023-01078-6.

[68] A.K. Chauhan, P. Kumar, S.N. Sharma, Perovskite solar cells: Past, present, and future, in: S. Sundaram, V. Subramaniam, R.P. Raffaelle, M.K. Nazeeruddin, A. Morales-Acevedo, M. Bernechea Navarro, A.F. Hepp (Eds.), Photovoltaics Beyond Silicon, Elsevier, Cambridge, MA, 2024, pp. 113–164.

[69] N.-G. Park, M. Grätzel, T. Miyasaka (Eds.), Organic-Inorganic Halide Perovskite Photovoltaics: From Fundamentals to Device Architectures, Springer International, Switzerland, 2016, 366 pp. Available from: https://link.springer.com/book/10.1007/978-3-319-35114-8.

[70] J. Bisquert, The Physics of Solar Cells: Perovskites, Organics, and Photovoltaic Fundamentals, CRC Press, Boca Raton, FL, USA (November 2017) 238 pp. https://www.taylorfrancis.com/books/mono/10.1201/b22380/physics-solar-cells-juan-bisquert.

[71] S. Thomas, A. Thankappan (Eds.), Perovskite Photovoltaics: Basic to Advanced Concepts and Implementation, Academic Press, London, UK, 2018, 581 pp. Available from: https://www.sciencedirect.com/book/9780128129159/perovskite-photovoltaics.

[72] D. Mulvaney, R.M. Richards, M.D. Bazilian, E. Hensley, G. Clough, S. Sridhar, Progress towards a circular economy in materials to decarbonize electricity and mobility, Renew. Sust. Energy Rev. 137 (2021) 110604. Available from: https://doi.org/10.1016/j.rser.2020.110604.

[73] L. McMillon-Brown, Biomimetic advances in photovoltaics with potential aerospace applications, in: V. Shyam, M. Eggermont, A.F. Hepp (Eds.), Biomimicry for Aerospace: Technology and Applications, Elsevier, Cambridge, MA, USA, 2022, pp. 291–329.

[74] W.-L. Min, A.P. Betancourt, P. Jiang, B. Jiang, Bioinspired broadband antireflection coatings on GaSb, Appl. Phys. Lett. 92 (2008) 141109. Available from: https://doi.org/10.1063/1.2908221.

[75] G. Shalev, S.W. Schmitt, H. Embrechts, G. Brönstrup, S. Christiansen, Enhanced photovoltaics inspired by the fovea centralis, Sci. Rep. 5 (2015) 8570. Available from: https://doi.org/10.1038/srep08570.

[76] D. Zhang, W. Zhang, J. Gu, T. Fan, Q. Liu, H. Su, et al., Inspiration from butterfly and moth wing scales: Characterization, modeling, and fabrication, Prog. Mater. Sci. 68 (2015) 67–96. Available from: https://doi.org/10.1016/j.pmatsci.2014.10.003.

[77] J. Ma, K.L. Man, S.-U. Guan, T.O. Ting, P.W.H. Wong, Parameter estimation of photovoltaic model via parallel particle swarm optimization algorithm, Int. J. Energy Res. 40 (3) (2016) 343–352. Available from: https://doi.org/10.1002/er.3359.

[78] K. Shanks, S. Senthilarasu, T.K. Mallick, Optics for concentrating photovoltaics: Trends, limits and opportunities for materials and design, Renew. Sustain. Energy Rev. 60 (2016) 394–407. Available from: https://doi.org/10.1016/j.rser.2016.01.089.

[79] C. Karam, C. Guerra-Nuñez, R. Habchi, Z. Herro, N. Abboud, et al., Urchin-inspired ZnO-TiO$_2$ core-shell as building blocks for dye sensitized solar cells, Mater. Des. 126 (2017) 314–321. Available from: https://doi.org/10.1016/j.matdes.2017.04.019.

[80] Z. Huang, C. Cai, L. Kuai, T. Li, M. Huttula, W. Cao, Leaf-structure patterning for antireflective and self-cleaning surfaces on Si-based solar cells, Sol. Energy 159 (2018) 733–741. Available from: https://doi.org/10.1016/j.solener.2017.11.020.

[81] M.V. da Rocha, L.P. Sampaio, S.A.O. da Silva, Comparative analysis of MPPT algorithms based on Bat algorithm for PV systems under partial shading condition, Sust. Energy Tech. Assess. 40 (2020) 100761. Available from: https://doi.org/10.1016/j.seta.2020.100761.

[82] X. Meng, Z. Cai, Y. Zhang, X. Hu, Z. Xing, et al., Bio-inspired vertebral design for scalable and flexible perovskite solar cells, Nat. Commun. 11 (2020) 3016. Available from: https://doi.org/10.1038/s41467-020-16831-3.

[83] T. Pickering, K. Shanks, S. Sundaram, Modelling technique and analysis of porous antireflective coatings for reducing wide angle reflectance of thin-film solar cells, J. Opt. 23 (2) (2021) 025901. Available from: https://doi.org/10.1088/2040-8986/abeaec.

[84] M.A. Sankar, Surface functionalization of bio-inspired nanostructures in organic photovoltaics and electronics based on feathers of avian species for inherent UV and IR reflection, Mater. Today: Proc. 44 (6) (2021) 4576–4582. Available from: https://doi.org/10.1016/j.matpr.2020.10.823.

[85] Y. Zhu, L. Shu, Q. Zhang, Y. Zhu, S. Poddar, et al., Moth eye-inspired highly efficient, robust, and neutral-colored semitransparent perovskite solar cells for building-integrated photovoltaics, EcoMat 3 (4) (2021) e12117. Available from: https://doi.org/10.1002/eom2.12117.

[86] D.T. Vu, N.M. Kieu, T.Q. Tien, T.P. Nguyen, H. Vu, S. Shin, et al., Solar concentrator bio-inspired by the superposition compound eye for high-concentration photovoltaic system up to thousands fold factor, Energies 15 (9) (2022) 3406. Available from: https://doi.org/10.3390/en15093406.

[87] G.E. Eperon, D. Bryant, J. Troughton, S.D. Stranks, M.B. Johnston, et al., Efficient, semitransparent neutral-colored solar cells based on microstructured formamidinium lead trihalide perovskite, J. Phys. Chem. Lett. 6 (1) (2015) 129–138. Available from: https://doi.org/10.1021/jz502367k.

[88] L.M. Terman, Spectral response of solar-cell structures, Solid-State Electron. 2 (1) (1961) 1–7. Available from: https://doi.org/10.1016/0038-1101(61)90050-8.

[89] J.G. Fossum, Computer-aided numerical analysis of silicon solar cells, Solid-State Electron. 19 (1976) 269–277.

[90] M. Burgelman, P. Nollet, S. Degrave, Modelling polycrystalline semiconductor solar cells, Thin Solid. Films 361 (2000) 527–532. Available from: https://doi.org/10.1016/S0040-6090(99)00825-1.

[91] P. Mahawela, S. Jeedigunta, C.S. Ferekides, D.L. Morel, Analysis of carrier transport phenomena in high band gap II−VI-based MIS photovoltaic devices, Thin Solid. Films 431 (2003) 461−465. Available from: https://doi.org/10.1016/S0040-6090(03)00168-8.

[92] T. Ma, H. Yang, L. Lu, Solar photovoltaic system modeling and performance prediction, Renew. Sustain. Energy Rev. 36 (2014) 304−315. Available from: https://doi.org/10.1016/j.rser.2014.04.057.

[93] O.K. Simya, A. Mahaboobbatcha, K. Balachander, A comparative study on the performance of kesterite based thin film solar cells using SCAPS simulation program, Superlattices Microstruct. 82 (2015) 248−261. Available from: https://doi.org/10.1016/j.spmi.2015.02.020.

[94] N. Khoshsirat, N.A.M. Yunus, M.N. Hamidon, S. Shafie, N. Amin, Analysis of absorber layer properties effect on CIGS solar cell performance using SCAPS, Optik 126 (2015) 681−686. Available from: https://doi.org/10.1016/j.ijleo.2015.02.037.

[95] H. Heriche, Z. Rouabah, N. Bouarissa, New ultra thin CIGS structure solar cells using SCAPS simulation program, Int. J. Hydrog. Energy 42 (2017) 9524−9532. Available from: https://doi.org/10.1016/j.ijhydene.2017.02.099.

[96] K. Kim, J. Gwak, S.K. Ahn, Y.-J. Eo, J.H. Park, et al., Simulations of chalcopyrite/c-Si tandem cells using SCAPS-1D, Sol. Energy 145 (2017) 52−58. Available from: https://doi.org/10.1016/j.solener.2017.01.031.

[97] M. Kumar, A. Raj, A. Kumar, A. Anshul, An optimized lead-free formamidinium Sn-based perovskite solar cell design for high power conversion efficiency by SCAPS simulation, Optical Mater. 108 (2020) 110213. Available from: https://doi.org/10.1016/j.optmat.2020.110213.

[98] S.S. Soly, A.D.D. Dwivedi, Numerical simulation and performance analysis of InGaP, GaAs, Ge single junction and InGaP/GaAs/Ge triple junction solar cells, Mater. Today: Proc. 39 (5) (2021) 2050−2055. Available from: https://doi.org/10.1016/j.matpr.2020.11.056.

[99] H.Q. Tan, X. Zhao, E. Birgersson, F. Lin, H. Xue, Optoelectronic modeling and sensitivity analysis of a four-terminal all-perovskite tandem solar cell - Identifying pathways to improve efficiency, Sol. Energy 216 (2021) 589−600. Available from: https://doi.org/10.1016/j.solener.2021.01.021.

[100] J. Smucker, J. Gong, A comparative study on the band diagrams and efficiencies of silicon and perovskite solar cells using wxAMPS and AMPS-1D, Sol. Energy 228 (2021) 187−199. Available from: https://doi.org/10.1016/j.solener.2021.09.066.

[101] X. Meng, T. Tang, R. Zhang, K. Liu, W. Li, et al., Optimization of germanium-based perovskite solar cells by SCAPS simulation, Optical Mater. 128 (2022) 112427. Available from: https://doi.org/10.1016/j.optmat.2022.112427.

[102] A.P. Shyma, R. Sellappan, Computational probing of tin-based lead-free perovskite solar cells: Effects of absorber parameters and various electron transport layer materials on device performance, Materials 15 (2022) 7859. Available from: https://doi.org/10.3390/ma15217859.

[103] F. Li, X. Peng, Z. Wang, Y. Zhou, Y. Wu, M. Jiang, et al., Machine learning (ML)-assisted design and fabrication for solar cells, Energy Environ. Mater. 2 (2019) 280−291. Available from: https://doi.org/10.1002/eem2.12049.

[104] F. Häse, L.M. Roch, P. Friederich, A. Aspuru-Guzik, Designing and understanding light-harvesting devices with machine learning, Nat. Commun. 11 (2020) 4587. Available from: https://doi.org/10.1038/s41467-020-17995-8.

[105] W. Sha, Y. Guo, Q. Yuan, S. Tang, X. Zhang, et al., Artificial intelligence to power the future of materials science and engineering, Adv. Intell. Syst. 2 (4) (2020) 1900143. Available from: https://doi.org/10.1002/aisy.201900143.

[106] Y. Liu, O.C. Asan, Z. Pan, L. An, Machine learning for advanced energy materials, Energy AI 3 (2021) 100049. Available from: https://doi.org/10.1016/j.egyai.2021.100049.
[107] Y. Juan, Y. Dai, Y. Yang, J. Zhang, Accelerating materials discovery using machine learning, J. Mater. Sci. Tech. 79 (2021) 178−190. Available from: https://doi.org/10.1016/j.jmst.2020.12.010.
[108] K. Choudhary, B. DeCost, C. Chen, A. Jain, F. Tavazza, et al., Recent advances and applications of deep learning methods in materials science, npj Comput. Mater. 8 (2022) 59. Available from: https://doi.org/10.1038/s41524-022-00734-6.
[109] H.F. Mateo Romero, M.Á. González Rebollo, V. Cardeñoso-Payo, V. Alonso Gómez, A. Redondo Plaza, R.T. Moyo, et al., Applications of artificial intelligence to photovoltaic systems: A review, Appl. Sci. 12 (2022) 10056. Available from: https://doi.org/10.3390/app121910056.
[110] T. Wu, J. Wang, Deep mining stable and nontoxic hybrid organic−inorganic perovskites for photovoltaics via progressive machine learning, ACS Appl. Mater. Interfaces 12 (52) (2020) 57821−57831. Available from: https://doi.org/10.1021/acsami.0c10371.
[111] S. Lu, Q. Zhou, Y. Ouyang, Q. Li, J. Wang, Accelerated discovery of stable lead-free hybrid organic-inorganic perovskites via machine learning, Nat. Commun. 9 (2018) 3405. Available from: https://doi.org/10.1038/s41467-018-05761-w.
[112] S. Sun, N.T.P. Hartono, Z.D. Ren, F. Oviedo, A.M. Buscemi, et al., Accelerated development of perovskite-inspired materials via high-throughput synthesis and machine-learning diagnosis, Joule 3 (6) (2019) 1437−1451. Available from: https://doi.org/10.1016/j.joule.2019.05.014.
[113] T. Wu, J. Wang, Global discovery of stable and non-toxic hybrid organic-inorganic perovskites for photovoltaic systems by combining machine learning method with first principle calculations, Nano Energy 66 (2019) 104070. Available from: https://doi.org/10.1016/j.nanoen.2019.104070.
[114] L. Zhang, M. He, S. Shao, Machine learning for halide perovskite materials, Nano Energy 78 (2020) 105380. Available from: https://doi.org/10.1016/j.nanoen.2020.105380.
[115] J. Kirman, A. Johnston, D.A. Kuntz, M. Askerka, Y. Gao, et al., Machine-learning-accelerated perovskite crystallization, Matter 2 (2020) 938−947. Available from: https://doi.org/10.1016/j.matt.2020.02.012.
[116] B. Yılmaz, R. Yıldırım, Critical review of machine learning applications in perovskite solar research, Nano Energy 80 (2020) 105546. Available from: https://doi.org/10.1016/j.nanoen.2020.105546.
[117] Y. Zhang, X. Xu, Machine learning lattice constants for cubic perovskite A_2XY_6 compounds, J. Solid. State Chem. 291 (2020) 121558. Available from: https://doi.org/10.1016/j.jssc.2020.121558.
[118] H. Park, R. Mall, A. Ali, S. Sanvito, H. Bensmail, F. El-Mellouhi, Importance of structural deformation features in the prediction of hybrid perovskite bandgaps, Comput. Mater. Sci. 184 (2020) 109858. Available from: https://doi.org/10.1016/j.commatsci.2020.109858.
[119] P. Omprakash, B. Manikandan, A. Sandeep, R. Shrivastava, P. Viswesh, D.B. Panemangalore, Graph representational learning for bandgap prediction in varied perovskite crystals, Comput. Mater. Sci. 196 (2021) 110530. Available from: https://doi.org/10.1016/j.commatsci.2021.110530.
[120] J. Chen, M. Feng, C. Zha, C. Shao, L. Zhang, L. Wang, Machine learning-driven design of promising perovskites for photovoltaic applications: A review, Surf. Interfaces 35 (2022) 102470. Available from: https://doi.org/10.1016/j.surfin.2022.102470.

[121] D. Padula, J.D. Simpson, A. Troisi, Combining electronic and structural features in machine learning models to predict organic solar cells properties, Mater. Horiz. 6 (2) (2019) 343−349. Available from: https://doi.org/10.1039/C8MH01135D.

[122] T. Wang, G. Kupgan, J.-L. Brédas, Organic photovoltaics: Relating chemical structure, local morphology, and electronic properties, Trends Chem. 2 (6) (2020) 535−554. Available from: https://doi.org/10.1016/j.trechm.2020.03.006.

[123] X. Rodríguez-Martínez, E. Pascual-San-José, M. Campoy-Quiles, Accelerating organic solar cell material's discovery: High-throughput screening and big data, Energy Environ. Sci. 14 (2021) 3301−3322. Available from: https://doi.org/10.1039/D1EE00559F.

[124] M. Rafiq, M. Salim, S. Noreen, R.A. Khera, S. Noor, U. Yaqoob, et al., End-capped modification of dithienosilole based small donor molecules for high performance organic solar cells using DFT approach, J. Mol. Liq. 345 (2022) 118138. Available from: https://doi.org/10.1016/j.molliq.2021.118138.

[125] P. Mahawela, G. Sivaraman, S. Jeediguna, J. Gaduputi, M. Ramalingam, et al., II−VI compounds as the top absorbers in tandem solar cell structures, Mater. Sci. Eng. B 116 (2005) 283−291. Available from: https://doi.org/10.1016/j.mseb.2004.05.054.

[126] H. Choubisa, P. Todorović, J.M. Pina, D.H. Parmar, Z. Li, O. Voznyy, et al., Interpretable discovery of semiconductors with machine learning, npj Comput. Mater. 9 (2023) 117. Available from: https://doi.org/10.1038/s41524-023-01066-9.

[127] I. Martin, K. Crowley, A.F. Hepp, Thin film solar cells and arrays for space power, in: R.P. Raffaelle, S.G. Bailey, D.C. Ferguson, S. Durbin, A.F. Hepp (Eds.), Space Photovoltaics: Materials, Missions and Alternative Technologies, Elsevier, Cambridge, MA USA, 2023, pp. 215−263.

[128] Y. Yang, M.T. Hoang, A. Bhardwaj, M. Wilhelm, S. Mathur, H. Wang, Perovskite solar cells based self-charging power packs: Fundamentals, applications and challenges, Nano Energy 94 (2022) 106910. Available from: https://doi.org/10.1016/j.nanoen.2021.106910.

[129] E. Wassweiler, A. Panda, T. Kadosh, T. Nguyen, W.-J. Hsu, et al., Design of a custom vapor transport co-deposition system for scalable production of perovskite solar cells, J. Vac. Sci. Technology A 41 (2023) 052801. Available from: https://doi.org/10.1116/6.0002668.

[130] G. Hoppe, E. Alvarez-Brito, G. Emanuel, J. Nekarda, M. Diehl, R. Preu, et al., Overcoming throughput limitations of laser systems in solar cell manufacturing via on-the-fly processing using polygon scanners, Energy Technology (Published online (2023) 17. Available from: https://doi.org/10.1002/ente.202300445.

[131] A. Gambhir, P. Sandwell, J. Nelson, The future costs of OPV − A bottom-up model of material and manufacturing costs with uncertainty analysis, Sol. Energy Mater. Sol. Cells 156 (2016) 49−58. Available from: https://doi.org/10.1016/j.solmat.2016.05.056.

[132] M.I. Hossain, A.M. Saleque, S. Ahmed, I. Saidjafarzoda, M. Shahiduzzaman, et al., Perovskite/perovskite planar tandem solar cells: A comprehensive guideline for reaching energy conversion efficiency beyond 30%, Nano Energy 79 (2021) 105400. Available from: https://doi.org/10.1016/j.nanoen.2020.105400.

[133] Z. Liu, N. Rolston, A.C. Flick, T.W. Colburn, Z. Ren, A. Sleugh, et al., Machine learning with knowledge constraints for process optimization of open-air perovskite solar cell manufacturing, Joule 6 (4) (2022) 834−849. Available from: https://doi.org/10.1016/j.joule.2022.03.003.

[134] W. Xu, Z. Liu, R.T. Piper, J.W.P. Hsu, Bayesian optimization of photonic curing process for flexible perovskite photovoltaic devices, Sol. Energy Mater. Sol. Cells 249 (2023) 112055. Available from: https://doi.org/10.1016/j.solmat.2022.112055.

[135] C.J. Mulligan, M. Wilson, G. Bryant, B. Vaughan, X. Zhou, et al., A projection of commercial-scale organic photovoltaic module costs, Sol. Energy Mater. Sol. Cells 120 (A) (2014) 9−17. Available from: https://doi.org/10.1016/j.solmat.2013.07.041.

[136] M.L. Chang, A.W.Y. Ho-Baillie, D. Vak, M. Gao, M.A. Green, R.J. Egan, Manufacturing cost and market potential analysis of demonstrated roll-to-roll perovskite photovoltaic cell processes, Sol. Energy Mater. Sol. Cells 174 (2018) 314−324. Available from: https://doi.org/10.1016/j.solmat.2017.08.038.

[137] N. Rolston, A. Sleugh, J.P. Chen, O. Zhao, T.W. Colburn, A.C. Flick, et al., Perspectives of open-air processing to enable perovskite solar cell manufacturing, Front. Energy Res. 9 (2021) 684082. Available from: https://doi.org/10.3389/fenrg.2021.684082.

[138] H.K.R. Rao, E. Gemechu, U. Thakur, K. Shankar, A. Kumar, Techno-economic assessment of titanium dioxide nanorod-based perovskite solar cells: From lab-scale to large-scale manufacturing, Appl. Energy 298 (2021) 117251. Available from: https://doi.org/10.1016/j.apenergy.2021.117251.

[139] R. Ichwani, S. Price, O.K. Oyewole, R. Neamtu, W.O. Soboyejo, Using machine learning for prediction of spray coated perovskite solar cells efficiency: From experimental to theoretical models, Mater. Design 233 (2023) 112161. Available from: https://doi.org/10.1016/j.matdes.2023.112161.

[140] L. Wang, Y. Zhang, M. Kim, M. Wright, R. Underwood, R.S. Bonilla, et al., Sustainability evaluations on material consumption for terawatt-scale manufacturing of silicon-based tandem solar cells, Prog. Photovolt. Res. Appl. (Published online (2023) 21. Available from: https://doi.org/10.1002/pip.3687.

[141] M.A. Green, The future of crystalline silicon solar cells, Prog. Photovolt. Res. Appl. 8 (2000) 127−139. add URL.

[142] L. El Chaar, L.A. lamont, N. El Zein, Review of photovoltaic technologies, Renewable and Sust. Energy Rev. 15 (5) (2011) 2165−2175. Available from: https://doi.org/10.1016/j.rser.2011.01.004.

[143] A. Goetzberger, J. Luther, G. Willek, Solar cells: Past, present, future, Solar Energy Mater. Solar Cells 74 (1) (2002). Available from: https://doi.org/10.1016/S0927-0248(02)00042-9.

[144] S. Guha, J. Yang, A. Banerjee, Amorphous silicon alloy photovoltaic research − present and future, Prog. Photovolt. Res. Appl. 8 (2000) 141−150.

[145] C. Battaglia, A. Cuevas, S. De Wolf, High-efficiency crystalline silicon solar cells: Status and perspectives, Energy and Environmental Science 9 (5) (2016) 1552−1576.

[146] T. Matsui, A. Bidiville, K. Maejima, H. Sai, T. Koida, et al., High-efficiency amorphous silicon solar cells: Impact of deposition rate on metastability, Appl. Phys. Lett 106 (52) (2015) 053901.

[147] U.S. Solar Panel Manufacturers. <https://www.solarpowerworldonline.com/u-s-solar-panel-manufacturers/> (accessed 30.05.23).

[148] J. Gwak, M. Lee, J.H. Yun, S. Ahn, A. Cho, et al., Selenium flux effect on Cu(In, Ga)Se$_2$ thin films grown by a 3-stage Co-evaporation process, Isr. J. Chem. 55 (10) (2015) 1115−1122. Available from: https://doi.org/10.1002/ijch.201500011.

[149] <https://www.pv-magazine.com/2019/01/21/solar-frontier-hits-new-cis-cell-efficiency-record/> (accessed 30.05.23).

[150] S. Bellani, A. Bartolotta, A. Agresti, G. Calogero, G. Grancini, et al., Solution-processed two-dimensional materials for next-generation photovoltaics, Chem. Soc. Rev. 50 (2021) 11870−11965. Available from: https://doi.org/10.1039/D1CS00106J.

[151] A.F. Hepp, J.S. McNatt, J.E. Dickman, D.L. Morel, C.S. Ferekides, et al., Thin-film solar cells on polymer substrates for space power, in: Proc. 2nd Int. Ener. Conv. Eng. Conf., Providence, RI, AIAA-2004−5537, August 2004, 18 pp.

[152] H.P. Mahabaduge, W.L. Rance, J.M. Burst, M.O. Reese, D.M. Meysing, et al., High-efficiency, flexible CdTe solar cells on ultra-thin glass substrates, Appl. Phys. Lett 106 (13) (2015) 133501.
[153] R. Carron, S. Nishiwaki, T. Feurer, R. Hertwig, E. Avancini, et al., Advanced alkali treatments for high-efficiency $Cu(In,Ga)Se_2$ solar cells on flexible substrates, Adv. Energy Mater. 9 (24) (2019) 1900408.
[154] Y. Kawano, J. Chantana, T. Nishimura, A. Mavlonov, T. Minemoto, Manipulation of [Ga]/([Ga] + [In]) profile in 1.4-μm-thick $Cu(In,Ga)Se_2$ thin film on flexible stainless steel substrate for enhancing short-circuit current density and conversion efficiency of its solar cell, Solar Energy 204 (2020) 231−237. Available from: https://doi.org/10.1016/j.solener.2020.04.069.
[155] L. Zortea, S. Nishiwaki, T.P. Weiss, S. Haass, J. Perrenoud, L. Greuter, et al., Cu (In,Ga)Se_2 solar cells on low cost mild steel substrates, Solar Energy 175 (2018) 25−30. Available from: https://doi.org/10.1016/j.solener.2017.12.057.
[156] J.C. Perrenoud, Low temperature grown CdTe thin film solar cells for the application on flexible substrates, Ph.D. Thesis, ETH Zürich, CH, No. 20460, 2012. <https://doi.org/10.3929/ethz-a-007339371>.
[157] S. Günes, H. Neugebauer, N.S. Sariciftci, Conjugated polymer-based organic solar cells, Chem. Rev. 107 (4) (2007) 1324−1338. Available from: https://doi.org/10.1021/cr050149z.
[158] M.S.A. Kamel, A. Al-jumaili, M. Oelgemöller, M.V. Jacob, Inorganic nanoparticles to overcome efficiency inhibitors of organic photovoltaics: An in-depth review, Renewable and Sustainable Energy Reviews 166 (2022) 112661. Available from: https://doi.org/10.1016/j.rser.2022.112661.
[159] A. Singh, M. Katiyar, A. Garg, Understanding the formation of PEDOT:PSS films by ink-jet printing for organic solar cell applications, RSC Advances 5 (96) (2015) 78677−78685. Available from: https://doi.org/10.1039/C5RA11032G.
[160] K.-J. Kim, C. Pan, S. Bansal, R. Malhotra, D.-H. Kim, C.-H. Chang, Scalably synthesized environmentally benign, aqueous-based binary nanoparticle inks for $Cu_2ZnSn(S,Se)_4$ photovoltaic cells achieving over 9% efficiency, Sustainable Energy and Fuels 1 (2) (2017) 267−274. Available from: https://doi.org/10.1039/C6SE00035E.
[161] B.S. Yadav, S.R. Day, S.R. Dhage, Effective ink-jet printing of aqueous ink for Cu (In, Ga) Se_2 thin film absorber for solar cell application, Solar Energy 179 (2019) 363−370. Available from: https://doi.org/10.1016/j.solener.2019.01.003.
[162] J. Li, J. Dagar, O. Shargaieva, M.A. Flatken, H. Köbler, et al., 20.8% slot-die coated $MAPbI_3$ perovskite solar cells by optimal DMSO-content and age of 2-ME based precursor inks, Adv. Energy Mater. 11 (10) (2021) 2003460. Available from: https://doi.org/10.1002/aenm.202003460.
[163] B. Wilk, S. Öz, E. Radicchi, F. Ünlü, T. Ahmad, et al., Green solvent-based perovskite precursor development for ink-jet printed flexible solar cells, ACS Sustainable Chem. Eng 10 (9) (2021) 3920−3930. Available from: https://doi.org/10.1021/acssuschemeng.0c09208.
[164] C.S. Pathak, G. Paramasivam, F. Mathies, K. Hirselandt, V. Schröder, et al., PTB7 as an ink-additive for spin-coated versus inkjet-printed perovskite solar cells, ACS Appl. Energy Mater 5 (4) (2022) 4085−4095. Available from: https://doi.org/10.1021/acsaem.1c03262.
[165] S. Mandati, R.K. Battula, G. Veerappan, E. Ramasamy, A promising scalable bar coating approach using a single crystal-derived precursor ink for high-performance large-area perovskite solar cells, Materials Today Chemistry 29 (2023) 101415. Available from: https://doi.org/10.1016/j.mtchem.2023.101415.
[166] The Materials Project, <https://next-gen.materialsproject.org/> (accessed 24.07.23).
[167] <https://www.nature.com/collections/egijhgcdcd> (accessed 24.07.23).

[168] <https://www.mrs.org/meetings-events/fall-meetings-exhibits/2023-mrs-fall-meeting> (accessed 24.07.23).
[169] <https://www.bnl.gov/compsci/c3d/> (accessed 24.07.23).
[170] <https://www.matse.psu.edu/computational-materials> (accessed 24.07.23).
[171] <https://sunthetics.io/> (accessed 24.07.23).
[172] <https://openlearning.mit.edu/> (accessed 24.07.23).
[173] <https://nanohub.org/resources/mseml> (accessed 24.07.23).
[174] <https://nanohub.org/> (accessed 24.07.23).
[175] K. Biradha, A. Ramanan, J.J. Vittal, Coordination polymers versus metal − organic frameworks, Cryst. Growth Des. 9 (7) (2009) 2969−2970. Available from: https://doi.org/10.1021/cg801381p.
[176] S.R. Batten, N.R. Champness, X.-M. Chen, J. Garcia-Martinez, S. Kitagawa, et al., Coordination polymers, metal−organic frameworks and the need for terminology guidelines, CrystEngComm 14 (9) (2012) 3001−3004. Available from: https://doi.org/10.1039/C2CE06488J.
[177] S.R. Batten, N.R. Champness, Coordination polymers and metal−organic frameworks: Materials by design, Phil. Trans. R. Soc. A 375 (2017) 20160032. Available from: https://doi.org/10.1098/rsta.2016.0032.
[178] E. Horowitz, M. Tryon, R.G. Christensen, T.P. Perros, Thermogravimetric study of some new transition metal−schiff base coordination polymers, J. Appl. Polym. Sci. 9 (7) (1965) 2321−2336.
[179] W.C. Grinonneau, P.L. Chapman, A.G. Menke, F. Walmsley, Transition metal coordination polymers of phenylphosphonic acid, J. Inorg. Nucl. Chem. 33 (9) (1971) 3011−3017.
[180] O.M. Yaghi, H. Li, Hydrothermal synthesis of a metal-organic framework containing large rectangular channels, J. Am. Chem. Soc. 117 (41) (1995) 10401−10402.
[181] L. Jiao, J.Y.R. Seow, W.S. Skinner, Z.L. Wang, H.-L. Jiang, Metal−organic frameworks: structures and functional applications, Mater. Today 27 (2019) 43−68. Available from: https://doi.org/10.1016/j.mattod.2018.10.038.
[182] G.E. Gomez, F. Roncaroli, Photofunctional metal-organic framework thin films for sensing, catalysis and device fabrication, Inorganica Chim. Acta 513 (2020) 119926. Available from: https://doi.org/10.1016/j.ica.2020.119926.
[183] O.M. Yaghi, J.L.C. Rowsell, Metal-organic frameworks: A new class of porous materials, Microporous Mesoporous Mater. 73 (1−2) (2004) 3−14.
[184] H. Furukawa, K.E. Cordova, M. O'Keeffe, O.M. Yaghi, The chemistry and applications of metal-organic frameworks, Science 341 (6149) (2013) 1230444.
[185] P.S. Sharanyakanth, M. Radhakrishnan, Synthesis of metal-organic frameworks (MOFs) and its application in food packaging: A critical review, Trends Food Sci. & Technol. 104 (2020) 102−116. Available from: https://doi.org/10.1016/j.tifs.2020.08.004.
[186] J.Y. Lu, Crystal engineering of Cu-containing metal-organic coordination polymers under hydrothermal conditions, Coord. Chem. Rev. 246 (1−2) (2003) 327−347.
[187] D.G. Madden, H.S. Scott, A. Kumar, K.-J. Chen, R. Sanii, et al., Flue-gas and direct-air capture of CO_2 by porous metal−organic materials, Phil. Trans. R. Soc. A 375 (2017) 20160025. Available from: https://doi.org/10.1098/rsta.2016.0025.
[188] S. Chuhadiya, H. Himanshu, D. Suthar, S.L. Patel, M.S. Dhaka, Metal organic frameworks as hybrid porous materials for energy storage and conversion devices: A review, Coord. Chem. Rev. 446 (2021) 214115.
[189] Y. Li, A. Pang, C. Wang, M. Wei, Metal-organic frameworks: Promising materials for improving the open circuit voltage of dye-sensitized solar cells, J. Mater. Chem. 21 (43) (2011) 17259−17264. Available from: https://doi.org/10.1039/C1JM12754C.
[190] S.-H. Hsu, C.-T. Li, H.-T. Chien, R.R. Salunkhe, N. Suzuki, et al., Platinum-free counter electrode comprised of metal-organic-framework (MOF)-derived cobalt

[218] R. Tang, R. Yin, S. Zhou, T. Ge, Z. Yuan, L. Zhang, et al., Layered MoS_2 coupled MOFs-derived dual-phase TiO_2 for enhanced photoelectrochemical performance, J. Mater. Chem. A 5 (2017) 4962−4971.

[219] A. Gu, W. Xiang, T. Wang, S. Gu, X. Zhao, Enhance photovoltaic performance of tris(2,2′-bipyridine) cobalt(II)/(III) based dye-sensitized solar cells via modifying TiO_2 surface with metal-organic frameworks, Sol. Energy 147 (2017) 126−132. Available from: https://doi.org/10.1016/j.solener.2017.03.045.

[220] A.S.A. Ahmed, W. Xiang, I.S. Amiinu, X. Zhao, Zeolitic-imidazolate-framework (ZIF-8)/PEDOT:PSS composite counter electrode for low cost and efficient dye-sensitized solar cells, N. J. Chem. 42 (2018) 17303−17310.

[221] J. Ou, B. Hu, S. He, W. Wang, Y. Han, Transparent $Cu_{2-x}Se$@N-doped carbon nanosheets as a cathode for Co(III/II)-mediated bifacial dye-sensitized solar cells, Sol. Energy 201 (2020) 693−700.

[222] A.S.A. Ahmed, W. Xiang, I.S. Amiinu, Z. Li, R. Yu, X. Zhao, ZnO-nitrogen doped carbon derived from a zeolitic imidazolate framework as an efficient counter electrode in dye-sensitized solar cells, Sustain. Energy Fuels 3 (2019) 1976−1987.

[223] A.N. Yang, J.T. Lin, C.T. Li, Electroactive and sustainable Cu-MOF/PEDOT composite electrocatalysts for multiple redox mediators and for high-performance dye-sensitized solar cells, ACS Appl. Mater. Interfaces 13 (2021) 8435−8444.

[224] X. Wang, Y. Xie, B. Bateer, K. Pan, X. Zhang, J. Wu, et al., $CoSe_2$/N-doped carbon hybrid derived from ZIF-67 as high-efficiency counter electrode for dye-sensitized solar cells, ACS Sustain, Chem. Eng. 7 (2019) 2784−2791.

[225] A.S.A. Ahmed, W. Xiang, Z. Li, I.S. Amiinu, X. Zhao, Yolk-shell m-SiO_2@ Nitrogen doped carbon derived zeolitic imidazolate framework high efficient counter electrode for dye-sensitized solar cells, Electrochim. Acta 292 (2018) 276−284.

[226] T.M.H. Nguyen, C.W. Bark, Synthesis of cobalt-doped $TiO2$ based on metal−organic frameworks as an effective electron transport material in perovskite solar cells, ACS Omega 5 (2020) 2280−2286. Available from: https://doi.org/10.1021/acsomega.9b03507.

[227] X. Hou, L. Pan, S. Huang, O.-Y. Wei, X. Chen, Enhanced efficiency and stability of perovskite solar cells using porous hierarchical TiO_2 nanostructures of scattered distribution as scaffold, Electrochim. Acta 236 (2017) 351−358. Available from: https://doi.org/10.1016/j.electacta.2017.03.192.

[228] Y.-N. Zhang, B. Li, L. Fu, Q. Li, L.-W. Yin, MOF-derived ZnO as electron transport layer for improving light harvesting and electron extraction efficiency in perovskite solar cells, Electrochim. Acta 330 (2020) 135280. Available from: https://doi.org/10.1016/j.electacta.2019.135280.

[229] U. Ryu, S. Jee, J.-S. Park, I.K. Han, J.H. Lee, M. Park, et al., Nanocrystalline titanium metal−organic frameworks for highly efficient and flexible perovskite solar cells, ACS Nano 12 (2018) 4968−4975. Available from: https://doi.org/10.1021/acsnano.8b02079.

[230] J. Wang, J. Zhang, Y. Yang, Y. Dong, W. Wang, et al., Li-TFSI endohedral metal-organic frameworks in stable perovskite solar cells for anti-deliquescent and restricting ion migration, Chem. Eng. J. 429 (2022) 132481. Available from: https://doi.org/10.1016/j.cej.2021.132481.

[231] X. Zhou, L. Qiu, R. Fan, H. Ye, C. Tian, S. Hao, et al., Toward high-efficiency and thermally-stable perovskite solar cells: A novel metal-organic framework with active pyridyl sites replacing 4-tert-butylpyridine, J. Power Sources 473 (2020) 228556. Available from: https://doi.org/10.1016/j.jpowsour.2020.228556.

[232] J. Zhang, S. Guo, M. Zhu, C. Li, J. Chen, et al., Simultaneous defect passivation and hole mobility enhancement of perovskite solar cells by incorporating anionic

metal-organic framework into hole transport materials, Chem. Eng. J. 408 (2021) 127328. Available from: https://doi.org/10.1016/j.cej.2020.127328.
[233] J. Wang, J. Zhang, Y. Yang, S. Gai, Y. Dong, et al., New insight into the Lewis basic sites in metal−organic framework doped hole transport materials for efficient and stable perovskite solar cells, ACS Appl. Mater. Interfaces 13 (2021) 5235−5244. Available from: https://doi.org/10.1021/acsami.0c19968.
[234] Y. Dong, J. Zhang, Y. Yang, L. Qiu, D. Xia, et al., Self assembly of hybrid oxidant POM@Cu-btc for enhanced efficiency and long-term stability of perovskite solar cells, Angew. Chem. Int. Ed. 58 (2019) 17610−17615. Available from: https://doi.org/10.1002/anie.201909291.
[235] C. Li, S. Guo, J. Chen, Z. Cheng, M. Zhu, et al., Mitigation of vacancy with ammonium salt-trapped ZIF-8 capsules for stable perovskite solar cells through simultaneous compensation and loss inhibition, Nanoscale Adv. 3 (2021) 3554−3562. Available from: https://doi.org/10.1039/D1NA00173F.
[236] J. Dou, C. Zhu, H. Wang, Y. Han, S. Ma, et al., Synergistic effects of Eu-MOF on perovskite solar cells with improved stability, Adv. Mater. 33 (2021) 2102947. Available from: https://doi.org/10.1002/adma.202102947.
[237] J. Ji, B. Liu, H. Huang, X. Wang, L. Yan, et al., Nondestructive passivation of the TiO_2 electron transport layer in perovskite solar cells by the PEIE-2D MOF interfacial modified layer, J. Mater. Chem. C. 9 (2021) 7057−7064. Available from: https://doi.org/10.1039/D1TC00036E.
[238] T.-H. Chang, C.-W. Kung, H.-W. Chen, T.-Y. Huang, S.-Y. Kao, et al., Planar heterojunction perovskite solar cells incorporating metal-organic framework nanocrystals, Adv. Mater. 27 (2015) 7229−7235. Available from: https://doi.org/10.1002/adma.201502537.
[239] S. Yuan, Y. Xian, Y. Long, A. Cabot, W. Li, J. Fan, Chromium-based metal−organic framework as A-site cation in $CsPbI_2Br$ perovskite solar cells, Adv. Funct. Mater. (2021) 2106233. Available from: https://doi.org/10.1002/adfm.202106233.
[240] X. Zhou, L. Qiu, R. Fan, J. Zhang, S. Hao, Y. Yang, Heterojunction incorporating perovskite and microporous metal−organic framework nanocrystals for efficient and stable solar cells, Nano-Micro Lett. 12 (2020) 80. Available from: https://doi.org/10.1007/s40820-020-00417-1.
[241] G. Kresse, J. Furthmüller, Efficiency of ab-initio total energy calculations for metals and semiconductors using a plane-wave basis set, Comput. Mater. Sci. 6 (1) (1996) 15−50. Available from: https://doi.org/10.1016/0927-0256(96)00008-0.
[242] S.S. Nagarkar, S.M. Unni, A. Sharma, S. Kurungot, S.K. Ghosh, Two-in-one: Inherent anhydrous and water-assisted high proton conduction in a 3D metal−organic framework, Angew. Chem. Int. Ed. 53 (10) (2014) 2638−2642. Available from: https://doi.org/10.1002/anie.201309077.
[243] M. Zhao, Y. Wang, Q. Ma, Y. Huang, X. Zhang, et al., Ultrathin 2D metal−organic framework nanosheets, Adv. Mater. 27 (45) (2015) 7372−7378. Available from: https://doi.org/10.1002/adma.201503648.
[244] Y. Zhao, L. Jiang, L. Shangguan, L. Mi, A. Liu, S. Liu, Synthesis of porphyrin-based two-dimensional metal−organic framework nanodisk with small size and few layers, J. Mater. Chem. A 6 (6) (2018) 2828−2833. Available from: https://doi.org/10.1039/C7TA07911G.
[245] L.M. Cavinato, E. Fresta, S. Ferrara, R.D. Costa, Merging biology and photovoltaics: How nature helps sun-catching, Adv. Energy Mater. 11 (43) (2021) 2100520. Available from: https://doi.org/10.1002/aenm.202100520.
[246] R.P. Raffaelle, A.F. Hepp, G.A. Landis, D.J. Hoffman, Mission applicability assessment of integrated power components and systems, Prog. Photovolt. Res. Appl. 10 (6) (2002) 391−397.

[247] A.F. Hepp, R.P. Raffaelle, I.T. Martin, Space photovoltaics: New technologies, environmental challenges, and missions, in: S. Sundaram, V. Subramaniam, R.P. Raffaelle, M.K. Nazeeruddin, A. Morales-Acevedo, M. Bernechea Navarro, A.F. Hepp (Eds.), Photovoltaics Beyond Silicon, Elsevier, Cambridge, MA, 2024, pp. 675−766.
[248] S. Bailey, R. Raffaelle, Chapter II-4-B − Operation of solar cells in a space environment, in: S. Kalogirou (Ed.), McEvoy's Handbook of Photovoltaics: Fundamentals and Applications, Third Edition, Academic Press, 2018, pp. 987−1003.
[249] G.A. Landis, A.F. Hepp, Applications of thin-film photovoltaics for space, in: Proc. 26th Intersoc. Energ. Conver. Eng. Conf., Vol. 2, August 4−9, 1991, pp. 256−261.
[250] D.J. Hoffman, T.W. Kerslake, A.F. Hepp, M.K. Jacobs, D. Ponnusamy, Thin-film photovoltaic solar array parametric assessment, in: Proc. 35th Intersoc. Energ. Conver. Eng. Conf., Vol. 1, July 24−28, 2000, pp. 670−680, AIAA-2000−2919, also: NASA/TM-2000−210342.
[251] N.G. Dhere, S.R. Ghongadi, M.B. Pandit, A.H. Jahagirdar, D. Scheiman, $CIGS_2$ thin-film solar cells on flexible foils for space power, Prog. Photovolt.: Res. Appl. 10 (6) (2002) 407−416.
[252] J.E. Dickman, A.F. Hepp, D.L. Morel, C.S. Ferekides, J.R. Tuttle, D.J. Hoffmann, et al., Utility of thin film solar cells on flexible substrates for space power, in: Proc. 1st Inter. Energ. Conv. Eng. Conf., August 2003, Portsmouth, VA, AIAA 2003−5922, 10 pp.
[253] R.P. Raffaelle, A.F. Hepp, J.E. Dickman, D.J. Hoffman, N. Dhere, J.R. Tuttle, et al., Thin-film solar cells on metal foil substrates for space power, in: Proc. 2nd Inter. Ener. Conver. Eng. Conf., Providence, RI, AIAA-2004−5735, August 2004, 18 pp.
[254] B.M. Başol, V.K. Kapur, A. Halani, C. Leidholm, Copper indium diselenide thin film solar cells fabricated on flexible foil substrates, Sol. Energy Mater. Sol. Cell 29 (2) (1993) 163−173.
[255] B.M. Başol, V.K. Kapur, C.R. Leidholm, A. Halani, K. Gledhill, Flexible and light weight copper indium diselenide solar cells on polyimide substrates, Sol. Energy Mater. Sol. Cell 43 (1) (1996) 93−98.
[256] N.G. Dhere, V.S. Gade, A.A. Kadam, A.H. Jahagirdar, S.S. Kulkarni, S.M. Bet, Development of $CIGS_2$ thin film solar cells, Mater. Sci. Eng. B 116 (3) (2005) 303−309.
[257] A.N. MacInnes, M.B. Power, A.F. Hepp, A.R. Barron, Indium tert-butylthiolates as single source precursors for indium sulfide thin films: Is molecular design enough? J. Organomet. Chem 449 (1−2) (1993) 95−104. Available from: https://doi.org/10.1016/0022-328X(93)80111-N.
[258] J.A. Hollingsworth, A.F. Hepp, W.E. Buhro, Spray CVD of copper indium disulfide films: Control of microstructure and crystallographic orientation, Chemical Vapor Deposition 5 (3) (1999) 105−108.
[259] K.K. Banger, J. Cowen, A.F. Hepp, Synthesis and characterization of the first liquid single-source precursors for the deposition of ternary chalcopyrite ($CuInS_2$) thin film materials, Chem. Mater. 13 (11) (2001) 3827−3829. Available from: https://doi.org/10.1021/cm010507o.
[260] K.K. Banger, M.H.-C. Jin, J.D. Harris, P.E. Fanwick, A.F. Hepp, A new facile route for the preparation of single-source precursors for bulk, thin-film, and nanocrystallite I-III-VI semiconductors, Inorg. Chem. 42 (24) (2003) 7713−7715. Available from: https://doi.org/10.1021/ic034802h.
[261] K.K. Banger, J.A. Hollingsworth, M.H.-C. Jin, J.D. Harris, E.W. Bohannan, et al., Ternary precursors for depositing I-III-VI_2 thin films for solar cells via spray CVD,

Thin Solid Films (2003) 63−67. Available from: https://doi.org/10.1016/S0040-6090(03)00196-2. 431−432.
[262] J.D. Harris, K.K. Banger, D.A. Scheiman, M.A. Smith, M.H.-C. Jin, A.F. Hepp, Characterization of $CuInS_2$ films prepared by atmospheric pressure spray chemical vapor deposition, Mater. Sci. Eng. B 98 (2) (2003) 150−155. Available from: https://doi.org/10.1016/S0921-5107(03)00041-2.
[263] J.E. Lau, J.D. Harris, D.G. Hehemann, O. Khan, N.V. Duffy, et al., Spray chemical vapor deposition of tris(bis(phenylmethyl)carbamo-dithioato-S,S') indium (III) for $CuInS_2$ polycrystalline thin-film photovoltaic devices, Mater. Sci. Eng. B 116 (3) (2005) 381−389. Available from: https://doi.org/10.1016/j.mseb.2004.08.016.
[264] A.F. Hepp, K.K. Banger, M.H.-C. Jin, J.D. Harris, J.S. McNatt, J.E. Dickman, Spray CVD of single-source precursors for chalcopyrite I-III-VI_2 thin-film materials, in: D.B. Mitzi (Ed.), Solution Processing of Inorganic Materials, John Wiley & Sons, Inc., 2008, pp. 157−198. Available from: https://doi.org/10.1002/9780470407790.ch6, https://ntrs.nasa.gov/citations/20090004419.
[265] J. Masnovi, K.K. Banger, P.E. Fanwick, A.F. Hepp, Structural characterization of copper−indium chalcopyrite precursors $(PPh_3)_2CuIn(ER)_4$ [R = CH_3, E = S and R = Ph, E = S and Se], Polyhedron 102 (2015) 246−252. Available from: https://doi.org/10.1016/j.poly.2015.09.038.
[266] P. Sheldon, Process integration issues in thin-film photovoltaics and their impact on future research directions, Prog. Photovolt. Res. Appl. 8 (2000) 77−91.
[267] H.-W. Schock, R. Noufi, CIGS-based solar cells for the next millennium, Prog. Photovolt. Res. Appl. 8 (2000) 151−160.
[268] D. Bonnet, P.V. Meyers, Cadmium telluride − material for thin film solar cells, J. Mater. Res 13 (1998) 2740−2753.
[269] J. Ramanujam, D.M. Bishop, T.K. Todorov, O. Gunawan, J. Rath, R. Nekovei, et al., Flexible CIGS, CdTe and a-Si:H based thin film solar cells: A review, Progress in Materials Science 110 (2020) 100619.
[270] T.D. Lee, A.U. Ebong, A review of thin film solar cell technologies and challenges, Renew. Sust. Energ. Rev. 70 (2017) 1286−1297. Available from: https://doi.org/10.1016/j.rser.2016.12.028.
[271] A. Romeo, E. Artegiani, CdTe-based thin film solar cells: Past, present and future, Energies 14 (6) (2021) 1684. Available from: https://doi.org/10.3390/en14061684.
[272] https://www.manchester.ac.uk/discover/news/obituary-professor-paul-obrien-cbe/, accessed August 1, 2023.
[273] M.A. Malik, N. Revaprasadu, P. O'Brien, Air-stable single-source precursors for the synthesis of chalcogenide semiconductor nanoparticles, Chem. Mater. 13 (3) (2001) 913−920. Available from: https://doi.org/10.1021/cm0011662.
[274] D. Fan, M. Afzaal, M.A. Mallik, C.Q. Nguyen, P. O'Brien, P.J. Thomas, Using coordination chemistry to develop new routes to semiconductor and other materials, Coord. Chem. Rev. 251 (13−14) (2007) 1878−1888. Available from: https://doi.org/10.1016/j.ccr.2007.03.021.
[275] M.A. Malik, M. Afzaal, P. O'Brien, Precursor chemistry for main group elements in semiconducting materials, Chem. Rev. 110 (7) (2010) 4417−4446. Available from: https://doi.org/10.1021/cr900406f.
[276] P. Kevin, D.J. Lewis, J. Raftery, M.A. Malik, P. O'Brien, Thin films of tin(II) sulphide (SnS) by aerosol-assisted chemical vapour deposition (AACVD) using tin(II) dithiocarbamates as single-source precursors, J. Crystal Growth 415 (2015) 93−99. Available from: https://doi.org/10.1016/j.jcrysgro.2014.07.019.
[277] G. Murtaza, M. Akhtar, M.A. Malik, P. O'Brien, N. Revaprasadu, Aerosol assisted chemical vapor deposition of Sb_2S_3 thin films: environmentally benign solar energy

material, Mater. Sci. Semicon. Proc. 40 (2015) 643−649. Available from: https://doi.org/10.1016/j.mssp.2015.07.038.
[278] S. Singh, A.K. Singh, A. Kumar, Single-source precursors for main group metal sulfides and solar cell applications, in: A. Apblett, A.R. Barron, A.F. Hepp (Eds.), Nanomaterials via Single-Source Precursors: Synthesis, Processing and Applications, Elsevier, Cambridge, 2022, pp. 357−387. Available from: https://doi.org/10.1016/B978-0-12-820340-8.00007-1.
[279] I.Y. Ahmet, M.S. Hill, A.L. Johnson, L.M. Peter, Polymorph-selective deposition of high purity SnS thin films from a single source precursor, Chem. Mater. 27 (22) (2015) 7680−7688. Available from: https://doi.org/10.1021/acs.chemmater.5b03220.
[280] I.Y. Ahmet, M. Guc, Y. Sánchez, M. Neuschitzer, V. Izquierdo-Roca, E. Saucedo, et al., Evaluation of AA-CVD deposited phase pure polymorphs of SnS for thin films solar cells, RSC Adv 9 (2019) 14899−14909. Available from: https://doi.org/10.1039/C9RA01938C.
[281] P. Fan, H. Lan, Z. Zheng, C. Lan, H. Peng, et al., Hysteresis-free two-dimensional perovskite solar cells prepared by single-source physical vapour deposition, Solar Energy 169 (2018) 179−186. Available from: https://doi.org/10.1016/j.solener.2018.04.051.
[282] Y. Kou, J. Bian, X. Pan, J. Guo, Enhancing photovoltaic performance and stability of perovskite solar cells through single-source evaporation and $CsPbBr_3$ quantum dots incorporation, Coatings 13 (5) (2023) 863. Available from: https://doi.org/10.3390/coatings13050863.
[283] S.A. Dowland, L.X. Reynolds, A. MacLachlan, U.B. Cappel, S.A. Haque, Photoinduced electron and hole transfer in CdS:P3HT nanocomposite films: Effect of nanomorphology on charge separation yield and solar cell performance, J. Mater. Chem. A 1 (2013) 13896−13901. Available from: https://doi.org/10.1039/C3TA12962D.
[284] M.A. Ehsan, M. Younas, A. Rehman, M. Altaf, M.Y. Khan, et al., Synthesis and utilization of platinum(II) dialkyldithiocarbamate precursors in aerosol assisted chemical vapor deposition of platinum thin films as counter electrodes for dye-sensitized solar cells, Polyhedron 166 (2019) 186−195. Available from: https://doi.org/10.1016/j.poly.2019.03.058.
[285] M. Wright, A. Uddin, Organic-inorganic hybrid solar cells: A comparative review, Sol. Energy Mater Sol. Cells 107 (2012) 87−111. Available from: https://doi.org/10.1016/j.solmat.2012.07.006.
[286] P.L. Musetha, N. Revaprasadu, G.A. Kolawole, R.V.S.R. Pullabhotla, K. Ramasamy, P. O'Brien, Homoleptic single molecular precursors for the deposition of platinum and palladium chalcogenide thin films, Thin Solid Films 519 (1) (2010) 197−202October. Available from: https://doi.org/10.1016/j.tsf.2010.07.101.
[287] Institute of Energy Conversion, <https://iec.udel.edu> (accessed 02.08.23).
[288] K.K. Banger, A.F. Hepp, J.D. Harris, M.H. Jin, S.L. Castro, Single-source precursors for ternary chalcopyrite materials, and methods of making and using the same, United States Patent 6,992,202 (B1), issued January 31, 2006.
[289] Nanotech Innovations LLC, <https://www.nanotech-innovations.com> (accessed 02.08.23).
[290] P. Sinsermsuksakul, L. Sun, S.W. Lee, H.H. Park, S.B. Kim, C. Yang, et al., Overcoming efficiency limitations of SnS-based solar cells, Adv. Energy Mater. 4 (15) (2014) 1400496. Available from: https://doi.org/10.1002/aenm.201400496.
[291] S. Kalogirou (Ed.), McEvoy's Handbook of Photovoltaics: Fundamentals and Applications, Third Edition, Academic Press, 2018, 1340 pp.
[292] V.K. Khanna (Author) (Ed.), Nano-Structured Photovoltaics: Solar Cells in the Nanotechnology Era, CRC Press, Boca Raton, FL, USA, 2022, 494 pp.

[293] C. Fradler, T. Rath, S. Dunst, I. Letofsky-Papst, R. Saf, et al., Flexible polymer/copper indium sulfide hybrid solar cells and modules based on the metal xanthate route and low temperature annealing, Sol. Energy Mater. Sol. Cells 124 (2014) 117−125. Available from: https://doi.org/10.1016/j.solmat.2014.01.043.
[294] F.A. Abou-Elfotouh, H. Moutinho, A. Bakry, T.J. Coutts, L.L. Kazmerski, Characterization of the defect levels in copper indium diselenide, Solar Cells 30 (1−4) (1991) 151−160. Available from: https://doi.org/10.1016/0379-6787(91)90048-T.
[295] J.C. Park, Y.S. Nam, Controlling surface defects of non-stoichiometric copper-indium-sulfide quantum dots, J. Colloid Interface Sci. 460 (2015) 173−180. Available from: https://doi.org/10.1016/j.jcis.2015.08.037.
[296] T.K. Das, P. Ilaiyaraja, P.S.V. Mocherla, G.M. Bhalerao, C. Sudakar, Influence of surface disorder, oxygen defects and bandgap in TiO_2 nanostructures on the photovoltaic properties of dye sensitized solar cells, Sol. Energy Mater. Sol. Cells 144 (2016) 194−209. Available from: https://doi.org/10.1016/j.solmat.2015.08.036.
[297] J.J.S. Scragg, J.K. Larsen, M. Kumar, C. Persson, J. Sendler, S. Siebentritt, et al., Cu−Zn disorder and band gap fluctuations in $Cu_2ZnSn(S,Se)_4$: theoretical and experimental investigations, Physica Status Solidi Rapid Research Letters 253 (2) (2016) 247−254. Available from: https://doi.org/10.1002/pssb.201552530.
[298] Y. Shao, Y. Yuan, J. Huang, Correlation of energy disorder and open-circuit voltage in hybrid perovskite solar cells, Nat. Energy 1 (2016) 15001. Available from: https://doi.org/10.1038/nenergy.2015.1.
[299] K. Ramki, N. Venkatesh, G. Sathiyan, R. Thangamuthu, P. Sakthivel, A comprehensive review on the reasons behind low power conversion efficiency of dibenzo derivatives based donors in bulk heterojunction organic solar cells, Organic Electronics 73 (2019) 182−204. Available from: https://doi.org/10.1016/j.orgel.2019.05.047.
[300] R. Nie, M. Hu, A.M. Risqi, Z. Li, S.I. Seok, Efficient and stable antimony selenoiodide solar cells, Adv. Sci. 8 (8) (2021) 2003172. Available from: https://doi.org/10.1002/advs.202003172.
[301] A.S.R. Bati, Y.L. Zhong, P.L. Burn, M.K. Nazeeruddin, P.E. Shaw, M. Batmunkh, Next-generation applications for integrated perovskite solar cells, Commun. Mater. 4 (2) (2023). Available from: https://doi.org/10.1038/s43246-022-00325-4.
[302] A. Gurung, Q. Qiao, Solar charging batteries: Advances, challenges, and opportunities, Joule 2 (7) (2018) 1217−1230. Available from: https://doi.org/10.1016/j.joule.2018.04.006.
[303] L. Fagiolari, M. Sampò, A. Lamberti, J. Amici, C. Francia, S. Bodoardo, et al., Integrated energy conversion and storage devices: Interfacing solar cells, batteries and supercapacitors, Energy Storage Materials 51 (2022) 400−434. Available from: https://doi.org/10.1016/j.ensm.2022.06.051.
[304] D. Lau, N. Song, C. Hall, Y. Jiang, S. Lim, I. Perez-Wurfl, et al., Hybrid solar energy harvesting and storage devices: The promises and challenges, Mater. Today Energy 13 (2019) 22−44. Available from: https://doi.org/10.1016/j.mtener.2019.04.003.
[305] V. Vega-Garita, L. Ramirez-Elizondo, N. Narayan, P. Bauer, Integrating a photovoltaic storage system in one device: A critical review, Prog. Photovolt. Res. Appl. 27 (4) (2019) 346−370. Available from: https://doi.org/10.1002/pip.3093.VVV.
[306] L.M. Wangatia, S. Yang, F. Zabihi, M. Zhu, S. Ramakrishna, Biomedical electronics powered by solar cells, Current Opinion in Biomedical Engineering 13 (2020) 25−31. Available from: https://doi.org/10.1016/j.cobme.2019.08.004.
[307] Y. Wu, C. Li, Z. Tian, J. Sun, Solar-driven integrated energy systems: State of the art and challenges, Journal of Power Sources 478 (2020) 228762. Available from: https://doi.org/10.1016/j.jpowsour.2020.228762.

[308] J. Xie, Y.-Z. Li, L. Yang, Y. Sun, M. Yuan, A review of the recent progress of stand-alone photovoltaic-battery hybrid energy systems in space and on the ground, Journal of Energy Storage 55 (Part C) (2022) 105735. Available from: https://doi.org/10.1016/j.est.2022.105735.

[309] T.K.N. Sweet, M.H. Rolley, W. Li, M.C. Paul, A. Johnson, et al., Design and characterization of hybrid III−V concentrator photovoltaic−thermoelectric receivers under primary and secondary optical elements, Applied Energy 226 (2018) 772−783. Available from: https://doi.org/10.1016/j.apenergy.2018.06.018.

[310] J. Bing, L. Granados Caro, H.P. Talathi, N.L. Chang, D.R. McKenzie, A.W.Y. Ho-Baillie, Perovskite solar cells for building integrated photovoltaics—Glazing applications, Joule 6 (7) (2022) 1446−1474. Available from: https://doi.org/10.1016/j.joule.2022.06.003.

[311] A. Ganguly, D. Misra, S. Ghosh, Modeling and analysis of solar photovoltaic-electrolyzer-fuel cell hybrid power system integrated with a floriculture greenhouse, Energy and Buildings 42 (11) (2010) 2036−2043. Available from: https://doi.org/10.1016/j.enbuild.2010.06.012.

[312] S.S. Kaleibari, Z. Yanping, S. Abanades, Solar-driven high temperature hydrogen production via integrated spectrally split concentrated photovoltaics (SSCPV) and solar power tower, International Journal of Hydrogen Energy 44 (5) (2019) 2519−2532. Available from: https://doi.org/10.1016/j.ijhydene.2018.12.039.

[313] T. Salameh, M.A. Abdelkareem, A.G. Olabi, E.T. Sayed, M. Al-Chaderchi, H. Rezk, Integrated standalone hybrid solar PV, fuel cell and diesel generator power system for battery or supercapacitor storage systems in Khorfakkan, United Arab Emirates, International Journal of Hydrogen Energy 46 (8) (2021) 6014−6027. Available from: https://doi.org/10.1016/j.ijhydene.2020.08.153.

[314] M. Rokonuzzaman, M.K. Mishu, N. Amin, M. Nadarajah, R.B. Roy, et al., Self-sustained autonomous wireless sensor network with integrated solar photovoltaic system for internet of smart home-building (IoSHB) applications, Micromachines 12 (6) (2021) 653. Available from: https://doi.org/10.3390/mi12060653.

[315] J. Wang, K.A. Al-attab, T.Y. Heng, Techno-economic and thermodynamic analysis of solid oxide fuel cell combined heat and power integrated with biomass gasification and solar assisted carbon capture and energy utilization system, Energy Conversion and Management 280 (2023) 116762. Available from: https://doi.org/10.1016/j.enconman.2023.116762.

[316] A. Kasaeian, M. Javidmehr, M.R. Mirzaie, L. Fereidooni, Integration of solid oxide fuel cells with solar energy systems: A review, Applied Thermal Engineering 224 (2023) 120117. Available from: https://doi.org/10.1016/j.applthermaleng.2023.120117.

[317] X.Z. Wang, H.L. Tam, K.S. Yong, Z.-K. Chen, F. Zhu, High performance optoelectronic device based on semitransparent organic photovoltaic cell integrated with organic light-emitting diode, Organic Electronics 12 (8) (2011) 1429−1433. Available from: https://doi.org/10.1016/j.orgel.2011.05.012.

[318] A.E. Ostfeld, A.M. Gaikwad, Y. Khan, A.C. Arias, High-performance flexible energy storage and harvesting system for wearable electronics, Sci. Rep. 6 (2016) 26122. Available from: https://doi.org/10.1038/srep26122.

[319] A. Nascetti, M. Mirasoli, E. Marchegiani, M. Zangheri, F. Costantini, A. Porchetta, et al., Integrated chemiluminescence-based lab-on-chip for detection of life markers in extraterrestrial environments, Biosens. Bioelectron. 123 (2019) 195−203. Available from: https://doi.org/10.1016/j.bios.2018.08.056.

[320] C. Li, S. Cong, Z. Tian, Y. Song, L. Yu, et al., Flexible perovskite solar cell-driven photo-rechargeable lithium-ion capacitor for self-powered wearable strain sensors, Nano Energy 60 (2019) 247−256. Available from: https://doi.org/10.1016/j.nanoen.2019.03.061.

[321] K. Song, J.H. Han, H.C. Yang, K.I. Nam, J. Lee, Generation of electrical power under human skin by subdermal solar cell arrays for implantable bioelectronic devices, Biosens. Bioelectron. 92 (2017) 364−371. Available from: https://doi.org/10.1016/j.bios.2016.10.095.

[322] C. García Núñez, L. Manjakkal, R. Dahiya, Energy autonomous electronic skin, npj Flex Electron 3 (1) (2019). Available from: https://doi.org/10.1038/s41528-018-0045-x.

[323] A. Scalia, F. Bella, A. Lamberti, S. Bianco, C. Gerbaldi, E. Tresso, et al., A flexible and portable powerpack by solid-state supercapacitor and dye-sensitized solar cell integration, J. Power Sources 359 (2017) 311−321. Available from: https://doi.org/10.1016/j.jpowsour.2017.05.072.

[324] T. Zhu, Y. Yang, Y. Liu, R. Lopez-Hallman, Z. Ma, L. Liu, et al., Wireless portable light-weight self-charging power packs by perovskite-organic tandem solar cells integrated with solid-state asymmetric supercapacitors, Nano Energy 78 (2020) 105397. Available from: https://doi.org/10.1016/j.nanoen.2020.105397.

[325] Y. Yang, L. Fan, N.D. Pham, D. Yao, T. Wang, Z. Wang, et al., Self-charging flexible solar capacitors based on integrated perovskite solar cells and quasi-solid-state supercapacitors fabricated at low temperature, J. Power Sources 479 (2020) 229046. Available from: https://doi.org/10.1016/j.jpowsour.2020.229046.

[326] Z. Song, J. Wu, L. Sun, T. Zhu, C. Deng, et al., Photocapacitor integrating perovskite solar cell and symmetrical supercapacitor generating a conversion storage efficiency over 20%, Nano Energy 100 (2022) 107501. Available from: https://doi.org/10.1016/j.nanoen.2022.107501.

[327] M.M. Rahman, A comprehensive review on perovskite solar cells integrated photo-supercapacitors and perovskites-based electrochemical supercapacitors energy conversion and storage devices: Interfacing solar cells, batteries and supercapacitors, Chemical Record (online August 29 (2023) e202300183. Available from: https://doi.org/10.1002/tcr.202300183.

[328] R.D. Andrés, T. Berestok, K. Shchyrba, A. Fischer, U. Würfel, A new figure of merit for solar charging systems: Case study for monolithically integrated photosupercapacitors composed of a large-area organic solar cell and a carbon double-layer capacitor, Solar RRL 6 (10) (2022) 2200614. Available from: https://doi.org/10.1002/solr.202200614.

[329] W. Song, X. Yin, D. Liu, W. Ma, M. Zhang, et al., A highly elastic self-charging power system for simultaneously harvesting solar and mechanical energy, Nano Energy 65 (2019) 103997. Available from: https://doi.org/10.1016/j.nanoen.2019.103997.

[330] D. Devadiga, M. Selvakumar, P. Shetty, M.S. Santosh, The integration of flexible dye-sensitized solar cells and storage devices towards wearable self-charging power systems: A review, Renew. Sustain. Energy Rev. 159 (2022) 112252. Available from: https://doi.org/10.1016/j.rser.2022.112252.

[331] K. Liu, L. Jiang, Bio-inspired design of multiscale structures for function integration, Nano Today 6 (2) (2011) 155−175. Available from: https://doi.org/10.1016/j.nantod.2011.02.002.

[332] Z. Tian, X. Tong, G. Sheng, Y. Shao, L. Yu, et al., Printable magnesium ion quasi-solid-state asymmetric supercapacitors for flexible solar-charging integrated units, Nat. Commun. 10 (2019) 4913. Available from: https://doi.org/10.1038/s41467-019-12900-4.

[333] A.S. Westover, K. Share, R. Carter, A.P. Cohn, L. Oakes, C.L. Pint, Direct integration of a supercapacitor into the backside of a silicon photovoltaic device, Appl Phys Lett 104 (21) (2014) 1−4. Available from: https://doi.org/10.1063/1.4880211.

[334] X. Xu, S. Li, H. Zhang, Y. Shen, S.M. Zakeeruddin, et al., A power pack based on organometallic perovskite solar cell and supercapacitor, ACS Nano 9 (2) (2015) 1782−1787. Available from: https://doi.org/10.1021/nn506651m.

[335] Y. Hu, Y. Bai, B. Luo, S. Wang, H. Hu, et al., A portable and efficient solar-rechargeable battery with ultrafast photo-charge/discharge rate, Adv. Energy Mater. 9 (28) (2019) 1900872. Available from: https://doi.org/10.1002/aenm.201900872.

[336] L.-C. Kin, Z. Liu, O. Astakhov, S.N. Agbo, H. Tempel, et al., Efficient area matched converter aided solar charging of lithium ion batteries using high voltage perovskite solar cells, ACS Appl. Energy Mater 3 (1) (2020) 431–439.

[337] T.L. Gibson, N.A. Kelly, Solar photovoltaic charging of lithium-ion batteries, J. Power Sources 195 (12) (2010) 3928–3932. Available from: https://doi.org/10.1016/j.jpowsour.2009.12.082.

[338] W. Li, H.-C. Fu, Y. Zhao 1, J.-H. He, S. Jin, 14.1% Efficient monolithically integrated solar flow battery, Chem 4 (11) (2018) 2644–2657. Available from: https://doi.org/10.1016/j.chempr.2018.08.023.

[339] A. Uddin, M.B. Upama, H. Yi, L. Duan, Encapsulation of organic and perovskite solar cells: A review, Coatings 9 (2) (2019) 65. Available from: https://doi.org/10.3390/coatings9020065.

[340] M. Yamaguchi, High-efficiency GaAs-based solar cells, in: M.M. Rahman, A.M. Asiri, A. Khan, I. Inamuddin, T. Tabbakh (Eds.), Post-Transition Metals, InTech Open, 2021. Available from: http://doi.org/10.5772/intechopen.94365.

[341] A. Sahu, A. Garg, A. Dixit, A review on quantum dot sensitized solar cells: Past, present and future towards carrier multiplication with a possibility for higher efficiency, Solar Energy 203 (2020) 210–239. Available from: https://doi.org/10.1016/j.solener.2020.04.044.

[342] K. Shanks, Concentrator and multijunction solar cells, in: S. Sundaram, V. Subramaniam, R.P. Raffaelle, M.K. Nazeeruddin, A. Morales-Acevedo, M. Bernechea Navarro, et al. (Eds.), Photovoltaics Beyond Silicon, Elsevier, Cambridge, MA, 2024, pp. 499–522.

[343] R.P. Raffaelle, An introduction to space photovoltaics: technologies, issues, and missions, in: S.G. Bailey, A.F. Hepp, D.C. Ferguson, R.P. Raffaelle, S.M. Durbin (Eds.), Space Photovoltaics: Materials, Missions and Alternative Technologies, Elsevier, Cambridge, MA, USA, 2022, pp. 3–27.

[344] S. Bailey, R. Raffaelle, Space solar cells and arrays. Decemberin: A. Luque, S. Hegedus (Eds.), Handbook of Photovoltaic Science and Engineering, John Wiley & Sons, Ltd, 2010, pp. 365–401. Available from: https://doi.org/10.1002/9780470974704.ch9.

[345] P. Pérez-Higueras, E. Muñoz, G. Almonacid, P.G. Vidal, High concentrator photovoltaics efficiencies: Present status and forecast, Renewable and Sustainable Energy Reviews 15 (4) (2011) 1810–1815. Available from: https://doi.org/10.1016/j.rser.2010.11.046.

[346] D.C. Law, R.R. King, H. Yoon, M.J. Archer, A. Boca, et al., Future technology pathways of terrestrial III–V multijunction solar cells for concentrator photovoltaic systems, Solar Energy Materials and Solar Cells 94 (8) (2010) 1314–1318. Available from: https://doi.org/10.1016/j.solmat.2008.07.014.

[347] N. Xu, J. Ji, W. Sun, L. Han, H. Chen, Z. Jin, Outdoor performance analysis of a 1090× point-focus Fresnel high concentrator photovoltaic/thermal system with triple-junction solar cells, Energy Conversion and Management 100 (2015) 191–200. Available from: https://doi.org/10.1016/j.enconman.2015.04.082.

[348] S. Akhil, S. Akash, A. Pasha, B. Kulkarni, M. Jalalah, et al., Review on perovskite silicon tandem solar cells: Status and prospects 2T, 3T and 4T for real world conditions, Materials & Design 211 (2021) 110138. Available from: https://doi.org/10.1016/j.matdes.2021.110138.

[349] M. Jošt, L. Kegelmann, L. Korte, S. Albrecht, Monolithic perovskite tandem solar cells: A review of the present status and advanced characterization methods toward

30% efficiency, Adv. Energy Mater. 10 (26) (2020) 1904102. Available from: https://doi.org/10.1002/aenm.201904102.

[350] M.I. Elsmani, N. Fatima, M.P.A. Jallorina, S. Sepeai, M.S. Su'ait, Recent issues and configuration factors in perovskite-silicon tandem solar cells towards large scaling production, Nanomaterials 11 (12) (2021) 3186. Available from: https://doi.org/10.3390/nano11123186.

[351] E. Raza, Z. Ahmad, Review on two-terminal and four-terminal crystalline-silicon/perovskite tandem solar cells; progress, challenges, and future perspectives, Energy Reports 8 (2022) 5820–5851. Available from: https://doi.org/10.1016/j.egyr.2022.04.028.

[352] F. Fu, J. Li, T.C.-J. Yang, H. Liang, A. Faes, et al., Monolithic perovskite-silicon tandem solar cells: From the lab to fab? Adv. Mater. 34 (24) (2022) 2106540. Available from: https://doi.org/10.1002/adma.202106540.

[353] O.M. Saif, A.H. Zekry, M. Abouelatta, A. Shaker, A comprehensive review of tandem solar cells integrated on silicon substrate: III/V vs perovskite, Silicon (May 2023) 23. Available from: https://doi.org/10.1007/s12633-023-02466-8.

[354] X. Chen, Z. Jia, Z. Chen, T. Jiang, L. Bai, et al., Efficient and reproducible monolithic perovskite/organic tandem solar cells with low-loss interconnecting layers, Joule 4 (7) (2020) 1594–1606. Available from: https://doi.org/10.1016/j.joule.2020.06.006.

[355] J. Zheng, C.F.J. Lau, H. Mehrvarz, F.-J. Ma, Y. Jiang, et al., Large area efficient interface layer free monolithic perovskite/homo-junction-silicon tandem solar cell with over 20% efficiency, Energy Environ. Sci. 11 (2018) 2432–2443. Available from: https://doi.org/10.1039/C8EE00689J.

[356] H. Shen, S.T. Omelchenko, D.A. Jacobs, S. Yalamanchili, Y. Wan, D. Yan, et al., In situ recombination junction between p-Si and TiO_2 enables high-efficiency monolithic perovskite/Si tandem cells, Sci. Adv. 4 (12) (2018) eaau9711. Available from: https://doi.org/10.1126/sciadv.aau971.

[357] Z. Li, Y. Zhao, X. Wang, Y. Sun, Z. Zhao, et al., Cost analysis of perovskite tandem photovoltaics, Joule 2 (8) (2018) 1559–1572. Available from: https://doi.org/10.1016/j.joule.2018.05.001.

[358] T. Druffela, R. Dharmadasa, B.W. Lavery, K. Ankireddy, Intense pulsed light processing for photovoltaic manufacturing, Sol. Energy Mater. Sol. Cells 174 (2018) 359–369. Available from: https://doi.org/10.1016/j.solmat.2017.09.010.

[359] W. Zi, Z. Jin, S. Liu, B. Xu, Flexible perovskite solar cells based on green, continuous roll-to-roll printing technology, J. Energy Chem. 27 (4) (2018) 971–989. Available from: https://doi.org/10.1016/j.jechem.2018.01.027.

[360] Y.Y. Kim, T.-Y. Yang, R. Suhonen, M. Välimäki, T. Maaninen, et al., Gravure-printed flexible perovskite solar cells: Toward roll-to-roll manufacturing, Adv. Sci. 6 (7) (2019) 1802094. Available from: https://doi.org/10.1002/advs.201802094.

[361] B.-J. Huang, C.-K. Guan, S.-H. Huang, W.-F. Su, Development of once-through manufacturing machine for large-area perovskite solar cell production, Solar Energy 205 (2020) 192–201. Available from: https://doi.org/10.1016/j.solener.2020.05.005.

[362] N. Mufti, T. Amrillah, A. Taufiq, S. Sunaryono, A. Aripriharta, et al., Review of CIGS-based solar cells manufacturing by structural engineering, Solar Energy 207 (2020) 1146–1157. Available from: https://doi.org/10.1016/j.solener.2020.07.065.

[363] N. Rolston, W.J. Scheideler, A.C. Flick, J.P. Chen, H. Elmaraghi, Rapid open-air fabrication of perovskite solar modules, Joule 4 (12) (2020) 2675–2692. Available from: https://doi.org/10.1016/j.joule.2020.11.001.

[364] J.W. Yoo, J. Jang, U. Kim, Y. Lee, S.-G. Ji, et al., Efficient perovskite solar minimodules fabricated via bar-coating using 2-methoxyethanol-based formamidinium

lead tri-iodide precursor solution, Joule 5 (9) (2021) 2420−2436. Available from: https://doi.org/10.1016/j.joule.2021.08.005.
[365] T.-Y. Yang, Y.Y. Kim, J. Seo, Roll-to-roll manufacturing toward lab-to-fab-translation of perovskite solar cells, APL Mater 9 (2021) 110901. Available from: https://doi.org/10.1063/5.0064073.
[366] V. Vohra, N.T. Razali, R. Wahi, L. Ganzer, T. Virgili, A comparative study of low-cost coating processes for green & sustainable organic solar cell active layer manufacturing, Opt. Mater. X 13 (2022) 100127. Available from: https://doi.org/10.1016/j.omx.2021.100127.
[367] L.W.T. Ng, S.W. Lee, D.W. Chang, J.M. Hodgkiss, D. Vak, Organic photovoltaics' new renaissance: advances toward roll-to-roll manufacturing of non-fullerene acceptor organic photovoltaics, Adv. Mater. Tech 7 (10) (2022) 2101556. Available from: https://doi.org/10.1002/admt.202101556.
[368] C. Kapnopoulos, A. Zachariadis, E. Mekeridis, S. Kassavetis, C. Gravalidis, A. Laskarakis, et al., On-the-fly short-pulse R2R laser patterning processes for the manufacturing of fully printed semitransparent organic photovoltaics, Materials 15 (22) (2022) 8218. Available from: https://doi.org/10.3390/ma15228218.
[369] S. Abbasi, X. Wang, P. Tipparak, C. Bhoomanee, P. Ruankham, Proper annealing process for a cost effective and superhydrophobic ambient-atmosphere fabricated perovskite solar cell, Mater. Sci. Semicon. Proc 155 (2023) 107241. Available from: https://doi.org/10.1016/j.mssp.2022.107241.
[370] M. Izu, T. Ellison, Roll-to-roll manufacturing of amorphous silicon alloy solar cells with in situ cell performance diagnostics, Sol. Energy Mater. Sol. Cells 78 (1−4) (2003) 613−626. Available from: https://doi.org/10.1016/S0927-0248(02)00454-3.
[371] S. Binetti, P. Garattini, R. Mereu, A. Le Donne, S. Marchionna, et al., Fabricating $Cu(In,Ga)Se_2$ solar cells on flexible substrates by a new roll-to-roll deposition system suitable for industrial applications, Semiconductor Science and Technology 30 (10) (2015) 105006. Available from: https://doi.org/10.1088/0268-1242/30/10/105006.
[372] J. Zheng, H. Mehrvarz, F.-J. Ma, C.F.J. Lau, M.A. Green, et al., 21.8% efficient monolithic perovskite/homo-junction-silicon tandem solar cell on 16 cm^2, ACS Energy Lett 3 (9) (2018) 2299−2300. Available from: https://doi.org/10.1021/acsenergylett.8b01382.
[373] Q.F. Han, Y.T. Hsieh, L. Meng, J.L. Wu, P.Y. Sun, et al., High-performance perovskite/$Cu(In, Ga)Se_2$ monolithic tandem solar cells, Science 361 (6405) (2018) 904−908. Available from: https://doi.org/10.1126/science.aat5055.
[374] A. Al-Ashouri, E. Köhnen, B. Li, A. Magomedov, H. Hempel, et al., Monolithic perovskite/silicon tandem solar cell with >29% efficiency by enhanced hole extraction, Science 370 (6522) (2020) 1300−1309. Available from: https://doi.org/10.1126/science.abd4016.
[375] B. Chen, P. Wang, R. Li, N. Ren, Y. Chen, et al., Composite electron transport layer for efficient n-i-p type monolithic perovskite/silicon tandem solar cells with high open-circuit voltage, Journal of Energy Chemistry 63 (2021) 461−467. Available from: https://doi.org/10.1016/j.jechem.2021.07.018.
[376] M. Jošt, E. Köhnen, A. Al-Ashouri, T. Bertram, Š. Tomšič, et al., Perovskite/CIGS tandem solar cells: from certified 24.2% toward 30% and beyond, ACS Energy Lett 7 (4) (2022) 1298−1307. Available from: https://doi.org/10.1021/acsenergylett.2c00274.
[377] Y. Ding, Q. Guo, Y. Geng, Z. Dai, Z. Wang, et al., A low-cost hole transport layer enables $CsPbI_2Br$ single-junction and tandem perovskite solar cells with record efficiencies of 17.8% and 21.4%, Nano Today 46 (2022) 101586. Available from: https://doi.org/10.1016/j.nantod.2022.101586.
[378] W. Chen, Y. Zhu, J. Xiu, G. Chen, H. Liang, Monolithic perovskite/organic tandem solar cells with 23.6% efficiency enabled by reduced voltage losses and

optimized interconnecting layer, Nat. Energy 7 (3) (2022) 229–237. Available from: https://doi.org/10.1038/s41560-021-00966-8.
[379] J. Luo, L. Tang, S. Wang, H. Yan, W. Wang, et al., Manipulating Ga growth profile enables all-flexible high-performance single-junction CIGS and 4 T perovskite/CIGS tandem solar cells, Chem. Eng. J. 455 (Part 2) (2023) 140960. Available from: https://doi.org/10.1016/j.cej.2022.140960.
[380] S. Wu, M. Liu, A.K.-Y. Jen, Prospects and challenges for perovskite-organic tandem solar cells, Joule 7 (3) (2023) 484–502. Available from: https://doi.org/10.1016/j.joule.2023.02.014.
[381] T. Nie, Z. Fang, X. Ren, Y. Duan, S.F. Liu, Recent advances in wide-bandgap organic–inorganic halide perovskite solar cells and tandem application, Nano-Micro Lett. 15 (2023) 70. Available from: https://doi.org/10.1007/s40820-023-01040-6.
[382] I.J. Park, H.K. An, Y. Chang, J.Y. Kim, Interfacial modification in perovskite-based tandem solar cells, Nano Convergence 10 (2023) 22. Available from: https://doi.org/10.1186/s40580-023-00374-6.
[383] L. Liu, H. Xiao, K. Jin, Z. Xiao, X. Du, et al., 4-Terminal inorganic perovskite/organic tandem solar cells offer 22% efficiency, Nano-Micro Lett 15 (2023) 23. Available from: https://doi.org/10.1007/s40820-022-00995-2.
[384] R. Wang, M. Han, Y. Wang, J. Zhao, J. Zhang, et al., Recent progress on efficient perovskite/organic tandem solar cells, Journal of Energy Chemistry 83 (2023) 158–172. Available from: https://doi.org/10.1016/j.jechem.2023.04.036.
[385] <https://www.csem.ch/press/new-world-records-perovskite-on-silicon-tandem-solar?pid = 172296> (accessed 23.06.23).
[386] M. Khalid, T.K. Mallick, S. Sundaram, Recent advances in perovskite-containing tandem structures, in: S. Sundaram, V. Subramaniam, R.P. Raffaelle, M.K. Nazeeruddin, A. Morales-Acevedo, M. Bernechea Navarro, A.F. Hepp (Eds.), Photovoltaics Beyond Silicon, Elsevier, Cambridge, MA, 2024, pp. 545–582.
[387] M. Khalid, T.S. Mallick, S. Sundaram, <https://www.nasa.gov/topics/aeronautics/features/trl_demystified.html> (accessed 15.08.23).
[388] <https://www.nasa.gov/pdf/458490main_TRL_Definitions.pdf> (accessed 15.08.23).
[389] Technology Readiness Assessment Guide: Best Practices for Evaluating the Readiness of Technology for Use in Acquisition Programs and Projects, U.S. General Accountability Office (Report GAO-20-48G), Washington, DC, 2020 149 pp.
[390] <http://dodmrl.com> (accessed 15.08.23).
[391] J.A. Fernandez, Contextual role of TRLs and MRLs in technology management, Sandia National Laboratories (Sandia Report, SAND2010–7595), Albuquerque, NM and Livermore, CA, 2010, 34 pp.
[392] <https://www.sciencedirect.com/topics/engineering/valley-of-death> (accessed 15.08.23).
[393] A. Jolink, Eva Niesten, Financing the energy transition: The role of public funding, collaboration and private equity, in: A. Rubino, A. Sapio, M. La Scala (Eds.), Handbook of Energy Economics and Policy: Fundamentals and Applications for Engineers and Energy Planners, Academic Press, 2021, pp. 521–547. Available from: https://doi.org/10.1016/B978-0-12-814712-2.00012-9.

CHAPTER ELEVEN

Green chemical synthesis of photovoltaic materials

O. Reyes-Vallejo[1,2], S. Torres-Arellano[3], J.L. Aleman-Ramirez[3] and P.J. Sebastian[2]

[1]Solid State Electronics Section (SEES) CINVESTAV-IPN, Gustavo A. Madero, Mexico City, Mexico
[2]Renewable Energy Institute IER-UNAM, Solar Materials Section, Temixco, Morelos, Mexico
[3]Renewable Energy Research and Innovation Institute IIIER-UNICACH, Tuxtla Gutiérrez, Chiapas, Mexico

11.1 Introduction

The great development of solar harvesting has motivated the production of diverse compounds through a variety of methods without considering the high impact they have on nature, not only the materials but also the methods of production because of their toxicity and tremendous energy consumption. Nowadays, it is an imperative necessity to take into consideration these aspects during the research and development in materials science.

The chemical methods for semiconductor production, such as hydrothermal and solvothermal chemical synthesis, electrodeposition, chemical bath deposition, and spin coating, just to mention some of them, involve the use of reactants, solvents, and auxiliary materials that can be considered hazardous pollutants by their own or by the mixing between themselves. Additionally, they also produce large volumes of waste. On the other hand, although physical methods such as sputtering or evaporation do not use many chemical compounds, solvents, or auxiliary agents nor do they produce large volumes of waste, they do use a great amount of energy so they cannot be considered out of focus when seeking the development of environmentally friendly processes.

Green chemistry, defined by Anastas and Warner [1] through the 12 principles (Fig. 11.1), has established parameters to develop new compounds and methods, taking into consideration the environment. The key principles on which green chemistry is based are the minimal use of toxic

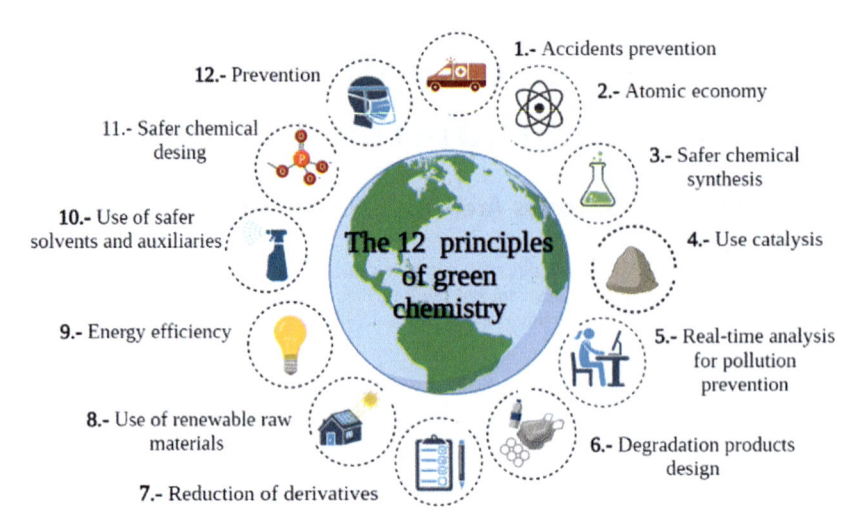

Figure 11.1 12 principles of green chemistry. *Adapted from P.T. Anastas, J.C. Warner, Green Chemistry: Theory and Practice, Oxford University Press, 1998, pp. 29–56.*

reactants and solvents, minimal production of wastes, minimization of by-products (Atomic economy), minimal toxicity of products, reduction in energy consumption, use of renewable raw matters, minimal use of auxiliary compounds or process, and maximization of catalyst and some others in the same direction. The application of these principles during the design of materials for solar harvesting is a promising approach to tackling environmental issues.

11.2 Green methods for synthesis of semiconductor materials

In recent times, to reduce the environmental impact in materials science, some greener methods have been developed and their results are notable and comparable to conventional methods. Examples of these methods are hydrothermal and solvothermal synthesis using green solvents, microwave-assisted chemical synthesis, mechanosynthesis, sonochemical synthesis, photo-assisted chemical synthesis (photocatalysis), magnetic field-assisted chemical synthesis, and biosynthesis processes using

Green synthesis processes

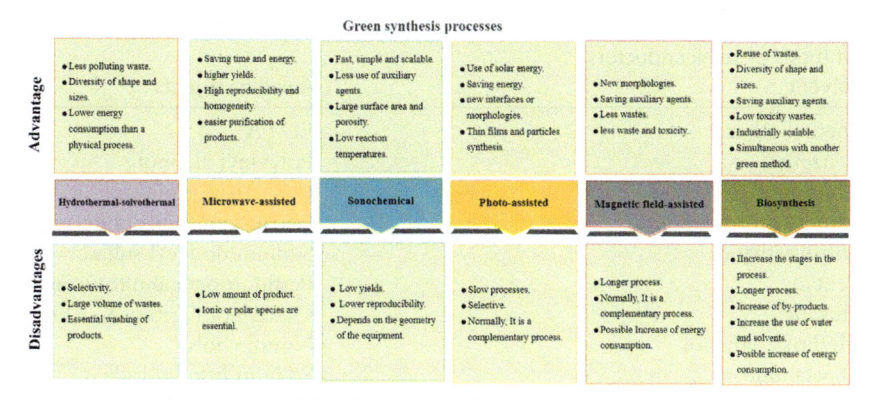

	Hydrothermal-solvothermal	Microwave-assisted	Sonochemical	Photo-assisted	Magnetic field-assisted	Biosynthesis
Advantages	• Less polluting waste. • Diversity of shape and sizes. • Lower energy consumption than a physical process.	• Saving time and energy. • higher yields. • High reproducibility and homogeneity. • easier purification of products.	• Fast, simple and scalable. • Less use of auxiliary agents. • Large surface area and porosity. • Low reaction temperatures.	• Use of solar energy. • Saving energy. • new interfaces or morphologies. • Thin films and particles synthesis.	• New morphologies. • Saving auxiliary agents. • Less wastes. • less waste and toxicity.	• Reuse of wastes. • Diversity of shape and sizes. • Saving auxiliary agents. • Low toxicity wastes. • Industrially scalable. • Simultaneous with another green method.
Disadvantages	• Selectivity. • Large volume of wastes. • Essential washing of products.	• Low amount of product. • Ionic or polar species are essential.	• Low yields. • Lower reproducibility. • Depends on the geometry of the equipment.	• Slow processes. • Selective. • Normally, It is a complementary process.	• Longer process. • Normally, It is a complementary process. • Possible Increase of energy consumption.	• Increase the stages in the process. • Longer process. • Increase of by-products. • Increase the use of water and solvents. • Possible increase of energy consumption.

Figure 11.2 Advantages and disadvantages of green synthesis processes.

microorganisms or organic wastes. In these methods, the use of reactants, solvents, or auxiliary chemicals less toxic than in conventional synthesis is considered and a better use of energy is promoted. Brilliant ideas for the crystallization and orientation of materials achieving diverse morphologies with great particle size control through the stimulation of light, sounds, magnetic fields or only by mixing reactants at high energy intensity. In Fig. 11.2, the advantages and disadvantages of every method are presented.

11.2.1 Green hydrothermal and solvothermal synthesis

Chemical methods for semiconductor synthesis, both powders and films, employ a variety of chemical products that perform functions such as metallic precursor, solvent, and auxiliary agent (complexing, surfactant, oxidizing, reducing, catalyst, and more). Many of these compounds can be toxic, hazardous, and difficult to handle in the laboratory, some of them are shown in Table 11.1, as are their wastes after the synthesis or processing.

Eco-friendly adaptations of these methods of synthesis have been proposed by using solvents and reactants less toxic and hazardous than those used for conventional methods. Prat et al. have proposed a less hazardous selection of solvents, following the 12 principles of green chemistry. Some examples of green solvents are listed in Table 11.2 [2].

Normally, the reactions are carried out in an autoclave reactor at high temperature and pressure (above normal boiling point), using water for the hydrothermal method or another green solvent for the solvothermal synthesis. These methods allow great control of the morphology and

Table 11.1 Solvents and auxiliary agents commonly used during the conventional synthesis of semiconductors.

Solvent	Auxiliary agents
Diethyl ether	EDTA
Benzene	Polyvinyl alcohol
Chloroform	Polyvinylpyrrolidone
Dimethylformamide	Cetrimonium bromide
Nitromethane	Sodium dodecyl sulfate
Pentane	Perfluorooctanesulfonic acid
Hexane	Octoxinol (Triton X-100)
Methoxy-ethanol	Nonoxynols
Oleylamine	Sodium borohydride
Pyridine	Hydrazine

Table 11.2 Alternative solvents for green synthesis.

Solvents	
Water	Acetone
Ethanol	Benzyl alcohol
Isopropanol	Sulfolane
Ethyl acetate	Acetic acid
Butyl acetate	Ethylene glycol
Isopropyl acetate	Cyclohexanone
Methoxybenzene	Acetic anhydride
2-Methyltetrahydrofuran	Methylcyclohexane

particle size, as well as the crystalline phase when ternary or quaternary compounds are required. Some disadvantages of these methods are the amount of waste produced and the long reaction times (days in some cases). The material synthesized needs a washing process involving the use of more water and solvents, and the use of auxiliary agents for the control of particle size, morphology, and functionalization is commonly required. Materials such as Cu_2ZnSnS_4, $Cu_2ZnSnSe_4$, TiO_2, and ZnO have been synthesized by using only water, ethylene glycol, ethanol, or mixtures of these, sometimes without the need for auxiliary agents [3–10].

11.2.2 Microwave-assisted chemical synthesis

This method is very similar to the hydrothermal and solvothermal methods, but the heating process is done mainly by radiation rather than conduction and convection. The chemical species of the ionic and polar reactants and solvents inside the reactor respond to the electromagnetic field by colliding and friction between them, which causes a superheating effect reaching a state in which the reactive processes happen much faster than in a conventional process. The materials that are synthesized in hours or days in an autoclave can be synthesized in minutes in a microwave, which also saves energy. Furthermore, the electromagnetic field causes a rearrangement of the atoms promoting morphologies different from those found by autoclaving synthesis. Compared with conventional heating this method is very reproducible, achieves higher yields, and produces materials with greater homogeneous size and shape distribution. Also, auxiliary agents can be employed to control the morphology, particle size, or functionalization of the material. Besides the similar disadvantages to hydrothermal synthesis, less production of material due to the size of reactors, it is essential to use the polar or ionic substances to heat and to promote reactions.

Nanostructures and thin films of materials such as Cu_3SbS_4, $CuInSe_2$, Cu_2ZnSnS_4, $Cu_2ZnSnSe_4$, Zn_2SnO_4, TiO_2, ZnO, and CdS have been synthesized and deposited using microwave radiation using green solvents [8,11−17]. In Fig. 11.3, octahedral particles of Zn_2SnO_4 are shown, which were synthesized by microwave-assisted chemical synthesis. $Zn(NO_3)_2 \cdot 6H_2O$ and $SnCl_2 \cdot 2H_2O$ were used as zinc and tin sources, NaOH as an oxidizing agent, and only water as a solvent. Well-defined octahedral particles are observed with a uniform size distribution, characteristic of this method. Detailed information can be found in the work of Reyes et al. [16].

Figure 11.3 Octahedral particles of Zn_2SnO_4 synthesized by microwave-assisted chemical synthesis. *Reproduced with permission from O. Reyes, M. Pal, J. Escorcia-García, R. Sánchez-Albores, P.J. Sebastian, Microwave-assisted chemical synthesis of Zn_2SnO_4 nanoparticles, Mater. Sci. Semicond. Proc. 108 (2020) 104878, copyright (2020) Elsevier.*

11.2.3 Mechanosynthesis: Ball milling

This is a very simple method and consists of ball milling of the reactants (most of the time elemental reagents), in which mechanical energy activates the molecules of reactants. These reactive molecules find each other randomly, synthesizing the desired material. This method is also named high energy ball milling to differentiate it from conventional grinding used for size reduction. Inert gases can be used to avoid oxidation or other undesirable reactions. Liquids can be used as lubricants or as auxiliary agents to promote conditions to achieve powders with specific characteristics, for example, ethanol can be used to obtain a smaller particle size. Some of the advantages of ball milling are low cost of equipment and reactants, no use of solvents nor auxiliary agents (or minimal use), suitability for continuous or batch operation, low energy consumption, ability to reach very small size particles (2–20 nm), minimal waste production, and no need to wash solids for purification and use. However, some of the disadvantages are low superficial area, wide particle size distribution, the amorphous shape of the nanomaterial, long reaction times (hours or days), possible contamination because of wear and tear, and an annealing process that is sometimes necessary (selenization and sulfurization in the case of chalcogenides). This method has been used for materials such as Cu_2ZnSnS_4, $Cu_2ZnSnSe_4$, $CH_3NH_3PbI_3$, CdS, and ZnO [18–21]. This method can also be used as a reagent mixture to improve the performance of other processes [19], which will be described in several sections next.

11.2.4 Sonochemical synthesis

This method is based on the activation of chemical species by ultrasound waves. Ultrasound means above the upper limit of human hearing (16 kHz–500 MHz), these waves are transmitted in an elastic medium (water) promoting chemical reactions. Chatel argues that optimized sonochemical processes, in most cases, follow the 12 principles of green chemistry [22], achieving better yield and selectivity, using water as a solvent, saving energy, and reducing reaction time. In general terms, the sonochemical reactions are often greener than those performed under silent conditions. More advantages of sonochemistry are the lower cost compared to other conventional methods (materials and equipment) and the use of less hazardous chemicals and environmentally friendly solvents. One of the limitations of sonochemistry is the problem of reproducibility; the geometry of equipment has a great impact on the materials synthesized so it makes it difficult for other research groups to reproduce. ZnO, Cu_2O, CuO, $NiSe_2$, TiO_2, and $AgInS_2$ are materials synthesized by this method [23−28].

11.2.5 Photo-assisted synthesis or photocatalysis

The effects of light activation on chemical species can be considered as green chemistry. This is mainly because sunlight is abundant, not polluting, and inexpensive; in fact, it is a source of renewable energy. The photochemical reactions are chemical reactions initiated or activated by the incidence of light into the chemical species, some examples are photoelectrochemistry, photocatalysis, and photodeposition. In the photoelectrochemical cells, the splitting of water occurs when semiconductors are illuminated producing hydrogen and oxygen at the electrodes. The stimulation by light is also observed in the photocatalysis process; when the catalyst is illuminated, the reaction takes place (or it is accelerated), and when illumination is off, the reaction stops. This process is also used for hydrogen production, CO_2 reduction, and more interesting processes. Also, light is used to heat a chemical process, which constitutes another section of photochemistry. In this process, solar concentration is applied to a photoreactor to promote desired reactions, for example, pyrolysis and thermal splitting of water. The main advantage of these examples is the use of solar light, which reduces or avoids the use of conventional energy. However, this technology is not well established, it is in the research stage; the high cost of materials in photoelectrochemistry and photocatalysis is another limitation. The photodeposition process is a method in which light activates the chemical species present in a solution to deposit on conductive surfaces. These surfaces could be or not be activated by light, forming thin films of the metal or metallic oxide when the solution is illuminated with a certain kind of light. This process has proved the controlled growth of very thin films of materials such as $BiVO_4$, ZnSe, CdSe, ZnSe, ZnO, MoS_2, and TiO_2 [29−34]. The photoactivation of organic perovskite precursors also has been reported to improve solar cell efficiency [35].

11.2.6 Magnetic field-assisted chemical synthesis

This novel method is an auxiliary strategy to control the morphology of the particles. The chemical system (solution) is confined and stimulated by a magnetic field before or during the reaction. This magnetic field does not activate or initiate the process but promotes the crystallization of the material in a certain order; this is a consequence of the orientation of the chemical species (ions) provoked by the magnetic fields. This is considered green because it avoids or limits the use of auxiliary agents such as surfactants and chelating, not only using fewer reactants but also avoiding waste production. Processes such as spin coating, solvothermal synthesis, thermal annealing, and plasma chemical vapor deposition have been assisted by magnetic field during thin-film deposition or synthesis of materials such as $CH_3NH_3PbI_3$, Fe_2O_3, Co_3O_4, and Fe_3O_4, which are of photovoltaic interest [36−41]. In Table 11.3, materials with photovoltaic interest [3,5,6,9,11,13,14,16−18,24,25,29,31−33,37−47] synthesized through nonbiological green methods are presented.

11.3 Biosynthesis

Biosynthesis is a group of synthesis in which reactants are converted into compounds through the use of microorganisms or plants (Fig. 11.4). In the case of microorganisms, during their metabolism, these living organisms catalyze substrates to produce compounds of interest. While the route using plants consists in replacing the auxiliary agents used in the conventional synthesis for the phytochemicals extracted from different parts of plants, fruits, or agro-wastes (Biomass extract).

11.3.1 Synthesis via microorganisms

The biological synthesis consists mainly of the manipulation of microorganisms such as bacteria, actinomycetes, fungi, yeast, algae, and viruses to produce chemical compounds of interest such as metallic particles and semiconductors. The catalysis of the substrates can take place in intracellular or extracellular mode, through enzymatic reactions during the metabolism of the microorganism. Smaller particles are obtained in the intracellular mode because the wall of the cell limits their growth. On the other hand, phytochemicals can be extracted by centrifugation, for example, from algae, and used in a similar way to biomass extract. In its case, the yeast has been reported as a template for the growth of TiO_2, promoting a porous thin film. Órdenes et al. biosynthesized TiO_2 for a dye-sensitized solar cell using the bacteria *Bacillus mycoide*. At the moment, this

Table 11.3 Materials with photovoltaic interest synthesized through nonbiological green methods.

Material	Method	Shape	Size	References
Cu_2ZnSnS_4	Hydrothermal	Irregular	50 nm	[5]
Cu_2ZnSnS_4		Spherical	29 nm (SEM)	[3]
Cu_2ZnSnS_4		Spherical	4–5 nm (TEM)	[42]
$Cu_2ZnSnSe_4$		Oval-like	80–120 (SEM)	[6]
ZnO		Nanoplates	500 × 30 nm (TEM)	[9]
Cu_2ZnSnS_4	Solvothermal	Spherical	20 nm	[43]
Cu_2O	Sonochemical	Cubic	400 nm (TEM)	[24]
Cu_2O		Spherical	80–150 nm (TEM)	[44]
CuO		Rectangular bars	0.6 × 1.8 μm (SEM)	[25]
CuO		Nanorods, spherical, and triangular	4–12 nm (XRD)	[45]
Cu_2ZnSnS_4	Mechanosynthesis	Irregular	6 nm (XRD)	[18]
$Cu_2ZnSnSe_4$		Irregular	30–40 nm (XRD)	[18]
Cu_2ZnSnS_4		Irregular	10 nm (XRD)	[46]
Cu_2ZnSnS_4		Irregular	100 nm (TEM)	[47]
Cu_3SbS_4	Microwave-assisted	Flower-like	1.5 μm (TEM)	[11]
$Cu_2ZnSnSe_4$		Irregular	6–8 nm (TEM)	[13]
Zn_2SnO_4		Octahedral	100 nm - 2 μm (SEM)	[16]
Cu_2ZnSnS_4		Spherical	1–2 μm (TEM)	[14]
CdS		Thin film	—	[17]
Co_3O_4	Magnetic field assisted	Chain	200 nm 10 μm (SEM)	[39]
Fe_2O_3		Thin film	—	[37]
Fe_2O_3		Thin film	—	[38]
Fe_3O_4		Spherical	2 nm (TEM)	[40]
Fe_3O_4		Cubic	—	[41]
$BiVO_4$	Photo-assisted	Thin film	—	[29]
ZnSe		Thin film	—	[31]
ZnO		Thin film	—	[32]
MoS_2		Thin film	—	[33]

is the unique report of semiconductor biosynthesis through the manipulation of microorganisms studied in solar cell technology [48]. Although these microorganisms have been manipulated for the synthesis of silver and gold metallic particles, mainly for biomedical and antimicrobiological applications, various semiconductors of photovoltaic interest have been synthesized in this way, listed in Table 11.4 [49–70], opening the

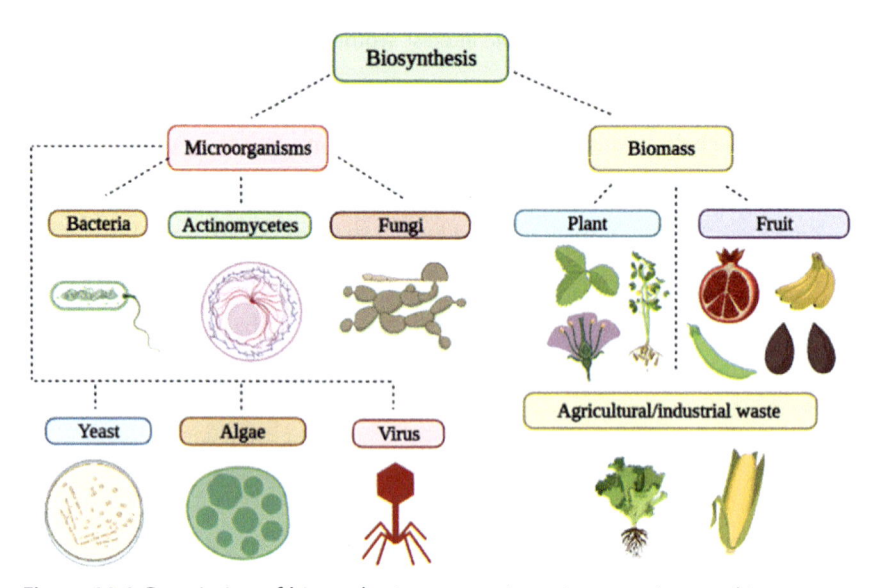

Figure 11.4 Description of biosynthesis routes using microorganisms or biomass.

possibility for their consideration in the dye-sensitized and quantum dot solar cells due to their size. Even these metallic gold and silver particles biosynthesized can be tested to improve electrical properties in these same types of solar cells.

11.3.2 Synthesis using biomass extracts

The extracts of plants, fruits, or agro-industrial wastes through the use of green solvents are employed as auxiliary agents during the synthesis of metallic and semiconductor materials. The compounds in the extracts such as flavonoids, terpenoids, steroids, carotenes, proteins, and more phytochemicals act as reducing, surfactant, capping, or stabilizing agents in a similar way an auxiliary agent does in a conventional hydrothermal or solvothermal synthesis process. This route of synthesis allows the production of particles with a great variety in size and shapes. Besides the full use of agri-waste (leaves, roots, tubers, flowers, barks of plants and trees, peels, pulp, seeds, and pods of fruits), some further advantages include the low toxicity of by-products, the low or zero use of auxiliary agents, low cost when raw materials of extracts are wastes, and the possible simultaneous use of other green processes such as the microwave or sonochemical synthesis. However, longer processes and greater use of water and solvents are required because of the extraction and purification processes. The

Table 11.4 Materials synthesized using microorganisms with photovoltaic interest.

Material	Microorganism	Name	Shape	Size	References
TiO$_2$	Bacteria	Lactobacillus sp	Spherical	25 nm (TEM)	[49]
ZnO		Aeromonas hydrophila	Spherical	<100 nm (SEM)	[50]
ZrO		Acinetobacter sp	Spherical	15 nm (TEM)	[51]
NiO		Microbacterium sp. Mrs-1	Flower-like	100–500 nm (SEM)	[52]
CuO	Actinomycetes	Actinomycete	Clusters	200 nm (SEM)	[53]
CdS	Fungi	Coriolus versicolor	Spherical	5–9 nm (TEM)	[54]
CeO$_2$		Aspergillus niger	Cubic and spherical	5–20 nm (TEM)	[55]
TiO$_2$		Aspergillus flavus	Hexagonal-spherical	60 nm (TEM)	[56]
CdS		Phanerochaete chrysosporium	Spherical	2 nm (TEM)	[57]
CdTe		Fusarium oxysporum	Spherical	15–20 nm (TEM)	[58]
ZnO	Yeast	Pichia kudriavzevii	Hexagonal	2–60 nm (TEM)	[59]
TiO$_2$		Saccharomyces cerevisae	Spherical	13 nm (TEM)	[49]
PbS		Rhodosporidium diobovatum	Spherical	2–5 nm (TEM)	[60]
CdS		Saccharomyces cerevisae	Spherical	2–5 nm (TEM)	[61]
ZnS		Saccharomyces cerevisae	Spherical	30–40 mm (TEM)	[62]
CuO/Cu$_2$O	Algae	Bifurcaria bifurcata	Spherical	5–45 nm (TEM)	[63]
Fe$_3$O$_4$		Ulva flexuosa	Cubo-spherical	12 nm (TEM)	[64]
ZnO		Sargassum muticum	Hexagonal	30–60 nm (SEM)	[65]
CdS		Chlamydomonas reinhardtii	Spherical	2–7 nm (TEM)	[66]
ZnO		Gracilaria edulis	Rods	60–100 nm (SEM)	[67]
CdS	Virus	M13 bacteriophage	Wire	3–5 nm (TEM)	[68]
ZnS		M13 bacteriophage	Spherical	20 nm (TEM)	[69]
ZnS		M13 bacteriophage	Wire	3–5 nm (TEM)	[68]
PbS		Tobacco Mosaic	Irregular	30 nm (TEM)	[70]

extracts of different parts of plants, fruit, and waste have been studied for the synthesis of materials with photovoltaic interest; some of them are enlisted in Table 11.5 [58,71−80].

The biosynthesis of Cu_2O using banana peel extract as a reducing agent, instead of chemical glucose, is a successful example of greener alternatives. The extraction of phytochemicals is processed with water; copper precursor is dissolved in water and mixed with the extract. Homogeneous and well-defined octahedral particles precipitate after heating. This is how simple a biosynthesis process can be, the detailed information on the process is in the reference [77] and the octahedral particles are presented in Fig. 11.5.

There is a great interest in the synthesis of materials for solar technology following the principles of green chemistry. It is true that solar technology such as solar cells, photoelectrochemical cells, or photocatalysts are respectful to the environment in their use but also the process that involves their creation, that is, the synthesis of the materials, and the methods to manufacture the structure need to be kind with nature.

Table 11.5 Material biosynthesized with plants or fruit extract with photovoltaic interest.

Material	Part	Name	Shape	Size	References
TiO_2	Plant extract (roots)	Glycyrrhiza glabra	Spherical	60−140 nm (EM)	[71]
TiO_2	Plant extract (flowers)	Caesalpinia pulcherrima	Spherical	20−25 nm (TEM)	[72]
TiO_2		Hibiscus Flower	Spherical	7 nm (XRD)	[58]
ZnO		Jacaranda mimosifolia	Spherical	2−4 nm (TEM)	[73]
TiO_2	Plant extract (stem)	Tinospora. cordifolia	Spherical	8−10 nm (XRD)	[74]
ZnO		Swertia chirayita	Spherical	10−40 nm (TEM)	[75]
ZnO		Euphorbia tirucalli	Spherical	20 nm (SEM)	[76]
Cu_2O	Fruit extract (peel)	Banana	Octahedral	1−4 μm (SEM)	[77]
TiO_2		Mango	Rice-grain-like	28 nm (TEM)	[78]
TiO_2		Mango	Spherical	17 nm (TEM)	[78]
CdS	Fruit extract (pulp)	Opuntia ficus-indica	Spherical	3−5 nm (TEM)	[79]
TiO_2		Aloe vera	Tetragonal	6−13 nm (TEM)	[80]

Figure 11.5 Octahedral particles of Cu_2O biosynthesized using banana peel extract as reducing agent. *Reproduced with permission from S. Torres-Arellano, O. Reyes-Vallejo, J. Pantoja Enriqueza, J.L. Aleman-Ramirez, A.M. Huerta-Flores, J. Moreira, J. Muñiz, L. Vargas-Estrada, P.J. Sebastian, Biosynthesis of cuprous oxide using banana pulp waste extract as reducing agent, Fuel 285 (2021) 119152, copyright (2021) Elsevier.*

Biosynthesized materials with possible photovoltaic interest are listed in Tables 11.4 and 11.5, including literature references. Although these particles have not been applied in solar cell technology, they do present promising characteristics for their use as absorber and window films, charge extractors, and porous or buffer layers, which can be deposited through techniques such as doctor blade, spray pyrolysis, spin coating, and drop casting for its study in solar cell technology. Fig. 11.6 shows a schematic representation of the solar cell fabrication through the biosynthesis process.

11.4 Photovoltaic materials through green synthesis

In Table 11.6, solar cells using materials synthesized by green methods described above are presented [4,7,8,10,12,19,23,26–28,30,32,34–36,81–92]. Both *p*- and *n*-type semiconductors are used in regular *p-n*, quantum dot-sensitized, dye-sensitized, inorganic electrolytic, and perovskite solar cells showing high efficiencies, even in some cases compared with conventional methods. Ramakrishnan et al. compared microwave-assisted and conventional solvothermal synthesis of TiO_2 nanoparticles for dye-sensitized solar cells, measuring efficiencies of 5.98% and 7.44%, respectively [8].

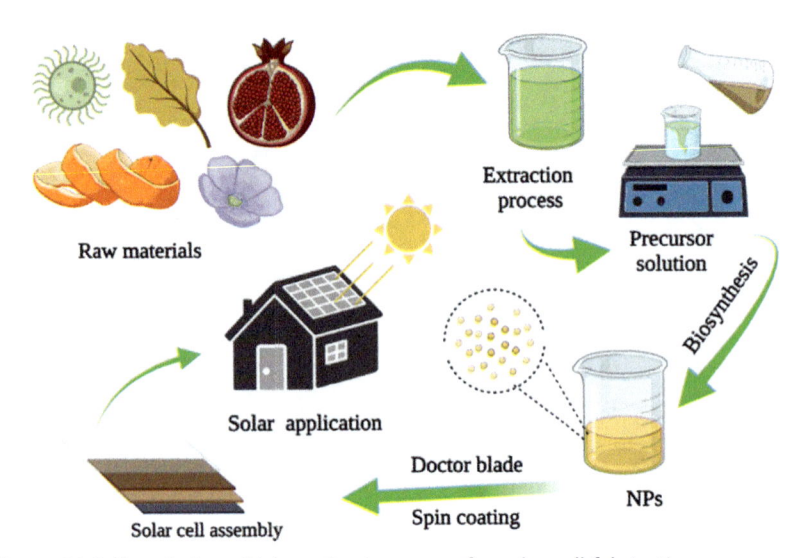

Figure 11.6 Description of biosynthesis process for solar cell fabrication.

The better performance of the microwave-assisted route is explained because of the higher superficial area and uniformity of the particles and film deposited, which allows better diffusion of electrolytes. Besides, higher electrical properties were measured, which improved the charge transference [8]. Prochowicz et al. found that a previous mechanosynthesis of precursors increased the efficiency of a perovskite solar cell from 8.0% to 8.9% compared with a similar solar cell without a mechanosynthesis step [19]. In a similar way, Lee et al. measured an increased efficiency when precursor formamidinium iodide (FAI) dissolved in isopropyl alcohol (IPA) is exposed to Light Emitting Diode (LED light), achieving efficiencies up to 22%, which is higher than without illumination. According to them, this is explained because light promotes higher ionization of precursor [35]. A similar increase in efficiency has been observed in solar cells when electrodeposition, sputtering, and chemical baths have been exposed to light [30,32,34]. Perovskite solar cells also have been assisted by a magnetic field, increasing their efficiency from 15.52% to 18.56% when the spin coating of perovskite is stimulated according to Wang [36]. These examples probe how eco-friendly methods can be used directly or indirectly to assist the synthesis and deposition of materials during the fabrication of solar cells.

On the other hand, materials produced via biosynthesis are less often studied. However, there is enthusiastic effort mainly in dye-sensitized solar cells and TiO_2 and ZnO materials, as can be observed in Table 11.7

Table 11.6 Solar cells using material synthesized by green methods.

Material	Method	Shape	Size	Type	η	References
SrTiO$_3$	Hydrothermal	Spherical	620 nm (TEM)	Dye-sensitized	–	[81]
TiO$_2$		Spherical	10 nm (TEM)	Dye-sensitized	4.37	[82]
TiO$_2$		Pseudo-cubic	18 nm (TEM)	Dye-sensitized	5.05	[4]
(CH$_3$NH$_3$)$_3$Bi$_2$I$_9$	Solvothermal/spin coating	Thin film	–	Perovskite	1.62	[83]
Cu$_2$ZnSnS$_4$	Solvothermal	Thin film	–	Dye-sensitized	5.65	[84]
Cu$_2$ZnSnS$_4$		Spherical	–	Perovskite	10.88	[7]
TiO$_2$		Spherical	6 nm (TEM)	Dye-sensitized	5.98	[8]
TiO$_2$		Nanowires	10 nm (TEM)	Dye-sensitized	7.78	[85]
TiO$_2$		Nanowires	10 nm (TEM)	Perovskite	8.52	[85]
AgInS$_2$	Sonochemical	Spherical	25–35 nm (TEM)	Inorganic electrolytic	0.12	[28]
CuIn$_{0.7}$Ga$_{0.3}$Se$_2$		–	20–30 nm (TEM)	p–n	0,16	[86]
NiSe$_2$		Spherical	6 nm (XRD)	Dye-sensitized	7.88	[26]
TiO$_2$		Spherical	50–200 nm (TEM)	Dye-sensitized	1.5	[27]
ZnO		Hierarchical	2–5 μm (TEM)	Dye-sensitized	6.42	[23]
CH$_3$NH$_3$PbI$_3$	Mechanosynthesis	Irregular	250–450 nm (SEM)	Perovskite	9.1	[19]
Cu$_2$SnS$_3$		–	–	p–n	1.1	[87]
Cu$_2$ZnSnS$_4$		–	–	p–n	1.52	[88]
Cu$_2$ZnSnSe$_4$		irregular	10–25 nm (TEM)	p–n	0.18	[89]

(*Continued*)

Table 11.6 (Continued)

Material	Method	Shape	Size	Type	η	References
Fe_3O_4		Spherical	400 nm (SEM)	p–n	1.53	[90]
$CuInSe_2$	Microwave-assisted	Plates, spherical	nm–µm	Inorganic electrolytic	–	[12]
Cu_2ZnSnS_4		Triangular round	6–22 nm (TEM)	p–n	4.48	[91]
TiO_2		Spherical	9 nm (TEM)	Dye-sensitized	7.44	[8]
TiO_2		Spherical	10 nm (TEM)	Dye-sensitized	6.6	[92]
ZnO		Flower-like (Hierarchical)	3–4 µm (TEM)	Dye-sensitized	4.12	[10]
$CH_3NH_3PbI_3$	Magnetic field-assisted	Thin film	–	Perovskite	18.56	[36]
CdSe	Photo-assisted	–	3–10 nm (TEM)	p–n	3.59	[30]
CH_5IN_2 (FAI-Precursor)		–	–	Perovskite	21	[35]
TiO_2		Thin film	–	Perovskite	13.6	[34]
ZnO		Thin film	–	p–n	14.7	[32]

Table 11.7 Solar cells using material biosynthesized by green methods.

Material	Type	Name	Shape	Size	Type	η	References
Co_3O_4	Plant extract (leave)	Calotropis gigantea	Spherical	50–60 nm (TEM)	Dye-sensitized	0.66	[95]
CuO		Calotropis gigantea	Spherical	20 nm (TEM)	Dye-sensitized	3.4	[96]
TiO_2		Phyllanthus emblica	Spherical	6 nm (TEM)	Dye-sensitized	5.42	[97]
ZnO		Azadirachta indica	Irregular	4 nm (TEM)	Dye-sensitized	2.1	[98]
ZnO		Vernonia amygdalina	Spherical	10 nm (XRD)	Dye-sensitized	0.63	[99]
ZnO		Tilia Tomentosa	Spherical	80 nm (SEM)	Dye-sensitized	1.97	[100]
ZnO		Carica papaya	Spherical	50 nm (SEM)	Dye-sensitized	1.6	[101]
ZnO		Albizia Amara	Flake-like	50–60 nm (SEM)	Perovskite	7.83	[94]
ZnO	Plant extract (roots)	Kniphofia schemperi	Spherical	10 nm (TEM)	Dye-sensitized	1.3	[102]
ZnO		Amorphophallus konjac	Rice shape	240 × 80 nm	Dye-sensitized	1.66	[15]
ZnO	Plant extract (tuber)	Colocasia esculenta	Nanorods	60 × 180 nm (SEM)	Dye-sensitized	1.6	[103]
ZrO		Gloriosa superba	Hexagonal	12 nm (XRD)	Dye-sensitized	1.6	[104]
CdO	Plant extract (flowers)	Hibiscus sabdariffa	Quasi-cuboidal	50–270 nm (SEM)	p–n	–	[105]
TiO_2	Plant extract (bark)	Terminalia arjuna	Spherical	5–10 nm (TEM)	Dye-sensitized	2.79	[106]
TiO_2		Phellinus linteus	Nanorod	25–150 nm (TEM)	Dye-sensitized	3.8	[107]
$CuInS_2$	Fruit extract (peel)	Pomegranate	Spherical	10–15 nm (SEM)	Dye-sensitized	0.37	[108]
TiO_2	Fruit extract (pulp)	Grape	Nanorods	>100 nm (SEM)	Dye-sensitized	4.33	[109]
TiO_2		Pineapple	Nanorods - spherical	>100 nm (SEM)	Dye-sensitized	4.11	[109]
TiO_2		Orange	Spherical	>100 nm (SEM)	Dye-sensitized	3.89	[109]
ZnO		Spondias mombin	Spherical	14–22 nm (TEM)	Dye-sensitized	0.63	[110]
TiO_2	Fruit extract (seed)	Bixa orellana	Spherical	13 nm (TEM)	Dye-sensitized	3.97	[93]
ZnO		Coffee seed	Spherical	80 nm (TEM)	p–n	3.12	[111]
TiO_2	Bacteria	Bacillus mycoides	Spherical	40–60 nm (TEM)	Quantum Dot Sensitized	–	[48]

[15,48,93−111]. Different parts of plants and fruits have been subjected to extraction processes using solvents such as water, ethanol, and ethylene glycol at temperatures lower than 100°C. The heating process has been reported by conventional and microwave-assisted methods. These extracts have been proven to work as an excellent capping and stabilizing agent because of their phytochemicals, producing well-defined particles and reduced size (nanometers). Which have provoked high efficiencies in solar cells, even higher compared with conventional ones. Maurya et al. synthesized TiO_2 using *Bixa orellana* seed extract and tested it in dye-sensitized solar cells measuring efficiencies of 2.97%, which is higher than the 1.0% obtained when the conventional route is followed. This behavior is partly explained by the increase in surface area and pore size when the particle size decreases in the extract synthesis pathway [93]. Pitchaiya et al. used *Albizia amara* extract for ZnO biosynthesis and used it in perovskite solar cells measuring efficiencies of 7.83%, which is higher than when ZnO is produced by regular chemical synthesis and reaching an efficiency of 6.81% [94].

11.5 Considerations for greener fabrication of solar cells

The performance of a solar cell does not have to be faced with a nature-friendly manufacturing process. Although green synthesis of materials has been shown in the sections earlier, a completely green process should be considered considering both the materials and the processes for their integration during the manufacture of the solar cell. There are some reports about how to perform a greener integration in various types of photovoltaics such as perovskite and dye-sensitized solar cells.

11.5.1 Perovskite solar cells

Detailed considerations have been published by Zhang et al. about the green processing of chemical solutions for the deposit of perovskite [112]. In addition to efforts to replace lead with less toxic materials in the perovskite molecule, it is well known that the solutions used to dissolve the reactants and to carry out better processing of the films are highly toxic and hazardous. *N,N*-dimethylformamide, dimethylacetamide, *N*-methyl-2-pyrrolidone, dimethyl sulfoxide, and γ-butyrolactone (GBL) are common solvents used in perovskite synthesis. On the other hand, toluene (TL), chlorobenzene, and diethyl ether (DE) are antisolvents used for improved perovskite crystallization processing. However, these solvents and antisolvents are toxic [112]. Alternative safer solvents have been tested for processing perovskite solar cells such as dimethylethanolamine; a mixture of water and isopropanol demonstrates poor performance [113]. As mixtures of hazardous and safer solvents until now are promising alternatives, Gardner et al. successfully tested mixtures of GBL with ethanol (EtOH), 1-propanol, IPA, and acetic acid that approached similar efficiencies achieved with toxic solvents [113]. Meanwhile, alternative antisolvents such as methoxybenzene (PhOMe), ethyl acetate, and tetraethyl orthosilicatehave been tested successfully replacing the toxic ones [112].

Molten salts are an alternative way of perovskite solar cell processing discovered by Zhou, in 2015, when $MAPbI_3$ was exposed to CH_3NH_2 (MA). The ability to transform solids into molten salts avoids the use of toxic solvents and antisolvents [114]. Safer solvents, for example, methylammonium acetate (MAAc) and a combination of methylamine and acetonitrile, have been probed with excellent results through these processing routes [115].

11.5.2 Dye-sensitized solar cells

The dye and the electrolyte are the most characteristic components that differentiate this type of cell from the others. In addition, they are key because the first one is responsible for most of the light absorption and the second one is responsible for the regeneration of the dye after the injection of electrons or holes (charge transference). There is a great variety of dyes that can be grouped into three groups: functionalized oligopyridine metal complexes, Zn-based, and fully organic. Its ecological impact is mainly due to its toxic residues during application and during its manufacture. In addition, it should be considered that in some dyes, rare metals such as ruthenium and osmium are usually used; as in N719, one of the most used. Mariotti et al. reviewed more ecological alternatives, classifying them as free of critical raw materials (CRM-free) and natural [116]. CRM-free is a great alternative due to its flexibility in molecular design, industrial scalability, and simple and inexpensive synthesis routes. The dyes YD-2, YD2-o-C8, and SM315 based on Zn are examples of this group of dyes, registering efficiencies higher than 11% in cells where they have been used [116]. The higher abundance of Zn in the Earth's crust with respect to Ru is significant, so it is a great point in favor of Zn.

The organic dyes R6 and AdekA-1 have shown a great response because cells in which it has been used reach efficiencies greater than 12%; however, they use cobalt-based electrolytes, which limits their use. These types of dyes are interesting due to the replacement of rare metals; however, issues remain to be solved during their manufacturing like reducing the number of step synthesis and avoiding rare metal-based catalysts. On the other hand, natural dyes are very friendly with nature in their use and manufacturing process, being nontoxic compounds, free of rare metals, biodegradable, and inexpensive. These pigments are obtained in a similar way to the biosynthesis processes, using plant and fruit parts such as flowers, seeds, leaves, and roots by extraction with green solvents such as water or ethanol. These pigments are usually classified into flavonoids, carotenoids, betalains, and chlorophylls. However, the efficiency of devices using natural colorants is still low between 1 and 4%, which is mainly due to their natural tendency to degrade because their absorption is limited to the region of 200–700 nm, leaving part of the spectrum without absorbing [116]. An excellent idea published by Maiaugree et al. proposes the fabrication of both the dye and counter electrode from waste mangosteen peel. The dye by extraction of mangosteen peel and the counter electrode from the carbonized peel. The device presented an efficiency of 2.6% [117].

The electrolyte should have high stability and high transparency in the visible region and appropriate redox potential with respect to the HOMO level of the sensitizer. The most used electrolytes are liquids because of high charge carrier diffusion and effective permeation in the photoelectrode porous structure, however, leakage is a great problem. The most common electrolytes are based on iodine or cobalt. Iodine electrolyte presents great dye regeneration and charge transference performance but is very corrosive with counter electrodes such as Pt. Also, I_2 is very volatile even in quasisolid electrolytes and absorbs part of the light limiting absorption by the dye. Meanwhile, cells with cobalt-based electrolytes have shown great efficiency; however, cobalt is a CRM, which increases its cost or limits its use from an industrial perspective [116]. Alternative electrolytes have been proposed, such as those developed based on selenium, and they have been tested registering similar performance to those based on iodine; however, selenium is scarce and a toxic material at high concentrations [118]. Iron-based electrolytes are interesting alternatives, considering their great abundance. However, many of its complexes are sensitive to air, so they must be processed in inert environments, limiting their industrial scaling. Some that are soluble in water are interesting because they are not corrosive, but their performance is still low compared to those based on iodine. This is explained by less regeneration of dye occurring and more recombination at TiO_2/electrolyte due to its low redox potential. Besides, problems of photodecomposition have been observed under UV and near UV light [119]. Meanwhile, copper-based electrolytes are very interesting because copper is nontoxic, is more abundant than cobalt, and its complexes are stable in air, so they are more scalable than iron-based electrolytes. The efficiencies that have been achieved are slightly lower than those recorded with iodine-based electrolytes but still remain competitive. However, these complexes use a variety of toxic substances during their production, such as TL and DE. Once this issue is solved, the copper-based electrolyte will be a great candidate.

Other catalysts based on vanadium, molybdenum, and tungsten are also under investigation, as well as totally organic and sulfur-based catalysts, which are friendly with nature. However, there are still some issues during their synthesis that have to be solved before being considered industrially scalable. Water-based electrolytes are very attractive from an ecological point of view because water is nontoxic as in the case of nitrile-based organics, but for a good performance, water must be ultrapure, so its cost and energy processing increase considerably. Electrolytes based on mixtures of water with organic solvents and electrolytes composed entirely of water with surfactants are a trend in electrolyte research, which undoubtedly represents an ecological improvement. Although these solar cells present low efficiencies, they are heading toward greener developments [120,121]. The transition toward solid and quasisolid electrolytes represents an important advance in solving stability and leakage problems; however, it does not represent an important advance in ecological matters due to the use of reactants, solvents, and additives similar to those used in liquid electrolytes [116]. Novel eco-friendly, sugar-based, and rice starch-based electrolytes have been proposed with low efficiencies but opening up new research areas [122,123].

A further drawback to be solved in DSSC cells is the use of platinum as a counter electrode, not only because of its cost and scarcity that limit its scale-up but also because it tends to degrade with iodine-based electrolytes. There are numerous investigations on this subject, such as carbonaceous materials from pomelo peels, orange fiber, silicone rubber, and humic acid. Materials based on transition metals, such as Co_3S_4, MoS_2, ZnS, FeS, FeS@C, NiS, TiS, $NiMoS_4$, Cu/FeSe, and Cu_2O, and their composites with carbon-based materials show outstanding results, in some cases, better than platinum. In contrast, toxic solvents often used during their synthesis, high processing temperatures needed, very long reaction times, and the use of catalysts based on scarce metals are disadvantages that must be solved because they go against a green process and its possible industrial scale-up [116].

11.6 Conclusion: Prospects for green synthesis of solar cells

The scientific community has made encouraging efforts to develop greener processes for solar cell processing. Many of them develop a single component of the structure, it is understandable that to show a comparative analysis with the best devices, these are taken as a basis, for example, testing TiO_2 particles developed by a biosynthesis process based on an extract of a plant in a DSSC built through typical processes and components that present their highest efficiency. It is time to consider a completely green and sustainable development of this type of device, from one contact to the other, as well as the origin of the reactants, solvents, and agents, taking into account the energy consumption on its manufacturing, the generation of waste, and its possible treatment.

The simultaneous use of green synthesis and deposition methods in order to avoid the use of auxiliary agents, toxic solvents, and rare and scarce metal catalysts and to reduce energy consumption and processing time is possible. The sonochemical, ball milling, photodeposition, microwave, and magnetic field can assist material processing and solar cell manufacturing, performing better structures, interfaces, and films developed by chemical bath, spin coating, spray pyrolysis, chemical vapor deposition, even with sputtering and thermal annealing, which leads to more efficient devices.

The use of biomass is another aspect to consider because it can be used for the development of materials of photovoltaic interest. Biosynthesis processes for semiconductor development are one of them and were previously described. However, biomass itself can be used for the development of photovoltaic semiconductors, materials such as carbon, graphene, graphene oxide, and its reduced oxide are clear examples, which can be synthesized from carbonaceous wastes like rice husk, glucose, hemp, paper cups, sugarcane bagasse, polyethylene terephthalate bottles, and chitosan, to mention some, but the sources are infinite [124,125]. In turn, the reduced graphene oxide can be processed by biosynthesis, through microorganisms like bacteria (*Shewanella, Escherichia coli*, and *Bacillussibtilus*) and through plant extracts like leaves, roots, rose water, and coconut water [126]. Materials like these present outstanding optical, electrical, electrocatalytic, and mechanical properties and have been tested in $p-n$ hetero-junction, dye-sensitized, quantum dot-sensitized, organic, and perovskite solar cells.

Carbon-based materials can be used as a transparent electrode, active interfacial layer, electron transport layer, hole transport layer, or electron/hole separation layer, which show very promising results most of the time [127–129]. Not only by themselves but also as a composite with semiconductors and metals, the carbon-based compounds present tremendous potential for their application. The development of carbon contacts is another field of application of biomass, by itself and as a composite. The fact that biomass is taken into consideration in photovoltaic development is very important and shows a clear trend toward the development of greener solar cell technology (Fig. 11.7). However, the processing of biomass must be reviewed, taking into consideration solvents and auxiliary agents used, as well as energy consumption and wastes produced.

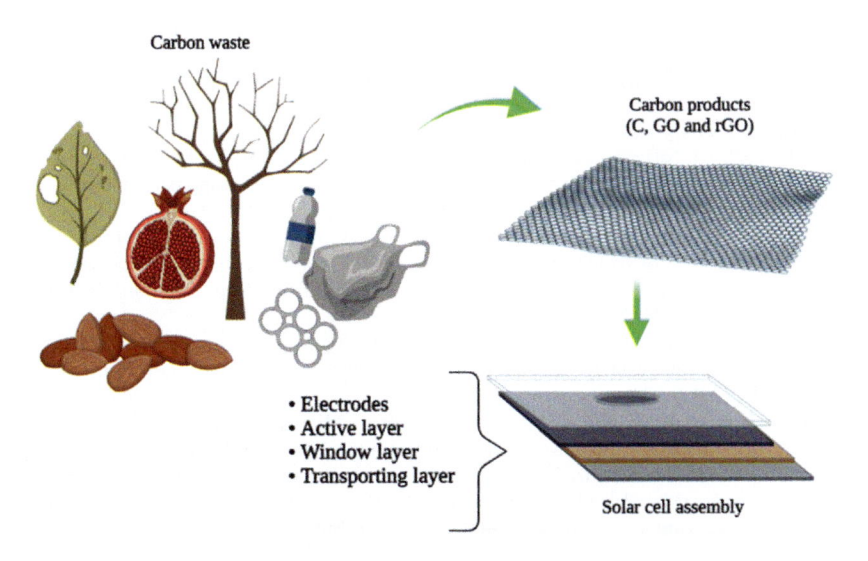

Figure 11.7 Carbon-based materials from wastes.

References

[1] P.T. Anastas, J.C.Warner Green, Chemistry: Theory and Practice, Oxford University Press, New York, 1998.
[2] D. Prat, J. Hayler, A. Wells, A survey of solvent selection guides, Green. Chem. 16 (10) (2014) 4546–4551.
[3] S.K. Aditha, A.D. Kurdekar, L.A. Chunduri, S. Patnaik, V. Kamisetti, Aqueous based reflux method for green synthesis of nanostructures: application in CZTS synthesis, MethodsX 3 (2016) 35–42.
[4] Y.X. Dong, X.L. Wang, E.M. Jin, S.M. Jeong, B. Jin, S.H. Lee, One-step hydrothermal synthesis of Ag decorated TiO_2 nanoparticles for dye-sensitized solar cell application, Renew. Energy 135 (2019) 1207–1212.
[5] S.A. Vanalakar, A.S. Kamble, S.W. Shin, S.S. Mali, G.L. Agawane, V.L. Patil, et al., Simplistic toxic to non-toxic hydrothermal route to synthesize Cu_2ZnSnS_4 nanoparticles for solar cell applications, Sol. Energy 122 (2015) 1146–1153.
[6] S.A. Vanalakar, G.L. Agawane, A.S. Kamble, P.S. Patil, J.H. Kim, The green hydrothermal synthesis of nanostructured $Cu_2ZnSnSe_4$ as solar cell material and study of their structural, optical and morphological properties, Appl. Phys. A 123 (12) (2017) 782.
[7] Z. Shadrokh, S. Sousani, S. Gholipour, Y. Abdi, Enhanced photovoltaic performance and stability of perovskite solar cells by interface engineering with poly (4-vinylpyridine) and Cu_2ZnSnS_4 & CNT, Sol. Energy 201 (2020) 908–915.
[8] V.M. Ramakrishnan, S. Pitchaiya, N. Muthukumarasamy, K. Kvamme, G. Rajesh, S. Agilan, et al., Performance of TiO_2 nanoparticles synthesized by microwave and solvothermal methods as photoanode in dye-sensitized solar cells (DSSC), Int. J. Hydrog. Energy 45 (51) (2020) 27036–27046.
[9] L. Zhu, Y. Li, W. Zeng, Hydrothermal synthesis of hierarchical flower-like ZnO nanostructure and its enhanced ethanol gas-sensing properties, Appl. Surf. Sci. 427 (2018) 281–287.
[10] R. Krishnapriya, S. Praneetha, A.V. Murugan, Investigation of the effect of reaction parameters on the microwave-assisted hydrothermal synthesis of hierarchical jasmine-flower-like ZnO nanostructures for dye-sensitized solar cells, N. J. Chem. 40 (6) (2016) 5080–5089.
[11] G. Chen, W. Wang, J. Zhao, W. Yang, S. Chen, Z. Huang, et al., Study on the synthesis and formation mechanism of flower-like Cu_3SbS_4 particles via microwave irradiation, J. Alloy. Compd. 679 (2016) 218–224.
[12] M. Sabet, M. Salavati-Niasari, D. Ghanbari, O. Amiri, N. Mir, M. Dadkhah, Synthesis and characterization of $CuInSe_2$ nanocrystals via facile microwave approach and study of their behavior in solar cell, Mater. Sci. Semicond. Proc. 25 (2014) 98–105.
[13] O.R. Vallejo, M. Sánchez, M. Pal, R. Espinal, J. Llorca, P.J. Sebastian, Synthesis and characterization of nanoparticles of CZTSe by microwave-assited chemical synthesis, Mater. Res. Express 3 (12) (2016) 125017. Available from: https://doi.org/10.1088/2053-1591/3/12/125017.
[14] R.S. Kumar, B.D. Ryu, S. Chandramohan, J.K. Seol, S.K. Lee, C.H. Hong, Rapid synthesis of sphere-like Cu_2ZnSnS_4 microparticles by microwave irradiation, Mater. Lett. 86 (2012) 174–177.
[15] P.N. Kumar, K. Sakthivel, V. Balasubramanian, Microwave assisted biosynthesis of rice shaped ZnO nanoparticles using Amorphophallus konjac tuber extract and its application in dye sensitized solar cells, Mater. Sci.-Poland 35 (1) (2017) 111–119.
[16] O. Reyes, M. Pal, J. Escorcia-García, R. Sánchez-Albores, P.J. Sebastian, Microwave-assisted chemical synthesis of Zn_2SnO_4 nanoparticles, Mater. Sci. Semicond. Proc. 108 (2020) 104878.

[17] M. Husham, Z. Hassan, A.M. Selman, N.K. Allam, Microwave-assisted chemical bath deposition of nanocrystalline CdS thin films with superior photodetection characteristics, Sens. Actuators A: Phys. 230 (2015) 9−16.
[18] T.S. Shyju, S. Anandhi, R. Suriakarthick, R. Gopalakrishnan, P. Kuppusami, Mechanosynthesis, deposition and characterization of CZTS and CZTSe materials for solar cell applications, J. Solid. State Chem. 227 (2015) 165−177.
[19] D. Prochowicz, M. Franckevičius, A.M. Cieślak, S.M. Zakeeruddin, M. Grätzel, J. Lewiński, Mechanosynthesis of the hybrid perovskite $CH_3NH_3PbI_3$: characterization and the corresponding solar cell efficiency, J. Mater. Chem. A 3 (41) (2015) 20772−20777.
[20] S. Patra, B. Satpati, S.K. Pradhan, Quickest single-step mechanosynthesis of CdS quantum dots and their microstructure characterization, J. Nanosci. Nanotechnol. 11 (6) (2011) 4771−4780.
[21] S. Dhara, P.K. Giri, Quick single-step mechanosynthesis of ZnO nanorods and their optical characterization: Milling time dependence, Appl. Nanosci. 1 (4) (2011) 165−171.
[22] G. Chatel, How sonochemistry contributes to green chemistry? Ultrason. Sonochem. 40 (B) (2018) 117−122.
[23] Y. Shi, C. Zhu, L. Wang, C. Zhao, W. Li, K.K. Fung, et al., Ultrarapid sonochemical synthesis of ZnO hierarchical structures: from fundamental research to high efficiencies up to 6.42% for quasi-solid dye-sensitized solar cells, Chem. Mater. 25 (6) (2013) 1000−1012.
[24] K. Kaviyarasan, S. Anandan, R.V. Mangalaraja, T. Sivasankar, M. Ashokkumar, Sonochemical synthesis of Cu_2O nanocubes for enhanced chemiluminescence applications, Ultrason. Sonochem. 29 (2016) 388−393.
[25] M.A. Ávila-López, E. Luevano-Hipolito, L.M. Torres-Martínez, CO_2 adsorption and its visible-light-driven reduction using CuO synthesized by an eco-friendly sonochemical method, J. Photochem. Photobiol. A: Chem. 382 (2019) 111933.
[26] M. Mousavi-Kamazani, M. Salavati-Niasari, M. Goudarzi, A. Gharehbaii, A facile novel sonochemical-assistance synthesis of $NiSe_2$ quantum dots to improve the efficiency of dye-sensitized solar cells, J. Inorg. Organomet. Polym. Mater. 26 (1) (2016) 259−263.
[27] Y.Q. Wang, S.G. Chen, X.H. Tang, O. Palchik, A. Zaban, Y. Koltypin, et al., Mesoporous titanium dioxide: sonochemical synthesis and application in dye-sensitized solar cells, J. Mater. Chem. 11 (2) (2001) 521−526.
[28] S.M. Hosseinpour-Mashkani, A. Sadeghinia, Z. Zarghami, K. Motevalli, $AgInS_2$ nanostructures: sonochemical synthesis, characterization, and its solar cell application, J. Mater. Sci.: Mater. Electron. 27 (1) (2016) 365−374.
[29] T.W. Kim, K.S. Choi, Improving stability and photoelectrochemical performance of $BiVO_4$ photoanodes in basic media by adding a $ZnFe_2O4$ layer, J. Phys. Chem. Lett. 7 (3) (2016) 447−451.
[30] X. Wang, H. Liu, W. Shen, Controllable in situ photo-assisted chemical deposition of CdSe quantum dots on ZnO/CdS nanorod arrays and its photovoltaic application, Nanotechnology 27 (8) (2016) 085605.
[31] D.D. Hile, H.C. Swart, S.V. Motloung, T.E. Motaung, K.O. Egbo, L.F. Koao, Comparative study of photo-and non-photo-assisted chemical bath deposition of zinc selenide thin films using different volumes of hydrazine hydrate, Superlatt. Microstruct. 134 (2019) 106222.
[32] F. Tsin, D. Hariskos, D. Lincot, J. Rousset, Photo-assisted electrodeposition of a ZnO front contact on ap/n junction, Electrochim. Acta 220 (2016) 176−183.
[33] W. Teng, Y. Wang, H. Huang, X. Li, Y. Tang, Enhanced photoelectrochemical performance of MoS_2 nanobelts-loaded TiO_2 nanotube arrays by photo-assisted electrodeposition, Appl. Surf. Sci. 425 (2017) 507−517.

[34] P. Lv, S.C. Chen, C. Xu, B. Peng, Photo-assisted deposited titanium dioxide for all-inorganic CsPbI$_2$Br perovskite solar cells with high efficiency exceeding 13.6%, Appl. Phys. Lett. 117 (9) (2020) 093902.
[35] D.G. Lee, D.H. Kim, J.M. Lee, B.J. Kim, J.Y. Kim, S.S. Shin, et al., High efficiency perovskite solar cells exceeding 22% via a photo-assisted two-step sequential deposition, Adv. Funct. Mater. 31 (9) (2021) 2006718.
[36] H. Wang, J. Lei, F. Gao, Z. Yang, D. Yang, J. Jiang, et al., Magnetic field-assisted perovskite film preparation for enhanced performance of solar cells, ACS Appl. Mater. Interf. 9 (26) (2017) 21756−21762.
[37] D. Lin, B. Deng, S.A. Sassman, Y. Hu, S. Suslov, G.J. Cheng, Magnetic field assisted growth of highly dense α-Fe$_2$O$_3$ single crystal nanosheets and their application in water treatment, RSC Adv. 4 (36) (2014) 18621−18626.
[38] M. Pyeon, V. Rauch, D. Stadler, M. Gürsoy, M. Deo, Y. Gönüllü, et al., Magnetic field-assisted control of phase composition and texture in photocatalytic hematite films, Adv. Eng. Mater. 21 (8) (2019) 1900195.
[39] X. Zhao, Z. Pang, M. Wu, X. Liu, H. Zhang, Y. Ma, et al., Magnetic field-assisted synthesis of wire-like Co$_3$O$_4$ nanostructures: electrochemical and photocatalytic studies, Mater. Res. Bull. 48 (1) (2013) 92−95.
[40] K. Zhang, W. Zhao, J.T. Lee, G. Jang, X. Shi, J.H. Park, A magnetic field assisted self-assembly strategy towards strongly coupled Fe$_3$O$_4$ nanocrystal/rGO paper for high-performance lithium ion batteries, J. Mater. Chem. A 2 (25) (2014) 9636−9644.
[41] W. Xiao, X. Liu, X. Hong, Y. Yang, Y. Lv, J. Fang, et al., Magnetic-field-assisted synthesis of magnetite nanoparticles via thermal decomposition and their hyperthermia properties, CrystEngComm 17 (19) (2015) 3652−3658.
[42] S. Kant Verma, V. Agrawal, K. Jain, R. Pasricha, S. Chand, Green synthesis of nanocrystalline Cu$_2$ZnSnS$_4$ powder Using hydrothermal route, J. Nanopart. 2013 (2013) 1−7.
[43] M. Karimi, M.J. Eshraghi, V. Jahangir, A facile and green synthetic approach based on deep eutectic solvents toward synthesis of CZTS nanoparticles, Mater. Lett. 171 (2016) 100−103.
[44] M.A. Bhosale, B.M. Bhanage, A simple approach for sonochemical synthesis of Cu$_2$O nanoparticles with high catalytic properties, Adv. Powder Technol. 27 (1) (2016) 238−244.
[45] T. Pandiyarajan, R. Saravanan, B. Karthikeyan, F. Gracia, H.D. Mansilla, M.A. Gracia-Pinilla, et al., Sonochemical synthesis of CuO nanostructures and their morphology dependent optical and visible light driven photocatalytic properties, J. Mater. Sci.: Mater. Electron. 28 (3) (2017) 2448−2457.
[46] Y. Wang, H. Gong, Cu$_2$ZnSnS$_4$ synthesized through a green and economic process, J. Alloy. Compd. 509 (40) (2011) 9627−9630.
[47] P. Baláž, M. Hegedus, M. Baláž, N. Daneu, P. Siffalovic, Z. Bujňáková, et al., Photovoltaic materials: Cu$_2$ZnSnS$_4$ (CZTS) nanocrystals synthesized via industrially scalable, green, one-step mechanochemical process, Prog. Photovolt.: Res. Appl. 27 (9) (2019) 798−811.
[48] N.A. Órdenes-Aenishanslins, L.A. Saona, V.M. Durán-Toro, J.P. Monrás, D.M. Bravo, J.M. Pérez-Donoso, Use of titanium dioxide nanoparticles biosynthesized by *Bacillus mycoides* in quantum dot sensitized solar cells, Microb. Cell Factories 13 (1) (2014) 1−10.
[49] A.K. Jha, K. Prasad, A.R. Kulkarni, Synthesis of TiO$_2$ nanoparticles using microorganisms, Colloids Surf. B: Biointerfaces 71 (2) (2009) 226−229.
[50] C. Jayaseelan, A.A. Rahuman, A.V. Kirthi, S. Marimuthu, T. Santhoshkumar, A. Bagavan, et al., Novel microbial route to synthesize ZnO nanoparticles using *Aeromonas hydrophila* and their activity against pathogenic bacteria and fungi, Spectrochimica Acta Part. A: Mol. Biomol. Spectrosc. 90 (2012) 78−84.

[51] S.P. Suriyaraj, G. Ramadoss, K. Chandraraj, R. Selvakumar, One pot facile green synthesis of crystalline bio-ZrO_2 nanoparticles using *Acinetobacter* sp. KCSI1 under room temperature, Mater. Sci. Eng.: C. 105 (2019) 110021.

[52] S. Sathyavathi, A. Manjula, J. Rajendhran, P. Gunasekaran, Extracellular synthesis and characterization of nickel oxide nanoparticles from *Microbacterium* sp. MRS-1 towards bioremediation of nickel electroplating industrial effluent, Bioresour. Technol. 165 (2014) 270−273.

[53] M.I. Nabila, K. Kannabiran, Biosynthesis, characterization and antibacterial activity of copper oxide nanoparticles (CuO NPs) from actinomycetes, Biocatal.s Agric. Biotechnol. 15 (2018) 56−62.

[54] R. Sanghi, P. Verma, A facile green extracellular biosynthesis of CdS nanoparticles by immobilized fungus, Chem. Eng. J. 155 (3) (2009) 886−891.

[55] K. Gopinath, V. Karthika, C. Sundaravadivelan, S. Gowri, A. Arumugam, Mycogenesis of cerium oxide nanoparticles using *Aspergillus niger* culture filtrate and their applications for antibacterial and larvicidal activities, J. Nanostruct. Chem. 5 (3) (2015) 295−303.

[56] G. Rajakumar, A.A. Rahuman, S.M. Roopan, V.G. Khanna, G. Elango, C. Kamaraj, et al., Fungus-mediated biosynthesis and characterization of TiO_2 nanoparticles and their activity against pathogenic bacteria, Spectrochimica Acta Part. A: Mol. Biomolecular Spectrosc. 91 (2012) 23−29.

[57] G. Chen, B. Yi, G. Zeng, Q. Niu, M. Yan, A. Chen, et al., Facile green extracellular biosynthesis of CdS quantum dots by white rot fungus *Phanerochaete chrysosporium*, Colloids Surf. B: Biointerfaces 117 (2014) 199−205.

[58] A. Syed, A. Ahmad, Extracellular biosynthesis of CdTe quantum dots by the fungus Fusarium oxysporum and their anti-bacterial activity, Spectrochimica Acta Part. A: Mol. Biomolecular Spectrosc. 106 (2013) 41−47.

[59] A.B. Moghaddam, M. Moniri, S. Azizi, R.A. Rahim, A.B. Ariff, W.Z. Saad, et al., Biosynthesis of ZnO nanoparticles by a new *Pichia* kudriavzevii yeast strain and evaluation of their antimicrobial and antioxidant activities, Molecules 22 (6) (2017) 872.

[60] S. Seshadri, K. Saranya, M. Kowshik, Green synthesis of lead sulfide nanoparticles by the lead resistant marine yeast, *Rhodosporidium diobovatum*, Biotechnol. Prog. 27 (5) (2011) 1464−1469.

[61] K. Prasad, A.K. Jha, Biosynthesis of CdS nanoparticles: a improved green and rapid procedure, J. Colloid Interface Sci. 342 (1) (2010) 68−72.

[62] J.G.S. Mala, C. Rose, Facile production of ZnS quantum dot nanoparticles by *Saccharomyces cerevisiae* MTCC 2918, J. Biotechnol. 170 (2014) 73−78.

[63] Y. Abboud, T. Saffaj, A. Chagraoui, A. El Bouari, K. Brouzi, O. Tanane, et al., Biosynthesis, characterization and antimicrobial activity of copper oxide nanoparticles (CONPs) produced using brown alga extract (*Bifurcaria bifurcata*), Appl. Nanosci. 4 (5) (2014) 571−576.

[64] S. Mashjoor, M. Yousefzadi, H. Zolgharnain, E. Kamrani, M. Alishahi, Organic and inorganic nano-Fe_3O_4: aAlga *Ulva flexuosa*-based synthesis, antimicrobial effects and acute toxicity to briny water rotifer Brachionus rotundiformis, Environ. Pollut. 237 (2018) 50−64.

[65] S. Azizi, M.B. Ahmad, F. Namvar, R. Mohamad, Green biosynthesis and characterization of zinc oxide nanoparticles using brown marine macroalga *Sargassum muticum* aqueous extract, Mater. Lett. 116 (2014) 275−277.

[66] M.D. Rao, G. Pennathur, Green synthesis and characterization of cadmium sulphide nanoparticles from *Chlamydomonas reinhardtii* and their application as photocatalysts, Mater. Res. Bull. 85 (2017) 64−73.

[67] R.I. Priyadharshini, G. Prasannaraj, N. Geetha, P. Venkatachalam, Microwave-mediated extracellular synthesis of metallic silver and zinc oxide nanoparticles using macro-algae (*Gracilaria edulis*) extracts and its anticancer activity against human PC3 cell lines, Appl. Biochem. Biotechnol. 174 (8) (2014) 2777−2790.

[68] C. Mao, C.E. Flynn, A. Hayhurst, R. Sweeney, J. Qi, G. Georgiou, et al., Viral assembly of oriented quantum dot nanowires, Proc. Natl Acad. Sci. 100 (12) (2003) 6946–6951.
[69] S.W. Lee, C. Mao, C.E. Flynn, A.M. Belcher, Ordering of quantum dots using genetically engineered viruses, Science 296 (5569) (2002) 892–895.
[70] W. Shenton, T. Douglas, M. Young, G. Stubbs, S. Mann, Inorganic–organic nanotube composites from template mineralization of tobacco mosaic virus, Adv. Mater. 11 (3) (1999) 253–256.
[71] M. Bavanilatha, L. Yoshitha, S. Nivedhitha, S. Sahithya, Bioactive studies of TiO_2 nanoparticles synthesized using *Glycyrrhiza glabra*, Biocatalysis Agric. Biotechnol. 19 (2019) 101131.
[72] S. Devikala, J.M. Abisharani, M. Bharath, Biosynthesis of TiO_2 nanoparticles from *Caesalpinia pulcherrima* flower extracts, Mater. Today: Proc. 40 (2021) S185–S188.
[73] D. Sharma, M.I. Sabela, S. Kanchi, P.S. Mdluli, G. Singh, T.A. Stenström, et al., Biosynthesis of ZnO nanoparticles using *Jacaranda mimosifolia* flowers extract: synergistic antibacterial activity and molecular simulated facet specific adsorption studies, J. Photochem. Photobiol. B: Biol. 162 (2016) 199–207.
[74] A. Maurya, P. Chauhan, A. Mishra, A.K. Pandey, Surface functionalization of TiO_2 with plant extracts and their combined antimicrobial activities against *E. faecalis* and *E. coli*, J. Res. Updates Polym. Sci. 1 (1) (2012) 43–51.
[75] R. Saha, K. Subramani, S. Sikdar, K. Fatma, S. Rangaraj, Effects of processing parameters on green synthesised ZnO nanoparticles using stem extract of *Swertia chirayita*, Biocatalysis Agric. Biotechnol. 33 (2021) 101968.
[76] S. Hiremath, C. Vidya, M.L. Antonyraj, M.N. Chandraprabha, P. Gandhi, A. Jain, et al., Biosynthesis of ZnO nano particles assisted by *Euphorbia tirucalli* (pencil cactus), Int. J. Curr. Eng. Technol. (2013) 176–179 (Special Issue 1).
[77] S. Torres-Arellano, O. Reyes-Vallejo, J. Pantoja Enriquez, J.L. Aleman-Ramirez, A.M. Huerta-Flores, J. Moreira, et al., Biosynthesis of cuprous oxide using banana pulp waste extract as reducing agent, Fuel 285 (2021) 119152.
[78] I.N. Isnaeni, D. Sumiarsa, I. Primadona, Green synthesis of different TiO_2 nanoparticle phases using mango-peel extract, Mater. Lett. 294 (2021) 129792.
[79] K. Kandasamy, M. Venkatesh, Y.S. Khadar, P. Rajasingh, One-pot green synthesis of CdS quantum dots using *Opuntia ficus-indica* fruit sap, Mater. Today: Proc. 26 (2020) 3503–3506.
[80] D. Hariharan, A.J. Christy, J. Mayandi, L.C. Nehru, Visible light active photocatalyst: hydrothermal green synthesized TiO_2 NPs for degradation of picric acid, Mater. Lett. 222 (2018) 45–49.
[81] P. Jayabal, V. Sasirekha, J. Mayandi, K. Jeganathan, V. Ramakrishnan, A facile hydrothermal synthesis of $SrTiO_3$ for dye sensitized solar cell application, J. Alloy. Compd. 586 (2014) 456–461.
[82] M.N. Asghar, A. Anwar, H.M.A. Rahman, S. Shahid, I. Nadeem, Green synthesis and characterization of metal ions-mixed titania for application in dye sensitized solar cells, Toxicol. Environ. Chem. 100 (8–10) (2018) 659–676.
[83] S.M. Jain, T. Edvinsson, J.R. Durrant, Green fabrication of stable lead-free bismuth based perovskite solar cells using a non-toxic solvent, Commun. Chem. 2 (1) (2019) 1–7.
[84] S. Chen, A. Xu, J. Tao, H. Tao, Y. Shen, L. Zhu, et al., In-situ and green method to prepare Pt-free Cu_2ZnSnS_4 (CZTS) counter electrodes for efficient and low cost dye-sensitized solar cells, ACS Sustain. Chem. & Eng. 3 (11) (2015) 2652–2659.
[85] L. Chu, J. Zhang, W. Liu, R. Zhang, J. Yang, R. Hu, et al., A facile and green approach to synthesize mesoporous anatase TiO_2 nanomaterials for efficient dye-sensitized and hole-conductor-free perovskite solar cells, ACS Sustain. Chem. & Eng. 6 (4) (2018) 5588–5597.

[86] A.C. Badgujar, R.O. Dusane, S.R. Dhage, Sonochemical synthesis of $CuIn_{0.7}Ga_{0.3}Se_2$ nanoparticles for thin film photo absorber application, Mater. Sci. Semicond. Proc. 81 (2018) 17−21.
[87] Q. Chen, X. Dou, Y. Ni, S. Cheng, S. Zhuang, Study and enhance the photovoltaic properties of narrow-bandgap Cu_2SnS_3 solar cell by p−n junction interface modification, J. Colloid Interface Sci. 376 (1) (2012) 327−330.
[88] Y.R. Lin, T.C. Chou, L.K. Liu, L.C. Chen, K.H. Chen, A facile and green synthesis of copper zinc tin sulfide materials for thin film photovoltaics, Thin Solid. Films 618 (2016) 124−129.
[89] Y. Liu, X. Xu, K. Liu, H. Liu, Synthesis of $Cu_2ZnSnSe_4$ thin-film solar cells from nanoparticles by a non-vacuum mechanical ball milling and rapid thermal processing, Micro Nano Lett. 15 (13) (2020) 887−891.
[90] F. Sadegh, A.R. Modarresi-Alam, M. Noroozifar, K. Kerman, A facile and green synthesis of superparamagnetic $Fe_3O_4@$ PANI nanocomposite with a core−shell structure to increase of triplet state population and efficiency of the solar cells, J. Environ. Chem. Eng. 9 (1) (2021) 104942.
[91] W.C. Chen, V. Tunuguntla, M.H. Chiu, L.J. Li, I. Shown, C.H. Lee, et al., Co-solvent effect on microwave-assisted Cu_2ZnSnS_4 nanoparticles synthesis for thin film solar cell, Sol. Energy Mater. Sol. Cell 161 (2017) 416−423.
[92] S.G. Ullattil, P. Periyat, Microwave-power induced green synthesis of randomly oriented mesoporous anatase TiO_2 nanoparticles for efficient dye sensitized solar cells, Sol. Energy 147 (2017) 99−105.
[93] I.C. Maurya, S. Singh, S. Senapati, P. Srivastava, L. Bahadur, Green synthesis of TiO_2 nanoparticles using *Bixa orellana* seed extract and its application for solar cells, Sol. Energy 194 (2019) 952−958.
[94] S. Pitchaiya, N. Eswaramoorthy, M. Natarajan, A. Santhanam, V.M. Ramakrishnan, V. Asokan, et al., Interfacing green synthesized flake like-ZnO with TiO_2 for bilayer electron extraction in perovskite solar cells, N. J. Chem. 44 (20) (2020) 8422−8433.
[95] J.K. Sharma, P. Srivastava, G. Singh, M.S. Akhtar, S.J.M.S. Ameen, Green synthesis of Co_3O_4 nanoparticles and their applications in thermal decomposition of ammonium perchlorate and dye-sensitized solar cells, Mater. Sci. Eng.: B 193 (2015) 181−188.
[96] J.K. Sharma, M.S. Akhtar, S. Ameen, P. Srivastava, G. Singh, Green synthesis of CuO nanoparticles with leaf extract of *Calotropis gigantea* and its dye-sensitized solar cells applications, J. Alloy. Compd. 632 (2015) 321−325.
[97] T. Solaiyammal, P. Murugakoothan, Green synthesis of Au and Au@ TiO_2 core−shell structure formation by hydrothermal method for dye sensitized solar cell applications, J. Mater. Sci.: Mater. Electron. 29 (1) (2018) 491−499.
[98] P. Saravanan, K. SenthilKannan, R. Divya, M. Vimalan, S. Tamilselvan, D. Sankar, A perspective approach towards appreciable size and cost-effective solar cell fabrication by synthesizing ZnO nanoparticles from *Azadirachta indica* leaves extract using domestic microwave oven, J. Mater. Sci.: Mater. Electron. 31 (5) (2020) 4301−4309.
[99] A.N. Ossai, S.C. Ezike, A.B. Dikko, Bio-synthesis of zinc oxide nanoparticles from bitter leaf (vernonia amygdalina) extract for dye-sensitized solar cell fabrication, J. Mater. Environ. Sci. 11 (3) (2020) 421−428.
[100] R. Shashanka, H. Esgin, V.M. Yilmaz, Y. Caglar, Fabrication and characterization of green synthesized ZnO nanoparticle based dye-sensitized solar cells, J. Sci.: Adv. Mater. Dev. 5 (2) (2020) 185−191.
[101] R. Rathnasamy, P. Thangasamy, R. Thangamuthu, S. Sampath, V. Alagan, Green synthesis of ZnO nanoparticles using *Carica papaya* leaf extracts for photocatalytic and photovoltaic applications, J. Mater. Sci.: Mater. Electron. 28 (14) (2017) 10374−10381.

[102] E.T. Bekele, E.A. Zereffa, N.S. Gultom, D.H. Kuo, B.A. Gonfa, F.K. Sabir, Biotemplated synthesis of titanium oxide nanoparticles in the presence of root extract of *Kniphofia schemperi* and its application for dye sensitized solar cells, Int. J. Photoenergy 2021 (2021) 6648325. Available from: https://doi.org/10.1155/2021/6648325.

[103] K. Sakthivel, V. Balasubramanian, D. Sengottaiyan, J. Suresh, Microwave assisted green synthesis of ZnO nanorods for dye sensitized solar cell application, Indian. J. Chem. Technol. (IJCT) 25 (4) (2019) 383−389.

[104] R. Vennila, P. Kamaraj, M. Arthanareeswari, M. Sridharan, G. Sudha, S. Devikala, et al., Biosynthesis of ZrO nanoparticles and its natural dye sensitized solar cell studies, Mater. Today: Proc. 5 (2) (2018) 8691−8698.

[105] N. Thovhogi, E. Park, E. Manikandan, M. Maaza, A. Gurib-Fakim, Physical properties of CdO nanoparticles synthesized by green chemistry via *Hibiscus Sabdariffa* flower extract, J. Alloy. Compd. 655 (2016) 314−320.

[106] K. Gopinath, S. Kumaraguru, K. Bhakyaraj, S. Thirumal, A. Arumugam, Eco-friendly synthesis of TiO_2, Au and Pt doped TiO_2 nanoparticles for dye sensitized solar cell applications and evaluation of toxicity, Superlatt. Microstruct. 92 (2016) 100−110.

[107] S.U. Ekar, G. Shekhar, Y.B. Khollam, P.N. Wani, S.R. Jadkar, M. Naushad, et al., Green synthesis and dye-sensitized solar cell application of rutile and anatase TiO_2 nanorods, J. Solid. State Electrochem. 21 (9) (2017) 2713−2718.

[108] M. Mousavi-Kamazani, M. Salavati-Niasari, M. Sadeghinia, Facile hydrothermal synthesis, formation mechanism and solar cell application of $CuInS_2$ nanoparticles using novel starting reagents, Mater. Lett. 142 (2015) 145−149.

[109] R. Senthamarai, V. Madurai Ramakrishnan, B. Palanisamy, S. Kulandhaivel, Synthesis of TiO_2 nanostructures by green approach as photoanodes for dye-sensitized solar cells, Int. J. Energy Res. 21 (2) (2021) 3089−3096.

[110] H.K. Reshma, B.S. Avinash, V.S. Chaturmukha, R.L. Ashok, H.S. Jayanna, Hog plum (*Spondias mombin*) assisted ZnO nanoparticles synthesis: characterization and its impact on the performance of dye-sensitized solar cells, Mater. Today: Proc. 37 (2021) 434−439.

[111] P. Sutradhar, M. Debbarma, M. Saha, Microwave synthesis of zinc oxide nanoparticles using coffee powder extract and its application for solar cell, Synth. React. Inorganic, Metal-Organic, Nano-Metal Chem. 46 (11) (2016) 1622−1627.

[112] M. Zhang, D. Xin, X. Zheng, Q. Chen, W.H. Zhang, Toward greener solution processing of perovskite solar cells, ACS Sustain. Chem. Eng. 8 (35) (2020) 13126−13138.

[113] K.L. Gardner, J.G. Tait, T. Merckx, W. Qiu, U.W. Paetzold, L. Kootstra, et al., Nonhazardous solvent systems for processing perovskite photovoltaics, Adv. Energy Mater. 6 (14) (2016) 1600386.

[114] Z. Zhou, Z. Wang, Y. Zhou, S. Pang, D. Wang, H. Xu, et al., Methylamine-gas-induced defect-healing behavior of $CH_3NH_3PbI_3$ thin films for perovskite solar cells, Angew. Chem. 127 (33) (2015) 9841−9845.

[115] N.K. Noel, S.N. Habisreutinger, B. Wenger, M.T. Klug, M.T. Hörantner, M.B. Johnston, et al., A low viscosity, low boiling point, clean solvent system for the rapid crystallisation of highly specular perovskite films, Energy Environ. Sci. 10 (1) (2017) 145−152.

[116] N. Mariotti, M. Bonomo, L. Fagiolari, N. Barbero, C. Gerbaldi, F. Bella, et al., Recent advances in eco-friendly and cost-effective materials towards sustainable dye-sensitized solar cells, Green. Chem. 22 (21) (2020) 7168−7218.

[117] W. Maiaugree, S. Lowpa, M. Towannang, P. Rutphonsan, A. Tangtrakarn, S. Pimanpang, et al., A dye sensitized solar cell using natural counter electrode and natural dye derived from mangosteen peel waste, Sci. Rep. 5 (1) (2015) 1−12.

[118] F. Bella, A. Sacco, G.P. Salvador, S. Bianco, E. Tresso, C.F. Pirri, et al., First pseudohalogen polymer electrolyte for dye-sensitized solar cells promising for in situ photopolymerization, J. Phys. Chem. C. 117 (40) (2013) 20421−20430.

[119] I.R. Perera, T. Daeneke, S. Makuta, Z. Yu, Y. Tachibana, A. Mishra, et al., Application of the tris (acetylacetonato) iron (III)/(II) redox couple in p-type dye-sensitized solar cells, Angew. Chem. Int. Ed. 54 (12) (2015) 3758−3762.

[120] Z. Hui, Y. Xiong, L. Heng, L. Yuan, W. Yu-Xiang, Explanation of effect of added water on dye-sensitized nanocrystalline TiO_2 solar cell: Correlation between performance and carrier relaxation kinetics, Chin. Phys. Lett. 24 (11) (2007) 3272−3275.

[121] Y.S. Jung, B. Yoo, M.K. Lim, S.Y. Lee, K.J. Kim, Effect of Triton X-100 in water-added electrolytes on the performance of dye-sensitized solar cells, Electrochim. Acta 54 (26) (2009) 6286−6291.

[122] C.L. Boldrini, N. Manfredi, F.M. Perna, V. Capriati, A. Abbotto, Eco-friendly sugar-based natural deep eutectic solvents as effective electrolyte solutions for dye-sensitized solar cells, ChemElectroChem 7 (7) (2020) 1707−1712.

[123] K.C. Yogananda, E. Ramasamy, S. Kumar, S.V. Kumar, M.N. Rani, D. Rangappa, Novel rice starch based aqueous gel electrolyte for dye sensitized solar cell application, Mater. Today: Proc. 4 (11) (2017) 12238−12244.

[124] N. Raghavan, S. Thangavel, G. Venugopal, A short review on preparation of graphene from waste and bioprecursors, Appl. Mater. Today 7 (2017) 246−254.

[125] M.T.U. Safian, U.S. Haron, M.M. Ibrahim, A review on bio-based graphene derived from biomass wastes, BioResources 15 (4) (2020) 9756.

[126] M. Agharkar, S. Kochrekar, S. Hidouri, M.A. Azeez, Trends in green reduction of graphene oxides, issues and challenges: a review, Mater. Res. Bull. 59 (2014) 323−328.

[127] T. Mahmoudi, Y. Wang, Y.B. Hahn, Graphene and its derivatives for solar cells application, Nano Energy 47 (2018) 51−65.

[128] S. Das, P. Sudhagar, Y.S. Kang, W. Choi, Graphene synthesis and application for solar cells, J. Mater. Res. 29 (3) (2014) 299−319.

[129] M.Z. Iqbal, A.U. Rehman, Recent progress in graphene incorporated solar cell devices, Sol. Energy 169 (2018) 634−647.

deposition, and crystallinity enabling structures and phases that might not be achievable in bulk [6].

Nonetheless, the best efficiency results for NC-based solar cells have been obtained with NCs containing lead [7], a toxic heavy metal that tends to bioaccumulate in soils, plants, and animals, harming the environment and human health [8−10]. Furthermore, several governments are already regulating the use of lead due to contamination issues during fabrication, use, and disposal. All these facts might compromise lead-based solar cell commercialization [11]. The toxicity issues might be less severe in solar cell farms or large-scale solar energy generation facilities but should be taken into consideration in wearable or indoor applications of solar cells, where NCs are making great progress and where health is primary [12]. Therefore special focus has been put on nontoxic and environmentally friendly NCs to be used as active layers in PV applications. In this chapter, we focus on the nontoxic compounds based on antimony (Sb), bismuth (Bi), tin (Sn), and iron (Fe) and their application in solution-processed solar cells including the progress using NCs.

12.2 Antimony compounds

Research on antimony-based solar cells has been driven by antimony's abundance, low cost, and low toxicity. These properties make antimony an attractive alternative to other common solar cell materials that contain lead, cadmium, tellurium, gallium, and arsenic [13,14]. Both binary compounds such as Sb_2S_3, Sb_2Se_3, and $Sb_2(S,Se)_3$, as well as ternary compounds, such as $CuSbS_2$ and $CuSbSe_2$, have exhibited promising efficiencies in heterojunction (HJ) and sensitized solar cells (SSCs) as shown in Table 12.1 [15−24].

In addition to offering low-cost production methods, solution-processed nanoparticle solar cells enable fine-tuning of electronic and optical properties such as bandgap, carrier type, and carrier concentration through control of synthesis conditions. Nanoparticle synthesis can also result in pure phases, hence preventing the formation of impurity phases that can trap charges in antimony chalcogenide solar cells [14]. Despite these advantages, solution-processed solar cells based on antimony-containing compounds remain largely unstudied. It is, therefore, not surprising that the few reports of solution-processed antimony-based solar

Table 12.1 Summary of the highest-performing antimony-based solar cells.

Compound	Type of material	Efficiency (%)	V_{oc} (mV)	J_{sc} (mA/cm^{-2})	FF (%)	Description	Ref.
Sb_2S_3	Bulk	7.5	711	16.1	65	SSC created by CBD on mesoporous TiO_2	[15]
	NCs	1.5	421	7.2	49	Dip coating of colloidal NCs onto mesoporous TiO_2	[16]
Sb_2Se_3	Bulk	9.2	400	32.6	70	Sublimation of Sb_2Se_3 nanorods to form a CdS/Sb_2Se_3 HJ	[17]
	NCs	1.2	570	4.2	52	SILAR growth of Sb_2Se_3 nanoparticles on mesoporous TiO_2	[18]
$Sb_2(S_xSe_{1-x})_3$	Bulk	10.5	664	23.8	66	Hydrothermal deposition of $Sb_2(S_xSe_{1-x})_3$ on CdS	[19]
$CuSbS_2$	Bulk	3.2	470	15.6	44	Spin coating of metal acetates then sulfurization	[20]
	NCs	1.9	380	9.9	50	Solvothermal growth of $CuSbS_2$ nanobricks on TiO_2 nanorods	[21]
$CuSbSe_2$	Bulk	4.7	336	26.3	53	Cosputtering of Cu_2Se and Sb_2Se_3	[22]
SbSeI	Bulk	4.1	473	14.8	59	Spin coating of SbI_3 on a thin layer of Sb_2Se_3	[23]

CBD, Chemical bath deposition; NCs, nanocrystals; SILAR, successive ionic layer adsorption and reaction; SSC, sensitized solar cells.

cells claim efficiencies much lower than solar cells fabricated with other solution- and vapor-deposition techniques, such as chemical bath deposition (CBD) [15], sublimation [17], hydrothermal deposition [19], and sputtering [22].

12.2.1 Binary antimony compounds and nanocrystals

Binary antimony compounds such as Sb_2S_3 and Sb_2Se_3 have desirable optical and electronic properties for PV applications such as high absorption coefficients (at peak, $\alpha > 10^5$/cm) [13,25–28]. The 1.1 eV bandgap of Sb_2Se_3 is close to ideal for a single-junction solar cell, and the 1.7 eV bandgap of Sb_2S_3 makes it a good candidate for the top cell of a tandem device on silicon with a theoretical efficiency above 40% [29]. Moreover, because Sb_2S_3 and Sb_2Se_3 form a homogenous solid solution, the sulfur and selenium content of $Sb_2(S_xSe_{1-x})_3$ can be controlled to finely tune material properties such as bandgap [13].

Savadogo et al. piqued research interest in Sb_2S_3 solar cells by reporting a photoresponse from Sb_2S_3 photoelectrochemical solar cells [30] and subsequently demonstrating n-Sb_2S_3/p-Si and n-Sb_2S_3/p-Ge HJ solar cell with power conversion efficiencies of 5.2% and 7.3%, respectively [31−33]. Further efficiency increases were brought about through improvements such as doping, increasing grain size, and nanostructuring [26,27]. The record Sb_2S_3 solar cell is a 7.5% efficient SSC fabricated by CBD of Sb_2S_3 on mesoporous TiO_2 [15]. This cell exhibited a high open-circuit voltage (V_{oc}) of 711 mV. Another recent study grew single-crystal Sb_2S_3 cuboids on a nanoparticle TiO_2 film to create a solar cell with a 5.12% certified efficiency [34].

The promise of Sb_2S_3 solar cells led researchers to explore Sb_2Se_3 PV. The bandgap of Sb_2Se_3 (1.1 eV) closely matches the optimal bandgap for single-junction solar cells calculated by Shockley and Queisser (1.1 eV), whereas Sb_2S_3 has a larger bandgap (1.7 eV) that results in lower theoretical single-junction efficiencies [2,13]. The first Sb_2Se_3 solar cell was created by the thermal decomposition of Sb_2Se_3 precursors spin-coated onto mesoporous TiO_2 and had a reported PV efficiency of 3.21% [35]. Further efficiency improvements have been achieved through interface engineering and control of the Sb_2Se_3 morphology [13]. The record Sb_2Se_3 efficiency of 9.2% with an impressive short-circuit current density (J_{sc}) of 32.6 mA/cm^{-2} was achieved by sublimation of core-shell Sb_2Se_3 nanorods on a Mo-coated glass substrate to form a CdS/Sb_2Se_3 HJ, but the V_{oc} remained modest at 400 mV [17]. The vertical nanorods orient the highly anisotropic crystal structure of Sb_2Se_3 to ensure efficient charge separation and charge transfer.

Significant effort has been expended to increase the V_{oc} of Sb_2Se_3 solar cells [36]. For example, a recent study substantially increased the V_{oc} to a record of 520 mV by suppressing nonradiative recombination using a rapid thermal annealing process, but the efficiency of these devices remains limited to 8.64% due to a lower J_{sc} [37]. Another study recently demonstrated a record efficiency of 9.19% with an increased V_{oc} of 461 mV enabled by an In_2S_3-CdS composite buffer layer that encouraged compact coating of a Sb_2Se_3 nanorod array [38]. Another approach to increasing the V_{oc} of Sb_2Se_3 solar cells is to combine Sb_2Se_3 with Sb_2S_3.

Sb_2S_3 and Sb_2Se_3 can form a homogeneous solid solution because they have the same crystal structure and similar unit-cell parameters

[13,39]. The advantages of these two materials can, therefore, be combined to form superior $Sb_2(S_xSe_{1-x})_3$ solar cells. For example, the high V_{oc} of Sb_2S_3 can be combined with the high J_{sc} of Sb_2Se_3 to form high-efficiency $Sb_2(S_xSe_{1-x})_3$ PV. Controlling the S:Se ratio also enables tuning of the bandgap from 1.1 to 1.7 eV [40]. Indeed, this fine-tuning has resulted in an efficiency of 10.5% for a solar cell fabricated by hydrothermal deposition of large-grained $Sb_2(S_xSe_{1-x})_3$ on CdS. The $Sb_2(S_xSe_{1-x})_3$ had a bandgap of 1.55 eV, and the solar cell exhibited reasonably high values for the V_{oc} (664 mV) and J_{sc} (23.8 mA/cm^{-2}) [19]. Fig. 12.1 shows the UV−vis absorption spectrum, a cross-sectional scanning electron microscope (SEM) image, the current−voltage curves, and the external

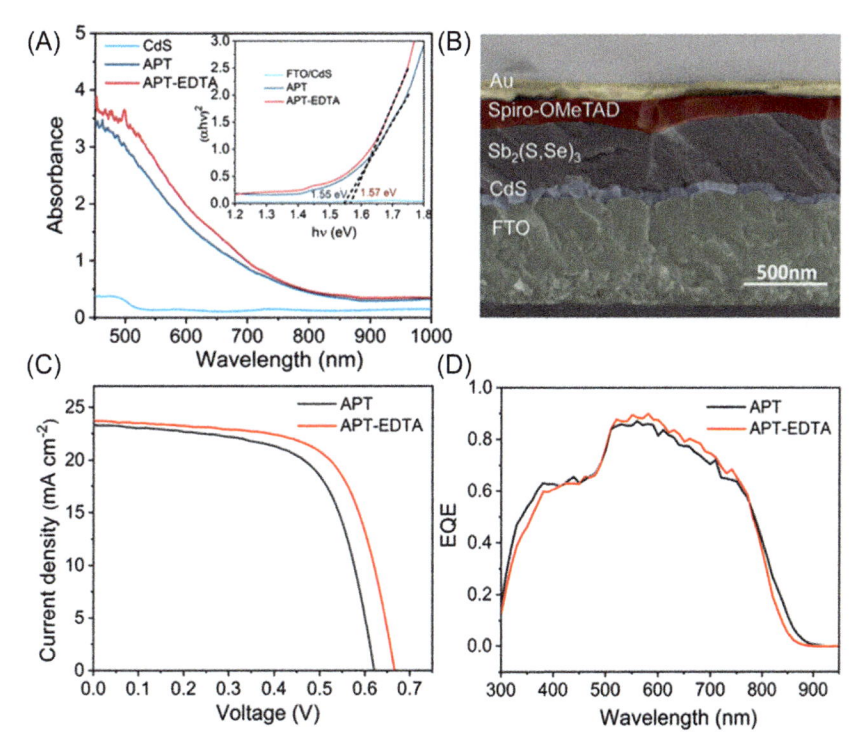

Figure 12.1 (A) UV−vis absorption spectrum of CdS and $Sb_2(S,Se)_3$ films synthesized with/without EDTA (ethylenediaminetetraacetic acid). Inset: The corresponding Tauc plots. (B) Cross-sectional SEM image of the solar cell with the configuration of FTO/CdS/$Sb_2(S,Se)_3$/Spiro-OMeTAD/Au. (C) $J-V$ curves of devices under AM 1.5 G (100 mW cm^{-2}) illumination conditions. (D) The external quantum efficiency of the devices made with antimony potassium tartrate (APT) and APT with EDTA. *Reproduced with permission from X. Wang, R. Tang, C. Jiang, W. Lian, H. Ju, et al., Manipulating the electrical properties of $Sb_2(S,Se)_3$ film for high-efficiency solar cell, Adv. Energy Mater. 10 (40) (Oct. 2020) 2002341, copyright (2020) John Wiley & Sons.*

quantum efficiency (EQE) of $Sb_2(S_xSe_{1-x})_3$ solar cells created in this study. The tunability of $Sb_2(S_xSe_{1-x})_3$ also enables the formation of multijunction solar cells by precisely tuning the bandgap of each layer [41].

Despite these promising efficiencies, there have been few studies of solution-processed nanoparticle solar cells made of antimony sulfides and selenides. Nanoparticles and nanowires (NWs) of Sb_2S_3 [42,43], Sb_2Se_3 [44,45], and $Sb_2(S_xSe_{1-x})_3$ [46] have been synthesized and investigated as components of batteries [47], supercapacitors [46], and photodiodes [48]. However, few studies have incorporated these nanoparticles into solution-processed solar cells.

In 2016 Abulikemu et al. reported a systematic study of hot-injection synthesis of colloidal Sb_2S_3 nanoparticles and their incorporation into a solar cell [16]. By changing conditions such as injection temperature, reaction time, and the reaction precursors, the researchers were able to synthesize nanoparticles with a variety of different sizes and shapes, including spherical nanoparticles, nanorods, and "urchinlike" nanostructures like those shown in Fig. 12.2. Solar cells created by dip coating of mesoporous TiO_2 with the colloidal nanoparticles exhibited a power conversion efficiency (PCE) of 1.5%. In another study, a 1.22% efficient solar cell was created through the growth of Sb_2Se_3 nanoparticles on mesoporous TiO_2 using successive ionic layer adsorption and reaction (SILAR) [18]. To our knowledge, there have been no reports of solution-processed $Sb_2(S_xSe_{1-x})_3$ solar cells.

Researchers have also begun incorporating antimony-based nanoparticles into solution-processed hybrid solar cells with limited success. Wang et al. used spin coating to deposit a film of colloidal Sb_2S_3 nanoparticles synthesized by hot injection and a film of poly(3-hexylthiophene) (P3HT) onto a compact TiO_2 film to create a hybrid solar cell with an efficiency of 1.5% [49]. Li et al. later demonstrated efficiencies up to 1.83% for hybrid solar cells created by hydrothermal growth of dendritic TiO_2, growth of Sb_2S_3 nanorods through thermal decomposition of precursors, and subsequent coating with a ternary polymer blend [50].

Solution-processed antimony sulfide and selenide solar cells continue to underperform antimony solar cells created with other methods. One paper identified defect states in Sb_2S_3 nanoparticles that can limit the efficiency of hybrid solar cells [51], but solution-processed Sb_2S_3 solar cells and possible avenues to their improvement remain largely unstudied. More substantial research in these areas is, therefore, expected to result in further efficiency increases.

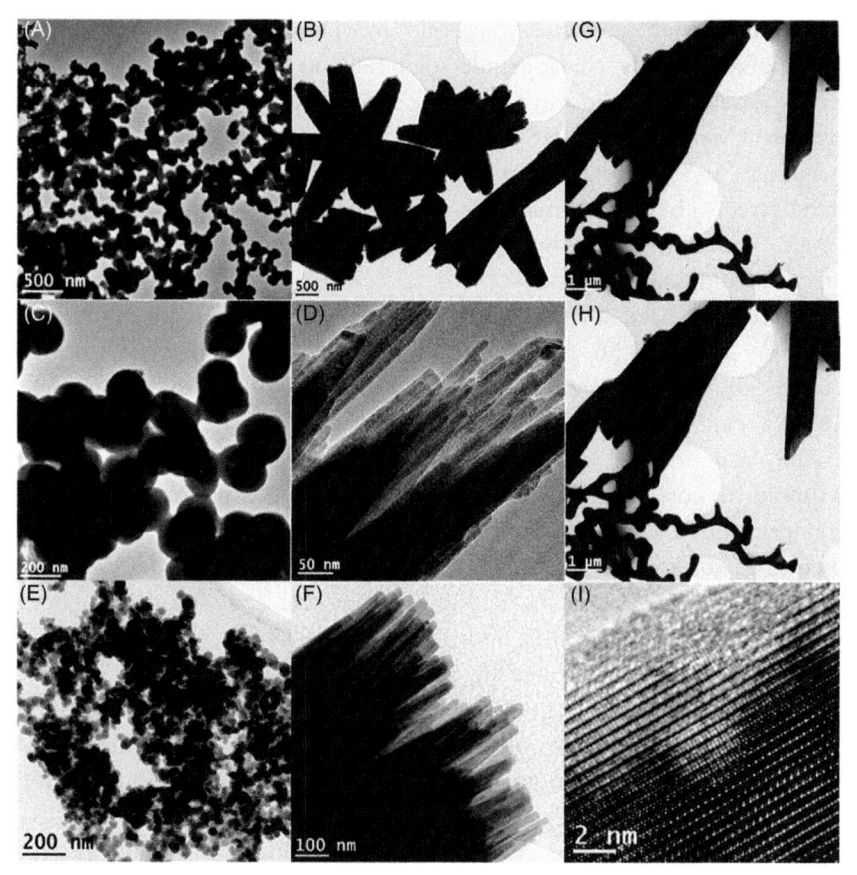

Figure 12.2 Transmission Electron Microscopy (TEM) and High-resolution Transmission Electron Microscopy (HRTEM) micrographs of Sb_2S_3 NCs with TMS solution injected at different temperatures and left for different reaction times. After the injection, the temperature of all was reduced to 100°C. (A) 100°C, 90 s; (B) 100°C, 3 h; (C) 140°C, 90 s; (D) 140°C, 2 h; (E) 180°C, 90 s; (F) 180°C, 90 min; (G) 230°C, 90 s; (H) and (I) 230°C, 90 min. *Reproduced with permission from M. Abulikemu, S. Del Gobbo, D.H. Anjum, M.A. Malik, O.M. Bakr, Colloidal Sb2S3 nanocrystals: Synthesis, characterization and fabrication of solid-state semiconductor sensitized solar cells, J. Mater. Chem. A 4 (2016) 6809–6814, copyright (2016) The Royal Society of Chemistry.*

12.2.2 Ternary antimony compounds and nanocrystals

Ternary antimony compounds have also been explored as solar absorbers. Although the success of silver-antimony chalcogenides [52–54] and sodium-antimony chalcogenides [55] has remained limited with reported efficiencies of less than 1%, copper-antimony chalcogenides, ternary compounds containing antimony and iodine, and antimony perovskites have demonstrated some promise. Peccerillo and Durose summarized progress on copper-antimony chalcogenide solar cells up to 2018 in an excellent review article [14]. The most-studied copper-antimony chalcogenides for solar are $CuSbS_2$ [20,56–58] and $CuSbSe_2$, but many other compounds including Cu_3SbS_3 [59,60], Cu_3SbS_4 [59], and $Cu_{12}Sb_4S_{13}$ have also been investigated. These bulk materials exhibit bandgaps ranging from 0.9 to 1.9 eV, with further bandgap tuning possible through the quantum size effect in nanoparticles.

There have been several reports of $CuSbS_2$ solar cells with efficiencies of 3.1%–3.2% [20,56–58]. While many of these studies deposit the $CuSbS_2$ absorption layer from the solution, none of the studies employs $CuSbS_2$ nanoparticles. The highest reported efficiency is 3.22% for $CuSbS_2$ deposited by sulfurization of spin-coated metal acetates [20], but it should be noted that this efficiency is within the experimental error of the next highest reported efficiencies [56–58]. The limited journal articles discussing solar cells with other copper-antimony sulfides such as Cu_3SbS_3, Cu_3SbS_4, and $Cu_{12}Sb_4S_{13}$ report power conversion efficiencies up to 1.49% [60], 0.46% [61], and 0.04% [62], respectively.

$CuSbSe_2$ solar cells created by cosputtering of Cu_2Se and Sb_2Se_3 have exhibited efficiencies up to 4.7% [22]. Just as Sb_2Se_3 tends to exhibit lower voltages and higher currents than Sb_2S_3, $CuSbSe_2$ solar cells often exhibit a lower V_{oc} and higher J_{sc} than $CuSbS_2$ solar cells as illustrated in Table 12.1. Combining $CuSbSe_2$ with $CuSbS_2$ to create $CuSb(Se_xS_{1-x})_2$, therefore, has the potential to create a solar cell with high V_{oc} and J_{sc} similar to what had been observed for $Sb_2(S_xSe_{1-x})_3$ solar cells [19]. Nevertheless, $CuSb(Se_xS_{1-x})_2$ solar cells remain largely unexplored with only one report of $CuSb(Se_{0.96}S_{0.04})_2$ solar cells with an efficiency of 2.70% [63].

One known problem with solar cells based on bulk copper-antimony sulfides is the existence of impurity phases that introduce defects and traps [14]. Nanoparticle synthesis can overcome these issues by selectively creating several individual phases including $CuSbS_2$, Cu_3SbS_3, Cu_3SbS_4, and $Cu_{12}Sb_4S_{13}$ [59,64,65]. Copper-antimony chalcogenide nanoparticles [21,59,66,67], nanorods [68], nanobricks [21], and nanoplates [64] have been synthesized with a variety of several methods including hot injection [59,64,66], solvothermal [21], and CBD [68]. Tuning the shape, size, composition, and surface chemistry of these ternary antimony nanoparticles offers control over important material properties such as bandgap [14].

Despite these advantages, there are few reports of solar cells fabricated from copper-antimony chalcogenide nanoparticles [21,65]. The first solution-processed $CuSbS_2$ nanoparticle solar cells had a modest PCE of only 0.01% [65]. Han et al. later reported an efficiency of 1.89% for solar cells comprised of $CuSbS_2$ nanobricks synthesized on TiO_2 nanorods by a solvothermal route [21]. Pulsed laser ablation in liquid has also been used to create colloidal $CuSbS_2$ nanoparticles that were incorporated into solar cells with a maximum efficiency of 0.65% [69]. Whereas $CuSbSe_2$ and $CuSb(Se_xS_{1-x})_2$ nanoparticles have been synthesized and tested as photocatalysts [67] and electrodes in supercapacitors [70], to our knowledge, there are no reports of solution-processed $CuSbSe_2$ or $CuSb(Se_xS_{1-x})_2$ nanoparticle solar cells. Copper-antimony chalcogenide solar cells fabricated with solution-processed nanoparticles currently underperform those created with alternate methods. However, this difference is likely due to the exceptionally small number of studies exploring solution-processed antimony-based nanoparticle PV, so further study is expected to result in substantial efficiency increases.

Antimony iodide compounds have also garnered interest as lead-free solar absorbers. For example, several groups have reported the solution deposition of thin films of SbSI [71,72], SbSeI [23,73], and $Sb(S_xSe_{1-x})I$ [40] with promising optoelectronic properties. Stoichiometric tuning of $Sb(S_xSe_{1-x})I$ enables precise tuning of important properties like bandgap [40]. A PCE of 4.1% was recently achieved for a SbSeI thin-film solar cell (TFSC) fabricated by spin coating SbI_3 solutions onto thin Sb_2Se_3 [23]. There have also been a few studies of antimony iodide perovskites with reported efficiencies up to 3.3% for both $(CH_3NH_3)_3Sb_2I_{9-x}Cl_x$ and $Cs_3Sb_2I_9$ solar cells [74—77]. Considering that research on these solar absorbers containing antimony and iodine remains in its infancy, more research into these compounds is expected to result in efficiency gains.

12.3 Bismuth compounds

Bismuth materials have gained attention in the last years due to their great optoelectronic properties, high defect tolerance, high dielectric constants, favorable bandgap, improved stability, relative abundancy in Earth's crust, and low and stable price. These properties coupled with the fact that bismuth is a nontoxic heavy metal make bismuth a potential substitute for lead in PV [78,79]. Binary (Bi_2S_3) and ternary compounds ($AgBiS_2$, $CuBiS_2$, $NaBiS_2$) have been studied as light harvester materials in HJ solar cells and QD solar cells and as cosensitizer in dye-SSCs (DSSC). Table 12.2 summarizes champion bismuth-based bulk and NC solar cells [80–87].

Table 12.2 Summary of highest-performing bulk and nanocrystals bismuth-based solar cells.

Compound	Type of material	Efficiency (%)	V_{oc} (mV)	J_{sc} (mA cm^{-2})	FF (%)	Description	Ref.
Bi_2S_3	Bulk	1.27	254	24.74	34.0	Chemical bath deposition with PbS and a CeO_2 buffer layer	[80]
	NCs	7.50	700	16.77	64.1	Cosensitizing with N719 dye in a DSSC	[81]
$CuBiS_2$	Bulk	1.70	320	18.20	30.3	Heterojunction solar cell with CdS QDs	[82]
	NCs	0.68	220	7.32	42.0	3 layer-by-layer $CuBiS_2$ NCs spin-coated onto TiO_2 and with P3HT on top	[83]
$AgBiS_2$	Bulk	2.87	500	13.27	43	4 SILAR cycles of bulk $AgBiS_2$ onto TiO_2 and with Co^{2+} doped P3HT on top	[84]
	NCs	9.2	495	27.1	68.4	Cation disorder-engineered $AgBiS_2$ NCs between SnO_2 and PTAA	[85]
$AgBiSe_2$	NCs	2.6	290	18.88	47.0	3 layer-by-layer deposited $AgBiSe_2$ NCs between ZnO and P3HT	[86]
$NaBiS_2$	NCs	0.07	638	0.18	56.0	4 layer-by-layer $NaBiS_2$ QDs deposited by SILAR between TiO_2 and Spiro-OMeTAD	[87]

DSSC, Dye sensitized solar cells; *NCs*, nanocrystals; *QDs*, quantum dots; *SILAR*, successive ionic layer adsorption and reaction.

12.3.1 Binary bismuth compounds and nanocrystals

The most widely studied bismuth binary compound is Bi_2S_3. It is a n-type semiconductor with a large absorption coefficient (in the range of $10^{-5}/$ cm) and reported bandgaps between 1.3 and 1.7 eV [79,88]. Bulk Bi_2S_3 TFSCs are scarcely reported. For instance, in 2010, Moreno-García et al. fabricated a TFSC device depositing PbS and Bi_2S_3 by CBD and achieving maximum efficiency of 0.5% [89]. This value was enhanced up to 1.27% by Ríos-Saldaña et al. when adding a CeO_2 buffer layer that increased the crystallinity of PbS thin film [80]. Song et al. reported a TFSC based on thermal deposition with structure ITO/NiO/Bi_2S_3/Au (where ITO = indium doped tin oxide) that yielded a 0.75% efficiency [90].

On the other hand, colloidal Bi_2S_3 NCs have received much more attention and have shown great promise for solution-processed solar cells with a record efficiency of 4.87% in HJ solar cells [91]. Bi_2S_3 (n-type) NCs have been combined with p-type semiconductors such as PbS QDs or an organic polymer, P3HT, to form a $p-n$ HJ. The first efficiencies reported for these materials were 1.61% [92] and 0.46% [93], respectively. In these studies, some efficiency limitations were identified such as high recombination, poor electron mobility, or high doping of the Bi_2S_3 layers. Later, the efficiency was increased up to 4.87% by mixing the PbS QDs and Bi_2S_3 NCs into a bulk HJ (BHJ) structure. This BHJ enhances the charge carrier transfer reducing recombination resulting in a higher short-circuit current [91]. A similar approach was studied for the P3HT, blending the organic p-type polymer with n-type Bi_2S_3 NCs to form a BHJ structure. The maximum efficiency yielded by these devices was 1% after applying a thiol treatment [94,95]. Furthermore, when preparing a BHJ with P3HT, Bi_2S_3, and Au nanoparticles, an efficiency of 2% was achieved [96].

Previous examples showed bismuth-based NCs randomly oriented in p-n or BHJ systems. Therefore a different strategy was adopted for these devices, orientating semiconductor Bi_2S_3 NWs to enhance performance. First, aligning Bi_2S_3 NWs vertically on a TiO_2 substrate and covered by an Ag_2S core—shell to decrease recombination showcased a maximum efficiency of 2.5% [97]. Furthermore, by interconnecting Bi_2S_3 NWs in a BHJ system with P3HT, the efficiency was increased up to 3.3% [98].

On the other hand, introducing Bi_2S_3 into DSSCs has produced impressive results. For example, CdS QDs and Bi_2S_3 QDs were grown using the SILAR method and were used to sensitize TiO_2 QDs. The best device with the structure $TiO_2/CdS/Bi_2S_3/ZnS$ provided an efficiency of 2.52% [99]. Furthermore, following this cosensitization strategy, by coating TiO_2 double-layer films with Bi_2S_3 QDs and N719 ruthenium dye, efficiency values as high as 7.5% were achieved [81].

The promise of Bi_2S_3 NCs is reflected not only by the numerous advantages of using colloidal semiconductors but also in their results when introduced in solution-processed solar cells. In fact, these devices provide better efficiencies than solar cells fabricated with bulk material using other deposition methods such as thermal or chemical deposition.

12.3.2 Ternary bismuth compounds and nanocrystals

Ternary bismuth compounds have received little attention in PV due to the more complex synthesis of ternary metal chalcogenides. Nonetheless, colloidal NCs of these ternary compounds offer several advantages such as low-cost, easy synthesis methods, tunable energy bandgap, large optical absorption coefficients, and suitable bandgaps for optoelectronic applications [83,100,101]. Most studies focus on modeling the potential applications of these materials; however, there are several reports where these NCs have been tested in devices. In fact, $CuBiS_2$, $AgBiS_2$, and $NaBiS_2$ NCs have been used as absorber layers in solar cells [82–85,87].

12.3.2.1 Copper bismuth sulfide

Copper bismuth sulfide ($CuBiS_2$) has been studied as a nontoxic candidate for solar cells as it was found that the calculated bandgap was 1.55 eV [102], close to the optimal value for PV. Additionally, $TiO_2/CuBiS_2$ films show low electron-hole recombination, indicating the potential of these HJs [103]. Despite its promising characteristics, $CuBiS_2$ compounds have been scarcely implemented in solar cell devices. To our knowledge, most $CuBiS_2$ thin-film studies reported in the literature are theoretical. Yang et al. managed to fabricate a $CdS/CuBiS_2$ HJ solar cell. The $CuBiS_2$ film was prepared by evaporation of the powdered bismuth and CuS precursors on a glass substrate, while the CdS layer was deposited using a CBD process. The maximum efficiency was 1.7%. In this study, the underperformance of the device was attributed to two reasons: (1) poor contact between CdS and $CuBiS_2$, which leads to carrier recombination reflected in low shunt resistance, low open-circuit voltage, and low fill factor, and (2) a poor contact between CdS and Mo, causing a Shockley barrier that limits carrier transport [82]. On the other hand, solution-processed solar cells based on $CuBiS_2$ NCs have yielded efficiencies lower than 1.7%. However, these studies do not include CdS in their structure.

In a study performed by Suriyawong et al., $CuBiS_2$ NCs (5–10 nm size) were grown on a mesoporous TiO_2 substrate using a CBD method to fabricate an SSC with polysulfide electrolyte. This device exhibited moderate short-circuit current density (J_{sc} = 6.87 mA/cm^2) and an efficiency of 0.62% [100]. These J_{sc} values might be limiting the performance of the device and were attributed to several factors. First, the measured bandgap of 2.1 eV is higher than the optimum for PV and second, the inhomogeneity observed for the deposition method of $CuBiS_2$ nanocrystals and the 2D layered structure [104] resulted in anisotropic carrier transport.

In 2017 Wang et al. managed to obtain a maximum efficiency of 0.68% using TiO_2 as the hole transport layer (HTL) and P3HT as an electron transport layer (ETL) in a QDs solution-processed solar cell. In this case, $CuBiS_2$ NCs were synthesized using a facile colloidal synthesis, unlike the previous case where a chemical deposition was used. Furthermore, a surface ligand passivation to substitute oleic acid ligands with EDT (1,2-ethanedithiol) shorter chains was applied, and a doping treatment with $InCl_3$ was also carried out [83]. Improving these contacts might be a good strategy for better-performing devices.

Despite the limitations detected and moderate efficiencies, $CuBiS_2$ NCs have shown promising properties as nontoxic alternatives for PV devices with only a few studies. However, they are still below the results achieved using bulk $CuBiS_2$ with CdS, which might be the reason behind this enhanced efficiency. In this sense, more research is needed regarding NC synthesis, contact optimization between layers, and surface ligand/passivation treatments.

12.3.2.2 Silver bismuth sulfide

Another ternary bismuth compound used in PV is silver bismuth sulfide ($AgBiS_2$). Thin-film $AgBiS_2$ has been deposited via spray pyrolysis onto ITO/ZnO supports to fabricate a solar cell device. Spiro-OMeTAD was used as HTL providing an efficiency of 1.5% [105]. Additionally, Calva-Yáñez et al. made a hybrid solar cell using $AgBiS_2$ deposited by the SILAR method as an absorber layer. The best device was obtained using mesoporous TiO_2 and Co^{2+} doped P3HT as ETL and HTL, respectively, yielding an efficiency of 2.87% [84]. Additionally, trap-assisted recombination was identified as the main limitation, which is enhanced by the absence of surface passivation treatments.

AgBiS$_2$ is another example where NCs have been more extensively explored and applied in solution-processed solar cells. Its nontoxicity, Earth abundance, and optimal bandgap (1.3 eV) [105,106], combined with the low-cost synthesis, easy processing, and ligand/surface passivation treatments make them ideal for PV applications. In fact, there are several studies with outstanding efficiencies reported when using AgBiS$_2$ NCs as the light-harvesting layer. One of the first and most promising results was achieved by Bernechea et al. in 2016. AgBiS$_2$ colloidal semiconductor NCs (4 nm) were synthesized using a simple and fast hot-injection method and were used as a light absorber layer in a solution-processed solar cell (Fig. 12.3A). Moreover, EDT and TMAI (tetramethyl ammonium iodide) ligand exchange treatments were applied to substitute the oleic acid ligands. A high short-circuit current was observed (22 mAh/cm^2) with a maximum certified efficiency of 6.3% with just 35 nm layer thickness in ITO/ZnO/AgBiS$_2$/PTB7/MoO$_3$/Ag structure. The NCs were sandwiched between a ZnO layer that acted as ETL and a thin PTB7 polymer layer that acted as HTL [106]. Despite the quite promising

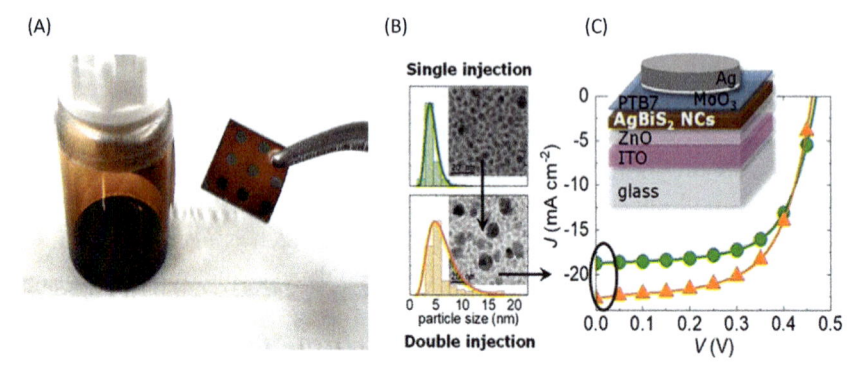

Figure 12.3 (A) Colloidal AgBiS$_2$ nanocrystals and the solar cell device fabricated with them using a single hot injection method. (B) TEM images for single injection and double injection AgBiS$_2$ nanocrystals with 4.3 and 6.5 nm size, respectively. (C) J–V curves for solar cell devices fabricated with both types of AgBiS$_2$ NCs. Double-injection nanocrystals show better performance. Scheme representing the structure of the device, ITO/ZnO/AgBiS$_2$/PTB7/MoO$_3$/Ag. Subfigure (A) reproduced with permission from M. Bernechea, N. Cates, G. Xercavins, D. So, A. Stavrinadis, G. Konstantatos, Solution-processed solar cells based on environmentally friendly AgBiS2 nanocrystals, Nat. Photonics 10 (2016) 521–525, copyright (2016) Springer Nature. Subfigures (B) and (C) reproduced with permission from I. Burgués-Ceballos, Y. Wang, M.Z. Akgul, G. Konstantatos, Colloidal AgBiS2 nanocrystals with reduced recombination yield 6.4% power conversion efficiency in solution-processed solar cells, Nano Energy 75, 104961 (2020), copyright (2020) Elsevier.

results obtained, several issues were identified, including inefficient extraction of carriers at higher light intensities, Shockley–Read–Hall trap-assisted recombination, poor carrier transport, and incomplete extraction before recombination [106]. Solutions that have been proposed to overcome these limitations are synthesis improvements, alternative surface/ligand passivation of the NCs that can reduce trap-assisted recombination, and nanostructuring.

Several strategies have been followed to synthesize $AgBiS_2$. For instance, an improved amine-based synthesis route was developed to enhance the quality of the NCs and their semiconducting properties. An efficiency of 4.3% was obtained in a $ZnO/AgBiS_2/P3HT/Au$ solar cell [108]. Furthermore, a room temperature and ambient conditions synthesis method was proposed for $AgBiS_2$ NCs. In this case, the best PV device showcased a 5.5% efficiency. This result is considered a significant step toward sustainable large-scale production [109]. In 2020 the efficiency of these devices was boosted up to 6.4% due to a photocurrent increase. This was achieved by modifying the synthesis method into a double-step hot-injection process (Fig. 12.3B), adding the sulfur precursor in two steps and, therefore, increasing the average size of the $AgBiS_2$ NCs (6.5 nm) [107]. The solar cell structure was similar to the one proposed by Bernechea et al. (Fig. 12.3C). In fact, when mixing these small-sized (4.7 nm) and big-sized (7.1 nm) $AgBiS_2$ NCs, thicker layers could be fabricated enhancing the charge transport and the efficiency up to 7.3% [110].

Other more environmentally friendly synthesis approaches have been explored by preparing $AgBiS_2$ aqueous inks. This was accomplished by performing an MPA (3-Mercaptopropionic acid) solution-phase ligand exchange. The maximum efficiency achieved with these inks was 7.3% [111]. Furthermore, cation-disorder-engineered $AgBiS_2$ colloidal NCs yielded a 9.2% record efficiency. It was demonstrated that cation disorder homogenization can enhance the absorption coefficient of the NCs, highlighting the importance of atomic configuration for PV applications [85].

These studies have demonstrated the potential of $AgBiS_2$ NCs as substitutes for lead in nontoxic solution-processed solar cells. Additionally, devices surpass TFSCs based on their bulk counterparts, especially because trap-assisted recombination was identified as one of the main problems that seem to be better controlled in NCs by ligand passivation treatments. Even so, several improvements regarding the synthesis of $AgBiS_2$ NCs have been reported recently trying to overcome the identified limitations of these devices. Upgrading the synthesis has been the main focus of these studies achieving promising efficiency results for low-cost and low-temperature methods. Nonetheless, other approaches should be explored such as different ligand exchange treatments of the $AgBiS_2$ NCs.

12.3.2.3 Silver bismuth selenide

Additionally, other chalcogenides have been explored for silver-bismuth compounds such as silver bismuth selenide ($AgBiSe_2$) colloidal NCs. A room-temperature and ambient conditions synthesis has been reported for $AgBiSe_2$ 6 nm spherical NCs, showcasing a maximum 2.6% efficiency when introduced in a solution-processed solar cell system [86]. In this case, $AgBiSe_2$ NCs solution was spin-coated on top of a 45 nm ZnO film. Afterward, P3HT as HTL and MoO_3 layers were deposited. Different active layer thicknesses were studied; nonetheless, 40 nm $AgBiSe_2$ NCs showed the best efficiency results with a PCE of 2.6%. This is claimed to be the first and only report of $AgBiSe_2$ NCs tested in solar cell devices. Furthermore, in this same paper, $AgBiS_xSe_y$ NCs have been synthesized and characterized demonstrating that the S/Se ratio can be modified for bandgap tuning. This suggests a promising route for selenide-based NCs toward optimization in PV.

12.3.2.4 Sodium bismuth sulfide

Sodium bismuth sulfide ($NaBiS_2$) and related compounds have received very little attention as semiconductor materials for PV as they have been mainly studied for photocatalytic applications. As a matter of fact, to the best of our knowledge, no TFSCs have been reported using bulk $NaBiS_2$ as an absorption layer. However, NCs seem to be more attractive for solar cells than bulk. The reported bandgap of $NaBiS_2$ NCs (1.1−1.4 eV) lies within the ideal range for PV applications and they crystallize in a disordered cubic structure similar to $AgBiS_2$ NCs, material that has provided outstanding results [112−114].

$NaBiS_2$ QDs with 4.5 nm size have been deposited onto a TiO_2 film using the SILAR method to fabricate an SSC with polysulfide electrolyte. The best efficiency achieved was a modest 0.05%. The observed low efficiency and instability of the cells suggest a defect-rich lattice and were attributed to the amorphous structure obtained. Additionally, slower hole-transfer has been identified when compared to Bi_2S_3 crystalline QDs [115].

Furthermore, $NaBiS_2$ NCs were synthesized at low temperatures and deposited layer-by-layer on top of mesoporous TiO_2. Afterward, a ligand exchange treatment with PbX_2 ($X = Br, I$) was applied to substitute the native insulating long-chain ligands, oleylamine, and neodecanoate, with Br^- and I^- ions. Spiro-OMeTAD was used as HTL. Finally, MoO_3 and Ag layers were deposited on top to complete the device. The maximum efficiency yielded was 0.07% and corresponded to four layer-by-layer $NaBiS_2$ NCs spin-coating deposition [87]. In spite of the promising optoelectronic properties of nontoxic Earth-abundant $NaBiS_2$ NCs, their application in PV devices is scarce and with limited efficiencies. However, the reported studies suggest that with further nanostructuring and ligand exchange optimization, $NaBiS_2$ NCs could become potential candidates for NC solar cells.

12.4 Tin compounds

Tin-based binary and ternary chalcogenide semiconductors such as SnSe, SnS, Cu-Sn-S system, or Cu_2ZnSnS are emerging hybrid absorber layers for low-cost and large-scale PV applications owing to their high optical absorption coefficient, optical bandgap, and nontoxic and Earth-abundant nature of the fundamental components [116,117]. The solution-based preparation of these compounds offers an encouraging route toward low-cost and large-scale production costs of green TFSCs [116]. Among them, hot injection synthesis methods are advantageous because they allow tuning of the size, shape, and composition by controlling synthesis conditions such as reaction time and temperature, solvent, precursors, and concentrations [118]. Table 12.3 provides a Summary of the highest-performing bulk and NCs tin-based binary and ternary chalcogenide solar cells [119–123].

12.4.1 Binary compounds and nanocrystals

The binary compounds in the Sn-S system have generated great interest among the scientific community and their fundamental properties have been investigated. SnS (orthorhombic), SnS_2 (trigonal), and Sn_2S_3 (rhombic) were the most closely studied [118,124–126]. Within this group, SnS is a p-type semiconductor with a bandgap of 1.16 eV [127], similar to silicon but with a higher optical absorption coefficient. Thus it has been explored as an absorber layer in PV applications [128]. Although the maximum theoretical efficiency of SnS TFSCs is 32% [2], the maximum

Table 12.3 Summary of the highest-performing bulk and NCs tin-based binary and ternary chalcogenide solar cells.

Compound	Type of material	Efficiency (%)	V_{oc} (mV)	J_{sc} (mA/cm^{-2})	FF (%)	Description	Ref.
SnS	Bulk	4.8	330	27.70	58.5	Solution processing and thermal annealing with additional SnCl$_2$ posttreatment SnS	[119]
	Bulk	4.36	372	22.60	58.0	ALD of purified SnS	[120]
	NCs	2.36	390	1.70	72.0	NCs of 3–4 nm placed on SiO$_2$ substrate in the form of film with the chitosan	[121]
Cu$_2$SnS$_3$	Bulk	6.73	442	26.60	57.1	Ge-doped Cu$_2$SnS$_3$ thin-film solar cells	[122]
Cu$_2$ZnSnS$_4$	NCs	7.23	430	31.20	53.9	as-synthesized CZTS nanocrystals	[123]

ALD, Atomic layer deposition; *CZTS*, Copper zinc tin sulfide; *NC*, nanocrystals.

efficiency actually achieved has been only 4.8% [119]. SnS$_2$ shows n-type conductivity and its presence influences the efficiency reached. Sn$_2$S$_3$ has a bandgap of 1.09 eV, similar to SnS, and an n-type behavior, although deviation from the stoichiometry can change the conduction [129]. Due to these promising properties, SnS has been included in TFSC assemblies with a number of electron- and hole-transporting materials. Unfortunately, the existence of SnS$_2$ and Sn$_2$S$_3$ phases with undesirable optoelectrical properties, several electrically active defects, and grain boundary recombination are generally recognized problems related to the quality of SnS absorber films [125].

Noguchi et al. developed one of the first solar cells based on SnS thin films deposited by the evaporation of a commercial SnS source [130]. The CdS/SnS planar heterojunction device showed a PCE of 0.29%. Since this report, the fabrication of SnS films has gained the attention of researchers; substantial efforts have been invested in obtaining them through diverse deposition methods. However, regardless of the deposition method used, secondary phases (SnS$_2$ and Sn$_2$S$_3$) are typically present in the final films limiting the efficiency [131–134].

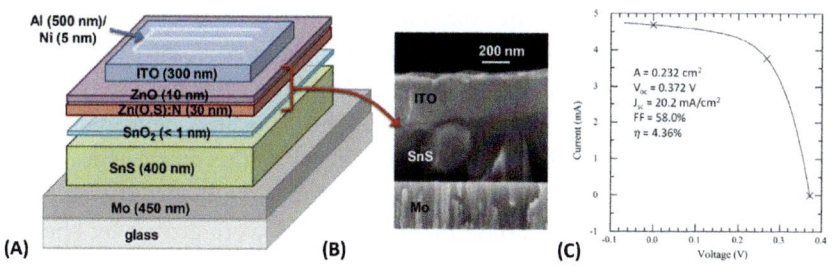

Figure 12.4 (A) A schematic diagram of a SnS-based solar cell. (B) A cross-sectional SEM image of an actual cell with SnS annealed in H_2S. (C) $J-V$ characteristics of the champion device with a certified (NREL) efficiency of 4.36%. *Images reproduced with permission from P. Sinsermsuksakul, L. Sun, S.W. Lee, H.H. Park, S.B. Kim, et al., Overcoming efficiency limitations of SnS-based solar cells, Adv. Energy Mater. 4 (2014) 1−7, copyright (2014) John Wiley & Sons, Inc.*

Up to now, most of the work has been focused on manufacturing single-phase SnS thin films through diverse approaches. Among the various deposition techniques, only atomic layer deposition (ALD) [120,135,136] and thermal evaporation (TE) of the purified SnS powder in vacuum [137] demonstrated the successful preparation of single-phase SnS films. Using these deposition methods, the US Department of Energy National Renewable Energy Laboratory (NREL) achieved a certified efficiency close to or higher than 4% in SnS-based solar cells [125].

In 2014 Sinsermsuksakul et al. reported a record efficiency of 4.36% for p-type SnS-based solar cells (Fig. 12.4) [120]. They achieved the successful preparation of single-phase SnS films by TE of the purified commercial SnS powder in a vacuum. Steinman and coworkers reported thin-film solar cell efficiency of 3.88% using TE as an alternative route for SnS film deposition using commercially available SnS powder and a congruent TE unit under high vacuum to produce phase-pure SnS films [137].

Later, Park et al. reported PV devices with efficiencies up to 2.9% through an improvement in the morphology via annealing of the SnS layers deposited using ALD and optimizing the composition of the buffer layer Zn(O,S), which helps in tuning the conduction band offset at the $p-n$ junction interface [136]. These efficiencies were achieved by simultaneously dealing with two technical issues: the use of single-phase SnS and the application of a Zn-based buffer layer. However, it is still unclear which is the main factor in reaching high efficiencies.

Studies utilizing diverse methods have focused on the selection of a suitable buffer layer for the SnS absorber, such as SnS/SnS_2 $p-n$ junction, with the goal to improve the SnS/buffer interface. Yue et al. fabricated doped SnS NW by chemical vapor deposition using B_2H_6 or PH_3 as dopant gases [138]. This strategy allowed combining n-type (SnS:P) and p-type (SnS:B) segments in a single NW. The fabricated devices showed an efficiency of 1.95%. Recently, Gedi and coworkers fabricated a SnS/SnS$_2$ solar cell by CBD (type-II heterostructure) with an efficiency of 0.51% [139]. In 2017 NW arrays of SnS/SnS$_2$ HJ were grown by chemical vapor transport catalyzed by Cu particles and assembled into devices yielding an efficiency of 1.4% [140]. Two years later, Yun et al. reported a SnCl$_2$ posttreatment (SPT) as an approach to improve the efficiency of the SnS/TiO$_2$ solar cells up to 4.8% [119]. This improvement is related to a higher density of the SnS thin film during SPT and the reduction of the defects associated with the introduction of Cl$^-$ into the lattice.

Moving from bulk to nanoscale, tin sulfide (SnS) NCs have attracted great interest in PV applications. This great potential is based on the precise control of shape, size, composition, and phase crystallinity, which can be achieved by controlling the synthetic conditions [116]. Rath and coworkers described a facile solution-based route for the preparation of nanostructured SnS layers [141], which consisted of a porous network of SnS nanoplates with a thickness of 60–80 nm and diameter from 0.1 to 1 μm. The solar cell with the architecture: ITO/planar-TiO$_2$/SnS/P3HT/MoO$_3$/Ag exhibited an efficiency of 1.2%. As an extension of this approach, Ding et al. fabricated SnS TFSCs based on mesoporous TiO$_2$/SnS/P3HT HJs with PCEs of up to 3.0%. Pure SnS layer (without secondary phases) was fabricated using a solution-based method that allows control of the film morphology and coverage tunability [142].

After initial attempts, Stavrinadis and coworkers demonstrated the feasibility of incorporating colloidal SnS NCs (sized 3.6 nm) into solar cells. The fabrication of SnS NCs/PbS NCs HJ devices yielded a PCE of 0.5% [143]. Dutta et al. demonstrated that solution-processed SnS solar cells made from QDs exhibit very low efficiencies (0.008%–0.066%) due to the increased number of grain boundaries and recombination losses [144]. More recently, Truong et al. reported the construction of SnS-based BHJ solar cells by combining SnS nanospheres with a mean size of 5–6 nm with conjugated polymers to build the photoactive layer. A moderate PCE of 0.71% was reached for the device with the structure of a SnS/PTB7 [145].

Nevertheless, Cheraghizade et al. found that sonochemical synthesis could be a suitable method to produce tin sulfide QDs. However, the structural studies showed a mixture of phases with an orthorhombic phase of SnS (48%) and Sn_2S_3 (33%) and hexagonal phase of SnS_2 (19%). While the particle size of QDs prepared was smaller with the sonication time. An exciting PCE of 2.36% was reached with QDs of 3—4 nm placed on SiO_2 substrate in the form of film with chitosan. The electrical characterization was performed by placing two Ag electrodes on the film side [121]. It is noteworthy that the PCE of tin-based solar cells has increased from 0.29% to 4.4% (for bulk thin films) and from 0.008 to 2.4 (for NC-based films). These results seem promising for future research and the use of these materials large-scale applications.

12.4.2 Ternary and quaternary compounds and nanocrystals

Since the 1990s, Cu_2SnS_3, Cu_3SnS_4, and Cu_4SnS_4 thin films have been deposited by various chemical and physical techniques but only Cu_2SnS_3 and Cu_4SnS_4 showed PV performance. Comparing Cu_2SnS_3 and Cu_4SnS_4, there was a notable increment of reports focusing on the physical properties and solar cell application of Cu_2SnS_3 in combination with CdS, In_2S_3, and ZnO as buffer layers [117].

In 1987 Titilayo et al. observed the PV behavior in Cu_2SnS_3 thin films in a Schottky-type solar cell with a reported efficiency of 0.11% [146]. After quite some time, the Katagiri group achieved a considerable enhancement in the Cu_2SnS_3 device efficiency [147—151]. Fabrication of Cu_2SnS_3 solar cells focused on various processes to improve their efficiencies such as electron beam evaporation, TE, sputtering, sulfurization, and posttreatment in various atmospheres. Specifically, the best efficiency of 4.29% was obtained for coevaporated Cu_2SnS_3 films (Cu, Sn, and sulfur were coevaporated and then annealed in a sulfur vapor atmosphere) [151].

In 2015 Nakashima et al. fabricated solar cells with doped Cu_2SnS_3 layers showing higher efficiency than those fabricated with the undoped films. Na-doped Cu_2SnS_3 solar cells delivered an efficiency of 4.63% [152]. The improvement in the efficiency of Na-doped Cu_2SnS_3 solar cell was related to the diminished carrier recombination. Later, a promising efficiency of 6.7% was reported by Umehara et al. al in 2016 employing a Ge-doped Cu_2SnS_3 film [122]. The thin films were deposited by annealing the cosputtered Cu—Sn precursors in S and GeS_2 vapors. These findings indicate that Cu_2SnS_3 can be a promising nontoxic material that can be used to develop cost-effective solar cells. However, some limiting factors in Cu_2SnS_3 solar cells have been identified: the poor quality of the

Cu_2SnS_3 films, issues in the back contact, and problems at the $p-n$ junction interface [117].

Copper zinc tin sulfide (Cu_2ZnSnS_4 or CZTS) is a low-cost, nontoxic material that is destined to become one of the most promising environmentally friendly and Earth-abundant absorber materials for solar cells. The PV performance was first demonstrated in 1988 by Ito et al. [153]. Even though CZTS solar cells have shown a record efficiency of 11% [154], the deposition of a single phase of Cu_2ZnSnS_4 thin films and the suppression of secondary phases (Cu−S, Zn−S, Sn−S, and Cu−Sn−S) are crucial. This is problematic due to its narrow window for single-phase formation in the phase diagram [155].

In this context, the investigation of CZTS NCs has drawn a lot of attention because it allows better control over the crystal phase, stoichiometry, and morphology of the materials previous to their processing into thin films, which is a key issue for building highly efficient solar cells based on quaternary CZTS. In 2009 Agrawal's group was the first to demonstrate the formation of kesterite CZTS NCs and their incorporation into solar cells [156]. Later, in 2010, the incorporation of as-synthesized CZTS NCs produced a PCE as high as 7.2% after light soaking, where the CZTS NCs ($Cu_{1.31 \pm 0.02}Zn_{0.91 \pm 0.03}Sn_{0.95 \pm 0.02}S_4$) were synthesized using a hot injection method [123].

As stated earlier, CZTS is a nontoxic semiconductor compound. However, there are other toxic components taking part in the CZTS solar cell structure such as Cd (CdS thin films are often used as the buffer layer) [155]. Recently a Cd-free Cu_2ZnSnS_4-Ag_2ZnSnS_4 HJ PV device was fabricated, where Cu_2ZnSnS_4 and Ag_2ZnSnS_4 layers were obtained by cosputtering method. The fabricated solar cell showed a promising efficiency of 4.51% [157].

12.5 Iron compounds

Iron-based materials are attractive for their integration into devices because iron is abundant, low-cost, and nontoxic. Iron is the fourth most abundant element in the Earth's crust and constitutes about a third of its entire mass. Moreover, in most of its forms, it can be considered nontoxic. Indeed, it is an essential trace element in living organisms and its deficiency can lead to cell damage, and eventually even death [158].

Table 12.4 Summary of the highest-performing bulk and nanocrystal iron-based solar cells.

Compound	Type of material	Efficiency (%)	V_{oc} (mV)	J_{sc} (mA cm^{-2})	FF (%)	Description	Ref.
FeS$_2$	Single crystal	2.8	187	42	50.0	In−Ga alloy/n-type FeS$_2$ single crystal/ I$^-$/I$_3^-$ electrolyte	[159]
	Bulk	1.98	570	6.55	53.0	Electrochemical deposition of FeS$_2$ ITO/FeS$_2$/ZnSe/Au	[160]
		5.42	830	10.71	61.0	ITO/3% Co-FeS$_2$/ ZnSe/Au	
	NCs	3.62	580	16.31	56.0	ITO/PEDOT:PSS/ FeS$_2$:P3HT:PCBM BHJ/Al	[161]
CuFeS$_2$	NCs	0.3	735	0.83	49.0	ITO/MEH-PPV: CuFeS$_2$ BHJ/Al	[162]
Cu$_2$FeSnS$_4$	Bulk	2.95	610	9.3	52.0	Cu$_2$FeSnS$_4$ thin films grown by SILAR Cu$_2$FeSnS$_4$/Bi$_2$S$_3$ junction	[163]

SILAR, Successive ionic layer adsorption and reaction.

In the area of PV, most of the research has focused on FeS$_2$ due to its attractive optoelectronic properties; however, experimental work has shown the difficulty of achieving these properties and high-efficiency devices. Ternary or quaternary-derived compounds could be an attractive alternative; nonetheless, very little research has been devoted to their introduction in solar cells. Table 12.4 summarizes some of the most relevant results of iron-based materials in PV devices [159−163].

12.5.1 Binary iron compounds and nanocrystals

Iron disulfide (FeS$_2$), also known as iron pyrite or fool's gold because of its golden aspect, crystallizes in a cubic structure, has a high absorption coefficient ($>10^5$/cm, two orders of magnitude higher than that of Si), and a suitable bandgap for PV (an indirect bandgap of 0.95 eV, similar to silicon, and a direct bandgap of 1.03 eV). Single crystals show *n*-type character while thin films appear to be *p*-type [126,155,159].

All of these properties, together with the significant natural abundance of iron, low cost of the material, and nontoxicity, make FeS_2 the ideal candidate for its use as an active material in solar cells. Indeed, it was suggested that a 4% efficient FeS_2 solar cell could produce electricity at the same price as a silicon solar cell with an efficiency of 19% [164]. This fact pushed research on iron pyrite in solar cells, a topic that has slowly declined, most likely because of the problems associated with this material. In FeS_2, sulfur is the form of disulfide anion (S_2^{2-}) instead of the more stable sulfide anion (S^{2-}). Therefore it is difficult to obtain a phase of pure material, and sulfur-deficient phases or polymorphs, with poor optoelectronic properties, are usually obtained together with FeS_2. Moreover, surface states, surface inversion, and bulk defects further limit optoelectronic FeS_2 thin-film properties [165–170].

In 1993 a PCE of 2.8% was first demonstrated in a single-crystal-based photoelectrochemical cell (*n*-type FeS_2/liquid electrolyte junction). Promising short-circuit current density (>30 mA/cm^{-2}) and good quantum efficiency were obtained, but the low open-circuit voltage (<0.2 V) limited the final efficiency [159]. Years later, FeS_2 nanoparticles were introduced as counter electrodes (CEs) in SSCs whether alone or combined with graphene delivering efficiencies of 8.39% [171], or 7.04% [172], respectively, higher than the analogous Pt-only CEs (8.2% and 6.63%, respectively).

Pure and cobalt-doped FeS_2 *p*-type bulk thin films were fabricated by electrochemical deposition and used in solar cells forming a HJ with ZnSe. The introduction of Co^{2+} does not affect the cubic structure, retains the original *p*-type behavior, and leads to an improved 5.42% efficiency as compared to the 1.98% efficiency of the undoped film [160].

There are several reports dealing with the synthesis of FeS_2 colloidal NCs, fabrication of thin films and providing some optoelectronic characterization, however, the successful incorporation of these NCs into solar cells is complicated [173–175]; in fact, these publications are scarce. As we commented above, the formation of surface defects related to sulfur vacancies affect the electrical properties, a problem that can be more prominent in NC devices due to higher number of surfaces and interfaces.

One of the first demonstrations on the use of FeS_2 NCs in solar cells was published in 2009 by Lin et al. FeS_2 NCs with an average diameter of approximately 10 nm and stabilized by 1,2-hexadecanediol and oleylamine were synthesized using a colloidal synthesis. A BHJ consisting on poly(3-hexylthiophene) (P3HT)/FeS$_2$ NCs was used as active layer delivering a PCE of 0.16%. Moreover, an extended red light harvesting up to 900 nm resulting from the NCs was observed [176].

In 2012 Kirkeminde et al. reported the synthesis of FeS$_2$ nanocubes via a seeded growth, involving a second injection of iron precursor into preformed FeS$_2$ NCs, and using octadecylamine as stabilizing ligand. Depending on the ageing time the size of the nanocubes could be varied from 80 to 120 nm. The 80 nm NCs were employed as the absorbing material in an all inorganic colloidal FeS$_2$:CdS BHJ device, with an active layer thickness of 500 nm, leading to a well-defined percolation network. The best device (1:1 FeS$_2$:CdS ratio) showed a significantly improved open-circuit voltage and an efficiency of 1.1% [177].

Some years later, oleylamine-capped FeS$_2$ NCs with an average diameter of approximately 60 nm were also used in a FeS$_2$:CdS BHJ device. In this study, using a similar solar cell architecture as in the previously commented report, the efficiency was 0.50%. The low efficiency can be attributed to film morphology since authors declare that most of the device suffers from short-circuit due to cracks in the film that could not be completely filled with subsequent coatings [178].

In 2014 the efficiency of FeS$_2$/polymer BHJ devices was further improved from 0.16% [176] to 3.62% [161] by introducing FeS$_2$ NCs into a 1:1 polymer matrix of poly(3-hexylthiophene-2,5-diyl):[6,6]-phenyl-C61-butyricacidmethylester (P3HT:PCBM). FeS$_2$ NCs had an average diameter of approximately 5 nm, were stabilized by dodecylamine, and were added with different wt.% loadings at the polymer mixture. The best device had a 20% (by weight) addition of NCs, leading to an efficiency of 3.62%, higher than the device without FeS$_2$ (2.32%). Interestingly, all the devices incorporating FeS$_2$ NCs showed a contribution to photocurrent in the 650–900 nm region of the EQE, consistent with the absorption of iron pyrite and the blends incorporating NCs (Fig. 12.5). More recently, a similar approach has been followed, adding FeS$_2$ to a PTB7:PCBM BHJ. In this case the PCE is enhanced from 5.69% to 6.47% when adding a small quantity of FeS$_2$ NCs (0.5 wt.%) [179].

In recent years FeS$_2$ NCs have been explored as HTL in CdTe or perovskite solar cells, showing improvements in the performance. For example, FeS$_2$ NCs were introduced as the copper-free back contact for CdTe solar cells, depositing the film using drop-casting coupled with a hydrazine treatment at ambient temperature and pressure. Copper-free

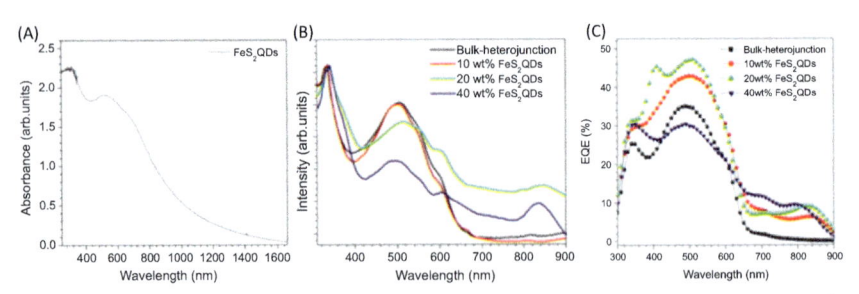

Figure 12.5 (A) UV–vis spectra of as-synthesized iron pyrite nanocrystals in dodecylamine. (B) UV–vis spectra of devices made from the controlled addition of FeS_2 nanocrystals in the active layer. (C) External quantum efficiency analysis of the devices. *Reproduced and adapted with permission from M.A. Khan, Y.-M. Kang, Synthesis and processing of strong light absorbent iron pyrite quantum dots in polymer matrix for efficiency enhancement of bulk-heterojunction solar cell, Mater. Lett. 132 (2014) 273–276, copyright (2014) Elsevier.*

solar cells based on the $CdS/CdTe/FeS_2$-NC/Au architecture exhibited device efficiencies 90% higher than that of a standard Cu/Au back contact device [180]. Similarly, p-type $Ni_{0.05}Fe_{0.95}S_2$ NCs were used as the HTL for CdTe solar cells with an average increase in efficiency of approximately 5% (from 11.2% to 12.1%) as compared to the laboratory standard copper/gold (Cu/Au) cell [181]. More recently, tri-octylphosphine oxide (TOPO)-caped FeS_2 NCs were deposited on top of a perovskite layer in a n-i-p configuration where the perovskite is an intrinsic semiconductor, TiO_2 is the ETL, and FeS_2 is the HTL. The PCE of the perovskite device using pyrite as HTL outperformed (11.2%) the cells without HTL (9.11%), and were close to those using spiro-OMeTAD (16.8%). Additionally, cost analysis of the pyrite HTL and spiro-OMeTAD indicated that the phase pure pyrite NCs are much cheaper to incorporate in large-scale devices than spiro-OMeTAD when materials cost was compared (>300 times less expensive) [182].

In general, the device performances for NC devices are low, compared with the single crystal or thin-film devices. Moreover, in most cases FeS_2 NCs are mainly incorporated as an additive, for example, to improve charge transport, but not as the active solar harvesting material. This could be due to the quality of the NCs and the scarcely explored device fabrication process. For example, the long-term maintained stability should be improved since FeS_2 NCs showed a tendency to aggregate, leading to films with a high surface roughness [178].

12.5.2 Ternary and quaternary iron compounds and nanocrystals

In view of the poor efficiencies shown by FeS_2, both as thin films or NCs, ternary or quaternary compounds, where sulfur is in the form of sulfide anion (S^{2-}), could be an interesting alternative. However, there are very few examples of the introduction of these materials in solar cells. Chalcopyrite ($CuFeS_2$) is a naturally occurring mineral with unusual optical, electrical and magnetic properties. In the bulk it has a small bandgap of about 0.5—0.6 eV, that can be tuned from 0.52 to 2 eV when in the form of NCs, making it attractive for PV applications [183].

Colloidal $CuFeS_2$ NCs prepared by thermal injection synthesis led to a phase-pure material with dimensions ranging from 10 to 40 nm, depending on the reaction time. Oleylamine-caped $CuFeS_2$ NCs films were used as HTL in CdTe solar cells. The best device delivered an efficiency of 11.9%, slightly higher than the device without HTL (11.8%) [184]. Like FeS_2 NCs, cubic $CuFeS_2$ have been used as CEs in SSC displaying a PCE of 8.10%, comparable to that of a cell with Pt as CE (7.74%) [185]. Additionally, $CuFeS_2$ NCs obtained following a hydrothermal strategy, were used as an acceptor in organic—inorganic hybrid solar cell where MEHPPV (poly[2-methoxy-5-(2-ethylhexyloxy)-1,4-phenylenevinylene]) acts as the donor and $CuFeS_2$ as the acceptor. A MEHPPV:$CuFeS_2$ BHJ device (1:1 weight ratio) was fabricated and the efficiency obtained was 0.3% [162].

$AgFeS_2$ is a ternary semiconductor with a bandgap between 0.88 and 1.2 eV [186]. Oleylamine-caped $AgFeS_2$ NCs were synthesized *via* a one-pot thermal decomposition route. The NCs showed an average diameter of 20 nm and the bandgap was reported to be 1.2 eV [187]. Later, $AgFeS_2$ NWs were obtained by reacting Ag NWs with iron chloride, sodium thiosulfate, and thioglycolic acid in dimethyl sulfoxide and water at 150°C showing a bandgap of 0.88 eV [186]. In spite of the interesting properties of $AgFeS_2$, including photocatalytic activity under solar light [188], its use in solar cells has not been reported.

$CoFe_2O_4$ NCs have been used with graphene nanosheets (GN) as CEs in DSSC leading to an improvement in performance from 6.54% of Pt to 9.04% of the $CoFe_2O_4$/GN hybrid [189]. Additionally, $CoFe_2O_4$, with an average diameter of 4—10 nm, was obtained through a sol-gel method and introduced into a P3HT:PCBM BHJ active layer. The large area (0.55 cm^2) flexible devices fabricated with these blends delivered an efficiency of 2.2%, for the active layer with $CoFe_2O_4$ NCs, higher than the reference device without (2.0%) [190]. More recently, $CoFe_2O_4$ NCs were used as interlayers between perovskite and the HTL in perovskite solar cells. The NCs were synthesized by a two-phase hydrothermal method and showed an average size of 30 nm. The best device incorporating a $CoFe_2O_4$ layer exhibited a PCE of 19.65%, higher than the reference without a $CoFe_2O_4$ layer (17.88%), attributed to enhanced hole extraction, reduced trap density, and better moisture resistance [191].

Cu_2FeSnS_4 (CFTS) is a very good candidate as a cheap absorber layer for PV due to its adequate bandgap (1.28–1.50 eV) and high absorption coefficient ($>10^4$ cm^{-1}). Cu_2FeSnS_4 thin films grown by SILAR were reported to show a p-type behavior and a bandgap of 1.5 eV. Several p-n junctions with CdS, Bi_2S_3, and Ag_2S were fabricated and sandwiched between ITO and aluminum electrodes for solar cell testing. CFTS/Bi_2S_3 exhibited a PCE of 2.95% with a high degree of reproducibility [163].

12.6 Conclusions

The use of Earth-abundant and nontoxic materials combined with solution processing offers the opportunity to reduce solar cell fabrication costs and the implementation of PV devices in contexts where human health or environmental protection are of special relevance. Thin-film active layers are preferably obtained using vapor-deposition techniques and among the solution-processed thin films, few of them use NCs to fabricate them. However, nontoxic Sb, Bi, Sn, and Fe-based chalcogenide NCs offer great optoelectronic properties, such as adequate bandgaps and high absorption coefficients, combined with the advantages offered by NCs such as simple and easy processing, production of pure phases, bandgap tuning, or modulation of band energy positions through ligand treatments.

In general, the efficiencies of bulk thin films surpass those of NC-based thin films, although it should be taken into account that the number of studies using NCs is smaller. Promising efficiencies have been obtained for bulk thin-film solar cells such as 10.5% for $Sb_2(S_xSe_{1-x})_3$, achieved using a solution-processed hydrothermal deposition, or 5.4% for an electrochemically deposited FeS_2 bulk film. The use of ternary and quaternary compounds seems to be an adequate approach to improve efficiencies, for example, bulk-Cu_2SnS_3 or Cu_2ZnSnS_4 NCs deliver efficiencies of 6.73% and 7.2%, respectively, higher than the efficiency obtained with SnS (4.8% for a bulk film posttreated with $SnCl_2$). Moreover, thin films based on $AgBiS_2$ NCs have shown outstanding results (9.2% efficiency), clearly surpassing the bulk counterpart (2.87%). In summary, solution-processed solar cells based on environmentally friendly materials have already shown promising efficiencies in spite of the limited number of studies. Among these, devices employing NCs have been even less studied and have delivered modest efficiencies, with the exception of $AgBiS_2$ NCs. More substantial research in these areas is therefore expected to result in further efficiency increases.

Acknowledgments

Authors acknowledge CIBER-BBN (financed by the Instituto de Salud Carlos III), the ICTS "NANBIOSIS," Agencia Estatal de Investigación-AEI (MCIN/AEI/10.13039/501100011033. Refs: PID2019−107893RB-I00 and PCI2019−103637), and Gobierno de Aragón (Ref: T57_20R) for financial support.

References

[1] M. Yuan, M. Liu, E.H. Sargent, Colloidal quantum dot solids for solution-processed solar cells, Nat. Energy 1 (2016). Available from: https://doi.org/10.1038/nenergy.2016.16.
[2] W. Shockley, H.J. Queisser, Detailed balance limit of efficiency of p-n junction solar cells, J. Appl. Phys. 32 (1961) 510−519. Available from: https://doi.org/10.1063/1.1736034.
[3] P.R. Brown, D. Kim, R.R. Lunt, N. Zhao, M.G. Bawendi, J.C. Grossman, et al., Energy level modification in lead sulfide quantum dot thin films through ligand exchange, ACS Nano 8 (2014) 5863−5872. Available from: https://doi.org/10.1021/nn500897c.
[4] M.C. Beard, R.J. Ellingson, Multiple exciton generation in semiconductor nanocrystals: toward efficient solar energy conversion, Laser Photonics Rev. 2 (2008) 377−399. Available from: https://doi.org/10.1002/lpor.200810013.
[5] H.W. Hillhouse, M.C. Beard, Solar cells from colloidal nanocrystals: fundamentals, materials, devices, and economics, Curr. Opin. Colloid Interface Sci. 14 (2009) 245−259. Available from: https://doi.org/10.1016/j.cocis.2009.05.002.
[6] A. Swarnkar, A.R. Marshall, E.M. Sanehira, B.D. Chernomordik, D.T. Moore, J.A. Christians, et al., Quantum dot−induced phase stabilization of α-$CsPbI_3$ perovskite for high-efficiency photovoltaics, Science. 354 (2016) 92−95. Available from: https://doi.org/10.1126/science.aag2700.
[7] M. Hao, Y. Bai, S. Zeiske, L. Ren, J. Liu, et al., Ligand-assisted cation-exchange engineering for high-efficiency colloidal $Cs_{1-x}FA_xPbI_3$ quantum dot solar cells with reduced phase segregation, Nat. Energy 5 (2020) 79−88. Available from: https://doi.org/10.1038/s41560-019-0535-7.
[8] J.W. Lee, H. Choi, U.K. Hwang, J.C. Kang, Y.J. Kang, K.I. Kim, et al., Toxic effects of lead exposure on bioaccumulation, oxidative stress, neurotoxicity, and immune responses in fish: a review, Environ. Toxicol. Pharmacol. 68 (2019) 101−108. Available from: https://doi.org/10.1016/j.etap.2019.03.010.
[9] D.A. Gidlow, Lead toxicity, Occup. Med. (Chic. Ill.) 54 (2004) 76−81. Available from: https://doi.org/10.1093/occmed/kqh019.

[10] W.C. Ma, Effect of soil pollution with metallic lead pellets on lead bioaccumulation and organ/body weight alterations in small mammals, Arch. Environ. Contam. Toxicol. 18 (1989) 617−622. Available from: https://doi.org/10.1007/BF01055030.

[11] A. Babayigit, H.-G. Boyen, B. Conings, Environment versus sustainable energy: the case of lead halide perovskite-based solar cells, MRS Energy Sustain. 5 (2018) 1−15. Available from: https://doi.org/10.1557/mre.2017.17.

[12] S.A. Hashemi, S. Ramakrishna, A.G. Aberle, Recent progress in flexible-wearable solar cells for self-powered electronic devices, Energy Environ. Sci. 13 (2020) 685−743. Available from: https://doi.org/10.1039/C9EE03046H.

[13] H. Lei, J. Chen, Z. Tan, G. Fang, Review of recent progress in antimony chalcogenide-based solar cells: materials and devices, Sol. RRL 3 (1990) 1900026. Available from: https://doi.org/10.1002/solr.201900026.

[14] E. Peccerillo, K. Durose, Copper—antimony and copper—bismuth chalcogenides—research opportunities and review for solar photovoltaics, MRS Energy & Sustainability. 5 (2018) E13. Available from: https://doi.org/10.1557/mre.2018.10.

[15] Y.C. Choi, D.U. Lee, J.H. Noh, E.K. Kim, S.I. Seok, Highly improved Sb_2S_3 sensitized-inorganic−organic heterojunction solar cells and quantification of traps by deep-level transient spectroscopy, Adv. Funct. Mater. 24 (2014) 3587−3592. Available from: https://doi.org/10.1002/adfm.201304238.

[16] M. Abulikemu, S. Del Gobbo, D.H. Anjum, M.A. Malik, O.M. Bakr, Colloidal Sb_2S_3 nanocrystals: synthesis, characterization and fabrication of solid-state semiconductor sensitized solar cells, J. Mater. Chem. A 4 (2016) 6809−6814. Available from: https://doi.org/10.1039/C5TA09546H.

[17] Z. Li, X. Liang, G. Li, H. Liu, H. Zhang, et al., 9.2%-efficient core-shell structured antimony selenide nanorod array solar cells, Nat. Commun. 10 (2019) 125. Available from: https://doi.org/10.1038/s41467-018-07903-6.

[18] B. Zhao, Z. Wan, J. Luo, F. Han, H.A. Malik, C. Jia, et al., Efficient Sb_2S_3 sensitized solar cells prepared through a facile SILAR process and improved performance by interface modification, Appl. Surf. Sci. 450 (2018) 228−235. Available from: https://doi.org/10.1016/j.apsusc.2018.04.152.

[19] X. Wang, R. Tang, C. Jiang, W. Lian, H. Ju, et al., Manipulating the electrical properties of $Sb_2(S,Se)_3$ film for high-efficiency solar cell, Adv. Energy Mater. 10 (40) (2020) 2002341. Available from: https://doi.org/10.1002/aenm.202002341.

[20] S. Banu, S.J. Ahn, S.K. Ahn, K. Yoon, A. Cho, Fabrication and characterization of cost-efficient $CuSbS_2$ thin film solar cells using hybrid inks, Sol. Energy Mater. Sol. Cell 151 (2016) 14−23. Available from: https://doi.org/10.1016/j.solmat.2016.02.013.

[21] M. Han, J. Jia, W.A. Wang, A novel $CuSbS_2$ hexagonal nanobricks @ TiO_2 nanorods heterostructure for enhanced photoelectrochemical characteristics, J. Alloy. Compd. 705 (2017) 356−362. Available from: https://doi.org/10.1016/j.jallcom.2017.02.185.

[22] A.W. Welch, L.L. Baranowski, H. Peng, H. Hempel, R. Eichberger, et al., Trade-offs in thin film solar cells with layered chalcostibite photovoltaic absorbers, Adv. Energy Mater. 7 (2017) 1601935. Available from: https://doi.org/10.1002/aenm.201601935.

[23] R. Nie, M. Hu, A.M. Risqi, Z. Li, S. Il Seok, Efficient and stable antimony selenoiodide solar cells, Adv. Sci. 8 (2021) 2003172. Available from: https://doi.org/10.1002/advs.202003172.

[24] A. Zakutayev, J.D. Major, X. Hao, A. Walsh, J. Tang, T.K. Todorov, et al., Emerging inorganic solar cell efficiency tables (version 2), J. Phys. Energy 3 (2021) 32003. Available from: https://doi.org/10.1088/2515-7655/abebca.

[25] C. Ghosh, B.P. Varma, Optical properties of amorphous and crystalline Sb_2S_3 thin films, Thin Solid. Films 60 (1979) 61−65. Available from: https://doi.org/10.1016/0040-6090(79)90347-X.

[26] Q. Wang, Z. Chen, J. Wang, Y. Xu, Y. Wei, et al., Sb_2S_3 solar cells: functional layer preparation and device performance, Inorg. Chem. Front. 6 (2019) 3381–3397. Available from: https://doi.org/10.1039/C9QI00800D.
[27] R. Kondrotas, C. Chen, J. Tang, Sb_2S_3 solar cells, Joule 2 (2018) 857–878. Available from: https://doi.org/10.1016/j.joule.2018.04.003.
[28] A. Mavlonov, T. Razykov, F. Raziq, J. Gan, J. Chantana, et al., A review of Sb_2S_3 photovoltaic absorber materials and thin-film solar cells, Sol. Energy 201 (2020) 227–246. Available from: https://doi.org/10.1016/j.solener.2020.03.009.
[29] A. De Vos, Detailed balance limit of the efficiency of tandem solar cells, J. Phys. D. Appl. Phys. 13 (1980) 839–846. Available from: https://doi.org/10.1088/0022-3727/13/5/018.
[30] O. Savadogo, K.C. Mandal, Characterizations of antimony tri-sulfide chemically deposited with silicotungstic acid, J. Electrochem. Soc. 139 (1992) L16–L18. Available from: https://doi.org/10.1149/1.2069211.
[31] O. Savadogo, K.C. Mandal, Low-cost technique for preparing n-Sb_2S_3/p-Si heterojunction solar cells, Appl. Phys. Lett. 63 (1993) 228–230. Available from: https://doi.org/10.1063/1.110349.
[32] O. Savadogo, K.C. Mandal, Fabrication of low-cost n-Sb_2S_3/p-Ge heterojunction solar cells, J. Phys. D. Appl. Phys. 27 (1994) 1070–1075. Available from: https://doi.org/10.1088/0022-3727/27/5/028.
[33] U.A. Shah, S. Chen, G.M.G. Khalaf, Z. Jin, H. Song, Wide bandgap Sb_2S_3 solar cells, Adv. Funct. Mater. 31 (2021) 2100265. Available from: https://doi.org/10.1002/adfm.202100265.
[34] J. Chen, J. Qi, R. Liu, X. Zhu, Z. Wan, et al., Preferentially oriented large antimony trisulfide single-crystalline cuboids grown on polycrystalline titania film for solar cells, Commun. Chem. 2 (2019) 121. Available from: https://doi.org/10.1038/s42004-019-0225-1.
[35] Y.C. Choi, T.N. Mandal, W.S. Yang, Y.H. Lee, S.H. Im, J.H. Noh, et al., Sb_2Se_3-sensitized inorganic–organic heterojunction solar cells fabricated using a single-source precursor, Angew. Chem. Int. Ed. 53 (2014) 1329–1333. Available from: https://doi.org/10.1002/ange.201308331.
[36] J. Dong, Y. Liu, Z. Wang, Y. Zhang, Boosting V_{oc} of antimony chalcogenide solar cells: a review on interfaces and defects, Nano Sel. 2 (2021) 1818–1848. Available from: https://doi.org/10.1002/nano.202000288.
[37] R. Tang, S. Chen, Z.H. Zheng, Z.H. Su, J.T. Luo, et al., Heterojunction annealing enabling record open-circuit voltage in antimony triselenide solar cells, Adv. Mater. 34 (2022) 2109078. Available from: https://doi.org/10.1002/adma.202109078.
[38] T. Liu, X. Liang, Y. Liu, X. Li, S. Wang, Y. Mai, et al., Conduction band energy-level engineering for improving open-circuit voltage in antimony selenide nanorod array solar cells, Adv. Sci. 8 (2021) 2100868. Available from: https://doi.org/10.1002/advs.202100868.
[39] R. Tang, X. Wang, W. Lian, J. Huang, Q. Wei, et al., Hydrothermal deposition of antimony selenosulfide thin films enables solar cells with 10% efficiency, Nat. Energy 5 (2020) 587–595. Available from: https://doi.org/10.1038/s41560-020-0652-3.
[40] K.-W. Jung, Y.C. Choi, Compositional engineering of antimony chalcoiodides via a two-step solution process for solar cell applications, ACS Appl. Energy Mater. 5 (2022) 5348–5355. Available from: https://doi.org/10.1021/acsaem.1c02676.
[41] Y. Cao, C. Liu, J. Jiang, X. Zhu, J. Zhou, et al., Theoretical insight into high-efficiency triple-junction tandem solar cells via the band engineering of antimony chalcogenides, Sol. RRL 5 (2021) 2000800. Available from: https://doi.org/10.1002/solr.202000800.

[42] R. Vogel, P. Hoyer, H. Weller, Quantum-sized PbS, CdS, Ag_2S, Sb_2S_3, and Bi_2S_3 particles as sensitizers for various nanoporous wide-bandgap semiconductors, J. Phys. Chem. 98 (1994) 3183−3188. Available from: https://doi.org/10.1021/j100063a022.

[43] Y. Liu, Y. Tang, Y. Zeng, X. Luo, J. Ran, et al., Colloidal synthesis and characterization of single-crystalline Sb_2S_3 nanowires, RSC Adv. 7 (2017) 24589−24593. Available from: https://doi.org/10.1039/C7RA03319B.

[44] W. Farfán, E. Mosquera, C. Marín, Synthesis and blue photoluminescence from naturally dispersed antimony selenide (Sb_2S_3) 0-D nanoparticles, Adv. Sci. Lett. 4 (2011) 85−88. Available from: https://doi.org/10.1166/asl.2011.1185.

[45] M. Loor, G. Bendt, J. Schaumann, U. Hagemann, M. Heidelmann, et al., Synthesis of Sb_2S_3 and Bi_2S_3 nanoparticles in ionic liquids at low temperatures and solid state structure of $[C4C1Im]_3[BiCl_6]$, Z. für Anorg. und Allg. Chem. 643 (2017) 60−68. Available from: https://doi.org/10.1002/zaac.201600325.

[46] M.D. Khan, S.U. Awan, C. Zequine, C. Zhang, R.K. Gupta, N. Revaprasadu, Controlled synthesis of $Sb_2(S_{1-x}Se_x)_3$ ($0 \leq x \leq 1$) solid solution and the effect of composition variation on electrocatalytic energy conversion and storage, ACS Appl. Energy Mater. 3 (2020) 1448−1460. Available from: https://doi.org/10.1021/acsaem.9b01895.

[47] K.V. Kravchyk, M.V. Kovalenko, M.I. Bodnarchuk, Colloidal antimony sulfide nanoparticles as a high-performance anode material for Li-ion and Na-ion batteries, Sci. Rep. 10 (2020) 2554. Available from: https://doi.org/10.1038/s41598-020-59512-3.

[48] D. Choi, Y. Jang, J. Lee, G.H. Jeong, D. Whang, S.W. Hwang, et al., Diameter-controlled and surface-modified Sb_2Se_3 nanowires and their photodetector performance, Sci. Rep. 4 (2014) 6714. Available from: https://doi.org/10.1038/srep06714.

[49] W. Wang, F. Strössner, E. Zimmermann, L. Schmidt-Mende, Hybrid solar cells from Sb_2S_3 nanoparticle ink, Sol. Energy Mater. Sol. Cell 172 (2017) 335−340. Available from: https://doi.org/10.1016/j.solmat.2017.07.046.

[50] Y. Li, Y. Wei, K. Feng, Y. Hao, J. Pei, Y. Zhang, et al., Introduction of PCPDTBT in P3HT:Spiro-OMeTAD blending system for solid-state hybrid solar cells with dendritic TiO_2/Sb_2S_3 nanorods composite film, J. Solid. State Chem. 276 (2019) 278−284. Available from: https://doi.org/10.1016/j.jssc.2019.05.020.

[51] D.U. Lee, S.W. Pak, S.G. Cho, E.K. Kim, S.I. Seok, Defect states in hybrid solar cells consisting of Sb_2S_3 quantum dots and TiO_2 nanoparticles, Appl. Phys. Lett. 103 (2013) 23901. Available from: https://doi.org/10.1063/1.4813272.

[52] J.G. Garza, S. Shaji, A.C. Rodriguez, T.K.D. Roy, B. Krishnan, $AgSbSe_2$ and $AgSb(S,Se)_2$ thin films for photovoltaic applications, Appl. Surf. Sci. 257 (2011) 10834−10838. Available from: https://doi.org/10.1016/j.apsusc.2011.07.115.

[53] Y.-R. Ho, M.-W. Lee, $AgSbS_2$ semiconductor-sensitized solar cells, Electrochem. Commun. 26 (2013) 48−51. Available from: https://doi.org/10.1016/j.elecom.2012.10.003.

[54] Y. Zhang, J. Tian, K. Jiang, J. Huang, F. Li, P. Wang, et al., $AgSbS_2$ thin film fabricated by in-situ gas-solid reaction and employed in solar cells as a light absorber, Mater. Lett. 232 (2018) 82−85. Available from: https://doi.org/10.1016/j.matlet.2018.08.081.

[55] W.W.W. Leung, C.N. Savory, R.G. Palgrave, D.O. Scanlon, An experimental and theoretical study into $NaSbS_2$ as an emerging solar absorber, J. Mater. Chem. C. 7 (2019) 2059−2067. Available from: https://doi.org/10.1039/C8TC06284F.

[56] S. Ikeda, Y. Iga, W. Septina, T. Harada, M. Matsumura, $CuSbS_2$-based thin film solar cells prepared from electrodeposited metallic stacks composed of Cu and Sb layers in 39th IEEE Photovoltaic Specialists Conference (PVSC), 2013, pp. 2598−2601. Available from: https://doi.org/10.1109/PVSC.2013.6745005.

[57] Y.C. Choi, E.J. Yeom, T.K. Ahn, S.I. Seok, $CuSbS_2$-sensitized inorganic−organic heterojunction solar cells fabricated using a metal−thiourea complex solution, Angew. Chem. Int. Ed. 54 (2015) 4005−4009. Available from: https://doi.org/10.1002/anie.201411329.

[58] W. Septina, S. Ikeda, Y. Iga, T. Harada, M. Matsumura, Thin film solar cell based on $CuSbS_2$ absorber fabricated from an electrochemically deposited metal stack, Thin Solid. Films 550 (2014) 700−704. Available from: https://doi.org/10.1016/j.tsf.2013.11.046.
[59] S. Ikeda, S. Sogawa, Y. Tokai, W. Septina, T. Harada, M. Matsumuraa, Selective production of $CuSbS_2$, Cu_3SbS_3, and Cu_3SbS_4 nanoparticles using a hot injection protocol, RSC Adv. 4 (2014) 40969−40972. Available from: https://doi.org/10.1039/C4RA07648F.
[60] S.A. Zaki, M.I. Abd-Elrahman, A.A. Abu-Sehly, N.M. Shaalan, M.M. Hafiz, Solar cell fabrication from semiconducting Cu_3SbS_3 on n-Si: parameters evolution, Mater. Sci. Semicond. Process. 115 (2020) 105123. Available from: https://doi.org/10.1016/j.mssp.2020.105123.
[61] N.D. Franzer, N.R. Paudel, C. Xiao, Y. Yan, Study of RF sputtered Cu_3SbS_4 thin-film solar cells, in 2014 IEEE 40th Photovoltaic Specialist Conference (PVSC), 2014, pp. 2326−2328. Available from: https://doi.org/10.1109/PVSC.2014.6925393.
[62] L. Wang, B. Yang, Z. Xia, M. Leng, Y. Zhou, et al., Synthesis and characterization of hydrazine solution processed $Cu_{12}Sb_4S_{13}$ film, Sol. Energy Mater. Sol. Cell 144 (2016) 33−39. Available from: https://doi.org/10.1016/j.solmat.2015.08.016.
[63] B. Yang, C. Wang, Z. Yuan, S. Chen, Y. He, et al., Hydrazine solution processed $CuSbSe_2$: temperature dependent phase and crystal orientation evolution, Sol. Energy Mater. Sol. Cell 168 (2017) 112−118. Available from: https://doi.org/10.1016/j.solmat.2017.04.030.
[64] K. Ramasamy, H. Sims, W.H. Butler, A. Gupta, Selective nanocrystal synthesis and calculated electronic structure of all four phases of copper−antimony−sulfide, Chem. Mater. 26 (2014) 2891−2899. Available from: https://doi.org/10.1021/cm5005642.
[65] S. Suehiro, K. Horita, M. Yuasa, T. Tanaka, K. Fujita, Y. Ishiwata, et al., Synthesis of copper−antimony-sulfide nanocrystals for solution-processed solar cells, Inorg. Chem. 54 (2015) 7840−7845. Available from: https://doi.org/10.1021/acs.inorgchem.5b00858.
[66] H.-I. Hsiang, C.-T. Yang, J.-H. Tu, Characterization of $CuSbSe_2$ crystallites synthesized using a hot injection method, RSC Adv. 6 (2016) 99297−99305. Available from: https://doi.org/10.1039/C6RA20692A.
[67] A.S. Kshirsagar, P.K. Khanna, $CuSbSe_2/TiO_2$: novel type-II heterojunction nano-photocatalyst, Mater. Chem. Front. 3 (2019) 437−449. Available from: https://doi.org/10.1039/C8QM00537K.
[68] C. An, Q. Liu, K. Tang, Q. Yang, X. Chen, J. Liu, et al., The influences of surfactant concentration on the quality of chalcostibite nanorods, J. Cryst. Growth 256 (2003) 128−133. Available from: https://doi.org/10.1016/S0022-0248(03)01297-1.
[69] S. Shaji, V. Vinayakumar, B. Krishnan, J. Johny, S.S. Kanakkillam, et al., Copper antimony sulfide nanoparticles by pulsed laser ablation in liquid and their thin film for photovoltaic application, Appl. Surf. Sci. 476 (2019) 94−106. Available from: https://doi.org/10.1016/j.apsusc.2019.01.072.
[70] K. Ramasamy, R.K. Gupta, S. Palchoudhury, S. Ivanov, A. Gupta, Layer-structured copper antimony chalcogenides ($CuSbSe_xS_{2-x}$): stable electrode materials for supercapacitors, Chem. Mater. 27 (2015) 379−386. Available from: https://doi.org/10.1021/cm5041166.
[71] R. Nie, H.S. Yun, M.J. Paik, A. Mehta, B.W. Park, Y.C. Choi, et al., Efficient solar cells based on light-harvesting antimony sulfoiodide, Adv. Energy Mater. 8 (2018) 1701901. Available from: https://doi.org/10.1002/aenm.201701901.
[72] Y.C. Choi, E. Hwang, D.-H. Kim, Controlled growth of SbSI thin films from amorphous Sb_2S_3 for low-temperature solution processed chalcohalide solar cells, APL. Mater. 6 (2018) 121108. Available from: https://doi.org/10.1063/1.5058166.
[73] Y.C. Choi, K.-W. Jung, One-step solution deposition of antimony selenoiodide films via precursor engineering for lead-free solar cell applications, Nanomaterials 11 (2021) 3206. Available from: https://doi.org/10.3390/nano11123206.

[74] K. Ahmad, S.M. Mobin, Recent progress and challenges in $A_3Sb_2X_9$-Based perovskite solar cells, ACS Omega 5 (2020) 28404−28412. Available from: https://doi.org/10.1021/acsomega.0c04174.
[75] Y. Yang, C. Liu, M. Cai, Y. Liao, Yong Ding, et al., Dimension-controlled growth of antimony-based perovskite-like halides for lead-free and semitransparent photovoltaics, ACS Appl. Mater. Interfaces 12 (2020) 17062−17069. Available from: https://doi.org/10.1021/acsami.0c00681.
[76] K.M. Boopathi, P. Karuppuswamy, A. Singh, C. Hanmandlu, L. Lin, et al., Solution-processable antimony-based light-absorbing materials beyond lead halide perovskites, J. Mater. Chem. A 5 (2017) 20843−20850. Available from: https://doi.org/10.1039/C7TA06679A.
[77] A. Singh, P.T. Lai, A. Mohapatra, C.Y. Chen, H.W. Lin, Y.J. Lu, et al., Panchromatic heterojunction solar cells for Pb-free all-inorganic antimony based perovskite, Chem. Eng. J. 419 (2021) 129424. Available from: https://doi.org/10.1016/j.cej.2021.129424.
[78] L.C. Lee, T.N. Huq, J.L. Macmanus-Driscoll, R.L.Z. Hoye, Research update: bismuth-based perovskite-inspired photovoltaic materials, APL. Mater. 6 (2018) 084502. Available from: https://doi.org/10.1063/1.5029484.
[79] N.C. Miller, M. Bernechea, Research update: bismuth based materials for photovoltaics, APL. Mater. 8 (2018) 084503. Available from: https://doi.org/10.1063/1.5026541.
[80] L.E. Ríos-Saldaña, V.D. Compeán-García, H. Moreno-García, A.G. Rodríguez, Improvement of the conversion efficiency of as-deposited Bi_2S_3/PbS solar cells using a CeO_2 buffer layer, Thin Solid. Films 670 (2019) 93−98. Available from: https://doi.org/10.1016/j.tsf.2018.12.017.
[81] J. Sun, H. Guo, L. Zhao, S. Wang, J. Hu, B. Dong, Co-sensitized efficient dye-sensitized solar cells with TiO_2 hollow sphere/nanoparticle double-layer film electrodes by Bi_2S_3 quantum dots and N719, Int. J. Electrochem. Sci. 12 (2017) 7941−7955. Available from: https://doi.org/10.20964/2017.09.01.
[82] Y. Yang, X. Xiong, H. Yin, M. Zhao, J. Han, Study of copper bismuth sulfide thin films for the photovoltaic application, J. Mater. Sci. Mater. Electron. 30 (2019) 1832−1837. Available from: https://doi.org/10.1007/s10854-018-0455-5.
[83] J.J. Wang, M.Z. Akgul, Y. Bi, S. Christodoulou, G. Konstantatos, Low-temperature colloidal synthesis of $CuBiS_2$ nanocrystals for optoelectronic devices, J. Mater. Chem. A 5 (2017) 24621−24625. Available from: https://doi.org/10.1039/C7TA08078F.
[84] J.C. Calva-Yáñez, O. Pérez-Valdovinos, E.A. Reynoso-Soto, G. Alvarado-Tenorio, O.A. Jaramillo-Quintero, M. Rincón, Interfacial evolution of $AgBiS_2$ absorber layer obtained by SILAR method in hybrid solar cells, J. Phys. D. Appl. Phys. 52 (2019) 125502. Available from: https://doi.org/10.1088/1361-6463/aafd88.
[85] Y. Wang, S.R. Kavanagh, I. Burgués-Ceballos, A. Walsh, D.O. Scanlon, G. Konstantatos, Cation disorder engineering yields $AgBiS_2$ nanocrystals with enhanced optical absorption for efficient ultrathin solar cells, Nat. Photonics 16 (2022) 235−241. Available from: https://doi.org/10.1038/s41566-021-00950-4.
[86] M.Z. Akgul, G. Konstantatos, $AgBiSe_2$ colloidal nanocrystals for use in solar cells, ACS Appl. Nano Mater. 4 (2021) 2887−2894. Available from: https://doi.org/10.1021/acsanm.1c00048.
[87] A.M. Medina-Gonzalez, B.A. Rosales, U.H. Hamdeh, M.G. Panthani, J. Vela, Surface chemistry of ternary nanocrystals: engineering the deposition of conductive $NaBiS_2$ films, Chem. Mater. 32 (2020) 6085−6096. Available from: https://doi.org/10.1021/acs.chemmater.0c01689.
[88] Y.-C. Lin, M.-W. Lee, Bi_2S_3 liquid-junction semiconductor-sensitized SnO_2 solar cells, J. Electrochem. Soc. 161 (2014) H1−H5. Available from: https://doi.org/10.1149/2.002401jes.

[89] H. Moreno-García, M.T.S. Nair, P.K. Nair, Chemically deposited lead sulfide and bismuth sulfide thin films and Bi_2S_3/PbS solar cells, Thin Solid. Films 519 (2011) 2287−2295. Available from: https://doi.org/10.1016/j.tsf.2010.11.009.

[90] H. Song, X. Zhan, D. Li, Y. Zhou, B. Yang, et al., Rapid thermal evaporation of Bi_2S_3 layer for thin film photovoltaics, Sol. Energy Mater. Sol. Cell 146 (2016) 1−7. Available from: https://doi.org/10.1016/j.solmat.2015.11.019.

[91] A.K. Rath, M. Bernechea, L. Martinez, F.P.G. de Arquer, J. Osmond, G. Konstantatos, Solution-processed inorganic bulk nano-heterojunctions and their application to solar cells, Nat. Photonics 6 (2012) 529−534. Available from: https://doi.org/10.1038/nphoton.2012.139.

[92] A.K. Rath, M. Bernechea, L. Martinez, G. Konstantatos, Solution-processed heterojunction solar cells based on p-type PbS quantum dots and n-type Bi_2S_3 nanocrystals, Adv. Mater. 23 (2011) 3712−3717. Available from: https://doi.org/10.1002/adma.201101399.

[93] L. Martinez, M. Bernechea, F.P.G. De Arquer, G. Konstantatos, Near IR-sensitive, non-toxic, polymer/nanocrystal solar cells employing Bi_2S_3 as the electron acceptor, Adv. Energy Mater. 1 (2011) 1029−1035. Available from: https://doi.org/10.1002/aenm.201100441.

[94] L. Martinez, A. Stavrinadis, S. Higuchi, S.L. Diedenhofen, M. Bernechea, K. Tajima, et al., Hybrid solution-processed bulk heterojunction solar cells based on bismuth sulfide nanocrystals, Phys. Chem. Chem. Phys. 15 (2013) 5482−5487. Available from: https://doi.org/10.1039/C3CP50599E.

[95] L. Martinez, S. Higuchi, A.J. MacLachlan, A. Stavrinadis, N. Cates, et al., Improved electronic coupling in hybrid organic-inorganic nanocomposites employing thiol-functionalized P3HT and bismuth sulfide nanocrystals, Nanoscale 6 (2014) 10018−10026. Available from: https://doi.org/10.1039/C4NR01679C.

[96] S.K. Saha, A.J. Pal, Schottky diodes between Bi_2S_3 nanorods and metal nanoparticles in a polymer matrix as hybrid bulk-heterojunction solar cells, J. Appl. Phys. 118 (2015) 014503. Available from: https://doi.org/10.1063/1.4923348.

[97] Y. Cao, M. Bernechea, A. Maclachlan, V. Zardetto, M. Creatore, S.A. Haque, et al., Solution processed bismuth sulfide nanowire array core/silver sulfide shell solar cells, Chem. Mater. 27 (2015) 3700−3706. Available from: https://doi.org/10.1021/acs.chemmater.5b00783.

[98] L. Whittaker-Brooks, J. Gao, A.K. Hailey, C.R. Thomas, N. Yao, Y.L. Loo, Bi_2S_3 nanowire networks as electron acceptor layers in solution-processed hybrid solar cells, J. Mater. Chem. C. 3 (2015) 2686−2692. Available from: https://doi.org/10.1039/C4TC02534B.

[99] D. Esparza, I. Zarazúa, T. López-Luke, R. Carriles, A. Torres-Castro, E. De la Rosa, Photovoltaic properties of Bi_2S_3 and CdS quantum dot sensitized TiO_2 solar cells, Electrochim. Acta 180 (2015) 486−492. Available from: https://doi.org/10.1016/j.electacta.2015.08.102.

[100] N. Suriyawong, B. Aragaw, J. Shi, J. Bin, M.W. Lee, Ternary $CuBiS_2$ nanoparticles as a sensitizer for quantum dot solar cells, J. Colloid Interface Sci. 473 (2016) 60−65. Available from: https://doi.org/10.1016/j.jcis.2016.03.062.

[101] J.T. Oh, S.Y. Bae, S.R. Ha, H. Cho, S.J. Lim, D.W. Boukhvalov, et al., Water-resistant $AgBiS_2$ colloidal nanocrystal solids for eco-friendly thin film photovoltaics, Nanoscale 11 (2019) 9633−9640. Available from: https://doi.org/10.1039/C9NR01192G.

[102] J.T.R. Dufton, A. Walsh, P.M. Panchmatia, L.M. Peter, D. Colombara, M.S. Islam, Structural and electronic properties of $CuSbS_2$ and $CuBiS_2$: potential absorber materials for thin-film solar cells, Phys. Chem. Chem. Phys. 14 (2012) 7229−7233. Available from: https://doi.org/10.1039/C2CP40916J.

[103] P.-C. Wang, F.Z. Li, Y. Zhang, H.C. Fan, X.Q. Zhou, Y.L. Song, et al., In situ deposition of $CuBiS_2$ on mesoporous TiO_2 film for light absorber in solar cells, J. Nanosci. Nanotechnol. 20 (2020) 7748−7752. Available from: https://doi.org/10.1166/jnn.2020.18618.

[104] D.J. Temple, A.B. Kehoe, J.P. Allen, G.W. Watson, D.O. Scanlon, Geometry, electronic structure, and bonding in CuMCh$_2$ (M = Sb, Bi; Ch = S, Se): alternative solar cell absorber materials? J. Phys. Chem. C. 116 (2012) 7334−7340. Available from: https://doi.org/10.1021/jp300862v.

[105] N. Pai, J. Lu, D.C. Senevirathna, A.S.R. Chesman, T. Gengenbach, et al., Spray deposition of AgBiS$_2$ and Cu$_3$BiS$_3$ thin films for photovoltaic applications, J. Mater. Chem. C. 6 (2018) 2483−2494. Available from: https://doi.org/10.1039/C7TC05711C.

[106] M. Bernechea, N. Cates, G. Xercavins, D. So, A. Stavrinadis, G. Konstantatos, Solution-processed solar cells based on environmentally friendly AgBiS$_2$ nanocrystals, Nat. Photonics 10 (2016) 521−525. Available from: https://doi.org/10.1038/nphoton.2016.108.

[107] I. Burgués-Ceballos, Y. Wang, M.Z. Akgul, G. Konstantatos, Colloidal AgBiS$_2$ nanocrystals with reduced recombination yield 6.4% power conversion efficiency in solution-processed solar cells, Nano Energy 75 (2020) 104961. Available from: https://doi.org/10.1016/j.nanoen.2020.104961.

[108] L. Hu, R.J. Patterson, Z. Zhang, Y. Hu, D. Li, et al., Enhanced optoelectronic performance in AgBiS$_2$ nanocrystals obtained via an improved amine-based synthesis route, J. Mater. Chem. C. 6 (2018) 731−737. Available from: https://doi.org/10.1039/C7TC05366E.

[109] M.Z. Akgul, A. Figueroba, S. Pradhan, Y. Bi, G. Konstantatos, Low-cost RoHS compliant solution processed photovoltaics enabled by ambient condition synthesis of AgBiS$_2$ nanocrystals, ACS Photonics 7 (2020) 588−595. Available from: https://doi.org/10.1021/acsphotonics.9b01757.

[110] I. Burgués-Ceballos, Y. Wang, G. Konstantatos, Mixed AgBiS$_2$ nanocrystals for photovoltaics and photodetectors, Nanoscale 14 (2022) 4987−4993. Available from: https://doi.org/10.1039/D2NR00589A.

[111] Y. Wang, L. Peng, Z. Wang, G. Konstantatos, Environmentally friendly AgBiS2 nanocrystal inks for efficient solar cells employing green solvent processing, Adv. Energy Mater. 12 (2022) 7−13. Available from: https://doi.org/10.1002/aenm.202200700.

[112] V.A. Öberg, M.B. Johansson, X. Zhang, E.M.J. Johansson, Cubic AgBiS$_2$ colloidal nanocrystals for solar cells, ACS Appl. Nano Mater. 3 (2020) 4014−4024. Available from: https://doi.org/10.1021/acsanm.9b02443.

[113] S. Suzuki, M. Tsuyama, Theoretical study of structural, electronic, and optical properties of ternary metal sulfides MBiS$_2$ (M = Ag, Na), Jpn. J. Appl. Phys. 59 (2020) 041002. Available from: https://doi.org/10.35848/1347-4065/ab7c93.

[114] S.N. Guin, S. Banerjee, D. Sanyal, S.K. Pati, K. Biswas, Origin of the order-disorder transition and the associated anomalous change of thermopower in AgBiS$_2$ Nanocrystals: a combined experimental and theoretical study, Inorg. Chem. 55 (2016) 6323−6331. Available from: https://doi.org/10.1021/acs.inorgchem.6b00997.

[115] I. Zumeta-Dubé, V.F. Ruiz-Ruiz, D. Díaz, S. Rodil-Posadas, A. Zeinert, TiO$_2$ sensitization with Bi$_2$S$_3$ quantum dots: the inconvenience of sodium ions in the deposition procedure, J. Phys. Chem. C. 118 (2014) 11495−11504. Available from: https://doi.org/10.1021/jp411516a.

[116] B. Pejjai, V.R. Minnam Reddy, S. Gedi, C. Park, Status review on earth-abundant and environmentally green Sn-X (X = Se, S) nanoparticle synthesis by solution methods for photovoltaic applications, Int. J. Hydrog. Energy 42 (2017) 2790−2831. Available from: https://doi.org/10.1016/j.ijhydene.2016.11.084.

[117] V.R.M. Reddy, M.R. Pallavolu, P.R. Guddeti, S. Gedi, K.K.Y.B. Reddy, et al., Review on Cu$_2$SnS$_3$, Cu$_3$SnS$_4$, and Cu$_4$SnS$_4$ thin films and their photovoltaic performance, J. Ind. Eng. Chem. 76 (2019) 39−74. Available from: https://doi.org/10.1016/j.jiec.2019.03.035.

[118] K.J. Norton, F. Alam, D.J. Lewis, A review of the synthesis, properties, and applications of bulk and two-dimensional tin (II) sulfide (SnS), Appl. Sci. 11 (2021) 2062. Available from: https://doi.org/10.3390/app11052062.

[119] H.S. Yun, B.W. Park, Y.C. Choi, J. Im, T.J. Shin, S.I. Seok, Efficient nanostructured TiO_2/SnS heterojunction solar cells, Adv. Energy Mater. 9 (2019) 1901343. Available from: https://doi.org/10.1002/aenm.201901343.
[120] P. Sinsermsuksakul, L. Sun, S.W. Lee, H.H. Park, S.B. Kim, C. Yang, et al., Overcoming efficiency limitations of SnS-based solar cells, Adv. Energy Mater. 4 (2014) 1–7. Available from: https://doi.org/10.1002/aenm.201400496.
[121] M. Cheraghizade, F. Jamali-Sheini, R. Yousefi, F. Niknia, M.R. Mahmoudian, M. Sookhakian, The effect of tin sulfide quantum dots size on photocatalytic and photovoltaic performance, Mater. Chem. Phys. 195 (2017) 187–194. Available from: https://doi.org/10.1016/j.matchemphys.2017.04.008.
[122] M. Umehara, S. Tajima, Y. Aoki, Y. Takeda, T. Motohiro, $Cu_2Sn_{1-x}Ge_xS_3$ solar cells fabricated with a graded bandgap structure, Appl. Phys. Express 9 (2016) 072301. Available from: https://doi.org/10.7567/APEX.9.072301.
[123] Q. Guo, G.M. Ford, W.C. Yang, B.C. Walker, E.A. Stach, H.W. Hillhouse, et al., Fabrication of 7.2% efficient CZTSSe solar cells using CZTS nanocrystals, J. Am. Chem. Soc. 132 (2010) 17384–17386. Available from: https://doi.org/10.1021/ja108427b.
[124] W. Albers, C. Haas, H.J. Vink, J.D. Wasscher, Investigations on SnS, J. Appl. Phys. 32 (1961) 2220–2225. Available from: https://doi.org/10.1063/1.1777047.
[125] D.G. Moon, S. Rehan, D.H. Yeon, S.M. Lee, S.J. Park, S. Ahn, et al., A review on binary metal sulfide heterojunction solar cells, Sol. Energy Mater. Sol. Cell 200 (2019) 109963. Available from: https://doi.org/10.1016/j.solmat.2019.109963.
[126] A. Le Donne, V. Trifiletti, S. Binetti, New earth-abundant thin film solar cells based on chalcogenides, Front. Chem. 7 (2019) 297. Available from: https://doi.org/10.3389/fchem.2019.00297.
[127] L.S. Price, I.P. Parkin, A.M.E. Hardy, R.J.H. Clark, T.G. Hibbert, K.C. Molloy, Atmospheric pressure chemical vapor deposition of tin sulfides (SnS, Sn_2S_3, and SnS_2) on glass, Chem. Mater. 11 (1999) 1792–1799. Available from: https://doi.org/10.1021/cm990005z.
[128] K. Kuepper, B. Schneider, V. Caciuc, M. Neumann, Electronic structure of $Sn_2P_2S_6$, Phys. Rev. B 67 (2003) 115101. Available from: https://doi.org/10.1103/PhysRevB.67.115101.
[129] L.A. Burton, D. Colombara, R.D. Abellon, F.C. Grozema, L.M. Peter, T.J. Savenije, et al., Synthesis, characterization, and electronic structure of single-crystal SnS, Sn_2S_3, and SnS_2, Chem. Mater. 25 (2013) 4908–4916. Available from: https://doi.org/10.1021/cm403046m.
[130] H. Noguchi, A. Setiyadi, H. Tanamura, T. Nagatomo, O. Omoto, Characterization of vacuum-evaporated tin sulfide film for solar cell materials, Sol. Energy Mater. Sol. Cell 35 (1994) 325–331. Available from: https://doi.org/10.1016/0927-0248(94)90158-9.
[131] G. Barone, T.G. Hibbert, M.F. Mahon, K.C. Molloy, L.S. Price, I.P. Parkin, et al., Deposition of tin sulfide thin films from tin(IV) thiolate precursors, J. Mater. Chem. 11 (2001) 464–468. Available from: https://doi.org/10.1039/B005888M.
[132] A. Sanchez-Juarez, A. Ortíz, Effects of precursor concentration on the optical and electrical properties of Sn_xS_y thin films prepared by plasma-enhanced chemical vapour deposition, Semicond. Sci. Technol. 17 (2002) 931–937. Available from: https://doi.org/10.1088/0268-1242/17/9/305.
[133] M. Khadraoui, M. Khadraoui, N. Benramdane, C. Mathieu, A. Bouzidi, R. Miloua, et al., Optical and electrical properties of Sn_2S_3 thin films grown by spray pyrolysis, Solid. State Commun. 150 (2010) 297–300. Available from: https://doi.org/10.1016/j.ssc.2009.10.032.
[134] N.R. Mathews, H.B.M. Anaya, M.A. Cortes-Jacome, C. Angeles-Chavez, J.A. Toledo-Antonio, Tin sulfide thin films by pulse electrodeposition: structural, morphological, and optical properties, J. Electrochem. Soc. 157 (2010) H337. Available from: https://doi.org/10.1149/1.3289318.

[135] P. Sinsermsuksakul, K. Hartman, S.B. Kim, J. Heo, L. Sun, et al., Enhancing the efficiency of SnS solar cells via band-offset engineering with a zinc oxysulfide buffer layer, Appl. Phys. Lett. 102 (2013) 53901. Available from: https://doi.org/10.1063/1.4789855.

[136] H.H. Park, R. Heasley, L. Sun, V. Steinmann, R. Jaramillo, et al., Co-optimization of SnS absorber and Zn(O,S) buffer materials for improved solar cells, Prog. Photovolt. Res. Appl. 23 (2015) 901−908. Available from: https://doi.org/10.1002/pip.2504.

[137] V. Steinmann, R. Jaramillo, K. Hartman, R. Chakraborty, R.E. Brandt, et al., 3.88% efficient tin sulfide solar cells using congruent thermal evaporation, Adv. Mater. 26 (2014) 7488−7492. Available from: https://doi.org/10.1002/adma.201402219.

[138] G. Yue, Y. Lin, X. Wen, L. Wang, D. Peng, SnS homojunction nanowire-based solar cells, J. Mater. Chem. 22 (2012) 16437−16441. Available from: https://doi.org/10.1039/C2JM32116E.

[139] S. Gedi, V.R.M. Reddy, B. Pejjai, C.W. Jeon, C. Park, K.T. R. Reddy, A facile inexpensive route for SnS thin film solar cells with SnS_2 buffer, Appl. Surf. Sci. 372 (2016) 116−124. Available from: https://doi.org/10.1016/j.apsusc.2016.03.032.

[140] A. Degrauw, R. Armstrong, A.A. Rahman, J. Ogle, L. Whittaker-Brooks, Catalytic growth of vertically aligned SnS/SnS_2 $p-n$ heterojunctions, Mater. Res. Express 4 (2017) 094002. Available from: https://doi.org/10.1088/2053-1591/aa8a37.

[141] T. Rath, L. Gury, I. Sánchez-Molina, L. Martínez, S.A. Haque, Formation of porous SnS nanoplate networks from solution and their application in hybrid solar cells, Chem. Commun. 51 (2015) 10198−10201. Available from: https://doi.org/10.1039/C5CC03125G.

[142] D. Ding, T. Rath, L. Lanzetta, J.M. Marin-Beloqui, S.A. Haque, Efficient hybrid solar cells based on solution processed mesoporous $TiO_2/Tin(II)$ sulfide heterojunctions, ACS Appl. Energy Mater. 1 (2018) 3042−3047. Available from: https://doi.org/10.1021/acsaem.8b00590.

[143] A. Stavrinadis, J.M. Smith, C.A. Cattley, A.G. Cook, P.S. Grant, A.A.R. Watt, SnS/PbS nanocrystal heterojunction photovoltaics, Nanotechnology 21 (2010) 185202. Available from: https://doi.org/10.1088/0957-4484/21/18/185202.

[144] P.K. Dutta, U.K. Sen, S. Mitra, Excellent electrochemical performance of tin monosulphide (SnS) as a sodium-ion battery anode, RSC Adv. 4 (2014) 43155−43159. Available from: https://doi.org/10.1039/C4RA05851H.

[145] N.T.N. Truong, H.H.T. Hoang, T.K. Trinh, V.T.H. Pham, R.P. Smith, C. Park, Effect of post-synthesis annealing on properties of SnS nanospheres and its solar cell performance, Korean J. Chem. Eng. 34 (2017) 1208−1213. Available from: https://doi.org/10.1007/s11814-016-0347-4.

[146] T.A. Kuku, O.A. Fakolujo, Photovoltaic characteristics of thin films of Cu_2SnS_3, Sol. Energy Mater. 16 (1987) 199−204. Available from: https://doi.org/10.1016/0165-1633(87)90019-0.

[147] J. Koike, K. Chino, N. Aihara, H. Araki, R. Nakamura, K. Jimbo, et al., Cu_2SnS_3 thin-film solar cells from electroplated precursors, Jpn. J. Appl. Phys. 51 (2012) 10NC34. Available from: https://doi.org/10.1143/JJAP.51.10NC34.

[148] N. Aihara, H. Araki, A. Takeuchi, K. Jimbo, H. Katagiri, Fabrication of Cu_2SnS_3 thin films by sulfurization of evaporated Cu-Sn precursors for solar cells, Phys. Status Solidi C. 10 (2013) 1086−1092. Available from: https://doi.org/10.1002/pssc.201200866.

[149] N. Aihara, A. Kanai, K. Kimura, M. Yamada, K. Toyonaga, H. Araki, et al., Sulfurization temperature dependences of photovoltaic properties in Cu_2SnS_3-based thin-film solar cells, Jpn. J. Appl. Phys. 53 (2014) 05FW13. Available from: https://doi.org/10.7567/JJAP.53.05FW13.

[150] A. Kanai, H. Araki, A. Takeuchi, H. Katagiri, Annealing temperature dependence of photovoltaic properties of solar cells containing Cu_2SnS_3 thin films produced by co-evaporation, Phys. Status Solidi 252 (2015) 1239−1243. Available from: https://doi.org/10.1002/pssb.201400297.

[151] A. Kanai, K. Toyonaga, K. Chino, H. Katagiri, H. Araki, Fabrication of Cu_2SnS_3 thin-film solar cells with power conversion efficiency of over 4%, Jpn. J. Appl. Phys. 54 (2015) 08KC06. Available from: https://doi.org/10.7567/JJAP.54.08KC06.

[152] M. Nakashima, J. Fujimoto, T. Yamaguchi, M. Izaki, Cu_2SnS_3 thin-film solar cells fabricated by sulfurization from NaF/Cu/Sn stacked precursor, Appl. Phys. Express 8 (2015) 042303. Available from: https://doi.org/10.7567/APEX.8.042303.

[153] K. Ito, T. Nakazawa, Electrical and optical properties of stannite-type quaternary semiconductor thin films, Jpn. J. Appl. Phys. 27 (1988) 2094–2097. Available from: https://doi.org/10.1143/JJAP.27.2094.

[154] C. Yan, J. Huang, K. Sun, S. Johnston, Y. Zhang, et al., Cu_2ZnSnS_4 solar cells with over 10% power conversion efficiency enabled by heterojunction heat treatment, Nat. Energy 3 (2018) 764–772. Available from: https://doi.org/10.1038/s41560-018-0206-0.

[155] H. Fu, Environmentally friendly and earth-abundant colloidal chalcogenide nanocrystals for photovoltaic applications, J. Mater. Chem. C. 6 (2018) 414–445. Available from: https://doi.org/10.1039/C7TC04952H.

[156] Q. Guo, H.W. Hillhouse, R. Agrawal, Synthesis of Cu_2ZnSnS_4 nanocrystal ink and its use for solar cells, J. Am. Chem. Soc. 131 (2009) 11672–11673. Available from: https://doi.org/10.1021/ja904981r.

[157] H. Guo, C. Ma, K. Zhang, X. Jia, Y. Li, N. Yuan, et al., The fabrication of Cd-free Cu_2ZnSnS_4-Ag_2ZnSnS_4 heterojunction photovoltaic devices, Sol. Energy Mater. Sol. Cell 178 (2018) 146–153. Available from: https://doi.org/10.1016/j.solmat.2018.01.022.

[158] C. Bolm, A new iron age, Nat. Chem. 1 (2009) 420. Available from: https://doi.org/10.1038/nchem.315.

[159] A. Ennaoui, S. Fiechter, Ch Pettenkofer, N. Alonso-Vante, K. Büker, M. Bronold, et al., Iron disulfide for solar energy conversion, Sol. Energy Mater. Sol. Cell 29 (1993) 289–370. Available from: https://doi.org/10.1016/0927-0248(93)90095-K.

[160] P. Prabukanthan, S. Thamaraiselvi, G. Harichandran, Single step electrochemical deposition of p-Type undoped and Co^{2+} Doped FeS_2 thin films and performance in heterojunction solid solar cells, J. Electrochem. Soc. 164 (2017) D581–D589. Available from: https://doi.org/10.1149/2.0991709jes.

[161] M.A. Khan, Y.-M. Kang, Synthesis and processing of strong light absorbent iron pyrite quantum dots in polymer matrix for efficiency enhancement of bulk-heterojunction solar cell, Mater. Lett. 132 (2014) 273–276. Available from: https://doi.org/10.1016/j.matlet.2014.06.106.

[162] S. Sil, A. Dey, S. Halder, J. Datta, P.P. Ray, Possibility to use hydrothermally synthesized $CuFeS_2$ nanocomposite as an acceptor in hybrid solar cell, J. Mater. Eng. Perform. 27 (2018) 2649–2654. Available from: https://doi.org/10.1007/s11665-018-3142-z.

[163] S. Chatterjee, A.J. Pal, A solution approach to p-type Cu_2FeSnS_4 thin-films and pn-junction solar cells: role of electron selective materials on their performance, Sol. Energy Mater. Sol. Cell 160 (2017) 233–240. Available from: https://doi.org/10.1016/j.solmat.2016.10.037.

[164] C. Wadia, A.P. Alivisatos, D.M. Kammen, Materials availability expands the opportunity for large-scale photovoltaics deployment, Environ. Sci. Technol. 43 (2009) 2072–2077. Available from: https://doi.org/10.1021/es8019534.

[165] L. Yu, S. Lany, R. Kykyneshi, V. Jieratum, R. Ravichandran, et al., Iron chalcogenide photovoltaic absorbers, Adv. Energy Mater. 1 (2011) 748–753. Available from: https://doi.org/10.1002/aenm.201100351.

[166] X. Zhang, T. Scott, T. Socha, D. Nielsen, M. Manno, et al., Phase stability and stoichiometry in thin film iron pyrite: impact on electronic transport properties, ACS Appl. Mater. Interfaces 7 (2015) 14130–14139. Available from: https://doi.org/10.1021/acsami.5b03422.

[167] M. Limpinsel, N. Farhi, N. Berry, J. Lindemuth, C.L. Perkins, Q. Linf, et al., An inversion layer at the surface of n-type iron pyrite, Energy Environ. Sci. 7 (2014) 1974−1989. Available from: https://doi.org/10.1039/C3EE43169J.
[168] M. Cabán-Acevedo, N.S. Kaiser, C.R. English, D. Liang, B.J. Thompson, et al., Ionization of high-density deep donor defect states explains the low photovoltage of iron pyrite single crystals, J. Am. Chem. Soc. 136 (2014) 17163−17179. Available from: https://doi.org/10.1021/ja509142w.
[169] B. Voigt, W. Moore, M. Maiti, J. Walter, B. Das, M. Manno, et al., Observation of an internal p-n junction in pyrite FeS_2 single crystals: potential origin of the low open circuit voltage in pyrite solar cells, ACS Mater. Lett. 2 (2020) 861−868. Available from: https://doi.org/10.1021/acsmaterialslett.0c00207.
[170] M. Rahman, G. Boschloo, A. Hagfeldt, T. Edvinsson, On the mechanistic understanding of photovoltage loss in iron pyrite solar cells, Adv. Mater. 32 (2020) 1905653. Available from: https://doi.org/10.1002/adma.201905653.
[171] S. Huang, Q. He, W. Chen, J. Zai, Q. Qiao, X. Qian, 3D hierarchical $FeSe_2$ microspheres: controlled synthesis and applications in dye-sensitized solar cells, Nano Energy 15 (2015) 205−215. Available from: https://doi.org/10.1016/j.nanoen.2015.04.027.
[172] B. Kilic, S. Turkdogan, Fabrication of dye-sensitized solar cells using graphene sandwiched 3D-ZnO nanostructures based photoanode and Pt-free pyrite counter electrode, Mater. Lett. 193 (2017) 195−198. Available from: https://doi.org/10.1016/j.matlet.2017.01.128.
[173] Y. Bi, Y. Yuan, C.L. Exstrom, S.A. Darveau, J. Huang, Air stable, photosensitive, phase pure iron pyrite nanocrystal thin films for photovoltaic application, Nano Lett. 11 (2011) 4953−4957. Available from: https://doi.org/10.1021/nl202902z.
[174] J. Puthussery, S. Seefeld, N. Berry, M. Gibbs, M. Law, Colloidal iron pyrite (FeS_2) nanocrystal inks for thin-film photovoltaics, J. Am. Chem. Soc. 133 (2011) 716−719. Available from: https://doi.org/10.1021/ja1096368.
[175] C. Steinhagen, T.B. Harvey, C.J. Stolle, J. Harris, B.A. Korgel, Pyrite nanocrystal solar cells: promising, or fool's gold? J. Phys. Chem. Lett. 3 (2012) 2352−2356. Available from: https://doi.org/10.1021/jz301023c.
[176] Y.Y. Lin, D.Y. Wang, H.C. Yen, H.L. Chen, C.C. Chen, C.M. Chen, et al., Extended red light harvesting in a poly(3-hexylthiophene)/iron disulfide nanocrystal hybrid solar cell, Nanotechnology 20 (2009) 405207. Available from: https://doi.org/10.1088/0957-4484/20/40/405207.
[177] A. Kirkeminde, R. Scott, R. Ren, All inorganic iron pyrite nano-heterojunction solar cells, Nanoscale 4 (2012) 7649−7654. Available from: https://doi.org/10.1039/C2NR32097E.
[178] M.A. Khan, J.C. Sarker, S. Lee, S.C. Mangham, M.O. Manasreh, Synthesis, characterization and processing of cubic iron pyrite nanocrystals in a photovoltaic cell, Mater. Chem. Phys. 148 (2014) 1022−1028. Available from: https://doi.org/10.1016/j.matchemphys.2014.09.013.
[179] O. Amargós-Reyes, J.L. Maldonado, O. Martínez-Alvarez, M.E. Nicho, J. Santos-Cruz, J. Nicasio-Collazo, et al., Nontoxic pyrite iron sulfide nanocrystals as second electron acceptor in PTB7:PC71BM-based organic photovoltaic cells, Beilstein J. Nanotechnol. 10 (2019) 2238−2250. Available from: https://doi.org/10.3762/bjnano.10.216.
[180] K.P. Bhandari, P. Koirala, N.R. Paudel, R.R. Khanal, A.B. Phillips, et al., Iron pyrite nanocrystal film serves as a copper-free back contact for polycrystalline CdTe thin film solar cells, Sol. Energy Mater. Sol. Cell 140 (2015) 108−114. Available from: https://doi.org/10.1016/j.solmat.2015.03.032.
[181] E. Bastola, K.P. Bhandari, R.J. Ellingson, Application of composition controlled nickel-alloyed iron sulfide pyrite nanocrystal thin films as the hole transport layer in cadmium telluride solar cells, J. Mater. Chem. C. 5 (2017) 4996−5004. Available from: https://doi.org/10.1039/C7TC00948H.

[182] A.J. Huckaba, P. Sanghyun, G. Grancini, E. Bastola, C.K. Taek, et al., Exceedingly cheap perovskite solar cells using iron pyrite hole transport materials, ChemistrySelect 1 (2016) 5316−5319. Available from: https://doi.org/10.1002/slct.201601378.
[183] B. Bhattacharyya, A. Pandey, $CuFeS_2$ quantum dots and highly luminescent $CuFeS_2$ based core/chell ctructures: synthesis, tunability, and photophysics, J. Am. Chem. Soc. 138 (2016) 10207−10213. Available from: https://doi.org/10.1021/jacs.6b04981.
[184] E. Bastola, K.P. Bhandari, I. Subedi, N.J. Podraza, R.J. Ellingson, Structural, optical, and hole transport properties of earth-abundant chalcopyrite ($CuFeS_2$) nanocrystals, Mater. Res. Soc. Commun. 8 (2018) 970−978. Available from: https://doi.org/10.1557/mrc.2018.117.
[185] Y. Wu, B. Zhou, C. Yang, S. Liao, W.H. Zhang, C. Li, $CuFeS_2$ colloidal nanocrystals as an efficient electrocatalyst for dye sensitized solar cells, Chem. Commun. 52 (2016) 11488−11491. Available from: https://doi.org/10.1039/C6CC06241E.
[186] B. Sciacca, A.O. Yalcin, E.C. Garnett, Transformation of Ag nanowires into semiconducting $AgFeS_2$ nanowires, J. Am. Chem. Soc. 137 (2015) 4340−4343. Available from: https://doi.org/10.1021/jacs.5b02051.
[187] S.K. Han, C. Gu, M. Gong, Z.M. Wang, S.H. Yu, Colloidal synthesis of ternary $AgFeS_2$ nanocrystals and their transformation to Ag_2S-Fe_7S_8 heterodimers, Small 9 (2013) 3765−3769. Available from: https://doi.org/10.1002/smll.201300268.
[188] X. Zheng, B. Sciacca, E.C. Garnett, L. Zhang, $AgFeS_2$-nanowire-modified $BiVO_4$ photoanodes for photoelectrochemical water splitting, Chempluschem 81 (2016) 1075−1082. Available from: https://doi.org/10.1002/cplu.201600095.
[189] B. Pang, S. Lin, Y. Shi, Y. Wang, Y. Chen, et al., Synthesis of $CoFe_2O_4$/graphene composite as a novel counter electrode for high performance dye-sensitized solar cells, Electrochim. Acta 297 (2019) 70−76. Available from: https://doi.org/10.1016/j.electacta.2018.11.170.
[190] M. De Sousa Pereira, F.A. de Sousa Lima, R. Queiros de Almeida, J.L. da Silva Martins, D. Bagnis, E. Bedê Barros, et al., Flexible, large-area organic solar cells with improved performance through incorporation of $CoFe_2O_4$ nanoparticles in the active layer, Mater. Res. 22 (2019) 20190417. Available from: https://doi.org/10.1590/1980-5373-MR-2019-0417.
[191] R. Li, Y. Liao, Y. Dou, D. Wang, G. Li, W. Sun, et al., $CoFe_2O_4$ nanocrystals for interface engineering to enhance performance of perovskite solar cells, Sol. Energy 220 (2021) 400−405. Available from: https://doi.org/10.1016/j.solener.2021.03.073.

CHAPTER THIRTEEN

Life cycle assessment of renewable energy from solar photovoltaic technologies

Annick Anctil
Michigan State University, East Lansing, MI, United States

13.1 Introduction to life cycle assessment

Life cycle assessment (LCA) is a comprehensive approach that considers all stages from raw material extraction (cradle) and ends when the material returns to the Earth (grave). There are established ISO standards to conduct LCA (i.e., ISO 14040 [1]). In addition, for photovoltaics (PV), there are recommended guidelines provided by the International Energy Agency (IEA) Photovoltaics Program (PVPS) [2]. The method provides standard assumptions such as the lifetime of the PV solar plant, annual degradation, and standard insolation conditions (1700 kWh/m^2 year) and recommends certain metrics to allow comparison between studies. In LCA, the cumulative environmental impacts from all stages of the product are included, allowing the evaluation of product and process selection trade-offs. Including all life cycle stages and multiple metrics prevents trading one problem for another. Fig. 13.1 illustrates the system boundary for solar PV LCA [3].

The stages considered include material extraction, material processing, product assembly, PV system assembly, product use, end-of-life, and disposal. The balance of systems (BOS) is similar to most PV modules and changes more slowly than the modules' technologies. Therefore, most publications focus on module manufacturing, PV operation under standard conditions, and the potential for recycling the module at end-of-life. The IEA PVPS program guides the LCA methodology but also the life cycle inventory for silicon and CdTe PV technologies, which, in the past, has been updated every five years [4]. However, with rapid changes in material and device structures, life cycle inventory data will need to be

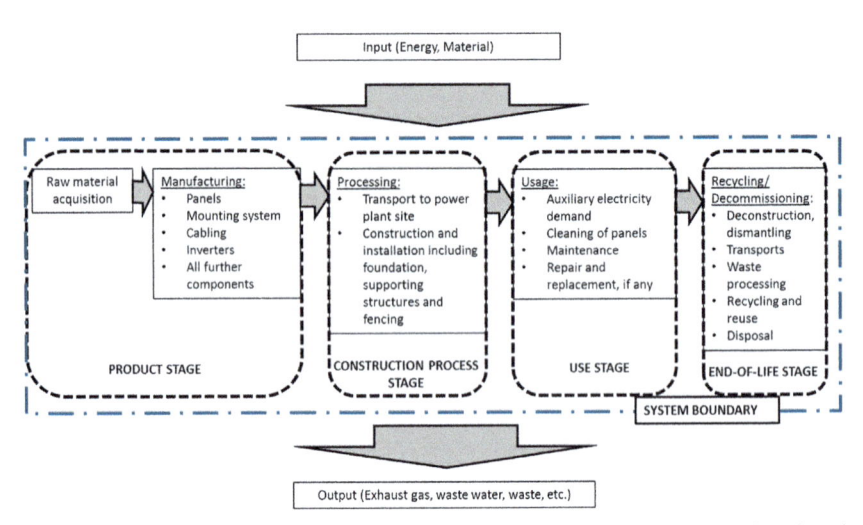

Figure 13.1 The system boundary of solar photovoltaic life cycle. *Reproduced with permission from N.A. Ludin, N.I. Mustafa, M.M. Hanafiah, M.A. Ibrahim, M.A.M. Teridi, et al., Prospects of life cycle assessment of renewable energy from solar photovoltaic technologies: A review, Renew. Sust. Energy Rev. 96 (November 2018) 11–28, copyright (2018) Elsevier.*

updated more regularly to represent newer PV technologies and manufacturing innovations.

Conducting LCA for module and inverter manufacturers is required to obtain an ecolabel from the Green Electronics Council's Electronic Product Environmental Assessment Tool (EPEAT) [5]. The product criteria are based on the NSF International standard #457: Sustainability Leadership for Photovoltaic Modules and Photovoltaic Inverters [6], which specifies metrics and inclusion. The LCA results must be reviewed by an external party and regularly updated. The objective is first to encourage manufacturers to conduct LCA of their product so they can identify opportunities for improvement. Therefore, low carbon certification requires a carbon footprint lower than 630 kg CO_2 eq/kW_p, while the threshold for ultra-low carbon is 400 kg CO_2 eq/kW_p.

For example, LCA is also used in international product category rules (PCRs) developed in Norway and Italy specifically for solar PV. The PCRs are used for environmental product declarations. Certification makes it easier for purchasers to select sustainable products and help identify greener products. LCA is also used by governments, corporate buyers, and international organizations to choose PV projects; for example, in a recent study commissioned by the United Nations [7], in France since

2011 through the simplified carbon methodology [8], and more recently in the Republic of Korea through the photovoltaic module carbon certification system [9]. While evaluating the carbon footprint of commercialized solar modules and establishing certification methods is valuable, there is a need to understand better other life cycle metrics that could be useful. In addition, current methods and certification focus on silicon and CdTe module manufacturing. At the same time, other benefits and concerns for both commercialized and emerging technologies need to be discussed. Finally, current LCA methods exclude the use phase (e.g., EPEAT) or assume ideal conditions for residential or utility-scale installations. There is a need to evaluate additional benefits solar can have in addition to generating electricity, for example, in building integrated PV and agrivoltaics where it affects building or crop temperature conditions. This chapter will provide an overview of LCA metrics, a summary of commercial and emerging solar technologies benefits and concerns throughout their life cycle, and an overview of future needs for LCA of PV.

13.2 Life cycle assessment metrics

Cumulative energy demand (CED) and greenhouse gases (GHGs) are the main impact categories for PV LCAs [9−11] because the main goal of PV is to replace fossil fuel energy production. The PV_{CED} is the cradle-to-grave primary energy required to produce the material, manufacture, install, operate, and dispose of the PV system [Eq. (13.1)]. In Eq. (13.1), PV_{CED} is the cradle-to-grave CED of PV; E_a is the annual PV electricity generation = Irradiance × module efficiency; E_{tot} is the total electricity produced over the system lifetime = Irradiance × module efficiency × lifetime; and η_{Grid} is grid efficiency.

The CED per Wh is used to calculate the energy payback time (EPBT) and the energy return on investment (EROI) as shown in Eqs. (13.1) and (13.2). The EPBT calculates how long the system needs to operate before it pays back the cradle-to-grave CED of the module. The EROI provides information on the ratio of energy returned to society in the form of a useful energy carrier over the total energy required or "invested" in finding, extracting, processing, and delivering the energy [12,13]. Energy systems should have an EROI higher than one, implying that the energy delivered is higher than the amount of energy required to manufacture the system. The carbon payback time is calculated similarly to the EPBT and uses the grid carbon intensity per Wh.

$$\text{EPBT (years)} = \frac{\text{PV}_{\text{CED}}}{\frac{E_d}{\eta_{\text{Grid}}}} \tag{13.1}$$

$$\text{EROI} = \frac{\frac{E_{\text{tot}}}{\eta_{\text{Grid}}}}{\text{PV}_{\text{CED}}} \tag{13.2}$$

In most cases, the EPBT and EROI use a static value for the grid efficiency and carbon footprint, which overestimate the benefit of PV with increasing renewables added to the grid. The effect is not important for the EPBT because it is usually less than a year [14], but it matters when estimating the avoided GHG emissions or cost-saving over the lifetime of the system. In an earlier study [14], Detroit, MI, Los Angeles, CA, Phoenix, AZ, and Honolulu, and HI were selected as examples of a diverse range of solar insolation conditions, climate zones, energy cost, and energy impact of electricity production that influence the benefit of both solar photovoltaic energy production and building energy balance (Fig. 13.2).

In addition to energy savings, the benefit of transparent PV in four locations in the US was evaluated based on the avoided cost and GHG for all scenarios (Fig. 13.3). For cities such as Los Angeles and Honolulu, the grid improvement cause a "curve" in Fig. 13.3B due to the decreasing amount of carbon being displaced by PV every year. Other common

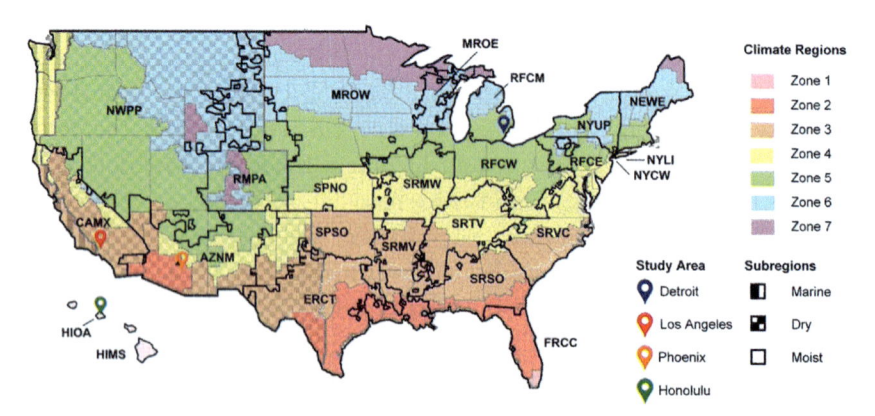

Figure 13.2 Summary of the climate zones, electricity grid regions, and locations considered. *Reproduced with permission from A. Anctil, E. Lee, R.R. Lunt, Net environmental and cost benefit of transparent organic solar cells in building-integrated applications, Appl. Energy 261 (March 2020) 114429, copyright (2020) Elsevier.*

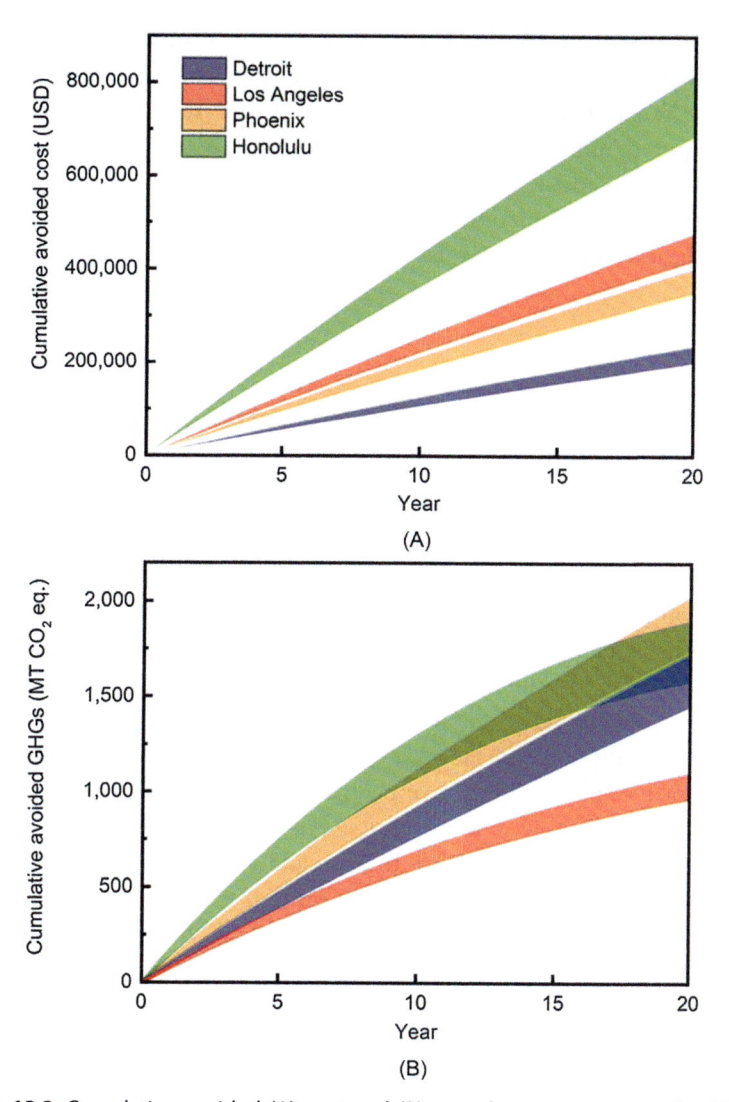

Figure 13.3 Cumulative avoided (A) cost and (B) greenhouse gases over the 20-year lifetime of glass transparent photovoltaic window ($\eta = 10\%$) for four cities considered, as discussed in the text; where the lowest value on the curve corresponds to application on the South, East, and West (SEW) direction of the building and the highest with the addition of the North face (SEWN). *Reproduced with permission from A. Anctil, E. Lee, R.R. Lunt, Net environmental and cost benefit of transparent organic solar cells in building-integrated applications, Appl. Energy 261 (March 2020) 114429, copyright (2020) Elsevier.*

metrics reported for PV LCA include land use, water use, human and environmental toxicity, and resource depletion. The IEA methodology guideline recommend specific indicators [2] and the NSF 457 specify eight required indicators: (1) CED; (2) Global Warming Potential; (3) Acidification potential; (4) Photochemical ozone creation potential (i.e., smog); (5) Ozone depletion potential; (6) Eutrophication potential; (7) Human health impact indicators; and (8) Ecotoxicity impact indicators).

13.3 Life cycle assessment of various photovoltaic technologies

The objective in most assessments is to ensure that the energy and carbon emissions associated with the manufacture and installation of PV are offset by its energy production and GHG over its lifetime (both energy and carbon payback time). PV LCA studies are often used to inform policy, usually by comparing solar with other energy technologies based on a single issue, such as water [15], land use [16], GHG [17], and cost [18]. In principle, the main benefit of conducting LCA studies is to prevent shifting issues by looking at all life stages and using multiple metrics. However, in the case of PV, most studies have been conducted to address specific concerns and include only a few metrics. For example, due to the general concern associated with cadmium toxicity and tellurium scarcity, the number of LCA studies for CdTe modules outnumber any other technologies, even though silicon represents 92% of the market [19].

For silicon, the active layer is an abundant and nontoxic material; for this reason, most studies have focused on the energy intensity to produce semiconductor-grade material, or lately, on the decreasing need for glass and aluminum from the frame [20]. Similarly, most work on III-V cells used in concentrators has focused on the concentrators rather than on the specific solar cell due to proprietary data from industry [21]. There are additional concerns with CIGS and perovskite results regarding toxicity because many metals such as gallium, selenium, and other intermediate chemicals are not characterized in LCA software, therefore, resulting in no impact. For organic photovoltaics, the focus of the studies has been on fabrication methods for large-scale production based on outdated material combinations, generally P3HT:C_{60}PCBM [22–25]. Table 13.1 summarizes the benefits or concerns associated with each LCA stage and related publications on the subject [22,26–93,95–97].

Table 13.1 Summary of commercial and emerging solar technologies benefits and concerns throughout their life cycle. *Assembled by author.*

		(1) Material extraction	(2) Material processing	(3) Product assembly	(4) Product use	(5) End-of-life	Cost
Crystalline Silicon	c-Si &mc-Si	Sand is considered abundant	Energy intensive to produce crystalline Si [11,26] Major reduction in energy consumption from crystalline Si in recent years [27,28]	Environmental impact depends on the module design (wafer size, bifacial, PERC, etc.) and manufacturing location [29–33]	Established technology with expected 30 yr lifetime. No toxicity concerns during the use phase [34,35]	Recyclability depends on amount and value of metals and glass [36–41] Environmental impact can be reduced by using recycled silicon [42] Low toxicity concerns from landfill disposal [43,44]	Can be reduced by reducing energy intensity of silicon manufacturing, reducing wafer loss, and switching from silver to cheaper contacts such as copper and aluminum
Thin Films	CdTe	Te scarcity [45–47]	Cadmium is toxic, but CdTe has lower toxicity [48,49]	Lower CdTe thickness and larger modules can reduce the environmental impact [50]	Low risk of cadmium contamination during fire [51] leaching [52] or from broken modules [53]	Established recycling process due to Te scarcity [39,41,54]	Dependent on tellurium prices and availability [55]

(Continued)

Table 13.1 (Continued)

		Life cycle stage				
		(1) Material extraction	(2) Material processing	(3) Product assembly	(4) Product use	(5) End-of-life	Cost
	CIGS	In, Ga, and Se scarcity [45]	Complex multimaterial system [56]	The thickness and type of material used in each layer vary [57]	Concerns about degradation [58]	Ga and In can be recycled in new modules [59]	Cost can be reduced by using recycled materials from old modules and increasing module area [55,60]
III-V	GaAs	Ga is a scarce material [61]	Energy intensive to produce crystalline material [62]	Lower purity material can be used without performance reduction [63]	Lightweights can be used for portable applications [64]	Small quantity of highly mixed high-value material. Can be recycled [65,66]	Expensive [63]
	Multi-Junctions	Scarce materials [67]	Expensive gaseous precursor for growth of high-quality solar cells [68]	Complex process to grow multiple junctions [69,70]	Used in concentrators for higher efficiency [69]	High value material [59,71]	Trade-off between higher efficiency and lowering cost [72]

Perovskites	Raw materials contribute to more than 80% of the CED [73]	Toxicity of Pb used in high-efficiency devices [74,75] Lead-free alternative tin iodine has lower toxicity and efficiency [76]	Environmental impact depends on material choice and efficiency [77,78]	Perovskites have low EPBT [73] Concerns about degradation [79–81] and lead toxicity from leaching [82]	Low residual value but toxic material is a regulatory and environmental concern [83] Incineration with energy recovery is better than landfill [77]	Similar or cheaper than existing PV [73,84,85]
Organics	Tunable material synthesized from petroleum feedstock [22]	Fullerenes are energy intensive [22]	Easy and fast roll-to-roll solution manufacturing [86,87]	Concerns about degradation [88,89] Lower energy and carbon payback time than Si PV [87] New applications: LED lamp [90,91] and BIPV [92]	Low residual value, unlikely to be recycled [93]	Expected to be extremely cheap but low efficiency [94]

13.4 Future need for life cycle assessment of photovoltaic technologies

To evaluate the impact of technologies currently under development, prospective or anticipatory LCA can be used either at a lab or pilot scale [94,98]. The objective is to identify potential concerns with new technologies and compare their environmental impact with existing technologies. In addition to evaluating new PV materials and device structures using prospective LCA, there is a need to account for the other benefits of the PV system in the assessment. In LCA, the functional unit refers to the unit of comparison where products perform the same way. Using the solar farm or rooftop application simplifies but also makes it possible to compare with other types of photovoltaics using Wp as a functional unit. For new applications where solar provides additional functions than electricity production, there is a need to redefine an adequate functional unit and method to adequately assess the potential of PV, beyond the traditional cost, efficiency, and lifetime approach. An alternative method based on LCA results but easier to interpret is the net environmental benefit (NEB). It has become popular in environmental engineering for remediation and wastewater treatment processes because it considers the no-action scenario as the baseline process. In wastewater treatment, not treating the water (no action) would have environmental consequences. Therefore, even though the treatment itself has environmental impacts due to energy or chemical consumption, as long as the total benefit is positive, then it is worth doing it.

The approach was demonstrated for transparent OPV in window applications [14,99]. In this case, the reference case corresponds to the energy consumption of the building without the OPV windows. In addition to electricity generation, the transparent OPV reduces the NIR transmission in the building, therefore, reducing the heat load of the building. While the benefit of this technology in warm climates might be beneficial to reduce the AC demand, without a complete analysis, there is a potential that the heating demand will increase during the winter in cold climates. Fig. 13.4 shows the window structure considered, the annual building cooling and heating demand for Lansing, MI and PV electricity production for windows covered with OPV on the south, east and west (SEW) and north (SEWN) directions (also see Fig. 13.3). The figure also shows the NEB of OPV for four scenarios including inclusion within the window (glass) or an additional encapsulated OPV inside the window for either SEWN and SEW directions.

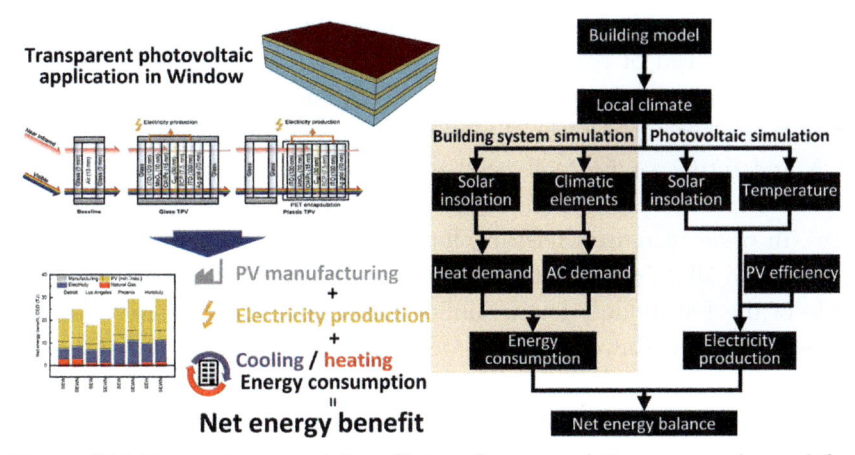

Figure 13.4 Net environmental benefit based on cumulative energy demand for transparent photovoltaics (PV) in buildings shows the importance of considering both the electricity for PV and the change in building energy demand. *Reproduced with permission from A. Anctil, E. Lee, R.R. Lunt, Net environmental and cost benefit of transparent organic solar cells in building-integrated applications, Appl. Energy 261 (March 2020) 114429, copyright (2020) Elsevier.*

Overall, for all cities considered, the net environmental impact was positive [14]. The approach suggested, therefore, allows for an easier interpretation of the results as well as a quantitative assessment of both the benefits associated with the PV module and the energy balance needed. Other applications that could benefit from the net benefit approach are all electronics-integrated applications such as in cell phones, tablets, laptops, and lamps. In those applications, the PV can reduce the energy consumption from the grid or reduce the size of the battery. In building- and car-integrated PV, PV can be used to help with heat/cooling management. In greenhouse applications, the goal is to absorb part of the solar spectrum that is not used by plants in order to produce solar energy while again playing a role in thermal management, reducing the cooling need during the summer and maintaining temperature during winter. LCA was used to evaluate the potential of OPV for this application.

13.5 Conclusion

In conclusion, LCA is becoming a standard method used for product selection through certification or preferred purchase for commercialized technologies. However, current methods are limited to the carbon footprint of the manufacturing stages. Existing LCA studies focus on areas that are perceived of particular concern (i.e., toxicity for perovskites and CdTe and energy consumption for silicon) and are often used to compare alternatives. Emerging technologies can be evaluated using prospective LCA and future studies will need to incorporate other benefits of PV beyond electricity production for product and building integrated applications. In order to remain relevant, future LCA methods and certifications will need to allow new PV technologies and comparison over the full life cycle of the PV module for nonutility applications for a variety of environmental metrics, beyond carbon footprint.

References

[1] ISO 14040, Environmental management — Life cycle assessment — Principles and framework; This standard was last reviewed and confirmed in 2022, therefore this version remains current. International Organization for Standardization, Geneva, Switzerland, 2006. <https://www.iso.org/standard/37456.html> (accessed 17.12.22).

[2] R. Frischknecht, P. Stolz, G. Heath, M. Raugei, P. Sinha, M. de Wild-Scholten, et al., Methodology Guidelines on Life Cycle Assessment of Photovoltaic Electricity, 4th edition. International Energy Agency Photovoltaic Power Systems Programme, Report IEA-PVPS T12−18:2020, April 2020, 34 pp. ISBN 978-3-906042-99-2.

[3] N.A. Ludin, N.I. Mustafa, M.M. Hanafiah, M.A. Ibrahim, M.A.M. Teridi, et al., Prospects of life cycle assessment of renewable energy from solar photovoltaic technologies: a review, Renew. Sust. Energy Rev. 96 (2018) 11−28. Available from: https://doi.org/10.1016/j.rser.2018.07.048.

[4] R. Frischknecht, G. Heath, M. Raugei, P. Sinha, M. de Wild-Scholten, Methodology Guidelines on Life Cycle Assessment of Photovoltaic Electricity, 3rd edition. International Energy Agency Photovoltaic Power Systems Programme, Report IEA-PVPS T12−08:2016, January 2016, 25 pp. ISBN 978-3-906042-39-8.

[5] EPEAT Program Continuous Monitoring Round Plan, Photovoltaic Modules and Inverters, GEC, PVMI-2021-01, March 26, 2021. <https://globalelectronicscouncil.org> (accessed 17.12.22).

[6] NSF International Standard/American National Standard, Sustainability Leadership Standard for Photovoltaic Modules and Photovoltaic Inverters, Ann Arbor, Michigan, USA, July 22, 2019, 57 pp.

[7] T. Gibon, Á.H. Menacho, M. Guiton, Life Cycle Assessment of Electricity Generation Options, United Nations Economic Commission for Europe (UNECE), UNECE Task Force on Carbon Neutrality, Geneva, Switzerland, 2021, 97 pp.

[8] Ultra Low-Carbon Solar Alliance. <https://ultralowcarbonsolar.org/blog/south-korea-implementing-carbon-footprint-assessment-regulations/> (accessed 17.12.22).

[9] H.C. Kim, V. Fthenakis, J.-K. Choi, D.E. Turney, Life cycle greenhouse gas emissions of thin-film photovoltaic electricity generation, J. Ind. Ecol. 16 (2012) S110−S121.

[10] D.D. Hsu, P. O'Donoughue, V. Fthenakis, G.A. Heath, H.C. Kim, P. Sawyer, et al., Life cycle greenhouse gas emissions of crystalline silicon photovoltaic electricity generation, J. Ind. Ecol. 16 (2012) S122−S135.

[11] J. Peng, L. Lu, H. Yang, Review on life cycle assessment of energy payback and greenhouse gas emission of solar photovoltaic systems, Renew. Sustain. Energy Rev. 19 (2013) 255−274.
[12] D.J. Murphy, C.A.S. Hall, Year in review-EROI or energy return on (energy) invested, Ann. N. Y. Acad. Sci. 1185 (1) (2010) 102−118.
[13] N. Kittner, S.H. Gheewala, D.M. Kammen, Energy return on investment (EROI) of mini-hydro and solar PV systems designed for a mini-grid, Renew. Energy 99 (2016) 410−419.
[14] A. Anctil, E. Lee, R.R. Lunt, Net environmental and cost benefit of transparent organic solar cells in building-integrated applications, Appl. Energy 261 (2020) 114429.
[15] J. Meldrum, S. Nettles-Anderson, G. Heath, J. Macknick, Life cycle water use for electricity generation: a review and harmonization of literature estimates, Environ. Res. Lett. 8 (1) (2013) 015031.
[16] S. Ong, C. Campbell, P. Denholm, R. Margolis, G. Heath, Land-use requirements for solar power plants in the United States, Technical Report NREL/TP-6A20-56290, National Renewable Energy Laboratory, Golden, CO, USA, June 2013, 39 pp. <https://www.nrel.gov/docs/fy13osti/56290.pdf>.
[17] V.M. Fthenakis, H.C. Kim, Greenhouse-gas emissions from solar electric- and nuclear power: a life-cycle study, Energy Policy 35 (4) (2017) 2549−2557.
[18] M. Ito, S. Lespinats, J. Merten, P. Malbranche, K. Kurokawa, Life cycle assessment and cost analysis of very large-scale PV systems and suitable locations in the world, Prog. Photovolt. Res. Appl. 24 (2) (2016) 159−174.
[19] S. Weckend, A. Wade, G. Heath, End-of-life management solar photovoltaic panels, International Energy Agency (IEA) Photovoltaic Power Systems Programme and International Renewable Energy Agency (IRENA), June 2016, 100 pp.
[20] M.J. De Wild-Scholten, Energy payback time and carbon footprint of commercial photovoltaic systems, Sol. Energy Mater. Sol. Cell 119 (2013) 296−305.
[21] V. Fthenakis, H.C. Kim, A.W. Bett, R.D. McConnell, G. Sala, F. Dimroth, Life cycle assessment of Amonix 7700 HCPV systems, AIP Conf. Proc. 1277 (2010) 260−263.
[22] N. Espinosa, R. García-Valverde, A. Urbina, F.C. Krebs, A life cycle analysis of polymer solar cell modules prepared using roll-to-roll methods under ambient conditions, Sol. Energy Mater. Sol. Cell 95 (5) (2011) 1293−1302.
[23] N. Espinosa, M. Hosel, D. Angmo, F.C. Krebs, Solar cells with one-day energy payback for the factories of the future, Energy Environ. Sci. 5 (1) (2012) 5117−5132.
[24] N. Espinosa, R. García-Valverde, A. Urbina, F. Lenzmann, M. Manceau, D. Angmo, et al., Life cycle assessment of ITO-free flexible polymer solar cells prepared by roll-to-roll coating and printing, Sol. Energy Mater. Sol. Cell 97 (2011) 3−13.
[25] S. Lizin, S. Van Passel, E. De Schepper, W. Maes, L. Lutsen, J. Manca, et al., Life cycle analyses of organic photovoltaics: a review, Energy Environ. Sci. 6 (11) (2013) 3136.
[26] C. Ramírez-Márquez, M.V. Otero, J.A. Vázquez-Castillo, M. Martín, J.G. Segovia-Hernández, Process design and intensification for the production of solar grade silicon, J. Clean. Prod. 170 (2018) 1579−1593.
[27] M. Woodhouse, B. Smith, A. Ramdas, R. Margolis, Crystalline silicon photovoltaic module manufacturing costs and sustainable pricing: 1H 2018 benchmark and cost reduction road map, Technical Report NREL/TP-6A20-72134, National Renewable Energy Laboratory, Golden, CO, USA, February 2020, 46 pp. <https://www.nrel.gov/docs/fy19osti/72134.pdf>.
[28] M.M. Lunardi, J.P. Alvarez-Gaitan, N.L. Chang, R. Corkish, Life cycle assessment on PERC solar modules, Sol. Energy Mater. Sol. Cell 187 (2018) 154−159.

[29] X. Jia, C. Zhou, Y. Tang, W. Wang, Life cycle assessment on PERC solar modules, Sol. Energy Mater. Sol. Cell 227 (2021) 111112.
[30] A. Müller, L. Friedrich, C. Reichel, S. Herceg, M. Mittag, D.H. Neuhaus, A comparative life cycle assessment of silicon PV modules: impact of module design, manufacturing location and inventory, Sol. Energy Mater. Sol. Cell 230 (2021) 111277.
[31] G. Hou, H. Sun, Z. Jiang, Z. Pan, Y. Wang, X. Zhang, et al., Life cycle assessment of grid-connected photovoltaic power generation from crystalline silicon solar modules in China, Appl. Energy 164 (2016) 882–890.
[32] W. Chen, J. Hong, X. Yuan, J. Liu, Environmental impact assessment of monocrystalline silicon solar photovoltaic cell production: a case study in China, J. Clean. Prod. 112 (2016) 1025–1032.
[33] P. Sinha, G. Heath, A. Wade, K. Komoto, Human Health Risk Assessment Methods for PV, Part 2: Breakage Risks, International Energy Agency Photovoltaic Power Systems Programme, Report IEA-PVPS T12-15:2019, September 2019,45 pp. ISBN 978-3-906042-87-9.
[34] P. Sinha, G.A. Heath, A. Wade, K. Komoto, Human Health Risk Assessment Methods for PV, Part 1: Fire Risks, International Energy Agency Photovoltaic Power Systems Programme, Report IEA-PVPS T12-14:2018, October 2018, 48 pp. ISBN 978-3-906042-78-7.
[35] C.E.L. Latunussa, F. Ardente, G.A. Blengini, L. Mancini, Life cycle assessment of an innovative recycling process for crystalline silicon photovoltaic panels, Sol. Energy Mater. Sol. Cell 156 (2016) 101–111.
[36] J.-K. Choi, V. Fthenakis, Crystalline silicon photovoltaic recycling planning: macro and micro perspectives, J. Clean. Prod. 66 (2014) 443–449.
[37] F. Corcelli, M. Ripa, S. Ulgiati, End-of-life treatment of crystalline silicon photovoltaic panels. an emergy-based case study, J. Clean. Prod. 161 (2017) 1129–1142.
[38] T. Maani, I. Celik, M.J. Heben, R.J. Ellingson, D. Apul, Environmental impacts of recycling crystalline silicon (c-Si) and cadmium telluride (CdTe) solar panels, Sci. Total. Environ. 735 (2020) 138827.
[39] V. Fiandra, L. Sannino, C. Andreozzi, G. Graditi, End-of-life of silicon PV panels: a sustainable materials recovery process, Waste Manag. 84 (2019) 91–101.
[40] F.C.S.M. Padoan, P. Altimari, F. Pagnanelli, Recycling of end of life photovoltaic panels: a chemical prospective on process development, Sol. Energy 177 (2019) 746–761.
[41] E. Klugmann-Radziemska, A. Kuczyńska-Łażewska, The use of recycled semiconductor material in crystalline silicon photovoltaic modules production - a life cycle assessment of environmental impacts, Sol. Energy Mater. Sol. Cell 205 (2020) 110259.
[42] M.K. Collins, A. Anctil, Implications for current regulatory waste toxicity characterisation methods from analysing metal and metalloid leaching from photovoltaic modules, Int. J. Sustain. Energy 35 (6) (2015) 531–544.
[43] P. Sinha, G. Heath, A. Wade, K. Komoto, Human Health Risk Assessment Methods for PV, Part 3: Module Disposal Risks, International Energy Agency Photovoltaic Power Systems Programme, Report IEA-PVPS T12-16:2020, May 2020, 46 pp. ISBN 978-3-906042-96-1.
[44] M. Woodhouse, A. Goodrich, R. Margolis, T.L. James, M. Lokanc, R. Eggert, Supply-chain dynamics of tellurium (Te), indium (In), and gallium (Ga) within the context of PV module manufacturing costs, in: Proc. 38th Photovoltaic Specialists Conference (PVSC) Part 2, 2012, pp. 1–6. Available from: https://doi.org/10.1109/PVSC-Vol2.2012.6656796.
[45] M.L. Bustamante, G. Gaustad, Challenges in assessment of clean energy supply-chains based on byproduct minerals: a case study of tellurium use in thin film photovoltaics, Appl. Energy 123 (2014) 397–414.

[46] V. Fthenakis, A. Anctil, Direct Te mining: resource availability and impact on cumulative energy demand of CdTe PV life cycles, IEEE J. Photovolt. 3 (1) (2013) 433−438.
[47] G. Genchi, M.S. Sinicropi, G. Lauria, A. Carocci, A. Catalano, The effects of cadmium toxicity, J. Environ. Res. Public. Health 17 (11) (2020) 3782.
[48] A.J. Rix, J.D.T. Steyl, M.J. Rudman, U. Terblanche, J.L. van Niekerk, First Solar's CdTe module technology − Performance, Life Cycle, Health and Safety Impact Assessment, Centre for Renewable and Sustainable Energy Studies, Stellenbosch University, Report No. 5004157, Stellenbosch, Western Cape, South Africa, December 2015, 32 pp.
[49] H. Kim, K. Cha, V.M. Fthenakis, P. Sinha, T. Hur, Life cycle assessment of cadmium telluride photovoltaic (CdTe PV) systems, Sol. Energy 103 (2014) 78−88.
[50] V.M. Fthenakis, M. Fuhrmann, J. Heiser, A. Lanzirotti, J. Fitts, W. Wang, Emissions and encapsulation of cadmium in CdTe PV modules during fires, Prog. Photovolt. Res. Appl. 13 (8) (2005) 713−723.
[51] P. Sinha, R. Balas, L. Krueger, A. Wade, Fate and transport evaluation of potential leaching risks from cadmium telluride photovoltaics, Environ. Toxicol. Chem. 31 (7) (2012) 1670−1675.
[52] W.D. Cyrs, H.J. Avens, Z.A. Capshaw, R.A. Kingsbury, J. Sahmel, B.E. Tvermoes, Landfill waste and recycling: use of a screening-level risk assessment tool for end-of-life cadmium telluride (CdTe) thin-film photovoltaic (PV) panels, Energy Policy 68 (2014) 524−533.
[53] M. Vellini, M. Gambini, V. Prattella, Environmental impacts of PV technology throughout the life cycle: importance of the end-of-life management for Si-panels and CdTe-panels, Energy 138 (2017) 1099−1111.
[54] M. Marwede, A. Reller, Estimation of life cycle material costs of cadmium telluride- and copper indium gallium diselenide-photovoltaic absorber materials based on life cycle material flows, J. Ind. Ecol. 18 (2) (2014) 254−267.
[55] N. Mufti, T. Amrillah, A. Taufiq, S. Sunaryono, A. Aripriharta, M. Diantoro, et al., Review of CIGS-based solar cells manufacturing by structural engineering, Sol. Energy 207 (2020) 1146−1157.
[56] S. Amarakoon, C. Vallet, M.A. Curran, P. Haldar, D. Metacarpa, D. Fobare, et al., Life cycle assessment of photovoltaic manufacturing consortium (PVMC) copper indium gallium (di)selenide (CIGS) modules, Int. J. Life Cycle Assess. 23 (4) (2018) 851−866.
[57] K. Sakurai, H. Tomita, D. Schmitz, S. Tokuda, K. Ogawa, H. Shibata, et al., Exploring suitable damp heat and potential induced degradation test procedures for Cu(In,Ga)(S,Se) photovoltaic modules, Jpn. J. Appl. Phys. 57 (8) (2018) 08RG02.
[58] A. Amato, F. Beolchini, End-of-life CIGS photovoltaic panel: a source of secondary indium and gallium, Prog. Photovolt. Res. Appl. 27 (3) (2019) 229−236.
[59] K.A.W. Horowitz, R. Fu, T. Silverman, M. Woodhouse, X. Sun, M.A. Alam, An Analysis of the Cost and Performance of Photovoltaic Systems as a Function of Module Area, Technical Report NREL/TP-6A20-67006, National Renewable Energy Laboratory, Golden, CO, USA, April 2017, 25 pp. <https://www.nrel.gov/docs/fy17osti/67006.pdf>.
[60] C.S. Tao, J. Jiang, M. Tao, Natural resource limitations to terawatt-scale solar cells, Sol. Energy Mater. Sol. Cell 95 (12) (2011) 3176−3180.
[61] N.J. Mohr, J.J. Schermer, M.A.J. Huijbregts, A. Meijer, L. Reijnders, Life cycle assessment of thin-film GaAs and GaInP/GaAs solar modules, Prog. Photovolt. Res. Appl. 15 (2) (2007) 163−179.
[62] J. Simon, C. Frank-Rotsch, K. Stolze, M. Young, M.A. Steiner, A.J. Ptak, GaAs solar cells grown on intentionally contaminated GaAs substrates, J. Cryst. Growth 541 (2020) 125668.
[63] A. Vijh, L. Washington, R.C. Parenti, High performance, lightweight GaAs solar cells for aerospace and mobile applications, in: Proc. 44th IEEE Photovoltaics Specialists Conference (PVSC), 2017, pp. 3520−3523. Available from: https://doi.org/10.1109/PVSC.2017.8366342.

[64] L. Zhan, F. Xia, Y. Xia, B. Xie, Recycle gallium and arsenic from GaAs-based E-wastes via pyrolysis-vacuum metallurgy separation: theory and feasibility, ACS Sustain. Chem. Eng. 6 (1) (2018) 1336−1342.

[65] S. Maneesuwannarat, A.S. Vangnai, M. Yamashita, P. Thiravetyan, Bioleaching of gallium from gallium arsenide by *Cellulosimicrobium funkei* and its application to semiconductor/electronic wastes, Process. Saf. Environ. Prot. 99 (2016) 80−87.

[66] P. Swart, J. Dewulf, H. Van Langenhove, K. Moonens, K. Dessein, C. Quaeyhaegens, Assessment of the overall resource consumption of germanium wafer production for high concentration photovoltaics, Resour. Conserv. Recycl. 55 (12) (2011) 1119−1128.

[67] B.L. Smith, C.W. Babbitt, K. Horowitz, G. Gaustad, S.M. Hubbard, Life cycle assessment of III-V precursors for photovoltaic and semiconductor applications, MRS Adv 3 (25) (2018) 1399−1404.

[68] J.F. Geisz, R.M. France, K.L. Schulte, M.A. Steiner, A.G. Norman, H.L. Guthrey, et al., Six-junction III−V solar cells with 47.1% conversion efficiency under 143 Suns concentration, Nat. Energy 5 (4) (2020) 326−335.

[69] S. Bagheri, R. Talebzadeh, B. Sardari, F. Mehdizadeh, Design and simulation of a high efficiency InGaP/GaAs multi junction solar cell with AlGaAs tunnel junction, Optik 199 (2019) 163315.

[70] Y. Xu, J. Li, Q. Tan, A.L. Peters, C. Yang, Global status of recycling waste solar panels: a review, Waste Manag. 75 (2018) 450−458.

[71] P.M. Rodrigo, A. Valera, E.F. Fernández, F.M. Almonacid, Performance and economic limits of passively cooled hybrid thermoelectric generator-concentrator photovoltaic modules, Appl. Energy 238 (2019) 1150−1162.

[72] J. Gong, S.B. Darling, F. You, Perovskite photovoltaics: life-cycle assessment of energy and environmental impacts, Energy Environ. Sci. 8 (7) (2015) 1953−1968.

[73] N. Espinosa, L. Serrano-Luján, A. Urbina, F.C. Krebs, Solution and vapour deposited lead perovskite solar cells: ecotoxicity from a life cycle assessment perspective, Sol. Energy Mater. Sol. Cell 137 (2015) 303−310.

[74] A. Babayigit, A. Ethirajan, M. Muller, B. Conings, Toxicity of organometal halide perovskite solar cells, Nat. Mater. 15 (3) (2016) 247−251.

[75] K.P. Goetz, A.D. Taylor, Y.J. Hofstetter, Y. Vaynzof, Sustainability in perovskite solar cells, ACS Appl. Mater. Interfaces 13 (1) (2021) 1−17.

[76] J. Zhang, X. Gao, Y. Deng, Y. Zha, C. Yuan, Comparison of life cycle environmental impacts of different perovskite solar cell systems, Sol. Energy Mater. Sol. Cell 166 (2017) 9−17.

[77] M.M. Lunardi, A. Wing, Y. Ho-Baillie, J.P. Alvarez-Gaitan, S. Moore, R. Corkish, A life cycle assessment of perovskite/silicon tandem solar cells, Prog. Photovolt. Res. Appl. 25 (8) (2017) 679−695.

[78] S.P. Dunfield, L. Bliss, F. Zhang, J.M. Luther, K. Zhu, M.F.A.M. van Hest, et al., From defects to degradation: a mechanistic understanding of degradation in perovskite solar cell devices and modules, Adv. Energy Mater. 10 (26) (2020) 1−35.

[79] J. Bisquert, E.J. Juarez-Perez, The causes of degradation of perovskite solar cells, J. Phys. Chem. Lett. 10 (19) (2019) 5889−5891.

[80] C.C. Boyd, R. Cheacharoen, T. Leijtens, M.D. McGehee, Understanding degradation mechanisms and improving stability of perovskite photovoltaics, Chem. Rev. 119 (5) (2019) 3418−3451.

[81] J. Li, H.-L. Cao, W.-B. Jiao, Q. Wang, M. Wei, I. Cantone, et al., Biological impact of lead from halide perovskites reveals the risk of introducing a safe threshold, Nat. Commun. 11 (1) (2020) 310.

[82] J.M. Kadro, A. Hagfeldt, The end-of-life of perovskite PV, Joule 1 (1) (2017) 29−46.

[83] Z. Li, Y. Zhao, X. Wang, Y. Sun, Z. Zhao, Y. Li, et al., Cost analysis of perovskite tandem photovoltaics, Joule 2 (8) (2018) 1559–1572.
[84] N.L. Chang, A.W.Y. Ho-Baillie, D. Vak, M. Gao, M.A. Green, R.J. Egan, Manufacturing cost and market potential analysis of demonstrated roll-to-roll perovskite photovoltaic cell processes, Sol. Energy Mater. Sol. Cell 174 (2018) 314–324.
[85] A. Anctil, C.W. Babbitt, R.P. Raffaelle, B.J. Landi, Cumulative energy demand for small molecule and polymer photovoltaics, Prog. Photovolt. Res. Appl. 21 (7) (2013) 1541–1554.
[86] M.P. Tsang, G.W. Sonnemann, D.M. Bassani, Life-cycle assessment of cradle-to-grave opportunities and environmental impacts of organic photovoltaic solar panels compared to conventional technologies, Sol. Energy Mater. Sol. Cell 156 (2016) 37–48.
[87] H. Cao, W. He, Y. Mao, X. Lin, K. Ishikawa, J.H. Dickerson, et al., Recent progress in degradation and stabilization of organic solar cells, J. Power Sources 264 (2014) 168–183.
[88] İ.V. Öner, E.Ç. Yilmaz, M.K. Yesilyurt, G. Ömeroglu, A.N. Özakin, Operational stability and degradation of organic solar cells, Period. Eng. Nat. Sci. 5 (2) (2017) 152–160.
[89] N. Espinosa, R. García-Valverde, F.C. Krebs, Life-cycle analysis of product integrated polymer solar cells, Energy Environ. Sci. 4 (5) (2011) 1547.
[90] N. Espinosa, F.O. Lenzmann, S. Ryley, D. Angmo, M. Hösel, R.R. Søndergaard, et al., OPV for mobile applications: an evaluation of roll-to-roll processed indium and silver free polymer solar cells through analysis of life cycle, cost and layer quality using inline optical and functional inspection tools, J. Mater. Chem. A 1 (24) (2013) 7037.
[91] D. Hengevoss, C. Baumgartner, G. Nisato, C. Hugi, Life cycle assessment and eco-efficiency of prospective, flexible, tandem organic photovoltaic module, Sol. Energy 137 (2016) 317–327.
[92] A. Anctil, V. Fthenakis, Critical metals in strategic photovoltaic technologies: abundance versus recyclability, Prog. Photovolt. Res. Appl. 21 (6) (2013) 1253–1259.
[93] K.A. Mazzio, C.K. Luscombe, The future of organic photovoltaics, Chem. Soc. Rev. 44 (1) (2015) 78–90.
[94] M.L. Parisi, S. Maranghi, L. Vesce, A. Sinicropi, A. Di Carlo, R. Basosi, Prospective life cycle assessment of third-generation photovoltaics at the pre-industrial scale: a long-term scenario approach, Renew. Sustain. Energy Rev. 121 (2020) 109703.
[95] J.H. Wong, M. Royapoor, C.W. Chan, Review of life cycle analyses and embodied energy requirements of single-crystalline and multi-crystalline silicon photovoltaic systems, Renew. Sustain. Energy Rev. 58 (2016) 608–618.
[96] D.H. Kim, J.B. Whitaker, Z. Li, M.F.A.M. van Hest, K. Zhu, Outlook and challenges of perovskite solar cells toward terawatt-scale photovoltaic module technology, Joule 2 (8) (2018) 1437–1451.
[97] S. Maranghi, M.L. Parisi, R. Basosi, A. Sinicropi, Environmental profile of the manufacturing process of perovskite photovoltaics: harmonization of life cycle assessment studies, Energies 12 (19) (2019) 3746.
[98] C.F. Blanco, S. Cucurachi, J.B. Guinée, M.G. Vijver, W.J.G.M. Peijnenburg, R. Trattnig, et al., Assessing the sustainability of emerging technologies: a probabilistic LCA method applied to advanced photovoltaics, J. Clean. Prod. 259 (2020) 120968.
[99] A. Anctil, E. Lee, J. Stephen, A. Munasinghe, C.J. Traverse, R.R. Lunt, Life cycle assessment of transparent organic photovoltaic for window and portable electronic applications, in Proc. 44th, IEEE Photovolt. Specialists Conf. (PVSC) (2017) 2124–2127. Available from: https://doi.org/10.1109/PVSC.2017.8366142.

SECTION V

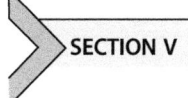

Concentrator and multijunction devices

CHAPTER FOURTEEN

Concentrator and multijunction solar cells

Katie Shanks
Solar Energy Research Group, Environment and Sustainability Institute, University of Exeter, Penryn Campus, Cornwall, United Kingdom

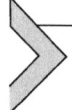

14.1 Introduction: Classifications of relevant technologies

Concentrator photovoltaic (CPV) technologies are systems made up of optical devices that focus light toward areas of photovoltaic (PV) material. Example systems are given in Fig. 14.1. Within a CPV system, there are concentrator optics, solar cell(s), and, if necessary, a passive or active cooling mount to manage the temperature of the solar cell. Within this chapter, we will be focusing on the solar cells used with concentrator optics, their differences from solar cells used in standard flat plate solar panels, and guidelines for matching concentrator optics to types of solar cells. When a solar cell has been designed specifically for concentrated light conditions, they are referred to as CPV or concentrator solar cells (CPV cells). Note that the term CPV in the literature can refer to the solar cell itself or the full system of concentrator optics, solar cells, and any supporting components such as cooling. Within this chapter, to aid clarification, we will include the appropriate component name when suitable. For example, we will use concentrator optics (CPV optics), CPV cells, or CPV system when discussing the optics, solar cells or the full system respectively.

14.1.1 Geometric concentration ratio

CPV cells and concentrator optics can be classified according to their geometrical concentration ratio. The geometric concentration ratio is the ratio of the input aperture area to the output aperture area or receiver area (see Fig. 14.1B and C). As we are discussing PV-based systems and not solar thermal-based systems (called concentrated power systems), our receiver area will be the solar cell area. Whether the output aperture area irradiance, are tested under lower irradiance conditions. Other 1 sun optimized cells may actually be more accurate for use as a nonconcentrating system equivalent. Typically, the irradiance source is set to 1000 W/m² (a standard testing condition) but the same comparison is also done with outdoor experiments where a concentrating module is compared to a nonconcentrating equivalent, both at the same orientations and hence experiencing the same incident starting irradiance, which varies depending on location, weather, time, and season [Eqs. (14.4)–(14.6)].

$$C_{\text{effect}} = \frac{\text{Power Output}_{\text{CPV}}}{\text{Power Output}_{\text{PV}}} \quad (14.3)$$

$$C_{\text{effect}} = C_{\text{opt}} \times \left(\frac{\eta_{\text{cell under concentration}}}{\eta_{\text{cell under 1 sun}}}\right) \quad (14.4)$$

$$C_{\text{effect}} = C_{\text{geo}} \times \eta_{\text{opt}} \times \eta_{\text{CellRatio}} \quad (14.5)$$

$$C_{\text{effect}} = \frac{\text{reciever area}^*}{\text{input aperture area}} \times \eta_{\text{opt}} \times \eta_{\text{CellRatio}} \quad (14.6)$$

where η_{opt} is the total optical efficiency of all the optical components within the system. For example, for the system shown in Fig. 14.1D, the optical efficiency of the system would include the reflectance of the large primary parabolic reflector, multiplied by the reflectance of the second smaller parabolic reflector and then multiplied by the transmittance of the receiver funnel optic.

Accurately measuring a solar cell's efficiency under concentrated light can be difficult, as it requires accurate calculation or measurement of the incident concentrated irradiance upon the cell without simply using the power output of the cell as an indicator (which of course will be affected by the cell efficiency). Methods of calculating the incidence irradiance upon the cell and hence the total incident power on the cell's surface area are often done using optical modeling and component optical tests; see, for example, the work of Shanks et al. [3]. The optical efficiency of the concentrator optics can be measured via transmittance, absorption, and reflectance measurements, so that all light loss can be accounted for.

14.1.3 Theoretical efficiency limits of concentrator and multijunction solar cells

There are many types of PV solar cells with varying record efficiencies. Single $p-n$ junction solar cells can only absorb one wavelength band of the incident irradiance. This limits their efficiency to approximately 33%, first calculated and named the Shockley-Queisser limit in 1961 [4]. Multijunction solar cells can utilize more wavelengths of the sun's radiation but are also more expensive [5]. At the time of writing, the record efficiency for a multijunction solar cell without concentrated light (1 sun) is 39.5% [6]. The number of junctions of a solar cell increases the theoretical maximum conversion efficiency possible of that cell as it is able to accept more of the solar spectrum [7]. According to Razykov et al. [8], the thermodynamic limit of a tandem (two junctions) solar cell is 55.6%, of three junctions 63.6%, and of four junctions 68.5% under unconcentrated sunlight conditions. Doeleman [9] took this further and calculated more realistic efficiency limits of approximately 46% for double junctions, 50% for triple, and 53.5% for 4-junction solar cells. Doeleman also suggests that an 8-junction solar cell can achieve around 60% efficiency and compares these efficiencies to those found in the literature where possible [9].

Concentrating light onto single and multijunction solar cells can also increase their efficiencies. On Earth, the maximum concentration level possible is 46,200 suns due to the law of etendue concerning the sun's diameter and its divergence of light (± 0.27 degrees) [10,11]. Assuming this maximum concentration ratio, the optimum theoretical efficiency of a single junction solar cell is extended to 45.1% [9]. For multijunction solar cells, there are a number of suggested theoretical maximum efficiencies such as the Carnot limit (95%) and the Landsberg limit (93.3%) [9]. The realistically achievable efficiencies will, however, be lower, especially when taking into account the monetary and carbon cost-effectiveness of such solar cells, but this shows the potential for solar concentrator technology. A timeline of CPV designs with predicted trends is illustrated in Fig. 14.2 [12].

14.2 Concentrator compared to one sun solar cells

Solar cells used within concentrator optical systems are often smaller than standard solar cell sizes for flat plate modules. Single junction cells can be used with low-concentration optics; this provides for a greater range of (performance and cost) options for a variety of applications. Finally, we address the utilization of emerging PV technologies; this area

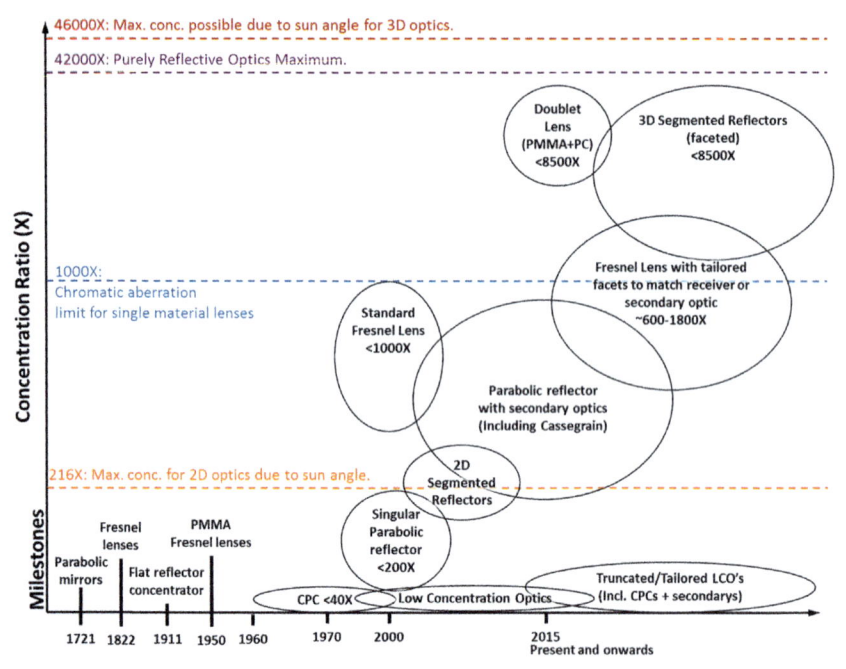

Figure 14.2 Timeline of concentrator photovoltaic designs and predicted future trends toward high and ultrahigh concentration ratios. *Reprinted with permission from K. Shanks, S. Senthilarasu, T.K. Mallick, Optics for concentrating photovoltaics: trends, limits and opportunities for materials and design, Renew. Sustain. Energy Rev. 60 (July 2016) 394–407, this article is distributed under the terms of the Creative Commons (CC-BY 4.0) license (http://creativecommons.org/licenses/by/4.0/), which permits unrestricted use.*

requires further study, especially for rapidly developing technologies such as perovskites and bifacial cells.

14.2.1 Concentrated light properties and cell design

High and ultrahigh concentrator designs in particular typically have solar PV cells of 1 cm^2 or less [13], so that the size of the primary optics can be more compact and easier to handle (e.g., lighter weight). Under concentrated sunlight, a PV cell will be exposed to an increased energy density of the solar spectrum. The exact intensity and shape of the spectrum incident on the solar cell(s) will depend on the concentrator optics and their efficiency at focusing each wavelength of sunlight. There has been various research into wavelength-selective optical designs and filtering materials within the field of CPV systems to match concentrated spectra to receiving solar cells [14–16] or for management of lighting and heating within building integrated systems [17–20].

This is one method of improving the performance of the CPV solar cells and the CPV system overall. There are previous literature reviews that also describe the solar cell structure design and electrode material choice and position (e.g., front contact vs. back contacts) [13], where back contact cell designs proved favorable due to reduced shading effects. Full solar cell structure redesign, however, is not necessary for concentrator systems, standard silicon solar cells can instead simply be metalized with a different pattern and cut to the required shape. Regardless of whether an optimized incident solar spectrum upon the PV material or not, there will be increased resistive losses due to the quantity of electrical current being generated in the solar cell [13]. This current needs to be carried away as quickly as possible and hence CPV cells are typically designed with a finer and less spaced metallization pattern, see Fig. 14.3.

The increased distribution of the solar cell fingers means electrical energy needs to travel less across the PV material before reaching the higher

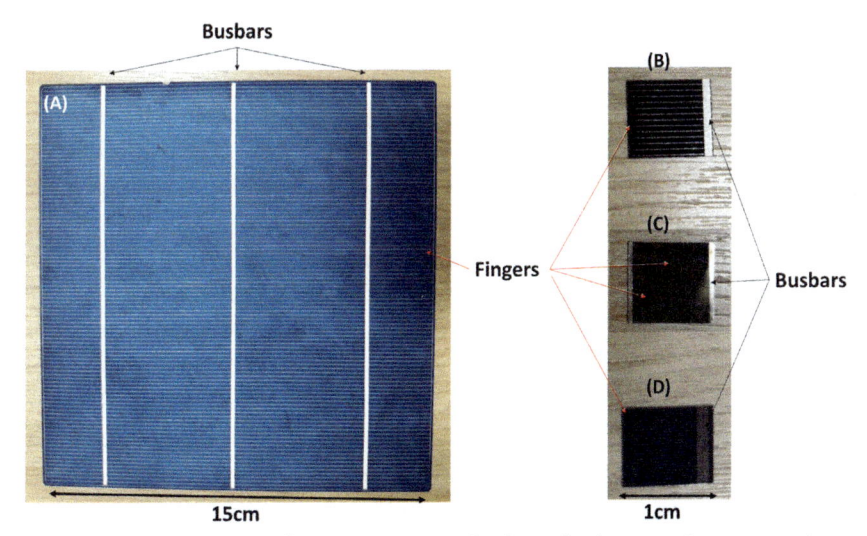

Figure 14.3 Example metallization patterns of solar cells showing the varying distribution of solar cell fingers, which carry collected current to busbars. (A) 15 cm by 15 cm Si solar cell with approximately 1.5 mm spaced fingers. (B) 1 cm by 1 cm silicon solar cell with approximately 0.9 mm spaced fingers. (C) 1 cm by 1 cm multijunction GaAs solar concentrator cell with vertical and horizontal fingers, horizontal fingers are <0.3 mm spaced. (D) 1 cm by 1 cm Si solar concentrator cell with <0.5 mm spaced fingers.

conducting channels [21]. There is, however, a compromise to be made here, as any metallization on the active surface of the solar cell will take up some of the exposed area and block the incident light from hitting the PV material. Hence there is an optimum amount of metallized area on the active surface of the solar cell depending on the energy density of the incident concentrated sunlight. The irradiance distribution on the solar cell will also have an impact on the exact optimum finger distribution [21], though this may be negligible at low concentration levels and simply not worthwhile to accommodate within scaled manufacturing operations. Higher solar concentration levels, however, will require a higher degree of metallization, such as that shown in Fig. 14.3C and D, compared to Fig. 14.3A and B. There is potential for further customized, irregularly spaced fingers to increase a multijunction concentrator cell's efficiency even further if optimized for a specific irradiance distribution and temperature profile.

Despite the above challenges of efficiently converting concentrated sunlight, it is viable to use solar cells designed for unconcentrated "1 sun" conditions within CPV systems (Fig. 14.3A and B). Typically, this is only done for low concentrator systems (<10 suns), where the gain in efficiency by customizing the solar cell may not outweigh the added cost in manufacturing/sourcing customized solar cells. At higher concentration levels, it will be required to have increased levels of passive or active cooling to keep the 1 sun-designed solar cell within its working temperature range compared to a specially designed concentrator solar cell. High-temperature materials are often chosen for CPV cells designed for high concentrations to ensure higher maximum working temperature limits [22].

Similarly, working with concentrator cells but exposing them to concentrated sunlight intensities significantly above their recommended working temperature range has also been done successfully by ensuring there is adequate cooling to maintain the solar cell within the working temperature range [23–26]. Again, this is done to optimize the cost/efficiency balance of the system. Hence, a variety of solar cells can be used with concentrator optics although best performance and lifetime results will be obtained by appropriately matching their properties and working conditions.

14.2.2 Single junction silicon cells and low concentrator optics

Examples of systems utilizing low concentrator optics ($<10 \times$ geometric concentration) and solar cells can be seen in building integrated PV systems and often as the final stage receiver optic in higher concentration systems (>10× geometric concentration) [12,27]. Geometric designs include reflective and refractive types, including but not limited to dome lens, V-trough, hyperbolic, and many variations of CPC optical funnels; see Fig. 14.4. As mentioned previously, these designs often attempt to match the solar cell shape and size, hence the elliptical entry aperture but square exit aperture; see Fig. 14.4F, for example.

Shanks et al. investigated 3.6 suns crossed compound parabolic concentrator (CCPC) made of plastic and glass with square-shaped concentrator single junction silicon cells of 1 cm^2 [3,17]. The resulting prototype achieved an effective concentration of 2.89× and an improved open circuit voltage compared to the cell under 1 sun conditions. Zhang et al. investigated an array of the same unit modules to be used as vertical windows [28]. The design allowed enhanced diffuse ambient lighting into the room, as well as thermal insulation properties, all whilst requiring less PV material [28]. Such concentrator systems that fulfill multiple functions for a building (e.g., lighting and thermal management) are of significant

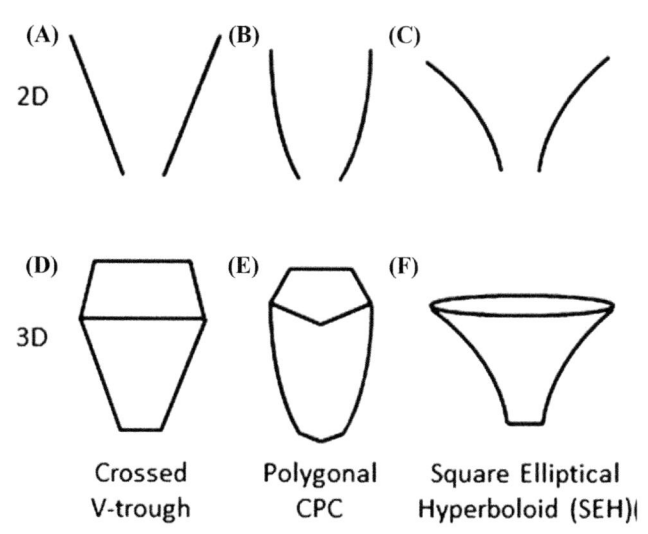

Figure 14.4 Example geometries of low concentration 2D (A–C) and 3D (D–F) optics for single junction solar cells: (A and D) crossed V-trough concentrator, (B and E) polygonal compound parabolic concentrator and (C and F) square elliptical hyperboloid concentrator. *Reproduced with permission from K. Shanks, S. Senthilarasu, T.K. Mallick, High-concentration optics for photovoltaic applications, in: P. Pérez-Higueras, E. F. Fernández, (Eds.), High Concentrator Photovoltaics: Fundamentals, Engineering and Power Plants, 1st edition. Springer International, 2015, pp. 85–113, copyright (2015) Springer Nature.*

interest for net zero building applications; see Fig. 14.5. In particular, there is demand to use this technology for high-rise office buildings where there is a vast area of vertical window space, which results in a high solar gain for the buildings in sunny locations or seasons. This then requires high AC power demands and hence is one of the target areas for technology development to transition to net zero.

14.2.3 Concentrator optics coupled with second- and third-generation photovoltaics

Sacco et. al have previously found that no detrimental damage was seen for dye-sensitized solar cells (DSSC) under solar concentration up to 6X, showing potential for low carbon footprint and esthetically pleasing building-integrated PV (BIPV) [29]. However, there has not been further interest in concentrator DSSC systems due to low efficiencies compared with silicon and MJ concentrator setups. Law et al. were one of the first few researchers to test perovskite solar cells (PSCs) under concentrated light intensities and also investigated concentrator DSSCs [30]. Promising results under low concentration conditions (<10 suns) were obtained, but above 50 suns, there was significant degradation of the cells. Sadhukhan et al. recently performed a useful review of perovskite potential with concentrator optics [30]. They conclude that PSCs could perform well in low-concentration systems with good stability and an impressive efficiency of 23.6% at 14 suns obtained in the reviewed literature [30]. Formamidinium-cesium-based hybrid perovskites appear to perform better than $MAPbI_3$ under solar

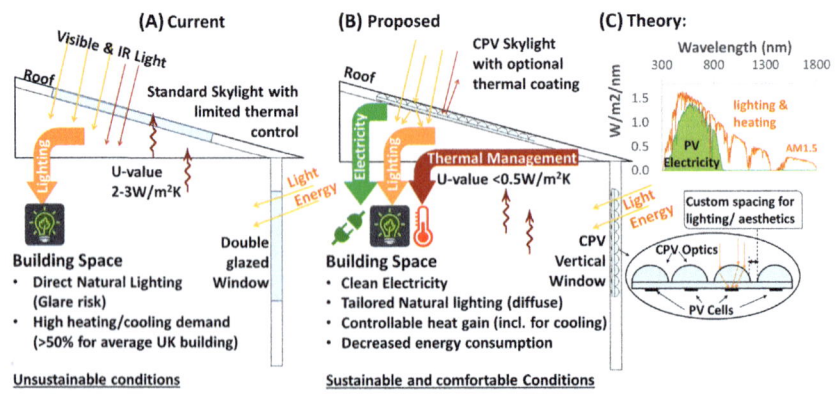

Figure 14.5 Example of multifunctional low-concentration PV systems for building integration showing customized management of natural lighting, thermal management (from solar gain and insulation properties), and power generation.

concentrator conditions, although far more research is required in this emerging field [30]. There are also literature reviews on nonconcentrator perovskites for building integrated applications [31].

Thin-film CdTe is a well-established technology [6], making up 5% of the global PV market, with First Solar, Inc. shipping the majority of this as 5.5 GW of CdTe modules in 2020 and an estimated 8 GW in 2021 (reports still to be confirmed at time of writing). CdTe-based concentrator PV (CdTe CPV) is a sparsely investigated concept with only a few references to be found in the literature [32,33]. Li et al. [32] investigated CdTe cells under increased irradiance intensities (1−26 suns concentration) and found that the short circuit current and open circuit voltage both increased with increased incident light intensity. While the fill factor decreased due to a significant increase in the electrical resistance of the front electrode [32].

There have been comparatively more studies into nonconcentrator BIPV CdTe than concentrator BIPV, although both can give interesting insights into the full building energy performance and balance of electrical, thermal, and lighting requirements. Alrashidi et al. investigated semitransparent CdTe for the electrical and thermal impact on an enclosure, showing that semitransparent CdTe can be used to reduce AC power demand for office buildings [34]. Sun and coworkers have carried out various BIPV integration considerations for a CdTe-based low concentrator system façade [18−20,28,35].

14.3 Multijunction solar cells for solar concentrator systems

Multijunction solar cells are PV cells that consist of multiple $p-n$ junctions made of different semiconductor materials. Each junction is tuned to efficiently convert a different wavelength range of light into electricity and hence reach far higher conversion efficiencies. There are a range of review articles within the literature detailing the progress, limitations, and material layering of multijunction solar cells [36−39]. Recent advances in fabrication techniques have led to new records in efficiency for cells with two, three, four, and six junctions.

For dual-junction (or tandem) solar cells, the record efficiency in 2020 was 32.9% held by Steiner et al. (National Renewable Energy Laboratory,

NREL) for a GaInAs/GaAsP quantum well solar cell under 1 sun illumination [6,40,41] and 35.5% under 38X concentrated sunlight [42]. This is an example where little adjustment was made to the cell for the concentrated scenario and still a higher efficiency was obtained. In June 2023, King Abdullah University of Science and Technology (KAUST) achieved a higher efficiency of 33.7% under the 1 sun conditions for a perovskite/silicon tandem cell [6,41]. This was announced after a 32.7% record efficiency only months before, highlighting the competitive field of perovskite research during this time [43].

For triple-junction solar cells, the record is 39.46% efficiency under 1 sun (AM1.5 global), set in 2021 by France et al. (NREL) for an optimized layering of GaInP top cell, GaInAs/GaAsP quantum well middle cell, and a lattice-matched GaInAs bottom cell [44]. The device achieves 34.2% efficiency under the AM0 space spectra (for space applications), which currently is the predominant use of triple junction solar cells [44]. All record cell efficiencies reported in NREL's best cell efficiency chart (Fig. 14.6 [6]) are tested at the standard AM1.5 spectra (1000 W/m^2) conditions under 1 sun or multiple suns concentration. It should be noted that different locations and environments experience a variety of incident spectra due to local weather (clouds, humidity, etc.), air pollution, and incidence of angle. Under concentrated sunlight, the record efficiency for triple junction solar cells is 44.4%, since 2013 by Sharp electronics Co.

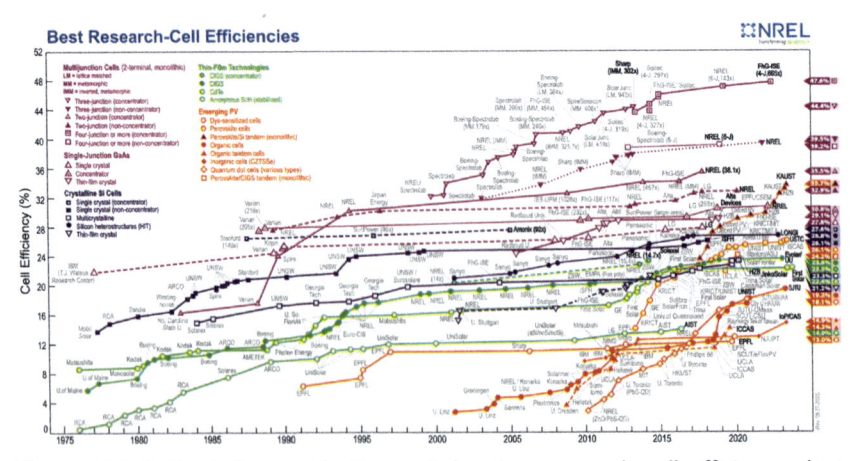

Figure 14.6 National Renewable Energy Laboratory, research cell efficiency chart, please see multijunction solar cells efficiency records under nonconcentrated and concentrated conditions. *From https://www.nrel.gov/pv/cell-efficiency.html, public domain as a US Government agency (Department of Energy) work product.*

[45], but there have not been any improvements to this in recent years despite the potential indicated by modeling papers [45].

The best efficiency single-junction solar cell under 1 sun is still held at 39.2%, since 2019 by NREL [6], which interestingly has now been surpassed by the 3-junction solar cell detailed above. This suggests there is further optimization required for a four-layer combination. The maximum efficiency record is 47.6% for a 4-junction solar cell at 665 suns concentration held by Fraunhofer [6]. This efficiency is even higher than the six-junction solar cell record of 47.1% efficiency under 146 suns, reported in 2020 by Geisz et al. (NREL) [46], showing the importance of careful subcell optimization and matching within multijunction layering. Lower concentration ratios are also easier to obtain, typically with lower costs and more consistent performance [12,47,48]. It is clear from these record cells that the GaInAs/GaAsP quantum well cell plays an important contribution as well as other combinations of these elements for multijunction cells. With the development of the perovskite/silicon tandem and its optimized performance, there are further combinations for 3- and 4-junction cells to be explored [37–39,45]. Despite these impressive efficiency values of almost 50%, multijunction solar cells are currently not as cost-effective as lower-efficiency tandem or 1-junction solar cells, and their main use is in applications where the area available for installation is limited or where the power-to-weight ratio needs to be a maximum such as in space applications.

Tandem solar cells are typically best matched to low and medium solar concentrator optics. Multijunction solar cells with three or more junctions can be used for high and ultrahigh solar concentrators, though care must be taken to ensure the temperature of the solar cell is maintained within the working temperature range. The maximum temperature experienced by a solar cell will depend on the location and local climate, as well as the specific CPV system design and the active or passive cooling components.

The ideal high-concentration PV system would consist of a high-efficiency multijunction solar cell that could handle the high flux intensity and thermal energy [49]. This may require a passive or active cooling mount in thermal contact with the solar cell. Reducing the size of the multijunction solar cell can greatly help reduce the temperature experienced by the cell and also allow overall a more compact solar concentrator system, as the geometrical concentration ratio is simply a multiple of the solar cell area. Hence a reduction in solar cell area results in a reduction in the required primary optic area for a specific concentration ratio [49].

One of the challenges of CPV systems is their low acceptance angle and requirement for sun tracking [50,51]. Low-concentration systems can be managed without solar tracking as they can have acceptance angles of approximately 30 degrees [52]. High and ultrahigh CPV systems require very accurate tracking systems, which can be costly [49]. Available high concentrator modules are often designed for approximately $500\times$ and the most common concentrator optics design involves a Fresnel lens (used within lighthouses), though reflective Cassegrain optics are another option (still used within telescopes) [53–55].

14.4 Matching solar cells and concentrator optics for optimum performance

Solar concentrator systems are capable of generating cost-effective electricity, especially when utilized in conjunction with high photovoltaic conversion efficiency (PCE) solar cells. High-efficiency MJ solar cells, most often comprised of III–V materials within flat plate modules or as solar concentrator cells within CPV systems, are so far the most effective method to achieve a high-power output per meter squared of space. Finally, we discuss the importance of a low carbon footprint. Until recently the carbon footprint, or carbon pay-back period, of solar PV devices has not been a dominant factor when comparing solar cells, panels, or CPV modules. Finally, Fig. 14.7 provides a visual connection between historical efficiency improvement (Fig. 14.6 [6]) and the evolution of PV materials [56].

14.4.1 Low cost

Due to the ever-decreasing cost of silicon and thin-film solar cells, CPV systems are most cost-competitive when utilizing multijunction solar cells under medium, high, and ultrahigh concentrated light. The cost of the optics, however, depends on the geometry, weight, material choice, and ease of manufacturing, so the viability also depends on these [57]. Table 14.1 from a 2018 study by Shanks et al. gives example costs for types of optics [57].

The effective concentration ratio can be used to estimate the required cost of the optics if the system is to compete with the nonconcentrator equivalent system [see Eq. (14.7) and (14.8) and for target criteria]. Here we assume that the cost of assembly of a PV panel is on a similar scale to the cost of assembly of a CPV system and that the supporting materials

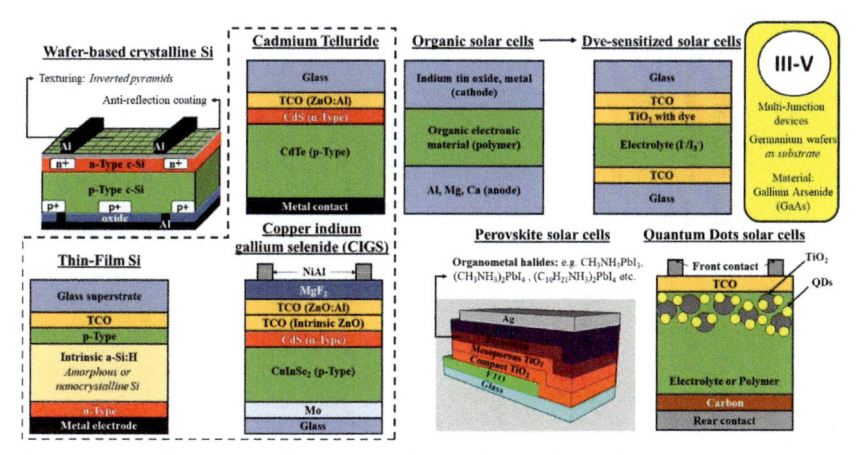

Figure 14.7 Progressive advances in photovoltaic solar cell materials. *Reproduced with permission from S.K. Verma, R. Kumar, M. Barthwal, D. Rakshit, A review on futuristic aspects of hybrid photovoltaic thermal systems (PV/T) in solar energy utilization: engineering and technological approaches, Sustain. Energy Technol. Assess. 53 (Part A) (October 2022) 102463, copyright (2022) Elsevier.*

(e.g., enclosure materials and tabbing wire) are also on a similar scale, especially if mass production processes lines are developed for each. This leaves the main cost difference between flat plate PV panels and CPV systems to be in the solar cell and concentrator optics. Hence, for:

$$\text{Cost}_{\text{Flat Plate PV panel}} = \text{Cost}_{\text{CPV system}} \tag{14.7}$$

$$\text{Cost}_{\text{PV Cells}} = \text{Cost}_{\text{CPV Cells}} + \text{Cost}_{\text{Concentrator Optics}} \tag{14.8}$$

If using the same PV cells in a CPV system and maintaining the final power output, then optics need to cost the same or less than the cost difference in PV cells required [Eqs. (14.9)–(14.12)]:

$$\text{Cost}_{\text{PV Cells}} = \frac{\text{Cost}_{\text{PV Cells}}}{C_{\text{effect}}} + \text{Cost}_{\text{Optics}} \tag{14.9}$$

$$\text{Cost}_{\text{Optics}} = \text{Cost}_{\text{PV Cells}} - \frac{\text{Cost}_{\text{PV Cells}}}{C_{\text{effect}}} \tag{14.10}$$

$$\text{Cost}_{\text{Optics}} = \text{Cost}_{\text{PV Cells}} \left(1 - \frac{1}{C_{\text{effect}}}\right) \tag{14.11}$$

$$\text{Cost}_{\text{Optics}} = \text{Cost}_{\text{PV Cells}} \left(1 - \frac{1}{C_{\text{geo}} \times \eta_{\text{opt}} \times \eta_{\text{CellRatio}}}\right) \tag{14.12}$$

Table 14.1 Example cost of different types of topics.

Optical Component	Processes	Materials	Initial cost (or <100 parts)	Cost thereafter (or >1000 parts)
SOG Fresnel lens (4 required for 1 system)	Injection molding	Plane glass silicon	<$140 (<$560 total)	<$90 (<£400 total)
Flat mirrors (4 large and 1 small for 1 system)	Polishing and/or coating	Sheet metal or plane glass with, e.g., silver/aluminum coating or mirror film	<$70 for small <$130 for large (<$570 total)	<$30 for small <$60 for large (<$250 total)
Tertiary 4-dome Central optic (low refractive index material)	Drilling and polishing	Aluminum casting material	<$90	~$8 per optic
Injection molded glass tertiary optics	Injection molding	Metal (usually) raw glass material	>$3500	$2–$10 per optic
Large curved mirrors (cassegrain system)	Metal spinning or CNC or coated glass molds	Metal, ABS plastic coated with aluminum/silver or raw glass and coating; see [53]	$400–$700 (for 21 × 21cm plastic mirrors)	~$140 each (vacuum metalized plastic mirrors)

Note that increased size and complexity results in higher costs but that these may scale favorably depending on manufacturing method and amount of bulk material required.

Source: Reproduced from K. Shanks, J.P. Ferrer-Rodriguez, E.F. Fernández, F. Almonacid, P. Pérez-Higueras, S. Senthilarasu, T. Mallick, A >3000 suns high concentrator photovoltaic design based on multiple Fresnel lens primaries focusing to one central solar cell, Sol. Energy (169) (2018) 457–467, this article is distributed under the terms of the Creative Commons (CC-BY 4.0) license (http://creativecommons.org/licenses/by/4.0/), which permits unrestricted use.

Eq. (14.12), compared to Eq. (14.11), reminds us that the optical quality, and hence optical efficiency of the optics, is important to ensure the effective concentration ratio is close to the geometric concentration ratio. Similarly, solar cells should work within the concentrated irradiance range otherwise the system could fail completely. Eq. (14.12) means that for low CPV systems, which typically utilize single-junction solar cells (taking current record single junction cell efficiency as 27.6%) and assuming 10% optical losses [Eqs. (14.13) and (14.14)]:

$$\text{Cost}_{\text{Optics}} = \text{Cost}_{\text{PV Cells}} \left(1 - \frac{1}{10 \times 0.9 \times 0.276}\right) \tag{14.13}$$

$$\text{Cost}_{\text{LCPV Optics}} \leq 0.9 \times \text{Cost}_{\text{PV Cells}} \tag{14.14}$$

For high CPV systems (HCPV), which achieve high effective concentration ratios (note that just because a system has a high geometric concentration ratio, the optical efficiency needs to be good enough so that the effective concentration ratio is also still high for these estimations), then the equation may become [Eqs. (14.15) and (14.16)]:

$$\text{Cost}_{\text{HCPV Optics}} = \text{Cost}_{\text{PV Cells}} \left(1 - \frac{1}{500 \times 0.85 \times 0.42}\right) \tag{14.15}$$

$$\text{Cost}_{\text{HCPV Optics}} \leq 178.5 \times \text{Cost}_{\text{PV Cells}} \tag{14.16}$$

In Eq. (14.15), we are taking the geometrical concentration ratio to be 500 (typical system), the optical efficiency to be 85% (high-quality optimized system), and a multijunction solar cell efficiency of 42% (realistic multijunction concentrator solar cell comparable to Azurespace cells). From these two calculation examples, it can be seen that higher concentration ratios with multijunction solar cells allow for a higher cost of the optics and are more likely to have higher cost savings over the flat plate solar PV counterpart. The optics, however, are often still too expensive, and achieving cost-effectiveness for scaled manufacturing and assembly of systems is a challenge due to the increased complexity with increasing concentration ratios of systems. Another cost factor is the need for a solar tracker, as medium, high, and ultrahigh concentration systems require constant alignment with the sun. For competitive cost-effectiveness with flat plate solar panels, the upper limit of the cost is defined by Eq. (14.17):

$$\text{Cost}_{\text{HCPV Optics}} + \text{Cost}_{\text{solar Tracker}} \leq 178.5 \times \text{Cost}_{\text{PV Cells}} \tag{14.17}$$

Perovskite PV coupled with low concentrator optics could be a competitive solution because of its low-cost manufacturing ability compared to the equivalent silicon concentrator system.

14.4.2 Space-limited high-efficiency photovoltaic technologies

When installation area is limited and cost is a secondary priority, then high-efficiency MJ PV technologies dominate. Space applications, such as for satellites, where there are other crucial practical factors limiting available installation area and weight, are hence very suitable for MJ and CPV technologies [58,59]. However, the recent development of lightweight copper indium gallium selenide (CIGS and cost-effective perovskite solar panels has resulted in these single junction technologies infiltrating this application area [58]. Shanks et al. investigated the power output and cost-effectiveness of an ultrahigh concentrator design with Azurespace MJ cells per meter squared area, showing that the cost and space per unit module is an important factor in optimization [60]. Although the system had a lower efficiency compared to other HCPV systems, the power output per area of PV cell utilized was much higher [57,60].

When calculating and comparing solar energy technologies for a limited space, the highest efficiency module will prevail. Care must, however, be taken to ensure that the cells and modules are closely packed. The total power output for the given array that fits within the space should be available from any providers and easily compared. Another factor is how much of the available area is utilized by the installed panels. Depending on the shape and size, there could be a best-fitting technology simply due to its smaller modularity and capability to fill more of the available space.

14.4.3 Low carbon footprint

The carbon footprint of a product is typically expressed in $kgCO_2$ eq, meaning the sum of greenhouse gas (GHG) emissions from the full life cycle of the product is translated into the equivalent amount of CO_2 emissions to simplify comparisons. It is good practice to report the initial carbon footprint or carbon "cost" of a renewable energy technology separate from its running cost or carbon savings (if positive) during its usage for an estimated number of years. The initial carbon footprint of a device will include all GHG emissions from material mining to final product delivery and sometimes includes the carbon cost of recycling or disposing of the materials at its end of life.

Polverini et al. provide a thorough review of kgCO$_2$-eq/kWh for various solar cell technologies [61]. The country of manufacture and country of use have a significant impact on the life cycle analysis of a product and its energy generation as any grid electricity that is displaced will be made up of different ratios of fossil fuels and renewables depending on the country [61–63]. Currently, evidence shows that CdTe solar cells produced by First Solar are currently the most carbon-friendly solar panels produced globally, although this varies on multiple factors, including the performance of a system (e.g., total kWh generated in lifetime) in a given application and site location [61–63]. CPV systems have a strong potential for lower carbon costs [64] due to the reduction in semiconductor material required, though the MJ solar cells will have a higher carbon footprint than single junction nonconcentrator cells. The carbon footprint of manufacturing high-quality optics also needs to be taken into account. Although the optical components will have minimum to zero toxic elements within them and depending on the material used (glass, plastic), there may be recycling options. Studies have shown that high concentrator systems are capable of as low as 50 gCO$_2$-eq/kWh and carbon payback periods of <1 year [64]. High concentrator systems are, however, not currently suitable for many applications due to their bulk and weight. Low-concentration building integrated systems are also good options for low GHG emissions [61,65,66], though there is much demand to understand can collate data on this topic more. Emerging PV technologies and CPV designs lack enough carbon life cycle information, but there is continued work on building the life cycle database for these technologies [63].

14.5 Conclusions

Various types and designs of solar concentrator optics and cells have been described in this chapter, introducing low, medium, high, and ultrahigh concentration ratios. Although concentrator multijunction solar cells are best matched for medium concentration ratios upward, single-junction solar cells can be used with low concentration ratios with little or zero alterations to cell design, such as the metallization grid layer. Solar cells can be used above their designed working range if adequately cooled but this may affect solar cell lifetime. Optimally matching solar cells with concentrator optics can result in high-efficiency designs that are very space-efficient and carbon-efficient but may not be the most cost-efficient option. Calculations were presented to help estimate these optimal designs; however, much more data is required for accurate life cycle analysis and comparison between materials and designs. CPV systems are still being developed with opportunities for customization depending on integration requirements such as for building integrated applications. CPV are capable of fulfilling multiple functions, including high power densities, which is why they are currently used in space applications. With the increasing urgency of global decarbonization targets, there is an opportunity for future CPV systems to be designed for agrivoltaic (greenhouse), vehicle-integrated (land, sea, and air), and marine-based (e.g., floating) PV applications.

References

[1] M. Alzahrani, A. Ahmed, K. Shanks, S. Sundaram, T. Mallick, Optical component analysis for ultrahigh concentrated photovoltaic system (UHCPV), Sol. Energy 227 (2021) 321−333. Available from: https://doi.org/10.1016/J.SOLENER.2021.09.019.

[2] D. Chemisana, T. Mallick, Building integrated concentrated solar systems, in: N. Enteria, A. Akbarzadeh (Eds.), Solar Energy Sciences and Engineering Applications, 1st Ed., CRC Press, 2014, pp. 545−587.

[3] K. Shanks, A. Knowles, A. Brierley, H. Baig, H. Orr, et al., An experimental analysis of the optical, thermal and power to weight performance of plastic and glass optics with AR coatings for embedded CPV windows, Sol. Energy Mater. Sol. Cell 200 (2019) 110027. Available from: https://doi.org/10.1016/j.solmat.2019.110027.

[4] W. Shockley, H.J. Queisser, Detailed balance limit of efficiency of p-n junction solar cells, J. Appl. Phys. 32 (1961) 510−519. Available from: https://doi.org/10.1063/1.1736034.

[5] E. Cartlidge, Bright outlook for solar cells, Phys. World 20 (7) (2007) 20. Available from: https://doi.org/10.1088/2058-7058/20/7/29.

[6] National Renewable Energy Laboratory (NREL), Research Cell Record Efficiency Chart. <https://www.nrel.gov/pv/cell-efficiency.html> (accessed 12.07.23).

[7] F.H. Alharbi, S. Kais, Theoretical limits of photovoltaics efficiency and possible improvements by intuitive approaches learned from photosynthesis and quantum coherence, Renew. Sustain. Energy Rev. 43 (2015) 1073−1089. Available from: https://doi.org/10.1016/J.RSER.2014.11.101.

[8] T.M. Razykov, C.S. Ferekides, D. Morel, E. Stefanakos, H.S. Ullal, H.M. Upadhyaya, Solar photovoltaic electricity: current status and future prospects, Sol. Energy 85 (8) (2011) 1580−1608. Available from: https://doi.org/10.1016/j.solener.2010.12.002.

[9] H. Doeleman, Limiting and realistic efficiencies of multi-junction solar cells, M.S. Thesis, AMOLF, Amsterdam, NL, 2012, 33 pp. <http://www.erbium.nl/wp-content/uploads/2016/08/Master-thesis-Hugo-Doeleman-2012.pdf>.

[10] R. Winston, J. Miñano, P. Benitez, Nonimaging Optics, Academic Press, 2005, p. 512. Available from: https://doi.org/10.1016/B978-0-12-759751-5.X5000-3.
[11] J. Chaves, Introduction to Nonimaging Optics, CRC Press, 2008, p. 786. Available from: https://doi.org/10.1201/b18785.
[12] K. Shanks, S. Senthilarasu, T.K. Mallick, Optics for concentrating photovoltaics: trends, limits and opportunities for materials and design, Renew. Sust. Energy Rev. 60 (2016) 394−407. Available from: https://doi.org/10.1016/J.RSER.2016.01.089.
[13] Y. Xing, P. Han, S. Wang, P. Liang, S. Lou, et al., A review of concentrator silicon solar cells, Renew. Sust. Energy Rev. 51 (2015) 1697−1708. Available from: https://doi.org/10.1016/j.rser.2015.07.035.
[14] K. Lu, B. Zhao, C. Xu, X. Li, G. Pei, A full-spectrum synergetic management strategy for passive cooling of solar cells, Sol. Energy Mater. Sol. Cell 245 (2022) 111860. Available from: https://doi.org/10.1016/J.SOLMAT.2022.111860.
[15] M.U. Sajid, Y. Bicer, Nanofluids as solar spectrum splitters: a critical review, Sol. Energy 207 (2020). Available from: https://doi.org/10.1016/j.solener.2020.07.009.
[16] H. Liang, F. Wang, L. Yang, Z. Cheng, Y. Shuai, H. Tan, Progress in full spectrum solar energy utilization by spectral beam splitting hybrid PV/T system, Renew. Sust. Energy Rev. 141 (2021). Available from: https://doi.org/10.1016/j.rser.2021.110785.
[17] K. Shanks, A. Knowles, A. Brierly, H. Baig, Y. Sun, Y. Wu, et al., Prototype optical modelling procedure and outdoor characterization of an embedded polyolefin crossed compound parabolic concentrator for integrated photovoltaic windows, AIP Conf. Proc. 2149 (1) (August 2019) 030005. <https://doi.org/10.1063/1.5124182>.
[18] Y. Sun, D. Liu, J.-F. Flor, Katie Shanks, H. Baig, et al., Analysis of the daylight performance of window integrated photovoltaics systems, Renew. Energy 145 (2020) 153−163. Available from: https://doi.org/10.1016/j.renene.2019.05.061.
[19] Y. Sun, K. Shanks, H. Baig, W. Zhang, X. Hao, et al., Integrated CdTe PV glazing into windows: energy and daylight performance for different window-to-wall ratio, Energy Procedia 158 (2019) 3014−3019. Available from: https://doi.org/10.1016/J.EGYPRO.2019.01.976.
[20] Y. Sun, K. Shanks, H. Baig, W. Zhang, X. Hao, et al., Integrated semi-transparent cadmium telluride photovoltaic glazing into windows: energy and daylight performance for different architecture designs, Appl. Energy 231 (2018) 972−984. Available from: https://doi.org/10.1016/J.APENERGY.2018.09.133.
[21] G. Li, Y. Lu, Q. Xuan, Y.G. Akhlaghi, G. Pei, J. Ji, et al., Small scale optimization in crystalline silicon solar cell on efficiency enhancement of low-concentrating photovoltaic cell, Sol. Energy 202 (2020) 316−325. Available from: https://doi.org/10.1016/J.SOLENER.2020.03.094.
[22] M.A. Steiner, E.E. Perl, J. Simon, AlGaInP/GaAs tandem solar cells for power conversion at 400°C and high concentration, 1881, 2017, p. 40007. <https://doi.org/10.1063/1.5001429>.
[23] L. Micheli, E.F. Fernandez, F. Almonacid, K.S. Reddy, T.K. Mallick, Enhancing ultra-high CPV passive cooling using least-material finned heat sinks, AIP Conf. Proc. 1679 (1) (September 2015) 130003. <https://doi.org/10.1063/1.4931563>.
[24] K. Djermane, S. Kadri, Nanofluid cooling optimization of high concentration photovoltaic panels, AIP Conf. Proc. 2149 (1) (August 2019) 020001. <https://doi.org/10.1063/1.5124171>.
[25] A. Radwan, M. Ahmed, S. Ookawara, Performance enhancement of concentrated photovoltaic systems using a microchannel heat sink with nanofluids, Energy Conv. Manag. 119 (2016) 289−303. Available from: https://doi.org/10.1016/j.enconman.2016.04.045.
[26] J. Siecker, K. Kusakana, B.P. Numbi, A review of solar photovoltaic systems cooling technologies, Renew. Sust. Energy Rev. 79 (2017) 192−203. Available from: https://doi.org/10.1063/1.5124182.

[27] K. Shanks, S. Senthilarasu, T.K. Mallick, High-concentration optics for photovoltaic applications, in: P. Pérez-Higueras, E.F. Fernández (Eds.), High Concentrator Photovoltaics: Fundamentals, Engineering and Power Plants, 1st Ed, Springer International, 2015, pp. 85–113. Available from: https://link.springer.com/chapter/10.1007/978-3-319-15039-0_4.

[28] W. Zhang, J. Li, L. Xie, X. Hao, T. Mallick, et al., Comprehensive analysis of electrical-optical performance and application potential for 3D concentrating photovoltaic window, Renew. Energy 189 (2022) 369–382. Available from: https://doi.org/10.1016/J.RENENE.2022.02.121.

[29] A. Sacco, M. Gerosa, S. Bianco, L. Mercatelli, R. Fontana, et al., Dye-sensitized solar cell for a solar concentrator system, Sol. Energy 125 (2016) 307–313. Available from: https://doi.org/10.1016/J.SOLENER.2015.11.026.

[30] P. Sadhukhan, A. Roy, P. Sengupta, S. Das, T.K. Mallick, et al., The emergence of concentrator photovoltaics for perovskite solar cells, Appl. Phys. Rev. 8 (4) (2021) 041324. Available from: https://doi.org/10.1063/5.0062671.

[31] A. Roy, A. Ghosh, S. Bhandari, S. Sundaram, T.K. Mallick, Perovskite solar cells for BIPV application: a review, Buildings 10 (7) (2020) 129. Available from: https://doi.org/10.3390/BUILDINGS10070129.

[32] W. Li, R. Yang, D. Wang, CdTe solar cell performance under high-intensity light irradiance, Sol. Energy Mater. Sol. Cell 123 (2014) 249–254. Available from: https://doi.org/10.1016/J.SOLMAT.2014.01.021.

[33] K. Shen, Q. Li, D. Wang, R. Yang, Y. Deng, M.-J. Jeng, et al., CdTe solar cell performance under low-intensity light irradiance, Sol. Energy Mater. Sol. Cell 144 (2016) 472–480. Available from: https://doi.org/10.1016/J.SOLMAT.2015.09.043.

[34] H. Alrashidi, W. Issa, N. Sellami, S. Sundaram, T. Mallick, Thermal performance evaluation and energy saving potential of semi-transparent CdTe in façade BIPV, Sol. Energy 232 (2022) 84–91. Available from: https://doi.org/10.1016/J.SOLENER.2021.12.037.

[35] R. Liang, Y. Sun, M. Aburas, R. Wilson, Y. Wu, An exploration of the combined effects of NIR and VIS spectrally selective thermochromic materials on building performance, Energy Build. 201 (2019) 149–162. Available from: https://doi.org/10.1016/J.ENBUILD.2019.05.061.

[36] M. Yamaguchi, High-efficiency GaAs-based solar cells, in: M.M. Rahman, A.M. Asiri, A. Khan, I. Inamuddin, T. Tabbakh (Eds.), Post-Transition Met, InTech Open, 2021. Available from: http://doi.org/10.5772/intechopen.94365.

[37] M. Yamaguchi, F. Dimroth, J.F. Geisz, N.J. Ekins-Daukes, Multi-junction solar cells paving the way for super high-efficiency, J. Appl. Phys. 129 (24) (2021) 240901. Available from: https://doi.org/10.1063/5.0048653/523212.

[38] I.M. Peters, C.D. Rodríguez Gallegos, L. Lüer, J.A. Hauch, C.J. Brabec, Practical limits of multijunction solar cells, Prog. Photovolt. 31 (10) (2023) 1006–1015. Available from: https://doi.org/10.1002/PIP.3705.

[39] A.S. Al-Ezzi, M.N.M. Ansari, Photovoltaic solar cells: a review, Appl. Syst. Innov. 5 (4) (2022) 67. Available from: https://doi.org/10.3390/ASI5040067.

[40] M.A. Steiner, R.M. France, J. Buencuerpo, J.F. Geisz, M.P. Nielsen, et al., High efficiency inverted GaAs and GaInP/GaAs solar cells with strain-balanced GaInAs/GaAsP quantum wells, Adv. Energy Mater. 11 (4) (2021) 2002874. Available from: https://doi.org/10.1002/AENM.202002874.

[41] M.A. Green, E.D. Dunlop, M. Yoshita, N. Kopidakis, K. Bothe, G. Siefer, et al., Solar cell efficiency tables (version 62), Prog. Photovolt. 31 (7) (2023) 651–663. Available from: https://doi.org/10.1002/pip.3726.

[42] N. Jain, K.L. Schulte, J.F. Geisz, D.J. Friedman, R.M. France, et al., High-efficiency inverted metamorphic 1.7/1.1 eV GaInAsP/GaInAs dual-junction solar cells, Appl. Phys. Lett. 112 (5) (2018) 053905. Available from: https://doi.org/10.1063/1.5008517.

[43] S. Mariotti, E. Köhnen, F. Scheler, K. Sveinbjörnsson, L. Zimmermann, et al., Interface engineering for high-performance, triple-halide perovskite-silicon tandem solar cells, Science 381 (6653) (2023) 63−69. Available from: https://doi.org/10.1126/science.adf5872.
[44] R.M. France, J.F. Geisz, T. Song, W. Olavarria, M. Young, A. Kibbler, et al., Triple-junction solar cells with 39.5% terrestrial and 34.2% space efficiency enabled by thick quantum well superlattices, Joule 6 (5) (2022) 1121−1135. Available from: https://doi.org/10.1016/J.JOULE.2022.04.024.
[45] D.N. Micha, R.T. Silvares Junior, The Influence of solar spectrum and concentration factor on the material choice and the efficiency of multijunction Solar Cells, Sci. Rep. 9 (1) (2019) 1−13. Available from: https://doi.org/10.1038/s41598-019-56457-0.
[46] J.F. Geisz, R.M. France, K.L. Schulte, M.A. Steiner, A.G. Norman, et al., Six-junction III−V solar cells with 47.1% conversion efficiency under 143 Suns concentration, Nat. Energy 5 (4) (2020) 326−335. Available from: https://doi.org/10.1038/s41560-020-0598-5.
[47] K. Shanks, S. Senthilarasu, T.K. Mallick, High-concentration optics for photovoltaic applications, Green. Energy Technol. 190 (2015) 85−113. Available from: https://doi.org/10.1007/978-3-319-15039-0_4.
[48] H. Singh, M. Sabry, D.A.G. Redpath, Experimental investigations into low concentrating line axis solar concentrators for CPV applications, Sol. Energy 136 (2016) 421−427. Available from: https://doi.org/10.1016/j.solener.2016.07.029.
[49] A.O.M. Maka, T.S. O'Donovan, A review of thermal load and performance characterisation of a high concentrating photovoltaic (HCPV) solar receiver assembly, Sol. Energy 206 (2020) 35−51. Available from: https://doi.org/10.1016/J.SOLENER.2020.05.022.
[50] H. Apostoleris, M. Stefancich, M. Chiesa, Tracking-integrated systems for concentrating photovoltaics, Nat. Energy 1 (4) (2016) 16018. Available from: https://doi.org/10.1038/nenergy.2016.18.
[51] M.A. Ceballos, P.J. Pérez-Higueras, E.F. Fernández, F. Almonacid, Tracking-integrated CPV technology: state-of-the-art and classification, Energies (Basel) 16 (15) (2023) 5605. Available from: https://doi.org/10.3390/en16155605.
[52] R.V. Parupudi, H. Singh, M. Kolokotroni, Low concentrating photovoltaics (LCPV) for buildings and their performance analyses, Appl. Energy 279 (2020) 115839. Available from: https://doi.org/10.1016/J.APENERGY.2020.115839.
[53] K. Shanks, H. Baig, N.P. Singh, S. Senthilarasu, K.S. Reddy, T.K. Mallick, Prototype fabrication and experimental investigation of a conjugate refractive reflective homogeniser in a Cassegrain concentrator, Sol. Energy 142 (2017) 97−108. Available from: https://doi.org/10.1016/j.solener.2016.11.038.
[54] V. Kumar, R.L. Shrivastava, S.P. Untawale, Fresnel lens: a promising alternative of reflectors in concentrated solar power, Renew. Sust. Energy Rev. 44 (2015) 376−390. Available from: https://doi.org/10.1016/j.rser.2014.12.006.
[55] M.J. O'Neill, Space photovoltaic concentrators for outer planet and near-Sun missions using ultralight Fresnel lenses, in: R.P. Raffaelle, S.G. Bailey, D.C. Ferguson, S. Durbin, A.F. Hepp (Eds.), Space Photovoltaics: Materials, Missions and Alternative Technologies, Elsevier, Cambridge, MA USA, 2022, pp. 411−432. Available from: https://doi.org/10.1016/B978-0-12-823300-9.00007-8.
[56] S.K. Verma, R. Kumar, M. Barthwal, D. Rakshit, A review on futuristic aspects of hybrid photovoltaic thermal systems (PV/T) in solar energy utilization: engineering and technological approaches, Sustain. Energy Technol. Assess. 53 (Part A) (2022) 102463. Available from: https://doi.org/10.1016/j.seta.2022.102463.
[57] K. Shanks, J.P. Ferrer-Rodriguez, E.F. Fernández, F. Almonacid, P. Peérez-Higueras, S. Senthilarasu, et al., A >3000 suns high concentrator photovoltaic design

based on multiple Fresnel lens primaries focusing to one central solar cell, Sol. Energy 169 (2018) 457−467. Available from: https://doi.org/10.1016/j.solener.2018.05.016.

[58] R.P. Raffaelle, An introduction to space photovoltaics: technologies, issues, and missions, in: R.P. Raffaelle, S.G. Bailey, D.C. Ferguson, S. Durbin, A.F. Hepp (Eds.), Photovoltaics for Space: Key Issues, Missions and Alternative Technologies, pp. 3−27. <https://doi.org/10.1016/B978-0-12-823300-9.00001-7>.

[59] P.T. Chiu, Space applications of III-V single- and multijunction solar cells, in: R.P. Raffaelle, S.G. Bailey, D.C. Ferguson, S. Durbin, A.F. Hepp (Eds.), Photovoltaics for Space: Key Issues, Missions and Alternative Technologies, January 2023, pp. 79−127. <https://doi.org/10.1016/B978-0-12-823300-9.00004-2>.

[60] K. Shanks, S. Senthilarasu, T. Mallick, Reliability investigation for a built ultrahigh concentrator prototype, AIP Conf. Proc. 1881 (1) (September 2017) 020011. <https://doi.org/10.1063/1.5001410>.

[61] D. Polverini, N. Espinosa, U. Eynard, E. Leccisi, F. Ardente, F. Mathieux, Assessing the carbon footprint of photovoltaic modules through the EU ecodesign directive, Sol. Energy 257 (2023) 1−9. Available from: https://doi.org/10.1016/J.SOLENER.2023.04.001.

[62] J. Lu, J. Tang, R. Shan, G. Li, P. Rao, N. Zhang, Spatiotemporal analysis of the future carbon footprint of solar electricity in the United States by a dynamic life cycle assessment, iScience 26 (3) (2023) 106188. Available from: https://doi.org/10.1016/J.ISCI.2023.106188.

[63] A.L. Carneiro, A.A. Martins, V.C.M. Duarte, T.M. Mata, L. Andrade, Energy consumption and carbon footprint of perovskite solar cells, Energy Rep. 8 (2022) 475−481. Available from: https://doi.org/10.1016/J.EGYR.2022.01.045.

[64] C. Lamnatou, D. Chemisana, Concentrating solar systems: life cycle assessment (LCA) and environmental issues, Renew. Sust. Energy Rev. 78 (2017) 916−932. Available from: https://doi.org/10.1016/J.RSER.2017.04.065.

[65] G. Li, Q. Xuan, G. Pei, Y. Su, Y. Lu, J. Ji, Life-cycle assessment of a low-concentration PV module for building south wall integration in China, Appl. Energy 215 (2018) 174−185. Available from: https://doi.org/10.1016/J.apenergy.2018.02.005.

[66] K. Menoufi, D. Chemisana, J.I. Rosell, Life cycle assessment of a building integrated concentrated photovoltaic scheme, Appl. Energy 111 (2013) 505−514. Available from: https://doi.org/10.1016/J.APENERGY.2013.05.037.

CHAPTER FIFTEEN

All perovskite tandem solar cells

Arunkumar Prabhakaran Shyma[1], Nandhakumar Eswaramoorthy[2], Raja Sellappan[3], Kamatchi Rajaram[2], Sowmya Sridharan[2] and Selvakumar Pitchaiya[4]

[1]Department of Physics, School of Advanced Science, VIT, Vellore, Tamil Nadu, India
[2]School of Mechanical Engineering, Vellore Institute of Technology, Vellore, Tamil Nadu, India
[3]Center for Nanotechnology Research, Vellore Institute of Technology, Vellore, Tamil Nadu, India
[4]Faculty of Engineering and Science, Western Norway University of Applied Sciences, Bergen, Norway

15.1 Introduction

The quantum leap in terms of energy consumption is going to constitute a significant challenge to the human race in the upcoming years. The proper utilization of carbon footprint-free resources is a mandate to troubleshoot the problems of climate change that are associated with the deployment of fossil fuels. Photovoltaics (PV) technology has always been a reliable and effective strategy for power generation with zero carbon footprint because the sun is a boundless source of energy. Presently, crystalline silicon (Si)-based solar cells (SCs) are dominating the PV market before their reliability and robust manufacturing process. However, the present commercially available SCs are based on a single junction, which is limited in terms of theoretical conversion efficiency of 33.8% due to the inherent auger recombination that has been widely seen in Si materials [1]. The two significant factors that limit the power conversion efficiency (PCE) of the single junction Si SCs are the fixed bandgap and thermalization losses. The inefficiency in absorbing low and high-energy photons that exceed the bandgap limit will result in the dissipation of excess heat, which in turn adversely affects the performance of the SCs [2].

Tandem SCs are an effective and commercially viable way of efficiently utilizing the solar spectrum by stacking materials with different bandgaps, which in turn absorb the maximum incoming photons [3]. Here each subcells that has stacked can absorb different parts of the solar spectrum, ruling out the possibility of thermalization losses. The hunt for an alternative absorber material has always been there among the PV research community. The implementation of perovskite material in SCs paved a new PV era that delivered a quantum leap in terms of PCE of single junction SCs.

Single junction perovskite SCs have already surpassed Si SCs and bottled near the Shockley−Queisser (SQ) limit [4]. The enhancement of PCE beyond SQ limit is possible by the integration of wide and low bandgap material in the form of tandem architecture. The two-terminal (2T) and four-terminal (4T) tandem architectures have already proposed a theoretical PCE of 43% in which tandem architecture with multiple cell stacks can go all the way up to 86% PCE [5]. Perovskite-based materials have gathered a significant amount of attention in the PV community to integrate within the tandem architecture, owing to their excellent bandgap tuning characteristics.

15.1.1 Perovskites and their unique stoichiometric flexibility

The first report on metal halide perovskite was delivered by Weber in 1978 [6]. Metal halide perovskites that are widely used in PSCs consist of a cuboctahedral geometry in the form of ABX_3 structure, which is depicted in Fig. 15.1A. It consists of a monovalent anion and two cations in which one is monovalent and the other is divalent [7]. The monovalent cation in the A position typically consists of methylammonium (MA^+), formamidinium (FA^+), cesium (Cs^+), and rubidium (Rb^+).

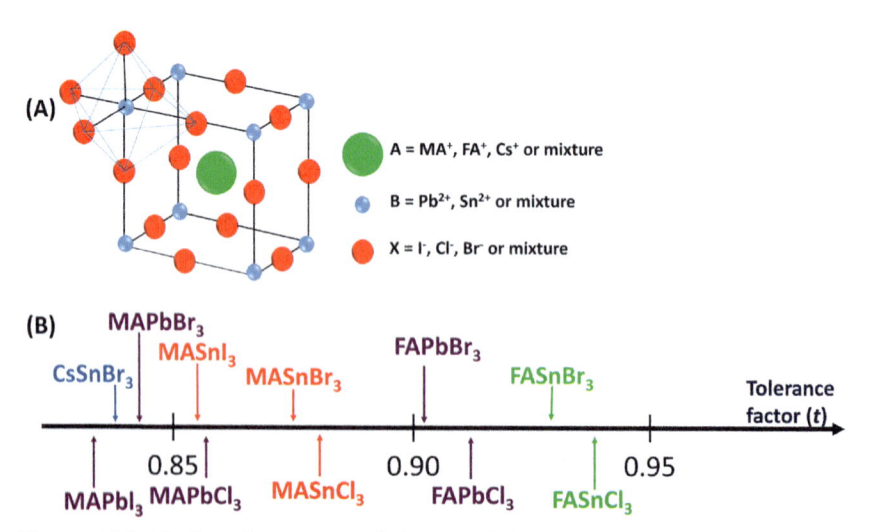

Figure 15.1 (A) Crystal structure of the metal halide-based perovskite materials. (B) Tolerance factor for different perovskite materials.

The substitution of more than one cation is also possible in the A position. Metal candidate such as Sn^{2+}, Pb^{2+}, and Ge^{2+} or their combinations are employed as divalent cations in the B position [8]. Halides such as Cl^-, I^-, Br^-, or their combination are employed as X-site anion [7]. The X anions as well as the B metallic cations form an octahedra (BX_6^{4-}) in which A cation resides with 12 coordination with the structure [1]. The tunable bandgap is one of the most significant characteristics of metal halide perovskites because it can be used in a wide variety of applications ranging from SCs to light-emitting diodes to photodetectors.

The structural tolerance of perovskite is also another prime factor considering the real-world applications. Goldschmidt has proposed the tolerance factor (t) for perovskites [Eq. (15.1)] to assess their structural tunability [9] as follows:

$$t = \frac{R_{A+}R_X}{\sqrt{2}(R_{B+}R_X)} \quad (15.1)$$

where, R_A, R_B, and R_X denote the ionic radii of A, B cations, and X anions, respectively. The values of t should fall between 0.813 and 1.107 to maintain a three-dimensional perovskite structure [10] (Fig. 15.1). The value of t higher than 1.0 leads to the formation of hexagonal structures and lower than 0.813 results in nonperovskite phases [11]. The interatomic distance between $A-X$ and $B-X$ is also considered a significant parameter according to the Goldschmidt factor. The idealized ratio between these two counterparts should be between 0.81 and 1.11 [12]. Another important parameter is the octahedral factor μ (R_B/R_X) of the BX_6 should be between 0.44 and 0.90 for the formation of the perovskite structure. Substitution of multiple halides in the X position is a viable method to control lattice parameters and the bandgap of the desired perovskite structure. Besides the structural tolerance, perovskites also exhibit electrical tolerance as well due to the formation of shallow levels as a result of intrinsic defects. The excellent carrier mobility (1.6–1.7 $cm^2/V/s$), as well as diffusion length (approximately 1 μm) of perovskite structures such as $MAPbI_3$, makes it an ideal candidate for PV applications [13].

15.1.2 Bandgap tuning and stability of perovskites

Methylammonium lead iodide ($MAPbI_3$) is one of the most explored absorber candidates for perovskite SCs. According to recent reports, $MAPbI_3$ managed to deliver PCEs beyond 22% [14]. The optimal bandgap of MAPbI$_3$ lies between 1.55 and 1.63 eV [15]. However, the long-term stability of the aforementioned absorber layer is still a concern to the PV community. The bandgap range of typical absorbers is mentioned in Fig. 15.2A. Many efforts have been carried out to optimize the bandgap via partial substitution of Sn along with Pb in the *B* cation sites and incorporation of multiple halides in the *X* position as well [16]. The change in halide composition is the most effective way to modify the bandgap because the bandgap tunability factor relies mainly on *X*–*B*–*X* bond angle and its length, as in Fig. 15.2B. The *A* position cation has no significant role in determining the bandgap of the perovskite [17]. The bandgap determination of any perovskite material solely depends upon the *s* and *p* orbitals of the *B* site cation and the *p* orbital of the *X* anion

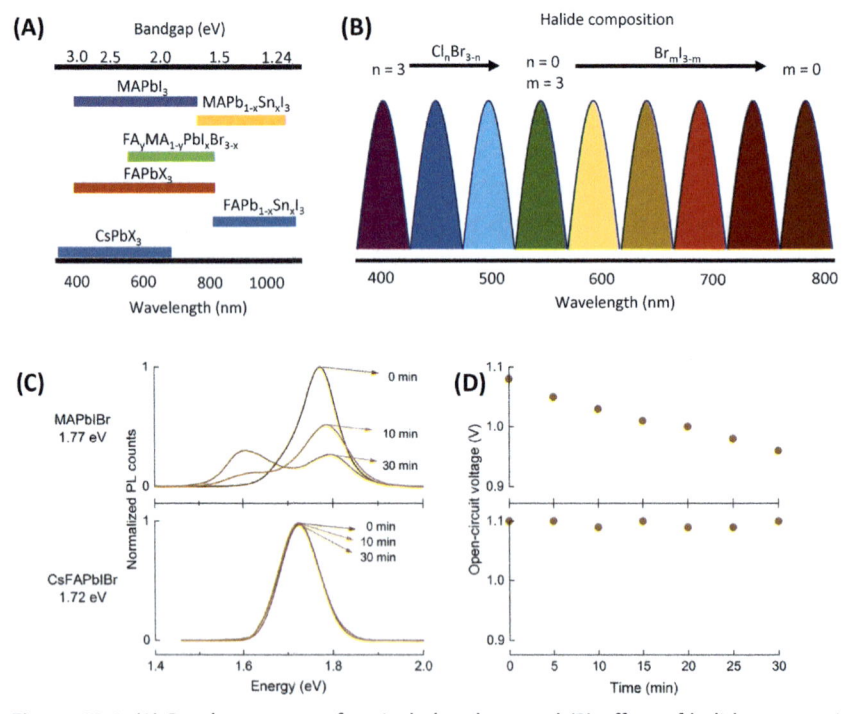

Figure 15.2 (A) Bandgap range of typical absorbers and (B) effect of halide composition on bandgap. (C) PL spectra of MAPbIBr and CsFAPbIBr thin films under continuous AM 1.5 G irradiation. (D) Open-circuit potential of MAPbIBr and CsFAPbIBr SCs under continuous illumination. *Reproduced with permission from M. Jaysankar, B.A.L. Raul, J. Bastos, C. Burgess, C. Weijtens, et al., Minimizing voltage loss in wide-bandgap perovskites for tandem solar cells, ACS Energy Lett. 4 (1) (2019) 259–264, copyright (2018) American Chemical Society.*

[18]. The increment in the perovskite bandgap depends upon the balance between antibonding and bonding levels. The narrower bandgap requires higher p orbitals of halide, which in turn increases the valence band maxima (VBM). The bandgap of the perovskite structure mainly depends upon the energy values of X site p orbitals [19]. Substitution of I/Br in the halide position is considered a good practice to achieve higher charge carrier mobility as well as stability toward ambient conditions [20].

The incorporation of FA with MA in the A site cation position is a viable and effective approach in terms of temperature stability [21]. Even though the A site cation does not have any direct influence on the bandgap, it indirectly impacts the bandgap via lattice contraction via second-order Jahn−Teller distortion [22]. The size of the A site cation plays a vital role in shaping the cuboctahedra structure. Smaller A cation triggers the bonding between B site cation and X site anion resulting in an increased bandgap. Thereby the bandgap of Pb/Sn-based halide perovskites increases with the incorporation of cations, which are smaller in size such as Cs and Rb, and decreases with the addition of cations such as FA, which is larger in size [23]. In fact, FA-based perovskites are more thermally stable, but lose their charm when coming to the shift in phase. The phase of $FAPb(I_{1-x}Br_x)_3$ transforms when I is incorporated with Br in the X position. An amorphous phase induces when $x = 0.3-0.6$ leads to low photon absorption and thermal loss [24]. Consequently, FA is employed at a low level along with MA to acquire improved thermal and structural stability. Partial Cs substitution in the A site, along with MA, also exhibits stability over a period of time as depicted in Fig. 15.2C and D [25].

15.1.3 All perovskite tandem solar cells

A typical tandem cell involves stacking of multiple $p-n$ junctions one on top of the other, which consists of a top semitransparent cell that absorbs high energy photons and a bottom cell that absorbs low energy photons, as shown in Fig. 15.3A and B; the typical absorption spectrum is given in Fig. 15.3C [5]. The fabrication of such tandem architectures is made in such a way that the wide bandgap semiconductors are employed in the top cell and low bandgap materials are employed in the bottom cell [26]. Typically, an all-perovskite tandem SC consists of a top cell that packs a wide bandgap absorber with ample transparency to receive the higher energy incoming photons and a bottom cell that contains a narrow bandgap absorber to ensure the capture of lower energy photons [27].

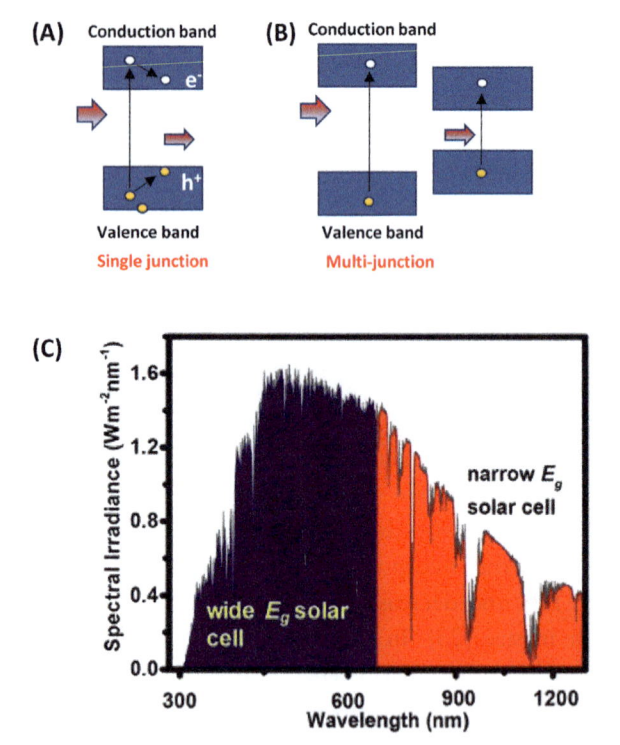

Figure 15.3 (A) Light trapping in single junction and (B) multijunction solar cell. (C) Absorption spectrum of a typical tandem solar cell. *Reproduced with permission from H. Li, W. Zhang, Perovskite tandem solar cells: From fundamentals to commercial deployment, Chem. Rev. 120 (18) (2020) 9835–9950, copyright (2020) American Chemical Society.*

This makes perovskite tandem SCs more charming with proper utilization of solar energy with minimal thermalization and spectral loss [28].

When compared with the existing absorber materials for tandem cells, perovskite materials exhibit far more enhanced performance, owing to their direct bandgap, long diffusion length, and high carrier mobility [29]. The availability of the perovskite materials is also comparable with the existing materials that are used in tandem structures in terms of cost-effectiveness and solvent processing. The bandgap of the halide perovskite can be tuned from 1.17 to 3.5 eV, which is a promising factor for using perovskite materials in both top and bottom cells [30]. Many tandem architectures are implemented incorporating perovskite as a top cell, such as perovskite/Si and perovskite/CIGS. However, all perovskite tandem structures that are realized by implementing perovskite/perovskite junctions serve an additional benefit of low material processing, easy fabrication, and cost-effectiveness. The main challenges that are faced by all perovskite tandem solar cells are semitransparency, voltage loss in the wide bandgap top cell, and optimization of the narrow bandgap cell for the bottom cell [31,32].

15.2 Light capture and device management

The perovskite tandem SCs can be effectively classified into two types based on the connectivity and light management as two terminal (2T) and four terminal (4T) devices. Here the significant deciding factor for a tandem structure to fall under either 2T or 4T tandem structure is the bandgap of the top as well as the bottom cells, which is depicted in Fig. 15.4A and B.

The bandgap as well as the semitransparency of the top cell is a must for photocurrent matching in between both subcells. The appropriate bandgap range for the proper light capture for the top cell lies between 1.75 and 1.85 eV [33]. However, additional adjustments in the spectrum are crucial for 2T configuration such as alloying with tin (Sn^{2+}) to lower the bandgap for the bottom cell compromising the unstable nature of Sn^{2+} (Fig. 15.4C) [34]. The proper control over the bandgap is the fundamental requirement for efficiently working all perovskite tandem SCs. Mixing $MAPbI_3$ with Br can effectively tune the bandgap from 1.5 to 2.3 eV. Sn^{2+} can reduce the bandgap as low as approximately 1.2 eV when alloyed with $MAPbI_3$ [35]. The thickness of the active layer is also a crucial parameter, which is a mandate to match the photocurrent between the top and the bottom cell, as shown in Fig. 15.4D.

15.2.1 Semitransparency of the top cell

Transparency of the top cell is a significant concern because the amount of photons that reach the bottom cell determines the total output of the tandem device. The replacement of conventional metal electrodes with transparent conductive oxides (TCOs) is the best way to address the aforementioned problem [36]. Many TCOs have been employed in tandem structures such as boron-doped ZnO (BZnO), Sn-doped indium oxide (Sn-In_2O_3), and Al-doped ZnO (Al-ZnO) [37]. Several transparent metal candidates are also employed, such as gold (Au) and silver (Ag) nanowires, by compromising the conductivity [38]. Indium tin oxide

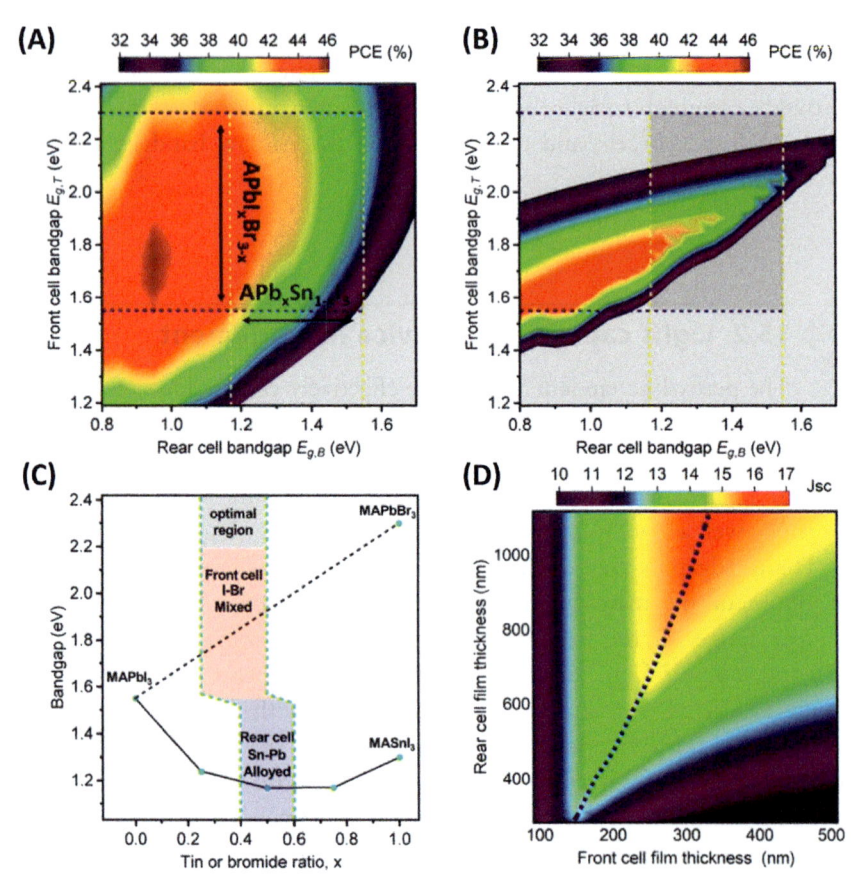

Figure 15.4 Bandgap arrangements for the (A) 4T and (B) 2T top and bottom (rear) tandem subcells. (C) Halide and metal tuning for tandem subcells. (D) Photogenerated current as a function of absorber layer thickness. *Reproduced with permission from X. Zheng, A.Y. Alsalloum, Y. Hou, E. Sargent, O. Bakr, All-perovskite tandem solar cells: A roadmap to uniting high efficiency with high stability, Acct. Mater. Res. 1 (1) (2020) 63−76, copyright (2020) American Chemical Society.*

(ITO) is the most commonly used one due to its wide bandgap (3.5−4.3 eV) and work function of 4.8 eV. Moreover, ITO exhibits the highest visible light transmittance with minimal electrical resistivity ($1.5-2.0 \times 10^{-4}$ Ω cm) [39−41]. ITO also helps PSCs gain a higher fill factor (FF) with ample band alignment. The performance of the semitransparent top cell is comparatively low with Au-deposited PSCs due to the charge deterioration that happens at the junctions [42]. To reduce the impact of the sputtering process on organic charge selective layers can be

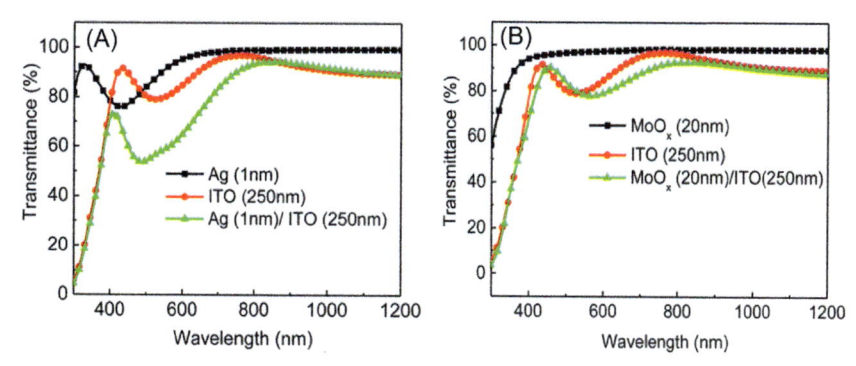

Figure 15.5 Transparency of the top ITO electrode with (A) 1 nm Ag and (B) 20 nm MoO_x. *Reproduced with permission from A. Guchhait, H.E. Dewi, S.W. Leow, H. Wang, G. Han, et al., Over 20% efficient CIGS–perovskite tandem solar cells, ACS Energy Lett. 2 (4) (2017) 807–812, copyright (2017) American Chemical Society.*

effectively employed by depositing a 30–50 nm layer of MoO_x, which possesses good hole extraction capabilities [36,43]. Additionally, MoO_x consists of a wide bandgap of around 2.9–3.3 eV with a higher work function of 6.7 eV and solution-processable compatibility [44]. This makes MoO_x the best protective layer candidate in semitransparent top cells, as depicted in Fig. 15.5 [45,46]. In the case of $MAPbI_3$, 50% of the photons below 800 nm will be transmitted via the top cell and used by the bottom cell, which possesses good near-infrared capabilities. The individual PCE of both cells will be higher than the overall tandem PCE due to the free carrier absorption and recombination that happens at the junctions.

Similarly, other researchers also tried protective layer and transparency optimization in the top cells. Yan and coworkers implemented a 10 nm of MoO_x in $FA_{0.8}Cs_{0.2}Pb(I_{0.7}Br_{0.3})_3$ and achieved a transparency of around 70%. The practice of applying paraffin oil between the top and the bottom cell is also an effective approach to minimize the optical losses [47]. In another work, Han et al. obtained a transparency of around 80% by implementing the ITO layer as the top contact along with an additional layer of MgF_2 in the 700–1300 nm region [48]. To date, the most effective material that has been used in the top cell is TCO due to its better electrical properties and optical transmittance.

15.2.2 Importance of low bandgap bottom cells

The effective way to achieve a low bandgap, high-performing bottom cell is to control the bandgap of the absorber near to 1.3 eV. Typically, the bottom cell in all perovskite tandem SCs consists of mixed A (FA, MA, and Cs) and B (Pb and Sn) site cations. The lowest bandgap that has been achieved without the presence of the Sn to date is 1.48 eV [49]. Sn plays a vital role in determining the bandgap of the perovskite because a bandgap of approximately 1.2 eV has been realized by employing Sn in $MAPb_{0.5}Sn_{0.5}I_3$ [50]. This can be noted that the aforementioned bandgap is lower than that of $MAPbI_3$ (1.55 eV) and $MASnI_3$ (1.3 eV) [19]. This is due to the variation in the ratio of Sn/Pb that leads to a nonlinear difference in the bandgap. The energy mismatch of the atomic orbitals of Pb and Sn forms a band edge that eventually transforms into alloy formation. However, here Sn acts as a double-edged sword [51]. Even though Sn enables the bandgap reduction in all perovskite SCs, it loses its charm when coming to the oxidation of Sn^{2+} to Sn^{4+}.

The number of various perovskite compositions for the bottom is higher than that of top wide bandgap cells. The first report on B site alloying of perovskite absorber layer dates back to 2014 and there on it made a consistent improvement in the efficiency of the perovskite SCs, reaching approximately 20% [52]. The recent record PCE of Sn-incorporated perovskite SC is 24.8% [53]. Lin et al. implemented a viable method of incorporating metallic Sn into the $MA_{0.3}FA_{0.7}Pb_{0.5}Sn_{0.5}I_3$ precursor solution that can effectively retard the oxidation mechanism of Sn^{2+} to Sn^{4+} [53]. They have employed the same composition in all perovskite tandem SCs and attained an efficiency of 24.8%. The studies that have been carried out on Sn/Pb alloying can be found elsewhere [13,54−58]. The ability to absorb maximum light from the near-infrared region is one of the critical activity markers for an all-perovskite tandem bottom cell. Therefore, it is noted that the absorber in the bottom cell should be fabricated in such a way that it is as thick as possible with good quality. The incorporation of additives such as CdI_2 and GuanSCN has achieved a thickness of ∼1 μm along with external quantum efficiency (EQE) of around 60%−90% [58]. Such types of modifications in Sn/Pb achieved a current density of 31 mA/cm^2.

15.2.3 Contact layers and absorbance loss

The contact layer (CL) between the top and the bottom cell is one of the significant parts when coming to the tandem structures. The processing method of CL also affects the overall efficiency of all perovskite tandem structures. The absorbance as a result of CL is considered as the potential shortage of light that is supposed to get absorbed by the perovskite active layer. Usually, this impacts the low bandgap bottom cell more than the wide bandgap cell because light must go through the CL between the top and the bottom cells. C_{60} and spiro-OMeTAD are generally employed as a second CL (CL2) in $p-i-n$ and $n-i-p$ architectures [59]. It has been reported that controlling the thickness of CL approximately 10 nm maintains less absorption than is accounted for CL [60]. Consequently, the less thickness of CL2 and CL3 can negatively impact the performance of the tandem structure. Most of the researchers are employing a thicker CL instead of a thinner one in practice [60].

15.3 All perovskite tandem architectures

All perovskite tandem SCs have garnered attention due to impressive PCE and low-cost solution processing fabrication methods. This is considered to be a great achievement compared to that of perovskite/Si and perovskite/CIGS tandems in which bottom cells typically require vacuum deposition methods. All perovskite tandem SCs have great viability for commercialization because the bandgap of the top and the bottom cell can be controlled. Another benefit is the compatibility of the precursor solution with the roll-to-roll process that can be effectively utilized in industrial manufacturing process. All perovskite tandem SCs mainly consist of two architectures, namely 2T and 4T.

15.3.1 Two terminal all perovskite tandem solar cells

As mentioned earlier, the bottom cell fabrication in perovskite/Si and perovskite/CIGS is employed via vacuum deposition technique. In the case of all perovskite tandem SCs, the fabrication of the top cell can cause dissolution of the bottom cell because the method employed is a solution process. To limit such issues, recombination layers with higher resistance to the perovskite solution should be implemented [61]. Additionally, high optical visibility and electrical transport should also be ensured. A 2T architecture is mainly prone to two main issues: nonradiative Shockley−Read−Hall recombination and the isolation of halide counterions, which leads to a detrimental effect in voltage [62]. Modification of bottom cell and bandgap engineering are the highest priorities in 2T architecture. Snaith and coworkers developed an absorber of $FA_{0.75}Cs_{0.25}Sn_{0.5}Pb_{0.5}I_3$ (1.2 eV) for bottom cell and clubbed it with a much higher wide bandgap absorber $FA_{0.83}Cs_{0.17}Pb(I_{0.5}Br_{0.5})_3$ (1.8 eV) for top cell and managed to obtain an efficiency of 17% along with 15.8 and 14.1 mA/cm^2 J_{sc} for bottom and top cells, respectively [63].

Good electronic properties also should be accompanied by good bandgap tunability because the absorber layer of the bottom cell requires comparatively thicker layers to ensure the maximum absorption of photons with minimal loss. Incorporation of Cl is considered as a viable option in Sn/Pb-based perovskites to obtain good quality films with increased grain size that can limit the electronic anomalies [31]. The Cl incorporation triggered the lifetime, as well as minority carrier mobility, resulting in a PCE of 18.4% [32]. The same bottom cell with an absorber layer thickness of 700 nm delivers EQE in the region between 700 and 900 nm, which is suitable for the performance of the tandem device [32]. The thickness of the top cell is also a quintessential parameter when coming to the overall performance of the tandem device since the current matching is a crucial element in 2T all perovskite SCs. One of the theoretical reports on 2T all perovskite tandem SCs suggests that the layer thickness of the top cell absorber should fall in between the 50 and 150 nm range for optimum performance [63]. They simulated a $MAPbI_3/MASn_{0.25}Pb_{0.75}I_3$ architecture and found the optimum thickness of layers to be 150 and 350 nm to obtain the maximum current density. The EQE measurement suggests lowering the bandgap further to obtain the spectral response beyond 1000 nm [53].

However, further increasing the Sn concentration is not a viable approach to minimize the bad gap further because the Sn/Pb relation exhibits no linearity at higher concentrations [64]. Cs is a robust and stable additive in the A cation position to get an even lower bandgap according to the desired process. Like in other perovskite 2T tandem architectures along with Si and CIGS, recombination layers (RLs) and tunnel junctions (TJs) play a major role in determining the output of the device. Along with TCOs that are already established in perovskite/Si and CIGS, some other alternatives are there as RL like PTAA and PEDOT:PSS/ITO. PTAA delivers ample carrier mobility of approximately 10^{-3} to 10^{-2} cm^2/V/s along with suitable energy levels and enhanced stability [65].

15.3.2 Multiple terminal all perovskite tandem solar cells via double-sided substrates

Yao et al. investigated the production of 2T, 3T, and 4T devices utilizing fabrication on both sides of a double-side-coated indium tin oxide (ITO) substrate [66]. The group reported efficiency of 21.17% for a 2T tandem device with a top cell of a wide (1.75 eV) bandgap perovskite ($Cs_{0.4}FA_{0.6}PbI_{1.95}Br_{1.05}$) and a bottom cell with a narrow (1.21 eV) bandgap perovskite (($FASnI_3)_{0.6}(MAPbI_3)_{0.4}$) absorber [66]; the champion device (see Fig. 15.6) achieved an open circuit voltage (V_{oc}) of 1.98 V and a J_{sc} of 13.12 mA/cm^2. Modifying the wiring method enabled the fabrication of 3T (PCE = 17.71%) and 4T (PCE = 21.55%) tandem devices. Due to the increased independence of individual devices, the stability over 45 days in an N_2-filled glovebox of 4T devices was found to be double and triple that found for 3T and 2T devices, respectively, similar to wide bandgap single-junction devices [66].

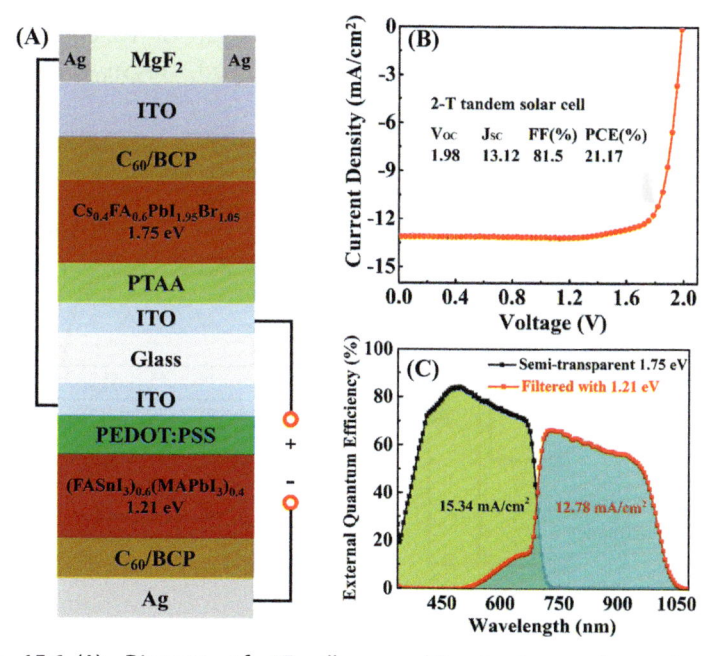

Figure 15.6 (A) Diagram of 2T all perovskite tandem solar cell with a $Cs_{0.4}FA_{0.6}PbI_{1.95}Br_{1.05}/(FASnI_3)_{0.6}(MAPbI_3)_{0.4}$ structure; (B) the J–V curve and (C) EQE spectra of the constructed cell on a double-side-coated indium tin oxide substrate. Reprinted with permission from Y. Yao, F. Lv, X. Zhao, X. Yang, B. Wu, P. Li, Q. Song, A monolithic all-perovskite tandem solar cell with 2-T, 3-T and 4-T architecture integrated, Sol. Energy Mater. Sol. Cells 259 (August 2023) 112376, copyright (2023) Elsevier.

15.3.3 Four terminal all perovskite tandem solar cells

The construction of a 4T (Fig. 15.7A) all-perovskite SCs is relatively less complex because the subcell compatibility is not a major concern here. Yan and coworkers developed a perovskite tandem device, which houses $MAPb_{1-x}Sn_xI_3$ absorber layer varying the B cation ratio ranging from 0.25 to 1.0, and observed that the bandgap of the absorber has considerably reduced from 1.6 to 1.2 eV [29]. They also found that increasing Sn content is gradually moving VBM in an upward direction. PEDOT:PSS is usually used here to ease the process of hole extraction in the perovskite/HTL interface. The incorporation of FA cations plays a vital role in alleviating Sn oxidation by delivering an efficiency of 14.19% [29]. The same group has clubbed this cell with $MAPbI_3$ top cell and attained an efficiency of 19.08%. This work shed light on the concept of utilizing compositional and interfacial engineering to realize high performing 4T all perovskite tandem SCs. The Sn substitution works in FA-based perovskites as well. Snaith and coworkers controlled the bandgap of $FAPbI_3$ with a 50% Sn/Pb ratio, which led to a bandgap of 1.2 eV [63]. The two absorber layers that they have developed, $FASn_{0.5}Pb_{0.5}I_3$ and

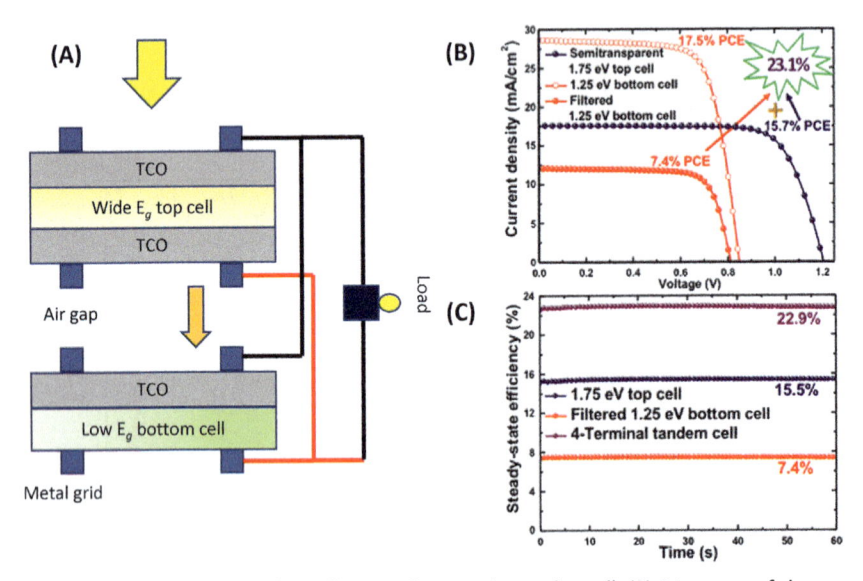

Figure 15.7 (A) Diagram of 4T all perovskite tandem solar cell, (B) IV curves of the top and the bottom cells, and (C) steady state efficiency of the top and the bottom cells. *Reproduced with permission from D. Zhao, C. Wang, Z. Song, Y. Yu, C. Chen, et al., Four-terminal all-perovskite tandem solar cells achieving power conversion efficiencies exceeding 23%, ACS Energy Lett. 3 (2) (2018) 305–306, copyright (2018) American Chemical Society.*

$FA_{0.75}Cs_{0.25}Sn_{0.5}Pb_{0.5}I_3$, exhibited enhanced stability and performance in 4T all perovskite tandem cell architecture. The constructed device achieved an efficiency of 20.3%. Yan and coworkers also implemented Sn in 4T all perovskite tandem architecture and managed to get an efficiency of 23%, as depicted in Figs. 15.7B and C [47]. They have employed $(FASnI_3)_{0.6}(MAPbI_3)_{0.4}$ with a bandgap of 1.25 eV as bottom cell and $FA_{0.8}Cs_{0.2}Pb(I_{0.7}Br_{0.3})_3$ with a bandgap of 1.75 eV as top cell. The wider bandgap top cell functions in a way that the lower wavelength photons from the IR region reach the bottom cell. This was the first time an all-perovskite tandem structure exhibited almost matching performance as a single junction perovskite SC.

Nejand et al. implemented a vacuum-based solution process to fabricate the top and the bottom cells [67]. By adopting the aforementioned method, an absorber with a good grain boundary, devoid of pinholes, and a long lifetime can be achieved. The practice of implementing the spectrum filters among the 4T all perovskite SCs helps to direct the desired wavelength for the maximum exposure by the absorber layer. Shen et al. proposed an $n-i-p$ all perovskite tandem structure, which possesses $MAPbBr_3$ as the top cell with a bandgap of 2.3 eV and $MAPbI_3$ as the bottom cell with a bandgap of 1.5 eV, and obtained a PCE of 13.4% in 4T all perovskite device architecture [68].

15.4 All perovskite flexible tandem solar cells

Flexible all perovskite tandem SCs deliver great hope for large-scale manufacturing from existing rigid substrate-based perovskite SCs. The flexible tandem devices open up a great avenue for low-cost manufacturing processes, which can potentially replace the existing Si-based devices. Palmstrom et al. developed a $DMA_{0.1}FA_{0.6}Cs_{0.3}PbI_{2.4}Br_{0.6}/FA_{0.75}Cs_{0.25}Sn_{0.5}Pb_{0.5}I_3$-based tandem perovskite SCs with A cation-based site engineering [60]. Later they applied the configuration onto a flexible polyethylene naphthalate (PEN) substrate and developed a 2T all perovskite architecture with an efficiency of 21.3%, which is a bit lesser than that of the same configuration on a glass substrate (24.8%). The main concern while playing with the flexible substrate is the processing temperature limitation because flexible substrates such as PEN cannot stand at more than 150°C [69]. Perovskite modules processed via low temperature

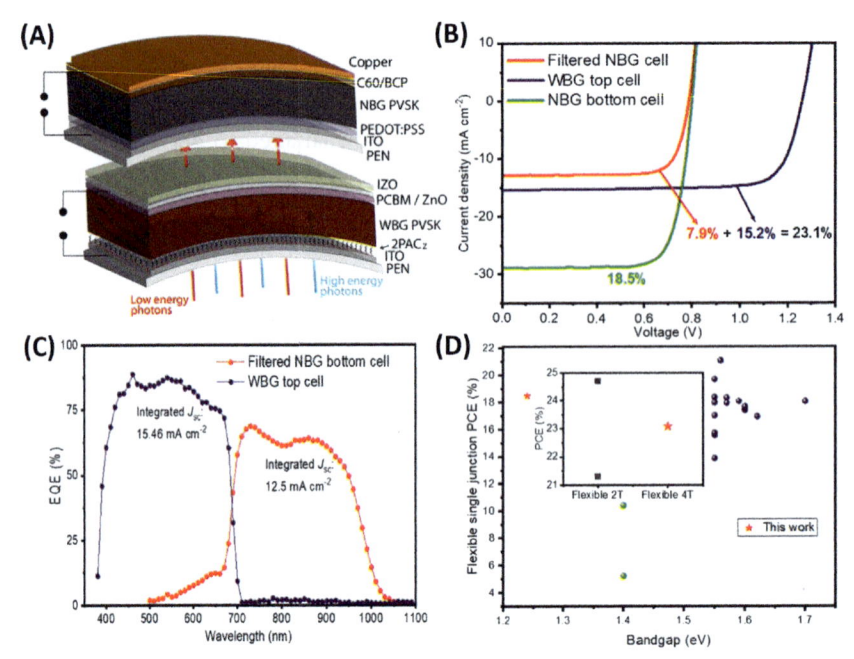

Figure 15.8 (A) Diagram of the PEN-based flexible all-perovskite 4T tandem solar cell, (B) I–V curves of the top and the bottom cells, (C) EQE of the top and the bottom cells, and (D) comparison with the reported single-junction perovskite solar cells. *Reproduced with permission from J.P. Pious, Y. Zwirner, H. Lai, S. Olthof, Q. Jeangros, et al., Revealing the role of tin fluoride additive in narrow bandgap Pb-Sn perovskites for highly efficient flexible all-perovskite tandem cells, ACS Appl. Mater. Interfaces 15 (7) (2023) 10150–10157, copyright (2023) American Chemical Society.*

can only be the suited candidates for flexible all-perovskite tandem SCs. Pious et al. investigated the scope of SnF_2 in the perovskite tandem architecture with PEN as the substrate and copper foil as the current collector (Fig. 15.8) [70]. The device showcased an efficiency of 23.1%.

15.5 Device stability

The stability of the all-perovskite tandem SCs is the collective stability of each subcell and tunnel, as well as the recombination junctions that have been used to construct the tandem structure. The light absorption instability of the top wide bandgap cell is a crucial issue in all perovskite tandem SCs [58]. Grain size engineering and the crystallinity of the absorber thin film are two viable options to eliminate the issues related to the light instability of the top cell [71]. Compositional engineering can also be considered as the best remedy to the aforementioned issue. Tuning of halide composition is an effective way because mixed I/Br halide perovskite showcases enhanced performance in terms of light absorption when kept under 10 suns [72]. Implementing triple halide perovskite also delivered 96% of its initial efficiency under a nitrogen atmosphere with maximum power point tracking at 60°C. Cho et al. further extended the work with Cl addition and found that Cl plays a phenomenal role in suppressing phase aggregation employing higher activation energy [73].

Device stability requires a dedicated approach for wide bandgap top cell and narrow bandgap bottom cell. Additive engineering in A and B cation sites is a best-suited approach concerning the wide bandgap top cell. Similarly, defect passivation can be applied to the narrow bandgap bottom cell since the formation of Sn vacancies can hinder the performance of the tandem device. The problem of Sn oxidation in narrow bandgap bottom cells is also a crucial issue that needs to be properly addressed. Prasanna et al. reported that the interaction of PEDOT:PSS with the Pb/Sn narrow bandgap perovskite adversely affects the charge extraction capability and the temperature stability [30]. The development of a hole transport layer-free structure can effectively resolve the issue with little compromise in the overall efficiency of the tandem device. They have put forward a hole transport layer-free structure narrow bandgap perovskite ($FA_{0.75}Cs_{0.25}Sn_{0.4}Pb_{0.6}I_3$) and observed stability of 1000 hours in ambient air without any loss of efficiency. They maintained the temperature at 85°C throughout the testing with moderate humidity. Employing other hole transport layers such as NiO and PTAA can also reduce the instability of the tandem device [74]. Palmstrom et al. reported an encapsulation strategy that can be applied to the tandem structure with an epoxy glass cover seal and sustained efficiency of 500 hours under constant illumination [60]. To conclude, multiple strategies can be applied for the stability of all perovskite tandem SCs such as site engineering, the addition of Cs and Sn, and encapsulation.

15.6 Conclusion

In summary, a comprehensive discussion on all perovskite tandem structures, their fabrication, stability, and recent advances has been executed.

The fundamental mechanism and strategy to improve the efficiency of all perovskite tandem devices are also included. Even though all perovskite tandem devices promise a bright future, further research and development are needed to cast out the hurdles before commercialization. Some points are to be noted for the future development of all perovskite tandem SCs to reach par with the proposed theoretical efficiency such as lower open circuit potential of wide bandgap top cell, limitation in scalable fabrication, and stability issues. The ongoing research in all perovskite tandem SC is expected to skyrocket in coming years with efficiency up to 30%—40%. The solution process-based fabrication can be considered a cost-effective production process even though it needs to be improved. Development of appropriate tunnel junctions and recombination layer is also a much-needed treatment in solution-processed all perovskite tandem SCs. The omission of the top metal electrode and the addition of more TCOs can resolve the transparency issues related to the wide bandgap top cell. Bridging the distance between cell-to-device efficiency is also a concern when coming to all perovskite tandem SCs. Here we could say the future development of all perovskite tandem SCs is not limited to efficiency enhancement but also will shed light on scalability, increased stability the lifetime of the perovskite absorber layer and cost effectiveness. The challenges that have been mentioned should be properly addressed to effectively develop all perovskite tandem SCs for large-scale manufacturing.

References

[1] P. Nayak, S. Mahesh, H. Snaith, D. Cahen, Photovoltaic solar cell technologies: analysing the state of the art, Nat. Rev. Mater. 4 (4) (2019) 269—285.
[2] M.A. Green, Limiting photovoltaic efficiency under new ASTM International G173-based reference spectra, Prog. Photovolt. 20 (8) (2012) 954—959.
[3] S.P. Bremner, C. Yi, I. Almansouri, A. Ho-Baillie, M.A. Green, Optimum band gap combinations to make best use of new photovoltaic materials, Sol. Energy 135 (2016) 750—757.
[4] L.C. Hirst, N.J. Ekins-Daukes, Fundamental losses in solar cells, Prog. Photovolt. 19 (3) (2011) 286—293.
[5] H. Li, W. Zhang, Perovskite tandem solar cells: from fundamentals to commercial deployment, Chem. Rev. 120 (18) (2020) 9835—9950.
[6] Y. Zhou, F. Lu, T. Fang, D. Gu, X. Feng, T. Song, et al., A brief review on metal halide perovskite photocatalysts: history, applications and prospects, J. Alloy. Compd. 911 (2022) 165062.
[7] M.A. Green, A. Ho-Baillie, H.J. Snaith, The emergence of perovskite solar cells, Nat. Photonics 8 (7) (2014) 506—514.
[8] G.E. Eperon, D.S. Ginger, B-site metal cation exchange in halide perovskites, ACS Energy Lett. 2 (5) (2017) 1190—1196.

[9] A.P. Shyma, A. Grace, V. Raghavan, G. Jacob, R. Sellappan, Tin as an emerging surrogate for lead-free perovskite solar cells, Nanosci. Nanotechnol. — Asia 11 (6) (2021) 3—16. Available from: https://doi.org/10.2174/2210681210999201120091911.

[10] W. Travis, E.N.K. Glover, H. Bronstein, D.O. Scanlon, R.G. Palgrave, On the application of the tolerance factor to inorganic and hybrid halide perovskites: a revised system, Chem. Sci. 7 (7) (2016) 4548—4556.

[11] J.S. Manser, J.A. Christians, P.V. Kamat, Intriguing optoelectronic properties of metal halide perovskites, Chem. Rev. 116 (21) (2016) 12956—13008.

[12] W.-J. Yin, J.-H. Yang, J. Kang, Y. Yan, S.-H. Wei, Halide perovskite materials for solar cells: a theoretical review, J. Mater. Chem. A 3 (17) (2015) 8926—8942.

[13] J. Tong, Z. Song, D.H. Kim, X. Chen, C. Chen, et al., Carrier lifetimes of >1 μs in Sn-Pb perovskites enable efficient all-perovskite tandem solar cells, Science 364 (6439) (2019) 475—479.

[14] F. Fu, T. Feurer, T. Jäger, E. Avancini, B. Bissig, et al., Low-temperature-processed efficient semi-transparent planar perovskite solar cells for bifacial and tandem applications, Nat, Commun. 6 (2015) 8932.

[15] L.A. Frolova, A.I. Davlethanov, N.N. Dremova, I. Zhidkov, A.F. Akbulatov, et al., Efficient and stable $MAPbI_3$-based perovskite solar cells using polyvinylcarbazole passivation, J. Phys. Chem. Lett. 11 (16) (2020) 6772—6778.

[16] A.P. Shyma, R. Sellappan, Computational probing of tin-based lead-free perovskite solar cells: effects of absorber parameters and various electron transport layer materials on device performance, Materials 15 (2022) 7859. Available from: https://doi.org/10.3390/ma15217859.

[17] C.C. Boyd, R. Cheacharoen, T. Leijtens, M.D. McGehee, Understanding degradation mechanisms and improving stability of perovskite photovoltaics, Chem. Rev. 119 (5) (2019) 3418—3451.

[18] D.B. Straus, R.J. Cava, Tuning the band gap in the halide perovskite $CsPbBr_3$ through Sr substitution, ACS Appl. Mater. Interfaces 14 (30) (2022) 34884—34890.

[19] R. Prasanna, A. Gold-Parker, T. Leijtens, B. Conings, A. Babayigit, et al., Band gap tuning via lattice contraction and octahedral tilting in perovskite materials for photovoltaics, J. Am. Chem. Soc. 139 (32) (2017) 11117—11124.

[20] C. Lin, S. Li, W. Zhang, C. Shao, Z. Yang, Effect of bromine substitution on the ion migration and optical absorption in $MAPbI_3$ perovskite solar cells: the first-principles study, ACS Appl. Energy Mater. 1 (3) (2018) 1374—1380.

[21] L.A. Muscarella, D. Petrova, R.J. Cervasio, A. Farawar, O. Lugier, Air-stable and oriented mixed lead halide perovskite (FA/MA) by the one-step deposition method using zinc iodide and an alkylammonium additive, ACS Appl. Mater. Interfaces 11 (19) (2019) 17555—17562.

[22] J.-H. Lee, N. Bristowe, J.H. Lee, S.-H. Lee, P. Bristowe, A. Cheetham, et al., Resolving the physical origin of octahedral tilting in halide perovskites, Chem. Mater. 28 (12) (2016) 4259—4266.

[23] R.M.I. Bandara, S.M. Silva, C.C.L. Underwood, K.D.G.I. Jayawardena, R.A. Sporea, S.R.P. Silva, Progress of Pb-Sn mixed perovskites for photovoltaics: a review, Energy Environ. Mater. 5 (2) (2022) 370—400.

[24] H. Lu, A. Krishna, S.M. Zakeeruddin, M. Grätzel, A. Hagfeldt, Compositional and interface engineering of organic-inorganic lead halide perovskite solar cells, iScience 23 (8) (2020) 101359.

[25] M. Jaysankar, B.A.L. Raul, J. Bastos, C. Burgess, C. Weijtens, et al., Minimizing voltage loss in wide-bandgap perovskites for tandem solar cells, ACS Energy Lett. 4 (1) (2019) 259—264.

[26] Y. Wang, M. Zhang, K. Xiao, R. Lin, X. Luo, Q. Han, et al., Recent progress in developing efficient monolithic all-perovskite tandem solar cells, J. Semiconductors 41 (5) (2020) 051201.
[27] M.T. Hörantner, T. Leijtens, M.E. Ziffer, G.E. Eperon, M.G. Christoforo, M.D. McGehee, et al., The potential of multijunction perovskite solar cells, ACS Energy Lett. 2 (10) (2017) 2506−2513.
[28] M. Chen, M.-G. Ju, H.F. Garces, A.D. Carl, L.K. Ono, et al., Highly stable and efficient all-inorganic lead-free perovskite solar cells with native-oxide passivation, Nat. Commun. 10 (1) (2019) 16.
[29] Z. Yang, A. Rajagopal, C.-C. Chueh, S.B. Jo, B. Liu, T. Zhao, et al., Stable low-bandgap Pb-Sn binary perovskites for tandem solar cells, Adv. Mater. 28 (40) (2016) 8990−8997.
[30] R. Prasanna, T. Leijtens, S.P. Dunfield, J.A. Raiford, E.J. Wolf, et al., Design of low bandgap tin−lead halide perovskite solar cells to achieve thermal, atmospheric and operational stability, Nat. Energy 4 (11) (2019) 939−947.
[31] D. Zhao, Y. Yu, C. Wang, W. Liao, N. Shrestha, et al., Low-bandgap mixed tin−lead iodide perovskite absorbers with long carrier lifetimes for all-perovskite tandem solar cells, Nat. Energy 2 (4) (2017) 17018.
[32] D. Zhao, C. Chen, C. Wang, M.M. Junda, Z. Song, et al., Efficient two-terminal all-perovskite tandem solar cells enabled by high-quality low-bandgap absorber layers, Nat. Energy 3 (12) (2018) 1093−1100.
[33] Y. Lin, B. Chen, Y. Fang, J. Zhao, C. Bao, et al., Excess charge-carrier induced instability of hybrid perovskites, Nat. Commun. 9 (1) (2018) 4981.
[34] X. Zheng, A.Y. Alsalloum, Y. Hou, E. Sargent, O. Bakr, All-perovskite tandem solar cells: a roadmap to uniting high efficiency with high stability, Acct. Mater. Res. 1 (1) (2020) 63−76.
[35] J.H. Noh, S.H. Im, J.H. Heo, T.N. Mandal, S.I. Seok, Chemical management for colorful, efficient, and stable inorganic−organic hybrid nanostructured solar cells, Nano Lett. 13 (4) (2013) 1764−1769.
[36] I. Irfan, H. Ding, Y. Gao, D.Y. Kim, J. Subbiah, F. So, Energy level evolution of molybdenum trioxide interlayer between indium tin oxide and organic semiconductor, Appl. Phys. Lett. 96 (7) (2010) 073304.
[37] P. Löper, S.-J. Moon, S. Martín de Nicolas, B. Niesen, M. Ledinsky, et al., Organic−inorganic halide perovskite/crystalline silicon four-terminal tandem solar cells, Phys. Chem. Chem. Phys. 17 (3) (2015) 1619−1629.
[38] C.D. Bailie, M.G. Christoforo, J.P. Mailoa, A.R. Bowring, E.L. Unger, et al., Semitransparent perovskite solar cells for tandems with silicon and CIGS, Energy Environ. Sci. 8 (3) (2015) 956−963.
[39] D. Zhang, M. Najafi, V. Zardetto, M. Dörenkämper, X. Zhou, et al., High efficiency 4-terminal perovskite/c-Si tandem cells, Sol. Energy Mater. Sol. Cell 188 (2018) 1−5.
[40] H. Shen, T. Duong, J. Peng, D. Jacobs, N. Wu, et al., Mechanically-stacked perovskite/CIGS tandem solar cells with efficiency of 23.9% and reduced oxygen sensitivity, Energy Environ. Sci. 11 (2) (2018) 394−406.
[41] I. Hamberg, C.G. Granqvist, K.-F. Berggren, B.E. Sernelius, L. Engström, Band-gap widening in heavily Sn-doped In_2O_3, Phys. Rev. B 30 (1984) 3240−3249.
[42] V. Bermudez, A. Perez-Rodriguez, Understanding the cell-to-module efficiency gap in $Cu(In,Ga)(S,Se)_2$ photovoltaics scale-up, Nat. Energy 3 (6) (2018) 466−475.
[43] B. Macco, M.F.J. Vos, N.F.W. Thissen, A.A. Bol, W.M.M. Kessels, Low-temperature atomic layer deposition of MoO_x for silicon heterojunction solar cells, Phys. Status. Solidi Rapid Res. Lett. 9 (7) (2015) 393−396.
[44] T. Yang, M. Wang, Y. Cao, F. Huang, L. Huang, et al., Polymer solar cells with a low-temperature-annealed sol−gel-derived MoO_x film as a hole extraction layer, Adv. Energy Mater. 2 (5) (2012) 523−527.

[45] Y. Zhao, A.M. Nardes, K. Zhu, Effective hole extraction using MoO_x-Al contact in perovskite $CH_3NH_3PbI_3$ solar cells, Appl. Phys. Lett. 104 (21) (2014) 213906.
[46] A. Guchhait, H.E. Dewi, S.W. Leow, H. Wang, G. Han, et al., Over 20% efficient CIGS−perovskite tandem solar cells, ACS Energy Lett. 2 (4) (2017) 807−812.
[47] D. Zhao, C. Wang, Z. Song, Y. Yu, C. Chen, et al., Four-terminal all-perovskite tandem solar cells achieving power conversion efficiencies exceeding 23, ACS Energy Lett. 3 (2) (2018) 305−306.
[48] Q. Han, Y.-T. Hsieh, L. Meng, J.-L. Wu, P. Sun, et al., High-performance perovskite/$Cu(In,Ga)Se_2$ monolithic tandem solar cells, Science 361 (6405) (2018) 904−908.
[49] W.S. Yang, J.H. Noh, N.J. Jeon, Y.C. Kim, S. Ryu, J. Seo, et al., High-performance photovoltaic perovskite layers fabricated through intramolecular exchange, Science 348 (6240) (2015) 1234−1237.
[50] F. Hao, C.C. Stoumpos, R.P.H. Chang, M.G. Kanatzidis, Anomalous band gap behavior in mixed Sn and Pb perovskites enables broadening of absorption spectrum in solar cells, J. Am. Chem. Soc. 136 (22) (2014) 8094−8099.
[51] A.K. Goyal, S. McKechnie, D. Pashov, W. Tumas, M. Schilfgaarde, V. Stevanović, Origin of pronounced nonlinear band gap behavior in lead−tin hybrid perovskite alloys, Chem. Mater. 30 (11) (2018) 3920−3928.
[52] Y. Ogomi, A. Morita, S. Tsukamoto, T. Saitho, N. Fujikawa, et al., $CH_3NH_3Sn_xPb_{(1-x)}I_3$ perovskite solar cells covering up to 1060 nm, J. Phys. Chem. Lett. 5 (6) (2014) 1004−1011.
[53] R. Lin, K. Xiao, Z. Qin, Q. Han, C. Zhang, et al., Monolithic all-perovskite tandem solar cells with 24.8% efficiency exploiting comproportionation to suppress Sn(II) oxidation in precursor ink, Nat. Energy 4 (10) (2019) 864−873.
[54] C. Wang, Z. Song, C. Li, D. Zhao, Y. Yan, Low-bandgap mixed tin-lead perovskites and their applications in all-perovskite tandem solar cells, Adv. Funct. Mater. 29 (47) (2019) 1808801.
[55] A. Rajagopal, Z. Yang, S.B. Jo, I.L. Braly, P.-W. Liang, H.W. Hillhouse, et al., Highly efficient perovskite−perovskite tandem solar cells reaching 80% of the theoretical limit in photovoltage, Adv. Mater. 29 (34) (2017) 1702140.
[56] G. Kapil, T. Bessho, C.H. Ng, K. Hamada, M. Pandey, et al., Strain relaxation and light management in tin−lead perovskite solar cells to achieve high efficiencies, ACS Energy Lett. 4 (8) (2019) 1991−1998.
[57] T. Jiang, Z. Chen, X. Chen, T. Liu, X. Chen, et al., Realizing high efficiency over 20% of low-bandgap Pb−Sn-alloyed perovskite solar cells by in situ reduction of Sn^{4+}, Sol. RRL 4 (3) (2020) 1900467.
[58] T. Leijtens, R. Prasanna, K.A. Bush, G.E. Eperon, J.A. Raiford, et al., Tin−lead halide perovskites with improved thermal and air stability for efficient all-perovskite tandem solar cells, Sustain. Energy Fuels 2 (11) (2018) 2450−2459.
[59] W.H. Nguyen, C.D. Bailie, E.L. Unger, M.D. McGehee, Enhancing the hole-conductivity of spiro-OMeTAD without oxygen or lithium salts by using spiro $(TFSI)_2$ in perovskite and dye-sensitized solar cells, J. Am. Chem. Soc. 136 (31) (2014) 10996−11001.
[60] A.F. Palmstrom, G.E. Eperon, T. Leijtens, R. Prasanna, S.N. Habisreutinger, et al., Enabling flexible all-perovskite tandem solar cells, Joule 3 (2019) 2193−2204. Available from: https://doi.org/10.1016/j.joule.2019.05.009.
[61] S.P. Bremner, M.Y. Levy, C.B. Honsberg, Analysis of tandem solar cell efficiencies under AM1.5G spectrum using a rapid flux calculation method, Prog. Photovolt. 16 (3) (2008) 225−233.
[62] Y. Hou, E. Aydin, M. De Bastiani, C. Xiao, F.H. Isikgor, et al., Efficient tandem solar cells with solution-processed perovskite on textured crystalline silicon, Science 367 (6482) (2020) 1135−1140.

[63] G.E. Eperon, T. Leijtens, K.A. Bush, R. Prasanna, T. Green, et al., Perovskite-perovskite tandem photovoltaics with optimized band gaps, Science 354 (2016) 861−865. Available from: https://doi.org/10.1126/science.aaf9717.

[64] J. Im, C.C. Stoumpos, H. Jin, A.J. Freeman, M.G. Kanatzidis, Antagonism between spin−orbit coupling and steric effects causes anomalous band gap evolution in the perovskite photovoltaic materials $CH_3NH_3Sn_{1-x}Pb_xI_3$, J. Phys. Chem. Lett. 6 (17) (2015) 3503−3509.

[65] Q. Zeng, L. Liu, Z. Xiao, F. Liu, Y. Hua, Y. Yuan, et al., A two-terminal all-inorganic perovskite/organic tandem solar cell, Sci. Bull. 64 (13) (2019) 885−887.

[66] Y. Yao, F. Lv, X. Zhao, X. Yang, B. Wu, P. Li, et al., A monolithic all-perovskite tandem solar cell with 2-T, 3-T and 4-T architecture integrated, Sol. Energy Mater. Sol. Cell 259 (2023) 112376. Available from: https://doi.org/10.1016/j.solmat.2023.112376.

[67] B.A. Nejand, I.M. Hossain, M. Jakoby, S. Moghadamzadeh, T. Abzieher, et al., Vacuum-assisted growth of low-bandgap thin films ($FA_{0.8}MA_{0.2}Sn_{0.5}Pb_{0.5}I_3$) for all-perovskite tandem solar cells, Adv. Energy Mater. 10 (5) (2020) 1902583.

[68] R. Sheng, A.W.Y. Ho-Baillie, S. Huang, M. Keevers, X. Hao, et al., Four-terminal tandem solar cells using $CH_3NH_3PbBr_3$ by spectrum splitting, J. Phys. Chem. Lett. 6 (19) (2015) 3931−3934.

[69] S. Khan, L. Lorenzelli, R.S. Dahiya, Technologies for printing sensors and electronics over large flexible substrates: a review, IEEE Sens. J. 15 (6) (2015) 3164−3185.

[70] J.P. Pious, Y. Zwirner, H. Lai, S. Olthof, Q. Jeangros, et al., Revealing the role of tin fluoride additive in narrow bandgap Pb-Sn perovskites for highly efficient flexible all-perovskite tandem cells, ACS Appl. Mater. Interfaces 15 (7) (2023) 10150−10157.

[71] M. Hu, C. Bi, Y. Yuan, Y. Bai, J. Huang, Stabilized wide bandgap $MAPbBr_xI_{3-x}$ perovskite by enhanced grain size and improved crystallinity, Adv. Sci. 3 (6) (2016) 1500301.

[72] J. Xu, C.C. Boyd, Z.J. Yu, A.F. Palmstrom, D.J. Witter, et al., Triple-halide wide-band gap perovskites with suppressed phase segregation for efficient tandems, Science 367 (6482) (2020) 1097−1104.

[73] J. Cho, P.V. Kamat, How chloride suppresses photoinduced phase segregation in mixed halide perovskites, Chem. Mater. 32 (14) (2020) 6206−6212.

[74] Q. Han, Y. Wei, R. Lin, Z. Fang, K. Xiao, et al., Low-temperature processed inorganic hole transport layer for efficient and stable mixed Pb-Sn low-bandgap perovskite solar cells, Sci. Bull. 64 (19) (2019) 1399−1401.

CHAPTER SIXTEEN

Recent advances in perovskite-containing tandem structures

Maria Khalid[1], Tapas K. Mallick[1] and Senthilarasu Sundaram[2]
[1]Solar Energy Research Group, Environment and Sustainability Institute, University of Exeter, Penryn Campus, Cornwall, United Kingdom
[2]School of Computing, Engineering and Digital Technologies, Teesside University, Tees Valley, Middlesbrough, United Kingdom

16.1 Introduction

In the evolution of photovoltaic (PV) technology, tandem solar cells are the next stage owing to their exceptional achievement of higher power conversion efficiency (PCE) than single-junction solar cells, which are inherently limited. The PCE of a single-junction solar cell is constitutionally restricted by the compromise between absorption losses, where photons owing lower energies than the bandgap are not absorbed, and thermalization losses, where hot carriers excited by high-energy photons quickly thermalize to the bandgap energy of the absorber, delivering the additional energy as heat. As needed by the principle of detailed balance in 1961, Shockley and Queisser determined the theoretical limit of PCE for PN-junction solar cells to be around 33%, scrutinizing only radiative recombination [1]. A feasible method to overcome the Shockley and Queisser limit for a single-junction PV device is stacking different solar cells with varying band gaps in a multijunction or tandem configuration.

Presently, Si-based single-junction solar cells dominate; as they reach their theoretical efficiency limits, performance improvement has become more challenging. On the other hand, different materials combined into a tandem solar cell architecture are emerging due to having two different absorber materials in their architecture sharing solar spectrum and higher efficiency. Perovskite solar cells (PSCs) have emerged recently as an exciting option that combines low-cost and increased efficiency over the past decade [1,2]. Perovskite materials are cheap and have fascinating optoelectronic properties for integration, such as the solar absorber and lower recombination defect chances at the interface, ease of fabrication, and a wide choice of fabrication approaches. Thus perovskites can be used as promising candidates exhibiting distinct properties, as mentioned earlier, to fulfill the requirement to be a perfect material for tandem solar cells. In all these respects, perovskite-based tandem solar cells can achieve the highest efficiencies in the near future [3–5].

In the tandem structure, a wide bandgap solar cell is stacked on the top of the solar cell with a narrow bandgap. Thus the top solar cell with a wide bandgap absorbs the high-energy photons with lower thermalization losses and transmits low-energy photons into the bottom solar cell with a narrow bandgap. Theoretical aspects of tandem devices are somewhat outside the purview of this chapter; the reader is encouraged to consult a recent source that includes a thorough background including the basic physics [5].

One key result from analyzing the basic physics of such devices is that the series resistance (R_s) of a tandem device is optimally as low as practicable due to its bulk contribution to the reduction of PV performance of the solar cells, mainly fill factor (FF) and short circuit current (I_{sc}). Series resistance involves contact resistance at the interface between each layer and sheet resistance of the transparent electrode of the device. The parallel resistance or shunt resistance (R_P) is caused by manufacturing faults that produce alternate pathways of current, resulting in reduced short circuit current density (J_{sc}) and open circuit–voltage (V_{oc}).

In addition to the parameters mentioned above, the bandgap of the absorbers can further limit the J_{sc} and open circuit voltage (V_{oc}) of the solar cell. The limit of Shockley–Queisser occurs in the absence of nonradiative recombination for a single p–n junction in a cell with a 1.1–1.4 eV bandgap is approximately 33%, indicating an optical loss of about 67% [1,5]. It also dictates that most of the energy coming from the source does not contribute to producing electricity. The primary sources of loss can be the transmission of photons with lower energies than the material's bandgap without absorption and the energy loss by photons of energy higher than the bandgap of the material via phonon emission by the vibrational and rotational energy levels. The idealized process is shown in Fig. 16.1.

The manufacturing of tandem PV devices with insignificant realized losses has become an effective methodology for overcoming the SQ limitation of the stand-alone solar cell. This concept has been successfully

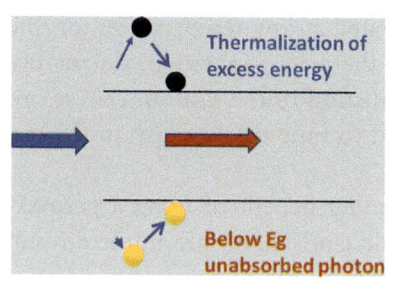

Figure 16.1 Thermalization process of the photo-induced electron and photon with lower energy than the bandgap (E_g) of the material.

realized utilizing expensive single-crystal III–V materials achieving an efficiency of up to 46% [6]. Polycrystalline inorganic thin-film solar cells have been used to develop low-cost tandems, despite the difficulty of producing tandems with a wide bandgap top solar cell [6]. Therefore the invention of the PSC has changed the research direction in this field due to its highly efficient and cost-effective thin-film PV [3–5].

Implementing low bandgap materials with hybrid perovskite material can boost the open-circuit voltage of both the subcells, leading to the achievement of high short circuit current by utilizing PV materials with higher absorption characteristics and simultaneously maintaining a broader solar spectrum. The unique architecture of a tandem solar cell by combining two, three, or more subcells by covering a wide spectral range to get more electricity with the top wide bandgap subcell and narrow bandgap for the bottom subcell. The increment in the number of stacking solar cells with varying band gaps significantly lifts the overall tandem device performance.

Organic–inorganic metal halide perovskites for solar cell applications have been the subject of intensive research due to their ease of fabrication and novel PV properties within the past decade. Organic–inorganic metal halide perovskites have some exceptional properties to compete with other PV systems:

1. These materials own extraordinary optoelectronic properties, effectively influencing PV active layer application and long carrier recombination lifetime, considerable carrier mobility, long carrier diffusion length, and defect tolerance resulting in an astonishing breakthrough.
2. These materials are Earth-abundant and offer cost-effective PV systems due to their recovery from other industrial products and materials.

3. They can be operated at low temperatures for spin coating processing, slot dies coating, thermal evaporation, doctor blading, and so on, to cut back the destruction to the bottom cells. Compositional engineering can be utilized to tune the bandgap in the range 1.17–3.10 eV to stabilize them.
4. The most outstanding benefit of using a perovskite layer for tandem structure is its facile fabrication onto numerous substrates due to defect tolerance and softness of the materials.

As additional breakthroughs in the potential of tandem configuration, PSCs are being fabricated and presented by researchers to maintain further improvement in enhancing the era by utilizing different processing and strategies to deposit the materials. Researchers have investigated several techniques, including solution-phase processing, hot casting, sequential deposition, vacuum-based deposition such as e-beam, RF-sputtering, and thermal evaporation, to fabricate different stacking of the tandem structure. The selection of the fabrication techniques depends on the materials chosen and their properties.

The perovskite-based tandem solar cells have effectively shown their prospect of converting sunlight into power more productively than stand-alone subcells. Notwithstanding, a great understanding of this fascinating technology and several challenges are needed for this class of materials, such as scaling up to large areas. Moreover, durability should be needed without compromising the system's performance to be operated at low temperatures for processing of spin coating, slot dies coating, thermal evaporation, and doctor blading to get to PCEs of more than 30%. Furthermore, the comprehension of this intriguing innovation must be widened to acknowledge the perception. According to Helmholtz–Zentrum Berlin (HZB), the realistic efficiency limit of the perovskite tandem configuration can have a bright future aspect to achieve over 30% PCE. Fig. 16.2 summarizes important milestones of perovskites in history and future direction with tandem configuration. In this chapter, the current status of different tandem architectures, perovskite/Si, perovskite/CIGS [$CuIn_{1-x}Ga_xSe2$ (also, copper indium gallium selenide)], and perovskite/perovskite, as well as other innovations in the field are, introduced and discussed.

16.2 An overview of tandem architectures

The important distinction for tandem solar cells is based on their device structure: (1) single package or two-terminal (2T), monolithic

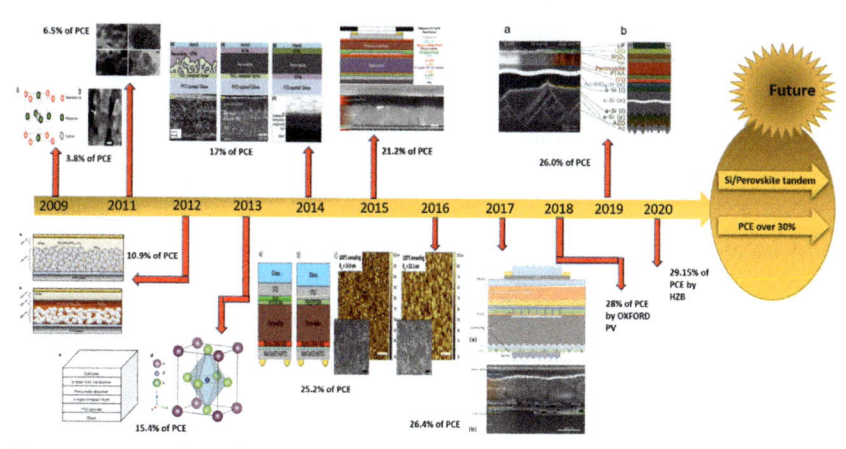

Figure 16.2 Timeline of important breakthroughs in perovskite solar cell technology.

design; (2) mechanically stacked cell separate subcells optically coupled (4T); or (3) three-terminal (3T). In 4T tandem solar cells, the top and the bottom cells are fabricated separately and mechanically placed upon one another. Multiple arrangements enable optimizing the efficiency of the tandem, but the integration cost due to processing complexity becomes high [7]. However, in 4T tandem solar cells, no electrical (and/or physical) coupling is involved between the subcells, which requires only the top cell to be semitransparent for optical coupling, making it easier to prototype. This is likely the reason why 4T tandem solar cells could achieve a high efficiency of 26.4% with the inclusion of efficient single-junction solar cells [8].

On the other hand, 2T tandem solar cell fabrication is more complicated often due to poor compatibility between every processing step within all interfaces and layers. The 2T tandems are fabricated on top of one another but need careful engineering of their components to achieve high efficiency. To utilize full theoretical efficiency, the subcells are integrated with the architecture required to connect in series with appropriate recombination layers. The function of the recombination layer is to make the electrical connection between both subcells in the series. It provides good selectivity for charge carriers and favorable energy levels for additional photovoltage and low series resistance without any losses and high transparency for bottom cells. For the fabrication of perovskite-based tandem solar cells, a recombination layer with highly doped transparent conducting oxide (TCO), silicon tunnel junction, and organic recombination layer can be used [9]. The recombination stack layer can act as a buffer layer to protect subshells from damage due to solvent and during the processing of the second subcell from surface treatment [10]. During the fabrication process, the involvement of temperature, solvent, and deposition techniques makes it complex. The incorporation of solvent resistance buffer layers is mandatory for all perovskite tandem solar cells due to the chances of damage by the solvent. This issue can be overcome by utilizing metal oxides either from solution or vacuum deposition processing [11,12]. Conformal deposition processes such as thermal evaporation and hybrid processes between thermal evaporation and infiltration are recommended to get fine surfaces of substrate used to fabricate 2T tandem solar cells [13].

In 2T tandem solar cells, subcells are interconnected in series, and these cells must show the same amount of photocurrent. However, the current mismatch can be compensated in part by an increase in FF [14]. The architecture of 2T tandem layers comprises two absorbing layers, (i.e., perovskite and Si) as well as several contact and charge extraction layers. Light-trapping schemes can enable the avoidance of parasitic losses; optical simulation can be utilized to discover such designs. Another essential aspect to consider is the tuning of optical bandgap by manipulating the composition of perovskite material, which gives captivating properties for different tandem cell applications [15,16]. The PCE ranging from 22% to 23.7% was reported by tuning the perovskite optical bandgap in several recent reports [17–19].

The parasitic absorption occurrence can complicate the realization of current matching, making it challenging to get high efficiency in 2T tandem solar cells. Several papers have reported fabricating tandem solar cells using different materials, including CIGS, GaAs, and CZTS [Cu_2ZnSnS_4 (also, copper zinc tin sulfide)]. At the same time, the incorporation of perovskite with Si PV can attain higher efficiency and trigger the low-cost PV technology to become favorable for large-scale commercialization. A 2T tandem solar cell with MoO_3 as a window layer and spiro-OMeTAD as a buffer layer was fabricated and achieved 16.8% efficiency [20].

To compare to inorganic hybrids, not involving a perovskite cell, Essig et al. fabricated 4T tandem solar cells by utilizing an n-Si absorber as the bottom subcell and InGaP, GaAs for top subcell, recording 32.8% overall efficiency [21]. A different approach used to fabricate a tandem cell involves utilizing Al_2O_3 as a surface passivating layer with an atomic layer deposition (ALD) method to obtain good results [22]. Wu et al. describe a tandem cell with Si as the bottom cell with the passivated front and rare rear surface, which helps maintain higher V_{oc}. An array of Cr/Pd/Ag metal stacks was applied, and they obtained a PCE of 22.5% on 1 cm^2 tandem solar cells [23]. Zheng et al. fabricated 2T tandem solar cells with a novel metal grid; the top cell was a perovskite (Cs$_{0.17}$FA$_{0.6}$Pb(Br$_{0.17}$I$_{0.7}$)$_3$) material and a homojunction silicon was used as the bottom subcell to obtain an efficiency of 21.8% [24]. Nanocrystalline silicon (nc-Si:H) was used to fabricate the recombination layer to integrate into the design of Si heterojunction tandem solar cell and obtained an efficiency of 22.7% [25]. A new configuration of Se/Ti/Se was utilized to fabricate tandem solar cells with CZTS as the top subcell and 30.2% efficiency was achieved [26]. Indium zinc oxide (IZO) has been employed in a tandem device to realize a J_{sc} above 19 mA/cm^2; tuning different layers of perovskite/Si tandems (mainly the perovskite absorber), a PCE of 26% could be obtained [12]. A comparison of different tandem architectures is shown in Fig. 16.3 [7].

Recently, researchers from the Swiss Center for Electronics and Microtechnology and École Polytechnique fédérale de Lausanne presented a study on a perovskite silicon tandem structure of 1 cm^2 area with an efficiency of 31.25% [27]. A hybrid vapor deposition technique was used in the research. The efficiency was certified by the US Department of Energy's National Renewable Energy Laboratory. However, the technical approach of the research was not disclosed by the team. Compared to the perovskite/Si tandem structure, perovskite/CIGS tandem structure shares a small market in PV field. The possible reason is the challenging integration of the material on the top of

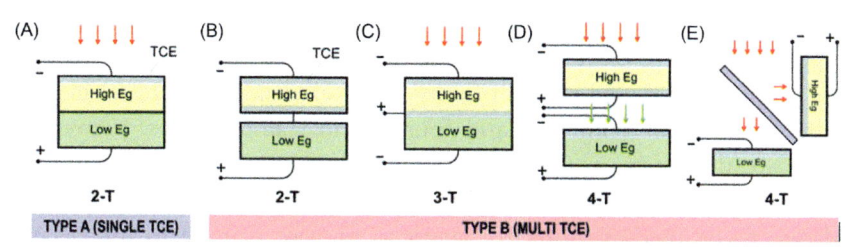

Figure 16.3 Different types of tandem solar cell configurations: (A) two-terminal (2T) with a single TCE (transparent conducting electrode), multi-TCE devices—(B) 2T mechanically stacked, (D) four-terminal (4T) mechanically stacked, (C) three-terminal monolithic stack, and (E) 4T spectrum-split. *Reproduced with permission from T. Todorov, O. Gunawan, S. Guha, A road towards 25% efficiency and beyond: Perovskite tandem solar cells, Mol. Syst. Des. Eng. 1 (2016) 370–376, copyright (2016) Elsevier.*

perovskite and the low PCE. The highest efficiency achieved by CIGS/perovskite tandem structure is 24.2%. These results describe the guidelines for optical measurements.

16.3 Challenges with tandem structures

In the implementation of tandem architectures with perovskites, several inherent problems are reported. One of the significant challenges is the health and safety issues of perovskite materials that contain lead (Pb). The hazardous nature of Pb affects commercial potential, especially in the presence of moisture. Moreover, high temperature is also not adequate for perovskite materials. The use of tin (Sn), an Earth-abundant nontoxic metal, within the perovskite layer, is emerging as a viable alternative to Pb. Another material, the solvent dimethyl formamide, also has been reported as highly harmful to human health. Therefore more progress and development of nontoxic materials as absorbing layers in tandem structures should be high on a researcher's priority list. The material's potential toxicity reduces its viability to compete on the commercial market with less hazardous alternatives such as silicon, thin film, and polymer-based solar cells.

Moreover, different layers in PSC, such as carrier transportation, buffer layers, and the interconnection of the independent layers, can also raise stability concerns. The instability of the PSC is enhanced in the presence of moisture, but it can also show self-degradation when fabricated in a controlled environment (inert gas environment). The phenomenon was explained by Back et al. whose work demonstrated that unique defects appear within perovskites by corrosion of metal electrodes [28].

The fabrication and choice of transparent electrodes are crucial in tandem structure. In both two and four-terminal architectures, perovskite layers are deposited on the top to allow lower-energy photons to reach the bottom subcell in the device. Barrier layers as sputter buffers are advised to be used as protective layers to prevent damage to underlying layers when processing the top layers. The average thickness of these layers (typically 5–30 nm) does not negatively interfere with electron (and hole) transport within the cell. The use of a MoO_x layer on the top of spiro-OMeTAD as a hole transport layer (HTL) is reported to achieve 26.4% efficiency for 4T perovskite/Si tandem structure [29]. Gold (Au) and Cu/Au are also used as thin metal layers with easy processing advantages [30]. However, strong parasitic absorption is a major disadvantage in the top subcell. Some suitable materials such as NiO_x and VO_x could be used as protective layers. Top cell/bottom cell interfaces are essential to minimize losses by depositing multiple layers stack such as antireflection coating to allow the refractive index to change from TCO layer to air. The use of a HTL and electron transport layer (ETL) with closely matched refractive index to perovskite can also reduce reflection losses. In 2T architecture, a recombination layer is mandatory, which allows current flow between the top and bottom subcells. The tandem set structure based on organic material processed by the solution method utilizes high-doped layers such as PEDOT:PSS or ZnO [31,32]. This approach was adopted to fabricate a perovskite/perovskite tandem structure.

Another solution is the use of TCO as a recombination layer such as indium tin oxide (ITO) or IZO. These highly conductive layers can limit the device's upscaling by making the carrier more disposed to direct the existing shunt pathway with the recombination layer. Some work has been reported to find solutions using n^+- and p^+-doped layers as recombination layers via the chemical vapor deposition (CVD) approach to fabricate large-area tandem solar cells [33]. The compatability during the processing of tandem solar cells are determined by the first subcell must be intact during the fabrication process of the second subcell. In Si/perovskite tandem architecture, Si is always the bottom or substrate cell, but Si can stay stable up to 500°C. Therefore the rest of the process for PSC can be performed using low-temperature processing. However, the perovskite/perovskite tandem structure is quite different to process due to its sensitivity to the high temperature. The use of the solution process to fabricate the perovskite layer is recommended to deposit the recombination layer or protective layer to avoid losses. Some vapor deposition techniques for perovskite layer processing could be exciting for upscaling and developing high-performance tandem solar cells.

Optical and electronic properties of perovskites need to be considered for optimization of cell performance and some important factors that hinder the device stability against degradation and carrier transportation. The bandgap of the PV material is known as an essential parameter. The maximum theoretical solar energy efficiency can be achieved based on the material's bandgap, according to SQ limit, and the quantum efficiency of the PV material can be measured rather than measuring the absorption coefficient. Maximum theoretical efficiency by a single-junction PSC can be 31% (SQ limit), while tandem architecture solar cell, with high bandgap PSC and lower bandgap of either CIGS or silicon, can be approximately 44%. Moreover, the density of defects within the bandgap of perovskite cells is another considerable parameter as the recombination mechanism within the perovskite film is dependent on the trap density of states. The absorption coefficient of perovskite film must be around 10^5-10^6/cm, which is higher than the c-Si solar cell, indicating high crystallinity of the PSC.

The high absorption coefficient of perovskite with a thin layer of about 100 nm—1 μm is significant enough to absorb light and give output efficiently. The dielectric constant is also a fundamental characteristic of a light-absorbing layer. The absorbing layer's exciton binding energy is responsible for whether a free hole and electron are drawn in a photon absorption event. The high carrier lifetime and mobility of the perovskite result in a high diffusion length of the PSC film and consequently high-efficiency output.

16.3.1 Parasitic absorption

In perovskite tandem cells, the absorbance of photons by nonphotoactive layers such as a transparent electrode, antireflective layers, ETL, HTL, and bottom electrode, which do not contribute to photocurrent, causes parasitic absorption. The experimental characterization of parasitic absorption loss can be estimated by the difference between the absorbance spectra and the external quantum efficiency (EQE). One of the primary parasitic absorption losses is in the transparent electrode, which is an unavoidable channel that can be mitigated by using low carrier concentration. However, low carrier concentration can lead to enhanced sheet resistance and consequently electrical loss. Another approach is thin ITO to reduce parasitic absorption loss [34]. Mechanically stacked tandem structures with three or four electrodes are recommended to mitigate parasitic loss and achieve higher efficiency.

16.3.2 Reflection losses

The mismatch in the refractive index (n) of the different layers within tandem structure reflection loss occurs, leading to power losses in tandem solar cells. The significant difference between the refractive index of air ($n = 1$), glass ($n = 1.52$), and transparent electrodes originates considerable reflection losses. This reflection loss is consequential at the front electrode.

Some light is reflected away from the higher layer without taking part in conversion to the photocurrent. The use of a nanocrystalline Si junction in Si/perovskite tandem solar cells can limit the parasitic absorption and reflection losses in the spectrum within red and infrared light regions, consequently producing more photocurrent in bottom Si solar cells.

In the 4T tandem design, two subcells are electrically isolated and physically separated. However, optical coupling helps to avoid reflection losses at the interface of the subcells. The coupling material is advised to have a refractive index close to the transparent electrode utilized at interfaces, and the material should be insulating and transparent. Another approach is the use of textured surfaces for light trapping to minimize reflection loss. However, the fabrication of PSC on the top of textured surface Si is still a challenge for scientists, especially in tandem structures. The antireflection coating is another effective strategy to reduce optical reflection loss by depositing the quarter wavelength film thickness rule. The deposition of the perovskite layer on the front side of textured Si would give low reflection loss with ultimately high photocurrent generation. In some cases, practically front-side texturing is challenging to do directly. In such cases, a micro-textured foil can be laminated on the front side of the electrode [20].

16.4 Power losses in perovskite tandem design

The understanding of power loss in a tandem device is crucial in the designing and fabrication process to improve performance. In tandem configuration, there are more power loss channels as compared to the single-junction solar cells. In 2T architecture, series resistance or electrical losses, such as current mismatching, are considerable, which results in low performance of the devices. Therefore the understanding of power loss and factors that give rise to losses are important to consider boosting tandem device performance.

The sheet resistance of the transparent electrode and charge carrier layers (CCLs) could play a pivotal role in determining the tandem devices' total PCE. The total electrical power loss by sheet resistance is proportional to the square of the current that flows through contacts and the sheet resistance. A trade-off relationship between sheet resistance and minimum carrier concentration is required for the suitable selection and design of the tandem solar cell. Reducing the thickness of transparent electrodes can reduce the parasitic absorption, but usually leads to increased sheet resistance increasing the FF loss in tandem devices.

Typically, the individual photocurrent of the unfiltered Si subcell and opaque top perovskite subcell is greater than the photocurrent of the tandem solar device. The smaller value of photocurrent in a tandem device provides flexibility to mitigate the electrical loss and gives more chance for the trade-off between parasitic absorption and electrical loss. In a monolithic tandem device, the current mismatch between two subcells leads to extensive power loss. The photocurrent of the tandem device is determined by the photocurrent of the subcell with a smaller value; therefore it is essential to match the current across the subcells to achieve an efficient performance of the device. The current match across the subcells of the tandem device is based on the interconnecting layers, including electron and HTLs, thereby cause shunt and series resistance in the device. To reduce power loss, it is critical to address material development and device configuration; carrier transport optimization also plays an important role. An effective method is to introduce a wide bandgap photoactive layer to maximize the transmission of light to the bottom subcell.

16.5 Tandem architecture(s) advantages

The introduction of tandem architecture to the PV field is due to the lack of a single-junction solar cell to effectively capture incident light and convert it to electricity. Some reasons reported by researchers are an inappropriate bandgap of the material or a mismatch of the photon energy.

In the perovskite tandem configuration, usually, the perovskite with a larger bandgap constitutes the upper subcell where incident light strikes first and transmits low-energy photons to the underlying bottom subcell with a lower bandgap, generating additional voltage to give high-performance solar cells. Perovskites as light absorbers are active up to 800 nm wavelength, whereas other semiconducting materials including germanium and Si stay active up to 1800 nm wavelength. Combining both short and long-wavelength absorbers enables a more efficient use of the solar spectrum resulting in better performance. The efficiency potential of tandem architecture makes them a preferable candidate to reduce the price of electricity.

16.6 Tandem configurations

Tandem solar cells can be configured in several architectures, such as 2T, 4T, and 3T. Each of these devices and architectures has its own individual advantages and disadvantages. These will be addressed in the remainder of this section.

16.6.1 Fabrication of four-terminal tandems

A typical four-tandem solar cell comprises two independent stacked subcells on top of each other, connected by an outer circuit. The sum of both the subcells determines the total output of the tandem device. These cells have the apparent benefits of an easy fabrication process, allowing optimal fabrication conditions to process simply by maintaining cell polarity, temperature, or substrate roughness. However, the stacking process results in challenges to minimize the losses within each subcell.

Most of the highly effective 4T perovskite tandem solar cells are fabricated with **Glass/ITO/HTL r**ecombination layer/perovskite absorber (MA, FA)/buffer layer/top ITO contact [11,35–37].

The 4T configuration is mechanically or optically stacked with voltage-matched perovskite with the bottom silicon cell. In this case, the efficiency of the device is limited by the lower voltage of the cell. As the open circuit voltage of the cell varies only logarithmically with exposure to light intensity and linear changes in short circuit current clearly indicate the advantage of mechanical or optical integration of the solar cell rather than monolithic integration of the solar cells. Therefore this type of module remains more stable than 2T against temperature variations, which are monolithically integrated, due to higher recombination losses. In addition, the 4T configuration can be fabricated and mechanically stacked by splitting spectrum. Mechanically stacked is explained earlier in which both subcells produce electrical output independently requiring for terminals to flow current. In a 4T spectral split device, a dichroic filter is needed for splitting the light spectrum. This type of system offers ease of fabrication of individual subcell to work and gives output as a stand-alone solar cell requiring only two terminals; not 3T like in standard 4T structure. The highest PCE for 4T terminal tandem configuration is predicted to be 46%, exceeding the SQ limit.

In this method, there are two ways to manipulate light. Firstly, the sunlight energy can be filtered above the solar cell, providing a response threshold to the target cell, where the rest of the light transmits to the next filter where the response of splitting is repeated. The second way is stacking of cells on one another with the cell giving the highest energy output on the top. A practical approach of optical splitting measurement of tandem solar cells was introduced in 2014 with 550 nm wavelength at 45 degrees position with both subcells. A shorter wavelength of 550 nm was reflected to the top subcell from the optical splitter, whereas light with a long wavelength was transmitted to the bottom silicon subcell, giving a total PCE of 28% for optically coupled perovskite/Si tandem solar cell [38]. The increment in cut-off wavelength was not found to be beneficial for performance of the device due to the smaller EQE of the top subcell as compared to the bottom subcell. Therefore it is recommended to enhance EQE of the top PSC by about 30% for optically coupled tandem devices. Moreover, the increment of V_{oc} in the top PSC cell can consequently lead to improved performance without compromising EQE and J_{sc}.

16.6.2 Performance of four-terminal tandems

A 28% PCE for 4T perovskite/CIGS with top perovskite subcell (E_g of 1.59 eV) and bottom CIGS subcell (E_g of 1.20 eV) using the optical splitting approach with 775 nm wavelength is reported as shown in Fig. 16.4. The perovskite/CIGS tandem configuration is quite more flexible than perovskite/Si. Optical splitting is not very popular, due to the additional cost of filters, and does not have commercial potential, but this work clearly indicates the superiority of CIGS to Si in tandem devices [39].

Reflective tandem can possibly be a good alternative to dealing with high costs. This concept offers great flexibility for combining with other PV technologies, such as solar thermal collectors. However, a drawback of the need of solar tracking where collection of diffuse light in the solar spectrum makes performance poor by module soiling. Therefore the focus of the researchers is usually on planar 4T stacked tandem devices. A summary of 4T perovskite tandem devices is shown in Table 16.1 [36,40-45].

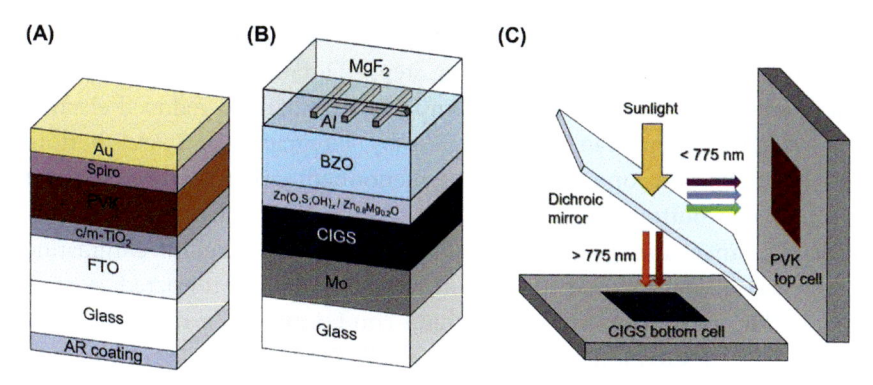

Figure 16.4 Schematic description of tandem solar cells: (A) perovskite top cell, (B) CIGS bottom cell, and (C) device with spectrum splitting system and a dichroic mirror. *Reproduced with permission from M. Nakamura, K. Tada, T. Kinoshita, T. Bessho, C. Nishiyama, et al., Perovskite/CIGS spectral splitting double junction solar cell with 28% power conversion efficiency, iScience 23 (2020) 101817, copyright (2020) the authors, published by Elsevier.*

Table 16.1 Example four-terminal perovskite tandem devices from the literature.

Device structure	Active area (cm^2)	PCE (%)	References
HIT c-Si: FTO/c-TiO$_2$/ MAPbI$_3$/spiro-OMETAD/CVD graphene/Au grid	0.16	13.2	[40]
ITO/PEDOT:PSS/FA$_{0.75}$Cs$_{0.25}$Sn$_{0.5}$Pb$_{0.5}$I$_3$/C$_{60}$/BCP/Ag	0.2	20.3	[41]
FACsPbI$_{3-x}$Br$_x$	0.09	19.8	[42]
MAPbI$_3$	0.25	20.1	[43]
HIT c-Si: ITO/SnO$_2$/PEIE/PCBM/ MAPbI$_3$/spiro-OMETAD/MoO$_x$/IO:H/ITO/Au grid	1	23.0	[36]
MAPbI$_3$	4	20.2	[44]
HIT c-Si: ITO/PTTA/MAPbI$_3$/ PCBM/C$_{60}$/BCP/Cu/Au	0.075	23.0	[45]

16.6.3 Two-terminal tandems

A 2T tandem device is the most common configuration; two subcells are optically stacked and connected electrically in series. The voltage of both subcells is added, and the photocurrent of both subcells is limited due to the low production of current. Therefore mismatch of current generated by changes in solar spectrum results in overall poor performance of the device. Compared to the 4T architecture, the 2T architecture requires only one transparent electrode, leading to low(er) manufacturing costs due to fewer materials and deposition steps. 2T tandems also have practical potential due to less parasitic absorption attributed to reduced electrodes. A 2T tandem can be fabricated by integrating top and bottom subcells with only one electrode, as mentioned earlier, or as a stack of two separate bottom and top cells connected in series having two terminals. The mechanically stacked tandem has the advantage of eliminating interfacial tunneling, and current matching is not as restricted as in a monolithic stack. The current matching condition can be realized by the adjustment of variation in areas of the cell. Strict current matching and compatibility, especially for the fabrication process, are needed in 2T configurations. Optimizing the absorber layer to make its current match in the spectrum is essential to mitigate the limitations due to lower current delivery to the subcell.

Moreover, due to the presence of silicon cells, low-temperature processing of the top perovskite cell is required because of the ITO layer on the silicon cell. Both subcell work functions need to be matched; therefore density and recombination layers need specific requirements. The first monolithic-based perovskite tandem structure was introduced in October 2014 with a CZTS bottom subcell with 4.4% efficiency [46]. This research was followed by the publication of a 4T configuration where both top and bottom subcells were connected separately with bottom cell silicon and CIGS [47]. However, the drawbacks of more electrodes than monolithic stacks include increased optical losses. Various 2T tandem structures have been reported to date. For example, a set of PSC/CIGS and PSC/CZTS-based tandem devices were constructed. A recombination layer of PEDOT:PSS/ITO was used in the configuration as shown in Fig. 16.5. Perovskite material (MAPb(I_xBr$_{1-x}$)$_3$) with a wide bandgap was used as the top subcell. The absorber layer was fabricated by utilizing a halide exchange reaction between the MAPbI$_3$ thin film and MABr vapor [48]. Individual cell PCE was reported as 11.4% for the combined CIGS top cell and PSC bottom cell. The monolithic tandem device was found to deliver 10.98% of PCE for PSC/CIGS, possibly limited by lower current density originated by optical losses in the top electrode. Furthermore, they estimated a PCE of 15.9% for a tandem solar device by assuming 100% transmission from the top electrode.

In 2T architecture, processing of temperature and solvent compatibility need to be considered carefully. High temperature is not suggested for the annealing process and both the top layer and bottom layer should not

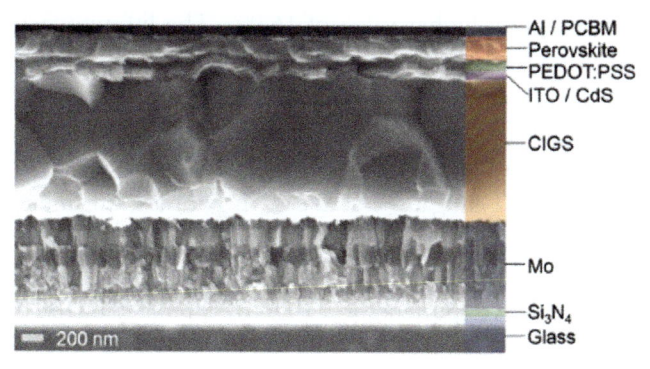

Figure 16.5 Scanning electron micrograph of a perovskite-CIGS tandem solar cell with device configuration Al/PCBM/perovskite/PEDOT:PSS/ITO/CdS/CIGS/Mo/Si$_3$N$_4$/glass. Reproduced with permission from C.D. Bailie, M.G. Christoforo, J.P. Mailoa, A.R. Bowring, E.L. Unger, et al., Semi-transparent perovskite solar cells for tandems with silicon and CIGS, Energy Environ. Sci. 8 (2015) 956–963, copyright (2015) Royal Society of Chemistry.

dissolve with each other. Some fabrication techniques such as physical vapor deposition, ALD, chemical vapor deposition, pulsed layer deposition, and thermal evaporation are reported to be used to cope with such interlayer problems of perovskite layer and fabrication of transparent electrode as well. ITO is one of the significant choices to fabricate transparent electrodes although it is not very popular in perovskite/perovskite tandem structures.

A polymer-based tandem structure with thermal treatment has also been reported. The fabrication of PSC on top of polymer-based cells requires low-temperature processing as postannealing can damage the polymer material. Therefore the use of narrow bandgap polymers to use as bottom subcell with excellent thermal sustainability is required to limit the damage. A polymer PBSeDTEG8 having a 1.31 eV bandgap was developed. The perovskite thin layer was crystallized rapidly using a solvent wash method followed by 100°C of postdeposition annealing to make it compatible with the IR polymer bottom subcell. A PCE of 10.23%, FF of 65%, V_{oc} of 1.52 V, and J_{sc} of 10.05 mA/cm^2 were measured for this PSC/polymer-based monolithic tandem device [11,35]. An alternative tandem device with a polymer-based cell as the top subcell was explored to avoid the postannealing process. The bandgap was similar for both the absorbers in the tandem cell of 1.55 eV. For an individual PSC architecture, an efficiency of 15.6% was measured; a tandem solar cell could achieve a slightly higher value of 16.0%. The limiting factor of low efficiency for polymer-based solar cells is relatively low EQE [11,35].

Recently, Snaith and coworkers presented $(FA_{0.83}Cs_{0.17}Pb(I_yBr_{1-y})_3)$ perovskite materials with n-butylammonium cations. The device showed 1.75% ± 1.3% for perovskite with a 1.61 eV bandgap and 15.8% ± 0.8% for perovskite with a bandgap of 1.72 eV. In contrast, an unencapsulated device retained 80% of PCE after 1000 h at the ambient environment and 80% PCE for encapsulation after 4000 h, as shown in Fig. 16.6 [49].

A 26.7% PCE was reported for 2T perovskite/Si tandem devices by Shin and coworkers. The initial PCE of the device was 20.7%, which was retained at 80% of it after continuous exposure to light for 1000 h [50]. The structural and electrical properties of the 2D passivated layer with PbI_2 addition were difficult to control during anion engineering. A summary of 2T perovskite tandem devices is shown in Table 16.2 [20,29,31,43,51].

16.6.4 Three-terminal tandems

Recently, 3T architecture has been introduced and sparked the interest of researchers. The purpose of this architecture is to overcome the drawbacks of the 2T tandem devices, specifically current mismatching and optical

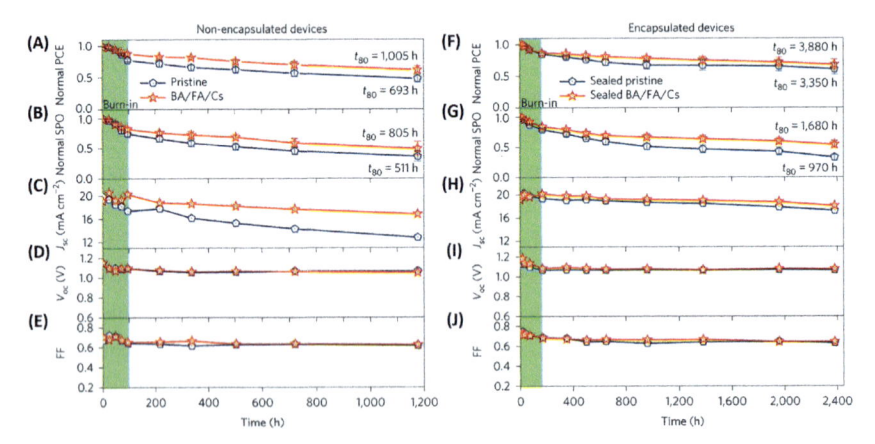

Figure 16.6 Comparison of stability of nonencapsulated and encapsulated high-performance devices: (A–E) nonencapsulated and (F–J) encapsulated devices. *Reproduced with permission from Z. Wang, Q. Lin, F.P. Chmiel, N. Sakai, L.M. Herz, H.J. Snaith, Efficient ambient-air-stable solar cells with 2D-3D heterostructured butylammonium-caesium-formamidinium lead halide perovskites, Nat. Energy 2 (2017) 1–10, copyright (2017) Springer Nature.*

Table 16.2 Summary of two-terminal perovskite tandem devices from the literature.

Device structure	Active area	J_{sc} (mA/cm^2)	V_{oc} (V)	Fill factor (%)	PCE (%)	References
Perovskite/perovskite (2T) illumination from bottom	10 mm^2	5.15	1.88	54	5.2	[31]
Perovskite/perovskite (2T) illumination from top	10 mm^2	6.61	1.89	56	7.0	[31]
FACsPbI$_{3-x}$Br$_x$	0.12 cm^2	14.0	1.78	79.5	18.1	[20]
MAPbI$_3$	0.25 cm^2	20	0.74	82	31	[43]
MAPbI$_3$	1.43 cm^2	36.19	1.14	80	25.57	[51]
MAPbI$_3$	0.25 cm^2	-	-	77	18.2	[29]

losses in 4T tandem devices. The concept of fabrication of tandem is an exciting step toward the betterment of the energy yield of tandem structures and accurate measuring of the PV parameters. These tandem solar cells operate in parallel connection with either cathode or anode. In this structure, the limitation of the voltage occurs by the lowest voltage, and the total current achieved is some of the individual currents by each subcell. The bottom subcell limits the V_{oc}, so it has a fixed value, and the total efficiency of this kind of tandem architecture is independent of the bandgap of the top cell. The 3T configuration shares common features with 2T and 4T devices, such as layer stack (2T) and independent electrically connected subcells (4T).

A numerical study of 3T perovskite/Si tandem solar cells was proposed by combining 2T and 4T architecture advantages. The tandem architecture was found to give the highest energy yield due to a larger photogenerated current as compared to 4T and better response to varying spectra as compared to 2T tandem indicating promising results [52]. Another 3T tandem was introduced by adding a thin recombination layer as the third contact to measure 2T perovskite/Si Tandem solar cell characterization for further optimization such as bandgap and optical absorption engineering was performed to reduce reflection losses and current matching. The champion efficiency of the device was 23.5%, with remarkable stability of maintaining 97% of its initial efficiency after 100 days [53]. A simulation study of multijunction based on 3T tandem balls was demonstrated to mitigate the limitations of 2T and 4T architecture. The 3T perovskite/Si tandem device exhibited a more promising potential than 2T and 4T tandem but was limited by $p-n/n-p$ configuration. The optical losses of the work are not analyzed in detail.

Another simulation study to demonstrate an optical model (GENPRO4) optimized 3T perovskite/Si tandem device. The tuning of the thickness of ITO, ETL, and HTL was performed to capture the maximum of incident light into the absorber layers. This study shows that if HTL is in front of a light-absorbing layer, it can originate parasitic absorption loss. If HTL is behind the light-absorbing layer, then parasitic absorption losses can be avoided, but reflection losses increase in this case. An interlayer of MgF_2 between top and bottom subcells was added to decrease reflection losses. The photocurrent density (J_{ph})of both subcells increased considerably to 1.6 mA/cm² for the top cell and 5.1 mA/cm² for the bottom cell resulting in total J_{ph} increase from 34.4 to 41.1 mA/cm². An equivalent circuit model was used to determine V_{oc} responding efficiency of the tandem device. The thicker perovskite layer showed a higher efficiency of around 32% as shown in Fig. 16.7 [54].

An interdigitated back contact 3T tandem concept was proposed and reported to achieve a stabilized PCE of 17.1%. Bias-dependent $J-V$ characterization was performed to investigate the mutual dependency of top and bottom subcells. By changing bias from V_{oc} to J_{sc}, a slight reduction of PV performance was observed. After optimization, 26.9% of efficiency was achieved [53].

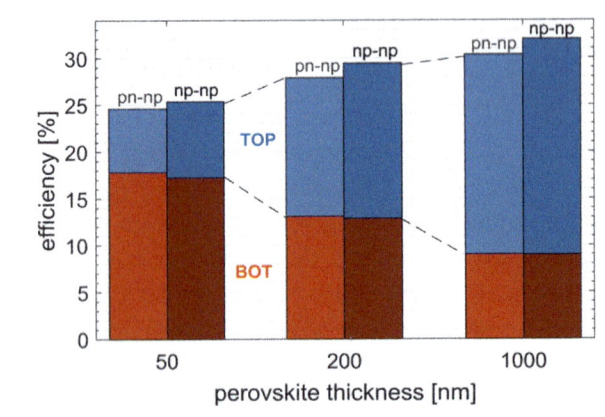

Figure 16.7 Power conversion efficiency of both top perovskite cell (blue) and bottom silicon cell (red) as a function of perovskite thickness. *Reproduced with permission from R. Santbergen, H. Uzu, K. Yamamoto, M. Zeman, Optimization of three-terminal perovskite/silicon tandem solar cells, IEEE J Photovoltaics 9 (2019) 446–451, copyright (2019) IEEE.*

16.7 Perovskite/Si tandem solar cells

After achieving excellent output from semitransparent PSCs with wide bandgap, promising light-harvesting properties, and high conductivity to mitigate the losses, tender architecture is a prerequisite and new challenge for researchers by combining suitable bandgap material to obtain considerable PCE. Among all the PV materials, Si's best solar cells are currently sharing 90% of the total global PV market with mature manufacturing and cost-effective strategies at the module level due to their excellent efficiency and high stability nature.

In a perovskite/Si random architecture, usually, the top subcell is perovskite with a wide bandgap to absorb high-energy photons and convert them at a high voltage. In contrast, the bottom Si subcell absorbs IR photons covering a broad spectrum to give the overall high output of the solar cell. The notable enhancement in tandem architecture is possible for several reasons:

1. Bandgap tunability of the perovskite subcell with a bandgap of about 1.55 eV to optimal bandgap that is needed for the top cell of 1.67−1.75 eV.
2. Light management of the device to minimize electrical and optical losses.
3. Appropriate configuration of both the subcells.
4. Optimization of the recombination layer in tandem architecture for effective subcell connection and to avoid optical losses and minimal shunt losses.
5. Compositional engineering to improve the stability of the device.
6. Modification of CCLs to reduce recombination losses between interface and CCL.

A PSC is suitable as the top subcell with both planar and inverted configuration with Si as the bottom cell. The regular ($n-i-p$) configuration of PSC at a high temperature of approximately 400°C can show decent PCE as the top subcell specifically in 4T tandem devices. The introduction of c-Si as as bottom cells to fabricate tandem devices is feasible for several reasons:

1. Bandgap energy of 1.1 eV, which is appropriate to match the current with a large bandgap of 1.75 eV of the top perovskite cell.
2. Si solar cell is an excellent substrate to deposit top PSC, and for the formation of heterojunction, it can easily be altered.
3. It is commercially available easily with promising efficiency.

At present, the highest tandem solar cell with a PCE of 29.1% is reported by HZB [55]. The first perovskite/Si tandem device was realized in 2015 by Mailoa et al. by utilizing n-type Si as the bottom cell [56]. The top and bottom cells were interconnected with tunnel junctions of highly doped c-Si layers, and flat TiO_2 was deposited using the ALD approach. Meso-TiO_2 was used as ETL in the configuration, followed by $MAPbI_3$ as a light-absorbing layer and top contact of spiro-OMeTAD and silver nanowires. The device achieved a PCE of 13.7% with a low V_{oc} of 1.58 V and J_{sc} of 11.5 mA/cm^2. The low value of V_{oc} and J_{sc} is a possible reason for the limited PCE of the device. The thickness of doped spiro-OMeTAD leads to parasitic absorption loss, limiting the J_{sc} of the device [56].

Several reports are presented using monolithic, 4T, and 3T configurations with perovskite/Si tandem devices concerning different aspects such as selecting TCO layer material and buffer layers to protect against damage and charge transportation within layers. Different buffer layers such as V_2O_5 and MoO_x as HTL can be deposited at low temperatures without further postannealing. A sputtering buffer layer of AZO nanoparticles was applied to enable ITO deposition using the solution process. The PSC was found to achieve a PCE of 12.3% with improved environment and thermal stability. After stacking with a c-Si solar cell in 4T architecture, the device showed 18% PCE. Some optical losses due to reflection were noticeable in planar tandem configuration [57].

Bush et al. reported a certified 2T perovskite/Si tandem device with an impressive PCE of 23.6% with an active area of 1 cm^2 for half an hour's continuous light exposure and maintaining performance. ALD was employed to deposit a SnO buffer layer; an ETL was directly deposited on the top of the perovskite layer in the regular $p-i-n$ configuration. The device, when measured, exhibited a FF of 79%, V_{oc} of 1.65 V, and J_{sc} of 18.1 mA/cm^2. This work observed the bottom Si subcell owing a high value of EQE of approximately 90% in the wavelength range of 800−875 nm and interlayer of ITO (20 nm) between the top and bottom subcells [58]. A 16 cm^2 large device with interconnecting layer of SnO_2 with top PSC and bottom Si-based tandem was demonstrated using a spin-coated process. A PCE of 17.6% with reverse scan and 17.1% of PCE based on maximum power point (MPP)-tracking were shown on a 16 cm^2 device. Whereas, for a 4 cm^2 active area device, a PCE of 21% for reverse scan, and finally, stabilized PCE of 20.5% was achieved [59]. The same group demonstrated a 21.8% PCE with a 16 cm^2 device area utilizing perovskite material $(FAPbI_3)_{0.83}(MAPbBr_3)_{0.17}$ as shown in Fig. 16.8 [24].

The parasitic absorption loss can be reduced with a thinner SnO_2 layer or by coupling it with any other p-type heterojunction contact. Implementation of the single high-quality transparent layer of graphene upon spiro-OMeTAD was laminated. The device enabled 6.2% of PCE without adding any auxiliary layer due to van der Waals forces that ensure the intimate and adhesion of graphene deposition on the spiro-OMeTAD layer. The device exhibited 13.2% of PCE when assembling as 4T tandem configuration, which was relatively higher than a single-junction c-Si solar cell (10.2%) with Au contacts [40]. To obtain promising performance of PSC on the top of Si solar cell using the spin coating process, the front

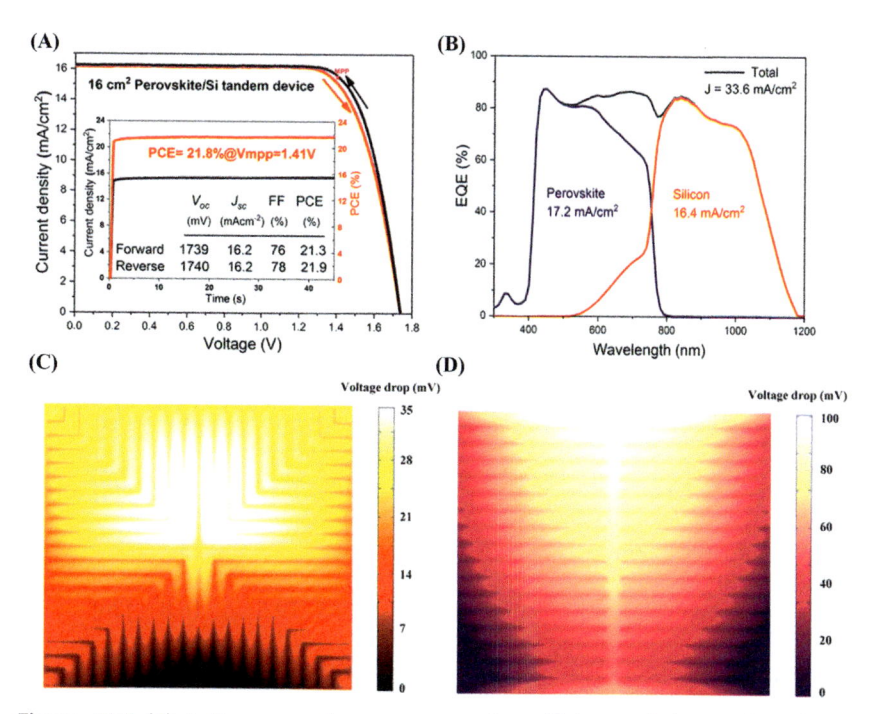

Figure 16.8 (A) J–V curve and power conversion efficiency of the tandem device with 16 cm² area. (B) EQE of the champion device. Simulated voltage drops across the perovskite solar cells with (C) new cell and (D) old cell with metal grid design. Reproduced with permission from J. Zheng, H. Mehrvarz, F.J. Ma, C.F.J. Lau, M.A. Green, et al., 21.8% efficient monolithic perovskite/homo-junction-silicon tandem solar cell on 16 cm², ACS Energy Lett. 3 (2018) 2299–2300, copyright (2018) American Chemical Society.

sides of the bottom Si cell are polished. Usually, SiN_x is used to polish the front side to better match the c-Si wafer's refractive index and to minimize the reflection losses. Whereas, the rear side of Si cell is textured polished with SiN_x having a low refractive index to improve light trapping. A considerable challenge to using SiN_x or other dielectric layers such as Al_2O_3 to polish or make textured Si solar cells is their good passivation, low reflectivity, and good insulation nature.

Few reports are present to circumvent this issue by replacing the dielectric layer with a recombination layer or removing it completely. For this purpose, the metal stack of Cr/Pd/Ag is used to contact opening and mitigate resistance between the Br-containing emitter and the ITO layer. For the rear contact opening, the same metal stack is used. The tandem device showed 22.5% stabilized PCE after 1200 s of light exposure [23]. The metal shading, reflection, and parasitic absorption losses contributed to lower J_{sc} being reduced by coping with these losses. A summary of perovskite/silicon tandem devices is shown in Table 16.3 [20,28,45,55,56,58,60−63]. A very recent review addresses current and future issues for improving 2T and 4T c-Si/perovskite tandem devices [64].

Table 16.3 A summary of perovskite/silicon tandem devices.

Device structure	Active area (cm^2)	J_{sc} (mA/cm^2)	V_{oc} (V)	Fill factor (%)	PCE (%)	References
Tandem perovskite/Si cell (4 T)	0.3	19.4	1.13	70	15.4	[29]
Perovskite/Si tandem device (2T)	1	18.1	1.65	79	23.6	[58]
Perovskite/Si tandem (4T)	4	39.0	1.08	77	23.0	[45]
LiF/Ag/spiro/perovskite/m-TiO$_2$/Si (2T)	1	11.5	1.58	75	13.7	[56]
Tandem perovskite/Si (2T)	−	11.8	1785	79.5	16.8	[20]
Tandem perovskite/Si (4T)	−	40.3	0.97	79	23.1	[60]
Perovskite/Si tandem solar cells (4T)	−	7.73	1.05	80	13.4	[61]
Perovskite/Si tandem (2T)	0.17	15.8	1692	79.9	21.4	[62]
Tandem perovskite/Si	1.10	19.0	1.79	74.6	25.4	[63]
Tandem perovskite/Si	1.06	−	−	−	29.02	[55]

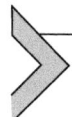

16.8 Perovskite/copper indium gallium selenide tandem solar cells

Apart from perovskite/Si, another interesting semiconducting material is CIGS for tandem devices with an optimal tunable bandgap of 1.08−1.15 eV, which is perfect for the bottom subcell. The CIGS-based solar cells share 2% of the PV market. CIGS has a higher tunable bandgap, even lowered to 1.1 eV better than PSC, and a large option coefficient of 10^5/cm makes it a promising candidate for tandem application. The direct bandgap of CIGS results in the reduction of absorber thickness and lowers the energy and material consumption consequently leading to relatively low cost. The junction of perovskite and CIGS enables the thin-film approach to achieve higher PCE. Both perovskite and CIGS are thin-film technology, which makes processing easy with high power-to-weight ratio. The bandgap requirement is almost the same as perovskite/Si tandem devices.

In CIGS tandem devices only, an inverted architecture can be used due to bandgap priority, specifically in 2T tandem devices. This is not an important issue as the inverted configuration is preferable in perovskite tandem solar cell fabrication due to low optical losses. In the literature, few reports can be found on 2T perovskite/CIGS-based tandem due to limitations of polarity and temperature, which restrict processing of the manufacturing of the device. An example of a successful perovskite/CIGS-based tandem device was reported to achieve 11% PCE on the first attempt [48]. In this work, the typical fabrication process of coevaporation or sputtering was not used; instead, a solution process was applied to reduce surface roughness. A thick layer of PEDOT:PSS was deposited between the top and bottom subcells to produce a smoother surface, although the insertion of a thin protective layer affects the overall performance of the tandem device by reducing V_{oc} [65]. A similar strategy for 2T tandem devices used electrodeposition or spin coating processes to fabricate CIS [CuInSe$_2$ (also, copper indium selenide)] to make the bottom cell surface smoother, producing a PCE of 11.0% [66] and 8.55% [67], respectively. An in situ bandgap tuning approach for MAPbX$_3$ was designed and the junction was used for both CIGS cells. They realized a PCE of only 10.9%; the poor PCE is attributed to weak contact between both subcells [48]. Further research was presented to solve this problem and found that for reliable contact between two subcells, smooth interconnecting layers are crucial to maintaining 2T tandem devices. They adopted a polishing method based on the chemical mechanical approach to smooth the ITO layer, which affected the mismatch work function between BZO and PTTA layers. This modification resulted in better ohmic contact for hole transportation, and 22.4% of PCE was achieved [68].

A different method to deposit HTL on the top of a rough CIGS surface rather than polishing it to get a smooth surface was presented [69]. The ALD process was used to grow this NiO_x-based HTL. This HTL prevented shunt loss of the top layer and obtained 18.0% of PCE. While, for the PTTA case, which did not cover all the rough peaks, depositing NiO_x underneath improved the performance and realized 21.6% of PCE for the 0.78 cm^2 area. Although the lower V_{oc} of 0.6 V of the device was the only downside as shown in Fig. 16.9. In Fig. 16.9B, the black curve shows PTAA, blue is for NiO_x, and red indicates behavior of NiO_x/PTAA. Whereas, solid lines indicate forward scanning and dashed lines indicate scan in the reverse direction.

Al-Ashouri et al. further demonstrated that conformal deposition of the the CIGS layer acting as HTL and joint to oxide layers such as ITO or ZnO could lead to certified PCE of 23.26% with improved current matching on an area of more than 1 cm^2 [70].

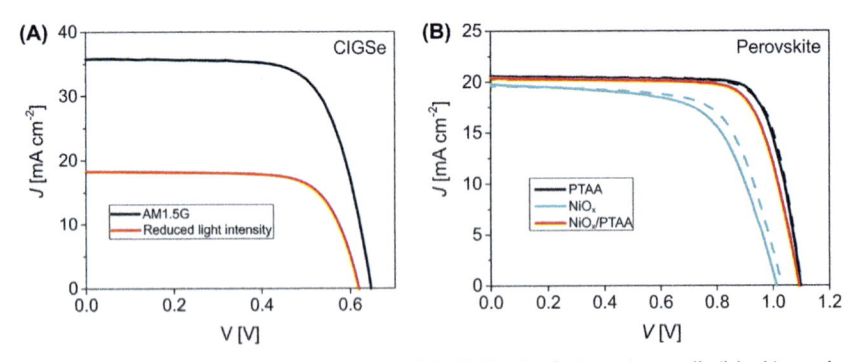

Figure 16.9 J−V characteristics curve (A) CIGS single-junction cell (black) under AM1.5G and reduced light intensity (red), (B) perovskite stand-alone cell with implementation of different HTLs under AM1.5G. *Reproduced with permission from M. Jošt, T. Bertram, D. Koushik, J.A. Marquez, M.A. Verheijen, et al., 21.6%-efficient monolithic perovskite/Cu(In, Ga)Se$_2$ tandem solar cells with thin conformal hole transport layers for integration on rough bottom cell surfaces, ACS Energy Lett. 4 (2019) 583−590, copyright (2019) American Chemical Society.*

A 4T perovskite/CIGS device does not need current matching, and both the subcells can work at the MPP point easily. In this tandem device, both regular and inverted configurations can be freely chosen. In PSC technology, light is illuminated from the substrate commonly. The superstate configuration restricts the selection of the substrate; usually, large bandgap and transparent substrates are preferable. A planar PSC with a feasible architecture was presented to achieve this goal. In the architecture, ZnO nanoparticles play an important role in protecting the perovskite layer from damage due to direct exposure during the AZO sputtering process of fabrication. The semitransparent PSC exhibited an average transmission of 80.4% in the 800–1200 nm wavelength range. The device showed a stabilized sum PCE of 22.1% and 20.9% for the champion device. The optimization of the bandgap of PSC is an approach to obtaining high performance of the tandem device.

Ballie et al. developed 4T perovskite/CIGS tandem devices with $MAPbI_3$ light-absorbing layer using Ag nanowires as the transparent electrode. They obtained 18.6% of PCE with the limitation of parasitic light absorption at 400–550 nm, resulting in loss of photocurrent in CIGS [47]. A well-defined blend of $Cs_x(MA_{0.17}FA_{0.83})_{1-x}(I_{0.83}Br_{0.17})_3$ perovskite with (five percent by weight of) Cs and the balance in the mixture of N-based cations (FA and MA) was employed to fabricate a 4T tandem device. The device obtained a PCE of 20.7%, which was higher than the stand-alone CIGS cell and PSC [71]. For perovskite/CIGS 4T, it is necessary to enhance the lifetime of the wide bandgap of perovskite material to reduce the defect density and high carrier mobility.

Kim et al. demonstrated approximately 20% of PCE of transparent PSC by using a bimolecular addition engineering strategy to incorporate PEAI and $Pb(SCN)_2$ additives in $FA_{0.65}MA_{0.20}Cs_{0.75}Pb_{2.4}I_{0.6}$ precursors. The coupling of additives maintains the challenge of both additives to improve the morphology and crystallinity, inhibiting the excess formation of PbI_2. It can limit the defect density and energetic disorder and enhance lifetime and carrier mobility. They obtained an efficiency of 25.9% for 4T tandem devices. The charge dynamic of the device is shown in Fig. 16.10 [72].

Fu et al. developed a 4T stacked tandem device by using hydrogenated indium oxide (In_2O_3:H) based transparent PSC processed at less than 50°C and CIGS as the bottom subcell. They achieved a PCE of 20.5% by combining both PSC and CIGS cells. The top cell individually offered PCE of 14.2% and 6.3% for filtered CIGS with low J_{sc} due to

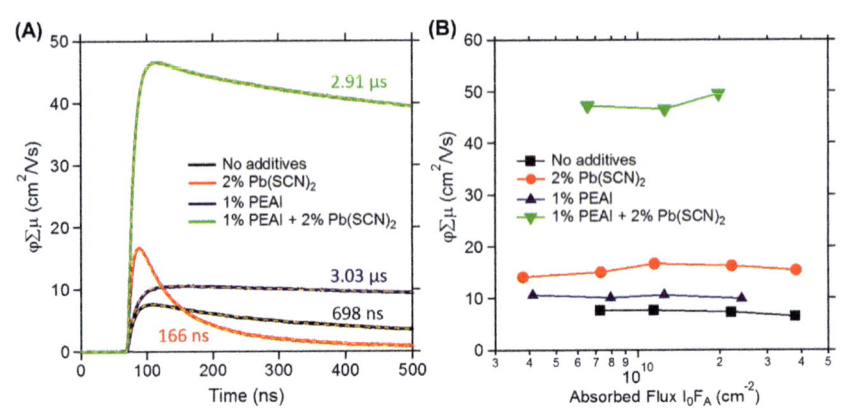

Figure 16.10 Charge-carrier dynamics mechanism of perovskite films: (A) Photoconductivity transient for perovskite films. (B) Excitation intensity dependence of the photoconductivity with I_0 FA in a specific absorbed flux range. *Reproduced with permission from D.H. Kim, C.P. Muzzillo, J. Tong, A.F. Palmstrom, B.W. Larson, et al., Bimolecular additives improve wide-band-gap perovskites for efficient tandem solar cells with CIGS, Joule 3 (2019) 1734–1745, copyright (2019) Elsevier.*

lack of light intensity [73]. Yang et al. reported a multilayer transparent top electrode with Au and Ag layers to give an efficiency of 11.5%. After junction with CIGS, it improved to 15.5% and showed 35% improvement as compared to PSC. The CIGS cell was fabricated using a solution process and realized a V_{oc} of 0.59 V, 29.8 mA/cm^2 for J_{sc}, an FF of 70.7%, and a 12.4% PCE [74]. Most of the tandem devices are based on rigid perovskite substrates restricting their cost and making them less viable in PV applications. Because the requirement of flexible tandem devices is to obtain promising high performance at a low cost is desirable. To achieve this goal, low-temperature processing on a flexible substrate is a viable approach; a recent review addresses methods to produce PSCs on flexible substrates [75]. Section 16.9 addresses all-perovskite tandems; another recent review addresses approaches to enable flexible tandems [11].

Finally, to enable improved perovskite/CIGS tandem devices, the following approaches are recommended: adjusting bandgap to optimal values, enhancement of electrical properties of PSC to limit nonradiative recombination, improvement of diffusion length and carrier lifetime, mitigation of optical losses, and selection of material for a feasible recombination layer. Both PSC and CIGS are thin-film devices, resulting in promising benefits as detailed above, it is therefore reasonable to promote

Table 16.4 Examples from the literature of perovskite/CIGS tandem devices.

Device structure	Active area (cm^2)	J_{sc} (mA/cm^2)	V_{oc} (V)	Fill factor (%)	PCE (%)	References
CZTSSe/perovskite tandem	—	5.6	1.353	60.4	4.6	[46]
Tandem Perovskite + GIGS	0.39	10.9	682	0.788	18.6	[47]
GIGS perovskite/ solar cell	0.108	10.2	0.56	69.6	4.0	[74]
Tandem CIGS/Perovskite solar cell (4 T)	0.108	10.2	0.56	69.6	15.5	[74]
Tandem solar cell (4 T)	0.213	12.7	667.4	74.9	20.5	[76]
Tandem perovskite/CIGS with Al contact	0.4	8.9	1315	68.9	8.0	[48]
Tandem perovskite/CIGS with Ca contact	0.4	12.7	1450	56.6	10.9	[48]
2T Tandem perovskite/CIGS	0.04	17.3	1.77	73.1	22.4	[68]
2T Tandem perovskite/CIGS	1.03	19.2	1.68	71.9	23.3	[68]

tandem technology. However, the difficulty of depositing PSC smooth films on CIGS rough films is another critical challenge. Because some optimization strategies, bandgap, tuning, and interface engineering can be helpful to grow this field. A summary of perovskite/CIGS tandem devices is listed in Table 16.4 [46–48,68,74,76].

16.9 Perovskite/perovskite tandem solar cells

The perovskite/perovskite tandem offers efficiency beyond the single-junction perovskite due to the junction of two transparent layers and efficient light absorbers. These tandems do not need a large budget or high cost of manufacturing. They can be processed at low temperatures and with flexible subcell, low environmental impact fabrication, presenting a route to highly efficient tandem devices. The stacking of devices with a wide bandgap front subcell and narrow bandgap bottom subcell is common in developing perovskite/perovskite tandem devices.

Replacement of Pb by Sn is a wide and successful method to fabricate narrower bandgap PSCs. A binary PSC based on the 4T was manufactured to realize a highly efficient device. The bandgap of Sn-based perovskite was found to decrease as Sn content increased from 0 to 75%, and finally, with 100% Sn, the bandgap of MASnI$_3$ was found to be 1.3 eV. Such Sn-based PSCs achieved a PCE of 14.35%. To improve the stability, oxidizing Sn^{2+} to Sn^{4+} in a hybrid absorber layer (MA$_{0.5}$FA$_{0.5}$-Pb$_{0.75}$Sn$_{0.25}$I$_3$) resulted in a 1.33 eV bandgap. A 4T device achieved a PCE of 19.08%, stacked with a MAPbI$_3$ front cell [77].

The mixed Sn-Pb hybrid perovskite presents another efficient route to developing a narrower bandgap than Pb-free or Sn-free perovskite. Wang et al. demonstrated Sn-Pb inverted PSC using PEDOT:PSS as HTL (FASnI$_3$)$_{0.6}$(MAPbI$_3$)$_{0.4}$ light absorber with a bandgap of 1.25 eV and SnF$_2$ as an additive. Both MAPbI$_3$ and FASnI$_3$ contain dot-sized grains compared to (FASnI$_3$)$_{0.6}$(MAPbI$_3$)$_{0.4}$ and exhibited 15.08% PCE with negligible hysteresis and good reproducibility. The grain size was increased by optimization of the precursor solution. The film thickness increased to 620 nm with a carrier lifetime of 225 ns. The PCE achieved with low bandgap perovskite was 17.5% and EQE was more than 70% in the NIR range. Whereas, for mechanically stacked 4T tandem devices, PCE was significantly enhanced [78].

The thickness of HTL plays a functional role in the performance of PSC or tandem configurations. The high thickness of the HTL can affect the hole extraction efficiency, leading to high recombination of holes with electrons at the HTL interface, consequently deteriorating the device's efficiency. While thin HTL can effectively perform hole transportation, it is challenging to fabricate thin HTL without additives to maintain adhesiveness between both subcells. Jiang et al. demonstrated two MAPbI$_3$-based subcells to fabricate tandem devices. The top cell followed the configuration glass-ITO/PEI/PCBM/MAPbI$_3$/spiro-OMeTAD/Ag and the bottom cell glass-FTO/c-TiO$_2$/m-TiO$_2$/MAPbI$_3$/spiro-OMeTAD/electrode. The top cell realized 8.3% with the addition of PEDOT:PSS, and with an Ag electrode, it reached 11.4%, while the bottom cell showed 10.1% and 11.7% of PCE, respectively. The tandem device obtained 7.0% of PCE with V_{oc} of 1.89 V [31].

Snaith et al. demonstrated an impressive PCE of 20.3% with an antisolvent immersion strategy to fabricate binary Pb-Sn perovskite film. In their work, they replaced FA$^+$ with Cs$^+$ and realized 14.1% of PCE for (FA$_{0.75}$Cs$_{0.25}$ Sn$_{0.15}$Pb$_{0.5}$I$_3$) perovskite layer [41]. Tin-based perovskite films with a smooth and pinhole-free structure are crucial to maintain. To demonstrate this point, the work of Abdollahi et al. should be considered [79]. By using the vacuum assistant growth method, large grain and

Table 16.5 Examples from the literature of perovskite/perovskite tandem devices.

Device structure	Active area	J_{sc} (mA/cm^2)	V_{oc} (V)	Fill factor (%)	PCE (%)	References
Top perovskite/perovskite bottom (2T)	1 cm^2	12.3	1.83	60	13.4	[41]
Top perovskite/perovskite bottom (4T)	1 cm^2	–	–	–	16.4	[41]
FTO/c-TiO$_2$/MAPbBr$_3$/P3HT/Au	0.096 cm^2	8.6	1.08	78	7.2	[80]
Perovskite/Perovskite (2T)	10 mm^2	6.61	1.89	56	7	[31]
ITO/PEDOT: PSS/MAPbI$_3$/PCBM/Au	0.096 cm^2	20.7	1.1	79	18	[80]
MAPbBr$_3$-MAPbI$_3$ Tandem cells	0.096 cm^2	8.4	1.95	66	10.8	[80]
FTO/c-TiO$_2$/MAPbBr$_3$/PTAA/Au	0.096 cm^2	8.4	1.3	77	8.4	[80]
2T (Sn-Pb) mixed-perovskite	0.105 cm^2	15.0	1.942	80.3	23.1	[68]

pinhole-free perovskites were obtained with improved carrier lifetime and a PCE of 18.2%. Furthermore, for stacked 4T devices, a 23% PCE was measured [79]. A summary of perovskite/perovskite tandem devices is shown in Table 16.5 [31,41,68,80].

16.10 Conclusion

Over the past few years, the combination of PSCs with crystalline Si solar cells, CIGS solar cells, and PSCs (or even polymer devices) in tandem configuration has shown tremendous promise. In the near future, tandem devices with high-efficiency, cost-effectiveness in converting sunlight into electricity, and operational stability can be anticipated. This chapter briefly introduced tandem structure configuration and important breakthroughs in PSC technology. Next, the significance of different fabrication approaches, such as 2T, 3T, and 4T was emphasized. Other challenges within the tandem structure in the presence of the perovskite are currently being addressed.

Lead in different layers in tandem structures, given the hazardous nature of Pb, can raise health and safety concerns. There has been significant effort over the past several years to find suitable and safer replacements that maintain performance to alleviate these concerns for commercial and other practical applications. An optimal combination of transparent electrode and barrier layers as sputter buffers is advised for use as protective layers to enhance stability. Aiming to highlight the development of different configurations of tandem structure, research has been the primary objective of this chapter. Clearly, each device configuration has its own unique opportunities and challenges.

References

[1] H.J. Snaith, Perovskites: the emergence of a new era for low-cost, high-efficiency solar cells, J. Phys. Chem. Lett. 4 (21) (2013) 3623−3630. Available from: https://pubs.acs.org/doi/10.1021/jz4020162.

[2] H.S. Kim, C.R. Lee, J.H. Im, K.B. Lee, T. Moehl, et al., Lead iodide perovskite sensitized all-solid-state submicron thin film mesoscopic solar cell with efficiency exceeding 9%, Sci. Rep. 2 (1) (2012) 7. Available from: https://doi.org/10.1038/srep00591.

[3] Z. Li, Y. Zhao, X. Wang, Y. Sun, Z. Zhao, et al., Cost analysis of perovskite tandem photovoltaics, Joule 2 (2018) 1559−1572. Available from: https://doi.org/10.1016/j.joule.2018.05.001.

[4] J. Werner, B. Niesen, C. Ballif, Perovskite/silicon tandem solar cells: marriage of convenience or true love story?—An overview, Adv. Mater. Interfaces 5 (2018) 1−19. Available from: https://doi.org/10.1002/admi.201700731.

[5] X. Zhao, H.Q. Tan, E. Birgersson, H. Xue, Elucidating the underlying physics in a two-terminal all-perovskite tandem solar cell: a guideline towards 30% power conversion efficiency, Sol. Energy 231 (2022) 716−731. Available from: https://doi.org/10.1016/j.solener.2021.11.029.

[6] Space photovoltaics: materials, in: S.G. Bailey, A.F. Hepp, D.C. Ferguson, R.P. Raffaelle, S.M. Durbin (Eds.), Missions and Alternative Technologies, Elsevier, Cambridge, MA, USA, 2022, 534 pp.

[7] T. Todorov, O. Gunawan, S. Guha, A road towards 25% efficiency and beyond: perovskite tandem solar cells, Mol. Syst. Des. Eng. 1 (2016) 370−376. Available from: https://doi.org/10.1039/c6me00041j.

[8] H. Shen, D. Walter, Y. Wu, K.C. Fong, D.A. Jacobs, et al., Monolithic perovskite/Si Tandem solar cells: pathways to over 30% efficiency, Adv. Energy Mater. 10 (2020) 1−18. Available from: https://doi.org/10.1002/aenm.201902840.

[9] J. Ávila, C. Momblona, P. Boix, M. Sessolo, M. Anaya, et al., High voltage vacuum-deposited $CH_3NH_3PbI_3$-$CH_3NH_3PbI_3$ tandem solar cells, Energy Environ. Sci. 11 (2018) 3292−3297. Available from: https://doi.org/10.1039/c8ee01936c.

[10] T. Leijtens, K.A. Bush, R. Prasanna, M.D. McGehee, Opportunities and challenges for tandem solar cells using metal halide perovskite semiconductors, Nat. Energy 3 (2018) 828−838. Available from: https://doi.org/10.1038/s41560-018-0190-4.

[11] A.F. Palmstrom, G.E. Eperon, T. Leijtens, R. Prasanna, S.N. Habisreutinger, et al., Enabling flexible all-perovskite tandem solar cells, Joule 3 (2019) 2193−2204. Available from: https://doi.org/10.1016/j.joule.2019.05.009.

[12] M.I. Hossain, A.M. Saleque, S. Ahmed, I. Saidjafarzoda, Md Shahiduzzaman, et al., Perovskite/perovskite planar tandem solar cells: a comprehensive guideline for reaching energy conversion efficiency beyond 30%, Nano Energy 79 (2021) 105400. Available from: https://doi.org/10.1016/j.nanoen.2020.105400.

[13] F. Sahli, J. Werner, B.A. Kamino, M. Bräuninger, R. Monnard, et al., Fully textured monolithic perovskite/silicon tandem solar cells with 25.2% power conversion efficiency, Nat. Mater. 17 (2018) 820–826. Available from: https://doi.org/10.1038/s41563-018-0115-4.

[14] E. Köhnen, M. Jošt, A.B. Morales-Vilches, P. Tockhorn, A. Al-Ashouri, et al., Highly efficient monolithic perovskite silicon tandem solar cells: analyzing the influence of current mismatch on device performance, Sustain. Energy Fuels 3 (2019) 1995–2005. Available from: https://doi.org/10.1039/c9se00120d.

[15] K. Jäger, L. Korte, B. Rech, S. Albrecht, Numerical optical optimization of monolithic planar perovskite-silicon tandem solar cells with regular and inverted device architectures, Opt. Express 25 (2017) A473. Available from: https://doi.org/10.1364/oe.25.00a473.

[16] M.T. Hörantner, H.J. Snaith, Predicting and optimising the energy yield of perovskite-on-silicon tandem solar cells under real world conditions, Energy Environ. Sci. 10 (2017) 1983–1993. Available from: https://doi.org/10.1039/c7ee01232b.

[17] A. Rajagopal, R.J. Stoddard, S.B. Jo, H.W. Hillhouse, A.K.Y. Jen, Overcoming the photovoltage plateau in large bandgap perovskite photovoltaics, Nano Lett. 18 (2018) 3985–3993. Available from: https://doi.org/10.1021/acs.nanolett.8b01480.

[18] S. Gharibzadeh, I.M. Hossain, P. Fassl, B.A. Nejand, T. Abzieher, et al., 2D/3D heterostructure for semitransparent perovskite solar cells with engineered bandgap enables efficiencies exceeding 25% in four-terminal tandems with silicon and CIGS, Adv. Funct. Mater. 30 (2020) 1909919. Available from: https://doi.org/10.1002/adfm.201909919.

[19] E.L. Unger, L. Kegelmann, K. Suchan, D. Sörell, L. Korte, S. Albrecht, Roadmap and roadblocks for the band gap tunability of metal halide perovskites, J. Mater. Chem. A 5 (2017) 11401–11409. Available from: https://doi.org/10.1039/c7ta00404d.

[20] S. Albrecht, M. Saliba, J.-P. Correa Baena, F. Lang, L. Kegelmann, et al., Monolithic perovskite/silicon-heterojunction tandem solar cells processed at low temperature, Energy Environ. Sci. 9 (2016) 81–88. Available from: https://doi.org/10.1039/c5ee02965a.

[21] S. Essig, C. Allebé, T. Remo, J.F. Geisz, M.A. Steiner, et al., Raising the one-sun conversion efficiency of III-V/Si solar cells to 32.8% for two junctions and 35.9% for three junctions, Nat. Energy 2 (2017) 17144. Available from: https://doi.org/10.1038/nenergy.2017.144.

[22] B. Hoex, J. Schmidt, R. Bock, P.P. Altermatt, M.C.M. Van De Sanden, W.M.M. Kessels, Excellent passivation of highly doped p-type Si surfaces by the negative-charge-dielectric Al_2O_3, Appl. Phys. Lett. 91 (2007) 1–4. Available from: https://doi.org/10.1063/1.2784168.

[23] Y. Wu, D. Yan, J. Peng, T. Duong, Y. Wan, et al., Monolithic perovskite/silicon-homojunction tandem solar cell with over 22% efficiency, Energy Environ. Sci. 10 (2017) 2472–2479. Available from: https://doi.org/10.1039/c7ee02288c.

[24] J. Zheng, H. Mehrvarz, F.J. Ma, C.F.J. Lau, M.A. Green, et al., 21.8% efficient monolithic perovskite/homo-junction-silicon tandem solar cell on 16 cm^2, ACS Energy Lett. 3 (2018) 2299–2300. Available from: https://doi.org/10.1021/acsenergylett.8b01382.

[25] Z. Song, A. Abate, S.C. Watthage, G.K. Liyanage, A.B. Phillips, et al., Perovskite solar cell stability in humid air: partially reversible phase transitions in the PbI_2-CH_3NH_3I-H_2O system, Adv. Energy Mater. 6 (2016) 1–7. Available from: https://doi.org/10.1002/aenm.201600846.

[26] H. Ferhati, F. Djeffal, Exceeding 30% efficiency for an environment-friendly tandem solar cell based on earth-abundant Se/CZTS materials, Phys. E Low. Dimens. Syst. Nanostruct. 109 (2019) 52−58. Available from: https://doi.org/10.1016/j.physe.2019.01.002.

[27] New world records: Perovskite-on-silicon-tandem solar cells, accessed March 23, 2024 https://actu.epfl.ch/news/new-world-records-perovskite-on-silicon-tandem-sol/

[28] H. Back, G. Kim, J. Kim, J. Kong, T.K. Kim, et al., Achieving long-term stable perovskite solar cells via ion neutralization, Energy Environ. Sci. 9 (2016) 1258−1263. Available from: https://doi.org/10.1039/c6ee00612d.

[29] T. Duong, Y.L. Wu, H. Shen, J. Peng, X. Fu, et al., Rubidium multication perovskite with optimized bandgap for perovskite-silicon tandem with over 26% efficiency, Adv. Energy Mater. 7 (2017) 1−11. Available from: https://doi.org/10.1002/aenm.201700228.

[30] H. Kanda, A. Uzum, H. Nishino, T. Umeyama, H. Imahori, et al., Interface optoelectronics engineering for mechanically stacked tandem solar cells based on perovskite and silicon, ACS Appl. Mater. Interfaces 8 (2016) 33553−33561. Available from: https://doi.org/10.1021/acsami.6b07781.

[31] F. Jiang, T. Liu, B. Luo, J. Tong, F. Qin, et al., A two-terminal perovskite/perovskite tandem solar cell, J. Mater. Chem. A 4 (2016) 1208−1213. Available from: https://doi.org/10.1039/c5ta08744a.

[32] C.-Y. Chang, L. Zuo, H.-L. Yip, C.-Z. Li, Y. Li, et al., Highly efficient polymer tandem cells and semitransparent cells for solar energy, Adv. Energy Mater. 4 (7) (2014) 1301645. Available from: https://doi.org/10.1002/aenm.201301645.

[33] F. Sahli, B.A. Kamino, J. Werner, M. Bräuninger, B. Paviet-Salomon, et al., Improved optics in monolithic perovskite/silicon tandem solar cells with a nanocrystalline silicon recombination junction, Adv. Energy Mater. 8 (2018) 1−8. Available from: https://doi.org/10.1002/aenm.201701609.

[34] S. Albrecht, M. Saliba, J.P. Correa-Baena, K. Jäger, L. Korte, et al., Towards optical optimization of planar monolithic perovskite/silicon-heterojunction tandem solar cells, J. Opt. 18 (2016) 064012. Available from: https://doi.org/10.1088/2040-8978/18/6/064012.

[35] M. Saliba, J.P. Correa-Baena, M. Grätzel, A. Hagfeldt, A. Abate, Perovskite solar cells: from the atomic level to film quality and device performance, Angew. Chem. Int. Ed. 57 (2017) 2554−2569. Available from: https://doi.org/10.1002/anie.201703226.

[36] J. Werner, L. Barraud, A. Walter, M. Bräuninger, F. Sahli, et al., Efficient near-infrared-transparent perovskite solar cells enabling direct comparison of 4-terminal and monolithic perovskite/silicon tandem cells, ACS Energy Lett. 1 (2016) 474−480. Available from: https://doi.org/10.1021/acsenergylett.6b00254.

[37] L. Peng, Y. Zhu, D. Chen, R.S. Ruoff, G. Yu, Two-dimensional materials for beyond-lithium-ion batteries, Adv. Energy Mater. 6 (2016) 1−21. Available from: https://doi.org/10.1002/aenm.201600025.

[38] H. Uzu, M. Ichikawa, M. Hino, K. Nakano, T. Meguro, et al., High efficiency solar cells combining a perovskite and a silicon heterojunction solar cells via an optical splitting system, Appl. Phys. Lett. 106 (2015) 013506. Available from: https://doi.org/10.1063/1.4905177.

[39] M. Nakamura, K. Tada, T. Kinoshita, T. Bessho, C. Nishiyama, et al., Perovskite/CIGS spectral splitting double junction solar cell with 28% power conversion efficiency, iScience 23 (2020) 101817. Available from: https://doi.org/10.1016/j.isci.2020.101817.

[40] F. Lang, M.A. Gluba, S. Albrecht, J. Rappich, L. Korte, et al., Perovskite solar cells with large-area CVD-graphene for tandem solar cells, J. Phys. Chem. Lett. 6 (2015) 2745−2750. Available from: https://doi.org/10.1021/acs.jpclett.5b01177.

[41] G.E. Eperon, T. Leijtens, K.A. Bush, R. Prasanna, T. Green, et al., Perovskite-perovskite tandem photovoltaics with optimized band gaps, Science 354 (2016) 861−865. Available from: https://doi.org/10.1126/science.aaf9717.

[42] D.P. McMeekin, G. Sadoughi, W. Rehman, G.E. Eperon, M. Saliba, et al., A mixed-cation lead mixed-halide perovskite absorber for tandem solar cells, Science 351 (2016) 151−155. Available from: https://www.science.org/doi/10.1126/science.aad5845.

[43] D.T. Grant, K.R. Catchpole, K.J. Weber, T.P. White, Design guidelines for perovskite/silicon 2-terminal tandem solar cells: an optical study, Opt. Express 24 (2016) A1454. Available from: https://doi.org/10.1364/oe.24.0a1454.

[44] L. Su, Y. Xiao, G. Han, L. Lu, H. Li, M. Zhu, Performance enhancement of perovskite solar cells using trimesic acid additive in the two-step solution method, J. Power Sources 426 (2019) 11−15. Available from: https://doi.org/10.1016/j.jpowsour.2019.04.024.

[45] B. Chen, Y. Bai, Z. Yu, T. Li, X. Zheng, et al., Efficient semitransparent perovskite solar cells for 23.0%-efficiency perovskite/silicon four-terminal tandem cells, Adv. Energy Mater. 6 (2016) 1−7. Available from: https://doi.org/10.1002/aenm.201601128.

[46] T. Todorov, T. Gershon, O. Gunawan, C. Sturdevant, S. Guha, Perovskite-kesterite monolithic tandem solar cells with high open-circuit voltage, Appl. Phys. Lett. 105 (2014) 173902. Available from: https://doi.org/10.1063/1.4899275.

[47] C.D. Bailie, M.G. Christoforo, J.P. Mailoa, A.R. Bowring, E.L. Unger, et al., Semi-transparent perovskite solar cells for tandems with silicon and CIGS, Energy Environ. Sci. 8 (2015) 956−963. Available from: https://doi.org/10.1039/c4ee03322a.

[48] T. Todorov, T. Gershon, O. Gunawan, Y.S. Lee, C. Sturdevant, et al., Monolithic perovskite-CIGS tandem solar cells via in situ band gap engineering, Adv. Energy Mater. 5 (2015) 1−6. Available from: https://doi.org/10.1002/aenm.201500799.

[49] Z. Wang, Q. Lin, F.P. Chmiel, N. Sakai, L.M. Herz, H.J. Snaith, Efficient ambient-air-stable solar cells with 2D-3D heterostructured butylammonium-caesium-formamidinium lead halide perovskites, Nat. Energy 2 (2017) 1−10. Available from: https://doi.org/10.1038/nenergy.2017.135.

[50] D. Kim, H.J. Jung, I.J. Park, B.W. Larson, S.P. Dunfield, et al., Efficient, stable silicon tandem cells enabled by anion-engineered wide-bandgap perovskites, Science 368 (2020) 155−160. Available from: https://doi.org/10.1126/science.aba3433.

[51] L. Zheng, J. Wang, Y. Xuan, M. Yan, X. Yu, et al., A perovskite/silicon hybrid system with a solar-to-electric power conversion efficiency of 25.5%, J. Mater. Chem. A 7 (2019) 26479−26489. Available from: https://doi.org/10.1039/c9ta10712f.

[52] F. Gota, M. Langenhorst, R. Schmager, J. Lehr, U.W. Paetzold, Energy yield advantages of three-terminal perovskite-silicon tandem photovoltaics, Joule 4 (2020) 2387−2403. Available from: https://doi.org/10.1016/j.joule.2020.08.021.

[53] P. Tockhorn, P. Wagner, L. Kegelmann, J.C. Stang, M. Mews, et al., Three-terminal perovskite/silicon tandem solar cells with top and interdigitated rear contacts, ACS Appl. Energy Mater. 3 (2020) 1381−1392. Available from: https://doi.org/10.1021/acsaem.9b01800.

[54] R. Santbergen, H. Uzu, K. Yamamoto, M. Zeman, Optimization of three-terminal perovskite/silicon tandem solar cells, IEEE J. Photovolt. 9 (2019) 446−451. Available from: https://doi.org/10.1109/JPHOTOV.2018.2888832.

[55] Helmholtz-Zentrum Berlin (H.Z.B.) Hits 29.15% efficiency with perovskite/silicon tandem cell. <https://renewablesnow.com/news/hzb-hits-2915-efficiency-with-perovskitesilicon-tandem-cell-685469/> (accessed 02.01.23).
[56] J.P. Mailoa, C.D. Bailie, E.C. Johlin, E.T. Hoke, A.J. Akey, et al., A 2-terminal perovskite/silicon multijunction solar cell enabled by a silicon tunnel junction, Appl. Phys. Lett. 106 (2015) 121105. Available from: https://doi.org/10.1063/1.4914179.
[57] K.A. Bush, C.D. Bailie, Y. Chen, A.R. Bowring, W. Wang, et al., Thermal and environmental stability of semi-transparent perovskite solar cells for tandems enabled by a solution-processed nanoparticle buffer layer and sputtered ITO electrode, Adv. Mater. 28 (2016) 3937−3943. Available from: https://doi.org/10.1002/adma.201505279.
[58] K.A. Bush, A.F. Palmstrom, Z.J. Yu, M. Boccard, R. Cheacharoen, et al., 23.6%-efficient monolithic perovskite/silicon tandem solar cells with improved stability, Nat. Energy 2 (2017) 1−7. Available from: https://doi.org/10.1038/nenergy.2017.9.
[59] J. Zheng, C.F.J. Lau, H. Mehrvarz, F.J. Ma, Y. Jiang, et al., Large area efficient interface layer free monolithic perovskite/homo-junction-silicon tandem solar cell with over 20% efficiency, Energy Environ. Sci. 11 (2018) 2432−2443. Available from: https://doi.org/10.1039/c8ee00689j.
[60] T. Duong, D. Grant, S. Rahman, A. Blakers, K.J. Weber, et al., Filterless spectral splitting perovskite-silicon tandem system with >23% calculated efficiency, IEEE J. Photovolt. 6 (2016) 1432−1439. Available from: https://doi.org/10.1109/JPHOTOV.2016.2600344.
[61] P. Löper, S.J. Moon, S. Martín De Nicolas, B. Niesen, M. Ledinsky, et al., Organic-inorganic halide perovskite/crystalline silicon four-terminal tandem solar cells, Phys. Chem. Chem. Phys. 17 (2015) 1619−1629. Available from: https://doi.org/10.1039/c4cp03788j.
[62] J. Werner, C.H. Weng, A. Walter, L. Fesquet, J.P. Seif, et al., Efficient monolithic perovskite/silicon tandem solar cell with cell area $>$ 1 cm^2, J. Phys. Chem. Lett. 7 (2016) 161−166. Available from: https://doi.org/10.1021/acs.jpclett.5b02686.
[63] L. Mazzarella, Y.H. Lin, S. Kirner, A.B. Morales-Vilches, L. Korte, et al., Infrared light management using a nanocrystalline silicon oxide interlayer in monolithic perovskite/silicon heterojunction tandem solar cells with efficiency above 25%, Adv. Energy Mater. 9 (2019) 1−9. Available from: https://doi.org/10.1002/aenm.201803241.
[64] E. Raza, Z. Ahmad, Review on two-terminal and four-terminal crystalline-silicon/perovskite tandem solar cells; progress, challenges, and future perspectives, Energy Rep. 8 (2022) 5820−5851. Available from: https://doi.org/10.1016/j.egyr.2022.04.028.
[65] L. Meng, J. You, T.F. Guo, Y. Yang, Recent advances in the inverted planar structure of perovskite solar cells, Acc. Chem. Res. 49 (2016) 155−165. Available from: https://doi.org/10.1021/acs.accounts.5b00404.
[66] A.R. Uhl, A. Rajagopal, J.A. Clark, A. Murray, T. Feurer, et al., Solution-processed low-bandgap $CuIn(S, Se)_2$ absorbers for high-efficiency single-junction and monolithic chalcopyrite-perovskite tandem solar cells, Adv. Energy Mater. 8 (2018) 1−8. Available from: https://doi.org/10.1002/aenm.201801254.
[67] Y.H. Jang, J.M. Lee, J.W. Seo, I. Kim, D.K. Lee, Monolithic tandem solar cells comprising electrodeposited $CuInSe_2$ and perovskite solar cells with a nanoparticulate ZnO buffer layer, J. Mater. Chem. A 5 (2017) 19439−19446. Available from: https://doi.org/10.1039/c7ta06163c.
[68] Q. Han, Y.T. Hsieh, L. Meng, J.L. Wu, P. Sun, et al., High-performance perovskite/ $Cu(In, Ga)Se_2$ monolithic tandem solar cells, Science 361 (6405) (2018) 904−908. Available from: https://doi.org/10.1126/science.aat5055.
[69] M. Jošt, T. Bertram, D. Koushik, J.A. Marquez, M.A. Verheijen, et al., 21.6%-efficient monolithic perovskite/$Cu(In, Ga)Se_2$ tandem solar cells with thin conformal

hole transport layers for integration on rough bottom cell surfaces, ACS Energy Lett. 4 (2019) 583–590. Available from: https://doi.org/10.1021/acsenergylett.9b00135.
[70] A. Al-Ashouri, A. Magomedov, M. Roß, M. Jošt, M. Talaikis, et al., Conformal monolayer contacts with lossless interfaces for perovskite single junction and monolithic tandem solar cells, Energy Environ. Sci. 12 (2019) 3356–3369. Available from: https://doi.org/10.1039/c9ee02268f.
[71] A. Guchhait, H.A. Dewi, S.W. Leow, H. Wang, G. Han, et al., Over 20% efficient CIGS-perovskite tandem solar cells, ACS Energy Lett. 2 (2017) 807–812. Available from: https://doi.org/10.1021/acsenergylett.7b00187.
[72] D.H. Kim, C.P. Muzzillo, J. Tong, A.F. Palmstrom, B.W. Larson, et al., Bimolecular additives improve wide-band-gap perovskites for efficient tandem solar cells with CIGS, Joule 3 (2019) 1734–1745. Available from: https://doi.org/10.1016/j.joule.2019.04.012.
[73] J. Fu, D. Sheng, X. Lu, Hydrogenation of levulinic acid over nickel catalysts supported on aluminum oxide to prepare γ-valerolactone, Catalysts 6 (1) (2015) 10. Available from: https://doi.org/10.3390/catal6010006.
[74] M.Y. Yang, Q. Chen, Y.T. Hsieh, T.-B. Song, N. De Marco, et al., Multilayer transparent top electrode for solution processed perovskite/Cu(In, Ga)(Se,S)$_2$ four terminal tandem solar cells, ACS Nano 9 (2015) 7714–7721. Available from: https://doi.org/10.1021/acsnano.5b03189.
[75] H.S. Jung, G.S. Han, N.-G. Park, M.J. Ko, Flexible perovskite solar cells, Joule 3 (8) (2019) 1850–1880. Available from: https://doi.org/10.1016/j.joule.2019.07.023.
[76] F. Fu, T. Feurer, T. Jäger, E. Avancini, B. Bissig, et al., Low-temperature-processed efficient semi-transparent planar perovskite solar cells for bifacial and tandem applications, Nat. Commun. 6 (1) (2015) 9. Available from: https://doi.org/10.1038/ncomms9932.
[77] Z. Yang, A. Rajagopal, C.C. Chueh, S.B. Jo, B. Liu, et al., Stable low-bandgap Pb−Sn binary perovskites for tandem solar cells, Adv. Mater. 28 (2016) 8990–8997. Available from: https://doi.org/10.1002/adma.201602696.
[78] C. Wang, D. Zhao, Y. Yu, N. Shrestha, C.R. Grice, et al., Compositional and morphological engineering of mixed cation perovskite films for highly efficient planar and flexible solar cells with reduced hysteresis, Nano Energy 35 (2017) 223–232. Available from: https://doi.org/10.1016/j.nanoen.2017.03.048.
[79] N.B. Abdollahi, I.M. Hossain, M. Jakoby, S. Moghadamzadeh, T. Abzieher, S. Gharibzadeh, et al., Vacuum-assisted growth of low-bandgap thin films (FA$_{0.8}$MA$_{0.2}$Sn$_{0.5}$Pb$_{0.5}$I$_3$) for all-perovskite tandem solar cells, Adv. Energy Mater. 10 (2020) 10. Available from: https://doi.org/10.1002/aenm.201902583.
[80] J.H. Heo, S.H. Im, CH$_3$NH$_3$PbBr$_3$−CH$_3$NH$_3$PbI$_3$ perovskite−perovskite tandem solar cells with exceeding 2.2V open circuit voltage, Adv. Mater. 28 (2016) 5121–5125. Available from: https://doi.org/10.1002/adma.201501629.

SECTION VI

Practical applications of photovoltaics

CHAPTER SEVENTEEN

Transparent photovoltaics: Overview and applications

Ananthakumar Soosaimanickam and Abhirami Murugavel
R&D Division, Intercomet S.L., Madrid, Spain

17.1 Transparent solar cells: An introduction

Thin-film solar cells (TFSCs) are generating increasing interest because of their light weight compared with traditional silicon (Si) solar cells. In this category, transparent solar cells (TSCs) have been intensively investigated in recent years due to their multifunctional applications [1−3]. When transparency is achieved in thin-film photovoltaic (PV) materials, PV devices may be useful for a broad range of applications. Most of these solar cells are wavelength-selective, especially in the UV-visible and near-infra red (NIR) region of the solar spectrum. It is predicted that the thermodynamic efficiency limit of the single-junction TSCs could reach up to 20% with an average visible transmission of 100% [4]. Transparent solar cells have been installed in recent years for building integrated photovoltaic (BIPV) applications with different colors and different transparency [5−7]. These solar cells that act as power-generating windows are greatly expected to be a breakthrough in the renewable energy sector. These power-generating windows improve the energy efficiency of the buildings and allow selected wavelengths to pass through. Factors such as haze ratio, coloring index, color perception, color temperature, color rendering, and esthetic elements are important for fabricating such colored windows hence fabrication of neutral-colored TSCs has much attracted in recent years [8−11]. In most of the literature, these solar cells are generally described as semitransparent solar cells (STSCs) because of the absorption/transmittance of the materials and different layers that are deposited during fabrication [12].

Because of their compatibility with a variety of devices, TSCs are also found to be useful in vehicles and screens in electronic components, to

Photovoltaics Beyond Silicon.
DOI: https://doi.org/10.1016/B978-0-323-90188-8.00018-X
© 2024 Elsevier Inc. All rights reserved, including those for text and data mining, AI training, and similar technologies.

give two examples. One of the recent breakthroughs in processing TSCs is employing roll-to-roll (R2R) processing, thus enabling advanced applications. TSCs are classified into two different types, namely wavelength-selective and nonselective [tend to have power conversion efficiencies (PCEs)]. Wavelength-selective TSCs generally function by harvesting UV-visible/NIR photons while admitting visible light. Thus, high transmittance in the visible region and also photon harvesting in the UV and NIR region are essential requirements while fabricating such TSCs. The transparency increase or decrease is dependent on several factors that include the nature of materials, thickness, number of layers, and other components in the device. These "see-through" solar cell technologies are highly desirable because of their diverse range of applications in different areas. At present, TSCs are fabricated in order to use two principles: firstly, a fraction of visible light is absorbed and the rest of the portion is transmitted and secondly, the light transmission is fixed at a particular region of the solar cell [13]. Another concept that is considered to be a breakthrough in TSC technology is the solar window (or) luminescent solar concentrator. In solar windows, nanocrystals (NCs) are mixed with a polymer; the mixture is used to produce transparent films [14]. When light crosses this film, the NCs undergo total internal reflection, transferring energy into the corner of the window where PV cells are placed. In other words, the NCs here act like a waveguide in collecting and transferring the photons, which ultimately produce photocurrent in the circuit. Due to the loss of photons during transmission, the TSCs are not as efficient as conventional opaque solar cells. However, because of their limited absorbing ability, TSCs could be directly used for BIPV purposes and so the wastage of huge surface area is avoided. The overall representation of materials and methodologies used for the BIPV is given in Fig. 17.1A [15]. In order to deliver high efficiency, the TSCs should be able to tackle the loss associated with the surface (reflective) and the loss related to internal factors such as recombination. The semitransparent or transparent notations mentioned in this chapter almost describe the same kind of material aspects of the layer (by simply changing the thickness, color, etc.); the readers should understand that the level of transparency is describing these aspects or properties of TSCs. Other than BIPV application, TSCs could be applied for roofs, shelters, PV curtains, wearable electronic appliances, and other products [21].

Transparency has been successfully achieved in different kinds of solar cells including organic solar cells (OSCs), Si solar cells, dye-sensitized solar cells (DSSCs), quantum-dot (sensitized) solar cells (QD(S)SCs), and

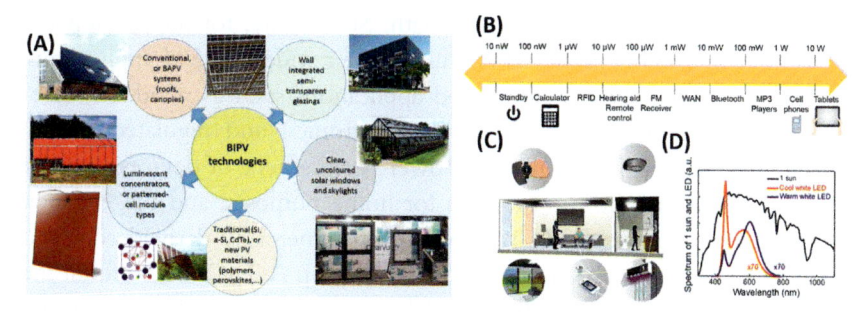

Figure 17.1 (A) different aspects of building integrated photovoltaic (PV) modules; (B) power requirement of devices for indoor application; (C) use of transparent PV devices for household purposes; (D) a comparison of emission spectra of the AM1.5 solar spectrum with a cool and warm white light emitting diode. *(A) Reprinted from M. Vasiliev, M. Nur-E-Alam, K. Alameh, Recent developments in solar energy harvesting technologies for building integration and distributed energy generation, Energies 12(6) 2019 1080, this article is distributed under the terms of the Creative Commons (CC-BY 4.0) license (http://creativecommons.org/licenses/by/4.0/), which permits unrestricted use.; (B–D) reprinted with permission from G. Kim, J.W. Lim, J. Kim, S.J. Yun, M.A. Park, Transparent Thin-Film Silicon Solar Cells for Indoor Light Harvesting with Conversion Efficiencies of 36% without Photodegradation, ACS Appl. Mater. Interfaces 12 (24) (2020) 27122–27130, copyright (2020) American Chemical Society.*

perovskite solar cells (PSCs) [12]. In these categories, concepts such as inverted and bifacial solar cells are implemented and considerable performance enhancement is observed. It is a general observation that both organic and inorganic materials have the capabilities of making transparency while depositing on a conducting substrate. To fulfill the requirements of semitransparency, it is necessary that the material should absorb in the region apart from a visible area (i.e., 380–800) and should make a high level of transparency in the visible region [16,17]. While fabricating TSCs, different conditions should be fulfilled, which include (i) higher PCE under the same transmittance at all angles, (ii) the quality of TSCs should be maintained for its purpose, and (iii) it should exhibit long-term stability [13]. Considering the efficiency, approximately 12%, currently crystalline silicon (c-Si) and perovskites-based TSCs dominate the research. Because the performance varies with the transmittance value, it is important to adjust the thickness of the film to improve the transmittance and performance. However, it should be remembered that when transmittance increases, the PCE of the device decreases and vice-versa. Apart from the solar cell, the concept of making transparency is expanded toward fabricating light-emitting diodes (LEDs), photodetectors, etc. Also, it is possible to govern the growth of plants under the closed atmosphere of TSCs by carefully admitting light of a particular wavelength. Specifically, for indoor applications, TSCs are highly preferred because of their wide-range of applications. It is also extended toward distributed applications such as windows, electronic device displays, automobiles, bioelectronics, and regulated neuromorphic computing [18–20]. The power requirement of devices used for indoor applications, the use of TSCs for home purposes, and the emission spectra of the air mass (AM) 1.5 with the commercially derived cool and warm LEDs are represented in Fig. 17.1B–D [21]. A TSC should fulfill the esthetic requirement of the needs and hence properties such as color and dimension are important to be taken under consideration while fabricating it. Such TSCs are also produced with different colors depending on the requirement for the BIPV applications [22,23]. For the introduction of colors in TSCs, approaches such as variation in the thickness of the antireflection coatings [24], different colored organic dyes [25], incorporation of a foreign compound as an additional layer [26], using a plasmonic color filter [27], and compositional and thickness modification [26] can be utilized. The detailed application aspects of TSCs are discussed in this chapter as a separate section. When TSCs are fabricated on a flexible substrate, such as poly-ethylene terephthalate (PET), they offer significant versatility in terms of applications. Thus, it is necessary to understand about TSCs and the different concepts used in the fabrication of various kinds of structures. This chapter highlights the recent advances existing in the different kinds of TSC categories and efforts undertaken toward optimization of stability and commercialization.

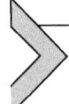

17.2 Solar cell parameters for (semi-)transparent applications

This section of the chapter defines several terms that will be used throughout, that are important to providing a consistent measure of effectiveness of TSCs. Solar cell efficiency, also referred to as PCE, is a universal metric for solar cells. The average visible transmittance (AVT) is particular to solar cells applied to systems where a measurable level of transparency is useful for multipurpose applications, such as BIPV.

17.2.1 Solar cell efficiency

The energy conversion efficiency (η) of a solar cell is defined as the ratio between the maximum power generated in the device and the incident power. This is generally measured using the following formula [13]:

$$\eta(\%) = P_{out}/P_{in} = V_{oc} \times J_{sc} \times FF/P_{in} \qquad (17.1)$$

where V_{oc} is the open-circuit voltage; J_{sc} is the short-circuit current density; FF is the fil factor. The V_{oc} is defined as a voltage across the circuit when there is a zero current, whereas J_{sc} is the maximum current in the circuit when there is zero voltage, and the fill factor (FF) is the ratio between maximum power output and product of V_{oc} and J_{sc}. Usually, the performance of solar cells is evaluated under air mass 1.5 (AM1.5) conditions. This condition is described as the path length of the sun through the atmosphere when it is zenith.

17.2.2 Average visible transmittance

To find the performance of TSCs, AVT is an important factor. It is defined as the average light transmittance in the visible range of the solar spectrum as observable with the human eye and this is numerically calculated by [Eq. (17.2)] [28]:

$$\text{AVT} = \int T(\lambda) \times V(\lambda) \times \text{AM1.5G}(\lambda)\, d(\lambda) / \int V(\lambda) \times \text{AM1.5G}(\lambda)\, d(\lambda). \qquad (17.2)$$

In Eq. (17.2), $T(\lambda)$ is the transparent spectrum of the semitransparent (or) TSC; $V(\lambda)$ is the photopic response of the human eye; AM1.5G(λ) is the photon flux under AM1.5 light illumination conditions. Typically, 25% of AVT is considered ideal for the TSCs for the window applications [29]. Also, the highest AVT requires the bandgap of the material should be larger than 2.75 eV (with respect to the wavelength 450 nm) [30]. The variation of AVT properties for the different kinds of PV technologies is given in Fig. 17.2 (a-f) [11].

17.2.3 Color rendering index

The color rendering index (CRI) is an important factor that defines the degree of variation between absorbed light and transmitted light. This property is generally expressed with the notation R_α. The higher value of CRI indicates a better color rendering property with a higher neutral color and this value typically varies from 0 to 100. According to the Commission Internationale de l'Eclairage (CIE), CRI is defined as the "effect of an illuminant on the appearance of objects by conscious or subconscious comparison with their color appearance under a reference

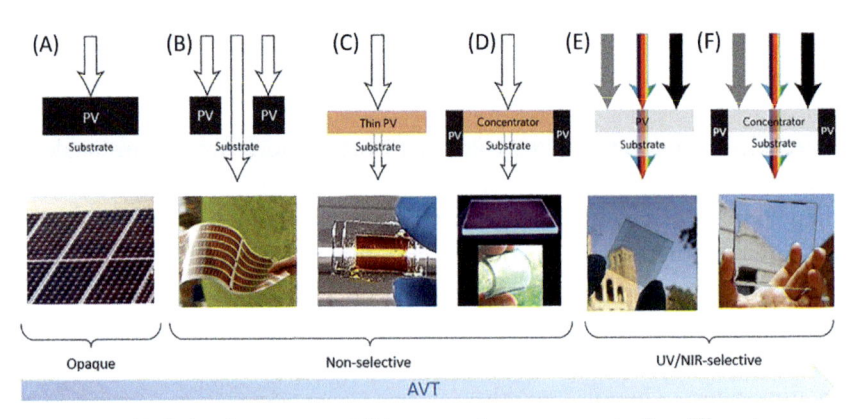

Figure 17.2 Variation in average visible transmittance property for different opaque, nonselective, and ultraviolet/near-infra red selective photovoltaic cell technologies. *Reprinted with permission from E. Pulli, E. Rozzi, F. Bella, Transparent photovoltaic technologies: Current trends towards upscaling, Energy Conv. Management 219 (2020) 112982, copyright (2020) Elsevier.*

illuminant." In the case of glazing type, the value of R_α can be 100 provided that the spectral transmission within the visible spectral range is absolutely constant [31]. The higher value of CRI denotes the transmitted light is sufficient in rendering the colors properly, whereas the lower value represents poor rendering properties. Through modulation of light coupling by the sandwich structure of dielectric material/metal with a low-bandgap organic polymer, the CRI value could be increased [32].

17.3 Electrodes in transparent solar cells

When we consider whether a complete device is transparent or semitransparent, the electrodes of the device should be transparent to optimize light transmittance. In the case of conventional organic or hybrid solar cells, the electrode is not transparent. Therefore, when the light passes through the active layer, the electrode helps to absorb more photons by doubling the optical path. While selecting the electrodes for TSCs, the conductivity and AVT of the material should be considered. An ideal electrode should possess low sheet resistance, high transmittance in the visible spectrum, and good mechanical flexibility. For large-scale applications, the use of indium tin oxide (ITO) is not recommended due to its high cost although it typically exhibits superior optical properties. Furthermore, the brittle nature of ITO restricts its use for flexible device applications. Thus, it is necessary to develop indium(In)-free electrodes for different kinds of applications. Therefore, other metal electrodes are suggested and studied for the TSCs. Specifically, solution-processed transparent metal electrodes have been getting more attention in recent years [33–38]. The comparison of optical transmittance and sheet resistance of the solution-processed electrodes is given in Fig. 17.3 [38].

When both front and back contact electrodes are transparent, in the case of tandem architecture for example, ideally it should minimize the

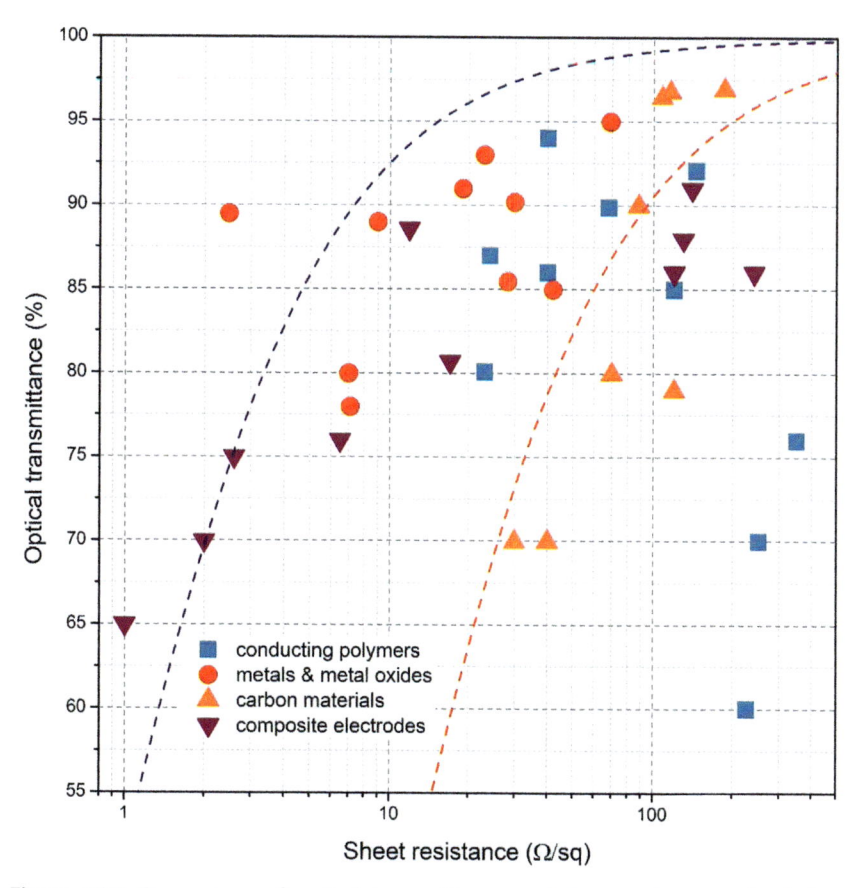

Figure 17.3 Comparison of optical transmittance and sheet resistance of solution-processed transparent electrodes. *Reprinted with permission from Y. Zhang, S-W. Ng, X. Lu, Z. Zheng, Solution-processed transparent electrodes for emerging thin-film solar cells, Chem. Rev. 120 (4) (2020) 2049–2122, copyright (2020) American Chemical Society.*

absorption and reflection in the 350−1200 nm spectral range [39]. When a metal electrode with reduced thickness is used, the AVT can be increased and thus tunable transparency can be achieved. Using this concept, ultrathin metals have been fabricated and their efficiency has been calculated. However, reducing thickness causes high sheet resistance, and hence materials with low resistance and high transmission are important for the electrode in TSCs. To learn more about the different kinds of metal electrodes that are used for future-generation PV, the readers are advised to go through some of the interesting review articles on this topic [40−42]. Because of the higher transmission over other electrodes, silver nanowires (Ag NWs) are seen as a gift to the TSCs industry [43−45]. Ag NWs have similar kinds of resistance and superior transmission properties to ITO electrodes. Importantly, the transmission of Ag NWs is reported more than 90% in many cases, and it indeed boosting up the effectiveness of TSCs. Moreover, while an ordered NCs network is favorable for good conducting ability, a random network of Ag NWs is found to be more efficient for the fabrication of electrodes. To achieve a highly transparent Ag NWs-based electrode, the density of the NWs should be low, which enhances the transparency. Through typical adjustments in the synthetic chemistry, it is possible to achieve such NWs, as reported in the literature [46,47]. Other than better light transmission, Ag NWs have superior conductivity and flexibility. Ag NWs are also more convenient for R2R processing methodologies and amenable to the use of different kinds of solvents. Together with Ag, several other metal electrodes could be fabricated with tunable thickness. These include copper (Cu) and gold (Au); their detailed aspects of the transparent electrode applications could be found elsewhere [34,48−50]. The addition of a buffer layer with the electrode material would eventually enhance the selective charge collection and also block the minority carriers. Besides, for the protection of Ag, the fabrication of multilayer electrodes using MoO_3 is suggested. In this case, the stacked architecture of layer-by-layer assembled $MoO_3/Ag/MoO_3$ transparent electrode provides long durability with good transmission of about 60%−85% in the visible region (400−700 nm). Here, MoO_3 is used to reduce the lattice mismatch of the Ag electrode and to reduce the optical reflection. The thickness of this antireflective MoO_3 layer is typically 30 nm to achieve transparency in the required region [51]. These features enable the deposition of these transparent electrodes on flexible substrates for TSC applications. To fabricate a TSC, it is important to concentrate to deposit transparent metal electrodes. Different approaches are followed including (a) reducing the thickness of the transparent electrode below 20 nm (b) fabricating metallic electrode fingers through screen-printing, and (c) printing semitransparent conductive electrode materials such as PEDOT or Ag NWs [52]. By choosing any one of these approaches, it is possible to achieve TSCs with good efficiency.

Other than MoO_3, materials such as TZO (tin-doped zinc oxide)/Ag/TZO [53], GZO (gallium doped zinc Oxide/Ag/GZO) [54], ZnO/Ag/CuSCN [55], Nb_2O_5/Ag/Nb_2O_5 [56], SnO_x/Ag/SnO_x [57], InZnSiO/Ag/InZnSiO [58], and ZrAcac/ polyetherimide (PEI)/Ag/Ta_2O_5 [59] structures are also used for multilayer electrode applications. Angmo et al. have developed a "flextrode," which consists of PET/Ag grid/PEDOT:PSS/ZnO, which is more suitable for devices with P3HT:PCBM architecture [60]. The authors also demonstrated that the use of Ag NWs/ZnO electrodes is more suitable for devices with tandem configuration. Amorphous oxide electrodes such as InZnSnO (IZTO) have also been used; their performance in solar cells has been evaluated [61]. The degradation of Ag NW electrodes can also be avoided by using an alloyed form of amorphous oxides with another transparent electrode, such as Au, as observed in the case of PSC [62]. Other than these inorganic compounds, carbon-based materials also show incredible performance for transparent electrode applications [63]. Because single-layer graphene exhibits more than 90% of transparency, graphene and graphene oxide electrodes are also used for the inverted type ST OSC with high transparency [64]. The important features of the graphene film such as high conductivity and transparency, high chemical and thermal stability, flexibility in transferring alternative substrates, and ultrasmooth surface. are admitting it to use window applications in solar cells [65].

When both anode and cathode electrodes are transparent, the resultant solar cell could absorb light from both directions. Such a kind of solar cell architecture is called as "bifacial solar cell" and electrodes play important role in this kind of assembly. It is possible to improve the performance of bifacial solar cells through electrode modifications. For example, when Ag NWs are combined with vanadium oxide (V_2O_5), the performance of the fabricated solar cell from Ag NWs side is improved from 5.45% to 8.91% [66]. Here, V_2O_5 increases the performance by reducing the reflection and improving the transmission of Ag NWs electrode. Properties such as mechanical properties, optical transparency, electrical conductivity, and thermal stability are the key factors in determining the effectivity of the transparent electrodes. As indicated earlier, the diameter of the NWs should be optimum, which provides compact films for the conductive pathways for efficient charge transport [67]. The comparison of optical transmission and sheet resistance of the solution-processed TSCs is represented in Fig. 17.3 [38]. Although Ag NWs are considered as most suitable for the cathode electrode application in TSCs, the high order of penetration of the active layer reduces the short-circuit current in the device, which may lower the efficiency considerably [68]. Also, it is possible to get degradation under ambient conditions owing to the higher aspect ratio of NWs. To avoid these issues, better encapsulation methods are essential to maintain efficiency and stability. Also, leakage current generated in the circuit during poor patterning will reduce the V_{oc} and FF of the device hence patterning carries important observations for the effectiveness of the device.

Patterning of the electrode is one of the important steps for the fabrication of TSC module systems. Efficient patterning of electrodes reduces the dead area of the device and creates a recombination network. Although several kinds of patterning methods are employed to fabricate the panels, laser patterning has been a well-established approach in recent decades. The salient features of laser patterning such as high-precision, high-throughput, and mask-free patterning processes are advancing the laser patterning approach for the fabrication of future-generation PV devices [69]. Laser patterning of the electrode is usually achieved through short pulse lasers, which form the efficient series connection between the subcells of the device. This method is often called as "laser ablation or laser scribing" technique and with respect to the energy of the laser beam, the removal of the active layers is achieved. The level of laser ablation of the layers in a thin-film Si solar cell is denoted with different notations, namely Pl, P2, and P3. The P1 laser scribing is carried out with the ITO-coated glass substrate using the laser with 1064 nm. Whereas the P2 pattern usually scribes the active layer with high energy (e.g., 532 nm) and after deposition of the contact electrode P3 pattern (i.e., electrical separation of the top electrode) is achieved. For patterning the substrate, compared with a normal laser beam, ultrafast laser pulses are highly preferred due to the precise ablation of the layers with minimum thermal influence. Therefore, the threshold fluence of electrodes should be determined before the electrodes are patterned. The collective representation of the laser scribing process with the general device architecture of a solar cell is given in Fig. 17.4 [13,70]. In general, a solar cell can be patterned through either during-patterning or postpatterning. For the postpatterning, a laser is used, and patterning during deposition is achieved through patch/intermittent coating or self-assembled coating [71].

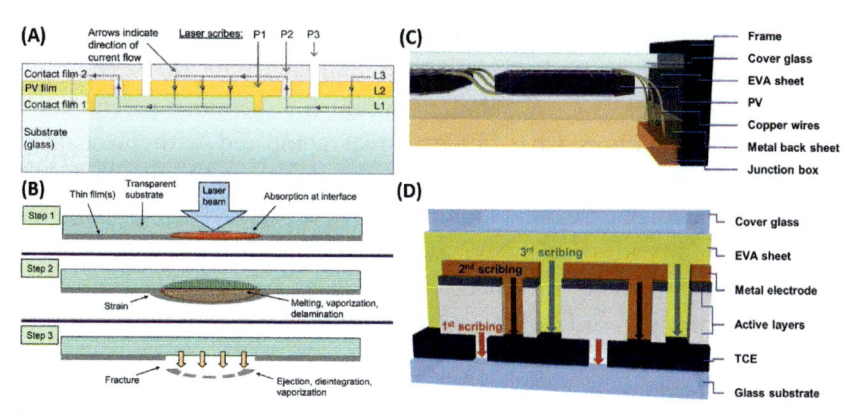

Figure 17.4 (A) Basic geometry of a thin-film solar cell with laser scribing, (B) removal of thin film through a laser beam and its effect through different stages. (C) Schematic diagram of the c-Si photovoltaic (PV) module and (D) a thin-film PV module fabricated through laser scribing process *(A) and (B) reprinted with permission from J. Bovatsek, A. Tamhankar, R.S. Patel, N.M. Bulgakova, J. Bonse, Thin film removal mechanisms in ns-laser processing of photovoltaic materials, Thin Solid Films 518 (10) (2010) 2897–2904, copyright (2010) Elsevier; (C) and (D) reprinted with permission from K. Lee, H.-D. Um, D. Choi, J. Park, N. Kim, H. Kim, K. Seo, The development of transparent photovoltaics, Cell Rep. Phys. Sci. 1 (8) (2020) 100143, copyright (2020) Elsevier.*

In the case of a flexible substrate, femtosecond and picosecond laser pulses are efficient in scribing for PV applications. It is found that femtosecond laser pulses induce minor damage to the ITO substrate [72]. For the monolithic tandem architecture, the interconnection between the two contact electrodes is a complex issue and much attention is required. Patterning of the electrode could be achieved either pre- or postprocessing of the active layers. While postprocessing, meticulous observation is required because the laser beam may affect the thin active layers and influence the device's efficiency. Also, a high level of laser ablation of the layers will affect the bottom electrode and may affect the device properties severely [69]. Hence, to achieve efficient laser ablation, a narrow laser pulse-width beam with a high repetition rate and peak power with the above material threshold is required. Besides, pulse-to-pulse stability and the quality of the laser beam are quite important to achieve a successful laser ablation process. Also to extract more amount of current, the inactive area of the solar cell (also called "dead cell") should be reduced with an efficient laser patterning approach. When the ultrafast pulse is used, the material damage of the active layer can be significantly avoided. It is demonstrated that through the ultrafast patterning approach, the FF in the resultant STSC module was significantly enhanced (up to approximately 63%) [69]. Rakocevic et al. used a picosecond laser with the wavelength 355 nm and made a PSC into the translucent (i.e., semitransparent) form [73]. Here, the authors have observed that compared with mechanical patterning, laser patterning is very efficient in the removal of layers without any residue. For flexible substrates such as PET, laser scribing has been carried out with the Ag NWs deposited PET [74]. In this analysis, it is found that by altering the energy of the laser, it is possible to remove Ag NWs from the surface of the PET efficiently. It is important to achieve a smooth and damage-free edge in the substrate to avoid the short-circuit in the device. Because surface deformation is hardly affecting the performance of the solar cell, much care should be taken when we use laser for the scribing of the substrate. From all these discussions, we can state that efficient laser scribing is essential to achieve good carrier transport in TSCs.

17.4 Types of transparent solar cells

As indicated in the introduction, transparency in organic and inorganic layers has led to the fabrication of future-generation TSCs for numerous kinds of applications. There are different varieties of solar cells with different configurations and almost TSCs have been fabricated in all these cases. Each category has some advantages with different aspects. The important experimental findings related to each category of the TSC are discussed below.

17.4.1 Organic transparent solar cells

17.4.1.1 Organic single-junction transparent solar cells

Organic semiconductors are well known for their high absorption coefficient values, which enable them to be deposited as a very thin active layer for TSC applications [75]. Also, their bandgap tunability extends from visible to NIR region, which helps to harvest a large number of photons. There are two different general architectures in OSCs for charge collection, namely bilayer heterojunction and bulk heterojunction. Fabrication of thin films in OSCs can be accomplished with different colors, which realize a high level of chromaticity. Also, by employing chemical routes, it is possible to synthesize the organic polymers with very narrow bandgap to harvest a large amount of photons up to the NIR region. In general, because of this narrow bandgap, semiconducting polymers could selectively absorb photons from the NIR region and allow the rest of the light to pass through. Importantly, by substituting electron-deficient acceptor molecules on their structure, it is possible to increase the intramolecular charge transfer process to obtain large transparency in the visible region. Besides, they are easily processed through solution-based low-cost deposition methods such as spin-coating and dip-coating. Also, sequential deposition of the layers has emerged as one of the potential routes to fabricate STSCs. Furthermore, large-scale processing methods such as slot-die deposition method and screen printing are possible, which help to fabricate organic polymers through the R2R processing approach. In the slot-die method, the material dissolved in a solution is dropped by a slot and deposited with squeezing.

Because of their versatility in printing large-area flexible conducting substrates, organic TSCs are often preferable for window applications. Concerning the photoactive layer, the color of the organic TSCs can be modified. When illuminated, an OSC undergoes three key processes, including the generation of exciton, dissociation at the interface, and efficient charge collection at their respective electrodes. Of two principal configurations of OSCs (bilayer and bulk-heterojunction), the bulk-heterojunction type is most-often studied and preferred due to efficient charge collection. The efficiency of the tandem-type organic TSCs has increased to 14.1% [76] due to the improved synthetic strategies to obtain narrow bandgap polymers. A wide-bandgap acceptor with high-lying lowest unoccupied molecular orbital (LUMO) is also interesting to improve the performance of an organic STSC [77]. Although ITO provides versatility in acting as a cathode layer, due to the high work-function, ITO-based electrodes are often surface-modified with potential inorganic compounds to reduce their work-function value. Because of their different functionality, organic polymers are identified as promising materials for surface modification. The first report on the transparent OSC was reported by Lunt and Bulovic in 2011 [78]. Here, the authors used chloroaluminum phthalocyanine (ClAlPc) as a donor and C_{60} as a molecular acceptor. Through optimizing the NIR interference, 1.70% ± 0.1% efficiency was achieved with the AVT 56% ± 2%. The schematic diagram of the fabricated structure, complex refractive index of the fabricated layers, current−voltage (IJV) measurement, and external quantum

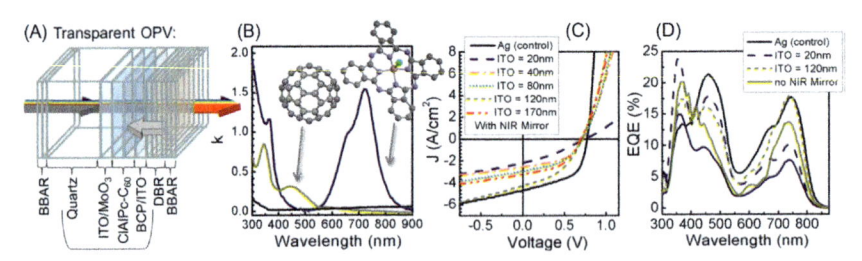

Figure 17.5 (A) Schematic diagram of the fabricated organic TSC device, (B) imaginary part of the complex refractive index of the fabricated active layers, (C) current—voltage (J—V), and (D) external quantum efficiency curves of the fabricated device. *Reprinted with permission from R. R. Lunt, V. Bulovic, Transparent, near-infrared organic photovoltaic solar cells for window and energy-scavenging applications, Appl. Phys. Lett. 98 2011 113305, copyright (2011) AIP Publishing.*

efficiency (EQE) of the fabricated device are given in Fig. 17.5. This investigation opened a gateway in the fabrication of TSCs research. Following this, by modifying an amphiphilic conjugated polymer PFPA-1, Tang et al. fabricated an STSC with two kinds of donors (P3TI and TQ1) [79]. In this case, a high internal quantum efficiency (approximately 90%) was achieved together with 50% transmittance. Effective bonding with ITO substrate through functional groups in PFPA-1 helps to achieve this kind of interesting result. Generally, a high CRI value often results in poor efficiency but when ternary organic components with suitable band alignment are used to fabricate a STSC, such a system achieves an efficiency of 9.73% with a CRI of nearly 100 [80].

The primary advantage of the organic semiconductor is that it is possible to modify the absorption spectrum by designing the molecular structure of the respective components presents in the structural moiety. Chen et al. designed a UV-NIR sensitive active layer donor compound, PBDTT-DPP, and fabricated STSCs using Ag NWs electrode [81]. Here, sol-gel-prepared TiO_2 nanoparticles were used to improve the adhesion of Ag NWs and to function as electron transporting layer. With the device structure of ITO/PEDOT:PSS/PBDTT-DPP:PCBM/TiO_2/Ag NWs, a PCE of 4% was achieved with 66% transparency at approximately 550 nm. Similar to the ITO, when Ag electrode is coupled with a capping (or) optical coupling layer like tellurium dioxide (TeO_2), the transmittance of Ag layer could be significantly improved [82—84]. Because of the high refractive index value (2.2), TeO_2 serves as the dielectric material improving the light transmission at the metal/dielectric interface [85]. The thickness of this TeO_2 layer is important because it is observed in the case of QDSSC that the optimum thickness required for the protection purpose is 20–30 nm [83]. Other than functioning as improving the transmittance, TeO_2 is serving as a back reflector of the light that comes from the ITO side [86]. Fabrication of organic semiconductor-based TSCs yet needs delicate control of the layers that are deposited. Although metal-based electrodes are widely used for solar cells, metal-free OSCs are also demonstrated with potential efficiency.

Zhou et al. used spin-coated, highly conductive PEDOT:PSS layers as top and bottom electrodes, and the bottom electrode was deposited on the glass substrate [87]. With P3HT:PCBM active layer, the authors achieved 1.8% efficiency under A.M 1.5 conditions. Despite low efficiency, this approach implies the possibility of replacing metal electrodes and reducing the fabrication cost. As already pointed out, the low bandgap of the semiconducting polymers is offering a large amount of photon absorption in the solar spectrum. Organic molecules that absorb in the NIR region are receiving more interest because they can block the NIR rays from the sunlight, which is beneficial for BIPV applications. There are several kinds of NIR absorbing organic acceptor molecules and their effect on the PV performance is evaluated [52,88,89]. Liu et al. used three kinds of IR-sensitive organic molecules namely IEICO-4F, F8IC, and FOIC as acceptors together with PTB7-Th as donor to construct a flexible type TSC [90]. With PTB7-Th/IEICO-4F combination, the cell efficiency reached ~10% with AVT 34%. However, the AVT values of most of these solar cells remain at the level of less than 40%. Li et al. fabricated an STSC (approximately 50% transparency) using a nonfullerene acceptor and achieved more than 10.8% efficiency [91]. Such kind of higher efficiency is also achieved by other groups using nonfullerene acceptors [92,93]. The use of nonfullerenes and fused ring acceptor molecules is interesting in TSCs because of their absorption in the NIR region, tailoring frontier energy levels, low synthetic cost, controlling crystallinity, etc. [94]. Because of their NIR absorption, OSCs with NIR acceptors mainly focused on windows and curtains applications. The ideal characteristics required for the windows and curtains using organic STSCs are listed in the Table. 17.1 [52].

Although much of the literature describes acceptors based on nonfullerene molecules, it is possible to fabricate an active layer (both donor and acceptor molecules) that is comprised of NIR active materials. Xie et al. have fabricated an OSC that consists of both donor and acceptor as NIR active components [95]. The fabricated device shows up to 61.5% transparency with an efficiency of 4.2%. Similarly, despite of low open-circuit

Table 17.1 Parameters required for the fabrication of solar windows using organic semitransparent solar cells.

Item	Target value
Chromaticity	Architects prefer blue, green, and neutral gray
Transparency	$\geq 30\%$
Geometrical fill factor (GFF)	$> 90\%$
Active area (AA)	> 1 cm^2

Source: Assembled by authors from data contained in T. Winkler, H. Schmidt, H. Flugge, F. Nikolayzik, I. Baumann, et al., Efficient large area semitransparent organic solar cells based on highly transparent and conductive ZTO/Ag/ZTO multilayer top electrodes, Org. Electr. 12 (10) (2011) 1612−1618.

voltage (V_{oc}) of organic semiconducting polymers, it is possible to produce more than 1 V of V_{oc} through thiophene copolymer for TSC applications [96]. Because NCs are also used for electron collection purposes, modification of the electron collection layer could lead to higher efficiency, over 12% [97]. Most of the laboratory-based OSCs are fabricated using a spin-coating method and hence controlling thickness is more feasible, and this helps to fabricate very thin active layers on a nanometer scale. However, it is proved that through the spray coating method, it is possible to achieve transparent layers for the fabrication of OSCs. Because of the selective tunability, organic compounds that absorb in the NIR region were also developed and delivered over 50% of AVT [51,98]. Compared with other technologies, through organic STSCs, it is possible to achieve a high value of the CRI, up to 100 [99]. For the large-scale deposition of TSCs, different methods such as slot-die coating, ink-jet printing, gravure printing, and screen printing are used. Each deposition process presents particular advantages and disadvantages; when properly integrated during manufacturing, TFSCs have been commercialized successfully. The required physical parameters for each method for the uniform coating are different and using one example, the viscosity requirement of each method for the uniform coating is given in Table 17.2. [100].

Inverted organic TSCs have also attracted considerable attention because of their potential for improved performance. By inserting a buffer layer of poly-ethylene oxide (PEO) between ITO and active layers, Zhou et al. have achieved fourfold enhancement in the performance of the resultant device [101]. In any case, the fabrication of transparent electrodes in the architecture is a challenging issue. For the hole extracting purpose, transition metal oxides such as molybdenum trioxide (MoO_3) and tungsten trioxide (WO_3) are generally used and thin layers of these compounds are

Table 17.2 Viscosity requirements for the different kinds of deposition techniques.

Deposition technique	Viscosity (mPa S)
Jet printing	1−40
Slot die coating	1−10,000
Gravure printing	15−200
Flexo printing	15−500
Screen printing	50−50,000

Source: Adapted from: Y. Galagan, R. Andriessen, Organic photovoltaics: Technologies and manufacturing, in: Third Generation Photovoltaics, Edited by Vasilis Fthenakis, 2012 Intech Publishers, distributed under the terms of the Creative Commons (CC-BY 4.0 (http://creativecommons.org/licenses/by/4.0/)) license, which permits unrestricted use.

quite feasible for carrier extraction. Other than functioning as a hole extraction layer, MoO_3 also functions as a protecting layer for the active layer from the sputter deposition of the contact electrode. Using this concept, an inverted solar cell with traditional active layer P3HT:PCBM has been fabricated and about 80% of transmission in the red region of the visible spectrum is realized [102]. The successful fabrication of organic TSCs also relies upon the fabrication of electrodes for charge transport. As discussed, the electrodes should possess high transparency, high conductivity, and appropriate work function. While depositing electrodes for the inverted solar cells, care should be taken because the transparent electrode is deposited on the active layer. Although all these features play a significant role in fabricating organic TSCs, demerits such as low-carrier mobility and less susceptibility to the atmosphere are pushing the search for other potential materials for the fabrication of stable, highly-efficient TSCs.

17.4.1.2 Organic tandem transparent solar cells

Compared with single-junction devices, a tandem architecture enables better light management, thereby reducing the total loss by reflection and parasitic absorption. Specifically, the typical thermalization loss that occurs in the narrow bandgap solar cell can be reduced when it is combined with a NIR TSC. Fortunately, because of the lower bandgap, organic molecules absorb NIR photons and are quite useful in the fabrication of tandem solar cells. In the case of tandem organic TSCs, where multiple junctions are coated on the substrate to harvest a large fraction of the solar spectrum, every layer of the device is important in providing transparency. Such tandem devices deliver impressive efficiency performance with good transparency. Significant developments in organic polymers, additives, electrodes, and solvent-assisted deposition technologies have led this research area into a new dimension. Here, two different bandgap layers of organic components are fabricated on the substrate, and carefully designed electrodes are applied to their respective structure. During fabrication, solvent-based deposition of every layer plays a key role in achieving higher performance. Using a NIR absorbing fused hexacyclic electron acceptor, Wang et al. fabricated a high-performance TSC that could deliver an efficiency of 9.77% [103]. Among the organic tandem TSCs, organic-perovskite tandem architecture has resulted in high performance in recent years. Especially over the four-terminal tandem approach, over 15% efficiency has been demonstrated [104]. Through solution-processed deposition approaches, it is possible to achieve highly crystalline, dense, and pin-hole-free perovskite layers as well as electron and hole-transporting layers. While fabricating organic/perovskite tandem solar cells, the challenge comes with the interconnecting layer. It is possible to frame these interconnecting layers through the thermal evaporation method, which results in higher efficiency. There is some better understanding through simulations-based investigation also available in this area [105,106].

Zuo et al. fabricated a $FAPbBr_{2.43}Cl_{0.57}$ perovskite layer; the single junction solar cell delivered efficiency of 7.5% with AVT 68% [107]; utilizing a device configuration of glass/ITO/NiO_x/PSS/$FAPbBr_{2.43}Cl_{0.57}$/$PC_{61}BM$/ZnO-NPs/LS-ITO/M-PEDOT:PSS/PTB7-Th:6TIC-4F/ZnONPs/ITO, the authors achieved 10.7% efficiency with AVT of 52.91%. [107]. Recently, Chen et al. have demonstrated that through passivating NiO hole-transporting layer using benzylphosphonic acid, the fabricated organic/perovskite tandem STSC could deliver 23.60% [108]. All these results confirm that organic tandem TSCs are promising for the future generation of solar cell architectures for BIPV and related applications.

17.4.2 Inorganic semiconductor-based transparent solar cells
17.4.2.1 Silicon-based transparent solar cells
As discussed in the introduction, the PCE achieved by crystalline Si in a traditional solar cell has turned out to be the same for TSC purposes. Achieving transparency in Si is a challenging task compared with other inorganic materials. Transparent silicon (Si) solar cells are highly interesting because of the high stability of Si over other kinds of inorganic semiconductor nanomaterials studied for PV applications. Silicon is normally opaque and achieving transparency was thought to be difficult. However, transparency has been achieved at a later stage by different approaches. Transparency in Si-wafers can be achieved through different methods,

including wet-chemical etching and laser. Through these methods, transparency is achieved through the selective formation of invisible holes on the Si-wafer, and due to these microholes, the resultant silicon wafer becomes transparent. The widely used silicon types namely, hydrogenated amorphous silicon (a-Si) and crystalline (c-Si) possess different band gap values (1.1 eV for c-Si and 1.5–2 eV for a-Si:H) and so the properties are entirely different to fabricate large-area solar cell devices. When reducing the thickness of crystalline-Si, care should be taken because Si has a brittle nature and any kind of instrumental process could break it easily. As discussed, to make transparency in Si wafers, chemical etching is one of the widely followed techniques. In this method, the thickness of the Si is reduced by a wide variety of etching chemical compounds (called "etchants"). With respect to the etching time, the light transmission in the visible region can be varied from 10 to 70% [109].

At the beginning of the 1990s, Takeoka et al. first reported on the fabrication of a see-through type a-Si solar cell through chemical etching as well as laser patterning [110]. Here, the fabricated submodule of the a-Si solar cell (35 cm × 76 cm) showed 3.1% efficiency with 10% transmittance. This investigation opened a new paradigm in this area to achieve transparent-Si through etching and other potential approaches. Lee et al. used the deep reactive ion etching method in order to make a light transmission window on the c-Si wafer with 200 μm thickness [111]. Through this, the authors achieved a transparent c-Si wafer that contains light transmission holes that are invisible to the eye. Also, with a p-type emitter layer, the transparent c-Si wafer-based solar cell (1 cm) showed 17.2% efficiency. The fabricated solar cell structure with other device parameters obtained through this analysis are represented in Fig. 17.6.

A detailed analysis of the etching of Si wafers using several kinds of chemical compounds is available elsewhere [112–114]. Park et al. used $HF/HNO_3/CH_3COOH$ (HNA) mixture to etch the c-Si wafer and the fabricated glass-like TSC with 25 cm^2 size showed 14.5% efficiency with 20% transmittance [109]. Although chemical etching has the potential to fabricate see-through type Si-TSCs, compared with the laser patterning approach this is less effective. Other than these two, plasma-enhanced chemical vapor deposition (PECVD) is followed to fabricate Si-TSCs. In this method, the Si layer is deposited on a nanometer scale and used for commercial purposes. For the industrial production of transparent (or) semitransparent-Si, this method is one of the widely followed approaches and different colors are achieved. Owing to the higher bandgap of the

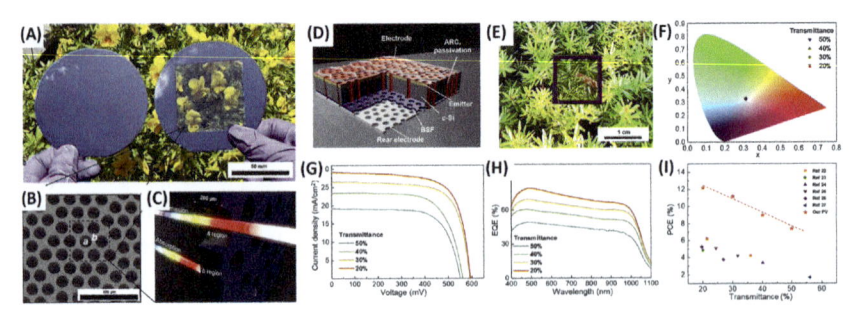

Figure 17.6 (A) photo of a polished c-Si wafer and neutral-colored c-Si wafer, (B) scanning electron microscopy image of the c-Si substrate (a = light transmission window and b = light absorption region), (C) illustration of the neutral-colored transparent c-Si substrate, (D) schematic representation of the c-Si based TSC, (E) photo of a TSC with the transmittance 50%, (F) color coordinate of the c-Si TSC with transmittance on the CIE 1931 chromaticity diagram, (G) J−V characteristics of the fabricated device, (H) external quantum efficiency of the device, and (I) comparison of the efficiency performance with the literature data. *Reprinted with permission from K. Lee, N. Kim, K. Kim, H.-D. Um, W. Jin, et al., Neutral-colored transparent crystalline silicon photovoltaics, Joule 4 (1) (2020) 235−246, copyright (2020) Elsevier.*

amorphous-silicon (a-Si), as compared to c-Si, the thickness can be controlled easily and hence this is used to fabricate TSCs through the sputtering technique. The drawback of Si-based TSCs is the need for a higher level of instrumental facility like PECVD and this is also consisting of critical experimental steps. It is hoped that future efforts in achieving transparent-Si through economically viable methods will solve difficulties associated with achieving Si-based TSCs. Patel et al. fabricated a TSC with the structure glass/FTO/nc-p-Si/i-Si/nc-n-Si/AZO/AgNW/TiO$_2$/Pt (nc = nanocrystalline; i = intrinsic; AZO = Al-doped ZnO) [115]. Here, the a-Si layer was deposited using PECVD, and AZO and TiO$_2$ layers were deposited using a sputtering method. The final TPV coupled with a transparent photoelectrochemical cell (TPEC) showed PCE of approximately 1.66%, with an impressive V_{oc} = 0.852. Kang et al. fabricated a TSC based on a free-standing n-silicon microwire array structure [116]. This structure is quite useful to fabricate a flexible type of TSC, which helps to reach an efficiency of up to 8%. Such flexible type transparent or STSCs are useful for diverse applications including power supply to contact lenses [117]. Advanced technologies such as transverse laser scribing and back-reflector screen printing are also found to be useful in fabricating see-through a-Si/microcrystalline(μ)-Si solar modules with different colors [118]. In the case of a-Si/μ-Si type tandem TSCs, modification in the

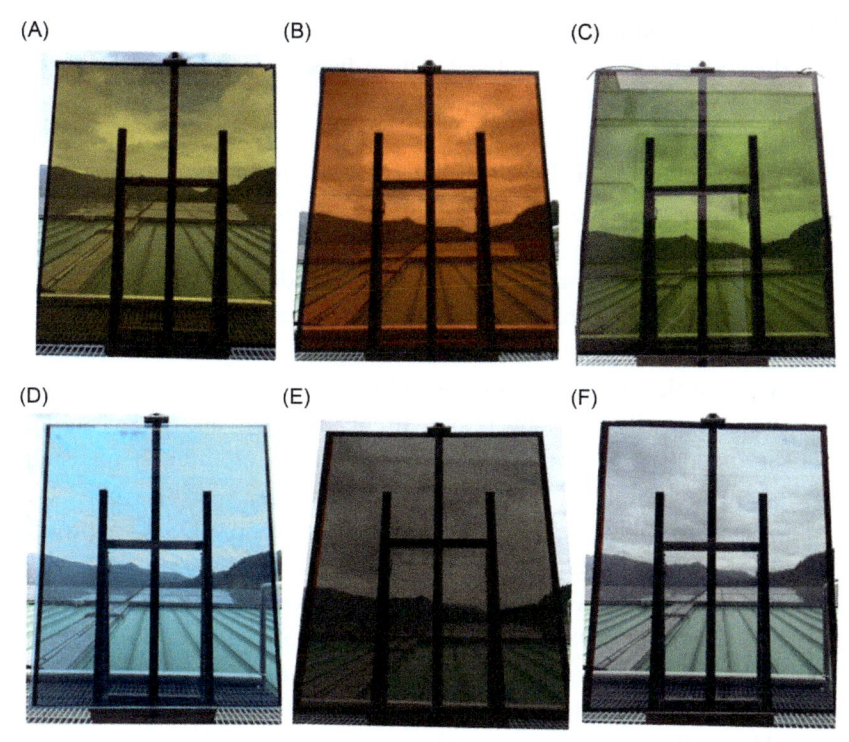

Figure 17.7 (A-F) Images of the a-Si:H semitransparent photovoltaic module fabricated through different approaches. *Reprinted with permission from S. Y. Myong, S. W. Jeon, Design of esthetic color for thin-film silicon semi-transparent photovoltaic modules, Solar Ener. Mater. Solar Cells 143 (2015) 442−449, copyright (2015) Elsevier.*

thickness of the bottom cell of hydrogenated microcrystalline Si (mc-Si:H) results over 8% efficiency [119].

Using a-Si:H, Kim et al. constructed a TSC through which the authors could achieve about 36% efficiency under the illumination of 3000 l × LED light [21]. This kind of highly efficient TSC is proposed for indoor light applications although high intensity light is required. Because the requirement usually consists of different colors, the color of the resultant semitransparent a-Si:H-based solar could be modified using the insertion of the intrinsic amorphous hydrogenated silicon (i-a-Si:H) together with the attachment of a back glass, simply inserting a colored back glass alone or insertion of a polyvinylbutyral (PVB) film with a back glass [120]. The images of the fabricated semitransparent a-Si:H solar module through this approach are given in Fig. 17.7. Through these kinds of technological developments, it can be stated that silicon-based TSCs could be made for commercialization in the near future.

17.4.2.2 Dye-sensitized transparent solar cells

Since their invention in 1991, DSSCs have been extensively studied with different aspects, and their promising avenues for different applications are explored [121−124]. The basic structure of DSSC consists of a mesoporous TiO_2 photoanode, a platinum(Pt) counter electrode, a sensitizing dye that absorbs a broad range of visible/NIR spectrum, and an electrolyte for charge transport. Most of the reports discuss the PV properties of ruthenium-based metal complexes, counter electrode materials, electrolytes, and modification of photoanodes. When light passes through a DSSC, the active dye molecules absorb and get excited. This molecular excitation transfers the carriers into the respective electrode through a redox electrolyte, which essentially generates photocurrent in the circuit. Several studies represent the different kinds of photoanode and photocathode materials, dye molecules, electrolytes, and device architectures [125−128].

Thus far, the highest efficiency achieved through the DSSC approach is over 15% through cosensitization process [129]; this can be further improved via novel materials and fabrication procedures. After the development of solid-state DSSCs, transparency in DSSCs is overlooked and analyzed by potential research groups. Zhou et al. used PEO-based gel electrolyte with TiO_2 nanoparticles/N719 for efficient charge transport [130]. Here, the use of quasisolid state polymeric electrolyte avoids leakage of electrolyte from the device and also improves the stability and conductivity. In this study, the fabricated STSC with 7 μm thickness of TiO_2 nanoparticles was shown 5.78%, and the same device with rear irradiation (bifacial type) was shown 3.76%. Moreover, the device has shown consistent efficiency performance for more than 1000 hours under unsealed conditions. This shows further investigation on atmospherically stable compounds will deliver a fruitful result in developing solid-state transparent DSSCs. Because counter electrodes play an important role in DSSC, a transparent counter electrode would be beneficial for the transportation of photons. The use of platinum, in this case, would be an expensive approach and so the development of transparent electrodes using Earth-abundant materials is highlighted [131]. Hussain et al. prepared large area transparent tungsten sulfide (WS_2) counter electrode through the radio frequency (RF) sputtering method and used it for the fabrication of DSSC [132]. Similarly, a transparent graphene electrode fabricated using a high-temperature method was also found to be useful for the fabrication of DSSC with high transparency [65]. For the sensitizers, different kinds of sensitizers are used to achieve semitransparency and their performance is evaluated. In most cases, NIR-based sensitizers are preferred because they are mostly colorless and thus provide a high level of transparency [133]. Therefore, squaraine, cyanine-based sensitizers are getting more important owing to their NIR absorbing ability. Recently, Baron et al. used pyrrolopyrrole cyanine dyes namely TB144 and TB207 to harvest NIR photons and achieved 4% efficiency with AVT 76% [134]. Here, the replacement of fluoride functional group by a phenyl group on boron is found to significantly improve the device performance. Such modifications in the functional group of the dyes will also be useful for developing transparency and other properties of the device. In the NIR dyes, aggregation is one of the major problems that affect the performance of the solar cells. However, the addition of additives, such as chenodeoxycholic acid (CDCA), could solve this problem. When NIR-selective heptamethine cyanine dyes are used to fabricate a TSC, the addition of CDCA is found to improve the performance of the device by reducing aggregation of the dye [135]. This kind of coadsorbents often improves the electron and hole injection of TiO_2 film and improves the efficiency [136]. In this case, the fabricated device showed 3.1% efficiency under extremely low temperatures ($-20°C$) with AVT 74%.

Through a multidye approach that connects the UV and NIR region, Zhang et al. achieved 3.4% efficiency with high transmittance (60.3%) using squaraine dyes (Y1 and HSQ5) [137]. Because of the limitation of such dyes, most studies are conducted using transparent electrodes and achieving highly efficient dye-sensitized TSCs [138,139]. Vesce et al. have reported the fabrication of a semitransparent DSSC module using a thiazole dye, TTZ5 [140]. Twenty modules (400 cm^2 each) of the prepared semitransparent DSSC panel showed a maximum efficiency of 2.7% under a tilt angle of 60 degrees. Also, the best module delivered 5.1% efficiency with a transparency of 35.7%. Besides, the module shows good stability records under thermal stress, which clearly emphasizes their possible applications for the BIPV sector. The fabrication steps and the $I-V$ curve of the best-performing module are given in Fig. 17.8. Despite DSSC TSCs suffering from the stability of the dyes and electrolyte-related issues, the use of novel dyes with long-term stability could deliver promising results in the near future.

17.4.2.3 Colloidal quantum dot-based transparent solar cells

Colloidally prepared semiconductor quantum dots (QDs) are getting much attention in the last few decades owing to their size and shape-dependent

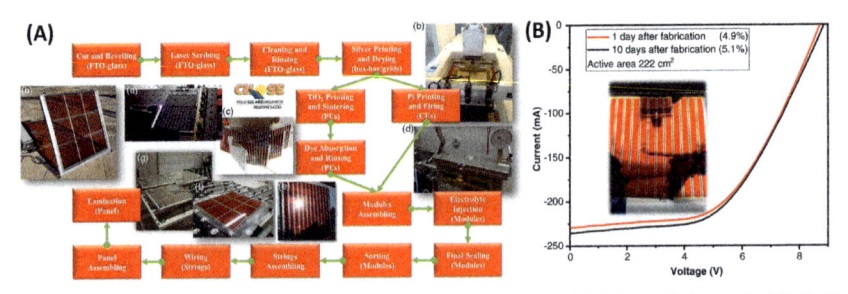

Figure 17.8 (A) Fabrication steps of a semitransparent DSSC module and (B) I–V curve of the best-performing module. *Reprinted with permission from L. Vesce, P. Mariani, M. Calamante, A. Dessi, A. Mordini, L. Zani, A. D. Carlo, Process engineering of semitransparent DSSC modules and panel incorporating an organic sensitizer, RRL Solar 6 (8) (2022) 2200403, copyright (2022) John Wiley & Sons, Inc.*

optical properties [141,142]. Metal chalcogenides (ex: CdTe, HgTe), metal phosphides (ex: InP), metals (ex: Si), and metal halide perovskites (ex: $CsPbBr_3$) QDs are prepared using high-temperature hot-injection method, as well as room temperature approach, and their excellent structural and optical properties are examined [143–147]. The presence of a strong quantum confinement effect and multiple exciton generation in QDs make them suitable candidates for PV-related applications. It is well known that QDs could harvest photons from UV to NIR region depending on their composition and size and hence fabrication of optoelectronic devices using these materials is much more attractive. Also, these materials could be deposited using cost-effective approaches such as spin-coating, spray-deposition, and doctor-blade deposition,. which makes them ideal for commercial purposes. Other than the above-discussed varieties of TSCs, colloidal QDs also be used for the fabrication of TSCs due to their outstanding optical properties. Zhang et al. first fabricated an STSC using PbS QDs as the active layer with a transparent Au electrode [148]. The fabricated solar cell, device assembly, J–V curve, and IPCE spectra of the device are given in Fig. 17.9 [148]. Using TeO_2 as a capping layer, Zhang et al. fabricated a QDSSC with the device configuration ITO/IZO/PbS QDs/Au/TeO_2 [83]. The authors achieved 7.3% efficiency with an AVT of 20.4% and this AVT value is found to increase when reducing the thickness of QD film. Previously, with the MoO_3/Au/MoO_3 transparent electrode, the authors were able to achieve 5.4% with AVT 24.1% [149]. This reveals the influence of thickness in modifying the AVT values. Other than QDs, bilayer chalcogenides-based heterojunction also seems to be a potential candidate for TSC applications. Cho et al. fabricated a bilayer structure based

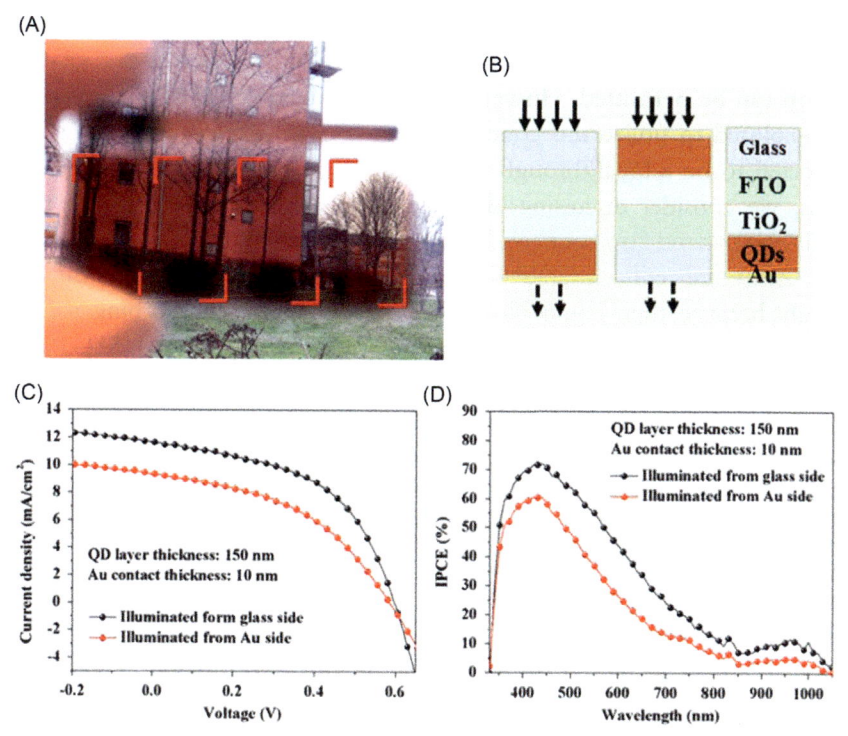

Figure 17.9 (A) Image of the fabricated transparent QDSSC, (B) schematic representation of the device structure with the direction of the illumination, (C) J–V curve of the device under different measurement conditions, and (D) IPCE spectra of the fabricated solar cell. *Reprinted with permission from X. Zhang, G. E. Eperon, J. Liu, E. M. J. Johansson, Semitransparent quantum dot solar cell, Nano Energy 22 (2016) 70–78, copyright (2016) Elsevier.*

on a two-dimensional (2D) WSe_2/MoS_2 junction on the glass substrate and deposited a layer of ITO on it (100 nm thickness) by RF sputtering method [150]. In this case, the authors achieved an impressive efficiency of approximately 10% with a high level of transparency (approximately 80%).

17.4.2.4 Metal-oxides-based transparent solar cells

Metal oxides are gaining attention due to their intriguing structural and electrical properties; several binary metal oxide semiconductors are being investigated to exploit them for PV applications. Also, compared with organic semiconductors, the stability of metal-oxides is quite high so they have significant potential for fabricating devices with good stability. By making a $p-n$ junction through the deposition of metal-oxide all these devices have shown long-term stability over months, which is higher than the organic, QDs, and halide perovskite-based TSCs discussed here. In view of expanding these metal oxides to the large area, the 25 cm^2 TPV with the structure FTO-coated glass/p-NiO/Si/n-ZnO/AZO/AgNWs (p-Si-n) produced 10.82 mW with 43% AVT [160]. These results provide a big hope for the metal-oxide heterojunction-based TSCs for the large-scale production and their use for building facades.

Chatterjee et al. fabricated a Cu_2O/SnO_2 heterojunction through the successive ionic layer adsorption and reaction (SILAR) method [161]. This (SILAR) method enables workers to sequentially deposit a thin layer of the semiconductor through precursors dissolved in solution and thickness can be precisely tuned through the number of cycles involved in the deposition. The authors deposited NiO and ZnO to block the electron and as a buffer layer using multiple-cycle deposition. The final device with $NiO/Cu_2O/ZnO/SnO_2$ structure showed about 1% efficiency, which indicates further optimization may produce higher photocurrent generation in this architecture. These kinds of multiple layer deposition methods are also useful to avoid recombination loss in the device.

Rana et al. fabricated a TSC with approximately 80% transparency, which comprised of the active layers $Cu_2O/ZnO/AZO$ [162]. Such a high level of transparency is also achieved by the same group using $Cu-Cu_2O/ZnO$ as an active layer [163]. All these layers are fabricated by means of the sputtering method and the fabricated device showed 0.7% efficiency. Although the performance of the device is not so high in this case, the obtained V_{oc} for this configuration (0.52 eV) shows that efficiency enhancement is still possible by varying the parameters. Because CuO and Cu_2O could generate different colors with respect to their composition, it is possible to tune them accordingly with different ratios of precursors in the synthesis. Recently, a TSC with $Cu_2O/CuO/SnO_2$ has been developed by the combination of thermal oxidation-sputtering method, and with approximately 75% of transparency, the fabricated device showed very good stability [164]. Here, the CuO serves as an intermediate layer to increase the charge injection of the carriers.

As we already stated, increasing transparency of the films leads to a decrease in their efficiency. Cho et al. fabricated a $p-n$ junction solar cell with the architecture GaTe/InGaZnO and achieved a higher level of approximately 90% transparency [165]. However, in this case, the efficiency was only 0.73%, which clearly indicates the necessity of a trade-off between efficiency and transparency. Together with gallium oxide

(Ga_2O_3), the heterojunction of Ga_2O_3/Cu_2O on the poly-ethylene naphthalate (PEN) substrate could block about 96% of UV light and can deliver 1.66% of efficiency [166]. The authors further investigated the heterojunction of n-Ga_2O_3/p-SnS for the TPV application and achieved approximately 3% efficiency [167,168]. These investigations reveal the importance of developing new materials that are useful in fabricating UV-blocking TSCs. Recently, the incorporation of MXenes with the heterojunction of TiO_2/NiO has been demonstrated for the TSCs [169] and it is expected that these kinds of efforts may deliver interesting results in the near future.

17.4.3 Organic-inorganic hybrid (including perovskite) transparent solar cells

With the general structural formula AX_3 (where A = Cs^+, methyl ammonium (MA), formamidinium (FA) etc., X = Cl, Br, and I), metal halide perovskites have concentrated much interest in recent years owing to their outstanding structural and optical properties [170,171]. Because of their simple processing approach, solution-processed organic–inorganic hybrid perovskites have shown incredible performance for solar cell applications [172–174]. Moreover, apart from their interesting optical and structural features, these materials are considered to be an alternative for silicon-based solar cells. Specifically, methyl ammonium lead halides ($CH_3NH_3PbX_3$ X = Cl, Br, and I) and formamidinium lead halides ($FAPbX_3$, X = Cl, Br, and I) are delivering remarkable efficiency with different configurations. The composition can be altered from hybrid to inorganic by replacing the cation using cesium (Cs^+) ions. Besides, halide perovskites generally tend to undergo modification of phase with respect to different conditions and this essentially allows them to be used for electrochromic switching applications [4]. The absorption coefficient of $MAPbI_3$ is relatively high (10^5/cm), which results in the perovskite film producing high photocurrent in the visible region. Also, their variable bandgap values with respect to composition allow them to produce different colored active layers, which is useful for BIPV applications [175,176]. In short, PSCs are showing an incredible efficiency record of over 25%, which has been achieved within a short period of time [177]. From the observed trend, metal halide perovskites can be solution processed through different variety of solvents and they can be fabricated as a film with more homogeneity and excellent transparency. Also, through vapor deposition approaches, film formation can be precisely controlled and the resultant solar cell can deliver higher performance.

Because of these features, fabrication of TSCs using hybrid perovskites is much thrived. Furthermore, compared with organic TSCs, a high transparency and efficiency can be achieved in this case. Also, hybrid perovskites are capable of adjusting themselves with different colors with respect to their composition and other experimental conditions [178], and it could be possible to obtain neutral-color devices [179−181]. Different nano [182] and microstructured TSCs of perovskites [183] are fabricated and the maximum level of AVT achieved using those devices remains less than 40%. Because the composition of the halide perovskites could be tuned in order to get absorption under different wavelength regimes, transparent or semitransparent PSCs are fabricated with different halide perovskite compounds.

Electrodes (i.e., Ag and Au), charge-transport layers, antireflective coatings, and deposition methods play an important role in the integration of each layer of the device. To maintain the cool conditions for indoor applications, it is necessary to prohibit the passage of NIR photons in the traditional TSCs, and in this aspect, ST-PSCs are found to be useful. In that case, the fabrication of selectively chosen stacked electrodes with a dielectric capping layer (ex: $Ag/MoO_3/ZnS$) serves as a thermal mirror and specifically blocks the photons from the NIR region [184]. Liu et al. used a series of lead halides (PbX_2, X = Cl, Br, I) to fabricate a UV-selective solar cell with the structure $ITO/PEDOT:PSS/PbX_2/C60/2,9$-dimethyl-4,7-diphenyl-1,10-phenanthroline (BCP)/Ag [185]. Here, by optimizing thickness, over 1% of efficiency with higher than 70% AVT was achieved for the mixed halide composition. It is possible to improve the transparency by reducing the thickness of the perovskite layer and using a highly transparent metal electrode [186,187]. The efficiency of the perovskite-based STSCs could excellently be modified by controlling the solvent-induced crystallization [188], modification in the composition of the perovskite [189], incorporation of additives [190], use of highly transparent electrode [191], etc. Bu et al. fabricated a fully air-processed semitransparent PSC using PEDOT:PSS as the top electrode [192]. Here, because PEDOT:PSS is a water-based dispersion, the transfer-lamination technique was followed to avoid the degradation of perovskite layer. The solar cell fabricated using this method delivered an efficiency of 10.1% (0.06 cm^2 area) with an average AVT of 7.3%. The images of the fabricated STSC in this study with different thicknesses of TiO_2, transmittance, and basic device structure are given in Fig. 17.11 [192].

Other approaches, such as the use of nanostructured and microstructured perovskite materials, carbon nanotubes, tandem architecture with

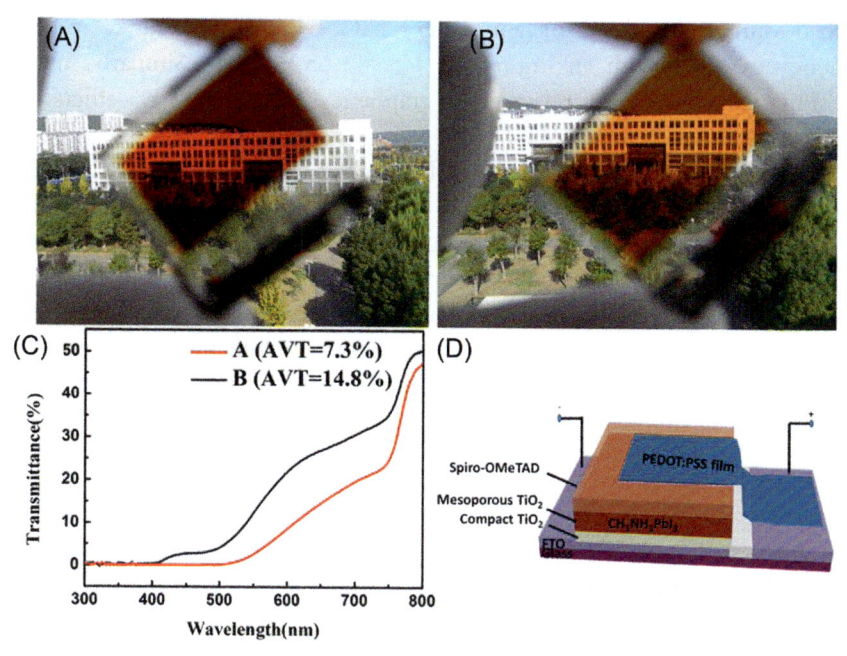

Figure 17.11 (A and B) Hybrid lead iodide perovskite STSCs with different thicknesses (240 and 120 nm), (C) transmittance of the fabricated solar cell with different thicknesses (240 and 120 nm), and (D) device structure of the fabricated solar cell. *Reprinted with permission from L. Bu, Z. Liu, M. Zhang, W. Li, A. Zhu, F. Cai, Z. Zhao, Y. Zhou, Semitransparent fully air processed perovskite solar cells, ACS Appl. Mater. Interfaces 7 (32) (2015) 17776–17781, copyright (2015) American Chemical Society.*

transparent silicon, and fabrication on the flexible substrates, are showing promising results in achieving semitransparent PSCs with high efficiency [193]. Specifically, through tandem architecture, over 25% efficiency has been achieved in the semi-TSCs [194,195]. In general, normally fabricated PSCs with Ag or Au electrode is opaque but transparency is attained through transparent electrodes. Bailie et al. fabricated a four-terminal tandem PSCs, which showed the semitransparent nature by processing transparent Ag NWs electrode [196]. Here, the authors combined the perovskite top cell with the Si and CIGS bottom cell through a tunnel junction and the resultant device delivered 18.6% efficiency. Practically, the electrodes are deposited either by spin coating or sputtering process. Although tandem architecture-based STSCs provide an efficiency over 20%, the complications associated with the fabrication of such structures in large areas should be rectified. Other than a bulk layer of halide perovskites, QDs of the inorganic counterpart, for example, $CsPbBr_3$ QDs, have been found to be useful for fabricating STSC, which could deliver an efficiency over 5% with an impressive $V_{oc} = 1.65$ V [197]. Similarly, using $CsPbI_3$ QDs as an active layer and graphene electrode, 4.95% efficiency was achieved with 53% AVT [198].

Despite several advantages, the efficiency of the perovskite TSCs is still lower than the opaque PSCs due to the following two reasons. First, the transparency of the absorber layer admits NIR photons due to the wide bandgap and so the efficiency loss occurs. Second, use of transparent electrode which does not reflect the light. However, because of the excellent absorption characteristics, efficiency, and device architecture development, it is expected that perovskite-based TSCs would lead the future generation solar cells industry.

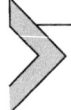

17.5 Applications of transparent solar cells in agriculture

The use of TSCs in different potential areas has been studied in recent years and different postulates have been made on such concepts [4,199,200]. It is demonstrated that solar cells can be used as a rooftop to control the growth of the plants and this process is generally defined as "agrivoltaics." Here, the electricity is generated using transparent solar windows and also the growth of the plants is maintained under a controlled atmosphere. One potential concept is to use TSCs for the greenhouse application. Here, because of the selective absorption, the TSCs with NIR absorbing materials are used for plant growth purposes in which NIR is not required [201,202]. This is because the photosynthesis process is sensitive to specific wavelengths of the solar spectrum and this is precisely controlled by admitting light through active layers. By preventing NIR region, the plant growth is accelerated and this helps in further improving other properties. Use of tinted semitransparent solar panels as rooftop for the growth of spinach has been found to provide a gross financial gain of more than 35% compared with the growth of sinach without the solar panel [199]. Parrish et al. found that the use of $CuInS_2$/ZnS core/shell structured QDs film affects the biomass accumulation in lettuce [203]. Furthermore, the use of TSCs that block UV-radiation also minimizes the use of pesticides in crop production because of the reduction of insects and fungal diseases [204]. Organic and DSSC-based STSCs are preferred for rooftop applications owing to their wavelength-selective, low weight, availing with different colors, better light control, and less intensive supporting structure [200]. There are several interesting studies on the use of organic TSCs for the rooftop applications but the use of organic-based modules has not resulted in commercialization owing to stability issues [205–207]. Wang et al. used a blend of two nonfullerene acceptors, one fullerene acceptor and an organic polymer donor, to fabricate an STSC, which resulted in plant growth factor of 26.3% [208]. Here, the fabrication was achieved using a nonhalogenated solvent (toluene), and the organic components used for the fabrication and the device assembly are listed in Fig. 17.12 [208].

At this time, greenhouse windows of this variety are quite popular; they have begun to be commercialized by several companies. Private industries such as UbiQDs and QuantumScience are actively involved in the industrial development of QDs-based solar windows for commercial purposes. Different categories of TSCs for greenhouse applications are studied including organic, DSSC, QDSSCs, and PSCs [209,210]. Yano et al. used an STSC that comprised silicon microcells (1.8 mm dia) and with a module of these solar cells, the authors achieved 4.5% efficiency

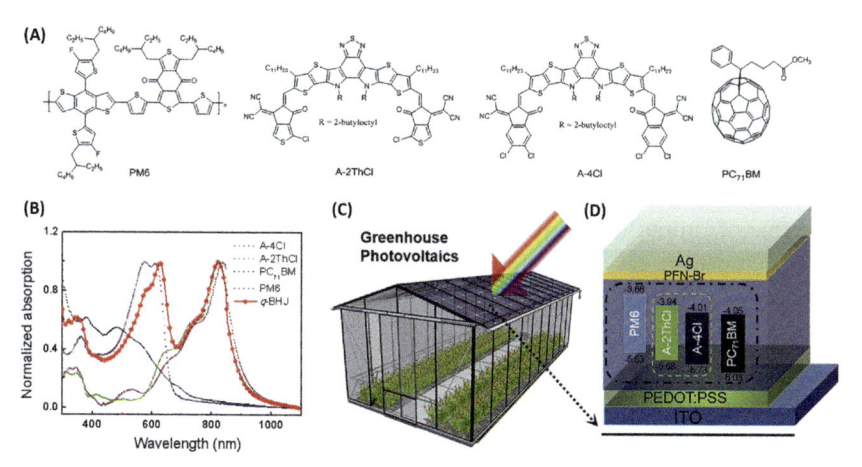

Figure 17.12 (A) Chemical structure of organic molecules used for the fabrication, (B) absorption spectra of the acceptor and donor molecules, (C) schematic representation of the semitransparent OSC for greenhouse PV application, and (D) Assembly of the device structure and energy level diagram of the molecules. *Reprinted with permission from D. Wang, H. Liu, Y. Li, G. Zhou, L. Zhan, et al., High-performance and eco-friendly semitransparent organic solar cells for greenhouse applications, Joule 5 (4) (2021) 945–957, copyright (2021) Elsevier.*

for the best one [211]. These kinds of solar cells are suitable for greenhouses in high-irradiation regions. In general, compared with other systems, c-Si and perovskite STSCs are found to be efficient for rooftop applications for plant growth.

17.6 Transparent luminescent solar concentrators

Transparent luminescent solar concentrators are a kind of approach to harvest solar energy based on redirecting the incident light to the solar panels in a concise arrangement. Here light strikes the transparent film consisting of fluorescent semiconductor NCs (or QDs) (or) dyes (generally called as "luminophore") and when the light undergoes a total internal reflection due to the scattering between the NCs (or) dye molecules, the photons are collected by the solar cells, which are embedded at the corner of the device.

For many decades, research and development around the globe have been aiming to attain higher PCEs at lower cost, which is the core aspect of cost-effective PV. Transparent luminescent solar concentrator (TLSC) is an optical device used to absorb invisible rays of light and convert them into grid-quality electricity, providing a source of clean power with minimal esthetic impact on energy-passive buildings. This transformation turns them into power generators that go hand in hand with a cost-effective approach and high PCE [212]. This paves an effortless way to large-area manufacturing with high defect tolerance, angle independent, invariable performance under different lighting conditions, good heat dissipation, and low-level energy cost. Hence, the cost will be reduced by means of replacing the high-cost solar cell with cheaper, eco-friendly, and efficient optical devices (plastic/glass waveguides with embedded species) [213]. The aesthetical inflexibility of opaque and semitransparent PV modules can be changed into transparent power-generating modules with efficient solar cell strips at the edges. The first generation luminescent solar concentrators (LSCs) were first suggested by Weber and Lambe in 1970 [214] and developed lately. However, until now, the technology has not delivered its full potential. The TLSCs are typically transparent in nature, and the level of transparency depends on various factors, including nature, thickness, and other physical and chemical properties of dyes, organic compounds, QDs, etc.

17.6.1 Principles of transparent luminescent solar concentrators

Transparent luminescent solar concentrators can optimize both power production (maximize PCE) and average visible transparency by selectively harvesting the invisible portion of the solar spectrum. It generally consists of transparent polymer sheets doped with luminescent species (or luminophores) [215] or QDs embedded within or on it. Incident sunlight (both direct and diffused) is absorbed by the luminescent species and re-emitted at a red-shifted wavelength, with high quantum efficiency. Those photons, which impact the substrate-air interface at greater than the critical angle, become trapped within the substrate due to the refractive index difference between the waveguide and the ambient environment. Then guided through the waveguide to the edge-mounted efficient solar cell by total internal reflection to convert it into electricity [213]. The Stokes shift which is the difference between the absorption onset and the emission wavelength has to be large enough to ensure that reabsorption during waveguide propagation is minimized. To have good conversion efficiency, the energy of the emitted photons (ideally) should be larger than the band gap of the attached solar cells. The most interesting part of this system is that the transparent or colorless LSCs look alike normal windows. TLSCs can comprise emitters with selective absorption in the UV, NIR, or both spectral wavelength regions. TLSCs are of special interest for BIPV or building-applied PV (BAPV) applications [216,217]. An alternate practical application is illustrated in Fig. 17.13: a schematic diagram of a near-infrared (NIR) TLSC that is integrated onto different surfaces of an automobile [218].

17.6.2 Material selection in transparent luminescent solar concentrators

The TLSC comprises two main components, one is the host material and the other one is the luminophore(s). Usually, the host matrix in TLSC is polymer [poly-methyl-methacrylate (PMMA)], polycorbante (PC), or glass that has a higher refractive index than that of the air [219]. The main criteria for selecting a good host material are to have a high refractive index, low absorption coefficient, transparency over the emission bandwidth of the luminophores, long outdoor lifetime, light weight, low cost, and readily available material. In addition, it should also allow the luminophore to have a high level of solubility. If the host material does not permit maximum

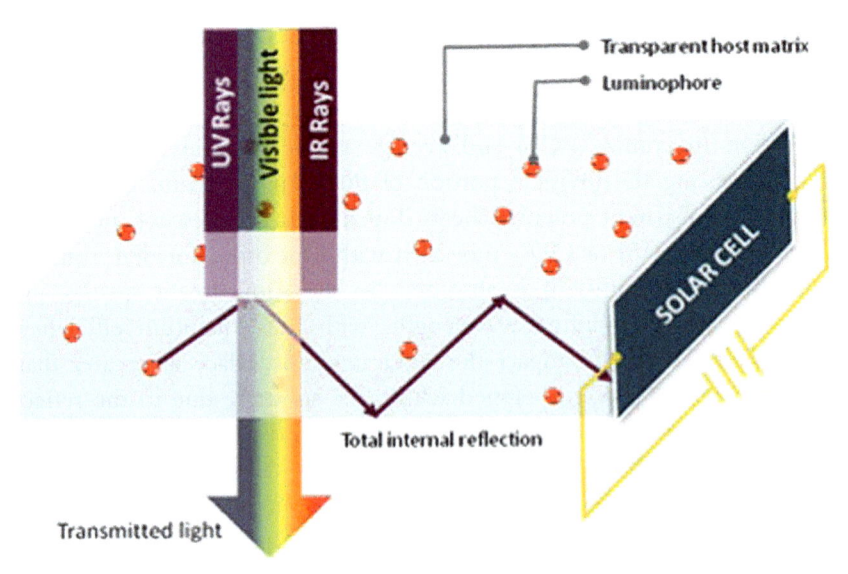

Figure 17.13 Conceptual schematic showing a near-infrared (NIR) harvesting transparent luminescent solar concentrator (TLSC) integrated onto the window ("Air Control") and nonwindow ("TLSC w/Low-n Film") parts of an automobile. While a black automobile is pictured, this could be applied to any color automobile without changing the architecture. With direct integration of LSC and TLSCs on such surfaces (included in the schematic as "Paint Control"), total internal reflection (TIR) is lost. *Reprinted with permission from C. Yang, D. Liu, A. Renny, P.S. Kuttipillai, R.R. Lunt, Integration of near-infrared harvesting transparent luminescent solar concentrators onto arbitrary surfaces, J. Lumin. 210 (June 2019) 230–246, copyright (2019) Elsevier.*

dissolution, material aggregation can lead to nonuniform distribution of the luminophore in the matrix and the formation of scattering centers. It should be compatible with the processing temperature of the given luminophore. Good optical coupling with PV cells by the host material is also necessary. The luminophores play a major role in TLSC as their concentration determines how many photons are absorbed and transferred to the edge. They should have a broad absorption spectrum combined with high absorption efficiency, high luminescence quantum efficiency, high Stokes shift, and long-term stability. The commonly used luminophores will be one of the following components—organic dyes [220], QDs [221], rare earth ions [222], inorganic nanoparticles [223], or phosphor [217]. All these luminophores exhibit some drawbacks such as high reabsorption losses, low absorption coefficients, photodegradability, narrow absorption bands, and so on. Engineering these parameters to achieve optimum efficiency opens up future scope to have TLSCs having higher efficiency over 10%.

The TLSCs can be stacked with different plates containing dyes with various absorption characteristics so as to break down the solar spectrum into segments and concentrate them at the same time leading to tandem TLSCs to improve the efficiency. The QD TLSC can be made into tandem TLSC with two QD materials, which are spectrally separated and accompanied by a large Stokes shift. The first layer should be UV-visible absorbing QDs and the second should be NIR-absorbing QDs so as to harvest to whole energy range of the material.

17.6.3 Efficiency of transparent luminescent solar concentrators

The efficiency of TLSCs is often misinterpreted with solar cell efficiency. The overall efficiency of the system is the combined efficiency of TLSC and the solar cell that is used in the edge. High-efficiency solar cell has to be used to obtain optimum output. The overall PCE of a TLSC system is the product of two components given by Eq. 17.3 [223]:

$$\eta_{LSC} = \eta_{opt} * \eta_{PV} \qquad (17.3)$$

where η_{PV} is the efficiency of edge mounted solar cell as normally calculated, taking into account the flux produced by the LSC, and η_{opt} is the optical efficiency of the TLSC, which is the ratio of photons reaching the edge-mounted solar cell to the incident photons on the TLSC. The optical efficiency (η_{opt}) of an LSC is expressed in Eq. (17.4) as outlined by Goetzberger [224–226]:

$$\eta_{opt} = (1 - R)PTIR \cdot \eta_{abs} \cdot \eta_{PLQY} \cdot \eta_{Stokes} \cdot \eta_{host} \cdot \eta_{TIR} \cdot \eta_{self} \qquad (17.4)$$

where R is the reflection of solar light from the waveguide surface, PTIR is the total internal reflection efficiency, η_{abs} is the solar spectrum absorption efficiency of luminophore from, η_{PLQY} is the photoluminescence quantum yield (PLQY) of the used luminophores, η_{Stokes} is the energy lost due to the heat generation during the absorption and emission event, η_{host} is the transport efficiency of the waveguided photons through the waveguide, η_{TIR} is the reflection efficiency of the waveguide determined by the smoothness of the waveguide surface, and η_{self} is the transport efficiency of the waveguided photons related to reabsorption of the emitted photons by another luminophore.

From the above equation, future research approaches exist for engineering the host material, luminophore, and the internal structure of concentrator waveguides. The PCE of the TLSC can be increased by choosing molecules, with a high Stokes shift, whose dipolar moment is higher in the excited state than in the ground state. Other PCE enhancements include high refractive index host material to ensure high transparency toward the reflected photons by luminophore, high absorption efficiency from the solar spectrum, and minimizing the reabsorption by adjacent dye molecules. Characterization of TLSCs is quite challenging as a greater range of possible measurement errors are encountered compared to normal TPV calculation. Practically, the overall conversion efficiency of the TLSC module system is limited to less than 20% due to the optical funneling, which reduces the PV area resulting in cost cut down [214,215,225,227].

17.6.4 Losses associated with transparent luminescent solar concentrators

Transparent luminescent solar concentrator losses occur in both the host material and the luminophore used. The main losses that occur are shaded or diffused light that is not absorbed properly, low PL quantum efficiency of the luminophores, low transparency of the host material to the reflected wavelength by the luminophore, and fluorescence quenching. The waveguide (or host) material should have a low absorption coefficient so that most of the re-emitted photons can propagate toward the LSC edges with minimum attenuation.

Surface and reabsorption losses are among the most significant parameters of device performance and scalability. Photons emitted by luminescent molecules inside the escape cone will be lost through the surfaces, which can be minimized by aligning the luminophores and choosing selective reflector material to reflect the wavelengths emitted by luminophores. The Stokes shift, which is the wavelength difference between the absorption and emission peak maxima for the same transition of luminophores, determines these losses. Several ways, such as core–shell QDs [228,229]; colloidal QDs [230]; rare earth ions, complexes, or materials [222,231,232]; metal ion doped QDs [233,234]; or ternary materials [235] are used to increase the Stokes shift by downshifting the radiative recombination. However, these QDs have the limitation of having continuous band-like absorption rather than selective absorption of invisible photons (UV and NIR). As a result, the transparency and esthetic quality, which are the most critical metrics of TLSCs, are reduced. To create emitters with high quantum yield, broad spectral absorption, narrow emission, large Stokes shifts, and wavelength selectivity to circumvent the losses, exigent material development with high availability and low cost is needed. If all these parameters gather together, highly efficient and exceptionally low-cost transparent solar harvesting systems can be developed. Waveguide roughness and optical transparency also have a major role as waveguides are scaled up to square meters, which can lead to reabsorption or scattering losses. There are many criteria that affect the solar concentrator efficiency as it deviates from ideal transparent solar concentrator conditions.

The transport losses (η_{self}), such as reabsorption of emission light by subsequent dye molecules and parasitic absorption of the emission light by the host waveguide material, imperfection in the waveguide, and even the presence of dust outside the waveguide also account for the performance of the TLSCs [236]. The loss will increase as the area of the TLSC increases due to the absorption tail of the luminophore. Even the unreacted monomers in the host material also have a great influence on the loss mechanism. Sometimes the additives used for the enhancement of host material properties can have some adverse effect on the overall output.

17.6.5 History of the development of transparent luminescent solar concentrators

Since the late 1970s, researchers have been developing TLSCs by focusing on enhancing the PCE by reducing the loss mechanisms. For instance, in the past few years, significant advances have been realized in the LSC system (both host material and luminophore). For example, in 2008, Slooff et al. achieved a PCE of 7.1% (26% increased efficiency) for a sample size of 5 cm × 5 cm × 0.5 cm using a configuration with four parallel-connected GaAs solar cells coupled to the edges of polymethylmethacrylate (PMMA) LSC sheet doped with a mixture of two organic dye luminophores, in which a diffuse reflector is used on the rear side of the sample [219]. In the same year, Currie et al. achieved a PCE of 6.8% using both single and tandem-waveguide organic solar concentrators without any reflection filter on the rear surface of the sample [237]. In 2009, Goldschmidt et al. achieved a PCE of 6.7% using four gallium indium phosphide (GaInP) solar cells in a multidye stacked configuration of two smaller plates of 20 mm × 20 mm × 3 mm and 5.6% for a single dye [238]. In 2014, Zhao et al. reported TLSC developed from luminophore blends of a cyanine and cyanine salt, combined with spectrally selective NIR harvesting with transparency of 86% and PCE of 0.4% [239].

Large-scale implementation requires long-term testing and adaptation. Several attempts have been already made to take advantage of design and fabrication freedom. In 2015, Aste et al. achieved PCE of 1.26% for a sample size of 50 cm × 50 cm × 6 cm TLSC with visible light transmittance (VLT) of approximately 40% with organic dyes [220]. The important results of this investigation are given in the Fig. 17.14 [220]. In 2017, Meinerdi et al. achieved η_{opt} of 2.85% for QDs LSC of 12 cm × 12 cm with VLT of 70% [240]. In the same year, Yang et al. reported the PCE of 6.9% and 20.6% for UV-only and UV- and NIR-selective TLSCs, respectively [241]. Originally organic dyes were used and now are shifting toward more stable inorganic phosphors and QDs as the stability of organic dyes is less.

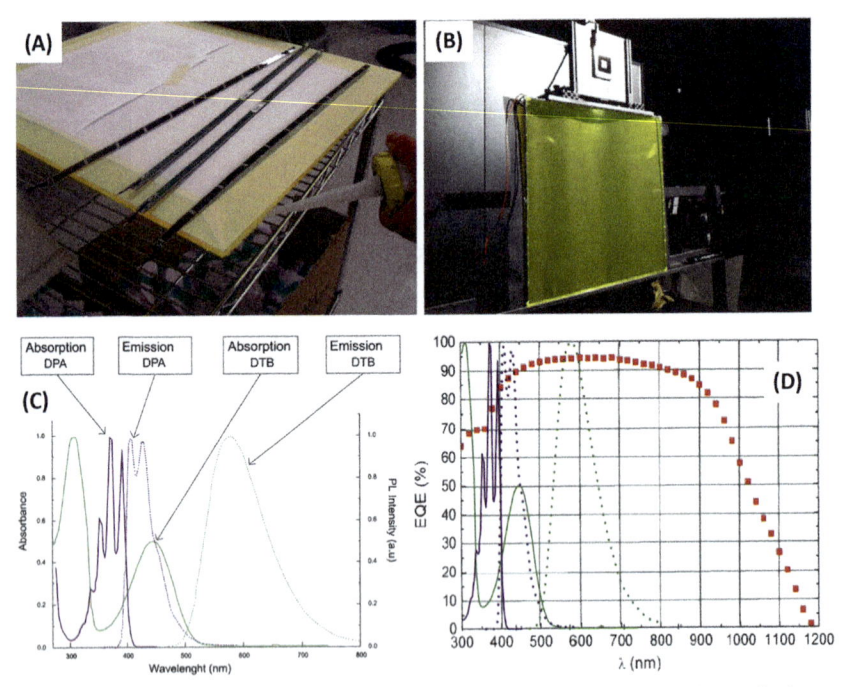

Figure 17.14 (A) assembly of the PMMA panel with photovoltaic cells, (B) flash test of the panel performed at SUPSI (University of Applied Sciences and Arts of Southern Switzerland) institute laboratories, (C) absorption and emission spectra of the dyes (DPA and DTB) in PMMA, and (D) external quantum efficiency of the fabricated LSC with the absorption and emission spectrum of the dyes. *Reprinted with permission from N. Aste, L.C. Tagliabue, C.D. Pero, D. Testa, R. Fusco, Performance analysis of a large-area luminescent solar concentrator module, Ren. Energy 76 (2015) 330–337, copyright (2015) Elsevier.*

In 2018, Bergren et al. achieved PCE of 2.18% ($\eta_{opt} = 8.1\%$) for a CuInS$_2$ QDs TLSC of 10 cm × 10 cm with VLT of 43.7% [221]. In the same year, Wu et al. achieved PCE of 3.1% for tandem QDs TLSC of 15.2 cm × 15.2 cm with VLT of nearly 30% [242]. Also during that year, Vasiliev et al. [217] reported PCE of 2.5% ($\eta_{opt} = 1.425\%$) for 10 cm × 75 cm of clear glass windows with a hybrid concentrator having phosphor luminophore and transparent NIR reflector coating at the back surface. The same author [243] reported PCE of 2.347% for 20 cm × 20 cm semitransparent diffraction-assisted TLSC with VLT greater than 60%. Recently, several studies have been carried out to find an alternative to the commonly used poly(methylmethacrylate) (PMMA) host matrix for the design of efficient luminescent solar concentrators (LSCs). Among them, Ostos et al. have reported a PCE of 11.4%, for 1.00 wt.% LR in poly(cyclohexylmethacrylate) (PCHMA), a 0.7% increase over devices utilizing PMMA [244]. The detailed discussion about the QDs-based TLCs is given in the next subsection of this chapter.

Several challenges are detailed for experimental devices with an efficiency greater than 6% with either single or stacked TLSC configuration for the total surface area of 14 cm × 14 cm [237,245]. In some cases, 4% efficiency was achieved for larger surfaces with the use of back reflectors. Applying wavelength-selective filters makes it possible to reach high-concentration factors. Otherwise, efficiency values remain significantly lower than the corresponding conventional solar PV due to a number of losses and the reported dimensions of the concentrators are still far below those of a commercial window. The development of suitable luminescent and host materials to generate solar energy with appreciable efficiency opens up future research. Ultimately, taking into account overall energy production, it is not only the individual TLSC efficiency that matters but when TLSCs are used over large areas, without aesthetical change and at low cost, when compared to practically used silicon or opaque solar cells, the efficiency of TLSCs enables their use in all buildings.

17.6.6 Emergence of quantum dots-based transparent luminescent solar concentrators

As discussed previously, colloidal QDs enable enhanced performance in future-generation solar cells due to their size-dependent properties. Because QDs could be processed through cost-effective approaches, fabrication of large-area QD-related devices is gaining greater interest. As the photostability of semiconductor QDs is higher than the organic dyes,

QDs are obviously useful for the LSCs [246]. For large-area device applications, methods such as electrophoretic deposition [247], nano-imprint lithography [248], and doctor-blade method [249] are employed. The spectral coverage of these QDs ranges from UV to NIR so different compound QDs and their core/shell structures are used to fabricate QD-based transparent solar concentrators [250,251]. Because of the size-dependent band gap tunability, the band gap energy and thus absorption onset of these QDs can be altered. Among the QDs, cadmium chalcogenides QDs (ex: CdTe, CdSe), Carbon QDs, lead chalcogenide QDs (PbS, PbSe), III-V group QDs (ex: InP), I-III-VI$_2$ QDs (CuInS$_2$, CuInSe$_2$), doped QDs and their core/shell assemblies are widely used and their performance is evaluated [252–256]. In order to reduce the toxicity-related issues with metal ions such as Pb^{2+} and Cd^{2+}, eco-friendly (or) less toxic and heavy metal-free QDs are concentrated and their performance for LSCs was investigated [257,258]. Most of the reported efficiency through different strategies in the TLSCs remains within 5%. We suggest several in-depth review articles that address basic principle, development, fabrication, and current trends of the QDs-based TLSCs for window applications [14,246,250].

Interesting results with different compound semiconductor QD/polymer composites are obtained for the fabrication of highly efficient solar windows. For example, Bergren et al. synthesized highly emissive NIR-emitting CuInS$_2$/ZnS QDs (PLQY approximately 96%) and embedded them into a polymer [221]. This polymeric layer is sandwiched between two iron-float glass with the dimensions 10×10 cm^2. Through this architecture, the authors achieved the maximum conversion efficiency of 2.94% with an optical efficiency of 8.1%. Several recent studies report investigating the fabrication of TLSCs with regard to their esthetics [255], minimizing scaling losses [259], and thermal effects [260]. Recently, due to the large range of absorption and high PLQY, metal halide perovskite NCs are also concentrated to fabricate transparent solar concentrators [261,262]. Because of their very low Stokes shift, doped-perovskite NCs are mainly used for the LSC applications to avoid reabsorption [263,264]. To avoid the toxicity associated with Pb^{2+}, lead-free perovskite NCs have emerged as the potential alternative for LSC applications [265]. Although the absorption and emission spectrum of the perovskite NCs typically show very small Stokes shift, it is possible to avoid the reabsorption effect through doping with elements like Mn^{2+}, using type-II core/shell architectures, and also modifying the composition of the halides [233,263,266,267].

Klimov et al. proposed a new concept, called a "quality factor," which is the ratio between absorption coefficients at the wavelengths of incident and reemitted light [268,269]. This actually helps to characterize the QDs light harvesting ability and self-induced reabsorption losses, also to balance the drop in PL quantum yield of QDs. In most cases, PMMA was used as the host polymer matrix to form the QDs/polymer nanocomposite due to its high transparency and high refractive index. However, other polymers such as poly-lauryl methacrylate (PLMA) [270,271], polyvinyl pyrrolidone (PVP) [272] are also used. Among all these, PVP exhibits high trapping efficiency owing to its relatively high refractive index value [273]. It should be noted that the compatibility of the QDs with the polymer matrix is dependent on several factors including reactivity of the QDs surface ligands. The prepared QDs should be agglomeration-free in order to disperse with the polymer-matrix uniformly. Also, the concentration of QDs within the polymer is important because increasing the QDs concentration will result in an increase in scattering coefficient, which results in scattering loss in the large-area devices. Because silicon covers visible spectrum and provides high efficiency, preparation of Si QDs, Si QDs/polymer nanocomposites, and depositing them for the fabrication of large-area transparent solar windows are also focussed intensively. Huang et al. used Si QDs for the fabrication of "quantum dot glass" and together with a polymer in a sandwiched structure, a surface area of 400 cm^2 was achieved [274]. This assembly demonstrated a 1.57% PCE (AVT of 84%) and a light-utilization efficiency performance (LUE) of 1.3%. The aesthetic quality of the transparent QDs glass with absorption, transmittance, reflectance, and haze spectrum of the fabricated QDs glass is given in Fig. 17.15 [274]. Such transparent windows are quite useful for BIPV applications and also for the energy-saving purposes. Although Si QDs have large Stokes shift, they possess lower absorption so the performance of Si QDs-based TLSCs is typically low. Similar to solar cells, tandem architecture also results in improved performance in TLSCs.

Zhao et al. proposed a novel concept of perovskite QDs and carbon-based tandem type solar concentrators; together with carbon QDs (approximately 2 nm)/PVP, $CsPb(Br_xI_{1-x})_3$ QDs/(PLMA-co-EGDA) were used to couple with the middle layer of $CsPb(Br_{1-x}Cl_x)_3$ QDs/ (PLMA-co-EGDA) [267]. This kind of tandem QDs solar concentrator is also demonstrated in other systems and interesting results are observed [242,275,276]. Moreover, coupling a QDs-based TLSC with a DSSC also results in high short-circuit photocurrent, which is another promising

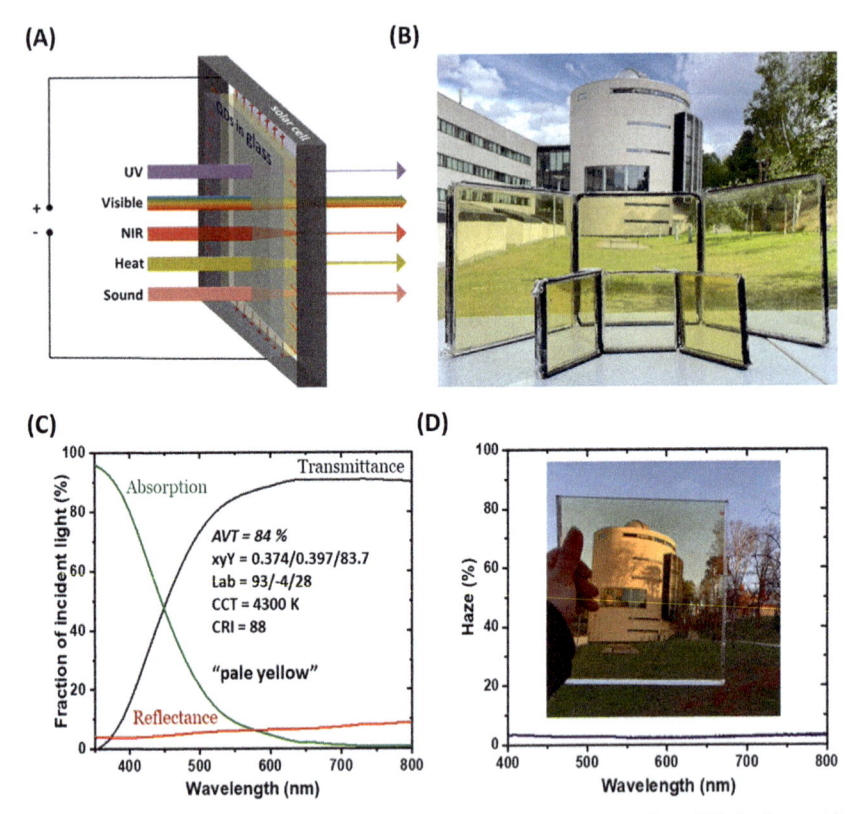

Figure 17.15 (A–C) Esthetic quality of the transparent quantum dots (QDs) glass with absorption, transmittance, and reflectance spectrum and (D) the haze spectrum of the fabricated QDs glass. *Reproduced with permission from J. Huang, J. Zhou, E. Jungstedt, A. Samanta, J. Linnros, L. A. Berglund, I. Sychugov, Large-area transparent "quantum dot glass" for building-integrated photovoltaics, ACS Photonics 9 (7) (2022) 2499–2509, this article is distributed under the terms of the Creative Commons (CC-BY 4.0) license (http://creativecommons.org/licenses/by/4.0/), which permits unrestricted use.*

direction [277]. Emergence of lead-free perovskites, particularly copper halide perovskites such as $Cs_3Cu_2I_5$ and $CsCu_2I_3$, which exhibit excellent optical properties, are also found to be suitable for fabricating TLSCs with impressive AVT [278]. Here, all these layers were prepared using a spin-coating method, and through this reabsorption-controlled architecture, the fabricated LSC showed an external optical efficiency of 2% with a geometrical gain factor of 45. Furthermore, the performance of these kinds of QD-based TLSCs can be enhanced through coupling with another nanomaterial, for example, carbon QDs with Ag nanoparticles [279]. Recent research on perovskite QDs with Au=nanoparticles demonstrates a reduction in self-absorption and improved efficiency [280]. Because high PLQY requires high-quality QDs, in general, core/shell assemblies of different compound semiconductor QDs are used to fabricate transparent LSCs. These results clearly emphasizing the importance of using QDs to develop TSLC architectures for smart windows applications.

17.7 Conclusion and future perspectives

The wide variety of device technologies and their disparate performance discussed in this chapter clearly demonstrate that materials selection and their processing technologies are key to achieving highly-efficient TSCs. These indeed could serve for future generations of power generators for different applications; further developments are anticipated for these intriguing devices. Fabrication methods of TPV should improve with advanced strategies, which are equal to the fabrication of silicon solar cells. Also, neutral-colored TSCs with a low haze transmission ratio could additionally strengthen the current trends in this area. The stability of fabricated TSCs is an important factor in considering device commercialization. It is important to achieve high stability by properly protecting the active layer materials from oxygen and moisture. Advanced level of device structures engineering through fabricating novel photoactive materials with transparency will result in better progress in this field. The difficulties associated with the deposition of transparent electrodes on a conducting substrate and active layers should be rectified to considerably improve the efficiency. Although, such innovative approaches for TSC processing are promising, the fabrication of these solar cells and achieving high PCE is a daunting challenge. The difficulties associated with the fabrication and integration of solar cell modules should be addressed in order to expand the commercial viability of the TSCs. In the case of dye-sensitized TSCs, the development of highly-efficient solid-state electrolytes is required to produce large-area devices. Cost-effective fabrication of large-area TSCs is required to facilitate commercialization, a common platform for further development is an example enabling appoarch. Further development in NIR absorbing inorganic and organic materials will ease the problems related to the absorption of NIR region. Metal oxide semiconductors such as ZnO, TiO_2, and NiO-based TSCs show promise for future generation BIPV applications, but further investigation is required to commercialize these device technologies.

Also, manufacturing solar cells using NCs with all these materials by low-cost fabrication approaches would be beneficial for selectively absorbing light in the visible spectrum. On the other hand, halide perovskite-based TSCs are showing promising directions for commercial production, and further research related to its long-term stability for BIPV application would be beneficial for a wide range of their applications. It is expected that future breakthroughs in terms of materials and device configurations will lead this field in a new direction.

Acknowledgments

The authors gratefully acknowledge the great support rendered by the management of Intercomet S.L. to write this chapter. Ananthakumar Soosaimanickam sincerely thanks the Ministerio de Ciencia e Innovación, Spain, for funding through the Torres-Quevedo program (PTQ2020-011398).

References

[1] L. Liu, K. Cao, S. Chen, W. Huang, Toward see-through optoelectronics: transparent light emitting diodes and solar cells, Adv. Opt. Mater. 8 (22) (2020) 2001122.
[2] S. Kim, H.V. Quy, C.W. Bark, Photovoltaic technologies for flexible solar cells: beyond silicon, Mater. Today Energy 19 (2021) 100583.
[3] G. Quesada, D. Rousse, Y. Dutil, M. Badache, S. Halle, A comprehensive review of solar facades. Transparent and translucent solar facades, Renew. Sust. Energy Rev. 16 (5) (2012) 2643–2651.
[4] H. Meddeb, M. Gotz-Kohler, N. Newgebohrn, U. Banik, N. Osterthun, et al., Tunable photovoltaics: Adapting solar cell technologies to versatile applications, Adv. Opt. Mater. 12 (28) (2022) 2200713.
[5] Y.T. Chae, J. Kim, H. Park, B. Shin, Building energy performance evaluation of building integrated photovoltaic (BIPV) window with semi-transparent solar cells, Appl. Energy 129 (2014) 217–227.
[6] A. Roy, A. Ghosh, S. Bhandari, S. Sundaram, T.K. Malik, Perovskite solar cells for BIPV application: a review, Buildings 10 (7) (2020) 129.
[7] A. Cannavale, F. Martellotta, F. Fiorito, U. Ayr, The challenge for building integration of highly transparent photovoltaics and photoelectrochromic devices, Energies 13 (8) (2020) 1929.
[8] A.A.F. Husain, W.Z.W. Hasan, S. Shafie, M.N. Hamidon, S.S. Pandey, A review of transparent solar photovoltaic technologies, Renew. Sust. Energy Rev. 94 (2018) 779–791.
[9] A.F. Nogueira, Shades of transparency, Nat. Energy 5 (2020) 428–429.
[10] B.H. Baby, M. Patel, K. Lee, J. Kim, Large-scale transparent photovoltaics for a sustainable energy future: review of inorganic transparent photovoltaics, Appl. Sci. Conver. Tech. 31 (1) (2022) 1–8.
[11] E. Pulli, E. Rozzi, F. Bella, Transparent photovoltaic technologies: current trends towards upscaling, Energy Conv. Manag. 219 (2020) 112982.
[12] Q. Tai, F. Yan, Emerging semitransparent solar cells: materials and device design, Adv. Mater. 29 (34) (2017) 1700192.
[13] K. Lee, H.-D. Um, D. Choi, J. Park, N. Kim, H. Kim, et al., The development of transparent photovoltaics, Cell Rep. Phys. Sci. 1 (8) (2020) 100143.

[14] L.R. Bradshaw, K.E. Knowles, S. McDowall, D.R. Gamelin, Nanocrystals for luminescent solar concentrators, Nano. Lett. 15 (2) (2014) 1315–1323.
[15] M. Vasiliev, M. Nur-E-Alam, K. Alameh, Recent developments in solar energy harvesting technologies for building integration and distributed energy generation, Energies 12 (6) (2019) 1080.
[16] N. Schopp, V.V. Brus, T.-Q. Nguyen, On optoelectronic processes in organic solar cells: From opaque to transparent, Adv. Opt. Mater. 9 (3) (2021) 2001481.
[17] N. Schopp, V.V. Brus, A review on the materials science and device physics of semitransparent organic photovoltaics, Energies 15 (13) (2022) 4639.
[18] C.J. Traverse, R. Pandey, M.C. Barr, R.R. Lunt, Emergence of highly transparent photovoltaics for distributed applications, Nat. Energy 2 (2017) 849–860.
[19] M. Kumar, S. Abbas, J. Kim, All-oxide-based highly transparent photonic synapse for neuromorphic computing, ACS Appl. Mater. Interfaces 10 (40) (2018) 34370–34376.
[20] P. Bhatnagar, M. Patel, T.T. Nguyen, S. Kim, J. Kim, Transparent photovoltaics for self-powered bioelectronics and neuromorphic applications, J. Phys. Chem. Lett. 12 (51) (2021) 12426–12436.
[21] G. Kim, J.W. Lim, J. Kim, S.J. Yun, M.A. Park, Transparent thin-film silicon solar cells for indoor light harvesting with conversion efficiencies of 36% without photodegradation, ACS Appl. Mater. Interfaces 12 (24) (2020) 27122–27130.
[22] Z. Li, T. Ma, H. Yang, L. Lu, R. Wang, Transparent and colored solar photovoltaics for building integration, RRL Sol. 5 (3) (2021) 2000614.
[23] H. Lee, H.-J. Song, Current status and perspective of colored photovoltaic modules, WIREs Energy Environ. 10 (6) (2021) e403.
[24] A. Royset, T. Kolas, B.P. Jelle, Colored building integrated photovoltaics: influence on energy efficiency, Energy Build. 208 (2020) 109623.
[25] W. Wu, J. Zhang, Y. Ren, Y. Zhang, Y. Yuan, Z. Shen, et al., Tuning the color palette of semi-transparent solar cells via later π-extension of polycyclic heteroaromatics of donor-acceptor dyes, ACS Appl. Energy Mater. 3 (5) (2020) 4549–4558.
[26] H. Wang, J. Li, H.A. Dewi, N. Mathews, S. Mhaisalkar, A. Bruno, Colorful perovskite solar cells: progress, strategies and potentials, J. Phys. Chem. Lett. 12 (4) (2021) 1321–1329.
[27] Y.S. Do, J.H. Park, B.Y. Hwang, S.-M. Lee, B.-K. Ju, K.C. Choi, Plasmonic color filter and it's fabrication for large-area applications, Adv. Opt. Mater. 1 (2) (2013) 133–138.
[28] Z. Hu, J. Wang, X. Ma, J. Gao, C. Xu, K. Yang, et al., A critical review on semitransparent organic solar cells, Nano Energy 78 (2020) 105376.
[29] K.-S. Chen, J.-F. Salinas, H.-L. Yip, L. Huo, J. Hou, A.K.-Y. Jen, Semi-transparent polymer solar cells with 6% PCE, 25% average visible transmittance and a color rendering index close to 100 for power generating window applications, Energy Environ. Sci. 5 (2012) 9551–9557.
[30] G. Liu, C. Wu, Z. Zhang, Z. Chen, L. Xiao, B. Qu, Ultraviolet-protective transparent photovoltaics based on lead-free double perovskites, RRL Sol. 4 (5) (2020) 2000056.
[31] N. Lynn, L. Mohanty, S. Wittkopf, Color rendering properties of semi-transparent thin-film PV modules, Build. Environ. 54 (2012) 148–158.
[32] S.-Y. Chang, P. Chang, G. Li, Y. Yang, Transparent polymer photovoltaics for solar energy harvesting and beyond, Joule 2 (6) (2018) 1039–1054.
[33] K. Ellmer, Past achievements and future challenges in the development of optically transparent electrode, Nat. Photonics 6 (2012) 809–817.
[34] M. Morales-Masis, S. De Wolf, R. Woods-Robinson, J.W. Ager, C. Ballif, Transparent electrodes for efficient optoelectronics, Adv. Elect. Mater. 3 (5) (2017) 1600529.

[35] J. Yun, Ultrathin metal films for transparent electrodes of flexible optoelectronic devices, Adv. Funct. Mater. 27 (18) (2017) 1606641.
[36] M.W. Rowell, M.D. McGehee, Transparent electrode requirements for thin film solar cell module, Energy Environ. Sci. 4 (2011) 131−134.
[37] D. Zhang, A.H. Alami, W.C.H. Choy, Recent progress on emerging transparent metallic electrodes for flexible organic and perovskite photovoltaics, RRL Sol. 6 (1) (2022) 2100830.
[38] Y. Zhang, S.-W. Ng, X. Lu, Z. Zheng, Solution-processed transparent electrodes for emerging thin-film solar cells, Chem. Rev. 120 (4) (2020) 2049−2122.
[39] G. Giuliano, A. Bonasera, G. Arrabito, B. Pignataro, Semitransparent perovskite solar cells for building integration and tandem photovoltaics: design strategies and chllanges, RRL Sol. 5 (12) (2021) 2100702.
[40] H. Lu, X. Ren, D. Ouyang, W.C.H. Choy, Emerging novel metal electrodes for photovoltaic applications, Small 14 (14) (2018) 1703140.
[41] J. Liu, D. Jia, J.M. Gardner, E.M.J. Johansson, X. Zhang, Metal nanowire networks: recent advances and challenges for new generation photovoltaics, Mat. Today Energy 13 (2019) 152−185.
[42] C. Zhang, C. Ji, Y.-B. Park, L. Guo, Thin-metal-film-based transparent conductors: material preparation, optical design, and device applications, Adv. Opt. Mater. 9 (3) (2021) 2001298.
[43] M.-R. Azani, A. Hassanpour, T. Torres, Benefits, problems and solutions of silver nanowire transparent conductive electrodes in Indium tin oxide (ITO)-free flexible solar cells, Adv. Energy Mater. 10 (48) (2020) 2002536.
[44] X. Wu, Z. Zhou, Y. Wang, J. Li, Syntheses of silver nanowires ink and printable flexible transparent conductive film: a review, Coatings 10 (9) (2020) 865.
[45] Y. Zhu, Y. Deng, P. Yi, L. Peng, X. Lai, Z. Lin, Flexible transparent electrodes based on silver nanowires: Material synthesis, fabrication, performance and applications, Adv. Mater. Tech. 4 (10) (2019) 1900413.
[46] X. Xiong, C.-L. Zou, X.-F. Ren, A.-P. Liu, Y.-X. Ye, F.-W. Sun, et al., Silver nanowires for photonics applications, Laser&Phot. Rev. 7 (6) (2013) 901−919.
[47] T. Muhmood, F. Ahmad, X. Hu, X. Yang, Silver nanowires: a focused review of their synthesis, properties and major factors limiting their commercialization, Nano Futures 6 (2022) 032006.
[48] T. Sannicolo, M. Legrange, A. Cabos, C. Celle, J.-P. Simonato, D. Bellet, Metallic nanowire-based transparent electrodes for next generation flexible devices: a review, Small 12 (44) (2016) 6052−6075.
[49] V.H. Nguyen, D.T. Papanastasiou, J. Resende, L. Bardet, T. Sannicolo, et al., Advances in flexible metallic transparent electrodes, Small 18 (19) (2022) 2106006.
[50] S. Ye, A.R. Rathmell, Z. Chen, I.E. Stewart, B.J. Wiley, Metal nanowire networks: the next generation of transparent conductors, Adv. Mater. 26 (39) (2014) 6670−6687.
[51] J. Lee, H. Cha, H. Yao, J. Hou, Y.-H. Suh, S. Jeong, et al., Toward visibly transparent organic photovoltaic cells based on a near-infrared harvesting bulk heterojunction blend, ACS Appl. Mater. Interfaces 12 (29) (2020) 32764−32770.
[52] E. Pascual-San-Jose, G. Sadoughi, L. Lucera, M. Stella, E. Martinez-Ferrero, et al., Towards photovoltaic windows: scalable fabrication of semitransparent modules based on non-fullerene acceptors via laser-patterning, J. Mater. Chem. A 8 (2020) 9882−9895.
[53] T. Winkler, H. Schmidt, H. Flugge, F. Nikolayzik, I. Baumann, et al., Efficient large area semitransparent organic solar cells based on highly transparent and conductive ZTO/Ag/ZTO multilayer top electrodes, Org. Electr. 12 (10) (2011) 1612−1618.

[54] H.-K. Park, J.-W. Kang, S.-I. Na, D.-Y. Kim, H.-K. Kim, Characteristics of indium-free GZO/Ag/GZO and AZO/Ag/AZO multilayer electrode grown by dual target DC sputtering at room temperature for low-cost organic photovoltaics, Sol. Energy Mat. Sol. Cell 93 (11) (2009) 1994–2002.

[55] Y. Ji, J. Yang, W. Luo, L. Tang, X. Bai, et al., Ultraflexible and high-performance multilayer transparent electrode based on ZnO/Ag/CuSCN, ACS Appl. Mater. Interfaces 10 (11) (2018) 9571–9578.

[56] A. Dhar, T.L. Alford, Optimization of Nb_2O_5/Ag/Nb_2O_5 multilayers as transparent composite electrode on flexible substrate with high figure of merit, J. Appl. Phys. 112 (2012) 103113.

[57] J.-D. Yang, S.-H. Cho, T.-W. Hong, D.I. Son, D.-H. Park, K.-W. Yoo, et al., Organic photovoltaic cells fabricated on a SnO_x/Ag/SnO_x multilayer transparent conducting electrode, Thin Solid. Films 520 (19) (2012) 6215–6220.

[58] H. Khachatryan, M. Kim, H.-J. Seok, H.-K. Kim, Fabrication of InZnSiO/Ag/InZnSiO transparence flexible heater on polymer substrate by continuous roll-to-roll sputtering advanced technology, Mater. Sci. Semi. Process. 99 (2019) 1–7.

[59] Z. Ying, W. Chen, Y. Lin, Z. He, T. Chen, et al., Supersmooth Ta_2O_5/Ag/polyetherimide film as the rear transparent electrode for high performance semitransparent perovskite solar cells, Adv. Opt. Mater. 7 (4) (2019) 1801409.

[60] D. Angmo, T.R. Andersen, J.J. Bentzen, M. Helgesen, R.R. Sondergaard, et al., Roll-to-roll printed silver nanowire semitransparent electrodes for fully ambient solution-processed tandem polymer solar cells, Adv. Func. Mater. 25 (28) (2015) 4539–4547.

[61] J.-G. Kim, J.-H. Lee, S.-I. Na, H.H. Lee, Y. Kim, H.-K. Kim, Semi-transparent perovskite solar cells with directly sputtered amorphous InZnSnO top cathode for building integrated photovoltaics, Org. Electron. 78 (2020) 105560.

[62] X. Dai, Y. Zhang, H. Shen, Q. Luo, X. Zhao, J. Li, et al., Working from both sides: composite metallic transparent top electrode for high performance perovskite solar cells, ACS Appl. Mater. Interfaces 8 (7) (2016) 4523–4531.

[63] A.I. Hofmann, E. Cloutet, G. Hadziioannou, Materials for transparent electrode: from metal oxides to organic alternatives, Adv. Electron. Mater. 4 (10) (2018) 1700412.

[64] Y.-Y. Lee, K.-H. Tu, C.-C. Yu, S.-S. Li, J.-Y. Hwang, et al., Top laminated graphene electrode in a semitransparent polymer solar cell by simultaneous thermal annealing/releasing method, ACS Nano 5 (8) (2011) 6564–6570.

[65] X. Wang, L. Zhi, K. Mullen, Transparent, conductive graphene electrodes for dye-sensitized solar cells, Nano Lett. 8 (1) (2008) 323–327.

[66] S. Pang, X. Li, H. Dong, D. Chen, W. Zhu, J. Chang, et al., Efficient bifacial semi-transparent perovskite solar cells using Ag/V_2O_5 as transparent anodes, ACS Appl. Mater. Interfaces 10 (15) (2018) 12731–12739.

[67] X. Li, P. Li, Z. Wu, D. Luo, H.-Y. Yu, Z.-H. Lu, Review and perspective of materials for flexible solar cells, Mater. Rep.: Energy 1 (1) (2021) 100001.

[68] D. Angmo, M. Hosel, F.C. Krebs, All solution processing of ITO-free organic solar cell modules directly on barrier foil, Sol. Energy Mat. Sol. Cell 107 (2012) 329–336.

[69] F. Guo, P. Kubis, T. Przybilla, E. Spiecker, A. Hollmann, et al., Nanowire interconnects for printed large-area semitransparent organic photovoltaic module, Adv. Energy Mat. 5 (12) (2015) 1401779.

[70] J. Bovatsek, A. Tamhankar, R.S. Patel, N.M. Bulgakova, J. Bonse, Thin film removal mechanisms in ns-laser processing of photovoltaic materials, Thin Solid. Films 518 (10) (2010) 2897–2904.

[71] Y. Galagan, I.G. de Vries, A.P. Langen, R. Andriessen, W.J.H. Verhees, S.C. Veenstra, et al., Technology development for roll-to-roll production of organic photovoltaics, Che. Eng. Process.: Process. Intens. 50 (2011) 454−461.
[72] C. McDonnel, D. Milne, C. Prieto, H. Chan, D. Rostohar, G.M. O'Connor, Laser patterning of very thin indium tin oxide thin films on PET substrate, Appl. Surf. Sci. 359 (2015) 567−575.
[73] L. Rakocevic, R. Gehlhaar, M. Jayasankar, W. Song, T. Aernouts, H. Fledderus, et al., Translucent, color-neutral and efficient perovskite thin film solar modules, J. Mater. Chem. C. 6 (2018) 3034−3041.
[74] C. Liang, X. Sun, J. Zheng, W. Su, Y. Hu, J. Duan, Surface ablation thresholds of femtosecond laser micropatterning silver nanowires network on flexible substrate, Micro. Eng. 232 (2020) 111396.
[75] K. Khandelwal, S. Biswas, A. Mishra, G.D. Sharma, Semitransparent organic solar cells: from molecular design to structure-performance relationships, J. Mater. Chem. C. 10 (2022) 13−43.
[76] L. Chen, Z. Gao, Y. Zheng, M. Cui, H. Yan, et al., 14.1% efficiency hybrid planar-Si/organic heterojunction solar cells with SnO_2 insertion layer, Sol. Energy 174 (2018) 549−555.
[77] P. Yin, Z. Yin, Y. Ma, Q. Zheng, Improving the charge transport of the ternary blend active layer for efficient semitransparent organic solar cells, Energy Environ. Sci. 13 (2020) 5177−5185.
[78] R.R. Lunt, V. Bulovic, Transparent, near-infrared organic photovoltaic solar cells for window and energy-scavenging applications, Appl. Phys. Lett. 98 (2011) 113305.
[79] Z. Tang, Z. George, Z. Ma, J. Bergqvist, K. Tvingstedt, et al., Semi-transparent tandem organic solar cells with 90% internal quantum efficiency, Adv. Energy Mater. 2 (12) (2012) 1467−1476.
[80] J. Zhang, G. Xu, F. Tao, G. Zeng, M. Zhang, Y. Yang, et al., Highly efficient semitransparent organic solar cells with color rendering index approaching 100, Adv. Mater. 31 (10) (2019) 1807159.
[81] C.-C. Chen, L. Dou, R. Zhu, C.-H. Chung, T.-B. Song, et al., Visibly transparent polymer solar cells produced by solution processing, ACS Nano 6 (8) (2012) 7185−7190.
[82] X. Zhang, C. Hagglund, M.B. Johansson, K. Sveinbjornsson, E.M.J. Johansson, Fine tuned nanolayered metal/metal oxide electrode for semitransparent colloidal quantum dot solar cells, Adv. Funct. Mater. 26 (12) (2016) 1921−1929.
[83] X. Zhang, C. Hagglund, E.M.J. Johansson, Highly efficient, transparent and stable semitransparent colloidal quantum dot solar cells: a combined numerical modelling and experimental approach, Energy Environ. Sci. 10 (2017) 216−224.
[84] Y. Li, C. He, L. Zuo, F. Zhao, L. Zhan, et al., High performance semi-transparent organic photovoltaic devices via improving absorbing selectivity, Adv. Energy Mater. 11 (11) (2021) 2003408.
[85] X. Li, R. Xia, K. Yan, H.-L. Yip, H. Chen, C.-Z. Li, Multifunctional semitransparent organic solar cells with excellent infrared photon rejection, Chin. Chem. Lett. 31 (6) (2020) 1608−1611.
[86] D. Chen, S. Pang, L. Zhou, X. Li, A. Su, et al., An efficient TeO_2/Ag transparent top electrode for 20%-efficiency bifacial perovskite solar cells with a bifaciality factor exceeding 80%, J. Mater. Chem. A 7 (2019) 15156−15163.
[87] Y. Zhou, H. Cheun, S. Choi, W.J. Potscavage, C. Fuentes-Hernandez, B. Kippelen, Indium tin oxide-free and metal-free semitransparent organic solar cells, Appl. Phys. Lett. 97 (2010) 153304.
[88] S. Dai, X. Zhan, Nonfullerene acceptors for semitransparent organic solar cells, Adv. Energy Mater. 8 (21) (2018) 1800002.

[89] J. Chen, G. Li, Q. Zhu, X. Guo, Q. Fan, W. Ma, et al., Highly efficient near-infrared and semitransparent polymer solar cells based on an ultra-narrow bandgap nonfullerne acceptor, J. Mater. Chem. A 7 (2019) 3745−3751.

[90] Y. Liu, P. Cheng, T. Li, R. Wang, Y. Li, et al., Unraveling sunlight by transparent organic semiconductors toward photovoltaic and photosynthesis, ACS Nano 13 (2) (2019) 1071−1077.

[91] Y. Li, X. Guo, Z. Peng, S.R. Forrest, Color-neutral, semitransparent organic photovoltaics for power window applications, Proc. Natl. Acad. Sci. USA 117 (35) (2020) 21147−21154.

[92] S. Chen, H. Yao, B. Hu, G. Zhang, L. Arunagiri, et al., A nonfullerene semitransparent tandem organic solar cell with 10.5% power conversion efficiency, Adv. Energy Mater. 8 (31) (2018) 1800529.

[93] Y. Zhang, D. Liu, T.-K. Lau, L. Zhan, D. Shen, et al., A novel wide-bandgap polymer with deep ionization potential enables exceeding 16% efficiency in ternary nonfullerene polymer solar cells, Adv. Funct. Mater. 30 (27) (2020) 1900466.

[94] G.P. Kini, S.J. Jeon, D.K. Moon, Latest progress on photoabsorbent materials for multifunctional semitransparent organic solar cells, Adv. Funct. Mater. 31 (15) (2021) 2007931.

[95] Y. Xie, R. Xia, T. Li, L. Ye, X. Zhan, H.-L. Yip, et al., Highly transparent organic solar cells with all-near-infrared photoactive materials, Small Methods 3 (12) (2019) 1900424.

[96] X. Wang, Y. Yao, X. Jing, F. Li, L. Yu, Y. Hao, et al., Thiophene copolymer for 1V high open-circuit voltage semitransparent photovoltaic devices, J. Mater. Chem. C. 7 (2019) 10868−10875.

[97] Y. Bai, C. Zhao, X. Chen, S. Zhang, S. Zhang, T. Hayat, et al., Interfacial engineering and optical coupling for multicolored semitransparent inverted organic photovoltaics with a record efficiency of over 12%, J. Mater. Chem. A 7 (2019) 15887−15894.

[98] J. Meiss, F. Holzmueller, R. Gresser, K. Leo, M. Riede, Near-infrared absorbing semitransparent organic solar cells, App. Phys. Lett. 99 (2011) 193307.

[99] W. Yu, X. Jia, M. Yao, L. Zhu, Y. Long, L. Shen, Semitransparent polymer solar cells with simultaneously improved efficiency and color rendering index, Phys. Chem. Chem. Phys. 17 (2015) 23732−23740.

[100] Y. Galagan, R. Andriessen, Organic photovoltaics: technologies and manufacturing, in: V. Fthenakis (ed.), Third Generation Photovoltaics, Intech Publishers, 2012. Available from: https://doi.org/10.5772/25901.

[101] Y. Zhou, F. Li, S. Barrau, W. Tian, O. Inganas, F. Zhang, Inverted and transparent polymer solar cells prepared with vacuum-free processing, Sol. Energy Mat. Sol. Cell 93 (4) (2009) 497−500.

[102] H. Schmidt, H. Flugge, T. Winkler, T. Bulow, T. Riedl, W. Kowalsky, Efficient semitransparent inverted organic solar cells with indium tin oxide top electrode, Appl. Phys. Lett. 94 (2009) 243302.

[103] W. Wang, C. Yan, T.-K. Lau, J. Wang, K. Liu, Y. Fan, et al., Fused hexacyclic nonfullerene acceptor with strong near-infrared absorption for semitransparent organic solar cells with 9.77% efficiency, Adv. Mater. 29 (31) (2017) 1701308.

[104] M. Nam, H.Y. Noh, J.-H. Kang, J. Cho, B.K. Min, J.W. Shim, et al., Semi-transparent quaternary organic blends for advanced photovoltaic applications, Nano Energy 58 (2019) 652−659.

[105] J. Mescher, S.W. Kettlitz, N. Christ, M.F.G. Klein, A. Puetz, et al., Design rules for semi-transparent organic tandem solar cells for window integration, Org. Electron. 15 (7) (2014) 1476−1480.

[106] D. Rossi, K. Forberich, F. Matteocci, M.A. der Maur, H.-J. Egelhaaf, C.J. Brabec, et al., Design of highly efficient semitransparent perovskite/organic tandem solar cells, RRL Sol. 6 (9) (2022) 2200242.

[107] L. Zuo, X. Shi, W. Fu, A.K.-Y. Jen, Highly efficient semitransparent solar cells with selective absorption and tandem architecture, Adv. Mater. 31 (36) (2019) 1901683.

[108] W. Chen, Y. Zhu, J. Xiu, G. Chen, H. Liang, et al., Monolithic perovskite/organic tandem solar cells with 23.6% efficiency enabled by reduced voltage losses and optimized interconnecting layer, Nat. Energy 7 (2022) 229−237.

[109] J. Park, K. Lee, K. Seo, 25cm^2 glass-like transparent crystalline silicon solar cells with an efficiency of 140.5%, Cell Rep. Phys. Sci. 3 (1) (2022) 100715.

[110] A. Takeoka, S. Kouzuma, H. Tanaka, H. Inoue, K. Murata, et al., Development and application of see-through a-Si solar cells, Sol. Energy Mater. Sol. Cell 29 (3) (1993) 243−252.

[111] K. Lee, N. Kim, K. Kim, H.-D. Um, W. Jin, et al., Neutral-colored transparent crystalline silicon photovoltaics, Joule 4 (1) (2020) 235−246.

[112] Z. Huang, N. Geyer, P. Werner, J. de Boor, U. Gosele, Metal-assisted chemical etching of silicon: a review, Adv. Mater. 23 (2) (2011) 285−308.

[113] H. Han, Z. Huang, W. Lee, Metal-assisted chemical etching of silicon and nanotechnology applications, Nano Today 9 (3) (2014) 271−304.

[114] X. Leng, C. Wang, Z. Yuan, Progress in metal-assisted chemical etching of silicon nanostructures, Procedia CIRP 89 (2020) 26−32.

[115] M. Patel, V.V. Satale, S. Kim, K. Lee, J. Kim, Transparent photovoltaic-based photocathodes for see-through energy systems, J. Power Sources 548 (2022) 232009.

[116] S.B. Kang, J.-H. Kim, M.H. Jeong, A. Sanger, C.U. Kim, C.-M. Kim, et al., Stretchable and colorless freestanding microwire arrays for transparent solar cells with flexibility, Light: Sci. & Appl. 8 (121) (2019) 1−13.

[117] E. Pourshaban, A. Banerjee, A. Deshpande, C. Ghosh, M.U. Kharkhanis, et al., Flexible and semi-transparent silicon solar cells as a power supply to smart contact lenses, ACS Appl. Electron. Mater. 4 (8) (2022) 4016−4022.

[118] C.-Y. Tsai, C.-Y. Tsai, See-through, light-through, and color modules for large-area tandem amorphous/microcrystalline silicon thin-film solar modules: technology development and practical considerations for building integrated photovoltaic applications, Renew. Energy 145 (2020) 2637−2646.

[119] J.-S. Cho, Y.H. Seo, A. Lee, S. Park, K. Kim, et al., Energy-harvesting photovoltaic transparent tandem devices using hydrogenated amorphous and microcrystalline silicon absorber layers for window applications, App. Surf. Sci. 589 (2022) 152936.

[120] S.Y. Myong, S.W. Jeon, Design of esthetic color for thin-film silicon semitransparent photovoltaic modules, Sol. Ener. Mater. Sol. Cell 143 (2015) 442−449.

[121] M. Gratzel, Dye-sensitized solar cells, J. Photochem. Photobiol. C: Photochem. Rev. 4 (2) (2003) 145−153.

[122] A.B. Munoz-Garcia, I. Benesperi, G. Boschloo, J.J. Concepcion, J.H. Delcamp, et al., Dye-sensitized solar cells strike back, Chem. Soc. Rev. 50 (2021) 12450−12550.

[123] M. Kokkonen, P. Talebi, J. Zhou, S. Asgari, S.A. Soomro, et al., Advanced research trends in dye-sensitized solar cells, J. Mater. Chem. A 9 (2021) 10527−10545.

[124] X. Wang, B. Zhao, W. Khan, Y. Xie, K. Pan, Review on low-cost counter electrode materials for dye-sensitized solar cells: effective strategy to improve photovoltaic performance, Adv. Mater. Inter. 9 (2) (2022) 2101229.

[125] C.-G. Wu, Ruthenium-based complex dyes for dye-sensitized solar cells, J. Chin. Chem. Soc. 69 (8) (2022) 1242−1252.

[126] A. Agrawal, S.A. Siddiqui, A. Soni, G.D. Sharma, Advancements, frontiers and analysis of metal oxide semiconductor, dye, electrolyte and counter electrode of dye-sensitized solar cell, Sol. Energy 233 (2022) 378–407.

[127] N.K. Farhana, N.M. Saidi, S. Bashir, S. Ramesh, K. Ramesh, Review on the revolution of polymer electrolytes for dye-sensitized solar cells, Energy Fuels 35 (23) (2021) 19320–19350.

[128] F. Odobel, Y. Pellegrin, E.A. Gibson, A. Hagfeldt, A.L. Smeigh, L. Hammarstrom, Recent advances and future directions to optimize the performance of p-type dye-sensitized solar cells, Coord. Chem. Rev. 256 (2012) 2414–2423.

[129] Y. Ren, D. Zhang, J. Suo, Y. Cao, F.T. Eickemeyer, et al., Hydroxamic acid pre-adsorption raises efficiency of cosensitized solar cells, Nature 613 (2023) 60–65.

[130] H. Zhou, J. Cui, J. Guo, S. Tao, X. Gao, et al., Semi-transparent and stable solar cells for building integrated photovoltaics: the confinement effects of the polymer gel electrolyte inside mesoporous films, ACS Omega 4 (12) (2019) 15097–15100.

[131] J. Briscoe, S. Dunn, The future of using earth-abundant elements in counter electrodes for dye-sensitized solar cells, Adv. Mater. 28 (20) (2016) 3802–3813.

[132] S. Hussain, S.F. Shaikh, D. Vikraman, R.S. Mane, O.-S. Joo, Mu Naushad, et al., Sputtering and sulfurization-combined synthesis of a transparent WS_2 counter electrode and its application to dye-sensitized solar cells, RSC Adv. 5 (2015) 103567–103572.

[133] F. Grifoni, M. Bonomo, W. Naim, N. Barbero, T. Alnasser, et al., Toward sustainable, colorless, and transparent photovoltaics: state of art and perspectives for the development of selective near-infrared dye-sensitized solar cells, Adv. Energy Mater. 11 (43) (2021) 2101598.

[134] T. Baron, W. Naim, I. Nikolinakos, B. Andrin, Y. Pellegrin, et al., Transparent and colorless dye-sensitized solar cells based on pyrrolopyrrole cyanine sensitizers, Angew. Chem. 134 (35) (2022) e202207459.

[135] W. Naim, V. Novelli, I. Nikolinakos, N. Barbero, I. Dzeba, et al., Transparent and colorless dye-sensitized solar cell exceeding 75% average visible transmittance, JACS Au 1 (4) (2021) 409–426.

[136] X. Xie, Y. Zhang, Y. Ren, L. He, Y. Yuan, J. Zhang, et al., Semitransparent dye-sensitized solar cell with 11% efficiency and photothermal stability, J. Phys. Chem. C. 126 (27) (2022) 11007–11015.

[137] K. Zhang, C. Qin, X. Yang, A. Islam, S. Zhang, H. Chen, et al., High-performance, transparent, dye-sensitized solar cells for see-through photovoltaic windows, Adv. Energy Mater. 4 (11) (2014) 1301966.

[138] Y. Duan, Q. Tang, B. He, R. Li, L. Yu, Transparent nickel selenide alloy counter electrodes for bifacial dye-sensitized solar cells exceeding 10% efficiency, Nanoscale 6 (2014) 12601–12608.

[139] J.-Y. Lin, J.-H. Liao, T.-C. Wei, Honeycomb-like CoS counter electrodes for transparent dye-sensitized solar cells, Electrochem. Solid-State Lett. 14 (2011) D41.

[140] L. Vesce, P. Mariani, M. Calamante, A. Dessi, A. Mordini, L. Zani, et al., Process engineering of semitransparent DSSC modules and panel incorporating an organic sensitizer, RRL Sol. 6 (8) (2022) 2200403.

[141] Y. Jang, A. Shapiro, M. Isarov, A. Rubin-Brusilovski, A. Safran, et al., Interface control of electronic and optical properties in IV-VI and II-VI core/shell colloidal quantum dots: a review, Chem. Commun. 53 (2017) 1002–1024.

[142] B. Guzelturk, P.L.H. Martinez, Q. Zhang, Q. Xiong, H. Sun, et al., Excitonics of semiconductor quantum dots and wires for lighting and displays, Laser Photonics Rev. 8 (1) (2014) 73–93.

[143] Y. Pu, F. Cai, D. Wang, J.-X. Wang, J.-F. Chen, Colloidal synthesis of semiconductor quantum dots toward large-scale production: a review, Ind. Eng. Chem. Res. 57 (6) (2018) 1790−1802.

[144] C.M. Donega, P. Liljeroth, D. Vanmaekelbergh, Physicochemical evaluation of the hot-injection method, a synthesis route for monodisperse nanocrystals, Small 1 (2) (2005) 1152−1162.

[145] D. Chen, X. Chen, Luminescent perovskite quantum dots: synthesis, microstructures, optical properties and applications, J. Mater. Chem. C. 7 (2019) 1413−1446.

[146] Q.V. Le, K. Hong, H.W. Jang, S.Y. Kim, Halide perovskite quantum dots for light-emitting diodes: properties, synthesis, applications and outlooks, Adv. Electr. Mater. 4 (12) (2018) 1800335.

[147] C. Greboval, A. Chu, N. Goubet, C. Livache, S. Ithurria, E. Lhuillier, Mercury chalcogenide quantum dots: material perspective for device integration, Chem. Rev. 121 (7) (2021) 3627−3700.

[148] X. Zhang, G.E. Eperon, J. Liu, E.M.J. Johansson, Semitransparent quantum dot solar cell, Nano Energy 22 (2016) 70−78.

[149] X. Zhang, C. Hagglund, M.B. Johansson, K. Sveinbjornsson, E.M.J. Johansson, Fine tuned metal/metal oxide electrode for semitransparent colloidal quantum dot solar cells, Adv. Funct. Mater. 26 (2) (2016) 1921−1929.

[150] A.-J. Cho, M.-K. Song, D.-W. Kang, J.-Y. Kwon, Two-dimensional WSe_2/MoS_2 p-n heterojunction-based transparent photovoltaic cell and its performance enhancement by fluoropolymer passivation, ACS Appl. Mater. Interfaces 10 (42) (2018) 35972−35977.

[151] M. Sadanand, N. Patel, W. Kumar, J. Lee, Kim, New concepts in all-metal-oxide-based ultraviolet transparent photovoltaics, IEEE Trans. Elec. Devices 69 (9) (2022) 5021.

[152] D.B. Patel, K.R. Chauhan, 50% transparent solar cells of CuO_x/TiO_2: device concepts, J. Alloy. Compd. 842 (2020) 155594.

[153] M. Patel, D.-K. Ban, A. Ray, J. Kim, Transparent all-oxide photovoltaics and broadband high-speed energy-efficient optoelectronics, Sol. En. Mat. Sol. Cell 194 (2019) 148−158.

[154] T.T. Nguyen, M. Patel, S. Kim, R.A. Mir, J. Yi, V.-A. Dao, et al., Transparent photovoltaic cells and self-powered photodetectors by TiO_2/NiO heterojunction, J. Power Sources 481 (2021) 228865.

[155] M. Patel, H.-S. Kim, J. Kim, J.-H. Yun, S.J. Kim, E.H. Choi, et al., Excitonic metal oxide heterojunction (NiO/ZnO) solar cells for all-transparent module integration, Sol. Energy Mater. Sol. Cell 170 (2017) 246−253.

[156] M. Patel, J. Song, D.-W. Kim, J. Kim, Carrier transport and working mechanism of transparent solar cells, Appl, Mater. Today 26 (2022) 101344.

[157] T.T. Naguyen, M. Patel, J.-W. Kim, W. Lee, J. Kim, Functional TiO_2 interlayer for all-transparent metal-oxide photovoltaics, J. Alloy. Compd. 816 (2020) 152602.

[158] A.J. Lopez-Garcia, O. Blazquez, C. Voz, J. Puigdollers, V. Izquierdo-Roca, A. Perez-Rodriguez, Ultrathin wide-bandgap a-Si:H based solar cells for transparent photovoltaic applications, RRL Sol. 6 (1) (2022) 2100909.

[159] A.L. Lopez-Garcia, A. Bauer, R.F. Rubio, D. Payno, Z.J. Li-Kao, et al., UV-selective optically transparent Zn(O,S)-based solar cells, RRL Sol. 4 (11) (2020) 2000470.

[160] S. Kim, M. Patel, T.T. Nguyen, N. Kumar, P. Bhatnagar, J. Kim, Highly transparent bidirectional transparent photovoltaics for on-site power generators, ACS Appl. Mater. Interfaces 14 (1) (2022) 706−716.

[161] S. Chatterjee, S.K. Saha, A.J. Pal, Formation of all-oxide solar cells in atmospheric condition based on Cu_2O thin-films grown through SILAR technique, Sol. Energy Mat. Sol. Cell 147 (2016) 17−26.

[162] A.K. Rana, D.-K. Ban, M. Patel, J.-H. Yun, J. Kim, A transparent photovoltaic device based on $Cu_2O/ZnO/AZO$ for see-through applications, Mater. Lett. 255 (2019) 126517.
[163] A.K. Rana, J.T. Park, J. Kim, C.-P. Wong, See-through metal oxide frameworks for transparent photovoltaics and broadband photodetectors, Nano Energy 64 (2019) 103952.
[164] Q. Yi, J. Pan, J. Mei, Z. Chen, P. Wang, et al., The $Cu_2O/CuO/SnO_2$ transparent pn junction film device towards photovoltaic enhancement with Cu^{2+} self-oxidation transition layer, J. Mater. Sci. 56 (2021) 5736−5747.
[165] A.-J. Cho, K. Park, S. Park, M.-K. Song, K.-B. Chung, J.-Y. Kwon, A transparent solar cell based on a mechanically exfoliated GaTe and InGaZnO p-n heterojunction, J. Mater. Chem. C. 5 (2017) 4327−4334.
[166] N. Kumar, M. Patel, J. Kim, C. Jeong, C.-P. Wong, Flexible transparent photovoltaics for ultra-UV photodetection and functional UV-shielding based on Ga_2O_3/Cu_2O heterojunction, App, Mater. Today 29 (2022) 101620.
[167] N. Kumar, M. Patel, D. Lim, K. Lee, J. Kim, Van der Waals semiconductor embedded transparent photovoltaic for broadband optoelectronics, Surf. Inter. 34 (2022) 102369.
[168] N. Kumar, U. Farva, M. Patel, W.-S. Cha, J. Lee, J. Kim, n-Ga_2O_3/p-SnS heterojunction thin-films based transparent photovoltaic device, J. Alloy. Compd. 921 (2022) 166177.
[169] T.T. Nguyen, G. Murali, M. Patel, S. Park, I. In, J. Kim, MXene-Integrated metal oxide transparent photovoltaics and self-powered photodetectors, ACS Appl. Energy Mater. 5 (6) (2022) 7134−7143.
[170] J.Y. Kim, J.-W. Lee, H.S. Jung, H. Shin, N.-G. Park, High-efficiency perovskite solar cells, Chem. Rev. 120 (15) (2020) 7867−7918.
[171] P. Roy, A. Ghosh, F. Barclay, A. Khare, E. Cuce, Perovskite solar cells: a review of the recent advances, Coatings 12 (8) (2022) 1089.
[172] P. Yan, D. Yang, H. Wang, S. Yang, Z. Ge, Recent advances in dopant-free organic hole-transporting materials for efficient, stable and low-cost perovskite solar cells, Energy Environ. Sci. 15 (2022) 3630−3669.
[173] D. Li, D. Zhang, K.-S. Lim, Y. Hu, Y. Rong, A. Mei, et al., A review on scaling up perovskite solar cells, Adv. Fun. Mater. 31 (2) (2021) 2008621.
[174] H.J. Snaith, Present status and future prospects of perovskite photovoltaics, Nat. Mater. 17 (2018) 372−376.
[175] T.M. Koh, H. Wang, Y.F. Ng, A. Bruno, S. Mhaisalkar, N. Mathews, Halide perovskite solar cells for building integrated photovoltaics: transforming building facades into power generators, Adv. Mater. 34 (25) (2022) 2104661.
[176] S. Rahmany, L. Etgar, Semitransparent perovskite solar cells, ACS Energy Lett. 5 (5) (2020) 1519−1531.
[177] K. Sveinbjornsson, B. Li, S. Mariotti, E. Jarzembowski, L. Kegelmann, et al., Monolithic perovskite/silicon tandem solar cell with 28.7% efficiency using industrial silicon bottom cells, ACS Energy Lett. 7 (8) (2022) 2654−2656.
[178] C.O.R. Quiroz, C. Bronnbauer, I. Levchuk, Y. Hou, C.J. Brabec, K. Forberich, Coloring semitransparent perovskite solar cells via dielectric mirrors, ACS Nano 10 (5) (2016) 5104−5112.
[179] S. Chen, B. Chen, X. Gao, B. Dong, H. Hu, et al., Neutral-colored semitransparent solar cells based on pseudohalide (SCN-)-doped perovskite, Sustain. Energy Fuels 1 (2017) 1034−1040.
[180] M. Mujahid, C. Chen, J. Zhang, C. Li, Y. Duan, Recent advances in semitransparent perovskite solar cells, InfoMat 3 (1) (2021) 101−124.
[181] G.E. Eperon, V.M. Burlakov, A. Goriely, H.J. Snaith, Neutral color semitransparent microstructured perovskite solar cells, ACS Nano 8 (1) (2014) 591−598.

[182] H.-C. Kwon, S. Ma, S.-C. Yun, G. Jang, H. Yang, J. Moon, A nanopillar-structured perovskite-based efficient semitransparent solar module for power-generating window applications, J. Mater. Chem. A 8 (2020) 1457−1468.

[183] G.E. Eperon, D. Bryant, J. Troughton, S.D. Stranks, M.B. Johnston, et al., Efficient, semitransparent neutral-colored solar cells based on microstructured formamidinium lead trihalide perovskite, J. Phys. Chem. Lett. 6 (1) (2015) 129−138.

[184] H. Kim, H.-S. Kim, J. Ha, N.-G. Park, S. Yoo, Solar cells: empowering semitransparent solar cells with thermal-mirror functionality, Adv. Mater. Mater. 6 (14) (2016) 1502466.

[185] D. Liu, C. Yang, P. Chen, M. Bates, S. Han, P. Askeland, et al., Lead halide ultraviolet-harvesting transparent photovoltaics with an efficiency exceeding 1, ACS Appl. Energy Mater. 2 (6) (2019) 3972−3978.

[186] E.D. Gaspera, Y. Peng, Q. Hou, L. Spiccia, U. Bach, J. Jasiniak, et al., Ultra-thin high semitransparent perovskite solar cells, Nano Energy 13 (2015) 249−257.

[187] K.-T. Lee, L.J. Guo, H.J. Park, Neutral- and multi-colored semitransparent perovskite solar cells molecules, Molecules 21 (4) (2016) 475.

[188] C.O.R. Quirroz, L. Levchuk, C. Bronnbauer, M. Salvador, K. Forberich, et al., Pushing efficiency limits for semitransparent perovskite solar cells, J. Mater. Chem. A 3 (2015) 24071−24081.

[189] L. Yuan, Z. Wang, R. Duan, P. Huang, K. Zhang, et al., Semi-transparent solar cells: unveiling the trade-off between transparency and efficiency, J. Mater. Chem. A 6 (2018) 19696−19702.

[190] O.M. Alkhudhari, A. Altujjar, M.Z. Mokhtar, B.F. Spencer, Q. Chen, et al., High efficiency semitransparent perovskite solar cells containing 2D nanopore arrays deposited in a single step, J. Mater. Chem. A 10 (2022) 10227−10241.

[191] Y. Xu, J. Wang, L. Sun, H. Huang, J. Han, H. Huang, et al., Top transparent electrodes for fabricating semitransparent organic and perovskite solar cells, J. Mater. Chem. C. 9 (2021) 9102−9123.

[192] L. Bu, Z. Liu, M. Zhang, W. Li, A. Zhu, F. Cai, et al., Semitransparent fully air processed perovskite solar cells, ACS Appl. Mater. Interfaces 7 (32) (2015) 17776−17781.

[193] M. Mujahid, C. Chen, W. Hu, Z.-K. Wang, Y. Duan, Progress of high-throughput and low-cost flexible perovskite solar cells, RRL Sol. 4 (8) (2020) 1900556.

[194] H.A. Dewi, H. Wang, J. Li, M. Thway, R. Sridharan, et al., Highly efficient semitransparent perovskite solar cells for four terminal perovskite-silicon tandems, ACS Appl. Mater. Interfaces 11 (37) (2019) 34178−34187.

[195] E. Raza, Z. Ahmad, Review on two-terminal and four-terminal crystalline-silicon/perovskite tandem solar cells: progress, challenges, and future perspectives, Energy Rep. 8 (2022) 5820−5851.

[196] C.D. Bailie, M.G. Christoforo, J.P. Mailoa, A.R. Bowring, E.L. Unger, et al., Semi-transparent perovskite solar cells for tandems with silicon and CIGS, Energy Environ. Sci. 8 (2015) 956−963.

[197] X. Zhang, Y. Qian, X. Ling, Y. Wang, Y. Zhang, et al., α-$CsPbBr_3$ perovskite quantum dots for application in semitransparent photovoltaics, ACS Appl. Mater. Interfaces 12 (24) (2020) 27307.

[198] M.M. Tavakoli, M. Nasilowski, J. Zhao, M.G. Bawendi, J. Kong, Efficient semitransparent $CsPbI_3$ quantum dots photovoltaics using a graphene electrode, Small Methods 3 (12) (2019) 1900449.

[199] E.P. Thompson, E.L. Bombelli, S. Shubham, H. Watson, A. Everard, et al., Tinted semi-transparent solar panels allow concurrent production of crops and electricity on the same cropland, Adv. Energy Mater. 10 (35) (2020) 2001189.

[200] S. Gorjian, E. Bousi, O.E. Ozdemir, M. Trommsdorff, N.M. Kumar, et al., Progress and challenges of crop production and electricity generation in agrivoltaic systems using semi-transparent photovoltaic technology, Renew. Sust. Energy Rev. 158 (2022) 112126.
[201] R.H.E. Hassanien, M. Li, F. Yin, The integration of semi-transparent photovoltaics on greenhouse roof for energy and plant production, Renew. Energy 121 (2018) 377–388.
[202] Y. Zhao, Y. Zhu, H.-W. Cheng, R. Zheng, D. Meng, Y. Yang, A review on semitransparent solar cells for agricultural applications, Mater. Today Energy 22 (2021) 100852.
[203] C.H. Parrish II, D. Hebert, A. Jackson, K. Ramasamy, H. McDaniel, G.A. Giacomelli, et al., Optimizing spectral quality with quantum dots to enhance crop yield in controlled environments, Comm. Bio. 4 (2021) 124.
[204] A. Mourtzikou, D. Sygkridou, T. Georgakopoulos, G. Katsagounos, E. Stathatos, Semi-transparent dye-sensitized solar panels for energy autonomous greenhouses, Int. J. Str. Constr. Eng. 14 (3) (2020) 1–6.
[205] M.F. Peretz, F. Geoola, I. Yehia, S. Ozer, A. Levi, et al., Testing organic photovoltaic modules for application as greenhouse cover or shading element, Bio. Eng. 184 (2019) 24–36.
[206] C.J.M. Emmott, J.A. Rohr, M. Campoy-quiles, T. Kirchartz, A. Urbina, N.J. Ekins-Daukes, et al., Organic photovoltaic greenhouses: a unique application for semi-transparent PV? Energy Environ. Sci. 8 (2015) 1317–1328.
[207] E. Ravishankar, R.E. Booth, C. Saravitz, H. Sederoff, H.W. Ade, B.T. O'Connor, Achieving net zero energy greenhouse by integrating semitransparent organic solar cells, Joule 4 (2) (2020) 490–506.
[208] D. Wang, H. Liu, Y. Li, G. Zhou, L. Zhan, et al., High-performance and eco-friendly semitransparent organic solar cells for greenhouse applications, Joule 5 (4) (2021) 945–957.
[209] H. Shi, R. Xia, G. Zhang, H.-L. Yip, Y. Cao, Spectral engineering semitransparent polymer solar cells for greenhouse applications, Adv. Energy Mater. 9 (5) (2019) 1803438.
[210] Z. Wang, X. Zhu, J. Feng, D. Yang, S. Liu, Semitransparent flexible perovskite solar cells for potential greenhouse applications, RRL Sol. 5 (8) (2021) 2100264.
[211] A. Yano, M. Onoe, J. Nakata, Prototype semi-transparent photovoltaic modules for greenhouse roof applications, Bio. Eng. 122 (2014) 62–73.
[212] M.G. Debije, Solar energy collectors with tunable transmission, Adv. Funct. Mater. 20 (9) (2010) 1498–1502.
[213] W. Sark, K. Barnham, L. Slooff, A. Chatten, A. Buchtemann, et al., Luminescent solar concentrators - a review of recent results, Opt. Expr. 16 (2008) 21773–21792.
[214] W.H. Weber, J. Lambe, Luminescent greenhouse collector for solar radiation, Appl. Opt. 15 (10) (1976) 2299.
[215] V. Wittwer, W. Stahl, A. Goetzberger, Fluorescent planar concentrators, Sol. Energy Mater. Sol. Cell 11 (3) (1984) 187–197.
[216] F. Meinardi, F. Bruni, S. Brovelli, Luminescent solar concentrators for building-integrated photovoltaics, Nat. Rev. Mater. 2 (12) (2017) 17072.
[217] M. Vasiliev, K. Alameh, M. Nur-E-Alam, Spectrally-selective energy-harvesting solar windows for public infrastructure applications, Appl. Sci. 8 (2018) 849.
[218] C. Yang, D. Liu, A. Renny, P.S. Kuttipillai, R.R. Lunt, Integration of near-infrared harvesting transparent luminescent solar concentrators onto arbitrary surfaces, J. Lumin. 210 (2019) 230–246. Available from: https://doi.org/10.1016/j.jlumin.2019.02.042.

[219] L.H. Slooff, E.E. Bende, A.R. Burgers, T. Budel, M. Pravettoni, et al., A luminescent solar concentrator with 7.1% power conversion efficiency, Phy. Stat. Solidi RRL 2 (2008) 257.
[220] N. Aste, L.C. Tagliabue, C.D. Pero, D. Testa, R. Fusco, Performance analysis of a large-area luminescent solar concentrator module, Renew. Energy 76 (2015) 330−337.
[221] M.R. Bergren, N.S. Makarov, K. Ramasamy, A. Jackson, R. Guglielmetti, H. McDaniel, High-performance $CuInS_2$ quantum dot laminated glass luminescent solar concentrators for windows, ACS Energy Lett. 3 (2018) 520−525.
[222] S. Marchionna, F. Meinardi, M. Acciarri, S. Binetti, A. Papagni, S. Pizzini, et al., Photovoltaic quantum efficiency enhancement by light harvesting of organolanthanide complexes, J. Lumin. 118 (2006) 325−329.
[223] P. Moraitis, R.E.I. Schropp, W.G.J.H.M. van Sark, Nanoparticles for luminescent solar concentrators - a review, Opt. Mater. 84 (2018) 636−645.
[224] C. Yang, R.R. Lunt, Limits of visibly transparent luminescent solar concentrators, Adv. Opt. Mater. 5 (2017) 1600851.
[225] A. Goetzberger, W. Greube, Solar energy conversion with fluorescent collectors, Appl. Phys. 14 (1977) 123−139.
[226] A. Goetzberger, V. Wittwer, Fluorescent planar collector-concentrators—a review, Sol. Cell 4 (1) (1981) 3−23.
[227] F. Galluzzi, E. Scafe, Spectrum shifting of sunlight by luminescent sheets: performance evaluation of photovoltaic applications, Sol. Energy 33 (1984) 501.
[228] I. Coropceanu, M.G. Bawendi, Core/shell QD based luminescent solar concentrators, Nano Lett. 14 (7) (2014) 4097−4101.
[229] Y. Zhou, D. Benneti, Z. Fan, H. Zhao, D. Ma, A.O. Govorov, et al., Near infrared, highly efficient luminescent solar concentrators, Adv. Energy Mater. 6 (11) (2016) 1501913.
[230] C.D.M. Donega, Synthesis and properties of colloidal heteronanocrystals, Chem. Soc. Rev. 40 (3) (2011) 1512−1546.
[231] V. Kataria, D.S. Mehta, Multispectral harvesting rare-earth oxysulphide based highly efficient transparent luminescent solar concentrator, J. Rare Earths 40 (1) (2022) 41−48.
[232] T. Wang, J. Zhang, W. Ma, Y. Luo, L. Wang, et al., Luminescent solar concentrator employing rare earth complex with zero self-absorption loss, Sol. Energy 85 (2011) 2571−2579.
[233] C.S. Erickson, L.R. Bradshaw, S. McDowall, J.D. Gilbertson, D.R. Gamelin, D.L. Patrick, Zero- reabsorption doped-nanocrystal luminescent solar concentrators, ACS Nano 8 (4) (2014) 3461−3467.
[234] M. Sharma, K. Gungor, A. Yeltik, M. Olutas, B. Guzelturk, Y. Kelestemur, et al., Near unity emitting copper doped colloidal semiconductor quantum wells, Adv. Mater. 29 (30) (2017) 1700821.
[235] C. Li, W. Chen, D. Wu, D. Quan, Z. Zhou, J. Hao, et al., Large stokes shift and high efficiency luminescent solar concentrator, Sci. Rep. 5 (2015) 17777.
[236] M.G. Debije, P.P.C. Verbunt, Thirty years of luminescent solar concentrator research: solar energy for the built environment, Adv. Energy Mater. 2 (2012) 12−35.
[237] M.J. Currie, J.K. Mapel, T.D. Heidel, S. Goffri, M.A. Baldo, High-efficiency organic solar concentrators for photovoltaics, Science 321 (5886) (2008) 226−228.
[238] J.C. Goldschmidt, M. Peters, A. Bosch, H. Helmers, F. Dimroth, S.W. Glunz, et al., Increasing the efficiency of fluorescent concentrator systems, Sol. Energy Mater. Sol. Cell 93 (2) (2009) 176−182.
[239] Y. Zhao, G.A. Meek, B.G. Levine, R.R. Lunt, Near-infrared harvesting transparent luminescent solar concentrators, Adv. Opt. Mater. 2 (7) (2014) 606−611.

[240] F. Meinardi, S. Ehrenberg, L. Dhamo, F. Carulli, M. Mauri, F. Bruni, et al., Highly efficient luminescent solar concentrators based on earth-abundant indirect-bandgap silicon quantum dots, Nat. Photonics 11 (2017) 177–185.
[241] C. Yang, D. Liu, R.R. Lunt, How to accurately report transparent luminescent solar concentrators, Joule 3 (2019) 1–6.
[242] K. Wu, H. Li, V.I. Klimov, Tandem luminescent solar concentrators based on engineered quantum dots, Nat. Photonics 12 (2018) 105–110.
[243] M. Vasiliev, K. Alameh, M.A. Badshah, S.-M. Kim, M. Nur-E-Alam, Semitransparent energy-harvesting solar concentrator windows employing infrared transmission-enhanced glass and large-area microstructured diffractive elements, Photonics 5 (3) (2018) 25.
[244] F.J. Ostos, G. Iasilli, M. Carlotti, A. Pucci, High-performance luminescent solar concentrators based on poly(cyclohexylmethacrylate) (PCHMA) films, Polymers 12 (12) (2020) 2898.
[245] L. Desmet, A.J.M. Ras, D.K.G. de Boer, M.G. Debije, Monocrystalline silicon photovoltaic luminescent solar concentrator with 42% power conversion efficiency, Opt. Lett. 37 (15) (2012) 3087–3089.
[246] R. Mazzaro, A. Vomiero, The renaissance of luminescent solar concentrators: the role of inorganic nanomaterials, Adv. Energy Mater. 8 (33) (2018) 1801903.
[247] J. Zhao, L. Chen, D. Li, Z. Shi, P. Liu, et al., Large-area patterning of full-color quantum dot arrays beyond 1000 pixels per inch by selective electrophoretic deposition, Nat. Comm. 12 (2021) 4603.
[248] S.-H. Shin, B. Hwang, Z.-J. Zhao, S.H. Jeon, J.Y. Jung, et al., Transparent displays utilizing nanopatterned quantum dot films, Sci. Rep. 8 (2018) 1–10.
[249] H. Li, K. Wu, J. Lim, H.-J. Song, V.I. Klimov, Doctor-blade deposition of quantum dots onto standard window glass for low-loss large-area luminescent solar concentrators, Nat. Energy 1 (2016) 16157.
[250] P. Moraitis, G. van Leeuwen, W. van Sark, Visual appearance of nanocrystal-based luminescent solar concentrators, Materials 12 (6) (2019) 885.
[251] J. Bomm, A. Buchtemann, A.J. Chatten, R. Bose, D.J. Farrell, et al., Fabrication and full characterization of state-of-the-art quantum dot luminescent solar concentrators, Sol. Energy Mat. Sol. Cell 95 (8) (2011) 2087–2094.
[252] F. Purcell-Milton, Y.K. Gun'ko, Quantum dots for luminescent solar concentrators, J. Mater. Chem. 22 (2012) 16687–16697.
[253] G. Liu, X. Wang, G. Han, J. Yu, H. Zhao, Earth abundant colloidal carbon quantum dots for luminescent solar concentrators, Mater. Adv. 1 (2020) 119–138.
[254] A. Kim, A. Hosseinmardi, P.K. Annamalai, P. Kumar, R. Patel, Review on colloidal quantum dots luminescent solar concentrators, ChemistrySelect 6 (20) (2021) 4948–4967.
[255] A.R.M. Velarde, E.R. Bartlett, N.S. Makarov, C. Castaneda, A. Jackson, K. Ramasamy, et al., Optimizing the aesthetics of high-performance $CuInS_2/ZnS$ quantum dot luminescent solar concentrator windows, ACS Appl. Energy Mater. 3 (9) (2020) 8159–8163.
[256] S. Sadeghi, H.B. Jalali, S.B. Srivastava, R. Melikov, I. Baylam, A. Sennaroglu, et al., High-performance, large-area, and eco-friendly luminescent solar concentrators using copper-doped InP quantum dots, iScience 23 (7) (2020) 101272.
[257] Z. Li, X. Zhao, C. Huang, X. Gong, Recent advances in green fabrication of luminescent solar concentrators using nontoxic quantum dots as fluorophores, J. Mater. Chem. C. 7 (2019) 12373–12387.
[258] W. Chen, J. Li, P. Liu, H. Liu, J. Xia, S. Li, et al., Heavy metal free nanocrystals with near infrared emission applying in luminescent solar concentrator, RRL Sol. 1 (6) (2017) 1700041.

[259] N.S. Marakov, D. Korus, D. Freppon, K. Ramasamy, D.W. Houck, et al., Minimizing scaling losses in high-performance quantum dot luminescent solar concentrators for large-area solar windows, ACS Appl. Mater. Interfaces 14 (26) (2022) 29679−29689.
[260] B. Liu, S. Ren, G. Han, H. Zhao, X. Huang, B. Sun, et al., Thermal effect on the efficiency and stability of luminescent solar concentrators based on colloidal quantum dots, J. Mater. Chem. C. 9 (2021) 5723−5731.
[261] J. Tong, J. Luo, L. Shi, J. Wu, L. Xu, J. Song, et al., Fabrication of highly emissive and highly stable perovskite nanocrystal-polymer slabs for luminescent solar concentrators, J. Mater. Chem. A 7 (2019) 4872−4880.
[262] Y. Zhang, W. Zhang, Y. Ye, K. Li, X. Gong, C. Liu, $CsPbBr_3$ nanocrystal-embedded glasses for luminescent solar concentrators, Sol. Energy Mat. Sol. Cell 238 (2022) 111619.
[263] F. Meinardi, Q.A. Akkerman, F. Bruni, S. Park, M. Mauri, Z. Dang, et al., Doped halide perovskite nanocrystals for reabsorption-free luminescent solar concentrators, ACS Energy Lett. 2 (10) (2017) 2368−2377.
[264] S.S. Bhosale, E. Jokar, Y.-T. Chiang, C.-H. Kuan, K. Khodakarami, et al., Mn-doped organic-inorganic perovskite nanocrystals for a flexible luminescent solar concentrator, ACS Appl. Energy Mater. 4 (10) (2021) 10565−10573.
[265] L. Zdrazil, S. Kalytchuk, M. Langer, R. Ahmad, J. Pospisil, et al., Transparent and low-loss luminescent solar concentrators based on self-trapped exciton emission in lead-free double perovskite nanocrystals, ACS Appl. Energy Mater. 4 (7) (2021) 6445−6453.
[266] Z. Krumer, S.J. Pera, R.J.A. van Dijk-Moes, Y. Zhao, A.F.P. de Brouwer, et al., Tackling self-absorption in luminescent solar concentrators with type-II colloidal quantum dots, Sol. Energy Mat. Sol. Cell 111 (2013) 57−65.
[267] H. Zhao, Y. Zhou, D. Benetti, D. Ma, F. Rosei, Perovskite quantum dots integrated in large-area luminescent solar concentrators, Nano Energy 37 (2017) 214−223.
[268] V.I. Klimov, T.A. Baker, J. Lim, K.A. Velizhanin, H. McDaniel, Quality factor of luminescent solar concentrators and practical concetration limits attainable with semiconductor quantum dots, ACS Photonics 3 (6) (2016) 1138−1148.
[269] K. Gungor, J. Du, V.I. Klimov, General trends in the performance of quantum dot luminescent solar concentrators (LSCs) revealed using the 'effective LSC quality factor, ACS Energy Lett. 7 (5) (2022) 1741−1749.
[270] N.D. Bronstein, Y. Yao, L. Xu, E. O'Brien, A.S. Powers, V.E. Ferry, et al., Quantum dot luminescent concentrator cavity exhibiting 30-fold concentration, ACS Photonics 2 (11) (2015) 1576−1583.
[271] H. Zhao, R. Sun, Z. Wang, K. Fu, X. Hu, Y. Zhang, Zero-dimensional perovskite nanocrystals for efficient luminescent solar concentrators, Adv. Fun. Mater. 29 (30) (2019) 1902262.
[272] H. Zhao, G. Liu, G. Han, High-performance laminated luminescent solar concentrators based on colloidal carbon quantum dots, Nanoscale Adv. 1 (2019) 4888−4894.
[273] G. Liu, R. Mazzaro, C. Sun, Y. Zhang, Y. Wang, H. Zhao, et al., Role of refractive index in highly efficient laminated solar concentrators, Nano Energy 70 (2020) 104470.
[274] J. Huang, J. Zhou, E. Jungstedt, A. Samanta, J. Linnros, L.A. Berglund, et al., Large-area transparent 'quantum dot glass' for building-integrated photovoltaics, ACS Photonics 9 (7) (2022) 2499−2509.
[275] G. Liu, H. Zhao, F. Diao, Z. Ling, Y. Wang, Stable tandem luminescent solar concentrators based on CdSe/CdS quantum dots and carbon dots, J. Mater. Chem. C. 6 (2018) 10059−10066.

[276] J. Chen, H. Zhao, Z. Li, X. Zhao, X. Gong, Highly efficient tandem luminescent solar concentrators based on eco-friendly copper iodide based hybrid nanoparticles and carbon dots, Energy Environ. Sci. 15 (2022) 799−805.
[277] L.J. Brennan, F. Purcell-Milton, B. McKenna, T.M. Watson, Y.K. Gun'ko, R.C. Evans, Large area quantum dot luminescent solar concentrators for use with dye-sensitized solar cells, J. Mater. Chem. A 6 (2018) 2671−2680.
[278] Y. Gu, X. Yao, H. Geng, G. Guan, M. Hu, M. Han, Highly transparent, dual-color emission, heterophase $Cs_3Cu_2I_5/CsCu_2I_3$ nanolayer for transparent luminescent solar concentrators, ACS Appl. Mater. Interfaces 13 (34) (2021) 40798−40805.
[279] X. Liu, D. Benetti, F. Rosei, Semi-transparent luminescent solar concentrators based on plasmon-enhanced carbon dots, J. Mater. Chem. A 9 (2021) 23345−23352.
[280] B. Mendewala, E.T. Vickers, K. Nikolaidou, A. DiBenedetto, W.G. Delmas, J.Z. Zhang, et al., Organo-metal halide perovskite quantum dots with plasmon enhancement, Adv. Opt. Mater. 9 (20) (2021) 2100754.

CHAPTER EIGHTEEN

Photovoltaic-powered vehicles: Current trends and future prospects

Jubaer Ahmed
Edinburgh Napier University, Edinburgh, United Kingdom

18.1 Introduction

To mitigate CO_2 emissions from the transport sector, electric vehicles (EVs) are being adopted globally as the future of transport infrastructure. The gradual transition process from shifting conventional fuel to electricity has already begun and successfully implemented in many places in the world. It is predicted by International Energy Agency (IEA) that 60% of the vehicle sold will be electric by 2030 [1]. Although EV have a low carbon footprint, it is vastly relying on conventional grid systems for charging. Because the conventional grid is still dominated by fossil fuels, the overall impact of the EV on achieving net zero is partly diminished. Besides, EV are highly dependent on battery size for running miles and charging infrastructure. Due to the lack of a strong charging infrastructure, EV may not be a favorable option for many. To counter these issues, the concept of PV being integrated into the body of vehicles is brought into the picture. In principle, PV being integrated into the body, EV could be charged on the go and will not be entirely dependent on the charging infrastructure. A simplified structure of a PV-powered EV (PV-EV) is presented in Fig. 18.1 [2]. As illustrated, PV modules can be integrated into the roof or any other part of the vehicle's body. The electricity from the PV modules is converted, using a DC/DC converter, to charge the battery responsible for driving the electric motor [2].

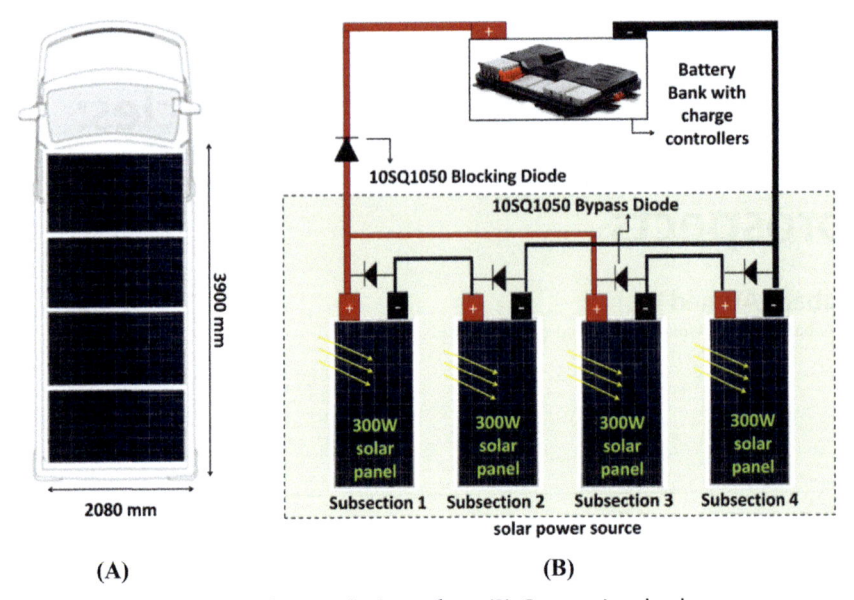

Figure 18.1 (A) Solar panels on vehicle rooftop. (B) Conventional solar power source architecture for electric vehicles. *Reproduced with permission from R.M. Prasad, A. Krishnamoorthy, Design, construction, testing and performance of split power solar source using mirror photovoltaic glass for electric vehicles, Energy 145 (February 2018) 374–387, copyright (2018) Elsevier.*

When batteries are being charged from the PV directly, it reduces the dependency on the fossil fuel-based grid supply. Therefore, the ideal concept was to make the EV self-sufficient and resilient. However, technologically at this current stage, it is quite impossible. The first challenge is the surface area available on the vehicle to install PV modules. For usual commercial vehicles, the surface area is too small to install a sufficient amount of PV panels to produce the energy needed. Besides, the surface area of the vehicles is often full of curvatures, which does not go hand in hand with the conventional PV panels. On top of that, PV panel efficiency is quite low to produce a sufficient amount to charge the EV batteries fully. Accepting these limitations, researchers and commercial developers endeavor into technologies where PV panels will be installed on the EVs as efficiently as possible to provide partial charging to the batteries. Depending on the availability of the sun, as much as possible, energy is expected to be captured to support the EV and reduce its dependency on the power grid. In the next section of this chapter, we will look into the current development that has been made in the field of EVs.

18.2 Recent trends in photovoltaic-powered vehicles

Along with the energy sector, it is a challenge for the transport sector to reduce its contribution to CO_2 emission and global warming. The proportion of CO_2 emission from road transport in the overall energy-related CO_2 emission in Japan, the USA, and the world are 15.9%, 27.8%, and 16.9%, respectively [3]. The rising usage of battery-powered electric vehicles (BEVs) has been contributing to that cause and has a very low carbon footprint compared to the hybrid EV, internal combustion engine, and fuel cell-powered vehicle. However, BEV is still reliable on the main power grid for charging and thus contributes to carbon emission passively. To reduce that even further, PV-EV are one of the best alternatives. A key finding of several recently reported studies was that PV-powered vehicles can reduce carbon emission by 85% compared to ICE and $\sim$ 65% lower than the conventional BEV as illustrated in Fig. 18.2 [4,5].

PV-powered EVs provide the opportunity to produce electricity on board with less dependency on the power grid. However, the surface area of the vehicles is very limited and thus poses a technological challenge to adopt PV-EV in the transport sector widely. In recent years, many R&D

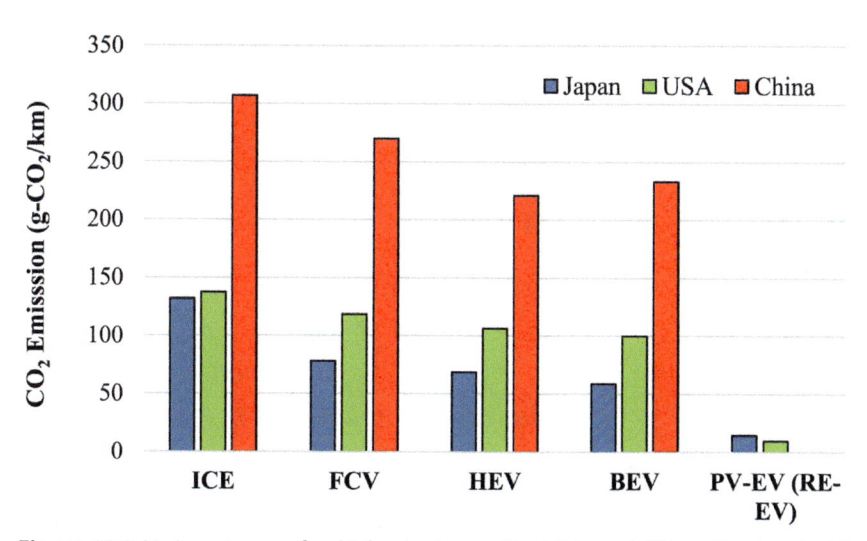

Figure 18.2 Various types of vehicles in Japan, the USA, and China. *Reprinted with permission from M. Yamaguchi, T. Masuda, K. Araki, Y. Ota, K. Nishioka, Impact and recent approaches of high-efficiency solar cell modules for PV-powered vehicles, Jpn. J. Appl. Phys. 61 (2022) SC0802, copyright (2022) The Japan Society of Applied Physics.*

projects have been initiated globally to overcome that challenge by improving PV cell efficiency and EV charging mechanisms. Commercial offers and R&D projects of PV-EV exist on different types of vehicles such as passenger vehicles, trucks, trains, camper vans, boats, planes, or spatial vehicles. In a recent IEA report [6], it is highlighted that there are around 50 R&D projects currently being conducted; it is clear that developing passenger cars is a priority over other types of vehicles as over half of the studies involve passenger cars [1,6]. Fig. 18.3 illustrates the dominance of PV-integrated passenger vehicles with an approximate number of PV-EV initiatives as a function of PV surface area.

Because there is a multibillion-dollar market anticipating PV-powered vehicles, several large automotive companies and start-ups are running R&D projects to develop efficient and cost-effective PV-powered vehicles [1,4–6]. Of the approximately 60 initiatives illustrated in Fig. 18.3, nearly

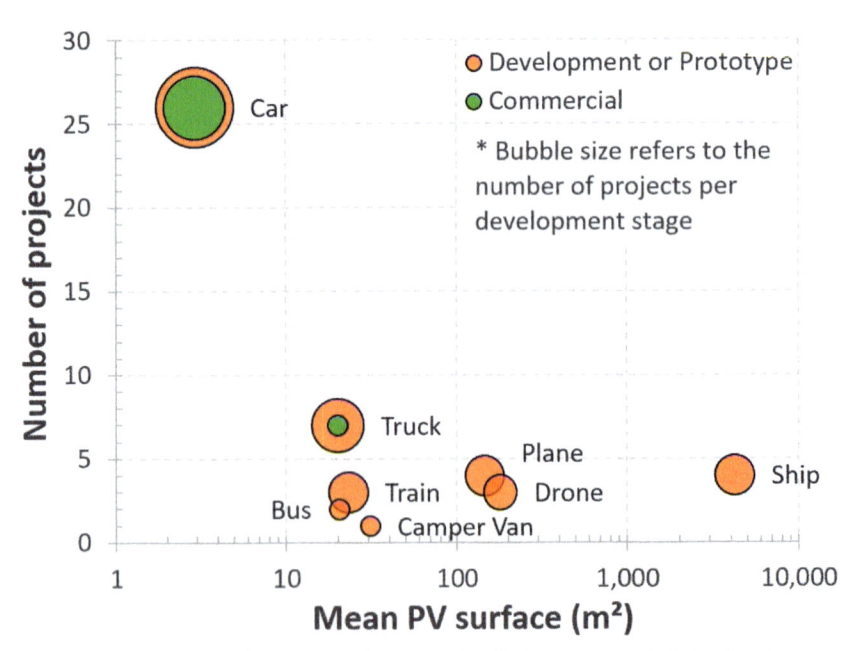

Figure 18.3 Number of initiatives for several vehicle types and their development stage as a function of photovoltaic surface area (m^2). Author's note: the commercial stage of passenger vehicles seems to be overstated. *Reproduced from B. Commault, T. Duigou, V. Maneval, J. Gaume, F. Chabuel, E. Voroshaz, Overview and perspectives for vehicle-integrated photovoltaics, Appl. Sci. 11 (24) (2021) 11598, this article is distributed under the terms of the Creative Commons (CC-BY 4.0) license (http://creativecommons. org/licenses/by/4.0/), which permits unrestricted use.*

half of these involve passenger vehicles. The type of vehicles varies: passenger vehicles, trains, buses, vans, boats, planes, and aerospace vehicles [7–17]. Several selected images of example developed/proof of concept vehicles for land, sea, and air use are presented in Fig. 18.4 [18–22].

The majority of the companies are primarily investing in passenger vehicles because it has the largest market at this moment. Thus, many of the developed passenger vehicles are commercially available while the rest of the vehicles are still in the process of development and prototype stage. The data provided in the IEA 2021 report suggests that almost 50% of PV-powered passenger vehicles are currently commercially available [6]. Meanwhile, some PV-powered trucks also can be seen commercially [12,13], while the rest of the types are still in the prototyping and development stages.

In Europe, Sono Motors [7] and Lightyear [8] are the companies pioneering the commercially available PV-EV; Aptera Motors, currently accepting orders, is based in the USA [9]. The Sono motor product "Sion" is expected to come to the market by the mid-2020s. It has integrated comparatively lightweight PV modules that can generate 1208 W_p (Watts

Figure 18.4 (A) Lightyear One solar car at Fully Charged Europe (2022); (B) Univ. of New South Wales Sunswift VI (Violet) (2017); (C) Planet Solar Tûranor in Miami (2010); (D) The solar-electric Helios Prototype (HP01) flying wing near the Hawaiian islands of Niihau and Lehua (2003); (E) Solar Impulse (HB-SIA) makes its first short flight (2009) in Dübendorf, Switzerland; (F) Artist's concept of Thales Alenia Stratobus. (A) Jan Ainali, CC 4.0 (BY-SA), https://creativecommons.org/licenses/by-sa/4.0; (B) CathyeeLiao, CC 4.0 (BY-SA); (C) Florence8787, CC 3.0 (BY-SA), https://creativecommons.org/licenses/by-sa/3.0; (D) Courtesy: NASA, (E) Matth1, CC 3.0 (BY-SA), (F) Source: Master Image Programs, CC 4.0 (BY-SA); Note: sub-images A-C, E, and F via Wikimedia Commons.

at peak production) [7]. It was estimated that Sion can go 5800 km/year using only solar energy and 34 km/day in Munich. A similar but more efficient product is Lightyear, which can run up to 70 km/day on solar depending on the favorable conditions [8]. In Japan, Toyota and Nissan have been powering PV-EVs using III-V multijunction solar cells, known to be used for space applications due to much higher (factor of approximately 1.75) efficiency but are quite expensive [23,24]. Such technologies will make PV-EVs more efficient but not be affordable soon [5].

Besides passenger vehicles, commercial transport like trucks or vans are also being equipped with PV modules to become more self-sufficient. Earlier 2021 Sono Motors and ARI added ultrathin chemically stressed solar panels to a 458 Box body truck roof. Under normal conditions in Munich, it was found to go 20 km daily without being plugged into the grid [25]. Sono Motors also successfully integrated PV modules into refrigerated trucks. According to their estimation, full solar integration on the roof and sides of a 40-ton semitruck trailer is expected to cover up to 50% of the cooling units' yearly average energy needs [25]. Researchers at Germany's Fraunhofer Institute for Solar Energy Systems have started testing the high-voltage solar modules they developed in partnership with industrial clients for heavy-duty trucks, as part of the Lade-PV project launched in April 2020 [26]. This heavy-duty truck is installed with 3.5KW PV on the entire roof. It is reported that the integrated PV system is able to cover 5%−10% of the energy needs of this truck depending on the location and the time of the year. The UK-based company SunSwap also invested in refrigeration trucks, which is in the process of being commercialized. They have branded their truck as a zero-emission vehicle and the solar panels are projected to save 580 L of fuel and reduce 1.5 metric tons of CO_2 emissions per vehicle each year [27,28].

Following the trend in land-based vehicles, PV panels are being integrated into water-based [14,29,30] and aerospace vehicles [15−17,23,24,31−33]. Millikan Boats, a company based in France, developed a marine boat in 2023. It is equipped with two sides solar wings of capacity 6 kWp. The boat is expected to run at 6 knots with full autonomy given that the sun is available [34]. Marlec Renewable Power, based in the UK, is also investing in PV-powered boats [35]. The largest PV-integrated boat so far, the Tûranor, is developed by Planet Solar, see Fig. 18.3C [20]. It was mounted with 825 solar modules with a peak performance of 93.5 kW [14]. Unlike land vehicles, marine vessels must be self-reliant while cruising. Integrated PV gives ships the potential to stay longer on the seas and rely less on burning fossil fuel while underway.

A recent study estimated a positive environmental impact [30]; Fig. 18.5 illustrates a top (top) and side view (middle) of a virtual ocean-going cargo vessel along with calculated reduction of sulfur oxides (SO_x), nitrogen oxides (NO_x), carbon dioxide (CO_2), and particulate matter.

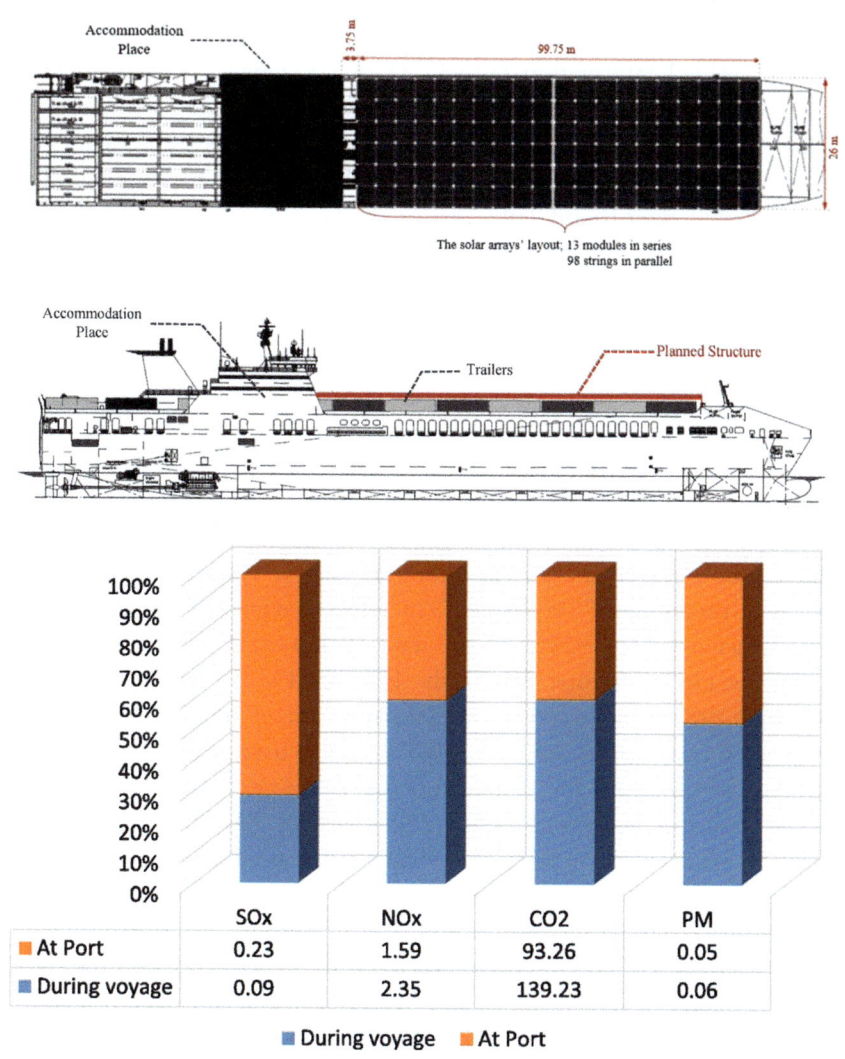

	SOx	NOx	CO2	PM
At Port	0.23	1.59	93.26	0.05
During voyage	0.09	2.35	139.23	0.06

Figure 18.5 (*Top*) Top view of a studied roll-on-roll-off (Ro-Ro) ocean-going cargo vessel and location of the solar array on the top deck. (*Middle*) The side view of the vessel and location of the planned durable structure. (*Bottom*) Amounts of reduction by pollutants and CO_2 through use of solar power for vessels; see reference for details. *Reproduced with permission from C. Karatuğ, Y. Durmuşoğlu, Design of a solar photovoltaic system for a Ro-Ro ship and estimation of performance analysis: A case study, Solar Energy 207 (September 2020) 1259–1268, copyright (2020) Elsevier.*

However, for a complex power system for such a large vessel, a complex control system is required. Tang describes a maximum power point tracking (MPPT) approach implemented for a green ocean-going vessel that relies primarily on PV power [30].

Given the significant literature concerning aerospace PV applications, the reader is directed to the next chapter and previously referenced literature for detailed information [15–17,23,24,31]. Finally, a representative list of the PV-integrated vehicles (PVIVs) is summarized in Table 18.1 [1,5–7,14,17,31,32,36–53]. For the sake of clarity, passenger vehicles and trucks will primarily refer to as PV-EVs, while the generic term, PVIV, will be used to refer to any land, sea, or air PV-integrated vehicle. As the emphasis of this book is PV beyond Si, as much as possible, PVIVs that utilize more advanced cell technologies are emphasized. However, given the need for economic viability, it is not surprising that conventional (Si-based) technologies continue to be the focus of prototype and commercial transportation platforms.

Except for the two commercial vehicles (bold) in Table 18.1, all of the vehicles could be classified as proof-of-concept or prototypes, although several of the early unmanned aerial vehicles (UAVs) or high-altitude airships (HAAs) could be more accurately described as developmental. The area-specific power (ASP) of solar arrays across the vehicle types is in the range of 160–200 W/m^2. This is expected given the very similar efficiency (22%–24%) of the state-of-the-art Si cells in the arrays. Several notable exceptions include the Hanergy PV-EV that uses flexible III-V ($\eta = 29\%$) cells, the train (SNCF), and camper van (Dethleffs) with low-performance arrays, and the UNSW racing team with access to the most recently-developed Si array technology.

The mass-specific power (MSP) will obviously be dominated by the vehicle weight; typical land and sea vehicles are approximately 1 W/kg, less for a lower-efficiency array. Given the nature of the competitive UNSW racing team, a fivefold increase (4.9 W/kg), mainly due to an exceedingly light platform, places the SunSwift solar racers in the same range as the early HAAs (2.4 and 8.7 W/kg), sponsored by the US Army, that utilized lightweight but lower-performance (1/2 of crystalline) amorphous Si arrays. The UAVs designed to carry heavier loads (Helios HP-03 and Zephyr S) and piloted solar airplanes ranged in MSP from ~ 10–20 W/kg. The early NASA UAVs and Stratobus HAA had the highest MSPs in the range of 40–60 W/kg. In fact, these types of vehicles are also described as high-altitude pseudo-satellites (HASPs), partly due to the very high altitudes

Table 18.1 Examples of photovoltaic-integrated vehicles.

Manufacturer[a]	Model/Project	Type	Year	Photovoltaics system metrics[b] (units as noted below)				Cell type[c]	Efficiency (η) (%)		PV use[d]	Refs.
				Power (kW_p)	ASP (W/m^2)	MSP (W/kg)			Cell	Module		
Planet Solar	Tûranor	Boat	2010	93.5	181	1.1		Silicon		18.1	P+	[14]
Energy Observer	Energy Observer	Boat	2017	21	162	0.84		Silicon		16.2	P+	[36]
Hyundai	**Sonata Hybrid**	Passenger vehicle	2020	0.204	157			Silicon	22.8	15.7	P+	[5,37]
Toyota	**Toyota - Prius IV**	Passenger vehicle	2017	0.18	200	0.13		Silicon		20.0	P	[5,38]
Toyota	Prius Prime	Passenger vehicle	2019	0.86				3 J III-V	34.0		P	[5,39]
SonoMotors	Sion	Passenger vehicle	2018	1.2	160	0.94		Silicon	24.0	16.0	P	[5,7]
Hanergy	Hanergy - Solar R. O. L and A	Passenger vehicle	2019	1–2	270[e]	0.83[e]		III-V flexible	29.0		P	[5,40]
UNSW	Sunswift solar racers	Passenger vehicle	1994–2017	0.8–1.8	220[e]	4.9[e]		Silicon	18–23		P	[41]
Dethleffs	E-home	Camper Van	2017	3	97	0.58		Silicon		9.7	P+	[42]
Solarbus.pro	FlixBus	Bus	2020	0.96				CIGS		17	A	[43]
Fast Concept Vehicle	Starter	Bus		3.4	172			Silicon	21.8	16.6	A	[44]
SNCF	TER	Train	2010	3.1	135			Silicon	21.0	13.5	A	[45]
Navistar	Catalist	Truck	2017	3.64	163			p-Silicon		16.3	A	[46]

(*Continued*)

Table 18.1 (Continued)

Manufacturer[a]	Model/Project	Type	Year	Photovoltaics system metrics[b] (units as noted below)			Cell type[c]	Efficiency (η) (%)		PV use[d]	Refs.
				Power (kW$_p$)	ASP (W/m^2)	MSP (W/kg)		Cell	Module		
NASA ERAST[f]	Centurian	UAV	1996–1998	31		58	Silicon			P+	[17,31,32,47]
NASA ERAST[f]	Pathfinder-Plus	UAV	1997–1998	12.5		40	Silicon			P+	
NASA ERAST[f]	Helios (HP-03)	UAV	2003	35	194	17.6	Silicon		19.4	P+	[32,48]
Airbus DS	Zephyr S	UAV	2008	0.9		14	IMM III-V	~30%		P+	[49]
Solar Stratos	Solar Stratos	Airplane	2017–Present	4.3[g]	195	10[g]	Silicon	23.0		P+	
Solar Impulse	Solar Impulse	Airplane	2004	52.5[g]	194	23[g]	Silicon	22.6		P+	[50]
U.S. Army	HiSentinel 80	HAA	2010	1.2		2.4	TF a-Si			P+	[32,51]
U.S. Army	HALE-D	HAA	2011	15		8.7	TF a-Si			P+	[32,52]
Thales Ariana	Stratobus	HAA	2018	200	200	40	Thin Si	24.0		P+	[15,53]

Source: Adapted from B. Commault, T. Duigou, V. Maneval, J. Gaume, F. Chabuel, E. Voroshaz, Overview and perspectives for vehicle-integrated photovoltaics, Appl. Sci. 11 (24) (2021) 11598, this article is distributed under the terms of the Creative Commons (CC-BY 4.0) license (http://creativecommons.org/licenses/by/4.0/), which permits unrestricted use, with additional information from numerous literature sources as discussed in text.

[a]All listed examples are prototypes, the two examples of commercial vehicles are bold.
[b]Units of peak power (kW$_p$), area specific power (ASP – W/m^2) and mass specific power (MSP – W/kg).
[c]Different types of crystalline silicon devices are employed but not specified here, except, amorphous (a–), polycrystalline (p–), and thin-Si (likely p-Si), triple junction (3 J) or flexible (IMM) III-V or copper indium gallium selenide (CIGS) have recently been employed, interested reader is encouraged to consult the original literature for more details.
[d]Photovoltaic use for propulsion (P), propulsion plus auxiliary (P+), or auxiliary only (A).
[e]Average of multiple vehicles.
[f]NASA-funded Environmental Research Aircraft and Sensor Technology program, unmanned aerial vehicles (UAVs) built by AeroVironment, Inc.
[g]Rough estimate assuming the PV arrays perform in a similar fashion to the Helios HP-03 UAV (drone) arrays.

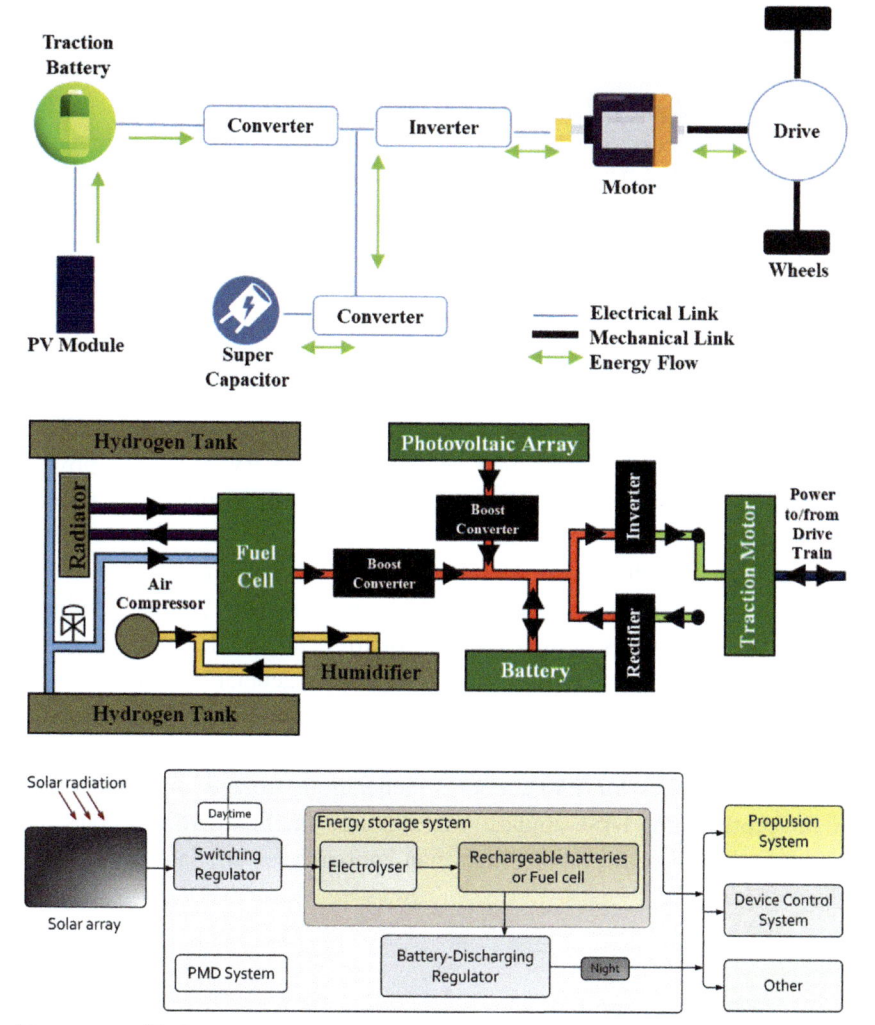

Figure 18.6 (*Top*) conceptual powertrain layout of a solar-assisted passenger electric vehicle; (*Middle*) schematic diagram of the powertrain of a PV/PEM fuel cell/NiCd battery hybrid surface vehicle. Color legends: (1) Blocks: green, primary components; beige, auxiliary fuel cell components; black, power conditioning units; and (2) Streams: purple, glycol/water mixture; light blue, hydrogen; yellow, air; red, DC power; light green, AC power; dark blue, mechanical power to/from the drive train. Black arrows indicate stream flow direction; (*Bottom*) Components of the energy system for a high-altitude airship. (*Top*) Reproduced with permission from N.R. Pochont, R. Sekhar, Recent trends in photovoltaic technologies for sustainable transportation in passenger vehicles — A review, Renew. Sustain. Energy Rev. 181 (July 2023) 113317, copyright (2023) Elsevier; (Middle) Reproduced from Z.S. Whiteman, P. Bubna, A. K. Prasad, B.A. Ogunnaike, Design, operation, control, and economics of a photovoltaic/fuel cell/battery hybrid renewable energy system for automotive applications, Processes 3(2) (2015) 452−470, this article is distributed under the terms of the Creative Commons (CC-BY 4.0) license (http://creativecommons.org/licenses/by/4.0/), which permits unrestricted use; (Bottom) Y. Xu, W. Zhu, J. Li, L. Zhang, Improvement of endurance performance for high-altitude solar-powered airships: A review, Acta Astronautica 167 (February 2020) 245−259, copyright (2020) Elsevier.

reached during flight [17,31−33], but also because their MSPs are in the lower range of current spacecraft (30−100 W/kg); however, near-term (150−200 W/kg) and mid- to far-term (200−250 W/kg) goals for MSP with multijunction III-V arrays [24] far exceed those of HASPs utilizing SOA Si technology. As expected, a typical ASP of 330−340 W/m^2 is approximately 60% higher for spacecraft employing MJ III-V arrays with efficiencies over 35% [24].

Finally, the issue of the power system for PV-EVs (or PVIVs) needs to be addressed. Three levels of utilization of PV-generated power are employed: propulsion and auxiliary functions, propulsion only, or auxiliary functions only. Sea- and air-based vehicles integrate PV-generated power into all functions; passenger vehicles typically focus on the use of PV power for propulsion, while heavier vehicles (i.e., buses, trucks, and trains) use PV power for auxiliary functions. In fact, there have been numerous in-depth studies and reviews for optimal power utilization for PVIVs that combine PV, fuel cells, batteries, and/or supercapacitors [1,2,17,24,30−33,51−63]. The first two technologies are utilized for power generation and the final three for energy storage as fuel cells can serve as both an energy generation and storage technology [17,32,33,54,58,61,63]. Fig. 18.6 illustrates three different approaches for PVIV energy generation, storage, and utilization for automotive and HASP applications, although the final two examples are quite generic and could define an overall approach for any land, sea, or air vehicle [33,54,63].

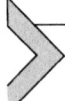

18.3 Solar cell technologies for photovoltaic-powered electric vehicles

Since the mid-1950s, solar cells and module technologies have demonstrated the potential for commercial applications [23]. Although initially it was dedicated to space technology [24,64], nowadays the applications have been expanded to numerous terrestrial power generations. Solar cells used in space technologies are higher in efficiency and price [23,24,64−66], unlike PV for terrestrial applications using cells with lower efficiency and cost. In space applications, PV cells are to be light, flexible, and reliable for reliable output in challenging environments. Thus, expensive III-V multijunction cells are typically deployed [23,24,64]. On the contrary, PV cells used for power generation on Earth are cheaper and

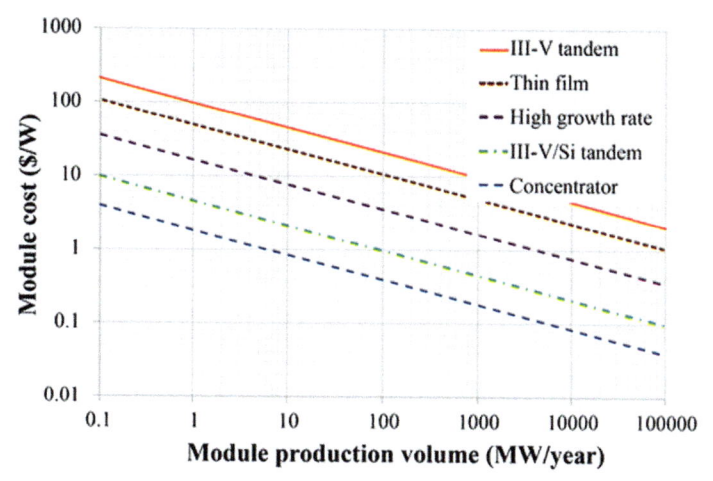

Figure 18.7 Comparison of estimated module cost for advanced PV (III-V tandem, III-V/Si tandem, and concentrator III-V/Si tandem solar cell) modules as a function of module production volume. *Reproduced from M. Yamaguchi, T. Masuda, K. Araki, D. Sato, K.-H. Lee, et al., Role of PV-powered vehicles in low-carbon society and some approaches of high-efficiency solar cell modules for cars, Energy and Power Engineering 12 (6) (June 2020) 375–395, this article is distributed under the terms of the Creative Commons (CC-BY 4.0) license (http://creativecommons.org/licenses/by/4.0/), which permits unrestricted use.*

lower efficiency mono or polycrystalline Si-based [1,4–6,54,61–63,67]; see Table 18.1. In between these two extremes, there is a huge gap in efficiency levels and production cost, which has been extensively explored in recent years [5,6,62–68]. Interestingly, Yamaguchi et al. provide a rough estimate (Fig. 18.7) of the potential reduced cost of recently-developed advanced PV technologies given greater production volume, a consequence of "economy of scale" [4]. However, it is not likely that market forces alone will drive the need for large manufacturing scale of advanced PV technologies. Hence, in this section, different cell technologies that are already commercialized and recent emerging technologies are described, and their merits are analyzed.

18.3.1 Correlation between vehicle surfaces and photovoltaic cell efficiency

In EVs, the available surface area is one of the challenging aspects of installing PV modules; primarily the area is limited and on top of that majority of the surfaces are full of curvatures [1,2,5,6,62,63,67]. Thus, conventional PV modules that are being used in the power generation in

PV plants are often not suitable for vehicle integration [69]. In Fig. 18.8, a correlation between PV panel area and module efficiency is presented [62,63,67]. In a study by Yamaguchi et al., the average available area on the roof of a generic car is considered to be less than 2 m^2 [67]. If both roof and engine hoods are used for PV installation, the available area could be raised up to 2.5 m^2. If an electric mileage of 17 km/kWh is considered (blue or bottom curve), only roof-covered EVs will need PV modules with an efficiency of 37%−38%; this is currently only available using multijunction III-V modules [23,24,64−69]. On the other hand, if both the roof and the hoods are covered with PV, the same performance could likely be delivered by PV modules with 25% efficiency, within the range of tandems, concentrators, and advanced single-junction devices [1,2,4−6,62,63,66−69]. These figures signify the challenges that need to

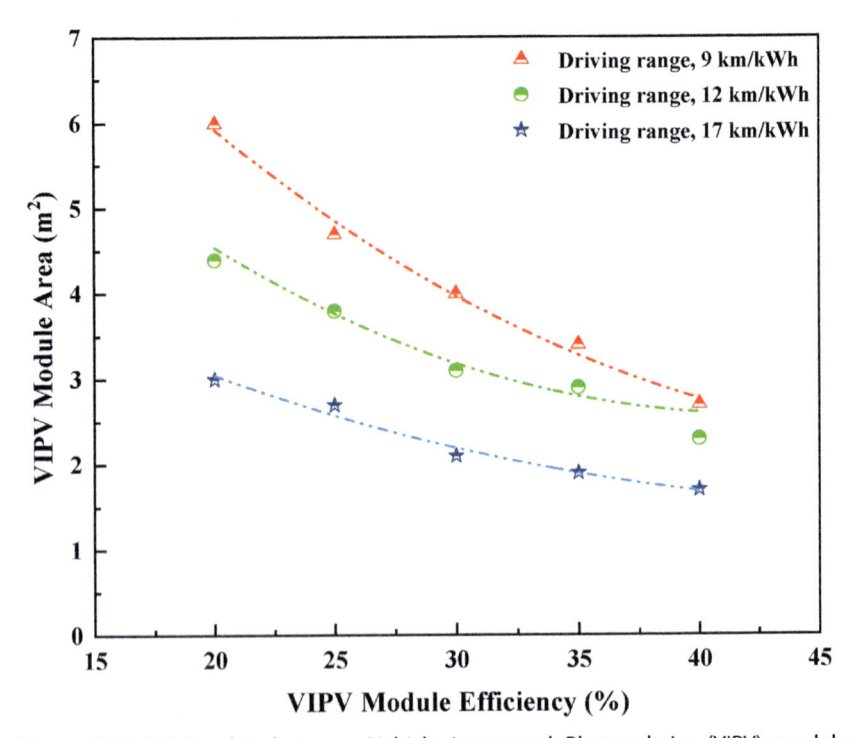

Figure 18.8 Relationship between Vehicle Integrated Photovoltaics (VIPV) module area, module efficiency, and the driving range. *Reproduced from S. Kim, M. Holz, S. Park, Y. Yoon, E. Cho, J. Yi, Future options for lightweight photovoltaic modules in electrical passenger cars, Sustainability 13 (5) (2021) 2532, this article is distributed under the terms of the Creative Commons (CC-BY 4.0) license (http://creativecommons.org/licenses/by/4.0/), which permits unrestricted use.*

be dealt with to develop PV-EVs; there is no alternative to developing high-efficiency solar cells if they are to be integrated with EVs [4–6,62,63,66–68].

18.3.2 Different types of solar cells for electric vehicle applications

A representative illustration of lower-cost higher-efficiency PV technologies that are used (or have potential use) in terrestrial applications is shown in Fig. 18.9; performance for various technologies is compared for individual cells and large-format (substrate-based) modules. Modules are either wafer-based $p-n$ junction (Si) solar cells connected in series and parallel combinations or semiconductor materials that are directly deposited on metals or glasses to produce large-format substrate PV modules. The latter technologies are more promising for PV-EVs because a significant portion of the body of the car could be used for constructing solar cells on top and that may result in better performance [1,2,4–6,62,63,66–69]. An extensive list of efficiencies of various PV technologies has recently been updated [69].

Materials composed of elements from Groups III and V from the periodic table [i.e., gallium arsenide (GaAs), indium phosphide (InP), and

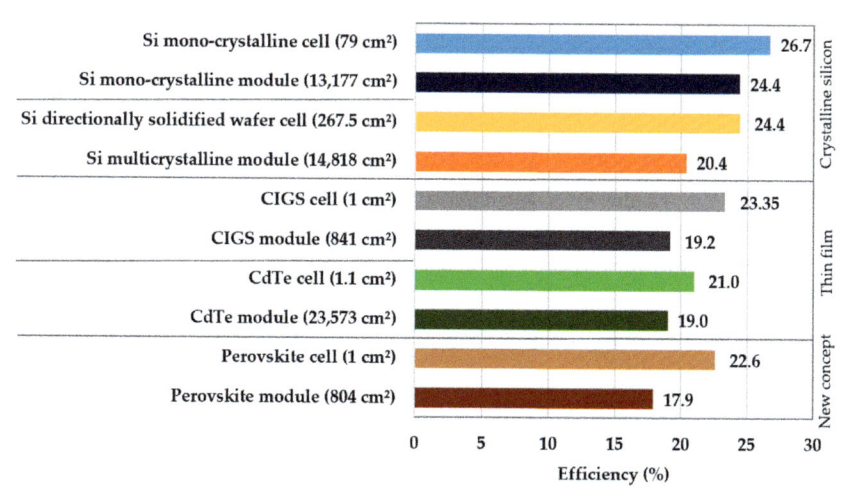

Technology	Cell/Module (area)	Efficiency (%)	Category
	Si mono-crystalline cell (79 cm²)	26.7	Crystalline silicon
	Si mono-crystalline module (13,177 cm²)	24.4	
	Si directionally solidified wafer cell (267.5 cm²)	24.4	
	Si multicrystalline module (14,818 cm²)	20.4	
	CIGS cell (1 cm²)	23.35	Thin film
	CIGS module (841 cm²)	19.2	
	CdTe cell (1.1 cm²)	21.0	
	CdTe module (23,573 cm²)	19.0	
	Perovskite cell (1 cm²)	22.6	New concept
	Perovskite module (804 cm²)	17.9	

Figure 18.9 Comparison of representative module and cell efficiencies for lower-cost, higher-efficiency single-junction photovoltaic technologies. *Reproduced from B. Commault, T. Duigou, V. Maneval, J. Gaume, F. Chabuel, E. Voroshaz, Overview and perspectives for vehicle-integrated photovoltaics, Appl. Sci. 11 (24) (2021) 11598, this article is distributed under terms of a Creative Commons (CC-BY 4.0) license (http://creativecommons.org/licenses/by/4.0/), which permits unrestricted use.*

gallium antimonide (GaSb)] are known collectively as III-V semiconductors [23,64−66,68]. These materials possess excellent electronic properties, making them ideal candidates for solar cell applications. Unlike traditional silicon PV cells, which are thick and rigid, III-V thin-film PV modules are characterized by their thin and flexible nature [23,24,64,65]; these are not to be confused with older inorganic thin film solar cells such as copper indium gallium selenide (CIGS) or cadmium telluride (CdTe) [23,64]. These solar cells are manufactured by molecular beam epitaxy and metalorganic chemical vapor deposition techniques that are used for epitaxial growth on a thin (or reusable) substrate [23,24,64,65]. Consequently, this reduces material usage, which makes them comparatively environmentally friendly. Also, III-V materials have a wider bandgap than Si and thus they are able to absorb a larger spectrum of light [23,24,64,65]. That results in exhibiting higher efficiency compared to traditional silicon-based PV modules. III-V multijunction cells are comparatively lightweight and flexible, which is highly desirable for EV applications. As discussed above, the MSPs goals for near- into mid- to far-term (150−250 W/kg) multijunction III-V arrays for spacecraft [24], albeit lighter platforms than terrestrial vehicles, far exceed those of HASPs utilizing SOA Si technology; see Table 18.1; a typical ASP of 330−340 W/m^2 is approximately 60% higher for spacecraft employing MJ III-V arrays with efficiencies over 35% [24].

Finally, III-V devices and modules are less sensitive to a variety of environmental conditions, such as terrestrial light-intensity variation, and thus practical for diffuse lights and variable weather conditions [23,24,64−66]. There are several reasons why they are rarely used in terrestrial applications: production cost is significantly high compared to silicon-based solar cells. However, as discussed above, higher production volume would reduce the price ($/W). Another, more intractable issue is that several elements that comprise III-V materials are not Earth-abundant elements, and their availability would certainly hamper increased production [70]. As silicon and production of a substrate, epitaxial deposition and metal contact formation represent the top three costs in fabricating the solar cells. Despite these concerns, several companies are actively investing in solar roofs made from III-V multijunction solar cells [1,4−6,39,40,63,67,71].

In contrast to several elements from III-V semiconductors, Si is the second most abundant element on Earth [70]. Thus, silicon-based solar cells are widely used for power generation, and currently, they constitute 94% of the solar market share [70,72]. In Si-based technologies, the cells are formed from wafers produced from either a single crystal ingot or from a multicrystalline cast block, which results in two different cell types, monocrystalline (mc-Si) and polycrystalline (p-Si) solar cells. Monocrystalline silicon PV modules are known for their high efficiency and uniform appearance. They are made from single crystal structures, which allow for efficient charge transport within the material. However, their production process involves cutting cylindrical ingots into thin wafers, leading to material wastage. Polycrystalline silicon PV modules are cost-effective alternatives to monocrystalline modules. While they are slightly less efficient, advancements in manufacturing have improved their performance over the years. Historically, p-Si has held a significant market share due to lower manufacturing costs but is less efficient [67,69,70,72]. However, intensive research in recent years [1,62,63,67] has extended the efficiency and reduced the cost of previous mc-Si cells. Premium n-type back contact mc-Si solar modules are presently achieving 21% in mass production and are projected to reach >23% [69,73]. Higher efficiency (>21.5%) silicon back contact cells were used to form the sun-facing body parts of the Lightyear One solar EV [8].

Despite being Earth-abundant and economically advantageous, Si-based solar cells are rigid in form which makes them unsuitable for EV (PVIV) application [1,4,5,62,63,67]. Recently, Si-based, polymer-encapsulated PV modules have been developed that can be laminated onto curved surfaces [74]. Such technologies make Si-based solar cells applicable for PVIVs. As discussed above, there is an ~50% efficiency advantage of III-V multijunction-based over Si-based PV technologies for PVIVs [1,4−6,24,62,63,67,69,70]. In two editions of the Toyota Prius, the differential performance of the application of these two different PV technologies is quite prominent; see Table 18.1 [1,5,39,40,63,67]. Efficiency wise MJ-III-V cells provide over 32% [64−66]. While Si-based cells are integrated into the roof, producing 180 W, the III-V multijunction cells are placed in the hood, rear, and doors along with the roof, producing 860 W; the rated output power from the III-V multijunction model is nearly 5 times higher than the Si-based Prius PHV [1,4−6,39,40,63,67]. Nevertheless, Si-based technologies are significantly less costly than MJ III-V cells; thus, the way forward could be investigating on the efficiency improvement of these technologies [1,4−6,61−63,66−70].

Other than the two prevailing solar cell production technologies, other technologies are also being investigated and developed [64−70]; see Fig. 18.9. Amorphous Si modules are Si films deposited on top of the glass and can be integrated into tandem cells [69,70]. Efficiency-wise, this technology only produces a power conversion efficiency (PCE) of 12%

[69,70]; this low efficiency is not very favorable for use on EV roofs and hoods [1,4−6,61−63,67]. However, these solar cells could be semitransparent, making them suitable for use on vehicle windows [75]. Chalcogenide thin-film modules account for 6% of global PV manufacturing and are composed of either CdTe or CIGS semiconductors deposited onto glass or metal foil [23,24,70,76]. Efficiency-wise they are very close to Si-based modules and on top of that they provide the opportunity to deposit semiconductor materials on the body of the vehicles directly [1,62,63,66−70]. Perovskite and organic PV modules share the benefit of (inorganic) chalcogenide thin films of enabling conformal coating of a wide range of substrate types with the added advantage they are solution processable at low temperatures [23,24,62−64,70].

A short list of PV modules that are suitable for EV applications is provided in Table 18.2 [1,7,8,40,63,77−87]; most of the projects in this list are recent, within the past five years. The majority of the listed projects considering Tables 18.1 and 18.2 involve Si-based technologies due to their trade-off between efficiency and cost-effectiveness. Efficiency-wise most of the modules (see Table 18.1, Fig. 18.9) vary between 15% and 22%, which is promising but apparently not yet commercially viable for mass EV application [4−6,63,67]. In fact, the technology readiness level (TRL), a heuristic scale developed by NASA [88] and adapted by other entities for a variety of projects and products [89,90], provides a rough guide for the development stage of a technology: a basic research result (1−2), laboratory proof-of-concept (3−5), road-ready prototype (6−7), or commercial ready (8−9). The most important performance metric, from the point-of-view of a driver, is a nominal range in km/day. Of the projects developed over the past five years, nearly all vehicles capable of over 20 km/day are flexible or MJ device technologies, including III-V/GaAs [40,81,86], CIGS [82], or a-Si [8,80,85].

The limited surface area of passenger vehicles (see Fig. 18.3, typically 2−3 m^2) increases the demand for high module ASP (W/m^2) and exploitation of highly curved surfaces; thus, manufacturers are more inclined to explore curved PV and flexible PV for EV applications [1,4−6,62,63,67]. Bendable PV modules employ a variety of different technologies, from amorphous- to thin Si-based cells to thin-film (i.e., CIGS) technologies or organic solar cells [1]. These modules are flexible and low-weight (0.7−6.7 kg/m^2, see Fig. 18.10) compared with the 11 kg/m^2 of a standard module. Their ASP (power density in W/m^2) varies widely due to the diversity of cell technologies used. However, module efficiencies close

Table 18.2 Different solar cells integrated into various VIPV technologies with their performance indication.

Year	PV cell[a]	Integrated platform[b]	Key Performance Indicator (km and/or kWh/day)	Vehicle Name	TRL[c]	Ref.
2010	Polycrystalline Si	Roof mounted	2 kWh/d	Piaggio porter van	4	[77]
2014	Monocrystalline Si	Roof with Fresnel lens	34 km/d and 8 kWh/d	C-max	7	[78]
2016	Monocrystalline Si	Roof	19 km/d and 1.5 kWh/d	P-mob	7	[79]
2018	a-Si thin-films	Roof	5.4 km/d and 1.6 kWh/d	Karma Revero	5	[80]
2018	Crystalline PV module	Roof	5.65 kWh/d	Wagon	2	[1]
2019	InGaP/GaAs/InGaAs	Roof and rear glass	44 km/d and 0.86 kWh/d	Prius Prime	7	[81]
2019	Thin GaAs	Roof and hood extended	20 km/d	Hanergy solar car	3	[40]
2019	CIGS thin-film	Roof	20 km/d and 1.6 kWh/d	K-Car	6	[82]
2020	Monocrystalline Si	Roof, hood and door panels	34 km/d and 4 kWh/d	Sion	6	[7]
2020	a-Si thin-films	Roof and hood extended	70 km/d and 5.8 kWh/d	Light year one	5	[8]
2020	Monocrystalline Si	Roof and hood	0.49–1.82 kWh/d	Fiat Punto	2	[83]
2020	a-Si thin-films	Roof	24 km/d	Tesla Model S & Cybertruck	3	[80]
2021	a-Si thin-films	Roof	3.5 km/d and 1 kWh/d	Sonata	8	[84]
2021	c-Si/a-Si heterojunction cells	Roof and compartment	36 km/d and 1.3 kWh/d	Light commercial vehicle	6	[85]
2021	III-V/Si 3J tandem	Lateral surfaces	30–50 km/d and 1 kWh/d	Electric vehicle	2	[86]
2022	a-Si thin-films	Roof	5 km/d and 1 kWh/d	Ocean	6	[87]

Source: Reproduced with permission from N.R. Pochont, R. Sekhar, Recent trends in photovoltaic technologies for sustainable transportation in passenger vehicles – A review, Renewable and Sustainable Energy Reviews 181 (July 2023) 113317, copyright (2023) Elsevier.

[a] Monocrystalline silicon (Si) is also known as single crystal, polycrystalline-Si is a less expensive (and less efficient) technology, multijunction devices, both III-V and III-V/Si are arranged with the higher-bandgap material being exposed to light first, allowing sub-bandgap photons to be absorbed by layers below, thin GaAs is a processing-driven PV device and not a traditional (CIGS, CdTe, or amorphous (a-)Si) thin-film technology.

[b] the actual area covered by PV modules are given in several literature sources [1,4-6], averaging ~2-5 m^2, see Fig. 19.3.

[c] TRL refers to technology readiness level and is discussed in the text, also in references [88-90], roughly 1-5 is applied research to proof-of-concept, 6-9 is a prototype to commercial vehicle.

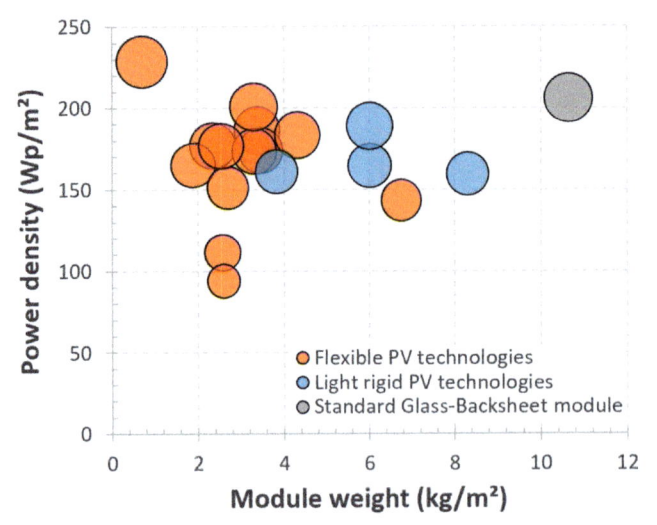

Figure 18.10 Number of flexible and light (nonflexible) modules (utilizing commercial brands) and their mass, where each bubble is a product; note: flexible modules do not include the typical 6 kg/m² metal sheet of a typical car body. *Reproduced from B. Commault, T. Duigou, V. Maneval, J. Gaume, F. Chabuel, E. Voroshaz, Overview and perspectives for vehicle-integrated photovoltaics, Appl. Sci. 11 (24) (2021) 11598, this article is distributed under terms of a Creative Commons (CC-BY 4.0) license (http://creativecommons.org/licenses/by/4.0/), which permits unrestricted use.*

to standard flat modules exist, and there are commercial vendors with available products [1]. Fig. 18.10 demonstrates the potential advantage to be realized by utilization of such technologies [1,4−6,63,67]; the highest ASP is within 50 W/m² of current space MJ modules [23,24].

18.4 Benefits of photovoltaic-powered vehicles and future directions

Achieving carbon neutrality in the automobile sector is quite a significant challenge due to the industry's dependency on fossil fuels directly or indirectly [3−6,61−63,67]. PV-integrated EVs (and all PVIV) will play an important role in reducing this dependency in the future. As documented in Table 18.1 and Table 18.2, this technology, on average, is still in the proof-of-concept to prototype stage; it is difficult to measure the impact that it might have in the future. In a general sense and based on the currently available data, the prospective benefits include: reduced

CO_2 emissions during driving, cost savings, reduction in the frequency of charging, and improvement in the battery health of PV-EVs [3–6,63,67].

In a recent study [5], Yamaguchi et al. determined the impact of PV-EVs on CO_2 emissions (Fig. 18.11). There is a clear correlation between increased PCE of PV modules with enhanced CO_2 reduction. We can extract from Fig. 18.11 that the annual CO_2 emission reduction of EVs with electric mileage (EM, a measure of efficiency) of 10 km/kWh is estimated to be 163 $kgCO_2$-eq/year (40% reduction), 237 $kgCO_2$-eq/year (58% reduction), and 282 $kgCO_2$-eq/year (70% reduction) for module efficiency of 20%, 30%, and 40%, respectively. In a study published by the IEA, six different driving patterns (heavy and light use for weekends, every day of the week, and weekday use only) were analyzed for PV-EVs; the details of the driving pattern can be found in the study [6]. The analysis results were quite promising. For five driving patterns, CO_2 emission is reduced by 150–250 kg-CO_2/year per vehicle. For daily short-distance

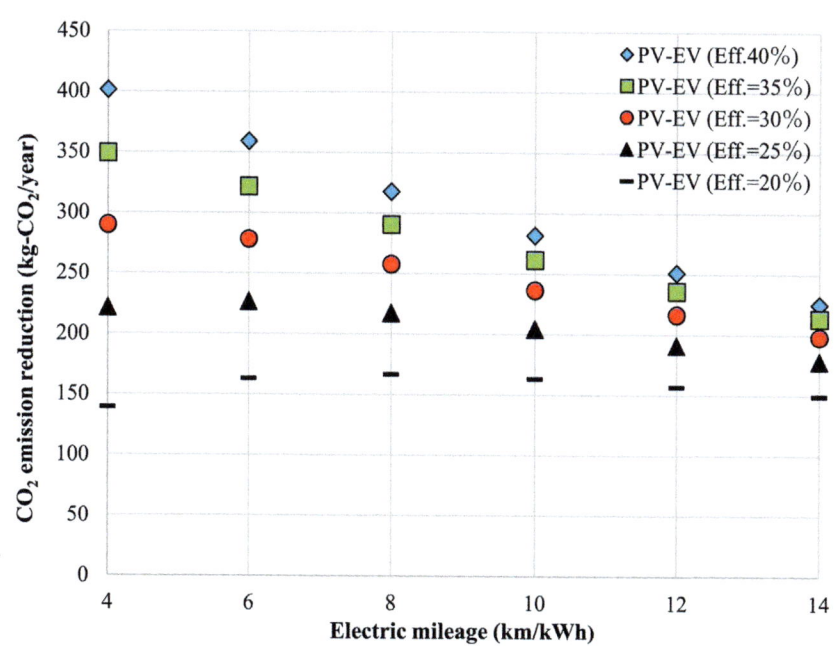

Figure 18.11 Calculated results for annual CO_2 emission reduction of PV-powered electric vehicles used in Japan, installed with solar cell modules with different efficiencies as a function of electric mileage. *Reprinted with permission from M. Yamaguchi, T. Masuda, K. Araki, Y. Ota, K. Nishioka, Impact and recent approaches of high-efficiency solar cell modules for PV-powered vehicles, Jpn. J. Appl. Phys. 61 (2022) SC0802, copyright (2022) The Japan Society of Applied Physics.*

journeys (light use), CO_2 emissions were calculated to be slightly higher due to the fact that the onboard PV capacity was not utilized.

Another significant impact found in the IEA study was the reduction in charging frequency [6]. The charging frequency is significantly reduced in some driving patterns. For light everyday use, only one charging was needed for the whole year. For the two other light use cases, there will be no charging required because the vehicles were driven for short distances and onboard PV nullified the requirement of grid charging over the span of the year. For the three heavy-use scenarios, the number of charges required was reduced from 20% (every day) to 33% (weekdays only) to 45% (weekends only) [6]. The study by Yamaguchi et al. translated the reduction in charging frequency to charging saving cost [5]; their findings for driving in Japan are presented in Fig. 18.12. It can be observed that high-efficiency solar cells significantly contribute to saving

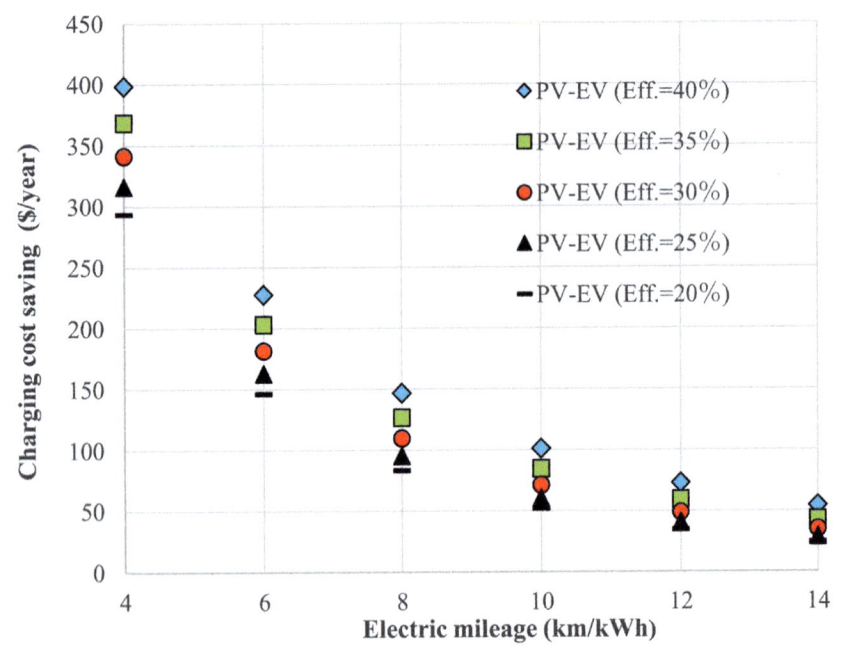

Figure 18.12 Charging electricity cost of PV-EVs as a function of electric millage by using cumulative frequency for daily mileage of Japanese passenger cars. *Reproduced with permission from M. Yamaguchi, T. Masuda, K. Araki, Y. Ota, K. Nishioka, Impact and recent approaches of high-efficiency solar cell modules for PV-powered vehicles, Jpn. J. Appl. Phys. 61 (2022) SC0802, copyright (2022) The Japan Society of Applied Physics.*

charging cost in EVs. For example, electricity cost saving is $398/year for the PV-EV module with a conversion efficiency of 40% and $293/year for the module with an efficiency of 20% in the case of electric mileage of 4 km/kWh. The amount of saving cost is $101/year for 40% efficiency and $52/year for 20% efficiency when the electric mileage is improved to 10 km/kWh as shown in Fig. 18.12. Monetarily it might not be highly lucrative for most of the consumers in the short term, however, in the long term, the accumulated savings could be significant.

18.5 Conclusions

To replace fossil fuel-based transport, PV-EVs are being intensively investigated as a significant future mode of transport in a number of countries. Several countries like Norway have moved into a majority use of EV-based vehicles in Europe; others have adopted policies to move in that direction. Along with the electrification of vehicles, it is imperative that the conventional power grid should increase renewable power generation to reduce CO_2 emissions to a minimum. However, it will take many decades to achieve this goal. In fact, there are many places where grid electricity is not feasible; alternative sources of electricity are required. To mitigate that issue, EVs are envisioned to be PV-powered so that they will be self-reliant, fully renewable, and with minimal carbon emissions. The trade-off between available space for PV integration on passenger vehicles and the efficiency of PV cells (and modules) is both a technological and economic challenge. At this time, it is clear that PV integration into a cross-section of land, sea, and air vehicles would be a supportive measure to reduce grid dependency for EVs.

As detailed above, PV is unable to provide full electric support to the vehicle at every moment of operation. Regardless of such limitations, numerous studies have demonstrated promising results of PV integration into a cross-section of vehicles. Depending on the country, PV-powered vehicles reduce CO_2 emissions significantly. Also, it provides on-the-go charging for vehicles enabling longer independent operation and less charging cycle for the batteries. Because of these promising aspects, multiple projects, consortia, and companies worldwide have been investing in PVIVs. As discussed in this book chapter, several commercial passenger vehicles are already on the market, as well as refrigeration trucks and vans.

Besides land-based vehicles, PV has been integrated into marine usage for boats and ships, as well as drones, planes, and HAAs, which have been tested successfully and put into specialized uses. To facilitate PV integration into the vehicles, other consortia are investing in specialized PV cells that are flexible, lightweight, transparent, and could be readily applied over the metal surfaces of the vehicles. As a result of the worldwide research and development efforts in advanced PV technologies, significant advances have been made in the fields of multijunction, concentrator, and hybrid (perovskite) solar cell technologies. Traditional thin-film cell materials such as amorphous Si and chalcogenides show promising output on vehicle windows and deposition on the metal frames of the car body. Such an ever-increasing range of technologies is expected to facilitate PV integration into a range of land, air, and sea vehicles; (near) grid-independent travel will be a reality in the future.

Acknowledgment

The author thanks Dr. A.F. Hepp, one of the coeditors of this book, for his advice and input that improved this chapter.

References

[1] B. Commault, T. Duigou, V. Maneval, J. Gaume, F. Chabuel, E. Voroshaz, Overview and perspectives for vehicle-integrated photovoltaics, Appl. Sci. 11 (24) (2021) 11598. Available from: https://doi.org/10.3390/app112411598.
[2] R.M. Prasad, A. Krishnamoorthy, Design, construction, testing and performance of split power solar source using mirror photovoltaic glass for electric vehicles, Energy 145 (2018) 374–387. Available from: https://doi.org/10.1016/j.energy.2017.12.131.
[3] O. Chen, V. García Tapia, A. Rogé, C. Fernandez Alvarez, P. Hugues M. Kueppers, et al., CO_2 emissions in 2022, International Energy Agency Report, March 2023, 19 pp. <https://www.iea.org/reports/co2-emissions-in-2022>.
[4] M. Yamaguchi, T. Masuda, K. Araki, D. Sato, K.-H. Lee, et al., Role of PV-powered vehicles in low-carbon society and some approaches of high-efficiency solar cell modules for cars, Energy Power Eng. 12 (6) (2020) 375–395. Available from: https://DOI.org/10.4236/epe.2020.126023.
[5] M. Yamaguchi, T. Masuda, K. Araki, Y. Ota, K. Nishioka, Impact and recent approaches of high-efficiency solar cell modules for PV-powered vehicles, Jpn. J. Appl. Phys. 61 (2022) SC0802. Available from: https://DOI.org/10.35848/1347-4065/ac461b.
[6] K. Araki, A.J. Carr, F. Chabuel, B. Commault, R. Derks, et al., State-of-the-art and expected benefits of PV-powered vehicles, International Energy Agency, Report IEA-PVPS T17-01:2021, April 2021, 155 pp. <https://iea-pvps.org/research-tasks/pv-for-transport/>.
[7] Sion Motors <https://sonomotors.com/en/sion/> (accessed 02.10.23).
[8] Lightyear, <https://lightyear.one/> (accessed 02.10.23).
[9] Aptera Solar, <https://aptera.us/> (accessed 02.10.23).
[10] Sunswift Racing, <https://www.sunswift.com/> (accessed 02.10.23).

[11] Toyota Prius Prime, <https://www.toyota.com/priusprime/> (accessed 02.10.23).
[12] Dethleffs E. HOME motorhome, <https://reiseziel-zukunft.dethleffs.de/en/reisemobil/> (accessed 02.10.23).
[13] <https://cleantechnica.com/2021/10/31/trucks-with-onboard-solar-are-becoming-a-thing/> (accessed 02.10.23).
[14] Planet Solar Tûranor, <https://www.planetsolar.swiss/en/world-premiere/boat/> (accessed 02.10.23).
[15] Thales Stratobus, <https://www.thalesgroup.com/en/worldwide/space/news/whats-stratobus> (accessed 02.10.23).
[16] NASA Solar Aircraft, <https://www.nasa.gov/centers-and-facilities/armstrong/sun-glider-builds-on-legacy-of-solar-aircraft/> (accessed 02.10.23).
[17] A.F. Hepp, P.N. Kumta, O.I. Velikokhatnyi, M.K. Datta, Batteries for aeronautics and space exploration: recent developments and future prospects, in: P.N. Kumta, A. F. Hepp, M.K. Datta, O.I. Velikokhatnyi (Eds.), Lithium-Sulfur Batteries: Advances in High-Energy Density Batteries, Elsevier, 2022, pp. 531–595. Available from: https://doi.org/10.1016/B978-0-12-819676-2.00011-6.
[18] <https://commons.wikimedia.org/wiki/File:Lightyear_one_Fully_Charged_Europe_2022_6.jpg> (accessed 02.10.23).
[19] <https://commons.wikimedia.org/wiki/File:Violet_Apperance.jpg> (accessed 02.10.23).
[20] <https://commons.wikimedia.org/wiki/File:PlanetSolar-In-Miami-Florida.jpg> (accessed 02.10.23).
[21] <https://www.dfrc.nasa.gov/Gallery/Photo/index.html> (accessed 02.10.23).
[22] <https://commons.wikimedia.org/wiki/File:Stratobus_artiste.jpg> (accessed 02.10.23).
[23] A.F. Hepp, R.P. Raffaelle, I.T. Martin, Space photovoltaics: New technologies, environmental challenges, and future missions, in: S. Sundaram, V. Subramaniam, R.P. Raffaelle, M.K. Nazeeruddin, A. Morales-Acevedo, M. Bernechea Navarro, A.F. Hepp (Eds.), Photovoltaics Beyond Silicon, Elsevier, Cambridge, MA, 2024, pp. 675–766.
[24] P.M. Beauchamp, J.A. Cutts, Solar power technologies for future planetary science missions, Jet Propulsion Laboratory, Report JPL D-101316, Pasadena, CA, USA, December 2017, 49 pp, full report available online.
[25] <https//sonomotors.com/> (accessed 02.10.23).
[26] <https://www.pv-magazine.com/2021/10/25/vehicle-integrated-pv-for-heavy-duty-trucks/> (accessed 02.10.23).
[27] <https://www.smmt.co.uk/2022/04/here-comes-the-sun-the-rising-use-of-solar-power-for-commercial-vehicles/> (accessed 02.10.23).
[28] <https://www.sunswap.co.uk/endurance/> (accessed 02.10.23).
[29] R. Tang, Large-scale photovoltaic system on green ship and its MPPT controlling, Sol. Energy 157 (2017) 614–628. Available from: https://doi.org/10.1016/j.solener.2017.08.058.
[30] C. Karatuğ, Y. Durmuşoğlu, Design of a solar photovoltaic system for a Ro-Ro ship and estimation of performance analysis: a case study, Sol. Energy 207 (2020) 1259–1268. Available from: https://doi.org/10.1016/j.solener.2020.07.037.
[31] X. Zhu, Z. Guo, Z. Hou, Solar-powered airplanes: a historical perspective and future challenges, Prog. Aerosp. Sci. 71 (2014) 36–53. Available from: https://doi.org/10.1016/j.paerosci.2014.06.003.
[32] J. Gonzalo, D. López, D. Domínguez, A. García, A. Escapa, On the capabilities and limitations of high altitude pseudo-satellites, Prog. Aerosp. Sci. 98 (2018) 37–56. Available from: https://doi.org/10.1016/j.paerosci.2018.03.006.
[33] Y. Xu, W. Zhu, J. Li, L. Zhang, Improvement of endurance performance for high-altitude solar-powered airships: a review, Acta Astronautica 167 (2020) 245–259. Available from: https://doi.org/10.1016/j.actaastro.2019.11.021.

[34] Millikan Boats https://millikan-boats.com/en/> (accessed 03.10.23).
[35] <https://www.marlec.co.uk/solar-power/boats-and-marine/> (accessed 03.10.23).
[36] <https://www.energy-observer.org> (accessed 10.10.23).
[37] <https://pv-magazine-usa.com/2022/03/24/hyundai-sonata-hybrid-is-equipped-with-a-solar-roof/> (accessed 10.10.23).
[38] <https://www.fleetnews.co.uk/news/manufacturer-news/2019/07/08/toyota-prius-solar-panels-to-generate-35-miles-of-charge-per-day> (accessed 10.10.23).
[39] <https://www.motortrend.com/features/the-2023-toyota-prius-primes-battery-could-take-three-weeks-to-recharge/> (accessed 10.10.23).
[40] <https://electricvehicles.in/hanergy-solar-powered-electric-car/> (accessed 10.10.23).
[41] <https://www.sunswift.com/our-vehicles> (accessed 10.10.23).
[42] <https://reiseziel-zukunft.dethleffs.de/en/reisemobil/> (accessed 10.10.23).
[43] <https://solarbus.pro/flixbus-florida/> (accessed 10.10.23).
[44] <https://www.fccbus.fr> (accessed 10.10.23) (translated to English).
[45] <https://innovationorigins.com/en/sun-powered-trains-sncf-unveils-green-initiative-with-new-solar-energy-subsidiary/> (accessed 10.10.23).
[46] <https://www.trucknews.com/transportation/comes-sun-solar-role-alternative-fuel/1003109612/> (accessed 10.10.23).
[47] <https://nasa.fandom.com/wiki/NASA_ERAST_Program#Pathfinder_Centurion_and_Helios> (accessed 11.10.23).
[48] <http://www.mldevices.com/index.php/product-services/photovoltaics> (accessed 11.10.23).
[49] <https://www.solarstratos.com/en/plane/> (accessed 11.10.23).
[50] < https://aroundtheworld.solarimpulse.com/adventure> (accessed 11.10.23).
[51] <https://lynceans.org/wp-content/uploads/2020/12/SwRI_HiSentinel.pdf> (accessed 11.10.23).
[52] <https://lynceans.org/wp-content/uploads/2020/12/Lockheed-Martin_HALE-D-1.pdf> (accessed 11.10.23).
[53] <https://lynceans.org/wp-content/uploads/2021/04/Thales-Alenia-Space_Stratobus-converted.pdf> (accessed 11.10.23).
[54] Z.S. Whiteman, P. Bubna, A.K. Prasad, B.A. Ogunnaike, Design, operation, control, and economics of a photovoltaic/fuel cell/battery hybrid renewable energy system for automotive applications, Proc 3 (2) (2015) 452−470. Available from: https://doi.org/10.3390/pr3020452.
[55] B.S.K. Reddy, A. Poondla, Performance analysis of solar powered unmanned aerial vehicle, Renew. Energy 104 (2017) 20−29. Available from: https://doi.org/10.1016/j.renene.2016.12.008.
[56] A. Makhsoos, H. Mousazadeh, S.S. Mohtasebi, M. Abdollahzadeh, Hamid Jafarbiglu, E. Omrani, et al., Design, simulation and experimental evaluation of energy system for an unmanned surface vehicle, Energy 148 (2018) 362−372. Available from: https://doi.org/10.1016/j.energy.2018.01.158.
[57] Q. Hoarau, Y. Perez, Interactions between electric mobility and photovoltaic generation: a review, Renew. Sustain. Energy Rev. 94 (2018) 510−522. Available from: https://doi.org/10.1016/j.rser.2018.06.039.
[58] J. Liao, Y. Jiang, J. Li, Y. Liao, H. Du, et al., An improved energy management strategy of hybrid photovoltaic/battery/fuel cell system for stratospheric airship, Acta Astronaut. 152 (2018) 727−739. Available from: https://doi.org/10.1016/j.actaastro.2018.09.007.
[59] W. Zhu, J. Li, Y. Xu, Optimum attitude planning of near-space solar powered airship, Aerosp. Sci. Tech. 84 (2019) 291−305. Available from: https://doi.org/10.1016/j.ast.2018.10.007.

[60] Z. Cobrane, D. Batool, J. Kim, K. Yoo, Design and simulation studies of battery-supercapacitor hybrid energy storage system for improved performances of traction system of solar vehicle, J. Energy Storage 32 (2020) 101943. Available from: https://doi.org/10.1016/j.est.2020.101943.
[61] C. Ogbonnaya, C. Abeykoon, A. Nasser, A. Turan, C.S. Ume, Prospects of integrated photovoltaic-fuel cell systems in a hydrogen economy: a comprehensive review, Energies 14 (20) (2021) 6827. Available from: https://doi.org/10.3390/en14206827.
[62] S. Kim, M. Holz, S. Park, Y. Yoon, E. Cho, J. Yi, Future options for lightweight photovoltaic modules in electrical passenger cars, Sustainability 13 (5) (2021) 2532. Available from: https://doi.org/10.3390/su13052532.
[63] N.R. Pochont, R. Sekhar, Recent trends in photovoltaic technologies for sustainable transportation in passenger vehicles — a review, Renew. Sustain. Energy Rev. 181 (2023) 113317. Available from: https://doi.org/10.1016/j.rser.2023.113317.
[64] R.P. Raffaelle, S.G. Bailey, D.C. Ferguson, S. Durbin, A.F. Hepp (Eds.), Space Photovoltaics: Materials, Missions and Alternative, Technologies, Elsevier, Cambridge, MA USA, 2023, p. 505.
[65] M. Yamaguchi, High-efficiency GaAs-based solar cells, in: M.M. Rahman, A.M. Asiri, A. Khan, I. Inamuddin, T. Tabbakh (Eds.), Post-Transition Metals, InTech Open, 2021. Available from: https://doi.org/10.5772/intechopen.94365.
[66] M. Yamaguchi, K.H. Lee, K. Araki, N. Kojima, H. Yamada, Y. Katsumata, Analysis for efficiency potential of high-efficiency and next generation solar cells, Prog. Photovol. 26 (2018) 543—552. Available from: https://doi.org/10.1002/pip.2955.
[67] M. Yamaguchi, T. Masuda, K. Araki, D. Sato, K.-H. Lee, et al., Development of high-efficiency and low-cost solar cells for PV-powered vehicles application, Prog. Photovolt. 29 (7) (2021) 684—693. Available from: https://doi.org/10.1002/pip.3343.
[68] M. Yamaguchi, H. Yamada, Y. Katsumata, K.-H. Lee, K. Araki, N. Kojima, Efficiency potential and recent activities of high-efficiency solar cells, J. Mater. Res. 32 (2017) 3445—3457. Available from: https://doi.org/10.1557/jmr.2017.335.
[69] M.A. Green, E.D. Dunlop, M. Yoshita, N. Kopidakis, K. Bothe, G. Siefer, et al., Solar cell efficiency tables (version 62), Prog. Photovolt. 31 (7) (2023) 651—663. Available from: https://doi.org/10.1002/pip.3726.
[70] A.F. Hepp, R.P. Raffaelle, Photovoltaics overview: historical background and current technologies, in: S. Sundaram, V. Subramaniam, R.P. Raffaelle, M.K. Nazeeruddin, A. Morales-Acevedo, M. Bernechea Navarro, A.F. Hepp (Eds.), Photovoltaics Beyond Silicon, Elsevier, Cambridge, MA, 2024, pp. 3—74.
[71] <https://www.pv-magazine-australia.com/2022/01/06/mercedes-newest-electric-car-comes-with-thin-film-solar-cells-on-the-roof/> (accessed 15.10.23).
[72] <https://www.ise.fraunhofer.de/content/dam/ise/de/documents/publications/studies/Photovoltaics-Report.pdf> (accessed 15.10.23).
[73] <https://www.vdma.org/international-technology-roadmap-photovoltaic> (accessed 15.10.23).
[74] S.K. Gaddam, R. Pothu, R. Boddula, Advanced polymer encapsulates for photovoltaic devices — a review, J. Materiomics 7 (5) (2021) 920—928. Available from: https://doi.org/10.1016/j.jmat.2021.04.004.
[75] M. Stuckelberger, R. Biron, N. Wyrsch, F.-J. Haug, C. Ballif, Review: progress in solar cells from hydrogenated amorphous silicon, Renew. Sust. Energ. Rev. 76 (2017) 1497—1523. Available from: https://doi.org/10.1016/j.rser.2016.11.190.
[76] M. Paire, S. Delbos, J. Vidal, N. Naghavi, J.F. Guillemoles, Chalcogenide thin-film solar cells, in: G. Conibeer, A. Willoughby (Eds.), Solar Cell Materials: Developing Technologies, John Wiley & Sons, Inc., 2014, pp. 145—215. Available from: https://doi.org/10.1002/9781118695784.ch7.

[77] G. Rizzo, I. Arsie, M. Sorrentino, Hybrid solar vehicles, in: R. Manyala (Ed.), Solar Collectors and Panels, InTech Open, 2010. Available from: https://doi.org/10.5772/10332.
[78] <https://media.ford.com/content/fordmedia/fna/us/en/news/2014/01/02/let-the-sun-in-ford-c-max-solar-energi-concept-goes-off-the-gri.html> (accessed 15.10.23).
[79] <https://www.greencarreports.com/news/1086233_electric-car-prototype-can-run-12-miles-on-solar-power-alone> (accessed 15.10.23).
[80] <https://electrek.co/2020/06/19/tesla-model-3-solar-roof-lightyear/> (accessed 15.10.23).
[81] <https://newatlas.com/toyota-prius-solar-roof/60461/> (accessed 15.10.23).
[82] <https://www.pv-magazine.com/2019/09/24/hanergy-touts-solar-powered-car-that-can-run-30-days-without-charging/> (accessed 15.10.23).
[83] F.A. Tiano, G. Rizzo, M. Marino, A. Monetti, Evaluation of the potential of solar photovoltaic panels installed on vehicle body including temperature effect on efficiency, eTransportation 5 (2020) 100067. Available from: https://doi.org/10.1016/j.etran.2020.100067.
[84] <https://www.greencarreports.com/news/1127957_2020-hyundai-sonata-hybrid-what-to-expect-from-its-mpg-boosting-solar-roof> (accessed 15.10.23).
[85] R. Peibst, H. Fischer, M. Brunner, A. Schießl, S. Wöhe, et al., Demonstration of feeding vehicle-integrated photovoltaic-converted energy into the high-voltage on-board network of practical light commercial vehicles for range extension, Sol. RRL 6 (5) (2022) 2100516. Available from: https://doi.org/10.1002/solr.202100516.
[86] M. Yamaguchi, K. Araki, Y. Ota, K. Nishioka, T. Takamoto, T. Masuda, et al., Potential of Si tandem solar cell modules for PV-powered vehicles, in: 2021 IEEE 48th Photovoltaic Specialists Conference (PVSC), 2021, pp. 120–122. <https://doi.org/10.1109/PVSC43889.2021.9518425>.
[87] <https://www.theverge.com/2020/1/5/21050802/fisker-ocean-electric-suv-solar-roof-range-price-ces-2020> (accessed 15.10.23).
[88] <https://www.nasa.gov/wp-content/uploads/2017/12/458490main_trl_definitions.pdf> (accessed 15.10.23).
[89] J.A. Fernandez, Contextual Role of TRLs and MRLs in Technology Management, Sandia Report, SAND2010-7595, Sandia National Laboratories, Albuquerque, NM and Livermore, CA, 2010, pp. 34. <https://www.osti.gov/biblio/1002093>
[90] A.F. Hepp, J.D. Harris, A.W. Apblett, A.R. Barron, Commercialization of single-source precursors: applications, intellectual property, and technology transfer, in: A.W. Apblett, A.R. Barron, A.F. Hepp (Eds.), Nanomaterials via Single-Source Precursors: Synthesis, Processing and Applications, Elsevier, Cambridge, MA USA, 2022, pp. 563–600. Available from: https://doi.org/10.1016/B978-0-12-820340-8.00008-3.

CHAPTER NINETEEN

Space photovoltaics: New technologies, environmental challenges, and missions

Aloysius F. Hepp[1], Ryne P. Raffaelle[2] and Ina T. Martin[3]
[1]Nanotech Innovations LLC, Oberlin, OH, United States
[2]Rochester Institute of Technology, Rochester, NY, United States
[3]Case Western Reserve University, Cleveland, OH, United States

19.1 A brief history of space solar cells

Photovoltaic (PV) arrays are self-contained, do not require replenishing with external fuel, and have no emissions, making them an ideal method for extraterrestrial power production. This introductory section begins with a brief history of the space age that gave rise to the modern era of PV. The early days of space power where solar cells were used for satellites and some interplanetary missions are described. The transition to higher power III-V-based technologies occurred with the transition to the twenty-first century and a string of successful missions to Mars and beyond.

19.1.1 The invention of the modern solar cell and the dawn of the "space race"

Previously, we detailed the history of the origins of the solar cell [1]. The modern PV era began in the Spring of 1954 when Bell Labs introduced their new invention, a "solar battery," or what we would today call a PV solar cell. They used these devices to power a child's toy and a radio transmitter solely from sunlight. These early cells had an impressive efficiency (η) or power conversion efficiency (PCE) for that time of 6% [2]. These devices were developed by Daryl Chapin, who was working on the development of a new power source to replace batteries for telephone systems in remote and humid areas, and Calvin Fuller and Gerald Pearson,

Photovoltaics Beyond Silicon.
DOI: https://doi.org/10.1016/B978-0-323-90188-8.00014-2
© 2024 Elsevier Inc. All rights reserved, including those for text and data mining, AI training, and similar technologies.

who were following in the footsteps of Ohl on doping and controlling the properties of semiconductors.

Chapin, Fuller, and Pearson created a single-crystal silicon (Si) $p-n$ junction solar cell and chained together multiple single-crystal silicon $p-n$ junction solar cells to form a solar panel [3]. The material systems that enabled these record results were arsenic and boron-doped Si. By 1958, small-area Si solar cells had reached a PCE of 14% under terrestrial sunlight [4]. Although Bell Labs saw commercial potential in the terrestrial use of solar energy, it would be several decades before any real terrestrial deployment at the scale of solar power systems would be realized. However, with the dawn of the "space race," a new driving force for the continued development of PV power was soon to arise: space solar power systems.

As part of the response to the former Soviet Union's launching of the Sputnik satellites, the United States launched the first solar-powered satellite, Vanguard 1, on March 17, 1958, see Fig. 19.1 [5]. Vanguard 1 was the second US satellite placed in orbit around the Earth and the first with solar power. It was the fourth satellite to be successfully launched, after the Sputnik 1 and 2 (launched October 4, 1957, and November 3, 1957, respectively) and the US Explorer 1 (January 31, 1958). Vanguard 1 has six body-mounted solar panels (see Fig. 19.1B), each with eighteen 2×0.5 cm p-on-n Si solar cells, with a 0.16 cm thick quartz coverglass. The cells were fabricated by Hoffman Electronics for the US Army Signal Research and Development Laboratory at Fort Monmouth New Jersey. The solar cells continued to power a radio signal until February 1965. Vanguard 1 continues as the oldest man-made object in orbit around Earth.

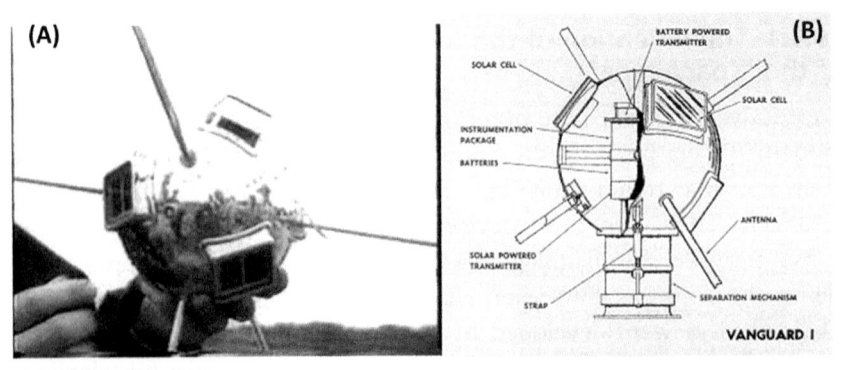

Figure 19.1 (A) Photograph and (B) schematic of Vanguard 1. *Courtesy: U.S. Naval Research Laboratory.*

19.1.2 Space photovoltaics during the 1960s to 1980s: Satellites and early science missions

Telstar is the name given to a series of over 20 communication satellites that began to be launched in the early 1960s and continue to the present day [6]. These satellites continued to grow in size, complexity, and power demands with each new generation. They have ranged from 77 and 79 kg for Telstar 1 and 2, respectively, to over 7000 kg for Telstar 19 V launched in 2018 aboard a SpaceX Falcon 9. Telstar 19 V became the heaviest commercial communications satellite ever launched. It can be argued that this series of satellites has been the greatest human achievement in communication since the telephone, or perhaps even the Guttenberg printing press. According to a US Information Agency poll, Telstar was better known in Great Britain in 1962 than Sputnik had been in 1957 [7].

Telstar 1 was part of a multinational agreement between NASA, commercial entities, and nongovernmental organizations in the United States, the United Kingdom, and France to establish satellite communications over the Atlantic Ocean. The almost identical, albeit slightly larger at 79 kg, Telstar 2 was launched less than a year later aboard a Delta B rocket. Although no longer functional, these satellites, like their predecessor Vanguard 1, continue to orbit the Earth. 21 Telstar satellites were deployed between 1962 and 2019, the most recent being Telstar 19 V, launched on July 22, 2018, and built by Space Systems Loral [8]. Bell Labs was responsible for developing the first two, Hughes for the next three, Lockheed Martin for the next three, and Space Systems Loral for the remaining 13.

The first satellites operated on less than a few hundred watts, and since the cost of the solar arrays was a small fraction of the overall satellite cost, the premium was put on reliability—as it still is today. However, as the size and complexity of the satellites continued to increase, so did their need for solar power. With launch costs approaching $10,000 or more per kilogram, considerable attention was paid to solar array-specific power (W/kg), and therefore the efficiency of the solar cells. In addition, many satellites were still using body-mounted arrays, which severely limited the array size and therefore required the highest efficiency cells available.

The initial development of space solar power systems focused on silicon, as it was relatively efficient and reliable. Theoretical predictions suggested that practical Si solar cells could be produced with efficiencies approaching

20%, although its bandgap was 0.4 eV below the optimum bandgap (approximately 1.5 eV) for terrestrial illumination [9]. Fundamental solar cell research was focused on understanding and mitigating the factors that limited cell efficiency (e.g., minority carrier lifetime, surface recombination velocity, series resistance, reflection of incident light, and nonideal diode behavior). However, new materials and cell designs were beginning to emerge. Examples were the use of III-V semiconductors or potential tandem or multijunction (MJ) cell designs. An optimized triple-junction (3J) cell was shown to have a theoretical efficiency of 37% [10].

19.1.3 Space photovoltaics in the 1990s and beyond: Transition to III-V materials

During the 1970s and 1980s, solar cell efficiencies rose dramatically [4,11]. While Si-based solar arrays still provided power for many missions, new III-V compound semiconductor materials were introduced [10,12]; these III-V semiconductors required new processing methods [13]. This subsection describes their introduction, new device structures, and early low-Earth orbit (LEO) missions.

Fig. 19.2 shows III-V semiconductor material devices and associated MJ cells [10]. As cell efficiencies continued to rise, the gap between theoretical efficiencies and experimental efficiencies for Si, gallium arsenide (GaAs), and indium phosphide (InP) diminished [14,15]. In addition, this

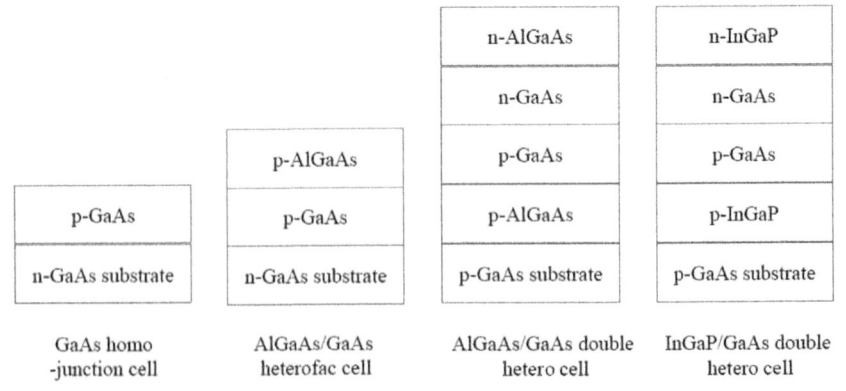

		n-AlGaAs	n-InGaP
		n-GaAs	n-GaAs
	p-AlGaAs	p-GaAs	p-GaAs
p-GaAs	p-GaAs	p-AlGaAs	p-InGaP
n-GaAs substrate	n-GaAs substrate	p-GaAs substrate	p-GaAs substrate
GaAs homo-junction cell	AlGaAs/GaAs heterofac cell	AlGaAs/GaAs double hetero cell	InGaP/GaAs double hetero cell

Figure 19.2 Device structures of GaAs solar cells developed historically. *GaAs, Gallium arsenide. Reproduced from M. Yamaguchi, High-efficiency GaAs-based solar cells, in: M.M. Rahman, A.M. Asiri, A. Khan, I. Inamuddin, T. Tabbakh (Eds.), Post-Transition Metals, InTech Open, 2021. DOI:10.5772/intechopen.94365 under the terms of the Creative Commons Attribution 3.0 Unported license (CC BY 3.0).*

period saw a host of other improvements to space solar cells such as the first use of shallow junction silicon cells for increased blue response and current output, the use of a back surface field, the low–high junction theory for increased silicon cell voltage output, and the development of wraparound contacts to enable automated array assembly and to reduce costs [16].

During this period, MJ III-V solar cells began to displace Si as the solar cell material system of choice for space power systems [10,11,15]. However, Si would continue to see some use, including some of the initial arrays on the largest space solar power system ever deployed, those of the International Space Station (ISS) launched on November 20, 1998. The ISS, as of 2021, had eight retractable solar array wings (SAW). The first pair was deployed in 1998 and the system was completed in March 2009 [17]. The arrays contain a total of 262,400 solar cells and cover an area of 2500 m^2. Altogether, the arrays are rated to generate about 240 kW of electrical power at 160 V in direct sunlight or an average of 84–120 kW cycling between sun and shade.

The array power output of the ISS (being in LEO) has been continuously degrading from radiation, primarily due to low-energy electrons trapped in the Van Allen radiation belts [18]; the ISS arrays have a 15-year life. NASA has replaced six of the eight existing power channels of the space station with new solar arrays. The arrays were delivered by the SpaceX Dragon cargo spacecraft during three resupply missions starting in June 2021, and completed in June 2023, during the writing of this book (Fig. 19.3).

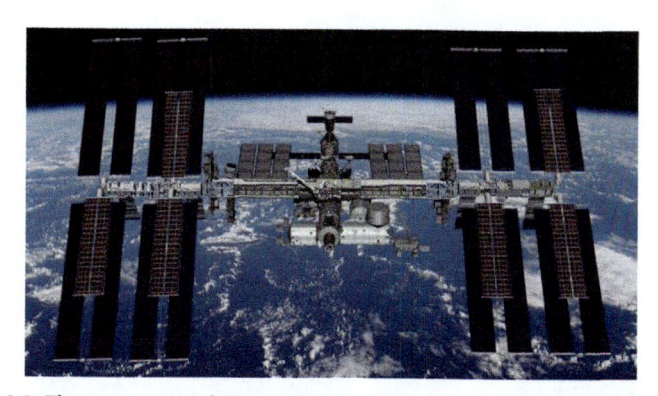

Figure 19.3 The International Space Station (ISS) showing both older solar array wing (SAW) and most recent ISS roll-out solar array (iROSA) photovoltaic arrays in the planned configuration in front of six of the eight existing power channels. *Courtesy: NASA.*

Boeing, NASA's prime contractor for space station operations provided the new arrays, which were manufactured by Deployable Space Systems (DSS) using Spectrolab Next Triple Junction (XTJ Prime) cells. The new arrays are a larger version of the roll-out solar array (ROSA) technology that was demonstrated aboard the space station in June 2017 [19]. These new arrays are smaller, due to the inclusion of high-efficiency Spectrolab triple-junction III-V solar cells. [12]. Positioned in front of six of the eight power channels (see Fig. 19.4), these new cells are estimated to provide up to a 30% increase in power for the ISS. An in-depth discussion of III-V single and MJ solar cells and arrays for space applications can be found in several excellent recently published sources [10,12]. To get a sense of the enormous size of the new arrays, Fig. 19.4 shows a NASA spacewalker riding a robotic arm carrying the ISS roll-out solar array (iROSA) to its installation site on the Starboard-4 truss segment.

Although it was launched before the ISS, the Mir modular space station that was assembled in LEO from 1986 to 1996 and operated until 2001 by the Soviet Union and later by Russia, actually used GaS solar cells (Fig. 19.5) [20]. Unfortunately, the deployment of its arrays, which occurred over 11 years, was slower than planned and the station continually suffered from a shortage of power as a result. Two of the final arrays were delivered by the Space Shuttle Atlantis during STS-74. Altogether, the solar arrays aboard Mir provided approximately 26 KW. The Mir

Figure 19.4 NASA spacewalker rides the Canadarm2 robotic arm carrying the ISS roll-out solar array to its installation site on the Starboard-4 truss segment in December 2022. *Courtesy: NASA.*

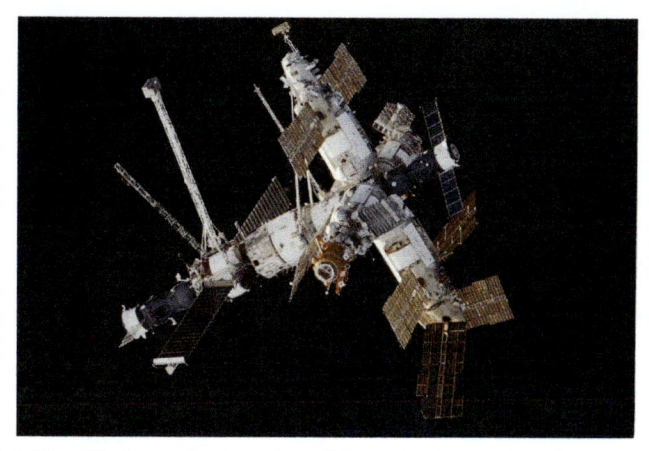

Figure 19.5 The Mir Space Station viewed from the Space Shuttle Endeavor during the SRS-89 rendezvous. *Courtesy: NASA.*

made history as the first continuously inhabited long-term research station. It provided a human presence in space for 3644 days, a record that was finally surpassed by the ISS on October 23, 2010 [21].

19.1.4 New devices and advanced materials technologies in the 21st century

Compared to the ISS, the transition from silicon to III-V and III-V MJ solar cells occurred more rapidly for most space power systems. During the 1990s and 2000s, satellites continued to grow in both size and power requirements, and innovative structures were designed to deploy ever-large solar arrays. The mass and fuel penalty for attitude control of very large arrays continued to be a huge driver for ever more efficient cells. III-V triple-junction solar cells on Ge or GaAs substrates became the *de facto* standard during this time. Spacecraft developers were still willing to pay a premium for the best cells, with satellite power systems continuing to be in $1000/W range, and thus there was a steady improvement in the efficiency of these cells as seen in the United States Department of Energy (DOE) National Renewable Energy Laboratory (NREL) efficiency chart (Fig. 19.6) [15,22−24].

State-of-practice (SOP) or state-of-the-art (SOA) solar arrays today have an AM0 (space solar spectrum, see below) efficiency of over 30% [23] and are typically triple-junction III-V cells grown on Ge or GaAs (Fig. 19.7) [14]; however, four-junction, five-junction, and six-junction cells are being developed and improved [10,12]; see below. Some of the more advanced techniques used to develop these stacks require them to

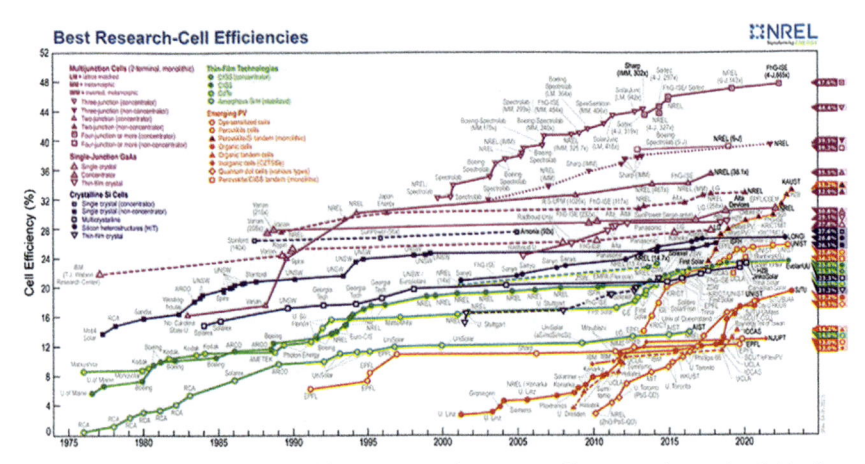

Figure 19.6 National Renewable Energy Laboratory efficiency chart. *Available from https://www.nrel.gov/pv/cell-efficiency.html. Public domain as U.S. Government-sponsored work product.*

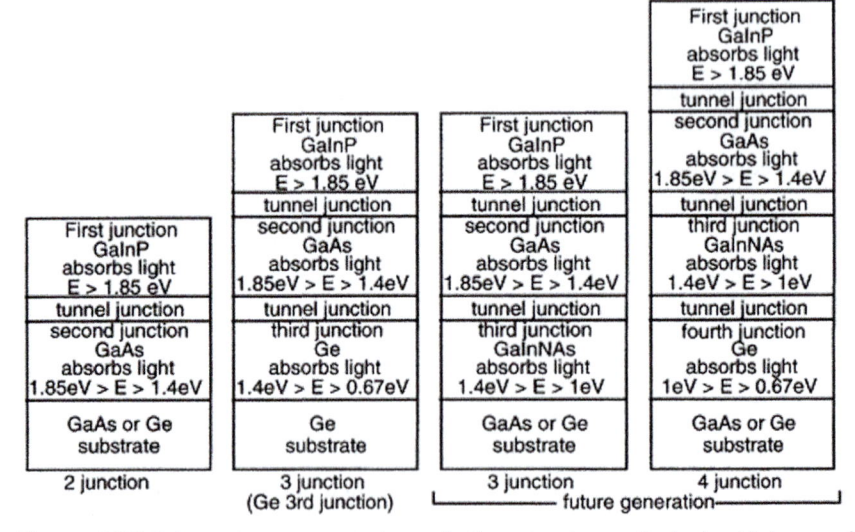

First junction GaInP absorbs light E > 1.85 eV	First junction GaInP absorbs light E > 1.85 eV	First junction GaInP absorbs light E > 1.85 eV	First junction GaInP absorbs light E > 1.85 eV
tunnel junction	tunnel junction	tunnel junction	tunnel junction
second junction GaAs absorbs light 1.85eV > E > 1.4eV	second junction GaAs absorbs light 1.85eV > E > 1.4eV	second junction GaAs absorbs light 1.85eV > E > 1.4eV	second junction GaAs absorbs light 1.85eV > E > 1.4eV
	tunnel junction	tunnel junction	tunnel junction
	third junction Ge absorbs light 1.4eV > E > 0.67eV	third junction GaInNAs absorbs light 1.4eV > E > 1eV	third junction GaInNAs absorbs light 1.4eV > E > 1eV
			tunnel junction
			fourth junction Ge absorbs light 1eV > E > 0.67eV
GaAs or Ge substrate	Ge substrate	GaAs or Ge substrate	GaAs or Ge substrate
2 junction	3 junction (Ge 3rd junction)	3 junction	4 junction

future generation

Figure 19.7 Schematic representation of the structure of dual-, triple-, and quadruple-junction solar cells. *Reprinted with permission from D.J. Friedman, J.F. Geisz, S.R. Kurtz, J.M. Olson, 1-eV solar cells with GaInNAs active layer, J. Cryst. Growth, 195 (1998) 409–415, copyright (1998) Elsevier.*

be grown metamorphically, with buffer layers that allow for relaxation of the conventional lattice-matched epitaxial growth, or by growing the structure in an inverted fashion, requiring substrate removal. Details of these particular structures and the various techniques used to achieve them have been covered in depth previously [10–14] and will be summarized in Section 19.3.

19.2 Space environment: Concerns, challenges, and issues for space photovoltaics

This section covers several key factors related to space versus terrestrial PV. The first is the difference in the solar spectrum. We detail the difference between space (air mass zero, AM0) and terrestrial (AM1.0 or AM1.5) solar radiation, which is a fundamental difference between standard measurements of cells used in terrestrial versus extraterrestrial applications. A discussion of key environmental challenges is the topic of this next section. Numerous thin-film PV technologies have been proposed for space; historically, most of this work has focused on inorganic thin films, but hybrid inorganic-organic perovskite materials have recently been the focus of extensive research. Finally, important considerations for space solar arrays are addressed. We summarize this subsection with some thoughts about future needs for new materials and devices in space.

19.2.1 Air mass standards for terrestrial and space photovoltaics

Without the absorptive properties of Earth's atmosphere, solar cells in space experience different levels of solar irradiation than terrestrial devices. Terrestrial solar cells are qualified using air mass (AM) 1.0 or 1.5, AM1.5 is typically used and quoted. Space efficiencies are qualified using AM0, see Fig. 19.8.

The air mass quantifies the absorption of sunlight by the Earth's atmosphere; the value is a ratio of the path length the light takes through the atmosphere relative to the shortest possible path length (when the sun is directly overhead). Thus AM1.0 is the solar constant at noon at the surface of the Earth on a perfectly clear day (1.000 kW/m^2). In comparison, AM1.5 corresponds to light passing through 1.5 atmospheres, which represents the annual average of the continental United States and other mid-latitude locations [25].

Fig. 19.8 shows that the lack of atmospheric absorption results in increased irradiance across the solar spectrum, particularly in the near UV

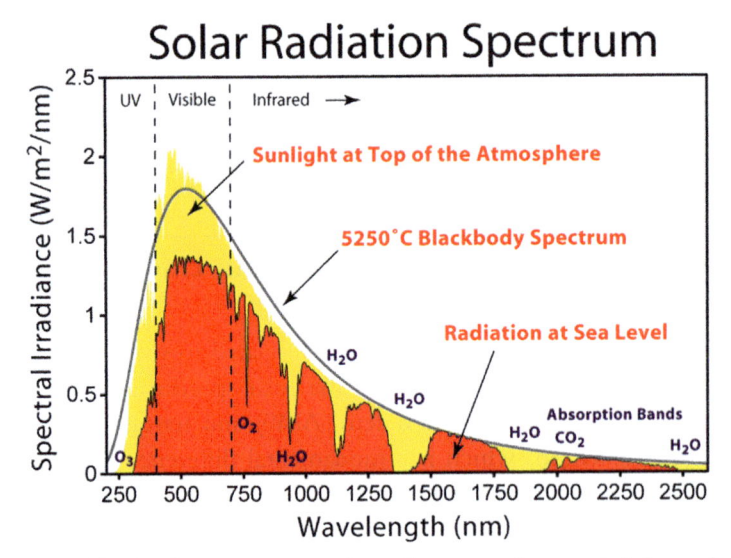

Figure 19.8 Solar irradiance spectrum above the atmosphere and at the surface. *Dr. R.A. Rohde, own work (CC BY-SA 3.0 (https://commons.wikimedia.org/w/index.php?curid = 2623187))*.

and short wavelength regions that are filtered out by Earth's atmosphere. The sum of the increased irradiance results in a higher power (1.367 kW/m^2) for AM0 conditions. Perhaps counterintuitively, this increased radiation often results in lower solar cell efficiencies. Solar cell efficiency is a ratio of power out to power in; thus devices with poorer performance at lower wavelengths commonly have a lower efficiency under AM0 conditions compared to AM1.5 [26,27]. Fig. 19.9 shows the predicted efficiency versus bandgap for PV materials for solar spectra in space (AM0) and on the surface of the Earth (AM1.5) at 300 K. Technology-dependent correction factors based on the spectral response of the cell can be used to estimate AM0 efficiency values from AM1.5 efficiency data. For example, a 15%—20% reduction in efficiency is reported for c-Si and chalcogenide solar cells (1—1.5 eV bandgaps absorbers).

19.2.2 Radiation challenges from the space environment

Space solar cell design complexity increased dramatically with the launching of Explorer I and the discovery of the Van Allen radiation belts [18] and damage due to electron and proton irradiation. This was made painfully evident with the US high-altitude nuclear weapon test "Starfish." When the Starfish Prime device was detonated on July 9, 1962, an

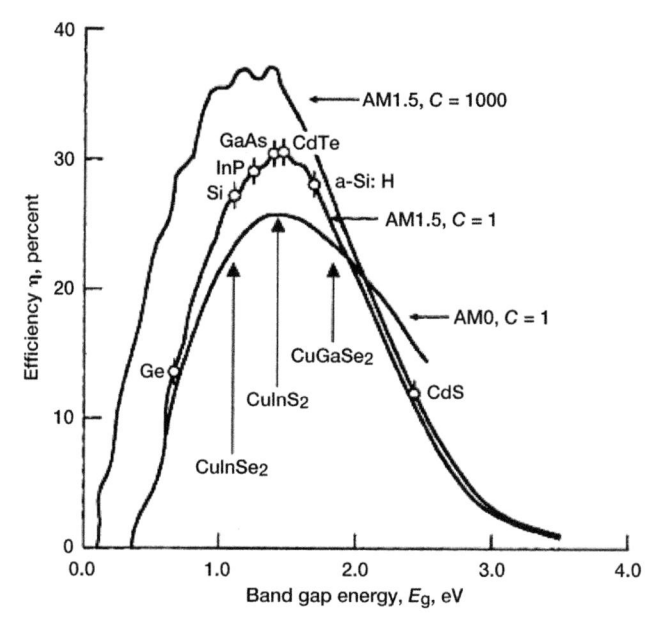

Figure 19.9 Predicted efficiency versus bandgap for photovoltaic materials for solar spectra in space (AM0) and on the surface of the Earth (AM1.5) at 300 K compared with unconcentrated (C = 1) and high concentration (C = 1000) sunlight. Thin-film I-III-VI$_2$ materials' bandgaps are indicated on the AM0 curve. *Courtesy: NASA.*

extremely large, and greater than expected, electromagnetic pulse was generated. The result of all the radiation that was dumped into the van Allen radiation belts was that several satellites ceased to function as their solar arrays failed due to the radiation damage. Even Telstar was negatively impacted, with a decreased solar power output following the Starfish detonation [28].

The lessons learned from Explorer I and Telstar prompted a surge of activity in radiation protection of space solar cells. The Naval Research Laboratories provided much-needed guidance to spacecraft designers on how to account for natural cell degradation due to radiation in space [29]. This influenced the design of silicon solar cells, which migrated away from the conventional *p*-on-*n* design to a *n*-on-*p* structure to provide better radiation resistance [30].

Solar cells, panels, or arrays will be exposed to different types and amounts of radiation depending on location. The Van Allen belts [18] and other solar wind regions are mostly composed of protons and electrons. High-energy protons, alpha particles, and heavier particles make up the galactic cosmic rays (GCR); solar flare activity results in eruptions of electromagnetic radiation and particle (electrons, protons, and heavier ions) acceleration. Generally, device performance will degrade as a function of the differential flux spectrum and the total ionizing dose [25–32]. A comparison of the three most commonly utilized cell technologies as a function of 1-MeV fluence is illustrated in Fig. 19.10. A recent comprehensive review provides a valuable analysis that can help spacecraft designers address radiation and other challenges (see below) presented by space environments during the development of PV assemblies for missions to various solar system destinations [31].

19.2.3 Thermal and other challenges of the space environment

Extraterrestrial environments provide unique stressors to PV devices, which vary with location [31]. As discussed above, the presence of a range of electromagnetic and ionizing radiation can result in displacement and ionization damage. Both radiation and thermal stressors can result in the interdiffusion of elements through the devices, resulting in changes to materials and device performance. Thermal cycling in space has greater extremes than those encountered by terrestrial installations, making delamination a particular concern. Atomic oxygen is an abundant reactive

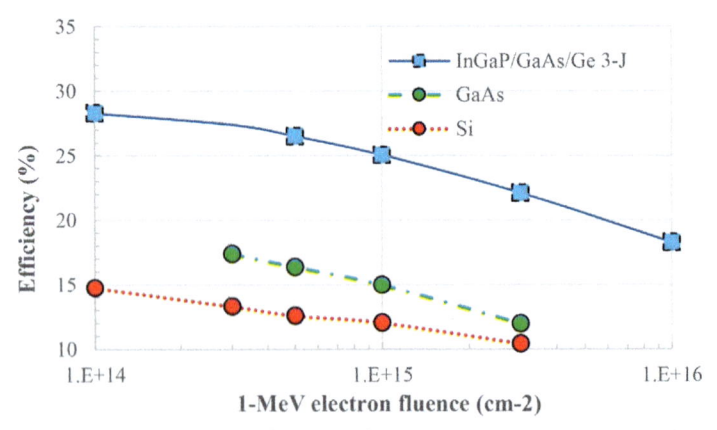

Figure 19.10 Changes in the efficiency of Si single-junction, GaAs single-junction, and InGaP/GaAs/Ge three-junction space solar cells as a function of 1-MeV electron fluence. *Reproduced from M. Yamaguchi, High-efficiency GaAs-based solar cells, in: M. M. Rahman, A.M. Asiri, A. Khan, I. Inamuddin, T. Tabbakh (Eds.), Post-Transition Metals, InTech Open, 2021. DOI:10.5772/intechopen.94365 under the terms of the Creative Commons Attribution 3.0 Unported license (CC BY 3.0).*

species in LEO that degrades both carbon- and silicon-based polymers. The former are susceptible to oxidation and thinning, while the latter are susceptible to oxidation and contamination from the resulting deposits [33]. Table 19.1 includes the multifaceted effects of the space environment on solar cells, which are key considerations for optimizing materials and device technology (Table 19.1) for future mission concepts and needs. A chapter by Ferguson et al. in a recently published book is suggested for a more in-depth treatment of space environmental effects [34].

Table 19.1 Summary of the impact of the space environment on solar cells.

Environment factors	Effects
Solar irradiance	Power conversion dependence
Temperature	Efficiency degradation
	Degradation mechanisms (carrier freeze-out and thermal barriers to conduction)
	Thermo-elastic stress cycles (e.g., cracks in solder joints of the interconnects)
	Electric resistances
Vacuum	Contamination (degassing)
	Pressure differentials (decompression)
Plasmas	Surface charging, electrostatic discharge, and dielectric breakdown
	Enhanced sputtering and reattraction of contamination
	Increased leakage current
Energetic particle radiation	Total ionizing dose effects (electronic degradation)
	Displacement damage
	Single-event effects (upset, latch-up, and burnout)
	Degradation in optical properties (e.g., coverglass and optics)
Electrically neutral particles	Mechanical effects (aerodynamic drag and physical sputtering)
	Chemical effects (ATOX and spacecraft flow)
Ultraviolet and X-ray radiation	Degradation of thermo-electric properties
	Degradation of optical properties (e.g., coverglass and optics)
	Structural damage(s)
Micrometeoroids and debris	Damage to cell active area and interconnects
	Damage to optical systems caused by hypervelocity impacts (coverglass, lenses, and mirrors)
	Increased cell shunt resistance

Source: Reproduced with permission from A. Bermudez-Garcia, P. Voarino, Olivier Raccurt, Environments, needs and opportunities for future space photovoltaic power generation: A review, Appl. Energy 290 (2021) 116757, copyright (2021) Elsevier.

Thermal issues are one of the fundamental stressors faced by space PV, leading to extreme thermal cycling results in interfacial stresses and delamination. A recent study by Banik et al. considered a potential solution to thermal issues confronting a TFSC (i.e., copper indium gallium selenide [CIGS]) operating in space [35]: a high emissivity space-proof coating. Compared to terrestrial applications, TFSCs operate at higher temperatures, based on both higher solar insolation and the lack of a medium for convection. At 3 K, the cosmic background is essentially a blackbody, making it a potential heat sink for any PV device capable of radiating out heat in the infrared. Therefore a practical solution to these thermal concerns would be the development of coatings with high transparency in the UV−VIS region of the electromagnetic spectrum, for efficient solar cell efficient operation, and high broadband emissivity in the mid-range infrared (MIR), to facilitate radiative cooling, see Fig. 19.11.

A more detailed understanding of the desirable coating characteristics comes from performing an energy balance. Due to the relatively low PCE of CIGS, only approximately 15% of the absorbed AM0 spectrum would be converted to electricity, while much of the remaining energy would be converted to heat. As radiative emission is the only way to shed this excess heat, an analysis is required to determine the integral emissive

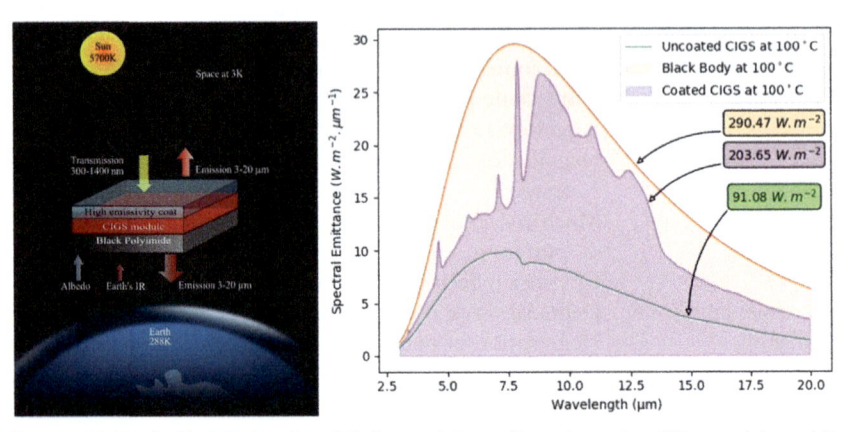

Figure 19.11 (*Left*) Schematic of light and heat flow into the PV panel in orbit. (*Right*) Total integral emittance comparison of a blackbody, coated, and uncoated module at 100°C. *Reproduced with permission from U. Banik, K. Sasaki, N. Reininghaus, K. Gehrke, M. Vehse, M. Sznajder, T. Sproewitz, C. Agert, Enhancing passive radiative cooling properties of flexible CIGS solar cells for space applications using single layer silicon oxycarbonitride films, Solar Energy Mater. Solar Cells 209 (2020) 110456, copyright (2020) the authors, published by Elsevier B.V.*

power gain from the coating. The desired properties of a coating designed for the best radiative cooling output from a solar cell are a very broad band emissivity in the 3–20 μm range and a high transmissivity in the 0.1–1.4 μm range. Assuming the modules are at 100°C in orbit, as discussed above, the thermal emittance capability of the coated and uncoated samples can be compared, see Fig. 19.9B. A theoretical black body at 100°C can emit an integral spectral emittance of 290.47 W/m^2 in the 3–20 μm wavelength range.

A successful demonstration of such a coating was achieved by the German Aerospace Center (DLR) group. Silicon oxycarbonitride (SiOCN) films made from organopolysilazane were deposited on CIGS cells on polyimide substrates; a single approximately 3.2 μm coating yielded the highest emissivity value of 0.72. Both the optical properties of SiOCN deposited by dip coating and the impact of the coating on device performance were investigated in detail [35]. Electrical characterization of the coated devices showed that the dip coating and postprocessing steps (including the low-temperature curing) had a negligible effect on the electrical performance. The 3.2 μm SiOCN-coated devices emitted 203.65 W/m^2 in comparison to an uncoated sample that could only emit 91.08 W/m^2, enhancing emissive power by 123%. The key is that the resonance peaks of the coating align well with the blackbody peak, enabling the stack to emit radiation with the highest intensity at orbit temperatures. These results coincide with an earlier study reported by M. Pscherer et al. on thin-film silazane/alumina high emissivity double-layer coatings for flexible CIGS solar cells [36]. Polysilazane-derived coatings may be considered a suitable alternative to cover glasses used on solar cells for space applications, due to their high ε, flexibility, and excellent thermal and electrical behavior.

19.2.4 Mass-specific power: Introduction to a critical metric for space photovoltaics

A significant consideration for the use of PV technologies in space is mass-specific power (MSP), or the power generated by a device divided by its weight (kW/kg). For example, the current launch costs of SpaceX's Falcon 9, which is regularly used to access the ISS, are approximately US $2700/kg (approximately $1200/lb) [37]. Minimizing the weight of the cells themselves is a crucial factor for the realization of advanced solar cell technology use in space. MSP costs motivate the development of solar cells on lightweight, flexible substrates.

In addition to the mass savings, flexible substrates can potentially benefit from roll-to-roll production techniques. These methods are used in the fabrication of other technologies, including flexible light emitting diodes; benefits traditionally associated with roll-to-roll production are increased cost-effectiveness and higher production rates, compared to, for example, wafer processing. [38]. Every gram of materials used in the solar module or array fabrication must be transported into orbit. Although the associated costs for this have decreased in recent years, they still remain a significant consideration. Because of this limiting factor, advanced technologies are being explored for higher MSP [1,11,22,27,39,40]. Note that the requirement for high specific power is also coupled with the need for stow-ability, or the power generated per unit volume (power density in W/m^3), and higher area-specific power (ASP in W/m^2). Here TFSCs also are at an advantage, as discussed in the following section.

19.3 Materials and devices: Connections to space exploration

In this section, we address classes of materials with relevance to space applications. The first subsection addresses III-V materials, the workhorse of future planetary exploration missions that are PV-powered. Thin-film materials were regarded as a potential competitor to III-V materials several decades ago. However, despite the great strides made with thin III-V multijunction devices, there may be niche applications where TFSC materials may be an optimal choice. We then examine the newest group of TFSC devices, including dye-sensitized solar cells (DSSCs), perovskite solar cells (PSCs), and organics photovoltaics (OPV). While the significant challenges imposed by space environmental issues remain, niche and specialized applications should be explored. Finally, we examine advanced device concepts such as PSC tandems and concentrator cells. We end this section with some overall assessments and a comparison of potential applications for each cell type.

19.3.1 Group III-V materials and device technologies

In 1955 an industry group was funded by the US Army Signal Corps, and later by the US Air Force (USAF), to work on the development of GaAs-based cells. Gallium arsenide has a nearly ideal direct bandgap of 1.42 eV for operation in a terrestrial solar spectrum [10,11]. It also has favorable thermal stability and radiation resistance as compared to silicon. However, it took a generation of efficiency improvements until the use of GaAs-based cells could be justified because of its much higher costs [1,10–12,15,22,41]. The lattice-matched heterojunction cells developed in the early 1970s led to the acceptance of GaAs as a viable PV material; see Fig. 19.2. The Air Force launched the Manufacturing Technology (ManTech) program for GaAs solar cells in 1982. This program was designed to develop metal-organic chemical vapor deposition (MOCVD) techniques necessary for the large-scale production of large quantities of GaAs solar cells [1,10,11]. The other significant development in GaAs technology was research into and use of alternative substrates.

The application in which this premium is most easily justified is in space utilization. Currently, a significant majority of solar cells being launched into space for power generation are MJ III-V cells [1,10–12]. In 1986 the USAF supported work by the Applied Solar Energy Corp. (ASEC), in which they developed GaAs cells grown on Ge substrates. This was possible due to the similarity in the lattice constants and thermal expansion coefficients of the two materials. This resulted in an improvement in the mechanical stability of the cells and a lowering of the production costs [1,10,11]. To summarize, MJ solar cells will be widely used in space because of their high conversion efficiency and good radiation resistance; see Figs. 19.7 and 19.10 [1,10–12,22,27,41–45].

The primary approach to increasing cell efficiency and improving upon the SOP or SOA MJ devices (Fig. 19.12A) is to optimize the bandgap of each subcell in an MJ stack. A significant source of energy loss is the mismatch between photon energies and junction bandgaps. Hence, to maximize efficiency, subcells must have bandgaps that nearly match, or are as close as feasible to, the photon's energy to convert photons while minimizing energy loss. Additionally, cell current in a stack of subcells in series is limited by the junction with the smallest photocurrent. Therefore subcells must also be current-matched to maximize efficiency. Ongoing developments are focused on growing materials and developing devices [12] that can simultaneously address both goals using a variety of approaches; this is shown schematically in Fig. 19.12B–E [42].

Inverted metamorphic MJ (IMM) solar cells include a 1.0 eV bandgap, as shown in Fig. 19.12B. The term "metamorphic" refers to a mismatch in the crystal lattices of different materials in the structure. In this case, the crystal lattice of the 1.0 eV material is not matched to the top two

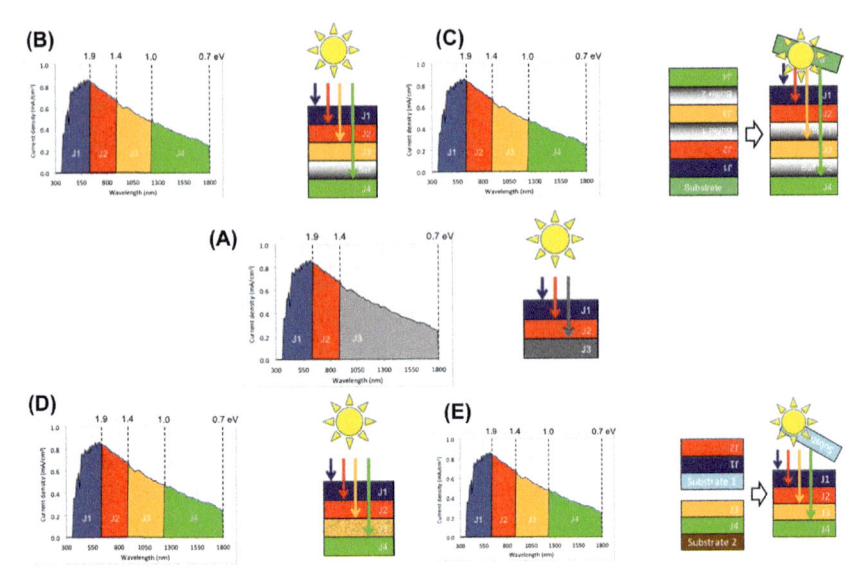

Figure 19.12 (A) Shortcomings of SOP solar cells. The bottom junction generates an excess current at a low potential due to its low bandgap (0.7 eV). (B) IMM solar cells. A 1.0 eV bandgap is added using the technique of inverted growth and subsequent removal of the growth substrate. (C) UMM solar cells. A 1.0 eV bandgap is added using lattice-mismatched materials and buffer layers to minimize the propagation of crystal defects. (D) Dilute nitride solar cells. A 1.0 eV bandgap is added using materials that are lattice-matched to GaAs. Nitrogen in the 1.0 eV material is used to fine-tune the lattice constant. (E) Semiconductor wafer bonding. Two separate wafers are bonded together to create a single cell. As a result, semiconductor quality is not affected by lattice mismatch. *Adapted from P.M. Beauchamp, J.A. Cutts, Solar Power Technologies for Future Planetary Science Missions, Jet Propulsion Laboratory, Report JPL D-101316, Pasadena, CA, USA, December 2017, 49 pp., Public Domain as a U.S. Government work product.*

junctions (GaInP$_2$ and GaInAs). Upright metamorphic MJ (UMM) materials generally include a 1.0 eV bandgap and at least four junctions. In contrast to IMM cells, UMM cells are fabricated by growing the structure in the same order as SOP cells, starting from the bottom subcell. The difficulty of lattice mismatch between the 1.0 eV material and the top two subcells is addressed by growing transparent buffer layers between the mismatched layers (Fig. 19.12C). The challenge of finding a material with a 1.0 eV bandgap with a crystal lattice matched to the SOP structure is solved by adding a small amount of nitrogen to the 1.0 eV material (dilute nitride materials—Fig. 19.12D). Hence, these cells provide optimized bandgaps without sacrificing crystal quality or introducing complex inverted growth techniques. Semiconductor wafer bonding technology (SBT) refers to mechanical connection of one semiconductor wafer on top of another. This approach enables two wafers that are grown separately, with different lattice constants, to be combined into a MJ stack (Fig. 19.12E). Finally, as mentioned above, five-junction cells are also now being produced (Fig. 19.13) [12].

19.3.2 Thin-film solar cell materials

Historically, classic crystalline technology has powered space exploration missions [1,10−12,22,27,42,46,47]. In space, as on Earth, cost, weight, and structural considerations provide ample motivation for developing thin-film technologies [22,46,48−51]. Amorphous or polycrystalline direct bandgap absorber materials can be lighter and deposited on light weight flexible substrates. Radiation test results are promising across thin-film technologies; however, materials whose performance depends on high deposition temperatures require thermally stable, flexible substrates. The recent comprehensive review referenced earlier in this chapter provided advice on appropriate PV technologies for specific solar system destinations, including optimal materials, device technologies, and key environmental challenges to overcome for future mission concepts and

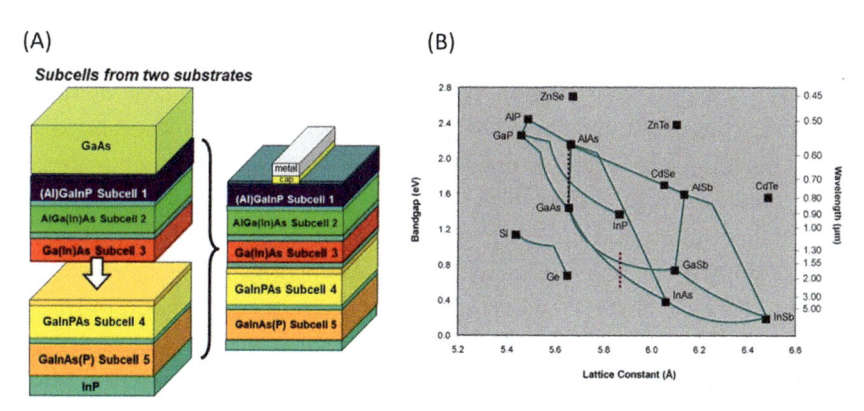

Figure 19.13 (A) Layer structure and process for a single-bonded (SBT) 5J cell. (B) Bandgap lattice constant diagram for the SBT 5J cell. Black line highlights the top three junctions' growth at the GaAs lattice constant. Dashed purple line highlights the bottom two junctions grown at the InP lattice constant. *Reproduced with permission from P.T. Chiu, Space applications of III-V single- and multi-junction solar cells, in: R.P. Raffaelle, S.G. Bailey, D.C. Ferguson, S. Durbin, A.F. Hepp (Eds.), Space Photovoltaics: Materials, Missions and Alternative Technologies, Elsevier, Cambridge, MA USA, 2023, copyright (2023) Elsevier.*

needs [31]. A significant finding concerning future missions to explore the outer solar system is that they will require high-power PV systems capable of functioning efficiently in low-intensity, low-temperature (LILT) conditions, and high radiation environments [31]; a detailed summary of the study by Sellers and coworkers is presented below [52]. This section details key environmental and spacecraft issues and goes into further details of past and present missions. It includes proposed technologies relevant to TFSCs and concludes with future space exploration challenges, technologies, and opportunities.

As discussed above, the original baseline PV arrays in space utilized crystalline solar cells [1,11,22,42,46,47]; this technology is being replaced by III-V-based MJ devices where higher PCE is essential [10−12,42,46,47]. Although crystalline solar cells continue to evolve, future missions requiring very high specific power and small launch stowed volumes may not be feasible using crystalline solar cells. Thin-film solar cell materials on lightweight substrates may be an enabling technology for future space missions by significantly increasing specific power while reducing launch stowed volume [38−42,46−52]. Inorganic TFSC technologies that have been widely studied over the last five decades include hydrogenated amorphous silicon (a-Si: H), CIGS (and related I-III-VI$_2$ materials), cadmium telluride (CdTe), and related materials. While these technologies have experienced significant advances over the years, it may take many years for them to reach the high AM0 efficiency required for many space applications [1,10,42,46−52].

Group 12−16 (II-VI)-based MJ, thin-film, solar cells have demonstrated the potential to achieve AM0 efficiencies of up to 25% and beyond [53]. Thin-film solar cells are grown layer-by-layer on a support material using methods that range from thermal evaporation to atmospheric MOCVD. Numerous architectures are used for TFSCs, [54−56] each of which comprise multiple layers with specific functions and constraints. In this section of the chapter, we provide an overview of materials, processes, and the impact of both processing and the space environment on device performance [55,56]. Fig. 19.14 provides illustrations of device structures of the three traditional single-junction inorganic TFSC material devices developed over the past several decades, including substrate and superstrate CdTe devices [54,56].

19.3.2.1 Amorphous silicon

For space applications, degradation studies show that c-Si devices have an acceptable radiation tolerance. Due to reduced thickness of the absorber

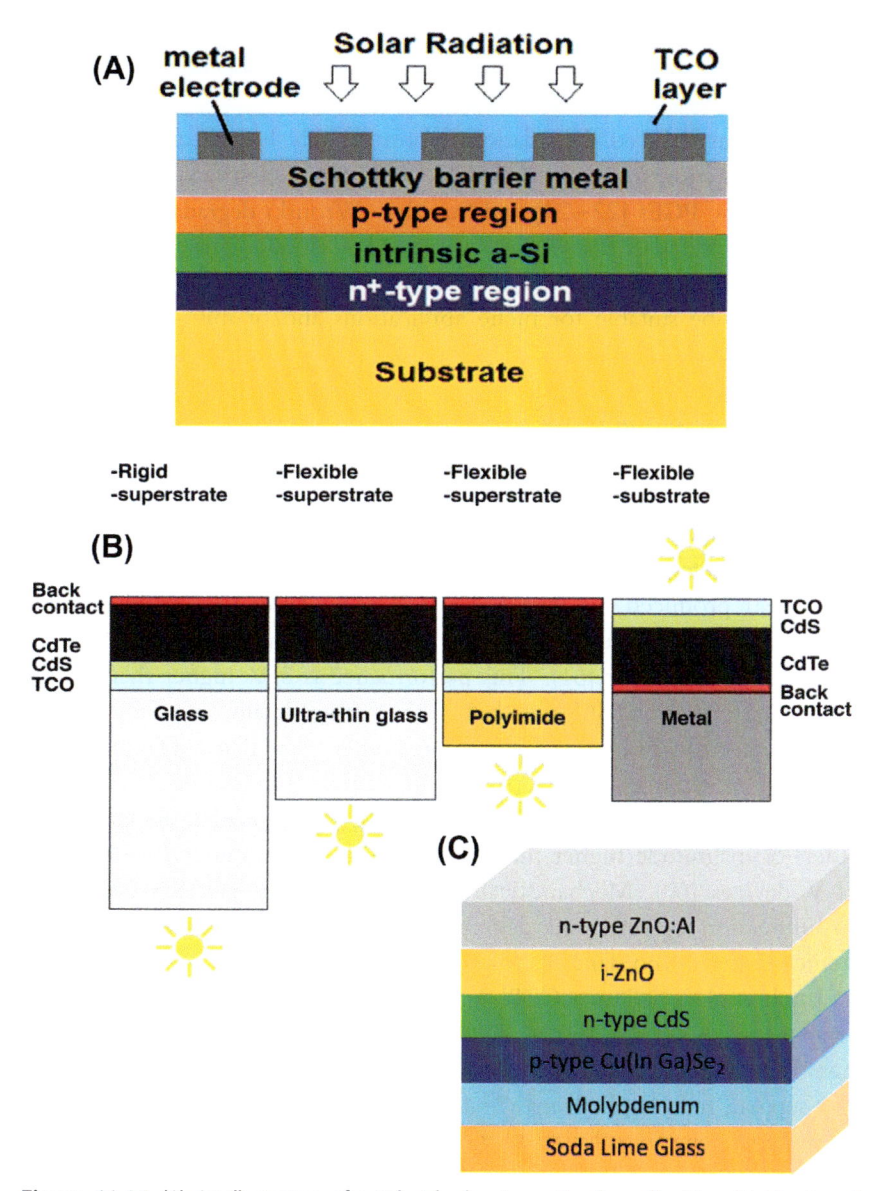

Figure 19.14 (A) An illustration of a Schottky barrier α-Si solar cell with a highly doped *p*-type region adjacent to the Schottky barrier high work function metal. (B) Schematic of superstrate and substrate CdTe devices (the substrates' width is not proportional but indicative of the differences between the devices). Both in substrate and superstrate configurations, light enters through wide bandgap window material (CdS). In the substrate structure, opaque substrates are used, and CdS need not undergo high-temperature processing. (C) Structure of a CIGS solar cell. *(A and C) Reproduced with permission from T.D. Lee, A.U. Ebong, A review of thin film solar cell technologies and challenges, Renew. Sust. Energ. Rev. 70 (April 2017) 1286–1297, copyright (2017) Elsevier; (B) Reproduced with permission from J. Ramanujam, D.M. Bishop, T.K. Todorov, O. Gunawan, J. Rath, R. Nekovei, E. Artegiani, A. Romeo, Flexible CIGS, CdTe and a-Si:H based thin film solar cells: A review, Prog. Mater. Sci. 110 (2020) 100619, copyright (2020) Elsevier.*

layers, proton-induced degradation is considerably lower for a-Si:H and nc-Si:H cells compared with c-Si devices [57,58]. Other factors, however, including low PCEs [22–24] and Staebler–Wronski degradation exclude thin-film Si materials from being viable candidates for missions requiring significant power [59]. As discussed below (Section 19.4.4), thin-film Si devices may be suitable for niche applications and/or missions requiring high MSP arrays.

19.3.2.2 Cadmium telluride

As an absorber layer, CdTe is a thermodynamically stable, strongly absorbing direct bandgap material, with a near-ideal 1.44 eV bandgap. Decades of research have resulted in the evolution of the device from a 4.1% CdTe/CdS heterojunction [54–56,60] to a >22% efficient device (19% large area) produced by First Solar [60]. The latter was enabled by a graded CdTe/CdSe$_x$Te$_{1-x}$ junction, and a high-transparency magnesium zinc oxide buffer layer [61]. This performance is 35% higher than a flex glass superstrate device [62]; see below for further details.

Both electron and proton irradiation studies show that CdTe solar cells are stable under high-energy particle irradiation. Romeo et al. observed degradation of CdTe/CdS heterojunction devices at particle fluences two orders of magnitude higher than those used to test monocrystalline Si or III-V devices [60]. Mechanistically, the high-energy particles generated recombination centers, and devices showed performance recovery that predicted little or no damage in space applications [63].

Lamb et al. investigated CdTe in a simulated space radiation environment using superstrate CdTe devices directly deposited onto cerium-doped cover glass. Deposition on the cover glass eliminated the need for an independent solid support, reducing both mass and cost [64]. Their work demonstrated that CdTe had a superior radiation hardness to protons compared to conventional MJ III-V solar cells. Device efficiency decreased by 5%, 18%, and 96% after exposure to fluences of 1×10^{12} cm^{-2}, 1×10^{13} cm^{-2}, and 1×10^{14} cm^{-2}, respectively. The lowest dose is estimated to represent a 20-year geostationary (or geosynchronous) Earth orbit (GEO) mission. The mechanism for the degradation of performance is proposed as the formation of interstitial hydrogen from proton irradiation, resulting in a shallow donor level and significantly decreased current. Annealing of the highest dose sample (100°C for 168 h in an inert environment) returned the device to 73% of its initial performance. Exposure to high temperatures in the space environment would provide conditions similar to those in the annealing experiment, allowing performance recovery.

In 2021 Lamb et al. detailed the performance of CdTe devices during 3 years in low Earth orbit onboard the AlSat-1N (3U) CubeSat, from 2016 to 2019 [32]. This work included the first $I-V$ curve data from CdTe solar cells in space and supported the predictions made from the laboratory studies [63,64]. Similar to the laboratory proton irradiation studies outlined above [64], TFSCs were deposited directly onto cerium-doped aluminosilicate glass in the following sequence: aluminum zinc oxide as the TCO, a ZnO buffer layer, *n*-type CdZnS, and arsenic-doped CdTe *p*-type layers, followed by a chlorine heat treatment to passivate grain boundaries, and evaporation of the Au back contact. Notably, there were no signs of delamination despite the thermal fluctuations between $-3.8°C$ and $51°C$. Photographs of the TFSC device architecture, the device, and experimental payload of the AlSat-1N (3U) CubeSat mission are included in Fig. 19.15 [32].

Current–voltage $(I-V)$ measurements were collected over 17,000 orbits. Over this time, the cells retained stable short circuit current and series resistance measurements, consistent with a stable front contact (aluminum zinc oxide for these devices). The V_{oc} values increased over time, which was attributed to lower overall cell temperatures in space, and light soaking effects (Fig. 19.16). Degradation of the fill factor (FF) values was observed due to a decrease in shunt resistance. The authors conjectured that the FF degradation was a result of diffusion of gold from the back contacts through the device, which could result in deep trap states in the CdTe absorber and micro-shunts between the contacts [32].

To summarize, laboratory [63,64] and GEO [32] data have shown that CdTe devices are suitable for GEO space applications. Notably, the device design incorporated materials unique to space applications. Given the current world record conversion efficiency for First Solar's thin-film CdTe cell of 22.1% (AM1.5) [60], it seems reasonable to target a CdTe solar cell for space applications that is radiation and thermally stable with 20% AM0 efficiency, a specific power of >1.5 kW/kg and a significantly lower production cost than SOA III-V MJ technology.

19.3.2.3 Copper indium gallium selenide and related materials

Along with CdTe, $CuIn_{1-x}Ga_xSe_2$ (CIGS) is the most studied chalcogenide TFSC technology. A quaternary material whose properties depend on the elemental ratios, it is commonly deposited using vacuum coevaporation of the individual constituents on a heated substrate ($400°C-600°C$) [56,65,66]. Alternate deposition methods also incorporate high-temperature steps to optimize device performance [50,67]. Table 19.2

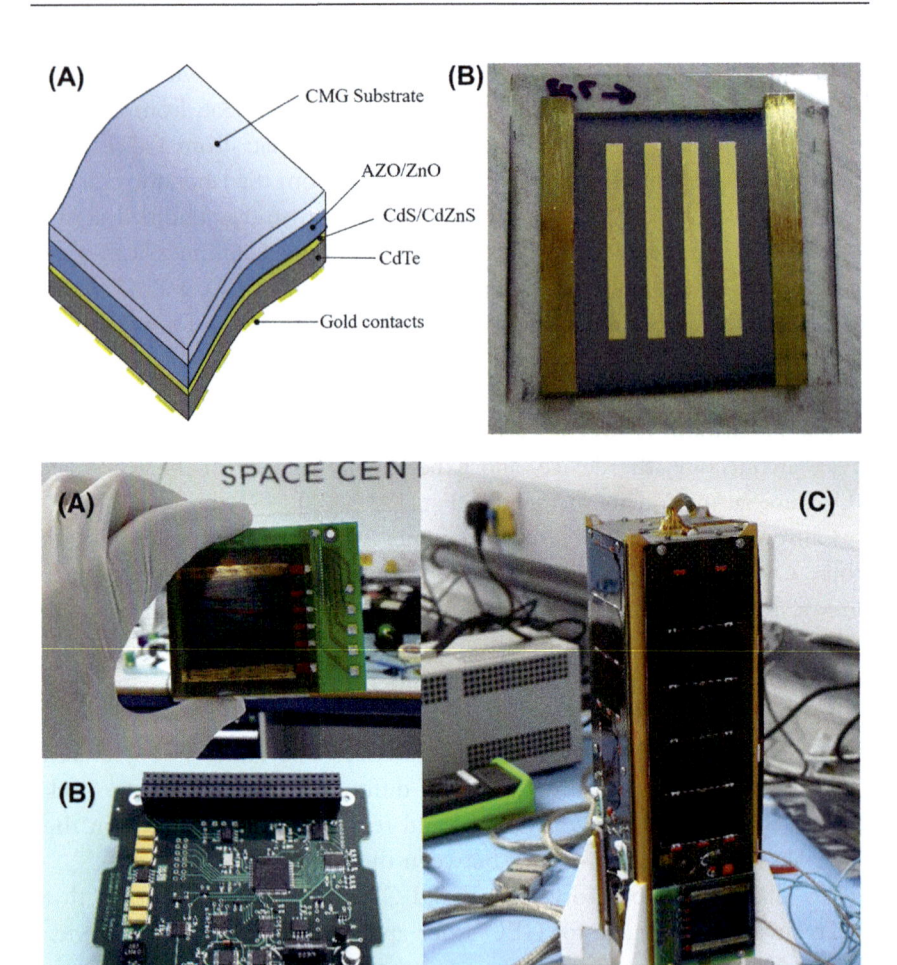

Figure 19.15 Photographs of the thin-film solar cells (TFSCs) and experiment payload of the AlSat-1N (3U) CubeSat mission: *Top*: (A) The CdTe device architecture on 100-micron cover glass; (B) A CdTe device deposited on to 60 × 60-mm 100-micron-thick cover glass; 4 × 1 cm² cells defined by gold contacts to the CdTe with two common contacts to the transparent conducting electrode. *Bottom*: (A) External PCB (printed circuit board) of the encapsulated TFSC payload. Gold contacts for the four cells and two common bus bars can be seen on the top and bottom of the glass. These are connected to the PCB electronic circuit via indium/tin solder and gold wires; (B) Internal PCB that, when commanded automatically, measures the four cells and the LM35 temperature sensor; (C) Assembled 3U CubeSat showing the face with the TFSC payload experiment attached alongside the conventional multijunction

(*Continued*)

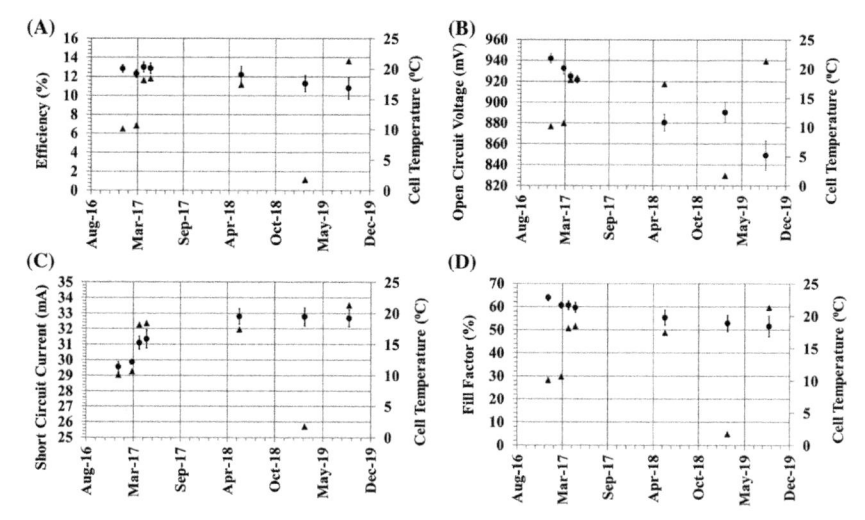

Figure 19.16 The mean (circles) and standard deviation (lines) $I-V$ parameters for surveys receiving solar flux above 115 mW/cm^2 over the duration of the mission. Cell temperature (triangles) for each survey is shown on the secondary y-axis. (A) Mean and standard deviation efficiency (%) for the four cells. (B) Mean and standard deviation V_{oc} (mV) for the four cells. (C) Intensity corrected mean and standard deviation I_{sc} (mA) for the four cells. (D) Mean and standard deviation fill factor (%) for the four cells. *Reproduced from D.A. Lamb, S.J.C. Irvine, M.A. Baker, C.I. Underwood, S. Mardhani, Thin film cadmium telluride solar cells on ultra-thin glass in low Earth orbit—3 years of performance data on the AlSat-1N CubeSat mission, Prog. Photovolt. Res. Appl. 29 (2021) 1000–1007, distributed under the terms of the Creative Commons (CC-BY 4.0) license (http://creativecommons.org/licenses/by/4.0/), which permits unrestricted use.*

provides representative examples of CIGS and CdTe devices on flexible substrates and superstrates, including PCEs and related metrics; devices on glass substrates, as well as several other TFSC technologies, are included for comparison [23,56,60,62,66,68–78]. The table includes examples of devices on glass that hold terrestrial efficiency records; for example, a polycrystalline CIGS thin–film absorber has now exceeded 23% (AM1.5) efficiency [66,68]. Solar Frontier, formerly a leading Japanese CIS company, in partnership with Japan's National Research and Development

solar cells that provide power to the spacecraft. *Reproduced from D.A. Lamb, S.J.C. Irvine, M.A. Baker, C.I. Underwood, S. Mardhani, Thin film cadmium telluride solar cells on ultra-thin glass in low Earth orbit—3 years of performance data on the AlSat-1N CubeSat mission, Prog. Photovolt. Res. Appl. 29 (2021) 1000–1007, this article is distributed under the terms of the Creative Commons (CC-BY 4.0) license (http://creativecommons.org/licenses/by/4.0/), which permits unrestricted use.*

Table 19.2 Representative CIGS and CdTe devices on flexible substrates (or superstrates[a]).

Material	Substrate or superstrate[a]	Eff. (h) (%)	V_{oc} (mV)	J_{sc} (mA/cm^2)	FF (%)	Year	Alkali metals/buffer layer	Ref.
CIGS	Glass	22.6	741	37.8	80.6	2016	KF + RbF	[68]
		22.9	746	38.5	79.7	2019	Cs	[66]
	Polyimide	20.4	736	35.1	78.9	2013	KF	[69]
		20.8	734	36.7	77.2	2019	RbF	[70]
	Stainless steel	17.5	601	40	72.5	2020	—	[71]
		18.0	692	34.5	75.5	2018	RbF + Ni/Cr	[71]
CdTe	Glass[a]	22.1	887	31.7	78.5	2015	—	[60]
	Flex glass[a]	16.4	831	25.5	77.4	2015	—	[62]
	Polyimide[a]	13.6	846	22.3	73.4	2012	—	[72]
	Mo foil	11.5	821	21.8	63.9	2013	—	[73]
a-Si	Glass	10.2	896	16.4	69.8	2015	—	[74]
Perovskite	Glass	24.8[b]	1060	26.0	75.1	2019	—	[75]
Organic	Glass	15.24	846.7	24.24	74.25	2021	—	[76]
DSSC	Glass	12.5	860	23.2	76.1	2020	—	[77]

Note: Efficiencies on glass substrates and others are included for comparison.
Source: Adapted from: J. Ramanujam, D.M. Bishop, T.K. Todorov, O. Gunawan, J. Rath, R. Nekovei, E. Artegiani, A. Romeo, Flexible CIGS, CdTe and a-Si:H based thin film solar cells: A review, Prog. Mater. Sci. 110 (2020) 100619 and other literature sources.
[a]Superstrate.
[b]Small area device aged 500 hours—large area (1 cm^2) device as prepared = 22.3%.

Agency's New Energy and Industrial Technology Development Organization announced a new record of 23.35% in 2019 [79], an increase of 0.35% over their previous record [66]. The use of metal foils and polymer substrates results in lower absolute PCEs than rigid substrates, but they are necessary to enhance mass-specific power, which is a critical metric for space PV systems [1,11,22,26,31,39,40,48—50,67,70—73,78]. CIGS industrial manufacturing is largely defunct as of the writing of this chapter [80,81], although it remains a topic of research at universities and other laboratories, such as a recent Belgian start-up [82]. At this time, it is unable to compete with the price points of silicon and CdTe technologies on a large scale.

The majority of devices included in Table 19.2 are on the order of 1 cm^2 active area. Obviously, small area cells (some much less than 1 cm^2), are optimal for breaking efficiency records [23]. However, this is not a practical solution for providing PV power for space applications. Further, AM0 efficiencies are on the order of 15%—20% lower as discussed above. In general, the typical efficiencies of III-V MJ modules or mini-modules are in the range of 30%—35%, and single-crystal Si and GaAs are in the 25%

efficiency range [23]. Perovskite tandem devices have also been demonstrated at over 30% efficiency (AM1.5), while perovskites, mc-Si, CdTe, and CIGS have demonstrated practical efficiencies in the range of 20%; finally, organics, tandem a-Si, and DSSC devices lag behind in the lower double-digit and upper single digit efficiencies [23]. However, as discussed above and below, efficiency at the cell level is only the beginning of the long road to adoption as a primary power source for space missions.

TFSCs are made of graded layers of amorphous and/or polycrystalline materials, ranging in thickness from nanometers to micrometers. During hot phases of growth and under illumination, heat, and current, diffusion between layers changes the composition and properties of the films, often leading to changes in performance over time. Space applications have unique requirements, and the changes in the processes induced by these requirements initially resulted in significantly lower PCEs. For example, growth on metal substrates with a process that is optimized for glass can lead to films with different microstructures and contamination from the metal substrate (e.g., Fe from stainless steel), leading to lower efficiency. This is consistent with the lower efficiency of CIGS and CdTe devices on metal foils, as shown in Table 19.2. Additionally, the devices need to be qualified for stressors unique to space. A research program that investigated CIGS deposited on titanium foil observed that thermal cycling consistent with a space environment resulted in delamination from the substrate at the interconnects [46].

A recent study by Sellers and coworkers examined the impact of LILT conditions near Mars and the outer planets Jupiter and Saturn on commercial CIGS (on stainless steel) devices [52]. They selected conditions consistent with variations in heating, bias, and/or illumination in two states of operation, *relaxed* and *metastable* states [52,83]. The relaxed state corresponds to a solar cell in the dark, at temperatures around 330 K for extended periods; in contrast, the metastable state is achieved by illuminating the cell, typically for 1 h with 1 sun power (AM0 or AM1.5) at room temperature [84]. The results are consistent with illumination saturating defect states at the interface, lowering the barrier to carrier transport, and reducing the corresponding resistance. At temperatures above 200 K, there is no difference between the performance of cells in the relaxed or metastable states. The metastability of the CIGS results in a barrier to minority carrier extraction. The presence of this barrier does not affect the performance in LILT conditions as CIGS solar cells operate with a higher PCE.

To assess the performance of flexible CIGS modules in the harsh radiation environments of deep space (particularly Jupiter), the effects of proton irradiation were examined to evaluate whether the previous predictions of higher radiation tolerance in CIGS [85] are transferred to LILT environments in flexible systems. Unencapsulated samples were irradiated with high-energy (1.5 MeV) protons at varying fluences to determine the solar cell regions most susceptible to damage. Fig. 19.17 shows a comparison of the light J-V measurements for the reference and irradiated CIGS solar cells under the LILT condition related to the planets of interest [52]. The solar cells were irradiated with 1.5 MeV protons at fluences of 1×10^{11}, 5×10^{11}, 1×10^{12}, and 1×10^{13} (H^+/cm^2) for samples designated as C, D, E, and F, respectively. Proton irradiation of the CIGS reduces the performance (Fig. 19.16) via defects. These results suggest

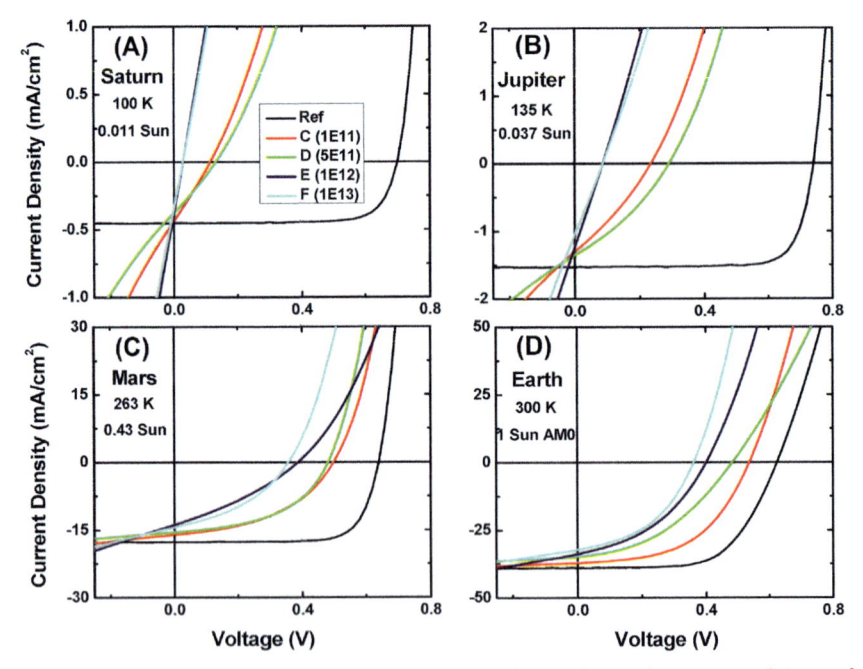

Figure 19.17 Light $J-V$ of reference and irradiated samples under LILT conditions of the planets. Samples C, D, E, and F are irradiated with 1.5 MeV protons and fluences of 1×10^{11}, 5×10^{11}, 1×10^{12}, and 1×10^{13} (H^+/cm^2), respectively. *Reproduced with permission from H. Afshari, B.K. Durant, C.R. Brown, K. Hossain, D. Poplavskyy, B. Rout, I.R. Sellers, The role of metastability and concentration on the performance of CIGS solar cells under Low-Intensity-Low-Temperature conditions, Solar Energy Mater. Solar Cells 212 (August 2020) 110571, copyright (2020) Elsevier.*

while CIGS performs well in LILT conditions, appropriate encapsulation would be required for practical applications in higher radiation environments such as the moons of Jupiter [31].

The Gossamer solar power array (GoSolAr) mission [86] by the DLR aims to demonstrate the reliability of TFSC technologies in powering spacecraft with applications that require high power such as space tugs and solar electric propulsion (SEP) [87,88]. The DLR's concept for a GoSolAr using TFSCs on the small satellite technology experimental platform (S^2TEP), a scalable microsatellite class (10−100 kg) bus, is currently under development (Fig. 19.18) [86]. The current focus of the GoSolAr program is to employ CIGS TFSCs on polyamide substrates. Currently, an assessment of available CIGS PV technologies is ongoing; two commercial manufacturers, one located in Europe and the other in the United States, are leading supplier candidates. Both have efficiencies of around 10% and

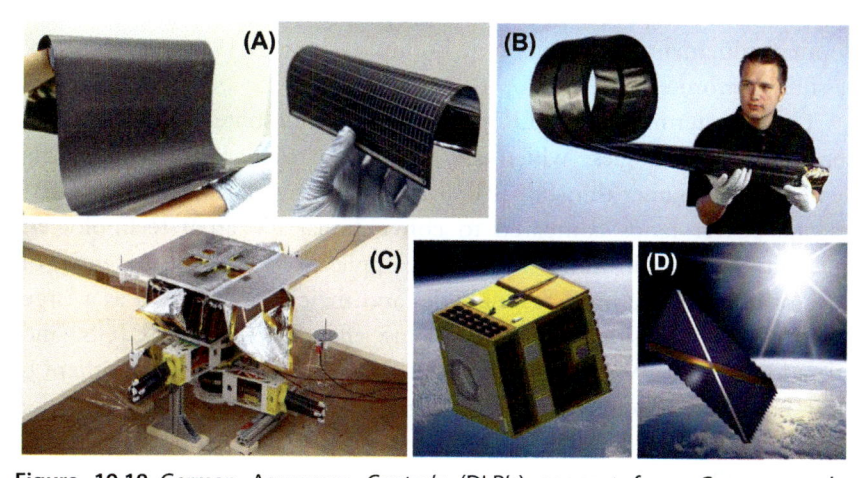

Figure 19.18 German Aerospace Center's (DLR's) concept for a Gossamer solar power array (GoSolAr) using thin-film photovoltaics on the small satellite technology experimental platform (S^2TEP), a scalable microsatellite class (10−100 kg) bus, currently under development at DLR (A) Flexible CIGS modules; *Left*: Flisom; *Right*: Ascent Solar. (B) DLR's tubular, reelable boom in partially deployed state. (C) Deployment test breadboard setup (two-boom deployment units with two booms each and one PV blanket stowed in storage box). (D) Artists view of GoSolAr on S^2TEP (*left*) and deployed GoSolAr demonstrator array (*right*). *Reproduced from T. Sproewitz, U. Banik, J.-T. Grundmann, F. Haack, M. Hillebrandt, H. Martens, S. Meyer, et al., Concept for a Gossamer solar power array using thin-film photovoltaics. CEAS Space J. 12 (2020) 125−135, article is distributed under the terms of the Creative Commons (CC-BY 4.0) license (http://creativecommons.org/licenses/by/4.0/), which permits unrestricted use.*

both manufacturers have good prospects for efficiency increase in the standard production process in the near future. While the use of thin GaAs technology is an alternative option, CIS on polymer has several inherent advantages. They are truly flexible down to a 25 mm roll radius (Fig. 19.18A), have a high mass-specific power (>0.75 kW/kg), and are relatively low cost (US$25/W) [86].

The study provides an overview of the concept: a large, lightweight, deployable Gossamer PV array based on thin-film PV, DLR's coilable carbon fiber reinforced plastic booms (Fig. 19.18B), and a two-dimensional array deployment (Fig. 19.18D) [86]. Together with the PV selection, the investigation of concepts to keep the PV within operating temperature ranges is ongoing, including layering the blanket, or using high-emitting coatings for thermal management [35]. One of the main practical issues is solid contact between PV and harness, and in between the harness itself, which is currently under investigation as a high priority. The first deployment tests with a prototype consisting of a PV blanket dummy and two boom deployment units with two booms each (Fig. 19.18C) were successfully performed [86].

The long-term goal of DLR's effort is to demonstrate several key concepts and technologies, which are as follows: the applicability of the deployable Gossamer thin-film PV array system; significantly higher mass-specific power ratios compared to conventional PV array technologies; suitability of the thin-film PV for space applications; and the potential of the scalability of the main technologies and their combined use as a large system. When considering the increasing efficiencies of the CIGS (and related thin-film) PV systems combined with Gossamer structures, there is clear potential of achieving a very high specific power value (1 kW/kg and beyond) exceeding that of conventional PV systems. Furthermore, the CIGS PV appears to be more radiation resistant and has already surpassed 23% efficiency in laboratories. Such efficiencies are expected to be achieved in the near future in a standard manufacturing process. However, flexible, thin-film GaAs cells are also under consideration for inclusion within GoSolAr [86].

19.3.3 Perovskite and organic photovoltaics

While this subsection will briefly consider organic and dye-sensitized solar cells, we will focus on PSCs and their suitability for use in space, due to their superior PCE and the level of interest and suitability of these materials. In a study presented nearly 10 years ago, the NASA Glenn Research Center (GRC) team made an initial assessment that cell technologies that contained a significant fraction of essential organic moieties were more suitable for terrestrial applications, while inorganic devices are more suitable for use in space exploration [89]; PSCs (both inorganic and hybrid) may be considered as an inorganic material for potential space applications.

Several recent studies have documented the significant loss of performance of dye-sensitized solar cells upon radiation [90−92]. The situation for OPV appears to be a bit more encouraging albeit preliminary and not comparable in scope and thoroughness with radiation damage studies for inorganic PV devices [1,10,12,22,25,26,51]. A recently published theoretical study assessed the effects of ionizing radiation on an OPV device (P3HT: PCBM) for total accumulated doses up to 1k Gy (SiO$_2$). The primary effect of irradiation on the OPV cell was to decrease the apparent V_{oc} [93]. The authors concluded that OPV would survive in space environments for 1-year missions under ISS orbit conditions, up to the commercial test standard (100 Gy (SiO$_2$)) [93].

A more practical test approach involved a high-altitude balloon flight experiment of PSC and OPV devices as a payload of a 35,000 m^3 stratospheric balloon, launched in October 2016 from the Esrange Space Center in northern Sweden [94]. The flight duration was limited to 5 hours, of which more than three were resident in the stratosphere, reaching an altitude of 32 km (20 miles) and tested both PSCs (32 MAPbI$_3$— not further discussed here) and 224 OPV. The primary issues on this flight would be temperature variations and ultraviolet radiation, related to the study of Sellers and coworkers discussed above (Section 19.3.2.3) [52]. The experiment remained at a temperature below 230 K for roughly 30 min, and reached temperatures in the range of 260−290 K within the following hour and throughout the phase of floatation at 32 km. Greater than 75% of the OPV were functional after flight, with an average 25% reduction to a slight increase in performance.

In 2020 an interesting report appeared on a rocket flight experiment with PSCs and OPV, reaching an altitude of 240 km (150 miles) [39]; the average altitude of the ISS is about 330 km (205 miles). The short flight experiment was launched on a suborbital sounding rocket from the Esrange Space Center in northern Sweden on the morning of June 13, 2019. The details surrounding the flight experiment, including cell deployment and measurements, are quite complex and are available in the original literature [39]. Therefore we will address only important details for discussing key results and lessons learned. The main measurement time (total of 6 minutes) began a minute after lift-off (LO) at an altitude above 70 km. The payload reached apogee at 240 km (150 miles) altitude after 251 s, and reentry mode detection occurred after 7 minutes at an altitude of 100 km (60 miles). The illumination states are distinguishable into three different main phases of incident illumination onto the solar cells. I, intense solar illumination from one side; II, no direct solar illumination onto the solar cells, reflected solar radiation from Earth's surface; III, intense solar illumination from a side opposite to phase I. For protection during ascent, each hatch was covered with a fused silica window that transmitted more than 90% of the light in the visible and UV wavelength range of the AM0 solar spectrum.

Four solar cell device types and in-flight champion performance $J-V$ plots of representative SOA single-junction PCSs and OPV are illustrated in Fig. 19.19 [39]: (1) mixed organic lead mixed halide perovskite of $n-i-p$ solar cells in planar (SnO_2) (Fig. 19.19A) and (2) mesoscopic (m-TiO_2) (Fig. 19.19C) architectures, both with a PCE of approximately 20% [23,24]; inverted organic bulk-heterojunctions of (3) narrow bandgap polymer:fullerene PTB7-Th:PC71BM type (Fig. 19.19B) and (4) nonfullerene PBDB-T: ITIC (Fig. 19.19D) with reported PCEs of approximately 8.4% [23,24].

During the active measurement time of 6 minutes, current-voltage characteristics were continuously recorded. Under strong solar irradiance (phases I and III), the solar cells performed efficiently; they also produced power with weak diffuse light (phase II) reflected from Earth's surface (albedo). This conforms with earlier flight data from an integrated power experiment (see Section 19.4.3.1). The measured temperatures ranged from 30°C to 60°C during the measurement time, and the temperature variation likely had little to no influence on the solar cell performance during this suborbital experiment. Solar cell damage due to radiation is also not considered to play a significant role in this short-term experiment. All four solar cell types demonstrated typical diode curves with a vertical offset due to the strong illumination; they also generated power in orbital altitudes under stable irradiance conditions under space conditions. During phases I and III (strong solar irradiation), PSCs and OSCs showed considerably efficient performance, exceeding power densities of 14 and 7 mW/cm^2, respectively. Both perovskite device types matched power densities measured in preflight characterization when the difference in solar spectra is taken into account (see Section 19.2.1). The highest measured J_{sc} exceeded 20 mA/cm^2 for TiO_2

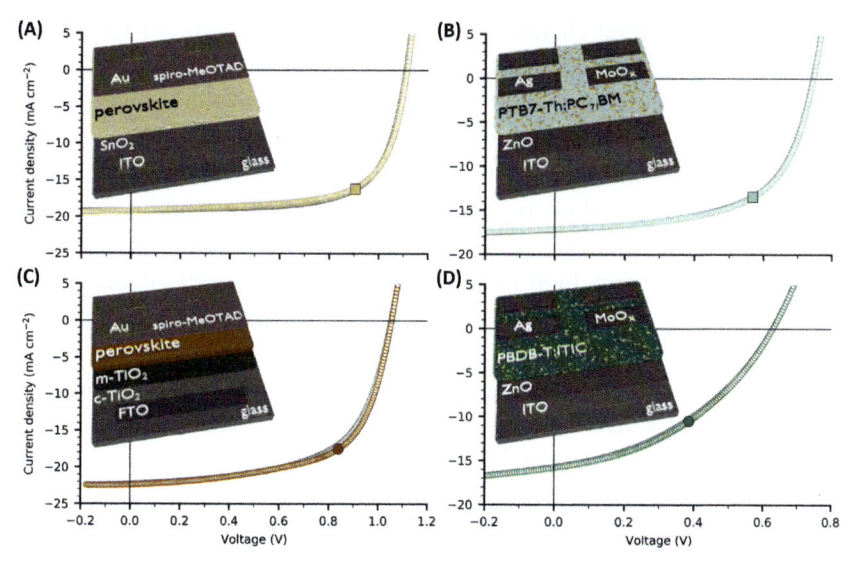

Figure 19.19 In-flight champion performance $J-V$ measurement cycles: (A) Planar perovskite with SnO_2 electron-transport layer, measurement begins 67.4 s after liftoff (LO); (B) Fullerene organic cell, measurement begins 87.9 s after LO; (C) Mesoscopic perovskite with TiO_2 electron-transport layer, measurement begins 67.4 s after LO; (D) Nonfullerene organic cell, measurement begins 353.5 s after LO. The voltage sweep forward is colored and backward grayed out; the maximum power points are indicated by marks that correspond to the respective segments. *Reprinted with permission from L.K. Reb, M. Böhmer, B. Predeschly, S. Grott, C.L. Weindl, et al., Perovskite and organic solar cells on a rocket flight, Joule 4 (9) (2020) 1880−1892, copyright (2020) Elsevier.*

perovskite cells (Fig. 19.19A); however, the planar PSC device (SnO_2) compensated with a better FF and higher V_{oc} (Fig. 19.19C). The nonfullerene OSC incurred a 25% decrease in performance, comparable to the balloon study above [94]. While a flight experiment of such short duration is hardly conclusive, several useful conclusions can be drawn. The ability to function in low-intensity conditions is relevant to missions to the outer planets (Section 19.4.4) [42]. Perovskite devices performed better than in the balloon flight experiment above [91]. The MSP of PSCs and OPV would be superior to inorganic devices, assuming comparable long-term durability in relevant environments. Lessons learned from in situ testing can inform further experiments. As with the high-altitude balloon experiment, orbital flight experiments are an excellent learning experience for students.

A recent irradiation study by Sun and coworkers employed a 200 MeV proton beam; measured radiation-induced electronic absorption or electronic structural changes were negligible to insignificant for several measured (P3HT, P3HT:PC60BM, DBA, and DBfA) polymers and their composite thin films up to 800 Rads [95]. The relatively small radiation effects of conjugated polymers were also observed in studies of irradiation with electrons, X-rays, and gamma rays [96−98]. The observed stability was ascribed by the authors to delocalization and high mobility of π electrons along the conjugated polymer main chains that can easily dissipate the energies of certain high-energy radiations. The results of these four studies [95−98] imply that conjugated polymer-based electronic and optoelectronic devices have the potential to be durable in space radiation environments. This is consistent with the many studies that demonstrate that thin-film cells have advantages in space applications because particles have enough energy to transit through the active area with negligible energy loss [26,27,32,42,48,50,57−60,63,64,83,85].

While theoretical studies and ground tests provide some initial guidance and allow for informed next steps, there is no substitute for flight experiments and real data. For example, the Materials International Space Station Experiment (MISSE) project, first launched in 2001, involves a series of spaceflight missions with experiments flown on the exterior of the ISS to test the performance and durability of materials and devices exposed to the LEO space environment [99−101]. Comparison of ground AO tests with actual space flight data resulted in large variations in predicted versus actual results by a factor of 1−37 depending upon the material(s). Utilization of specialized test facilities such as those located at NASA GRC [102] can improve the validity of the space simulation, but there is no substitute for flight experiments, even those involving CubeSats, suborbital flights, or high-altitude balloons [32,39,94].

With this caveat in mind, a recent study by Kirmani et al. proposed a framework for performing radiation-testing experiments on PSCs; the authors' overall approach is illustrated in Fig. 19.20 [103]. Proton fluence trends demonstrate that low-energy protons (approximately 0.05 MeV) have an order of magnitude higher fluence than high-energy protons ($>$50 MeV) for a majority of Earth orbits. Fig. 19.20D illustrates that electrons have a relatively lower energy upper limit ($<$10 MeV) for Earth orbits; electron fluence falls steeply for energies $>$1 MeV [103]. The ISS orbit is relatively benign, having a minimal radiation threat, given its approximately 420 km altitude above Earth. The orbit of the near-Jupiter Juno mission is of particular interest for radiation-tolerant materials; extremely energetic electrons and protons (approximately 1 GeV) are

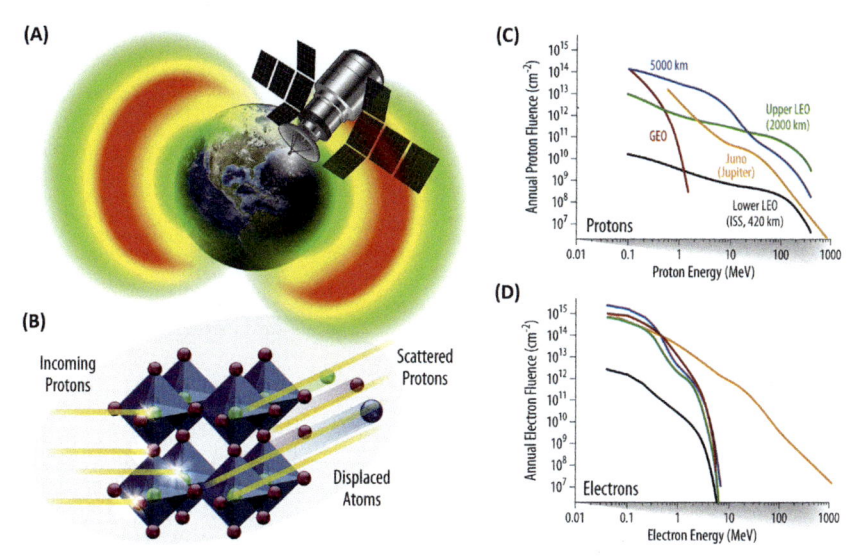

Figure 19.20 Radiation effects in space orbits: (A) artist conception of a satellite powered by perovskite solar panels in an Earth orbit and the surrounding radiation field; (B) schematic of nuclear displacements caused by protons resulting in atomic vacancies; (C) simulated annual fluences from the Space Environment Information System (SPENVIS) as a function of proton energy; (D) electron energies for ISS (black), upper-LEO (green), GEO (red), 5000 km (blue), and Juno (yellow) orbits. *Note*: (C) and (D) show integral fluences; see: https://www.spenvis.oma.be/ for differential spectra. *Reproduced with permission from A.R. Kirmani, B.K. Durant, J. Grandidier, N.M. Haegel, M.D. Kelzenberg, Y.M. Lao, M.D. McGehee, et al., Countdown to perovskite space launch: Guidelines to performing relevant radiation-hardness experiments, Joule 6 (5) (2022) 1015–1031, copyright (2022) Elsevier.*

present, albeit at very low fluences. For missions to the inner planets, including a Venus flyby, cells would need to function at high temperatures [31,42]. As interplanetary missions are expected to involve several years of flight and exploration, high-energy particles are expected to make a significant impact on performance, posing unique challenges to solar panels [22,26,29–31,42].

Prior proton irradiation studies summarized in this study [103] are reproduced in Table 19.3 [104–110]. Although early demonstrations have been used to promote the radiation-tolerant aspect of perovskite PV, the relatively low PCEs of solar cells employed may undermine this claim. This is because low-PCE solar cells likely already have a significant concentration of defect states, and the interaction of charged particles with defect-ridden perovskites can result in defect healing, masking the true effect of radiation damage and

Table 19.3 Summary of literature reports on proton irradiation of perovskite solar cells.

Absorber & substrate	Facility & flux (cm^{-2}/s)	Irrad. direction	Perovskite bandgap (eV)	Proton energy (MeV)*	Fluence (cm^{-2})	PCE$_0$/PCE$_{irrad}$	Year	Ref.
FAPI (p-i-n) quartz	Aerospace Corporation, US	Top metal	1.47	0.050[a]	1E12	12.3/12.5	2017	[105]
Triple-cation (n-i-p) quartz	Wakasa Wan Energy Research Japan; 3E11	Top metal	1.62	0.050[a]	1E12	4.4/4.4	2018	[106]
					1E13	4.4/4.4		
					1E14	4.4/3.3		
					1E15	4.4/2.4		
MAPI (p-i-n) PET	National Institute for Quantum and Radiological Science and Technology (QST), Japan; 3.5E9	Top metal	1.54	0.100[a]	3E10	11/11	2020	[107]
					3E12	11/8.8		
Triple-cation (n-i-p) quartz	Surrey Ion Beam Centre, UK; 3E11	top metal	1.62	0.150[a]	1E13	15/12	2019	[108]
					1E14	15/3		
					1E15	15/0.0		
Triple-cation (p-i-n) quartz	Helmholtz-Zentrum Berlin; 9E8	Substrate	1.62	20[b]	1E12	17.0/17.0	2019	[110]
				68[b]	1E12	18.8/17.9		
Triple-cation/CIGS (Tandem)	Helmholtz-Zentrum Berlin; 7E8	top IZO	1.62	68[b]	2E12	18/14.9	2020	[111]
Triple-cation/Si (Tandem)					2E12	21.1/0.18		
Triple-halide (p-i-n) quartz	University of North Texas; E9–7E10	top ITO	1.70	0.050[a]	1E12	8.5/6.3	2021	[111]
				0.085[a]	1E12	8.4/8.5		
				0.300[b]	1E12	9.1/10.1		
				0.650[b]	1E12	8.7/9.4		
				1.500[b]	1E12	7.9/8.5		
				2.500[b]	1E12	8.3/8.4		

Source: Reproduced with permission from A.R. Kirmani, B.K. Durant, J. Grandidier, N.M. Haegel, M.D. Kelzenberg, Y.M. Lao, M.D. McGehee, et al., Countdown to perovskite space launch: Guidelines to performing relevant radiation-hardness experiments, Joule 6 (5) (2022) 1015–1031, copyright (2022) Elsevier.

*Proton energies are marked with a symbol footnote ([a]Recommended or [b]Not recommended) for assessing radiation hardness (as discussed in the text). Perovskite absorbers are approximately 500 nm thick. Irradiation is considered in a vacuum of approximately 1E-5–1E-6 mbar, and the cells have been irradiated for approximately 1–100 s depending on the target proton fluence.

dosage on the materials and device parameters. This same concern is relevant for other lower-efficiency devices such as OPV and inorganic thin-film devices [22–24,26,32,39,41,48–50,52,60,83,84,89,94–98]. Ideally, stable, high-PCE devices should be used for investigating the radiation hardness of PSCs. Recently, single- and MJ solar cells involving perovskites and PCEs >15% have been utilized for some studies; see Table 19.3. However, as pointed out by Kirmani et al. [103], some studies employ proton energy ranges not representative of the space environment [108,110].

Interestingly, the authors also simulated and compared the vacancy density of perovskites with current crystalline PV technologies, focusing on proton interactions and subsequent vacancy creation; the following hypothetical device architecture was employed: Au (100 nm)/Spiro (100 nm)/absorber (250 nm)/SnO$_2$ (50 nm)/ITO (150 nm)/glass (70 nm). Simulations were carried out for three perovskite absorbers CsPbI$_2$Br, CH$_3$NH$_3$PbI$_3$, and (Cs$_{0.05}$(MA$_{0.17}$FA$_{0.83}$)$_{0.95}$Pb(I$_{0.83}$Br$_{0.17}$)$_3$ and two current crystalline PV materials with 0.05 MeV protons. Conventional device materials were compared directly using the same thickness and architecture for a materials comparison with perovskites, although it must be emphasized that Si and InGaP use significantly different device architectures and absorber thicknesses [1,4,10–13,23,41]; results are shown in Fig. 19.21 [103].

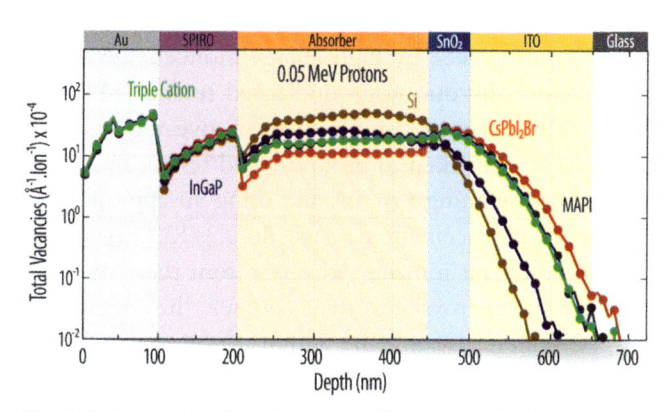

Figure 19.21 Defect creation in simulated solar cell technologies. Comparison of vacancies formed in hypothetical device stacks with perovskite absorbers (Cs$_{0.05}$(MA$_{0.17}$FA$_{0.83}$)$_{0.95}$Pb(I$_{0.83}$Br$_{0.17}$)$_3$, CH$_3$NH$_3$PbI$_3$, and CsPbI$_2$Br) and conventional PV absorbers (Si and InGaP), upon irradiation with 0.05 MeV protons and assuming identical thickness absorber. *Reproduced with permission from: A.R. Kirmani, B.K. Durant, J. Grandidier, N.M. Haegel, M.D. Kelzenberg, Y.M. Lao, M.D. McGehee, et al., Countdown to perovskite space launch: guidelines to performing relevant radiation-hardness experiments, Joule 6 (5) (2022) 1015–1031, copyright (2022) Elsevier.*

Silicon undergoes maximum displacements, resulting in the most vacancies due to the low displacement energy of Si as well as its low atomic number, followed by InGaP. In fact, this trend, based on simulations, agrees with experimentally observed PCE degradation of irradiated Si and InGaP solar cells [10,43−45,111−113]. Because active layer thicknesses in conventional solar cells can be several microns with Si PV having a 100-mm-thick absorber, the higher energy proton spectrum is "redshifted" to lower energies while passing through; this will further contribute to displacements and device damage. The greater thicknesses of conventional PV technologies result in greater radiation damage; however, there are continuing efforts to improve the performance and radiation damage resistance of MJ III-V PV arrays [1,10,12−14,43−45].

Low-energy proton fluences of approximately 10^{13} cm^{-2}, typically encountered for deep-space missions, [1,10,22,31,43] can affect Si, InGaP, and other III-V solar cells, if these panels are not heavily shielded, as has been previously shown [44,45]. Using encapsulation, such as a cover glass, is a way to mitigate these effects but increases solar array weight and reduces the specific power of the panels. Interactions of protons and other charged particles with III-V and Si solar cells have been extensively studied in the past [10,22,43]; Messenger et al. established protocols for ground-based testing of conventional cells [111−113]. It was found that an omnidirectional polyenergetic proton flux creates a uniform damage profile inside these devices that can be mimicked by high-energy monoenergetic (1−10 MeV) protons incident normally during ground-based testing [113]. Space radiation creates a uniform damage profile in the active regions of III-V solar cells that can be best mimicked in ground-based testing by irradiation with high-energy, fully penetrating protons and using an appropriate fluence to achieve appropriate displacement damage dose [113,114]. Recent studies indicate that the dominant ionizing processes from these protons may lead to healing effects in perovskites that counter the detrimental defects [109,115,116]. For these reasons, the authors propose that PSCs require a fresh set of radiation-testing conventions [103].

Kirmani et al. demonstrated that PSCs made with a approximately 500-nm-thick metal halide layer are most effectively tested with protons that have an energy of 0.05−0.15 MeV [103]. These protons create a space-representative uniform damage profile in the cells, producing approximately three orders of magnitude more damage than >10 MeV protons. If a cover glass is employed, the proton energy should be increased to maximize damage even after passing through the glass. Furthermore, the thinness of these devices, their ability to withstand exceptionally high defect densities, and their tendency to self-heal upon irradiation from high-energy particles set them apart from conventional PV materials that tend to be tested at higher energy or with more accessible electron sources. Considering particle fluences found in space orbits of interest and nonionizing energy loss (NIEL), it is suggested that proton irradiation be the focus of near-term experiments. The initial encouraging results for PSC use in space must be seen in the light of extensive testing and space flown track record required by mission planners and directors. This will form the basis of the discussion throughout the final section of this chapter.

19.3.4 Approaches to enhance device performance: Multijunction and concentrators

Given the ability to grow new III-V materials epitaxially or metamorphically [1,10−12,41,42], as outlined above (Section 19.3.1), space cell and array manufacturers now have an unprecedented ability to tailor MJ devices to succeed in specific illumination and environmental conditions. Conventional arrays have reached panel-level specific power of over 100 W/kg and pathways now exist for even higher specific power levels (150−500 W/kg), allowing manufacturers to meet the new demands for SEP systems [87,88], which can be used to explore the solar system and neighboring planets, or systems which may have harsh or challenging environments (i.e., low solar intensity and low temperature [LILT], high solar intensity and high temperature, and high-radiation exposure) [51]. In the future, space applications, including challenging missions, may also be enabled via other advanced materials systems that utilize lightweight substrates [39,40,48,50,56,94,116]; see discussions in Sections 19.3.2 and 19.3.3. The reader is also encouraged to consult further details in several chapters on TFSC and/or PSC materials in our recently published book on PV for space [117−119].

Another way to decrease array cost and/or enhance performance is by the use of concentrator arrays. Concentrator arrays use either refractive or reflective optics to direct concentrated sunlight onto a smaller active area of solar cells. Deep Space 1 (DS1), launched by NASA in October 1998, was the first spacecraft to rely upon concentrator arrays; see Fig. 19.22 [120]. It had two such arrays with each capable of producing 2.5 kW at 100 V (DC). The Solar Concentrator Array with Refractive Linear Element Technology (SCARLET) arrays was developed by AEC-ABLE Engineering, Inc. through support provided by the US Ballistic Missile Defense Organization. These arrays performed flawlessly in this inaugural demonstration.

Figure 19.22 (A) The Deep Space 1 satellite with SCARLET solar concentrator arrays in the stowed position and (B) the SCARLET array deployed. *Courtesy: NASA.*

Unfortunately, not all of the concentrator arrays used in the 1990s fared as well. The original Boeing Space Systems (BSS-702) bus used for communication satellites exhibited more rapid than expected degradation [121]. The problem was isolated to the concentrator reflector surfaces, which degraded after becoming coated with contaminants outgassing from the array because of its increased temperature. Although concentrator arrays are not currently in production for flight systems, concentrator technology previously developed for flight and technologies currently under development have potential importance for planetary exploration [42,51,122]. The concentration of sunlight on solar cells alleviates the performance losses associated with LILT operation at the outer planets because concentration increases the effective solar irradiance on the cells. A thorough, practical overview of Fresnel lenses for concentrator technologies for space is provided by O'Neill in a recently published book on space PV [122].

While device stability, environmental durability, and MSP are important considerations, efficiency projections are certainly a consideration going forward to assess the future viability of a device technology to serve as the primary power source. Fig. 19.23 summarizes the efficiency potential of single-junction and MJ solar cells, calculated by using a theoretical approach outlined by Yamaguchi [10,123,124], in comparison with experimentally realized efficiencies under 1-sun illumination. This model

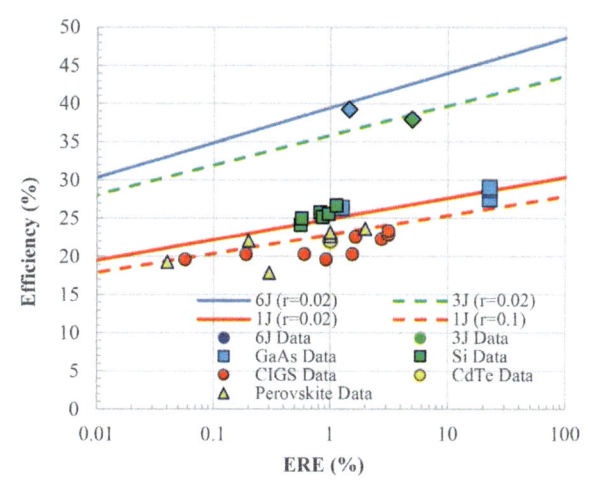

Figure 19.23 Calculated conversion efficiencies of various single-junction, three-junction, and six-junction solar cells, calculated by using a procedure as outlined in the text, in comparison with experimentally realized efficiencies under 1-sun illumination. *Reproduced from M. Yamaguchi, High-efficiency GaAs-based solar cells, in: M.M. Rahman, A.M. Asiri, A. Khan, I. Inamuddin, T. Tabbakh (Eds.), Post-Transition Metals, InTech Open, 2021. DOI:10.5772/intechopen.94365 under the terms of the Creative Commons Attribution 3.0 Unported license (CC BY 3.0).*

considers efficiency losses such as nonradiative recombination and resistance because conventional solar cells often have a minimal optical loss these assumptions are warranted.

The nonradiative recombination loss is characterized by external radiative efficiency (ERE), which is the ratio of radiatively recombined carriers against all recombined carriers; the ERE = 1 at the Shockley–Queisser limit [125]. Further analysis of EREs of several solar cell technologies can be found in several in-depth references [123,126–128]. Although single-junction solar cells have potential efficiencies of less than 32%, three-junction and six-junction solar cells have potential efficiencies of 42% and 46%, respectively (Fig. 19.23). In fact, four- and six-junction III-V cells under concentration have demonstrated efficiencies over 47.2% (see Fig. 19.6), with predicted efficiencies approaching 50% [10,129] (Fig. 19.24).

We have discussed several actual space applications of PV above but have mainly focused on background, history, and PV technologies. In the next section of the chapter, we will examine past, present, and future space missions [1,10,12,22,27,42,47], and the suitability of various PV technologies given the environmental challenges [31] of different regions

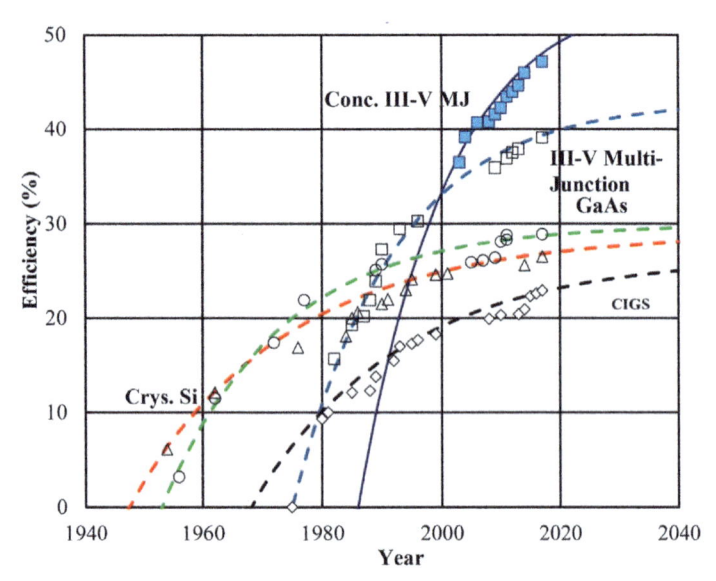

Figure 19.24 Historical record-efficiency of III-V multijunction (MJ) and concentrator MJ solar cells in comparison with 1-sun efficiencies of GaAs and crystalline Si solar cells, along with their extrapolations. *Reproduced from M. Yamaguchi, High-efficiency GaAs-based solar cells, in: M.M. Rahman, A.M. Asiri, A. Khan, I. Inamuddin, T. Tabbakh (Eds.), Post-Transition Metals, InTech Open, 2021. DOI:10.5772/intechopen.94365 under the terms of the Creative Commons Attribution 3.0 Unported license (CC BY 3.0).*

and localities in our solar system. Some mission-enabling improvements over SOA, but not all, can be achieved with solar cell and solar array optimization for the particular environment [31,42]. This is discussed in more detail below (Section 19.4.4). Clearly, III-V-related cells, modules, and arrays will dominate the space power generation landscape for the foreseeable future; however, there will likely be niche applications where other materials and technologies [26,32,35,39–41,46,48–57,60,63,64,67,83–89,93–95,103–110,115–120] will be enabling; we will examine those possibilities below (Section 19.4.4).

19.4 Space applications and exploration missions: Past, present, and future

This final section of the chapter loosely adheres to a chronological approach to the examination of space PV and the exploration of the solar system. The first subsection examines a variety of mission destinations, the

PV technology(ies) employed, and the solar cell, module, or array configurations for a cross-section of exemplary missions. As we have discussed throughout this chapter, the transition from Si to III-V device technology occurred during the 1980s while beginning-of-life (BOL) efficiency has essentially tripled during the 50 years from 1965 to 2015 (Fig. 19.25) [42]. Notable space missions and technology demonstrations are the main topics of the second subsection. The section ends with a summary and discussion of space PV and exploration of the solar system into the future. We conclude the section with an informed prediction of the potential for employment of new PV technologies going forward.

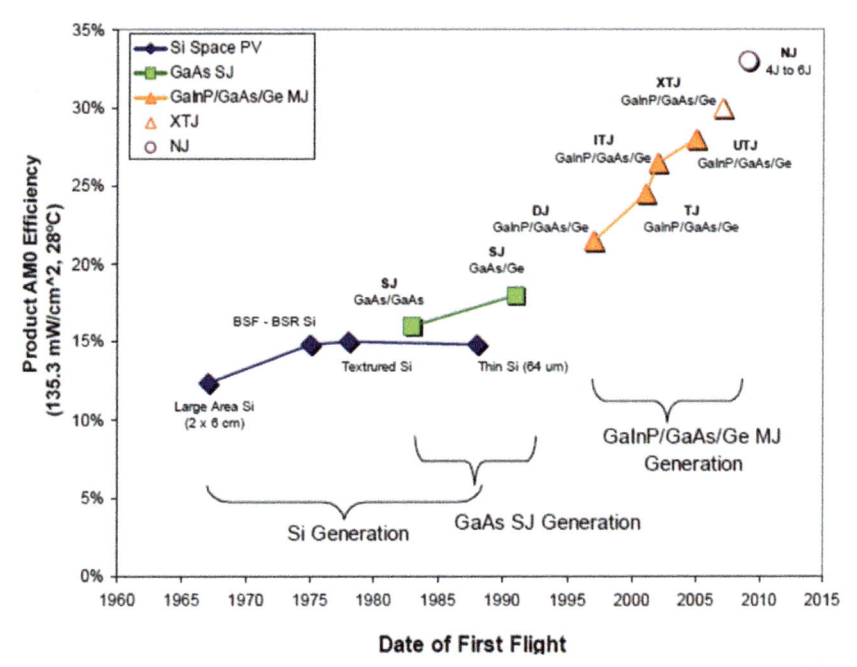

Figure 19.25 Historical product efficiency of space solar cells against the date of first flight. Open points are for planned products with estimated flight dates, as of the date of this report's publication. Abbreviations: single-junction (SJ), dual-junction (DJ), triple-junction (TJ), and Spectrolab (https://www.spectrolab.com/photovoltaics.html, accessed May 21, 2023) improved triple-junction (ITJ), ultra triple-junction (UTJ), neXt triple-junction (XTJ), and 4–6 junction (NJ) cells. *Adapted from P.M. Beauchamp, J.A. Cutts, Jet Propulsion Laboratory, Solar Power Technologies for Future Planetary Science Missions, JPL D-101316, Pasadena, CA, USA, December 2017, 49 pp. Courtesy: NASA.*

19.4.1 Space photovoltaic technologies for exemplary missions

As discussed above and detailed in Fig. 19.25, space solar cell and solar array technology has advanced significantly since the first solar-powered satellite in 1958 with solar cell PCE increasing by a factor of three. Arrays have grown in power from milliwatts to over 20 kW on spacecraft; the new arrays on the ISS will produce 120 kW. In addition, the remaining uncovered solar array pair and partially uncovered original arrays will continue to generate approximately 95 kilowatts of power for a total of up to 215 kW of power available to support the ISS, which is currently slated to remain in operation until at least 2030 [130].

The efficiency of SOP triple-junction cells, designed and optimized for Earth orbital missions, is typically approximately 30% under standard test conditions (1 AU, 28°C). The specific power of the solar arrays has also improved from 30 W/kg to 100 W/kg during the past 25 years and has enabled several Mars (orbital and surface), small body (flyby and orbital), and inner planetary (flyby and orbital) missions. The entries in Table 19.4 provide a representative sample of missions [131−145] during a 20-year period from 1996 to 2016, to a variety of destinations, employing one of three solar cell configurations, using MJ III-V technology exclusively, after 2001, with a power capability (at 1 AU) ranging from approximately 300 W for a lunar mission (LCROSS [135]) to 14 kW for a Jupiter mission (Juno [142]). The reader is encouraged to consult several NASA sources for a rather large amount of detailed information on past, present, and future space missions, if interested [146−148]. In the next subsection, we will review in more detail the diversity of solar cell and array technology configurations employed in space exploration missions across the solar system.

19.4.2 Notable space missions, technology demonstrations, and array configurations

19.4.2.1 Classic gimbaled wings: Dawn mission (2007−18)

The gimbaled wings configuration may be considered the "classic design"; however, specific details of solar array designs are set by specific mission requirements. To reduce costs, several later missions rigidly affixed their solar arrays to the spacecraft bus instead of providing gimbaled sun-tracking. Fig. 19.26 includes images of various examples of solar cell, module, or array configurations. The Dawn mission, launched in 2007,

Table 19.4 Solar arrays on example NASA planetary science missions.

Mission class	Mission[a]	Destination	Launch date	Solar cell technology	Solar array technology	Power capability at 1 AU (W)	Mission details Ref.
Inner planetary system	MESSENGER	Mercury	3-Aug-04	Triple junction	Deployable rigid	450	[132]
Lunar Probes	Lunar Reconnaissance Orbiter	Moon	18-Jun-09	Triple junction	Body-mounted	600	[133]
	GRAIL	Moon	10-Sep-11	Triple junction	Deployable rigid	1850	[134]
	LADEE	Moon	7-Sep-13	Triple junction	Deployable rigid	763	[135]
	LCROSS	Moon	18-Jun-09	Triple junction	Body-mounted	295	[136]
Mars	Mars Global Surveyor	Mars	7-Nov-96	GaAs/Ge and Si	Deployable rigid	2100	[137]
	Mars Odyssey	Mars	7-Apr-01	GaAs/Ge	Deployable rigid	2092	[138]
	Mars Exploration Rover (2 rovers)	Mars surface	10-Jun-03 7-Jul-03	Triple junction	Deployable rigid	390	[139]
	Mars Reconnaissance Orbiter	Mars	12-Aug-05	Triple junction	Deployable rigid	6000	[140]
	Phoenix	Mars surface	4-Aug-07	Triple junction	Flexible fold-out	1255	[141]
	MAVEN	Mars	18-Nov-13	Triple junction	Deployable rigid	3165	[142]
Outer planets	Juno	Jupiter	5-Aug-11	Triple junction	Deployable rigid	14000	[143]
Asteroids/comets	Deep Impact/EPOXI	Tempel-1 Hartley-2	12-Jan-05	Triple junction	Body-mounted	620	[144]
	Dawn (with solar electric propulsion)	Vesta, Ceres	27-Sep-07	Triple junction	Deployable rigid	10300	[145]
	OSIRIS-REx	Bennu	8-Sep-16	Triple junction	Deployable rigid	3000	[146]

Source: P.M. Beauchamp, J.A. Cutts, Jet Propulsion Laboratory, Solar Power Technologies for Future Planetary Science Missions, JPL D-101316, Pasadena, CA, USA, December 2017, 49 pp.
[a]MESSENGER—Mercury Surface, Space Environment, Geochemistry, and Ranging; GRAIL—Gravity Recovery and Interior Laboratory; LCROSS—Lunar Crater Observation & Sensing Satellite; LADEE—Lunar Atmosphere Dust & Environment Explorer; MAVEN—Mars Atmosphere & Volatile Evolution; EPOXI—Extrasolar Planet Observation & Characterization Investigation (EPOCh) + Deep Impact Extended Investigation (DIXI); OSIRIS-REx—Origins, Spectral Interpretation, Resource Identification, Security-Regolith Explorer.

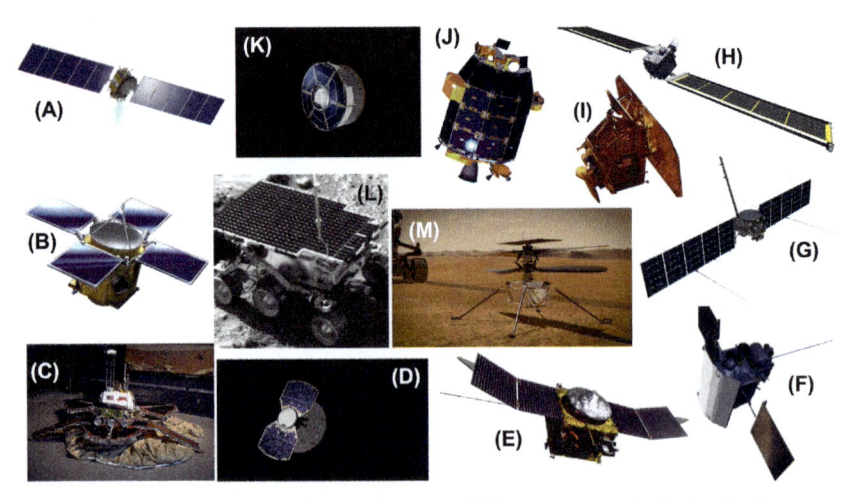

Figure 19.26 Artist's conceptions or images of (from counter-clockwise in the upper left): (A) Dawn; (B) NEAR Shoemaker; (C) Pathfinder lander; (D) Mars InSight cruiser; (E) MAVEN; (F) MESSENGER; (G) Europa Clipper; (H) DART; (I) Deep Impact (EPOXI); (J) LADEE; (K) Mars 2020 cruiser; (L) Mars Pathfinder Sojourner rover; and (M) Mars 2020 Ingenuity Mars helicopter. *Courtesy: NASA.*

visited Ceres and Vesta on a 4.3 billion-mile (6.9 billion kilometers) journey. Propelled by ion engines, the spacecraft used a dual solar array gimbaled about the long axis to control its orientation with respect to the Sun. Dawn achieved many firsts until its extended mission concluded on October 31, 2018 (Fig. 19.26A) [144].

19.4.2.2 Fixed wings or paddles: Small bodies and Mars missions

In the early 1990s, NASA established the Discovery program to reduce the time and cost to develop planetary science missions, using competitively selected missions. Three types of solar-powered spacecraft or vehicles of this design can be articulated. An example of a deep space enterprise, the Near-Earth Asteroid Rendezvous (NEAR) Shoemaker mission was the first of the Discovery program selected [149]. As mentioned above, this spacecraft had solar arrays rigidly attached to the spacecraft bus instead of providing gimbaled sun-tracking to reduce costs (Fig. 19.26B). NASA's NEAR Shoemaker, launched in February 1996, was the first spacecraft to orbit an asteroid (Eros) and also was the first to land on one. Interestingly, NEAR, not designed as a lander, survived touchdown on the asteroid and returned valuable data for about 2 weeks until contact was lost to the spacecraft on February 28, 2001, having likely succumbed to the extreme cold.

The Mars Pathfinder mission, launched in December 1996, delivered a rover (Sojourner) to Mars, the first robotic rover to another planet [150]. The Pathfinder, lander, and rover all used solar arrays for primary power. The Mars Pathfinder lined the petals with solar cells to power a communications relay between Sojourner and Earth (Fig. 19.26C). Note that the cruiser stage for Mars Pathfinder used body-mounted solar panels mounted directly to the aft end of the transit vehicle, providing power for avionics and communications during transit in an economical design. The cruiser stage for another Mars mission, Mars InSight [151], employed twin paddles, extending just slightly beyond the diameter of the spacecraft, as shown in Fig. 19.26D. Another notable aspect of this mission, which will be examined below (Section 19.4.3), is the key contribution of interplanetary CubeSats [152,153].

19.4.2.3 Unique solar arrays: Fixed and gimbaled

Beyond generating electrical power using solar energy, solar arrays on various deep space missions have also been innovatively used for aerobraking, tilted for thermal management and collision avoidance, and other mission-enabling applications. The Mars Atmosphere and Volatile Evolution (MAVEN) mission (Fig. 19.26E), part of the Mars Scout Program, includes eight instruments to enable the study of the planet's upper atmosphere, ionosphere, and interactions with the Sun and solar wind [141]. The solar panels provide 1.2 kW of power and aerobraking to assist in maneuvering and adjusting its orbit around Mars. The spacecraft is still operating. It was originally intended for 2 years but has been extended several times; MAVEN has enough fuel to operate until 2030. In late November and early December 2015, MAVEN made a series of close flybys of Mars' largest moon, Phobos. MAVEN has produced a wealth of information about Mars that has been published over the past several years; the numerous websites provide some guidance to locating it.

The Mercury Surface, Space Environment, Geochemistry, and Ranging (MESSENGER) spacecraft, launched in August 2004, is an example of a solar array with a tilt-back configuration, specifically designed for its 2004 mission to orbit Mercury [131]. The MESSENGER mission's primary goal was to study Mercury's geology, magnetic field, and chemical composition; on March 18, 2011, MESSENGER made history by becoming the first spacecraft ever to orbit Mercury. Santo et al. detail a number of technological advances developed for the MESSENGER spacecraft to address the environmental challenges presented by a mission to Mercury (extreme temperature(s), intense solar radiation [up to 11 Earth-equivalent suns], and vacuum) [154]. The solar array, composed of 30% cells (GaAs/Ge) and 70% mirrors, was one of the few elements exposed to the Sun. The solar arrays and thermally conductive composite face sheets were used together with off-pointing by folding the arrays back along the spacecraft body (Fig. 19.26F). This allowed for operations at temperatures up to 150°C and survivability up to 275°C [155]. MESSENGER's mission, originally planned as a 1-year orbital mission, was completed on April 30, 2015, by impact on the surface of the planet.

There are several recent and upcoming deep-space science missions [Lucy (2021) [156], Psyche (2023) [157], and Europa Clipper (2024) [158] (Fig. 19.26G)] with unique solar arrays. Another example is the SEP-driven (see Section 19.4.4.1) Double Asteroid Redirect (DART) mission, launched in November 2021 aboard a Falcon 9 rocket [159]. DART, managed by the Planetary Defense Coordination Office [160], is the first-ever mission dedicated to investigating and demonstrating asteroid deflection by changing an asteroid's motion in space through kinetic impact. The DART mission was designed to study a deflection technique that may be used should a future near-Earth asteroid [161] classified as a potentially hazardous asteroid (PHA) arrive on a trajectory and with a mass that could seriously harm life on Earth [160,161]. This technology demonstration was designed to minimize mass and stowed volume for potential flight as a rideshare. DART uses a ROSA (Fig. 19.26H) that employs a flexible substrate for the solar cells, substantially reducing the solar array mass and stowed volume. In addition to being the first mission to fly these novel arrays as a primary power source, several reflector-based solar concentrators will be incorporated into the array to demonstrate 2X concentration on IMM solar cells for significantly improved performance on LILT missions to the outer planets [45,162]. The DART mission was successfully concluded on September 26, 2022; analysis of the data indicates that kinetic impact is a viable planetary defense method [163]. As with the Mars Insight mission discussed above (Section 19.4.2.2), DART included a CubeSat [164] that played a key role in documenting the successful impact on Dimorphos; this will also be discussed below (Section 19.4.3).

19.4.2.4 Deployable single panels: Cost savings and multipurpose arrays

While most spacecraft use symmetrically located solar arrays to reduce the need for navigation corrections caused by asymmetric solar pressure, several planetary science missions found that even if power loads require the use of deployable solar arrays, using a single array can substantially reduce mission costs. The cost savings can be realized simply because of the reduced number of deployment mechanisms, yokes, and gimbals, but also because the reuse of prior Earth-orbiting designs can reduce development costs (such as for the Lunar Reconnaissance Orbiter [LRO] [132] and Mars Observer [137]). Also, a single wing can serve as an effective drag shield for aerobraking. The Deep Impact EPOXI (EPOXI—Extrasolar Planet Observation and Characterization Investigation [**EPO**Ch] + Deep Impact Extended Investigation [DI**XI**]) mission was launched in January 2005 [143] (Fig. 19.26I). The triple-junction array provided 600 W of power; this was sufficient for two missions before termination in September 2013. The initial (and primary mission) of Deep Impact was to probe beneath the surface of a comet. The spacecraft delivered an impactor into the path of Tempel 1 to investigate the internal composition and structure of a comet. Because a significant quantity of fuel remained after the successful completion of its primary mission, Deep Impact was redirected toward Comet 103 P/Hartley (or Hartley 2), starting with an engine burn on November 1, 2007. The "mission of opportunity," called EPOXI, set Deep Impact on three consecutive Earth flybys, spread over 2 years (December 2007–June 2010) before the final leg of this circuitous voyage to meet Comet Hartley 2 [165].

19.4.2.5 Missions using body-mounted solar arrays

The initial phase of US solar-powered solar system exploration concluded with the 1978 launches of the Pioneer 12 and 13 to Venus [47,148,166]. Both spacecraft utilized body-mounted solar cells [45,47]. Pioneer 13 dropped four probes that descended through the dense Venusian atmosphere to the surface of Venus, while Pioneer 12 served as an orbiting data relay station [148,166]. The Lunar Atmosphere and Dust Environment Explorer (LADEE) space science mission [134] launched from Wallops Flight Research Facility in Virginia in September 2013 also employs body-mounted solar arrays; see Table 19.4 and Fig. 19.26J. LADEE, the first mission in the Lunar Quest series, was designed to orbit the Moon and study its thin atmosphere and the lunar dust environment. Specifically, the goal was to collect data on the global density, composition, and time variability of the exosphere.

The design of spacecraft used to deliver landed assets is often relatively simple. The cruise stages for the Mars Pathfinder [150], Mars Exploration Rover [138], Mars Science Laboratory [167], and Mars 2020 [168] used body-mounted solar panels mounted directly to the aft end of the transit vehicle, as shown in Fig. 19.26K, providing power for avionics and communications during transit in an economical design. The tiny Sojourner rover of the Mars Pathfinder mission, the first to operate on another planet, was designed to be low-cost; therefore its design was relatively simple (Fig. 19.26L). A single solar panel affixed to the entire upper surface of the vehicle accommodated over 200 GaAs/Ge cells on the 0.22 m^2 panel to generate 35 W of electrical power at noon on Mars, roughly equivalent to the power required by a modest incandescent lightbulb. While not traversing, excess electrical energy from the solar array was used to resistively heat a warm electronics box, although three radioisotope heater units provided the main source of nighttime warmth. Several experiments on the Sojourner rover were assembled and monitored by researchers at NASA GRC [169,170].

The Mars 2020 mission, launched in July 2020 [168], included a rover [Perseverance [171]) and a helicopter flyer (Ingenuity [172] (Fig. 19.26M)] to demonstrate the first powered flight on Mars. Ingenuity rode to Mars attached to the belly of the Perseverance rover. The solar panel for Ingenuity was manufactured in SolAero's SOA production facility in Albuquerque, NM [173]. The panel is populated with SolAero's industry-leading, 33.0% efficient IMM (see Fig. 19.12) class of solar cells; the panels are approximately 425 × 165 mm and are used to charge six helicopter lithium-ion batteries. As we discussed previously (Section 19.3.1), growing the cell structure in an inverted fashion with the removal of the substrate can dramatically improve the specific power; the IMM cells are more than 40% lighter than typical space-grade solar cells [1,12,45]. The combination of higher efficiency and significantly lower mass was a critical factor that was deemed mission-enabling for Ingenuity's successful flight. As of this writing, since taking flight in April 2021, Ingenuity has had more than 50 flights, covered >seven miles (approximately 12 km), reaching altitudes as high as 59 ft (18.0 m) [172]. The success of Ingenuity further extends SolAero's long history of powering satellites and spacecraft for NASA and JPL, including the Mars 2020 [168] cruise stage, the Interior Exploration using Seismic Investigations, Geodesy and Heat Transport (InSight) lander [151] that operated on the Mars surface from late November 2018 to late December 2022, the DAWN asteroid mission [144], the Parker Solar Probe to the Sun [174], the LADEE [134] mission, and many others.

19.4.3 SmallSats, CubeSats, and rideshare missions: Opportunities for technology demonstrations

Another class of planetary science missions includes small satellites (SmallSats) and CubeSats [175–177], as well as rideshare missions—secondary missions that share a launch vehicle with another, primary, mission [47]. SmallSats, CubeSats (typically 6U; 1U = 10 cm × 10 cm × 10 cm [175–177]), and rideshare missions have the potential to dramatically change the way planetary science missions are conducted. As the small spacecraft industry evolves [177], the use of commercial-off-the-shelf components, including solar arrays, could reduce mission costs and allow for more frequently launched small planetary science missions. While it is likely that the most efficient solar cells will still be used to reduce overall costs, lower-efficiency cells with much lower manufacturing costs and high(er) MSPs could contribute to reduced mission costs to enable (very) low-cost deep-space (rideshare) missions. The example missions examined below flew in both the Earth–Moon region of the Solar System (Starshine 3, LCROSS, and Artemis 1 CubeSats), and deep space (MarCO and LICIACube); the MarCO twins stand out as unique CubeSat spacecraft, being free flyers for most of the journey to Mars [152,153]. The reader is directed to several recent sources that introduce smaller spacecraft from picosatellites (0.01–1 kg) to minisatellites (100–180 kg), including CubeSats [175–177].

19.4.3.1 Starshine 3: Demonstration of integrated micropower technology

The PV group at NASA GRC and its partners investigated the viability of MJ thin-film solar cells to potentially drive the AM0 efficiency of these devices toward 25% and beyond [50,119,178]. At the same time, we were investigating carbon nanotubes and eventually, Si-based materials for use as anodes in Li-ion batteries [179,180]. We recognized that we could develop a related device structure of a solar cell integrated with a battery and subsequently initiated an effort to develop an integrated (micro-) power source or system (IMPS or IPS) [181–185].

Integrated power devices (or systems), combining three traditionally separate power system components (energy conversion, storage, and control), would be ideal for scientific payloads, experiments, and smaller satellites [182,184]. The volume formerly required by traditionally separately located chemical batteries frees up valuable space for other systems or an increased payload. Another benefit of using an I(M)PS as a distributed power system is a reduction in spacecraft complexity, especially with respect to power distribution wiring, simplifying spacecraft integration [177,182]. Integrated power systems could also be used in specialized instances, serving as decentralized (or distributed) power sources or uninterruptible power supplies for discrete components. Of course, this requires solar cells (or arrays) to be located such that they have a view of the Sun for at least during some portion of the orbit or flight to a mission locale. Finally, with future improvement in thin-film power generation [185] and energy storage [115−119], IMPS should also find application as a main power system for upcoming missions using constellations of very small spacecraft: microsatellites, nanosatellites, picosatellites [176,186], or CubeSats [176,177].

In 1999 two NASA GRC power experiments [185,187,188] were late additions to Starshine 3, the third in a series of small satellite launches intended to engage students and amateur radio enthusiasts [189]. The primary mission of Starshine 3 was to measure atmospheric density as a function of altitude. Starshine 3 was essentially a passive satellite. There was no use of electrical power to assist the orbital tracking. This made Starshine 3 an excellent platform to test new power technologies without the burden of mission success depending on the power system. The Starshine 3 satellite (Fig. 19.27B (top)) was 1 meter in diameter with a mass of 88 kg, thus it was a microsatellite [176]. Its surface is covered with 1000 student-polished mirrors, 31 laser retro-reflectors, 48 triple-junction 2×2 cm commercial (EMCORE Corp.) solar cells, and five IMPS assembled at NASA GRC (top of Fig. 19.25).

This would be the first in-space demonstration of an IMPS, given the short lead time, we chose low-risk and/or commercial off-the-shelf (COTS) technologies to successfully demonstrate proof-of-concept: an in-house produced GaAs monolithically integrated module (MIM) 1 cm^2 solar cell array, consisting of solar cells connected in series, with an output of approximately 7 V and 3 mA, a commercial 3 V Li-ion rechargeable (Panasonic ML2020 coin cell) battery with a nominal capacity of 45 mA-h and fool-proof (COTS) control electronics (a micropower voltage regulator (MAXIM 1726EUK) and a blocking diode). The voltage regulator kept the battery from charging above 3 V; the blocking diode prevented current from flowing back through the array when it was in the dark. The load side included two p-type MOSFETS that shut off the load from the IMPS below 2.3 V. [183,185,187]. The MIM used in this flight demonstration experiment had more than enough voltage and current to both charge a Li-ion battery and power an equivalent load.

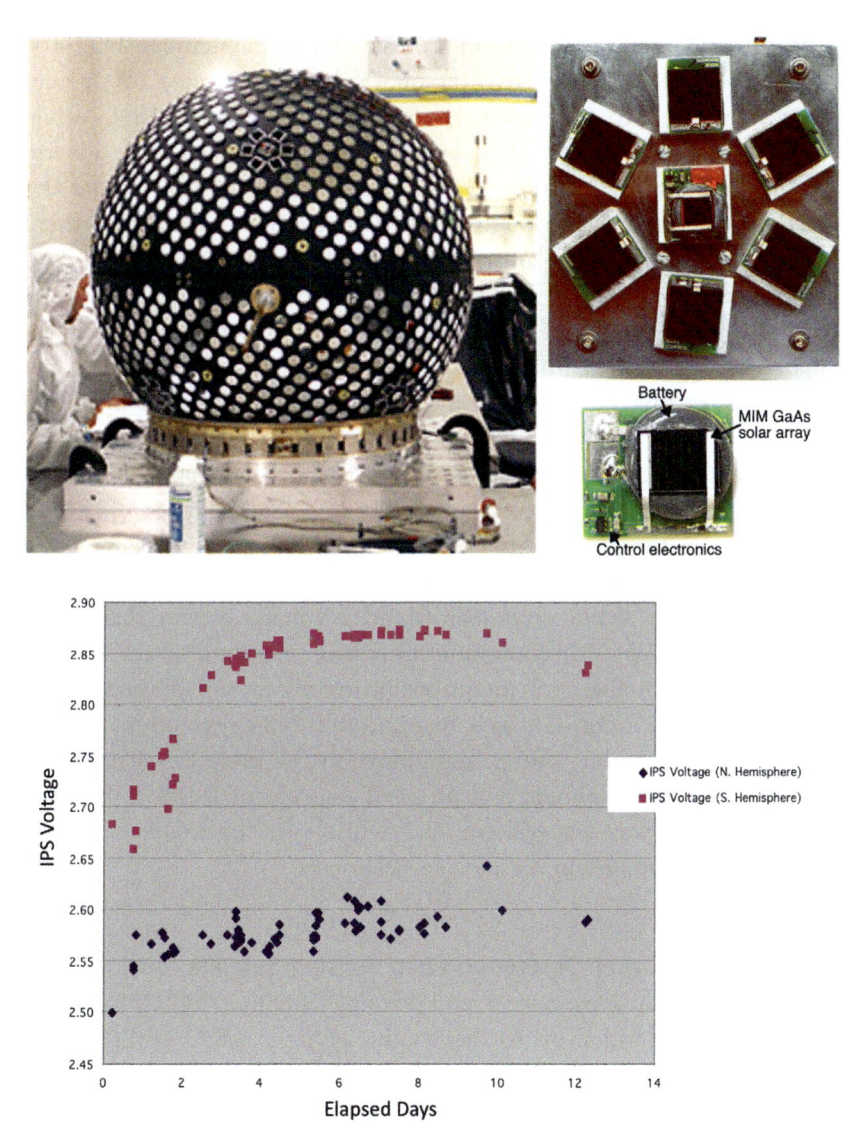

Space photovoltaics: New technologies, environmental challenges, and missions 727

Figure 19.27 (*Top*) Photos clockwise from left: Starshine 3 small satellite; expanded view of a constellation (one of five on satellite) of a seven-junction, 1 cm^2 monolithically interconnected GaAs module (MIM) surrounding an integrated (micro-)power device; detail of an IPS showing a single MIM GaAs solar array, commercial Li-ion battery, and control electronics. (*Bottom*) Data from five integrated micro power supplies (IMPS) designed and fabricated at NASA GRC that operated onboard Starshine 3 from late-September to mid-October 2001. Four of the IMPS units had particularly challenging operating conditions due to a severe initial solar beta angle that provided mostly Earth albedo (light reflected off the Earth) charging and an average operating temperature of −10°C. *Courtesy: NASA.*

Starshine 3 was launched from Kodiak, AK on September 29, 2001; it was deployed to a circular orbit of 475 km (LEO) at a 67 degrees inclination with a fixed rotational velocity of 5 degrees/s [189]. While in orbit, the five independent power supplies experienced quite adverse operating conditions; Starshine 3 was deployed in an orbit where the solar beta angle passed through ±90 degrees. This meant that there were periods when the solar array on an IMPS did not "see" the Sun for days at a time. Any charging would have to come from light reflected off the surface of the Earth (albedo). Furthermore, when illuminated by only the Earth's albedo, the operating temperature of the IMPS sank as low as −18°C; only 2°C above the battery's operating range.

Despite these adverse conditions, the IMPS on the upper (Northern) hemisphere still managed to charge under albedo lighting only [183,185,187]. The IMPS in the lower (Southern) hemisphere quickly charged (by direct solar insolation) to its maximum voltage of 2.9 V. During the first 100 days, all five IMPS units maintained a voltage greater than 2.5 V (bottom Fig. 19.25). All communication from Starshine 3 ceased after 100 days. The satellite was intentionally deorbited and burned up on January 21, 2003, after 479 days in orbit and 7434 revolutions around Earth, with one orbit lasting approximately 90 minutes. Starshine 3 marked the first deployment and proof-of-concept of IMPS technology via a successful space experiment. We will revisit the topic of power-integrated technologies in the next subsection (19.4.4).

19.4.3.2 Lunar Crater Observation and Sensing Satellite mission
The Lunar Crater Observation and Sensing Satellite (LCROSS [135]) mission flew as a rideshare to the Moon with the LRO [132] mission launched in June 2009. LCROSS's innovative design incorporated the launch vehicle's secondary payload adapter ring as its primary structure, with its fuel tank mounted within the ring, and all its subsystems mounted to the ring's six ports. One of these ports was allocated for the solar panel; the efficient design of the spacecraft enabled low-power consumption—the array produced less than 300 W. This single wing was oriented to the Sun during the cruise phase. The primary objective of the rideshare LCROSS was to observe ejecta created by the impact of the launch vehicle's upper stage into the Moon to determine the presence or absence of water ice in permanently shadowed regions [190].

19.4.3.3 Mars CubeSat One (MarCO) participation in the Mars Insight mission

The Mars Cube One twin (MarCO-A and B, nicknamed Eve and Wall-E, respectively) 6U microsatellites are the first CubeSats to successfully complete an interplanetary mission (Fig. 19.28 bottom) [152,153,191,192]. The NASA Interior Exploration using Seismic Investigations, Geodesy and Heat

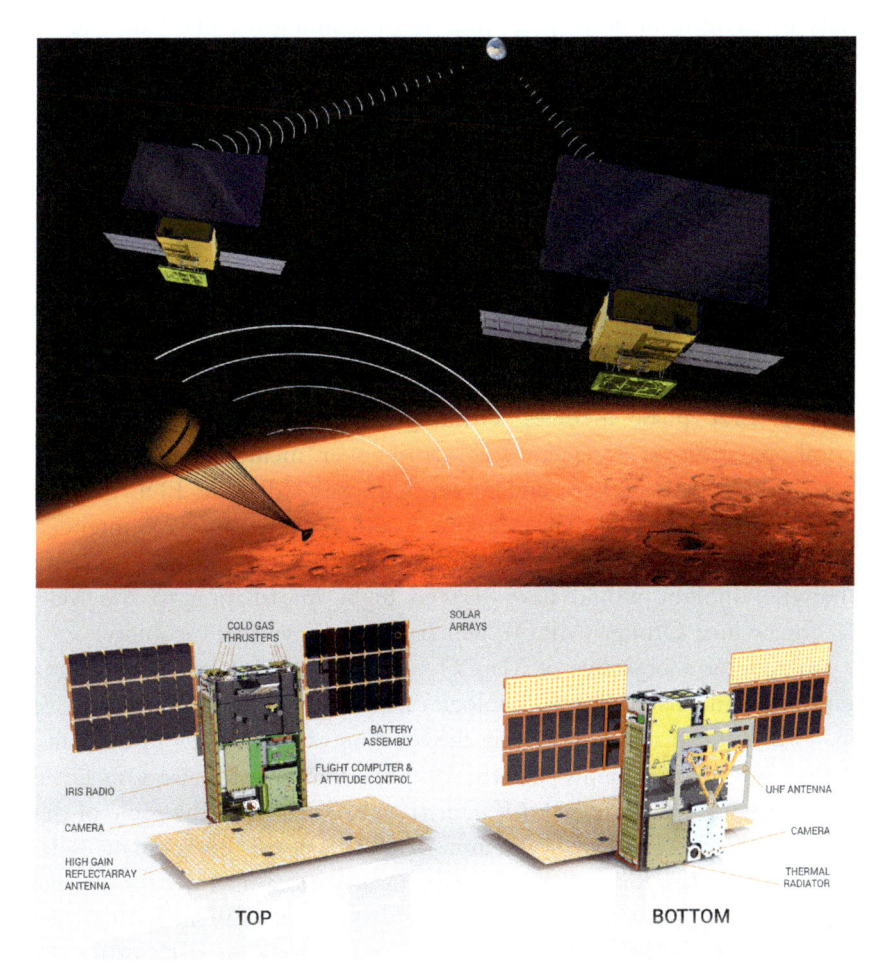

Figure 19.28 *Top*: MarCO A and B monitoring the InSight landing (artist's concept). *Bottom*: illustration of the Mars Cube One spacecraft, indicating the location of key components. *Reprinted with permission from F.C. Krause, J.A. Loveland, M.C. Smart, E.J. Brandon, R.V. Bugga, Implementation of commercial Li-ion cells on the MarCO deep space CubeSats, Journal of Power Sources 449 (February 2020) 227544, copyright (2020) Elsevier.*

Transport (InSight) mission landed on Elysium Planitia, an equatorial site on Mars, on November 26, 2018 [151]; the InSight lander's goal is to study the crust mantle and core of Mars. InSight (with Eve and Wall-E) was launched on May 5, 2018. One of several unique features of the InSight mission was the addition of the redundant MarCO 6U CubeSats. Although launched together with InSight, the two MarCO spacecraft separated soon after launch to fly their own trajectories to Mars to test their endurance and ability to navigate in deep space, thus also becoming the first CubeSats to fly in deep space. The two spacecraft were kept approximately 10,000 km (6200 miles) away from InSight at either flank during flight to protect the main spacecraft from potential mishap; this distance was reduced to 3500 km as the three spacecraft approached Mars [152,185,191].

The primary objectives of MarCO were to demonstrate new miniaturized communication and navigation technologies in the CubeSat form factor in a deep space environment, and more importantly, provide a real-time communications relay during the InSight lander's entry, descent, and landing (EDL) phase, the most critical and uncertain phase of any surface mission (Fig. 19.26 top) [152,191,192]. Although only lasting 7–10 minutes, the EDL phase involves activating several pyrotechnic devices for critical operations to decelerate a lander or rover, including opening parachutes, firing retro-rockets, and so on; Curiosity's difficult EDL is available to download [193]. The MarCO-A and B CubeSats were intended to provide real-time data relay, with minimal cost to the overall mission; otherwise, InSight would have relayed flight information to the Mars Reconnaissance Orbiter (MRO), which did not have that capability.

An additional objective of MarCO was to test new miniaturized communication and navigation technologies in a deep space environment; see Fig. 19.26 for a diagram indicating the location of several key instruments; a listing of key components [152,153,191] of a MarCO CubeSat is included in a prior book chapter [185]. Of interest and relevance to the topic of this chapter is the successful demonstration of COTS power components (solar cells and batteries) for an interplanetary (Mars) mission. Two three-panel solar arrays of 21 3J-GaAs solar cells panels provided 35 W near Earth and 17 W near Mars [185,194]. While simulated LEO testing [195] of COTS batteries for Earth-orbiting CubeSats discovered shortcomings [196], this technology proved to be successful for the MarCO mission [153]. A critical difference between LEO simulation and the MarCO mission conditions is the absence of significant solar array shadowing during the transit to and short-term mission performance around Mars.

The MarCO 6-month mission was successfully completed on November 26, 2018. Overall, MarCO-A successfully transmitted 93% of the InSight EDL data and MarCO-B sent 97%. The first image InSight took of Mars was relayed to Earth by the MarCO CubeSats. Over the following days, MarCO-B transmitted several images of Mars taken by its wide field-of-view camera, including the landing site; the poles of Mars, white from the ice caps; several volcanoes; even two small dim images of Phobos; and finally, a "farewell" image of Mars [191]. Due to the loss of communication, the primary MarCO mission was declared complete in February 2019.

19.4.3.4 LICIACube: Witness to the DART impact on Dimorphos

The DART mission, discussed above (Section 19.4.2.3), visited a binary asteroid system: asteroid Didymos is orbited by its small moonlet, Dimorphos [197]. The binary asteroids, not a threat to Earth, were chosen as the target for DART because they are relatively close to Earth. The DART mission included a rideshare 6U (10 × 20 × 30 cm) CubeSat (mass ≈ 14 kg), provided by the Italian Space Agency (ASI). The Light Italian Cubesat for Imaging of Asteroid (LICIACube) (Fig. 19.29) [164] main scientific payloads, LEIA (a narrow-angle camera) and LUKE (a wide-angle imager with an RGB filter), investigated the nature of a binary NEA for the first time. While details of the PV arrays are minimal [164], it is reasonable to estimate that the total power output of the two (likely 3J-GaAs and MarCO [153]) solar panels to power the 6U CubeSat would be 30−40 W, provided by a commercial vendor [177,185,194].

The LICIACube was deployed from a spring-loaded box on September 11, 2022, in the asteroid Didymos system and followed 3 minutes behind after a separation maneuver [198]; see Fig. 19.29 (bottom) for an overview of the concept of operations (CONOPS). DART intercepted the smaller moonlet asteroid on September 26, 2022. DART impacted Dimorphos at high speed—about 4 miles (or 6.6 km)/s (14,000 mph). Dimorphos was about 6.8 million miles (11 million km) from Earth at the time of DART's impact [197]. LICIACube performed an autonomous fly-by of the Didymos system probing the DART impact of Dimorphos; it downlinked images directly to Earth after the fly-by [199]. Analysis of data by the DART investigation team shows that DART's kinetic impact with its target asteroid (Dimorphos) successfully altered its orbit. This marks the full-scale demonstration of asteroid deflection technology [200]. Analysis of data from the first kinetic impact test, as well as the study of the nature and the

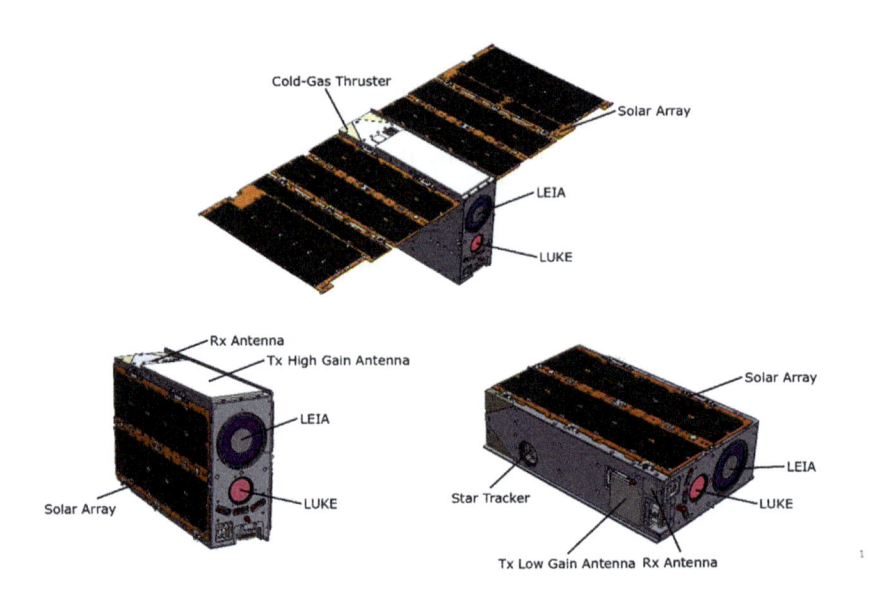

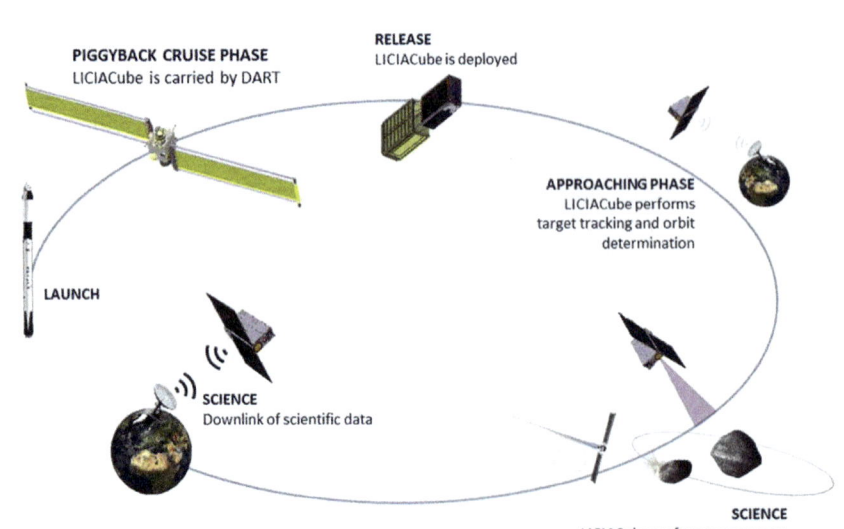

Figure 19.29 Three-dimensional view (also with deployed Solar Array, *top*) of the LICIACube spacecraft, with the two payloads, LEIA and LUKE, onboard (see text). (*bottom*) Sketch of the LICIACube nominal mission phases or CONOPS. *Reproduced with permission from E. Dotto, V. Della Corte, M. Amoroso, I. Bertini, J.R. Brucato, et al., LICIACube - The Light Italian Cubesat for Imaging of Asteroids in support of the NASA DART mission towards asteroid (65803) Didymos, Planetary Space Sci. 199 (May 2021) 105185, copyright (2021) Elsevier.*

evolution of the dust plume produced by the DART and LICIACube teams, has produced a wealth of information on the physics of the impact as well as the composition and the structure of the material composing a small NEA (Dimorphos) [200−203].

19.4.3.5 Rideshare CubeSats on Artemis: Profiles in CubeSat mission success

Artemis I (Exploration Mission-1) launched from NASA Kennedy Space Center, Cape Canaveral, Florida on November 16, 2022 [204]. The mission lasted 25.5 days, and traveled a total distance of 1.4 million miles (2.25 million km); the splashdown was on December 11, 2022. It was the first integrated flight test of the Orion spacecraft and Space Launch System (SLS) rocket. The primary operations goal of the mission was to ensure a safe crew module entry, descent, splashdown, and recovery. In addition to sending Orion on its journey around the Moon, SLS carried ten 6U CubeSats, each with its own science and technology investigations; a summary of the current status of these CubeSat missions is summarized in brief in Table 19.5 [205−223].

Table 19.5 identifies the four successful missions; two deep-space CubeSats (BioSentinel [205] and EQUULEUS [209]) are still operating, as of this writing. It is important to note that the selection process itself was arduous: thirteen CubeSats were selected for the mission, but three were not ready by the deadline for launch [224]. As CubeSats often involve educational institutions and/or students, the teams that made the final cut and were included in the Artemis 1 mission were already successful from an educational perspective, training the next generation of aerospace leaders.

One of us (AFH) was involved in a rigorous design process [225] and subsequent proposal of a deep space asteroid rendezvous 6U CubeSat mission, Diminutive Asteroid Visitor using Ion Drive (DAVID). While the mission survived the preliminary review rounds and was selected for further technology development [226], DAVID was not selected for inclusion in Artemis 1. The selected NEA Scout CubeSat [215] proposed a much more intriguing mission. Solar sail technology and missions represent an opportunity to utilize advanced PV technologies and will be discussed in more detail below (Section 19.4.4). One positive postscript to the three CubeSats not included on Artemis 1 is the success of Lunar Flashlight [227,228] launched from a SpaceX Falcon 9 rocket on December 11, 2022.

Table 19.5 Summary of ride-along 6U CubeSats released during the Artemis 1 Mission in November 2022.

CubeSat name[a,b]	Mission/Goal(s)	CubeSat team	Status	Ref.(s)
ArgoMoon	To observe the SLS interim cryogenic propulsion stage (ICPS) with advanced optics and software imaging systems.	ArgoTec, Italian Space Agency (ASI), ESA	Successfully deployed: Images of Earth and Moon returned but not ICPS (primary object).	[222,223]
BioSentinel	Develop a biosensor instrument to detect and measure the impact of space radiation on living organisms over long durations beyond low-Earth orbit (LEO).	NASA Ames Research Center (ARC)	Successfully deployed; overcame initial tumbling issue and is currently orbiting the Sun.	[205,206]
CuSP	"Space weather" mission to measure solar particles and magnetic fields.	Southwest Research Institute	Successfully deployed after initial hours of transmission, contact lost, not reestablished.	[207,208]
EQUULEUS	Will travel to Earth–Moon Lagrange Point 2 (LP-2), Earth–Moon orbit where the gravitational pull of the Earth and Moon equals the force required for a small object to move with them will demonstrate trajectory control techniques within Sun–Earth–Moon region and image Earth's plasmasphere.	Japan Aerospace Exploration Agency (JAXA) and Univ. of Tokyo	Successfully deployed and currently headed to LP-2. Successfully photographed Comet ZTF (Comet C/2022 E3) from space on February 23, 2023.	[209,210]

(Continued)

Table 19.5 (Continued)

CubeSat name[a,b]	Mission/Goal(s)	CubeSat team	Status	Ref.(s)
LunaH-Map	Will enter a polar orbit around the Moon with a low altitude (5–12 km) centered on the lunar South Pole. Carries two neutron spectrometers that will produce maps of near-surface hydrogen (H).	Led by Arizona State University, includes team members from industry, NASA ARC & JPL	Successfully deployed; collected images for several weeks but the mission ended due to propulsion issues.	[211,212]
Lunar IceCube	Technology demonstrator developed by Morehead State Univ. and partners. Purpose: look for surface ice in permanently shadowed regions near lunar south pole and test new spacecraft technologies.	Morehead State Univ. w/NASA Goddard Space Flight Center (SFC) and the Busek Co.	Successfully deployed, but no communications established.	[213]
LunIR	Perform a lunar flyby, taking surface images to appraise the Moon's thermal environment. After flyby, conduct technology demonstrations: maneuvering and deep-space operations for future Mars missions.	Lockheed Martin	Successfully deployed but no communications established.	[214]

(*Continued*)

Table 19.5 (Continued)

CubeSat name[a,b]	Mission/Goal(s)	CubeSat team	Status	Ref.(s)
NEA Scout	The purpose of the mission is twofold: one, to demonstrate solar sail deployment; and two, to demonstrate solar sail navigation by rendezvousing (and characterizing) the near-Earth asteroid (NEA) 2020 GE.	NASA Marshal SFC, & NASA JPL, w/NASA Goddard SFC, Johnson SC & Langley Research Center.	Successfully deployed but communication with NEA Scout never successfully established.	[215,216]
OMOTENASHI	Will make a semihard survivable landing on the Moon with the primary objective of testing the technologies and trajectory maneuvers that allow for such a landing. It will also be measuring the radiation environment beyond LEO.	JAXA	Successfully deployed; but began to tumble. It could not be stabilized in time to conduct the landing mission or any backup missions.	[217,218]
Team Miles	Designed to travel about 60 million km from Earth on a trajectory toward Mars. Employs plasma iodine thrusters that generate low-frequency electromagnetic waves as propulsion.	Miles Space (FL-based aerospace company)	Successfully deployed but contact was never established.	[219]

Source: Numerous websites and literature as noted in the table and text.
[a] See references for acronyms of CubeSats.
[b] Successful mission names are bold.

All ten CubeSats were successfully deployed. However, communications and/or power issues terminated the six unsuccessful missions in Table 19.5; discharged batteries that could not be recharged in a timely fashion are a likely cause [153,229,230]. CubeSats of all sizes have failed for a variety of reasons; a valuable website provides a wealth of detailed information about all aspects of CubeSats [231]. Information gleaned from the details of the various CubeSat missions is that power to be expected from the PV arrays ranged from 30 to 50 W at 1 AU. However, it must be pointed out that PV is but one component of the power system [177,185].

ArgoMoon [222], in important aspects, was similar to the highly successful LICIACube [164] rideshare mission on DART: a simple mission with reliable components and an experienced team. Luna-H Map also had a straightforward mission with an extensive and experienced mission team. While the initial goals of the mission were successful, propulsion issues shortened Luna-H Map's mission [212]; Lunar Flashlight had similar challenges [228]. Despite a complex set of mission objectives, EQUULEUS, one of two CubeSats from the Japan Aerospace Exploration Agency (JAXA), is currently headed toward the second Earth-Moon Lagrange Point (LP-2), a key location that is one million (1.5 million km) from Earth [232].

NASA Ames Research Center (ARC) has a long and impressive history of space biology, particularly bio-science-themed CubeSats [233]. BioSentinel (Fig. 19.30), the sole biological experiment included, aims to

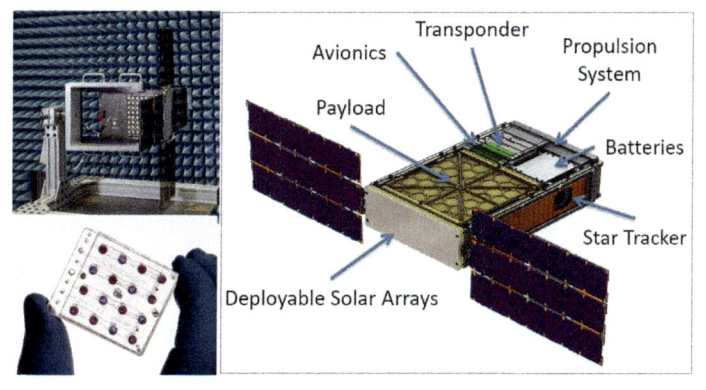

Figure 19.30 NASA BioSentinel 6U CubeSat: (*upper left*) BioSentinel flight unit with deployed solar array after completion of an electromagnetic compatibility test procedure; (*lower left*) BioSentinel's microfluidics card, designed at NASA Ames, will be used to study the impact of interplanetary space radiation on yeast. Here, pink wells contain actively growing yeast cells that have reduced the metabolic dye from blue to pink in color. (*right*) BioSentinel 6U CubeSat general system configuration—exploded view. *Courtesy: NASA.*

develop a biosensor to detect and measure the impact of space radiation on living organisms over long durations beyond LEO. Identifying and characterizing biological radiation effects using Earth-based facilities has made significant progress; however, as discussed above for PV materials, a terrestrial source cannot fully simulate the radiation environment encountered in deep space. BioSentinel had an issue with tumbling after deployment but the experienced team at NASA ARC was able to overcome this challenge; as of this writing, BioSentinel is currently in a heliocentric orbit. BioSentinal carried both an experiment with living yeast cells, and an onboard radiation detector. The initial mission phase ended in April 2023, but was extended up to another 18 months, to continue the collection of deep space radiation data outside of LEO. The mission will likely be completed (late-2024) before the publication of this book.

19.4.4 Photovoltaics technologies into the future: Opportunities for exploration of the solar system

Power systems are one of the most conservatively designed systems on a spacecraft, making it a challenge to introduce new technologies into the power system design. Any new technology must be flight-proven before it can be flown, which can greatly reduce the pace of innovation. This is demonstrated visually by examining the timeline of cell technologies for primary power sources for space missions (Fig. 19.23). Mission planners must look for capabilities and performance from solar arrays as well as the challenging environments of the mission destinations [31,45].

Despite numerous advances in SOP (SOA) MJ III-V solar arrays, their operational capabilities in extreme environments (Table 19.6) [31] are still limited. These include LILT environments on the outer planets, high temperatures (high or low solar irradiance) near the inner planets, corrosive environments on Venus, and dusty conditions on Mars [31,45]. In view of these challenges, the SOA solar power systems need improvements to successfully complete future outer planet, inner planet, and Mars missions under consideration for near future (into the early 2030s) planetary science missions, including reductions in mass and volume to power for SEP (or solar sail) missions to small bodies and outer planet destinations. Other niche applications, power-integrated technologies, and mission-enabling opportunities for the employment of advanced PV technologies are briefly addressed below.

Table 19.6 Solar power system needs for planetary exploration.

General destination	Mission type	Mission	Performance capability	Needs
Inner Planets	Orbiter	Venus and Mercury		• Low–Medium Temperature Operation • High Solar Intensities at Mercury
	Aerial (high to mid altitudes)	Venus		• Medium–High Temperature Operation (0°C–300°C) • Venus Solar Spectrum Operation • Operation in Corrosive Environments
Mars	Orbiter	Mars Missions	• Low Mass (>3× lower than SOP) • Low Volume (>3× lower than SOP)	• Long Life (15 years) • High Reliability • Mars Solar Spectrum Operation
	Landers/Rovers		• Low Mass (>3× lower than SOP) • Low Volume (>3× lower than SOP) • High Power Density (50% higher than SOP)	• High-Efficiency Cells • Dust Removal Capability • High Reliability
	Aerial Vehicle		• Low Mass (>4× lower than SOP) • High Power Density (50% higher than SOP)	• Mars Solar Spectrum Operation • Dust Tolerance
Outer Planets	Orbiters/Flyby	Jupiter* Saturn Europa* Titan Enceladus	• LILT Capability (η > 38% at 10 AU and T < −140°C) • High Voltage (>100 V) • High Power (>50 kW at 1 AU) • Low Mass (3× lower than SOP) • Low Volume (3× lower than SOP)	• Long Life (>15 years) • High Reliability • Radiation Tolerance (6×10^{15} 1 MeV e^-/cm^2)*

(Continued)

Table 19.6 (Continued)

General destination	Mission type	Mission	Performance capability	Needs
Small body	Orbiters/Flyby	Asteroids and comets	• High-efficiency solar cells ($\eta > 38\%$) • High voltage ($>100–200$ V) • High Power (>20 kW) • Low Mass ($>3\times$ lower than SOP) • Low Volume ($>3\times$ lower than SOP)	• Long Life (15 years) • High Reliability
	Surface Missions		• Low Mass ($>3\times$ lower than SOP) • Low volume ($>3\times$ lower than SOP) • High power density (50% higher than SOP)	• High Reliability

Source: Adapted from P.M. Beauchamp, J.A. Cutts, Jet Propulsion Laboratory, Solar Power Technologies for Future Planetary Science Missions, JPL D-101316, Pasadena, CA, USA, December 2017, 49 pp.

19.4.4.1 An introduction to solar sails and solar electric propulsion

As discussed above, the NEA Scout 6U CubeSat mission was to rendezvous with a small asteroid propelled by a solar sail measuring 925 ft^2 (86 m^2) [215,234]. The mirror-like thin aluminized polymer sail, deployed by stainless steel alloy booms [235], would utilize sunlight to generate thrust by reflecting solar photons. Sailing on light is highly efficient; momentum is generated by reflecting solar photons from membrane sails, allowing vehicles to be propelled indefinitely in space, reach and maintain novel orbits otherwise inaccessible, and conduct orbital plane changes more efficiently compared to spacecraft using conventional chemical propulsion [236–238]. Thus solar sails are a potentially disruptive enabling technology for nonnuclear exploration of the outer solar system; given the potential for spacecraft to achieve remarkable speeds.

Unfortunately, the NEA Scout team was unable to overcome the initial challenges encountered after CubeSat's successful deployment (Section 19.4.3.5). However, the solar sail concept and technology are being actively pursued by a number of space exploration entities [238–242]; in fact, several references [235,241,242] are included in an entire recent issue of the journal *Advances in Space Research*. Images and a diagram of two successful solar sail missions are displayed in Fig. 19.31; the missions include NanoSail-D2 (Fig. 19.31A and B) [238] and IKAROS (Fig. 19.29C and D) [239,240].

SEP is a related but competing technology discussed above (Sections 19.3.2.3 and 19.3.4) [84,85,243,244]. SEP utilizes the electricity generated by solar arrays to power electric propulsion engines. Several NASA missions have already demonstrated the efficacy of SEP for the exploration of small bodies: Dawn [144], DART [159], DS1 [245], and Psyche [246]. The specific power of current sheets yields 500 W/kg; there are designs utilizing IMMs under development that could potentially deliver even higher (i.e., 700 W/kg) mass-specific power [247]; this cell technology has been proposed for SEP missions to the outer solar system [85,122,236].

19.4.4.2 Photovoltaics for solar sails and solar electric propulsion

An interesting study, published in 2006, evaluated and compared the performance of these two (solar sails and SEP) advanced propulsion concepts with traditional (liquid) rocket engines [248]. Mission scenarios considered to compare the performance of these upper-stage systems included Mars, the Asteroid Belt, and Jupiter. The goal of the study was to help inform mission planners considering these three propulsion systems for future deep space missions. The analysis determined that solar sails and SEP

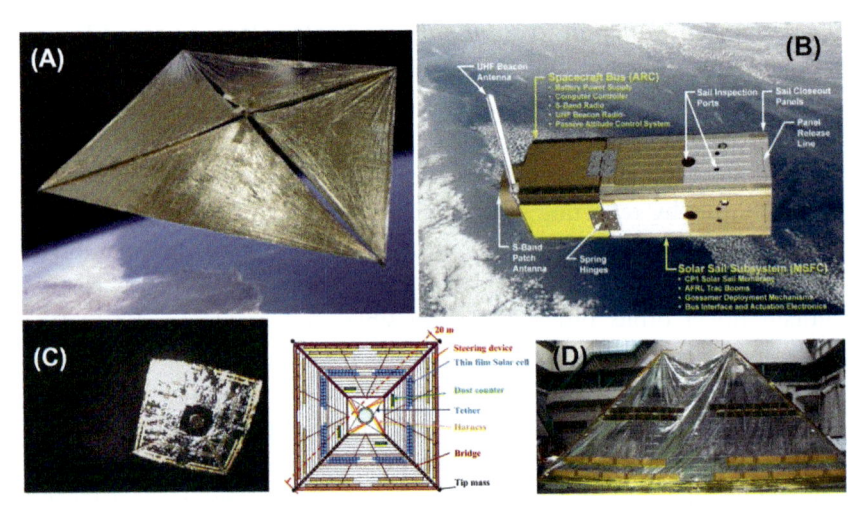

Figure 19.31 (A) NanoSail-D2 with 10 m² sail fully deployed in early 2011; (B) On-orbit view of the CubeSat before sail deployment in December 2010; the right-hand 1U of the 3U CubeSat holds NanoSail-D2; (C) Solar-sailing IKAROS in the interplanetary field, captured by Deployable Camera (DCAM) on June 14, 2010; (D) IKAROS solar sail (left: Overall configuration—a-Si cells are blue in the middle, right: One of four petals of the sail flight model). *(A and B) Courtesy: NASA; (C and D) Reproduced with permission from Y. Tsuda, O. Mori, R. Funase, H. Sawada, T. Yamamoto, et al., Achievement of IKAROS—Japanese deep space solar sail demonstration mission, Acta Astronaut. 82 (2) (2013) 183–188, copyright (2013) Elsevier.*

technologies fared reasonably well when compared to traditional rockets [248]. Solar sails may prove to be effective. In general, the benefits of electric engines over solar sails relate to launch and deployment considerations; the benefits of solar sails become apparent for longer, farther missions to the outer planets because they require little propellant. The most reasonable take-home lesson of this study is that the benefit(s) of a propulsion system varies according to specific missions.

Since this study was published, both types of PV technologies, MJ III-V and TFSC/PSCs/OPV, have improved [12,40–45,48,50]. However, several events have occurred that tend toward NASA favoring SEP, which are as follows: (1) two more SEP missions to small bodies have been successful [144,157]; (2) over the past 5 years, NASA GRC's small spacecraft electric propulsion (SSEP) project has further developed high-performance subkilowatt (<1 kW) Hall-effect thruster and power processing technologies to enable smaller spacecraft missions [249]; (3) NASA canceled a major solar sail demonstration mission, the Solar Cruiser in June 2022, due to serious budgetary issues [250]; (4) the NEA Scout CubeSat mission was not successful (however, Luna-H Map had electric propulsion issues that terminated the mission early); and (5). several solar sail missions have been successful over the past 15 years [238−241].

From the perspective of choice of PV technology, SEP has utilized MJ III-V arrays [12,13,43−45,120,144,244,245], while solar sails [236−242] would likely employ high-MSP materials on lightweight substrates such as TFSCs, PSCs, and OPV [40−42,48,50,56,91,116]. In fact, the JAXA Interplanetary Kite-craft Accelerated by Radiation Of the Sun (IKAROS) mission utilized a-Si solar arrays, attached to the sail membrane (14 m $\times$ 14 m) to generate electricity [239,240,251]. A hybrid SEP-capable JAXA solar power sail spacecraft, Oversize Kite-craft for Exploration and AstroNautics in the Outer Solar system (OKEANOS) (see bottom of Fig. 19.32) is a mission concept under development to rendezvous with a Jupiter Trojan asteroid [252], deploy a lander for in situ measurements, and even feasibly return samples to the Earth [242]. The estimated 17,000 TFSCs to be integrated onto the 39.7 m $\times$ 39.7 m sail (approximately eight times the area of the IKAROS solar sail) are currently slated to be commercial CIGS modules. A recent JAXA study investigated combining a hybrid solar sail/SEP spacecraft with a novel gravity assist maneuver to facilitate deep space exploration by small spacecraft [88]; the space solar sheets utilized for this concept include 32% efficient inverted 3J III-V cells [253]. To summarize the utility of the two fundamental types of cell technologies, high-MSP (kW/kg) arrays of thin-film (inorganic, hybrid, or organic) materials would seem to be better suited for large area solar sail spacecraft while high-ASP (kW/m^2) MJ (likely III-V) arrays are better suited for (small) SEP spacecraft.

19.4.4.3 Novel applications: Integrated power, CubeSats, and in situ resource utilization

We began a prior Section 19.4.3.1 with a description of a successful demonstration of integrated micropower technology [181−185,187,188] on a small satellite, Starshine 3 [189]. Starshine 3 predated the beginnings of the CubeSat format by several years [175−177,254]. The unique opportunity to fly several power experiments in LEO was unusually fortuitous as we were not providing the primary power source as the microsatellite was unpowered. Also, we had a rare opportunity to test a low-technology readiness level (TRL) technology [255,256] that had been on-ground tested with no prior flight heritage [181,183,188]. The IMPS data

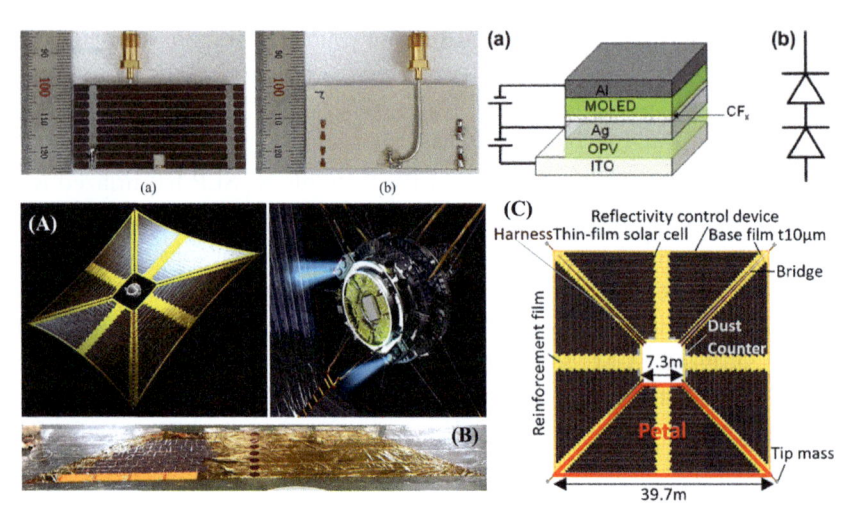

Figure 19.32 Images or diagrams of a variety of PV-integrated technologies. (*Top left*) Photograph of the fabricated prototype of an integrated solar cell/antenna device: (A) top view and (B) bottom view. (*Top right*) Device structure and circuitry diagram of the device components. (A) Schematic diagram of an integrated OPV/MOLED device, with a layer configuration of glass/ITO/ZnPc:C_{60} (35 nm)/C_{60} (25 nm)/Bphen (7 nm)/Ag (30 nm)/CF_x/NPB (55 nm)/Alq_3:C545 (25 nm)/Alq_3 (25 nm)/LiF (0.7 nm)/Al (150 nm). (B) A schematic equivalent circuitry of the integrated device. (*Bottom*) (A) Artist's conception of OKEANOS power sail (*left*) and SEP drive (*right*), (B) Photograph of the fabricated prototype, (C) Design of the OKEANOS power sail. (*Top left*) Reproduced from A. Ali, H. Wang, J. Lee, Y.H. Ahn, I. Park, Ultra-low profile solar-cell-integrated antenna with a high form factor, Sci. Rep. 11 (2021) 20918, distributed under the terms of the Creative Commons (CC-BY 4.0) license (http://creativecommons.org/licenses/by/4.0/), which permits unrestricted use; (*Top right*) Reproduced with permission from X.Z. Wang, H.L. Tam, K.S. Yong, Z.-K. Chen, F. Zhu, High performance optoelectronic device based on semitransparent organic photovoltaic cell integrated with organic light-emitting diode, Organic Electronics 12 (8) (August 2011) 1429−1433, copyright (2011) Elsevier; (*Bottom*) Reproduced from M. Matsushita, T. Chujo, J. Matsumoto, O. Mori, R. Yokota, et al., Solar power sail membrane prototype for OKEANOS mission, Advances in Space Research 67 (9) (May 2021) 2899−2911, distributed under the terms of the Creative Commons (CC-BY 4.0) license (http://creativecommons.org/licenses/by/4.0/), which permits unrestricted use.

downloaded during the mission [182,185] was facilitated by the other two NASA GRC experiments [187].

Recently, we reviewed the status of other power-integrated technologies developed over the past several decades [119,185]. The literature in power-integrated systems for energy conversion and storage [257−259], as well as other novel applications such as powered sensors [260−262], has grown dramatically over the past several decades. The PV technologies employed include a broad spectrum of materials and device structures [119,185,257−264]. For possible space applications, several instruments have been described in the recent literature, including an OPV-organic light emitting diode sensor [260], a CIGS solar sensor [262], an integrated lab-on-chip device for detection of life markers in extraterrestrial environments [263], and an integrated antenna/PV device [264]. In the previous subsection (19.4.4.2), we described several examples of JAXA spacecraft and/or studies that integrated various PV into solar sails: a-Si—IKAROS [240], CIGS—OKEANOS [242], and a 2021 study that models a mission using 3 J III-V cells [88]. Along with providing power in concert with solar sails, PV-integrated technologies are a potential opportunity for utilizing advanced PV technologies beyond Si [11,48−50,67,86,117−120,178,185,257−264]; see Fig. 19.32 for depictions of example integrated technologies [242,260,264], as well as an earlier example of an IMPS on Starshine 3 (Fig. 19.27 [181−183]).

As discussed above (Section 19.3.2.2), the AlSat-1N 3U CubeSat mission provided 3 years (2016−19) of performance data on thin-film CdTe solar cells on ultra-thin glass in LEO [32], the initial data to support the predictions made from earlier laboratory studies [63,64]. To date, this is a singular in-space test of a thin-film technology. The GoSolAR platform being developed by the DLR is currently planning to use either CIGS or thin, flexible GaAs cells [83]. The high-altitude balloon flight experiment of OPV and PSCs described above (Section 19.3.3.3) is, to date, the closest to space-related testing of lightweight cell technologies in situ [91]. The recent extensive Small Spacecraft Technology report lists approximately 10 commercial manufacturers of CubeSat solar cells (or panels or arrays) with MSPs ranging from 40 to 160 W/kg, although most commonly in the range of 75−100 W/kg [177]. These tend to be high (TRL 7−9 [255]) MJ III-V technology or less frequently Si technologies, including the Powerfilm module flown on IKAROS [251]. As CubeSat mission planners tend to view power as part of the baseline infrastructure, it is not realistic to expect novel PV technologies that are not reliable, and flight qualified to provide critical mission power [177]. Disruptive technologies such as solar cell-integrated devices are more likely to be selected for future flight opportunities [119,177,181−185,257−264].

The reality of off-world settlements or bases on the Moon, Mars, and beyond is still many decades away [265−270]. However, NASA (and many others) have conducted paper studies and mission analyses since the late 1970s [271−276]. A foundational paradigm and related portfolio of technologies revolve around in situ resource utilization (ISRU) where, as much as possible, essential raw materials are obtained from available local resources [277–282]. In fact, there have been terrestrial tests of relevant materials processing technologies by NASA on Mauna Kea (Hawaii) in 2010 [281] and 2012 [282]; many of the technologies tested involve smaller, mobile platforms that will likely be more complex, capable versions of current PV-powered rovers [138]. Coordinating all of the processing and materials handling, storage, and transport for mining operations, as well as integration into building the basic infrastructure for an off-world outpost, is illustrated in Fig. 19.33 [281]. The prevalence of Si on the surface of the Moon and Mars makes it clear that PV manufacturing off-world will primarily produce Si solar cells or arrays [272,273,275,280]. Although alternative solar cell types (pyrite) have been considered [283], Si solar cells are most likely to dominate the product mix.

The significant and complex infrastructure that must be established to "live off the land" on a nearby or deep space off-world surface is the subject of numerous studies published over the past 40-plus years [265–279]. Most likely, early settlements will rely on critical supplies from Earth and will use crude solutions to survivability challenges; for example, to protect personnel radiation, one solution is piling regolith on top of simple structures to serve as habitats

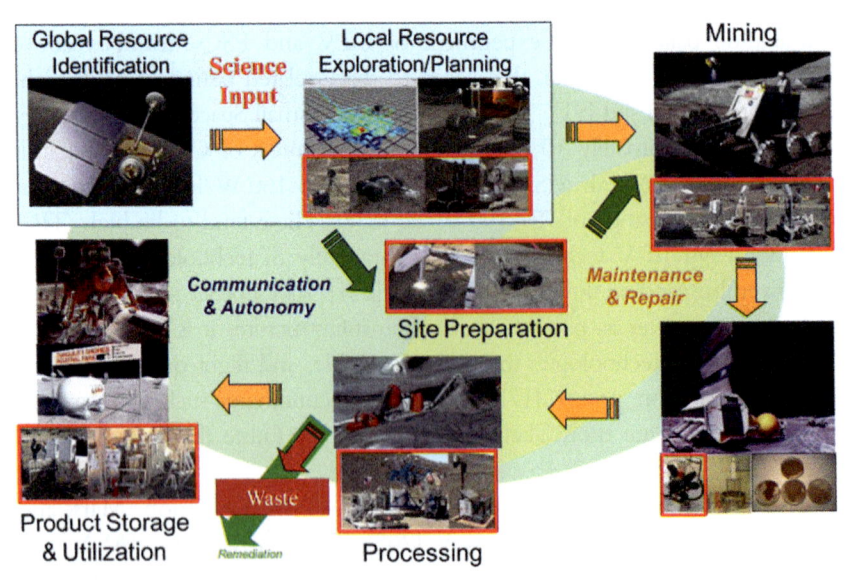

Figure 19.33 Mining operations as part of the development of infrastructure for a lunar outpost. *Reprinted with permission from G.B. Sanders, W.E. Larson, Integration of in-situ resource utilization into Lunar/Mars exploration through field analogs, Adv. Space Res. 47 (1) (Jan. 2011) 20–29, copyright (2011) Elsevier.*

[281,284]. Even simple settlements will require significant logistics, including preliminary cargo missions. An artist's conception of a lunar settlement from a 1989 study commissioned by NASA JSC would likely take many years to realize (Fig. 19.34) [265,268,270,275,281,282,284].

This is a further rationale for utilizing space stations at key locations such as LEO [19], Lagrange points [232,285], and orbiting the Moon itself [286,287]. The Gateway Program is an international effort led by NASA to build a small, human-tended space station orbiting the Moon, which is envisioned to be 1/8 the volume of the ISS and will provide extensive capabilities to support the Artemis campaign (Fig. 19.35) [227,287]. Gateway's capabilities for supporting sustained exploration and research in deep space include docking ports for a variety of visiting spacecraft, space for the crew to live and work, and onboard science investigations. Gateway will be a critical platform for developing technology and capabilities to support Moon and Mars exploration in the coming years.

Of most relevance to PV technologies is the power and propulsion element (PPE), a high-power (60 kW) SEP spacecraft that will provide power [using

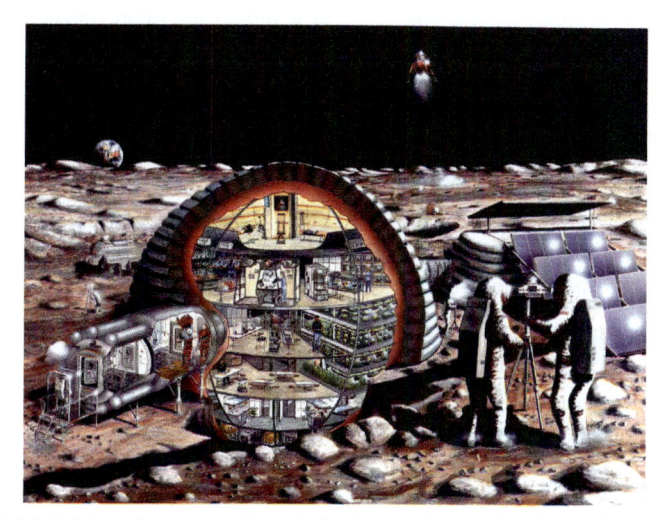

Figure 19.34 A 16 m diameter inflatable habitat is depicted and could accommodate the needs of a dozen astronauts living and working on the surface of the Moon. Illustrated are power and mining infrastructure (outside), astronauts exercising, a base operations center, a pressurized lunar rover, a small clean room, a fully equipped life sciences laboratory, a lunar lander, selenological work, hydroponic gardens, a wardroom, private crew quarters, dust-removing devices for lunar surface work, and an airlock. The top level of the habitat shows joggers required to run with their bodies almost parallel to the floor because of the low gravity. *Courtesy: NASA.*

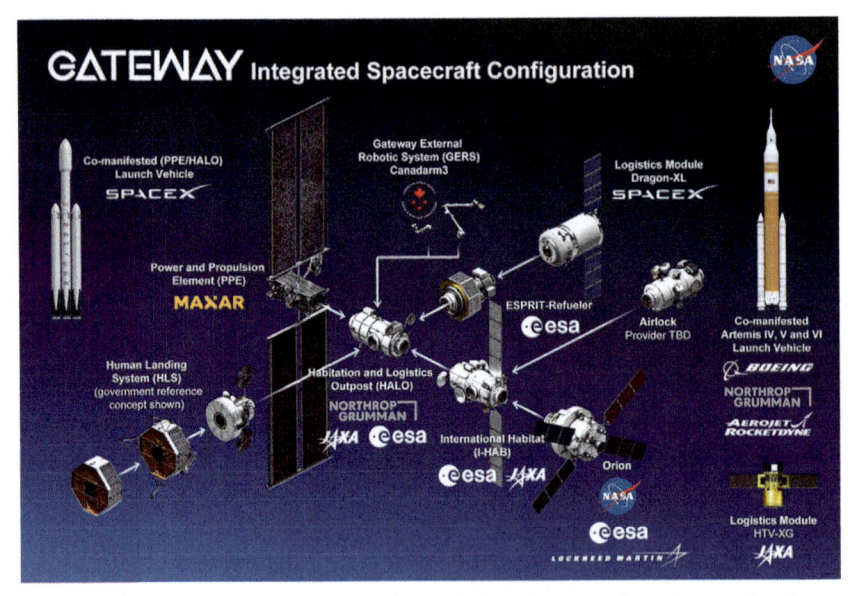

Figure 19.35 A full view of Gateway that includes elements from international partners. Built with commercial and international partners, the Gateway is critical to sustainable lunar exploration and will serve as a model for future missions to Mars. *Courtesy: NASA.*

updated ROSA arrays [12,19] (Section 19.1.3)], high-rate communications, attitude control, and orbital maintenance and transfer capabilities for the Lunar-orbiting space station. An important aspect of building Gateway is commercial partnerships; for example, the PPE is being developed and built by Maxar Technologies of Westminster, Colorado, and is managed by the NASA GRC in Cleveland, Ohio [288]. Clearly, the use of proven platforms and technologies such as space stations, rovers, SEP, and MJ III-V PV will serve as the foundation of space exploration in the future. There is certainly room for novel, disruptive technologies to contribute when a clear advantage or mission-enabling impact is demonstrated.

19.5 Conclusions

This chapter presented a succinct history of space PV, from the early days of solar cell development through the current space exploration missions that utilize advanced PV technologies. We have emphasized the impact that PV technology continues to have on the exploration and development of space. The history of space PV parallels, in many ways, the history of PV. The development of space solar power systems drove much of the development of early PV solar cells and arrays; indeed, the first PV specialist conference (PVSC) was held in 1961 at NASA headquarters. The solar power industry today owes a tremendous debt of gratitude to the space power scientists and engineers who developed the foundational technology that is still being used in terrestrial power systems. The previous chapters in this book outline many intriguing details of PV beyond Si. The reader is also encouraged to consult several sources for information related to space PV technologies for future science missions, space exploration, and environment needs and opportunities for space PV [31,45,47,84,177,267,286].

The second section mentions several key factors related to space versus terrestrial PV, including the solar spectrum itself. The efficiency of the solar cell is sensitive to the power and energy of the radiative light source; thus different light sources are used for standard measurements of cells used in terrestrial versus extraterrestrial applications. We clearly outlined the numerous challenges involved with operating in space or the vicinity of off-world locales [31]. High-energy protons, alpha particles, and heavier particles make up GCR, and solar flare activity results in eruptions of electromagnetic radiation and particle (electrons, protons, and heavier ions) acceleration. Generally, device performance will degrade as a function of the differential flux spectrum and the total ionizing dose [25–32]. While numerous PV technologies have been proposed for space, inorganic materials (especially MJ III-V devices) have a demonstrated track record for the ability to perform under adverse conditions.

In the third section, we addressed classes of materials with relevance to space applications and environmental challenges, especially radiation tolerance. Clearly, the workhorse technology of future PV-powered planetary exploration missions that is will be MJ III-V materials. We examined three major types of inorganic thin-film materials; CdTe and CIGS were regarded as a potential competitor to III-V materials several decades ago. However, great strides have been made with MJ III-V devices, including thin cells and concentrator-enhanced devices. We also examined the newest group (third generation) of materials that include DSSCs, PSCs, and OPV. While the significant challenges imposed by space environmental issues and performance metrics remain, as with TFSCs, niche and specialized applications for these next-generation materials should be explored.

As solar cell technologies evolve and adapt, even those that reach technological maturity may not become established successes in terms of production over time. However, the knowledge gained can live on in applications well beyond those for which they were developed. For example, robust and effective TCOs created for use in a-SiH solar cells resulted in materials and methods that are broadly used in display and lighting technology. Indeed, the lessons of a-Si:H formation and stability are now being used in advanced Si architectures, including silicon heterojunction solar cells. Domain knowledge from DSSCs and planar OPV devices accelerated the development of perovskite solar cells, from the device architectures to the materials specifications and constraints of the charge transport materials. We hope that the reader will be able to apply some of the lessons learned in this chapter to their own unique PV research problems.

The achievements of the spectacular missions described in the fourth section of this chapter are grounded in the solar array designs of the past. The earliest spacecraft progressed from simple body-mounted cells to fixed fold-out panels or paddles, evolving in little over a decade into the deployable and gimbaled multipanel wings that are often used today. While this gimbaled design is ubiquitous in satellites, unique mission requirements drive unique solutions for deep-space planetary science missions. The early array designs persist; all robotic US planetary science missions have employed some variant on these designs, employing unique variations as mission requirements demanded. Solar arrays have enabled the robotic exploration of the solar system and the space environment; the reach of solar-powered missions continues to increase.

It is especially difficult to introduce new technologies into a power system design due to the necessity that any new technology must be flight-proven before it can be utilized in a mission. This reality of space technology development can greatly reduce the pace of innovation. Section 19.4 concluded with a consideration of power-integrated technologies, CubeSats, and local resource utilization to support off-world surface operations. Solar sails and SEP are complementary technologies that provide both propulsion and power for a spacecraft, particularly when PV technology is integrated into the sail. Currently, SEP is powered by MJ III-V arrays while high mass-specific power lightweight solar cell technologies have the potential for enabling integrated PV and propulsion with solar sails. CubeSats (and SmallSats) present a viable platform for flight testing novel and disruptive technologies involving lightweight inorganic, hybrid, and organic PV technologies. Surface operations on the Moon, Mars, and beyond have the potential to be supported by local resources. However, a substantial infrastructure must be employed, including a resilient concept of operations and mission architecture. PV technologies can be mission-enabling for small mobile materials processing platforms. Clearly, the use of proven platforms and technologies such as space stations, rovers, SEP, and MJ III-V PV will serve as the foundation of space exploration in the future. There is certainly room for novel, disruptive technologies to contribute when a clear advantage or mission-enabling impact is demonstrated.

Finally, we note several take-home lessons for the reader when considering the technologies and missions discussed in this chapter: (1) the importance for academics to partner with industry and government labs, especially those with prior spaceflight experience; (2) the spaceflight and mission landscape is quite competitive, several attempts (proposals) to be a part of this endeavor may be required; (3) CubeSats and instruments are a very good entry into this enterprise as lower TRL technologies can be selected for further development or flight; (4) in our experience, student projects and theses involving PV or space exploration are very popular and often result in quite positive impacts on student engagement and their future careers.

References

[1] R.P. Raffaelle, An introduction to space photovoltaics: technologies, issues, and missions, in: S.G. Bailey, A.F. Hepp, D.C. Ferguson, R.P. Raffaelle, S.M. Durbin (Eds.), Space Photovoltaics: Materials, Missions and Alternative Technologies, Elsevier, Cambridge, MA, USA, 2023, pp. 3−27. Available from: https://www.sciencedirect.com/science/article/pii/B9780128233009000017.

[2] D.M. Chapin, C.S. Fuller, G.L. Pearson, New silicon p-n junction photocells for conversion of solar radiation into electrical power, J. Appl. Phys. 25 (1954) 676−677.

[3] D.M. Chapin, C.S. Fuller, G.L. Pearson, Solar energy converting apparatus, U.S. Patent 2,780,765, 1957.

[4] A. Goetzberger, C. Hebling, H.-W. Schock, Photovoltaic materials, history, status and outlook, Mater. Sci. Eng.: R: Rep. 40 (1) (2003) 1−46. Available from: https://doi.org/10.1016/S0927-796X(02)00092-X.

[5] R. Easton, M. Votaw, I. Vanguard, I.G.Y. Satellite, (1958 Beta), Rev. Sci. Instrum. 30 (2) (1959) 70−75.

[6] D.F. Hoth, E.F. O'Neill, I. Welber, The Telstar satellite system, Bell Syst. Tech. J. 42 (4) (1963) 765−799.

[7] D.R. Glover, NASA experimental communications satellites, 1958−1995, in: A.J. Butrica (Ed.), Beyond the Ionosphere: Fifty Years of Satellite Communication SP-4217, Washington, D.C., 1997. <https://history.nasa.gov/SP-4217/ch6.htm>.

[8] Telstar I., Vol. 1: Telstar satellite design, construction, ground facilities, and uses, NASA SP-32, NASA Goddard Space Flight Center, Greenbelt, MD USA, June 1963. <https://ntrs.nasa.gov/citations/19640000959> (accessed 26.1121).

[9] J.J. Loeferski, Theoretical considerations governing the choice of the optimum semiconductor for photovoltaic solar energy conversion, J. Appl. Phys. 27 (7) (1956) 777−785.

[10] M. Yamaguchi, High-efficiency GaAs-based solar cells, in: M.M. Rahman, A.M. Asiri, A. Khan, I. Inamuddin, T. Tabbakh (Eds.), Post-Transition Metals, InTech Open, 2021. Available from: 10.5772/intechopen.94365.

[11] A.F. Hepp, S.G. Bailey, R.P. Raffaelle, Inorganic photovoltaic materials and devices: past, present, and future, in: S.-S. Sun, N.S. Sariciftci (Eds.), Organic Photovoltaics: Mechanisms, Materials and Devices, CRC Press, Boca Raton, FL, USA, 2005, pp. 19−36.

[12] P.T. Chiu, Space applications of III-V single- and multi-junction solar cells, in: R.P. Raffaelle, S.G. Bailey, D.C. Ferguson, S. Durbin, A.F. Hepp (Eds.), Space Photovoltaics: Materials, Missions and Alternative Technologies, Elsevier, Cambridge, MA USA, 2023, pp. 79−127. Available from: https://www.sciencedirect.com/science/article/pii/B9780128233009000042.

[13] Y. Okada, N. Miyashita, Inverted lattice-matched GaInP/GaAs/GaInNAsSb triple junction solar cells: epitaxial lift-off thin film devices and potential space applications, in: R.P. Raffaelle, S.G. Bailey, D.C. Ferguson, S. Durbin, A.F. Hepp (Eds.), Space Photovoltaics: Materials, Missions and Alternative Technologies, Elsevier, Cambridge, MA USA, 2023, pp. 265−291. Available from: https://www.sciencedirect.com/science/article/pii/B9780128233009000133.

[14] D.J. Friedman, J.F. Geisz, S.R. Kurtz, J.M. Olson, 1-eV solar cells with GaInNAs active layer, J. Cryst. Growth 195 (1998) 409−415. Available from: https://doi.org/10.1016/S0022-0248(98)00561-2.

[15] S.G. Bailey, R. Raffaelle, K. Emery, Space and terrestrial photovoltaics: synergy and diversity, Prog. Photovolt.: Res. Appl. 10 (6) (2002) 399−406.

[16] AIAA Solar cells and solar panels committee on standards, qualification and quality requirements for space solar cells, in: AIAA S-111A-2014 (Revision of AIAA S-111−2005), American Institute of Aeronautics and Astronautics, Reston, VA USA, June 2014, 25 pp. Available from: https://doi.org/10.2514/4.102806.

[17] L.M. Hague, K.J. Metcalf, G.M. Shannon, R.C. Hill, C.-Y. Lu, Performance of international space station electric power system during station assembly, in: Proceedings of the 31st Intersociety Energy Conversion Engineering Conference, Part 1, August 11−16, 1996, pp. 154−159. Available from: https://doi.org/10.1109/IECEC.1996.552863.

[18] J.A. Van Allen, L.A. Frank, Radiation around the earth to a radial distance of 107,400 km, Nature 183 (4659) (1959) 430−434.

[19] New solar arrays to power NASA's International Space Station research. <https://www.nasa.gov/feature/new-solar-arrays-to-power-nasa-s-international-space-station-research> (accessed 12.1121).

[20] P. Bouisset, V.D. Nguyen, Y.A. Akatov, M. Siegrist, N. Parmentier, V.V. Archangelsky, et al., Quality factor and dose equivalent investigations aboard the Soviet space station MIR, Adv. Space Res. 12 (2−3) (1992) 363−367.

[21] M.G. Smith, M. Kelley, M. Basner, A brief history of spaceflight from 1961 to 2020: an analysis of missions and astronaut demographics, Acta Astronaut. 175 (2020) 290−299.

[22] S. Bailey, R. Raffaelle, Chapter II-4-B−operation of solar cells in a space environment, in: S. Kalogirou (Ed.), McEvoy's Handbook of Photovoltaics (Third Edition) Fundamentals and Applications, Academic Press, 2017, pp. 987−1003.

[23] M.A. Green, E.D. Dunlop, J. Hohl-Ebinger, M. Yoshita, N. Kopidakis, et al., Solar cell efficiency tables (Version 60), Prog. Photovolt. 30 (7) (2022) 687−701. Available from: https://doi.org/10.1002/pip.3595.
[24] NREL Best Research-Cell Efficiency Chart. <https://www.nrel.gov/pv/cell-efficiency.html> (accessed 27.04.24).
[25] C.A. Gueymard, D. Myers, K. Emery, Proposed reference irradiance spectra for solar energy systems testing, Sol. Energy 73 (6) (2003) 443−467.
[26] J.R. Woodyard, G.A. Landis, Radiation resistance of thin-film solar cells for space photovoltaic power, Sol. Cell (1991) 297−329. Available from: https://doi.org/10.1016/0379-6787(91)90103-V.
[27] S.G. Bailey, D.J. Flood, Space photovoltaics, Prog. Photovolt. Res. Appl. 6 (1998) 1−14.
[28] H.S. Rauschenbach, Solar Cell Array Design Handbook, JPL SP43-38, Vol. 1, NASA CR-149364, 1976. <https://ntrs.nasa.gov/citations/19770007250>.
[29] D.J. Curtin, R.L. Statler, Review of radiation damage to silicon solar cells, IEEE Trans. Aerosp. Electron. Syst. AES 11 (4) (1975) 499−513.
[30] R.L. Crabb, Solar cell radiation damage, Radiat. Phys. Chem. 43 (1−2) (1994) 93−103. Available from: https://doi.org/10.1016/0969-806X(94)90204-6.
[31] A. Bermudez-Garcia, P. Voarino, O. Raccurt, Environments, needs and opportunities for future space photovoltaic power generation: a review, Appl. Energy 290 (2021) 116757. Available from: https://doi.org/10.1016/j.apenergy.2021.116757.
[32] D.A. Lamb, S.J.C. Irvine, M.A. Baker, C.I. Underwood, S. Mardhani, Thin film cadmium telluride solar cells on ultra-thin glass in low earth orbit—3 years of performance data on the AlSat-1N CubeSat mission, Prog. Photovolt. Res. Appl. 29 (2021) 1000−1007. Available from: https://doi.org/10.1002/pip.3423.
[33] B.A. Banks, S.K. Miller, K.K. de Groh, D.L. Waters, Lessons learned from atomic oxygen interaction with spacecraft materials in low earth orbit, AIP Conf. Proc. 1087 (2009) 312−328. Available from: https://doi.org/10.1063/1.3076845.
[34] D.C. Ferguson, D.P. Engelhart, R.C. Hoffmann, V.J. Murray, E.A. Plis, Space solar arrays and spacecraft charging, in: R.P. Raffaelle, S.G. Bailey, D.C. Ferguson, S. Durbin, A.F. Hepp (Eds.), Space Photovoltaics: Materials, Missions and Alternative Technologies, Elsevier, Cambridge, MA USA, 2022, pp. 29−50. Available from: https://www.sciencedirect.com/science/article/pii/B9780128233009000029.
[35] U. Banik, K. Sasaki, N. Reininghaus, K. Gehrke, M. Vehse, M. Sznajder, et al., Enhancing passive radiative cooling properties of flexible CIGS solar cells for space applications using single layer silicon oxycarbonitride films, Sol. Energy Mater. & Sol. Cell 209 (2020) 110456. Available from: https://doi.org/10.1016/j.solmat.2020.110456.
[36] M. Pscherer, M. Günthner, C.A. Kaufmann, A. Rahm, G. Motz, Thin-film silazane/alumina high emissivity double layer coatings for flexible Cu(In, Ga)Se2 solar cells, Sol. Energy Mater. Sol. Cell. 132 (2015) 296−302. Available from: https://doi.org/10.1016/j.solmat.2014.09.015.
[37] https://www.spacex.com/media/Capabilities&Services.pdf (accessed16.03.23).
[38] K. Keränen, P. Korhonen, J. Rekilä, O. Tapaninen, T. Happonen, P. Makkonen, et al., Roll-to-roll printed and assembled large area LED lighting element, Int. J. Adv. Manuf. Technol. 81 (1) (2015) 529−536. Available from: https://doi.org/10.1007/s00170-015-7244-6.
[39] L.K. Reb, M. Böhmer, B. Predeschly, S. Grott, C.L. Weindl, et al., Perovskite and organic solar cells on a rocket flight, Joule 4 (9) (2020) 1880−1892. Available from: https://doi.org/10.1016/j.joule.2020.07.004.
[40] Y. Hu, T. Niu, Y. Liu, Y. Zhou, Y. Xia, et al., Flexible perovskite solar cells with high power-per-weight: progress, application, and perspectives, ACS Energy Lett. 6 (8) (2021) 2917−2943. Available from: https://doi.org/10.1021/acsenergylett.1c01193.

[41] A.K.-W. Chee, On current technology for light absorber materials used in highly efficient industrial solar cells, Renew. Sustain. Energy Rev. 173 (2023) 113027. Available from: https://doi.org/10.1016/j.rser.2022.113027.

[42] P.M. Beauchamp, J.A. Cutts, Solar power technologies for future planetary science missions, Jet Propulsion Laboratory, Report JPL D-101316, Pasadena, CA, USA, December 2017, 49 pp., full report available online.

[43] J.M. Raya-Armenta, N. Bazmohammadi, J.C. Vasquez, J.M. Guerrero, A short review of radiation-induced degradation of III−V photovoltaic cells for space applications, Sol. Energy Mater. Sol. Cell 233 (2021) 111379. Available from: https://doi.org/10.1016/j.solmat.2021.111379.

[44] T. Sumita, M. Imaizumi, S. Matsuda, T. Ohshima, A. Ohi, H. Itoh, Proton radiation analysis of multi-junction space solar cells, Nucl. Instrum. Methods Phys. Res. B: Beam Interact. Mater. At. 206 (2003) 448−451. Available from: https://doi.org/10.1016/S0168-583X(03)00791-2.

[45] M. Imaizumi, T. Nakamura, T. Takamoto, T. Ohshima, M. Tajima, Radiation degradation characteristics of component subcells in inverted metamorphic triple-junction solar cells irradiated with electrons and protons, Prog. Photovolt.: Res. Appl. 25 (2017) 161−174. Available from: https://doi.org/10.1002/pip.2840.

[46] R. Zwanenburg, M. Kroon, Requirements for thin film solar arrays and the development status of a new solar cell blanket concept, Eur. Space Agency (Special Publication) 589 (2005) 299−304 (Seventh European Space Power Conference, 9 May−13 May 2005).

[47] C.R. Mercer, Solar array designs for deep-space science missions, in: S.G. Bailey, A. F. Hepp, D.C. Ferguson, R.P. Raffaelle, S.M. Durbin (Eds.), Photovoltaics for Space: Key Issues, Missions and Alternative Technologies, Elsevier, 2022, pp. 349−378. Available from: https://www.sciencedirect.com/science/article/pii/B9780128233009000029.

[48] D.J. Hoffman, T.W. Kerslake, A.F. Hepp, M.K. Jacobs, D. Ponnusamy, Thin-film photovoltaic solar array parametric assessment, in: Proceedings of the 35th Intersociety Energy Conversion Engineering Conference (1), AIAA-2000−2919, July 24−28, 2000, pp. 670−680, also: NASA/TM-2000−210342. <https://ntrs.nasa.gov/citations/20000081753>.

[49] S.G. Bailey, A.F. Hepp, R.P. Raffaelle, Thin film photovoltaics for space applications, in: Proceedings of the 36th Intersociety Energy Conversion Engineering Conference (1), July 29−August 2, 2001, pp. 235−238.

[50] A.F. Hepp, J.S. McNatt, S.G. Bailey, R.P. Raffaelle, B.J. Landi, S.-S. Sun, et al., Ultra-lightweight space power from hybrid thin-film solar cells, IEEE Aerosp. Elec. Syst. 23 (9) (2008) 31−41.

[51] S. Bailey, R. Raffaelle, Space solar cells and arrays, in: A. Luque, S. Hegedus (Eds.), Handbook of Photovoltaics Science and Engineering, John Wiley & Sons, 2011, pp. 365−401. Available from: https://doi.org/10.1002/9780470974704.ch9.

[52] H. Afshari, B.K. Durant, C.R. Brown, K. Hossain, D. Poplavskyy, B. Rout, et al., The role of metastability and concentration on the performance of CIGS solar cells under low-intensity-low-temperature conditions, Sol. Energy Mater. Sol. Cell 212 (2020) 110571. August.

[53] P. Mahawela, G. Sivaraman, S. Jeedigunta, J. Gaduputi, M. Ramalingam, S. Subramanian, et al., II-VI compounds as the top absorbers in tandem solar cell structures, Mater. Sci. Eng. B 116 (2005) 283−291.

[54] T.D. Lee, A.U. Ebong, A review of thin film solar cell technologies and challenges, Renew. Sustain. Energy Rev. 70 (2017) 1286−1297. Available from: https://doi.org/10.1016/j.rser.2016.12.028.

[55] G. Han, S. Zhang, P.P. Boix, L.H. Wong, L. Sun, S.-Y. Lien, Towards high efficiency thin film solar cells, Prog. Mater. Sci. 87 (2017) 246—291. Available from: https://doi.org/10.1016/j.pmatsci.2017.02.003.
[56] J. Ramanujam, D.M. Bishop, T.K. Todorov, O. Gunawan, J. Rath, R. Nekovei, et al., Flexible CIGS, CdTe and a-Si:H based thin film solar cells: a review, Prog. Mater. Sci. 110 (2020) 100619.
[57] S.-I. Sato, H. Sai, T. Ohshima, M. Imaizumi, K. Shimazaki, M. Kondo, Temperature influence on performance degradation of hydrogenated amorphous silicon solar cells irradiated with protons, Prog. Photovolt. Res. Appl. 21 (2013) 1499—1506.
[58] A.D. Verkerk, J.K. Rath, R.E.I. Schropp, Degradation of thin film nanocrystalline silicon solar cells with 1 MeV protons, Energy Procedia 2 (1) (2010) 221—226. Available from: https://doi.org/10.1016/j.egypro.2010.07.032.
[59] H. Fritzsche, Development in understanding and controlling the Staebler-Wronski effect in a-Si:H, Annu. Rev. Mater. Sci. 31 (2001) 47—79.
[60] A. Romeo, E. Artegiani, CdTe-based thin film solar cells: past, present and future, Energies 14 (6) (2021) 1684. Available from: https://doi.org/10.3390/en14061684.
[61] F. Bittau, C. Potamialis, M. Togay, A. Abbas, P.J.M. Isherwood, J.W. Bowers, et al., Analysis and optimization of the glass/TCO/MZO stack for thin film CdTe solar cells, Sol. Energy Mater. Sol. Cell 187 (2018) 15—22.
[62] H.P. Mahabaduge, W.L. Rance, J.M. Burst, M.O. Reese, D.M. Meysing, C.A. Wolden, et al., High-efficiency, flexible CdTe solar cells on ultra-thin glass substrates, Appl. Phys. Lett. 106 (13) (2015) 133501.
[63] A. Romeo, D.L. Bätzner, H. Zogg, A.N. Tiwari, Potential of CdTe thin film solar cells for space applications, in: Proceedings of the 17th European Photovoltaic Conference and Exhibition, Munich, Germany, 22—26 October 2001, pp. 2183—2186.
[64] D.A. Lamb, C.I. Underwood, V. Barrioz, R. Gwilliam, J. Hall, M.A. Baker, et al., Proton irradiation of CdTe thin film photovoltaics deposited on cerium-doped space glass, Prog. Photovolt.: Res. Appl. 25 (12) (2017) 1059—1067.
[65] J. Gwak, M. Lee, J.H. Yun, S. Ahn, A. Cho, S. Ahn, et al., Selenium flux effect on Cu(In, Ga)Se$_2$ thin films grown by a 3-stage Co-evaporation process, Isr. J. Chem. 55 (10) (2015) 1115—1122. Available from: https://doi.org/10.1002/ijch.201500011.
[66] T. Kato, J.-L. Wu, Y. Hirai, H. Sugimoto, V. Bermudez, Record efficiency for thin-film polycrystalline solar cells Up to 22.9% achieved by Cs-treated Cu(In, Ga)(Se,S)$_2$, IEEE J. Photovolt. 9 (1) (2019) 325—330. Available from: https://doi.org/10.1109/JPHOTOV.2018.2882206.
[67] R.P. Raffaelle, A.F. Hepp, J.E. Dickman, D.J. Hoffman, N. Dhere, J.R. Tuttle, et al., Thin-film solar cells on metal foil substrates for space power, in: Proceedings of 2nd International Energy Conversion Engineering Conference, Providence, RI, AIAA-2004—5735, August 2004, 18 pp.
[68] P. Jackson, R. Wuerz, D. Hariskos, E. Lotter, W. Witte, M. Powalla, Effects of heavy alkali elements in Cu(In, Ga)Se$_2$ solar cells with efficiencies up to 22.6%, Phys. Status Solidi Rapid Res. Lett. 10 (8) (2016) 583—586. Available from: https://doi.org/10.1002/pssr.201600199.
[69] A. Chirilă, P. Reinhard, F. Pianezzi, P. Bloesch, A.R. Uhl, C. Fella, et al., Potassium-induced surface modification of Cu(In, Ga)Se$_2$ thin films for high-efficiency solar cells, Nat. Mater. 12 (2013) 1107—1111.
[70] R. Carron, S. Nishiwaki, T. Feurer, R. Hertwig, E. Avancini, J. Löckinger, et al., Advanced alkali treatments for high-efficiency Cu(In, Ga)Se$_2$ solar cells on flexible substrates, Adv. Energy Mater. 9 (24) (2019) 1900408.
[71] L. Zortea, S. Nishiwaki, T.P. Weiss, S. Haass, J. Perrenoud, L. Greuter, et al., Cu(In, Ga)Se$_2$ solar cells on low cost mild steel substrates, Sol. Energy 175 (2018) 25—30. Available from: https://doi.org/10.1016/j.solener.2017.12.057.

[72] J.C. Perrenoud, Low temperature grown CdTe thin film solar cells for the application on flexible substrates, Ph.D. Thesis, ETH Zürich, CH, No. 20460, 2012. Available from: https://doi.org/10.3929/ethz-a-007339371.

[73] L. Kranz, C. Gretener, J. Perrenoud, R. Schmitt, F. Pianezzi, F. La Mattina, et al., Doping of polycrystalline CdTe for high-efficiency solar cells on flexible metal foil, Nat. Commun. 4 (2013) 2306. Available from: https://doi.org/10.1038/ncomms3306.

[74] T. Matsui, A. Bidiville, H. Sai, et al., High-efficiency amorphous silicon solar cells: impact of deposition rate on metastability, Appl. Phys. Lett. 106 (5) (2015) 053901. Available from: https://doi.org/10.1063/1.4907001.

[75] M. Jeong, I.W. Choi, E.M. Go, Y. Cho, M. Kim, et al., Stable perovskite solar cells with efficiency exceeding 24.8% and 0.3-V voltage loss, Science 369 (6511) (2020) 1615−1620. Available from: https://doi.org/10.1126/science.abb7167.

[76] U. Würfel, J. Herterich, M. List, et al., A 1 cm^2 organic solar cell with 15.2% certified efficiency: detailed characterization and identification of optimization potential, Sol. RRL 5 (4) (2021) 2000802. Available from: https://doi.org/10.1002/solr.202000802.

[77] P. Nithya, C. Roumana, Unexpected high efficient dye sensitized solar cell based $NiWO_4$ decorated bio activated carbon nanosheets hybrid photoanodes by one-pot facile hydrothermal approach, Inorg. Chem. Commun. 118 (2020) 108039.

[78] Y. Kawano, J. Chantana, T. Nishimura, A. Mavlonov, T. Minemoto, Manipulation of [Ga]/([Ga] + [In]) profile in 1.4-μm-thick Cu(In, Ga)Se_2 thin film on flexible stainless steel substrate for enhancing short-circuit current density and conversion efficiency of its solar cell, Sol. Energy 204 (2020) 231−237. Available from: https://doi.org/10.1016/j.solener.2020.04.069.

[79] <https://www.pv-magazine.com/2019/01/21/solar-frontier-hits-new-cis-cell-efficiency-record/> (accessed 22.08.23).

[80] <https://www.pv-magazine.com/2021/11/01/solar-frontier-abandons-thin-film-launches-250-w-monocrystalline-panel/> (accessed 22.08.23).

[81] <https://www.pv-magazine.com/2023/03/15/cigs-module-manufacturer-announces-switch-to-perovskite/> (accessed 22.08.23).

[82] <https://www.pv-magazine.com/2023/07/25/belgian-startup-offers-custom-cigs-solar-foils/> (accessed 22.08.23).

[83] C. Morioka, K. Shimazaki, S. Kawakita, M. Imaizumi, H. Yamaguchi, T. Takamoto, et al., First flight demonstration of film-laminated InGaP/GaAs and CIGS thin-film solar cells by JAXA's small satellite in LEO, Prog. Photovolt. Res. Appl. 19 (2011) 825−833. Available from: https://doi.org/10.1002/pip.1046.

[84] C.R. Brown, V.R. Whiteside, D. Poplavskyy, K. Hossain, M.S. Dhoubhadel, I.R. Sellers, Flexible Cu(In, Ga)Se_2 solar cells for outer planetary missions: investigation under low-intensity low-temperature conditions, IEEE J. Photovolt. 9 (2019) 552−558. Available from: https://doi.org/10.1109/JPHOTOV.2018.2889179.

[85] A. Jasenek, U. Rau, K. Weinert, I.M. Kötschau, G. Hanna, et al., Radiation resistance of Cu(In, Ga)Se_2 solar cells under 1-MeV electron irradiation, Thin Solid. Films 387 (2001) 228−230. Available from: https://doi.org/10.1016/S0040-6090(00)01847-2.

[86] T. Sproewitz, U. Banik, J.-T. Grundmann, F. Haack, M. Hillebrandt, H. Martens, et al., Concept for a Gossamer solar power array using thin-film photovoltaics, CEAS Space J. 12 (2020) 125−135. Available from: https://doi.org/10.1007/s12567-019-00276-6.

[87] F. Chang Díaz, J. Carr, L. Johnson, W. Johnson, G. Genta, P.F. Maffione, Solar electric propulsion for human Mars missions, Acta Astronaut. 160 (2019) 183−194.

[88] Y. Takao, O. Mori, M. Matsushita, A.K. Sugihara, Solar electric propulsion by a solar power sail for small spacecraft missions to the outer solar system, Acta Astronaut. 181 (2021) 362−376.

[89] A.F. Hepp, S.G. Bailey, J.S. McNatt, M.V.S. Chandrashekhar, J.D. Harris, et al, Novel materials, processing, and device technologies for space exploration with potential dual-use applications, in: Propulsion and Energy Forum 2014, American Institute of Aeronautics and Astronautics, Cleveland, Ohio, USA July 28−30, 2014, AIAA−2014−3464, also available as NASA/TM- 2015−218866, December 2015, 21 pp.

[90] M. Liu, X. Tang, Y. Liu, Z. Xu, H. Wang, M. Fang, et al., A study on the variation of dye-sensitized solar cell parameters under γ irradiation, J. Radioanalytical Nucl. Chem. 308 (2) (2016) 631−637. May.

[91] Y. Liu, X. Tang, M. Liu, Z. Xu, Z. Zhang, M. Fang, Effect of 10 MeV electron irradiation on dye-sensitized solar cells, Radiat. Eff. Defects Solids 172 (3−4) (2017) 342−353.

[92] Z. Zhang, X. Tang, Y. Liu, Z. Xu, K. Liu, Z. Yuan, et al., A study on the degradation of dye-sensitized solar cells irradiated by two different dose rates of γ-rays, J. Radioanalytical Nucl. Chem. 312 (3) (2017) 609−614.

[93] I. Piña-López, K.M. García-Ruiz, C. Barrueta-Flores, C. Amador-Bedolla, et al., Theoretical study of the open circuit voltage decay on organic photovoltaic (OPV) solar cells based on space radiation ionizing damage, J. Phys.: Conf. Ser. 1723 (2021) 012017. Available from: https://doi.org/10.1088/1742-6596/1723/1/012017.

[94] I. Cardinaletti, T. Vangerven, S. Nagels, R. Cornelissen, D. Schreurs, Organic and perovskite solar cells for space applications, Sol. Energy Mater. Sol. Cell 182 (2018) 121−127.

[95] H.O. Lee III, M. Hasib, S.-S. Sun, Proton radiation studies on conjugated polymer thin films, MRS Adv. 2 (51) (2017) 2967−2972. Available from: https://doi.org/10.1557/adv.2017.389.

[96] H. Ahn, D.W. Oblas, J.E. Whitten, Electron irradiation of poly(3-hexylthiophene) films, Macromolecules 37 (9) (2004) 3381−3387.

[97] K. Kambour, N. Rosen, C. Kouhestani, D. Nguyen, C. Mayberry, et al., Modeling of the X-irradiation response of the carrier relaxation time in P3HT:PCBM organic-based photocells, IEEE Trans. Nucl. Sci. 59 (6) (2012) 2902−2906.

[98] M.S. Kim, J.H. Han, C.E. Lee, Electron mobility in poly(3-hexylthiophene) enhanced by gamma-ray irradiation, Curr. Appl. Phys. 15 (1) (2015) 25−28.

[99] K.K. de Groh, B.A. Banks, Atomic oxygen erosion data from the MISSE 2−8 missions, NASA/TM-2019−219982, May 2019, 32 pp. <https://ntrs.nasa.gov/citations/20190025445>.

[100] K.K. de Groh, B.A. Banks, MISSE-Flight Facility: Polymers and Composites Experiment 1−4 (PCE 1−4), NASA/TM-20205008863, February 2021, 90 pp. <https://ntrs.nasa.gov/citations/20205008863>.

[101] <https://www1.grc.nasa.gov/space/iss-research/misse/> (accessed 23.04.23).

[102] <https://www1.grc.nasa.gov/facilities/> (accessed 23.04.23).

[103] A.R. Kirmani, B.K. Durant, J. Grandidier, N.M. Haegel, M.D. Kelzenberg, et al., Countdown to perovskite space launch: guidelines to performing relevant radiation-hardness experiments, Joule 6 (5) (2022) 1015−1031.

[104] J. Huang, M.D. Kelzenberg, P. Espinet-González, C. Mann, D. Walker, et al., Effects of electron and proton radiation on perovskite solar cells for space solar power application, in: Proceedings of the 2017 IEEE 44th Photovoltaic Specialist Conference (PVSC), 2017, pp. 1248−1252.

[105] Y. Miyazawa, M. Ikegami, H.W. Chen, T. Ohshima, M. Imaizumi, K. Hirose, et al., Tolerance of perovskite solar cell to high-energy particle irradiations in space environment, iScience 2 (2018) 148−155. Available from: https://doi.org/10.1016/j.isci.2018.03.020.

[106] O. Malinkiewicz, M. Imaizumi, S.B. Sapkota, T. Ohshima, S. Öz, Radiation effects on the performance of flexible perovskite solar cells for space applications, Emerg. Mater. Res. 3 (2020) 9–14. Available from: https://doi.org/10.1007/s42247-020-00071-8.

[107] J. Barbé, D. Hughes, Z. Wei, A. Pockett, H.K.H. Lee, et al., Radiation hardness of perovskite solar cells based on aluminum-doped zinc oxide electrode Under proton irradiation, Sol. RRL 3 (2019) 1900219. Available from: https://doi.org/10.1002/solr.201900219.

[108] F. Lang, M. Jošt, J. Bundesmann, A. Denker, S. Albrecht, et al., Efficient minority carrier detrapping mediating the radiation hardness of triple-cation perovskite solar cells under proton irradiation, Energy Environ. Sci. 12 (2019) 1634–1647. Available from: https://doi.org/10.1039/C9EE00077A.

[109] B.K. Durant, H. Afshari, S. Singh, B. Rout, G.E. Eperon, I.R. Sellers, Tolerance of perovskite solar cells to targeted proton irradiation and electronic ionization induced healing, ACS Energy Lett. 6 (2021) 2362–2368. Available from: https://doi.org/10.1021/acsenergylett.1c00756.

[110] F. Lang, M. Jošt, K. Frohna, E. Köhnen, A. Al-Ashouri, et al., Proton radiation hardness of perovskite tandem photovoltaics, Joule 4 (2020) 1054–1069. Available from: https://doi.org/10.1016/j.joule.2020.03.006.

[111] S.R. Messenger, E.A. Burke, G.P. Summers, M.A. Xapsos, R.J. Walters, E.M. Jackson, et al., Nonionizing energy loss (NIEL) for heavy ions, IEEE Trans. Nucl. Sci. 46 (1999) 1595–1602. Available from: https://doi.org/10.1109/23.819126.

[112] S.R. Messenger, G.P. Summers, E.A. Burke, R.J. Walters, M.A. Xapsos, Modeling solar cell degradation in space: a comparison of the NRL displacement damage dose and the JPL equivalent fluence approaches, Prog. Photovolt.: Res. Appl. 9 (2001) 103–121. Available from: https://doi.org/10.1002/pip.357.

[113] S.R. Messenger, E.A. Burke, R.J. Walters, J.H. Warner, G.P. Summers, T.L. Morton, Effect of omnidirectional proton irradiation on shielded solar cells, IEEE Trans. Nucl. Sci. 53 (2006) 3771–3778. Available from: https://doi.org/10.1109/TNS.2006.886220.

[114] G.P. Summers, R.J. Walters, M.A. Xapsos, E.A. Burke, S.R. Messenger, P. Shapiro, et al., A new approach to damage prediction for solar cells exposed to different radiations, in: Proceedings of IEEE 1st World Conference on Photovoltaic Energy Conversion-WCPEC (A Joint Conference of PVSC, PVSEC and PSEC), 1994, pp. 2068–2075.

[115] B.K. Durant, H. Afshari, S. Sourabh, V. Yeddu, M.T. Bamidele, S. Singh, et al., Radiation stability of mixed tin–lead halide perovskites: implications for space applications, Sol. Energy Mater. Sol. Cell 230 (2021) 111232.

[116] F. Lang, N.H. Nickel, J. Bundesmann, S. Seidel, A. Denker, S. Albrecht, et al., Radiation hardness and self-healing of perovskite solar cells, Adv. Mater. 28 (2016) 8726–8731.

[117] L. McMillon-Brown, T.J. Peshek, Prospects for perovskites in space, in: R.P. Raffaelle, S.G. Bailey, D.C. Ferguson, S. Durbin, A.F. Hepp (Eds.), Space Photovoltaics: Materials, Missions and Alternative Technologies, Elsevier, Cambridge, MA USA, 2023, pp. 129–156. Available from: https://www.sciencedirect.com/science/article/pii/B9780128233009000182.

[118] B.K. Durant, I.R. Sellers, Perovskite solar cells on the horizon for space power systems, in: R.P. Raffaelle, S.G. Bailey, D.C. Ferguson, S. Durbin, A.F. Hepp (Eds.), Space Photovoltaics: Materials, Missions and Alternative Technologies, Elsevier, Cambridge, MA USA, 2023, pp. 175–195. Available from: https://www.sciencedirect.com/science/article/pii/B978012823300900011X.

[119] I.T. Martin, K. Crowley, A.F. Hepp, Thin film solar cells and arrays for space power, in: R.P. Raffaelle, S.G. Bailey, D.C. Ferguson, S. Durbin, A.F. Hepp (Eds.), Space Photovoltaics: Materials, Missions and Alternative Technologies, Elsevier, Cambridge, MA USA, 2023, pp. 215−263. Available from: https://www.sciencedirect.com/science/article/pii/B9780128233009000157.

[120] M.J. O'Neil, M.F. McDanal, M.F. Piszczor, D.L. Edwards, M.I. Eskenazi, H.W. Brandhorst, Recent technology advances for the stretched lens array (SLA), a space solar array offering state of the art performance at low cost and ultra-light mass, in: Conference Record of the Thirty-first IEEE Photovoltaic Specialists Conference, 2005, pp. 810−813. doi:10.1109/PVSC.2005.1488256.

[121] G.A. Landis, Tabulation of power-related satellite failure causes, Ain: IAA 2013−3736, 11th International Energy Conversion Engineering Conference, July 14−17, 2013, San Jose, CA, USA. doi.org/10.2514/6.2013-3736.

[122] M.J. O'Neill, Space PV concentrators for outer planet and near-sun missions using ultra-light Fresnel lenses, in: R.P. Raffaelle, S.G. Bailey, D.C. Ferguson, S. Durbin, A.F. Hepp (Eds.), Space Photovoltaics: Materials, Missions and Alternative Technologies, Elsevier, Cambridge, MA USA, 2023, pp. 411−432. Available from: https://www.sciencedirect.com/science/article/pii/B9780128233009000078.

[123] M. Yamaguchi, H. Yamada, Y. Katsumata, K.-H. Lee, K. Araki, N. Kojima, Efficiency potential and recent activities of high-efficiency solar cells, J. Mater. Res. 32 (2017) 3445−3457. Available from: https://doi.org/10.1557/jmr.2017.335.

[124] M. Yamaguchi, K.H. Lee, K. Araki, N. Kojima, H. Yamada, Y. Katsumata, Analysis for efficiency potential of high-efficiency and next generation solar cells, Prog. Photovol. 26 (2018) 543−552. Available from: https://doi.org/10.1002/pip.2955.

[125] W. Shockley, H.J. Queisser, Detailed balance limit of efficiency of p-n junction solar cells, J. Appl. Phys. 32 (1961) 510−519. Available from: https://doi.org/10.1063/1.1736034.

[126] O. Dupré, R. Vaillon, M.A. Green, Physics of the temperature coefficients of solar cells, Sol. Energy Mater. Sol. Cell 140 (2015) 92−100.

[127] D. Jariwala, A.R. Davoyan, J. Wong, H.A. Atwater, Van der Waals materials for atomically-thin photovoltaics: promise and outlook, ACS Photonics 4 (12) (2017) 2962−2970. Available from: https://pubs.acs.org/doi/10.1021/acsphotonics.7b01103.

[128] M. Ochoa, S. Buecheler, A.N. Tiwari, R. Carron, Challenges and opportunities for an efficiency boost of next generation Cu(In, Ga)Se$_2$ solar cells: prospects for a paradigm shift, Energy Environ. Sci. 13 (2020) 2047−2055. Available from: https://doi.org/10.1039/D0EE00834F.

[129] J.F. Geisz, R.M. France, K.L. Schulte, M.A. Steiner, A.G. Norman, et al., Six junction III-V solar cells with 47.1% conversion efficiency under 143 suns concentration, Nat. Energy 5 (2020) 326. Available from: https://doi.org/10.1038/s41560-020-0598-5.

[130] <https://www.nasa.gov/feature/new-solar-arrays-to-power-nasa-s-international-space-station-research> (accessed 29.04.23).

[131] <https://www.nasa.gov/mission_pages/messenger/main/index.html> (accessed 29.04.23).

[132] <https://www.nasa.gov/mission_pages/LRO/main/index.html> (accessed 01.05.23).

[133] <https://solarsystem.nasa.gov/missions/grail/in-depth/> (accessed 01.05.23).

[134] <https://solarsystem.nasa.gov/missions/ladee/in-depth/> (accessed 01.05.23).

[135] <https://www.nasa.gov/mission_pages/LCROSS/overview/index.html> (accessed 01.05.23).

[136] <https://mars.nasa.gov/mars-exploration/missions/mars-global-surveyor/> (accessed 01.05.23).

[137] <https://solarsystem.nasa.gov/missions/mars-odyssey/in-depth/> (accessed 01.05.23).

[138] <https://mars.nasa.gov/mer/> (accessed 01.05.23).
[139] <https://mars.nasa.gov/mro/> (accessed 01.05.23).
[140] <https://mars.nasa.gov/mars-exploration/missions/phoenix/> (accessed 01.05.23).
[141] <https://mars.nasa.gov/maven/>, accessed on May 1, 2023.
[142] <https://www.jpl.nasa.gov/missions/juno>, accessed on May 1, 2023.
[143] <https://solarsystem.nasa.gov/missions/deep-impact-epoxi/in-depth/> (accessed 01.05.23).
[144] <https://solarsystem.nasa.gov/missions/dawn/overview/> (accessed 01.05.23).
[145] <https://www.nasa.gov/osiris-rex> (accessed 01.05.23).
[146] <https://www.nasa.gov/content/solar-missions-list> (accessed 01.05.23).
[147] <https://nssdc.gsfc.nasa.gov/planetary/upcoming.html> (accessed 01.05.23)..
[148] A.A. Siddiqi, Beyond Earth: A Chronicle of Deep Space Exploration, 1958–2016, 2nd Ed., The NASA History Series, National Aeronautics and Space Administration, Office of Communications, NASA History Division, Washington, DC, 2018, NASA SP2018–4041. <https://solarsystem.nasa.gov/resources/1060/beyond-earth-a-chronicle-of-deep-space-exploration/> (accessed 01.05.23).
[149] <https://solarsystem.nasa.gov/missions/near-shoemaker/in-depth/> (accessed 01.05.23).
[150] <https://www.nasa.gov/mission_pages/mars-pathfinder> (accessed 01.05.23).
[151] <https://mars.nasa.gov/insight/> (accessed 01.05.23).
[152] <https://www.jpl.nasa.gov/missions/mars-cube-one-marco> (accessed 01.05.23).
[153] F.C. Krause, J.A. Loveland, M.C. Smart, E.J. Brandon, R.V. Bugga, Implementation of commercial Li-ion cells on the MarCO deep space CubeSats, J. Power Sources 449 (2020) 227544. Available from: https://doi.org/10.1016/j.jpowsour.2019.227544.
[154] A.G. Santo, R.E. Gold, R.L. McNutt Jr., S.C. Solomon, C.J. Ercol, et al., The MESSENGER mission to Mercury: spacecraft and mission design, Planet. Space Sci. 49 (14–15) (2001) 1481–1500.
[155] C.J. Ercol, G. Dakermanji, B. Le, The MESSENGER spacecraft power subsystem thermal design and early mission performance, in: 4th International Energy Conversion Engineering Conference and Exhibit (IECEC), June 26–29, 2006, San Diego, CA, USA, AIAA 2006–4144, 14pp. <https://messenger.jhuapl.edu/Resources/Publications/Ercol.et.al.2006.pdf>.
[156] <https://solarsystem.nasa.gov/missions/lucy/in-depth/> (accessed 01.05.23).
[157] <https://solarsystem.nasa.gov/missions/psyche/overview/> (accessed 01.05.23).
[158] <https://europa.nasa.gov/> (accessed 01.05.23).
[159] <https://dart.jhuapl.edu/> (accessed 01.05.23).
[160] <https://www.nasa.gov/planetarydefense> (accessed 01.05.23).
[161] <https://cneos.jpl.nasa.gov/about/neo_groups.html> (accessed 01.05.23).
[162] E. Gaddy, Transformational solar array final report, NASA Contract NNC16CA19C. <https://ntrs.nasa.gov/api/citations/20170010684/downloads/20170010684.pdf> (accessed 01.05.23).
[163] <https://dart.jhuapl.edu/News-and-Resources/article.php?id = 20230301> (accessed 01.05.23).
[164] E. Dotto, V. Della Corte, M. Amoroso, I. Bertini, J.R. Brucato, et al., LICIACube – The light Italian Cubesat for imaging of asteroids in support of the NASA DART mission towards asteroid (65803) Didymos, Planet. Space Sci. 199 (2021) 105185.
[165] <https://epoxi.astro.umd.edu> (accessed 02.05.23).
[166] <https://nssdc.gsfc.nasa.gov/planetary/pioneer_venus.html> (accessed 02.05.23).
[167] <https://mars.nasa.gov/msl/home/> (accessed 02.05.23).
[168] <https://mars.nasa.gov/mars2020/> (accessed 02.05.23).

[169] D.C. Ferguson, D.M. Wilt, A.F. Hepp, J.C. Kolecki, M.W. Siebert, P.P. Jenkins, et al., The Mars Pathfinder wheel abrasion experiment, Mater. Des. 22 (7) (2001) 555−564.
[170] G.A. Landis, P.P. Jenkins, Measurement of the settling rate of atmospheric dust on Mars by the MAE instrument on Mars Pathfinder, J. Geophys. Research: Planets 105 (E1) (2000) 1855−1857. Available from: https://doi.org/10.1029/1999JE001029.
[171] <https://mars.nasa.gov/files/mars2020/Mars2020_Fact_Sheet.pdf> (accessed 02.05.23).
[172] <https://mars.nasa.gov/technology/helicopter/#Quick-Facts> (accessed 02.05.23).
[173] <https://www.rocketlabusa.com/space-systems/solar/> (accessed 02.05.23).
[174] <https://solarsystem.nasa.gov/missions/parker-solar-probe/in-depth/> (accessed 02.05.23).
[175] C. Cappelletti, S. Battistini, B.K. Malphrus (Eds.), Cubesat Handbook: From Mission Design to Operations, Academic Press, 2021, p. 498. Available from: https://www.sciencedirect.com/book/9780128178843/cubesat-handbook.
[176] <https://www.nasa.gov/content/what-are-smallsats-and-cubesats, Accessed May 4, 2023.
[177] B. Yost, S. Weston, C. Burkhard, W. Czakon, A. Dono, et al., State-of-the-Art Small Spacecraft Technology, NASA/TP—2022−0018058, January 2023, NASA Ames Research Center, Moffett Field, CA, USA, 366 pp. <https://www.nasa.gov/smallsat-institute/sst-soa> (accessed 04.05.23).
[178] A.F. Hepp, M.A. Smith, J.H. Scofield, J.E. Dickman, G.B. Lush, D.L. Morel, et al., Multi-Junction Thin-Film Solar Cells on Flexible Substrates for Space Power, NASA TM/2002−211834, October 2002, 12 pp.
[179] J.P. Maranchi, A.F. Hepp, P.N. Kumta, High-capacity, reversible silicon thin-film anodes for lithium ion batteries, Electrochem. Solid. State Lett. 6 (9) (2003) A198−A201.
[180] R.P. Raffaelle, B.J. Landi, J.D. Harris, S.G. Bailey, A.F. Hepp, Carbon nanotubes for power applications, Mater. Sci. Eng. B 116 (3) (2005) 233−243.
[181] R.P. Raffaelle, J. Underwood, D. Scheiman, J. Cowen, P. Jenkins, A.F. Hepp, et al., Integrated solar power systems, in: Conference Record of the 28th IEEE Photovoltaic Specialists Conference, Anchorage, AK, USA, 2000, pp. 1370−1373.
[182] R.P. Raffaelle, A.F. Hepp, G.A. Landis, D.J. Hoffman, Mission applicability assessment of integrated power components and systems, Prog. Photovoltaics: Res. Appl. 10 (6) (2002) 391−397.
[183] D.M. Wilt, A.F. Hepp, M. Moran, P.P. Jenkins, D.A. Scheiman, R.P. Raffaelle, Integrated micro-power system (IMPS) development at NASA Glenn Research Center, in: Electrochemical Society Proceedings 2002-25, 2003, pp. 194−202.
[184] E.J. Simburger, J.H. Matsumoto, P.A. Gierow, A.F. Hepp, Integrated thin film battery and circuit module, U.S. Patent 7,045,246, May 16, 2006.
[185] A.F. Hepp, P.N. Kumta, O. Velikokhatnyi, R.P. Raffaelle, Batteries for integrated power and CubeSats: recent developments and future prospects, in: A.F. Hepp, P. N. Kumta, O. Velikokhatnyi, M.K. Datta (Eds.), Silicon Anode Systems for Lithium-Ion Batteries, Elsevier, New York, 2021, pp. 457−508.
[186] J. Bouwmeester, J. Guo, Survey of worldwide pico- and nanosatellite missions, distributions and subsystem technology, Acta Astronaut. 67 (7−8) (2010) 854−862.
[187] P. Jenkins, T. Kerslake, D. Scheiman, D. Wilt, R. Button, et al., First results from the Starshine 3 power technology experiment, in: Conference Record of the Twenty-Ninth IEEE Photovoltaic Specialists Conference, May 19−24, 2002, New Orleans, LA, USA, pp. 788−791. doi:10.1109/PVSC.2002.1190689.
[188] R.P. Raffaelle, P. Jenkins, D. Scheiman, M.A. Smith, D. Wilt, T. Kerslake, et al., Microsat power supplies, in: Proc. 40th AIAA Aerospace Sciences Meeting & Exhibit, Reno, NV, USA, 2002, AIAA-2002-0719, 4 pp. https://doi.org/10.2514/6.2002-719.

[189] <http://www.azinet.com/starshine/> (accessed 04.05.23).
[190] K. Ennico, M. Shirley, A. Colaprete, L. Osetinsky, The Lunar crater observation and sensing satellite (LCROSS) payload development and performance in flight, Space Sci. Rev. 167 (2012) 23−69. Available from: https://doi.org/10.1007/s11214-011-9753-4.
[191] <https://solarsystem.nasa.gov/missions/mars-cube-one/in-depth/> (accessed 05.05.23).
[192] K. Oudrhiri, O..Yang; D. Buccino; D. Kahan; P. Withers, et al., MarCO radio occultation: how the first interplanetary Cubesat can help improve future missions, in: 2020 IEEE Aerospace Conference, Big Sky, MT, USA, 2020, pp. 1−10. doi:10.1109/AERO47225.2020.9172734.
[193] <https://www.jpl.nasa.gov/infographics/infographic.view.php?id=10776> (accessed 05.05.23).
[194] <https://mmadesignllc.com/products/solar-arrays/> (accessed 05.05.23).
[195] R.W. Cook, L.G. Swan, K.P. Plucknett, Failure mode analysis of lithium ion batteries operated for low earth orbit CubeSat applications, J. Energy Storage 31 (2020) 101561.
[196] Panasonic, NCR18650B Data Sheet. <https://www.batteryspace.com/prod-specs/NCR18650B.pdf> (accessed 05.05.23).
[197] <https://solarsystem.nasa.gov/asteroids-comets-and-meteors/asteroids/didymos/in-depth/> (accessed 05.05.23).
[198] <https://nssdc.gsfc.nasa.gov/nmc/spacecraft/display.action?id=2021-110C> (accessed 05.05.23).
[199] <https://www.ssdc.asi.it/liciacube/> (accessed 05.05.23).
[200] <https://www.nasa.gov/feature/nasa-dart-imagery-shows-changed-orbit-of-target-asteroid> (accessed 05.05.23).
[201] T.S. Statler, S.D. Raducan, O.S. Barnouin, M.E. DeCoster, S.R. Chesley, After DART: using the first full-scale test of a kinetic impactor to inform a future planetary defense mission, Planet. Sci. J. 3 (2022) 244. Available from: https://doi.org/10.3847/PSJ/ac94c1.
[202] J.Y. Li, M. Hirabayashi, T.L. Farnham, J.M. Sunshine, M.M. Knight, et al., Ejecta from the DART-produced active asteroid Dimorphos, Nature 616 (7957) (2023) 452−456. Available from: https://doi.org/10.1038/s41586-023-05811-4.
[203] L. Shestakova, A. Serebryanskiy, G. Aimanova, Observations of alkaline emissions NaI, KI and LiI during first minutes after DART probe impact on Dimorphos, Icarus 401 (2023) 115595.
[204] <https://www.nasa.gov/artemis-1> (accessed 05.05.23).
[205] <https://www.nasa.gov/centers/ames/engineering/projects/biosentinel.html> (accessed 06.05.23).
[206] <https://eyes.nasa.gov/apps/solar-system/#/sc_biosentinel> (accessed 06.05.23).
[207] <https://cusp.space.swri.edu/> (accessed 06.05.23).
[208] <https://spaceref.com/newspace-and-tech/artemis-i-payload-cusp-cubesat-has-apparently-failed/> (accessed 06.05.23).
[209] <https://www.space.t.u-tokyo.ac.jp/equuleus/en/> (accessed 06.05.23).
[210] <https://www.eoportal.org/satellite-missions/equuleus#overview> (accessed 06.05.23).
[211] <https://lunahmap.asu.edu/mission> (accessed 06.05.23).
[212] <https://spacenews.com/artemis-1-cubesat-nearing-end-of-mission> (accessed 06.05.23).
[213] <https://nssdc.gsfc.nasa.gov/nmc/spacecraft/display.action?id=L-ICECUBE> (accessed 06.05.23).
[214] <https://nssdc.gsfc.nasa.gov/nmc/spacecraft/display.action?id=LUNIR> (accessed 06.05.23).
[215] <https://www.nasa.gov/content/nea-scout> (accessed 06.05.23).

[216] <https://www.nasa.gov/centers/marshall/news/2022/nea-scout-status-update.html> (accessed 06.05.23).
[217] <https://nssdc.gsfc.nasa.gov/nmc/spacecraft/display.action?id = OMOTENASH> (accessed 06.05.23).
[218] <https://space.skyrocket.de/doc_sdat/omotenashi.htm> (accessed 06.05.23).
[219] <https://nssdc.gsfc.nasa.gov/nmc/spacecraft/display.action?id = TEAMMILES> (accessed 06.05.23).
[220] <https://www.space.com/nasa-artemis-1-moon-mission-cubesats#section-the-cubesats-and-their-missions> (accessed 06.05.23).
[221] B.K. Malphrus, A. Freeman, R. Staehle, A.T. Klesh, R. Walker, Interplanetary CubeSat missions, in: C. Cappelletti, S. Battistini, B.K. Malphrus (Eds.), Cubesat Handbook: From Mission Design to Operations, Academic Press, 2021, pp. 85−121. Available from: https://doi.org/10.1016/B978-0-12-817884-3.00004-7.
[222] <https://www.eoportal.org/satellite-missions/argomoon#eop-quick-facts-section> (accessed 06.05.23).
[223] <https://www.argotecgroup.com/artemis-1-argomoon-captured-photos-of-the-earth-and-moon/> (accessed 06.05.23).
[224] <https://www.nasaspaceflight.com/2022/11/artemis-i-cubesats/> (accessed 06.05.23).
[225] <https://www1.grc.nasa.gov/facilities/compass-lab/> (accessed 06.05.23).
[226] <https://www.nasa.gov/feature/nasa-s-space-cubes-small-satellites-provide-big-payoffs> (accessed 06.05.23).
[227] <https://www.jpl.nasa.gov/missions/lunar-flashlight> (accessed 06.05.23).
[228] <https://blogs.nasa.gov/smallsatellites/2023/03/23/team-troubleshoots-propulsion-for-nasas-lunar-flashlight/> (accessed 06.05.23).
[229] <https://payloadspace.com/artemis-i-cubesats-fail-to-power-up/> (accessed 06.05.23).
[230] <https://spaceflightnow.com/2022/08/23/cubesats-on-artemis-1-to-pursue-bold-missions-in-deep-space/> (accessed 06.05.23).
[231] Prof. M. Swartwout's (Assoc. Professor of Aerospace & Mech. Engineering, Saint Louis University) CubeSat Database. <https://sites.google.com/a/slu.edu/swartwout/cubesat-database?authuser = 0> (accessed 09.05.23).
[232] <https://www.esa.int/Science_Exploration/Space_Science/Herschel/L2_the_second_Lagrangian_Point> (accessed 06.05.23).
[233] L. Zea, S.R. Santa Maria, A.J. Ricco, CubeSats for microbiology and astrobiology research, in: C. Cappelletti, S. Battistini, B.K. Malphrus (Eds.), Cubesat Handbook: From Mission Design to Operations, Academic Press, 2021, pp. 147−162. Available from: https://doi.org/10.1016/B978-0-12-817884-3.00007-2.
[234] J. Pezent, R. Sood, A. Heaton, High-fidelity contingency trajectory design and analysis for NASA's near-earth asteroid (NEA) Scout solar sail mission, Acta Astronaut. 159 (2019) 385−396.
[235] L.T. Hibbert, H.W. Jordaan, Considerations in the design and deployment of flexible booms for a solar sail, Adv. Space Res. 67 (9) (2021) 2716−2726. Available from: https://doi.org/10.1016/j.asr.2020.01.019.
[236] Y. Li, K.H. Cheah, Solar sail as propellant-less micropropulsion, in: K.H. Cheah (Ed.), Space Micropropulsion for Nanosatellites: Progress, Challenges and Future, Elsevier, 2022, pp. 273−284. Available from: https://doi.org/10.1016/B978-0-12-819037-1.00008-6.
[237] B. Fu, E. Sperber, F. Eke, Solar sail technology—a state of the art review, Prog. Aerosp. Sci. 86 (2016) 1−19. Available from: https://doi.org/10.1016/j.paerosci.2016.07.001.
[238] L. Johnson, R. Young, E. Montgomery, D. Alhorn, Status of solar sail technology within NASA, Adv. Space Res. 48 (11) (2011) 1687−1694. Available from: https://doi.org/10.1016/j.asr.2010.12.011.

[239] Y. Tsuda, O. Mori, R. Funase, H. Sawada, T. Yamamoto, et al., Flight status of IKAROS deep space solar sail demonstrator, Acta Astronauti. 69 (9–10) (2011) 833–840. Available from: https://doi.org/10.1016/j.actaastro.2011.06.005.
[240] Y. Tsuda, O. Mori, R. Funase, H. Sawada, T. Yamamoto, et al., Achievement of IKAROS—Japanese deep space solar sail demonstration mission, Acta Astronaut. 82 (2) (2013) 183–188. Available from: https://doi.org/10.1016/j.actaastro.2012.03.032.
[241] D.A. Spencer, B. Betts, J.M. Bellardo, A. Diaz, B. Plante, J.R. Mansell, The LightSail 2 solar sailing technology demonstration, Adv. Space Res. 67 (9) (2021) 2878–2889. Available from: https://doi.org/10.1016/j.asr.2020.06.029.
[242] M. Matsushita, T. Chujo, J. Matsumoto, R. O.Mori, Yokota, et al., Solar power sail membrane prototype for OKEANOS mission, Adv. Space Res. 67 (9) (2021) 2899–2911. Available from: https://doi.org/10.1016/j.asr.2020.10.007.
[243] J.R. Brophy, M. Noca, Electric propulsion for solar system exploration, J. Propuls. Power 14 (5) (1998) 700–707.
[244] A.A. Quarta, G. Mengali, Minimum-time space missions with solar electric propulsion, Aerosp. Sci. Technol. 15 (5) (2011) 381–392.
[245] <https://solarsystem.nasa.gov/missions/deep-space-1/in-depth/> (accessed 10.05.23).
[246] <https://www.jpl.nasa.gov/news/solar-electric-propulsion-makes-nasas-psyche-spacecraft-go> (accessed 10.05.23).
[247] J.T. Howell, M.J. O'Neill, J.C. Mankins, High-voltage array ground test for direct-drive solar electric propulsion, Acta Astronaut. 59 (1–5) (2006) 206–215.
[248] B.J. Duffy, A.P. Mazzoleni, Advanced propulsion concepts for Interplanetary exploration: Solar sails, rockets, and electric propulsion, in: Proceedings of the 47th AIAA/ASME/ASCE/AHS/ASC Structures, Structural Dynamics, and Materials Conference, May 1–4, 2006, Newport, RI, USA, 2006-1702, 17 pp. https://doi.org/10.2514/6.2006-1702.
[249] <https://www.nasa.gov/feature/glenn/2022/small-spacecraft-electric-propulsion-opens-new-deep-space-opportunities> (accessed 11.05.23).
[250] NASA cancels the Solar Cruiser mission in June 2022 after a review deemed it unrealistic within its proposed budget. <https://www.planetary.org/space-missions/solar-cruiser> (accessed 11.05.23).
[251] <https://www.powerfilmsolar.com/custom-solutions/case-studies/space-yacht> (accessed 11.05.23).
[252] <https://solarsystem.nasa.gov/asteroids-comets-and-meteors/asteroids/in-depth/> (accessed 11.05.23).
[253] H. Yamaguchi, R. Ijichi, Y. Suzuki, S. Ooka, K., Shimada; et al., Development of space solar sheet with inverted triple-junction cells, in: 2015 IEEE 42nd Photovoltaic Specialist Conference (PVSC), New Orleans, LA, USA, 2015, pp. 1–5. <https://ieeexplore.ieee.org/document/7356138>.
[254] A. Poghosyan, A. Golkar, CubeSat evolution: analyzing CubeSat capabilities for conducting science missions, Prog. Aerosp. Sci. 88 (2017) 59–83.
[255] <https://www.nasa.gov/pdf/458490main_TRL_Definitions.pdf> (accessed 15.05.23).
[256] J. Straub, In search of technology readiness level (TRL) 10, Aerosp. Sci. Technol. 46 (2015) 312–320. Available from: https://doi.org/10.1016/j.ast.2015.07.007.
[257] A. Gurung, Q. Qiao, Solar charging batteries: advances, challenges, and opportunities, Joule 2 (7) (2018) 1217–1230.
[258] V. Vega-Garita, L. Ramirez-Elizondo, N. Narayan, P. Bauer, Integrating a photovoltaic storage system in one device: a critical review, Prog. Photovolt. 27 (4) (2019) 346–370.
[259] D. Lau, N. Song, C. Hall, Y. Jiang, S. Lim, I. Perez-Wurfl, et al., Hybrid solar energy harvesting and storage devices: the promises and challenges, Mater. Today Energy 13 (2019) 22–44.

[260] X.Z. Wang, H.L. Tam, K.S. Yong, Z.-K. Chen, F. Zhu, High performance optoelectronic device based on semitransparent organic photovoltaic cell integrated with organic light-emitting diode, Org. Electron. 12 (8) (2011) 1429–1433. Available from: https://doi.org/10.1016/j.orgel.2011.05.012.

[261] Y. Wang, L. Zhang, K. Cui, C. Xu, H. Li, H. Liu, et al., Solar driven electrochromic photoelectrochemical fuel cells for simultaneous energy conversion, storage and self-powered sensing, Nanoscale 10 (2018) 3421–3428. Available from: https://doi.org/10.1039/c7nr09275j.

[262] T. Böhnke, M. Edoff, Copper indium gallium diselenide thin films for sun angle detectors in space applications, Thin Solid. Films 517 (6) (2009) 2063–2068. Available from: https://doi.org/10.1016/j.tsf.2008.10.028.

[263] A. Nascetti, M. Mirasoli, E. Marchegiani, M. Zangheri, F. Costantini, A. Porchetta, et al., Integrated chemiluminescence-based lab-on-chip for detection of life markers in extraterrestrial environments, Biosens. Bioelectron. 123 (2019) 195–203. Available from: https://doi.org/10.1016/j.bios.2018.08.056.

[264] A. Ali, H. Wang, J. Lee, Y.H. Ahn, I. Park, Ultra-low profile solar-cell-integrated antenna with a high form factor, Sci. Rep. 11 (2021) 20918. Available from: https://doi.org/10.1038/s41598-021-00461-w.

[265] B. Sherwood, Decadal opportunities for space architects, Acta Astronaut. 81 (2012) 600–609. Available from: https://doi.org/10.1016/j.actaastro.2012.07.021.

[266] P.T. Metzger, A. Muscatello, R.P. Mueller, J. Mantovani, Affordable, rapid bootstrapping of space industry and solar system civilization, J. Aerosp. Eng. 26 (1) (2013) 18–29January. Available from: https://ascelibrary.org/doi/epdf/10.1061/%28ASCE%29AS.1943-5525.0000236.

[267] P. Ehrenfreund, C. McKay, J.D. Rummel, B.H. Foing, C.R. Neal, T. Masson-Zwaan, et al., Toward a global space exploration program: a stepping stone approach, Adv. Space Res. 49 (1) (2012) 2–48.

[268] S. Lim, V.L. Prahbu, M. Anand, L.A. Taylor, Extra-terrestrial construction processes – advancements, opportunities and challenges, Adv. Space Res. 60 (7) (2017) 1413–1429. Available from: https://doi.org/10.1016/j.asr.2017.06.038.

[269] S.O. Starr, A.C. Muscatello, Mars in situ resource utilization: a review, Planet. Space Sci. 182 (2020) 104824. Available from: https://doi.org/10.1016/j.pss.2019.104824.

[270] M.F. Palos, P. Serra, S. Fereres, K. Stephenson, R. González-Cinca, Lunar ISRU energy storage and electricity generation, Acta Astronautica 170 (2020) 412–420. Available from: https://doi.org/10.1016/j.actaastro.2020.02.005.

[271] R.L. Ash, W.L. Dowler, G. Varsi, Feasibility of rocket propellant production on Mars, Acta Astronaut. 5 (9) (1978) 705–724. Available from: https://doi.org/10.1016/0094-5765(78)90049-8.

[272] G.A. Landis, S.G. Bailey, D.J. Brinker, D.J. Flood, Photovoltaic power for a Lunar base, Acta Astronaut. 22 (1990) 197–203. Available from: https://doi.org/10.1016/0094-5765(90)90021-C.

[273] A.F. Hepp, D.L. Linne, G.A. Landis, M.F. Wade, J.E. Colvin, Production and use of metals and oxygen for lunar propulsion, J. Propuls. Power 10 (6) (1994) 834–840.

[274] B.K. Breedlove, G.M. Ferrence, J. Washington, C.P. Kubiak, A photoelectrochemical approach to splitting carbon dioxide for a manned mission to Mars, Mater. Des. 22 (7) (2001) 577–584. Available from: https://doi.org/10.1016/S0261-3069(01)00018-8.

[275] G.A. Landis, Materials refining on the moon, Acta Astronaut. 60 (10-11) (2007) 906–915. Available from: https://doi.org/10.1016/j.actaastro.2006.11.004.

[276] H. Williams, E. Butler-Jones, Additive manufacturing standards for space resource utilization, Addit. Manuf. 28 (2019) 676–681. Available from: https://doi.org/10.1016/j.addma.2019.06.007.

[277] P. Fleith, A. Cowley, A.C. Pou, A.V. Lozano, R. Frank, P.L. Córdoba, et al., In-situ approach for thermal energy storage and thermoelectricity generation on the moon: modelling and simulation, Planet. Space Sci. 181 (2020) 104789. Available from: https://doi.org/10.1016/j.pss.2019.104789.

[278] B. Kading, J. Straub, Utilizing in-situ resources and 3D printing structures for a manned Mars mission, Acta Astronautica 107 (2015) 317–326. Available from: https://doi.org/10.1016/j.actaastro.2014.11.036.

[279] H. Chen, T.S. du Jonchay, L. Hou, K. Ho, Integrated in-situ resource utilization system design and logistics for Mars exploration, Acta Astronautica 170 (2020) 80–92. Available from: https://doi.org/10.1016/j.actaastro.2020.01.031.

[280] G.H. Heiken, D.T. Vaniman, B.M. French (Eds.), Lunar Sourcebook, Cambridge University Press, Cambridge, UK, 1991book available at. Available from: https://www.lpi.usra.edu/publications/books/lunar_sourcebook/.

[281] G.B. Sanders, W.E. Larson, Integration of in-situ resource utilization into Lunar/Mars exploration through field analogs, Adv. Space Res. 47 (1) (2011) 20–29. Available from: https://doi.org/10.1016/j.asr.2010.08.020.

[282] E. Reid, P. Iles, J. Muise, N. Cristello, B. Jones, et al., The Artemis Jr. rover: mobility platform for lunar ISRU mission simulation, Adv. Space Res. 55 (10) (2015) 2472–2483.

[283] K. Kristmann, T. Raadik, M. Altosaar, M. Grossberg-Kuusk, J. Krustok, Pyrite as promising monograin layer solar cell absorber material for in-situ solar cell fabrication on the Moon, Acta Astronaut. 199 (2022) 420–424. Available from: https://doi.org/10.1016/j.actaastro.2022.07.043.

[284] C.P. Spedding, W.J. Nuttall, S. Lim, Energy requirements of a thermally processed ISRU radiation shield for a lunar habitat, Adv. Space Res. 65 (11) (2020) 2467–2474. Available from: https://doi.org/10.1016/j.asr.2020.03.015.

[285] K. Yazdi, E. Messerschmid, A lunar exploration architecture using lunar libration point one, Aerosp. Sci. Technol. 12 (3) (2008) 231–240. Available from: https://doi.org/10.1016/j.ast.2007.06.001.

[286] <https://www.nasa.gov/gateway/overview> (accessed 15.05.23).

[287] S. Fuller, E. Lehnhardt, C. Zaid, K. Halloran, Gateway program status and overview, J. Space Saf. Eng. 9 (4) (2022) 625–628. Available from: https://doi.org/10.1016/j.jsse.2022.07.008.

[288] <https://investor.maxar.com/investor-news/press-release-details/2021/Maxar-Completes-Power-and-Propulsion-Element-Preliminary-Design-Review/default.aspx> (accessed 15.05.23).

Index

Note: Page numbers followed by "*f*" and "*t*" refer to figures and tables, respectively

A

Absorbance loss, 532–533
Absorption, 186, 188
  coefficient, 553–554
Abstract, title, and keywords (ATK), 337
Acceptor light absorption, 258
Acetic acid (CH_3COOH), 307–308
Actinomycetes, 412–414
Additives, 142–143
Agriculture, applications of transparent solar cells in, 616–618, 617*f*
AI. *See* Artificial intelligence (AI)
Air mass (AM), 12, 228–229, 683
  standards for terrestrial and space photovoltaics, 683–684
Air mass 1 (AM1), 13–15
Al-doped ZnO (Al-ZnO), 529–531
Albizia amara, 418–422
ALD method. *See* Atomic layer deposition method (ALD method)
Algae, 412–414
Aluminum (Al), 6–7
Aluminum gallium arsenide (AlGaAs), 20–21
Aluminum oxide (Al_2O_3), 127–128
AM. *See* Air mass (AM)
AM1. *See* Air mass 1 (AM1)
American Society for Testing and Materials, The, 228–229
Ammonium (NH_3), 307–308
Amorphous silicon solar cells (a-Si solar cells), 114, 602–605, 694–696
AMPS. *See* Analysis of microelectronic and photonic structures (AMPS)
Analysis of microelectronic and photonic structures (AMPS), 338
Annealing process, 410, 560–561
Anode electrodes, 593–594
Antimony (Sb), 6–7, 129–130, 438
  antimony-based nanoparticles, 442
  antimony-based solar cells, 438
  compounds, 438–445
    highest-performing antimony-based solar cells, 439*t*
  nanocrystals
    binary antimony compounds and, 439–442
    ternary antimony compounds and, 443–445
Antimony iodide compounds, 445
Antimony selenoiodide (SbSeI), 362–363
Antireflection coatings, 247
Antireflective coated GaAs (ARC GaAs), 42–43
Antisolvents, 422–423
  engineering approaches, 129–130
  immersion strategy, 574–575
Applied Solar Energy Corp. (ASEC), 691
ARC. *See* NASA Ames Research Center (ARC)
ARC GaAs. *See* Antireflective coated GaAs (ARC GaAs)
Area-specific power (ASP), 654, 690
ArgoMoon, 737
Array fabrication, 690
Arsenic (As), 6–7
Artemis, Rideshare CubeSats on, 733–738
Artemis I, 733
Artificial intelligence (AI), 325–326, 337
Artificial photosynthesis, 225–226
Arylamine-based ligands, 212
ASEC. *See* Applied Solar Energy Corp. (ASEC)
ASI. *See* Italian Space Agency (ASI)
ASP. *See* Area-specific power (ASP)
Asteroid Didymos system, 731–733
ATK. *See* Abstract, title, and keywords (ATK)
Atmosphere-assisted CVD (AACVD), 358
Atomic layer deposition method (ALD method), 140–142, 455, 550–551
Atomic oxygen, 686–687

767

Average reflectance, 189–190
Average transmittance, 188
Average upper-layer transmittance, 190
Average visible transmittance (AVT), 334, 588–589
AVT. *See* Average visible transmittance (AVT)
Azadipyrromethenes, 271–272

B

Bacillus mycoide, 412–414
Bacteria, 412–414
Ball milling, 410
Bandgap
  energy, 545
  materials, 547
  solar cell, 546
  tuning and stability of perovskites, 525–527
BAPV. *See* Building-applied PV (BAPV)
Bathocuproine (BCP), 142
Batteries, 647–648
Battery-powered electric vehicles (BEVs), 649
BCP. *See* Bathocuproine (BCP)
Becquerel effect, 4–5
Beginning-of-life (BOL), 716–717
Benzodithiazole (BT), 265–266
Benzodithiophene-based copolymers, 263
BEVs. *See* Battery-powered electric vehicles (BEVs)
BHJ. *See* Bulk heterojunction (BHJ)
1,1-bi-2-naphthol central scaffold, 211–212
Bias-dependent J–V characterization, 564
Bifacial solar cell, 593–594
Binary antimony compounds, 439–442
Binary bismuth compounds, 447–448
Binary compounds, 453–457
  highest-performing bulk, 454t
  NCs tin-based binary, 454t
  SnS-based solar cell, 455f
  ternary chalcogenide solar cells, 454t
Binary iron compounds, 459–462
Binary oxides, 245–246
Binding energy, 554
Bio-inspired approaches, 332–337

Bio-inspired solar cell devices, 365–366
Biomass, 426–427
  synthesis using biomass extracts, 414–417
Biomedical electronics, 366
Biomimetics, 326
Biomimicry, 325–337
  background and overview, 326–328
  green materials and processing, 328–332
  lessons learned from nature, 332–337
Bionics, 325–326
Biphenyl derivatives, 201, 203–205
Biphenyl-based molecules, 201
Biphenylamino groups, 205–206
BIPV. *See* Building-integrated photovoltaics (BIPV)
Bis(trifluoromethane) sulfonamide lithium (Li-TFSI), 142–143
Bismuth (Bi), 6–7, 129–130, 438
  compounds, 445–453
  highest-performing bulk and nanocrystals bismuth-based solar cells, 446t
  nanocrystals
    binary bismuth compounds and, 447–448
    ternary bismuth compounds and, 448–453
Bixa orellana, 418–422
Body-mounted solar arrays, missions using, 723–724
Boeing Space Systems, 714
BOL. *See* Beginning-of-life (BOL)
Boron (B), 6–7
Boron difluoride (BF_2^+), 271–272
Boron-doped ZnO (BZnO), 529–531
BT. *See* Benzodithiazole (BT)
Buffer layer in perovskite solar cells, 145
Building-applied PV (BAPV), 619
Building-integrated photovoltaics (BIPV), 75, 102–103, 334, 508–509, 585
Bulk film perovskite cells, 176–177
Bulk heterojunction (BHJ), 257, 447
Bulk thin films, 437–438

C

c-TiO$_2$ layer, 197
Cadmium selenide (CdSe), 93–94

Index

Cadmium telluride (CdTe), 9, 303–304, 661–662, 694, 696–697
  representative CIGS and CdTe devices on flexible substrates, 700t
  solar cells, 114
CAGR. *See* Compound annual growth rate (CAGR)
Carbazole, 199
  carbazole-based copolymers, 263
  carbazole-based HTMs, 199–200
  carbazole-based organic HTMs, 203–205
  carbazole-based scaffolds, 199, 203–205
  derivatives, 199–201, 203–205
  2,7 carbazole *bis*(4-methoxy phenyl) amine groups, 211–212
  3,6 carbazole *bis*(4-methoxy phenyl) amine groups, 211–212
Carbazole core-based hole-transporting materials, 199–208
  chemical structure
    of star-shaped carbazole core-based organic HTM, 200f
    of synthesized carbazole-based HTMs, 202f
    of two carbazole-based HTMs, 199f
  molecular design feature of target HTMs, 204f
  molecular structure
    of facile donor (D)-π-D triphenylamine-based organic hole-transporting material, 203f
    of indolo(3,2-b)carbazole-based HTMs, 206f
    of investigated linear and starburst-shaped hole-transporting materials, 207f
    of triphenylamine-and carbazole-based HTMs, 205f
    of two hole-transporting materials with triphenylene core, 208f
  structures of hole-transporting materials with fused aromatic rings, 204f
Carbon dioxide (CO_2), 303–304, 653–654
  emissions, 647, 649
Carbon emissions, 484, 649
Carbon footprint, 480, 512, 516–517
Carbon ink, 146–147
Carbon monoxide (CO), 303–304
Carbon nanotubes (CNT), 146–147, 614–616
Carbon neutrality, 666–667
Carbon-based materials, 427
Carotenes, 414–416
Catalysts, 425
Cathode electrodes, 593–594
Cations, 527
CB. *See* Conduction band (CB)
CBD. *See* Chemical bath deposition (CBD)
CBE. *See* Conduction band edge (CBE)
CC. *See* Current constraint (CC)
CCLs. *See* Charge carrier layers (CCLs)
CCPC. *See* Crossed compound parabolic concentrator (CCPC)
CDCA. *See* Chenodeoxycholic acid (CDCA)
CED. *See* Cumulative energy demand (CED)
Cells, 676
  design, 504–506
  performance within concentrator photovoltaic systems, 501–502
  photocurrent density, 188
  technologies, 658–659
Cerium-doped aluminosilicate glass, 697
Cerium(IV) oxide (CeO_2)
  CeO_2-based scattering approach, 244–245
  CeO_2-TiO_2-based hybrid photoanode DSSC, 244–245
CEs. *See* Counter electrodes (CEs)
CFL. *See* Compact fluorescent lamp (CFL)
CFTS. *See* Cu_2FeSnS_4 (CFTS)
Chalcogenide thin-film modules, 663–664
Chalcogenide thin-film solar cells, 23–27
Chalcopyrite, 463
Characterization techniques, 135
Charge carrier layers (CCLs), 555–556
Charge recombination processes, 120–121, 234–235
Charge transfer (CT), 90–91, 197, 258
Chemical bath deposition (CBD), 305–308, 405, 438–439

Chemical capacitance, 128
Chemical deposition processes, 127
Chemical methods
 for semiconductor production, 405
 for semiconductor synthesis, 407
Chemical vapor deposition (CVD), 553
Chenodeoxycholic acid (CDCA), 606–607
Chloroaluminum phthalocyanine (ClAlPc), 596–597
Chlorobenzene, 289
CIS. *See* Copper indium diselenide (CIS)
CL. *See* Contact layers (CL)
ClAlPc. *See* Chloroaluminum phthalocyanine (ClAlPc)
Classic crystalline technology, 693–694
Classic gimbaled wings, 718–720
CNT. *See* Carbon nanotubes (CNT)
Cobalt, 90–91, 424–425
$CoFe_2O_4$ NCs, 463–464
Colloidal quantum dot-based transparent solar cells, 607–609, 609f
Color rendering index (CRI), 589–590
 comparison of optical transmittance and sheet resistance, 591f
Commercial off-the-shelf (COTS), 726
Commission Internationale de l'Eclairage (CIE), 589–590
Compact fluorescent lamp (CFL), 147
Compositional engineering, 548
Compound annual growth rate (CAGR), 93–94
Computer-aided approaches to improved photovoltaics, 337–344
Computer-aided insights, 325–326
Concentrated light irradiation on dye-sensitized solar cells, 230
Concentrated light properties, 504–506
Concentrated power systems, 499–500
Concentrator optics, 499–500, 502
 coupled with second-and third-generation photovoltaics, 508–509
Concentrator photovoltaic technologies (CPV technologies), 19, 230, 499
Concentrator solar cells, 3–4, 34–37, 499
 classifications of relevant technologies, 499–503, 500f
  geometric concentration ratio, 499–501
  measuring optical and cell performance within concentrator photovoltaic systems, 501–502
  theoretical efficiency limits, 502–503
 concentrator compared to one sun solar cells, 503–509
  concentrated light properties and cell design, 504–506, 505f
  concentrator optics coupled with second-and third-generation photovoltaics, 508–509
  single junction silicon cells and low concentrator optics, 506–508, 507f
 matching solar cells and concentrator optics for optimum performance, 512–517
Concept of operations (CONOPS), 731–733
Conduction band (CB), 227
Conduction band edge (CBE), 233–234
Conjugated polymer main chains, 707–708
Conjugated polymer-based electronic devices, 707–708
Conjugated polymer-based optoelectronic devices, 707–708
CONOPS. *See* Concept of operations (CONOPS)
Contact layers (CL), 532–533
Conventional device materials, 711
Conventional lattice-matched epitaxial growth, 681–683
Conventional meso-structured PSCs, 197
Conventional $n-i-p$ perovskite cell, 190–191
Conventional perovskite, 199–200
Conventional solar cells, 93–94, 712
Copper (Cu), 129, 146, 311, 461–462, 591–593
Copper bismuth sulfide ($CuBiS_2$), 448–449
Copper chloride ($CuCl_2$), 307–308
Copper halide perovskites, 627–629
Copper indium diselenide (CIS), 303–304

Copper indium gallium selenide (CIGS),
    114, 303–304, 661–662, 688, 697–700
    tandem solar cells, 569–573
Copper oxide (CuOx), 144, 609–610
Copper sulfate (CuSO$_4$), 307–308
Copper thiocyanate (CuSCN), 144
Copper zinc tin sulfide (CZTS), 304, 458
    comparison of performance of vacuum-
        and nonvacuum-based solar cells,
        316–317
        copper zinc tin sulfide nonvacuum-
            processed solar cells, 317
        copper zinc tin sulfide vacuum-
            processed solar cells, 316–317
    copper zinc tin sulfide-based thin-film
        solar cell deposition techniques,
        309–315
        vacuum-based approaches, 310–315
    future scope, 318
    nonvacuum-processed solar cells, 317
    structure, 304–305
    synthesis methods of copper zinc tin
        sulfide nanoparticles and thin films,
        305–309
        CBD, 307–308
        sol–gel method, 306–307
        solvothermal method, 309
        spin-coating, 308
        spray pyrolysis, 308
        thermal evaporation technique, 309
    vacuum-processed solar cells, 316–317
Copper-antimony chalcogenide
    nanoparticles, 444
    solar cells, 445
Correlation between vehicle surfaces and
    photovoltaic cell efficiency, 659–661
Cosputter deposition technique, 316–317
COTS. See Commercial off-the-shelf
    (COTS)
Coumarin, 199
Counter electrodes (CEs), 81, 227, 348, 460
CPV technologies. See Concentrator
    photovoltaic technologies (CPV
    technologies)
CRI. See Color rendering index (CRI)
Crossed compound parabolic concentrator
    (CCPC), 507–508

Crystalline silicon (c-Si), 15, 303–304,
    325, 586–588, 602–603
    crystalline silicon-based solar cells,
        91–92, 523
Crystalline solar cells, 694
CT. See Charge transfer (CT)
Cu$_2$FeSnS$_4$ (CFTS), 464
CubeSats missions, 725–738, 743–748
    profiles in, 733–738
Cumulative energy demand (CED), 481
Cupric chloride, 308
Current constraint (CC), 38–40
Current–voltage measurements (I–V
    measurements), 697
CuSbSe$_2$ solar cells, 444
CVD. See Chemical vapor deposition
    (CVD)
2-cyanoacrylic acid-4-(bis-
    dimethylfluoreneaniline)dithiophene,
    233–234
Cyclopentadithiophene-based copolymers,
    263

D

D-A molecular HTMs.
    See Donor–acceptor molecular HTMs
    (D-A molecular HTMs)
Dark nonradiative recombination,
    173–174
Dark radiative recombination in thin cells,
    172–173
DART impact on Dimorphos, witness to,
    731–733
DART mission. See Double Asteroid
    Redirect mission (DART mission)
DARWIN. See Deep adaptive regressive
    weighted intelligent network
    (DARWIN)
DAVID. See Diminutive Asteroid Visitor
    using Ion Drive (DAVID)
Dawn mission, 718–720
DC materials. See Down-conversion
    materials (DC materials)
Deep adaptive regressive weighted
    intelligent network (DARWIN), 342
Deep Impact Extended Investigation
    mission (DIXI mission), 722–723

Deep learning (DL), 337
Deep Space 1 (DS1), 713
Delafossite-based HTLs, 144–145
Density functional theory (DFT), 325–326, 338–339
Deployable Gossamer thin-film PV array system, 704
Deployable single panels, 722–723
Deployable Space Systems (DSS), 680
Deposition methods, 123–124, 127, 309–310, 455, 599–600, 697–700
Device fabrication for photovoltaics, 358–375
Device stability, 538–539
DFT. See Density functional theory (DFT)
Di-substituted hole-transporting materials, molecular structures of, 209f
Di[anisyl]amino substituents, 203–205
Dibenzo [a,c]–carbazole (DBC), 201–203
Dielectric constant, 554
Diffusion limited semiconductors, 165
1,8-diiodooctane (DIO), 263–264
4,4-dimenthyltriphenylamine, 208–209
4,4′-dimethoxy triphenylamine, 212–214
Dimethoxydiphenylamine, 209–210
4,4-dimethoxydiphenylamine, 209–210
Dimethoxytriphenylamine, 201, 208–209
Dimethyl sulfoxide, 422–423
9-dimethyl-9H-fluoren-2-amine, 201–203
Dimethylacetamide, 422–423
Dimethylethanolamine, 422–423
Diminutive Asteroid Visitor using Ion Drive (DAVID), 733
Dipping process, 127
Direct radiation, 13–15
Distributed power system, 725–726
Dithiocarbamates (DTCs), 358–359
DL. See Deep learning (DL)
DOE. See United States Department of Energy (DOE)
DOE NREL. See US Department of Energy's National Renewable Energy Laboratory (DOE NREL)
Donor light absorption, 258
Donor–acceptor blend, 286
Donor–acceptor molecular HTMs (D-A molecular HTMs), 199–200

Dopant-free organic HTMs, 205–206
Double Asteroid Redirect mission (DART mission), 722
Down-conversion materials (DC materials), 29, 42
DS1. See Deep Space 1 (DS1)
DSS. See Deployable Space Systems (DSS)
DSSCs. See Dye-sensitized solar cells (DSSCs)
DTCs. See Dithiocarbamates (DTCs)
"Dual-junction" cell, 18
Dual-layer molybdenum contacts, 312
Dual-source thermal evaporation, 123–124
Dye-metal oxide complex formation, 245–246
Dye-sensitized solar cells (DSSCs), 3–4, 27–29, 75–76, 90–93, 195, 224, 226–228, 328, 422–426, 445–446, 508–509, 586–588, 690, 704–705
  brief history of, 225–226
  concentrated light irradiation on, 230
  device performance, 245–247
  nanostructured materials for performance improvement of, 241–243
  overall relative scope of performance improvement of, 239–241
  use of metal-organic framework materials in, 349–352
Dye-sensitized transparent solar cells, 606–607, 608f

E

EA. See Ethyl ammonium (EA)
Earth abundance
  AgBiS$_2$, 450–451
  of elements, 56, 305f
Earth-abundant
  absorber materials, 458
  CZTS, 304–305, 316, 329
  element(s), 23–25, 56, 377
  materials, 3, 327t, 328–329, 377, 464, 606–607
  nontoxic metal, 450–451
  NaBiS$_2$, 452–453
  silicon(-based solar cells), 663–664
Earth orbital missions, 718
Earth-orbiting designs, 722–723

Index

Earth's atmosphere, solar cells, 683
Economic assessment and market status of third-generation photovoltaics, 83–104
 DSSCs, 90–93
 OSCs, 83–90
 perovskite solar cells, 96–104
 quantum-dot-sensitized solar cells, 93–96
Edge mounted solar cell, 621
EDL phase. *See* Entry, descent, and landing phase (EDL phase)
Effective concentration ratio, 512–513
EIS. *See* Electrochemical impedance spectroscopy (EIS)
Electric field vectors, 185
Electric mileage (EM), 667–668
Electric vehicles (EVs), 647
 applications
  annual CO_2 emission reduction of PV-powered electric vehicles, 667f
  different types of solar cells for, 661–666
  flexible and light modules, 666f
Electrical current, 505
Electrical power, 721, 726
Electrically independent 4-T devices, 49–50
Electricity, 616–617
 generation, 488
Electrochemical approach, 233
Electrochemical impedance spectroscopy (EIS), 128
Electrodeposition, 314, 405
Electrodes, 593–594, 614–616
 for perovskite solar cells, 145–147
Electrolyte, 424–425
Electromagnetic radiation, 685–686
Electron transport layers (ETL), 120, 449
 buffer layer in perovskite solar cells, 145
 different ranges of, 140–145
Electron transport materials (ETMs), 140, 195
Electronic Product Environmental Assessment Tool (EPEAT), 480
Electrons, 140, 197, 227–228
 donors optimized for fullerene-based acceptors, 261–266
 irradiation, 696
 transfer process, 247–248

EM. *See* Electric mileage (EM)
Encapsulation on perovskite solar cells, effect of, 147–148
Energy, 113, 303–304
 consumption, 523
 efficiency, 588–589
 emissions, 484
 loss, 691
 quantum concept, 5
 sector, 649
 from space race to energy crisis, 7–9
 systems, 481–482
Energy payback time (EPBT), 481–482
Energy return on investment (EROI), 481–482
Enhanced solar cells, metal-organic framework materials for, 344–357
Entry, descent, and landing phase (EDL phase), 730
EPBT. *See* Energy payback time (EPBT)
EPEAT. *See* Electronic Product Environmental Assessment Tool (EPEAT)
EQE. *See* External quantum efficiency (EQE)
Equivalent circuit, 241
ERE. *See* External radiative efficiency (ERE)
EROI. *See* Energy return on investment (EROI)
ETA. *See* Extremely thin absorber (ETA)
Etchants, 602–603
Ethanol, 424
Ethyl acetate, 422–423
Ethyl ammonium (EA), 129–130
ETL. *See* Electron transport layers (ETL)
ETMs. *See* Electron transport materials (ETMs)
Evaporation, 311–312, 405
EVs. *See* Electric vehicles (EVs)
Exploration missions
 notable space missions, technology demonstrations, and array configurations, 718–724
  classic gimbaled wings, 718–720
  deployable single panels: cost savings and multipurpose arrays, 722–723

Exploration missions (*Continued*)
  fixed wings or paddles, 720–721
  missions using body-mounted solar arrays, 723–724
  unique solar arrays, 721–722
  photovoltaics technologies into future, 738–748
  SmallSats, CubeSats, and rideshare missions, 725–738
    LICIACube, 731–733, 734t
    lunar crater observation and sensing satellite mission, 728
    Mars CubeSat One participation in Mars Insight mission, 729–731
    Rideshare CubeSats on Artemis, 733–738
    starshine 3, 725–728
  space applications and, 716–748
  space photovoltaic technologies for exemplary missions, 718, 719t
External quantum efficiency (EQE), 440–442, 532, 596–597
External radiative efficiency (ERE), 15, 715
Extrasolar Planet Observation and Characterization Investigation (EPOCh), 722–723
Extraterrestrial environments, 686–687
Extremely thin absorber (ETA), 122–123

F

FA. *See* Formamidinium (FA)
Fabrication, 527–528, 596–597
  approaches, 245–247, 484, 548–550, 557
  of four-terminal tandems, 557–558
FBTD. *See* Fluorine substituted benzothiadiazole (FBTD)
FDSSC. *See* Flexible DSSC (FDSSC)
FF. *See* Fill factor (FF)
Fill factor (FF), 133, 175, 230–231, 529–531, 546, 588, 697
Film morphology, 201–203
First-generation silicon solar cells, 168
First-generation solar cells, 224–225
Fixed wings, 720–721
Flavonoids, 414–416

Flexible DSSC (FDSSC), 365–366
Flexible perovskite solar cells, 100
Flexible substrates, 690
Fluorene-based molecules, 201
Fluorinated ITIC, 275–276
Fluorine, 201
  atoms, 199–200
Fluorine substituted benzothiadiazole (FBTD), 212–214
Fluorine-doped tin oxide (FTO), 125, 227
FOE. *See* Fuorophen-oxyethylammonium (FOE)
Formamidinium (FA), 613–614
  based perovskite production, 139
  formamidinium-cesium-based hybrid perovskites, 508–509
Formamidinium lead iodide (CH(NH_2)$_2$PbI$_3$), 116
Formamidinium tin iodide (FASnI$_3$), 129–130
Förster resonance energy transfer (FRET), 233
Fossil fuels, 195, 303–304
Four terminal all perovskite tandem solar cells, 536–537
Four terminal devices (4T devices), 529
  tandem architectures, 524
Fresnel coefficients, 186
Fresnel equations, 185–186
Fresnel lens, 35–37, 500–501
FRET. *See* Förster resonance energy transfer (FRET)
FTO. *See* Fluorine-doped tin oxide (FTO)
Fullerene-based acceptors, electron donors optimized for, 261–266
Fullerene-based OPV, 275–276
Fullerene-derivatives, 266–267
Fungi, 412–414
Fuorophen-oxyethylammonium (FOE), 129–130
Fused aromatic rings, structures of hole-transporting materials with, 204f
Fused ring acceptor molecules, 599

G

Galactic cosmic rays (GCR), 685–686
Gallium (Ga), 6–7

Gallium antimonide (GaSb), 661−662
Gallium arsenide (GaAs), 9−10, 661−662, 678−679
  GaAs-based cells, 10
  and III−V materials, 17−22
Gallium doped Si (Ga-doped Si), 7
Gallium doped zinc oxide (GZO), 593
Gallium indium phosphide (GaInP), 623
Gallium oxide (Ga_2O_3), 612−613
γ-butyrolactone (GBL), 422−423
GBL. *See* γ-butyrolactone (GBL)
GCR. *See* Galactic cosmic rays (GCR)
GEO mission. *See* Geostationary Earth orbit mission (GEO mission)
Geometric concentration ratio, 499−501
Geostationary Earth orbit mission (GEO mission), 696
Geothermal energy, 303−304
Germanium (Ge), 129
GHG emissions. *See* Greenhouse gas emissions (GHG emissions)
Gimbaled wings configuration, 718−720
Glenn Research Center (GRC), 358
Global warming, 649
Gold (Au), 4−5, 145−146, 461−462, 552−553, 591−593
  nanoparticles, 627−629
  nanowires, 529−531
Gossamer solar power array mission (GoSolAr mission), 703−704
Gradual transition process, 647
Graphene oxide (GO), 125
Graphene QDs, 233
GRC. *See* Glenn Research Center (GRC)
Green chemical synthesis of photovoltaic materials
  biosynthesis, 412−417
    synthesis using biomass extracts, 414−417
    synthesis via microorganisms, 412−414
  considerations for greener fabrication of solar cells, 422−426
    dye-sensitized solar cells, 423−426
    perovskite solar cells, 422−423
  green methods for synthesis of semiconductor materials, 406−412

  green hydrothermal and solvothermal synthesis, 407−408
  magnetic field-assisted chemical synthesis, 411−412
  mechanosynthesis, 410
  microwave-assisted chemical synthesis, 408−409
  photo-assisted synthesis or photocatalysis, 411
  sonochemical synthesis, 410−411
  photovoltaic materials through green synthesis, 417−422
  prospects for green synthesis of solar cells, 426−427
Green materials, 328
  and processing, 328−332
Green technologies, 326−337
  background and overview, 326−328
  green materials and processing, 328−332
  lessons learned from nature, 332−337
Greenhouse gas emissions (GHG emissions), 481, 516−517

H

HAAs. *See* High-altitude airships (HAAs)
Halides, 524−525
  perovskite, 528−529
HASPs. *See* High-altitude pseudo-satellites (HASPs)
HCPV. *See* High CPV systems (HCPV)
Helmholtz−Zentrum Berlin (HZB), 548
Hemispherical tilted spectrum, 228−229
Heterojunction (HJ), 438, 610−611
High concentrator systems, 517
High CPV systems (HCPV), 34, 515
High-altitude airships (HAAs), 654
High-altitude pseudo-satellites (HASPs), 654−658
High-purity crystalline silicon solar cells, 165
High-throughput open-air processing method, 371
Highest occupied molecular orbital (HOMO), 126, 197, 227, 257
HJ. *See* Heterojunction (HJ)
HOIP. *See* Hybrid organic−inorganic perovskite (HOIP)

HOISC. See Hybrid organic—inorganic solar cell (HOISC)
Hole selective layer (HSL), 183—184
Hole transport layers (HTL), 83—84, 120, 286—288, 449, 539
  buffer layer in perovskite solar cells, 145
  different ranges of, 140—145
Hole-transporting materials (HTM), 117, 195—196
  impact of design of HTM on performance of perovskite solar cells, 196f
  structures of HTM with fused aromatic rings, 204f
  with triphenylene core, 208f
HOMO. See Highest occupied molecular orbital (HOMO)
Hot injection synthesis methods, 453
HSL. See Hole selective layer (HSL)
HTL. See Hole transport layers (HTL)
HTM. See Hole-transporting materials (HTM)
Hybrid inorganic-organic perovskite materials, 683
Hybrid organic—inorganic perovskite (HOIP), 338—339
Hybrid organic—inorganic solar cell (HOISC), 359
Hybrid perovskites, 170, 175
  materials, 176, 547
Hybrid vapor technique, 551—552
Hydrogenated amorphous silicon (a-Si:H), 693—694
Hydrogenated indium oxide (In_2O_3:H), 571—572
Hydrothermal chemical synthesis, 405
Hydrothermal methods, 309, 408—409
HZB. See Helmholtz—Zentrum Berlin (HZB)

I

i-EX. See Intra-moiety excimer (i-EX)
i-FoM. See Industrial figure of merit (i-FoM)
IB. See Intermediate band (IB)
IBSC. See Intermediate bandgap solar cells (IBSC)
ICBA. See Indene-C_{60} bis-adduct (ICBA)
IDT. See Indacenodithiophene (IDT)
IEA. See International Energy Agency (IEA)
IKAROS. See Interplanetary Kite-craft Accelerated by Radiation Of Sun (IKAROS)
Imidazole, 199
Implantable PV (IPV), 366
IMPS. See Integrated micro-power source (IMPS)
In situ bandgap, 569—570
In situ resource utilization (ISRU), 743—748
Incident photon to current efficiency (IPCE), 231—232
Indacenodithiophene (IDT), 275—276
Indene-C_{60} bis-adduct (ICBA), 261
Indium (In), 6—7
Indium gallium phosphide (InGaP), 20—21
Indium phosphide (InP), 19, 661—662, 678—679
Indium tin oxide (ITO), 529—531, 535, 590—591
Indium zinc oxide (IZO), 550—551
Indium-doped tin oxide, 227
Indolo[3,2-b]carbazole-based small molecules (ICZ-based small molecules), 205—206
Industrial figure of merit (i-FoM), 281
Infrared range (IR range), 228—229
InGaP. See Indium gallium phosphide (InGaP)
Inorganic coordination dyes, 231—232
Inorganic metal complex dyes, 90—91
Inorganic semiconductor-based transparent solar cells, 602—613
  colloidal quantum dot-based transparent solar cells, 607—609, 609f
  dye-sensitized transparent solar cells, 606—607, 608f
  metal-oxides-based transparent solar cells, 609—613, 611f
  organic-inorganic hybrid transparent solar cells, 613—616, 615f
  silicon-based transparent solar cells, 602—605, 604f

Inorganic thin-film solar cells, 661–662
InP. *See* Indium phosphide (InP)
Integrated micro-power source (IMPS), 363, 725
Integrated micropower technology, demonstration of, 725–728
Integrated photocapacitor (IPC), 366–368
Integrated power, 743–748
Integrated power sources (IPSs), 328, 363–370, 725
Interfacial electron kinetics, strategies to improve, 230–234
Interhalogen-based binary-redox couples, 233
Intermediate band (IB), 29–31
Intermediate bandgap (concentrator) solar cells (IB(C)SCs), 3–4
Intermediate bandgap solar cells (IBSC), 33–34, 37–40
Internal spectral transmittance, 188
International Energy Agency (IEA), 479, 647
International Space Station (ISS), 679
International Summit on Organic Photovoltaic Stability (ISOS), 147
Interplanetary Kite-craft Accelerated by Radiation Of Sun (IKAROS), 743
Intra-moiety excimer (i-EX), 279
Intramolecular charge transfer process, 596–597
Intrinsic amorphous hydrogenated silicon (i-a-Si:H), 605
Inverted cells, 183–184
Inverted $p-i-n$ solar cell, 183–184, 190–191
Inverted PSCs, 126
Inverted solar cell, 190
IPA. *See* Isopropyl alcohol (IPA)
IPC. *See* Integrated photocapacitor (IPC)
IPCE. *See* Incident photon to current efficiency (IPCE)
IPSs. *See* Integrated power sources (IPSs)
IPV. *See* Implantable PV (IPV)
IR range. *See* Infrared range (IR range)
Iron (Fe), 438
    compounds, 458–464
    engines, 718–720
    highest-performing bulk and nanocrystal iron-based solar cells, 459t
    iron-based electrolytes, 424–425
    iron-based materials, 458
    nanocrystals
        binary iron compounds and, 459–462
        ternary and quaternary iron compounds and, 463–464
Iron disulfide (FeS$_2$), 459
Iron pyrite, 459
iROSA. *See* ISS roll-out solar array (iROSA)
Isopropyl alcohol (IPA), 417–418
ISOS. *See* International Summit on Organic Photovoltaic Stability (ISOS)
ISRU. *See* In situ resource utilization (ISRU)
ISS. *See* International Space Station (ISS)
ISS roll-out solar array (iROSA), 680
IT-4F, 275–276
Italian Space Agency (ASI), 731
ITO. *See* Indium tin oxide (ITO)
IZO. *See* Indium zinc oxide (IZO)

J

Japan Aerospace Exploration Agency (JAXA), 737
JAXA. *See* Japan Aerospace Exploration Agency (JAXA)

K

Kelvin probe force microscopy, 135, 610–611
Kesterite (CZTS(e)-based) phase solar cells, 328–329

L

Lactuca sativa, 330
LADEE. *See* Lunar Atmosphere and Dust Environment Explorer (LADEE)
Laser ablation technique, 594
Laser scribing technique, 594
Lattice-matched heterojunction cells, 10
Layered system, 185
LCA. *See* Life cycle assessment (LCA)
LCPV. *See* Low concentration (LCPV)

LCROSS. *See* Lunar Crater Observation and Sensing Satellite (LCROSS)
Lead (Pb), 552
Lead halide perovskite solar cells, 165
Lead sulfide (PbS), 93–94
Lead-free perovskite solar cells, 129–130
Learning curve, 8–9
LEDs. *See* Light-emitting diodes (LEDs)
LEO. *See* Low-Earth orbit (LEO)
LIB. *See* Lithium-ion battery (LIB)
Life cycle assessment (LCA), 479
  metrics, 481–484
    commercial and emerging solar technologies, 485t
  of photovoltaic technologies, 484–489
  system boundary of solar photovoltaic life cycle, 480f
Light absorption, 538–539
Light capture and device management, 529–533
  bandgap arrangements, 530f
  contact layers and absorbance loss, 532–533
  low bandgap bottom cells, 531–532
  semitransparency of top cell, 529–531
Light conditioning techniques, 230
Light Italian Cubesat for Imaging of Asteroid (LICIACube), 731–733, 734t
Light scattering, effect of, 243–245
Light transmission holes, 603
Light-emitting diodes (LEDs), 586–588
Light-induced phase segregation, 132–133
Light-trapping schemes, 550
Light-utilization efficiency (LUE), 627
LILT conditions. *See* Low-intensity, low-temperature conditions (LILT conditions)
Linear HTM, 207
Linear-shaped hole-transporting materials, molecular structure of investigated, 207f
Liquid phase electrolyte, 195
Lithium (Li), 6–7
Lithium bis (trifluoromethanesulfonyl) imide (Li-TFSI), 135
Lithium-doped Si, 7
Lithium-ion battery (LIB), 363
LMWF. *See* Low molecular weight fraction (LMWF)

Low bandgap bottom cells, 531–532
Low carrier concentration, 554
Low concentration (LCPV), 34
Low concentrator optics, 506–508
Low molecular weight fraction (LMWF), 279–280
Low solar intensity and low temperature (LILT), 713
Low-bandgap organic semiconducting polymers, 83–84
Low-Earth orbit (LEO), 678
Low-energy electrons, 679
Low-intensity, low-temperature conditions (LILT conditions), 693–694
Lowest unoccupied molecular orbital (LUMO), 227, 255–257, 596–597
LSCs. *See* Luminescent solar concentrators (LSCs)
LUE. *See* Light-utilization efficiency (LUE)
Luminescent solar concentrators (LSCs), 618, 625
LUMO. *See* Lowest unoccupied molecular orbital (LUMO)
Lunar Atmosphere and Dust Environment Explorer (LADEE), 723
Lunar Crater Observation and Sensing Satellite (LCROSS), 728
Lunar Reconnaissance Orbiter (LRO), 722–723

M

MA. *See* Methyl ammonium (MA)
MAAc. *See* Methylammonium acetate (MAAc)
Machine learning (ML), 90, 325–326
Magnetic field vectors, 185
Magnetic field-assisted chemical synthesis, 411–412
ManTech program. *See* Manufacturing Technology program (ManTech program)
Manufacturing Technology program (ManTech program), 690–691
(MAPbI$_3$) perovskite solar cells, 183–184
MarCO. *See* Mars CubeSat One (MarCO)
Marlec Renewable Power, 652

Mars Atmosphere and Volatile Evolution mission (MAVEN mission), 721
Mars CubeSat One (MarCO), 729–731
Mars Insight mission, 729–731
Mars missions, 720–721
Mars Pathfinder mission, 721
Mars Reconnaissance Orbiter (MRO), 730
Mass-specific power (MSP), 22, 654–658, 689–690
Massachusetts Institute of Technology, 94
Material selection in transparent luminescent solar concentrators, 619–621
Material systems, 259
Materials International Space Station Experiment (MISSE) project, 708
Materials Project, The, 343–344
Materials Research Society, The, 343–344
Mathematical models, 183
Matrix multiplication, 187–188
MAVEN mission. See Mars Atmosphere and Volatile Evolution mission (MAVEN mission)
Maximum efficiency point (MEP), 366–368
Maximum power point tracking approach (MPPT approach), 653–654
MBE. See Molecular beam epitaxy (MBE)
MCPs. See Metal coordination polymers (MCPs)
MCPV. See Medium concentration (MCPV)
MEA. See Monoethanolamine (MEA)
Mechanosynthesis, 410
Medium concentration (MCPV), 34
MEG. See Multiple exciton generation (MEG)
MEHPPV. See Poly[2-methoxy-5-(2-ethylhexyloxy)-1,4-phenylenevinylene] (MEHPPV)
MEIS. See Moth eye-inspired structure (MEIS)
MEP. See Maximum efficiency point (MEP)
Mercury Surface, Space Environment, Geochemistry, and Ranging spacecraft (MESSENGER spacecraft), 721–722

Mesoscopic PSCs, 123
MESSENGER spacecraft. See Mercury Surface, Space Environment, Geochemistry, and Ranging spacecraft (MESSENGER spacecraft)
Metal coordination polymers (MCPs), 328
Metal electrode (Au), 195
Metal halide layer, 712–713
Metal halide perovskites, 119, 197, 613–614
Metal oxides, 609–610
Metal-organic chemical vapor deposition techniques (MOCVD techniques), 10, 690–691
Metal-organic frameworks (MOFs), 3, 328
  details and photovoltaic applications, 348–349
  materials for enhanced solar cells, 344–357
  synthesis, 345–348
  use of metal-organic framework materials in dye-sensitized solar cells, 349–352
  use of metal-organic framework materials in perovskite solar cells, 352–357
Metal-oxide semiconductors, 609–610
Metal-oxides-based transparent solar cells, 609–613, 611f
Metallization, 505–506
Metals, 484
Methoxybenzene (PhOMe), 422–423
2-methoxyethanol (2-ME), 306–307
Methoxyl-arylamine terminal units, 208
Methyl ammonium (MA), 613–614
Methylammonium acetate (MAAc), 423
Methylammonium lead iodide (MAPbI$_3$), 115–116, 525–527
Methylammonium lead iodide chloride (CH$_3$NH$_3$PbI$_{3-x}$Clx), 116
Microorganisms, synthesis via, 412–414
Microstructured perovskite materials, 614–616
Microwave-assisted chemical synthesis, 408–409
Mid-range infrared (MIR), 688
Millikan Boats, 652

MIM. See Monolithically integrated module (MIM)
MIR. See Mid-range infrared (MIR)
Mission-enabling improvements, 715−716
Mixed organic lead mixed halide perovskite, 706
ML. See Machine learning (ML)
MOCVD techniques. See Metal-organic chemical vapor deposition techniques (MOCVD techniques)
MOF-808, 355−357
MOFs. See Metal-organic frameworks (MOFs)
Molecular beam epitaxy (MBE), 10
Molecular electron acceptors, 267
Molecular-designed HTMs, 199−200
Molten salts, 423
Molybdenum, 425
Molybdenum trioxide (MoO_3), 600−601
Mono-substituted HTM, 208
  molecular structures of, 209f
Monocrystalline silicon (mc-Si), 662−663
Monoethanolamine (MEA), 306−307
Monolithic tandem device, 556
Monolithically integrated module (MIM), 726
Moth eye-inspired structure (MEIS), 334
MPPT approach. See Maximum power point tracking approach (MPPT approach)
MRO. See Mars Reconnaissance Orbiter (MRO)
MSP. See Mass-specific power (MSP)
Multicrystalline-Si (mc-Si), 15−17
Multifunctional PEDOT:EVA layer, 336−337
Multijunction (MJ) devices, 3−4, 44−55, 75
Multijunction solar cells, 502−503
  classifications of relevant technologies, 499−503, 500f
    geometric concentration ratio, 499−501
    measuring optical and cell performance within concentrator photovoltaic systems, 501−502
    theoretical efficiency limits, 502−503
  concentrator compared to one sun solar cells, 503−509
  matching solar cells and concentrator optics for optimum performance, 512−517
  for solar concentrator systems, 509−512, 510f
Multilayered systems, 186−188
Multiple exciton generation (MEG), 437−438
Multiple photovoltaics technologies for higher efficiency, 34−37

N

N,N-di-p-methoxyphenylamine-substituted pyrene derivatives, 214
N,N-dimethylformamide, 422−423
N-[4-methoxyphenyl]-9, 201−203
N-ethylcarbazole, 208−209
N-methyl-2-pyrrolidone, 422−423
n-propylammonium iodide (PAI), 129−130
n-type semiconductors, 127
Nanobelt, 242−243
Nanobricks, 444
Nanocomb, 242−243
Nanocrystalline silicon (nc-Si:H), 550−551, 554−555
Nanocrystalline thin films, 437−438
Nanocrystals (NCs), 43−44, 437−438, 585−586
  binary antimony compounds and, 439−442
  binary bismuth compounds and, 447−448
  binary compounds and, 453−457
    highest-performing bulk, 454t
    NCs tin-based binary, 454t
    SnS-based solar cell, 455f
    ternary chalcogenide solar cells, 454t
  binary iron compounds and, 459−462
  ternary and quaternary compounds and, 457−458
    iron compounds and, 463−464
  ternary antimony compounds and, 443−445
  ternary bismuth compounds and, 448−453

Index 781

AgBiS$_2$, 449–452
AgBiSe$_2$, 452
CuBiS$_2$, 448–449
NaBiS$_2$, 452–453
Nanofibers, 127
Nanoflower, 242–243
Nanoparticle SnO$_2$ product (np-SnO$_2$), 124–125
Nanoparticles, 233, 242–243, 438–439, 442, 444
Nanoplates, 444
Nanorods, 127, 444
Nanoscale, 456
Nanostructured materials for performance improvement of dye-sensitized solar cells, 241–243
Nanostructured perovskite materials, 614–616
Nanotechnology, 75
Nanowires (NWs), 127, 442
Naphthalene, 199
Naphthol core-based small molecule hole-transporting materials, 211–214, 211f
Naphthol-core HTMs, 211–212
Naphthothiadiazole (NTz), 265–266
Narrow bandgap solar cell, 601–602
NASA. See National Aeronautics and Space Administration (NASA)
NASA Ames Research Center (ARC), 737–738
National Aeronautics and Space Administration (NASA), 720–721
National Renewable Energy Laboratory (NREL), 9, 94, 681
Nature-inspired insights, 325–326
NCs. See Nanocrystals (NCs)
NEAR Shoemaker mission. See Near-Earth Asteroid Rendezvous Shoemaker mission (NEAR Shoemaker mission)
Near-Earth Asteroid Rendezvous Shoemaker mission (NEAR Shoemaker mission), 720–721
Near-infra red region (NIR region), 585
Near-infrared TLSC (NIR TLSC), 619
NEB. See Net environmental benefit (NEB)
Net benefit approach, 489–490
Net environmental benefit (NEB), 488
New Energy and Industrial Technology Development Organization, 697–700
Next Triple Junction Prime cells (XTJ Prime cells), 680
NFA. See Nonfullerene acceptors (NFA)
Nickel oxide (NiOx), 140, 609–610
NIEL. See Nonionizing energy loss (NIEL)
NIR region. See Near-infra red region (NIR region)
NIR TLSC. See Near-infrared TLSC (NIR TLSC)
Nitrogen oxides (NOx), 303–304, 653–654
Nitrogen-filled glove box, 134
Nonfullerene acceptors (NFA), 258–259
Nonfullerene electron acceptors, 266–281
Nonionizing energy loss (NIEL), 712–713
Nonplanar acceptors, 275–276
Nonvacuum-based approaches, 313
Nonvacuum-based solar cells, comparison of performance of, 316–317
NREL. See National Renewable Energy Laboratory (NREL)
NTz. See Naphthothiadiazole (NTz)
NWs. See Nanowires (NWs)

O

OKEANOS. See Oversize Kite-craft for Exploration and AstroNautics in Outer Solar system (OKEANOS)
1D nanostructures, 241–242
Open circuit voltage, 175
Optical concentration ratio, 501
Optical coupling, 619–621
Optical design of perovskite solar cells multilayered systems, 186–188
  optical calculations for perovskite solar cells, 188–192, 192f
    average reflectance as function, 190f
    front layer thickness of TiO$_2$ and PEDOT, 191f
    optical properties of cell materials, 189f
  optical modeling, 184–186
    electromagnetic fields, 185f
  perovskite solar cell structures, 184f

Optical matrix (M_{opt}), 187
Optical performance within concentrator photovoltaic systems, 501–502
Optical splitting, 558
Optimization strategies, 140–142
Optoelectronic devices, 183
OPV. See Organic photovoltaics (OPV)
Organic dyes, 90–91, 225–226
Organic hole-transporting materials, 195–196, 198–214
   carbazole and triphenylamine core-based hole-transporting materials, 199–208
   device architecture and mechanism of perovskite solar cells, 197
   further perspectives, 215–216
   naphthol, thiophene, and pyrene core-based small molecule hole-transporting materials, 211–214
   molecular structures of star-shaped HTMs, 213f
   perovskite solar cells, 195–197
   phenothiazine core-based organic hole-transporting materials, 208–211
Organic HTLs, 144
Organic layers, 596
Organic molecules, 599
Organic photovoltaics (OPV), 10–11, 75, 125, 255, 328, 484, 690, 704–713
Organic semiconductors, 126, 255–257, 596–599
Organic solar cells (OSCs), 10–11, 27–29, 83–90, 90f, 328, 596–597, 599–600, 664–666, 704–705
Organic transparent solar cells, 596–602
   organic single-junction transparent solar cells, 596–601, 601t
   organic tandem transparent solar cells, 601–602
Organic-inorganic hybrid transparent solar cells, 613–616, 615f
Organic–inorganic metal halide perovskites, 547–548
Organo-metal halide perovskite semiconductors, 117
OSCsOrganic solar cells (OSCs)
Osmium, 423
   complex dyes, 90–91

Oversize Kite-craft for Exploration and AstroNautics in Outer Solar system (OKEANOS), 743
Oxide layers, 570

P

P3HT. See Poly(3-hexylthiophene-2,5-diyl) (P3HT)
Paddles, 720–721
PAI. See n-propylammonium iodide (PAI)
Parabolic mirrors, 35–37
Parasitic absorption, 552–554
   loss, 564, 570
   occurrence, 550
Passenger vehicles, 651
Passive scaffold devices, 128
PC. See Polycarbonate (PC)
PCBM. See [6,6]-phenyl C61 butyric acid methyl ester (PCBM)
PCEs. See Power conversion efficiencies (PCEs)
PCRs. See Product category rules (PCRs)
PDIs. See Perylene diimides (PDIs)
PDMS. See Polydimethylsiloxane (PDMS)
PEA. See Phenethyl ammonium (PEA)
PECVD. See Plasma-enhanced chemical vapor deposition (PECVD)
PEDOT:EVA. See Poly(3,4-ethylenedioxythiophene):poly(ethylene-co-vinyl acetate) (PEDOT:EVA)
PEIE. See Polyethyleneimine ethoxylated (PEIE)
Perovskite QDs solar cells (PQDSCs), 90–91
Perovskite solar cells (PSCs), 3–4, 27–29, 77, 96–104, 114–118, 115f, 165, 184, 195–197, 328, 422–423, 508–509, 545–546, 690
   buffer layer in, 145
   crystal structure of example perovskite material, 116f
   impact of design of hole-transporting materials on performance of, 196f
   device architectures of, 122–129
   and mechanism of, 197
   mesoscopic architecture of perovskite solar cells, 126–129

planar architecture of perovskite solar
    cells, 123–126
  schematic illustration of common
    device architecture of PeSCs, 198f
  electrodes for, 145–147
  effect of encapsulation on, 147–148
  flexible, 100
  modeling
    results for thin-film perovskite cells,
      176–180
    thick vs. thin solar cells, 166–169
    thin perovskite solar cells, 169–176
  optical calculations for, 188–192, 192f
  overview of perovskite single-junction
    and tandem cells, 103–104
  perovskite-perovskite tandem solar cells,
    102–103
  perovskite-silicon tandem solar cells, 102
  schematic structure of typical, 116f
  single-junction perovskite solar cells,
    99–100
  stability study of, 133–135
  tandem, 100–101
  total current density in, 175–176
  use of metal-organic framework
    materials in, 352–357
  working mechanism of, 118–121
Perovskite tandem solar cells, 527–529,
  528f, 554, 697–700
  all perovskite flexible tandem solar cells,
    537–538, 538f
  all perovskite tandem architectures,
    533–537
    four terminal all perovskite tandem
      solar cells, 536–537
    multiple terminal all perovskite
      tandem solar cells via double-sided
      substrates, 535, 535f
    two terminal all perovskite tandem
      solar cells, 533–534
  bandgap tuning and stability of
    perovskites, 525–527
  device stability, 538–539
  light capture and device management,
    529–533
  perovskites and unique stoichiometric
    flexibility, 524–525

Perovskites, 119, 135, 422, 704–713
  challenges with tandem structures,
    552–555
    parasitic absorption, 554
    reflection losses, 554–555
  commercially available easily with
    promising efficiency, 565, 568t
  film degradation processes, 148
  film fabrication processes, 136–139
    printing method, 139
    spin coating method, 136–137
    spray coating method, 138–139
    vacuum thermal deposition methods,
      137–138
  materials, 523–524, 545–546
  perovskite solar cell technology, 549f
  perovskite-based solar cells, 183–184
  perovskite-based tandem solar cells, 132,
    548
  perovskite-perovskite tandem solar cells,
    102–103
  perovskite-silicon solar, tandem
    architecture of cells, 130–133
  perovskite-silicon tandem solar cells, 102
  perovskite/copper indium gallium
    selenide tandem solar cells, 569–573
  perovskite/perovskite tandem solar cells,
    573–575, 573t
  perovskite/Si tandem solar cells,
    565–568
  photovoltaics, 119
  power losses in perovskite tandem
    design, 555–556
  single-junction and tandem cells,
    103–104
  tandem architectures, 548–552, 551f,
    556–557
  tandem configurations, 556–557
    fabrication of four-terminal tandems,
      557–558
    performance of four-terminal
      tandems, 558, 559f
    three-terminal tandems, 562–564
    two-terminal tandems, 559–562, 562f
  thermalization process of photo-induced
    electron and photon, 547f
Perylene diimides (PDIs), 268–269

PET. *See* Poly(ethylene terephthalate) (PET)
PHA. *See* Potentially hazardous asteroid (PHA)
Phenanthrene group, 201–203
Phenethyl ammonium (PEA), 129–130
Phenothiazine (Pz), 199, 208–211
  core comprising carbazole, 208–209
  phenothiazine core-based organic hole-transporting materials, 208–211
    chemical structure of three promising hole-transporting materials, 210f
    molecular structures of mono-and di-substituted hole-transporting materials, 209f
  phenothiazine-based PSCs, 208
[6,6]-phenyl C_{61} butyric acid methyl ester (PCBM), 119, 140, 258–260, 461
[6,6]-phenyl C71-butyric acid methyl ester (PC71BM), 261
Phosphorus (P), 6–7
Photo-assisted synthesis, 411
Photo-generated current, 174–175
Photoanode improvement of state-of-the-art dye-sensitized solar cells, 234–235
Photocatalysis, 411
Photocurrent, 11–12, 606
  density, 174–175
Photodeposition process, 411
Photoelectrochemical technology, 226–227
Photoelectrode, 81–82
Photoluminescence, 135
Photoluminescence quantum yield (PLQY), 621
Photon flux, 192
Photons, 188, 546, 622–623
Photosensitizers, 236–237
Photosynthesis process, 616–617
Photovoltaics (PV), 3, 479. *See also* Third-generation photovoltaics
  advanced device concepts and structures, 29–44
    concentrator solar cells, 34–37
    intermediate bandgap solar cells, 37–40
    multijunction devices, 44–55
    quantum dot solar cells, 29–34
    rare earth-containing up-and down-conversion materials to improve solar cell efficiency, 41–44
  arrays, 675
  background of photovoltaic materials and devices, 10–55
    dye-sensitized, organic, and perovskite solar cells, 27–29
    gallium arsenide and III–V materials, 17–22
    silicon solar cells, 11–17
    thin-film solar cells, 22–27
  computer-aided approaches to improved, 337–344
  correlation between vehicle surfaces and photovoltaic cell efficiency, 659–661
  devices, 113–114
  effect, 4–6, 113–114
  into future, 738–748
    novel applications, 743–748
    photovoltaics for solar sails and solar electric propulsion, 741–743
    solar sails and solar electric propulsion, 741
  industry, 303–304, 437–438
  market, 165
  materials, 499, 585
    through green synthesis, 417–422
  modules, 647
    carbon certification system, 480–481
  novel materials and device fabrication for, 358–375
    low-cost manufacturing, 370–375
    photovoltaics-integrated devices and integrated power sources, 363–370
    single-source precursor routes to solar cell materials, 358–363
  performance, 136
  photovoltaic-powered vehicles
    benefits of photovoltaic-powered vehicles and future directions, 666–669
    photovoltaic-integrated vehicles, 655t
    recent trends in, 649–658, 649f, 665t
    solar cell technologies for photovoltaic-powered electric vehicles, 658–666

photovoltaics-integrated devices, 363–370
short history of, 4–10
first century of, 4–6
new materials and new challenges, 9–10, 11f
from practical solar cell device to dawn of space race, 6–7
from space race to energy crisis, 7–9
technologies, 75, 195, 325
life cycle assessment of, 484–489
Phthalocyanin, 199
Physical deposition processes, 127
π-conjugated thiophene bridges, 212
π-linker, 201
π–π interactions, 278–279
Planar PSCs, 122–123
Plasma-enhanced chemical vapor deposition (PECVD), 603–605
Platinum (Pt), 4–5, 91, 606
PLD. See Pulsed laser deposition (PLD)
PLQY. See Photoluminescence quantum yield (PLQY)
PMSAs. See Polymerizing small molecule acceptors (PMSAs)
Poly-ethylene oxide (PEO), 600–601
Poly-ethylene terephthalate (PET), 148, 586–588
Poly-lauryl methacrylate (PLMA), 627
Poly-methyl-methacrylate (PMMA), 619–621, 623
Poly(3-hexylthiophene-2,5-diyl) (P3HT), 461
Poly(3-hexylthiophene) (P3HT), 119, 124, 135, 143–144, 258–259, 442, 460–461
Poly(3,4-ethylenedioxythiophene) (PEDOT), 140
Poly(3,4-ethylenedioxythiophene):poly (ethylene-co-vinyl acetate) (PEDOT: EVA), 334–336
Poly(cyclohexylmethacrylate) (PCHMA), 625
Poly(ethylene terephthalate) (PET), 361–362
Poly(methylmethacrylate) (PMMA), 625
Poly[2-methoxy-5-(2-ethylhexyloxy)-1,4-phenylenevinylene] (MEHPPV), 463

Poly[bis(4-phenyl)(2,4,6-trimethylphenyl)] (PTAA), 135, 140, 143–144
Polycarbonate (PC), 143–144, 619–621
Polycrystalline inorganic thin-film solar cells, 546–547
Polycrystalline materials, 701
Polycrystalline silicon (p-Si), 15–17, 303–304, 662–663
Polycrystalline thin-film devices, 9
Polydimethylsiloxane (PDMS), 143–144
Polyetherimide (PEI), 593
Polyethylene naphthalate (PEN), 537–538, 612–613
Polyethyleneimine ethoxylated (PEIE), 142
Polymer design, 265
Polymer donors, 281–282
Polymer-based cell, 561–562
Polymer-based OPV, 255
Polymer-fullerene BHJ, 285
Polymeric electron acceptors, 267
Polymerizing small molecule acceptors (PMSAs), 280–281
Polyphenylene vinylene (PPV)-type polymer donor, 260
Polysilazane-derived coatings, 689
Polythiophene (PT), 126
Polyvinyl alcohol (PVA), 315
Polyvinyl pyrrolidone (PVP), 627
Polyvinylbutyral (PVB), 605
Portable electronics, 102–103
Potentially hazardous asteroid (PHA), 722
Power and propulsion element (PPE), 747–748
Power conversion efficiencies (PCEs), 3–4, 113–114, 195, 224–225, 258–259, 442, 523, 545, 586–588, 663–664, 675
Power losses in perovskite tandem design, 555–556
Power systems, 738
Power-integrated systems, 744–745
PPE. See Power and propulsion element (PPE)
PQDSCs. See Perovskite QDs solar cells (PQDSCs)
Primary optics, 504–505
Printing method, 139
Product category rules (PCRs), 480–481

Propulsion systems, 741–742
Proteins, 414–416
Proton irradiation, 696
PSCs. See Perovskite solar cells (PSCs)
PT. See Polythiophene (PT)
PTQ10, 281–282
Pulsed laser deposition (PLD), 310–313
PVA. See Polyvinyl alcohol (PVA)
PV. See Photovoltaics (PV)
Pyrene, 199
   chemical structure of pyrene-based hole-transporting materials, 214f
   core-based small molecule hole-transporting materials, 211–214
   derivatives, 214
Pyrrolopyrrole cyanine dyes, 606–607

Q

QDs. See Quantum dots (QDs)
QDSCs. See Quantum dot solar cells (QDSCs)
QDSSCs. See Quantum dot sensitized solar cells (QDSSCs)
Quantum dot sensitized solar cells (QDSSCs), 76, 90–91, 93–96, 246–247
Quantum dot solar cells (QDSCs), 3–4, 28–34, 94–95
Quantum dots (QDs), 75, 224–225, 437–438, 607–609
   quantum dots-based transparent luminescent solar concentrators, 625–629, 628f
Quaternary compounds, 457–458
Quaternary iron compounds, 463–464
Quaternary-derived compounds, 459

R

R&D. See Research and development (R&D)
R2R. See Roll-to roll (R2R)
Radiation challenges from space environment, 684–686
Radiative emission, 688
Radio frequency (RF), 606–607
Rapid spray plasma processing (RSPP), 371
Rare earth-containing up-and down-conversion materials to improve solar cell efficiency, 41–44
Rare metals, 423
Rare-earth-doped upconversion nanoparticles, 233
Re-emitted photons, 622
Reabsorption losses, 622–623
Recombination layers (RLs), 534
Redox flow batteries (RFBs), 365–366
Reflection losses, 554–555
Refractive index, 554–555
Renewable energy, 303–304
Research and development (R&D), 78
RF. See Radio frequency (RF)
RFBs. See Redox flow batteries (RFBs)
Rideshare CubeSats on Artemis, 733–738
Rideshare missions, 725–738
RLs. See Recombination layers (RLs)
Roll-out solar array (ROSA), 680
Roll-to roll (R2R), 115–116
   fabrication, 115–116
   manufacturing, 371–373
   process, 91–92, 533
   production, 371
ROSA. See Roll-out solar array (ROSA)
RSPP. See Rapid spray plasma processing (RSPP)
Ruthenium, 90–91, 423
   bipyridine analogs, 231–232

S

Satellites, 516, 675
SAW. See Solar array wings (SAW)
SBT. See Semiconductor wafer bonding technology (SBT)
SBUs. See Secondary building units (SBUs)
Scanning electron microscope (SEM), 440–442
SCARLET. See Solar Concentrator Array with Refractive Linear Element Technology (SCARLET)
Screen printing of TiO_2, 139
SCs. See Solar cells (SCs)
Second-generation photovoltaics, concentrator optics coupled with, 508–509

Secondary building units (SBUs), 344
Selenide solar cells, 442
Selenium (Se), 4−5
SEM. *See* Scanning electron microscope (SEM)
Semiconductor wafer bonding technology (SBT), 691−693
Semiconductors, 128
  materials, 678
    green methods for synthesis of, 406−412
  QDs, 233
Semitransparency of top cell, 529−531, 531*f*
Semitransparent polymer solar cell (ST-PSC), 89, 334
Semitransparent solar cells (STSCs), 585
Sensitized solar cells (SSCs), 438
Sensitizers, 236, 606
SEP. *See* Solar electric propulsion (SEP)
SERI. *See* Solar Research Institute (SERI)
Shockley equations, 166−167
Shockley−Queisser limit (SQ limit), 524
Shockley−Read−Hall recombination, 533−534
Shunt resistance, 241
SILAR method. *See* Successive ionic layer adsorption and reaction method (SILAR method)
Silicon (Si), 4, 15, 114, 165, 325, 484, 602−603, 712
  PV, 15
  Si-based single-junction solar cells, 545−546
  silicon-based solar arrays, 678
  silicon-based transparent solar cells, 602−605
  solar cells, 11−17, 224, 505, 662−663
  tandem solar cells, 565−568
Silicon oxycarbonitride (SiOCN), 689
Silver (Ag), 145−146, 149
Silver bismuth selenide ($AgBiSe_2$), 452
Silver bismuth sulfide ($AgBiS_2$), 449−452, 463
Silver iodide (AgI), 145−146
Silver nanowires (Ag NWs), 529−531, 571, 591−593
Single junction silicon cells, 506−508

Single p−n junction solar cells, 502−503
Single-crystal silicon solar cells, 114
Single-junction perovskite solar cells, 99−100
Single-junction solar cells, 100−101, 545
Single-source precursors (SSPs), 328
  routes to solar cell materials, 358−363
Single-wall carbon nanotubes (SWNTs), 143−144
SLS rocket. *See* Space Launch System rocket (SLS rocket)
Small bodies, 720−721
Small organic molecules, 196−197
Small satellite technology experimental platform (S^2TEP), 697−700
Small satellites missions (SmallSats missions), 725−738
Small spacecraft electric propulsion (SSEP), 742−743
Small Spacecraft Technology, 745
SmallSats missions. *See* Small satellites missions (SmallSats missions)
Sn-doped indium oxide ($Sn-In_2O_3$), 529−531
$SnCl_2$ posttreatment (SPT), 455−456
SOA. *See* State-of-the-art (SOA)
SOC. *See* Spin-orbit coupling (SOC)
Sodium bismuth sulfide ($NaBiS_2$), 452−453
Solar array wings (SAW), 679
Solar arrays (SOA), 677, 738
Solar battery, 7, 675
Solar cells (SCs), 118, 183−184, 188−189, 438, 444, 456, 460−461, 503−504, 523, 585, 616−617, 675
  classification of, 224−225, 225*f*
  and concentrator optics for optimum performance, 512−517
    low carbon footprint, 516−517
    low cost, 512−516, 514*t*
    progressive advances in photovoltaic solar cell materials, 513*f*
    space-limited high-efficiency photovoltaic technologies, 516
  considerations for greener fabrication of, 422−426
  device technologies, 165

Solar cells (SCs) (*Continued*)
  efficiency, 683–684
    rare earth-containing up-and down-conversion materials to improve, 41–44
    materials, single-source precursor routes to, 358–363
    parameters for (semi-)transparent applications, 588–590
      average visible transmittance, 589
      color rendering index, 589–590
      solar cell efficiency, 588–589
    performance, 501–502
    production technologies, 663–664
    prospects for green synthesis of, 426–427
    technologies for photovoltaic-powered electric vehicles, 658–666
      correlation between vehicle surfaces and photovoltaic cell efficiency, 659–661
      different types of solar cells for electric vehicle applications, 661–666
Solar Concentrator Array with Refractive Linear Element Technology (SCARLET), 713
Solar concentrator systems, 501, 512
  multijunction solar cells for, 509–512, 510*f*
Solar electric propulsion (SEP), 703–704, 741
  photovoltaics for, 741–743
Solar energy, 113–114, 303–304, 489, 651–652
  systems, 652
  technologies, 516
Solar irradiation, 228–230
Solar module, 690
Solar panels, 721
Solar photons, 741
Solar radiation, 683
Solar Research Institute (SERI), 9
Solar sails, 741
  photovoltaics for, 741–743
Solar spectrum, 15, 589
Solar system, 725
  exploration of, 738–748
Solar tracking, 558
Solar tree, 89
Sol–gel
  deposition technique, 314–315
  method, 306–307
  spin-coating process, 308
Solid state HTMs, 195–196
Solution process spin-coating, 123–124
Solution-processable organic photovoltaics
  considerations for commercialization, 281–289
  electron donors optimized for fullerene-based acceptors, 261–266
  highest certified efficiencies of organic photovoltaics, 259*f*
  layer structure of organic solar cell, 258*f*
  nonfullerene electron acceptors, 266–281
  solution-processed bulk heterojunction organic photovoltaics, 260–261
Solution-processed antimony sulfide, 442
Solution-processed deposition approaches, 601–602
Solution-processed planar PSCs, 201
Solution-processed solar cells, 437–439
Solvent-based deposition, 601–602
Solvothermal chemical synthesis, 405
Solvothermal methods, 309, 315, 408–409
Sonochemical synthesis, 410–411
SOP. *See* State-of-practice (SOP)
South Korea Research Institute of Chemical Technology, 117
South Korea Ulsan National Institute of Science and Technology (UNIST), 117
Space Launch System rocket (SLS rocket), 733
Space photovoltaics, 675. *See also* Photovoltaics (PV)
  history of space solar cells, 675–683
    invention of modern solar cell and dawn of space race, 675–676
    new devices and advanced materials technologies in 21st century, 681–683
    space photovoltaics during 1960s to 1980s, 677–678
    space photovoltaics in 1990s, 678–681, 739*t*

materials and devices, 690—716
  group III-V materials and device
    technologies, 690—693
  thin-film solar cell materials, 693—704
space applications and exploration
    missions, 716—748
space environment, 683—690, 687t
  air mass standards for terrestrial and
    space photovoltaics, 683—684
  mass-specific power, 689—690
  radiation challenges from space
    environment, 684—686
  thermal and challenges of space
    environment, 686—689
  technologies for exemplary missions, 718
Space power, 675
Space race
  to energy crisis, 7—9
  from practical solar cell device to dawn
    of, 6—7
Space radiation, 13—15, 712
Space solar cells, 678—679
Space solar power systems, 677—678
Space technology development, 738
Space-limited high-efficiency photovoltaic
    technologies, 516
Spacecraft developers, 681
Spectral reflectance, 184
Spin-coating, 127, 136—137, 308,
    314—315, 405, 599—600, 607—609
Spin-orbit coupling (SOC), 129
Spiro-OMeTAD, 142—143, 183—184,
    196, 199—200, 205—206, 214, 449
9-spirobifluorene, 206—207
Spray coating method, 138—139
Spray pyrolysis, 127, 308, 313
Spray-deposition, 607—609
SPT. See $SnCl_2$ posttreatment (SPT)
Sputnik 1, 7
Sputnik 2, 7
Sputtering, 127, 311, 405, 612
SQ limit. See Shockley—Queisser limit (SQ
    limit)
SSC. See Symmetrical supercapacitor (SSC)
SSCs. See Sensitized solar cells (SSCs)
SSEP. See Small spacecraft electric
    propulsion (SSEP)

SSPs. See Single-source precursors (SSPs)
ST-PSC. See Semitransparent polymer solar
    cell (ST-PSC)
Stacking
  process, 557
  sequences, 307—308
Staebler—Wronski degradation, 694—696
Staebler—Wronski effect, 23
Standard testing protocols, 148
Stannic chloride ($SnCl_4$), 313
Stannous chloride ($SnCl_2$), 313
Star-shaped carbazole core-based organic
    HTM, 200
  chemical structure of, 200f
Star-shaped HTMs, molecular structures of,
    213f
Starburst shaped HTMs, 207
  molecular structure of investigated, 207f
Starfish Prime device, 684—685
Starshine 3, 725—728
State-of-practice (SOP), 681—683
State-of-the-art (SOA), 681—683
  dye-sensitized solar cells
    counter-electrodes, improvements
      over, 238—239
    photoanode improvement of,
      234—235
    sensitizers, improvements over, 236—237
Steroids, 414—416
Stokes shift, 619, 622—623
STSCs. See Semitransparent solar cells
    (STSCs)
Successive ionic layer adsorption and
    reaction method (SILAR method), 442,
    612
Sulfur, 308
Sulfur oxides (SOx), 303—304, 653—654
Sunlight, 506, 510—511
  absorption, 118—119
  energy, 558
Surface losses, 622—623
Sustainability, 326—337
  background and overview, 326—328
  green materials and processing,
    328—332
  lessons learned from nature, 332—337
Sustainable materials, 328

Sustainable solution-processed solar cells
  antimony compounds, 438–445
  bismuth compounds, 445–453
  iron compounds, 458–464
  tin compounds, 453–458
SWNTs. See Single-wall carbon nanotubes (SWNTs)
Symmetrical supercapacitor (SSC), 366–368
Synthesis methods of copper zinc tin sulfide nanoparticles and thin films, 305–309
Synthesized carbazole-based HTMs, chemical structure of, 202f
Synthetic organic chemistry, 198

T

Tandem perovskite solar cells, 100–101
Tandem PV devices, 546–547
Tandem solar cells, 511, 545, 547, 557
TBTIT, 265–266
TCOs. See Transparent conductive oxides (TCOs)
TE technique. See Thermal evaporation technique (TE technique)
Technology readiness level (TRL), 375, 664, 743–744
Tellurium dioxide (TeO_2), 598–599
Telstar, 677
Telstar I, 8
TENG. See Triboelectric nanogenerator (TENG)
Terawatt energy generation challenges, 78–80
Ternary antimony compounds, 443–445
Ternary bismuth compounds, 448–453
  $AgBiS_2$, 449–452
  $AgBiSe_2$, 452
  $CuBiS_2$, 448–449
  $NaBiS_2$, 452–453
Ternary chalcogenide semiconductors, 453
Ternary compounds, 457–458
Ternary iron compounds, 463–464
Ternary oxides, 245
Ternary-derived compounds, 459
Terpenoids, 414–416
Terrestrial solar cells, 683

Tertbutyl alcohol (tBuOH), 139
4-tertbutylpyriding, 142–143
Tetraethyl orthosilicate, 422–423
2,2′,7,7′-tetrakis-(N,N-dimethoxyphenylamine)-9, 206–207
Tetrakis(4-carboxyphenyl) porphyrin (TCPP), 357
TFSCs. See Thin-film solar cells (TFSCs)
Thermal and challenges of space environment, 686–689
Thermal cycling, 701
Thermal evaporation technique (TE technique), 137, 309, 455
Thermal oxidation-sputtering method, 612
Thick solar cells, 166–169
Thieno-[3,2-b]thiophene (TbT), 212
Thin cells, dark radiative recombination in, 172–173
Thin-film $AgBiS_2$, 449
Thin-film perovskite cells, 176–177
  modeling results for, 176–180
Thin-film solar cells (TFSCs), 22–27, 313, 358, 585
  chalcogenide thin-film solar cells, 23–27
  materials, 693–704
    amorphous silicon, 694–696
    approaches to enhance device performance, 713–716
    cadmium telluride, 696–697
    perovskite and organic photovoltaics, 704–713, 711f
  thin-film silicon solar cells, 22–23
Thin perovskite solar cells, 169–176
  dark nonradiative recombination, 173–174
  dark radiative recombination in thin cells, 172–173
  photo-generated current, 174–175
  total current density in perovskite solar cells, 175–176
Thin solar cells, 166–169
Thin-film approaches, 9
Thin-film PV technologies, 683
Thin-film silicon solar cells, 22–23
Thin-film solar cell, 453–454
Thioacetamide (C_2H_5NS), 307–308

Thiophene, 212–214
  core-based small molecule hole-transporting materials, 211–214
  molecular structure of thiophene bridged hole-transporting materials, 214f
Thiourea, 308
Third-generation photovoltaics, 75, 76f, 239
  concentrator optics coupled with, 508–509
  economic assessment and market status of, 83–104
  features of third-generation solar cells, 77–78
  overview of, 80–82
  terawatt energy generation challenges, 78–80
Third-generation tandem solar cells, 44–55
Three-dimension (3D), 234
  3D-shaped TiO_2, 242–243
  network, 234
Three-terminal (3T), 548–549
  power conversion efficiency, 564f
  tandems, 562–564
Tin (Sn), 129, 311, 438
  compounds, 453–458
  nanocrystals
    binary compounds and, 453–457
    ternary and quaternary compounds and, 457–458
  precursors, 313
  tin-based binary, 453
  tin-based perovskites material, 129
Tin sulfide (SnS), 456
Tin sulfide TFSCs, 361–362
Tin-chloride ($SnCl_2$), 307–308
Tin-doped Zinc oxide (TZO), 593
Tin(IV)oxide (SnO_2), 140
Titanium (Ti), 129
Titanium dioxide (TiO_2), 190–191, 225–226, 609–610
  nanocrystals, 127
  nanoparticles, 122, 606
  nanotubes, 242
TJs. See Tunnel junctions (TJs)
TLSC. See Transparent luminescent solar concentrators (TLSC)

TOPO. See Tri-octylphosphine oxide (TOPO)
Total average reflectance, 184
Total current density in perovskite solar cells, 175–176
TPEC. See Transparent photoelectrochemical cell (TPEC)
Traditional inorganic materials, 22–27
Transfer matrix model, 184–185
Transparent conductive oxides (TCOs), 227, 529–531
Transparent electrodes, 554–556
Transparent luminescent solar concentrators (TLSC), 618–629
  efficiency of, 621–622
  emergence of quantum dots-based, 625–629, 628f
  history of development of, 623–625
  losses associated with, 622–623
  material selection in, 619–621
  principles, 619
Transparent photoelectrochemical cell (TPEC), 603–605
Transparent photovoltaics, 482–484
  in agriculture, 616–618, 617f
  solar cell parameters for (semi-)transparent applications, 588–590
  transparent luminescent solar concentrators, 618–629
  transparent solar cells, 585–588
  types of transparent solar cells, 596–616
    inorganic semiconductor-based transparent solar cells, 602–613
    organic transparent solar cells, 596–602
Transparent solar harvesting systems, 622–623
Tri-octylphosphine oxide (TOPO), 461–462
Triboelectric nanogenerator (TENG), 89, 365–366
Triphenylamine, 199, 208–209
  components, 201
  groups, 201
  triphenylamine-based donor (D)-π-D HTMs, 201
  units, 201

Triphenylamine core-based hole-transporting materials, 199–208
  chemical structure
    of star-shaped carbazole core-based organic HTM, 200f
    of synthesized carbazole-based HTMs, 202f
    of two carbazole-based HTMs, 199f
  molecular design feature of target HTMs, 204f
  molecular structure
    of facile donor (D)-π-D triphenylamine-based organic hole-transporting material, 203f
    of indolo[3, 2-b]carbazole-based HTMs, 206f
    of investigated linear and starburst-shaped hole-transporting materials, 207f
    of triphenylamine-and carbazole-based HTMs, 205f
    of two hole-transporting materials with triphenylene core, 208f
  structures of hole-transporting materials with fused aromatic rings, 204f
Triphenylbenzene (TPB), 209–210
Triphenylene core, molecular structures of two hole-transporting materials with, 208f
Triple-junction cell (3J cell), 510–511, 677–678
1,3,5-tris(2-((N,N-di(4 methoxyphenyl) amino)phenyl))benzene (TPB(2-MeOTAD)), 209–210
1,3,5-tris(2-(N-phenothiazylo)phenyl) benzene (TPB(2-TPTZ)), 209–210
TRL. See Technology readiness level (TRL)
Tunable bandgaps, 437–438
Tungsten, 425
Tungsten disulfide (WS_2), 91–92
Tungsten sulfide, 606–607
Tungsten trioxide (WO_3), 600–601
Tunnel junctions (TJs), 534
Two terminal (2T), 49–50, 529, 548–550, 559–560
  all perovskite tandem solar cells, 533–534
  devices, 529
  monolithic tandem solar cell, 132
  tandems, 559–562
    architectures, 524
    two-terminal perovskite tandem devices, 563t
Two-dimension (2D), 607–609
  2D-shaped TiO_2, 242–243
  polymers, 265
  triphenylene core, 208

U

UAVs. See Unmanned aerial vehicles (UAVs)
UiO-66, 355–357
Ultrasound (US), 410–411
  solar-powered solar system, 723
Ultraviolet (UV), 124–125
  UV-curable epoxy resin, 148
  UV-visible region, 585
UMM materials. See Upright metamorphic MJ materials (UMM materials)
UNIST. See South Korea Ulsan National Institute of Science and Technology (UNIST)
United States Air Force (USAF), 9–10, 690–691
United States Department of Energy (DOE), 681
Unmanned aerial vehicles (UAVs), 654
Up-conversion materials, 42
Up/down conversion materials, 3–4
Upright metamorphic MJ materials (UMM materials), 691–693
US Army Signal Corps, 9–10
US Ballistic Missile Defense Organization, 713
US Department of Energy's National Renewable Energy Laboratory (DOE NREL), 9, 551–552
USAF. See United States Air Force (USAF)
UV. See Ultraviolet (UV)

V

Vacuum thermal deposition methods, 137–138

Vacuum-based approaches, 310−315
  electrodeposition technique, 314
  evaporation techniques, 311−312
  nonvacuum-based approaches, 313
  pulsed laser deposition methods, 312−313
  sol−gel deposition technique, 314−315
  solvothermal method, 315
  spray pyrolysis deposition technique, 313
  sputtering techniques, 311
Vacuum-based solar cells, comparison of performance of, 316−317
Vacuum-processed CZTS solar cell, 316−317
Valence band (VB), 37
Valence band maxima (VBM), 525−527
Van Allen radiation belts, 684−685
Vanadium, 425
Vanadium oxide (V_2O_5), 593−594
Vanguard 1, 8
VASP. See Vienna Ab Initio Simulation Package (VASP)
VB. See Valence band (VB)
VBM. See Valence band maxima (VBM)
Venus flyby, 708−709
Vertical optical cavity design, 247
Vienna Ab Initio Simulation Package (VASP), 355
Viruses, 412−414
Visible light transmittance (VLT), 624
VLT. See Visible light transmittance (VLT)

W

Wastewater treatment, 488
Water, 424
  water-based electrolytes, 425
Weighted average reflectance, 189−190
Wide-band gap semiconducting metal oxides, 609−610
Wulff construction, 242−243

X

X-ray diffraction (XRD), 134
XRD. See X-ray diffraction (XRD)
XTJ Prime cells. See Next Triple Junction Prime cells (XTJ Prime cells)

Y

Yeast, 412−414

Z

Zinc (Zn), 311
Zinc oxide (ZnO), 124, 609−610
  film electrode, 225−226
  nanoparticles, 571
Zinc sulfide (ZnS), 32−33, 93−94, 307−308